Political Map of the World

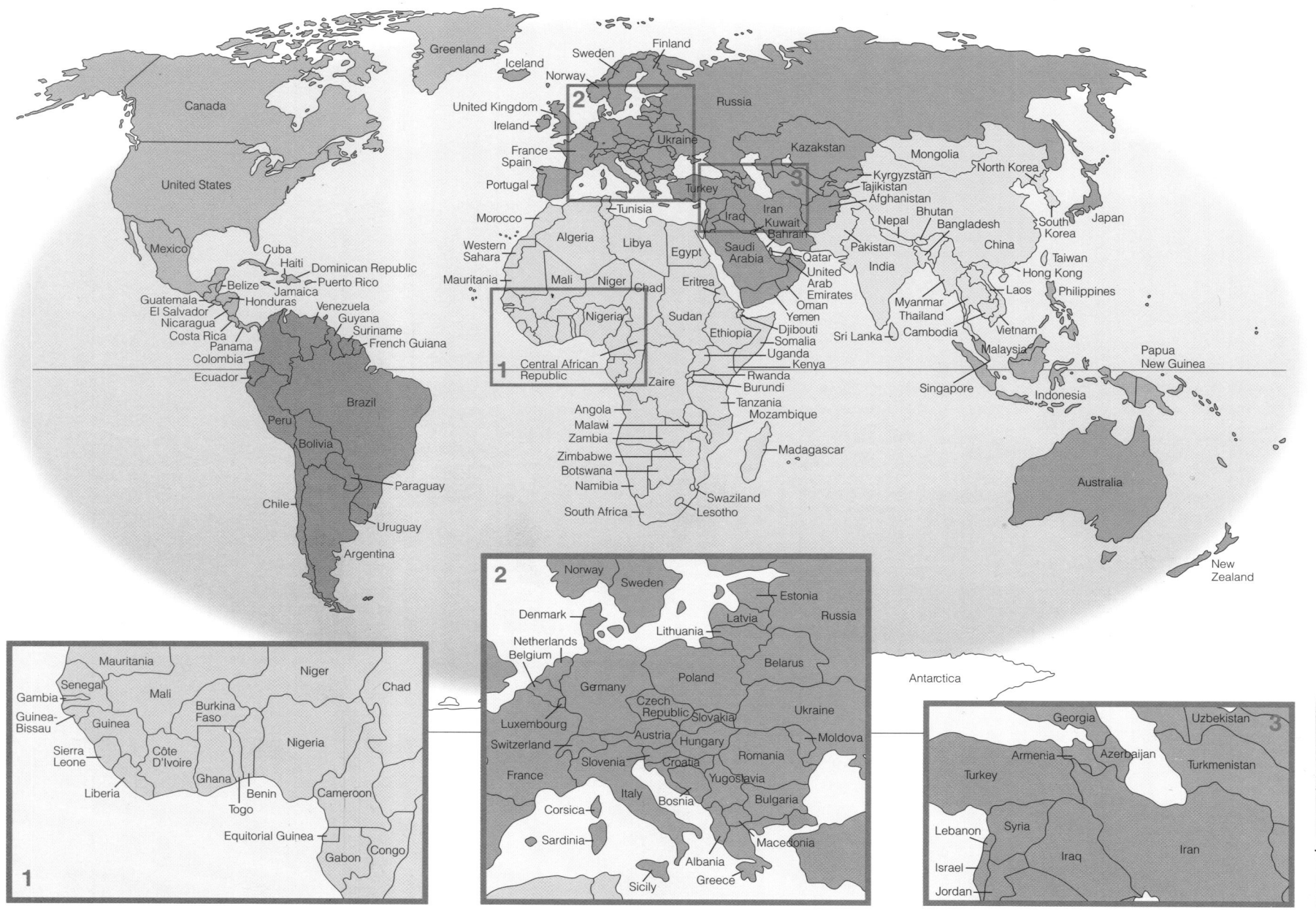

Sociology in Our Times

FIFTH EDITION

Diana Kendall
Baylor University

WADSWORTH

★

THOMSON LEARNING ™

Australia • Canada • Mexico • Singapore • Spain • United Kingdom • United States

To Harold Osborne, professor emeritus of sociology at Baylor University, who not only has inspired me in the writing of this text, but also has challenged me to take on new and exciting teaching and research endeavors.

WADSWORTH
THOMSON LEARNING

Sociology Editor: Robert Jucha
Development Editor: Shelley Murphy
Assistant Editor: Stephanie Monzon
Editorial Assistant: Melissa Walter
Technology Project Manager: Dee Dee Zobian
Marketing Manager: Matthew Wright
Marketing Assistant: Tara Pierson
Advertising Project Manager: Linda Yip
Project Manager, Editorial Production: Cheri Palmer
Print/Media Buyer: Judy Inouye
Permissions Editor: Kiely Sexton

Production Service: Greg Hubit Bookworks
Text Designer: Lisa Delgado, delgadoandcompany, inc.
Photo Researcher: Linda Rill
Copy Editor: Donald Pharr
Illustrator: Thompson Type
Cover Designer: Bill Stanton
Cover Image: © Romilly Lockyer/Getty Images
Cover Printer: Quebecor World/Dubuque
Compositor: Thompson Type
Printer: Quebecor World/Dubuque

Printed in the United States of America
2 3 4 5 6 7   08 07 06 05 04

For more information about our products, contact us at:
Thomson Learning Academic Resource Center
1-800-423-0563
For permission to use material from this text or product, submit a request online at:
http://www.thomsonrights.com
Any additional questions about permissions can be submitted by email to:
thomsonrights@thomson.com

Library of Congress Control Number: 2003116252

Student Edition: ISBN 0-534-62685-8

Instructor's Edition: ISBN 0-534-62686-6

0-534-62710-2

Thomson Wadsworth
10 Davis Drive
Belmont, CA 94002-3098
USA

Asia
Thomson Learning
5 Shenton Way #01-01
UIC Building
Singapore 068808

Australia/New Zealand
Thomson Learning
102 Dodds Street
Southbank, Victoria 3006
Australia

Canada
Nelson
1120 Birchmount Road
Toronto, Ontario M1K 5G4
Canada

Europe/Middle East/Africa
Thomson Learning
High Holborn House
50/51 Bedford Row
London WC1R 4LR
United Kingdom

Latin America
Thomson Learning
Seneca, 53
Colonia Polanco
11560 Mexico D.F.
Mexico

Spain/Portugal
Paraninfo
Calle Magallanes, 25
28015 Madrid, Spain

BRIEF CONTENTS

CONTENTS

BOXES

PREFACE

Welcome to the fifth edition of *Sociology in Our Times*! The twenty-first century offers unprecedented challenges and opportunities for each of us as individuals and for our larger society and world. In the United States, we can no longer take for granted the peace and economic prosperity that many—but far from all—people were able to enjoy in previous decades. However, even as some things change, others remain the same, and among the things that have not changed are the significance of education and the profound importance of understanding how and why people act the way they do. It is also important to analyze how societies grapple with issues such as economic hardship and the threat of terrorist attacks and war, and to gain a better understanding of why many of us seek stability in our social institutions—including family, religion, education, government, and media—even if we believe that some of these institutions might benefit from certain changes.

Like previous editions of this widely read text, the fifth edition of *Sociology in Our Times* is a cutting-edge book that highlights the relevance of sociology. It does this in at least two ways: (1) by including a diversity of classical and contemporary theory, interesting and relevant research, and lived experiences that accurately mirror the diversity in society itself, and (2) by showing students that sociology involves important questions and issues that they confront both personally and vicariously (for example, through the media). This text speaks to a wide variety of students and captures their interest by taking into account their concerns and perspectives. The research used in this text includes the best work of classical and established contemporary sociologists—including many white women and people of color—and it weaves an inclusive treatment of *all* people into the examination of sociology in *all* chapters. Through the use of the latest theorizing and research, *Sociology in Our Times* not only provides students with the most relevant information about sociological thinking but also helps students consider the significance of the interlocking nature of class, race, and gender in all aspects of social life.

I would encourage you to read a chapter in the book and judge for yourself the writing style, which I have sought to make both accessible and engaging for students and instructors. Concepts and theories are presented in a straightforward and understandable way, and the wealth of concrete examples and lived experiences woven throughout the chapters makes the relevance of sociological theory and research abundantly clear to students.

ORGANIZATION OF THIS TEXT

Sociology in Our Times, fifth edition, contains twenty carefully written, well-organized chapters to introduce students to the best of sociological thinking. **Chapter 1** introduces students to the sociological imagination and traces the development of sociological thinking. The chapter sets forth the major theoretical perspectives used by sociologists in analyzing compelling social issues such as the problem of credit card abuse and hyperconsumerism among college students and others. **Chapter 2** focuses on how sociologists conduct research. This chapter provides a thorough description of both quantitative and qualitative methods of sociological research, showing how these approaches have been used from the era of Emile Durkheim to the present to study social concerns such as suicide.

The next five chapters focus on the nature of social life and core sociological concepts. In **Chapter 3,** culture is spotlighted as either a stabilizing force or a force that can generate discord, conflict, and even violence in societies. Cultural diversity is discussed as a contemporary issue, and unique coverage is given to popular culture and leisure and to divergent perspectives on popular culture. **Chapter 4** looks at positive and negative aspects of socialization and presents an innovative analysis of gender socialization, racial–ethnic socialization, and issues associated with recent immigration. **Chapter 5** examines society, social structure, and social interaction in detail, using homelessness as a sustained example of the dynamic interplay of structure and interaction in society. Unique to this chapter are discussions of the sociology of emotions and of personal space as viewed through the lenses of race, class, gender, and age. **Chapter 6** analyzes groups and organizations, including innovative forms of social organization and ways in which organizational structures may differentially affect people based on race, class, gender, and age. **Chapter 7** examines diverse perspectives on deviance, crime, and the criminal justice system. Key issues are dramatized for students through an analysis of recent research on peer cliques and gangs.

The next five chapters examine social differences and social inequality, looking at issues of class, race/ethnicity, and sex/gender. **Chapter 8** addresses systems of global stratification and examines differences in wealth and poverty in rich and poor nations around the world. **Chapter 9** looks at social class in the United States, including the causes and consequences of inequality and poverty as well as the accessibility of the American Dream. Explanations for these differences are discussed. The focus of **Chapter 10** is race and ethnicity, and the chapter uses as an illustration the historical relationship (or lack of it) between sports and upward mobility by persons from diverse racial–ethnic groups. A thorough analysis of prejudice, discrimination, theoretical perspectives, and the experiences of racial and ethnic groups is presented, along with global racial and ethnic issues in the twenty-first century. **Chapter 11** examines sex and gender, with special emphasis on gender stratification in historical perspective. Linkages between gender socialization and contemporary gender inequality are described and illustrated by lived experiences and perspectives on body image. **Chapter 12** provides a cutting-edge analysis of aging, including theoretical perspectives and inequalities experienced by people across the life course.

Next are six chapters that examine social institutions, making students more aware of the importance of social institutions and showing how problems in one can have a significant impact on others. The economy and work are explored in **Chapter 13,** which examines global economic systems, the social organization of work in the United States, unemployment, and worker resistance and activism. The chapter concludes with a discussion of the global economy in the future. **Chapter 14** discusses the intertwining nature of politics, government, and the media. Political systems are examined in global perspective, and politics and government in the United States are analyzed with attention to gov-

ernmental bureaucracy and the military–industrial complex. Families and intimate relationships are explored in **Chapter 15,** which focuses on families in global perspective and on the diversity found in U.S. families today. **Chapter 16** investigates the history of education in the United States and contrasts it with systems of education in other nations. In the process, the chapter highlights issues of race, class, and gender inequalities in current U.S. education. In **Chapter 17,** religion is examined in global perspective, including a survey of world religions and an analysis of how religious beliefs affect other aspects of social life. Current trends in U.S. religion are also explored, including various sociological explanations of how and why religion is a means by which people seek purpose and meaning in everyday life. **Chapter 18** analyzes health, health care, and disability in the United States and worldwide. This chapter is unique in that it contains one of the only thorough discussions in any introductory sociology text of mental illness and disability as factors in health and health care delivery.

The final two chapters focus on social dynamics and social change. **Chapter 19** examines population, urbanization, and migration, looking at demography, global population change, and the process and consequences of urbanization. **Chapter 20** discusses collective behavior, social movements, and social change. Environmental activism is used as a sustained example to help students grasp the importance of collective behavior and social movements in producing social change. The concluding section takes a final look at the physical environment, population, technology, social institutions, and change in the future.

DISTINCTIVE FEATURES OF THIS TEXT

The following special features are specifically designed to reflect the themes of relevance and diversity in *Sociology in Our Times,* as well as to support students' learning.

Interesting and Engaging Lived Experiences Throughout Chapters

Authentic first-person accounts are used as opening vignettes and throughout each chapter to create interest and give concrete meaning to the topics being discussed. Lived experiences provide opportunities for students to examine social life beyond their own experiences and for instructors to systematically incorporate into lectures and discussions an array of interesting and relevant topics demonstrating to students the value of applying sociology to their everyday lives. Some examples of the lived experiences include the following:

- Kurt Cobain's suicide note, describing how a person about to commit suicide might feel, shows how such beliefs can be linked to sociological theories and research on pressing social issues such as suicide (Chapter 2, "Sociological Research Methods").
- George Pataki, New York governor, discussing his first trip to the World Trade Center disaster site following the 2001 terrorist attacks (Chapter 6, "Groups and Organizations").
- Greg Penhaligon, a former senior front-end developer for a dotcom company, recalling what it was like when the bubble burst on the information technology (IT) industry and many remaining jobs were relocated to other nations (Chapter 13, "The Economy and Work in Global Perspective").
- John Cronin and Robert F. Kennedy, Jr., environmental activists, describing their work to save the Hudson River from environmental degradation (Chapter 20, "Collective Behavior, Social Movements, and Social Change").

Focus on the Relationship Between Sociology and Everyday Life

Each chapter has a brief quiz that relates the sociological perspective to the pressing social issues presented in the opening vignette. (Answers are provided on a subsequent page.) Topics such as these will pique students' interest:

- "How Much Do You Know About Homeless Persons?" (Chapter 5, "Society, Social Structure, and Interaction")
- "How Much Do You Know About Privacy in Groups and Organizations?" (Chapter 6, "Groups and Organizations")
- "How Much Do You Know About Body Image and Gender?" (Chapter 11, "Sex and Gender")
- "How Much Do You Know About Migration?" (Chapter 19, "Population and Urbanization")

Emphasis on the Importance of a Global Perspective

Sociology in Our Times analyzes our interconnected world and reveals how the sociological imagination extends beyond national borders. Global implications of all topics are examined throughout each chapter and in the Sociology in Global Perspective box found in many chapters. Here are a few examples:

- "Stay-at-Home Dads: Socialization for Kids and Resocialization for Parents" (Chapter 4, "Socialization")
- "Opposition, Resistance, and the Women of Afghanistan" (Chapter 11, "Sex and Gender")
- "The European Union: Transcending National Borders and Governments" (Chapter 14, "Politics and Government in Global Perspective")
- "'Flash Mobs': Collective Behavior in the Information Age" (Chapter 20, "Collective Behavior, Social Movements, and Social Change")

Focusing on "Changing Times: Media and Technology" to Encourage Critical Thinking

A significant benefit of a sociology course is encouraging critical thinking about such things as how media and technological changes influence our daily lives. Here are the topics of a few Media/Technology boxes that will foster critical-thinking skills:

- "Online Shopping and Your Privacy: The Changing Nature of Social Life" (Chapter 1, "The Sociological Perspective")
- "*American Idol* and the Diffusion of Popular Culture" (Chapter 2, "Sociological Research Methods")
- "Can We Form Social Groups and True Communities on the Internet?" (Chapter 6, "Groups and Organizations")
- "'On the Road Again': Older People and the Irrationality of Stereotypes" (Chapter 12, "Aging and Inequality Based on Age")
- "Islamophobia and the Media" (Chapter 17, "Religion")

Applying the Sociological Imagination to Social Policy

The Sociology and Social Policy boxes in selected chapters help students understand the connection between sociology and social policy issues in society. Here are a few of the topics in these interesting and informative boxes:

- "Computer Privacy in the Workplace" (Chapter 6, "Groups and Organizations")
- "The U.S. Supreme Court and the Continuing Debate Over Affirmative Action" (Chapter 10, "Race and Ethnicity")

- "The Electoral College: Is School Over for This Body?" (Chapter 14, "Politics and Government in Global Perspective")
- "The Ongoing Debate Over School Vouchers" (Chapter 16, "Education")

"You Can Make a Difference" Helps Get Students Involved in Each Chapter

The You Can Make a Difference boxes address the ways in which students can find out how the chapter theme affects their lives. For example:

- "One Person's Trash May Be Another Person's Treasure: Recycling for Good Causes" (Chapter 5) makes students aware of programs—some of which may be located on campus—that gather discarded items to benefit local charities.
- "Creating Small Communities of Our Own Within Large Organizations" (Chapter 6) passes on ideas for building microcommunities of informal friendships and small groups within larger organizations.
- "Global Networking to Reduce World Hunger and Poverty" (Chapter 8) and "Feeding the Hungry" (Chapter 9) address the issue of hunger and how everyday people can contribute in some small way to alleviating the suffering of others.
- "Creating Access to Information Technologies for Workers with Disabilities" (Chapter 13) describes how cost-efficient changes in the workplace can make it easier for some persons with a disability to be employed. Some students will eventually become owners or supervisors who are in a position to make decisions that provide greater access to meaningful employment for all people.
- "Recycling for Tomorrow" (Chapter 20) discusses how we can make a difference by reusing or recycling products and the packaging in which they come rather than creating more excess waste than already exists.

INNOVATIONS IN THE FIFTH EDITION

The fifth edition of *Sociology in Our Times* builds on the best from previous editions while providing students with new insights for their times.

A new **Writing in Sociology** feature has been added to each chapter. This exciting feature asks a critical-thinking essay question at the end of one of the boxes (e.g., Social Policy, Global Perspective, or Changing Times: Media and Technology) in each chapter. It calls on students to use the resources of their school library, the online resources offered with the text (including *InfoTrac College Edition, Micro-Case,* and the *Opposing Viewpoints Resource Center*), and their own writing skills to develop an answer. Students gain valuable skills in formulating and organizing a coherent response to the question. Thus, students learn how to write a research paper, learn how to use and evaluate online content, and develop useful skills that will assist them for the rest of their lives.

Census Profiles, a feature in most chapters, provide recent data from the U.S. Census as applied to the specific chapter in which they are located. Topics include "Consumer Spending" (Chapter 1), "Languages Spoken in U.S. Households" (Chapter 3), "Age of the U.S. Population" (Chapter 4), "Computer and Internet Access in U.S. Households" (Chapter 5), "Single Mothers with Children Under 18" (Chapter 11), and "Disability and Employment Status" (Chapter 18).

Chapter 1 ("The Sociological Perspective") provides a new and interesting approach for teaching sociological theory. Throughout the chapter, consumerism is used as an example of a research topic that sociologists pursue. The chapter reinforces the significance of theory in understanding a diversity of social issues, including how shopping, spending, and credit card debt have become a major problem for some people. Chapter 1 features more recent theoretical insights, including postmodernist, not typically found in other introductory texts. It also includes expanded coverage of George Herbert Mead's contributions to sociology. There is a new Census Profile on consumer spending and a new Media and Technology box examining online shopping and the issue of privacy.

Chapter 2 ("Sociological Research Methods") provides a chapter-length analysis of sociological research methods. This discussion of how sociologists do social research will provide students with new insights on how problems such as suicide can be viewed from a social, rather than a purely individualistic, perspective. The chapter features a new Changing Times: Media and Technology box, "Using the Internet for Research," that helps students understand how and why to avoid plagiarism in their work.

Chapter 3 ("Culture") has a new opening lived experience describing the day when Saddam Hussein's statue was pulled down in Baghdad and discusses the cultural significance of why many people were kicking the statue with their feet. This chapter includes a new Concept Table on analysis of culture and a new Changing Times: Media and Technology box that analyzes *American Idol* and the diffusion of popular culture.

In **Chapter 4** ("Socialization"), the opening lived experience features Drew Barrymore's discussion about her first encounter with her father. This discussion has been extremely popular with students and professors because it calls attention to the fact that although some people may grow up in a home where there are problems, it is still possible to overcome such concerns and live a productive life. A new Sociology in Global Perspective box, "Stay-at-Home Dads: Socialization for Kids and Resocialization for Parents," describes how more men are becoming actively involved in the socialization of their children.

The linkage of society, social structure, interaction, and the social problem of homelessness in **Chapter 5** ("Society, Social Structure, and Interaction"), which was so well-received in earlier editions, has been updated in the fifth edition. A section on "Societies, Technology, and Sociocultural Change" discusses types of societies (hunting and gathering, horticultural and pastoral, industrial, and postindustrial). A new You Can Make a Difference box, "One Person's Trash May Be Another Person's Treasure," describes why it is important to recycle for good causes.

Chapter 6 ("Groups and Organizations") has an interesting new opening lived experience dealing with the challenge that organizations in New York City faced as they worked to get back up and running after the 2001 terrorist attacks. This chapter applies the sociological imagination to an examination of groups, organizations, and privacy issues in contemporary societies. All of the boxes in the chapter are new: "How Much Do You Know About Privacy in Groups and Organizations?" (Sociology in Everyday Life), "Can We Form Social Groups and True Communities on the Internet?" (Changing Times: Media and Technology), "Computer Privacy in the Workplace" (Sociology and Social Policy), and "Creating Small Communities of Our Own Within Large Organizations" (You Can Make a Difference).

Chapter 7 ("Deviance and Crime") includes the latest available crime statistics, and an opening lived experience features an interview of a gang member by the sociologist Felix M. Padilla. There is a discussion of how terrorist attacks are a form of violent deviant behavior, and a revised section on the criminal justice system. A new Changing Times: Media and Technology box, "Child's Play? Extreme Violence and Video Games," asks students to think about the possible relationship between play and violent behavior.

Chapter 8 focuses on global stratification. This chapter examines wealth and poverty in global perspective, and it includes new opening lived experiences in which people in various African and Eastern European nations describe poverty as a pressing prob-

lem that pervades all aspects of their daily life. This chapter features a new Sociology in Global Perspective box that discusses "Poverty and the Curse of Civil War in Africa." A revised Changing Times: Media and Technology box ("A Wired World?") highlights the difference between the "haves" and the "have-nots" in regard to new technologies such as the Internet and the World Wide Web.

Chapter 9 focuses on the U.S. system of social stratification and takes into account the downturn in the U.S. economy in the early 2000s. A new opening lived experience highlights how three young men achieved what, in the eyes of many people, would be "the American Dream" when they achieved their education goals and were named among the forty most influential African Americans by *Essence* magazine.

Chapter 10 ("Race and Ethnicity") examines the relationship between racial–ethnic groups and sports throughout U.S. history. A revised Sociology and Social Policy box invites students to consider the continuing debate over affirmative action in colleges and universities. A new Sociology in Global Perspective box describes racism and anti-racism in European football.

Chapter 11 ("Sex and Gender") begins with an opening lived experience (about weight and appearance) in which a young woman describes her persistent problem with anorexia nervosa. The chapter distinguishes between sex—the biological dimension—and gender—the cultural dimension. The chapter features a new Sociology in Global Perspective box on "Oppression, Resistance, and the Women of Afghanistan" and a revised Media and Technology box on "*Red Jack: Revenge of the Brethren, Barbie Super Sports,* and Gendered Play."

Chapter 12 focuses on aging and problems associated with age. It features a section on "Age in Global Perspective" regarding how people are treated based on age in hunting and gathering, horticultural, pastoral and agrarian, industrial, and postindustrial societies. Two new boxes focus on issues relevant to the aging U.S. population. In the Media and Technology box, "'On the Road Again': Older People and the Irrationality of Stereotypes," the popularity of RVing is discussed. The Sociology and Social Policy box looks at "Driving While Elderly: Policies Pertaining to Age and Driving."

Chapter 13, regarding the economy and work, takes into account recent changes in the U.S. and global economies. New opening lived experiences focus on the loss of U.S. jobs due to labor practices such as off-shoring, near-shoring, and hiring temporary immigrant workers. A popular Media and Technology box has been revised to look at how the economy may influence the amount of traffic in a particular region, in this case California's Silicon Valley. The chapter also features a new discussion of blogging and privacy issues in the workplace.

Chapter 14 focuses on the effect that the intertwining of politics and the media has on the United States and other nations. A new opening lived experience demonstrates how the practice of embedding the media with U.S. troops during the War in Iraq turned some members of the media into "insiders" in government operations. This chapter has been revised to include up-to-date accounts of the wars in Afghanistan and Iraq. Two new boxes in this chapter are Sociology in Global Perspective, which examines "The European Union: Transcending National Borders and Governments," and You Can Make a Difference, which explains to students how they can keep an eye on the media.

Chapter 15 ("Families and Intimate Relationships") has opening lived experiences that call attention to the changing nature of families and how family members view their lives, particularly when a father explains how divorce and joint custody affected his son, Nick. Students are then able to hear from Nick (a number of years later) about how he thinks the joint custody arrangement influenced his life. All figures and tables have been updated with the latest available information on such topics as household composition. A new Sociology in Global Perspective box, "*Fukugan Shufu:* Changing Terminology and Family Life in Japan," shows how the changing role of women has influenced families in Japan.

Chapter 16 has been completely revised to focus on pressing issues in education. The chapter highlights ways in which race, class, and gender differentially affect people's access to education and its outcomes. New sections include "Bullying, Teasing, and Sexual Harassment"; "School Vouchers"; "Charter Schools and 'For Profit' Schools"; and "Homeschooling." A new Sociology and Social Policy box encourages students to think about "The Ongoing Debate Over School Vouchers."

Chapter 17 starts with a discussion of separation of church and state in the United States. It also contrasts the sociology of religion with theological perspectives on religion. It contains a balanced discussion of the world's religions, including how they originated, what their central teachings are, and what forms of social conflict have been found in each. In addition, the chapter examines the relationship between U.S. religion and social inequality in central-city and suburban churches. A new Changing Times: Media and Technology box, "Islamophobia and the Media," encourages students to analyze media images of people who some might consider to be "different" based on their religion or culture.

Chapter 18 ("Health, Health Care, and Disability") features an opening lived experience in which a first-year college student describes her experience with chronic illness. The Census Profile highlights disability and employment status, and new material discusses how the neighborhood in which a person lives might affect his or her chances of dying in the next year. Chapter 18 also features a unique discussion of the availability and costs of health care in rural areas of the United States.

Chapter 19 ("Population and Urbanization") looks at the issue of global migration as it affects population change. It opens with a lived experience from an immigrant worker in the United States and features a Sociology in Everyday Life box on migration. The chapter features new sections on "The Gated Community in the Capitalist Economy" and "Rural Community Issues in the United States." A new Concept Table has been added on perspectives on urbanism and the growth of cities.

The final chapter, **Chapter 20** ("Collective Behavior, Social Movements, and Social Change"), features an opening lived experience in which John Cronin and Robert F. Kennedy, Jr., discuss the importance of the Hudson Riverkeepers project. This chapter explains why sociologists study collective behavior and includes innovative sections on "Social Constructionist Theory: Frame Analysis" and "New Social Movement Theory." The chapter includes new "hot topics" such as blogging and the 2003 major power outage in the United States and Canada. A new Sociology in Global Perspective box discusses "'Flash Mobs': Collective Behavior in the Information Age" and shows how information technologies make it possible for organizations to bring together a large number of people in a relatively short period of time.

SUPPLEMENTS

Supplements for the Instructor

Instructor's Edition of *Sociology in Our Times,* Fifth Edition. An Instructor's Edition (IE) of this text containing several useful features for instructors is available. Found in the IE is the Resources Integration Guide, a 20-page chart that correlates the key instructor and student supplements by chapter. The IE also contains the Visual Preface, a walk-through of the several themes and many features of *Sociology in Our Times,* along with a complete listing of available bundles for this text. To obtain a copy of the Instructor's Edition, contact your Thomson sales representative.

Instructor's Resource Manual. This manual contains lecture outlines, chapter summaries, student learning objectives (grouped according to Bloom's taxonomy), key terms, discussion questions, lecture suggestions, student activities, video suggestions, Internet and *InfoTrac* exercises, and a list of suggested resources for instructors. Also included is a table of contents for the *CNN Today* Sociology Video Series and concise user guides for both *InfoTrac* and *WebTutor*.

Instructor's Resource Manual Notebook (with the ***Multimedia Manager*** CD-ROM). This enhanced instructor's manual provides instructors with the opportunity to integrate their personal notes into a three-hole punched and tabbed notebook containing a wealth of comprehensive teaching resources. The manual contains lecture outlines, chapter summaries, student learning objectives (grouped according to Bloom's taxonomy), key terms, discussion questions, lecture suggestions, student activities, video suggestions, Internet and *InfoTrac* exercises, a list of suggested resources for instructors, a table of contents for the *CNN Today* Sociology Video Series, and concise user guides for both *InfoTrac* and *WebTutor*. Also included is Jerry M. Lewis's helpful booklet *Tips for Teaching Introductory Sociology* as well as *Wadsworth's Sociology Online Resources and Writing Companion*, containing *InfoTrac*, *MicroCase*, and *Opposing Viewpoints Resource Center* exercises that can be assigned to students to help them build essential research and writing skills in sociology. The *Instructor's Resource Manual Notebook* is packaged with an all-new *Multimedia Manager* CD-ROM. This new instructor resource includes book-specific PowerPoint Lecture Slides, graphics from the book itself, the IRM Word documents, the Test Bank, CNN Video Clips, and links to many of Wadsworth's important sociology resources. All of your media teaching resources are in one place!

Test Bank. The *Test Bank* consists of 75–100 multiple-choice questions with question type indicated (fact, concept, or concept application) and 20–25 true/false questions per chapter, all with answers and page references. The *Test Bank* also includes 10–15 short-answer and 5–10 essay questions per chapter. This test bank is also available electronically on the *ExamView Computerized Testing CD-ROM* as well as on the *Multimedia Manager Resource CD-ROM* that comes packaged in the *Instructor's Resource Manual Notebook*.

ExamView Computerized Testing. Create, deliver, and customize tests and study guides (both print and online) in minutes with this easy-to-use assessment and tutorial system. *ExamView* offers both a Quick Test Wizard and an Online Test Wizard that guide you step-by-step through the process of creating tests. The test appears on screen exactly as it will print or display online. Using *ExamView's* complete word processing capabilities, you can enter an unlimited number of new questions or edit existing questions included with *ExamView*.

Wadsworth's Introduction to Sociology 2005 Transparency Acetates. A set of four-color acetates consisting of tables and figures from Wadsworth's introductory sociology texts is available to help prepare lecture presentations. Free to qualified adopters.

Videos. Adopters of *Sociology in Our Times* have several different video options available with the text.

Wadsworth's Lecture Launchers for Introductory Sociology. An exclusive offering jointly created by Wadsworth/Thomson Learning and DALLAS TeleLearning, this video contains a collection of video highlights taken from the *Exploring Society: An Introduction to Sociology* telecourse (formerly the *Sociological Imagination*). Each 3–6-minute video segment has been chosen to enhance and enliven class lectures and discussion of twenty key topics covered in any introductory sociology text. Accompanying the video is a brief written description of each clip, along with suggested discussion questions to help effectively incorporate the material into the classroom.

Sociology: Core Concepts. An exclusive offering jointly created by Wadsworth/Thomson Learning and DALLAS TeleLearning, this video contains a collection of video highlights taken from the *Exploring Society: An Introduction to Sociology* telecourse (formerly the *Sociological Imagination*). Each 15–20-minute video segment will enhance student learning of the essential concepts in the introductory course and can be used to initiate class lectures, discussion, and review. The video covers topics such as the sociological imagination, stratification, race and ethnic relations, and social change.

CNN® Today Sociology Video Series, Volumes I–VII. Illustrate the relevance of sociology to everyday life with this exclusive series of videos for the introduction to sociology course. Jointly created by Wadsworth and CNN, each video consists of approximately 45 minutes of footage originally broadcast on CNN and specifically selected to illustrate important sociological concepts. The videos are broken into short, 2- to 7-minute segments, perfect for use as lecture launchers or as illustrations of key sociological concepts. Each video includes an annotated table of contents, descriptions of the segments, and suggestions on their use within the course.

Wadsworth Sociology Video Library. Bring sociological concepts to life with videos from Wadsworth's Sociology Video Library, which includes thought-provoking offerings from Films for Humanities as well as from other excellent educational video sources. This extensive collection illustrates important sociological concepts covered in many sociology courses. Certain adoption conditions apply.

Supplements for the Student

Study Guide. Author Diana Kendall and Kathryn Sinast Mueller of Baylor University created this supplement to give students further opportunities to think sociologically and master course material. Each chapter of this student study tool includes a brief chapter outline, a chapter summary, learning objectives, key terms and people, a detailed chapter outline, student projects, and all new *InfoTrac College Edition* and Internet exercises. The *Study Guide* also includes practice tests consisting of 25–30 multiple-choice and 10–15 true/false questions, all with answers and page references, as well as five short-answer/essay questions for each chapter of the text.

Practice Tests. This booklet now contains more test questions, with 50–60 multiple-choice questions and 15–20 true/false questions for each chapter of the text (with page references) to help students test their knowledge of chapter concepts.

Wadsworth's Sociology Online Resources and Writing Companion for Kendall's *Sociology in Our Times,* Fifth Edition. This valuable guide shows students how they can use Wadsworth's exclusive online resources—*InfoTrac College Edition,* the *Opposing Viewpoints Resource Center (OVRC),* and *MicroCase Online*—to assist them in their study of sociology and build essential research and writing skills. Part One provides informative user guides that introduce each of these powerful research tools, while Part Two contains chapter-by-chapter directed exercises designed to develop research and critical thinking proficiency for each of the core topics in sociology. Part Three encourages students to "put it all together" and apply their newly acquired research skills to respond to the specific sociological questions posed in the "Writing in Sociology" feature in the Kendall text. Part Four then provides an overview of some of the research and writing tools available on-line, such as *InfoWrite* and the *OVRC Research Guide,* and shows students how they can effectively integrate their research findings into class assignments.

***SocCoach* CD-ROM for Kendall's *Sociology in Our Times,* Fifth Edition.** The new, interactive *SocCoach* CD-ROM is automatically packaged for free with each new copy of the text and can be located inside the back cover. This tutorially driven CD-ROM is firmly grounded in sociology. It enables students to review chapter content, conduct online research, think critically about sociology statistics, watch well-known sociologists discussing important concepts, and complete book-specific quizzes all on one easy-to-use CD-ROM! The new Study Plan feature prompts students to take a diagnostic chapter quiz, then generates a personalized study plan that shows students exactly what they need to review further. Students can then access study material for each concept, including material from the book itself, illustrative graphs, videos, and statistics that help students to better understand each concept.

Readers. Along with several other excellent introductory sociology readers (see the Instructor's Edition's Visual Preface), Wadsworth publishes the following reader especially matched to this text: *Classic Readings in Sociology,* Third Edition, edited by Eve Howard. This series of classic articles written by key sociologists will complement any introductory sociology textbook. This reader serves as a touchstone where students can read original works that teach the fundamental ideas of sociology.

Online Resources

Wadsworth's Virtual Society: The Wadsworth Sociology Resource Center (**http://www.wadsworth.com/sociology**). Here you will find a wealth of sociology resources such as Census 2000: A Student Guide for Sociology, Breaking News in Sociology, a Guide to Researching Sociology on the Internet, and Sociology in Action. Contained on the home page is the text-specific site for *Sociology in Our Times.*

Companion Web Site for *Sociology in Our Times,* Fifth Edition (http://sociology.wadsworth.com/kendall/times5e). On the book's companion site, you will find the Web research projects and Web interactive exercises identified by the icons in the text's margins. In addition, you will be able to access useful learning resources for each chapter of the book:

- Tutorial Practice Quizzes that can be scored and e-mailed to the instructor

- Internet Exercises and Web Links
- Video Exercises
- Periodical Exercises from *InfoTrac College Edition*
- Flashcards of the Text's Glossary
- Crossword Puzzles
- Essay Questions
- Learning Objectives
- *MicroCase Online* Data Exercises
- Virtual Explorations
- And much more!

WebTutor™ Advantage on WebCT and Blackboard.
This Web-based software for students and instructors takes a course beyond the classroom to an anywhere, anytime environment. Students gain access to a full array of study tools, including chapter outlines, chapter-specific quizzing material, interactive games and maps, and videos. With *WebTutor Advantage,* instructors can provide virtual office hours, post syllabi, track student progress with the quizzing material, and even customize the content to suit their needs. Instructors can also use the communication tools to do such things as set up threaded discussions and conduct "real-time" chats, as well as bring the latest developments from the field into the classroom using *NewsEdge,* an authoritative news source that delivers customized news feeds daily. "Out of the box" or customized, WebTutor Advantage provides powerful tools for instructors and students alike.

InfoTrac® College Edition.   With each purchase of a new copy of the text comes a free four-month passcode to *InfoTrac College Edition,* the online library that gives students anytime, anywhere access to reliable resources. This fully searchable database offers twenty years' worth of full-text articles from almost 5,000 diverse sources, such as academic journals, newsletters, and up-to-the-minute periodicals including *Time, Newsweek, Science, Forbes,* and *USA Today.* This incredible depth and breadth of material—available twenty-four hours a day from any computer with Internet access—makes conducting research so easy that your students will want to use it to enhance their work in every course! Through *InfoTrac's InfoWrite,* students now also have instant access to critical-thinking and paper-writing tools. Both adopters and their students receive unlimited access for four months.

Opposing Viewpoints Resource Center (OVRC).
Newly available from Wadsworth, this online center presents varying perspectives on today's most compelling issues. *OVRC* draws on Greenhaven Press's acclaimed Social Issues series as well as core reference content from other Gale and Macmillan Reference USA sources. The result is a dynamic online library of current event topics—the facts as well as the arguments of each topic's proponents and detractors. Special sections focus on critical thinking—walking students through the steps involved in critically evaluating point–counterpoint arguments—and researching and writing papers.

Online Chapters.   Two online chapters are available to augment coverage of two important topics in sociology today—school violence and the environment. Visit the text's companion web site for additional information on accessing these two online chapters and bundling a printed version of the chapters with the text.

Suburban Youth and School Violence explores the concept of youth culture and the various sociological and criminological theories that postulate why youth become deviant.

Environmental Sociology examines such topics as the development of environmental sociology, social theory and the environment, environmental issues and problems, and the future of the environment.

ACKNOWLEDGMENTS

Sociology in Our Times would not have been possible without the insightful critiques of these colleagues, who have reviewed some or all of this book. My profound thanks to each one for engaging in this time-consuming process:

Dennis L. R. Anderson
Butler County Community College

Kay Boston
Bossier Parish Community College

Christopher Bradley
Bowling Green State University

Cynthia Calhoun
Southwest Tennessee Community College

Evandro Camara
Emporia State University

Verna M. Cavey
Community College of Denver

Vanessa Eslinger-Brown
Strayer University

Jan Fiola
Minnesota State University–Moorhead

Jianjun Ji
University of Wisconsin–Eau Claire

Charles S. Koeber
Wichita State University

Dan Muhwezi
Butler County Community College

Luis Salinas
Houston Community College and
University of Houston

Leslie Stanley-Stevens
Tarleton State University

James Walter
Southern New Hampshire University

Previous Reviewers of *Sociology in Our Times*

Jan Abu-Shakrah, Portland Community College; Bonni Korn Ach, Chapman University; Ted Alleman, Penn State University; Ginna Babcock, University of Idaho; John Bandy, Texas Lutheran University; Kay Barches, Skyline College; Sampson Lee Blair, University of Oklahoma; Barbara Bollmann, Community College of Denver; Robert L. Boyd, Mississippi State University; Joni Boye-Beaman, Wayne State College; Valerie Brown, Case Western Reserve University; William D. Camp, Luzerne County Community College; Deborah Carter, Johnson C. Smith University; Ralph Cherry, Purdue University Calumet; Lillian Daughaday, Murray State University; Mary Davidson, Columbia–Greene Community College; Kevin Delaney, Temple University; Judith DiIorio, Indiana University–Purdue University; Kevin Early, Oakland University; Don Ecklund, Lincoln Land Community College; Julian Thomas Euell, Ithaca College; Grant Farr, Portland State University; Joe R. Feagin, University of Florida; Kathryn M. Feltey, University of Akron; William Finlay, University of Georgia; Mark Foster, Johnson County Community College; Richard Gale, University of Oregon; DeAnn K. Gauthier, University of Southwestern Louisiana; Michael Goslin, Tallahassee Community College; Anna Hall, Delgado Community College; Gary D. Hampe, University of Wyoming; Carole Hill, Shelton State Community College; Kinko Ito, University of Arkansas at Little Rock; Irwin Kantor, Middlesex County College; William Kelly, University of Texas, Austin; Bud Khleif, University of New Hampshire; Michael B. Kleiman, University of South Florida;

Janet Koenigsamen, College of St. Benedict; Monica Kumba, Buffalo State University; Yechiel Lahavy, Atlantic Community College; Dwight Landua, Southeastern Oklahoma State University; Abraham Levine, El Camino College; Diane Levy, University of North Carolina, Wilmington; Stephen Lilley, Sacred Heart University; Michael Lovaglia, University of Iowa; Richard J. Lundman, Ohio State University; Errol M. Magidson, Richard J. Daley College; Akbar Mahdi, Ohio Wesleyan University; M. Cathey Maze, Franklin University; Jane McCandless, State University of West Georgia; John Moland, Alabama State University; Martin Monto, University of Portland; David C. Moore, University of Nebraska at Omaha; Betty Morrow, Florida International University; Tina Mougouris, San Jacinto College; Hart M. Nelsen, Pennsylvania State University; Dale Parent, Southeastern Louisiana University; Ellen Rosengarten, Sinclair Community College; Phil Rutledge, University of North Carolina, Charlotte; Marcia Texler Segal, Indiana University Southeast; Dorothy Smith, State University of New York–Plattsburg; Jackie Stanfield, Northern Colorado University; K. Stewart-Cain, Trident Technical College; Mary Texeira, California State University–San Bernardino; Gary Tiedeman, Oregon State University; Steven Vassar, Minnesota State University, Mankato; Henry Walker, Cornell University; Susan Waller, University of Central Oklahoma; John Wilson, Duke University; Mary Lou Wylie, James Madison University; John Zipp, University of Wisconsin–Milwaukee

I deeply appreciate the energy, creativity, and dedication of the many people responsible for the development and production of *Sociology in Our Times*. I wish to thank Wadsworth Publishing Company's Susan Badger, Eve Howard, Bob Jucha, and Shelley Murphy for their enthusiasm and insights throughout the development of this text. Many other people worked hard on the production of the fifth edition of *Sociology in Our Times*, especially Greg Hubit and Donald Pharr. I am extremely grateful to them. My most profound thanks go to my husband, Terrence Kendall, who made invaluable contributions to *Sociology in Our Times*.

I invite you to send your comments and suggestions about this book to me in care of:

Wadsworth Publishing Company
10 Davis Drive
Belmont, CA 94002

ABOUT THE AUTHOR

Diana Kendall received a Ph.D. from the University of Texas at Austin, where she was invited to membership in Phi Kappa Phi Honor Society. Her areas of specialization and primary research interests are sociological theory, race/class/gender studies, and the sociology of medicine. In addition to *Sociology in Our Times*, she is the author of *The Power of Good Deeds: Privileged Women and the Social Reproduction of the Upper Class* (Rowman & Littlefield, 2002) and—along with Jane Lothian Murray and Rick Linden—of *Sociology in Our Times: Third Canadian Edition* (Nelson Thomson Learning, 2004). Her articles and presented papers primarily focus on the scholarship of teaching and on an examination of U.S. women of the upper classes across racial and ethnic groups.

Diana Kendall is currently a sociology professor at Baylor University, where she has taught a variety of courses, including Introduction to Sociology, Sociological Theory (undergraduate and graduate), Sociology of Medicine, and Race, Class, and Gender. Previously, she enjoyed many years of teaching sociology and serving as chair of the Social and Behavioral Science Division at Austin Community College.

Professor Kendall is actively involved in national and regional sociological associations, including the American Sociological Association, Sociologists for Women in Society, the Society for the Study of Social Problems, and the Southwestern Sociological Association.

Sociology in Our Times

CHAPTER 1

The Sociological Perspective

I got my first credit card, an American Express, a few weeks before the mall trip. . . . And let me confess up front that I have always loved shopping and buying things for myself, even when I couldn't afford them. *Especially* when I couldn't afford them. So my story of the mall trip is a story of pleasure, but also a story about denial—for I spent money I did not have, as usual, and I could only enjoy that while denying I was doing it. . . . I looked long and lovingly at a red and white plaid flannel dress—it would have been perfect with a white t-shirt and black docmartin boots. But it was $75.00, two weeks worth of groceries or several much-needed university press books. So I knew I could not buy any new clothes. [Instead,] I bought a video. . . .

Using my credit card was really easy. I just handed it over to the young man at the cash register, who had me sign a slip of paper. Then, the video was mine. . . . I think my urge to call this "losing my virginity" is just one more way to deny what actually happened to me. . . . Denial is a fairly predictable process—whatever it is that you repress to get your pleasure always comes back to haunt you. Credit card denial works like that: first you buy something for "free," then your monthly statement comes with an itemized list which tells you exactly what you'll have to pay for your pleasure. . . .

■ According to sociologists, our consumer society continues to grow as more people shop at home via the telephone and Internet.

L ike millions of college students in the United States and other high-income nations, Annalee quickly learned both the liberating and constraining aspects of living in a "consumer society" where credit cards play a crucial role in everyday life. Sociologists are interested in studying the *consumer society,* which refers to a society, such as ours, in which discretionary consumption is a mass phenomenon among people across diverse income categories. In the consumer society, purchasing goods and services is not in the exclusive province of the rich or even the middle classes; people in all but the lowest income categories may spend extensive amounts of time, energy, and money shopping, while amassing larger credit card debts in the process (see Baudrillard, 1998/1970; Ritzer, 1995; Schor, 1999). Did you know that recent surveys show that people in this country go to shopping centers more often than they go to church? Are you aware that U.S. teenagers spend more time at malls than anywhere besides school or home? (Twitchell, 1999).

According to sociologists, shopping and consumption—in this instance, the money that people spend on goods and services—are processes that extend beyond our individual choices and are rooted in larger structural conditions in the social, political, and economic order in which we live. Our ability to engage in twenty-four-hour-a-day consumption has been aided by the growth of "cathedrals of consumption," which include mega-shopping malls, Disney World-like theme parks, fast-food restaurants, home shopping television networks, cybermalls on the Internet, and the omnipresent credit card industry (see Baudrillard, 1998/1970; Gottdiener, 1997; Ritzer, 1999). This increase in consumption has also produced—even among people under the age of twenty-five—a dramatic increase in bankruptcy filings (Sullivan, Warren, and Westbrook, 2000).

Why have shopping, spending, credit card debt, and bankruptcy become major problems for some people? How are social relations and social meanings shaped by what people in a given society produce and how they consume? What national and worldwide social processes shape the production and consumption of goods, services, and information? In this chapter, we see how the sociological perspective helps us examine complex questions such as these, and we wrestle with some of the difficulties of attempting to study human behavior. Before reading on, take the quiz in Box 1.1, which lists a number of commonsense notions about shopping, consumption, and consumer credit.

QUESTIONS AND ISSUES

Chapter Focus Question: How does sociology add to our knowledge of human societies and of social issues such as consumerism?

What is the sociological imagination?

Why were early thinkers concerned with social order and stability?

Why were later social thinkers concerned with change?

What are the assumptions behind each of the contemporary theoretical perspectives?

PUTTING SOCIAL LIFE INTO PERSPECTIVE

Sociology **is the systematic study of human society and social interaction.** It is a *systematic* study because sociologists apply both theoretical perspectives and research methods (or orderly approaches) to examinations of social behavior. Sociologists study human societies and their social interactions to develop theories of how human behavior is shaped by group life and how, in turn, group life is affected by individuals.

Why Study Sociology?

Sociology helps us gain a better understanding of ourselves and our social world. It enables us to see how behavior is largely shaped by the groups to which we belong and the society in which we live.

Most of us take our social world for granted and view our lives in very personal terms. Because of our culture's emphasis on individualism, we often do not consider the complex connections between our own lives and the larger, recurring patterns of the society and world in which we live. Sociology helps us look beyond our personal experiences and gain insights into society and the larger world order. A *society* **is a large social grouping that shares the same geographical territory and is subject to the same political authority and dominant cultural expectations,** such as the United States, Mexico, or Nigeria. Examining the world order helps us understand that each of us is affected by *global interdependence*—a relationship in which the lives of all people are intertwined closely and any one nation's problems are part of a larger global problem.

Individuals can make use of sociology on a more personal level. Sociology enables us to move beyond established ways of thinking, thus allowing us to gain new insights into ourselves and to develop a greater awareness of the connection between our own "world" and that of other people. According to the sociologist Peter Berger (1963: 23), sociological inquiry helps us see that "things are not what they seem." Sociology provides new ways of approaching problems and making decisions in everyday life. Sociology promotes understanding and tolerance by enabling each of us to look beyond our personal experiences (see Figure 1.1 on page 7).

Many of us rely on intuition or common sense gained from personal experience to help us understand our daily lives and other people's behavior. *Commonsense knowledge* guides ordinary conduct in everyday life. We often rely on common sense—or "what everybody knows"—to answer key questions about behavior: Why do people behave the way they do? Who makes the rules? Why do some people break rules and other people follow rules?

Box 1.1 SOCIOLOGY AND EVERYDAY LIFE

How Much Do You Know About Consumption and Credit Cards?

True	False	
T	F	1. Less than half of all undergraduate students at four-year colleges have at least one credit card.
T	F	2. The average debt owed on undergraduate college students' credit cards is more than $2,000.
T	F	3. In the United States, it is illegal to offer incentives such as free T-shirts and Frisbees to encourage students to apply for credit cards.
T	F	4. Some consumer activist groups want Congress to adopt a measure requiring people under age 21 to get parental approval or show that they have sufficient income prior to obtaining a credit card.
T	F	5. More than one million people in this country file for bankruptcy each year.
T	F	6. Banks and credit card companies lose less than $500,000 on uncollected credit card debts each year.
T	F	7. If we added up everyone's credit card balances in the United States, we would find that the total amount owed is more than $555 billion.
T	F	8. More than 100,000 people under the age of 25 file for personal bankruptcy each year.
T	F	9. The sociology of consumption focuses primarily on how individuals determine what they are going to purchase and how much they will spend.
T	F	10. Overspending is primarily a problem for people in the higher-income brackets in the United States and other affluent nations.

Answers on page 6.

Many commonsense notions are actually myths. A *myth* is a popular but false notion that may be used, either intentionally or unintentionally, to perpetuate certain beliefs or "theories" even in the light of conclusive evidence to the contrary. For example, one widely held myth is that "money can buy happiness." By contrast, sociologists strive to use scientific standards, not popular myths or hearsay, in studying society and social interaction. They use systematic research techniques and are accountable to the scientific community for their methods and the presentation of their findings. Although some sociologists argue that sociology must be completely value free—without distorting subjective (personal or emotional) bias—others do not think that total objectivity is an attainable or desirable goal when studying human behavior. However, all sociologists attempt to discover patterns or commonalities in human behavior. For example, when they study shopping behavior or credit card abuse, sociologists look for recurring patterns of behavior and for larger, structural factors that contribute to people's behavior. Women's studies scholar Juliet B. Schor, who wrote *The Overspent American* (1999: 68), refers to consumption as the "see–want–borrow–buy" process, which she believes is a comparative process in which desire is structured by what we see around us. As sociologists examine patterns such as these, they begin to use the sociological imagination.

The Sociological Imagination

Sociologist C. Wright Mills (1959b) described sociological reasoning as the ***sociological imagination— the ability to see the relationship between individual experiences and the larger society.*** This awareness enables us to understand the link between our personal experiences and the social contexts in which they occur. The sociological imagination helps us distinguish between personal troubles and social

Box 1.1 SOCIOLOGY AND EVERYDAY LIFE

Answers to the Sociology Quiz on Consumption and Credit Cards

1. **False.** About 78 percent of college students have at least one credit card, and 32 percent have four or more cards.

2. **True.** The average debt on undergraduate college students' credit cards in 2001 was about $2,748.

3. **False.** Aggressive marketing of credit cards to college students and others is not illegal; the credit card industry routinely pays colleges and universities fees to rent tables for campus solicitations, and alumni groups offer "affinity" cards linked to the schools.

4. **True.** Organizations such as the Consumer Federation of America, the Consumers Union, and the U.S. Public Interest Research Group have pointed out the problem of credit card debt among college students and encouraged Congress to adopt a measure requiring age or income requirements on the issuance of credit cards.

5. **True.** About 1.5 million people in the United States file for bankruptcy each year.

6. **False.** Despite aggressive collection efforts, banks and credit card companies have to write off approximately $17 billion in uncollected credit card debts each year. However, they continue to make large profits on the interest (and sometimes annual fees) that are paid by those who do not default on their debt.

7. **True.** The total would be about $555 billion, with $450 billion of that amount being owed on general-use cards such as Visa, MasterCard, Discover, and American Express.

8. **True.** One study estimated, for example, that 120,000 people under the age of 25 filed for personal bankruptcy in the year 2000.

9. **False.** The sociology of consumption does not focus on individual decisions about purchasing products as much as it offers us a way to learn more about society as a whole, including how consumption affects people's social identities and the society in which we live.

10. **False.** Recent studies by social scientists in various disciplines have generally shown that people in middle- and lower-income brackets are also engaging in overspending, partially due to the availability of credit, aggressive advertising campaigns, and new means of consumption such as cybermalls and mega-shopping centers.

Sources: Based on Associated Press, 1999; Hoover, 2001; Schor, 1999; and Sullivan, Warren, and Westbrook, 2000.

(or public) issues. *Personal troubles* are private problems that affect individuals and the networks of people with which they associate regularly. As a result, those problems must be solved by individuals within their immediate social settings. For example, one person being unemployed or running up a high credit card debt could be identified as a personal trouble. *Public issues* are problems that affect large numbers of people and often require solutions at the societal level. Widespread unemployment and massive, nationwide consumer debt are examples of public issues. The sociological imagination helps us place seemingly personal troubles, such as losing one's job or overspending on credit cards, into a larger social context, where we can distinguish whether and how personal troubles may be related to public issues.

Overspending as a Personal Trouble Although the character of the individual can contribute to social problems, some individual experiences are

Figure 1.1 **Fields That Use Social Science Research**

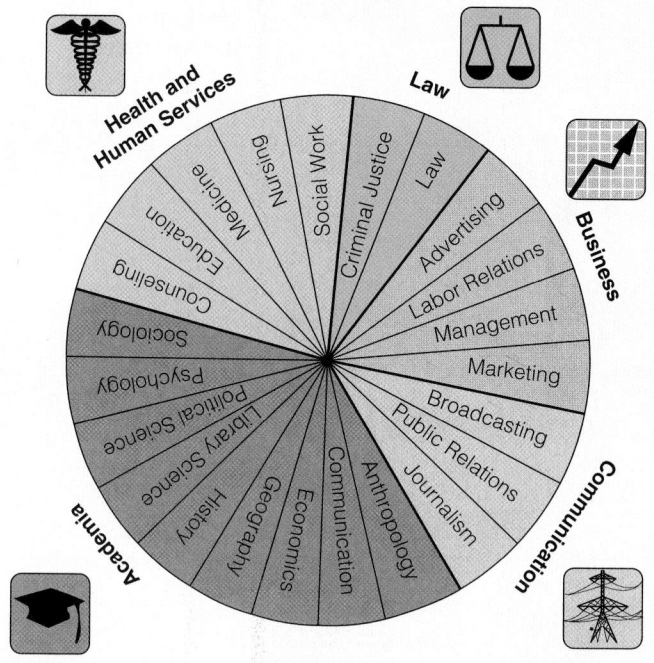

In many careers, including jobs in academia, business, communication, health and human services, and law, the ability to analyze social science research is an important asset.

Source: Based on Katzer, Cook, and Crouch, 1991.

largely beyond the individual's control. They are influenced and in some situations determined by the society as a whole—by its historical development and its organization. In everyday life, we often blame individuals for "creating" their own problems. If a person sinks into debt due to overspending or credit card abuse, many people consider it to be the result of his or her own personal failings. However, this approach overlooks debt among people who are in low income brackets, having no way other than debt to gain the basic necessities of life. By contrast, at middle- and upper-income levels, overspending takes on a variety of other meanings.

At the individual level, people may accumulate credit cards and spend more than they can afford, thereby affecting all aspects of their lives, including health, family relationships, and employment stability. Sociologist George Ritzer (1999: 29) suggests that people may overspend through a gradual process in which credit cards "lure people into consumption by easy credit and then entice them into still further consumption by offers of 'payment holidays,' new cards, and increased credit limits." A classic example of Ritzer's description is Chip H., who describes how he has had problems with overspending on credit cards

since his freshman year of college but still carries around seven of them:

> I was pretty good for about a year. Then I bought $200 speakers for my car and that put me over the top. I figured if I'm a little in the hole, then why not spend more. Then if you don't keep track, you end up just paying the interest, and you don't get anywhere. They get you, though. They bump you up another $500 or $600 because of your "outstanding credit." Then you charge more and go into deeper debt. It's like gambling, once you start you can't stop. (qtd. in McDonald, 1997: 4–5)

Chip, like millions of others, remains heavily in debt, recently transferring $5,500 of debt from his Master-Card and Visa accounts to an AT&T Universal card that offered him lower interest, at least initially (McDonald, 1997).

Overspending as a Public Issue We can use the sociological imagination to look at the problem of overspending and credit card debt as a public issue—a societal problem. For example, Ritzer (1998) suggests that the relationship between credit card debt

Is this a familiar-looking scene on your college campus? A major source of new credit card customers is college students.

© Lyntha Scott Eiler/Stock Boston

and the relatively low *savings rate* in the United States constitutes a public issue. Between 1990 and 2000, credit card debt tripled in the United States while savings diminished. Since savings is money that governments, businesses, and individuals can borrow for expansion, lack of savings may create problems for future economic growth. The increasing rate of bankruptcies in this country is a problem both for financial institutions and the government. As corporations "write off" bad debt from those who declare bankruptcy or simply do not pay their bills, all consumers pay either directly or indirectly for that debt. Finally, poverty is forgotten as a social issue when more-affluent people are having a spending holiday and consuming all, or more than, they can afford to purchase.

Some practices of the credit card industry are also a public issue (Ritzer, 1998). In a recent study of credit card use among college students, the sociologist Robert D. Manning (1999) found that students are aggressively targeted through marketing campaigns by credit card companies even though it is an accepted fact that some of the students will ruin their credit while still in college. Taking a walking tour of most large university campuses at registration time provides ample evidence of aggressive credit card marketing to students (and sometimes to faculty and

staff). Card offers are on or near campus, tucked into school newspapers, and distributed by area bookstores that sell texts and school supplies. The cards offer "identification" in the form of a picture of the cardholder set against a backdrop of a historical university building or the school mascot. Some offer airline frequent-flier points for routine purchases. Solicitation letters begin arriving in students' mailboxes early in the freshman year. As graduation time approaches, seniors are regaled with letters of "congratulations" from the credit card industry and one last appeal to get a "lifestyle" credit card that benefits the school's alumni association with a small contribution when purchases are charged on the card. The problem has grown so great that some in the credit card industry have acknowledged it is a problem. As one bank vice president stated, "I've seen customers who've had as many as 13 Visas and MasterCards. . . . Banks are guilty in that they make credit too easily available" (*ABA Banking Journal,* 1990: 42).

As these examples show, Mills's *The Sociological Imagination* (1959b) remains useful for examining issues in the twenty-first century because it helps integrate microlevel (individual and small group) troubles with compelling public issues of our day. Recently, his ideas have been applied at the global level as well.

THE IMPORTANCE OF A GLOBAL SOCIOLOGICAL IMAGINATION

Although existing sociological theory and research provide the foundation for sociological thinking, we must reach beyond past studies that have focused primarily on the United States to develop a more comprehensive *global* approach for the future. In the twenty-first century, we face important challenges in a rapidly changing nation and world. The world's ***high-income countries* are nations with highly industrialized economies; technologically advanced industrial, administrative, and service occupations; and relatively high levels of national and personal income.** Examples include the United States, Canada, Australia, New Zealand, Japan, and the countries of Western Europe.

As compared with other nations of the world, many high-income nations have a high standard of living and a lower death rate due to advances in nutrition and medical technology. However, everyone living in a so-called high-income country does not necessarily have a high income or an outstanding quality of life. Even among middle- and upper-income people, problems such as personal debt may threaten economic and social stability. In the economic boom of the late 1990s in the United States, the personal bankruptcy rate soared. For example, more than 1.5 million people in this country filed for bankruptcy in 2002, almost double the number of filings in 1990, and preliminary figures show an even larger number of bankruptcies being filed in 2003 (American Bankruptcy Institute, 2003). More than 97 percent of all U.S. bankruptcies were filed by consumers. Figure 1.2 shows the number of bankruptcies in the United States per year over the past decade.

In contrast, ***middle-income countries* are nations with industrializing economies, particularly in urban areas, and moderate levels of national and personal income.** Examples of middle-income countries include the nations of Eastern Europe and many Latin American countries, where nations such as Brazil and Mexico are industrializing rapidly. ***Low-income countries* are primarily agrarian nations with little industrialization and low levels of national and personal income.** Examples of low-income countries include many of the nations of Africa and Asia, particularly the

Figure 1.2	U.S. Bankruptcies, 1993–2002

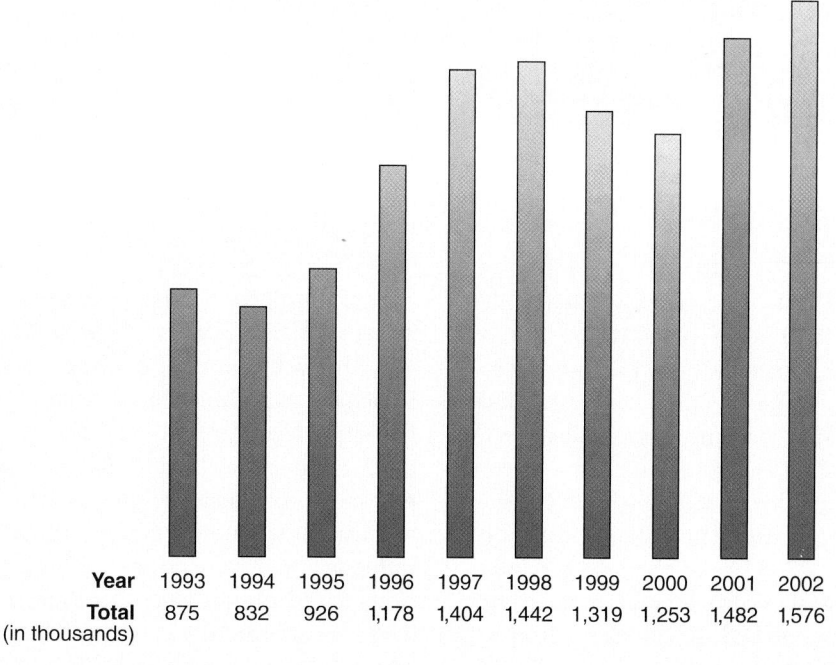

Year	1993	1994	1995	1996	1997	1998	1999	2000	2001	2002
Total (in thousands)	875	832	926	1,178	1,404	1,442	1,319	1,253	1,482	1,576

Source: Based on American Bankruptcy Institute, 2003.

Box 1.2 SOCIOLOGY IN GLOBAL PERSPECTIVE

Charge! The New Global Rallying Cry?

- Who would have guessed that the plump envelopes dropping through British mailboxes offering potential European customers the chance to attend the World Cup soccer match and get a low interest rate on a MasterCard had been issued by Capital One in Glen Allen, Virginia?
- Who would have guessed that China is MasterCard's second-largest market after the United States?

When you go to the college bookstore, open your mail, or read the newspaper, are you swamped by credit card issuers who want to make you one of their new credit card holders? If so, you are among the billions of people not only in the United States but also worldwide who are bombarded by advertisements from U.S.-based credit card companies. The global reach of these companies continues to expand in the twenty-first century. An example is the recently formed partnership of U.S. banking giant Citigroup with Shanghai Pudong Development Bank (a commercial bank in China) that enables Citigroup to enter China's credit card market (BBC, 2003).

What difference does it make if U.S. corporations enter the global credit card business? Sociologist George Ritzer suggests that credit cards, like Big Macs and Cokes, constitute U.S. inroads into the way of life of people in other nations. According to Ritzer, credit cards are a mechanism that aids our ability to consume. The explosive growth of mega-shopping malls, cybermalls, and other available means for con-

■ Consumption of U.S.-based brands such as Coca-Cola occurs on a worldwide basis. In Beijing, a man creates a sculpture of Paris's Arc de Triomphe with Coca-Cola cans.

suming goods and services has been fueled in part by credit cards because the cards lure customers into buying more than they otherwise might. As a result, consumers often go deeper into debt, find it difficult to extricate themselves from what they owe, and pay high interest rates on balances for years, if not decades.

People's Republic of China and India, where people typically work the land and are among the poorest in the world. However, generalizations are difficult to make because there are wide differences in income and standards of living within many nations (see Chapter 8, "Global Stratification").

Credit cards and other forms of global consumerism such as Disney amusement parks, fast-food restaurants, and U.S.-origin soft drinks are sold around the world. McDonald's is an example of the globalizing influence of U.S.-based megacorporations. Although McDonald's originated in 1955 as a small fast-food restaurant in Des Plaines, Illinois, the corporation now has an empire made up of franchises

around the globe, ranging from Rovaniemi, Finland, in the north to Invercargill, New Zealand, in the south, with the easternmost and westernmost McDonald's franchises straddling the international date line, in Gisborne, New Zealand, and Western Samoa, respectively (Grimes, 1999: 11). As these examples show, new means of consumption such as fast-food restaurants and international credit cards are largely a U.S. influence on other nations. Although such innovations have brought with them many benefits, they have also affected the everyday lives of many people around the world (see Box 1.2).

Throughout this text, we will continue to develop our sociological imaginations by examining social life

Not only in the United States but also in Britain and China, credit card companies are aggressively seeking new customers. For example, in the United States about 3.5 billion letters soliciting new credit card holders are sent out annually by credit card companies. However, as the potential for credit card growth has leveled off in the United States, U.S. lenders have begun to aim full-scale marketing campaigns toward people living in high-income nations such as Britain or toward the more affluent in middle- and low-income nations such as China, where a flourishing middle class has developed in recent years.

How might credit cards affect people's daily lives in other nations? First, revolving credit cards may influence how much money people spend. Many people in Britain have come to accept the idea of *revolving credit*—an agreement under which a person may make a minimum payment on the total balance each month, pay interest on the unpaid balance, and purchase additional items within a preestablished credit limit. However, in continental Europe and China, people primarily use *debit cards*—cardholders must deposit money into their accounts, and purchases are automatically deducted from these funds. In the past, most people in China did not even possess U.S.-style credit cards. Of China's 1.3 billion consumers, about 20 million today have credit cards, and many of these can be used only locally (BBC, 2003). However, some analysts believe that this will change as credit card companies aggressively market their products in the name of economic development and "modernization." Since China's emerging middle class already possesses Jeeps, Fords, BMWs, cell phones, and computers, it is likely that credit cards will contribute to their rapid accumulation of material things in the future.

Young people around the world are particularly viewed as new consumer markets, and they are continually bombarded with ads for clothes, makeup, music, and other products and services (BBC, 2002). Purchasing these items is much easier when credit cards are readily available, and the motivation is there for such purchases when high levels of consumption are associated with making a better life for themselves, as the ads often suggest.

While consumerism and credit card debt may be major problems for some people, this is not a universal issue for all of the world's peoples. Even as global capitalism grows and the U.S.-based credit card industry has new opportunities to attract cardholders worldwide, issues such as credit card ownership or interest rates are of no consequence for those people around the globe who still do not have adequate economic resources to acquire even the basic necessities of life.

Sources: Based on BBC, 2002, 2003; *The Economist*, 1999; Ritzer, 1995, 1999; and Rosenthal, 1998.

WRITING IN SOCIOLOGY ASSIGNMENT

What are the major arguments *for* and *against* sociologist George Ritzer's statement that credit cards, just like Big Macs and Cokes, con-  stitute U.S. inroads into the way of life of people in other nations?

in the United States and other nations. The future of our nation is deeply intertwined with the future of all other nations of the world on economic, political, environmental, and humanitarian levels. We buy many goods and services that were produced in other nations, and we sell much of what we produce to the people of other nations. Peace in other nations is important if we are to ensure peace within our borders. Famine, unrest, and brutality in other regions of the world must be of concern to people in the United States. Moreover, fires, earthquakes, famine, or environmental pollution in one nation typically has an adverse influence on other nations as well. Global problems contribute to the large influx of immigrants who arrive in the United States annually. These immigrants bring with them a rich diversity of language, customs, religions, and previous life experiences; they also contribute to dramatic population changes that will have a long-term effect on this country.

Throughout this book, we will look at data from the U.S. Census Bureau, which—among other things—reflect how the United States is changing. According to the Census Bureau, for example, more than 32 million (11.5 percent of the total population) of the people living in the United States in 2002 were not U.S. citizens at birth but now reside in this country (Schmidley, 2003). With the increasing diversity of the population, what issues do you think are most important for all of us to think about? Examples might include what educational and employment opportunities are available; how people from diverse social, cultural, and religious backgrounds will interact with one another; and what part social institutions will play in increasing harmony—or sometimes, unfortunately, producing greater discord—among people. For each of us, developing a better understanding of diversity and tolerance for people who are different from us is important for our personal, social, and economic well-being.

Whatever your race/ethnicity, class, sex, or age, are you able to include in your thinking the perspectives of people who are quite different from you in experiences and points of view? Before you answer this question, a few definitions are in order. *Race* is a term used by many people to specify groups of people distinguished by physical characteristics such as skin color; in fact, there are no "pure" racial types, and the concept of race is considered by most sociologists to be a social construction people use to justify existing social inequalities. *Ethnicity* refers to the cultural heritage or identity of a group and is based on factors such as language or country of origin. *Class* is the relative location of a person or group within the larger society, based on wealth, power, prestige, or other valued resources. *Sex* refers to the biological and anatomical differences between females and males. By contrast, *gender* refers to the meanings, beliefs, and practices associated with sex differences, referred to as *femininity* and *masculinity* (Scott, 1986: 1054).

In forming your own global sociological imagination and in seeing the possibilities for sociology in the twenty-first century, it will be helpful for you to understand the development of the discipline.

THE ORIGINS OF SOCIOLOGICAL THINKING

Throughout history, social philosophers and religious authorities have made countless observations about human behavior, but the first systematic analysis of society is found in the philosophies of early Greek philosophers such as Plato (c. 427–347 B.C.E.) and Aristotle (384–322 B.C.E.). For example, Aristotle was concerned with developing a system of knowledge, and he engaged in theorizing and the empirical analysis of data collected from people in Greek cities regarding their views about social life when ruled by

kings or aristocracies or when living in democracies (Collins, 1994). However, early thinkers such as Plato and Aristotle provided thoughts on what they believed society *ought* to be like, rather than describing how society actually *was*.

Social thought began to change rapidly in the seventeenth century with the scientific revolution. Like their predecessors in the natural sciences, social thinkers sought to develop a scientific understanding of social life, believing that their work might enable people to reach their full potential. The contributions of Isaac Newton (1642–1727) to modern science, including the discovery of the laws of gravity and motion and the development of calculus, inspired social thinkers to believe that similar advances could be made in systematically studying human behavior. As Newton advanced the cause of physics and the natural sciences, he was viewed by many as the model of a true scientist. Moreover, his belief that the universe is an orderly, self-regulating system strongly influenced the thinking of early social theorists.

Sociology and the Age of Enlightenment

The origins of sociological thinking as we know it today can be traced to the scientific revolution in the late seventeenth and mid-eighteenth centuries and to the Age of Enlightenment. In this period of European thought, emphasis was placed on the individual's possession of critical reasoning and experience. There was also widespread skepticism regarding the primacy of religion as a source of knowledge and heartfelt opposition to traditional authority. A basic assumption of the Enlightenment was that scientific laws had been designed with a view to human happiness and that the "invisible hand" of either Providence or the emerging economic system of capitalism would ensure that the individual's pursuit of enlightened self-interest would always be conducive to the welfare of society as a whole.

In France, the Enlightenment (also referred to as the *Age of Reason*) was dominated by a group of thinkers referred to collectively as the *philosophes*. The philosophes included such well-known intellectuals as Charles Montesquieu (1689–1755), Jean-Jacques Rousseau (1712–1778), and Jacques Turgot (1727–1781). They defined a *philosophe* as one who, trampling on prejudice, tradition, universal consent, and authority—in a word, all that enslaves most minds—dares to think for himself, to go back and search for the clearest general principles, and to admit nothing except on the testimony of his experience and reason (Kramnick, 1995). For the most part, these men were optimistic about the future, believing that human society could be improved through scientific discoveries. In this view, if people were free from the ignorance and superstition of the past, they could create new forms of political and economic organization such as democracy and capitalism, which would eventually produce wealth and destroy aristocracy and other oppressive forms of political leadership.

Although women were categorically excluded from much of public life in France because of the sexism of the day, some women strongly influenced the philosophes and their thinking through their participation in the *salon*—an open house held to stimulate discussion and intellectual debate. Salons provided a place for intellectuals and authors to discuss ideas and opinions and for women and men to engage in witty repartee regarding the issues of the day, but the "brotherhood" of philosophes typically viewed the women primarily as good listeners or mistresses more than as intellectual equals, even though the men sometimes later adopted the women's ideas as if they were their own. However, the writings of Mary Wollstonecraft (1759–1797) reflect the Enlightenment spirit, and her works have recently received recognition for influencing people's thoughts on the idea of human equality, particularly as it relates to social equality and women's right to education.

For women and men alike, the idea of observing how people lived in order to find out what they thought, and doing so in a systematic manner that could be verified, did not take hold until sweeping political and economic changes in the late eighteenth and early nineteenth centuries caused many people to realize that several of the answers provided by philosophers and theologians to some very pressing questions no longer seemed relevant. Many of these questions concerned the social upheaval brought above by the age of revolution, particularly the American Revolution of 1776 and the French Revolution of 1789, and the rapid industrialization and urbanization that occurred first in Britain, then in Western Europe, and later in the United States.

Sociology and the Age of Revolution, Industrialization, and Urbanization

Several types of revolution that took place in the eighteenth century had a profound influence on the origins of sociology. The Enlightenment produced an *intellectual revolution* in how people thought about social change, progress, and critical thinking. The optimistic views of the philosophes and other social thinkers regarding progress and equal opportunity (at least for some people) became part of the impetus for *political* and *economic revolutions,* first in America and then in France. The Enlightenment thinkers had emphasized a sense of common purpose and hope for human progress; the French Revolution and its aftermath replaced these ideals with discord and overt conflict (see Schama, 1989; Arendt, 1973).

During the nineteenth and early twentieth centuries, another form of revolution occurred: the *Industrial Revolution.* **Industrialization is the process by which societies are transformed from dependence on agriculture and handmade products to an emphasis on manufacturing and related industries.** This process first occurred during the Industrial Revolution in Britain between 1760 and 1850, and was soon repeated throughout Western Europe. By the mid-nineteenth century, industrialization was well under way in the United States. Massive economic, technological, and social changes occurred as machine technology and the factory system shifted the economic base of these nations from agriculture to manufacturing. A new social class of industrialists emerged in textiles, iron smelting, and related industries. Many people who had labored on the land were forced to leave their tightly knit rural communities and sacrifice well-defined social relationships to seek employment as factory workers in the emerging cities, which became the centers of industrial work.

Urbanization accompanied modernization and the rapid process of industrialization. **Urbanization is the process by which an increasing proportion of a population lives in cities rather than in rural areas.** Although cities existed long before the Industrial Revolution, the development of the factory system led to a rapid increase in both the number of cities and the size of their populations. People from very diverse backgrounds worked together in the same factory. At the same time, many people shifted from being *producers* to being *consumers.* For example, families living in the cities had to buy food with their wages because they could no longer grow their own crops to consume or to barter for other resources. Similarly, people had to pay rent for their lodging because they could no longer exchange their services for shelter.

These living and working conditions led to the development of new social problems: inadequate housing, crowding, unsanitary conditions, poverty, pollution, and crime. Wages were so low that entire families—including very young children—were forced to work,

As the Industrial Revolution swept through the United States in the nineteenth century, sights like this became increasingly common. The Remington Arms Works of Bridgeport, Connecticut, is symbolic of the factory system that shifted the base of the U.S. economy from agriculture to manufacturing. What new technologies are transforming the U.S. economy in the twenty-first century?

Hulton/Archive/Getty Images

often under hazardous conditions and with no job security. As these conditions became more visible, a new breed of social thinkers turned its attention to trying to understand why and how society was changing.

THE DEVELOPMENT OF MODERN SOCIOLOGY

At the same time that urban problems were growing worse, natural scientists had been using reason, or rational thinking, to discover the laws of physics and the movement of the planets. Social thinkers started to believe that by applying the methods developed by the natural sciences, they might discover the laws of human behavior and apply these laws to solve social problems. Historically, the time was ripe for such thoughts because the Age of Enlightenment had produced a belief in reason and humanity's ability to perfect itself.

Early Thinkers: A Concern with Social Order and Stability

Early social thinkers—such as Auguste Comte, Harriet Martineau, Herbert Spencer, and Emile Durkheim—were interested in analyzing social order and stability, and many of their ideas had a dramatic influence on modern sociology.

Auguste Comte The French philosopher Auguste Comte (1798–1857) coined the term *sociology* from the Latin *socius* ("social, being with others") and the Greek *logos* ("study of") to describe a new science that would engage in the study of society. Even though he never actually conducted sociological research, Comte is considered by some to be the "founder of sociology." Comte's theory that societies contain *social statics* (forces for social order and stability) and *social dynamics* (forces for conflict and change) continues to be used, although not in these exact terms, in contemporary sociology.

Drawing heavily on the ideas of his mentor, Count Henri de Saint-Simon, Comte stressed that the methods of the natural sciences should be applied to the objective study of society. Saint-Simon's primary interest in studying society was social reform, but Comte sought to unlock the secrets of society so that intellectuals like himself could become the new secular (as contrasted with religious) "high priests" of society (Nisbet, 1979). For Comte, the best policies involved order and authority. He envisioned that a new consensus would emerge on social issues and that the new science of sociology would play a significant part in the reorganization of society (Lenzer, 1998).

Comte's philosophy became known as ***positivism*—a belief that the world can best be understood through scientific inquiry.** Comte believed that objective, bias-free knowledge was attainable only through the use of science rather than religion. However, scientific knowledge was "relative knowledge," not absolute and final. Comte's positivism had two

The Granger Collection, New York

Auguste Comte

dimensions: (1) methodological—the application of scientific knowledge to both physical and social phenomena—and (2) social and political—the use of such knowledge to predict the likely results of different policies so that the best one could be chosen.

What did Comte actually mean by positivism? Some recent scholars have suggested that Comte's ideas regarding positivism are often misinterpreted. They believe he was arguing that we should limit what we consider to be "valid knowledge" to those testable statements that have proved to be true. In this view, Comte was not advocating that the social sciences should merely imitate the methods of the natural sciences; rather, he was developing a historical and differential theory of science based upon distinguishing levels of complexity (Heilbron, 1995).

The ideas of Saint-Simon and Comte regarding the objective, scientific study of society are deeply embedded in the discipline of sociology. Of particular importance is Comte's idea that the nature of human thinking and knowledge passed through several stages as societies evolved from simple to more complex. Comte described how the idea systems and their corresponding social structural arrangements changed in what he termed the *law of the three stages:* the theological, metaphysical, and scientific (or positivistic) stages. Comte believed that knowledge began in the *theological stage*—explanations were based on religion and the supernatural. Next, knowledge moved to the *metaphysical stage*—explanations were based on abstract philosophical speculation. Finally, knowledge would reach the *scientific* or *positive stage*—explanations are based on systematic observation, experimentation, comparison, and historical analysis. Shifts in the forms of knowledge in societies were linked to changes in the structural systems of society. In the theological stage, kinship was the most prominent unit of society; however, in the metaphysical stage, the state became the prominent unit, and control shifted from small groups to the state, military, and law. In the scientific or positive stage, industry became the prominent structural unit in society, and scientists become the spiritual leaders, replacing in importance the priests and philosophers of the previous stages of knowledge. For Comte, this progression through the three stages constituted the basic law of social dynamics, and, when coupled with the laws of statics (which emphasized social order and stability), the new science of sociology could bring about positive social change.

Social analysts have praised Comte for his advocacy of sociology and his contributions to positivism. His insights regarding linkages between the social structural elements of society (such as family, religion, and government) and social thinking in specific historical epochs were useful to later sociologists. However, a number of contemporary sociologists argue that Comte, among others, brought about an overemphasis on the "natural science model" that has been detrimental to sociology (Vaughan, Sjoberg, and Reynolds, 1993). Still others state that sociology, while claiming to be "scientific" and "objective," has focused on the experiences of a privileged few, to the exclusion by class, gender, race, ethnicity, and age of all others (Harding, 1986; Collins, 1990).

Harriet Martineau Comte's works were made more accessible for a wide variety of scholars through the efforts of the British sociologist Harriet Martineau (1802–1876). Until recently, Martineau received no recognition in the field of sociology, partly because she was a woman in a male-dominated discipline and society. Not only did she translate and condense Comte's work, but she was also an active sociologist in her own right. Martineau studied the social customs of Britain and the United States, and analyzed the consequences of industrialization and capitalism. In *Society in America* (1962/1837), she examined religion, politics, child rearing, slavery, and immigration in the United States, paying special attention to social distinctions based

Harriet Martineau

on class, race, and gender. Her works explore the status of women, children, and "sufferers" (persons who are considered to be criminal, mentally ill, handicapped, poor, or alcoholic).

Based on her reading of Mary Wollstonecraft's *A Vindication of the Rights of Women* (1974/1797), Martineau advocated racial and gender equality. She was also committed to creating a science of society that would be grounded in empirical observations and widely accessible to people. She argued that sociologists should be impartial in their assessment of society but that it is entirely appropriate to compare the existing state of society with the principles on which it was founded (Lengermann and Niebrugge-Brantley, 1998).

Recently, some scholars have argued that Martineau's place in the history of sociology should be as a founding member of this field of study, not just as the translator of Auguste Comte's work (Hoecker-Drysdale, 1992; Lengermann and Niebrugge-Brantley, 1998). Others have highlighted her influence in spreading the idea that societal progress could be brought about by the spread of democracy and the growth of industrial capitalism (Polanyi, 1944). Martineau believed that a better society would emerge if women and men were treated equally, enlightened reform occurred, and cooperation existed among people in all social classes (but led by the middle class).

In keeping with the sociological imagination, Martineau not only analyzed large-scale social structures in society, but she also explored how these factors influenced the lives of people, particularly women, children, and those who were marginalized by virtue of being criminal, mentally ill, disabled, poor, or alcoholic (Lengermann and Niebrugge-Brantley, 1998). She remained convinced that sociology, the "true science of human nature," could bring about new knowledge and understanding, enlarging people's capacity to create a just society and live heroic lives (Hoecker-Drysdale, 1992).

Herbert Spencer Unlike Comte, who was strongly influenced by the upheavals of the French Revolution, the British social theorist Herbert Spencer (1820–1903) was born in a more peaceful and optimistic period in his country's history. Spencer's major contribution to sociology was an evolutionary perspective on social order and social change. Although the term *evolution* has various meanings, evolutionary theory should be taken to mean "a theory to explain the mechanisms of organic/social change" (Haines, 1997: 81). According to Spencer's Theory of General Evolution, society, like a biological organism, has various interdependent parts (such as the family, the economy, and the government) that work to ensure the stability and survival of the entire society.

Spencer believed that societies developed through a process of "struggle" (for existence) and "fitness" (for survival), which he referred to as the "survival of the fittest." Because this phrase is often attributed to Charles Darwin, Spencer's view of society is known as *social Darwinism*—**the belief that those species of animals, including human beings, best adapted to their environment survive and prosper, whereas those poorly adapted die out.** Spencer equated this process of *natural selection* with progress, because only the "fittest" members of society would survive the competition, and the "unfit" would be filtered out of society. Based on this belief, he strongly opposed any social reform that might interfere with the natural selection process and, thus, damage society by favoring its least-worthy members.

Critics have suggested that many of his ideas contain serious flaws. For one thing, societies are not the same as biological systems; people are able to create and transform the environment in which they live. Moreover, the notion of the survival of the fittest can easily be used to justify class, racial–ethnic, and gender inequalities and to rationalize the lack of action to eliminate harmful practices that contribute to such inequalities. Not surprisingly, Spencer's "hands-off" view was applauded by many wealthy industrialists of his day. John D. Rockefeller, who gained monopolistic control of much of the U.S. oil industry early in the twentieth century, maintained that the growth of giant businesses was merely the "survival of the fittest" (Feagin and Feagin, 1997).

Social Darwinism served as a rationalization for some people's assertion of the superiority of the white race. After the Civil War, it was used to justify the repression and neglect of African Americans as well as the policies that resulted in the annihilation of Native American populations. Although some social reformers spoke out against these justifications, "scientific" racism continued to exist (Turner, Singleton, and Musick, 1984). In both positive and negative ways, many of Spencer's ideas and concepts have been deeply embedded in social thinking and public policy for over a century.

Emile Durkheim French sociologist Emile Durkheim (1858–1917) was an avowed critic of some of Spencer's views while incorporating others into his own writing. Durkheim stressed that people are the product of their social environment and that behavior cannot be fully understood in terms of *individual* biological and psychological traits. He believed that the limits of human potential are *socially* based, not *biologically* based. As Durkheim saw religious traditions evaporating in his society, he searched for a scientific, rational way to provide for societal integration and stability (Hadden, 1997).

In *The Rules of Sociological Method* (1964a/1895), Durkheim set forth one of his most important contributions to sociology: the idea that societies are built on social facts. **Social facts are patterned ways of acting, thinking, and feeling that exist *outside* any one individual but that exert social control over each person.** Durkheim believed that social facts must be explained by other social facts—by reference to the social structure rather than to individual attributes.

Durkheim was concerned with social order and social stability because he lived during the period of rapid social changes in Europe resulting from industrialization and urbanization. His recurring question was this: How do societies manage to hold together? In *The Division of Labor in Society* (1933/1893), Durkheim concluded that preindustrial societies were held together by strong traditions and by members' shared moral beliefs and values. As societies industrialized, more specialized economic activity became the basis of the social bond because people became interdependent on one another.

Emile Durkheim

Durkheim observed that rapid social change and a more specialized division of labor produce *strains* in society. These strains lead to a breakdown in traditional organization, values, and authority and to a dramatic increase in **anomie—a condition in which social control becomes ineffective as a result of the loss of shared values and of a sense of purpose in society.** According to Durkheim, anomie is most likely to occur during a period of rapid social change. In *Suicide* (1964b/1897), he explored the relationship between anomic social conditions and suicide, as discussed in Chapter 2.

Durkheim's contributions to sociology are so significant that he has been referred to as "*the* crucial figure in the development of sociology as an academic discipline [and as] one of the deepest roots of the sociological imagination" (Tiryakian, 1978: 187). He has long been viewed as a proponent of the scientific approach to examining social facts that lie outside individuals. He is also described as the founding figure of the functionalist theoretical tradition. Recently, scholars have acknowledged Durkheim's influence on contemporary social theory, including the structuralist and postmodernist schools of thought. Like Comte, Martineau, and Spencer, Durkheim emphasized that sociology should be a science based on observation and the systematic study of social facts rather than on individual characteristics or traits.

Although they acknowledge Durkheim's important contributions, some critics note that his emphasis on societal stability, or the "problem of order"—how society can establish and maintain social stability and cohesiveness—obscures the *subjective meaning* that individuals give to social phenomena such as religion, work, and suicide. In this view, overemphasis on *structure* and the determining power of "society" resulted in a corresponding neglect of *agency,* the beliefs and actions of the actors involved, in much of Durkheim's theorizing (Zeitlin, 1997).

Can Durkheim's ideas be applied to our ongoing analysis of credit cards? Durkheim was interested in examining the "social glue" that could hold contemporary societies together and provide people with a "sense of belonging." Ironically, the credit card industry has created what we might call a "pseudo-sense of belonging" through the creation of "affinity cards," like those designed to encourage members of an organization (such as a university alumni association) or people who share interests and activities (such as dog owners and skydiving enthusiasts) to possess a particular card. In later chapters, we examine Durkheim's theoretical contributions to diverse subjects ranging from suicide and deviance to education and religion.

Differing Views on the Status Quo: Stability Versus Change

Together with Karl Marx, Max Weber, and Georg Simmel, Durkheim established the course for modern sociology. We will look first at Marx's and Weber's divergent thoughts about conflict and social change in societies, and then at Georg Simmel's analysis of society.

Karl Marx In sharp contrast to Durkheim's focus on the stability of society, German economist and philosopher Karl Marx (1818–1883) stressed that history is a continuous clash between conflicting ideas and forces. He believed that conflict—especially class conflict—is necessary in order to produce social change and a better society. For Marx, the most important changes were economic. He concluded that the capitalist economic system was responsible for the overwhelming poverty that he observed in London at the beginning of the Industrial Revolution (Marx and Engels, 1967/1848).

In the Marxian framework, *class conflict* is the struggle between the capitalist class and the working

Karl Marx

class. The capitalist class, or *bourgeoisie,* comprises those who own and control the means of production—the tools, land, factories, and money for investment that form the economic basis of a society. The working class, or *proletariat,* is composed of those who must sell their labor because they have no other means to earn a livelihood. From Marx's viewpoint, the capitalist class controls and exploits the masses of struggling workers by paying less than the value of their labor. This exploitation results in workers' *alienation*—a feeling of powerlessness and estrangement from other people and from themselves. Marx predicted that the working class would become aware of its exploitation, overthrow the capitalists, and establish a free and classless society.

Can Marx provide useful insights on the means of consumption? Although Marx primarily analyzed the process of production, he linked production and consumption in his definition of *commodities* as products that workers produce. Marx believed that commodities have a use value and an exchange value. *Use value* refers to objects that people produce to meet their personal needs or the needs of those in their immediate surroundings. By contrast, *exchange value* refers to the value that a commodity has when it is exchanged for money in the open market. In turn, this money is used to acquire other use values, and the cycle continues. According to Marx, commodities play a central role in capitalism, but the workers who give value to the commodities eventually fail to see this fact. Marx coined

"I totally agree with you about capitalism, neo-colonialism, and globalization, but you really come down too hard on shopping."

the phrase the *fetishism of commodities* to describe the situation in which workers fail to recognize that their labor gives the commodity its value and instead come to believe that a commodity's value is based on the natural properties of the thing itself. By extending Marx's idea in this regard, we might conclude that the workers did not rebel against capitalism for several reasons: (1) they falsely believed that what capitalists did was in their own best interests as well, (2) they believed that the products they produced had a value in the marketplace that was independent of anything the workers did, and (3) they came to view ownership of the commodities as a desirable end in itself and to work longer hours so that they could afford to purchase more goods and services.

Although Marx's ideas on exploitation of workers cannot be fully developed into a theory of consumer exploitation, it has been argued that a form of exploitation does occur when capitalists "devote increasing attention to getting consumers to buy more goods and services" (Ritzer, 1995: 19). The primary ways by which capitalists can increase their profits are cutting costs and selling more products. To encourage continual increases in spending (and thus profits), capitalists have created mega-shopping malls, cable television shopping networks, and online shopping in order to provide consumers with greater opportunities to purchase more goods (see Box 1.3), thus increasing the consumers' credit card debt and forcing them to continue to work in order to pay their bills. Perhaps the ultimate agents of consumption are the industries that produce *desire* and make it possible for people to consume beyond their means. These include the credit card, advertising, and marketing industries, which encourage consumers to spend more money, in many cases far beyond their available cash, on goods and services (Ritzer, 1995, 1999).

Marx's theories provide a springboard for neo-Marxist analysts and other scholars to examine the economic, political, and social relations embedded in production and consumption in historical and contemporary societies. But what is Marx's place in the history of sociology? Marx is regarded as one of the most profound sociological thinkers, one who combined ideas derived from philosophy, history, and the social sciences into a new theoretical configuration. However, his social and economic analyses have also inspired heated debates among generations of social scientists. Central to his view was the belief that society should not just be studied but should also be changed, because the *status quo* (the existing state of society) involved the oppression of most of the population by a small group of wealthy people. Those who believe that sociology should be value free are uncomfortable with Marx's advocacy of what some perceive to be radical social change. Scholars who examine society through the lens of race, gender, and class believe that his analysis places too much emphasis on class relations, often to the exclusion of issues regarding race/ethnicity and gender. In regard to power differences between women and men, Marx believed that the primary form of oppression was rooted in *class divisions.* He justified his explanation by noting that gender and class divisions were not present in the earliest forms of human society. Instead, men first came into power over women when class divisions emerged. Then the institution of marriage established that women were a form of "private property" to be owned by men. According to Marx, women would become free from this bondage only when class divisions were overcome. Some scholars have criticized this notion, arguing that it does not fully reflect the nature of men's and women's power relations (see Hartmann, 1981). Marx also viewed power differentials across racial–ethnic lines as being primarily *class* divisions. Critics disagree with the notion that racial–ethnic inequalities can be reduced to class divisions, meaning that racism can be eradicated only after the class struggle has been won (Martin and Cohen, 1980).

In recent decades, scholars have shown renewed interest in Marx's *social theory,* as opposed to his radical ideology (see Postone, 1997; Lewis, 1998). Throughout this text, we will continue to explore Marx's various contributions to sociological thinking.

Box 1.3 CHANGING TIMES: MEDIA AND TECHNOLOGY

Online Shopping and Your Privacy: The Changing Nature of Social Life

Motorcycle jacket for kid brother on the Internet—$300

Monogrammed golf balls for dad on the Internet—$50

Vintage smoking robe for husband on the Internet—$80

Not having to hear "attention shoppers"—not even once—priceless.

The way to pay on the Internet and everywhere else you see the MasterCard logo: MasterCard.

—MasterCard advertisement (qtd. in Manning, 2000: 114)

This advertisement taps into a vital source of revenue for companies that issue credit cards: Online customers are an increasing percentage of those persons who use credit cards to make daily purchases. Some analysts estimate that online shopping generates more than $6 billion a year in revenues, and much of that $6 billion is based on credit (or debit) card purchases.

Earlier in this chapter, we mentioned that industrialization and urbanization were important historical factors that brought about significant changes in social life. Today, however, social life continues to change rapidly as computers and the Internet increasingly become an integral part of our daily lives, influencing how we communicate with others, how we go about seemingly mundane activities such as shopping, and how we view our privacy.

This raises some important sociological questions: Who is watching your online shopping activity? How far are companies willing to go in "snooping" on those who visit their Web sites? At the time of this writing, companies that sell products or services on the Internet are not required to respect the privacy of shop-

■ How many screens similar to this one have you seen on your computer? Do you care if outsiders watch your online shopping activities?

pers who order from their sites. According to the American Bar Association (2003), "This means the seller may collect data on which site pages you visit, which products you buy, when you buy them, and where you ship them. Then, the seller may share the information with other companies or sell it to them." Some Web sites have privacy policies posted but still insert "cookies" onto the hard drive of your computer. These cookies help the site's owner know where you go and what you do on the site. In some cases, the site owner records your e-mail address and begins sending you e-mail messages (known as "spam") about that company's products, whether you want to receive them or not.

Contrast this lack of privacy with the "good old days," when your grandparents or great-grandparents shopped at "mom and pop" stores where they knew the people they were buying from and often developed a level of trust with them. They thought that any information they provided the shop owners and clerks in face-to-face encounters was safe from the prying eyes of people who had no right to know about their business dealings. By contrast, in the twenty-first century, shopping—like other aspects of life—has become increasingly impersonal, and technology offers new opportunities for surveillance, particularly as millions of people now shop online and provide personal information and credit card data to faceless online retailers.

To offset people's fears of invasion of privacy or abuse of their credit card information, corporations in the online sales business have sought to reassure customers that they are not being tracked and that it is safe to give out personal information online. However, organizations such as the American Bar Association (ABA) business law section advise caution in Internet interactions. According to the ABA, consumers using a credit card for an online purchase should ask whether or not their credit card number will be kept on file by the seller for automatic use in future orders. Some people prefer not to have their credit card number kept on file (American Bar Association, 2003). Online shoppers should also find out what the seller's privacy policy is. It is especially important to know what information the seller is gathering about you, how the seller will use this information, and whether you can "opt out" of having this information gathered on you (American Bar Association, 2003).

Although these issues did not arise until recently, the foundation for analyzing some of them can be found in the ideas of some of the early social theorists discussed in this chapter. If these theorists were alive today, how might they address these issues? Karl Marx might relate contemporary Internet issues to the prevalence of capitalism and its exploitative nature. By contrast, Max Weber might view these privacy issues as problems associated with our acceptance of the impersonal nature of bureaucratic organizational structure. Georg Simmel might apply his study of money to consumerism on the Internet and demonstrate how people may get so carried away with the things that money can buy that they are willing to give up a great deal of themselves, including highly personal information, in order to acquire the products and services they so desire.

Contemporary sociologists might focus on how technology and consumerism have changed our way of life. For example, when we go to the mall, we are aware that we are potential customers. However, when we surf the Internet, we may be less aware that our home has become what sociologist George Ritzer refers to as a means of consumption:

> Traditional and new means of consumption have been battering so hard and so often at our doors that many of us simply have given up and welcomed them all in. Why not? Spread before our eyes and ears is a cornucopia of goods and services. All that is required is a telephone call, a keystroke, and a credit card number. In this way our homes have become means of consumption. It is one thing to be trapped at the mall, but quite another thing to be trapped at home. No matter how trapped one is at the mall, one must eventually leave. However, most people do not have the option of leaving a home that has become commercialized. (Ritzer, 1999: 150)

As Ritzer suggests, technology has narrowed the gap between the public and private aspects of our lives, and the choice remains for each of us to determine the extent to which we want to "share" our privacy with those who seek to sell us goods and services.

Can sociology enhance our knowledge of social life through a study of consumption? According to some social analysts, studying consumption not only offers us a good way to learn more about the larger society but also to find out more about its varying aspects, including how people live, what becomes a fad or fashion, and how consumption affects our social identities. For example, consider these questions: How are your shopping habits influenced by other people? By the college or university you attend? By the larger society in which you live?

Sources: Based on American Bar Association, 2003; Ritzer, 1999.

Max Weber German social scientist Max Weber (pronounced VAY-ber) (1864–1920) was also concerned about the changes brought about by the Industrial Revolution. Although he disagreed with Marx's idea that economics is *the* central force in social change, Weber acknowledged that economic interests are important in shaping human action. Even so, he thought that economic systems are heavily influenced by other factors in a society. As we will see in Chapter 17 ("Religion"), one of Weber's most important works, *The Protestant Ethic and the Spirit of Capitalism* (1976/1904–1905), evaluated the role of the Protestant Reformation in producing a social climate in which capitalism could exist and flourish.

Unlike many early analysts, who believed that values could not be separated from the research process, Weber emphasized that sociology should be *value free*—research should be conducted in a scientific manner and should exclude the researcher's personal values and economic interests (Turner, Beeghley, and Powers, 2002). However, Weber realized that social behavior cannot be analyzed by the objective criteria that we use to measure such things as temperature or weight. Although he recognized that sociologists cannot be totally value free, Weber stressed that they should employ *verstehen* (German for "understanding" or "insight") to gain the ability to see the world as others see it. In contemporary sociology, Weber's idea has been incorporated into the concept of the sociological imagination (discussed earlier in this chapter).

Weber was also concerned that large-scale organizations (bureaucracies) were becoming increasingly oriented toward routine administration and a specialized division of labor, which he believed were destructive to human vitality and freedom. According to Weber, rational bureaucracy, rather than class struggle, was the most significant factor in determining the social relations among people in industrial societies. In this view, bureaucratic domination can be used to maintain powerful (capitalist) interests in society. As discussed in Chapter 6 ("Groups and Organizations"), Weber's work on bureaucracy has had a far-reaching impact.

What might Weber's work contribute to a contemporary study of consumerism and the credit card industry? One of Weber's most useful concepts in this regard is *rationalization*—"the process by which the modern world has come to be increasingly dominated by structures devoted to efficiency, calculability, predictability, and technological control" (Ritzer, 1995: 21). According to Ritzer, the credit card industry has contributed to the rationalization process by the *effi-*

Max Weber

ciency with which it makes loans and deals with consumers. For example, prior to the introduction of credit cards, the process of obtaining a loan was slow and cumbersome. Today, the process of obtaining a credit card is highly efficient. It may take only minutes from the time a brief questionnaire is filled out until credit records are checked by computer and the application is approved or disapproved. *Calculability* is demonstrated by scorecards that allow lenders to score potential borrowers based on prior statistics of other people's performance in paying their bills. Factors that are typically calculated include home ownership versus renting, length of time with present employer, and current bank and/or credit card references (see Figure 1.3). Calculability is also reflected in such things as cardholders' monthly statements, which show when and where charges were made; available credit limits; and current interest rates.

The *predictability* of credit cards is easy to see. If the cardholder is current on paying bills and the merchant accepts that particular kind of card, the person knows that he or she will not be turned down on a purchase. Even the general appearance of the cards is highly predictable, as Ritzer (1998: 107) points out: "Whatever company issues them, they are likely to be made out of the same material, to feel the same, to be the same shape, to include similar information in similar places, and to do just about the same things." Finally, the use of *technological control* in the contemporary rationalization process is apparent in the credit card industry. These technologies range from the computerized system that determines whether or not

Figure 1.3	Typical Credit Report "Scorecard"

The more points you score, the more likely you are to receive credit.
Factors include the following: **Score**

☐ Other credit cards? (Too many cards may lose points; having no cards may lose even more points.)	
☐ Credit card payment history. (Delinquencies of 30 days may lose points; 60 days or more may lose even more points.)	
☐ Lawsuits and bankruptcies? (Having filed bankruptcy *really* loses points.)	
☐ Stability on the job and at place of residence. (Frequent changes loses points; being a homeowner adds points.)	
☐ Occupation and income. (Prestigious job and high income gain points.)	
TOTAL:	

Sources: Based on Ritzer, 1998; United College Marketing Services, 1997.

a new credit card will be issued to cards embedded with computer chips, ATM machines, and online systems that permit instantaneous transfers of funds. As Ritzer's application of the concept of rationalization to the credit card industry shows, many of Weber's ideas have served as the springboard for contemporary sociological theories and research.

Weber made significant contributions to modern sociology by emphasizing the goal of value-free inquiry and the necessity of understanding how others see the world. He also provided important insights on the process of rationalization, bureaucracy, religion, and many other topics. In his writings, Weber was more aware of women's issues than were many of the scholars of his day. Perhaps his awareness at least partially resulted from the fact that his wife, Marianne Weber, was an important figure in the women's movement in Germany in the early twentieth century (Roth, 1988).

Georg Simmel At about the same time that Durkheim was developing the field of sociology in France, the German sociologist Georg Simmel (pronounced ZIM-mel) (1858–1918) was theorizing about society as a web of patterned interactions among people. The main purpose of sociology, according to Simmel, should be to examine these social interaction processes within groups. In *The Sociology of Georg Simmel* (1950/1902–1917), he analyzed how social interactions vary depending on the size of the social group. He concluded that interaction patterns differed between a *dyad*, a social group with two members, and a *triad*, a social group with three members. He developed *formal sociology*, an approach that focuses attention on the universal recurring social forms that underlie the varying content of social interaction. Simmel referred to these forms as the "geometry of social life." He also distinguished between the *forms* of social interaction (such as cooperation or conflict) and the *content* of social interaction in different contexts (for example, between leaders and followers).

Like the other social thinkers of his day, Simmel analyzed the impact of industrialization and urbanization on people's lives. He concluded that class conflict was becoming more pronounced in modern industrial societies. He also linked the increase in individualism, as opposed to concern for the group, to the fact that people now had many cross-cutting "social spheres"—membership in a number of organizations and voluntary associations—rather than

According to the sociologist Georg Simmel, society is a web of patterned interactions among people. If we focus on the behavior of individuals in isolation, such as any one of the members of this women's rowing team, we may miss the underlying forms that make up the "geometry of social life."

Richard Pasley/Stock Boston

having the singular community ties of the past. Simmel also assessed the costs of "progress" on the upper-class city dweller, who, he believed, had to develop certain techniques to survive the overwhelming stimulation of the city. Simmel's ultimate concern was to protect the autonomy of the individual in society.

The Philosophy of Money (1990/1907), one of Simmel's most insightful studies, sheds light on the issue of consumerism. According to Simmel, money takes on a life of its own as people come to see money and the things that it can purchase as an end in themselves. Eventually, everything (and everybody) is seen as having a price, and people become blasé, losing the ability to differentiate between what is really of value and what is not. If money increases imprudence in consumption, credit cards afford even greater opportunities for people to spend money they do not have for things they do not need and, in the process, to sink deeper into debt (Ritzer, 1995). An example is Diane Curran, a teacher in Syracuse, New York, who accumulated $27,452 of debt on a dozen credit cards and other loans (Frank, 1999). Even when her monthly credit card payments equaled her take-home pay, she was still acquiring new credit cards, which she used to keep up with the other credit card payments. After Curran was forced to file for bankruptcy, she stated that "I wish somebody had cut me off 10 years earlier" (qtd. in Hays, 1996: B1, B6).

Simmel's perspective on money is only one of many possible examples of how his writings provide insights into social life. Simmel's contributions to sociology are significant. He wrote more than thirty books and numerous essays on diverse topics, leading some critics to state that his work was fragmentary and piecemeal. However, his thinking has influenced a wide array of sociologists, including the members of the "Chicago School" in the United States.

The Beginnings of Sociology in the United States

From Western Europe, sociology spread in the 1890s to the United States, where it thrived as a result of the intellectual climate and the rapid rate of social change. The first departments of sociology in the United States were located at the University of Chicago and at Atlanta University, then an African American school.

The Chicago School The first department of sociology in the United States was established at the University of Chicago, where the faculty was instrumental in starting the American Sociological Society (now known as the American Sociological Association). Robert E. Park (1864–1944), a member of the Chicago faculty, asserted that urbanization had a disintegrating influence on social life by producing an increase in the crime rate and in racial and class antagonisms that contributed to the segregation and isolation of neighborhoods (Ross, 1991). George Herbert Mead (1863–1931), another member of the faculty at Chicago, founded the symbolic interaction perspective, which is discussed later in this chapter. Mead made many significant contributions to sociology. Among these were his emphasis on the importance of studying the group ("the social") rather than starting with separate individuals. Mead also called our attention to the importance of shared communication among people based on language and gestures. As discussed in Chapter 4 ("Socialization"), Mead gave us important insights on how we develop our self-concept through interaction with those persons who are the most significant influences in our lives.

Jane Addams

Jane Addams (1860–1935) is one of the best-known early women sociologists in the United States, because she founded Hull House, one of the most famous settlement houses, in an impoverished area of Chicago. Throughout her career, she was actively engaged in sociological endeavors: She lectured at numerous colleges, was a charter member of the American Sociological Society, and published a number of articles and books. Addams was one of the authors of *Hull-House Maps and Papers,* a groundbreaking book that used a methodological technique employed by sociologists for the next forty years (Deegan, 1988). She was also awarded a Nobel Prize for her assistance to the underprivileged.

W. E. B. Du Bois and Atlanta University

The second department of sociology in the United States was founded by W. E. B. Du Bois (1868–1963) at Atlanta University. He created a laboratory of sociology, instituted a program of systematic research, founded and conducted regular sociological conferences on research, founded two journals, and established a record of valuable publications. His classic work, *The Philadelphia Negro: A Social Study* (1967/1899), was based on his research into Philadelphia's African American community and stressed the strengths and weaknesses of a community wrestling with overwhelming social problems. Du Bois was one of the first scholars to note that a dual heritage creates conflict for people of color. He called this duality *double-consciousness*—the identity conflict of being a black and an American. Du Bois pointed out that although people in this country espouse such values as democracy, freedom, and equality, they also accept racism and group discrimination. African Americans are the victims of these conflicting values and the actions that result from them (Benjamin, 1991).

UPI/Corbis

W. E. B. Du Bois

CONTEMPORARY THEORETICAL PERSPECTIVES

Given the many and varied ideas and trends that influenced the development of sociology, how do contemporary sociologists view society? Some see it as basically a stable and ongoing entity; others view it in terms of many groups competing for scarce resources; still others describe it based on the everyday, routine interactions among individuals. Each of these views represents a method of examining the same phenomena. Each is based on general ideas as to how social life is organized and represents an effort to link specific observations in a meaningful way. Each uses a *theory*—**a set of logically interrelated statements that attempts to describe, explain, and (occasionally) predict social events.** Each theory helps interpret reality in a distinct way by providing a framework in which observations may be logically ordered. Sociologists refer to this theoretical framework as a *perspective*—an overall approach to or viewpoint on some subject. Three major theoretical perspectives have emerged in sociology: the functionalist, conflict, and symbolic interactionist perspectives. Other perspectives, such as postmodernism, have emerged and gained acceptance among some social thinkers more recently. Before turning to the specifics of these perspectives, however, we should note that

Shopping malls are a reflection of a consumer society. A manifest function of a shopping mall is to sell goods and services to shoppers; however, a latent function may be to provide a communal area in which people can visit with friends and eat. For this reason, food courts have proven to be a boon in shopping malls around the globe.

Bob Daemmrich/The Image Works

some theorists and theories do not neatly fit into any of these perspectives.

Functionalist Perspectives

Also known as *functionalism* and *structural functionalism*, **functionalist perspectives are based on the assumption that society is a stable, orderly system.** This stable system is characterized by *societal consensus*, whereby the majority of members share a common set of values, beliefs, and behavioral expectations. According to this perspective, a society is composed of interrelated parts, each of which serves a function and (ideally) contributes to the overall stability of the society. Societies develop social structures, or institutions, that persist because they play a part in helping society survive. These institutions include the family, education, government, religion, and the economy. If anything adverse happens to one of these institutions or parts, all other parts are affected and the system no longer functions properly. As Durkheim noted, rapid social change and a more specialized division of labor produce *strains* in society that lead to a breakdown in these traditional institutions and may result in social problems such as an increase in crime and suicide rates.

Talcott Parsons and Robert Merton   Talcott Parsons (1902–1979), perhaps the most influential contemporary advocate of the functionalist perspective, stressed that all societies must provide for meet-

ing social needs in order to survive. Parsons (1955) suggested, for example, that a division of labor (distinct, specialized functions) between husband and wife is essential for family stability and social order. The husband/father performs the *instrumental tasks*, which involve leadership and decision-making responsibilities in the home and employment outside the home to support the family. The wife/mother is responsible for the *expressive tasks*, including housework, caring for the children, and providing emotional support for the entire family. Parsons believed that other institutions, including school, church, and government, must function to assist the family and that all institutions must work together to preserve the system over time (Parsons, 1955).

Functionalism was refined further by Robert K. Merton (1910–2003), who distinguished between manifest and latent functions of social institutions. **Manifest functions are intended and/or overtly recognized by the participants in a social unit.** In contrast, **latent functions are unintended functions that are hidden and remain unacknowledged by participants.** For example, a manifest function of education is the transmission of knowledge and skills from one generation to the next; a latent function is the establishment of social relations and networks. Merton noted that all features of a social system may not be functional at all times; *dysfunctions* are the undesirable consequences of any element of a society. A dysfunction of education in the United States is the perpetuation of gender, racial, and class inequalities. Such dysfunctions may threaten the capacity of a society to adapt and survive (Merton, 1968).

Applying a Functional Perspective to Shopping and Consumption

How might functionalists analyze shopping and consumption? In examining the part-to-whole relationships of contemporary society in high-income nations, it immediately becomes apparent that each social institution depends on the others for its well-being. For example, a booming economy benefits other social institutions, including the family (members are gainfully employed), religion (churches, mosques, synagogues, and temples receive larger contributions), and education (school taxes are higher when property values are higher). A strong economy also makes it possible for more people to purchase more goods and services. Due to the significance of the strength of the economy, the U.S. Census Bureau conducts surveys (for the Bureau of Labor Statistics) to determine how people are spending their money (see the "Census Profiles" feature). If people have "extra" money to spend and can afford leisure time away from work, they are more likely to dine out, take trips, and purchase things they might otherwise forgo.

Clearly, the manifest functions of shopping and consumption include purchasing necessary items such as food, clothing, household items, and sometimes transportation. In contemporary societies, purchasing entertainment and information is another function of shopping, both in actual stores and in virtual stores online. But what are the latent functions of shopping malls, for example? Many teens go to the mall to "hang out," visit with friends, maybe buy a T-shirt, and eat lunch at the food court. People of all ages go shopping for pleasure, relaxation, and perhaps to enhance their feelings of self-worth. ("If I buy this product, I'll look younger/beautiful/handsome/sexy, etc.!") As one scholar noted, "Shopping entails the joy of going into a safe spot filled with things to look at where [shoppers] are treated deferentially. Although no one has hooked up a turbo lie detector to a shopper out for fun, if they did, the machine would register increased arousal, heightened involvement, perceived freedom, and fantasy fulfillment" (Twitchell, 1999: 243).

However, shopping and consuming may also produce problems or dysfunctions. Some people are "shopaholics" or "credit card junkies" who cannot stop spending money; others are kleptomaniacs, who steal products rather than paying for them. In the end, however, the typical functionalist approach to

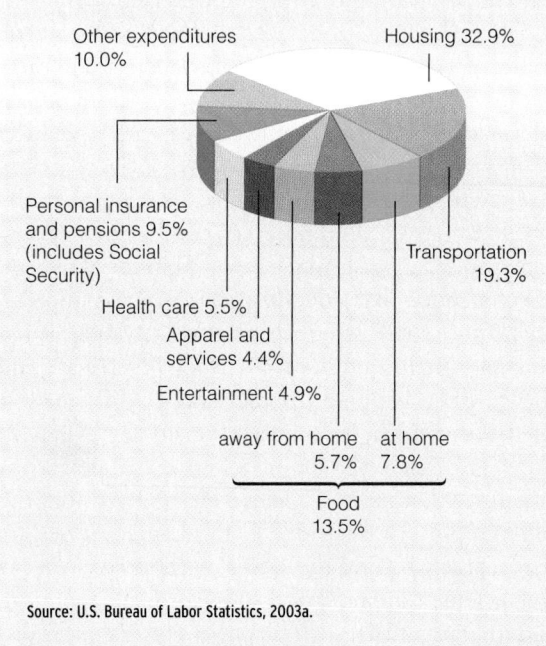

consumerism is shown in this comment by one scholar: "Let's face it, the idea that consumerism creates artificial desires rests on a wistful ignorance of history and human nature, on the hazy, romantic feeling that there existed some halcyon era of noble savages with purely natural needs. Once fed and sheltered, our needs have always been cultural, not natural. Until there is some other system to codify and satisfy those needs and yearnings, capitalism—and the culture it carries with it—will continue not just to thrive but to triumph" (Twitchell, 1999: 283).

Conflict Perspectives

According to **conflict perspectives, groups in society are engaged in a continuous power struggle for control of scarce resources.** Conflict may take the form of politics, litigation, negotiations, or family discussions about financial matters. Simmel, Marx, and Weber contributed significantly to this perspective by focusing on the inevitability of clashes between social groups. Today, advocates of the conflict perspective view social life as a continuous power struggle among competing social groups.

Max Weber and C. Wright Mills

As previously discussed, Karl Marx focused on the exploitation and oppression of the proletariat (the workers) by the bourgeoisie (the owners or capitalist class). Max Weber recognized the importance of economic conditions in producing inequality and conflict in society but added *power* and *prestige* as other sources of inequality. Weber (1968/1922) defined *power* as the ability of a person within a social relationship to carry out his or her own will despite resistance from others, and *prestige* as a positive or negative social estimation of honor (Weber, 1968/1922).

C. Wright Mills (1916–1962), a key figure in the development of contemporary conflict theory, encouraged sociologists to get involved in social reform. He contended that value-free sociology was impossible because social scientists must make value-related choices—including the topics they investigate and the theoretical approaches they adopt. Mills encouraged everyone to look beneath everyday events in order to observe the major resource and power inequalities that exist in society. He believed that the most important decisions in the United States are made largely behind the scenes by the *power elite*—a small clique composed of the top corporate, political, and military officials. Mills's power elite theory is discussed in Chapter 14 ("Politics and Government in Global Perspective").

As one of the wealthiest and most-beloved entertainers in the world, Oprah Winfrey is an example of Max Weber's concept of *prestige*—a positive estimate of honor.

The conflict perspective is not one unified theory but rather encompasses several branches. One branch is the neo-Marxist approach, which views struggle between the classes as inevitable and as a prime source of social change. A second branch focuses on racial–ethnic inequalities and the continued exploitation of members of some racial–ethnic groups. A third branch is the feminist perspective, which focuses on gender issues (Feagin and Feagin, 1997).

The Feminist Approach

A feminist approach (or "feminism") directs attention to women's experiences and the importance of gender as an element of social structure. This approach is based on the belief that "women and men are equal and should be equally valued as well as have equal rights" (Basow, 1992). According to feminists (including many men as well as women), we live in a patriarchy, a system in which men dominate women and in which things that are considered to be "male" or "masculine" are more highly valued than those considered to be "female" or "feminine." The feminist perspective assumes that gender is socially created, rather than determined by one's biological inheritance, and that change is essential in order for people to achieve their human potential without limits based on gender. It also assumes that society reinforces social expectations through social learning, which is acquired through social institutions such as education, religion, and the political and economic structure of society. Some feminists argue that women's subordination can end only after the patriarchal system becomes obsolete. Note, however, that feminism is not one single, unified approach. Rather, there are several feminist perspectives, which are discussed in Chapter 11 ("Sex and Gender").

Applying Conflict Perspectives to Shopping and Consumption

How might advocates of a conflict approach analyze the process of shopping and consumption? A contemporary conflict analysis of consumption might look at how inequalities based on racism, sexism, and income differentials affect people's ability to acquire the things they need and want. It might also look at inequalities regarding the issuance of credit cards and access to "cathedrals of consumption" such as mega-shopping malls and tourist resorts (see Ritzer, 1999: 197–214). However, one of the earliest social theorists to discuss the relationship between social class and consumption patterns was the U.S. social scientist Thorstein Veblen (1857–1929). In *The Theory of the Leisure Class* (1967/1899), Veblen described early wealthy U.S. industrialists as engaging in *conspicuous consumption*— the continuous public display of one's wealth and status through purchases such as expensive houses, clothing, motor vehicles, and other consumer goods. According to Veblen, the leisurely lifestyle of the upper classes typically does not provide them with adequate opportunities to show off their wealth and status. In order to attract public admiration, the wealthy often engage in consumption and leisure activities that are both highly visible and highly wasteful. Examples of conspicuous consumption range from Cornelius Vanderbilt's 8 lavish mansions (including one with 137 rooms) and 10 major summer estates in the Gilded Age (about 1890 to the beginning of World War I) to the 2,400 pairs of shoes owned by Imelda Marcos, wife of the late President Ferdinand Marcos of the Philippines (Frank, 1999; Twitchell, 1999). However, as Ritzer (1999) points out, some of today's wealthiest people engage in

inconspicuous consumption, perhaps to maintain a low public profile or out of fear for their own safety.

According to contemporary social analysts, conspicuous consumption has become more widely acceptable at all income levels, and some middle- and lower-income individuals and families now use as their frame of reference the lifestyles of the more affluent in their communities. As a result, many families live on credit in order to purchase the goods and services that they would like to have or that keep them on the competitive edge with their friends, neighbors, and co-workers (Schor, 1999). However, others may decide not to overspend, instead seeking to make changes in their lives and encouraging others to do likewise (see Box 1.4).

Living in a society that overemphasizes consumption is particularly difficult for people in low-income categories, as women's studies scholar Juliet B. Schor (1999: 39) states: "For many low-income individuals, the lure of consumerism is hard to resist. When the money isn't there, however, feelings of deprivation, personal failure, and deep psychic pain result. In a culture where consuming means so much, not having money is a profound social disability. For parents, faced with the desires of their children, the failure can feel overwhelming." According to conflict theorists, the economic gains of the upper classes are often at the expense of those in the lower classes who may have had to struggle (sometimes unsuccessfully) to have adequate food, clothing, and shelter for themselves and their children. Chapter 8 ("Global Stratification") and Chapter 9 ("Social Class in the United States") discuss contemporary conflict perspectives on class-based inequalities.

Symbolic Interactionist Perspectives

The conflict and functionalist perspectives have been criticized for focusing primarily on macrolevel analysis. A **macrolevel analysis examines whole societies, large-scale social structures, and social systems** instead of looking at important social dynamics in individuals' lives. Our third perspective, symbolic interactionism, fills this void by examining people's day-to-day interactions and their behavior in groups. Thus, symbolic interactionist approaches are based on a **microlevel analysis, which focuses on small groups rather than on large-scale social structures.**

We can trace the origins of this perspective to the Chicago school, especially George Herbert Mead and the sociologist Herbert Bloomer (1900–1986), who is credited with coining the term *symbolic interactionism.* According to **symbolic interactionist perspectives, society is the sum of the interactions of**

Box 1.4 YOU CAN MAKE A DIFFERENCE

Taking a Stand Against Overspending

I never like what's already in my closet. I only like what I'm about to buy.... I'm thinking of converting my guest bedroom into one big closet. Who needs friends when you have clothes?

—actor Jennifer Tilly (quoted in *People,* 1999: 106-107)

My little girl had a friend visit her who was really into Guess jeans. My nine-year-old didn't even know what Guess jeans were. Well, after that kid left, that was all she talked about. She had to have a pair of Guess jeans.

—a mother (quoted in Schor, 1999: 68)

People's consumer desires and the urge to overspend are often fueled by others. Shopping and spending are routinely glorified in the media. For example, a *People* magazine cover story touted "Hollywood's Super Shoppers," including celebrities such as Jennifer Tilly (quoted above), Arnold Schwarzenegger, and Charles Oakley, who like to "shop till they drop." Although celebrities are frequently used to "pitch" products in commercials, our family members and friends (such as the nine-year-old whose friend "sold" her on the Guess jeans) can be even stronger advocates for certain consumer goods and services.

Is it possible to take a stand against overconsumption and overspending? Recently, some people have been choosing *voluntary simplicity* or *downshifting* in an effort to simplify their lives, including reconsidering where they work, what they purchase, and how much they spend. Downshifting has been defined as

Jim McHugh/Corbis/Outline

■ During a visit to the Prada boutique in Beverly Hills, film star Jennifer Tilly admits that she views shopping as a mission. Does the behavior of celebrities such as Tilly encourage consumerism in all of us?

"the practice of replacing busyness with business. It is the deliberate act of opting out of conspicuous consumption and possession of possessions which end up

---

individuals and groups. Theorists using this perspective focus on the process of *interaction*—defined as immediate reciprocally oriented communication between two or more people—and the part that *symbols* play in giving meaning to human communication. A *symbol* is anything that meaningfully represents something else. Examples of symbols include signs, gestures, written language, and shared values. Symbolic interaction occurs when people communicate through the use of symbols; for example, a gift of food—a cake or a casserole—to a newcomer in a neighborhood is a symbol of welcome and friendship. But symbolic communication occurs in a variety of forms, including facial gestures, posture, tone of voice,

and other symbolic gestures (such as a handshake or a clenched fist).

Symbols are instrumental in helping people derive meanings from social situations. In social encounters, each person's interpretation or definition of a given situation becomes a *subjective reality* from that person's viewpoint. We often assume that what we consider to be "reality" is shared by others; however, this assumption is often incorrect. Subjective reality is acquired and shared through agreed-upon symbols, especially language. If a person shouts "Fire!" in a crowded movie theater, for example, that language produces the same response (attempting to escape) in all of those who hear and understand it. When people

possessing you, and rejection of non-job sacrifices such as missing reading the children a story at night because of the pursuit of job promotions in the career rat race" (Buckingham, 1999). Many reassessments of life go into such decisions, including seeking a less stressful lifestyle, wanting more time to spend with family and friends, and having concern about the environment. However, most "voluntary downshifters" also decide that they want to reduce the "clutter" (an excess of material possessions) in their lives and seek ways to decide to forgo excessive consumerism. According to the economist Juliet Schor (1999), there are several things that downshifters and others can do to avoid overspending, including (1) *controlling desire* by gaining knowledge of the process of consumption and its effect on people; (2) *helping make exclusivity uncool* by demystifying the belief that people are "better" for having purchased extraordinarily expensive items; (3) *exercising voluntary restraints on competitive consumption* by encouraging friends and acquaintances to decelerate spending on presents and other purchases; (4) *learning to share,* particularly expensive items such as a lawn mower or a boat; (5) *becoming an educated consumer;* and (6) *avoiding the use of shopping as a form of therapy.* Since overspending is often linked with credit card use, other social analysts suggest that a person should have no more than one credit card and that the entire balance on the card should be paid off each month.

Can an individual make a difference using these suggestions? Certainly, these ideas may change an individual's shopping and overspending habits. However, if we apply C. Wright Mills's (1959b) *sociological imagination,* including the distinction between personal troubles and public issues (as discussed in this chapter), we see that these suggestions focus exclusively on what *individuals* can do to change their own behavior. As a result, this approach may be somewhat useful but may still overlook the larger structural factors that contribute to people's overspending. According to the sociologist George Ritzer (1999), individual actions in this regard are likely to fail as long as there are no changes in the larger society, particularly in the cathedrals of consumption, advertisers, credit card companies, and other businesses that have a vested interest in promoting hyperconsumption. Some organizations suggest that the only way to reduce overconsumption and credit card debt is through activism, such as getting the age limit raised at which people can be issued their first credit card. In this view, those who want to make a difference could also become involved in advocating social change. If you would like to know more about the simplicity or downshifting movements, here are several organizations to contact:

- Center for a New American Dream, 6930 Carroll Avenue, Suite 900, Takoma Park, MD 20912.
- The Media Foundation, 1243 W. 7th Avenue, Vancouver, BC V6H 1B7, Canada. Online:

http://www.adbusters.org

- Co-op America, 1612 K Street, Washington, DC 20006. Online:

http://www.coopamerica.org

- Living Lightly on the Earth:

http://www.scn.org/earth/lightly

Sources: Based on Buckingham, 1999; *People,* 1999; Ritzer, 1999; and Schor, 1999.

in a group do not share the same meaning for a given symbol, however, confusion results; for example, people who did not know the meaning of the word *fire* would not know what the commotion was about. How people *interpret* the messages they receive and the situations they encounter becomes their subjective reality and may strongly influence their behavior.

Symbolic interactionists attempt to study how people make sense of their life situations and the way they go about their activities, in conjunction with others, on a day-to-day basis (Prus, 1996). How do people develop the capacity to think and act in socially prescribed ways? According to symbolic interactionists, our thoughts and behavior are shaped by our social interactions with others. Early theorists such as Charles H. Cooley and George Herbert Mead explored how individual personalities are developed from social experience and concluded that we would not have an identity, a "self," without communication with other people. This idea is developed in Cooley's notion of the "looking-glass self" and Mead's "generalized other," as discussed in Chapter 4 ("Socialization"). From this perspective, the attainment of language is essential not only for the development of a "self" but also for establishing common understandings about social life.

How do symbolic interactionists view social organization and the larger society? According to symbolic

Sporting events are a prime location for seeing how college students use symbols to convey shared meanings. From the colors of clothing to hand gestures, students show pride in their school.

© Bill Aron/PhotoEdit

interactionists, social organization and society are possible only through people's everyday interactions. In other words, group life takes its shape as people interact with one another (Blumer, 1986/1969). Although macrolevel factors such as economic and political institutions constrain and define the forms of interaction that we have with others, the social world is dynamic and always changing. Chapter 5 ("Society, Social Structure, and Interaction") explores two similar approaches—rational choice and exchange theories—that focus specifically on how people rationally try to get what they need by exchanging valued resources with others.

As we attempt to present ourselves to others in a particular way, we engage in behavior that the sociologist Erving Goffman (1959) referred to as "impression management." Chapter 5 also presents some of Goffman's ideas, including *dramaturgical analysis,* which envisions that individuals go through their life somewhat like actors performing on a stage, playing out their roles before other people. Symbolic interac-

tionism involves both a theoretical perspective and specific research methods, such as observation, participant observation, and interviews, that focus on the individual and small group behavior (see Chapter 2, "Sociological Research Methods").

Applying Symbolic Interactionist Perspectives to Shopping and Consumption Sociologists applying a symbolic interactionist framework to the study of shopping and consumption would primarily focus on a microlevel analysis of people's face-to-face interactions and the roles that people play in society. In our efforts to interact with others, we define any situation according to our own subjective reality. This theoretical viewpoint applies to shopping and consumption just as it does to other types of conduct. For example, when a customer goes into a store to make a purchase and offers a credit card to the cashier, what meanings are embedded in the interaction process that takes place between the two of them? The roles that the two people play are based on their histories of interaction in previous situations. They bring to the present encounter symbolically charged ideas, based on previous experiences. Each person also has a certain level of emotional energy available for each interaction. When we are feeling positive, we have a high level of emotional energy, and the opposite is also true. Each time we engage in a new interaction, the situation has to be negotiated all over again, and the outcome cannot be known beforehand (Collins, 1987).

In the case of the shopper–cashier interaction, how successful will the interaction be for each of them? The answer to this question depends on a kind of social marketplace in which such interactions can either raise or lower one's emotional energy (Collins, 1987). If the customer's credit card is rejected, he or she may come away with lower emotional energy. If the customer is angry at the cashier, he or she may attempt to "save face" by reacting in a haughty manner regarding the rejection of the card. ("What's wrong with you? Can't you do anything right? I'll never shop here again!") If this type of encounter occurs, the cashier may also come out of the interaction with a lower level of emotional energy, which may affect the cashier's interactions with subsequent customers. Likewise, the next time the customer uses a credit card, he or she may say something like "I hope this card isn't over its limit. Sometimes I lose track," even if the person knows that the card's credit limit has not been exceeded. This is only one of many ways in which the rich tradition of symbolic interactionism might be used to examine shopping and consumption. Other areas of interest might include the social nature of the shopping experience, social interaction patterns in families regarding credit card debts, and why we might spend money to impress others.

Postmodern Perspectives

According to *postmodern perspectives,* **existing theories have been unsuccessful in explaining social life in contemporary societies that are characterized by postindustrialization, consumerism, and global communications.** Postmodern social theorists reject the theoretical perspectives we have previously discussed, as well as how those thinkers created the theories (Ritzer, 1996). These theorists oppose the grand narratives that characterize modern thinking and believe that boundaries should not be placed on academic disciplines—such as philosophy, literature, art, and the social sciences—where much could be learned by sharing ideas.

Just as functionalist, conflict, and symbolic interactionist perspectives emerged in the aftermath of the Industrial Revolution, postmodern theories emerged after World War II (in the late 1940s) and reflected the belief that some nations were entering a period of postindustrialization. Postmodern (or "postindustrial") societies are characterized by an *information explosion* and an economy in which large numbers of people either provide or apply information, or they are employed in professional occupations (such as lawyers and physicians) or service jobs (such as fast-food servers and health care workers). There is a corresponding *rise of a consumer society* and the emergence of a *global village* in which people around the world communicate with one another by electronic technologies such as television, telephone, fax, e-mail, and the Internet.

Jean Baudrillard, a well-known French social theorist, is one of the key figures in postmodern theory, even though he would dispute this label. Baudrillard has extensively explored how the shift from production of goods (such as in the era of Marx and Weber) to consumption of information, services, and products in contemporary societies has created a new form of social control. According to Baudrillard's approach, capitalists strive to control people's shopping habits, much like the output of factory workers in industrial economies, to enhance their profits and to keep everyday people from rebelling against social inequality (1998/1970). How does this work? When consumers are encouraged to purchase more than they need or can afford, they often sink deeper in debt and must keep working to meet their monthly payments.

Instead of consumption being related to our needs, it is based on factors such as our "wants" and the need we feel to distinguish ourselves from others. We will look at this idea in more detail in the next section, where we apply a postmodern perspective to shopping and consumption. We will also return to Baudrillard's general ideas on postmodern societies in Chapter 3 ("Culture").

Today, postmodern theory remains an emerging perspective in the social sciences. How influential will this approach be? It remains to be seen what influence postmodern thinkers will have on the social sciences. Although this approach opens up broad new avenues of inquiry by challenging existing perspectives and questioning current belief systems, it also tends to ignore many of the central social problems of our time—such as inequalities based on race, class, and gender, and global political and economic oppression (Ritzer, 1996).

Applying Postmodern Perspectives to Shopping and Consumption According to some social theorists, the postmodern society is a consumer society. The focus of the capitalist economy has shifted from production to consumption. Today, the emphasis is on getting people to consume more and to own a greater variety of things. As previous discussed, credit cards may encourage people to spend more money than they should, and often more than they can afford (Ritzer, 1998). Television shopping networks and cybermalls make it possible for people to shop around the clock without having to leave home or encounter "real" people. As Ritzer (1998: 121) explains, "So many of our interactions in these settings . . . are simulated, and we become so accustomed to them, that in the end all we have are simulated interactions; there are no more 'real' interactions. The entire distinction between the simulated and the real is lost; simulated interaction *is* the reality" (also see Baudrillard, 1983). Similarly, Ritzer (1998: 121) points out that a credit card is a simulation:

Any given credit card is a simulation of all other cards of the same brand; there was no "original" card from which all others are copied; there is no "real" credit card. Furthermore, credit cards can be seen as simulations of simulations. That is, they simulate currency, but each bill is a simulation, a copy, of every other bill and, again, there was never an original bill from which all others have been copied. But currencies, in turn, can be seen as simulations of material wealth, or of the faith one has in the Treasury, or whatever one imagines to be the "real" basis of wealth. Thus, the credit card shows how we live in a world characterized by a never-ending spiral of simulation built upon simulation.

As this example suggests, postmodern theorists do not focus on actors (human agents) as they go about their everyday lives, but instead offer more-abstract conceptions of what constitutes "reality." For postmodernists, social life is not an objective reality waiting for us to discover how it works. Rather, what we experience as social life is actually nothing more or less than how we think about it, and there are many diverse ways of doing that. According to a postmodernist perspective, the Enlightenment goal of intentionally creating a better world out of some knowable truth is an illusion. Although some might choose to dismiss postmodernist approaches, they do give us new and important questions to think about regarding the nature of social life.

Concept Table 1.A reviews all four of these perspectives. Throughout this book, we will be using these perspectives as lenses through which to view our social world.

COMPARING SOCIOLOGY WITH OTHER SOCIAL SCIENCES

In this chapter, we have discussed how sociologists examine social life. We have focused on shopping and consumption as an example of the many topics studied by sociologists and other social scientists. Let's briefly examine how other social sciences investigate human relationships and then compare each discipline to sociology.

Anthropology

Anthropologists and sociologists are interested in studying human behavior; however, there are differences between the two disciplines. Anthropology seeks to understand human existence over geographic space and evolutionary time (American Anthropological Association, 2001), whereas sociology seeks to understand contemporary social organization, relations, and change. Some anthropologists focus on the beginnings of human history, millions of years ago, whereas others primarily study contemporary societies. Anthropology is divided into four main subfields: sociocultural, linguistic, archaeological, and biological anthropology. Cultural anthropologists focus on culture and its many manifestations, including art, religion, and politics. Linguistic anthropologists primarily study language because culture itself depends on language. Archaeologists are interested in discovering and analyzing material artifacts—such as cave paintings, discarded stone tools, and abandoned baskets—from which they piece together a record of social life in earlier societies and assess what this information adds to our knowledge of contemporary cultures. Biological (or physical) anthropologists study the biological origins, evolutionary development, and genetic diversity of primates, including Homo Sapiens (human beings). Clearly, some cultural anthropologists would be interested in the issue we have looked at in this chapter—why people consume items of material culture and how their behavior as consumers is interwoven with all other aspects of social life in this society.

Psychology

Psychology is the systematic study of behavior and mental processes—what occurs in the mind. Psychologists focus not only on behavior that is directly observable, such as talking, laughing, and eating, but also on mental processes that cannot be directly observed, such as thinking and dreaming. For psychologists, behavior and mental processes are interwoven; therefore, to understand behavior, they examine the emotions that underlie people's actions. For example, a psychologist interested in studying why some individuals have excessive credit card debt might identify the specific emotions that a person has when purchasing an expensive item that is well beyond his or her budget.

Psychology is a diverse field. Some psychologists work in clinical settings, where they diagnose and treat psychological disorders; others practice in schools, where they are concerned with the intellectual, social, and emotional development of schoolchildren. Still other psychologists work in business, industry, and other work-related settings. Another branch of psychology—social psychology—is similar to sociology in that it emphasizes how social conditions affect individual behavior. Social psychological perspectives on human development are useful to sociologists who study the process of socialization (see Chapter 4, "Socialization"). However, a distinction between psychology and sociology is the extent to which most psychological studies focus on internal factors relating to the individual in their explanations of human behavior, whereas sociological research

Concept Table 1.A	THE MAJOR THEORETICAL PERSPECTIVES	
Perspective	**Analysis Level**	**View of Society**
Functionalist	Macrolevel	Society is composed of interrelated parts that work together to maintain stability within society. This stability is threatened by dysfunctional acts and institutions.
Conflict	Macrolevel	Society is characterized by social inequality; social life is a struggle for scarce resources. Social arrangements benefit some groups at the expense of others.
Symbolic Interactionist	Microlevel	Society is the sum of the interactions of people and groups. Behavior is learned in interaction with other people; how people define a situation becomes the foundation for how they behave.
Postmodernist	Macrolevel/Microlevel	Societies characterized by postindustrialization, consumerism, and global communications bring into question existing assumptions about social life and the nature of reality.

examines the effects of groups, organizations, and social institutions on social life.

Economics

Unlike the other social sciences we have discussed, economics concentrates primarily on a single institution in society—the economy (see Chapter 13, "The Economy and Work in Global Perspective"). Economists attempt to explain how the limited resources of a society are allocated among competing demands. Economics is divided into two different branches. Macroeconomics looks at such things as the total amount of goods and services produced by a society; microeconomics studies such things as decisions made by individual businesses. Thus, consumerism and credit card debt would be issues of interest to economists because such topics can be analyzed at global, national, and individual levels. However, a distinction between economics and sociology is that economists focus on the complex workings of economic systems (such as monetary policy, inflation, and the national debt), whereas sociologists focus on a number of social institutions, one of which is the economy.

Political Science

Political science is the academic discipline that studies political institutions such as the state, government, and political parties (see Chapter 14, "Politics and Government in Global Perspective"). Political scientists study power relations and seek to determine how power is distributed in various types of political systems. Some political scientists focus primarily on international relations and similarities and differences in the political institutions across nations. Other political scientists look at the political institutions in a particular country. An example is a political scientist interested in studying consumerism in the United States who systematically examines how the political process—such as the efforts of lobbyists and interest groups to influence governmental policies—affects credit card interest rates and consumer spending in this country. Political scientists concentrate on political institutions, whereas sociologists study these institutions within the larger context of other social institutions such as families, religion, education, and the media.

In Sum

Clearly, the areas of interest and research in the social sciences overlap in that the goal of scholars, teachers, and students is to learn more about human behavior, including its causes and consequences. In applied sociology, there is increasing collaboration among researchers across disciplines to develop a more holistic, integrated view of how human behavior and social life take place in societies. Join me as, throughout this book, we examine specific ways in which sociology creates its own realm of knowledge yet benefits from the other social sciences in order to provide new information and personal insights that we can use in our everyday lives.

CHAPTER REVIEW

■ **What is sociology, and how can it help us to understand ourselves and others?**

Sociology is the systematic study of human society and social interaction. We study sociology to understand how human behavior is shaped by group life and, in turn, how group life is affected by individuals. Our culture tends to emphasize individualism, and sociology pushes us to consider more-complex connections between our personal lives and the larger world.

■ **What is the sociological imagination, and why is it important to have a global sociological imagination?**

According to C. Wright Mills, the sociological imagination helps us understand how seemingly personal troubles, such as suicide, are actually related to larger social forces. It is the ability to see the relationship between individual experiences and the larger society. It is important to have a global sociological imagination because the future of this nation is deeply intertwined with the future of all nations of the world on economic, political, and humanitarian levels.

■ **What factors contributed to the emergence of sociology as a discipline?**

Industrialization and urbanization increased rapidly in the late eighteenth century, and social thinkers began to examine the consequences of these powerful forces. Auguste Comte coined the term *sociology* to describe a new science that would engage in the study of society.

■ **What are the major contributions of early sociologists such as Durkheim, Marx, and Weber?**

The ideas of Emile Durkheim, Karl Marx, and Max Weber helped lead the way to contemporary sociology. Durkheim argued that societies are built on social facts, that rapid social change produces strains in society, and that the loss of shared values and purpose can lead to a condition of anomie. Marx stressed that within society there is a continuous clash between the owners of the means of production and the workers who have no choice but to sell their labor to others. According to Weber, sociology should be value free, and people should become more aware of the role that bureaucracies play in daily life.

■ **How did Simmel's perspective differ from that of other early sociologists?**

Whereas other sociologists primarily focused on society as a whole, Simmel explored small social groups and argued that society was best seen as a web of patterned interactions among people.

■ **What are the major contemporary sociological perspectives?**

Functionalist perspectives assume that society is a stable, orderly system characterized by societal consensus. Conflict perspectives argue that society is a continuous power struggle among competing groups, often based on class, race, ethnicity, or gender. Interactionist perspectives focus on how people make sense of their everyday social interactions, which are made possible by the use of mutually understood symbols. From an alternative perspective, postmodern theorists believe that entirely new ways of examining social life are needed and that it is time to move beyond functionalist, conflict, and interactionist approaches.

KEY TERMS

anomie 17
conflict perspectives 28
functionalist perspectives 26
high-income countries 9
industrialization 13
latent functions 26
low-income countries 9
macrolevel analysis 29
manifest functions 26
microlevel analysis 29
middle-income countries 9
positivism 14
postmodern perspectives 33
social Darwinism 16
social facts 17
society 4
sociological imagination 5
sociology 4
symbolic interactionist perspectives 29
theory 25
urbanization 13

QUESTIONS FOR CRITICAL THINKING

1. What does C. Wright Mills mean when he says the sociological imagination helps us "to grasp history and biography and the relations between the two

within society" (Mills, 1959b: 6)? How might this idea be applied to today's consumer society?

2. As a sociologist, how would you remain objective yet see the world as others see it? Would you make subjective decisions when trying to understand the perspectives of others?

3. Early social thinkers were concerned about stability in times of rapid change. In our more global world, is stability still a primary goal? Or is constant conflict important for the well-being of all humans? Use the conflict and functionalist perspectives to bolster your analysis.

4. Some social analysts believe that college students relate better to commercials and advertising culture than they do to history, literature, or probably anything else (Twitchell, 1996). How would you use the various sociological perspectives to explore the validity of this assertion in regard to students on your college campus?

RESOURCES ON THE INTERNET

Chapter-Related Web Sites

The following Web sites have been selected for their relevance to the topics in this chapter. These sites are among the more stable, but please note that Web site addresses change frequently. For an updated list of chapter-related Web sites with URL links, please visit the *Sociology in Our Times* Web site (**www.wadsworth.com/KendallSIOT**).

Dead Sociologists Index
http://www2.pfeiffer.edu/~lridener/DSS /INDEX. HTML

Visit this Web site to learn more about the sociologists discussed in this chapter. Click on the name of an individual you wish to study, and access biographical information, summaries of key ideas, and selections from original works.

A Sociological Tour Through Cyberspace
http://www.trinity.edu/~mkearl/index.html

Professor Michael C. Kearl of Trinity University in San Antonio has developed a comprehensive gateway site to a number of Internet resources in the field of sociology. For Chapter 1, click on "General Sociological Resources" and "Sociological Theory."

American Sociological Society (ASA)
http://www.asanet.org

The American Sociological Association (ASA) is a national organization for sociologists. Its home page provides information on annual meetings, resources available for sociological research, information on careers in sociology, special reports on current activities in social policy, and information on how students can get involved in this sociological organization, including the Minority Fellowship Program.

ONLINE STUDY AND RESEARCH TOOLS

Accompanying this text are many *free* powerful online study tools that will help you master the material in this chapter, help increase your depth of understanding, and help you make the grade!

SocCoach CD-ROM

Use the SocCoach CD-ROM enclosed with this text to help you formulate a customized study plan for this chapter. After you take the Diagnostic Quiz, SocCoach will generate a customized study plan just for you! It will identify sections of the chapter that you should review and will provide videos, charts, graphs, and excerpts from the text to supplement your studies and enhance your understanding. You'll also find fun, interactive activities such as Virtual Explorations and Map the Stats to apply what you've learned and stretch your sociological imagination.

The Companion Web Site for Sociology in Our Times, *Fifth Edition*
www.wadsworth.com/KendallSIOT

Gain an even better grasp on this chapter by going to the companion Web site to take one of the Tutorial Quizzes, use the Flash Cards to master key terms, or check out the many other study aids you'll find there. You'll also find special features such as GSS Data and Census 2000 information that'll put data and resources at your fingertips to help you with that special project or help you as you do some research on your own.

In this chapter, when you see the icon on the left, it alerts you to a specific exercise found in *Wadsworth's Sociology Online Resources and Writing Companion*. This valuable guide shows you how to use Wadsworth's exclusive online resources—*InfoTrac College Edition*, the *Opposing Viewpoints Resource Center*, and *MicroCase Online*—to assist you in your study of sociology and to build essential research and writing skills.

Sociological Research Methods

This note should be pretty easy to understand. All the wording's from the Punk Rock 101. . . . I haven't felt the excitement of listening to as well as creating music, along with reading and writing for too many years now. . . . I've tried everything that's in my power to appreciate it, and I do. . . . I'm too sensitive. I need to be slightly numb in order to regain the enthusiasm I had as a child. . . . There's good in all of us and I simply love people too much. So much that it makes me feel too . . . sad. . . . I have it good, very good, and I'm grateful. But, since the age of seven, I've become hateful toward all humans in general. . . . I'm too much of an erratic, moody baby! I don't have passion anymore, and so remember, it's better to burn out than to fade away.

—Peace, love, empathy, Kurt Cobain (qtd. in Etkind, 1997: 38–39; originally appeared in *Rolling Stone*, June 2, 1994)

When the fire goes out you better learn to fake it.
It's better to rise than fade away.

—After Cobain's suicide, Courtney Love responded to his death with these lyrics in *Celebrity Skin,* an album that she and her band recorded.

T he first example set forth here is a suicide statement left by Kurt Cobain—founder of the band Nirvana, which is credited with inventing the grunge sound—who apparently found himself trapped in a downward spiral of depression and drug addiction. Why did Cobain commit suicide? It appeared that he had everything to live for. Millions of Nirvana albums had been sold, he was married to celebrity Courtney Love, and wealth and success were his. In the aftermath of his death, Love issued her reply to his feeling that "it's better to burn out than to fade away" by suggesting that people should go on with their lives even if "the fire goes out" for a period of time. According to Love's philosophy, "It's better to rise than fade away," and apparently she has done this as she has continued with her best-selling albums and successful career.

Although Cobain and Love are only two people in a world comprising more than six billion people, the issues that are raised by his death are of importance in many social contexts. They bring us to a larger sociological question: Why does anyone commit suicide? Is suicide purely an individual phenomenon, or is it related to the social environments and societies in which people live?

In this chapter, we examine how sociological theories and research can help us understand the seemingly individualistic act of taking one's own life. We

Paul Morse/Corbis

■ Suicides of celebrities such as Kurt Cobain (pictured here with his wife, Courtney Love) call our attention to significant social facts related to suicide rates in contemporary societies. Sociologists ask questions such as why highly successful people, such as Cobain, commit suicide.

will see how sociological theory and research methods might be used to answer complex questions, and we will wrestle with some of the difficulties that sociologists experience as they study human behavior.

QUESTIONS AND ISSUES

Chapter Focus Question: How do sociological theory and research add to our knowledge of human societies and social issues such as suicide?

What is the relationship between theory and research?

What are the steps in the conventional research process?

What can qualitative methods add to our understanding of human behavior?

Why is it important to have a variety of research methods available?

What has research contributed to our understanding of suicide?

Why is a code of ethics for sociological research necessary?

WHY IS SOCIOLOGICAL RESEARCH NECESSARY?

Sociologists obtain their knowledge of human behavior through research, which results in a body of information that helps us move beyond guesswork and common sense in understanding society. The sociological perspective incorporates theory and research to arrive at a more accurate understanding of the "hows" and "whys" of human social interaction. Once we have an informed perspective about social issues, such as who commits suicide and why, we are in a better position to find solutions and make changes. Social research, then, is a key part of sociology.

Common Sense and Sociological Research

Most of us have commonsense ideas about suicide. Common sense, for example, may tell us that people who threaten suicide will not commit suicide. Sociological research indicates that this assumption is frequently incorrect: People who threaten to kill themselves are often sending messages to others and may indeed attempt suicide. Common sense may also tell us that suicide is caused by despair or depression. However, research suggests that suicide is sometimes used as a means of lashing out at friends and relatives because of real or imagined wrongs. Before reading on, take the quiz in Box 2.1, which lists a number of commonsense notions about suicide.

Historically, the commonsense view of suicide was that it was a sin, a crime, and a mental illness (Evans and Farberow, 1988). Emile Durkheim refused to accept these explanations. In what is probably the first sociological study to use scientific research methods, he related suicide to the issue of cohesiveness (or lack

David Keeler/Getty Images

Psychics and astrologers such as Jacqueline Stallone now have access to a wider array of potential clients through television advertising, the Internet, and lucrative book contracts. What problems are associated with relying on such individuals for counseling and advice?

of cohesiveness) in society instead of viewing suicide as an isolated act that could be understood only by studying individual personalities or inherited tendencies. In *Suicide* (1964b/1897), Durkheim documented his contention that a high suicide rate was symptomatic of large-scale societal problems. In the process, he developed an approach to research that influences researchers to this day. As we discuss sociological research, we will use the problem of suicide to demonstrate the research process.

Box 2.1 SOCIOLOGY AND EVERYDAY LIFE

How Much Do You Know About Suicide?

True	False	
T	F	1. For people thinking of suicide, it is difficult, if not impossible, to see the bright side of life.
T	F	2. People who talk about suicide don't do it.
T	F	3. Once people contemplate or attempt suicide, they must be considered suicidal for the rest of their lives.
T	F	4. In the United States, suicide occurs on the average of one every eighteen minutes.
T	F	5. Accidents and injuries sustained by teenagers and young adults may indicate suicidal inclinations.
T	F	6. Alcohol and drugs are outlets for anger and thus reduce the risk of suicide.
T	F	7. Older women have lower rates of both attempted and completed suicide than do older men.
T	F	8. Children don't know enough to be able to intentionally kill themselves.
T	F	9. Suicide rates for African Americans are higher than for white Americans.
T	F	10. Suicidal people are fully intent upon dying.

Answers on page 42.

Since much of sociology deals with everyday life, we might think that common sense, our own personal experiences, and the media are the best sources of information. However, our personal experiences are subjective, and much of the information provided by the media comes from sources seeking support for a particular point of view. The content of the media is also influenced by the continual need for audience ratings.

We need to be able to evaluate the information we receive. This is especially true because the quantity—but, in some instances, not the quality—of information available has grown dramatically as a result of the information explosion brought about by computers and by the telecommunications industry.

Sociology and Scientific Evidence

In taking this course, you will be studying social science research and may be asked to write research reports or read and evaluate journal articles. If you attend graduate or professional school in fields that use sociological research, you will be expected to evaluate existing research and perhaps do your own. Hopefully, you will find that social research is relevant to the practical, everyday concerns of the real world.

Sociology involves *debunking*—the unmasking of fallacies (false or mistaken ideas or opinions) in the everyday and official interpretations of society (Mills, 1959b). Since problems such as suicide involve threats to existing societal values, we cannot analyze these problems without acknowledging the values involved. For example, should assisted suicide for terminally ill patients who wish to die be legal? We often answer questions like this by using either the normative or the empirical approach. The *normative approach* uses religion, customs, habits, traditions, and law to answer important questions. It is based on strong beliefs about what is right and wrong and what "ought to be" in society. Issues such as assisted suicide are often answered by the normative approach. From a legal standpoint, the consequences of assisting in another person's suicide may be severe.

Although these issues are immediate and profound, some sociologists discourage the use of the normative approach in their field and advocate the use of the empirical approach instead. The *empirical approach* attempts to answer questions through systematic collection and analysis of data. This approach is referred to as the conventional model, or the "scientific method," and is based on the assumption that knowledge is best gained by direct, systematic observation.

Box 2.1 SOCIOLOGY AND EVERYDAY LIFE

Answers to the Sociology Quiz on Suicide

1. **True.** To people thinking of suicide, an acknowledgment that there is a bright side only confirms and conveys the message that they have failed; otherwise, they, too, could see the bright side of life.

2. **False.** Some people who talk about suicide do kill themselves. Warning signals of possible suicide attempts include talk of suicide, the desire not to exist anymore, despair, and hopelessness.

3. **False.** Most people think of suicide for only a limited amount of time. When the crisis is over and the problems leading to suicidal thoughts are resolved, people usually cease to think of suicide as an option.

4. **True.** A suicide occurs on the average of every eighteen minutes in the United States; however, this rate differs with respect to the sex, race/ethnicity, and age of the individual. For example, men are four times more likely to kill themselves than are women.

5. **True.** Accidents, injuries, and other types of life-threatening behavior may be signs that a person is on a course of self-destruction. One study concluded that the incidence of suicide was twelve times higher among adolescents and young adults who had been previously hospitalized because of an injury.

6. **False.** Excessive use of alcohol or drugs may enhance a person's feelings of anger and frustration, making suicide a greater possibility. This risk appears to be especially high for men who abuse alcohol or drugs.

7. **True.** In the United States, as in other countries, suicide rates are the highest among men over age seventy. One theory of why this is true asserts that older women may have a more flexible and diverse coping style than do older men.

8. **False.** Children do know how to intentionally hurt or kill themselves. They may learn the means and methods from television, movies, and other people. However, the National Center for Health Statistics (the agency responsible for compiling suicide statistics) does not recognize suicides under the age of ten; they are classified as accidents, despite evidence that young children have taken their own lives.

9. **False.** Suicide rates are much higher among white Americans than African Americans. For example, in 2000 the overall U.S. suicide rate was 10.7 per 100,000 population. For white males, the rate was 19.1; for African American males, it was 9.8. For white females, it was 4.5; for African American females, the rate was 1.8.

10. **False.** Suicidal people often have an ambivalence about dying—they want to live and to die at the same time. They want to end the pain or problems they are experiencing, but they also wish that something or someone would remove the pain or problem so that life could continue.

Sources: Based on American Association of Suicidology, 2003; Leenaars, 1991; Levy and Deykin, 1989; Patros and Shamoo, 1989; and Wickett, 1989.

Many sociologists believe that two basic scientific standards must be met: (1) scientific beliefs should be supported by good evidence or information and (2) these beliefs should be open to public debate and critiques from other scholars, with alternative interpretations being considered (Cancian, 1992).

Sociologists typically use two types of empirical studies: descriptive and explanatory. *Descriptive studies* attempt to describe social reality or provide facts about some group, practice, or event. Studies of this type are designed to find out what is happening to whom, where, and when. For example, a descriptive study of suicide might attempt to determine the number of people who recently thought about committing suicide. On other topics, well-known descriptive studies include the U.S. Census and the FBI's Uniform Crime Reports. However, it is important to note that even studies which are considered to be "objective" have certain biases because of the limitations inherent in doing certain types of research, as discussed in this chapter and in Chapter 7 ("Deviance and Crime"). By contrast, *explanatory studies* attempt to explain cause-and-effect relationships and to provide information on why certain events do or do not occur. In an explanatory study of suicide, we might ask questions such as these: Why do African American men over age sixty-five have a significantly lower suicide rate than white males in the same age bracket? Why are women more likely to attempt suicide than men? Sociologists engage in theorizing and conducting research in order to describe, explain, and sometimes predict how and why people will act in certain situations.

The Theory and Research Cycle

The relationship between theory and research has been referred to as a continuous cycle, as shown in Figure 2.1 (Wallace, 1971). You will recall that a *theory* is a set of logically interrelated statements that attempts to describe, explain, and (occasionally) predict social events. A theory attempts to explain why something is the way it is. *Research* is the process of systematically collecting information for the purpose of testing an existing theory or generating a new one. The theory and research cycle consists of deductive and inductive approaches. In the *deductive approach,* the researcher begins with a theory and uses research to test the theory. This approach proceeds as follows: (1) theories generate hypotheses, (2) hypotheses lead to observations (data gathering), (3) observations lead to the formation of generalizations, and (4) generalizations are used to support the theory, to suggest modifications to it, or to refute it. To illustrate, if we use the deductive method to determine why people commit suicide, we start by formulating a theory about the "causes" of suicide and then test our theory by collecting and analyzing data (for example, vital statistics on suicides or surveys to determine whether adult church members view suicide differently from nonmembers).

In the *inductive approach,* the researcher collects information or data (facts or evidence) and then generates theories from the analysis of that data. Under the inductive approach, we would proceed as follows: (1) specific observations suggest generalizations, (2) generalizations produce a tentative theory, (3) the theory is tested through the formation of hypotheses,

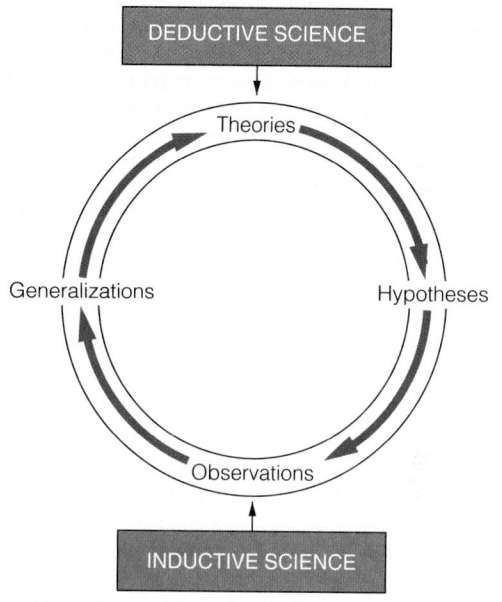

Figure 2.1 | **The Theory and Research Cycle**

The theory and research cycle can be compared to a relay race: Although all participants do not necessarily start or stop at the same point, they share a common goal—to examine all levels of social life.

Source: Adapted from Walter Wallace, *The Logic of Science in Sociology.* New York: Aldine de Gruyer, 1971.

and (4) hypotheses may provide suggestions for additional observations. Using the inductive approach to study suicide, we might start by simultaneously collecting and analyzing data related to suicidal behavior and then generate a theory (see Glaser and Strauss, 1967; Reinharz, 1992). Researchers may break into the cycle at different points depending on what they want to know and what information is available.

Theory gives meaning to research; research helps support theory. For example, data collected from interviews with 25 women aged 15 to 24 who recently attempted suicide will not give us an explanation of *why* women are more likely than men to *attempt* to take their own lives. Similarly, theories unsupported by data are meaningless. Suppose, for instance, that we made the following assertions: Women are more likely to attempt suicide because of problems in their personal relationships whereas men are more likely to be suicidal when they have economic difficulties, are unemployed, or experience a severe physical illness (Canetto, 1992). Our assertions are unsupported because we have not tested their validity.

Research helps us question such assumptions about suicide and other social concerns. Sociologists

Understanding Statistical Data Presentations

Are men or women more likely to commit suicide? Are suicide rates increasing or decreasing? Such questions may be answered in numerical terms. Sociologists often use statistical tables as a concise way to present data because such tables convey a large amount of information in a relatively small space; Table 1 gives an example. To understand a table, follow these steps:

1. *Read the title.* The title indicates the topic. From the title, "U.S. Suicides, by Sex and Method Used, 1970 and 2000," we learn that the table shows relationships between two variables: sex and method of suicide used. It also indicates that the table contains data for two different time periods: 1970 and 2000.
2. *Check the source and other explanatory notes.* In this case, the sources are the *U.S. Census Bureau, 2002,* and *American Association of Suicidology, 2003.* Checking the source helps determine its reliability and timeliness. The first footnote indicates that the table includes only people who reside in the United States. The next two footnotes provide more information about exactly what is included in each category.
3. *Read the headings for each column and each row.* The main column headings in Table 1 are "Method," "Males," and "Females." These latter two column headings are divided into two groups: 1970 and 2000. The columns present information (usually numbers) arranged vertically. The rows present information horizon-

tally. Here, the row headings indicate suicide methods.
4. *Examine and compare the data.* To examine the data, determine what units of measurement have been used. In Table 1, the figures are numerical counts (for example, the total number of reported female suicides by poisoning in 2000 was 2,067) and percentages (for example, in 2000, poisoning accounted for 36.1 percent of all female suicides reported). A *percentage,* or proportion, shows how many of a given item there are in every one hundred. Percentages allow us to compare groups of different sizes. For example, percentages show the proportion of people who used each method, thus giving a more meaningful comparison.
5. *Draw conclusions.* By looking for patterns, some conclusions can be drawn from Table 1.
   a. *Determining the increase or decrease.* Between 1970 and 2000, reported male suicides increased from 16,629 to 23,618–an increase of 6,989 (but keep in mind that the overall male population also increased!)–while female suicides decreased by 1,119 (even as the female population increased). This represents a *total* increase (for males) and decrease (for females) in suicides for the two years being compared. The *amount* of increase or decrease can be stated as a percentage: Total male suicides were about 42 percent higher in 2000, calculated by dividing the total increase (6,989) by the

suggest that a healthy skepticism (a feature of science) is important in research because it keeps us open to the possibility of alternative explanations. Some degree of skepticism is built into each step of the research process. With that in mind, let's explore the steps in the sociological research process.

THE SOCIOLOGICAL RESEARCH PROCESS

Not all sociologists conduct research in the same manner. Some researchers primarily engage in quantitative research whereas others engage in qualitative

research. With *quantitative research,* the goal is scientific objectivity, and the focus is on data that can be measured numerically. Quantitative research typically emphasizes complex statistical techniques. Most sociological studies on suicide have used quantitative research. They have compared rates of suicide with almost every conceivable variable, including age, sex, race/ethnicity, education, and even sports participation (see Lester, 1992). For example, researchers in one study examined the effects of church membership, divorce, and migration on suicide rates in the United States and concluded that suicide rates are typically higher where divorce and migration rates are higher and church membership is lower (Breault, 1986). (The box above, "Understand-

earlier (lower) number. Total female suicides were about 16.3 percent lower in 2000, calculated by dividing the total decrease (1,119) by the earlier (higher) number.

b. *Drawing appropriate conclusions.* The number of female suicides by firearms increased about 3 percent between 1970 and 2000; the number for poisoning dropped by about 37 percent. We might conclude that more women preferred firearms over poisoning as a means

of killing themselves in 2000 than in 1970. Does that mean more women gained access to guns? That taking one's life with a firearm became more acceptable? Such generalizations do not take into account that we are looking at *fatal* suicides. If we had accurate data for *nonfatal* suicidal behavior (attempted suicide), we would likely find a higher rate for poisoning than for firearms.

Table 1 U.S. SUICIDES, BY SEX AND METHOD USED, 1970 AND 2000[a]

METHOD	MALES		FEMALES	
	1970	2000	1970	2000
Total	16,629	23,618	6,851	5,732
Firearms	9,704	14,454	2,068	2,132
(% of total)	(58.4)	(61.2)	(30.2)	(37.2)
Poisoning[b]	3,299	2,792	3,285	2,067
(% of total)	(19.8)	(11.8)	(48.0)	(36.1)
Hanging/strangulation[c]	2,422	4,733	831	955
(% of total)	(14.6)	(20.0)	(12.1)	(16.7)
Other	1,204	1,639	667	578
(% of total)	(7.2)	(6.9)	(9.7)	(10.0)

[a]Excludes deaths of nonresidents of the United States.
[b]Includes solids, liquids, and gases.
[c]Includes suffocation.
Sources: U.S. Census Bureau, 2002; American Association of Suicidology, 2003.

ing Statistical Data Presentations," explains how to read numerical tables, how to interpret the data and draw conclusions, and how to calculate ratios and rates.)

With *qualitative research,* interpretive description (words) rather than statistics (numbers) is used to analyze underlying meanings and patterns of social relationships. An example of qualitative research is a study in which the researcher systematically analyzed the contents of the notes of suicide victims to determine recurring themes, such as a feeling of despair or failure. Through this study, the researcher hoped to determine if any patterns could be found that would help in understanding why people might kill themselves (Leenaars, 1988).

The "Conventional" Research Model

Research models are tailored to the specific problem being investigated and the focus of the researcher. Both quantitative research and qualitative research contribute to our knowledge of society and human social interaction, and both involve a series of steps, as shown in Figure 2.2. We will now trace the steps in the "conventional" research model, which focuses on quantitative research. Then we will describe an alternative model that emphasizes qualitative research.

1. *Select and define the research problem.* When you engage in research, the first step is to select and clearly define the research topic. Sometimes, a specific experience such as having known someone

Figure 2.2 Steps in Sociological Research

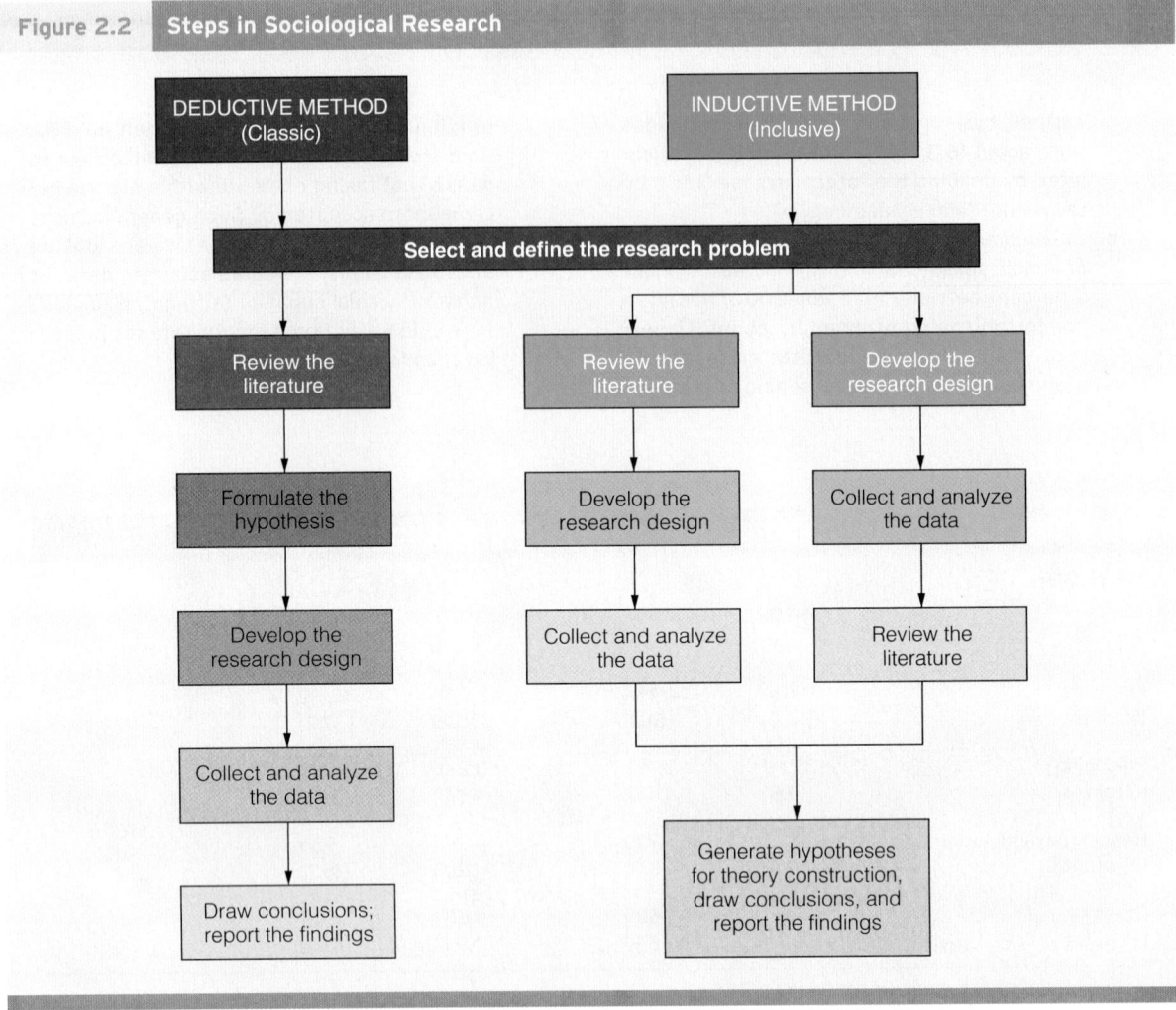

who committed suicide can trigger your interest in a topic. Other times, you might select topics to fill gaps or challenge misconceptions in existing research or to test a specific theory (Babbie, 2001). Emile Durkheim selected suicide because he wanted to demonstrate the importance of *society* in situations that might appear to be arbitrary acts by individuals. Suicide was a suitable topic because it was widely believed that suicide was a uniquely individualistic act. However, Durkheim emphasized that *suicide rates* provide better explanations for suicide than do *individual acts* of suicide. He reasoned that if suicide were purely an individual act, then the rate of suicide (the relative number of people who kill themselves each year) should be the same for every group regardless of culture and social structure (see Box 2.2 on page 48 for a current example). Moreover, Durkheim wanted to know why there were different rates of suicide—whether factors such as religion, marital status, sex, and age had an effect on social cohesion.

2. *Review previous research.* Before you begin your research, it is important to review the literature to see what others have written about the topic. Analyzing what previous researchers have found helps to clarify issues and focus the direction of your own research. But when Durkheim began his study, very little sociological literature existed for him to review other than the works of Henry Morselli (1975/1881), who concluded that suicide was a part of an evolutionary process whereby "weak-brained" individuals were sorted out by insanity and voluntary death.

3. *Formulate the hypothesis (if applicable).* You may formulate a **hypothesis**—a statement of the relationship between two or more concepts. Concepts are the abstract elements representing some aspect of the world in simplified form (such as "social integration" or "loneliness"). As you

Figure 2.3	Hypothesized Relationships Between Variables

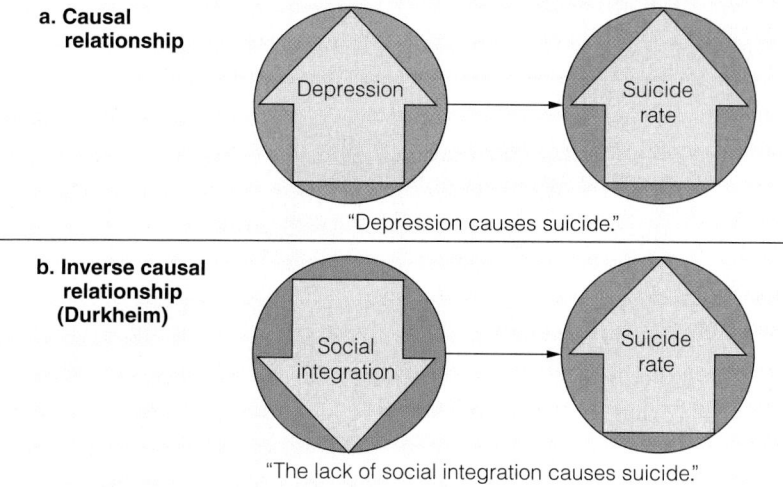

a. Causal relationship

"Depression causes suicide."

b. Inverse causal relationship (Durkheim)

"The lack of social integration causes suicide."

A causal hypothesis connects one or more independent (causal) variables with a dependent (affected) variable. The diagram illustrates three hypotheses about the causes of suicide. To test these hypotheses, social scientists would need to operationalize the variables (define them in measurable terms) and then investigate whether the data support the proposed explanation.

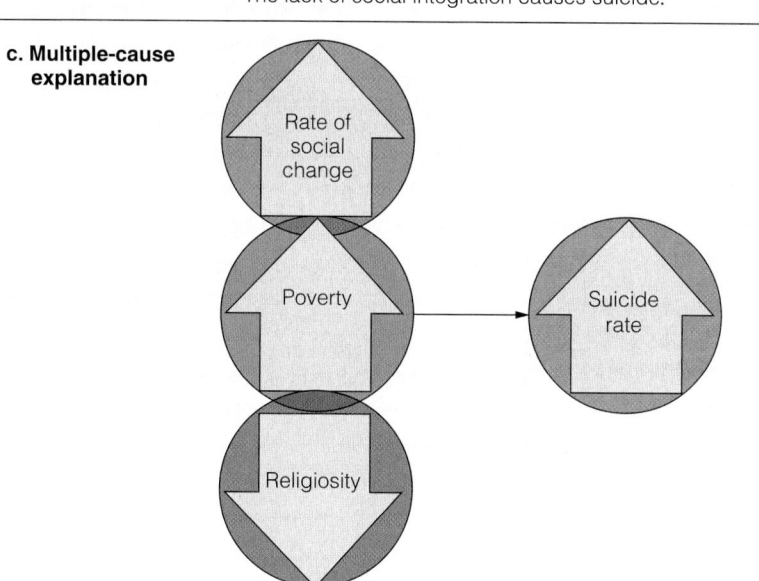

c. Multiple-cause explanation

"Many factors interact to cause suicide."

formulate your hypothesis about suicide, you may need to convert concepts to variables. A *variable* is any concept with measurable traits or characteristics that can change or vary from one person, time, situation, or society to another. Variables are the observable and/or measurable counterparts of concepts. For example, "suicide" is a concept; the "rate of suicide" is a variable.

The most fundamental relationship in a hypothesis is between a dependent variable and one or more independent variables (see Figure 2.3). The ***independent variable* is presumed to cause or determine a dependent variable.** Age, sex,

race, and ethnicity are often used as independent variables. The ***dependent variable* is assumed to depend on or be caused by the independent variable(s)** (Babbie, 2001). Durkheim used the degree of social integration in society as the independent variable to determine its influence on the dependent variable, the rate of suicide.

Whether a variable is dependent or independent depends on the context in which it is used. To use variables in the contemporary research process, sociologists create operational definitions. An *operational definition* is an explanation of an abstract concept in terms of observable features

Box 2.2 SOCIOLOGY IN GLOBAL PERSPECTIVE

Comparing Suicide Statistics from Different Nations

We're often told that this is the age of convergence. As global optimists see it, free-market capitalism, democratic norms, and maybe even a better appreciation for the sanctity of life are gaining ascendancy across the borders. In the rich world at least, you get the sense that a country's unique way of looking at the world will eventually be submerged by these big global trends.

There's some truth to that. But every once in a while here in Japan, I'm abruptly reminded that some things about this remarkable culture I'll never begin to fathom. To my mind the most fundamental one is the prevalence of suicide in a nation that boasts some of the highest living standards and longest life expectancies in the world.

–Brian Bremner (2000), Tokyo bureau chief for *Business Week,* stating his concern about the "suicide epidemic" in Japan

When sociologists select and define a research problem, they may look at a current social phenomenon such as statistical trends in suicides or how the rates of suicide compare across nations. If we look at the rates of suicide per 100,000 people in various countries, will these rates differ? In fact, the answer to this question is "Yes." There is a wide disparity among suicide rates in various nations. For example, Lithuania has a suicide rate of 48 per 100,000 people, as compared with the United States, which has a suicide rate of 12 per 100,000 people. Can any patterns be identified in regard to Lithuanian suicides? One that has been identified is that the suicide rate has risen in Lithuania over the past three decades as that nation has gone through a lengthy period of economic, political, and social upheaval.

© 2003 AP/Wide World Photos

■ Global economic woes and worries about the health of banks in Japan have contributed to heightening concern among many Japanese workers. Sociologists continue to explore the relationship between economic problems and suicide rates in nations such as Japan, where the recent suicide rate has been twice that of the United States.

We might ask this: Is Lithuania's suicide rate increase due to that social upheaval? Japan has also had an upswing in the number of reported suicides, culminating in the late 1990s, when there was an un-

that are specific enough to measure the variable. For example, suppose that your goal is to earn an *A* in this course. Your professor may have created an operational definition by defining an *A* as earning an exam average of 90 percent or above (Babbie, 2001).

Events such as suicide are too complex to be caused by any one variable. Therefore, they must be explained in terms of *multiple causation*—that is, an event occurs as a result of many factors operating in combination. What *does* cause suicide? Social scientists cite multiple causes, including rapid social change, economic conditions, hope-

less poverty, and lack of religiosity (the degree to which an individual or group feels committed to a particular system of religious beliefs). Usually, no one factor will cause a person to commit suicide. Rather, other factors must combine with a factor such as poverty to cause a person to commit suicide. Sociologists cannot produce an equation (such as poverty + homelessness = suicide) to predict a social occurrence. Not all social research makes use of hypotheses.

4. *Develop the research design.* In developing the research design, you must first consider the units of analysis and the time frame of the study. A

Comparative Suicide Rates

Lithuania	48
Hungary	38
Japan	24
U.S.	12

Suicides per 100,000 People

Source: *BBC News*, 1999.

precedented 35-percent surge in suicides. The suicide rate in Japan (24 per 100,000 people) is half of that of Lithuania but twice that of the United States. To put the suicide rate in Japan in perspective, that country has three times as many reported suicide victims as it does traffic fatalities. In 1999, men accounted for 71 percent of all suicides in Japan, with 40 percent of the men being in their forties and fifties, and many of them having recently experienced major financial difficulties.

Using theory as the foundation for our research on Japanese and Lithuanian suicides, we might reflect on Emile Durkheim's types of suicides and test whether or not some of these suicides might be best described as *anomic suicides,* which may be brought about by rapid social change. Japan has experienced a prolonged economic slump that has particularly affected the employment rates of middle-aged men. In a society where people are accustomed to full employment (meaning that most people who wanted a job could have one), many people have found themselves unemployed for the first time. Recent statistics showed that 47 percent of those who killed themselves in Japan were unemployed. Business failures and inability to meet basic living costs were two of the major reasons cited for the upswing in deaths among Japanese men in the late 1990s. An example was Masaaki Kobayashi, age fifty-one, who left a suicide note asking his firm's accountant to use Kobayashi's $3.14 million in life insurance to raise money to head off the bankruptcy of his company.

Some social analysts have suggested that additional factors come into play in explaining the high rates of suicide in any nation. Although there is no consensus on the factors associated with suicide in Japan, some analysts attribute the problem to "cultural factors" such as a belief that the group is more important than the individual and that, under certain conditions, suicide is an honorable deed. Examples include the samurai warriors who committed hara-kiri and the kamikaze pilots of World War II. Other analysts believe that the two main Japanese religions—Shintoism and Buddhism—are related to the high rate of suicides because these religions have no moral prohibition on self killing. However, it is important for us to realize that these views about cultural differences as factors remain nothing more than assumptions until more social scientists conduct systematic research on these issues and are able to more conclusively demonstrate that cause-and-effect relationships exist. Which of Durkheim's types of suicide do you think might be applicable in a study of suicide in Japan?

Sources: Based on *BBC News*, 1999; Bremner, 2000; Lamar, 2000; and Lev, 1998.

unit of analysis is *what* or *whom* is being studied (Babbie, 2001). In social science research, individuals are the most typical unit of analysis. Social groups (such as families, cities, or geographic regions), organizations (such as clubs, labor unions, or political parties), and social artifacts (such as books, paintings, or weddings) may also be units of analysis. Durkheim's unit of analysis was social groups, not individuals, because he believed that the study of individual cases of suicide would not explain the rates of suicide.

After determining the unit of analysis for your study, you must select a time frame for study:

cross-sectional or longitudinal. *Cross-sectional studies* are based on observations that take place at a single point in time; these studies focus on behavior or responses at a specific moment. *Longitudinal studies* are concerned with what is happening over a period of time or at several different points in time; they focus on processes and social change. Some longitudinal studies are designed to examine the same set of people each time, whereas others look at trends within a general population. Using longitudinal data, Durkheim was able to compare suicide rates over a period of time in France and other European nations.

An operational definition is an explanation of an abstract concept in terms of observable features that are specific enough to measure the variable. The operational definition of an *A* may be an exam average of 90 percent or above, for example. After college professors have established the grading requirements for a course, students seek to meet those expectations by performing well on examinations.

Gary Conner/PhotoEdit

5. *Collect and analyze the data.* Your next step is to collect and analyze data. You must decide which population—persons about whom we want to be able to draw conclusions—will be observed or questioned. Then it is necessary to select a sample of people from the larger population to be studied. It is important that the sample accurately represent the larger population. For example, if you arbitrarily selected five students from your sociology class to interview, they probably would not be representative of your school's total student body. However, if you selected five students from the total student body by a random sample, they might be closer to being representative (although a random sample of five students would be too small to yield much useful data). In *random sampling,* **every member of an entire population being studied has the same chance of being selected.** You would have a more representative sample of the total student body, for example, if you placed all the students' names in a rotating drum and conducted a drawing. By contrast, in *probability sampling,* **participants are deliberately chosen because they have specific characteristics,** possibly including such factors as age, sex, race/ethnicity, and educational attainment.

In addition to problems with sampling, sociologists must maintain the validity and reliability of the data they collect. *Validity* **is the extent to which a study or research instrument accurately measures what it is supposed to measure.** For example, sociologists who analyze the relationship between religious beliefs and suicide must determine whether "church membership" is

an accurate indicator of a person's religious beliefs. In fact, one person may be very religious but not belong to a specific church, whereas another person may be a member of a church yet not hold any deep religious convictions. To maintain validity, some sociologists study the relationship between suicide and religion not only in terms of people's specific behaviors (e.g., frequency of attendance at church services) but also as a set of values, beliefs, or attitudes (Breault, 1986). *Reliability* **is the extent to which a study or research instrument yields consistent results** when applied to different individuals at one time or to the same individuals over time. An important issue in reliability is the fact that sociologists have found that the characteristics of interviewers and how they ask questions may produce different answers from the people being interviewed. As a result, different studies of college students who have contemplated suicide may arrive at different conclusions. Problems of validity are also linked to how data is analyzed. *Analysis* is the process through which data are organized so that comparisons can be made and conclusions drawn. Sociologists use many techniques to analyze data. The process for each type of research method is discussed later in this chapter.

In Durkheim's study, he collected data from vital statistics for approximately 26,000 suicides. He classified them separately according to age, sex, marital status, presence or absence of children in the family, religion, geographic location, calendar date, method of suicide, and a number of other variables. As Durkheim analyzed his data, four distinct categories of suicide emerged: egoistic, anomic, altruistic, and fatalistic. *Egoistic suicide* occurs among people who are isolated from any social group. For example, Durkheim concluded that suicide rates were relatively high in Protestant countries in Europe because Protestants believed in individualism and were more loosely tied to the church than were Catholics. Single people had proportionately higher suicide rates than married persons because they had a low degree of social integration, which contributed to their loneliness. In contrast, *altruistic suicide* occurs among individuals who are excessively integrated into society. An example is military leaders who kill themselves after defeat in battle because they have so strongly identified themselves with their cause that they believe they cannot live with defeat. According to Durkheim, people are more likely to kill themselves when social cohesion is either very weak or very strong.

Durkheim further observed that degree of social integration is not the only variable that influences suicide rates. Rapid social change and shifts in moral values make it difficult for people to know what is right and wrong. *Anomic suicide* results from a lack of shared values or purpose and from the absence of social regulation. By contrast, excessive regulation and oppressive discipline may contribute to *fatalistic suicide,* as in the suicides of slaves.

6. *Draw conclusions and report the findings.* After analyzing the data, your first step in drawing conclusions is to return to your hypothesis or research objective to clarify how the data relate both to the hypothesis and to the larger issues being addressed. At this stage, you note the limitations of the study, such as problems with the sample, the influence of variables over which you had no control, or variables that your study was unable to measure.

At the end of your research, it is important to report your findings. This report usually includes a review of each step you took so that others can replicate your work in substantially the same way that it was originally conducted. Social scientists generally present their findings in papers at professional meetings and publish them in academic journals and scholarly books. Durkheim reported his findings in his book *Suicide* (1964b/1897), in which he concluded that suicide is an indicator of the moral condition of a society and that the suicide rate reflects such factors as the presence or absence of strong social bonds among people, and shifting standards of behavior.

We have traced the steps in the "conventional" research process (based on quantitative research). But what steps might be taken in an alternative approach based on qualitative research?

A Qualitative Research Model

Although the same underlying logic is involved in both quantitative and qualitative sociological research, the *styles* of these two models are very different (King, Keohane, and Verba, 1994). As previously stated, qualitative research is more likely to be used when the research question does not easily lend itself to numbers and statistical methods. As compared to a quantitative model, a qualitative approach often involves a different type of research question and a smaller number of cases. As a result, the outcome of a qualitative study is a complex, more holistic picture of some particular social phenomenon or human problem (King, Keohane, and Verba, 1994; Creswell, 1998).

How might qualitative research be used to study suicidal behavior? In studying different rates of suicide among women and men, for example, the social psychologist Silvia Canetto (1992) questioned whether existing theories and quantitative research provided an adequate explanation for gender differences in suicidal behavior and decided that she would explore alternate explanations. As a result, Canetto redefined the concept of suicidal behavior to focus on outcome ("fatal" versus "nonfatal") rather than in terms of intent ("completed" versus "attempted"). Analyzing previous research, Canetto learned that most studies linked suicidal behavior in women to problems in their personal relationships, particularly with members of the opposite sex, whereas men's suicides most often were linked to performance pressure, especially when their self-esteem and independence were threatened. However, from her analysis of existing research, Canetto believed that gender differences in suicidal behavior are more closely associated with beliefs about and expectations for men and women in a particular culture rather than purely interpersonal crises (Canetto, 1992).

As in Canetto's case, researchers using a qualitative approach may engage in *problem formulation* to clarify the research question and to develop questions of concern and interest to the research participants (Reinharz, 1992). To create a research design for Canetto's study, we might start with the proposition that most studies may have attributed women's and men's suicidal behavior to the wrong causes. Next, we might decide to interview people who have attempted

Recently, sociological research on suicide has begun to look at issues such as what social factors might motivate suicide bombers. Some researchers might ask why suicide bomber Raed Abdel-Hameed Misk (shown here with his children) would take his own life in the process of committing a terrorist attack.

suicide by using a collaborative approach in which the participants suggest avenues of inquiry that the researcher should explore (Reinharz, 1992).

Although Canetto did not gather data in her study, she made an important contribution to our knowledge about gender differences in suicidal behavior by suggesting that there may be a relationship between suicide and feelings of fear, especially in cases of domestic violence. She also pointed out that cultural norms often encourage nonfatal suicide in women and fatal suicide in men (e.g., "real men" don't fail when they take their own life). Canetto concluded that most researchers do not explore social structure factors such as the effect of low income or restricted job mobility on women's suicidal behavior. Similarly, men's suicidal behavior tends to be linked to the lack of relationships with other people and the loss of social privilege (such as might occur at retirement).

In a qualitative approach, the next step is to collect and analyze data to assess the validity of the starting proposition. Qualitative researchers typically gather data in natural settings, such as where people live or work, rather than in a laboratory or other research setting. In this environment, the researcher can play a background rather than a foreground role, and the data analysis frequently uses the language of the people being studied, not the researcher. Often, this approach generates new theories and innovative research that incorporate the perspectives of people previously excluded on the basis of race, class, gender, sexual orientation, or other attributes.

Although the qualitative approach follows the conventional research approach in presenting a problem,

asking a question, collecting and analyzing data, and seeking to answer the question, it also has several unique features (based on Creswell, 1998, and Kvale, 1996):

1. *The researcher begins with a general approach rather than a highly detailed plan.* Flexibility is necessary because of the nature of the research question. The topic needs to be explored so that we can know "how" or "what" is going on, but we may not be able to explain "why" a particular social phenomenon is occurring.

2. *The researcher has to decide when the literature review and theory application should take place.* Initial work may involve redefining existing concepts or reconceptualizing how existing studies have been conducted. The literature review may take place at an early stage, before the research design is fully developed, or it may occur after development of the research design, and after the data collection has already occurred. Many of us who teach sociological theory would like to see greater use of theory to inform both qualitative and quantitative studies because this approach provides a framework for interpreting the data collected (see also Kvale, 1996).

3. *The study presents a detailed view of the topic.* Qualitative research usually involves a smaller number of cases and many variables, whereas quantitative researchers typically work with a few variables and many cases (Creswell, 1998).

4. *Access to people or other resources that can provide the necessary data is crucial.* Unlike the quantitative researcher, who often uses existing databases,

Qualitative researchers often generate their own data by using research methods such as the focus group. In this photo, the focus group is being observed through a one-way mirror. What are the strengths of such an approach? What are the limitations?

Spencer Grant/Stock Boston

many qualitative researchers generate their own data. As a result, it is necessary to have access to people and build rapport with them.

5. *Appropriate research method(s) are important for acquiring useful qualitative data.* Qualitative studies are often based on field research such as observation, participant observation, case studies, ethnography, and unstructured interviews, as discussed in the next section.

RESEARCH METHODS

How do sociologists know which research method to use? Are some approaches better than others? Which method is best for a particular problem? **Research methods are specific strategies or techniques for systematically conducting research.** The methods should be acceptable to a larger community of scholars and nonacademic researchers who routinely engage in research endeavors (see Box 2.3). Qualitative researchers frequently attempt to study the social world from the point of view of the people they are studying. By contrast, quantitative researchers generally use surveys, secondary analyses of existing statistical data, and experimental designs. We will now look at these research methods.

Survey Research

A *survey* is a poll in which the researcher gathers facts or attempts to determine the relationships among facts. Surveys are often done when the re-

searcher wants to describe, compare, and predict knowledge, attitudes, and behavior. For example, a community survey might describe and compare such things as income, educational level, and type of employment in regard to people's attitudes about a juvenile curfew ordinance that prohibits adolescents from being out on the streets at certain nighttime hours.

Researchers frequently select a representative sample (a small group of respondents) from a larger population (the total group of people) to answer questions about their attitudes, opinions, or behavior. For example, if the larger population consists of 10,000 people, 51 percent of whom are female, with 30 percent over age 35, a representative sample will have fewer people (perhaps 1,000) but must still consist of 51 percent females, with 30 percent over age 35 (Fink, 1995). ***Respondents* are persons who provide data for analysis through interviews or questionnaires.** The Gallup and Harris polls are among the most widely known large-scale surveys; however, government agencies such as the U.S. Census Bureau conduct a variety of surveys as well. Unlike many polls that use various methods of gaining a representative sample of the larger population, the Census Bureau attempts to gain information from all persons in the United States. The decennial census occurs every 10 years, in the years ending in "0." The purpose of this census is to count the population and housing units of the entire United States. The population count determines how seats in the U.S. House of Representatives are apportioned; however, census figures are also used in formulating public policy and in planning and decision making in the private sector. The Census Bureau attempts to survey the *entire* U.S. population by using two forms—a "short form" of

Box 2.3 CHANGING TIMES: MEDIA AND TECHNOLOGY

Using the Internet for Research

Need Term Paper Help . . . *FAST?!?*

OUR GUARANTEE–You <u>WILL</u> find a model research paper on THIS site or *we'll write one as FAST as you need!* There are more than 100,000 example term papers listed at Fastpapers.com– available for <u>same day delivery</u> via email, fax or Federal Express! Search our model term papers by keyword or topic. *FAST!!!* THERE ARE NO SIGN-UP FEES & NO MEMBERSHIP IS REQUIRED! YOU PAY ONLY FOR THOSE PAPERS <u>YOU</u> CHOOSE TO ORDER!! (The Paper Store Enterprises, 2003)

I found this ad and a number of similar ones on the Internet after a student asked me, "Did you know that your textbook is used as a reference in online research papers that are being sold on the Internet?" Although a disclaimer on a number of the sites indicates that the papers being sold are for reference only, and that students should write their own paper, this is not what actually happens in many instances: Students simply hand in a paper they obtained from such a source (possibly with a few minor changes), hoping that no one will notice.

We have access to a wealth of useful information in doing research today. Radio, television, newspapers, magazines, CDs, DVDs, and Internet sites provide us with instant access to more ideas and information than were available to most people in previous generations during a lifetime. One negative aspect of this information explosion is the fact that it has become very easy to use the work of other people without acknowledging their role in producing what we claim as our own. At best, that is unintentional plagiarism; at worst, it is outright fraud. On work that you do in connection with a course, it may result in you receiving a failing grade or even being expelled from college.

According to the *Code of Ethics* of the American Sociological Association (1997), in order to avoid plagiarism, we should identify, credit, and reference the author of any material taken from another person's work, and we should never present others' work as our own, regardless of whether or not that other work is published. As the sociologist Earl Babbie (1998) concisely states in an online article, "Plagiarism is the presentation of another's words or ideas as your own. It is a **bad** thing. Don't do it." To show Babbie's excellent example, I am now going to quote directly from his article:

> Let's suppose you were assigned to write a book review of Theodore M. Porter's book, *Trust in Numbers: The Pursuit of Objectivity in Science and Public Life* (Princeton, NJ: Princeton University Press, 1995). In preparing to write your paper, you come across a book review by Lisa R. Staffen, published in *Contemporary Sociology* (March 1996, Vol. 25, No. 2, pp. 154-156). Staffen's review begins as follows:
>> It has become fashionable to reject the notion of absolute objectivity on the grounds that objectivity is simply unattainable or, even if attainable, is undesirable.

Based on this opening, Babbie points out that to simply change Staffen's statement (in your own paper) to read "*I feel* it has become fashionable to reject the notion of absolute objectivity on the grounds that objectivity is simply unattainable" is blatant plagia-

---

questions asked of *all* respondents, and a "long form" that contains additional questions asked of a *representative sample* of about one in six respondents. Statistics from the Census Bureau provide information that sociologists use in their research. An example is shown in the Census Profiles feature: "How People in the United States Self-Identify as to Race." Note that because of recent changes in the methods used to collect data by the Census Bureau, information on race from the 2000 census is not directly comparable with data from earlier censuses.

Surveys are the most widely used research method in the social sciences because they make it possible to study things that are not directly observable—such as people's attitudes and beliefs—and to describe a population too large to observe directly (Babbie, 2001). Let's take a brief look at the most frequently used types of surveys.

Types of Surveys Survey data are collected by using self-administered questionnaires, face-to-face interviews, and/or telephone interviews. A ***questionnaire*** **is a printed research instrument containing a series of items to which subjects respond.** Items are often in the form of statements with which the respondent is asked to "agree" or "disagree." Question-

rism. Likewise, to change the passage to read "*I feel it has become stylish to reject the idea of absolute objectivity on the grounds that objectivity cannot be achieved*" (italicized words reflecting changes from Staffen's original comment) would also constitute plagiarism: "the idea expressed, along with many of the phrases, have been taken from someone else, without acknowledging that fact" (Babbie, 1998).

As Babbie correctly notes, even in instances where only a few of the original words remain in a passage, incorporating the thoughts of another person into your own work without attributing them to that person still constitutes plagiarism. Based upon the same passage from Staffen's review, Babbie gives this example of this form of plagiarism: "*Many people today have rejected the idea that there is such a thing as absolute objectivity since they do not believe that it can be achieved.*"

By now, you may be wondering how to use other people's words and ideas in your own papers, and the answer lies—as Babbie (1998) suggests—in acknowledgment and citation. Here are several examples he gives of proper attribution of Staffen's article:

1. Lisa Staffen (1996:154) begins her review of Porter's book by suggesting "It has become fash-ionable to reject the notion of absolute objectivity on the grounds that objectivity is simply unattainable or, even if attainable, is undesirable."
2. In her review of Porter's book, Lisa Staffen (1996:154) says the idea of absolute objectivity is now commonly rejected as "simply unattainable or, even if attainable, [as] undesirable."
3. According to Lisa Staffen (1996:154), it has become fashionable to reject the idea of absolute objectivity altogether.

Babbie further notes that the bibliography or references at the end of the paper should include the following information:

Lisa R. Staffen, "Featured Essays," *Contemporary Sociology,* March 1996, Vol. 25, No. 2, pp. 154-156.

There is a lot of information available—from traditional sources such as books and scholarly journals and from newer resources such as the Internet—to assist you in conducting research and writing papers. By all means, you should use it. However, don't simply copy someone else's work and pretend that it is your own. As Earl Babbie (1998) states in his article, plagiarism is lying, and it is an insult to your instructor and your fellow students.

Source: Based on Babbie (1998), used with permission. To read Earl Babbie's full article, "Plagiarism," go to http://www.csubak.edu/ssric/Modules/Other/plagiarism.htm.

WRITING IN SOCIOLOGY ASSIGNMENT

Discuss the problem of plagiarism at colleges and universities. Explain how technology has  made plagiarism easier for students but has also made plagiarism easier to detect.

naires may be administered by interviewers in face-to-face encounters or by telephone, but the most commonly used technique is the *self-administered questionnaire.* The questionnaires are typically mailed or delivered to the respondents' homes; however, they may also be administered to groups of respondents gathered at the same place at the same time.

Sociologist Kevin E. Early (1992), for example, conducted a survey regarding the lower rates of suicide among African Americans than whites in the United States. Early collected and analyzed survey data to test his hypothesis that "the black church's influence is an essential factor in ameliorating and buffering social forces that otherwise would lead to suicide." A self-administered questionnaire was completed by congregation members in conjunction with services at six black churches considered representative of the thirty-seven black churches in Gainesville, Florida.

Self-administered questionnaires have certain strengths. They are relatively simple and inexpensive to administer, they allow for rapid data collection and analysis, and they permit respondents to remain anonymous (an important consideration when the questions are of a personal nature). A major disadvantage is the low response rate. Mailed surveys sometimes

Conducting surveys and polls is an important means of gathering data from respondents. Some surveys take place on street corners; however, increasingly, such surveys are done by telephone, Internet, or other means.

© HIRB/IndexStock Imagery

have a response rate as low as 10 percent—and a 50-percent response rate is considered by some to be minimally adequate (Babbie, 2001). The response rate is usually somewhat higher if the survey is handed out to a group that is asked to fill it out on the spot. Moreover, for surveys involving ethnically diverse or international respondents, the questionnaire must be available in languages other than English (Fink, 1995).

Survey data may also be collected by interviews. An **interview is a data-collection encounter in which an interviewer asks the respondent questions and records the answers.** Survey research often uses *structured interviews,* in which the interviewer asks questions from a standardized questionnaire. Structured interviews tend to produce uniform or replicable data that can be elicited time after time by different interviews. For example, in addition to surveying congregation members, Early (1992) conducted interviews with pastors of African American churches, using a series of open-ended questions. Next, he read four vignettes (stories about people) relating to suicide to the pastors and then asked questions designed to determine the pastors' opinions and attitudes concerning the behavior displayed in the vignettes. His goal was to learn the extent to which the African American church reinforces attitudes, values, beliefs, and norms that discourage suicide.

Unlike the open-ended questions used in Early's more qualitative approach, closed-ended questions may be used when researchers want to have a large number of respondents and to generate standardized answers to questions. For example, in a study involv-

ing forty-nine hospital accident and emergency departments and psychiatric services in the United Kingdom, researchers developed closed-ended questions to be asked of patients who had attempted suicide by poisoning. The purpose of the study was to determine if media representation of suicide and deliberate self-harm encouraged suicidal behavior in vulnerable individuals. Specifically, the researchers wanted to know if the rate of self-poisoning and the choice of overdose drugs was influenced by a television drama, *Casualty,* which portrayed an acetaminophen overdose in one of its episodes. Questionnaires were completed by more than 1,000 self-poisoning patients during the three-week periods before and after the program was broadcast. Was there a direct link between viewing the episode and the person's decision to take an overdose, choice of drug, and speed with which he or she arrived at the hospital? According to the researchers, there was a 17-percent increase in the number of hospital patients who reported that they engaged in self-poisoning in the week after the broadcast, and a 9-percent increase in the second week. Moreover, the rate of poisonings by acetaminophen increased more than that of any other drug. In fact, 20 percent of the patients interviewed indicated that the program had influenced their decision to overdose, and 17 percent said it had influenced their drug choice (Hawton, Simkin, Deeks, et al., 1999).

Interviews have specific advantages. They are usually more effective in dealing with complicated issues and provide an opportunity for face-to-face communication between the interviewer and the respondent. When open-ended questions are used, the researcher may gain new perspectives. The pastors interviewed in Early's study distinguished between suicide (which is "unthinkable for black people" because it is a "white thing, not a black thing") and alcohol abuse, drug addiction, and homicide (which are wrong and "sinful" but are "an understandable response to the socioeconomic and political conditions of blacks in the United States") (Early, 1992: 79). When closed-ended questions are used, it is easier for interviewers to code responses and for researchers to compare individuals' responses across categories of interest. For example, researchers in the overdose study were able to compare such variables as sex, age, choice of overdose drug, history of taking overdoses, and whether the choice of substance was influenced by television programs. Based on their findings, the researchers argued that media portrayals of self-poisoning or self-injury on popular television shows may contribute to self-harming behavior and choice of method used, and thus should be of concern to the general public and to media producers as well (Hawton, Simkin, Deeks, et al., 1999). As this and other research studies show, interviews provide a wide variety of useful information; however, a major disadvantage is the cost and time involved in conducting the interviews and analyzing the results. Also, one weakness of interviews is that people may be influenced by the interviewer's race, age, sex, size, or other attributes in responding to the questions asked.

A quicker method of administering questionnaires is the *telephone survey,* which is becoming an increasingly popular way to collect data. Telephone surveys save time and money as compared to self-administered questionnaires or face-to-face interviews. Some respondents may be more honest than when they are facing an interviewer. Telephone surveys also give greater control over data collection and provide greater personal safety for respondents and researchers than do personal encounters. In *computer-assisted telephone interviewing* (sometimes called CATI), the interviewer uses a computer to dial random telephone numbers, reads the questions shown on the video monitor to the respondent, and then types the responses into the computer terminal. The answers are immediately stored in the central computer, which automatically prepares them for data analysis. Although use of the CATI system overcomes the problem of unlisted telephone numbers by randomly dialing numbers, it is limited by people's widespread use of answering machines, voice mail, and caller ID to filter their incoming telephone calls.

CENSUS ★ PROFILES

How People in the United States Self-Identify as to Race

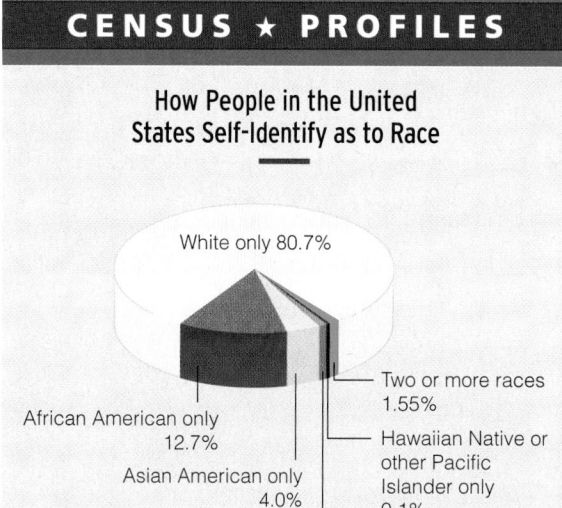

White only 80.7%

African American only 12.7%

Asian American only 4.0%

Two or more races 1.55%

Hawaiian Native or other Pacific Islander only 0.1%

Native American only 0.95%

Beginning with Census 2000, the U.S. Census Bureau has made it possible for people responding to census questions regarding their race to mark more than one racial category. Although the vast majority of respondents select only one category (see above), the Census Bureau reports that in 2002 approximately 4.2 million people (1.55 percent of the population) in the United States self-identified as being of more than one race. As a result, if you look at the figures set forth below, they total more than 100 percent of the total population. How can this be? Simply stated, some individuals are counted at least twice, based on the number of racial categories they listed.

Race	Percentage of Total Population
White alone or in combination with one or more other races	81.9
African American alone or in combination with one or more other races	13.3
Asian American alone or in combination with one or more other races	4.5
Native American alone or in combination with one or more other races	1.5
Native Hawaiian or other Pacific Islander alone or in combination with one or more other races	0.3
Total	101.5

Source: U.S. Census Bureau, 2002.

Computer-assisted telephone interviewing is an easy and cost-efficient method of conducting research. The widespread use of answering machines, voice mail, and caller ID may make this form of research more difficult in the twenty-first century.

© Spencer Grant/PhotoEdit

Strengths and Weaknesses of Surveys Survey research has several important strengths. First, it is useful in describing the characteristics of a large population without having to interview each person in that population. Second, survey research enables the researcher to search for causes and effects and to assess the relative importance of a number of variables. In recent years, computer technology has enhanced our ability to do *multivariate analysis*—research involving more than two independent variables. For example, to assess the influence of religion on suicidal behavior among African Americans, a researcher might look at the effects of age, sex, income level, and other variables all at once to determine which of these independent variables influences suicide the most or least and how influential each variable is relative to the others. Third, survey research can be useful in analyzing social change or in documenting the existence of a social problem. Contemporary scholars have used survey research to provide information about such problems as racial discrimination, sexual harassment, and sex-based inequality in employment by documenting the fact that they are more widespread than previously thought (Reinharz, 1992).

Survey research also has weaknesses. One is that the use of standardized questions tends to force respondents into categories in which they may or may not belong. Another weakness concerns validity. People's opinions on issues seldom take the form of a standard response ranging from "strongly agree" to "strongly disagree." Moreover, as in other types of research, people may be less than truthful, especially on emotionally charged issues such as suicide, thus making reliance on self-reported attitudes problematic.

Some scholars have also criticized the way survey data are used. They believe that survey data do not always constitute the "hard facts" that other analysts may use to justify changes in public policy or law. For example, survey statistics may over- or underestimate the extent of a problem and work against some categories of people more than others, as shown in Table 2.1.

Secondary Analysis of Existing Data

In ***secondary analysis,*** **researchers use existing material and analyze data that were originally collected by others.** Existing data sources include public records, official reports of organizations and government agencies, and surveys conducted by researchers in universities and private corporations. Research data gathered from studies are available in data banks, such as the Inter-University Consortium for Political and Social Research, the National Opinion Research Center (NORC), and the Roper Public Opinion Research Center. Other sources of data for secondary analysis are books, magazines, newspapers, radio and television programs, and personal documents. Secondary analysis is referred to as *unobtrusive research* because it has no impact on the people being studied. In Durkheim's study of suicide, for example, his analysis of existing statistics on suicide did nothing to increase or decrease the number of people who *actually* committed suicide.

Analyzing Existing Statistics Secondary analysis may involve obtaining *raw data* collected by other researchers and undertaking a statistical analysis of the data, or it may involve the use of other researchers'

Table 2.1 STATISTICS: WHAT WE KNOW (AND DON'T KNOW)

	TOPIC		
	Homelessness in the United States	Gay Men in the United States	Suicide in the United States
Research Finding	At least 250,000 people in this country are homeless.	At least 1.2 million American men are exclusively homosexual.	At least 30,575 Americans committed suicide in 1998.
Possible Problem	Does that badly under-estimate the total number of homeless people?	That is about 1 percent of the male population— is the actual figure ten times that high?	Are suicide rates different for some categories of U.S. citizens?
Explanation	The homeless are difficult to count, frequently attempting to avoid interviews with census takers. Critics of the census figures assert that the actual number may be 3 million and that the government intentionally undercounts the homeless.	As one analyst noted, many people lie about their sexuality more than anything else; people are often afraid to say they are gay. This may result in estimates being too low; however, gay rights organizations may overstate percentages to gain political clout.	U.S. census data place Latino/as in the category of whites. Other than African Americans, all other people of color are listed as "nonwhite-other." Thus, census data on specific cate-gories are not available.

existing statistical analyses. In analysis of existing statistics, the unit of analysis is often *not* the individual. Most existing statistics are *aggregated:* They describe a group. Durkheim wanted to determine whether Protestants or Catholics were more likely to commit suicide; however, none of the available records indicated the religion of those who committed suicide. Although Durkheim suggested that Protestants were more likely to commit suicide than Catholics, it was impossible for him to determine that from the existing data.

In a contemporary study of suicide, K. D. Breault (1986) analyzed secondary data collected by government agencies to test Durkheim's hypothesis that religion and social integration provide protection from suicide. Using suicide as the dependent variable and church membership, divorce, unemployment, and female labor force participation as several of his independent variables, Breault performed a series of sophisticated statistical analyses and concluded that the data supported Durkheim's views on social integration and his theory of egoistic suicide. He also found support for Durkheim's proposition that Catholics are less likely to commit suicide than are Protestants. However, it should be noted that Durkheim did not attribute lower rates of suicide among Catholics to the role of church beliefs as much as to the tendency of Catholicism to promote social integration through rituals and regulation of standards of

faith and moral conduct (Ellison, Burr, and McCall, 1997).

Numerous other studies have used secondary data to examine the relationship between religious factors and rates of suicide. For example, in a recent study, researchers used data from sources including the National Center for Health Statistics and a large survey of religious denominations to examine the extent to which religious homogeneity—how well community residents adhere to a single religion or a small number of faiths—is associated with lower suicide rates (Ellison, Burr, and McCall, 1997). Using larger categories such as conservative Protestant, moderate Protestant, Catholic, Mormon, Orthodox, and Jewish, the researchers concluded that religious homogeneity was linked with lower suicide rates, particularly in the northeastern and southern United States (Ellison, Burr, and McCall, 1997).

Analyzing Content *Content analysis* **is the systematic examination of cultural artifacts or various forms of communication to extract thematic data and draw conclusions about social life.** *Cultural artifacts* are products of individual activity, social organizations, technology, and cultural patterns (Reinharz, 1992). Among the materials studied are *written records,* such as diaries, love letters, poems, books, and graffiti, and *narratives and visual texts,* such as movies, television programs, advertisements,

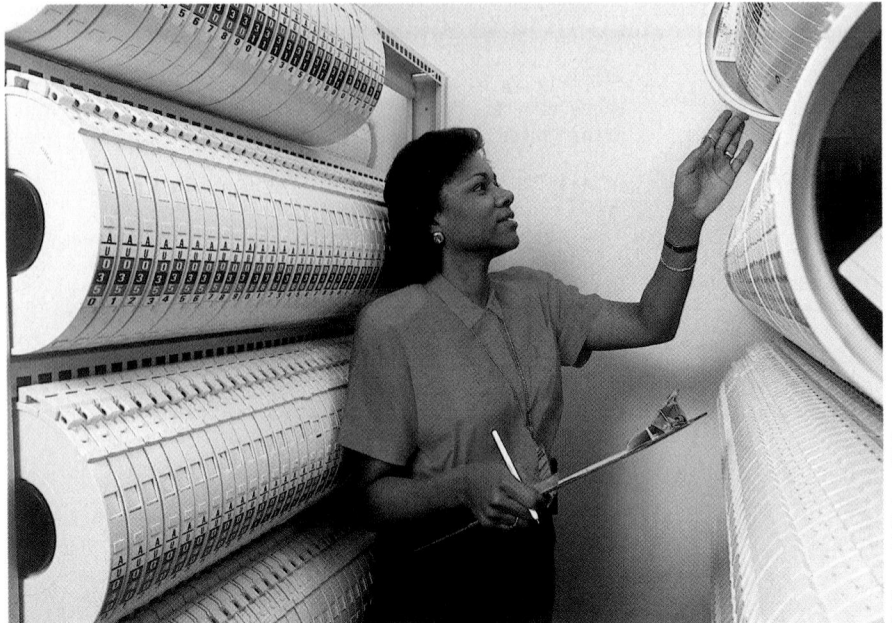

Today, many researchers conduct secondary analysis of existing data that have been generated by sources such as the U.S. Census Bureau. Regional data banks, such as this one in Austin, Texas, collect and process data for the Census Bureau.

Bob Daemmrich/The Image Works

and greeting cards. Also studied are *material culture,* such as music, art, and even garbage, and *behavioral residues,* such as patterns of wear and tear on the floors in front of various exhibits at museums to determine which exhibits are the most popular (see Webb, 1966). Harriet Martineau stated that more could be learned about a society in a day by studying "things" than by talking with individuals for a year (Martineau, 1988/1838). Researchers may look for regular patterns, such as frequency of suicide as a topic on television talk shows. They may also examine subject matter to determine how it has been handled, such as how the mass media handle "celebrity" suicides.

Content analysis provides objective coding procedure for analyzing written material (see Berg, 1998; Manning and Cullum-Swan, 1994). It also allows for the counting and arranging of data into clearly identifiable categories (manifest coding) and provides for the creation of analytically developed categories (latent or open coding). Using latent or open coding, it is possible to identify general themes, create generalizations, and develop "grounded theoretical" explanations (Glaser and Strauss, 1967). As this explanation suggests, researchers use both qualitative and quantitative procedures in content analysis.

How might a social scientist use content analysis in research on why people commit suicide? Suicide notes and diaries are useful forms of cultural artifacts. Suicide notes have been subjected to extensive analysis because they are "ultrapersonal documents" that are not solicited by others and frequently are written just before the person's death (Leenaars, 1988: 34).

Many notes provide new levels of meaning regarding the *individuality* of the person who committed or attempted suicide. Suicide notes and diaries often reveal that people committing suicide consider their death as a "passing on to another world" or simply "escaping this world." Some notes indicate that people may want to get revenge and make other people feel guilty or responsible for their suicide: "Now you'll be sorry for what you did" or "It's all your fault!" Thus, suicide notes may be a valuable starting point for finding patterns of suicidal behavior and determining the characteristics of people who are most likely to commit suicide (Leenaars, 1988). Today, researchers analyze the suicide notes of both women and men. However, earlier studies of suicide notes primarily focused only on those written by men, even though women have been found to leave notes more often than do men (Lester, 1988, 1992).

Strengths and Weaknesses of Secondary Analysis
One strength of secondary analysis is that data are readily available and inexpensive. Another is that, because the researcher often does not collect the data personally, the chances of bias may be reduced. In addition, the use of existing sources makes it possible to analyze longitudinal data to provide a historical context within which to locate original research. However, secondary analysis has inherent problems. For one thing, the data may be incomplete, unauthentic, or inaccurate. A second issue is that the various data from which content analysis is done may not be strictly comparable with one another (Reinharz,

Field research takes place in a wide variety of settings. For example, how might sociologists study the ways in which parents and their college-age children cope with change when the students first leave home and move into college housing?

1992), and *coding* this data—sorting, categorizing, and organizing them into conceptual categories—may be difficult (Babbie, 2001).

Field Research

Field research **is the study of social life in its natural setting: observing and interviewing people where they live, work, and play.** Some kinds of behavior can be studied best by "being there"; a fuller understanding can be developed through observations, face-to-face discussions, and participation in events. Researchers use these methods to generate *qualitative* data: observations that are best described verbally rather than numerically. Although field research is less structured and more flexible than the other methods we have discussed, it still places many demands on the researcher. To engage in field research, sociologists must select the method or combination of methods that will best reveal what they want to know. For example, they must decide how to approach the target group, whether to identify themselves as researchers, and whether to participate in the events they are observing.

Participant Observation Sociologists who are interested in observing social interaction as it occurs may use participant observation. *Participant observation* **refers to the process of collecting data while being part of the activities of the group that the researcher is studying.** As this definition states, the researcher gains insight into some aspect of social life

by participating in what is going on while observing what is taking place.

Let's assume that you wanted to study how volunteers at a suicide prevention center learned how to counsel people by telephone. You might become a volunteer-in-training and attend the orientation sessions for volunteers, taking notes on how others responded to the information being provided and how various volunteers interacted with one another. Then you might serve as a "hot line" volunteer, not only seeking to help the people who called but also observing how the volunteers interacted with callers. Throughout this process, of course, you would have to be aware of issues relating to ethics and with conducting research with human subjects, as discussed later in this chapter.

What are the strengths of participant observation research? Participant observation generates more "inside" information than simply asking questions or observing from the outside. For example, to learn more about how coroners make a ruling of "suicide" in connection with a death and to analyze what, if any, effect such a ruling has on the accuracy of "official" suicide statistics, sociologist Steve Taylor (1982) engaged in participant observation at a coroner's office over a six-month period. As he followed a number of cases from the initial report of death through the various stages of investigation, Taylor found that it was important to "be around" so that he could listen to the discussion of particular cases and ask the coroners questions. According to Taylor, intuition and guesswork play a much larger part in coroners' decisions than they are willing to acknowledge.

What are the limitations of participant observation research? This type of research requires time and expertise on the part of the researcher, who must be able to become a participant while still maintaining some distance from those being observed. Interpreting the results of this type of research also requires that the researchers distance themselves from people with whom they may have spent a great deal of time and with whom they developed personal relationships during the research project.

Case Studies Most participant observation research takes the form of a *case study,* which is often an in-depth, multifaceted investigation of a single event, person, or social grouping (Feagin, Orum, and Sjoberg, 1991). However, case studies may also involve multiple cases and then be referred to as a *collective case study* (Stake, 1995). Whether the case is single or collective, most case studies require detailed, in-depth data collection involving multiple sources of rich information such as documents and records and the use of methods such as participant observation, unstructured or in-depth interviews, and life histories (Creswell, 1998). As they collect extensive amounts of data, the researchers seek to develop a detailed description of the case, to analyze the themes or issues that emerge, and to interpret or create their own assertions about the case (Stake, 1995).

When do social scientists decide to do case studies? Initially, some researchers have only a general idea of what they wish to investigate. In other cases, they literally "back into" the research. They may find themselves close to interesting people or situations. For example, the anthropologist Elliot Liebow "backed into" his study of single, homeless women living in emergency shelters by becoming a volunteer at a shelter. As he got to know the women, Liebow became fascinated with their lives and survival strategies. Prior to Liebow's research, most studies of the homeless focused primarily on men. These studies typically asked questions like "How many homeless are there?" and "What proportion of the homeless are chronically mentally ill?" By contrast, Liebow wanted to know more about the homeless women themselves, wondering such things as "What are they carrying in those [shopping] bags?" (Coughlin, 1993: A8). Liebow spent the next four years engaged in participant observation research that culminated in his book *Tell Them Who I Am* (1993).

In participant observation studies, the researcher must decide whether to let people know they are being studied. After Liebow decided that he would like to take notes on informal conversations and conduct interviews with the women, he asked the shelter director and the women for permission and told them that he would like to write about them. Liebow's findings are discussed in Chapter 5 ("Society, Social Structure, and Interaction"). Although some social scientists gain permission from their subjects, others fear that people will refuse to participate or will change their behavior if they know they are being observed. On the one hand, researchers who do not obtain consent from their subjects may be acting unethically. On the other hand, when subjects know they are being observed, they risk succumbing to the Hawthorne effect (discussed later in this chapter).

The next step is to gain the trust of participants. Liebow had previous experience in blending in with individuals he wanted to observe when he gained the trust of young, lower-class African American men who talked and passed time on an inner-city street corner in Washington, D.C., in the 1960s. In his classic study *Tally's Corner* (1967), Liebow described how he (as a thirty-seven-year-old white anthropology graduate student) played pool and drank beer with his subjects. While interacting with the men, Liebow gathered a large volume of data that led him to conclude that his subjects had created their own "society" after being unable to find a place in the existing one. Liebow found "insiders" to help him gain the trust of other participants in his research. In a participant observation study, you may wish to identify possible *informants*—individuals who introduce you to others, give suggestions about how to "get around" in the natural setting, and provide you with essential insider information on what you are observing. Informants are especially useful in the community study/ethnography.

Ethnography An **ethnography is a detailed study of the life and activities of a group of people by researchers who may live with that group over a period of years** (Feagin, Orum, and Sjoberg, 1991). Although this approach is similar in some ways to participant observation, these studies typically take place over much longer periods of time. In fact, ethnography has been referred to as "the study of the way of life of a group of people" (Prus, 1996). For example, *Middletown* and *Middletown in Transition* describe the sociologists Robert Lynd and Helen Lynd's (1929, 1937) study in Muncie, Indiana. The Lynds, who lived in this midwestern town for a number of years, applied ethnographic research to the daily lives of residents, conducting interviews and reading newspaper files in order to build a historical base for their own research. The Lynds showed how a dominant family "ruled" the city and how the working class developed as a result of industry moving into Muncie. They concluded that the people had strong

Sociologist Elijah Anderson's fourteen-year study of two Philadelphia neighborhoods—one populated by low-income African Americans, the other racially mixed but becoming increasingly middle- to upper-income and white—is an example of ethnographic research.

beliefs about the importance of religion, hard work, self-reliance, and civic pride. When a team of sociologists returned to Muncie in the late 1970s, they found that the people there still held these views (Bahr and Caplow, 1991).

In another classic study, *Street Corner Society,* the sociologist William F. Whyte (1988/1943) conducted long-term participant observation studies in Boston's low-income Italian neighborhoods. Whereas "outsiders" generally regarded these neighborhoods as disorganized slums with high crime rates, Whyte found the residents to be hardworking people who tried to take care of one another. More recently, the sociologist Elijah Anderson (1990) conducted a study in two Philadelphia neighborhoods—one populated by low-income African Americans, the other racially mixed but becoming increasingly middle- to upper-income and white. Over the course of fourteen years, Anderson spent numerous hours on the streets, talking and listening to the people (Anderson, 1990: ix). In this longitudinal study, Anderson was able to document the changes brought about by drug abuse, loss of jobs, decreases in city services despite increases in taxes, and the eventual exodus of middle-income people. As these examples show, ethnographic work involves not only immersing oneself into the group or community that the researcher studies but also engaging in dialogue to learn more about social life through ongoing interaction with others (Burawoy, 1991).

Unstructured Interviews An ***unstructured interview*** **is an extended, open-ended interaction between an interviewer and an interviewee.** This type

of interview is referred to as an *unstructured,* or *nonstandardized, interview* because few predetermined or standardized procedures are established for conducting this type of interview. Since many decisions have to be made during the interview, this approach requires that the researcher have a high level of skill in interviewing and extensive knowledge regarding the interview topic (Kvale, 1996). Here, the interviewer has a general plan of inquiry but not a specific set of questions that must be asked, as is often the case with surveys. Unstructured interviews are essentially conversations in which interviewers establish the general direction by asking open-ended questions, to which interviewees may respond flexibly. Interviewers have the ability to "shift gears" to pursue specific topics raised by interviewees, because answers to one question are used to suggest the next question or new areas of inquiry.

Sociologist Joe Feagin's (1991) study of middleclass African Americans is an example of research that used in-depth interviews to examine public discrimination and victims' coping strategies. No specific questions were asked regarding discrimination in public accommodations or other public places. Rather, discussion of discrimination was generated by answers to general questions about barriers to personal goals or in digressions in answers to specific questions about employment, education, and housing (Feagin, 1991).

Even in unstructured interviews, researchers must prepare a few general or "lead-in" questions to get the interview started. Following the interviewee's initial responses, the interviewer may wish to ask additional questions on the same topic, probe for more information (by using questions such as "In what ways?" or "Anything else?"), or introduce a new line of inquiry. At all points in the interview, *careful listening* is essential. It provides the opportunity to introduce new questions as the interview proceeds while simultaneously keeping the interview focused on the research topic. It also enables the interviewer to envision the interviewees' experiences and to glean multiple levels of meaning.

The Interview and Sampling Process

Before conducting in-depth interviews, researchers must make a number of decisions, including how the people to be interviewed will be selected. Respondents for unstructured interviews are often chosen by "snowball sampling." In *snowball sampling,* the researcher interviews a few individuals who possess a certain characteristic; these interviewees are then asked to supply the names of others with the same characteristic. The process continues until the sample has "snowballed" into an acceptable size and no new information of any significance is being gained.

Researchers must make other key decisions. Will people be interviewed more than once? If so, how long will the interviews be? Are there a specific number and order of questions to be followed? Will the interviewees have an opportunity to question the interviewer? Where will the interview take place? How will information be recorded? Who should do the interviewing? Who should be present at the interview (Reinharz, 1992; Kvale, 1996)? Unstructured, open-ended interviews do not mean that the researcher simply walks into a room, has a conversation with someone, and the research is complete. Planning and preparation are essential. Similarly, the follow-up, analysis of data, and write-up of the study must be carefully designed and carried out.

Interviews and Theory Construction

In-depth interviews, along with participant observation and case studies, are frequently used to develop theories through observation. The term *grounded theory* was developed by sociologists Barney Glaser and Anselm Strauss (1967) to describe this inductive method of theory construction. Researchers who use grounded theory collect and analyze data simultaneously. For example, after in-depth interviews with 106 suicide attempters, researchers in one study concluded that half of the individuals who attempted suicide wanted *both* to live *and* to die at the time of their attempt. From these unstructured interviews, it became obvious that ambivalence led about half of "se-

rious" suicidal attempters to "literally gamble with death" (Kovacs and Beck, 1977, quoted in Taylor, 1982: 144). After asking their initial unstructured questions of the interviewees, Kovacs and Taylor decided to widen the research question from "Why do people kill themselves?" to a broader question: "Why do people engage in acts of self damage which may result in death?" In other words, uncertainty of outcome is a common feature of most suicidal acts. In previous studies, researchers had simply assumed that in "dangerous attempts" the individual really wanted to die whereas in "moderate" attempts the person was ambivalent (Taylor, 1982: 160).

Strengths and Weaknesses of Field Research

Participant observation research, case studies, ethnography, and unstructured interviews provide opportunities for researchers to view from the inside what may not be obvious to an outside observer. They are useful when attitudes and behaviors can be understood best within their natural setting or when the researcher wants to study social processes and change over a period of time. They provide a wealth of information about the reactions of people and give us an opportunity to generate theories from the data collected (Whyte, 1989). For example, through unstructured interviews, researchers gain access to "people's ideas, thoughts, and memories in their own words rather than in the words of the researcher" (Reinharz, 1992: 19). Research of this type is important for the study of race, ethnicity, and gender because it often includes those who have been previously excluded from studies and provides information on them.

Social scientists who believe that quantitative research methods (such as survey research) provide the most scientific and accurate means of measuring attitudes, beliefs, and behavior are often critical of data obtained through field research. They argue that what is learned from a specific group or community cannot be generalized to a larger population. They also suggest that the data collected in natural settings are descriptive and do not lend themselves to precise measurement. Researchers who want to determine cause and effect or to test a theory emphasize that it is impossible to demonstrate such relationships from participant observation studies. For these reasons and others, some qualitative researchers (particularly ethnographers) use computer-assisted qualitative data analysis (CAQDA) programs. Such programs make it easier for researchers to enter, organize, annotate, code, retrieve, count, and analyze data (Dohan and Sanchez-Jankowski, 1998). However, other ethnographers and field researchers do not use CAQDA pro-

Not all experiments occur in laboratory settings. Natural experiments may be conducted when some unforeseen event occurs. From a sociological perspective, what adaptation strategies did these people use during the massive 2003 power outage in the northeastern United States?

© 2003 AP/Wide World Photos

grams in their research (see Charmaz and Olesen, 1997; Horowitz, 1997; and Morrill and Fine, 1997).

Experiments

An *experiment* **is a carefully designed situation in which the researcher studies the impact of certain variables on subjects' attitudes or behavior.** Experiments are designed to create "real-life" situations, ideally under controlled circumstances, in which the influence of different variables can be modified and measured.

Types of Experiments Conventional experiments require that subjects be divided into two groups: an experimental group and a control group. The *experimental group* **contains the subjects who are exposed to an independent variable** (the experimental condition) to study its effect on them. The *control group* **contains the subjects who are not exposed to the independent variable.** The members of the two groups are matched for similar characteristics so that comparisons may be made between the groups. In the simplest experimental design, subjects are (1) pretested (measured) in terms of the dependent variable in the hypothesis, (2) exposed to a stimulus representing an independent variable, and (3) post-tested (remeasured) in terms of the dependent variable. The experimental and control groups are then compared to see if they differ in relation to the dependent variable, and the hypothesis stating the relationship of the two variables is confirmed or rejected.

In a *laboratory experiment,* subjects are studied in a closed setting so that researchers can maintain as much control as possible over the research. For example, if you wanted to examine the influence of the media on attitudes regarding suicide, you might decide to use a laboratory experiment. Sociologist Arturo Biblarz and colleagues (1991) designed a laboratory study to investigate the effects of the media on people's attitudes toward suicide. Researchers showed one group of subjects a film about suicide, showed a second group a film about violence, and showed a third a film containing neither suicide nor violence. Some evidence was found that media exposure to suicidal acts or violence may arouse an emotional state favorable to suicidal behavior, especially in those persons already "at risk" for suicide.

Not all experiments occur in laboratory settings. *Natural experiments* are real-life occurrences such as floods and other disasters that provide researchers with "living laboratories." Sociologist Kai Erikson (1976) studied the consequences of a deadly 1972 flood in Buffalo Creek, West Virginia, and found that extensive disruption of community ties occurred. Natural experiments cannot be replicated because it is impossible to re-create the exact conditions, nor would we want to do so.

Demonstrating Cause-and-Effect Relationships Researchers may use experiments when they want to demonstrate that a cause-and-effect

Figure 2.4   Correlation Versus Causation

A study might find that exposure to a suicide hot line is associated (correlated) with a change in attitude toward suicide. But if some of the students who were exposed to the hot line also received psychiatric counseling, the counseling may be the "hidden" cause of the observed change in attitude. In general, correlations alone do not prove causation.

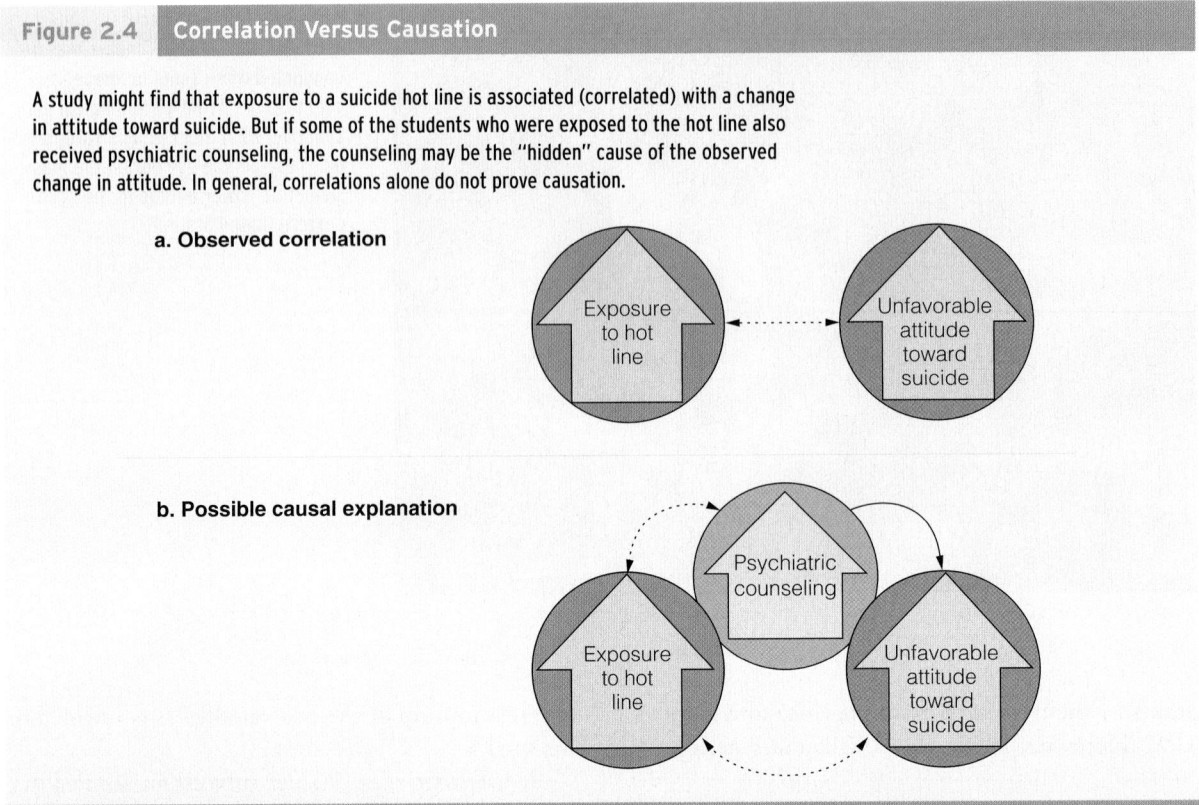

a. Observed correlation

b. Possible causal explanation

relationship exists between variables. In order to show that a change in one variable causes a change in another, these three conditions must be fulfilled:

1. *You must show that a correlation exists between the two variables.* **Correlation *exists when two variables are associated more frequently than could be expected by chance*** (Hoover, 1992). For example, suppose that you wanted to test the hypothesis that the availability of a crisis intervention center with a twenty-four-hour counseling "hot line" on your campus causes a change in students' attitudes toward suicide (see Figure 2.4). To demonstrate correlation, you would need to show that the students had different attitudes toward committing suicide depending on whether they had any experience with the crisis intervention center.

2. *You must ensure that the independent variable preceded the dependent variable.* If differences in students' attitudes toward suicide were evident before the students were exposed to the intervention center, exposure to the center could not be the cause of these differences.

3. *You must make sure that any change in the dependent variable was not due to an extraneous variable*—one outside the stated hypothesis. If some of the students receive counseling from off-

campus psychiatrists, any change in attitude that they experience could be due to this third variable and not to the hot line. This is referred to as a *spurious correlation*—the association of two variables that is actually caused by a third variable and does not demonstrate a cause-and-effect relationship.

Strengths and Weaknesses of Experiments

The major advantage of the controlled experiment is the researcher's control over the environment and the ability to isolate the experimental variable. Since many experiments require relatively little time and money and can be conducted with limited numbers of subjects, it is possible for researchers to replicate an experiment several times by using different groups of subjects. Replication strengthens claims about the validity and generalizability of the original research findings (Babbie, 2001).

Perhaps the greatest limitation of experiments is that they are artificial. Social processes that occur in a laboratory setting often do not occur in the same way in real-life settings. For example, social scientists frequently rely on volunteers or captive audiences. As a result, the subjects of most experiments may not be representative of a larger population, and the findings cannot be generalized to other groups.

Experiments have several other limitations. First, the rigid control and manipulation of variables de-

Multiple research methods are often used to gain information about important social concerns. Which methods might be most effective in learning more about the problems of homeless persons, such as these in India?

© Rob Crandall/The Image Works

manded by experiments do not allow for a more communal approach to data gathering. Second, biases can influence each of the stages in an experiment, and research subjects may become the objects of sex/class/race biases. Third, the unnatural characteristics of laboratory experiments and of group competition in such settings have a negative effect on subjects (Reinharz, 1992).

Researchers acknowledge that experiments have the additional problem of *reactivity*—the tendency of subjects to change their behavior in response to the researcher or to the fact that they know they are being studied. This problem was first noted in a study conducted between 1927 and 1932 by the social psychologist Elton Mayo, who used a series of experiments to determine how worker productivity and morale might be improved at Western Electric's Hawthorne plant. To identify variables that tend to increase worker productivity, Mayo separated one group of women (the experimental group) from the other workers and then systematically varied factors in that group's work environment while closely observing them. Meanwhile, the working conditions of the other workers (the control group) were not changed. The researchers tested a number of hypotheses, including one stating that an increase in the amount of lighting would raise the workers' productivity. Much to the researchers' surprise, the level of productivity rose not only when the lighting was brightened but also when it was dimmed.

Indeed, all of the changes increased productivity. Mayo concluded that the subjects were trying to please the researchers because of the interest being shown in the subjects (Roethlisberger and Dickson, 1939). Thus, the *Hawthorne effect* **refers to changes in the subject's behavior caused by the researcher's presence or by the subject's awareness of being studied.** Other aspects of this study are discussed in Chapter 6 ("Groups and Organizations").

Multiple Methods: Triangulation

What is the best method for studying a particular topic? How can we get accurate answers to questions about suicide and other important social concerns? Concept Table 2.A compares the various social research methods. There is no one best research method because of the "complexity of social reality and the limitations of all research methodologies" (Snow and Anderson, 1991: 158).

Many sociologists believe that it is best to combine multiple methods in a given study. *Triangulation* is the term used to describe this approach (Denzin, 1989). Triangulation refers not only to research methods but also to multiple data sources, investigators, and theoretical perspectives in a study. Multiple data sources include persons, situations, contexts, and time (Snow and Anderson, 1991). For example, in a study of "unattached homeless men and women living in and passing through Austin, Texas, in the mid-1980s," sociologists David Snow and Leon Anderson (1991: 158) used as their primary data sources "the homeless themselves and the array of settings, agency personnel, business proprietors, city officials, and neighborhood activities relevant to the routines of the homeless." Snow and Anderson gained a detailed portrait of the homeless and their experiences and institutional contacts by tracking more than seven hundred homeless individuals through a network of seven institutions with which they had varying degrees of contact.

The study also tracked a number of the individuals over a period of time and used a variety of methods,

Concept Table 2.A   STRENGTHS AND WEAKNESSES OF SOCIAL RESEARCH METHODS

Research Method	Strengths	Weaknesses
Experiments (Laboratory, Field, Natural)	Control over research Ability to isolate experimental factors Relatively little time and money required Replication possible, except for natural experiments	Artificial by nature Frequent reliance on volunteers or captive audiences Ethical questions of deception
Survey Research (Questionnaire, Interview, Telephone Survey)	Useful in describing features of a large population without interviewing everyone Relatively large samples possible Multivariate analysis possible	Potentially forced answers Respondent untruthfulness on emotional issues Data that are not always "hard facts" presented as such in statistical analyses
Secondary Analysis of Existing Data (Existing Statistics, Content Analysis)	Data often readily available, inexpensive to collect Longitudinal and comparative studies possible Replication possible	Difficulty in determining accuracy of some of the data Failure of data gathered by others to meet goals of current research Questions of privacy when using diaries, other personal documents
Field Research (Participant Observation, Case Study, Ethnography, Unstructured Interview)	Opportunity to gain insider's view Useful for studying attitudes and behavior in natural settings Longitudinal/comparative studies possible Documentation of important social problems of excluded groups possible Access to people's ideas in their words Forum for previously excluded groups Documentation of need for social reform	Problems in generalizing results to a larger population Nonprecise data measurements Inability to demonstrate cause/effect relationship or test theories Difficult to make comparisons because of lack of structure Not a representative sample

including "participant observation and informal, conversational interviewing with the homeless; participant and nonparticipation observation, coupled with formal and informal interviewing in street agencies and settings; and a systematic survey of agency records" (Snow and Anderson, 1991: 158–169). This study is discussed in depth in Chapter 5 ("Society, Social Structure, and Interaction").

Multiple methods and approaches provide a wider scope of information and enhance our understanding of critical issues. Many researchers also use multiple methods to validate or refine one type of data by use of another type. For example, both quantitative and qualitative methods of analysis were employed in a study that sought to use one of Durkheim's types of altruistic suicide—namely, "heroic suicide"—to better understand actual cases of combat suicide where military personnel were killed in the line of duty and had received the Congressional Medal of Honor (Riemer, 1998), the highest award for valor in action against an enemy force that can be presented to a per-

son serving in the U.S. armed services. Although Durkheim did not deal specifically with suicide in military combat situations, contemporary researchers have found that his concepts are applicable to cases of "heroic suicide" (Riemer, 1998). Four criteria were used for heroic suicide: (1) the act occurred during military combat, (2) the act involved the sacrifice of one's life for one's comrades, (3) death was certain by choosing the act, and (4) death immediately resulted from the act (Riemer, 1998). Among the data and methods used in the study were content analysis of government documents, interviews with the director of the Congressional Medal of Honor Headquarters, a comparison of 3,408 cases (recipients of the medal) to determine which ones met the criteria for heroic suicides, and extensive investigation of the 125 cases that were identified as heroic suicides. Most of the cases involved military personnel who placed their body over an explosive device such as a hand grenade to save their comrades. This research supported Durkheim's assertion that heroic suicides would be more frequent among noncommissioned officers than commissioned officers, among elite troops than those who display less cohesion, and among those in "leadership roles" than those who are not. It also provides an example of how qualitative and quantitative analyses may be combined to provide information and insights that might otherwise be lacking if either element were neglected (Riemer, 1998).

ETHICAL ISSUES IN SOCIOLOGICAL RESEARCH

The study of people ("human subjects") raises vital questions about ethical concerns in sociological research. Beginning in the 1960s, the U.S. government set up regulations for "the protection of human subjects." Because of scientific abuses in the past, researchers are now mandated to weigh the societal benefits of research against the potential physical and emotional costs to participants. Researchers are required to obtain written "informed consent" statements from the persons they study. However, these guidelines have produced many new questions. What constitutes "informed consent"? What constitutes harm to a person? How do researchers protect the identity and confidentiality of their sources?

The ASA *Code of Ethics*

The American Sociological Association (ASA) *Code of Ethics* (1997) sets forth certain basic standards that sociologists must follow in conducting research:

1. Researchers must endeavor to maintain objectivity and integrity in their research by disclosing their research findings in full and including all possible interpretations of the data (even those interpretations that do not support their own viewpoints).
2. Researchers must safeguard the participants' right to privacy and dignity while protecting them from harm.
3. Researchers must protect confidential information provided by participants, even when this information is not considered to be "privileged" (legally protected, as is the case between doctor and patient and between attorney and client) and legal pressure is applied to reveal this information.
4. Researchers must acknowledge research collaboration and assistance they receive from others and disclose all sources of financial support.

Sociologists are obligated to adhere to this code and to protect research participants; however, many ethical issues arise that cannot be easily resolved. Ethics in sociological research is a difficult and often ambiguous topic. But ethical issues cannot be ignored by researchers, whether they are sociology professors, graduate students conducting investigations for their dissertations, or undergraduates conducting a class research project. Sociologists have a burden of "self-reflection"—of seeking to understand the role they play in contemporary social processes while at the same time assessing how these social processes affect their findings (Gouldner, 1970).

How honest do researchers have to be with potential participants? Let's look at a specific case in point. Where does the "right to know" end and the "right to privacy" begin in this situation?

The Zellner Research

Sociologist William Zellner (1978) sought to interview the family, friends, and acquaintances of persons killed in single-car crashes that he thought might have been "autocides." Zellner wondered if some automobile "accidents" were actually suicides—instances in which the individual wished to protect other people and perhaps make it easier for them to collect insurance benefits that might not be paid if the death was a suicide. By interviewing people who knew the victims, Zellner hoped to obtain information that would help determine if the deaths were accidental or intentional. To recruit respondents, he suggested that their participation in his study might reduce the number of accidents in the future; however, he did not mention that he suspected autocide. In each interview, he asked if the deceased had recently talked about suicide or about himself or herself in a negative manner.

From the data he collected, Zellner concluded that at least 12 percent of the fatal single-occupant crashes were suicides. He also learned that in a number of the crashes, other people (innocent bystanders) were killed or critically injured. Was Zellner's research unethical because he misrepresented the reasons for his study? In this situation, does the right to know outweigh the right to privacy? Using methods that did not involve misrepresentation, other researchers have attempted to learn more about "parasuicides" in single-occupant car crashes (Peck and Warner, 1995). *Parasuicide* refers to what is similar to, or resembles, an intentional act of suicide. For example, a one-vehicle crash may sometimes show unconscious intention to commit suicide (Peck and Warner, 1995). However, information on such car crashes is limited to secondary sources such as official medical examiner records. In one study, researchers found that law enforcement officials routinely classified a single-occupant crash as an "accident" unless a suicide note was found because the law requires that suicidal intent must be proven. As a result, single-car crashes may be a form of parasuicide that is unacknowledged in official records and largely unconfirmed by researchers who are unable to gain access to data, other than anecdotal information, to prove that a parasuicide has taken place (Peck and Warner, 1995).

The Humphreys Research

Laud Humphreys (1970), then a sociology graduate student, decided to study homosexual conduct as a topic for his doctoral dissertation. His research focused on homosexual acts between strangers meeting in "tearooms," public restrooms in parks. He did *not* ask permission of his subjects, nor did he inform them that they were being studied. Instead, Humphreys showed up at public restrooms that were known to be tearooms and offered to be the lookout while others engaged in homosexual acts. Then he systematically recorded the encounters that took place.

Humphreys was interested in the fact that the tearoom participants seemed to live "normal" lives apart from these encounters, and he decided to learn more about their everyday lives. To determine who they were, he wrote down their auto license numbers and tracked down their names and addresses. Later, he arranged for these men to be included in a medical survey so that he could go out and interview them personally. He wore different disguises and drove a different car so that they would not recognize him (Henslin, 1997). From these interviews, he collected personal information and determined that most of the men were married and lived very conventional lives.

Would Humphreys have gained access to these subjects if he had identified himself as a researcher? Probably not—nevertheless, the fact that he did not do so produced widespread criticism from sociologists and journalists. Despite the fact that his study, *Tearoom Trade* (1970), won an award for its scholarship, the controversy surrounding his project was never fully resolved.

The Scarce Research

As a graduate student in 1993, the sociologist Rik Scarce was jailed for more than five months because he refused to testify before a grand jury investigating break-ins by animal-rights activists at a Washington State University laboratory. Although Scarce was not a suspect in the break-ins, he had previously interviewed activists in the environmental movement for his book, *Eco-Warriors: Understanding the Radical Environmental Movement* (1990). Investigators believed that a person who had been house-sitting for Scarce had been involved in a 1991 raid that resulted in $150,000 in damage to the university lab and the release of 23 animals. The prosecutor in the federal grand jury investigation of the incident claimed that he wanted to ask Scarce questions about an acquaintance who was supposed to have participated in the raid, not about his research. The judge ruled that the "government had a clear need for the information, and that no blanket privilege protected academics, or even journalists, from testifying" (Monaghan, 1993: A10). Scarce said that he was taking his stand because he refused to breach guarantees of confidentiality he had made to his subjects. He was also "concerned that activists within the environmental movement may refuse to speak to me if I testify." Scarce noted that social scientists in the future might be less willing to go out and do research that requires them to promise confidentiality to their respondents if they could be required to testify (Monaghan, 1993). Based on the ASA *Code of Ethics,* many sociologists supported Scarce's stand (Monaghan, 1993); however, other analysts argued that social scientists should not be granted an exemption from testifying in court if they have information relevant to possible crimes (Comarow, 1993). As can be seen from these examples, ethical issues in research may be difficult to resolve.

In this chapter, we have looked at the research process and the methods used to pursue sociological knowledge. We have also critiqued many of the existing approaches and suggested alternate ways of pursuing research. The important thing to realize is that research is the "lifeblood" of sociology. Theory provides the framework for an analysis, and research takes us beyond common sense and provides opportunities for us to use our sociological imagination to generate new knowledge. For example, as we have seen in this chapter, suicide cannot be explained by common sense or a few isolated variables. In answering questions such as "Why do people commit suicide?" we have to take into account many aspects of personal choice and social structure that are related to one another in extremely complex ways. Research can help us unravel the complexities of social life if sociologists observe, talk to, and interact with people in real-life situations (Feagin, Orum, and Sjoberg, 1991).

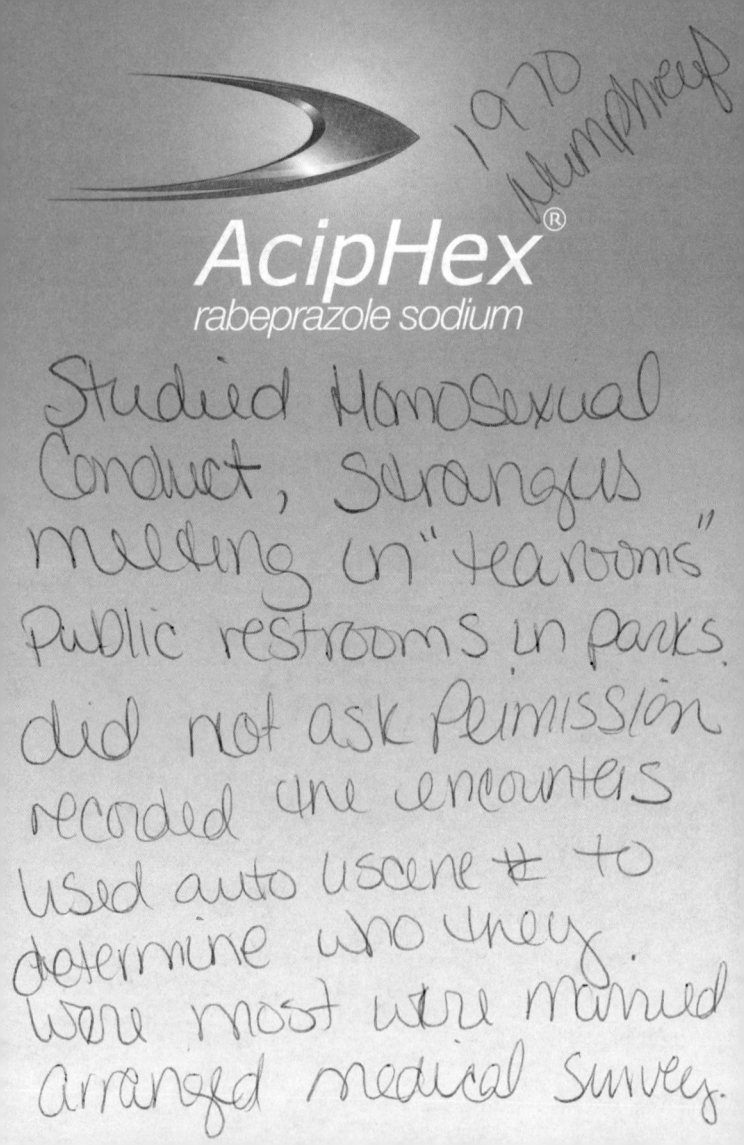

AcipHex®
rabeprazole sodium

1970 Humphreys

Studied Homosexual Conduct, strangers meeting in "tearooms" public restrooms in parks. did not ask permission recorded the encounters used auto license # to determine who they were. most were married arranged medical survey.

Box 2.4 YOU CAN MAKE A DIFFERENCE

Responding to a Cry for Help

■ Can a suicide crisis center prevent a person from committing suicide? People who understand factors that contribute to suicide may be able to better counsel those who call for help.

Chad felt that he knew Frank quite well. After all, they had been roommates for two years at State U. As a result, Chad was taken aback when Frank became very withdrawn, sleeping most of the day and mumbling about how unhappy he was. One evening, Chad began to wonder whether he needed to do something since Frank had begun to talk about "ending it all" and saying things like "the world will be better off without me." If you were in Chad's place, would you know the warning signs that you should look for? Do you know what you might do to help someone like Frank?

The American Suicide Foundation, a national nonprofit organization dedicated to funding research, education, and treatment programs for depression and suicide prevention, suggests that each of us should be aware of these warning signs of suicide:

- *Talking about it.* Be alert to such statements as "Everyone would be better off without me." Sometimes, individuals who are thinking about suicide speak as if they are saying good-bye.
- *Making plans.* The person may do such things as giving away valuable items, paying off debts, and otherwise "putting things in order."
- *Showing signs of depression.* Although most depressed people are not suicidal, most suicidal people are depressed. Serious depression tends to be expressed as a loss of pleasure or withdrawal from activities that a person has previously enjoyed. It is especially important to note whether five of the following symptoms are present almost every day for several weeks: change in appetite or weight, change in sleeping patterns, speaking or moving with unusual speed or slowness, loss of interest in usual activities, decrease in sexual drive, fatigue, feelings of guilt or worthlessness, and indecisiveness or inability to concentrate.

The possibility of suicide must be taken seriously: Most people who commit suicide give some warning to family members or friends. Instead of attempting to argue the person out of suicide or saying "You have so much to live for," let the person know that you care and understand, and that his or her problems can be solved. Urge the person to see a school counselor, a physician, or a mental health professional immediately. If you think the person is in imminent danger of committing suicide, you should take the person to an emergency room or a walk-in clinic at a psychiatric hospital. It is best to remain with the person until help is available.

For more information about suicide prevention, contact either of the following agencies:

- American Suicide Foundation, 1045 Park Avenue, New York, NY 10028 (800-531-4477 or 212-410-1111).
- SOS (Survivors of Suicide), The Link Counseling Center, 348 Mt. Vernon Highway, NE, Sandy Springs, GA 30328 (404-256-9797).

On the Internet:

- Suicide Awareness Voices of Education (**http://www.save.org**) is a resource index with links to other valuable resources, such as "Questions Most Frequently Asked on Suicide," "Symptoms of Depression and Danger Signs of Suicide," and "What to Do If Someone You Love Is Suicidal."

Our challenge today is to find new ways to integrate knowledge and action and to include all people in the research process in order to help fill the gaps in our existing knowledge about social life and how it is shaped by gender, race, class, age, and the broader social and cultural contexts in which everyday life occurs (Cancian, 1992). Each of us can and should find new ways to integrate knowledge and action into our daily lives (see Box 2.4).

CHAPTER REVIEW

■ **How does sociological research differ from commonsense knowledge?**

Sociological research provides a factual and objective counterpoint to commonsense knowledge and ill-informed sources of information. It is based on an empirical approach that answers questions through a direct, systematic collection and analysis of data.

■ **What is the relationship between theory and research?**

Theory and research form a continuous cycle that encompasses both deductive and inductive approaches. With the deductive approach, the researcher begins with a theory and then collects and analyzes research to test it. With the inductive approach, the researcher collects and analyzes data and then generates a theory based on that analysis.

■ **How does quantitative research differ from qualitative research?**

Quantitative research focuses on data that can be measured numerically (comparing rates of suicide, for example). Qualitative research focuses on interpretive description (words) rather than statistics to analyze underlying meanings and patterns of social relationships.

■ **What are the key steps in the conventional research process?**

A conventional research process based on deduction and the quantitative approach has these key steps: (1) selecting and defining the research problem; (2) reviewing previous research; (3) formulating the hypothesis, which involves constructing variables; (4) developing the research design; (5) collecting and analyzing the data; and (6) drawing conclusions and reporting the findings.

■ **What steps are often taken by researchers using the qualitative approach?**

A researcher taking the qualitative approach might (1) formulate the problem to be studied instead of creating a hypothesis, (2) collect and analyze the data, and (3) report the results.

■ **What are the major types of research methods?**

The main types of research methods are surveys, secondary analysis of existing data, field research, and experiments. Surveys are polls used to gather facts about people's attitudes, opinions, or behaviors; a representative sample of respondents provides data through questionnaires or interviews. In secondary analysis, researchers analyze existing data, such as a government census, or cultural artifacts, such as a diary. In field research, sociologists study social life in its natural setting through participant observation, case studies, unstructured interviews, and ethnography. Through experiments, researchers study the impact of certain variables on their subjects.

■ **What ethical issues are involved in sociological research?**

Since sociology involves the study of people ("human subjects"), researchers are required to obtain the informed consent of the people they study; however, in some instances what constitutes "informed consent" may be difficult to determine.

KEY TERMS

content analysis 59
control group 65
correlation 66
dependent variable 47
ethnography 62
experiment 65
experimental group 65
field research 61
Hawthorne effect 67
hypothesis 46
independent variable 47
interview 56
participant observation 61
probability sampling 50
questionnaire 54
random sampling 50
reliability 50
research methods 53
respondents 53
secondary analysis 58
survey 53
unstructured interview 63
validity 50

QUESTIONS FOR CRITICAL THINKING

1. The agency that funds the local suicide clinic has asked you to study the clinic's effectiveness in preventing suicide. What would you need to measure?

What can you measure? What research method(s) would provide the best data for analysis?

2. Recent studies have suggested that groups with high levels of *suicide acceptability* (holding the belief that suicide is an acceptable way to end one's life under certain circumstances) tend to have a higher than average suicide risk (Stack and Wasserman, 1995; Stack, 1998). What implications might such findings have on public policy issues such as the legalization of physician-assisted suicide and euthanasia? What implications might the findings have on an individual who is thinking about committing suicide? Analyze your responses using a sociological perspective.

3. In high-income nations, computers have changed many aspects of people's lives. Thinking about the various research methods discussed in this chapter, which approaches do you believe would be most affected by greater reliance on computers for collecting, organizing, and analyzing data? What are the advantages and limitations of conducting sociological research via the Internet?

RESOURCES ON THE INTERNET

Chapter-Related Web Sites

The following Web sites have been selected for their relevance to the topics in this chapter. These sites are among the more stable, but please note that Web site addresses change frequently. For an updated list of chapter-related Web sites with URL links, please visit the *Sociology in Our Times* Web site (**www.wadsworth.com /KendallSIOT**).

The Gallup Organization
http://www.gallup.com

Perhaps best known for its Gallup poll, the Gallup Organization specializes in gauging and understanding human attitudes and behaviors. Visit this site and click on "Gallup Poll News Service" to access public opinion surveys on a wide variety of topics. You can look through an alphabetical index of poll topics or conduct a keyword search.

Research Methods and Statistics
http://www.trinity.edu/mkearl/methods .html#ms

Visit Professor Michael C. Kearl's research methods and statistics Web site to access resources on a wide range of topics, including strategies for data collection, tactics for improving statistical and methodological skills, and information on how to write a research paper.

ASA Code of Ethics
http://www.asanet.org/ecoderev.htm

Learn more about the principles and ethical standards that sociologists must follow in conducting research by visiting the complete online version of the American Sociological Association's *Code of Ethics* (1997).

ONLINE STUDY AND RESEARCH TOOLS

Accompanying this text are many *free* powerful online study tools that will help you master the material in this chapter, help increase your depth of understanding, and help you make the grade!

SocCoach CD-ROM

Use the SocCoach CD-ROM enclosed with this text to help you formulate a customized study plan for this chapter. After you take the Diagnostic Quiz, SocCoach will generate a customized study plan just for you! It will identify sections of the chapter that you should review and will provide videos, charts, graphs, and excerpts from the text to supplement your studies and enhance your understanding. You'll also find fun, interactive activities such as Virtual Explorations and Map the Stats to apply what you've learned and stretch your sociological imagination.

The Companion Web Site for Sociology in Our Times, *Fifth Edition*
www.wadsworth.com/KendallSIOT

Gain an even better grasp on this chapter by going to the companion Web site to take one of the Tutorial Quizzes, use the Flash Cards to master key terms, or check out the many other study aids you'll find there. You'll also find special features such as GSS Data and Census 2000 information that'll put data and resources at your fingertips to help you with that special project or help you as you do some research on your own.

In this chapter, when you see the icon on the left, it alerts you to a specific exercise found in *Wadsworth's Sociology Online Resources and Writing Companion*. This valuable guide shows you how to use Wadsworth's exclusive online resources—*InfoTrac College Edition*, the *Opposing Viewpoints Resource Center*, and *MicroCase Online*—to assist you in your study of sociology and to build essential research and writing skills.

Culture

First the crowd of Iraqis sledgehammer the sculpture of Saddam Hussein. With the help of American soldiers, they yank it to the ground and drag the severed head through the streets. And then, in a gesture of pure cultural insult, they begin pounding the stew out of the statue with their shoes.

A mustachioed man whacks the icon with his sandal. Another presses his laced sneaker into the huge forehead. A boy pummels the statue with pink shoes.

It is one of the strongest insults in the Arab world—sticking the sole of your shoe in somebody's face, in a culture where the foot is considered the dirtiest part of the body. And flogging someone with your footwear, says Georgetown University Arab studies professor Samer Shehata, is a certain-sure symbol of disrespect. (Weeks, 2003)

A s one of the many media images of Operation Iraqi Freedom shown around the world in 2003, the sight of people belligerently kicking the statue of Saddam Hussein, Iraq's deposed leader, became a significant cultural statement that had a variety of meanings for those who viewed this occasion. How people interpreted the kicking of the statue was related to factors such as what they thought about Saddam Hussein's Iraqi regime; how they viewed the involvement of the United States, Britain, and other coalition forces in a war in the Middle East; and other issues regarding economics, politics, and religion.

What people in Iraq and millions of television and Internet viewers around the world had observed as they saw angry men kicking Hussein's statue was an enactment of one of the strongest insults that can be heaped upon another person, particularly in the Middle East. According to the anthropologist David B. Givens, "Getting an insult through words is not as insulting fundamentally as getting something visually" (qtd. in Weeks, 2003). Activities such as kicking others, showing the sole of your shoe (while sitting down), or writing a person's name on the bottom of your shoe and then walking on it are considered to be among the worst insults in some cultures because of these cultures' negative view of human feet: "In ancient times, men would etch the silhouette of their enemies on the bottom of their sandals. So they could walk on them" (qtd. in Weeks, 2003).

As our world appears to grow increasingly smaller because of rapid transportation, global communications, and international business transactions and political alliances—and sometimes because of hostility, terrorism, and warfare—learning about cultural diversity, within our own nation and globally, is extremely important for our individual and col-

■ How people demonstrate admiration or scorn varies from one culture to another, as shown by these Iraqis kicking the toppled statue of Saddam Hussein, who was later captured by the United States.

lective well-being. Although the world's population shares a common humanity—and perhaps some components of culture—cultural differences pose crucial barriers to our understanding of others. Sociology provides us with a framework for examining and developing a greater awareness of culture and cultural diversity, and how cultures change over time and place.

What is culture? Why is it so significant to our personal identities? What happens when others are intolerant of our culture? **Culture is the knowledge, language, values, customs, and material objects that are passed from person to person and from one generation to the next in a human group or society.** As previously defined, a *society* is a large social grouping that occupies the same geographic territory and is subject to the same political authority and dominant cultural expectations.

Whereas a society is composed of people, a culture is composed of ideas, behaviors, and material possessions. Society and culture are interdependent; neither could exist without the other.

If we look within our own nation, culture can be an enormously stabilizing force for a society, and it can provide a sense of continuity. However, culture can also be a force that generates discord, conflict, and even violence. How people view culture is intricately related to their location in society with regard to their race/ethnicity, class, sex, and age. From one perspective, U.S. culture does not condone intolerance or *hate crimes*—attacks against people because of their race, religion, color, disability, sexual orientation, national origin, or ancestry (Levin and McDevitt, 1993). Most states have laws against such behavior, and persons apprehended and convicted of hate crimes may be punished. From another perspective, however, intolerance and hatred may be the downside of some "positive" cultural val-

ues—such as individualism, competition, and consumerism—found in U.S. society. Just as attitudes of love and tolerance may be embedded in societal values and teachings, beliefs that reinforce acts of intolerance and hatred may also be embedded in culture.

If we look across the cultures of various nations, we may see opportunities for future cooperation based on our shared beliefs, values, and attitudes, or we may see potential for lack of understanding, discord, and conflict based on divergent ideas and world views.

In this chapter, we examine society and culture, with special attention to the components of culture and the relationship between cultural change and diversity. We also analyze culture from functionalist, conflict, symbolic interactionist, and postmodern perspectives. Before reading on, test your knowledge of the relationship between culture and intolerance toward others in the United States by answering the questions in Box 3.1.

QUESTIONS AND ISSUES

Chapter Focus Question: What part does culture play in shaping people and the social relations in which they participate?

What are the essential components of culture?

To what degree are we shaped by popular culture?

How do subcultures and countercultures reflect diversity within a society?

How do the various sociological perspectives view culture?

CULTURE AND SOCIETY IN A CHANGING WORLD

Understanding how culture affects our lives helps us develop a sociological imagination. When we meet someone from a culture vastly different from our own, or when we travel in another country, it may be easier to perceive the enormous influence of culture on people's lives. However, as our society has become more diverse, and communication among members of international cultures more frequent, the need to appreciate diversity and to understand how people in other cultures view their world has also increased (Samovar and Porter, 1991b). For example, many international travelers and businesspeople have learned the importance of knowing what gestures mean in various nations (see Fig-

ure 3.1). Although the "hook 'em Horns" sign—the pinky and index finger raised up and the middle two fingers folded down—is used by fans to express their support for University of Texas at Austin sports teams, for millions of Italians the same gesture means "Your spouse is being unfaithful." In Argentina, rotating one's index finger around the front of the ear means "You have a telephone call," but in the United States it usually suggests that a person is "crazy" (Axtell, 1991). Similarly, making a circle with your thumb and index finger indicates "OK" in the United States, but in Tunisia it means "I'll kill you!" (Samovar and Porter, 1991a).

The Importance of Culture

How important is culture in determining how people think and act on a daily basis? Simply stated, culture

Box 3.1 SOCIOLOGY AND EVERYDAY LIFE

How Much Do You Know About Culture and Intolerance Toward Others?

True	False	
T	F	1. Core values in the United States are opposed to racism and a belief in the superiority of one's own group.
T	F	2. As a form of popular culture, some rap music has antiviolence and antidrug themes.
T	F	3. Some individuals view the Confederate flag as a racist symbol associated with slavery.
T	F	4. Individuals can do very little to reduce or eliminate intolerance in society.
T	F	5. As the rate of immigration into the United States has increased rapidly in recent years, anti-immigrant feelings have also risen.
T	F	6. The U.S. Constitution designates English as the official language of this country.
T	F	7. Individuals have been physically attacked for speaking Spanish or other non-English languages in public places in the United States.
T	F	8. As the United States is increasing in diversity, most dominant-group members (middle- and high-income white Anglo-Saxon Protestants) are becoming more tolerant of social and cultural diversity.

Answers on page 78.

is essential for our individual survival and for our communication with other people. We rely on culture because we are not born with the information we need to survive. We do not know how to take care of ourselves, how to behave, how to dress, what to eat, which gods to worship, or how to make or spend money. We must learn about culture through interaction, observation, and imitation in order to participate as members of the group (Samovar and Porter, 1991a). Sharing a common culture with others simplifies day-to-day interactions. However, we must also understand other cultures and the world views therein.

Just as culture is essential for individuals, it is also fundamental for the survival of societies. Culture has been described as "the common denominator that makes the actions of individuals intelligible to the group" (Haviland, 1993: 30). Some system of rule making and enforcing necessarily exists in all societies. What would happen, for example, if *all* rules and laws in the United States suddenly disappeared? At a basic level, we need rules in order to navigate our bicycles and cars through traffic. At a more abstract level, we need laws to establish and protect our rights.

In order to survive, societies need rules about civility and tolerance toward others. We are not born

knowing how to express kindness or hatred toward others, although some people may say "Well, that's just human nature" when explaining someone's behavior. Such a statement is built on the assumption that what we do as human beings is determined by *nature* (our biological and genetic makeup) rather than *nurture* (our social environment)—in other words, that our behavior is instinctive. An *instinct* is an unlearned, biologically determined behavior pattern common to all members of a species that predictably occurs whenever certain environmental conditions exist. For example, spiders do not learn to build webs. They build webs because of instincts that are triggered by basic biological needs such as protection and reproduction.

Humans do not have instincts. What we most often think of as instinctive behavior can actually be attributed to reflexes and drives. A *reflex* is an unlearned, biologically determined involuntary response to some physical stimuli (such as a sneeze after breathing some pepper in through the nose or the blinking of an eye when a speck of dust gets in it). *Drives* are unlearned, biologically determined impulses common to all members of a species that satisfy needs such as sleep, food, water, and sexual

Box 3.1 SOCIOLOGY AND EVERYDAY LIFE

Answers to the Sociology Quiz on Culture and Intolerance Toward Others

1. **False.** Among the core American values identified by sociologists is the belief that one's own racial or ethnic group should be valued above all others. Inherent in this belief may be the assumption of racism—that members of racial-ethnic categories other than one's own are somehow inferior.

2. **True.** Although some people associate all rap music with "gangsta rap," which glorifies violence, drug use, and hostility toward women, other forms of rap music discourage such behavior and draw attention to the severe economic barriers that increasingly divide poor African Americans in central cities from middle- and upper-middle-class African Americans.

3. **True.** The Confederate flag has become an emotional symbol that continues to divide whites and African Americans in the South.

4. **False.** Hatred for members of other racial and ethnic groups—like other forms of prejudice and discrimination—is learned from the individuals with whom we associate. It is not rooted in human biology. Thus, individuals can help reduce or eliminate intolerance in society.

5. **True.** Polls show that high rates of immigration are related to an increase in anti-immigrant sentiment.

6. **False.** Although the desire to protect the supremacy of the English language in the United States dates back to colonial times, the framers of the Constitution chose in 1780 *not* to establish a national language. They thought that a national language was inconsistent with the cultural composition of the new nation.

7. **True.** Individuals speaking languages other than English have been the victims of verbal and physical abuse by individual bigots and members of hate groups, who view such persons as "outsiders who should go back home."

8. **False.** Recent polls have shown that as the United States has increased in diversity, most dominant-group members are *not* becoming more tolerant. Three examples demonstrate this point: recent demands in this country that immigration laws be more strictly enforced, renewed interest in establishing English as the "official" language of the United States, and pressure to eliminate affirmative action programs that might otherwise benefit minority-group members.

Sources: Based on Dyson, 1993; *Harvard Law Review*, 1987; Herek and Berrill, 1992; and Levin and McDevitt, 1993.

gratification. Reflexes and drives do not determine how people will behave in human societies; even the expression of these biological characteristics is channeled by culture. For example, we may be taught that the "appropriate" way to sneeze (an involuntary response) is to use a tissue or turn our head away from others (a learned response). Similarly, we may learn to sleep on mats or in beds. Most contemporary sociologists agree that culture and social learning, not nature, account for virtually all of our behavior patterns.

Since humans cannot rely on instincts in order to survive, culture is a "tool kit" for survival. According to the sociologist Ann Swidler (1986: 273), culture is a "tool kit of symbols, stories, rituals, and world views, which people may use in varying configurations to solve different kinds of problems." The tools we choose will vary according to our own personality and the situations we face. We are not puppets on a string; we make choices from among the items in our own "tool box."

Material Culture and Nonmaterial Culture

Our cultural tool box is divided into two major parts: material culture and nonmaterial culture (Ogburn,

Figure 3.1 Hand Gestures with Different Meanings in Other Societies

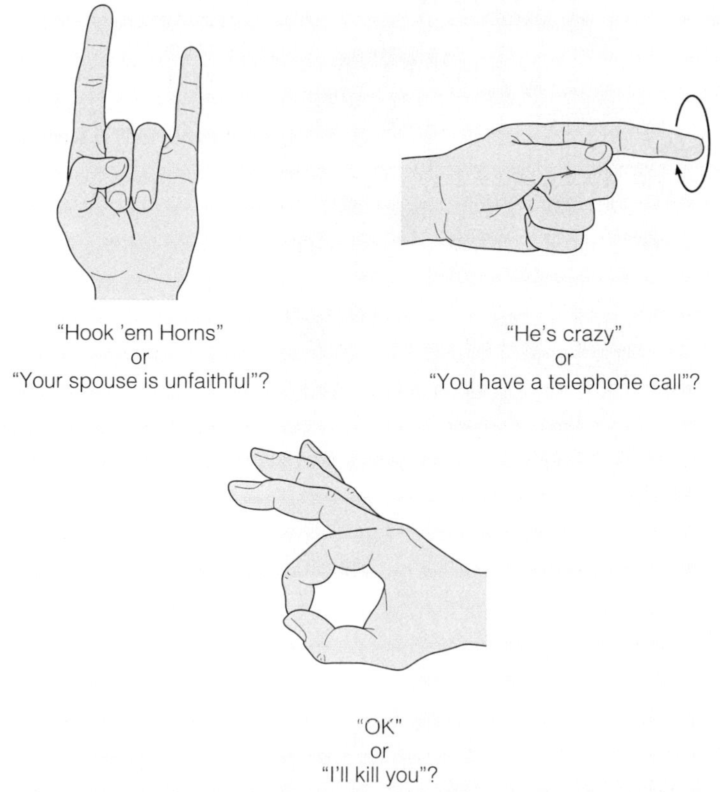

As international travelers and businesspeople have learned, hand gestures may have very different meanings in different cultures.

"Hook 'em Horns"
or
"Your spouse is unfaithful"?

"He's crazy"
or
"You have a telephone call"?

"OK"
or
"I'll kill you"?

1966/1922). *Material culture* **consists of the physical or tangible creations that members of a society make, use, and share.** Initially, items of material culture begin as raw materials or resources such as ore, trees, and oil. Through technology, these raw materials are transformed into usable items (ranging from books and computers to guns and tanks). Sociologists define *technology* as the knowledge, techniques, and tools that make it possible for people to transform resources into usable forms, and the knowledge and skills required to use them after they are developed. From this standpoint, technology is both concrete and abstract. For example, technology includes a pair of scissors and the knowledge and skill necessary to make them from iron, carbon, and chromium (Westrum, 1991). At the most basic level, material culture is important because it is our buffer against the environment. For example, we create shelter to protect ourselves from the weather and to provide ourselves with privacy. Beyond the survival level, we make, use, and share objects that are both interesting and important to us. Why are you wearing the particular clothes that you have on today? Perhaps you're communicating something about yourself, such as

where you attend school, what kind of music you like, or where you went on vacation.

Nonmaterial culture **consists of the abstract or intangible human creations of society that influence people's behavior.** Language, beliefs, values, rules of behavior, family patterns, and political systems are examples of nonmaterial culture. A central component of nonmaterial culture is *beliefs*—the mental acceptance or conviction that certain things are true or real. Beliefs may be based on tradition, faith, experience, scientific research, or some combination of these. Faith in a supreme being and trust in another person are examples of beliefs. We may also have a belief in items of material culture. When we travel by airplane, for instance, we believe that it is possible to fly at 33,000 feet and to arrive at our destination even though we know that we could not do this without the airplane itself.

Cultural Universals

Because all humans face the same basic needs (such as for food, clothing, and shelter), we engage in similar activities that contribute to our survival. Anthropologist George Murdock (1945: 124) compiled a list

© Phil Banka (Stone)/Getty Images

© James Nelson (Stone)/Getty Images

Shelter is a universal type of material culture, but it comes in a wide variety of shapes and forms. What might be some of the reasons for the similarities and differences you see in these cross-cultural examples?

© Jean-Marie Truchet (Stone)/Getty Images

of over seventy *cultural universals*—**customs and practices that occur across all societies.** His categories included appearance (such as bodily adornment and hairstyles), activities (such as sports, dancing, games, joking, and visiting), social institutions (such as family, law, and religion), and customary practices (such as cooking, folklore, gift giving, and hospitality). These general customs and practices may be present in all cultures, but their specific forms vary from one group to another and from one time to another within the same group. For example, although telling jokes may be a universal practice, what is considered to be a joke in one society may be an insult in another.

How do sociologists view cultural universals? In terms of their functions, cultural universals are useful because they ensure the smooth and continual operation of society (Radcliffe-Brown, 1952). A society must meet basic human needs by providing food, shelter, and some degree of safety for its members so that they will survive. Children and other new members (such as immigrants) must be taught the ways of the group. A society must also settle disputes and deal with people's emotions. All the while, the self-interest of individuals must be balanced with the needs of society as a whole. Cultural universals help fulfill these important functions of society.

From another perspective, however, cultural universals are not the result of functional necessity; these practices may have been *imposed* by members of one society on members of another. Similar customs and practices do not necessarily constitute cultural univer-

Spencer Grant/PhotoEdit

© Mark Richards/PhotoEdit

© Michael Greenlar/The Image Works

The customs and rituals associated with weddings are one example of nonmaterial culture. What can you infer about beliefs and attitudes concerning marriage in the societies represented by these photographs?

sals. They may be an indication that a conquering nation used its power to enforce certain types of behavior on those who were defeated (Sargent, 1987). Sociologists might ask questions such as "Who determines the dominant cultural patterns?" For example, although religion is a cultural universal, the traditional religious practices of indigenous peoples (those who first live in an area) have often been repressed and even stamped out by subsequent settlers or conquerors who have gained political and economic power over them. However, many people believe there is cause for optimism in the United States because the democratic ideas of this nation provide more guarantees of religious freedom than might be found in some other nations. For example, Josif Shrayman, aged sixty-two, who recently became a U.S. citizen, believes that he has found freedom here that he never previously had as a Jew in Ukraine: "For Jews it's very good here. God has helped me. America is a free country. I celebrate the Jewish holidays. I go to synagogue every day. I eat kosher" (qtd. in Dugger, 1997: 27).

COMPONENTS OF CULTURE

Even though the specifics of individual cultures vary widely, all cultures have four common nonmaterial cultural components: symbols, language, values, and norms. These components contribute to both harmony and strife in a society.

Symbols

A *symbol* **is anything that meaningfully represents something else.** Culture could not exist without symbols because there would be no shared meanings among people. Symbols can simultaneously produce loyalty and animosity, and love and hate. They help us communicate ideas such as love or patriotism because they express abstract concepts with visible objects.

For example, flags can stand for patriotism, nationalism, school spirit, or religious beliefs held by

© 2002 AP/Wide World Photos

Controversy continues over the use of symbols such as the Confederate flag. Members of the South Carolina Council of Conservative Citizens, shown here, were very displeased when the flag was removed from atop the dome of the state capitol in Columbia, South Carolina. Other groups applauded the removal of the flag, which they believed was a racist symbol. How would sociologists explain the divergent meanings of such a symbol?

members of a group or society. They also can be a source of discord and strife among people, as evidenced by recent controversies over the Confederate flag, the banner of the southern states during the Civil War. To some people, this flag symbolizes the "Old South" or "American history." To others, such as members of the Ku Klux Klan, it symbolizes "white supremacy." To still others, it symbolizes slavery. In 1993, when the United Daughters of the Confederacy asked Congress to renew the design patent on their logo (which is a laurel wreath encircling the Confederate flag), they asserted that the flag was part of their cultural heritage, not a racist symbol. Former U.S. Senator Carol Moseley-Braun, an African American, delivered the following argument against renewing the patent:

> On this issue there can be no consensus. It is an outrage. It is an insult.
>
> It is absolutely unacceptable to me and to millions of Americans, black or white, that we would put the imprimatur of the United States Senate on a symbol of this kind of idea. . . .
>
> This is no small matter. This is not a matter of little old ladies walking around doing good deeds.

There is no reason why these little old ladies cannot do good deeds anyway. If they choose to wave the Confederate flag, that certainly is their right. . . .

> [But] a flag that symbolized slavery should not be underwritten, underscored, adopted, approved by the United States Senate. (qtd. in Clymer, 1993: A10)

Other conflicts over use of the Confederate flag have occurred in Columbia, South Carolina, where it has been removed from the dome of the state capitol, where it flew for four decades, and at the University of Mississippi, where one supporter described the symbolic nature of the flag as follows:

> [L]et us freely admit that all symbols can be (and usually are) many things to many people. To adopt the notion that any and all displays of the Confederate flag are symbols of racism and slavery is not only ridiculous, it also displays an abysmal ignorance of U.S. history and a lack of understanding of human behavior. . . .
>
> What then is the Confederate flag? It is many things to many people. It has been used (and shall continue to be used) by those who espouse evil endeavors. It has been used by racists, South and North, but it is also recognized by thousands of others as a symbol of the American South, a short-lived government, an era now "Gone With the Wind," and a cause for which thousands of their ancestors died. With our 20th-century eyeglasses, we may look back upon this time, with its scourge of human slavery, rhetoric about state sovereignty, federalism, etc., and easily judge a few of the issues. But we dishonor the memories and sacrifices (on both sides) of this time if we allow simplistic flapdoodle to obscure the real issues of that great conflict that divided but also forged us together as a great nation. (Ezell, 1993: B2)

To both of these people, the Confederate flag is a meaningful symbol of some value, belief, or institution. In an effort to compromise on this emotional issue, political leaders in several states have decided to move the Confederate flag from highly visible locations such as atop the state capitol to less prominent locations on or around other public buildings.

Symbols can stand for love (a heart on a valentine), peace (a dove), or hate (a Nazi swastika), just as words can be used to convey these meanings. Symbols can also transmit other types of ideas. A siren is a symbol that denotes an emergency situation and sends the message to clear the way immediately. Gestures are also a symbolic form of communication—a movement of the head, body, or hands can express our ideas or feelings to others. For example, in the United States, pointing toward your chest with your thumb or finger is a symbol for "me."

Symbols affect our thoughts about class. For example, how a person is dressed or the kind of car that he or she drives is often at least subconsciously used as a measure of that individual's economic standing or position. With regard to clothing, although many people wear casual clothes on a daily basis, where the clothing was purchased is sometimes used as a symbol of social status. Were the items purchased at K-Mart, Old Navy, Abercrombie & Fitch, or Saks Fifth Avenue? What indicators are there on the items of clothing—such as the Nike *swoosh,* some other logo, or a brand name—that say something about the status of the product? Automobiles and their logos are also symbols that have cultural meaning beyond the shopping environment in which they originate.

Symbols may also affect our beliefs about race and ethnicity. Although black and white are not truly colors at all, the symbolic meanings associated with these labels permeate society and affect everyone. English-language scholar Alison Lurie (1981: 184) suggests that it is incorrect to speak of "whites" and "blacks." She notes that "pinkish-tan persons . . . have designated themselves the 'White race' while affixing the term 'Black' [to people] whose skin is some shade of brown or gold." The result of this "semantic sleight of hand" has been the association of pinkish-tan skin with virtue and cleanliness, and "brown or golden skin with evil, dirt and danger" (Lurie, 1981: 184).

Language

Language **is a set of symbols that expresses ideas and enables people to think and communicate with one another.** Verbal (spoken) language and nonverbal (written or gestured) language help us describe reality. One of our most important human attributes is the ability to use language to share our experiences, feelings, and knowledge with others. Language can create visual images in our head, such as "the kittens look like little cotton balls" (Samovar and Porter, 1991a). Language also allows people to distinguish themselves from outsiders and to maintain group boundaries and solidarity (Farb, 1973).

Language is not solely a human characteristic. Other animals use sounds, gestures, touch, and smell to communicate with one another, but they use signals with fixed meanings that are limited to the immediate situation (the present) and cannot encompass

past or future situations. For example, chimpanzees can use elements of Standard American Sign Language and manipulate physical objects to make "sentences," but they are not physically endowed with the vocal apparatus needed to form the consonants required for oral language. As a result, nonhuman animals cannot transmit the more complex aspects of culture to their offspring. Humans have a unique ability to manipulate symbols to express abstract concepts and rules, and thus to create and transmit culture from one generation to the next.

Language and Social Reality Does language *create* or simply *communicate* reality? Anthropological linguists Edward Sapir and Benjamin Whorf have suggested that language not only expresses our thoughts and perceptions but also influences our perception of reality. According to the *Sapir–Whorf hypothesis,* **language shapes the view of reality of its speakers** (Whorf, 1956; Sapir, 1961). If people are able to think only through language, then language must precede thought.

If language actually shapes the reality we perceive and experience, then some aspects of the world are viewed as important and others are virtually neglected, because people know the world only in terms of the vocabulary and grammar of their own language. For example, the Eskimo language has more than twenty words associated with snow, making it possible for people to make subtle distinctions regarding different types of snowfalls. In the United States, people perceive time as something that can be kept, saved, lost, or wasted; therefore, "being on time" or "not wasting time" is important. Many English words divide time into units (years, months, weeks, days, hours, minutes, seconds, and milliseconds) and into the past, present, and future (yesterday, today, and tomorrow) (Samovar and Porter, 1991a). By contrast, according to Sapir and Whorf, the Hopi language does not contain past, present, and future tenses of verbs or nouns for times, days, or years (Carroll, 1956); however, scholars have recently argued that this assertion is incorrect (see Edgerton, 1992).

If language does create reality, are we trapped by our language? Many social scientists agree that the Sapir–Whorf hypothesis overstates the relationship between language and our thoughts and behavior patterns. Although they acknowledge that language has many subtle meanings and that words used by people reflect their central concerns, most sociologists contend that language may *influence* our behavior and interpretation of social reality but not *determine* them.

Does the language that people use influence their perception of reality? According to social scientists, our word choices strongly influence our relationships with other people. What kind of interaction is taking place in this photo?

© Joe Pololillo Photography

Language and Gender   What is the relationship between language and gender? What cultural assumptions about women and men does language reflect? Scholars have suggested several ways in which language and gender are intertwined:

- The English language ignores women by using the masculine form to refer to human beings in general (Basow, 1992). For example, the word *man* is used generically in words like *chairman* and *mankind,* which allegedly include both men and women. However, *man* can mean either "all human beings" or "a male human being" (Miller and Swift, 1991: 71).

- Use of the pronouns *he* and *she* affects our thinking about gender. Pronouns show the gender of the person we *expect* to be in a particular occupation. For instance, nurses, secretaries, and schoolteachers are usually referred to as *she,* but doctors, engineers, electricians, and presidents are referred to as *he* (Baron, 1986).

- Words have positive connotations when relating to male power, prestige, and leadership; when relating to women, they carry negative overtones of weakness, inferiority, and immaturity (Epstein, 1988: 224). Table 3.1 shows how gender-based language reflects the traditional acceptance of men and women in certain positions, implying that the jobs are different when filled by women rather than men.

- A language-based predisposition to think about women in sexual terms reinforces the notion that women are sexual objects. Women are often described by terms such as *fox, broad, bitch, babe,* or *doll,* which ascribe childlike or even petlike characteristics to them. By

contrast, men have performance pressures placed on them by being defined in terms of their sexual prowess, such as *dude, stud,* and *hunk* (Baker, 1993).

Gender in language has been debated and studied extensively in recent years, and some changes have occurred. The preference of many women to be called *Ms.* (rather than *Miss* or *Mrs.* in reference to their marital status) has received a degree of acceptance in public life and the media (Maggio, 1988). Many organizations and publications have established guidelines for the use of nonsexist language and have changed titles such as *chairman* to *chair* or *chairperson.* "Men Working" signs in many areas have been replaced with "People Working" (Epstein, 1988). Some occupations have been given "genderless" titles, such as *firefighter* or *flight attendant* (Maggio, 1988). To develop a more inclusive and equitable society, many scholars suggest that a more inclusive language is needed (see Basow, 1992). Yet many people resist change, arguing that the English language is being ruined (Epstein, 1988).

Language, Race, and Ethnicity   Language may create and reinforce our perceptions about race and ethnicity by transmitting preconceived ideas about the superiority of one category of people over another. Let's look at a few images conveyed by words in the English language in regard to race/ethnicity:

- Words may have more than one meaning and create and/or reinforce negative images. Terms such as *blackhearted* (malevolent) and expressions such as *a black mark* (a detrimental fact) and *Chinaman's chance of success* (unlikely to

Table 3.1 LANGUAGE AND GENDER

MALE TERM	FEMALE TERM	NEUTRAL TERM
Teacher	Teacher	Teacher
Chairman	Chairwoman	Chair, chairperson
Congressman	Congresswoman	Representative
Policeman	Policewoman	Police officer
Fireman	Lady fireman	Firefighter
Airline steward	Airline stewardess	Flight attendant
Race car driver	Woman race car driver	Race car driver
Wrestler	Lady/woman wrestler	Wrestler
Professor	Female/woman professor	Professor
Doctor	Lady/woman doctor	Doctor
Bachelor	Spinster/old maid	Single person
Male prostitute	Prostitute	Prostitute
Welfare recipient	Welfare mother	Welfare recipient
Worker/employee	Working mother	Worker/employee
Janitor/maintenance man	Maid/cleaning lady	Custodial attendant

Sources: Adapted from Korsmeyer, 1981: 122; and Miller and Swift, 1991.

succeed) associate the words *black* or *Chinaman* with negative associations and derogatory imagery. By contrast, expressions such as *that's white of you* and *the good guys wear white hats* reinforce positive associations with the color white.

- Overtly derogatory terms such as *nigger, kike, gook, honkey, chink, spic,* and other racial–ethnic slurs have been "popularized" in movies, music, comic routines, and so on. Such derogatory terms are often used in conjunction with physical threats against persons.
- Words are frequently used to create or reinforce perceptions about a group. For example, Native Americans have been referred to as "savages" and "primitive," and African Americans have been described as "uncivilized," "cannibalistic," and "pagan."
- The "voice" of verbs may minimize or incorrectly identify the activities or achievements of people of color. For example, the use of the passive voice in the statement "African Americans *were given* the right to vote" ignores how African Americans *fought* for that right. Active-voice verbs may also inaccurately attribute achievements to people or groups. Some historians argue that cultural bias is shown by the

very notion that "Columbus discovered America"—given that America was already inhabited by people who later became known as Native Americans (see Stannard, 1992; Takaki, 1993).

- Adjectives that typically have positive connotations can have entirely different meanings when used in certain contexts. Regarding employment, someone may say that a person of color is "qualified" for a position when it is taken for granted that whites in the same position *are* qualified (see Moore, 1992).

In addition to these concerns about the English language, problems also arise when more than one language is involved. Consider an incident in Florida described by journalist Carlos Alberto Montaner, a native of Cuba:

I was walking quietly with my wife on a sidewalk in Miami Beach. We were speaking Spanish, of course, because that is our language. Suddenly, we were accosted by a spry little old lady, wearing a baseball cap and sneakers, who told us: "Talk English. You are in the United States." She continued on her way at once, without stopping to see our reaction. The expression on her face, curiously, was not that of somebody performing a

Rapid changes in language and culture in the United States are reflected in this street scene. How do functionalist and conflict theorists' views regarding language differ?

Michael Newman/PhotoEdit

rude action, but of someone performing a sacred patriotic duty. (Montaner, 1992: 163)

Why was this woman upset about a conversation between two people she did not even know? Montaner explains why he believes she reacted as she did:

[T]he truth is that the lady in question was not an eccentric madwoman. Thousands, millions of people are mortified that in their country there is a vast minority that constantly speaks a language that they do not understand. It disturbs them to hear Spanish prattle in shops, at work, in restaurants. They are irritated when conversations that they do not understand are held in their presence. . . .

[One] of the key elements in the configuration of a nation is its language. A monolingual American who suddenly finds himself on Miami's Calle Ocho [in Little Havana] or in San Francisco's Chinatown has the feeling that he is not in his own country. And when one is not in one's own country, one feels endangered. (Montaner, 1992: 163)

In recent decades, the United States has experienced rapid changes in language and culture. Recent data gathered by the U.S. Census Bureau (see "Census Profiles: Languages Spoken in U.S. Households") indicate that although more than 82 percent of the people in this country speak only English at home, more than 17 percent speak a language other than English, with the largest portion (over 10 percent of the U.S. population) speaking Spanish at home. French, German, Italian, Polish, Russian, Arabic, Chinese, Tagalog, Korean, and Vietnamese are among the top languages other than English and Spanish spoken at home in U.S. households.

From the functionalist perspective, a shared language is essential to a common culture; language is a stabilizing force in society. This view was expressed by Mauro E. Mujica (2002), chairman of the organization called U.S. English:

I immigrated to the United States from Chile in 1965 to study architecture at Columbia University. While English was not my first language, I am perfectly bilingual today. Learning English was never an option nor was it something to which I objected or feared. It was required for success if I wanted to enjoy a prosperous life in the U.S.

Now, I am chairman of U.S.ENGLISH, the nation's oldest and largest organization fighting to make our common language, English, the official language of government at the federal and state levels. . . .

Let me be clear: Encouraging immigrants to learn English is not about bigotry or exclusion. On the contrary, teaching newcomers English is one of the strongest acts of inclusion to our society our government can provide. The whole notion of a melting pot culture is threatened if immigrants aren't encouraged to adopt the common language of this country.

Language is an important means of cultural transmission. Through language, children learn about their cultural heritage and develop a sense of personal identity in relationship to their group. For example, Latinos/as in New Mexico and south Texas use *dichos*—proverbs or sayings that are unique to the Spanish language—as a means of expressing themselves and as a reflection of their cultural heritage. Examples of *dichos* include *Anda tu camino sin ayuda de vecino* ("Walk your own road without the help of a neighbor") and *Amor de lejos es para pendejos* ("A long-distance romance is for fools"). *Dichos* are passed from generation to generation as a priceless verbal tradition whereby people can give advice or teach a lesson (Gandara, 1995).

Conflict theorists view language as a source of power and social control; it perpetuates inequalities between people and between groups because words are used (whether or not intentionally) to "keep people in their place." As the linguist Deborah Tannen (1993: B5) has suggested, "The devastating group hatreds that result in so much suffering in our own country and around the world are related in origin to the small intolerances in our everyday conversations—our readiness to attribute good intentions to ourselves and bad intentions to others." Language, then, is a reflection of our feelings and values.

Values

Values **are collective ideas about what is right or wrong, good or bad, and desirable or undesirable in a particular culture** (Williams, 1970). Values do not dictate which behaviors are appropriate and which ones are not, but they provide us with the criteria by which we evaluate people, objects, and events. Values typically come in pairs of positive and negative values, such as being brave or cowardly, hardworking or lazy. Since we use values to justify our behavior, we tend to defend them staunchly (Kluckhohn, 1961).

Core American Values Do we have shared values in the United States? Sociologists disagree about the extent to which all people in this country share a core set of values. Functionalists tend to believe that shared values are essential for societies and have conducted most of the research on core values. Sociologist Robin M. Williams, Jr. (1970) has identified ten core values as being important to people in the United States:

1. *Individualism.* People are responsible for their own success or failure. Individual ability and hard work are the keys to success. Those who do not succeed have only themselves to blame because of their lack of ability, laziness, immorality, or other character defects.

CENSUS ★ PROFILES

Languages Spoken in U.S. Households

> a. Does this person speak a language other than English at home?
> ☐ Yes
> ☐ No → *Skip to 12*
> b. **What is this language?**
> ☐☐☐☐☐☐☐☐☐☐☐☐☐☐☐☐☐☐☐
> *(For example: Korean, Italian, Spanish, Vietnamese)*
> c. **How well does this person speak English?**
> ☐ Very well
> ☐ Well
> ☐ Not well
> ☐ Not at all

Shown above is one of the questions from the 2000 census. For a number of years, the U.S. Census Bureau has gathered data on the languages spoken in U.S. households. People who speak a language other than English at home are asked not only to indicate which other languages they speak but also how well they speak English. Approximately 43 percent of people who speak a language other than English at home report that they speak English "less than well." The languages that are most frequently spoken at home are listed below. Do you think that changes in the languages spoken in this country will bring about other significant changes in U.S. culture? Why or why not?

Language	Total Estimated Number of Speakers
English only	215,423,557
Spanish	28,101,052
Chinese	2,022,143
French	1,643,838
German	1,383,442
Tagalog	1,224,241
Vietnamese	1,009,627
Italian	1,008,370
Korean	894,063
Russian	706,242
Polish	667,414
Arabic	614,582

Source: U.S. Census Bureau, 2003a.

Baseball star Barry Bonds continues to break home-run records. Which core American values are reflected in sports such as baseball?

© Lou Dematteis/Reuters New Media, Inc./Corbis

2. *Achievement and success.* Personal achievement results from successful competition with others. Individuals are encouraged to do better than others in school and to work in order to gain wealth, power, and prestige. Material possessions are seen as a sign of personal achievement.

3. *Activity and work.* People who are industrious are praised for their achievement; those perceived as lazy are ridiculed. From the time of the early Puritans, work has been viewed as important. Even during their leisure time, many people "work" in their play. Think, for example, of all the individuals who take exercise classes, run in marathons, garden, repair or restore cars, and so on in their spare time.

4. *Science and technology.* People in the United States have a great deal of faith in science and technology. They expect scientific and technological advances ultimately to control nature, the aging process, and even death.

5. *Progress and material comfort.* The material comforts of life include not only basic necessities (such as adequate shelter, nutrition, and medical care) but also the goods and services that make life easier and more pleasant.

6. *Efficiency and practicality.* People want things to be bigger, better, and faster. As a result, great value is placed on efficiency ("How well does it work?") and practicality ("Is this a realistic thing to do?").

7. *Equality.* Since colonial times, overt class distinctions have been rejected in the United States. However, "equality" has been defined as "equality of *opportunity*"—an assumed equal chance to achieve success—not as "equality of *outcome*."

8. *Morality and humanitarianism.* Aiding others, especially following natural disasters (such as floods or hurricanes), is seen as a value. The notion of helping others was originally a part of religious teachings and tied to the idea of morality. Today, people engage in humanitarian acts without necessarily perceiving that it is the "moral" thing to do.

9. *Freedom and liberty.* Individual freedom is highly valued in the United States. The idea of freedom includes the right to private ownership of property, the ability to engage in private enterprise, freedom of the press, and other freedoms that are considered to be "basic" rights.

10. *Racism and group superiority.* People value their own racial or ethnic group above all others. Such feelings of superiority may lead to discrimination; slavery and segregation laws are classic examples. Many people also believe in the superiority of their country and that "the American way of life" is best.

As you can see from this list, some core values may contradict others.

Value Contradictions All societies, including the United States, have value contradictions. *Value contradictions* are values that conflict with one another or are mutually exclusive (achieving one makes it difficult, if not impossible, to achieve another). Core values of morality and humanitarianism may conflict with values of individual achievement and success. In the 1990s, for example, humanitarian values reflected in welfare and other government aid programs came into conflict with values emphasizing hard work and personal achievement. Many people are more ambivalent about helping the poor or homeless than about helping victims of natural disasters such as floods, hurricanes, or earthquakes in this or other countries. In the aftermath of the earthquake in Kobe, Japan, in January 1995, for instance, some people in the United States viewed economic aid to Japan more favorably than welfare benefits for unwed mothers under the age of eighteen (Pear, 1995).

Ideal Versus Real Culture   What is the relationship between values and human behavior? Sociologists stress that a gap always exists between ideal culture and real culture in a society. *Ideal culture* refers to the values and standards of behavior that people in a society profess to hold. *Real culture* refers to the values and standards of behavior that people actually follow. For example, we may claim to be law-abiding (ideal cultural value) but smoke marijuana (real cultural behavior), or we may regularly drive over the speed limit but think of ourselves as "good citizens."

Most of us are not completely honest about how well we adhere to societal values. In a University of Arizona study known as the "Garbage Project," household waste was analyzed to determine the rate of alcohol consumption in Tucson, Arizona. People were asked about their level of alcohol consumption, and in some areas of the city, they reported very low levels of alcohol use. However, when their garbage was analyzed, researchers found that over 80 percent of those households consumed some beer, and more than half discarded eight or more empty beer cans a week (Haviland, 1993). Obviously, this study shows a discrepancy between ideal cultural values and people's actual behavior.

The degree of discrepancy between ideal culture and real culture is relevant to sociologists investigating social change. Large discrepancies provide a foothold for demonstrating hypocrisy (pretending to be what one is not or to feel what one does not feel). These discrepancies are often a source of social problems; if the discrepancy is perceived, leaders of social movements may use it to point out people's contradictory behavior. For example, preserving our natural environment may be a core value, but our behavior (such as littering highways and lakes) contributes to its degradation, as discussed in Chapter 20 ("Collective Behavior, Social Movements, and Social Change").

Norms

Values provide ideals or beliefs about behavior but do not state explicitly how we should behave. Norms, on the other hand, do have specific behavioral expectations. **Norms are established rules of behavior or standards of conduct.** *Prescriptive norms* state what behavior is appropriate or acceptable. For example, persons making a certain amount of money are expected to file a tax return and pay any taxes they owe. Norms based on custom direct us to open a door for a person carrying a heavy load. By contrast, *proscriptive norms* state what behavior is inappropriate or unacceptable. Laws that prohibit us from driving over the speed limit and "good manners" that preclude you from reading a newspaper during class are examples. Prescriptive and proscriptive norms operate at all levels of society, from our everyday actions to the formulation of laws.

Formal and Informal Norms   Not all norms are of equal importance; those that are most crucial are formalized. *Formal norms* are written down and involve specific punishments for violators. Laws are the most common type of formal norms; they have been codified and may be enforced by sanctions. **Sanctions are rewards for appropriate behavior or penalties for inappropriate behavior.** Examples of *positive sanctions* include praise, honors, or medals for conformity to specific norms. *Negative sanctions* range from mild disapproval to the death penalty. In the case of law, formal sanctions are clearly defined and can be administered only by persons in certain official positions (such as police officers and judges), who are given the authority to impose the sanctions.

Norms considered to be less important are referred to as *informal norms*—unwritten standards of behavior understood by people who share a common identity. When individuals violate informal norms, other people may apply informal sanctions. *Informal sanctions* are not clearly defined and can be applied by any member of a group (such as frowning at someone or making a negative comment or gesture).

Folkways   Norms are also classified according to their relative social importance. **Folkways are informal norms or everyday customs that may be violated without serious consequences within a particular culture** (Sumner, 1959/1906). They provide rules for conduct but are not considered to be essential to society's survival. In the United States, folkways include using underarm deodorant, brushing our teeth, and wearing appropriate clothing for a specific occasion. Often, folkways are not enforced; when they are enforced, the resulting sanctions tend to be informal and relatively mild.

Folkways are culture specific; they are learned patterns of behavior that can vary markedly from one society to another. In Japan, for example, where the walls of restroom stalls reach to the floor, folkways dictate that a person should knock on the door before entering a stall (you cannot tell if anyone is there without knocking). However, people in the United States find it disconcerting when someone knocks on the door of the stall (A. Collins, 1991).

Mores   Other norms are considered to be highly essential to the stability of society. ***Mores* are a particular culture's strongly held norms with moral and ethical connotations that may not be violated without serious consequences.** Since mores (pronounced MOR-ays) are based on cultural values and are considered to be crucial for the well-being of the group, violators are subject to more severe negative sanctions (such as ridicule, loss of employment, or imprisonment) than are those who fail to adhere to folkways. The strongest mores are referred to as taboos. ***Taboos* are mores so strong that their violation is considered to be extremely offensive and even unmentionable.** Violation of taboos is punishable by the group or even, according to certain belief systems, by a supernatural force. The incest taboo, which prohibits sexual or marital relations between certain categories of kin, is an example of a nearly universal taboo.

Folkways and mores provide structure and security in a society. They make everyday life more predictable and provide people with some guidelines for appearance and behavior. As individuals travel in countries other than their own, they become aware of cross-cultural differences in folkways and mores. For example, women from the United States traveling in Muslim nations quickly become aware of mores, based on the Sharia (the edicts of the Qur'an), that prescribe the dominance of men over women. In Saudi Arabia, for instance, women are not allowed to mix with men in public. Banks have branches with only women tellers—and only women customers. In hospitals, female doctors are supposed to tend only to children and other women (Alireza, 1990; Ibrahim, 1990).

Laws   ***Laws* are formal, standardized norms that have been enacted by legislatures and are enforced by formal sanctions.** Laws may be either civil or criminal. *Civil law* deals with disputes among persons or groups. Persons who lose civil suits may encounter negative sanctions such as having to pay compensation to the other party or being ordered to stop certain conduct. *Criminal law,* on the other hand, deals with public safety and well-being. When criminal laws are violated, fines and prison sentences are the most likely negative sanctions, although in some states the death penalty is handed down for certain major offenses.

As with material objects, all of the nonmaterial components of culture—symbols, language, values, and norms—are reflected in the popular culture of contemporary society.

TECHNOLOGY, CULTURAL CHANGE, AND DIVERSITY

Cultures do not generally remain static. There are many forces working toward change and diversity. Some societies and individuals adapt to this change, whereas others suffer culture shock and succumb to ethnocentrism.

Cultural Change

Societies continually experience cultural change at both material and nonmaterial levels. Changes in technology continue to shape the material culture of society. ***Technology* refers to the knowledge, techniques, and tools that allow people to transform resources into usable forms and the knowledge and skills required to use what is developed.** Although most technological changes are primarily modifications of existing technology, *new technologies* are changes that make a significant difference in many people's lives. Examples of new technologies include the introduction of the printing press more than 500 years ago and the advent of computers and electronic communications in the twentieth century. The pace of technological change has increased rapidly in the past 150 years, as contrasted with the 4,000 years prior to that, during which humans advanced from digging sticks and hoes to the plow.

All parts of culture do not change at the same pace. When a change occurs in the material culture of a society, nonmaterial culture must adapt to that change. Frequently, this rate of change is uneven, resulting in a gap between the two. Sociologist William F. Ogburn (1966/1922) referred to this disparity as ***cultural lag*—a gap between the technical development of a society and its moral and legal institutions.** In other words, cultural lag occurs when material culture changes faster than nonmaterial culture, thus creating a lag between the two cultural components. For example, at the material cultural level, the personal computer and electronic coding have made it possible to create a unique health identifier for each person in the United States. Based on available technology (material culture), it would be possible to create a national data bank that included everyone's individual medical records from birth to death. Using this identifier, health providers and insurance companies could rapidly transfer medical records around the globe, and researchers could access unlimited data on people's diseases, test results, and treatments. However, the availability of this technology does not mean that it will be accepted by people who believe (nonmaterial culture) that such a national data bank would constitute an invasion of privacy and could easily be abused by others. The failure of nonmaterial culture to keep pace with material culture is linked to social conflict and societal problems. As in the above example, such changes are often set in motion by discovery, invention, and diffusion.

Discovery **is the process of learning about something previously unknown or unrecognized.** Historically, discovery involved unearthing natural elements or existing realities, such as "discovering" fire or the true shape of the earth. Today, discovery most often results from scientific research. For example, the discovery of a polio vaccine virtually eliminated one of the major childhood diseases. A future discovery of a cure for cancer or the common cold could result in longer and more productive lives for many people.

As more discoveries have occurred, people have been able to reconfigure existing material and nonmaterial cultural items through invention. *Invention* **is the process of reshaping existing cultural items into a new form.** Guns, video games, airplanes, and First Amendment rights are examples of inventions that positively or negatively affect our lives today.

When diverse groups of people come into contact, they begin to adapt one another's discoveries, inventions, and ideas for their own use. *Diffusion* **is the transmission of cultural items or social practices from one group or society to another** through such means as exploration, military endeavors, the media, tourism, and immigration. To illustrate, piñatas can be traced back to the twelfth century, when Marco Polo brought them back from China, where they were used to celebrate the springtime harvest, to Italy, where they were filled with costly gifts in a game played by the nobility. When the piñata traveled to Spain, it became part of Lenten traditions. In Mexico, it was used to celebrate the birth of the Aztec god Huitzilopochtli (Burciaga, 1993). Today, children in many countries squeal with excitement at parties as they swing a stick at a piñata. In today's "shrinking

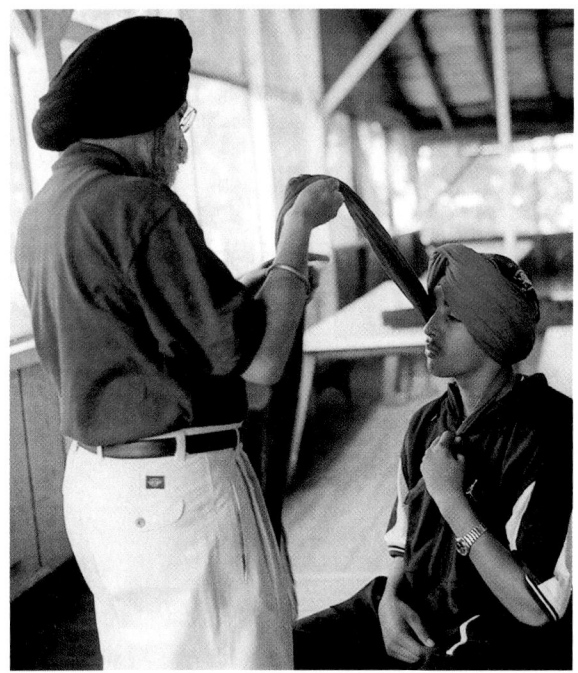

Nancy Sisel/NYT Pictures

In heterogeneous societies such as the United States, people from very diverse cultures encourage their children to learn about their heritage. At Camp Robin Hood, a Sikh camp in Pennsylvania, children learn turban-tying and other practices of their religious tradition.

globe," cultural diffusion moves at a very rapid pace as countries continually seek new markets for their products.

Cultural Diversity

Cultural diversity refers to the wide range of cultural differences found between and within nations. Cultural diversity between countries may be the result of natural circumstances (such as climate and geography) or social circumstances (such as level of technology and composition of the population). Some nations—such as Sweden—are referred to as *homogeneous societies,* meaning that they include people who share a common culture and who are typically from similar social, religious, political, and economic backgrounds. By contrast, other nations—including the United States—are referred to as *heterogeneous societies,* meaning that they include people who are dissimilar in regard to social characteristics such as religion, income, or race/ethnicity (see Figure 3.2).

Immigration contributes to cultural diversity in a society. Throughout its history, the United States has been a nation of immigrants. Over the past 185 years, more than 60 million "documented" (legal) immigrants have arrived here; innumerable people have also

Figure 3.2   Heterogeneity of U.S. Society

Throughout history, the United States has been heterogeneous. Today, we represent a wide diversity of social categories, including our religious affiliations, income levels, and racial-ethnic categories.

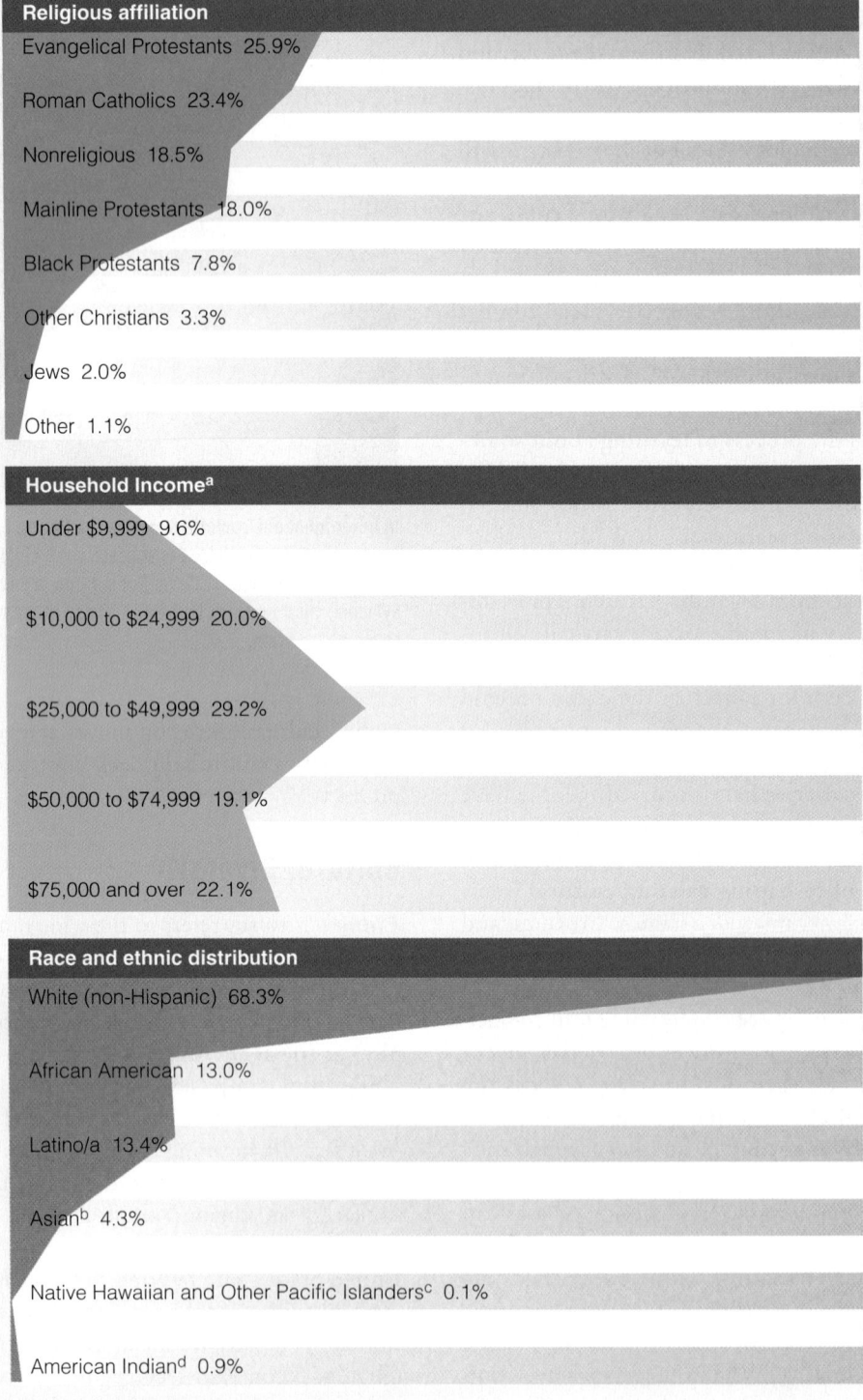

Religious affiliation

- Evangelical Protestants 25.9%
- Roman Catholics 23.4%
- Nonreligious 18.5%
- Mainline Protestants 18.0%
- Black Protestants 7.8%
- Other Christians 3.3%
- Jews 2.0%
- Other 1.1%

Household Income[a]

- Under $9,999 9.6%
- $10,000 to $24,999 20.0%
- $25,000 to $49,999 29.2%
- $50,000 to $74,999 19.1%
- $75,000 and over 22.1%

Race and ethnic distribution

- White (non-Hispanic) 68.3%
- African American 13.0%
- Latino/a 13.4%
- Asian[b] 4.3%
- Native Hawaiian and Other Pacific Islanders[c] 0.1%
- American Indian[d] 0.9%

[a]In Census Bureau terminology, a household consists of people who occupy a housing unit.

[b]Includes Chinese, Filipino, Japanese, Asian Indian, Korean, Vietnamese, and other Asians.

[c]Includes Native Hawaiian, Guamanian or Chamorro, Samoan, and other Pacific Islanders.

[d]Includes American Indians, Eskimos, and Aleuts.

Source: U.S. Census Bureau, 2002.

entered the country as undocumented immigrants. Immigration can cause feelings of frustration and hostility, especially in people who feel threatened by the changes that large numbers of immigrants may produce. Often, people are intolerant of those who are different from themselves. When societal tensions rise, people may look for others on whom they can place blame—or single out persons because they are the "other," the "outsider," the one who does not "belong." Ronald Takaki, an ethnic studies scholar, described his experience of being singled out as an "other":

> I had flown from San Francisco to Norfolk and was riding in a taxi to my hotel to attend a conference on multiculturalism. . . . My driver and I chatted about the weather and the tourists. . . . The rearview mirror reflected a white man in his forties. "How long have you been in this country?" he asked. "All my life," I replied, wincing. "I was born in the United States." With a strong southern drawl, he remarked: "I was wondering because your English is excellent!" Then, as I had many times before, I explained: "My grandfather came here from Japan in the 1880s. My family has been here, in America, for over a hundred years." He glanced at me in the mirror. Somehow I did not look "American" to him; my eyes and complexion looked foreign. (Takaki, 1993: 1)

Have you ever been made to feel like an "outsider"? Each of us receives cultural messages that may make us feel good or bad about ourselves or may give us the perception that we "belong" or "do not belong." Can people overcome such feelings in a culturally diverse society such as the United States? Some analysts believe it is possible to communicate with others despite differences in race, ethnicity, national origin, age, sexual orientation, religion, social class, occupation, leisure pursuits, regionalism, and so on (see Box 3.2). People who differ from the dominant group may also find reassurance and social support in a subculture or a counterculture.

Subcultures
A *subculture* **is a category of people who share distinguishing attributes, beliefs, values, and/or norms that set them apart in some significant manner from the dominant culture.** Emerging from the functionalist tradition, this concept has been applied to distinctions ranging from ethnic, religious, regional, and age-based categories to those categories presumed to be "deviant" or marginalized from the larger society. In the broadest use of the concept, thousands of categories of people residing in the United States might be classified as par-

Take a close look at the boy's feet as he follows the horse-drawn carriage of his Old Order Amish family. Are modernization and consumerism a threat to the way of life of subcultures such as the Amish? Why or why not?

ticipants in one or more subcultures, including Native Americans, Muslims, Generation Xers, and motorcycle enthusiasts. However, to see how subcultural participants interact with the dominant U.S. culture, many sociological studies of subcultures have limited the scope of inquiry to more visible, distinct subcultures such as the Old Order Amish and ethnic enclaves in large urban areas.

The Old Order Amish Having arrived in the United States in the early 1700s, members of the Old Order Amish have fought to maintain their distinct identity. Today, over 75 percent of the more than 100,000 Amish live in Pennsylvania, Ohio, and Indiana, where they practice their religious beliefs and remain a relatively closed social network. According to sociologists, this religious community is a subculture because its members share values and norms that differ significantly from those of people who primarily identify with the dominant culture. The Amish have a strong faith in God and reject worldly concerns. Their core values include the joy of work, the primacy of the home, faithfulness, thriftiness, tradition, and humility. The Amish hold a conservative view of the family, believing that women are subordinate to men, birth control is unacceptable, and wives should remain at home. Children (about seven per family) are cherished and seen as an economic asset: They help with

Box 3.2 YOU CAN MAKE A DIFFERENCE

Understanding People from Other Cultures

Why study chemistry if one can-not effectively tell an audience how to use various chemicals responsibly? Or why study geography and history if one cannot link the environment and the past to people's current ways of life? . . . [W]hy learn about culture and fail to appreciate differences among people and communicate the same message to others? (Kabagarama, 1993: vi–vii)

Daisy Kabagarama, a U.S. college professor who was born in Uganda, poses these questions in her provocative book *Breaking the Ice* (1993), in which she explains how to further the cross-cultural understanding needed with rapid population changes and globalization. We can help others communicate across cultures by passing Kabagarama's techniques on to them:

- *Get acquainted.* Show genuine interest, have a sense of curiosity and appreciation, feel empathy for others, be nonjudgmental, and demonstrate flexibility.

- *Ask the right questions.* Ask general questions first and specific ones later, making sure that questions are clear and simple and are asked in a relaxed, nonthreatening manner.

- *Consider visual images.* Use compliments carefully; it is easy to misjudge other people based on their physical appearance alone, and appearance norms differ widely across cultures.

- *Deal with stereotypes.* Overcome stereotyping and myths about people from other cultures through sincere self-examination, searching for knowledge, and practicing objectivity.

- *Establish trust and cooperation.* Be available when needed. Give and accept criticism in a positive manner and be spontaneous in interactions with others, but remember that rules regarding spontaneity are different for each culture.

Electronic systems now link people around the world, making it possible for us to communicate with people from diverse racial–ethnic backgrounds and cultures without even leaving home or school. Try these Web sites on the Internet for interesting information on multicultural issues and cultural diversity:

- *Multicultural Pavilion* provides resources on racism, sexism, and classism in the United States, as well as access to multicultural news-groups, essays, and a large list of multicultural links on the Web:

http://www.edchange.org/multicultural

- *Multiworld* is a bilingual (Chinese and English) e-zine that includes information on culture, people, art, and nature, plus sites about nations such as the United States, Canada, Ireland, China, Belgium, and Brazil:

http://sunsite.nus.edu.sg/mw

WRITING IN SOCIOLOGY ASSIGNMENT

If you wanted to learn more about the cultures of people in the United States from backgrounds different from your own, where would you obtain the information? Based on the communities with which you are most familiar, such as where your  home is located and/or where you attend college, which groups do you think you already know the most about? Which groups do you know the least about? Why?

the farming and other work. Many of the Old Order Amish speak Pennsylvania Dutch (a dialect of German) as well as English. They dress in traditional clothing, live on farms, and rely on the horse and buggy for transportation.

The Amish are aware that they share distinctive values and look different from other people; these differences provide them with a collective identity and make them feel close to one another (Kephart and Zellner, 1994). The belief system and group cohesive-ness of the Amish remain strong despite the intrusion of corporations and tourists, the vanishing farmlands, and increasing levels of government regulation in their daily lives (Kephart and Zellner, 1994).

Ethnic Subcultures Some people who have unique shared behaviors linked to a common racial, language, or national background identify themselves as members of a specific subculture, whereas others do not. Examples of ethnic subcultures include African

Even as global travel and the media make us more aware of people around the world, the distinctiveness of the Yanomamö in South America remains apparent. Are people today more or less likely than those in the past to experience culture shock upon encountering diverse groups of people such as these Yanomamö?

© Gerhard Hurner/Contrast/Gamma Press

Americans, Latinos/Latinas (Hispanic Americans), Asian Americans, and Native Americans. Some analysts include "white ethnics" such as Irish Americans, Italian Americans, and Polish Americans. Others also include Anglo Americans (Caucasians).

Although people in ethnic subcultures are dispersed throughout the United States, a concentration of members of some ethnic subcultures is visible in many larger communities and cities. For example, Chinatowns, located in cities such as San Francisco, Los Angeles, and New York, are one of the more visible ethnic subcultures in the United States. In San Francisco, over 100,000 Chinese Americans live in a twenty-four-block "city" within a city, which is the largest Chinese community outside of Asia. Traditionally, the core values of this subculture have included loyalty to others and respect for one's family. Obedience to parental authority, especially the father's, is expected, and sexual restraint and control over one's emotions in public are also highly valued.

By living close to one another and clinging to their original customs and language, first-generation immigrants can survive the abrupt changes they experience in material and nonmaterial cultural patterns. In New York City, for example, Korean Americans and Puerto Rican Americans constitute distinctive subcultures, each with its own food, music, and personal style. In San Antonio, Mexican Americans enjoy different food and music than do Puerto Rican Americans or other groups. Subcultures provide opportunities for expression of distinctive lifestyles, as well as sometimes helping people adapt to abrupt cultural change. Subcultures can also serve as a buffer against the discrimination experienced by many ethnic or religious groups in the United States. However, some people may be forced by economic or social disadvantage to remain in such ethnic enclaves.

Countercultures Some subcultures actively oppose the larger society. A ***counterculture* is a group that strongly rejects dominant societal values and norms and seeks alternative lifestyles** (Yinger, 1960, 1982). Young people are most likely to join countercultural groups, perhaps because younger persons generally have less invested in the existing culture. Examples of countercultures include the beatniks of the 1950s, the flower children of the 1960s, the drug enthusiasts of the 1970s, and members of nonmainstream religious sects, or cults. Some countercultures (such as the Ku Klux Klan, militias, neo-Nazi skinheads, and the Nation of Islam) engage in revolutionary political activities.

Culture Shock

***Culture shock* is the disorientation that people feel when they encounter cultures radically different from their own** and believe they cannot depend on their own taken-for-granted assumptions about life. When people travel to another society, they may not know how to respond to that setting. For example, Napoleon Chagnon (1992) described his initial shock at seeing the Yanomamö (pronounced yah-noh-MAH-mah) tribe of South America on his first trip in 1964.

The Yanomamö (also referred to as the "Yanomami") are a tribe of about 20,000 South American Indians who live in the rain forest. Although Chagnon traveled in a small aluminum motorboat for three days to reach these people, he was not prepared for the sight that met his eyes when he arrived:

> I looked up and gasped to see a dozen burly, naked, sweaty, hideous men staring at us down the shafts of their drawn arrows. Immense wads of green tobacco were stuck between their lower teeth and lips, making them look even more hideous, and strands of dark-green slime dripped from their nostrils—strands so long that they reached down to their pectoral muscles or drizzled down their chins and stuck to their chests and bellies. We arrived as the men were blowing ebene, a hallucinogenic drug, up their noses. . . . I was horrified. What kind of welcome was this for someone who had come to live with these people and learn their way of life—to become friends with them? But when they recognized Barker [a guide], they put their weapons down and returned to their chanting, while keeping a nervous eye on the village entrances. (Chagnon, 1992: 12–14)

The Yanomamö have no written language, system of numbers, or calendar. They lead a nomadic lifestyle, carrying everything they own on their backs. They wear no clothes and paint their bodies; the women insert slender sticks through holes in the lower lip and through the pierced nasal septum. In other words, the Yanomamö—like the members of thousands of other cultures around the world—live in a culture very different from that of the United States.

Ethnocentrism and Cultural Relativism

When observing people from other cultures, many of us use our own culture as the yardstick by which we judge their behavior. Sociologists refer to this approach as ***ethnocentrism—the practice of judging all other cultures by one's own culture*** (Sumner, 1959/1906). Ethnocentrism is based on the assumption that one's own way of life is superior to all others. For example, most schoolchildren are taught that their own school and country are the best. The school song, the pledge to the flag, and the national anthem are forms of *positive ethnocentrism*. However, *negative ethnocentrism* can also result from constant emphasis on the superiority of one's own group or nation. Negative ethnocentrism is manifested in derogatory stereotypes that ridicule recent immigrants whose customs, dress, eating habits, or religious beliefs are markedly different from those of dominant-group members. Long-term U.S. residents

who are members of racial and ethnic minority groups, such as Native Americans, African Americans, and Latinas/os, have also been the target of ethnocentric practices by other groups.

An alternative to ethnocentrism is ***cultural relativism—the belief that the behaviors and customs of any culture must be viewed and analyzed by the culture's own standards.*** For example, the anthropologist Marvin Harris (1974, 1985) uses cultural relativism to explain why cattle, which are viewed as sacred, are not killed and eaten in India, where widespread hunger and malnutrition exist. From an ethnocentric viewpoint, we might conclude that cow worship is the cause of the hunger and poverty in India. However, according to Harris, the Hindu taboo against killing cattle is very important to their economic system. Live cows are more valuable than dead ones because they have more important uses than as a direct source of food. As part of the ecological system, cows consume grasses of little value to humans. Then they produce two valuable resources—oxen (the neutered offspring of cows) to power the plows and manure (for fuel and fertilizer)—as well as milk, floor covering, and leather. As Harris's study reveals, culture must be viewed from the standpoint of those who live in a particular society.

Cultural relativism also has a downside. It may be used to excuse customs and behavior (such as cannibalism) that may violate basic human rights. Cultural relativism is a part of the sociological imagination; researchers must be aware of the customs and norms of the society they are studying and then spell out their background assumptions so that others can spot possible biases in their studies. However, according to some social scientists, issues surrounding ethnocentrism and cultural relativism may become less distinct in the future as people around the globe increasingly share a common popular culture. Others, of course, disagree with this perspective. Let's see what you think.

■ A GLOBAL POPULAR CULTURE?

Before taking this course, what was the first thing you thought about when you heard the term *culture*? In everyday life, culture is often used to describe the fine arts, literature, and classical music. When people say that a person is "cultured," they may mean that the individual has a highly developed sense of style or aesthetic appreciation of the "finer" things.

People of all ages are spending many hours each week using computers, playing video games, and watching television. How is this behavior different from the ways in which people enjoyed popular culture in previous generations?

© Eric Fowke/PhotoEdit

High Culture and Popular Culture

Some sociologists use the concepts of high culture and popular culture to distinguish among different cultural forms. *High culture* consists of classical music, opera, ballet, live theater, and other activities usually patronized by elite audiences, composed primarily of members of the upper-middle and upper classes, who have the time, money, and knowledge assumed to be necessary for its appreciation. In the United States, high culture is often viewed as being international in scope, arriving in this country through the process of diffusion, because many art forms originated in European nations or other countries of the world. By contrast, much of U.S. popular culture is often thought of as "homegrown." **Popular culture consists of activities, products, and services that are assumed to appeal primarily to members of the middle and working classes.** These include rock concerts, spectator sports, movies, and television soap operas and situation comedies. Although we will distinguish between "high" and "popular" culture in our discussion, it is important to note that some social analysts believe that the rise of a consumer society in which luxury items have become more widely accessible to the masses has greatly reduced the huge divide between activities and possessions associated with wealthy people or a social elite (see Huyssen, 1984; Lash and Urry, 1994).

However, most sociological examinations of high culture and popular culture focus primarily on the link between culture and social class. French sociologist Pierre Bourdieu's (1984) *cultural capital theory* views high culture as a device used by the dominant class to exclude the subordinate classes. According to Bourdieu, people must be trained to appreciate and understand high culture. Individuals learn about high culture in upper-middle- and upper-class families and in elite education systems, especially higher education. Once they acquire this trained capacity, they possess a form of cultural capital. Persons from poor and working-class backgrounds typically do not acquire this cultural capital. Since knowledge and appreciation of high culture are considered a prerequisite for access to the dominant class, its members can use their cultural capital to deny access to subordinate-group members and thus preserve and reproduce the existing class structure (but see Halle, 1993).

Forms of Popular Culture

Three prevalent forms of popular culture are fads, fashions, and leisure activities. A *fad* is a temporary but widely copied activity followed enthusiastically by large numbers of people. Most fads are short-lived novelties (Garreau, 1993). According to the sociologist John Lofland (1993), fads can be divided into four major categories. First, *object fads* are items that people purchase despite the fact that they have little use or intrinsic value. Recent examples include Beanie Babies, Furbys, and Pokémon games, toys, trading cards, clothing, cartoons, and snack foods. Second, *activity fads* include pursuits such as body piercing, "surfing" the Internet, and "blade nights" in New York City, where thousands of in-line skaters swarm down the city's streets in processions that are several

Activity fads such as flash mobs are particularly popular with young people. Why are fads (such as this umbrella-carrying flash mob in London) often short-lived?

© Scott Barbour/Getty Images

blocks long. Third are *idea fads,* such as New Age ideologies. Fourth are *personality fads,* such as those surrounding celebrities such as Jennifer Lopez, Tiger Woods, Eminem, and Brad Pitt.

A *fashion* is a currently valued style of behavior, thinking, or appearance that is longer lasting and more widespread than a fad. Examples of fashion are found in many areas, including child rearing, education, arts, clothing, music, and sports. Soccer is an example of a fashion in sports. Until recently, only schoolchildren played soccer in the United States. Now it has become a popular sport, perhaps in part because of immigration from Latin America and other areas of the world where soccer is widely played.

Like soccer, other forms of popular culture move across nations. In fact, popular culture is the United States' second largest export (after aircraft) to other nations (Rockwell, 1994). Of the world's 100 most-attended films in the 1990s, for example, 88 were produced by U.S.-based film companies. Likewise, music, television shows, novels, and street fashions from the United States have become a part of many other cultures. In turn, people in this country continue to be strongly influenced by popular culture from other nations (see Box 3.3 for an example). For example, contemporary music and clothing in the United States reflect African, Caribbean, and Asian cultural influences, among others.

Will the spread of popular culture produce a homogeneous global culture? Critics argue that the world is not developing a global culture; rather, other cultures are becoming westernized. Political and religious leaders in some nations oppose this process, which they view as **cultural imperialism—the extensive infusion of one nation's culture into other nations** (see Box 3.4). For example, some view the widespread infusion of the English language into countries that speak other languages as a form of cultural imperialism. On the other hand, the concept of cultural imperialism may fail to take into account various cross-cultural influences. For example, cultural diffusion of literature, music, clothing, and food has occurred on a global scale. A global culture, if it comes into existence, will most likely include components from many societies and cultures.

SOCIOLOGICAL ANALYSIS OF CULTURE

Sociologists regard culture as a central ingredient in human behavior. Although all sociologists share a similar purpose, they typically see culture through somewhat different lenses as they are guided by different theoretical perspectives in their research. What do these perspectives tell us about culture?

Functionalist Perspectives

As previously discussed, functionalist perspectives are based on the assumption that society is a stable, orderly system with interrelated parts that serve specific functions. Anthropologist Bronislaw Malinowski (1922) suggested that culture helps people meet their *biological needs* (including food and procreation), *instrumental needs* (including law and education), and *integrative needs* (including religion and art). Societies in which people share a common language and core values are more likely to have consensus and harmony.

How might functionalist analysts view popular culture? According to many functionalist theorists, popular culture serves a significant function in society in

Box 3.3 CHANGING TIMES: MEDIA AND TECHNOLOGY

American Idol and the Diffusion of Popular Culture

The hot topic at this year's Passover Seder was *American Idol.* It was Wednesday night, and the show was down to the final six. Every 10 minutes, the somber proceedings of the annual ritual were interrupted to check in with Ryan, Paula, Randy, and Simon [the show's judges]. Sometime between the 4 questions and the 10 plagues [in the Passover Seder], the week's loser was announced and dinner continued. Our voices were speaking the Haggadah [the text that is recited at the Seder], but our minds were still reeling: *Who would be next? I'm so glad Kimberly was kicked off and Clay is gonna be back. Josh's performance was horrible!*
    —Russell Brown (2003), writing for *The Simon,* an Internet commentary on current issues

■ These U.S. contestants during a recent season of *American Idol* were not alone in their pursuit of fame and fortune in the entertainment industry. Similar "Idol" programs are popular with television audiences in many other countries.

In this personal commentary about his family's celebration of the Jewish Passover, Internet columnist Russell Brown describes the fascination that he and millions of other viewers in the United States have with *American Idol,* the Fox television series that searches for a new national solo pop idol. The competition allows viewers to be involved in the process of selecting the next "star" because (in addition to celebrity judges who evaluate each contestant and render an opinion about their performance) television viewers call in and vote for their favorite contestant. According to Brown (2003), it is this illusion of choice and of supposedly having the power to help decide who might be the next new musical star that makes *American Idol* a hit with viewers.

Although the success of *American Idol* is an interesting sociological study in itself, the prominence of this TV show is actually part of a larger story of cultural diffusion—the transmission of cultural items or social practices from one group or society to another. In fact, the story of *American Idol* does not start in the United States: The program originated in the United Kingdom as *Pop Idol.* After that show received high viewer ratings in Britain, versions of *Idol* were adapted for the United States, Germany, the Netherlands, Poland, and South Africa. Among adults aged sixteen to forty-nine in these countries, the various *Idols* swept the TV ratings week after week. In the United States, more than 30 million viewers watched the final show of *American Idol*'s second season, and interest has been sustained by competition for the next season and the release of the movie *From Justin to Kelly,* a teen musical starring the winner and runner-up from the first *American Idol* (Stein, 2003). However, our story does not end there: By 2003, *Idols* had popped up in other countries, including *Superstar* on Lebanon-based TV, *Idool 2003*

in Belgium, *A la Recherche de la Nouvelle Star* in France, *Idol* in Norway, *Canadian Idol* in Canada, and *Idols* in Finland, to name only a few.

Is this show highly successful across nations strictly because of corporate marketing strategies used by the various television networks? Although the show's hype in various countries no doubt plays a part in generating contestants and millions of viewers, there is a larger cultural appeal to the program that cannot be ignored. Brown (2003) believes that part of the appeal of *Idol* is that even though the program is produced by powerful media industries, viewers believe that they can take culture into their own hands, defy what the judges think, and help determine the outcome. Perhaps a more enduring cultural appeal of the show around the world is that viewers believe that they are connecting with other people—that they are developing new sets of friends and that they are choosing how they will be entertained.

How long will the *Idols* be popular in various nations? These shows may just be fads that temporarily catch some people's interest, or they may produce offshoots (contests involving younger would-be-stars, for example) that help such reality-based programming remain popular for some more extended period of time.

What kinds of television programming do you believe will have the widest appeal to diverse viewers around the world? Which U.S. programs are least likely to survive the process of cultural diffusion? Why?

Box 3.4 SOCIOLOGY IN GLOBAL PERSPECTIVE

Popular Culture, the Internet, and Cultural Imperialism

- In Hanoi, a photographer in Lenin Park charges the equivalent of 50 cents (U.S. currency) for taking a person's picture standing beside a human-size Mickey Mouse doll. ("Mouse Makes the Man," 1995)
- Euro Disney, outside Paris, is the single largest tourist attraction in France, far surpassing the Eiffel Tower and the Louvre. (Kraft, 1994)
- During Moscow's December holiday season, parents can buy their children nonalcoholic champagne in bottles decorated with Disney's Pocahontas and the Lion King. (Hilsenrath, 1996)
- Ninety percent of worldwide traffic on the Internet is in English; the Internet is anchored in the United States, and the vast majority of Web sites are based in this country. (Harvard Law School Seminar, 1996)

Are these examples of cultural imperialism and the "Americanization" of the world's cultures? According to Richard Kuisel (1993), *Americanization* is not the central issue when people around the globe go to U.S.-inspired amusement parks or surf the Internet. As Kuisel (1993: 4) states, "Although the phenomenon is still described as Americanization, it has become increasingly disconnected from America. Perhaps it would be better described as the coming of the consumer society." However, other global analysts disagree with Kuisel's assertion and suggest that we should think about how we would feel if we were in this situation:

> What if one day you woke up, turned on the radio and could not find an American song? What if you went to the movies and the only films were foreign? What if you wanted to buy a book and you found that the only American works were located in a small "Americana" section of the store? Furthermore, what if you came home to find your kids glued to the television for back-to-back reruns of a French soap opera and two German police dramas? It sounds foreign, even silly to an American. Ask a Canadian, a German, or a Greek the same question, however, and you will get a different reaction. (Harvard Law School Seminar, 1996: 1)

In nations such as France, which is extremely proud of its cultural identity, many people believe that the effects of cultural imperialism are evident in the types of entertainment and other cultural products available to them in everyday life. According to this view, a nation's *cultural products,* including its books, films, television programs, and other modes of communication, define its *identity.* As a result, some believe that "the

AP/Wide World Photos

■ What is the location of the Sleeping Beauty castle shown in this photo? Although it looks similar to the castles at Disneyland and Disney World, this particular tourist attraction is at Euro Disneyland, outside of Paris. Is this an example of a global culture? Why or why not?

Internet will have a huge impetus in the Anglification of the world and that English, currently the unofficial language of world commerce, will become that of world culture" (Harvard Law School Seminar, 1996: 5). Leaders in some nations have become concerned that the Internet constitutes a new tool of U.S. cultural imperialism and have formed groups such as France's "La Francophonie," an organization created to preserve the use of French in cyberspace.

Although the ideas and products of many nations have also permeated U.S. culture and influenced our consumption patterns, many social analysts believe that U.S. culture has had a greater influence on other nations' cultures than other nations' cultures have had on ours. Some social analysts believe that American culture is likely to become the "second culture" of people around the globe (Kuisel, 1993). Is the commercialized diffusion of popular culture beneficial to people in other nations? Or is it a form of cultural imperialism? What do you think?

that it may be the "glue" that holds society together. Regardless of race, class, sex, age, or other characteristics, many people are brought together (at least in spirit) to cheer teams competing in major sporting events such as the Super Bowl or the Olympic Games. Television and the Internet help integrate recent immigrants into the mainstream culture, whereas longer-term residents may become more homogenized as a result of seeing the same images and being exposed to the same beliefs and values (Gerbner et al., 1987).

However, functionalists acknowledge that all societies have dysfunctions that produce a variety of societal problems. When a society contains numerous subcultures, discord results from a lack of consensus about core values. In fact, popular culture may undermine core cultural values rather than reinforce them (Christians, Rotzoll, and Fackler, 1987). For example, movies may glorify crime, rather than hard work, as the quickest way to get ahead. According to some analysts, excessive violence in music videos, movies, and television programs may be harmful to children and young people (Medved, 1992). From this perspective, popular culture can be a factor in antisocial behavior as seemingly diverse as hate crimes and fatal shootings in public schools.

A strength of the functionalist perspective on culture is its focus on the needs of society and the fact that stability is essential for society's continued survival. A shortcoming is its overemphasis on harmony and cooperation. This approach also fails to fully account for factors embedded in the structure of society—such as class-based inequalities, racism, and sexism—that may contribute to conflict among people in the United States or to global strife.

Conflict Perspectives

Conflict perspectives are based on the assumption that social life is a continuous struggle in which members of powerful groups seek to control scarce resources. According to this approach, values and norms help create and sustain the privileged position of the powerful in society while excluding others. As early conflict theorist Karl Marx stressed, ideas are *cultural creations* of a society's most powerful members. Thus, it is possible for political, economic, and social leaders to use *ideology*—an integrated system of ideas that is external to, and coercive of, people—to maintain their positions of dominance in a society. As Marx stated,

> The ideas of the ruling class are in every epoch the ruling ideas, i.e., the class which is the ruling material force in society, is at the same time, its ruling

intellectual force. The class, which has the means of material production at its disposal, has control at the same time over the means of mental production. . . . The ruling ideas are nothing more than the ideal expression of the dominant material relationships, the dominant material relationships grasped as ideas. (Marx and Engels, 1970/1845–1846: 64)

Many contemporary conflict theorists agree with Marx's assertion that ideas, a nonmaterial component of culture, are used by agents of the ruling class to affect the thoughts and actions of members of other classes.

How might conflict theorists view popular culture? Some conflict theorists believe that popular culture, which originated with everyday people, has been largely removed from their domain and has become nothing more than a part of the capitalist economy in the United States (Gans, 1974; Cantor, 1980, 1987). From this approach, media conglomerates like Time Warner, Disney, and Viacom create popular culture, such as films, television shows, and amusement parks, in the same way that they would produce any other product or service. Creating new popular culture also promotes consumption of *commodities*—objects outside ourselves that we purchase to satisfy our human needs or wants (Fjellman, 1992). Recent studies have shown that moviegoers spend more money for popcorn, drinks, candy, and other concession-stand food than they do for tickets to get into the theater. Similarly, park-goers at Disneyland and Walt Disney World spend as much money on merchandise—such as Magic Kingdom pencils, Mickey Mouse hats, kitchen accessories, and clothing—as they do on admission tickets and rides (Fjellman, 1992).

From this perspective, people come to believe that they *need* things they ordinarily would not purchase. Their desire is intensified by marketing techniques that promote public trust in products and services provided by a corporation such as the Walt Disney Company. Sociologist Pierre Bourdieu (1984: 291) refers to this public trust as *symbolic capital*: "the acquisition of a reputation for competence and an image of respectability and honourability." Symbolic capital consists of culturally approved intangibles—such as honor, integrity, esteem, trust, and goodwill—that may be accumulated and used for tangible (economic) gain. Thus, people buy products at Walt Disney World (and Disney stores throughout the country) because they believe in the trustworthiness of the item ("These children's pajamas are bound to be flame retardant; they came from the Disney Store") and the integrity of the company ("I can trust Disney; it has been around for a long time").

In recent years, there has been a significant increase in the number of immigrants who have become U.S. citizens. However, an upsurge in anti-immigrant sentiment has put pressure on the Border Patrol and the Immigration and Naturalization Service, which are charged with enforcing immigration laws.

Other conflict theorists examine the intertwining relationship among race, gender, and popular culture. According to the sociologist K. Sue Jewell (1993), popular cultural images are often linked to negative stereotypes of people of color, particularly African American women. Jewell believes that cultural images depicting African American women as mammies or domestics—such as those previously used in Aunt Jemima Pancake ads and recent resurrections of films like *Gone with the Wind*—affect contemporary black women's economic prospects in profound ways (Jewell, 1993).

A strength of the conflict perspective is that it stresses how cultural values and norms may perpetuate social inequalities. It also highlights the inevitability of change and the constant tension between those who want to maintain the status quo and those who desire change. A limitation is its focus on societal discord and the divisiveness of culture.

Symbolic Interactionist Perspectives

Unlike functionalists and conflict theorists, who focus primarily on macrolevel concerns, symbolic interactionists engage in a microlevel analysis that views society as the sum of all people's interactions. From this perspective, people create, maintain, and modify culture as they go about their everyday activities. Symbols make communication with others possible because they provide us with shared meanings.

According to some symbolic interactionists, people continually negotiate their social realities. Values and norms are not independent realities that automatically determine our behavior. Instead, we reinterpret them in each social situation we encounter. However, the classical sociologist Georg Simmel warned that the larger cultural world—including both material culture and nonmaterial culture—eventually takes on a life of its own apart from the actors who daily re-create social life. As a result, individuals may be more controlled by culture than they realize. Simmel (1990/1907) suggested that money is an example of how people may be controlled by their culture. According to Simmel, people initially create money as a means of exchange, but then money acquires a social meaning that extends beyond its purely economic function. Money becomes an end in itself, rather than a means to an end. Today, we are aware of the relative "worth" not only of objects but also of individuals. Many people revere wealthy entrepreneurs and highly paid celebrities, entertainers, and sports figures for the amount of money they make, not for their intrinsic qualities. According to Simmel (1990/1907), money makes it possible for us to *relativize* everything, including our relationships with other people. When social life can be reduced to money, people become cynical, believing that anything—including people, objects, beauty, and truth—can be bought if we can pay the price. Although Simmel acknowledged the positive functions of money, he believed that the social interpretations people give to money often produce individual feelings of cynicism and isolation.

A symbolic interactionist approach highlights how people maintain and change culture through their interactions with others. However, interactionism does not provide a systematic framework for analyzing how we shape culture and how it, in turn, shapes us. It also does not provide insight into how shared meanings are developed among people, and it does not take into account the many situations in which there is disagreement on meanings. Whereas the functional and conflict approaches tend to overemphasize the macrolevel workings of society, the interactionist viewpoint often fails to take these larger social structures into account.

Postmodernist Perspectives

Postmodernist theorists believe that much of what has been written about culture in the Western world is Eurocentric—that it is based on the uncritical assumption that European culture (including its dispersed versions in countries such as the United States, Australia, and South Africa) is the true, universal culture in which all the world's people ought to believe (Lemert, 1997). By contrast, postmodernists believe that we should speak of *cultures,* rather than *culture.*

However, Jean Baudrillard, one of the best-known French social theorists, believes that the world of culture today is based on *simulation,* not reality. According to Baudrillard, social life is much more a spectacle that simulates reality than reality itself. Many people gain "reality" from the media or cyberspace. For example, consider the many U.S. children who, upon entering school for the first time, have already watched more hours of television than the total number of hours of classroom instruction they will encounter in their entire school careers (Lemert, 1997). Add to this the number of hours that some will have spent playing computer games or surfing the Internet. Baudrillard refers to this social creation as *hyperreality*—a situation in which the *simulation* of reality is more real than the thing itself. For Baudrillard, everyday life has been captured by the signs and symbols generated to represent it, and we ultimately relate to simulations and models as if they were reality.

Baudrillard (1983) uses Disneyland as an example of a simulation that conceals the reality that exists outside rather than inside the boundaries of the artificial perimeter. According to Baudrillard, Disney-like theme parks constitute a form of seduction that substitutes symbolic (seductive) power for real power, particularly the ability to bring about social change. From this perspective, amusement park "guests" may feel like "survivors" after enduring the rapid speed and gravity-defying movements of the roller coaster rides or see themselves as "winners" after surviving fights with hideous cartoon villains on the "dark rides"—when they have actually experienced the substitution of an *appearance* of power over their lives for the *absence* of real power. Similarly, the anthropologist Stephen M. Fjellman (1992) studied Disney World in Orlando, Florida, and noted that people may forget, at least briefly, that the outside world can be threatening while they stroll Disney World's streets without fear of crime or automobiles. Although this freedom may be temporarily empowering, it also may lull people into accepting a "worldview that presents an idealized United States as heaven. . . . How nice if they could all be like us—with kids, a dog, and General Electric appliances—in a world whose only problems are avoiding Captain Hook, the witch's apple, and Toad Hall weasels" (Fjellman, 1992: 317).

In their examination of culture, postmodernist social theorists make us aware of the fact that no single perspective can grasp the complexity and diversity of the social world. They also make us aware that reality may not be what it seems. According to the postmodernist view, no one authority can claim to know social reality, and we should deconstruct—take apart and subject to intense critical scrutiny—existing beliefs and theories about culture in hopes of gaining new insights (Ritzer, 1997).

Although postmodern theories of culture have been criticized on a number of grounds, we will examine only three. One criticism is postmodernism's lack of a clear conceptualization of ideas. Another is the tendency to critique other perspectives as being "grand narratives," whereas postmodernists offer their own varieties of such narratives. Finally, some analysts believe that postmodern analyses of culture lead to profound pessimism about the future.

Concept Table 3.A reviews the components of culture as well as how the four major perspectives view culture.

CULTURE IN THE FUTURE

As we have discussed in this chapter, many changes are occurring in the United States. Increasing cultural diversity can either cause long-simmering racial and ethnic antagonisms to come closer to a boiling point or result in the creation of a truly "rainbow culture" in which diversity is respected and encouraged.

In the future, the issue of cultural diversity will increase in importance, especially in schools. Multicultural education that focuses on the contributions of a wide variety of people from different backgrounds will continue to be an issue of controversy from kindergarten through college. In the Los Angeles school district, for example, students speak more than 114 different languages and dialects. Schools will face the challenge of embracing widespread cultural diversity while conveying a sense of community and national identity to students.

Concept Table 3.A   ANALYSIS OF CULTURE

Components of Culture	Symbol	Anything that meaningfully represents something else.
	Language	A set of symbols that expresses ideas and enables people to think and communicate with one another.
	Values	Collective ideas about what is right or wrong, good or bad, and desirable or undesirable in a particular culture.
	Norms	Established rules of behavior or standards of conduct.
Sociological Analysis of Culture	Functionalist Perspectives	Culture helps people meet their biological, instrumental, and expressive needs.
	Conflict Perspectives	Ideas are a cultural creation of society's most powerful members and can be used by the ruling class to affect the thoughts and actions of members of other classes.
	Symbolic Interactionist Perspectives	People create, maintain, and modify culture during their everyday activities; however, cultural creations can take on a life of their own and end up controlling people.
	Postmodern Perspectives	Much of culture today is based on simulation of reality (e.g., what we see on television) rather than reality itself.

Technology will continue to have a profound effect on culture. Television and radio, films and videos, and electronic communications will continue to accelerate the flow of information and expand cultural diffusion throughout the world. Global communication devices will move images of people's lives, behavior, and fashions instantaneously among almost all nations (Petersen, 1994). Increasingly, computers and cyberspace will become people's window on the world and, in the process, promote greater integration or fragmentation among nations. Integration occurs when there is a widespread acceptance of ideas and items—such as democracy, rock music, blue jeans, and McDonald's hamburgers—among cultures. By contrast, fragmentation occurs when people in one culture disdain the beliefs and actions of other cultures. As a force for both cultural integration and fragmentation, technology will continue to revolutionize communications, but most of the world's population will not participate in this revolution (Petersen, 1994).

From a sociological perspective, the study of culture helps us not only understand our own "tool kit" of symbols, stories, rituals, and world views but also expand our insights to include those of other people of the world, who also seek strategies for enhancing their own lives. If we understand how culture is used by people, how cultural elements constrain or further certain patterns of action, what aspects of our cultural heritage have enduring effects on our actions, and what specific historical changes undermine the validity of some cultural patterns and give rise to others, we can apply our sociological imagination not only

to our own society but to the entire world as well (see Swidler, 1986).

Bill Bachmann/The Image Works

New technologies have made educational opportunities available to a wider diversity of students, including persons with a disability. How will global communications technologies continue to change culture and social life in the future?

CHAPTER REVIEW

■ **What is culture?**

Culture is the knowledge, language, values, and customs passed from one generation to the next in a human group or society. Culture can be either material or nonmaterial. Material culture consists of the physical creations of society. Nonmaterial culture is more abstract and reflects the ideas, values, and beliefs of a society.

■ **What are cultural universals?**

Cultural universals are customs and practices that exist in all societies and include activities and institutions such as storytelling, families, and laws. Specific forms of these universals vary from one cultural group to another, however.

■ **What are the four nonmaterial components of culture that are common to all societies?**

These components are symbols, language, values, and norms. Symbols express shared meanings; through them, groups communicate cultural ideas and abstract concepts. Language is a set of symbols through which groups communicate. Values are a culture's collective ideas about what is acceptable or not acceptable. Norms are the specific behavioral expectations within a culture.

■ **What are the main types of norms?**

Folkways are norms that express the everyday customs of a group, whereas mores are norms with strong moral and ethical connotations and are essential to the stability of a culture. Laws are formal, standardized norms that are enforced by formal sanctions.

■ **What are high culture and popular culture?**

High culture consists of classical music, opera, ballet, and other activities usually patronized by elite audiences. Popular culture consists of the activities, products, and services of a culture that appeal primarily to members of the middle and working classes.

■ **What causes cultural change in societies?**

Cultural change takes place in all societies. Change occurs through discovery and invention, and through diffusion, which is the transmission of culture from one society or group to another.

■ **How is cultural diversity reflected in society?**

Cultural diversity is reflected through race, ethnicity, age, sexual orientation, religion, occupation, and so forth. A diverse culture also includes subcultures and countercultures. A subculture has distinctive ideas and behaviors that differ from the larger society to which it belongs. A counterculture rejects the dominant societal values and norms.

■ **What are culture shock, ethnocentrism, and cultural relativism?**

Culture shock refers to the anxiety that people experience when they encounter cultures radically different from their own. Ethnocentrism is the assumption that one's own culture is superior to others. Cultural relativism views and analyzes another culture in terms of that culture's own values and standards.

■ **How do the major sociological perspectives view culture?**

A functionalist analysis of culture assumes that a common language and shared values help produce consensus and harmony. According to some conflict theorists, culture may be used by certain groups to maintain their privilege and exclude others from society's benefits. Symbolic interactionists suggest that people create, maintain, and modify culture as they go about their everyday activities. Postmodern thinkers believe that there are many cultures within the United States alone. In order to grasp a better understanding of how popular culture may simulate reality rather than being reality, postmodernists believe that we need a new way of conceptualizing culture and society.

KEY TERMS

counterculture 95
cultural imperialism 98
cultural lag 90
cultural relativism 96
cultural universals 80
culture 75
culture shock 95
diffusion 91
discovery 91
ethnocentrism 96
folkways 89
invention 91
language 83
laws 90
material culture 79

QUESTIONS FOR CRITICAL THINKING

1. Would it be possible today to live in a totally separate culture in the United States? Could you avoid all influences from the mainstream popular culture or from the values and norms of other cultures? How would you be able to avoid any change in your culture?

2. Do fads and fashions reflect and reinforce or challenge and change the values and norms of a society? Consider a wide variety of fads and fashions: musical styles, computer and video games and other technologies, literature, and political, social, and religious ideas.

3. You are doing a survey analysis of neo-Nazi skinheads to determine the effects of popular culture on their views and behavior. What are some of the questions you would use in your survey?

RESOURCES ON THE INTERNET

Chapter-Related Web Sites

The following Web sites have been selected for their relevance to the topics in this chapter. These sites are among the more stable, but please note that Web site addresses change frequently. For an updated list of chapter-related Web sites with URL links, please visit the *Sociology in Our Times* Web site (**www.wadsworth.com /KendallSIOT**).

"What Is Culture?"
http://www.wsu.edu:8001/vcwsu/commons /topics/culture/culture-index.html

This site explores the concept of human culture. In addition to providing a baseline definition of culture, the site examines pivotal discussions as well as debates about culture and provides various student interpretations of culture, interesting links, and many other resources.

Cultural Studies Central
http://www.culturalstudies.net/index.html

Discuss, read about, and enjoy contemporary culture at this site, which features interactive commentary, informative links, and interactive Web projects on topics such as art crimes, urban legends, and gender borders.

Center for the Study of Popular Culture
http://www.cspc.org

Maintained by the Los Angeles-based Center for the Study of Popular Culture, this Web site will connect you to resources related to politics, media, and entertainment culture in addition to providing exposure to an array of opinions about the state of Western culture.

ONLINE STUDY AND RESEARCH TOOLS

Accompanying this text are many *free* powerful online study tools that will help you master the material in this chapter, help increase your depth of understanding, and help you make the grade!

SocCoach CD-ROM

Use the SocCoach CD-ROM enclosed with this text to help you formulate a customized study plan for this chapter. After you take the Diagnostic Quiz, SocCoach will generate a customized study plan just for you! It will identify sections of the chapter that you should review and will provide videos, charts, graphs, and excerpts from the text to supplement your studies and enhance your understanding. You'll also find fun, interactive activities such as Virtual Explorations and Map the Stats to apply what you've learned and stretch your sociological imagination.

The Companion Web Site for Sociology in Our Times, *Fifth Edition*
www.wadsworth.com/KendallSIOT

Gain an even better grasp on this chapter by going to the companion Web site to take one of the Tutorial Quizzes, use the Flash Cards to master key terms, or check out the many other study aids you'll find there. You'll also find special features such as GSS Data and Census 2000 information that'll put data and resources at your fingertips to help you with that special project or help you as you do some research on your own.

 In this chapter, when you see the icon on the left, it alerts you to a specific exercise found in *Wadsworth's Sociology Online Resources and Writing Companion*. This valuable guide shows you how to use Wadsworth's exclusive online resources—*InfoTrac*

College Edition, the *Opposing Viewpoints Resource Center,* and *MicroCase Online*—to assist you in your study of sociology and to build essential research and writing skills.

CHAPTER 4

Socialization

I think my mom and dad were boyfriend and girlfriend for a couple of years, but they were apart by the time I was born. . . . The earliest memory I have of my father isn't pleasant. I was three years old. It was afternoon, and I was wearing a pair of jeans and these cute Mickey Mouse suspenders, a favorite play outfit. My mom and I were standing in the kitchen, doing the laundry. . . . Suddenly the door swung open and there was this man standing there. I yelled, "Daddy!" Even though I didn't know what he looked like, I just automatically knew it was him.

He paused in the doorway, like he was making a dramatic entrance, and I think he said something, but he was so drunk, it was unintelligible. It sounded more like a growl. We stood there, staring at him. I was so excited to see him. I was just coming to the age where I noticed that I didn't have a father like everyone else, and I wanted one. I didn't really know what my dad was like, but I learned real fast.

In a blur of anger he roared into the room and threw my mom down on the ground. Then he turned on me. I didn't know what was happening. I was still excited to see him, still hearing the echo of my gleeful yell, "Daddy!" when he picked me up and threw me into the wall. Luckily, half of my body landed on a big sack of laundry, and I wasn't hurt. But my dad didn't even look back at me.

> He turned and grabbed a bottle of tequila, shattered a bunch of glasses all over the floor, and then stormed out of the house.... And that was it. That was the first time I remember seeing my dad.
>
> —Actress Drew Barrymore, describing her first encounter with her father (Barrymore, 1994: 185–187)

© AP/Wide World Photos

■ Actress Drew Barrymore's description of childhood maltreatment in her family makes us aware of the importance of early socialization in all our lives. As an adult, she has experienced many happier times.

The process of socialization is of major significance to sociologists. Although most children are nurtured, trusted, and loved by their parents, Barrymore's experience is not an isolated incident: Large numbers of children suffer maltreatment at the hands of family members or other caregivers such as baby-sitters or day-care workers. Child maltreatment includes physical abuse, sexual abuse, physical neglect, and emotional mistreatment of children and young adolescents. Such maltreatment is of interest to sociologists because it has a serious impact on a child's social growth, behavior, and self-image—all of which develop within the process of socialization. By contrast, children who are treated with respect by their parents are more likely to develop a positive self-image and learn healthy conduct because their parents provide appropriate models of behavior.

In this chapter, we examine why socialization is so crucial, and we discuss both sociological and social psychological theories of human development. We look at the dynamics of socialization—how it occurs and what shapes it. Throughout the chapter, we focus on positive and negative aspects of the socialization process. Before reading on, test your knowledge of socialization and child care by taking the quiz in Box 4.1.

QUESTIONS AND ISSUES

Chapter Focus Question: What happens when children do not have an environment that supports positive socialization?

What purpose does socialization serve?

How do individuals develop a sense of self?

How does socialization occur?

Who experiences resocialization?

WHY IS SOCIALIZATION IMPORTANT AROUND THE GLOBE?

Socialization **is the lifelong process of social interaction through which individuals acquire a self-identity and the physical, mental, and social skills needed for survival in society.** It is the essential link between the individual and society. Socialization enables each of us to develop our human potential and to learn the ways of thinking, talking, and acting that are necessary for social living.

Socialization is essential for the individual's survival and for human development. The many people who met the early material and social needs of each of us were central to our establishing our own identity. During the first three years of our life, we begin to develop both a unique identity and the ability to manipulate things and to walk. We acquire sophisticated cognitive tools for thinking and for analyzing a wide variety of situations, and we learn effective communication skills. In the process, we begin a relatively long socialization process that culminates in our integration into a complex social and cultural system (Garcia Coll, 1990).

Socialization is also essential for the survival and stability of society. Members of a society must be socialized to support and maintain the existing social structure. From a functionalist perspective, individual conformity to existing norms is not taken for granted; rather, basic individual needs and desires must be balanced against the needs of the social structure. The socialization process is most effective when people conform to the norms of society because they believe that this is the best course of action. Socialization enables a society to "reproduce" itself by passing on its culture from one generation to the next.

Although the techniques used to teach newcomers the beliefs, values, and rules of behavior are somewhat similar in many nations, the *content* of socialization differs greatly from society to society. How people walk, talk, eat, make love, and wage war are all functions of the culture in which they are raised. At the same time, we are also influenced by our exposure to subcultures of class, race, ethnicity, religion, and gender. In addition, each of us has unique experiences in our families and friendship groupings. The kind of human being that we become depends greatly on the particular society and social groups that surround us at birth and during early childhood. What we believe about ourselves, our society, and the world does not spring full-blown from inside ourselves; rather, we learn these things from our interactions with others.

Human Development: Biology and Society

What does it mean to be "human"? To be human includes being conscious of ourselves as individuals with unique identities, personalities, and relationships with others. As humans, we have ideas, emotions, and values. We have the capacity to think and to make rational decisions. But what is the source of "humanness"? Are we born with these human characteristics, or do we develop them through our interactions with others?

When we are born, we are totally dependent on others for our survival. We cannot turn ourselves over, speak, reason, plan, or do many of the things that are associated with being human. Although we can nurse, wet, and cry, most small mammals can also do those things. As discussed in Chapter 3, we humans differ from nonhuman animals because we lack instincts and must rely on learning for our survival. Human infants have the potential for developing human characteristics if they are exposed to an adequate socialization process.

Every human being is a product of biology, society, and personal experiences—that is, of heredity and environment or, in even more basic terms, "nature" and "nurture." How much of our development can be explained by socialization? How much by our genetic heritage? Sociologists focus on how humans design their own culture and transmit it from generation to

Box 4.1 SOCIOLOGY AND EVERYDAY LIFE

How Much Do You Know About Early Socialization and Child Care?

True	False	
T	F	1. In the United States, full-day child care often costs as much per year as college tuition at a public college or university.
T	F	2. The cost of child care is a major problem for many U.S. families.
T	F	3. After-school programs have greatly reduced the number of children who are home alone after school.
T	F	4. The average annual salary of a child-care worker is less than the average yearly salaries for funeral attendants or garbage collectors.
T	F	5. All states require teachers in child-care centers to have training in their field and to pass a licensing examination.
T	F	6. In a family in which child abuse occurs, all the children are likely to be victims.
T	F	7. It is against the law to fail to report child abuse.
T	F	8. Some people are "born" child abusers whereas others learn abusive behavior from their family and friends.

Answers on page 112.

generation through socialization. By contrast, sociobiologists assert that nature, in the form of our genetic makeup, is a major factor in shaping human behavior. ***Sociobiology*** **is the systematic study of how biology affects social behavior** (Wilson, 1975). According to the zoologist Edward O. Wilson, who pioneered sociobiology, genetic inheritance underlies many forms of social behavior such as war and peace, envy and concern for others, and competition and cooperation. Most sociologists disagree with the notion that biological principles can be used to explain all human behavior. Obviously, however, some aspects of our physical makeup—such as eye color, hair color, height, and weight—are largely determined by our heredity.

How important is social influence ("nurture") in human development? There is hardly a single behavior that is not influenced socially. Except for simple reflexes, most human actions are social, either in their causes or in their consequences. Even solitary actions such as crying and brushing our teeth are ultimately social. We cry because someone has hurt us. We brush our teeth because our parents (or dentist) told us it was important. Social environment probably has a greater effect than heredity on the way we develop and the way we act. However, heredity does provide the basic material from which other people help to mold an individual's human characteristics.

Our biological needs and emotional needs are related in a complex equation. Children whose needs are met in settings characterized by affection, warmth, and closeness see the world as a safe and comfortable place and see other people as trustworthy and helpful. By contrast, infants and children who receive less-than-adequate care or who are emotionally rejected or abused often view the world as hostile and have feelings of suspicion and fear.

Problems Associated with Social Isolation and Maltreatment

Social environment, then, is a crucial part of an individual's socialization. Even nonhuman primates such as monkeys and chimpanzees need social contact with others of their species in order to develop properly. As we will see, appropriate social contact is even more important for humans.

Isolation and Nonhuman Primates Researchers have attempted to demonstrate the effects of social isolation on nonhuman primates raised without contact with others of their own species. In a series of laboratory experiments, the psychologists Harry and Margaret Harlow (1962, 1977) took infant

Box 4.1 SOCIOLOGY AND EVERYDAY LIFE

Answers to the Sociology Quiz on Early Socialization and Child Care

1. **True.** Full-day child care typically costs between $4,000 and $10,000 per child per year, which is as much or more than tuition at many public colleges and universities.

2. **True.** Child care outside the home is a major financial burden, particularly for the one out of every four families with young children but with an income of less than $25,000 a year.

3. **False.** Although after-school programs have slightly reduced the number of children at home alone after school, nearly seven million school-age children are alone each week while their parents work.

4. **True.** The average salary for a child-care worker is only $15,430 per year, which is less than the yearly salaries for people in many other employment categories.

5. **False.** Although all states require hairdressers and manicurists to have about 1,500 hours of training at an accredited school, only 11 states require child-care providers to have any early childhood training prior to taking care of children.

6. **False.** In some families, one child may be the victim of repeated abuse whereas others are not.

7. **True.** In the United States, all states have reporting requirements for child maltreatment; however, there has been inconsistent compliance with these legal mandates. Some states have requirements that everyone who suspects abuse or neglect must report it. Other states mandate reporting only by certain persons, such as medical personnel and child-care providers.

8. **False.** No one is "born" to be an abuser. People learn abusive behavior from their family and friends.

Source: Based on Children's Defense Fund, 2002.

rhesus monkeys from their mothers and isolated them in separate cages. Each cage contained two nonliving "mother substitutes" made of wire, one with a feeding bottle attached and the other covered with soft terry cloth but without a bottle. The infant monkeys instinctively clung to the cloth "mother" and would not abandon it until hunger drove them to the bottle attached to the wire "mother." As soon as they were full, they went back to the cloth "mother" seeking warmth, affection, and physical comfort.

The Harlows' experiments show the detrimental effects of isolation on nonhuman primates. When the young monkeys were later introduced to other members of their species, they cringed in the corner. Having been deprived of social contact with other monkeys during their first six months of life, they never learned how to relate to other monkeys or to become well-adjusted adults—they were fearful of or hostile toward other monkeys (Harlow and Harlow, 1962, 1977).

Because humans rely more heavily on social learning than do monkeys, the process of socialization is even more important for us.

Isolated Children Of course, sociologists would never place children in isolated circumstances so that they could observe what happened to them. However, some cases have arisen in which parents or other caregivers failed to fulfill their responsibilities, leaving children alone or placing them in isolated circumstances. From analysis of these situations, social scientists have documented cases in which children were deliberately raised in isolation. A look at the lives of two children who suffered such emotional abuse provides important insights into the importance of a positive socialization process and the negative effects of social isolation.

Anna Born in 1932 to an unmarried, mentally impaired woman, Anna was an unwanted child. She was

Bill Aron/PhotoEdit

Bob Daemmrich/The Image Works

What are the consequences to children of isolation and physical abuse, as contrasted with social interaction and parental affection? Sociologists emphasize that social environment is a crucial part of an individual's socialization.

kept in an attic-like room in her grandfather's house. Her mother, who worked on the farm all day and often went out at night, gave Anna just enough care to keep her alive; she received no other care. Sociologist Kingsley Davis (1940) described Anna's condition when she was found in 1938:

> [Anna] had no glimmering of speech, absolutely no ability to walk, no sense of gesture, not the least capacity to feed herself even when the food was put in front of her, and no comprehension of cleanliness. She was so apathetic that it was hard to tell whether or not she could hear. And all of this at the age of nearly six years.

When she was placed in a special school and given the necessary care, Anna slowly learned to walk, talk, and care for herself. Just before her death at the age of ten, Anna reportedly could follow directions, talk in phrases, wash her hands, brush her teeth, and try to help other children (Davis, 1940).

Genie Almost four decades later, Genie was found in 1970 at the age of thirteen. She had been locked in a bedroom alone, alternately strapped down to a child's

potty chair or straitjacketed into a sleeping bag, since she was twenty months old. She had been fed baby food and beaten with a wooden paddle when she whimpered. She had not heard the sounds of human speech because no one talked to her and there was no television or radio in her room (Curtiss, 1977; Pines, 1981). Genie was placed in a pediatric hospital, where one of the psychologists described her condition:

> At the time of her admission she was virtually unsocialized. She could not stand erect, salivated continuously, had never been toilet-trained and had no control over her urinary or bowel functions. She was unable to chew solid food and had the weight, height and appearance of a child half her age. (Rigler, 1993: 35)

In addition to her physical condition, Genie showed psychological traits associated with neglect, as described by one of her psychiatrists:

> If you gave [Genie] a toy, she would reach out and touch it, hold it, caress it with her fingertips, as though she didn't trust her eyes. She would rub it against her cheek to feel it. So when I met her and

she began to notice me standing beside her bed, I held my hand out and she reached out and took my hand and carefully felt my thumb and fingers individually, and then put my hand against her cheek. She was exactly like a blind child. (Rymer, 1993: 45)

Extensive therapy was used in an attempt to socialize Genie and develop her language abilities (Curtiss, 1977; Pines, 1981). These efforts met with limited success: In the 1990s, Genie was living in a board-and-care home for retarded adults (see Angier, 1993; Rigler, 1993; Rymer, 1993).

Why do we discuss children who have been the victims of maltreatment in a chapter that looks at the socialization process? The answer lies in the fact that such cases are important to our understanding of the socialization process because they show the importance of this process and reflect how detrimental social isolation and neglect can be to the well-being of people.

Child Maltreatment What do the terms *child maltreatment* and *child abuse* mean to you? When asked what constitutes child maltreatment, many people first think of cases that involve severe physical injuries or sexual abuse. However, neglect is the most frequent form of child maltreatment (Dubowitz et al., 1993). Child neglect occurs when children's basic needs—including emotional warmth and security, adequate shelter, food, health care, education, clothing, and protection—are not met, regardless of cause (Dubowitz et al., 1993: 12). Neglect often involves acts of omission (where parents or caregivers fail to provide adequate physical or emotional care for children) rather than acts of commission (such as physical or sexual abuse). Of course, what constitutes child maltreatment differs from society to society.

SOCIAL PSYCHOLOGICAL THEORIES OF HUMAN DEVELOPMENT

Over the past hundred years, a variety of psychological and sociological theories have been developed not only to explain child abuse but also to describe how a positive process of socialization occurs. Let's look first at several psychological theories that focus primarily on how the individual personality develops.

Freud and the Psychoanalytic Perspective

The basic assumption in Sigmund Freud's (1924) psychoanalytic approach is that human behavior and personality originate from unconscious forces within individuals. Sigmund Freud (1856–1939), who is known as the founder of psychoanalytic theory, developed his major theories in the Victorian era, when biological explanations of human behavior were prevalent. It was also an era of extreme sexual repression and male dominance when compared to contemporary U.S. standards. Freud's theory was greatly influenced by these cultural factors, as reflected in the importance he assigned to sexual motives in explaining behavior. For example, Freud based his ideas on the belief that people have two basic tendencies: the urge to survive and the urge to procreate.

According to Freud (1924), human development occurs in three states that reflect different levels of the personality, which he referred to as the *id, ego,* and *superego.* The ***id* is the component of personality that includes all of the individual's basic biological drives and needs that demand immediate gratification.** For Freud, the newborn child's personality is all id, and from birth the child finds that urges for self-gratification—such as wanting to be held, fed, or changed—are not going to be satisfied immediately. However, id remains with people throughout their life in the form of *psychic energy,* the urges and desires that account for behavior. By contrast, the second level of personality—the ego—develops as infants discover that their most basic desires are not always going to be met by others. The ***ego* is the rational, reality-oriented component of personality that imposes restrictions on the innate, pleasure-seeking drives of the id.** The ego channels the desire of the id for immediate gratification into the most advantageous direction for the individual. The third level of personality—the superego—is in opposition to both the id and the ego. The ***superego,* or conscience, consists of the moral and ethical aspects of personality.** It is first expressed as the recognition of parental control and eventually matures as the child learns that parental control is a reflection of the values and moral demands of the larger society. When a person is well adjusted, the ego successfully manages the opposing forces of the id and the superego. Figure 4.1 illustrates Freud's theory of personality.

Although subject to harsh criticism, Freud's theory made people aware of the importance of early childhood experiences, including abuse and neglect. His theories have also had a profound influence on con-

| Figure 4.1 | Freud's Theory of Personality |

This illustration shows how Freud might picture a person's internal conflict over whether to commit an antisocial act such as stealing a candy bar. In addition to dividing personality into three components, Freud theorized that our personalities are largely unconscious—hidden from our normal awareness. To dramatize his point, Freud compared conscious awareness (portions of the ego and superego) to the visible tip of an iceberg. Most of personality—including the id, with its raw desires and impulses—lies submerged in our subconscious.

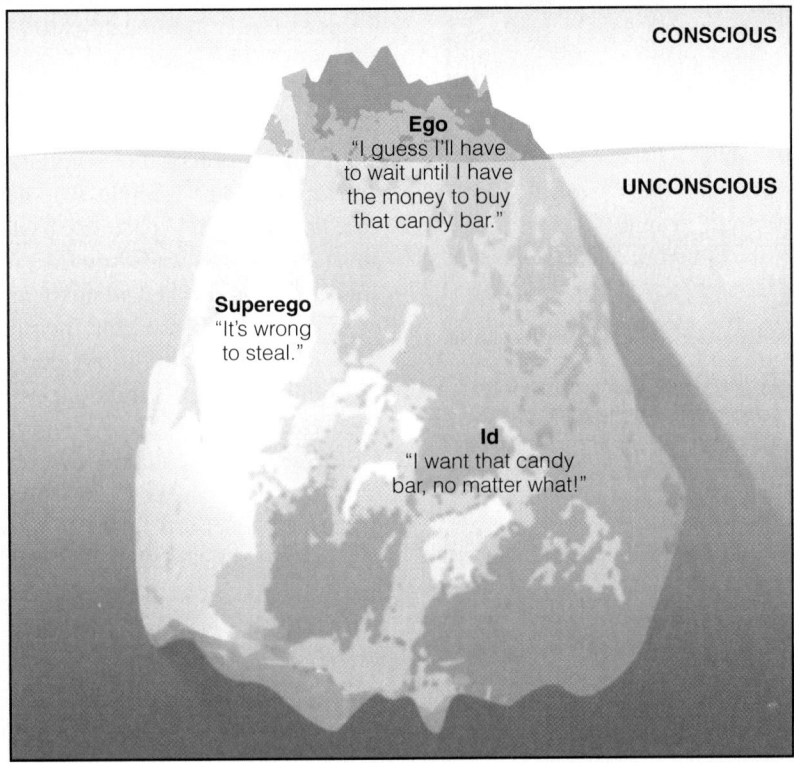

temporary mental health practitioners and on other human development theories.

Erikson and Psychosocial Development

Erik H. Erikson (1902–1994) drew from Freud's theory and identified eight psychosocial stages of development. According to Erikson (1980/1959), each stage is accompanied by a crisis or potential crisis that involves transitions in social relationships:

1. *Trust versus mistrust* (birth to age one). If infants receive good care and nurturing (characterized by emotional warmth, security, and love) from their parents, they will develop a sense of trust. If they do not receive such care, they will become mistrustful and anxious about their surroundings.

2. *Autonomy versus shame and doubt* (age one to three). As children gain a feeling of control over their behavior and develop a variety of physical and mental abilities, they begin to assert their independence. If allowed to explore their environment, children will grow more autonomous. If parents disapprove of or discourage them, children will begin to doubt their abilities.

3. *Initiative versus guilt* (age three to five). If parents encourage initiative during this stage, children will develop a sense of initiative. If parents make children feel that their actions are bad or that they are a nuisance, children may develop a strong sense of guilt.

4. *Industry versus inferiority* (age six to eleven). At this stage, children want to manipulate objects and learn how things work. Adults who encourage children's efforts and praise the results—both

© Stuart Cohen/Index Stock Imagery

Erik H. Erikson's theory of psychosocial development states that individuals go through eight stages of development. One of these is a period of intimacy versus isolation, which covers courtship and early family life. The college years are often a time when young people seek to establish permanent relationships with others.

at home and at school—produce a feeling of industry in children. Feelings of inferiority result when parents or teachers appear to view children's efforts as silly or as a nuisance.

5. *Identity versus role confusion* (age twelve to eighteen). During this stage, adolescents attempt to develop a sense of identity. As young people take on new roles, the new roles must be combined with the old ones to create a strong self-identity. Role confusion results when individuals fail to acquire an accurate sense of personal identity.

6. *Intimacy versus isolation* (age eighteen to thirty-five). The challenge of this stage (which covers courtship and early family life) is to develop close and meaningful relationships. If individuals establish successful relationships, intimacy ensues. If they fail to do so, they may feel isolated.

7. *Generativity versus self-absorption* (age thirty-five to fifty-five). Generativity means looking beyond oneself and being concerned about the next generation and the future of the world in general. Self-absorbed people may be preoccupied with their own well-being and material gains or be overwhelmed by stagnation, boredom, and interpersonal impoverishment.

8. *Integrity versus despair* (maturity and old age). Integrity results when individuals have resolved previous psychosocial crises and are able to look back at their life as having been meaningful and personally fulfilling. Despair results when previous crises remain unresolved and individuals view their life as a series of disappointments, failures, and misfortunes.

Erikson's psychosocial stages broaden the framework of Freud's theory by focusing on social and cultural forces and by examining development throughout the life course. The psychosocial approach encompasses the conflicts that coincide with major changes in a person's social environment and describes how satisfactory resolution of these conflicts results in positive development. For example, if adolescents who experience an identity crisis are able to determine who they are and what they want from life, they may be able to achieve a positive self-identity and acquire greater psychological distance from their parents.

Critics have pointed out that Erikson's research was limited to white, middle-class respondents from industrial societies (Slugoski and Ginsburg, 1989). However, other scholars have used his theoretical framework to examine racial–ethnic variations in the process of psychosocial development. Most of the studies have concluded that all children face the same developmental tasks at each stage but that children of color often have greater difficulty in obtaining a positive outcome because of experiences with racial prejudice and discrimination in society (Rotheram and Phinney, 1987). Although establishing an identity is difficult for most adolescents, one study found that it was especially problematic for children of recent Asian American immigrants who had experienced high levels of stress related to immigration (Huang and Ying, 1989).

Piaget and Cognitive Development

Unlike psychoanalytic approaches, which focus primarily on personality development, cognitive approaches emphasize the intellectual (cognitive) development of children. The Swiss psychologist Jean Piaget (1896–1980) was a pioneer in the field of cognitive development. Cognitive theorists are interested in how people obtain, process, and use information—that is, in how we think. Cognitive development relates to changes over time in how we think.

According to Piaget (1954), in each stage of human development (from birth through adolescence), children's activities are governed by their perception of the world around them. His four stages of cognitive development are organized around specific tasks that, when mastered, lead to the acquisition of new mental capacities, which then serve as the basis for the next level of development. Thus, development is a continuous process of successive changes in which the child must go through each stage in the sequence before moving on to the next one. However, Piaget believed that the length of time each child remained in a specific stage would vary based on the child's individual attributes and the cultural context in which the development process occurred.

1. *Sensorimotor stage* (birth to age two). Children understand the world only through sensory contact and immediate action; they cannot engage in symbolic thought or use language. Children gradually comprehend *object permanence*—the realization that objects exist even when the items are placed out of their sight.

2. *Preoperational stage* (age two to seven). Children begin to use words as mental symbols and to form mental images. However, they have limited ability to use logic to solve problems or to realize that physical objects may change in shape or appearance but still retain their physical properties.

3. *Concrete operational stage* (age seven to eleven). Children think in terms of tangible objects and actual events. They can draw conclusions about the likely physical consequences of an action without always having to try the action out. Children begin to take the role of others and start to empathize with the viewpoints of others.

4. *Formal operational stage* (age twelve through adolescence). Adolescents have the potential to engage in highly abstract thought and understand places, things, and events they have never seen. They can think about the future and evaluate different options or courses of action.

Using this cognitive model, Piaget (1932) also investigated moral development. In one study, he told stories and asked children to judge how "good" or "bad" the characters were. One story involved a child who *accidentally* broke fifteen cups while another *deliberately* broke one cup. Piaget asked the children in his study if they thought one child's behavior was worse than the other's. From his research, Piaget concluded that younger children (lasting until about age eight or ten) believe that it is more evil to break a large number of cups (or steal large sums of money) than to break one cup (or steal small sums of money) for whatever reason. In contrast, older children (beginning at about age eleven) are more likely to consider principles, including the intentions and motives behind people's behavior.

Psychologist Jean Piaget identified four stages of cognitive development, including the preoperational stage, in which children have limited ability to realize that physical objects may change in shape or appearance. To prove his point, Piaget showed children two different-sized beakers, each containing the same amount of water, and asked the children to determine which one contained the greater quantity.

Tony Freeman/PhotoEdit

Piaget's stages of cognitive development provide us with useful insights on children's logical thinking and how children invent or construct the rules that govern their understanding of the world. His views on moral development show that children move from greater external influence, such as parental and other forms of moral authority, to being more autonomous, based on their own moral judgments about behavior. However, critics have pointed out that his theory says little about children's individual differences, including how gender or culture may influence children's beliefs and actions.

Kohlberg and the Stages of Moral Development

Lawrence Kohlberg (b. 1927) elaborated on Piaget's theories of cognitive reasoning by conducting a series of studies in which children, adolescents, and adults were presented with moral dilemmas that took the form of stories. Based on his findings, Kohlberg (1969, 1981) classified moral reasoning into three sequential levels:

1. *Preconventional level* (age seven to ten). Children's perceptions are based on punishment and

Social psychological theories of human development typically suggest that individuals go through stages of cognitive and moral development. According to Lawrence Kohlberg, during the conventional level of moral development, people are most concerned with how they are perceived by their peers.

obedience. Evil behavior is that which is likely to be punished; good conduct is based on obedience and avoidance of unwanted consequences.

2. *Conventional level* (age ten through adulthood). People are most concerned with how they are perceived by their peers and on how one conforms to rules.

3. *Postconventional level* (few adults reach this stage). People view morality in terms of individual rights; "moral conduct" is judged by principles based on human rights that transcend government and laws.

Although Kohlberg presents interesting ideas about the moral judgments of children, some critics have challenged the universality of his stages of moral development. They have also suggested that the elaborate "moral dilemmas" he used are too abstract for children. In one story, for example, a husband contemplates stealing for his critically ill wife medicine that he cannot afford. When questions are made simpler, or when children and adolescents are observed in natural (as opposed to laboratory) settings, they often demonstrate sophisticated levels of moral reasoning (Darley and Shultz, 1990; Lapsley, 1990).

Gilligan's View on Gender and Moral Development

Psychologist Carol Gilligan (b. 1936) is one of the major critics of Kohlberg's theory of moral development. According to Gilligan (1982), Kohlberg's model was developed solely on the basis of research with male respondents, and women and men often have divergent views on morality based on differences in socialization and life experiences. Gilligan believes that men become more concerned with law and order but that women analyze social relationships and the social consequences of behavior. For example, in Kohlberg's story about the man who is thinking about stealing medicine for his wife, Gilligan argues that male respondents are more likely to use *abstract standards* of right and wrong, whereas female respondents are more likely to be concerned about what *consequences* his stealing the drug might have for the man and his family. Does this constitute a "moral deficiency" on the part of either women or men? Not according to Gilligan.

To correct what she perceived to be a male bias in Kohlberg's research, Gilligan (1982) examined morality in women by interviewing twenty-nine pregnant women who were contemplating having an abortion. Based on her research, Gilligan concluded that Kohlberg's stages do not reflect the ways that many women think about moral problems. As a result, Gilligan identified three stages in female moral development. In stage 1, the woman is motivated primarily by selfish concerns ("This is what I want . . . this is what I need"). In stage 2, she increasingly recognizes her responsibility to others. In stage 3, she makes a decision based on her desire to do the greatest good for both herself and for others. Gilligan argued that men are socialized to make moral decisions based on a justice perspective ("What is the fairest thing to do?"), whereas women are socialized to make such decisions on a care and responsibility perspective ("Who will be hurt least?").

Subsequent research that directly compared women's and men's reasoning about moral dilemmas has supported some of Gilligan's assertions but not others. For example, some other researchers have not found that women are more compassionate than men (Tavris, 1993). Overall, however, Gilligan's argument that people make moral decisions according to both abstract principles of justice and principles of compassion and care is an important contribution to our knowledge about moral reasoning. Her book *In a Different Voice* (1982) also made social scientists more aware that the same situation may be viewed quite differently by men and by women.

SOCIOLOGICAL THEORIES OF HUMAN DEVELOPMENT

Although social scientists acknowledge the contributions of psychoanalytic and psychologically based explanations of human development, sociologists believe that it is important to bring a sociological perspective to bear on how people develop an awareness of self and learn about the culture in which they live. According to a sociological perspective, we cannot form a sense of self or personal identity without intense social contact with others. The self represents the sum total of perceptions and feelings that an individual has of being a distinct, unique person—a sense of who and what one is. When we speak of the "self," we typically use words such as *I, me, my, mine,* and *myself* (Cooley, 1998/1902). This sense of self (also referred to as *self-concept*) is not present at birth; it arises in the process of social experience. **Self-concept is the totality of our beliefs and feelings about ourselves.** Four components make up our self-concept: (1) the physical self ("I am tall"), (2) the active self ("I am good at soccer"), (3) the social self ("I am nice to others"), and (4) the psychological self ("I believe in world peace"). Between early and late childhood, a child's focus tends to shift from the physical and active dimensions of self toward the social and psychological aspects. Self-concept is the foundation for communication with others; it continues to develop and change throughout our lives.

Our *self-identity* is our perception about what kind of person we are. As we have seen, socially isolated children do not have typical self-identities; they have had no experience of "humanness." According to symbolic interactionists, we do not know who we are until we see ourselves as we believe that others see us. We gain information about the self largely through language, symbols, and interaction with others. Our

Throughout life, our self-concept is influenced by our interactions with others.

interpretation and evaluation of these messages are central to the social construction of our identity. However, we are not just passive reactors to situations, programmed by society to respond in fixed ways. Instead, we are active agents who develop plans out of the pieces supplied by culture and attempt to execute these plans in social encounters (McCall and Simmons, 1978).

Cooley, Mead, and Symbolic Interactionist Perspectives

Social constructionism is a term that is applied to theories that emphasize the socially created nature of social life. This perspective is linked to symbolic interactionist theory, and its roots can be traced to the Chicago school and early theorists such as Charles Horton Cooley and George Herbert Mead.

Cooley and the Looking-Glass Self
According to the sociologist Charles Horton Cooley (1864–1929), the *looking-glass self* **refers to the way in which a person's sense of self is derived from the perceptions of others.** Our looking-glass self is not

Figure 4.2 How the Looking-Glass Self Works

who we actually are or what people actually think about us; rather, it is based on our perception of *how* other people think of us (Cooley, 1998/1902). Cooley asserted that we base our perception of who we are on how we think other people see us and on whether this opinion seems good or bad to us.

As Figure 4.2 shows, the looking-glass self is a self-concept derived from a three-step process:

1. We imagine how our personality and appearance will look to other people. We may imagine that we are attractive or unattractive, heavy or slim, friendly or unfriendly, and so on.
2. We imagine how other people judge the appearance and personality that we think we present. This step involves our perception of how we think they are judging us. We may be correct or incorrect!
3. We develop a self-concept. If we think the evaluation of others is favorable, our self-concept is enhanced. If we think the evaluation is unfavorable, our self-concept is diminished. (Cooley, 1998/1902)

According to Cooley, we use our interactions with others as a mirror for our own thoughts and actions; our sense of self depends on how we interpret what others do and say. Consequently, our sense of self is not per-

manently fixed; it is always developing as we interact with others in the larger society. For Cooley, self and society are merely two sides of the same coin: "Self and society go together, as phases of a common whole. I am aware of the social groups in which I live as immediately and authentically as I am aware of myself" (Cooley, 1963/1909: 8–9). Accordingly, the self develops only through contact with others, just as social institutions and societies do not exist independently of the interaction of acting individuals (Schubert, 1998). By developing the idea of the looking-glass self, Cooley made us aware of the mutual interrelationship between the individual and society—namely, that society shapes people and people shape society.

Mead and Role-Taking George Herbert Mead (1863–1931) extended Cooley's insights by linking the idea of self-concept to ***role-taking*—the process by which a person mentally assumes the role of another person or group in order to understand the world from that person's or group's point of view.** Role-taking often occurs through play and games, as children try out different roles (such as being mommy, daddy, doctor, or teacher) and gain an appreciation of them. First, people come to take the role of the other (role-taking). By taking the roles of others, the individual hopes to ascertain the intention or direction of the acts of others. Then the person begins to construct his or her own roles (role-making) and to anticipate other individuals' responses. Finally, the person plays at her or his particular role (role-playing) (Marshall, 1998).

According to Mead (1934), in the early months of life, children do not realize that they are separate from others. However, they do begin early on to see a mirrored image of themselves in others. Shortly after birth, infants start to notice the faces of those around them, especially the significant others, whose faces start to have meaning because they are associated with experiences like feeding and cuddling. ***Significant others* are those persons whose care, affection, and approval are especially desired and who are most important in the development of the self.** Gradually, we distinguish ourselves from our caregivers and begin to perceive ourselves in contrast to them. As we develop language skills and learn to understand symbols, we begin to develop a self-concept. When we can represent ourselves in our minds as objects distinct from everything else, our self has been formed.

Mead (1934) divided the self into the "I" and the "me." The "I" is the subjective element of the self and represents the spontaneous and unique traits of each person. The "me" is the objective element of the self, which is composed of the internalized attitudes and demands of other members of society and the individual's awareness of those demands. Both the "I" and the "me" are needed to form the social self. The unity of the two constitutes the full development of the individual. According to Mead, the "I" develops first, and the "me" takes form during the three stages of self development:

1. During the *preparatory stage,* up to about age three, interactions lack meaning, and children largely imitate the people around them. At this stage, children are preparing for role-taking.
2. In the *play stage,* from about age three to five, children learn to use language and other symbols, thus enabling them to pretend to take the roles of specific people. At this stage, they begin to see themselves in relation to others, but they do not see role-taking as something they have to do.
3. During the *game stage,* which begins in the early school years, children understand not only their own social position but also the positions of others around them. In contrast to play, games are structured by rules, are often competitive, and involve a number of other "players." At this time, children become concerned about the demands

and expectations of others and of the larger society. Mead used the example of a baseball game to describe this stage because children, like baseball players, must take into account the roles of all the other players at the same time. Mead's concept of the ***generalized other* refers to the child's awareness of the demands and expectations of the society as a whole or of the child's subculture.**

Is socialization a one-way process? No, according to Mead. Socialization is a two-way process between society and the individual. Just as the society in which we live helps determine what kind of individuals we will become, we have the ability to shape certain aspects of our social environment and perhaps even the larger society.

How useful are symbolic interactionist perspectives such as Cooley's and Mead's in enhancing our understanding of the socialization process? Certainly, this approach contributes to our understanding of how the self develops. Cooley's idea of the looking-glass self makes us aware that our perception of how we think others see us is not always correct. Mead extended Cooley's ideas by emphasizing the cognitive skills acquired through role-taking. His concept of the generalized other helps us see that the self is a social creation. According to Mead (1934: 196), "Selves can only exist in definite relations to other selves. No hard-and-fast line can be drawn between our own selves and the selves of others." However, the viewpoints of symbolic interactionists such as Cooley and Mead have certain limitations. Sociologist Anne Kaspar (1986) suggests that Mead's ideas about the social self may be more applicable to men than to women because women are more likely to experience inherent conflicts between the meanings they derive from their personal experiences and those they take from the culture, particularly in regard to balancing the responsibilities of family life and paid employment.

Recent Symbolic Interactionist Perspectives

The symbolic interactionist approach emphasizes that socialization is a collective process in which children are active and creative agents, not just passive recipients of the socialization process. From this view, childhood is a *socially constructed* category (Adler and Adler, 1998). Children are capable of actively constructing their own shared meanings as they acquire language skills and accumulate interactive experiences (Qvortrup, 1990). According to the sociologist William A. Corsaro's (1985, 1997) "orb web model," children's cultural knowledge

According to sociologist George Herbert Mead, the self develops through three stages. In the preparatory stage, children imitate others; in the play stage, children pretend to take the roles of specific people; and in the game stage, children become aware of the "rules of the game" and the expectations of others.

reflects not only the beliefs of the adult world but also the unique interpretations and aspects of the children's own peer culture. Corsaro (1992: 162) states that *peer culture* is "a stable set of activities or routines, artifacts, values, and concerns that children produce and share." This peer culture emerges through interactions as children "borrow" from the adult culture but transform it so that it fits their own situation. Based on ethnographic studies of U.S. and Italian preschoolers, Corsaro found that very young children engage in predictable patterns of interaction. For example, when playing together, children often permit some children to gain access to their group and play area while preventing others from becoming a part of their group. Children also play "approach–avoidance" games in which they alternate between approaching a threatening person or group and then running away. In fact,

Corsaro (1992) believes that the peer group is the most significant public realm for children. (Peer groups as agents of socialization are discussed later in the chapter.) This approach contributes to our knowledge about human development because it focuses on group life rather than individuals. Researchers using this approach "look at social relations, the organization and meanings of social situations, and the collective practices through which children create and recreate key constructs in their daily interactions" (Adler and Adler, 1998: 10; see also Thorne, 1993; Eder, 1995).

Ecological Perspectives

Another approach that emphasizes cultural or environmental influences on human development is the ecological perspective. One of the best-known ecological approaches is developmental psychologist Urie Bronfenbrenner's (1989) *ecological systems theory.* The ecological systems in this theory consist of the interactions a child has with other people, as well as how those interactions are influenced by still other people and situations. The four ecological systems are as follows, starting with the one closest to the child: the *microsystem,* the *mesosystem,* the *exosystem,* and the *macrosystem.* In the microsystem, a child is engaged in immediate face-to-face interaction with the child's parents, siblings, and other immediate family members. By contrast, in the mesosystem, the child's interactions with family members are influenced by the interactions of those family members. For example, how the mother reacts to her son is influenced by how she is getting along with the father. The exosystem relates to how the immediate family members are influenced by another setting, such as the mother's job. Finally, the macrosystem involves how interaction with the child is affected by all the components of the larger society, including public policy, such as child-care legislation.

Bronfenbrenner's ecological perspective provides interesting insights on the overall context in which child development occurs. However, research using this approach is somewhat difficult because of the complex nature of the systems approach that he suggests. As a result, many sociological studies have focused on specific agents of socialization rather than the larger societal context in which child development occurs.

AGENTS OF SOCIALIZATION

***Agents of socialization* are the persons, groups, or institutions that teach us what we need to know in order to participate in society.** We are exposed to many agents of socialization throughout our lifetime; in turn, we have an influence on those socializing agents and organizations. Here, we look at the most pervasive ones in childhood—the family, the school, peer groups, and the mass media.

The Family

The family is the most important agent of socialization in all societies. From infancy, our families transmit cultural and social values to us. As discussed later in this book, families vary in size and structure. Some families consist of two parents and their biological children, whereas others consist of a single parent and one or more children. Still other families reflect changing patterns of divorce and remarriage, and an increasing number are made up of same-sex partners and their children. Over time, patterns have changed in some two-parent families so that fathers, rather than mothers, are the primary daytime agents of socialization for their young children (see Box 4.2).

Theorists using a functionalist perspective emphasize that families serve important functions in society because they are the primary locus for the procreation and socialization of children. Most of us form an emerging sense of self and acquire most of our beliefs and values within the family context. We also learn about the larger dominant culture (including language, attitudes, beliefs, values, and norms) and the primary subcultures to which our parents and other relatives belong.

Families are also the primary source of emotional support. Ideally, people receive love, understanding, security, acceptance, intimacy, and companionship within families (Benokraitis, 2002). The role of the family is especially significant because young children have little social experience beyond the family's boundaries; they have no basis for comparing or evaluating how they are treated by their own family.

To a large extent, the family is where we acquire our specific social position in society. From birth, we are a part of the specific racial, ethnic, class, religious, and regional subcultural grouping of our family. Studies show that families socialize their children somewhat differently based on race, ethnicity, and class (Kohn, 1977; Kohn et al., 1990; Harrison et al., 1990). For example, sociologist Melvin Kohn (1977; Kohn et al., 1990) has suggested that social class (as measured by parental occupation) is one of the strongest influences on what and how parents teach their children. On the one hand, working-class parents, who are closely supervised and expected to follow orders at work, typically emphasize to their children the importance of obedience and conformity. On the other hand, parents from the middle and professional classes, who have more freedom and flexibility at work, tend to give their children more freedom to make their own decisions and to be creative. Kohn concluded that differences in parents' occupations were a better predictor of child-rearing practices than was social class itself.

Whether or not Kohn's findings are valid today, the issues he examined make us aware that not everyone has the same family experiences. Many factors—including our cultural background, nation of origin, religion, and gender—are important in determining how we are socialized by family members and others who are a part of our daily life.

Box 4.2 | SOCIOLOGY IN GLOBAL PERSPECTIVE

Stay-at-Home Dads: Socialization for Kids and Resocialization for Parents

"You want to do what?!" came the reply from my boss, when told of my plans to leave work and stay home with my boys. I had been employed with the company for some six years as a Buyer and was assured a successful career, in the fullness of time. All this was about to change! As my wife Lucy was due to return to work following her Maternity Leave, we set about investigating our options for childcare and quickly discovered that in real terms we didn't have any. Childcare placements for two babies were scarce, expensive and not very flexible in relation to the unpredictability of two working parents. The decision was obvious. . . . it was time to join the new breed of Homedads!

—Karl Betschwar (2002), a father in the United Kingdom, describing his boss's reaction when he quit his job to become the family's primary daytime caregiver

■ Many fathers around the world are assuming full-time responsibility for their children. In some families, this is a matter of choice, but in others it is a matter of necessity. How do changes in family life affect our perceptions of the roles of men and women?

Recent media reports about the growing number of men who are assuming full-time child-care responsibilities rather than being employed outside the home (Horsburgh, 2003), along with the release of the Eddie Murphy film *Daddy Day Care,* have made us aware that more fathers in two-parent families have now assumed the role of primary caregivers for their young children. In 2002, about 105,000 stay-at-home dads in the United States were caring for 189,000 children under age 15 (Horsburgh, 2003). In the United Kingdom, the numbers were similar: About 100,000 dads were bringing up their children, and it was estimated that this number had more than doubled in recent years (Cavender, 2001). What contributes to this trend? Recent accounts suggest that a man is more likely to become a stay-at-home dad when his wife has a high-paying job that she really

likes, when the husband's job pays less and is viewed as less psychologically rewarding, or when the husband is laid off from his job (Horsburgh, 2003; Tyre and McGinn, 2003).

What do children learn from having a father as their primary caregiver? First, children may learn that much housework is gender neutral, meaning that neither women nor men are necessarily "better" at performing certain tasks at home. As Brian Fogg, a stay-at-home dad, told a journalist, "I think I'm a better housekeeper than any mother out there." According to Fogg, he does a great job of dusting, ironing, and cooking fettuccine Alfredo, but he doesn't do a good job of styling hair for his four daughters (qtd. in Horsburgh, 2003: 79). Second, children may learn that men's and women's paid employment roles

Conflict theorists stress that socialization contributes to false consciousness—a lack of awareness and a distorted perception of the reality of class as it affects all aspects of social life. As a result, socialization reaffirms and reproduces the class structure in the next generation rather than challenging the con-

ditions that presently exist. For example, children in low-income families may be unintentionally socialized to believe that acquiring an education and aspiring to lofty ambitions are pointless because of existing economic conditions in the family (Ballantine, 2001). By contrast, middle- and upper-income families typi-

can be flexible: In some families, the mother leaves home for work while the father remains at home; in others, the father leaves for work, and the mother remains at home; and in still other families, both parents go off to work. (Of course, in single-parent families, all the work within the home and in the paid workplace typically falls most heavily on the parent who lives at home with the children.) Finally, if children who are cared for by their fathers receive adequate nurturing and have a variety of positive learning experiences, there is no indication that the children experience any detrimental effects—and may even experience many positive effects—from having their father present in their everyday lives.

In regard to socialization, perhaps the greatest adjustments are not for the children but rather for the mothers and fathers. Some fathers have to resocialize themselves for the role of stay-at-home dad, and some mothers have to resocialize themselves for assuming the role of the family's only breadwinner. This is because many men have been taught from an early age that they are supposed to be their family's primary breadwinner and because many women view themselves as possibly one of the wage earners in the family but not necessarily the only breadwinner (Tyre and McGinn, 2003). Further confounding the problem of adjusting are the lag in public perception and the lag in public accommodations: Some people do not believe that a woman should be the primary wage earner in a family (Tyre and McGinn, 2003), and many public accommodations are not geared to the role of men as child-care providers. Stay-at-home dads are particularly frustrated when they need to change children's diapers in shopping and leisure centers. As one stay-at-home father in the United Kingdom commented, "Many shops in leisure centres still don't cater for men with young children. It's amazing how many baby changing facilities are found only in the ladies' toilets, so dads end up having to change nappies on the floor in the gents (never a pleasant situation) or in the back of the car." But, as he concluded, "Being a man out with young children does have its advantages, however. Older men will open doors for me, older women will stand aside and let me go first, and I always get my bags packed at Waitrose [a grocery store]" (qtd. in Cavender, 2001).

Because they view their situation as unique, some stay-at-home dads around the world have joined online support groups. Web sites such as the United Kingdom's "HomeDads" (**http://www.HomeDad.org.uk**), the United States's Slowlane (**http://www.Slowlane.com**), and "House Fathers" in Japan provide information on child rearing and food preparation, as well as offering online discussions where the men can communicate about child-care issues and personal concerns.

In the future, will there be more flexible roles for women and men in the socialization of infants and young children? According to some analysts, today's college students are typically more open than their parents are in regard to roles in marriage and the family. Economic conditions, including the recent downturn in employment opportunities and problematic family finances, may contribute to a belief that greater flexibility will be necessary in work and family roles in the future (Kantrowitz, 2003). It is also possible that as today's infants and children see their parents assuming more flexible roles, including that of "Home Dad," they will not view these roles as unique.

Sources: Based on Betschwar, 2002; Cavender, 2001; Horsburgh, 2003; Kantrowitz, 2003; and Tyre and McGinn, 2003.

WRITING IN SOCIOLOGY ASSIGNMENT

How does an increase in the number of stay-at-home dads relate to other, larger trends in the family and the economy?

cally instill ideas of monetary and social success in children while encouraging them to think and behave in "socially acceptable" ways.

The social constructionist/symbolic interactionist perspective helps us recognize that children affect their parents' lives and change the overall household environment. When we examine the context in which family life takes place, we also see that grandparents and other relatives have a strong influence on how parents socialize their children. In turn, the children's behavior may have an effect on how parents, siblings, and grandparents get along with one another. For example, in

As this birthday celebration attended by three generations of family members illustrates, socialization enables society to "reproduce" itself.

Michael Newman/PhotoEdit

families where there is already intense personal conflict, the birth of an infant may intensify the stress and discord, sometimes resulting in child maltreatment, spousal battering, or elder abuse. By contrast, in families where partners feel happiness and personal satisfaction, the birth of an infant may contribute to the success of the marriage and bringing about positive interpersonal communications among relatives.

The School

As the amount of specialized technical and scientific knowledge has expanded rapidly and as the amount of time that children are in educational settings has increased, schools continue to play an enormous role in the socialization of young people. For many people, the formal education process is an undertaking that lasts up to twenty years.

As the number of one-parent families and families in which both parents work outside the home has increased dramatically, the number of children in day-care and preschool programs has also grown rapidly. Currently, about 60 percent of all U.S. preschool children are in day care, either in private homes or institutional settings, and this percentage continues to climb (Children's Defense Fund, 2002). Generally, studies have found that quality day-care and preschool programs have a positive effect on the overall socialization of children. These programs provide children with the opportunity to have frequent interactions with teachers and to learn how to build their language and literacy skills. High-quality programs also have a positive effect on the academic performance of chil-

dren, particularly those from low-income families. For example, several states with pre-kindergarten programs reported an increase in children's math and reading scores, school attendance records, and parents' involvement in their children's education (Children's Defense Fund, 2002).

Although schools teach specific knowledge and skills, they also have a profound effect on children's self-image, beliefs, and values. As children enter school for the first time, they are evaluated and systematically compared with one another by the teacher. A permanent, official record is kept of each child's personal behavior and academic activities. From a functionalist perspective, schools are responsible for (1) socialization, or teaching students to be productive members of society; (2) transmission of culture; (3) social control and personal development; and (4) the selection, training, and placement of individuals on different rungs in the society (Ballantine, 2001).

In contrast, conflict theorists assert that students have different experiences in the school system depending on their social class, their racial–ethnic background, the neighborhood in which they live, their gender, and other factors. According to the sociologists Samuel Bowles and Herbert Gintis (1976), much of what happens in school amounts to teaching a hidden curriculum in which children learn to be neat, to be on time, to be quiet, to wait their turn, and to remain attentive to their work. Thus, schools do not socialize children for their own well-being but rather for their later roles in the work force, where it is important to be punctual and to show deference to supervisors. Students who are destined for leadership or

Day-care centers have become important agents of socialization for increasing numbers of children. Today, about 60 percent of all U.S. preschool children are in day care of one kind or another.

elite positions acquire different skills and knowledge than those who will enter working-class and middle-class occupations (see Cookson and Persell, 1985).

Symbolic interactionists examining socialization in the school environment might focus on how daily interactions and practices in schools affect the construction of students' beliefs regarding such things as patriotism, feelings of aggression or cooperation, and gender practices as they influence girls and boys. For example, recent studies have shown that the school environment often fosters a high degree of gender segregation, including having boys and girls line up separately to participate in different types of extracurricular activities in middle schools and high schools (Eder, 1995; Thorne, 1993).

Peer Groups

As soon as we are old enough to have acquaintances outside the home, most of us begin to rely heavily on peer groups as a source of information and approval about social behavior. A *peer group* **is a group of people who are linked by common interests, equal social position, and (usually) similar age.** In early childhood, peer groups are often composed of classmates in day care, preschool, and elementary school. Recent studies have found that preadolescence—the latter part of the elementary school years—is an age period in which children's peer culture has an important effect on how children perceive themselves and how they internalize society's expectations (Adler and Adler, 1998). In adolescence, peer groups are typically made up of people with similar interests and social activities. As adults, we continue to participate in peer groups of people with whom we share common

interests and comparable occupations, income, and/or social position.

Peer groups function as agents of socialization by contributing to our sense of "belonging" and our feelings of self-worth. As early as the preschool years, peer groups provide children with an opportunity for successful adaptation to situations such as gaining access to ongoing play, protecting shared activities from intruders, and building solidarity and mutual trust during ongoing activities (Corsaro, 1985; Rizzo and Corsaro, 1995). Unlike families and schools, peer groups provide children and adolescents with some degree of freedom from parents and other authority figures (Corsaro, 1992). Although peer groups afford children some degree of freedom, they also teach cultural norms such as what constitutes "acceptable" behavior in a specific situation. Peer groups simultaneously reflect the larger culture and serve as a conduit for passing on culture to young people. As a result, the peer group is both a product of culture and one of its major transmitters (Elkin and Handel, 1989).

Is there such a thing as "peer pressure"? Individuals must earn their acceptance with their peers by conforming to a given group's norms, attitudes, speech patterns, and dress codes. When we conform to our peer group's expectations, we are rewarded; if we do not conform, we may be ridiculed or even expelled from the group. Conforming to the demands of peers frequently places children and adolescents at cross-purposes with their parents. Sociologist William A. Corsaro (1992) notes that children experience strong peer pressure even during their preschool years. For example, children are frequently under pressure to obtain certain valued material possessions (such as toys, videotapes, clothing, or athletic shoes); they then pass this pressure on to their parents through emotional pleas to purchase the desired items. In this way, adult caregivers learn about the latest fads and fashions from children, and they may contribute to the peer culture by purchasing the items desired by the children (Corsaro, 1992). Sociologists Patricia A. Adler and Peter Adler (1998: 215) summarized their research regarding the power of peer culture on preadolescent youth as follows:

> From their subculture, children learned the culturally acceptable behavioral guidelines and the consequences of violating them. They saw the positive and negative sanctions, social status and ridicule, applied to individuals based on their power, position, attitudes, and behavior. They learned and experienced the identity outcomes of their placement in relation to the peer culture.

The pleasure of participating in activities with friends is one of the many attractions of adolescent peer groups. What groups have contributed the most to your sense of belonging and self-worth?

Their peer culture divided and unified them. It divided them by stratifying and setting them against each other. But it also forged them together in relation to older and younger groups in society. It took what was culturally appropriate (by preadolescent standards) out of the adult culture and translated it into preadolescent ways of thinking and communicating. It deconstructed and reconstructed elements of adult culture into preadolescent cultural forms. It supported age-related standards that stood in defiance of adult standards, that gave preadolescents their own beliefs and code.

When we consider that preadolescence, as defined by Adler and Adler (1998), refers to children below the age of eleven, we can see how influential early peer groups are in the socialization process. Throughout our lives, we are influenced by our close friends and casual associates.

Mass Media

An agent of socialization that has a profound impact on both children and adults is the *mass media,* composed of large-scale organizations that use print or electronic means (such as radio, television, film, and the Internet) to communicate with large numbers of people. The media function as socializing agents in several ways: (1) they inform us about events; (2) they introduce us to a wide variety of people; (3) they provide an array of viewpoints on current issues; (4) they make us aware of products and services that, if we purchase them, will supposedly help us to be accepted by others; and (5) they entertain us by providing the opportunity to live vicariously (through other peo-

ple's experiences). Although most of us take for granted that the media play an important part in contemporary socialization, we frequently underestimate the enormous influence this agent of socialization may have on children's attitudes and behavior.

Recent studies have shown that U.S. children, on average, are spending more time each year in front of TV sets, computers, and video games. According to the Annenberg Public Policy Center (University of Pennsylvania) study on media in the home, "The introduction of new media continues to transform the environment in American homes with children. . . . Rather than displacing television as the dominant medium, new technologies have supplemented it, resulting in an aggregate increase in electronic media penetration and use by America's youth" (qtd. in Dart, 1999: A5). It is estimated that U.S. children spend 2.5 hours per day watching television programs and about 2 hours with computers, video games, or a VCR, which adds up to about 1,642 hours per year (Dart, 1999). By contrast, U.S. children spend about 1,000 hours per year in school. Considering television-watching time alone, by the time that students graduate from high school, they will have spent more time in front of the television set than in the classroom (American Academy of Child and Adolescent Psychiatry, 1997; Dart, 1999). Perhaps it is no surprise that the Annenberg researchers found that 93 percent of children between the ages of 10 and 17 know that Homer, Bart, and Maggie are characters on the animated Fox series *The Simpsons* whereas only 63 percent could name the current vice president of the United States (Dart, 1999).

Parents, educators, social scientists, and public officials have widely debated the consequences of young people watching that much television. Television has been praised for offering numerous positive experiences to children. Some scholars suggest that television (when used wisely) can enhance children's development by improving their language abilities, concept-formation skills, and reading skills and by encouraging prosocial development (Winn, 1985). However, other studies have shown that children and adolescents who spend a lot of time watching television often have lower grades in school, read fewer books, exercise less, and are overweight (American Academy of Child and Adolescent Psychiatry, 1997).

Of special concern to many people is the issue of television violence. It is estimated that the typical young person who watches 28 hours of television a week will have seen 16,000 simulated murders and 200,000 acts of violence by the time he or she reaches age 18. A report by the American Psychological Association states that about 80 percent of all television programs contain acts of violence and that commercial television for children is 50 to 60 times more violent than prime-time television for adults. For example, some cartoons average more than 80 violent acts per hour (APA Online, 2000). The violent content of media programming and the marketing and advertising practices of mass media industries that routinely target children under age 17 have come under the scrutiny of government agencies such as the Federal Trade Commission due to concerns raised by parents and social analysts (see Box 4.3 for further discussion).

In addition to concerns about violence in television programming, motion pictures, and electronic games, television shows have been criticized for projecting negative images of women and people of color. Although the mass media have changed some of the roles that they depict women as playing (such as showing Xena, Warrior Princess, who is able to vanquish everything that stands in her way), even these newer images tend to reinforce existing stereotypes of women as sex symbols because of the clothing they wear in their action adventures. Throughout this text, we will look at additional examples of how the media—ranging from advertising and television programs to video games and the Internet—socialize all of us, particularly when we are young, in ways that we may or may not realize. For example, cultural studies scholars and some postmodern theorists believe that "media culture" has in recent years dramatically changed the socialization process for very young children.

GENDER AND RACIAL-ETHNIC SOCIALIZATION

Gender socialization is the aspect of socialization that contains specific messages and practices concerning the nature of being female or male in a specific group or society. Gender socialization is important in determining what we *think* the "preferred" sex of a child should be and in influencing our beliefs about acceptable behaviors for males and females. In some families, gender socialization starts before birth. Parents who learn the sex of the fetus through ultrasound or amniocentesis often purchase color-coded and gender-typed clothes, toys, and nursery decorations in anticipation of their daughter's or son's arrival. After the child has been born, parents may respond differently toward male and female infants;

they often play more roughly with boys and talk more lovingly to girls. Throughout childhood and adolescence, boys and girls are typically assigned different household chores and given different privileges (such as how late they may stay out at night).

When we look at the relationship between gender socialization and social class, the picture becomes more complex. Although some studies have found less-rigid gender stereotyping in higher-income families (Seegmiller, Suter, and Duviant, 1980; Brooks-Gunn, 1986), others have found more (Bardwell, Cochran, and Walker, 1986). One study found that higher-income families are more likely than low-income families to give "male-oriented" toys (which develop visual/spatial and problem-solving skills) to children of both sexes (Serbin et al., 1990). Working-class families tend to adhere to more-rigid gender expectations than do middle-class families (Canter and Ageton, 1984; Brooks-Gunn, 1986).

We are limited in our knowledge about gender socialization practices among racial–ethnic groups because most studies have focused on white, middle-class families. In a study of African American families, the sociologist Janice Hale-Benson (1986) found that children typically are not taught to think of gender strictly in "male–female" terms. Both daughters and sons are socialized toward autonomy, independence, self-confidence, and nurturance of children (Bardwell, Cochran, and Walker, 1986). Sociologist Patricia Hill Collins (1990) has suggested that "othermothers" (women other than a child's biological mother) play an important part in the gender socialization and motivation of African American children, especially girls. Othermothers often serve as gender role models and encourage women to become activists on behalf of their children and community (Collins, 1990). By contrast, studies of Korean American and Latino/a families have found more traditional gender socialization (Min, 1988), although some evidence indicates that this pattern may be changing (Jaramillo and Zapata, 1987).

Like the family, schools, peer groups, and the media contribute to our gender socialization. From kindergarten through college, teachers and peers reward gender-appropriate attitudes and behavior. Sports reinforce traditional gender roles through a rigid division of events into male and female categories. The media are also a powerful source of gender socialization; starting very early in childhood, children's books, television programs, movies, and music provide subtle and not-so-subtle messages about "masculine" and "feminine" behavior. Gender socialization is discussed in more depth in Chapter 11 ("Sex and Gender").

Box 4.3 SOCIOLOGY AND SOCIAL POLICY

The Debate Over Regulation of Entertainment Products with Violent Content

Do the motion picture, music recording, and electronic game industries promote products that they themselves acknowledge warrant parental caution in situations where children make up a substantial percentage of the audience? If so, are the advertisements for those products intended to attract children and teenagers?

In a report issued in September 2000, the Federal Trade Commission ("FTC") answered "yes" to both those questions. The report found that, despite criticism, these media industries still routinely target children under age seventeen in marketing products that their own ratings systems deem inappropriate for such audiences or that warrant parental caution due to violent content: Advertisements for violent video games, movies, and music were aired during television shows for younger audiences and printed in teen magazines. The FTC report also noted that children under age seventeen are frequently able to buy tickets to R-rated movies without being accompanied by an adult or to purchase music recordings and electronic games that contain parental advisory labels or are supposed to be sold only to an older audience.

Many parents are quite concerned about the amount of violence on television, in films, and in other forms of entertainment. They are especially concerned about how the media target young audiences when the media are advertising these products. Parents and social analysts assert that ratings systems—such as those for movies and television programs—should be strengthened and enforced in order to protect young audiences from gratuitous depictions of violence.

Although advocates of change argue that "the government" should do something about the situation, the FTC report did not suggest this. The report recommends only that the motion picture, music recording, and electronic games industry—voluntarily—try to do something about this problem. To many critics, this is like "having a fox guard the chicken house." Why do media marketers target young audiences? The reason is clear: Young people are the ones who spend money on these media products, and those who create media products hope to give these audiences what they think the young consumers want.

Why did the FTC report stop short of recommending that "the government" do something to reduce or eliminate inappropriate content targeted at younger audiences? The First Amendment to the United States

Television and other forms of media may have a positive influence on young children, introducing them to a world beyond their home and family. However, the media may also have a negative influence on children, especially when they are exposed to many hours of violent content.

Constitution prohibits the government from making any law or regulation that would ban or indirectly tend to suppress—to "chill"—free speech or expression. This provision in the nation's Bill of Rights has been interpreted by the courts over the history of this nation in a manner intended to protect individuals from government attempts to suppress political, ideological, or scientific ideas or information. However, there obviously have to be limits on anyone's right to free speech: A person should not be allowed to yell "fire" in a crowded theater, for example. Generally speaking, the courts allow the government to place limits on free speech only to the extent that the government proves that the restriction promotes a compelling government interest and is narrowly tailored to promote that interest. If a less-restrictive alternative will serve the government's purpose, the government must use that alternative.

Is the violent content of media programming and the marketing and advertising practices of mass media industries that routinely target children under age seventeen a compelling government interest? What are the best alternatives to government regulation of the media? What do you think?

Source: Based on Federal Trade Commission, 2000.

In addition to gender-role socialization, we receive racial socialization throughout our lives. ***Racial socialization* is the aspect of socialization that contains specific messages and practices concerning the nature of one's racial or ethnic status** as it relates to (1) personal and group identity, (2) intergroup and interindividual relationships, and (3) position in the social hierarchy. Racial socialization includes direct statements regarding race, modeling behavior (wherein a child imitates the behavior of a parent or other caregiver), and indirect activities such as exposure to specific objects, contexts, and environments that represent one's racial–ethnic group (Thornton et al., 1990).

The most important aspects of our racial identity and attitudes toward other racial–ethnic groups are passed down in our families from generation to generation. As discussed in Chapter 3, some of the core values of U.S. society may support racist beliefs. As sociologist Martin Marger (1994: 97) notes, "Fear of, dislike for, and antipathy toward one group or another is learned in much the same way that people learn to eat with a knife or fork rather than with their bare hands or to respect others' privacy in personal matters." These beliefs can be transmitted in subtle and largely unconscious ways; they do not have to be taught directly or intentionally. Scholars have found that ethnic values and attitudes begin to crystallize among children as young as age four (Goodman, 1964; Porter, 1971). By this age, the society's ethnic hierarchy has become apparent to the child (Marger, 2003). Some minority parents feel that racial socialization is essential because it provides children with the skills and abilities they will need to survive in the larger society (Hale-Benson, 1986). For example, Chuck Hayashi, a Japanese American, describes how his parents helped him:

> [Racism] never tore me up, but what helped me was just little things my parents used to tell me. Like, yes, you are different, but the people that really count will overlook things like that. My dad would say things like that. But he would also say, look, you are a minority, and you will have to compete for jobs and things with the majority. And he would tell me. I know it was to make me work harder because he would say we had to do twice as good to get the jobs. He told me that several times. (qtd. in Tuan, 1998: 69)

Scholars may be hesitant to point out differences in socialization practices among diverse racial–ethnic and social class groupings because such differences have typically been interpreted by others to be a sign of inadequate (or inferior) socialization practices.

© AP/Wide World Photos

Do you believe that what this child is learning here will have an influence on her actions in the future? What other childhood experiences might offset early negative racial socialization?

SOCIALIZATION THROUGH THE LIFE COURSE

Why is socialization a lifelong process? Throughout our lives, we continue to learn. Each time we experience a change in status (such as becoming a college student or getting married), we learn a new set of rules, roles, and relationships. Even before we achieve a new status, we often participate in ***anticipatory socialization*—the process by which knowledge and skills are learned for future roles.** Many societies organize social activities according to age and gather data regarding the age composition of the people who live in that society. For example, the U.S. Census Bureau gathers and maintains those data in the United States (see "Census Profile: Age of the U.S. Population"). Some societies have distinct *rites of passage,* based on age or other factors, that publicly dramatize and validate changes in a person's status. In the United States

CENSUS ★ PROFILES

Age of the U.S. Population

❹ **What is this person's age and what is this person's date of birth?**

Age on April 1, 2000

| | |

Print numbers in boxes.

Month   Day   Year of Birth

| | | | | | | | | |

Just as age is a crucial variable in the socialization process, the U.S. Census Bureau gathers data about people's age so that the government and other interested parties will know how many individuals residing in this country are in different age categories. This chapter examines how a person's age is related to socialization and one's life experiences. Shown below is a depiction of the nation's population in the year 2000, separated into three broad age categories.

Below age 25   Age 25 to 54   Age 55 and above

Can age be a source of social cohesion among people? Why might age differences produce conflict among individuals in different age groups? What do you think?

Source: U.S. Census Bureau, 2002.

and other industrialized societies, the most common categories of age are infancy, childhood, adolescence, and adulthood (often subdivided into young adulthood, middle adulthood, and older adulthood).

Infancy and Childhood

Some social scientists believe that a child's sense of self is formed at a very early age and that it is difficult to change this self-perception later in life. Symbolic interactionists emphasize that during infancy and early childhood, family support and guidance are crucial to a child's developing self-concept. In some families, children are provided with emotional warmth, feelings of mutual trust, and a sense of security. These families come closer to our ideal cultural belief that childhood should be a time of carefree play, safety, and freedom from economic, political, and sexual responsibilities. However, other families reflect the discrepancy between cultural ideals and reality—children grow up in a setting characterized by fear, danger, and risks that are created by parental neglect, emotional maltreatment, or premature economic and sexual demands (Knudsen, 1992).

Abused children often experience low self-esteem, an inability to trust others, feelings of isolationism and powerlessness, and denial of their feelings. However, the manner in which parental abuse affects children's ongoing development is subject to much debate and uncertainty. For example, some scholars and therapists assert that the intergenerational hypothesis—the idea that abused children will become abusive parents—is valid, but others have found little support for this hypothesis (Knudsen, 1992).

According to the developmental psychologist Urie Bronfenbrenner (1990), mutual interaction with a caring adult—and preferably a number of nurturing adults—is essential for the child's emotional, physical, intellectual, and social growth. However, Bronfenbrenner also states that at the macrosystem level, it is necessary for communities and the major economic, social, and political institutions of the entire society to provide the public policies and practices that support positive child-rearing activities on the part of families.

Adolescence

In industrialized societies, the adolescent (or teenage) years represent a buffer between childhood and adulthood. In the United States, no specific rites of passage exist to mark children's move into adulthood; therefore, young people have to pursue their own routes to self-identity and adulthood (Gilmore, 1990). Anticipatory socialization is often associated with adolescence, during which many young people spend much of their time planning or being educated for future roles they hope to occupy. However, other adolescents (such as eleven- and twelve-year-old mothers) may have to plunge into adult responsibilities at this time. Adolescence is often characterized by emotional and social unrest. In the process of developing their own identities, some young people come into conflict with parents, teachers, and other authority figures who attempt to restrict their freedom. Adolescents may also find themselves caught between the

An important rite of passage for many Latinas is the quince-añera—a celebration of their fifteenth birthday and their passage into womanhood. Can you see how this occasion might also be a form of anticipatory socialization?

Richard Lord/The Image Works

demands of adulthood and their own lack of financial independence and experience in the job market. The experiences of individuals during adolescence vary according to race, class, and gender. Based on their family's economic situation, some young people move directly into the adult world of work. However, those from upper-middle-class and upper-class families may extend adolescence into their late twenties or early thirties by attending graduate or professional school and then receiving additional advice and financial support from their parents as they start their own families, careers, or businesses.

Adulthood

One of the major differences between child socialization and adult socialization is the degree of freedom of choice. If young adults are able to support themselves financially, they gain the ability to make more choices about their own lives. In early adulthood (usually until about age forty), people work toward their own goals of creating meaningful relationships with others, finding employment, and seeking personal fulfillment. Of course, young adults continue to be socialized by their parents, teachers, peers, and the media, but they also learn new attitudes and behaviors. When we marry or have children, for example, we learn new roles as partners or parents. Adults often learn about fads and fashions in clothing, music, and language from their children. Parents in one study indicated that they had learned new attitudes and behaviors about drug use,

sexuality, sports, leisure, and racial–ethnic issues from their college-age children (Peters, 1985).

Workplace (occupational) socialization is one of the most important types of adult socialization. Sociologist Wilbert Moore (1968) divided occupational socialization into four phases: (1) career choice, (2) anticipatory socialization (learning different aspects of the occupation before entering it), (3) conditioning and commitment (learning the "ups" and "downs" of the occupation and remaining committed to it), and (4) continuous commitment (remaining committed to the work even when problems or other alternatives may arise). This type of socialization tends to be most intense immediately after a person makes the transition from school to the workplace; however, this process continues throughout our years of employment. In the future, many people will experience continuous workplace socialization as a result of individuals having more than one career in their lifetime (Lefrançois, 1999).

Between the ages of forty and sixty-five, people enter middle adulthood, and many begin to compare their accomplishments with their earlier expectations. This is the point at which people either decide that they have reached their goals or recognize that they have attained as much as they are likely to achieve.

In older adulthood (age sixty-five and over), some people are quite happy and content; others are not. Erik Erikson noted that difficult changes in adult attitudes and behavior occur in the last years of life, when people experience decreased physical ability, lower prestige, and the prospect of death. Older adults in industrialized societies may experience ***social devaluation—wherein a person or group is considered to have less social value than other persons or groups.*** Social devaluation is especially acute when people are leaving roles that have defined their sense of social identity and provided them with meaningful activity.

It is important to note that not everyone goes through passages or stages of a life course at the same age. Sociologist Alice Rossi (1980) suggests that human experience is much more diverse than life-course models suggest. She also points out that young people growing up today live in a different world, with a different set of opportunities and problems, than did the young people of previous generations (Epstein, 1988). Rossi further suggests that women's and men's experiences are not identical throughout the life course and that the life course of women today is remarkably different from that of their mothers and grandmothers because of changing societal roles and expectations. Life-course patterns are strongly influenced by race/ethnicity and social class as well.

■ RESOCIALIZATION

***Resocialization* is the process of learning a new and different set of attitudes, values, and behaviors from those in one's background and previous experience.** Resocialization may be voluntary or involuntary. In either case, people undergo changes that are much more rapid and pervasive than the gradual adaptations that socialization usually involves.

Voluntary Resocialization

Resocialization is voluntary when we assume a new status (such as becoming a student, an employee, or a retiree) of our own free will. Sometimes, voluntary resocialization involves medical or psychological treatment or religious conversion, in which case the person's existing attitudes, beliefs, and behaviors must undergo strenuous modification to a new regime and a new way of life. For example, resocialization for adult survivors of emotional/physical child abuse includes extensive therapy in order to form new patterns of thinking and action, somewhat like Alcoholics Anonymous and its twelve-step program, which has become the basis for many other programs dealing with addictive behavior (Parrish, 1990).

Involuntary Resocialization

Involuntary resocialization occurs against a person's wishes and generally takes place within a ***total institution*—a place where people are isolated from the rest of society for a set period of time and come under the control of the officials who run the institution** (Goffman, 1961a). Military boot camps, jails and prisons, concentration camps, and some mental hospitals are total institutions. In these settings, people are totally stripped of their former selves—or depersonalized—through a degradation ceremony (Goffman, 1961a). For example, inmates entering prison are required to strip, shower, and wear assigned institutional clothing. In the process, they are searched, weighed, fingerprinted, photographed, and given no privacy even in showers and restrooms. Their official identification becomes not a name but a number. In this abrupt break from their former existence, they must leave behind their personal possessions and their family and friends. The depersonalization process continues as they are required to obey rigid rules and to conform to their new environment.

After stripping people of their former identities, the institution attempts to build a more compliant person. A system of rewards and punishments (such

Larry Kolvoord/The Image Works

People in military training are resocialized through extensive, grueling military drills and maneuvers. What new values and behaviors are learned in marching drills such as this?

as providing or withholding television or exercise privileges) encourages conformity to institutional norms. Some individuals may be rehabilitated; others become angry and hostile toward the system that has taken away their freedom. Although the assumed purpose of involuntary resocialization is to reform persons so that they will conform to societal standards of conduct after their release, the ability of total institutions to modify offenders' behavior in a meaningful manner has been widely questioned. In many prisons, for example, inmates may conform to the norms of the prison or of other inmates, but little relationship exists between those norms and the laws of society.

■ SOCIALIZATION IN THE FUTURE

In the future, the family is likely to remain the institution that most fundamentally shapes and nurtures personal values and self-identity. However, parents may increasingly feel overburdened by this responsibility, especially without societal support—such as high-quality, affordable child care—and more education in parenting skills. Some analysts have suggested that there will be an increase in known cases of child maltreatment and in the number of children who experience delayed psychosocial development, learning difficulties, and emotional and behavioral problems (see Box 4.4 for suggestions on how to prevent child maltreatment). They attribute these increases to the dramatic changes occurring in the size, structure, and economic stability of families.

A central value-oriented issue facing parents and teachers as they attempt to socialize children is the growing dominance of the mass media and other forms

Box 4.4 YOU CAN MAKE A DIFFERENCE

Helping a Child Reach Adulthood

After Tina—one of your best friends—moves into a large apartment complex near her university, she keeps hearing a baby cry at all hours of the day and night. Although the crying is coming from the apartment next to Tina's, she never sees anyone come or go from it. On several occasions, she knocks on the door, but no one answers. At first Tina tries to ignore the situation, but eventually she can't sleep or study because the baby keeps crying. Tina decides she must take action and asks you, "What do you think I ought to do?" What advice could you give Tina?

Like Tina, many of us do not know if we should get involved in other people's lives. We also do not know how to report child maltreatment. However, social workers and researchers suggest that bystanders must be willing to get involved in cases of possible abuse or neglect to save a child from harm by others. They also note the importance of people knowing how to report incidents of maltreatment:

- *Report child maltreatment.* Cases of child maltreatment can be reported to any social service or law enforcement agency.
- *Identify yourself to authorities.* Although most agencies are willing to accept anonymous reports, many staff members prefer to know your name, address, telephone number, and other basic information so that they can determine that you are not a self-interested person such as a hostile relative, ex-spouse, or vindictive neighbor.
- *Follow up with authorities.* Once an agency has validated a report of child maltreatment, the agency's first goal is to stop the neglect or abuse of that child, whose health and safety are paramount concerns. However, intervention also has long-term goals. Sometimes, the situation can be improved simply by teaching the parents different values about child rearing or by pointing them to other agencies and organizations that can provide needed help. Other times, it may be necessary to remove the child from the parents' custody and place the child in a foster home, at least temporarily. Either way, the situation for the child will be better than if he or she had been left in an abusive or neglectful home environment.

So the best advice for Tina—or anyone else who has reason to believe that child maltreatment is occurring—is to report it to the appropriate authorities. In most telephone directories, the number can be located in the government listings section. Here are some other resources for help:

- Child Help USA, which offers a 24-hour crisis hot line, national information, and referral network for support groups and therapists and for reporting suspected abuse: 6463 Independence Avenue, Woodland Hills, CA 91367. (800) 422-4453.
- Child Welfare League of America, a Washington, D.C., association of nearly 800 public and private nonprofit agencies, serves as an advocacy group for children who have experienced maltreatment: 440 First Street, NW, Suite 310, Washington, DC 20001-2085. (202) 638-2952.
- The National Committee to Prevent Child Abuse (NCPCA), a nonprofit organization that seeks the prevention of child abuse and the promotion of public education about the causes and consequences of child maltreatment. For the location of the chapter nearest you, contact P.O. Box 2866, Chicago, IL 60690. (800) 556-2722.

On the Internet:

- The National Center for Missing and Exploited Children provides brochures about child safety and child protection upon request:

http://www.missingkids.org/index.html

---

of technology. For example, interactive television and computer networking systems will enable children to experience many things outside their own homes and schools and to communicate regularly with people around the world. If futurists are correct in predicting that ideas and information and access to them will be the basis for personal, business, and political advancement in the twenty-first century, persons without access to computers and other information technology will become even more disadvantaged. This prediction raises important issues about the effects of social class and race on the socialization process. Socialization—a lifelong learning process—can no longer be viewed as a "glance in the rearview mirror" or as a reaction to some previous experience. With the rapid pace of technological change, we must not only learn about the past but also learn how to anticipate the future—and consider its consequences (Westrum, 1991).

CHAPTER REVIEW

■ **What is socialization, and why is it important for human beings?**

Socialization is the lifelong process through which individuals acquire their self-identity and learn the physical, mental, and social skills needed for survival in society. The kind of person we become depends greatly on what we learn during our formative years from our surrounding social groups and social environment.

■ **How much of our unique human characteristics comes from heredity and how much from our social environment?**

As individual human beings, we have unique identities, personalities, and relationships with others. Individuals are born with some of their unique physical characteristics; other characteristics and traits are gained during the socialization process. Each of us is a product of two forces: (1) heredity, referred to as "nature," and (2) the social environment, referred to as "nurture." Whereas biology dictates our physical makeup, the social environment largely determines how we develop and behave.

■ **Why is social contact essential for human beings?**

Social contact is essential in developing a self, or self-concept, which represents an individual's perceptions and feelings of being a distinct or separate person. Much of what we think about ourselves is gained from our interactions with others and from what we perceive others think of us.

■ **How do sociologists believe that we develop a self-concept?**

According to Charles Horton Cooley's concept of the looking-glass self, we develop a self-concept as we see ourselves through the perceptions of others. Our initial sense of self is typically based on how our family perceives and treats us. George Herbert Mead suggested that we develop a self-concept through role-taking and learning the rules of social interaction. According to Mead, the self is divided into the "I" and the "me." The "I" represents the spontaneous and unique traits of each person. The "me" represents the internalized attitudes and demands of other members of society.

■ **What are the main psychological theories on human development?**

According to Sigmund Freud, the self emerges from three interrelated forces: the id, the ego, and the superego. When a person is well adjusted, the three forces act in balance. Jean Piaget identified four cognitive stages of development: (1) sensorimotor, in which children understand the world only through sensory contact and immediate action; (2) preoperational, the ability to use words as mental symbols; (3) concrete operational, in which children think in terms of tangible objects and actual events; and (4) formal operational, the ability to engage in abstract thought.

■ **What is Lawrence Kohlberg's perspective on moral development?**

Lawrence Kohlberg classified moral development into stages; certain levels of cognitive development are essential before corresponding levels of moral reasoning may occur. By contrast, Carol Gilligan suggested that there are male–female differences regarding morality and identified three stages in female moral development.

■ **What are the primary agents of socialization?**

The agents of socialization include the family, schools, peer groups, the media, and the workplace. Our families, which transmit cultural and social values to us, are the most important agents of socialization in all societies, serving these functions: (1) procreating and socializing children, (2) providing emotional support, and (3) assigning social position. Schools primarily teach knowledge and skills but also have a profound influence on the self-image, beliefs, and values of children. Peer groups contribute to our sense of belonging and self-worth, and are a key source of information about acceptable behavior. The media function as socializing agents by (1) informing us about world events, (2) introducing us to a wide variety of people, and (3) providing an opportunity to live vicariously through other people's experiences.

■ **Are socialization practices universal?**

No, social class, gender, and race are all determining factors in socialization practices. Social class is one of the strongest influences on what and how parents teach their children. Gender socialization strongly influences what we believe to be acceptable behavior for females and males. Racial socialization concerns messages and practices about the nature of racial status as it relates to personal and group identity, relationships within groups and between individuals, and position in the social hierarchy.

■ **When does socialization end?**

Socialization is ongoing throughout the life course. We learn knowledge and skills for future roles through anticipatory socialization. Parents are socialized by their own children, and adults learn through workplace socialization. Resocialization is the process of learning new attitudes, values, and behaviors, either voluntarily or involuntarily.

KEY TERMS

agents of socialization 123
anticipatory socialization 131
ego 114
gender socialization 129
generalized other 122
id 114
looking-glass self 119
peer group 127
racial socialization 131
resocialization 134
role-taking 120
self-concept 119
significant other 121
social devaluation 133
socialization 110
sociobiology 111
superego 114
total institution 134

QUESTIONS FOR CRITICAL THINKING

1. Consider the concept of the looking-glass self. How do you think others perceive you? Do you think most people perceive you correctly?
2. What are your "I" traits? What are your "me" traits? Which ones are stronger?
3. What are some different ways that you might study the effect of toys on the socialization of children? How could you isolate the toy variable from other variables that influence children's socialization?
4. Is the attempted rehabilitation of criminal offenders—through boot camp programs, for example—a form of socialization or resocialization?

RESOURCES ON THE INTERNET

Chapter-Related Web Sites

The following Web sites have been selected for their relevance to the topics in this chapter. These sites are among the more stable, but please note that Web site addresses change frequently. For an updated list of chapter-related Web sites with URL links, please visit the *Sociology in Our Times* Web site (**www.wadsworth.com/KendallSIOT**).

Social Psychology Index

http://www.trinity.edu/mkearl/socpsy.html#in

From Michael Kearl's award-winning sociology Web gateway, this page will link you to a number of resources on topics discussed in this chapter, including the nature versus nurture debate, the looking-glass self, agents of socialization, and much more.

Center for Social Informatics

http://www.slis.lib.indiana.edu/CSI

The Center for Social Informatics at Indiana University conducts research on the influence of technology and computerization on society. Its Web site will link you to journals, papers, workshops, and other resources that examine the social aspects of computerization.

ONLINE STUDY AND RESEARCH TOOLS

Accompanying this text are many *free* powerful online study tools that will help you master the material in this chapter, help increase your depth of understanding, and help you make the grade!

SocCoach CD-ROM

Use the SocCoach CD-ROM enclosed with this text to help you formulate a customized study plan for this chapter. After you take the Diagnostic Quiz, SocCoach will generate a customized study plan just for you! It will identify sections of the chapter that you should review and will provide videos, charts, graphs, and excerpts from the text to supplement your studies and enhance your understanding. You'll also find fun, interactive activities such as Virtual Explorations and Map the Stats to apply what you've learned and stretch your sociological imagination.

The Companion Web Site for Sociology in Our Times, *Fifth Edition*

www.wadsworth.com/KendallSIOT

Gain an even better grasp on this chapter by going to the companion Web site to take one of the Tutorial Quizzes, use the Flash Cards to master key terms, or check out the many other study aids you'll find there. You'll also find special features such as GSS Data and Census 2000 information that'll put data and resources at your fingertips to help you with that special project or help you as you do some research on your own.

In this chapter, when you see the icon on the left, it alerts you to a specific exercise found in *Wadsworth's Sociology Online Resources and Writing Companion*. This valuable guide shows you how to use Wadsworth's exclusive online resources—*InfoTrac College Edition*, the *Opposing Viewpoints Resource Center*, and *MicroCase Online*—to assist you in your study of sociology and to build essential research and writing skills.

CHAPTER 5

Society, Social Structure, and Interaction

I began Dumpster diving [scavenging in a large garbage bin] about a year before I became homeless. . . . The area I frequent is inhabited by many affluent college students. I am not here by chance; the Dumpsters in this area are very rich. Students throw out many good things, including food. In particular they tend to throw everything out when they move at the end of a semester, before and after breaks, and around midterm, when many of them despair of college. So I find it advantageous to keep an eye on the academic calendar.

I learned to scavenge gradually, on my own. Since then I have initiated several companions into the trade. I have learned that there is a predictable series of stages a person goes through in learning to scavenge.

At first the new scavenger is filled with disgust and self-loathing. He is ashamed of being seen and may lurk around, trying to duck behind things, or he may dive at night. (In fact, most people instinctively look away from a scavenger. By skulking around, the novice calls attention to himself and arouses suspicion. Diving at night is ineffective and needlessly messy.) . . . That stage passes with experience. The scavenger finds a pair of running shoes that fit and look and smell brand-new. . . . He begins to understand: People throw away perfectly good stuff, a lot of perfectly good stuff.

At this stage, Dumpster shyness begins to dissipate. The diver, after all, has the last laugh. He is finding all manner of good things that are his for the taking. Those who disparage his profession are the fools, not he.

–Author Lars Eighner recalls his experiences as a Dumpster diver while living under a shower curtain in a stand of bamboo in a public park. Eighner became homeless when he was evicted from his "shack" after being unemployed for about a year. (Eighner, 1993: 111–119)

■ All activities in life—including scavenging in garbage bins and living "on the streets"—are social in nature.

Eighner's "diving" activities reflect a specific pattern of social behavior. All activities in life—including scavenging in garbage bins and living "on the streets"—are social in nature. Homeless persons and domiciled persons (those with homes) live in social worlds that have predictable patterns of social interaction. *Social interaction* **is the process by which people act toward or respond to other people** and is the foundation for all relationships and groups in society. In this chapter, we look at the relationship between social structure and social interaction. In the process, homelessness is used as an example of how social problems occur and how they may be perpetuated within social structures and patterns of interaction.

Social structure **is the complex framework of societal institutions (such as the economy, politics, and religion) and the social practices (such as rules and social roles) that make up a society and that organize and establish limits on people's behavior.** This structure is essential for the survival of society and for the well-being of individuals because it provides a social web of familial support and social relationships that connects each of us to the larger society. Many homeless people have lost this vital linkage. As a result, they often experience a loss of personal dignity and a sense of moral worth because of their "homeless" condition (Snow and Anderson, 1993).

Who are the homeless? Before reading on, take the quiz on homelessness in Box 5.1. The characteristics of the homeless population in the United States vary widely. Among the homeless are single men, single women, and families. In recent years, families with children have accounted for 41 percent of the homeless population (U.S. Conference of Mayors, 2002). Further, people of color are overrepresented among the homeless. In 2002, African Americans made up 50 percent of the homeless population, whites (Caucasians) 35 percent, Latinas/os (Hispanics) 12 percent, Native Americans 2 percent, and Asian Americans

1 percent (U.S. Conference of Mayors, 2002). These percentages obviously vary across communities and different areas of the country.

Homeless persons come from all walks of life. They include undocumented workers, parolees, runaway youths and children, Vietnam veterans, and the elderly. They live in cities, suburbs, and rural areas. Contrary to popular myths, most of the homeless are not on the streets by choice or because they were deinstitutionalized by mental hospitals. Not all of the homeless are unemployed. About 22 percent of homeless people hold full- or part-time jobs but earn too little to find an affordable place to live (U.S. Conference of Mayors, 2002).

QUESTIONS AND ISSUES

Chapter Focus Question: How is homelessness related to the social structure of a society?

How do societies change over time?

What are the components of social structure?

Why do societies have shared patterns of social interaction?

How are daily interactions similar to being onstage?

Do positive changes in society occur through individual efforts or institutional efforts?

SOCIAL STRUCTURE: THE MACROLEVEL PERSPECTIVE

Social structure provides the framework within which we interact with others. This framework is an orderly, fixed arrangement of parts that together make up the whole group or society (see Figure 5.1). As defined in Chapter 1, a *society* is a large social grouping that shares the same geographical territory and is subject to the same political authority and dominant cultural expectations. At the macrolevel, the social structure of a society has several essential elements: social institutions, groups, statuses, roles, and norms.

Functional theorists emphasize that social structure is essential because it creates order and predictability in a society (Parsons, 1951). Social structure is also important for our human development. As we saw in Chapter 4, we develop a self-concept as we learn

the attitudes, values, and behaviors of the people around us. When these attitudes and values are part of a predictable structure, it is easier to develop that self-concept.

Social structure gives us the ability to interpret the social situations we encounter. For example, we expect our families to care for us, our schools to educate us, and our police to protect us. When our circumstances change dramatically, most of us feel an acute sense of anxiety because we do not know what to expect or what is expected of us. For example, newly homeless individuals may feel disoriented because they do not know how to function in their new setting. The person is likely to ask questions: "How will I survive on the streets?" "Where do I go to get help?" "Should I stay at a shelter?" "Where can I get a job?" Social structure helps people make sense out of their environment, even when they find themselves on the streets. As sociologists David Snow and Leon Anderson (1993) suggest in their study of unat-

tached, homeless men, survival strategies are the product of the interplay between the resourcefulness and ingenuity of the homeless and local political and ecological constraints.

In addition to providing a map for our encounters with others, social structure may limit our options and place us in arbitrary categories not of our own choosing. Conflict theorists maintain that there is more to social structure than is readily visible and that we must explore the deeper, underlying structures that determine social relations in a society. Karl Marx suggested that the way economic production is organized is the most important structural aspect of any society. In capitalistic societies, where a few people control the labor of many, the social structure reflects a system of relationships of domination among categories of people (for example, owner–worker and employer–employee).

Social structure creates boundaries that define which persons or groups will be the "insiders" and which will be the "outsiders." *Social marginality* is the state of being part insider and part outsider in the social structure. Sociologist Robert Park (1928) coined this term to refer to persons (such as immigrants) who simultaneously share the life and traditions of two distinct groups. Social marginality results in stigmatization. A *stigma* is any physical or social attribute or sign that so devalues a person's social identity that it disqualifies that person from full social acceptance (Goffman, 1963b). A convicted criminal, wearing a prison uniform, is an example of a person who has

been stigmatized; the uniform says that the person has done something wrong and should not be allowed unsupervised outside the prison walls.

COMPONENTS OF SOCIAL STRUCTURE

The social structure of a society includes its social positions, the relationships among those positions, and the kinds of resources attached to each of the positions. Social structure also includes all the groups that make up society and the relationships among those groups (Smelser, 1988). We begin by examining the social positions that are closest to the individual.

Status

A *status* **is a socially defined position in a group or society characterized by certain expectations, rights, and duties.** Statuses exist independently of the specific people occupying them (Linton, 1936); the statuses of professional athlete, rock musician, professor, college student, and homeless person all exist exclusive of the specific individuals who occupy these social positions. For example, although thousands of new students arrive on college campuses each year to occupy the status of first-year student, the status of

Box 5.1 SOCIOLOGY AND EVERYDAY LIFE

Answers to the Sociology Quiz on Homeless Persons

1. **False.** Less than 6 percent of all homeless people are that way by choice.

2. **False.** Most homeless persons did not inflict upon themselves the conditions that produced their homelessness. Some are the victims of child abuse or violence.

3. **False.** Many homeless people are among the working poor. Minimum-wage jobs do not pay enough for an individual to support a family or pay inner-city rent.

4. **False.** Most homeless people are not mentally ill; estimates suggest that about one-fourth of the homeless are emotionally disturbed.

5. **False.** Many homeless persons panhandle to pay for food, a bed at a shelter, or other survival needs.

6. **False.** Most homeless people are not heavy drug users. Estimates suggest that about one-third of the homeless are substance abusers. Many of these are part of the one-fourth of the homeless who are mentally ill.

7. **False.** Although an encounter with a homeless person occasionally ends in tragedy, most homeless persons are among the least threatening members of society. They are often the victims of crime, not the perpetrators.

8. **True.** Scholars have found that homelessness has always existed in the United States. However, the number of homeless persons has increased or decreased with fluctuations in the national economy.

9. **True.** Families with children are the fastest growing category of homeless persons in the United States. The number of such families nearly doubled between 1984 and 1989, and continues to do so. Many homeless children are alone. They may be runaways or "throwaways" whose parents do not want them to return home.

10. **True.** Some homeless persons have attended college and graduate school, and many have completed high school.

Sources: Based on Kroloff, 1993; Liebow, 1993; Snow and Anderson, 1993; U.S. Conference of Mayors, 2002; Vissing, 1996; and Waxman and Hinderliter, 1996.

college student and the expectations attached to that position have remained relatively unchanged for the past one hundred years.

Does the term *status* refer only to high-level positions in society? No, not in a sociological sense. Although many people equate the term *status* with high levels of prestige, sociologists use it to refer to all socially defined positions—high rank and low rank. For example, both the position of director of the Department of Health and Human Services in Washington, D.C., and that of a homeless person who is paid about five dollars a week (plus bed and board) to clean up the dining room at a homeless shelter are social statuses (see Snow and Anderson, 1993).

Take a moment to answer the question "Who am I?" To determine who you are, you must think about your social identity, which is derived from the statuses you occupy and is based on your status set. A status set comprises all the statuses that a person occupies at a given time. For example, Maria may be a psychologist, a professor, a wife, a mother, a Catholic, a school volunteer, a Texas resident, and a Mexican American. All of these socially defined positions constitute her status set.

Ascribed and Achieved Status Statuses are distinguished by the manner in which we acquire them. An ***ascribed status*** **is a social position conferred at birth or received involuntarily later in life,** based on attributes over which the individual has little or no control, such as race/ethnicity, age, and gender. For example, Maria is a female born to Mexican American parents; she was assigned these statuses at birth. She is an adult and—if she lives long enough—

Figure 5.1   Social Structure Framework

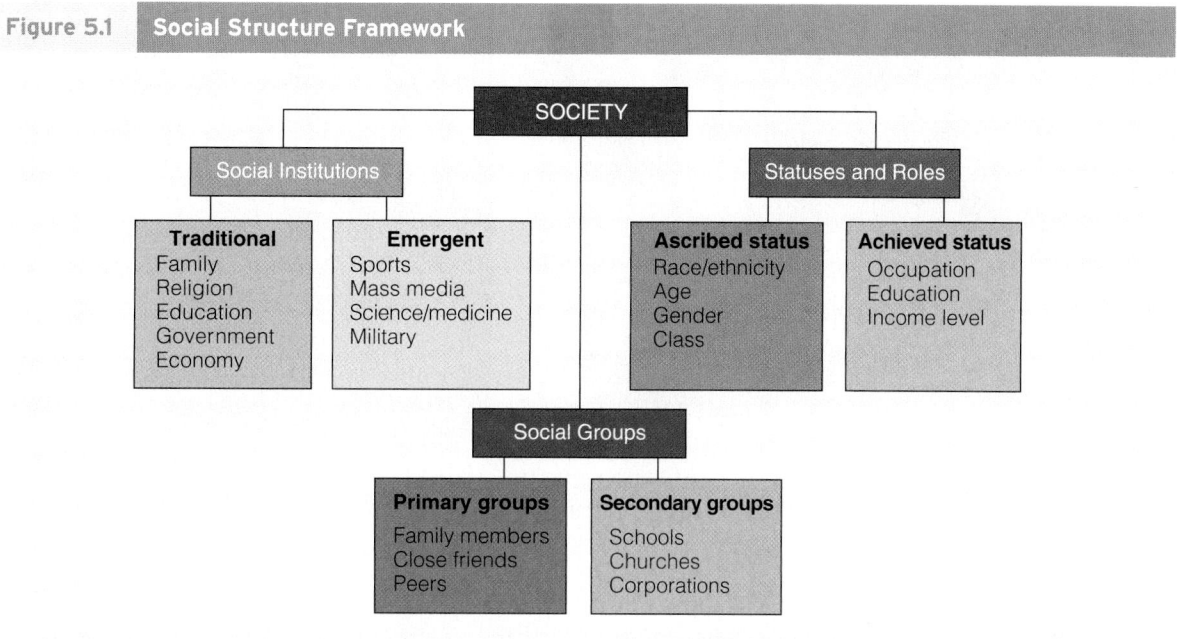

will someday become an "older adult," which is an ascribed status received involuntarily later in life. An *achieved status* **is a social position that a person assumes voluntarily as a result of personal choice, merit, or direct effort.** Achieved statuses (such as occupation, education, and income) are thought to be gained as a result of personal ability or successful competition. Most occupational positions in modern societies are achieved statuses. For instance, Maria voluntarily assumed the statuses of psychologist, professor, wife, mother, and school volunteer. However, not all achieved statuses are positions that most people would want to attain; for example, being a criminal, a drug addict, or a homeless person is a negative achieved status.

Ascribed statuses have a significant influence on the achieved statuses we occupy. Race/ethnicity, gender, and age affect each person's opportunity to acquire certain achieved statuses. Those who are privileged by their positive ascribed statuses are more likely to achieve the more prestigious positions in a society. Those who are disadvantaged by their ascribed statuses may more easily acquire negative achieved statuses.

Master Status If we occupy many different statuses, how can we determine which is the most important? Sociologist Everett Hughes has stated that societies resolve this ambiguity by determining master statuses. A *master status* **is the most important status a person occupies;** it dominates all of the indi-

vidual's other statuses and is the overriding ingredient in determining a person's general social position (Hughes, 1945). Being poor or rich is a master status that influences many other areas of life, including health, education, and life opportunities. Historically, the most common master statuses for women have related to positions in the family, such as daughter, wife, and mother. For men, occupation has usually been the most important status, although occupation is increasingly a master status for many women as well. "What do you do?" is one of the first questions many people ask when meeting another. Occupation provides important clues to a person's educational level, income, and family background. An individual's race/ethnicity may also constitute a master status in a society in which dominant-group members single out members of other groups as "inferior" on the basis of real or alleged physical, cultural, or nationality characteristics (see Feagin and Feagin, 2003).

Master statuses are vital to how we view ourselves, how we are seen by others, and how we interact with others. Justice Ruth Bader Ginsburg is both a U.S. Supreme Court justice and a mother. Which is her master status? Can you imagine how she would react if attorneys arguing a case before the Supreme Court treated her as if she were a mother rather than a justice? Lawyers wisely use "justice" as her master status and act accordingly.

Master statuses confer high or low levels of personal worth and dignity on people. Those are not characteristics that we inherently possess; they are

Sociologists believe that being rich or poor may be a master status in the United States. How do the lifestyles of these two women differ based on their master statuses?

derived from the statuses we occupy. For those who have no residence, being a homeless person readily becomes a master status regardless of the person's other attributes. Homelessness is a stigmatized master status that confers disrepute on its occupant because domiciled people often believe a homeless person has a "character flaw." The circumstances under which someone becomes homeless determine the extent to which that person is stigmatized. For example, individuals who become homeless as a result of natural disasters (such as a hurricane or a brush fire) are not seen as causing their homelessness or as being a threat to the community. Thus, they are less likely to be stigmatized. However, in cases in which homeless persons are viewed as the cause of their own problems, they are more likely to be stigmatized and marginalized by others. Snow and Anderson (1993: 199) observed the effects of homelessness as a master status:

> It was late afternoon, and the homeless were congregated in front of [the Salvation Army shelter] for dinner. A school bus approached that was packed with Anglo junior high school students being bused from an eastside barrio school to their upper-middle and upper-class homes in the city's northwest neighborhoods. As the bus rolled by, a fusillade of coins came flying out the windows, as the students made obscene gestures and shouted, "Get a job." Some of the homeless gestured back, some scrambled for the scattered coins—mostly pennies—others angrily threw the coins at the bus, and a few seemed oblivious to the encounter. For the passing junior high schoolers, the exchange was harmless fun, a way to work off the restless energy built up in school; but for the homeless it was a stark reminder of their stig-

matized status and of the extent to which they are the objects of negative attention.

Status Symbols   When people are proud of a particular social status that they occupy, they often choose to use visible means to let others know about their position. ***Status symbols* are material signs that inform others of a person's specific status.** For example, just as wearing a wedding ring proclaims that a person is married, owning a Rolls-Royce announces that one has "made it." As we saw in Chapter 3, achievement and success are core U.S. values. For this reason, people who have "made it" frequently want to display symbols to inform others of their accomplishments.

In our daily lives, status symbols both announce our statuses and aid our interactions with others. In hospitals affiliated with medical schools, the length and color of a person's uniform coat (worn over street clothing) indicates the individual's status within the medical center. Physicians wear longer white coats, medical students wear shorter white coats, laboratory technicians wear short blue coats, and so forth.

Status symbols for the domiciled and for the homeless may have different meanings. Among affluent persons, a full shopping cart in the grocery store and bags of merchandise from expensive department stores indicate a lofty financial position. By contrast, among the homeless, bulging shopping bags and overloaded grocery carts suggest a completely different status. Carts and bags are essential to street life; there is no other place to keep things, as shown by this description of Darian, a homeless woman in New York City:

> The possessions in her postal cart consist of a whole house full of things, from pots and pans to books,

shoes, magazines, toilet articles, personal papers and clothing, most of which she made herself. . . .

Because of its weight and size, Darian cannot get the cart up over the curb. She keeps it in the street near the cars. This means that as she pushes it slowly up and down the street all day long, she is living almost her entire life directly in traffic. She stops off along her route to sit or sleep for awhile and to be both stared at as a spectacle and to stare back. Every aspect of her life including sleeping, eating, and going to the bathroom is constantly in public view. . . . [S]he has no space to call her own and she never has a moment's privacy. Her privacy, her home, is her cart with all its possessions. (Rousseau, 1981: 141)

For homeless women and men, possessions are not status symbols as much as they are a link with the past, a hope for the future, and a potential source of immediate cash. As Snow and Anderson (1993: 147) note, selling personal possessions is not uncommon among most social classes; members of the working and middle classes hold garage sales, and those in the upper classes have estate sales. However, when homeless persons sell their personal possessions, they do so to meet their immediate needs, not because they want to "clean house."

Roles

Role is the dynamic aspect of a status. Whereas we occupy a status, we play a role. A *role is a set of behavioral expectations associated with a given status.* For example, a carpenter (employee) hired to remodel a kitchen is not expected to sit down uninvited and join the family (employer) for dinner.

Role expectation is a group's or society's definition of the way that a specific role ought to be played. By contrast, *role performance is how a person actually plays the role.* Role performance does not always match role expectation. Some statuses have role expectations that are highly specific, such as that of surgeon or college professor. Other statuses, such as friend or significant other, have less-structured expectations. The role expectations tied to the status of student are more specific than those of being a friend. Role expectations are typically based on a range of acceptable behavior rather than on strictly defined standards.

Our roles are relational (or complementary); that is, they are defined in the context of roles performed by others. We can play the role of student because someone else fulfills the role of professor. Conversely, to perform the role of professor, the teacher must have one or more students.

Role ambiguity occurs when the expectations associated with a role are unclear. For example, it is not always clear when the provider–dependent aspect of the parent–child relationship ends. Should it end at age eighteen or twenty-one? When a person is no longer in school? Different people will answer these questions differently depending on their experiences and socialization, as well as on the parents' financial capability and psychological willingness to continue contributing to the welfare of their adult children.

Role Conflict and Role Strain Most people occupy a number of statuses, each of which has numerous role expectations attached. For example, Charles is a student who attends morning classes at the university, and he is an employee at a fast-food restaurant, where he works from 3:00 to 10:00 P.M. He is also Stephanie's boyfriend, and she would like to see him more often. On December 7, Charles has a final exam at 7:00 P.M., when he is supposed to be working. Meanwhile, Stephanie is pressuring him to take her to a movie. To top it off, his mother calls, asking him to fly home because his father is going to have emergency surgery. How can Charles be in all these places at once? Such experiences of role conflict can be overwhelming.

Role conflict **occurs when incompatible role demands are placed on a person by two or more statuses held at the same time.** When role conflict occurs, we may feel pulled in different directions. To deal with this problem, we may prioritize our roles and first complete the one we consider to be most important. Or we may compartmentalize our lives and "insulate" our various roles (Merton, 1968). That is, we may perform the activities linked to one role for part of the day and then engage in the activities associated with another role in some other time period or elsewhere. For example, under routine circumstances, Charles would fulfill his student role for part of the day and his employee role for another part of the day. In his current situation, however, he is unable to compartmentalize his roles.

Role conflict may occur as a result of changing statuses and roles in society. Research has found that women who engage in behavior that is gender-typed as "masculine" tend to have higher rates of role conflict than those who engage in traditional "feminine" behavior (Basow, 1992). According to the sociologist Tracey Watson (1987), role conflict can sometimes be attributed not to the roles themselves but to the pressures people feel when they do not fit into culturally

© Lisette Le Bon/SuperStock

What are the competing demands of working parents in contemporary societies? What sociological term best describes this situation?

prescribed roles. In her study of women athletes in college sports programs, Watson found role conflict in the traditionally incongruent identities of being a woman and being an athlete. Even though the women athletes in her study wore makeup and presented a conventional image when they were not on the basketball court, their peers in school still saw them as "female jocks," thus leading to role conflict.

Whereas role conflict occurs between two or more statuses (such as being homeless and being a temporary employee of a social services agency), role strain takes place within one status. **Role strain occurs when incompatible demands are built into a single status that a person occupies** (Goode, 1960). For example, many women experience role strain in the labor force because they hold jobs that are "less satisfying and more stressful than men's jobs since they involve less money, less prestige, fewer job openings, more career roadblocks, and so forth" (Basow, 1992: 192). Similarly, married women may experience more role strain than married men because of work overload, marital inequality with their spouse,

exclusive parenting responsibilities, unclear expectations, and lack of emotional support.

Recent social changes may have increased role strain in men. In the family, men's traditional position of dominance has eroded as more women have entered the paid labor force and demanded more assistance in child-rearing and homemaking responsibilities. Role strain may occur among African American men who have internalized North American cultural norms regarding masculinity yet find it very difficult (if not impossible) to attain cultural norms of achievement, success, and power because of racism and economic exploitation (Basow, 1992).

Sexual orientation, age, and occupation are frequently associated with role strain. Lesbians and gay men often experience role strain because of the pressures associated with having an identity heavily stigmatized by the dominant cultural group (Basow, 1992). Women in their thirties may experience the highest levels of role strain; they face a large amount of stress in terms of role demands and conflicting work and family expectations (Basow, 1992). Dentists, psychiatrists, and police officers have been found to experience high levels of occupation-related role strain, which may result in suicide. (The concepts of role expectation, role performance, role conflict, and role strain are illustrated in Figure 5.2.)

Individuals frequently distance themselves from a role they find extremely stressful or otherwise problematic. *Role distancing* occurs when people consciously foster the impression of a lack of commitment or attachment to a particular role and merely go through the motions of role performance (Goffman, 1961b). People use distancing techniques when they do not want others to take them as the "self" implied in a particular role, especially if they think the role is "beneath them." While Charles is working in the fast-food restaurant, for example, he does not want people to think of him as a "loser in a dead-end job." He wants them to view him as a college student who is working there just to "pick up a few bucks" until he graduates. When customers from the university come in, Charles talks to them about what courses they are taking, what they are majoring in, and what professors they have. He does not discuss whether the bacon cheeseburger is better than the chili burger. When Charles is really involved in role distancing, he tells his friends that he "works there but wouldn't eat there."

Role Exit *Role exit* **occurs when people disengage from social roles that have been central to their self-identity** (Ebaugh, 1988). Sociologist Helen Rose Fuchs Ebaugh studied this process by interview-

Figure 5.2 Role Expectation, Performance, Conflict, and Strain

When playing the role of "student," do you sometimes personally encounter these concepts?

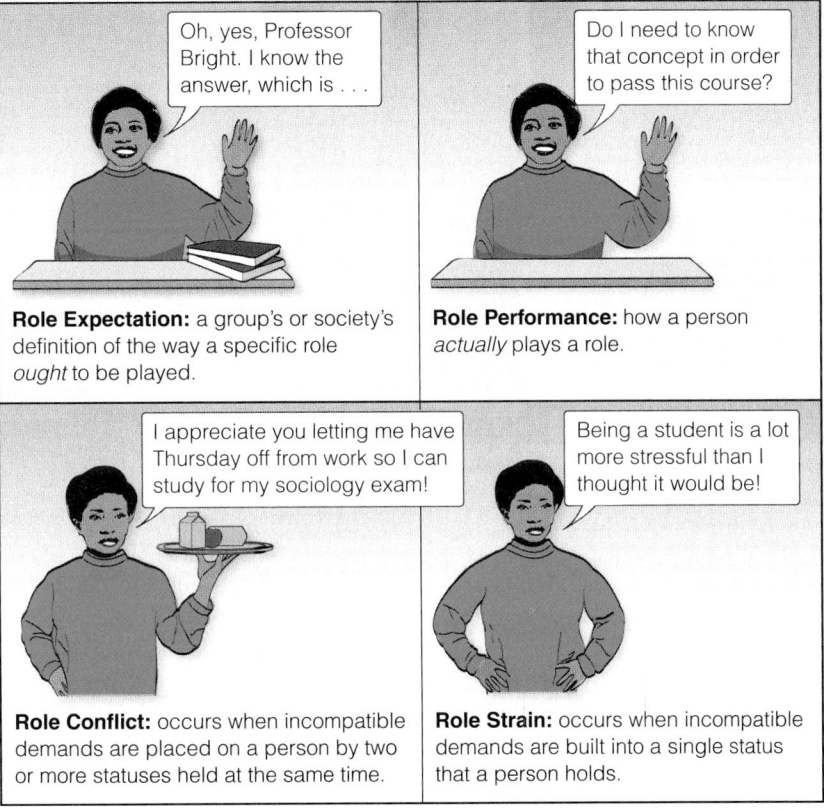

Role Expectation: a group's or society's definition of the way a specific role *ought* to be played.

Role Performance: how a person *actually* plays a role.

Role Conflict: occurs when incompatible demands are placed on a person by two or more statuses held at the same time.

Role Strain: occurs when incompatible demands are built into a single status that a person holds.

ing ex-convicts, ex-nuns, retirees, divorced men and women, and others who had exited voluntarily from significant social roles. According to Ebaugh, role exit occurs in four stages. The first stage is doubt, in which people experience frustration or burnout when they reflect on their existing roles. The second stage involves a search for alternatives; here, people may take a leave of absence from their work or temporarily separate from their marriage partner. The third stage is the turning point, at which people realize that they must take some final action, such as quitting their job or getting a divorce. The fourth and final stage involves the creation of a new identity.

Exiting the "homeless" role is often very difficult. The longer a person remains on the streets, the more difficult it becomes to exit this role. Personal resources diminish over time. Possessions are often stolen, lost, sold, or pawned. Work experience and skills become outdated, and physical disabilities that prevent individuals from working are likely to develop. However, a number of homeless people are able to exit this role. For example, Christopher, a former

crack addict who had lived in New York subway stations, was able to become a domiciled person after completing a drug rehabilitation program and receiving assistance from a community service agency:

> I felt like I had reached the end of my life. I felt awful talking to people with dirty tennis shoes and ripped pants, smelling bad. I couldn't look anybody in the eye. Once you're taking drugs, nobody respects you, especially you. (qtd. in R. Kennedy, 1993: A15)

Of course, many of the homeless do not beat the odds and exit this role. Instead, they shift their focus from role exiting to survival on the streets.

Groups

Groups are another important component of social structure. To sociologists, a ***social group*** **consists of two or more people who interact frequently and share a common identity and a feeling of interdependence.** Throughout our lives, most of us participate

For many years, capitalism has been dominated by powerful "old-boy" social networks. Professional women have increasingly created their own social networks to enhance their business opportunities.

in groups: our families and childhood friends, our college classes, our work and community organizations, and even society.

Primary and secondary groups are the two basic types of social groups. A **primary group is a small, less specialized group in which members engage in face-to-face, emotion-based interactions over an extended period of time.** Primary groups include our family, close friends, and school- or work-related peer groups. By contrast, a **secondary group is a larger, more specialized group in which members engage in more impersonal, goal-oriented relationships for a limited period of time.** Schools, churches, and corporations are examples of secondary groups. In secondary groups, people have few, if any, emotional ties to one another. Instead, they come together for some specific, practical purpose, such as getting a degree or a paycheck. Secondary groups are more specialized than primary ones; individuals relate to one another in terms of specific roles (such as professor and student) and more-limited activities (such as course-related endeavors). Primary and secondary groups are further discussed in Chapter 6 ("Groups and Organizations").

Social solidarity, or cohesion, relates to a group's ability to maintain itself in the face of obstacles. Social solidarity exists when social bonds, attractions, or other forces hold members of a group in interaction over a period of time (Jary and Jary, 1991). For example, if a local church is destroyed by fire and congregation members still worship together in a makeshift setting, then they have a high degree of social solidarity.

Many of us build social networks that involve our personal friends in primary groups and our acquaintances in secondary groups. A *social network* is a series of social relationships that links an individual to others. Social networks work differently for men and women, for different races/ethnicities, and for members of different social classes. Traditionally, people of color and white women have been excluded from powerful "old-boy" social networks. At the middle- and upper-class levels, individuals tap social networks to find employment, make business deals, and win political elections. However, social networks typically do not work effectively for poor and homeless individuals. Snow and Anderson (1993) found that homeless men have fragile social networks that are plagued with instability. Homeless men often do not even know one another's "real" names.

Sociological research on the homeless has noted the social isolation experienced by people on the streets. Sociologist Peter H. Rossi (1989) found that a high degree of social isolation exists because the homeless are separated from their extended family and former friends. Rossi noted that among the homeless who did have families, most either did not wish to return or believed that they would not be welcome. Most of the avenues for exiting the homeless role and acquiring housing are intertwined with the large-scale, secondary groups that sociologists refer to as formal organizations.

A **formal organization is a highly structured group formed for the purpose of completing certain tasks or achieving specific goals.** Many of us

As a formal organization, the Salvation Army completes certain tasks and achieves certain goals that otherwise might not be fulfilled in contemporary societies. People waiting in line to be admitted into Salvation Army facilities for food and bedding have become an all-too-familiar sight in many cities.

AP Photo/Mike Wintroath

spend most of our time in formal organizations, such as colleges, corporations, or the government. In Chapter 6 ("Groups and Organizations"), we analyze the characteristics of bureaucratic organizations; however, at this point we should note that these organizations are a very important component of social structure in all industrialized societies. We expect such organizations to educate us, solve our social problems (such as crime and homelessness), and provide work opportunities.

Today, many formal organizations are referred to as "people-processing" organizations. For example, the Salvation Army and other caregiver groups provide services for the homeless and others in need. However, these organizations must work with limited monetary resources and at the same time maintain some control over their clientele. This control is necessary in order to provide their services in an orderly and timely fashion, according to a major at the Salvation Army:

> I'll sleep and feed almost anybody, but such help requires that they be deserving. Some people would say I'm cold-hearted, but I rule with an iron hand. I have to because these guys need to respect authority. . . . The experience of working with these guys has taught us the necessity of rules in order to avoid problems. (qtd. in Snow and Anderson, 1993: 81)

Because of its rules and policies, the "Sally" (as the Salvation Army is sometimes called) tends to close its doors to those who are currently inebriated, are chronic drunks, or are viewed as "troublemakers." Likewise, a number of women's shelters have restrictions and regulations that some of the women feel deprive them of their personhood. One shelter used to require a compulsory gynecological examination of its residents (Golden, 1992). Another required that the women be out of the building by 7:00 A.M. and not return before 7:00 P.M. Fearful of violence among shelter residents or between residents and staff, many shelters use elaborate questionnaires and interviews to screen out potentially disruptive clients. Those who are supposed to benefit from the services of such shelters often find the experience demeaning and alienating. Nevertheless, organizations such as the Salvation Army and women's shelters do help people within the limited means they have available.

Social Institutions

At the macrolevel of all societies, certain basic activities routinely occur—children are born and socialized, goods and services are produced and distributed, order is preserved, and a sense of purpose is maintained (Aberle et al., 1950; Mack and Bradford, 1979). Social institutions are the means by which these basic needs are met. A *social institution* **is a set of organized beliefs and rules that establishes how a society will attempt to meet its basic social needs.** In the past, these needs have centered around five basic social institutions: the family, religion, education, the economy, and the government or politics. Today, mass media, sports, science and medicine, and the military are also considered to be social institutions.

What is the difference between a group and a social institution? A group is composed of specific, identifiable people; an institution is a standardized way of doing something. The concept of "family" helps to distinguish between the two. When we talk about "your family" or "my family," we are referring to a specific family. When we refer to the family as a social institution, we are talking about ideologies and standardized patterns of behavior that organize family life. For example, the family as a social institution contains certain statuses organized into well-defined relationships, such as husband–wife, parent–child, and brother–sister. Specific families do not always conform to these ideologies and behavior patterns.

Functional theorists emphasize that social institutions exist because they perform five essential tasks:

1. *Replacing members.* Societies and groups must have socially approved ways of replacing members who move away or die. The family provides the structure for legitimated sexual activity—and thus procreation—between adults.
2. *Teaching new members.* People who are born into a society or move into it must learn the group's values and customs. The family is essential in teaching new members, but other social institutions educate new members as well.
3. *Producing, distributing, and consuming goods and services.* All societies must provide and distribute goods and services for their members. The economy is the primary social institution fulfilling this need; the government is often involved in the regulation of economic activity.
4. *Preserving order.* Every group or society must preserve order within its boundaries and protect itself from attack by outsiders. The government legitimates the creation of law enforcement agencies to preserve internal order and some form of military for external defense.
5. *Providing and maintaining a sense of purpose.* In order to motivate people to cooperate with one another, a sense of purpose is needed.

Although this list of functional prerequisites is shared by all societies, the institutions in each society perform these tasks in somewhat different ways depending on their specific cultural values and norms.

Conflict theorists agree with functionalists that social institutions are originally organized to meet basic social needs. However, they do not believe that social institutions work for the common good of everyone in society. For example, the homeless lack the power and resources to promote their own interests when they are opposed by dominant social groups. This is a problem not only in the United States but also for homeless people, especially children and youth, throughout the world (see Box 5.2). From the conflict perspective, social institutions such as the government maintain the privileges of the wealthy and powerful while contributing to the powerlessness of others (see Domhoff, 2002). For example, U.S. government policies in urban areas have benefited some people but exacerbated the problems of others. Urban renewal and transportation projects have caused the destruction of low-cost housing and put large numbers of people "on the street" (Katz, 1989). Similarly, the shift in governmental policies toward the mentally ill and welfare recipients has resulted in more people struggling—and often failing—to find affordable housing. Meanwhile, many wealthy and privileged bankers, investors, developers, and builders have benefited at the expense of the low-income casualties of those policies.

SOCIETIES, TECHNOLOGY, AND SOCIOCULTURAL CHANGE

As we think about homeless people today, it is difficult to realize that for people in some societies being without a place of residence is a way of life. Where people live and the mode(s) of production they use to generate a food supply are related to *subsistence technology*—the methods and tools that are available for acquiring the basic needs of daily life. Social scientists have identified five types of societies based on various levels of subsistence technology: hunting and gathering, horticultural and pastoral, agrarian, industrial, and postindustrial societies. The first three of these—hunting and gathering, horticultural and pastoral, and agrarian—are also referred to as preindustrial societies. According to the social scientists Gerhard Lenski and Jean Lenski, societies change over time through the process of *sociocultural evolution,* the changes that occur as a society gains new technology (see Nolan and Lenski, 1999). However, not all anthropologists and sociologists agree on the effects of new technology.

Box 5.2 SOCIOLOGY IN GLOBAL PERSPECTIVE

Homelessness as an International Problem: The Plight of the "Street Children"

The eyes of the homeless, those who really need help, are the same eyes in any country—eyes that can see every door is closed to them.

>   —Valery Sokolov, a journalist and president of the Nochlyazhka Charitable Foundation in St. Petersburg, Russia, at the time he made this statement, knows what it means to be homeless: During six years of homelessness, he slept in a railroad station and sometimes in the forest (qtd. in Spence, 1997).

Homelessness is a widespread problem around the globe. In cities such as St. Petersburg, Vancouver, Tokyo, Paris, Sydney, and New York, some homeless individuals and families are quite visible to others as they sleep in train stations, on the sidewalks, in the doorways, and near warm air vents. But not all homeless people are so highly visible: Some reside in temporary shelters, sleep on the spare beds or sofas of friends and relatives, or are the "hidden homeless" who, in nations like Australia, camp in caravans or live in cars.

Who are the homeless in other nations? A partial answer is that homeless people share many commonalities regardless of where they reside. According to recent reports from a variety of nations, the faces of homelessness have changed in recent decades. As a study of homelessness in Australia stated,

>   The old, derelict wino on the park bench has been joined by younger men, unemployed, and hopeless; by the confused and mentally ill, frightened by the pace of activity surrounding them; by women and children, desperate to escape violent and destructive domestic situations; by young people, cast off by families who can't cope or don't care. (Harrison, 2001)

As this statement suggests, people from all walks of life may be affected by homelessness. A particularly pressing issue in many countries is the growing number of families with children and the "chronically homeless children," typically between the ages of twelve and eighteen, who reside on the streets or move between shelters, street life, and squatting (Harrison, 2001).

Let's specifically focus our attention on the plight of the so-called "street children." The World Health Organization's Street Children Project identified four categories of young people who fit the definition:

1. Children living on the streets, whose immediate concerns are survival and shelter.
2. Children who are detached from their families and living in temporary shelter, such as abandoned houses and other buildings (hostels/refuges/shelters), or moving about between friends.
3. Children who remain in contact with their families but because of poverty, overcrowding, or sexual or physical abuse within the family will spend some nights, or most days, on the streets.
4. Children who are in institutional care, who come from a situation of homelessness and are at risk of returning to a homeless existence.

If our children are the future of our nations, what can be said for those children who do not have a home to call their own? Some analysts believe that the plight of "street children" is especially dire in that they experience alienation and exclusion from society's mainstream structures and systems, including schools, churches, and other organizations that may provide stability and hope to youth. In addition, "street children" may participate in significant antisocial and self-destructive behavior, and have a profound distrust for social services and welfare systems that might, in some cases, make it possible for them to leave a life of homelessness. Whether in England, Australia, Russia, Norway, Canada, the United States, or other nations of the world, homelessness, particularly of children and youth, marks the failure of existing social institutions to adequately provide for the next generation and puts people in a devalued master status in which they lack the power and resources to promote their own interests and in which survival becomes their primary objective.

WRITING IN SOCIOLOGY ASSIGNMENT

Do nations have a responsibility to take care of homeless people within their borders? What about homeless children? Is there a difference  between homeless adults and homeless children when we talk about social responsibility?

In contemporary hunting and gathering societies, women contribute to the food supply by gathering plants and sometimes hunting for small animals. These women of the Kalahari in Botswana gather and share edible roots.

© Jason Laure/The Image Works

Hunting and Gathering Societies

At present, fewer than 250,000 people support themselves solely through hunting, fishing, and gathering wild plant foods (Haviland, 1999). However, from the origins of human existence (several million years ago) until about 10,000 years ago, hunting and gathering societies were the only type of human society that existed. ***Hunting and gathering societies* use simple technology for hunting animals and gathering vegetation.** The technology in these societies is limited to tools and weapons that are used for basic subsistence, including spears, bows and arrows, nets, traps for hunting, and digging sticks for plant collecting. All tools and weapons are made of natural materials such as stone, bone, and wood.

In hunting and gathering societies, the basic social unit is the kinship group or family. People do not have private households or residences as we think of them. Instead, they live in small groups of about twenty-five to forty people. Kinship ties constitute the basic economic unit through which food is acquired and distributed. With no stable food supply, hunters and gatherers continually search for wild animals and edible plants. As a result, they remain on the move and seldom establish a permanent settlement (Nolan and Lenski, 1999).

Hunting and gathering societies are relatively egalitarian. Since it is impossible to accumulate a surplus of food, there are few resources upon which individuals or groups can build a power base. Some specialization (division of labor) occurs, primarily based on age and sex. Young children and older people are expected to contribute what they can to securing the food supply, but healthy adults of both sexes are expected to obtain most of the food. In some societies, men hunt for animals and women gather plants; in others, both women and men gather plants and hunt for wild game, with women more actively participating when smaller animals are nearby (Lorber, 1994; Volti, 1995).

In these societies, education, religion, and politics are not formal social institutions. Instead, their functions take place on an informal basis in the kinship group, which is responsible for teaching children basic survival skills such as how to hunt and gather food. Religion is based on *animism,* the belief that spirits inhabit virtually everything in the world. There is no organized religious body; the *shaman,* or religious leader, exercises some degree of leadership but receives no material rewards for his duties and is expected to work like everyone else to obtain food (Nolan and Lenski, 1999). Contemporary hunting and gathering societies are located in relatively isolated geographical areas. However, some analysts predict that these groups will soon cease to exist, as food producers with more dominating technologies usurp the geographic areas from which these groups have derived their food supply (Nolan and Lenski, 1999).

What would life be like if we had to find our own food supply rather going to the grocery store? Consider how the Netsilik, an Inuit community in northern Canada, move from place to place by sledge (a type of sled, sometimes pulled by dogs, that is used for carrying loads), searching for food:

Traveling and moving camps [is] a very arduous task. Lack of dog feed severely [limits] the keeping of dogs to only one or two per family. The heavy sledges [must] be pulled or pushed by both men and women. Only very small children [are] allowed to sit on the sledge. . . . Seal hunting [involves] a motionless watch on the flat ice under intense cold maintained for many hours. . . . Beating the fast running caribou [a form of reindeer] over great distances in the tundra [is] an exhausting task. Stalking the caribou with the bow and arrow [requires] endless pursuits across the tundra, the hunters lying on the wet grass and trying to approach the game while hiding behind tufts of moss. . . . Hunting [is] a never-ceasing pursuit, the game [must] be brought to camp at all cost, and the hunter [must] stay out until a successful kill. (Balikci, 1968: 81)

Although not all hunting and gathering societies have this much difficulty securing food, the Netsilik adapted to harsh environmental conditions and survived by using basic subsistence technology.

Horticultural and Pastoral Societies

The period between 13,000 and 7,000 B.C.E. marks the beginning of horticultural and pastoral societies. During this period, there was a gradual shift from *collecting* food to *producing* food, a change that has been attributed to three factors: (1) the depletion of the supply of large game animals as a source of food, (2) an increase in the size of the human population to feed, and (3) dramatic weather and environmental changes that probably occurred by the end of the Ice Age (Ferraro, 1992).

Why did some societies become horticultural while others became pastoral? Whether horticultural activities or pastoral activities became a society's primary mode of food production was related to water supply, terrain, and soils. **Pastoral societies are based on technology that supports the domestication of large animals to provide food** and emerged in mountainous regions and areas with low amounts of annual rainfall. Pastoralists—people in pastoral societies—typically remain nomadic as they seek new grazing lands and water sources for their animals. **Horticultural societies are based on technology that supports the cultivation of plants to provide food.** These societies emerged in more fertile areas that were better suited for growing plants through the use of hand tools.

The family is the basic unit in horticultural and pastoral societies. Because they typically do not move as often as hunter-gatherers or pastoralists, horticulturalists establish more permanent family ties and create complex systems for tracing family lineage. Some social analysts believe that the invention of a hoe with a metal blade was a contributing factor to the less nomadic lifestyle of the horticulturalists. Unlike the digging stick, use of the metal-blade hoe made planting more efficient and productive. Horticulturists using a hoe are able to cultivate the soil more deeply, and crops can be grown in the same area for longer periods. As a result, people become more *sedentary,* remaining settled for longer periods in the same location.

Unless there are fires, floods, droughts, or environmental problems, herding animals and farming are more reliable sources of food than hunting and gathering. When food is no longer in short supply, more infants are born, and children have a greater likelihood of surviving. When people are no longer nomadic, children are viewed as an economic asset: They can cultivate crops, tend flocks, or care for younger siblings.

Division of labor increases in horticultural and pastoral societies. As the food supply grows, not everyone needs to be engaged in food production. Some people can pursue activities such as weaving cloth or carpets, crafting jewelry, serving as priests, or creating the tools needed for building the society's structure. Horticultural and pastoral societies are less egalitarian than hunter-gatherers. Even though land is initially communally controlled (often through an extended kinship group), the idea of property rights emerges as people establish more-permanent settlements. At this point, families with the largest surpluses not only have an economic advantage but also gain prestige and power, including the ability to control others. Slavery is a fairly common practice, and being a slave is a hereditary status in some pastoral societies.

In simple horticultural societies, a fairly high degree of gender equality exists because neither sex controls the food supply. Women contribute to food production because hoe cultivation is compatible with child care (Basow, 1992). In contemporary horticultural societies, women still do most of the farming while men hunt game, clear land, work with arts and crafts, make tools, participate in religious and ceremonial activities, and engage in war (Nielsen, 1990). Gender inequality is greater in pastoral societies because men herd the large animals and women contribute relatively little to subsistence production. In some herding societies, women's primary value is seen as their ability to produce male offspring so that the family lineage can be preserved and a sufficient number of males are available to protect the group against enemy attack (Nielsen, 1990).

Education, religion, and politics remain relatively informal in horticultural and pastoral societies. Boys

Early in the twenty-first century, most people around the globe reside in agrarian societies that are in various stages of industrialization. These Balinese women are selling food at an open-air market.

Michelle Garrett/Corbis

learn how to plant and harvest crops, domesticate large animals, and fight. Girls learn how to do domestic chores, care for younger children, and, sometimes, cultivate the land. In horticultural societies, religion is based on ancestor worship; in pastoral societies, religion is based on belief in a god or gods, who are believed to take an active role in human affairs. Politics is based on a simple form of government that is backed up by military force.

Although these societies have existed throughout the world, most horticultural and pastoral societies of today are located in Africa, Asia, and South America. Horticultural societies include the Gururumba of New Guinea, the Yanamamö of Brazil and Venezuela, the Truk of the Caroline Islands in the Western Pacific, and the Udu of Nigeria. Few truly pastoral societies still exist, but modified forms of pastoralism are found among cattle herders in eastern and southern Africa, camel herders in northern Africa and the Arabian peninsula, reindeer herders in the subarctic areas of eastern Europe and Siberia, and pastoralists who engage in mixed herding in various parts of Europe and Asia.

Agrarian Societies

About five to six thousand years ago, agrarian (or agricultural) societies emerged, first in Mesopotamia and Egypt and slightly later in China. *Agrarian societies use the technology of large-scale farming, including animal-drawn or energy-powered plows and equipment, to produce their food supply.* Farming made it possible for people to spend their entire lives in the same location, and food surpluses made it possible for people to live in cities where they were not directly involved in food production. Unlike the digging sticks and hoes that had previously been used in farming, the use of animals to pull plows made it possible for people to generate a large surplus of food. In agrarian societies, land is cleared of all vegetation and cultivated with the use of the plow, a process that not only controls the weeds that might kill crops but also helps maintain the fertility of the soil. The land can be used more or less continuously because the plow turns the topsoil, thus returning more nutrients to the soil. In some cases, farmers reap several harvests each year from the same plot of land.

In agrarian societies, social inequality is the highest of all preindustrial societies in terms of both class and gender. The two major classes are the landlords and the peasants. The landlords own the fields and the harvests produced by the peasants. Inheritance becomes important as families of wealthy landlords own the same land for generations. By contrast, the landless peasants enter into an agreement with the landowners to live on and cultivate a parcel of land in exchange for part of the harvest or other economic incentives. Over time, the landlords grow increasingly wealthy and powerful as they extract labor, rent, and taxation from the landless workers. Politics is based on a feudal system controlled by a political–economic elite made up of the ruler, his royal family, and members of the landowning class. Peasants have no political power and may be suppressed through the use of force or military power.

Gender-based inequality grows dramatically in agrarian societies. Men gain control over both the disposition of the food surplus and the kinship system (Lorber, 1994). Because agrarian tasks require more labor and greater physical strength than horticultural ones, men become more involved in food production.

Women may be excluded from these tasks because they are seen as too weak for the work or it is believed that their child-care responsibilities are incompatible with the full-time labor that the tasks require (Nielsen, 1990). As more people own land or businesses, the rules pertaining to marriage become stronger, and women's lives become more restricted. Men demand that women practice premarital virginity and marital fidelity so that "legitimate" heirs can be produced to inherit the land and other possessions (Nielsen, 1990). This belief is supported by religion, which is a powerful force in agrarian societies. In simple agrarian societies, the gods are seen as being concerned about the individual's moral conduct. In advanced agrarian societies, monotheism (belief in one god) replaces a belief in multiple gods. Today, gender inequality continues in agrarian societies; the division of labor between women and men is very distinct in areas such as parts of the Middle East. Here, women's work takes place in the private sphere (inside the home), and men's work occurs in the public sphere, providing men with more recognition and greater formal status.

Overall, the agricultural revolution that began five to six thousand years ago has been referred to as the "the dawn of civilization" because it brought about many sociocultural changes, including the invention of the wheel for pottery making and transportation, the development of writing and mathematics for communication and scientific progress, and the emergence of the first major cities (Nolan and Lenski, 1999). The potential size of agricultural societies today is much greater than in the past and can run as high as several million people. Most of the world's population lives in agrarian societies that are in various stages of industrialization.

Industrial Societies

Industrial societies are based on technology that mechanizes production. Originating in England during the Industrial Revolution, this mode of production dramatically transformed predominantly rural and agrarian societies into urban and industrial societies. Chapter 1 describes how the revolution first began in Britain and then spread to other countries, including the United States.

Industrialism involves the application of scientific knowledge to the technology of production, thus making it possible for machines to do the work previously done by people or animals. New technologies, such as the invention of the steam engine and fuel-powered machinery, stimulated many changes. Before the invention of the steam engine, machines were run by natural power sources (such as wind or water mills) or harnessed power (either human or animal power). The steam engine made it possible to produce goods by machines powered by fuels rather than undependable natural sources or physical labor.

As inventions and discoveries build upon one another, the rate of social and technological change increases. For example, the invention of the steam engine brought about new types of transportation, including trains and steamships. Inventions such as electric lights made it possible for people to work around the clock without regard to whether it was daylight or dark outside. Take a look around you: Most of what you see would not exist if it were not for industrialization. Cars, computers, electric lights, stereos, telephones, and virtually every other possession we own is a product of an industrial society.

Industrialism changes the nature of subsistence production. In countries such as the United States, large-scale agribusinesses have practically replaced small, family-owned farms and ranches. However, large-scale agriculture has produced many environmental problems while providing solutions to the problem of food supply.

In industrial societies, a large proportion of the population lives in or near cities. Large corporations and government bureaucracies grow in size and complexity. The nature of social life changes as people come to know one another more as statuses than as individuals. In fact, a person's occupation becomes a key defining characteristic in industrial societies, whereas his or her kinship ties are most important in preindustrial societies. Although time is freed up for leisure activities, many people still work long hours or multiple jobs.

Social institutions are transformed by industrialism. The family diminishes in significance as the economy, education, and political institutions grow in size and complexity. Although the family is still a major social institution for the care and socialization of children, it loses many of its other production functions to businesses and corporations. The family is now a consumption unit, not a production unit. In advanced industrial societies such as the United States, families take on many diverse forms, including single-parent families, single-person families, and stepfamilies (see Chapter 15, "Families and Intimate Relationships"). Although the influence of traditional religion is diminished in industrial societies, religion remains a powerful institution. Religious organizations are important in determining what moral issues will be brought to the forefront (e.g., unapproved drugs, abortion, and violence and sex in the media) and in trying to influence lawmakers to pass laws regulating

In postindustrial economies, many service- and information-based jobs are located in countries far removed from where a corporation's consumers actually live. These call center employees in India are helping customers around the world.

© Alyssa Banta/Getty Images

people's conduct. Politics in industrial societies is usually based on a democratic form of government. As nations such as South Korea, the People's Republic of China, and Mexico have become more industrialized, many people in these nations have intensified their demands for political participation.

Although the standard of living rises in industrial societies, social inequality remains a pressing problem. As societies industrialize, the status of women tends to decline further. For example, industrialization in the United States created a gap between the nonpaid work performed by women at home and the paid work that was increasingly performed by men and unmarried girls. The division of labor between men and women in the middle and upper classes also became much more distinct: Men were responsible for being "breadwinners"; women were seen as "homemakers" (Amott and Matthaei, 1996). This gendered division of labor increased the economic and political subordination of women. Likewise, although industrialization was a source of upward mobility for many whites, most people of color were left behind (Lorber, 1994).

In short, industrial societies have brought about some of the greatest innovations in all of human history, but they have also maintained and perpetuated some of the greatest problems, including violence; race-, class-, and gender-based inequalities; and environmental degradation.

Postindustrial Societies

A **postindustrial society** is one in which technology supports a service- and information-based economy. As discussed in Chapter 1, postmodern (or "postindustrial") societies are characterized by an *information explosion* and an economy in which large numbers of people either provide or apply information or are employed in service jobs (such as fast-food server or health care worker). For example, banking, law, and the travel industry are characteristic forms of employment in postindustrial societies, whereas producing steel or automobiles is representative of employment in industrial societies. There is a corresponding *rise of a consumer society* and the emergence of a *global village* in which people around the world communicate with one another by electronic technologies such as television, telephone, fax, e-mail, and the Internet.

Societies do not make a clear transition from industrialism to postindustrialism. According to the sociologist Daniel Bell (1973, 1976), the most advanced industrial nations, such as the United States, Japan, Australia, New Zealand, Germany, and Switzerland, are in the process of evolving from industrial to postindustrial societies.

Postindustrial societies produce knowledge that becomes a commodity. This knowledge can be leased or sold to others, or it can be used to generate goods, services, or more knowledge. In the previous types of societies we have examined, machinery or raw materials are crucial to how the economy operates. In postindustrial societies, the economy is based on involvement with people and communications technologies such as the mass media, computers, and the World Wide Web. For example, recent information from the U.S. Census Bureau indicates that more than half of all U.S. households have at least one computer (see "Census Profile: Computer and Internet Access in U.S. Households"). Some analysts refer to postindustrial societies as "service economies," based on the assumption that many workers provide services for others. Examples include home health care workers and airline flight attendants. However, most of the new service occupations pay relatively low wages and offer limited opportunities for advancement.

Previous forms of production, including agriculture and manufacturing, do not disappear in postindustrial societies. Instead, they become more efficient through computerization and other technological innovations. Work that relies on manual labor is often shifted to less technologically advanced societies, where workers are paid low wages to produce profits for corporations based in industrial and postindustrial societies.

Knowledge is viewed as the basic source of innovation and policy formulation in postindustrial societies. As a result, education becomes one of the most important social institutions (Bell, 1973). Formal education and other sources of information become crucial to the success of individuals and organizations. Scientific research becomes institutionalized, and new industries—such as computer manufacturing and software development—come into existence that would not have been possible without the new knowledge and technological strategies. (The features of the different types of societies, distinguished by technoeconomic base, are summarized in Table 5.1.) Throughout this text, we will examine key features of postindustrial societies as well as the postmodern theoretical perspectives that have come to be associated with the process of postindustrialism.

STABILITY AND CHANGE IN SOCIETIES

How do societies maintain some degree of social solidarity in the face of the changes we have described? As you may recall from Chapter 1, theorists using a functionalist perspective focus on the stability of societies and the importance of equilibrium even in times of rapid social change. By contrast, conflict perspectives highlight how societies go through continuous struggles for scarce resources and how innovation, rebellion, and conquest may bring about social change. Sociologists Emile Durkheim and Ferdinand Tönnies developed typologies to explain the processes of stability and change in the social structure of societies. A *typology* is a classification scheme containing two or more mutually exclusive categories that are used to compare different kinds of behavior or types of societies.

CENSUS ★ PROFILES

Computer and Internet Access in U.S. Households

The U.S. Census Bureau collects extensive data on U.S. households in addition to the questions it used for Census 2000. For example, Current Population Survey data, collected from about 50,000 U.S. households during 2000, show an increase in the percentage of homes with computers and access to the Internet, as the following figure illustrates:

Computers and Internet Access in the Home: 1984 to 2000
(civilian noninstitutional population)

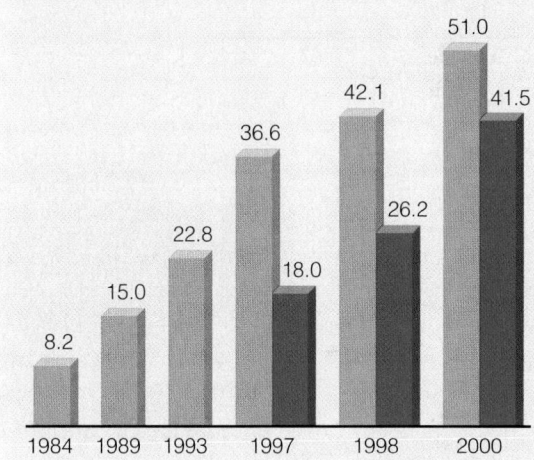

- Percentage of households with a computer
- Percentage of households with Internet access

Note: Data on Internet access were not collected before 1997.

Since 1984, the first year in which the Census Bureau collected data on computer ownership and use, there has been more than a fivefold increase in the percentage of households with computers. However, in Chapter 9 ("Social Class in the United States"), we will see that computer ownership varies widely by income and educational level.

Source: Newburger, 2001.

Durkheim: Mechanical and Organic Solidarity

Emile Durkheim (1933/1893) was concerned with the question "How do societies manage to hold together?" He asserted that preindustrial societies are held together by strong traditions and by the members' shared moral beliefs and values. As societies industrialized and

Table 5.1　TECHNOECONOMIC BASES OF SOCIETY

	HUNTING AND GATHERING	HORTICULTURAL AND PASTORAL	AGRARIAN
Change from Prior Society	–	Use of hand tools, such as digging stick and hoe	Use of animal-drawn plows and equipment
Economic Characteristics	Hunting game, gathering roots and berries	Planting crops, domestication of animals for food	Labor-intensive farming
Control of Surplus	None	Men begin to control societies	Men who own land or herds
Inheritance	None	Shared—patrilineal and matrilineal	Patrilineal
Control over Procreation	None	Increasingly by men	Men—to ensure legitimacy of heirs
Women's status	Relative equality	Decreasing in move to pastoralism	Low

Holton Collection/SuperStock

Alison Wright/The Image Works

Source: Adapted from Lorber, 1994: 140.

developed more specialized economic activities, social solidarity came to be rooted in the members' shared dependence on one another. From Durkheim's perspective, social solidarity derives from a society's social structure, which, in turn, is based on the society's division of labor. *Division of labor* refers to how the various tasks of a society are divided up and performed. People in diverse societies (or in the same society at different points in time) divide their tasks somewhat differently, based on their own history, physical environment, and level of technological development.

To explain social change, Durkheim categorized societies as having either mechanical or organic solidarity. *Mechanical solidarity* **refers to the social cohesion of preindustrial societies, in which there is minimal division of labor and people feel united by shared values and common social bonds.** Durk-

heim used the term *mechanical solidarity* because he believed that people in such preindustrial societies feel a more or less automatic sense of belonging. Social interaction is characterized by face-to-face, intimate, primary-group relationships. Everyone is engaged in similar work, and little specialization is found in the division of labor.

Organic solidarity **refers to the social cohesion found in industrial (and perhaps postindustrial) societies, in which people perform very specialized tasks and feel united by their mutual dependence.** Durkheim chose the term *organic solidarity* because he believed that individuals in industrial societies come to rely on one another in much the same way that the organs of the human body function interdependently. Social interaction is less personal, more status oriented, and more focused on specific goals

Table 5.1 (continued)		
	INDUSTRIAL	**POSTINDUSTRIAL**
Change from Prior Society	Invention of steam engine	Invention of computer and development of "high-tech" society
Economic Characteristics	Mechanized production of goods	Information and service economy
Control of Surplus	Men who own means of production	Corporate shareholders and high-tech entrepreneurs
Inheritance	Patrilineal	Patrilineal
Control over Procreation	Men—but less so in later stages	Mixed
Women's status	Low	Varies by class, race, and age

and objectives. People no longer rely on morality or shared values for social solidarity; instead, they are bound together by practical considerations. Which of Durkheim's categories most closely describes the United States today?

Tönnies: *Gemeinschaft* and *Gesellschaft*

Sociologist Ferdinand Tönnies (1855–1936) used the terms *Gemeinschaft* and *Gesellschaft* to characterize the degree of social solidarity and social control found in societies. He was especially concerned about what happens to social solidarity in a society when a "loss of community" occurs.

The **Gemeinschaft (guh-MINE-shoft) is a traditional society in which social relationships are based on personal bonds of friendship and kinship and on intergenerational stability.** These relationships are based on ascribed rather than achieved status. In such societies, people have a commitment to the entire group and feel a sense of togetherness. Tönnies (1963/1887) used the German term *Gemeinschaft* because it means "commune" or "community"; social solidarity and social control are maintained by the community. Members have a strong sense of belonging, but they also have very limited privacy.

By contrast, the **Gesellschaft (guh-ZELL-shoft) is a large, urban society in which social bonds are based on impersonal and specialized relationships, with little long-term commitment to the group or consensus on values.** In such societies, most people are "strangers" who perceive that they have very little in common with most other people. Consequently,

Figure 5.3 *Gemeinschaft* and *Gesellschaft* Societies

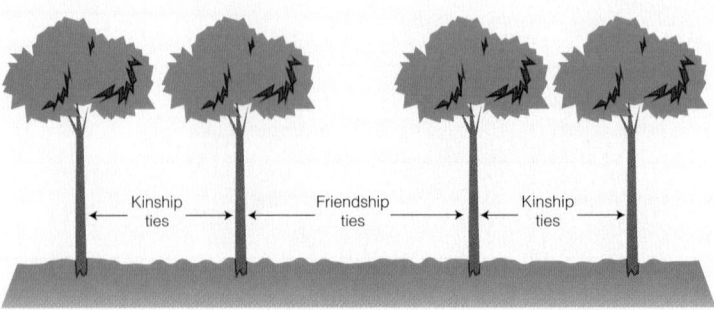

If we compare societies to trees, a *Gemeinschaft* society would be made up of the various family trees (clans or kinship groups) and how they are related to one another.

By contrast, a *Gesellschaft* society would be made up of clumps of trees (kinship/friendship ties) that are important, but each of them has a more specialized relationship and may not be committed to the others, just as an individual tree and the pot it is growing in could be moved to somewhere else.

self-interest dominates, and little consensus exists regarding values. Tönnies (1963/1887) selected the German term *Gesellschaft* because it means "association"; relationships are based on achieved statuses, and interactions among people are both rational and calculated. See Figure 5.3.

Social Structure and Homelessness

In *Gesellschaft* societies such as the United States, a prevailing core value is that people should be able to take care of themselves. Thus, many people view the homeless as "throwaways"—as beyond help or as having already had enough done for them by society. Some argue that the homeless made their own bad decisions, which led them into alcoholism or drug ad-

diction, and should be held responsible for the consequences of their actions. In this sense, homeless people serve as a visible example to others to "follow the rules" lest they experience a similar fate.

Alternative explanations for homelessness in *Gesellschaft* societies have been suggested. Elliot Liebow (1993) notes that homelessness is rooted in poverty; overwhelmingly, homeless people are poor people who come from poor families. Homelessness is a "social class phenomenon, the direct result of a steady, across-the-board lowering of the standard of living of the working class and lower class" (Liebow, 1993: 224). As the standard of living falls, those at the bottom rungs of society are plunged into homelessness. The problem is exacerbated by a lack of jobs. Of those who find work, a growing number work full time, year-round, but remain poor because of substandard wages. Half of the households living below the poverty line pay more than 70 percent of their income for rent—if they are able to find accommodations that they can afford at all (Roob and McCambridge, 1992). Clearly, there is no simple answer to the question about what should be done to help the homeless. Nor, as discussed in Box 5.3, is there any consensus on what rights the homeless have in public spaces, such as parks and sidewalks. The answers we derive as a society and as individuals are often based on our social construction of this reality of life.

SOCIAL INTERACTION: THE MICROLEVEL PERSPECTIVE

So far in this chapter, we have focused on society and social structure from a macrolevel perspective, seeing how the structure of society affects the statuses we occupy, the roles we play, and the groups and organizations to which we belong. Functionalist and conflict perspectives provide a macrosociological overview because they concentrate on large-scale events and broad social features. For example, sociologists using the macrosociological approach to study the homeless might analyze how social institutions have operated to produce current conditions. By contrast, the symbolic interactionist perspective takes a microsociological approach, asking how social institutions affect our daily lives. We will now look at society from the microlevel perspective, which focuses on social interactions among individuals, especially face-to-face encounters.

Social Interaction and Meaning

When you are with other people, do you often wonder what they think of you? If so, you are not alone! Because most of us are concerned about the meanings that others ascribe to our behavior, we try to interpret their words and actions so that we can plan how we will react toward them (Blumer, 1969). We know that others have expectations of us. We also have certain expectations about them. For example, if we enter an elevator that has only one other person in it, we do not expect that individual to confront us and stare into our eyes. As a matter of fact, we would be quite upset if the person did so.

Social interaction within a given society has certain shared meanings across situations. For instance,

our reaction would be the same regardless of *which* elevator we rode in *which* building. Sociologist Erving Goffman (1963b) described these shared meanings in his observation about two pedestrians approaching each other on a public sidewalk. He noted that each will tend to look at the other just long enough to acknowledge the other's presence. By the time they are about eight feet away from each other, both individuals will tend to look downward. Goffman referred to this behavior as *civil inattention*—the ways in which an individual shows an awareness that another is present without making this person the object of particular attention. The fact that people engage in civil inattention demonstrates that interaction does have a pattern, or *interaction order,* which regulates the form and processes (but not the content) of social interaction.

Does everyone interpret social interaction rituals in the same way? No. Race/ethnicity, gender, and social class play a part in the meanings we give to our interactions with others, including chance encounters on elevators or the street. Our perceptions about the meaning of a situation vary widely based on the statuses we occupy and our unique personal experiences. For example, sociologist Carol Brooks Gardner (1989) found that women frequently do not perceive street encounters to be "routine" rituals. They fear for their personal safety and try to avoid comments and propositions that are sexual in nature when they walk down the street. African Americans may also feel uncomfortable in street encounters. A middle-class African American college student described his experiences walking home at night from a campus job:

> So, even if you wanted to, it's difficult just to live a life where you don't come into conflict with others. . . . Every day that you live as a black person you're reminded how you're perceived in society. You walk the streets at night; white people cross the streets. I've seen white couples and individuals dart in front of cars to not be on the same side of the street. Just the other day, I was walking down the street, and this white female with a child, I saw her pass a young white male about 20 yards ahead. When she saw me, she quickly dragged the child and herself across the busy street. . . . [When I pass,] white men tighten their grip on their women. I've seen people turn around and seem like they're going to take blows from me. . . . So, every day you realize [you're black]. Even though you're not doing anything wrong; you're just existing. You're just a person. But you're a black person perceived in an unblack world. (qtd. in Feagin, 1991: 111–112)

Box 5.3 SOCIOLOGY AND SOCIAL POLICY

Homeless Rights Versus Public Space

SANTA MONICA, CALIF.–At precisely 11:55 P.M., Bernardo Discensio pops his pup tent out of a backpack, places it on the curb beneath a streetlight, and crawls inside.

"I'm in bed by midnight and up by 5 A.M. when the street sweepers become my alarm clock," says Mr. Discensio, a former aerospace worker now unemployed. Up and down the Third Street Promenade–a premier tourist area of shops, movie houses, and upscale restaurants–doorways and benches are filled with homeless people huddled under blankets. (Wood, 2002)

This scene has become an all-too-common sight in many cities. Record numbers of homeless individuals and families continue to seek refuge on the streets and in public parks because they have nowhere else to go. However, this seemingly individualistic problem is actually linked to larger social concerns in our nation, including high rates of unemployment, lack of job training and education, lack of affordable housing, and recent major cutbacks in social service agency budgets. The problem of homelessness also raises significant social policy issues, including the extent to which cities can make it illegal for people to remain for extended periods of time in public spaces.

Should homeless persons be allowed to sleep on streets, in parks, and in other public areas? This issue has been the source of controversy in a number of cities, including San Francisco, Portland, Seattle, Baltimore, and Santa Monica, Ca. As cities have sought to improve their downtown areas and public spaces, they have taken measures to enforce city ordinances controlling loitering (standing around or sleeping in public spaces) and disorderly conduct. For example, Santa Monica passed a law that makes it illegal for a person to occupy the doorway of a business between the hours of 11 P.M. and 7 A.M. if the owner has posted a sign to that effect (Wood, 2002).

Advocates for the homeless and civil liberties groups have filed lawsuits in several cities claiming that the rights of the homeless are being violated by the enforcement of these laws. The lawsuits assert that the homeless have a right to sleep in parks because no affordable housing is available for them. Advocates also argue that panhandling is a legitimate means of livelihood for some of the homeless and is protected speech under the First Amendment. In addition, they accuse public and law enforcement officials of seeking to punish the homeless on the

■ Contrary to a popular myth that most homeless people are single drifters, an increasing number of families are now homeless.

basis of their "status," a cruel and unusual punishment prohibited by the Eighth Amendment.

The "homeless problem" is not a new one for city governments. Of the limited public funding that is designated for the homeless, most has been spent on shelters that are frequently overcrowded and otherwise inadequate. Officials in some cities have given homeless people a one-way ticket to another city. Still others have routinely run them out of public spaces. The issue has become more pressing for homeless advocates because cities such as Santa Monica and San Francisco, which previously tolerated the homeless, have now grown weary of the individuals who beg on the streets and live in public spaces.

What responsibility does society have to the homeless? Are laws restricting the hours that public areas or parks are open to the public unfair to homeless persons? Should city workers remove cardboard boxes, blankets, and other "makeshift" homes created by the homeless in parks? Some critics have argued that if the homeless and their advocates win these lawsuits, what they have won (at best) is the right for the homeless to live on the street, to slowly freeze to death, and to drink themselves into oblivion with the option of continuing to forgo seeking the help they need. Others have disputed this assertion and note that if society does not make available affordable housing and job opportunities, the least it can do is stop harassing homeless people who are getting by as best they can. What do you think? What rights are involved? Whose rights should prevail?

Sources: Based on Kaufman, 1996; Teir, 1994; and Wood, 2002.

Sharply contrasting perceptions of the same reality are evident in these people's views regarding the war in Iraq.

As this passage indicates, social encounters have different meanings for men and women, whites and people of color, and individuals from different social classes. Members of the dominant classes regard the poor, unemployed, and working class as less worthy of attention, frequently subjecting them to subtle yet systematic "attention deprivation" (Derber, 1983). The same can certainly be said about how members of the dominant classes "interact" with the homeless.

The Social Construction of Reality

If we interpret other people's actions so subjectively, can we have a shared social reality? Some symbolic interaction theorists believe that there is very little shared reality beyond that which is socially created. Symbolic interactionists refer to this as the *social construction of reality*—**the process by which our perception of reality is largely shaped by the subjective meaning that we give to an experience** (Berger and Luckmann, 1967). This meaning strongly influences what we "see" and how we respond to situations.

When you watch a football game, do you "see" the same game as everyone else? The answer is no, according to researchers who asked Princeton and Dartmouth students to watch a film of a recent game between their two schools. The students were instructed to watch for infractions of the rules by each team. Although both groups saw the same film, the Princeton students saw twice as many rule infractions involving the Dartmouth team as the Dartmouth students saw. The researchers noted that one version of what transpired at the game was just as "real" to one person as another (entirely different) version was to another person (Hastorf and Cantril, 1954). When we see what we want or expect to see, we are engaged in the social construction of reality.

As discussed previously, our perceptions and behavior are influenced by how we initially define situations: We act on reality as we see it. Sociologists describe this process as the *definition of the situation,* meaning that we analyze a social context in which we find ourselves, determine what is in our best interest, and adjust our attitudes and actions accordingly. This can result in a *self-fulfilling prophecy*—**a false belief or prediction that produces behavior that makes the originally false belief come true** (Merton, 1968). An example would be a person who has been told repeatedly that she or he is not a good student; eventually, this person might come to believe it to be true, stop studying, and receive failing grades.

People may define a given situation in very different ways, a tendency demonstrated by the sociologist Jacqueline Wiseman (1970) in her study of "Pacific City's" skid row. She wanted to know how people who live or work on skid row (a run-down area found in all cities) felt about it. Wiseman found that homeless persons living on skid row evaluated it very differently from the social workers who dealt with them there.

On the one hand, many of the social workers "saw" skid row as a smelly, depressing area filled with men who were "down-and-out," alcoholic, and often physically and mentally ill. On the other hand, the men who lived on skid row did not see it in such a negative light. They experienced some degree of satisfaction with their "bottle clubs [and a] remarkably indomitable and creative spirit"—at least initially (Wiseman, 1970: 18). As this study shows, we define situations from our own frame of reference, based on the statuses that we occupy and the roles that we play.

Dominant-group members with prestigious statuses may have the ability to establish how other people define "reality" (Berger and Luckmann, 1967: 109). Some sociologists have suggested that dominant groups, particularly higher-income white males in powerful economic and political statuses, perpetuate their own world view through ideologies that are frequently seen as "social reality." For example, the sociologist Dorothy E. Smith (1999) points out that the term "Standard North American Family" (meaning a heterosexual two-parent family) is an ideological code promulgated by the dominant group to identify how people's family life *should* be arranged. According to Smith (1999), this code plays a powerful role in determining how people in organizations such as the government and schools believe that a family should be. Likewise, the sociologist Patricia Hill Collins (1998) argues that "reality" may be viewed differently by African American women and other historically oppressed groups when compared to the perspectives of dominant-group members. Collins (1998) gives the example of how groups view economic and social justice. If the views of marginalized "outsiders" were taken into account by members of the dominant group, new angles of vision on problems such as race- and class-based injustice might bring about new social realities. However, according to Collins (1998), mainstream, dominant-group members sometimes fail to realize how much they could learn about "reality" from "outsiders." As these theorists state, social reality and social structure are often hotly debated issues in contemporary societies.

Ethnomethodology

How do we know how to interact in a given situation? What rules do we follow? Ethnomethodologists are interested in the answers to these questions. ***Ethnomethodology* is the study of the commonsense knowledge that people use to understand the situations in which they find themselves** (Heritage, 1984: 4). Sociologist Harold Garfinkel (1967) initi-

ated this approach and coined the term: *ethno* for "people" or "folk" and *methodology* for "a system of methods." Garfinkel was critical of mainstream sociology for not recognizing the ongoing ways in which people create reality and produce their own world. Consequently, ethnomethodologists examine existing patterns of conventional behavior in order to uncover people's background expectancies—that is, their shared interpretation of objects and events—as well as their resulting actions. According to ethnomethodologists, interaction is based on assumptions of shared expectancies. For example, when you are talking with someone, what expectations do you have that you will take turns? Based on your background expectancies, would you be surprised if the other person talked for an hour and never gave you a chance to speak?

To uncover people's background expectancies, ethnomethodologists frequently break "rules" or act as though they do not understand some basic rule of social life so that they can observe other people's responses. In a series of *breaching experiments,* Garfinkel assigned different activities to his students to see how breaking the unspoken rules of behavior created confusion. In one experiment, when students participating in the study were asked "How are you?" by persons not in the study, they were instructed to respond with very detailed accounts of their health and personal problems, as in this example:

ACQUAINTANCE: How are you?

STUDENT: How am I in regard to what? My health, my finances, my school work, my peace of mind, my . . .

ACQUAINTANCE (red in the face and suddenly out of control): Look! I was just trying to be polite. Frankly, I don't give a damn how you are. (Garfinkel, 1967: 44)

In this encounter, the acquaintance expected the student to use conventional behavior in answering the question. By acting unconventionally, the student violated background expectancies and effectively "sabotaged" the interaction.

The ethnomethodological approach contributes to our knowledge of social interaction by making us aware of subconscious social realities in our daily lives. However, a number of sociologists regard ethnomethodology as a frivolous approach to studying human behavior because it does not examine the impact of macrolevel social institutions—such as the economy and education—on people's expectancies. Women's studies scholars suggest that ethnomethodologists fail to do what they claim to do: look at how social realities are created. Rather, they take ascribed

According to the sociologist Erving Goffman, our day-to-day interactions have much in common with a dramatic production. How do these characters from the Broadway play *Hairspray* show impression management?

© 2003 AP/Wide World Photos

statuses (such as race, class, gender, and age) as "givens," not as *socially created* realities. For example, in the experiments that Garfinkel assigned to his students, he did not account for how gender affected their experiences. When Garfinkel asked students to reduce the distance between themselves and a nonrelative to the point that "their noses were almost touching," he ignored the fact that gender was as important to the encounter as was the proximity of the two persons. Scholars have recently emphasized that our expectations about reality are strongly influenced by our assumptions relating to gender, race, and social class (see Bologh, 1992).

Dramaturgical Analysis

Erving Goffman suggested that day-to-day interactions have much in common with being on stage or in a dramatic production. ***Dramaturgical analysis* is the study of social interaction that compares everyday life to a theatrical presentation.** Members of our "audience" judge our performance and are aware that we may slip and reveal our true character (Goffman, 1959, 1963a). Consequently, most of us attempt to play our role as well as possible and to control the impressions we give to others. ***Impression management (presentation of self)* refers to people's efforts to present themselves to others in ways that are most favorable to their own interests or image.**

For example, suppose that a professor has returned graded exams to your class. Will you discuss the exam and your grade with others in the class? If you are like most people, you probably play your student role differently depending on whom you are talking to and what grade you received on the exam. Your "presentation" may vary depending on the grade earned by the other person (your "audience"). In one study, students who all received high grades ("Ace–Ace encounters") willingly talked with one another about their grades and sometimes engaged in a little bragging about how they had "aced" the test. However, encounters between students who had received high grades and those who had received low or failing grades ("Ace–Bomber encounters") were uncomfortable. The Aces felt as if they had to minimize their own grade. Consequently, they tended to attribute their success to "luck" and were quick to offer the Bombers words of encouragement. On the other hand, the Bombers believed that they had to praise the Aces and hide their own feelings of frustration and disappointment. Students who received low or failing grades ("Bomber–Bomber encounters") were more comfortable when they talked with one another because they could share their negative emotions. They often indulged in self-pity and relied on face-saving excuses (such as an illness or an unfair exam) for their poor performances (Albas and Albas, 1988).

In Goffman's terminology, *face-saving behavior* refers to the strategies we use to rescue our performance when we experience a potential or actual loss of face. When the Bombers made excuses for their low scores, they were engaged in face-saving; the Aces attempted to help them save face by asserting that the test was unfair or that it was only a small part of the final grade.

Why would the Aces and Bombers both participate in face-saving behavior? In most social interactions, all role players have an interest in keeping the "play" going so that they can maintain their overall definition of the situation in which they perform their roles.

Goffman noted that people consciously participate in *studied nonobservance,* a face-saving technique in which one role player ignores the flaws in another's performance to avoid embarrassment for everyone involved. Most of us remember times when we have failed in our role and know that it is likely to happen again; thus, we may be more forgiving of the role failures of others.

Social interaction, like a theater, has a front stage and a back stage. The *front stage* is the area where a player performs a specific role before an audience. The *back stage* is the area where a player is not required to perform a specific role because it is out of view of a given audience. For example, when the Aces and Bombers were talking with each other at school, they were on the "front stage." When they were in the privacy of their own residences, they were in "back stage" settings—they no longer had to perform the Ace and Bomber roles and could be themselves.

The need for impression management is most intense when role players have widely divergent or devalued statuses. As we have seen with the Aces and Bombers, the participants often play different roles under different circumstances and keep their various audiences separated from one another. If one audience becomes aware of other roles that a person plays, the impression being given at that time may be ruined. For example, homeless people may lose jobs or the opportunity to get them when their homelessness becomes known. One woman had worked as a receptionist in a doctor's office for several weeks but was fired when the doctor learned that she was living in a shelter (Liebow, 1993). However, the homeless do not passively accept the roles into which they are cast. For the most part, they attempt—as we all do—to engage in impression management in their everyday life.

The dramaturgical approach helps us think about the roles we play and the audiences who judge our presentation of self. Like all other approaches, it has its critics. Sociologist Alvin Gouldner (1970) criticized this approach for focusing on appearances and not the underlying substance. Others have argued that Goffman's work reduces the self to "a peg on which the clothes of the role are hung" (see Burns, 1992) or have suggested that this approach does not place enough emphasis on the ways in which our everyday interactions with other people are influenced by occurrences within the larger society. For example, if some members of Congress belittle the homeless as being lazy and unwilling to work, it may become easier for people walking down a street to do likewise. Goffman's defenders counter that he captured the essence of society because social interaction "turns out to be not only where most of the world's work gets done, but where the solid buildings of the social world are in fact constructed" (Burns, 1992: 380). Goffman's work was influential in the development of the sociology of emotions, a relatively new area of theory and research.

The Sociology of Emotions

Why do we laugh, cry, or become angry? Are these emotional expressions biological or social in nature? To some extent, emotions are a biologically given sense (like hearing, smell, and touch), but they are also social in origin. We are socialized to feel certain emotions, and we learn how and when to express (or not express) those emotions (Hochschild, 1983).

How do we know which emotions are appropriate for a given role? Sociologist Arlie Hochschild (1983) suggests that we acquire a set of *feeling rules* that shapes the appropriate emotions for a given role or specific situation. These rules include how, where, when, and with whom an emotion should be expressed. For example, for the role of a mourner at a funeral, feeling rules tell us which emotions are required (sadness and grief, for example), which are acceptable (a sense of relief that the deceased no longer has to suffer), and which are unacceptable (enjoyment of the occasion expressed by laughing out loud) (see Hochschild, 1983: 63–68).

Feeling rules also apply to our occupational roles. For example, the truck driver who handles explosive cargos must be able to suppress fear. Although all jobs place some burden on our feelings, *emotional labor* occurs only in jobs that require personal contact with the public or the production of a state of mind (such as hope, desire, or fear) in others (Hochschild, 1983). With emotional labor, employees must display only certain carefully selected emotions. For example, flight attendants are required to act friendly toward passengers, to be helpful and open to requests, and to maintain an "omnipresent smile" in order to enhance the customers' status. By contrast, bill collectors are encouraged to show anger and make threats to customers, thereby supposedly deflating the customers' status and wearing down their presumed resistance to paying past-due bills. In both jobs, the employees are expected to show feelings that are often not their true ones (Hochschild, 1983).

Emotional labor may produce feelings of estrangement from one's "true" self. C. Wright Mills (1956) suggested that when we "sell our personality" in the course of selling goods or services, we engage in a seriously self-alienating process. In other words, the "commercialization" of our feelings may dehumanize our work-role performance and create alienation and contempt that spill over into other aspects of our life (Hochschild, 1983; Smith and Kleinman, 1989).

Those who are unemployed and homeless are also required to engage in emotional labor. Governmental agencies and nonprofit organizations that function as caregivers to the homeless sometimes require emotional labor (such as feelings of gratitude or penitence) from their recipients. Homeless people have been denied social services even when they were eligible and have been asked to leave shelters when they did not show the appropriate deference and gratitude toward staff members (Liebow, 1993).

Do all people experience and express emotions the same way? It is widely believed that women express emotions more readily than men; as a result, very little research has been conducted to determine the accuracy of this belief. In fact, women and men may differ more in the way they express their emotions than in their actual feelings. Differences in emotional expression may also be attributed to socialization, for the extent to which men and women have been taught that a given emotion is appropriate (or inappropriate) to their gender no doubt plays an important part in their perceptions.

Social class is also a determinant in managed expression and emotion management. Emotional labor is emphasized in middle- and upper-class families. Since middle- and upper-class parents often work with people, they are more likely to teach their children the importance of emotional labor in their own careers than are working-class parents, who tend to work with things, not people (Hochschild, 1983). Race is also an important factor in emotional labor. People of color spend much of their life engaged in emotional labor because racist attitudes and discrimination make it continually necessary to manage one's feelings.

Clearly, Hochschild's contribution to the sociology of emotions helps us understand the social context of our feelings and the relationship between the roles we play and the emotions we experience. However, her thesis has been criticized for overemphasizing the cost of emotional labor and the emotional controls that exist outside the individual (Wouters, 1989). The context in which emotions are studied and the specific emotions examined are important factors in determining the costs and benefits of emotional labor.

Nonverbal Communication

In a typical stage drama, the players not only speak their lines but also convey information by nonverbal communication. In Chapter 3, we discussed the importance of language; now we will look at the messages we communicate without speaking. ***Nonverbal communication* is the transfer of information between persons without the use of words.** It includes not only visual cues (gestures, appearances) but also vocal features (inflection, volume, pitch) and environmental factors (use of space, position) that affect meanings (Wood, 1999). Facial expressions, head movements, body positions, and other gestures carry as much of the total meaning of our communication with others as our spoken words do (Wood, 1999).

Nonverbal communication may be intentional or unintentional. Actors, politicians, and salespersons may make deliberate use of nonverbal communication to convey an idea or "make a sale." We may also send nonverbal messages through gestures or facial expressions or even our appearance without intending to let other people know what we are thinking.

Functions of Nonverbal Communication

Nonverbal communication often supplements verbal communication (Wood, 1999). Head and facial movements may provide us with information about other people's emotional states, and others receive similar information from us (Samovar and Porter, 1991a). We obtain first impressions of others from various kinds of nonverbal communication, such as the clothing they wear and their body positions.

Our social interaction is regulated by nonverbal communication. Through our body posture and eye contact, we signal that we do or do not wish to speak to someone. For example, we may look down at the sidewalk or off into the distance when we pass homeless persons who look as if they are going to ask for money.

Nonverbal communication establishes the relationship among people in terms of their responsiveness to and power over one another (Wood, 1999). For example, we show that we are responsive toward or like another person by maintaining eye contact and attentive body posture and perhaps by touching and standing close. By contrast, we signal to others that we do not wish to be near them or that we dislike them by refusing to look them in the eye or stand near them. We can even express power or control over others through nonverbal communication. Goffman (1956) suggested that *demeanor* (how we behave or conduct ourselves) is relative to social power. People

Nonverbal communication is influenced by many factors, including a person's race, class, and gender. What messages are conveyed by the manner in which the people shown here greet each other?

in positions of dominance are allowed a wider range of permissible actions than are their subordinates, who are expected to show deference. *Deference* is the symbolic means by which subordinates give a required permissive response to those in power; it confirms the existence of inequality and reaffirms each person's relationship to the other (Rollins, 1985).

Facial Expression, Eye Contact, and Touching
Deference behavior is important in regard to facial expression, eye contact, and touching. This type of nonverbal communication is symbolic of our relationships with others. Who smiles? Who stares? Who makes and sustains eye contact? Who touches whom? All these questions relate to demeanor and deference; the key issue is the status of the person who is doing the smiling, staring, or touching relative to the status of the recipient (Goffman, 1967).

Facial expressions, especially smiles, also reflect gender-based patterns of dominance and subordination in society. Typically, white women have been so-

cialized to smile and frequently do so even when they are not actually happy (Halberstadt and Saitta, 1987). Jobs held predominantly by women (including flight attendant, secretary, elementary schoolteacher, and nurse) are more closely associated with being pleasant and smiling than are "men's jobs." In addition to smiling more frequently, many women tend to tilt their heads in deferential positions when they are talking or listening to others. By contrast, men tend to display less emotion through smiles or other facial expressions and instead seek to show that they are reserved and in control (Wood, 1999).

Women are more likely to sustain eye contact during conversations (but not otherwise) as a means of showing their interest in and involvement with others. By contrast, men are less likely to maintain prolonged eye contact during conversations but are more likely to stare at other people (especially men) in order to challenge them and assert their own status (Pearson, 1985).

Eye contact can be a sign of domination or deference. For example, in a participant observation study of domestic (household) workers and their employers, the sociologist Judith Rollins (1985) found that the domestics were supposed to show deference by averting their eyes when they talked to their employers. Deference also required that they present an "exaggeratedly subservient demeanor" by standing less erect and walking tentatively.

Touching is another form of nonverbal behavior that has many different shades of meaning. Gender and power differences are evident in tactile communication from birth. Studies have shown that touching has variable meanings to parents: Boys are touched more roughly and playfully, whereas girls are handled more gently and protectively (Condry, Condry, and Pogatshnik, 1983). This pattern continues into adulthood, with women touched more frequently than men. Sociologist Nancy Henley (1977) attributed this pattern to power differentials between men and women and to the nature of women's roles as mothers, nurses, teachers, and secretaries. Clearly, touching has a different meaning to women than to men. Women may hug and touch others to indicate affection and emotional support, but men are more likely to touch others to give directions, assert power, and express sexual interest (Wood, 1999). The "meaning" we give to touching is related to its "duration, intensity, frequency, and the body parts touching and being touched" (Wood, 1994: 162).

Personal Space Physical space is an important component of nonverbal communication. Anthropologist Edward Hall (1966) analyzed the physical distance between people speaking to each other and found that the amount of personal space that people prefer varies from one culture to another. ***Personal space is the immediate area surrounding a person that the person claims as private.*** Our personal space is contained within an invisible boundary surrounding our body, much like a snail's shell. When others invade our space, we may retreat, stand our ground, or even lash out, depending on our cultural background (Samovar and Porter, 1991a). Hall (1966) observed that people in the United States have different "distance zones":

1. Intimate distance (contact to about 18 inches): reserved for spouses, lovers, and close friends, for purposes of lovemaking, comforting, and protecting.
2. Personal distance (18 inches to 4 feet): reserved for friends and acquaintances, for purposes of ordinary conversation, card playing, and similar activities.
3. Social distance (4 to 12 feet): marks impersonal or formal relationships, such as in job interviews and business transactions.
4. Public distance (beyond 12 feet): marks an even more formal relationship and makes interpersonal communication nearly impossible. This distance often denotes a status difference between dignitaries or speakers and their audience or the general public.

Hall based his observations primarily on middle-class adults living in the northeastern United States. He emphasized that people from different cultures have different distance zones. In Latin American countries, for example, people communicate while closer to each other than do whites in the United States (Hall, 1966).

Age, gender, kind of relationship, and social class are important factors in the allocation of personal space. Power differentials between people (including adults and children, men and women, and dominant-group members and people of color) are reflected in personal space and privacy issues. With regard to age, adults generally do not hesitate to enter the personal space of a child (Thorne, Kramarae, and Henley, 1983). Similarly, young children who invade the personal space of an adult tend to elicit a more favorable response than do older uninvited visitors (Dean, Willis, and la Rocco, 1976). The need for personal space appears to increase with age (Baxter, 1970; Aiello and Jones, 1971) although it may begin to decrease at about age forty (Heshka and Nelson, 1972).

For some people, the idea of privacy or personal space is an unheard-of luxury afforded only to those in the middle and upper classes. As we have seen in this chapter, homeless bag ladies may have as their only personal space the bags they carry or the shopping carts they push down the streets. Some of the homeless may try to "stake a claim" on a heat grate or the same bed in a shelter for more than one night, but such claims have dubious authenticity in a society in which the homeless are assumed to own nothing and to have no right to lay claim to anything in the public domain.

In sum, all forms of nonverbal communication are influenced by gender, race, social class, and the personal contexts in which they occur. Although it is difficult to generalize about people's nonverbal behavior, we still need to think about our own nonverbal communication patterns. Recognizing that differences in social interaction exist is important. We should be wary of making value judgments—the differences are simply differences. Learning to understand and respect alternative styles of social interaction enhances

Box 5.4 YOU CAN MAKE A DIFFERENCE

One Person's Trash May Be Another Person's Treasure: Recycling for Good Causes

The average college student tosses out 640 pounds of trash through the year, and almost 30 percent of that is at the end of the year.

–Lisa K. Heller, founder of Dump & Run, a national nonprofit organization that helps colleges collect discarded items and organize sales for local charities and campus groups (qtd. in Crawford, 2003: A6)

At the beginning of this chapter, Lars Eighner (1993) pointed out that in his years as a Dumpster diver, he was able to find many useful things in large garbage bins because the area he "worked" was inhabited by many affluent college students. As you will recall, he stated that students are prone to throw away valuable items at certain times in the school year—particularly when they move out at the end of a semester. Although Eighner's Dumpster diving was a random process, Lisa K. Heller has organized a systematic way in which discarded items can be turned into cash to benefit organizations such as soup kitchens and Head Start programs.

Heller's Dump & Run program has been successful at schools such as Brown University and has inspired similar programs such as "Clean Sweep" at Bates Col-

lege. Students at still other schools, including Texas Christian University and Wilkes University, do not turn trash into cash but instead recycle unwanted items such as furniture and half-full bottles of detergent so that these items may be used at food banks, soup kitchens, and resale shops (Crawford, 2003).

Would you like an opportunity to participate in a similar program in your community? If none currently exists, perhaps you could start a group to take on a similar project. Even if there is no large college or university campus near you, no doubt many people throw away items that might be useful to others. In some cities, people put pieces of furniture, used household goods, and items of clothing out by their trash cans to be recycled. Some of these items may not be as "fancy" as the George Foreman grills, designer clothing, blenders, and cappuccino machines that Ms. Heller and others have found near college campuses through the Dump & Run program; however, even less flashy items can benefit many individuals and families.

To learn more about Dump & Run, visit its Web site (**http://www.dumpandrun.org**), or write to Dump & Run, Inc., P.O. Box 397, Brookfield, MA 01506.

our personal effectiveness by increasing the range of options we have for communicating with different people in diverse contexts and for varied reasons (Wood, 1999).

FUTURE CHANGES IN SOCIETY, SOCIAL STRUCTURE, AND INTERACTION

The social structure in the United States has been changing rapidly in recent decades. Currently, there are more possible statuses for persons to occupy and roles to play than at any other time in history. Although achieved statuses are considered very important, ascribed statuses still have a significant effect on the options and opportunities that people have.

Ironically, at a time when we have more technological capability, more leisure activities and types of entertainment, and more quantities of material goods available for consumption than ever before, many people experience high levels of stress, fear for their lives because of crime, and face problems such as homelessness. In a society that can send astronauts into space to perform complex scientific experiments, is it impossible to solve some of the problems that plague us here on Earth?

Individuals and groups often show initiative in trying to solve some of our pressing problems (see Box 5.4). For example, Ellen Baxter has single-handedly tried to create housing for hundreds of New York City's homeless by reinventing well-maintained, single-room-occupancy residential hotels to provide cheap lodging and social services (Anderson, 1993). However, individual initiative alone will not solve all our social problems in the future. Large-scale, formal organizations must become more responsive to society's needs.

At the microlevel, we need to regard social problems as everyone's problem; if we do not, they have a way of becoming everyone's problem anyway. When we think about "the homeless," for example, we are thinking in a somewhat misleading manner. "The homeless" suggests a uniform set of problems and a single category of poor people. Jonathan Kozol (1988: 92) emphasizes that "their miseries are somewhat uniform; the squalor is uniform; the density of living space is uniform. [However, the] uniformity is in their mode of suffering, not in themselves."

What can be done about homelessness in the future? Martha R. Burt, director of the Urban Institute's 1987 national study of urban homeless shelter and soup kitchen users, notes that we must first become dissatisfied with explanations that see personal problems as the cause of homelessness. Many people in the past have suffered from poverty, mental illness, alcoholism, physical handicaps, and drug addiction, but they have not become homeless. Only structural changes can explain why we have a larger homeless population today than in earlier times. As Burt suggests,

> To undo the effects of changing structural factors we will have to address the factors themselves, not the vulnerabilities of the people caught by changing times. We can take a short-term approach, raising benefit levels and expanding eligibility to cover those most vulnerable to homelessness. Such actions would prevent homelessness rather than ameliorate it, and are therefore preferable to building emergency shelters. They would not, however, change the underlying conditions, and the need for public support would be likely to continue indefinitely. A far better approach is to fulfill our commitments to support people, such as those with severe mental illness, who cannot be expected to support themselves, and also address simultaneously the employer and the employee requirements for increasing productivity, by reshaping the work environment and improving education and training. If we succeed at this much larger agenda, we will solve not only the problem of homelessness, but also the problem of declining living standards for a much broader spectrum of American workers. (Burt, 1992: 225–226)

In sum, the future of this country rests on our collective ability to deal with major social problems at both the macrolevel and the microlevel of society.

CHAPTER REVIEW

■ **How does social structure shape our social interactions?**

The stable patterns of social relationships within a particular society make up its social structure. Social structure is a macrolevel influence because it shapes and determines the overall patterns in which social interaction occurs. Social structure provides an ordered framework for society and for our interactions with others.

■ **What are the main components of social structure?**

Social structure comprises statuses, roles, groups, and social institutions. A status is a specific position in a group or society and is characterized by certain expectations, rights, and duties. Ascribed statuses, such as gender, class, and race/ethnicity, are acquired at birth or involuntarily later in life. Achieved statuses, such as education and occupation, are assumed voluntarily as a result of personal choice, merit, or direct effort. We occupy a status, but a role is the set of behavioral expectations associated with a given status. A social group consists of two or more people who interact frequently and share a common identity and sense of interdependence. A formal organization is a highly structured group formed to complete certain tasks or achieve specific goals. A social institution is a set of organized beliefs and rules that establishes how a society attempts to meet its basic needs.

■ **What are the functionalist and conflict perspectives on social institutions?**

According to functionalist theorists, social institutions perform several prerequisites of all societies: replace members; teach new members; produce, distribute, and consume goods and services; preserve order; and provide and maintain a sense of purpose. Conflict theorists suggest that social institutions do not work for the common good of all individuals. Institutions may enhance and uphold the power of some groups but exclude others, such as the homeless.

■ **What are the major types of societies?**

Social scientists have identified five types of societies. Three of these are referred to as preindustrial societies—hunting and gathering, horticultural and pastoral, and agrarian societies. The other two are industrial and post-industrial societies. Industrial societies are characterized by mechanized production of goods. Postindustrial societies are based on technology that supports an information-based economy in which providing services is based on knowledge more than on the production of goods.

■ How do societies maintain stability in times of social change?

According to Emile Durkheim, although changes in social structure may dramatically affect individuals and groups, societies manage to maintain some degree of stability. People in preindustrial societies are united by mechanical solidarity because they have shared values and common social bonds. Industrial societies are characterized by organic solidarity, which refers to the cohesion that results when people perform specialized tasks and are united by mutual dependence.

■ How do *Gemeinschaft* and *Gesellschaft* societies differ in social solidarity?

According to Ferdinand Tönnies, the *Gemeinschaft* is a traditional society in which relationships are based on personal bonds of friendship and kinship and on intergenerational stability. The *Gesellschaft* is an urban society in which social bonds are based on impersonal and specialized relationships, with little group commitment or consensus on values.

■ Is all social interaction based on shared meanings?

Social interaction within a society, particularly face-to-face encounters, is guided by certain shared meanings of how we should behave. All meanings may not be shared—race/ethnicity, gender, and social class often influence people's perceptions of meaning.

■ What is the dramaturgical perspective?

According to Erving Goffman's dramaturgical analysis, our daily interactions are similar to dramatic productions. Presentation of self refers to efforts to present our own self to others in ways that are most favorable to our own interests or self-image.

■ Why are feeling rules important?

Feeling rules shape the appropriate emotions for a given role or specific situation. Our emotions are not always private, and specific emotions may be demanded of us on certain occasions.

KEY TERMS

achieved status 143
agrarian societies 154
ascribed status 142
dramaturgical analysis 165
ethnomethodology 164
formal organization 148
Gemeinschaft 159
Gesellschaft 159
horticultural societies 153
hunting and gathering societies 152
impression management (presentation of self) 165
industrial societies 155
master status 143
mechanical solidarity 158
nonverbal communication 167
organic solidarity 158
pastoral societies 153
personal space 169
postindustrial societies 156
primary group 148
role 145
role conflict 145
role exit 146
role expectation 145
role performance 145
role strain 146
secondary group 148
self-fulfilling prophecy 163
social construction of reality 163
social group 147
social institution 149
social interaction 139
social structure 139
status 141
status symbol 144

QUESTIONS FOR CRITICAL THINKING

1. Think of a person you know well who often irritates you or whose behavior grates on your nerves (it could be a parent, friend, relative, or teacher). First, list that person's statuses and roles. Then analyze the person's possible role expectations, role performance, role conflicts, and role strains. Does anything you find in your analysis help to explain the irritating behavior? How helpful are the concepts of social structure in analyzing individual behavior?

2. Are structural problems responsible for homelessness, or are homeless individuals responsible for their own situation? Use functionalist, conflict, symbolic interactionist, and postmodernist theoretical perspectives as tools for analyzing this issue.

3. You are conducting field research on gender differences in nonverbal communication styles. How are you going to account for variations among age, race, and social class?

4. When communicating with other genders, races, and ages, is it better to express and acknowledge different styles or to develop a common, uniform style?

RESOURCES ON THE INTERNET

Chapter-Related Web Sites

The following Web sites have been selected for their relevance to the topics in this chapter. These sites are among the more stable, but please note that Web site addresses change frequently. For an updated list of chapter-related Web sites with URL links, please visit the *Sociology in Our Times* Web site (**www.wadsworth.com/KendallSIOT**).

The Emile Durkheim Archive
http://durkheim.itgo.com/main.html

Visit this Web site to learn more about Emile Durkheim and his contributions to sociology. Click on "solidarity" for additional information about Durkheim's concepts of mechanical and organic solidarity.

The National Coalition for the Homeless
http://www.nationalhomeless.org

The mission of this organization is to end homelessness. Its Web site provides facts about homelessness, information on legislation and policy, personal stories, and additional resources pertinent to the study of homelessness.

The Psychology of Cyberspace
http://www.rider.edu/users/suler/psycyber /psycyber.html

This site features an online hypertext book that explores social interaction and cyberspace, with links to additional sites that investigate the implications of cyberspace for interpersonal interaction.

ONLINE STUDY AND RESEARCH TOOLS

Accompanying this text are many *free* powerful online study tools that will help you master the material in this chapter, help increase your depth of understanding, and help you make the grade!

SocCoach CD-ROM

Use the SocCoach CD-ROM enclosed with this text to help you formulate a customized study plan for this chapter. After you take the Diagnostic Quiz, SocCoach will generate a customized study plan just for you! It will identify sections of the chapter that you should review and will provide videos, charts, graphs, and excerpts from the text to supplement your studies and enhance your understanding. You'll also find fun, interactive activities such as Virtual Explorations and Map the Stats to apply what you've learned and stretch your sociological imagination.

The Companion Web Site for Sociology in Our Times, *Fifth Edition*
www.wadsworth.com/KendallSIOT

Gain an even better grasp on this chapter by going to the companion Web site to take one of the Tutorial Quizzes, use the Flash Cards to master key terms, or check out the many other study aids you'll find there. You'll also find special features such as GSS Data and Census 2000 information that'll put data and resources at your fingertips to help you with that special project or help you as you do some research on your own.

In this chapter, when you see the icon on the left, it alerts you to a specific exercise found in *Wadsworth's Sociology Online Resources and Writing Companion.* This valuable guide shows you how to use Wadsworth's exclusive online resources—*InfoTrac College Edition,* the *Opposing Viewpoints Resource Center,* and *MicroCase Online*—to assist you in your study of sociology and to build essential research and writing skills.

Groups and Organizations

It's just incomprehensible to see what it was like down there. You know, I remember seeing one of these Cold War movies [that showed what things might be like] after the nuclear attacks with the Hollywood portrayal of a nuclear winter. It looked worse than that in downtown Manhattan, and it wasn't some grade "B" movie. It was life. It was real.

–New York Governor George Pataki describing his first trip to the World Trade Center disaster site after September 11, 2001 (qtd. in Nacos, 2002: 13)

Following the terrorist attacks of 2001, the U.S. Senate Judiciary Committee conducted hearings in an effort to learn what information about terrorist activities and possible targets in this country had been available to federal agencies such as the Federal Bureau of Investigation and the Central Intelligence Agency in advance of September 11, and why the government had not acted on any available information in a manner that might have prevented the attacks from occurring on that fateful day. On June 6, 2002, the Judiciary Committee interviewed FBI agent Coleen Rowley, who had written a letter setting forth information that had been known by the FBI; she discussed her belief that the "culture" of that agency had prevented it from acting on what it knew (*New York Times,* 2002b):

AGENT ROWLEY: We have a culture in the FBI that there's a certain pecking order, and it's pretty strong. And it's very rare that someone picks up the phone and calls a rank or two above themselves. It would have to be only on the strongest reasons. Typically, you would have to . . . pick up the phone and talk to somebody who is at your rank. So when you have an item that requires review by a higher level, it's incumbent for you to go to a higher-level person in your office and then for that person to make a call. . . .

SENATOR GRASSLEY: In your letter [to the FBI director], you mention a culture of fear, especially a fear of taking action, and the problem of careerism. Could you talk about how this hurts investigations in the field, what the causes are, and what you think might fix these problems?

AGENT ROWLEY: [W]hen I looked up the definition [of careerism], I really said [it's] unbelievable how appropriate that is. I think the FBI does have a problem with that. And if I remember right, it means, "promoting one's career over integrity." So, when people make decisions, and it's basically so that [they] can get to the next level and not rock—either it's not rock the boat or do what a boss says without question. And either way that works, if you're making a decision to try to get to the next level, but you're not making that decision for the real right reason, that's a problem. . . .

AP/Wide World Photos

■ The terrorist attacks on New York City and Washington, D.C., raised serious questions about how groups and organizations relay critical information that might prevent such occurrences.

According to sociologists, we need groups and organizations—just like we need culture and socialization—in order to live and to participate in societies as we know them; however, there are both positive and negative aspects of groups, regardless of their size. For example, a small group may be made up only of people who are good friends and enjoy being with one another, yet the group may disband if one or more of its members feel left out and cease to participate. A large group may be able to accomplish tasks that a smaller number of people could never achieve, yet the larger group may become unwieldy due to its size, with the result that some members come to feel that it is no longer focused on its original goals. In each instance, part of the problem may be inadequate or ineffective communication.

When we encounter a major problem with an existing organization, we may wonder whether the group should be restructured or possibly replaced by some new entity that we believe might meet our needs more efficiently. Such has been the case in the aftermath of the terrorist attacks on the United States: The Federal Bureau of Investigation, the Central Intelligence Agency, and other governmental organizations have faced criticism about how they used information before these attacks—not only how cru-

cial information flowed or failed to flow within the particular organization but also why such information was not more adequately conveyed among supposedly cooperating agencies. This is an area in which sociologists have applied the sociological imagination by studying how bureaucracies and technology contribute to (or impede) information flow within and between large-scale organizations.

As we shall see in this chapter, bureaucratic structure is the most common organizational form in governments, businesses, schools, and other institutions around the globe. Although there are many benefits from this organizational model, there are also significant limitations, including loss of personal privacy. Before we take a closer look at social groups and bureaucratic organizations, take the quiz in Box 6.1 on privacy in groups and organizations.

QUESTIONS AND ISSUES

Chapter Focus Question: Why is it important for groups and organizations to enhance communication among participants and facilitate the flow of information while protecting the privacy of individuals?

What constitutes a social group?

How are groups and their members shaped by group size, leadership style, and pressures to conform?

What is the relationship between information and social organizations in societies such as ours?

What purposes does bureaucracy serve?

What alternative forms of organization exist as compared with the most widespread forms today?

■ SOCIAL GROUPS

Three strangers are standing at a street corner waiting for a traffic light to change. Do they constitute a group? Five hundred women and men are first-year graduate students at a university. Do they constitute a group? In everyday usage, we use the word *group* to mean any collection of people. According to sociologists, however, the answer to these questions is no; individuals who happen to share a common feature or to be in the same place at the same time do not constitute social groups.

Groups, Aggregates, and Categories

As we saw in Chapter 5, a *social group* is a collection of two or more people who interact frequently with

one another, share a sense of belonging, and have a feeling of interdependence. Several people waiting for a traffic light to change constitute an *aggregate*—**a collection of people who happen to be in the same place at the same time but share little else in common.** Shoppers in a department store and passengers on an airplane flight are also examples of aggregates. People in aggregates share a common purpose (such as purchasing items or arriving at their destination) but generally do not interact with one another, except perhaps briefly. The first-year graduate students, at least initially, constitute a *category*—**a number of people who may never have met one another but share a similar characteristic** (such as education level, age, race, or gender). Men and women make up categories, as do Native Americans and Latinos/as, and victims of sexual or racial harassment. Categories are not social groups because the people in them usu-

Box 6.1 SOCIOLOGY AND EVERYDAY LIFE

How Much Do You Know About Privacy in Groups and Organizations?

True	False	
T	F	1. A fast-food restaurant can legally require all employees under the age of twenty-one to submit to periodic, unannounced drug testing.
T	F	2. Members of a high school football team can be required to submit to periodic, unannounced drug testing.
T	F	3. Parents of students at all U.S. colleges and universities are entitled to obtain a transcript of their children's college grades, regardless of the student's age.
T	F	4. A company has the right to keep its employees under video surveillance at all times while they are at the company's place of business—even in the company's restrooms.
T	F	5. A private club has the right to require an applicant for membership to provide his or her Social Security number as a condition of membership.
T	F	6. If a person applies for a job in a workplace that has more than twenty-five employees, the employer can require that person to provide medical information or take a physical examination prior to offering him or her a job.
T	F	7. Students at a church youth group meeting who hear one member of the group confess to an illegal act can be required to divulge what that member said.
T	F	8. A student's privacy is protected when using a computer, even if it is owned by the college or university, because deleting an e-mail or other document from a computer prevents anyone else from examining that document.

Answers on page 178.

---

ally do not create a social structure or have anything in common other than a particular trait.

Occasionally, people in aggregates and categories form social groups. For instance, people within the category known as "graduate students" may become an aggregate when they get together for an orientation to graduate school. Some of them may form social groups as they interact with one another in classes and seminars, find that they have mutual interests and concerns, and develop a sense of belonging to the group. Information technology raises new and interesting questions about what constitutes a group. For example, some people question whether we can form a social group on the Internet (see Box 6.2 on page 180).

Types of Groups

As you will recall from Chapter 5, groups have varying degrees of social solidarity and structure. This structure is flexible in some groups and more rigid in others. Some groups are small and personal; others are large and impersonal. We more closely identify with the members of some groups than we do with others.

Cooley's Primary and Secondary Groups

Sociologist Charles H. Cooley (1963/1909) used the term *primary group* to describe a small, less specialized group in which members engage in face-to-face, emotion-based interactions over an extended period of time. We have primary relationships with other individuals in our primary groups—that is, with our *significant others,* who frequently serve as role models.

In contrast, you will recall, a *secondary group* is a larger, more specialized group in which the members engage in more impersonal, goal-oriented relationships for a limited period of time. The size of a secondary group may vary. Twelve students in a graduate seminar may start out as a secondary group but eventually

Box 6.1 SOCIOLOGY AND EVERYDAY LIFE

Answers to the Sociology Quiz on Privacy

1. True. In all but a few states, an employee in the private sector of the economy can be required to submit to a drug test even where nothing about the employee's job performance or history suggests illegal drug use. An employee who refuses can be terminated without legal recourse.

2. True. The U.S. Supreme Court has ruled that schools may require students to submit to random drug testing as a condition to participating in extracurricular activities such as sports teams, the school band, the future homemakers' club, the cheerleading squad, and the choir.

3. False. The Family Educational Right to Privacy Act, which allows parents of a student under age eighteen to obtain their child's grades, requires the student's consent once he or she has attained age eighteen; however, that law applies only to institutions that receive federal educational funds.

4. False. An employer may not engage in video surveillance of its employees in situations where they have a reasonable right of privacy. At least in the absence of a sign warning of surveillance, employees have such a right in company restrooms.

5. True. Although the Privacy Act of 1974 makes it illegal for federal, state, and local governmental agencies to deny rights, privileges, or benefits to individuals who refuse to provide their Social Security number unless disclosure is required by law, no federal law extends this prohibition to private groups and organizations.

6. False. The Americans with Disabilities Act prohibits employers in workplaces with more than twenty-five employees from asking job applicants about medical information or requiring a physical examination prior to employment.

7. True. Although confidential communications made privately to a minister, priest, rabbi, or other religious leader (or to an individual the person reasonably believes to hold such a position) generally cannot be divulged without the consent of the person making the communication, this does not apply when other people are present who are likely to hear the statement.

8. False. Deleting an e-mail or other document from a computer does not actually remove it from the computer's memory. Until other files are entered that actually write over the space where the document was located, experts can retrieve the document that was deleted.

---

become a primary group as they get to know one another and communicate on a more personal basis. Formal organizations are secondary groups, but they also contain many primary groups within them. For example, how many primary groups do you think there are within the secondary-group setting of your college?

Sumner's Ingroups and Outgroups All groups set boundaries by distinguishing between insiders who are members and outsiders who are not. Sociologist William Graham Sumner (1959/1906) coined the terms *ingroup* and *outgroup* to describe people's feelings toward members of their own and other groups. An ***ingroup* is a group to which a person belongs and with which the person feels a sense of identity.** An ***outgroup* is a group to which**

a person does not belong and toward which the person may feel a sense of competitiveness or hostility. Distinguishing between our ingroups and our outgroups helps us establish our individual identity and self-worth. Likewise, groups are solidified by ingroup and outgroup distinctions; the presence of an enemy or hostile group binds members more closely together (Coser, 1956).

Group boundaries may be formal, with clearly defined criteria for membership. For example, a country club that requires an applicant for membership to be recommended by four current members, to pay a $25,000 initiation fee, and to pay $1,000 per month in membership dues has clearly set requirements for its members. The club may even post a sign at its entrance that states "Members Only" and use security

© Tony Roberts/Corbis

Sometimes the distinction between what constitutes an ingroup and an outgroup is subtle. Other times, it is not subtle at all. Consider, for example, this "members only" sign at a club in Aberdeen, Scotland. We have many similar signs through the United States.

personnel to ensure that nonmembers do not encroach on its grounds. Boundary distinctions are often reflected in symbols such as emblems or clothing. Members of the country club are given membership cards to gain access to the club's facilities or to charge food to their account. They may wear sun visors and shirts with the country club's logo on them. All these symbols denote that the bearer/wearer is a member of the ingroup; they are status symbols. However, group boundaries are not always that formal. For example, friendship groups usually do not have clear guidelines for membership; rather, the boundaries tend to be very informal and vaguely defined.

Ingroup and outgroup distinctions may encourage social cohesion among members, but they also may promote classism, racism, sexism, and ageism. Ingroup members typically view themselves positively and members of outgroups negatively. These feelings of group superiority, or *ethnocentrism,* are somewhat inevitable. However, members of some groups feel more free than others to act on their beliefs. If groups are embedded in larger groups and organizations, the large organization may discourage such beliefs and their consequences (Merton, 1968). Conversely, organizations may covertly foster these ingroup/outgroup distinctions by denying their existence or by failing to take action when misconduct occurs.

Reference Groups Ingroups provide us not only with a source of identity but also with a point of reference. A *reference group* **is a group that strongly influences a person's behavior and social attitudes, regardless of whether that individual is an actual member.** When we attempt to evaluate our appearance, ideas, or goals, we automatically refer to the standards of some group. Sometimes, we will refer to

our membership groups, such as family or friends. Other times, we will rely on groups to which we do not currently belong but that we might wish to join in the future, such as a social club or a profession. We may also have negative reference groups. For many people, the Ku Klux Klan and neo-Nazi skinheads are examples of negative reference groups because most people's racial attitudes compare favorably with such groups' blatantly racist behavior.

Reference groups help explain why our behavior and attitudes sometimes differ from those of our membership groups. We may accept the values and norms of a group with which we identify rather than one to which we belong. We may also act more like members of a group we want to join than members of groups to which we already belong. In this case, reference groups are a source of anticipatory socialization. Many people have more than one reference group and often receive conflicting messages from these groups about how they should view themselves. For most of us, our reference group attachments change many times during our life course, especially when we acquire a new status in a formal organization.

Networks A *network* **is a web of social relationships that links one person with other people and, through them, with other people they know.** Frequently, networks connect people who share common interests but who otherwise might not identify and interact with one another. For example, if A is tied to B, and B is tied to C, then a network is more likely to

Recently laid-off individuals hope that their networks will provide them with other job opportunities. High-tech layoffs resulted in gatherings such as this dot-com "pink slip" party, where job hunters got together to share leads regarding new employment opportunities.

Bob Daemmrich/The Image Works

Box 6.2   CHANGING TIMES: MEDIA AND TECHNOLOGY

Can We Form Social Groups and True Communities on the Internet?

Meeting new friends,
Imagining smiles . . .
Across the networks
Spanning the miles. . . .

From all walks of life
We come to the net.
A community of friends
Who have never met.
−from "Thoughts of Internet Friendships" by Jamie Wilkerson (1996)

As this excerpt from a poem posted on the Internet suggests, many people believe that they can make new friends and establish a community online. From this perspective, the "Internet community" is "a body of people looking for similar information, dealing with similar conditions, and abiding by the same 'general rules'" (thewritemarket.com, 2003). However, as you study sociology, you might ask whether this form of "community" is actually a true community.

In our discussion of societies in Chapter 5, we defined the *Gemeinschaft* as a traditional society in which social relationships are based on personal bonds of friendship and kinship and on intergenerational stability, and we defined a *social group* as a collection of two or more people who interact frequently with one another, share a sense of belonging, and

■ Chat rooms and other forms of communication on the Internet are extremely popular with millions of people; however, some sociologists question whether we can actually form social groups and true communities on the Internet. Is cyber chat different from our face-to-face interactions with others?

have a feeling of interdependence. Both of these definitions suggest that people must have a sense of place (be in the same place at the same time at least part of the time) in order to establish a true social group or community. However, these definitions were developed before introduction of the computer and the Internet. With the rapid communications that link

---

be formed among individuals A, B, and C. If this seems a little confusing at first, let's assume that Alice knows of Dolores and Eduardo only through her good friends Bill and Carolyn. For almost a year, Alice has been trying (without success) to purchase a house she can afford. Since large numbers of people are moving into her community, the real estate market is "tight," and houses frequently sell before a "for sale" sign goes up in the yard. However, through her friends Bill and Carolyn, Alice learns that their friends—Dolores and Eduardo—are about to put their house up for sale. Bill and Carolyn call Dolores and Eduardo to set up an appointment for Alice to see the house before it goes on the real estate market. Thanks to Alice's network, she is able to purchase the house before other people learn that it is for sale. Although Alice had not previously met Dolores and Eduardo, they are part of her network through her friendship with Bill and Carolyn. Scarce resources (in

this case, the number of affordable houses available) are unequally distributed, and people often must engage in collaboration and competition in their efforts to deal with this scarcity. Another example of the use of networks to help overcome scarce resources is recent college graduates who seek help from friends and acquaintances in order to find a good job.

What are your networks? For a start, your networks consist of all the people linked to you by primary ties, including your relatives and close friends. Your networks also include your secondary ties, such as acquaintances, classmates, professors, and—if you are employed—your supervisor and co-workers. However, your networks actually extend far beyond these ties to include not only the people that you *know,* but also the people that you *know of*—and who know of you—through your primary and secondary ties. In fact, your networks potentially include a pool of between 500 and 2,500 acquaintances, if you

people around the world today, are we able to form groups and establish communities with people whom we have never actually met?

Although some social scientists believe that virtual communities established on the Internet constitute true communities (see Wellman, 2001), the sociologists Robyn Bateman Driskell and Larry Lyon (2002) examined existing theories and research on this topic and concluded that true communities cannot be established in the digital environment of cyberspace. According to Driskell and Lyon, although a virtual community provides two core elements of community—common ties and social interaction—there is a lack of identification with place (a specific geographical location in which people get together and have extended personal relationships with one another), which is a key component of true community. Driskell and Lyon (2002: 1) state that if community includes "the close, emotional, holistic ties of *Gemeinschaft,* then the virtual community is not true community."

Although the Internet provides us with the opportunity to share interests with others whom we have not met (such as through chat groups) and to communicate with people we already know (such as by e-mail and instant messaging), the original concept of community which "emphasized local place, common ties, and social interaction that is intimate, holistic, and all-encompassing" is lacking (Driskell and Lyon, 2002: 6). Virtual communities on the Internet (so-called "communities in cyberspace") do not have geographic and social boundaries, are limited in their scope to specific areas of interest, are psychologically detached from close interpersonal ties, and have only limited concern for their "members" (Driskell and Lyon, 2002: 6). In fact, the Internet may reduce community, rather than enhance it, if we spend many hours in social isolation doing impersonal searches for information.

Clearly, the Internet provides a greater means of communication among people who have computers and Internet access, but this may not add up to the establishment of true social groups and true communities in the traditional sociological sense of these terms. However, it is possible that the Internet will create a "weak community replacement" for people based on a virtual community of specialized ties developed by e-mail correspondence and chat room discussions (Driskell and Lyon, 2002).

WRITING IN SOCIOLOGY ASSIGNMENT

Do your communications on the Internet give you the same feeling of connectivity to groups and a community as your "real world" interac-  tions with family, friends, other students, and co-workers? Why or why not?

count the connections of everyone in your networks (Milgram, 1967). Today, the term *networking* is widely used to describe the contacts that people make to find jobs or other opportunities; however, sociologists have studied social networks for many years in an effort to learn more about the linkages between individuals and their group memberships.

GROUP CHARACTERISTICS AND DYNAMICS

What purpose do groups serve? Why are individuals willing to relinquish some of their freedom to participate in groups? According to functionalists, people form groups to meet instrumental and expressive needs. *Instrumental,* or task-oriented, needs cannot always be met by one person, so the group works cooperatively to fulfill a specific goal. For example, think of how hard it would be to function as a one-person football team or to single-handedly build a skyscraper. Groups help members do jobs that are impossible to do alone or that would be very difficult and time-consuming at best. In addition to instrumental needs, groups also help people meet their *expressive,* or emotional, needs, especially those involving self-expression and support from family, friends, and peers.

Although not disputing that groups ideally perform such functions, conflict theorists suggest that groups also involve a series of power relationships whereby the needs of individual members may not be equally served. Symbolic interactionists focus on how the size of a group influences the kind of interaction that takes place among members. To many postmodernists,

According to the sociologist Georg Simmel, interaction patterns change when a third person joins a dyad—a group composed of two members. How might the conversation between this man and woman change when another person arrives to talk with them?

groups and organizations—like other aspects of postmodern societies—are generally characterized by superficiality and depthlessness in social relationships (Jameson, 1984). One postmodern thinker who focuses on this issue is the literary theorist Fredric Jameson, whose works continue to have a significant influence on contemporary sociological theorizing. According to Jameson's (1984) analysis, not only are postmodern organizations (and societies as a whole) characterized by superficial relations and lack of depth; people also experience a waning of emotion because the world, and the people in it, have become more fragmented (Ritzer, 1997). For example, the sociologist George Ritzer (1997) examined fast-food restaurants using a postmodern approach and concluded that both restaurant employees and customers interact in extremely superficial ways that are largely scripted by large-scale organizations: The employees learn to follow scripts in taking and filling customers' orders (Leidner, 1993), whereas customers also respond with their own "recipied" action (Schutz, 1967/1932). According to Ritzer (1997: 226), "[C]ustomers are mindlessly following what they consider tried-and-true social recipes, either learned or created by them previously, on how to deal with restaurant employees and, more generally, how to work their way through the system associated with the fast-food restaurant."

We will now look at certain characteristics of groups, such as how size affects group dynamics.

Group Size

The size of a group is one of its most important features. Interactions are more personal and intense in a *small group,* **a collectivity small enough for all members to be acquainted with one another and to interact simultaneously.**

Sociologist Georg Simmel (1950/1902–1917) suggested that small groups have distinctive interaction patterns that do not exist in larger groups. According to Simmel, in a *dyad*—**a group composed of two members**—the active participation of both members is crucial for the group's survival. If one member withdraws from interaction or "quits," the group ceases to exist. Examples of dyads include two people who are best friends, married couples, and domestic partnerships. Dyads provide members with a more intense bond and a sense of unity not found in most larger groups.

When a third person is added to a dyad, a *triad,* **a group composed of three members,** is formed. The nature of the relationship and interaction patterns changes with the addition of the third person. In a triad, even if one member ignores another or declines to participate, the group can still function. In addition, two members may unite to create a coalition that can subject the third member to group pressure to conform. A *coalition* is an alliance created in an attempt to reach a shared objective or goal. If two members form a coalition, the other member may be seen as an outsider or intruder. Like dyads, triads can exist as separate entities or be contained within formal organizations.

As the size of a group increases beyond three people, members tend to specialize in different tasks, and everyday communication patterns change. For instance, in groups of more than six or seven people, it becomes increasingly difficult for everyone to take part in the same conversation; therefore, several conversations will probably take place simultaneously.

Figure 6.1 | **Growth of Possible Social Interaction Based on Group Size**

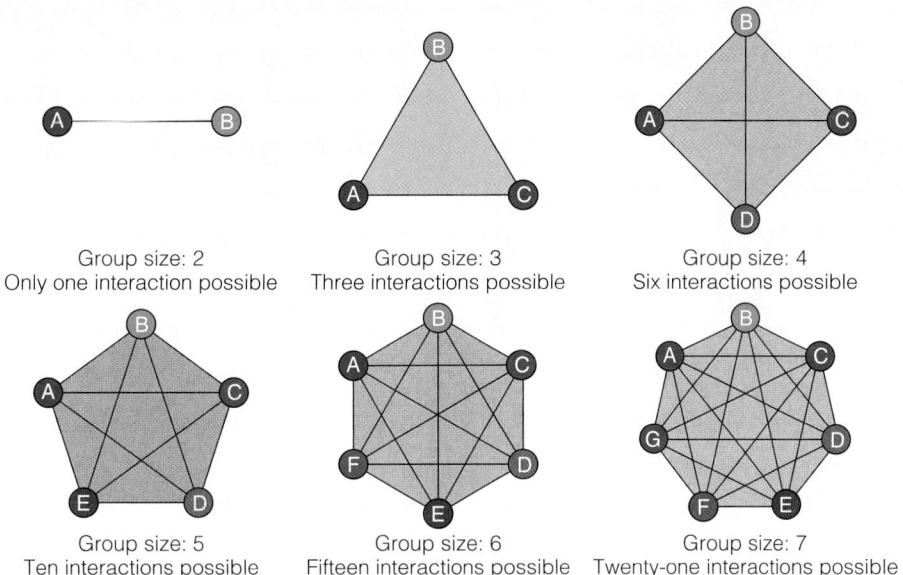

Group size: 2
Only one interaction possible

Group size: 3
Three interactions possible

Group size: 4
Six interactions possible

Group size: 5
Ten interactions possible

Group size: 6
Fifteen interactions possible

Group size: 7
Twenty-one interactions possible

Members are also likely to take sides on issues and form a number of coalitions. In groups of more than ten or twelve people, it becomes virtually impossible for all members to participate in a single conversation unless one person serves as moderator and guides the discussion. As shown in Figure 6.1, when the size of the group increases, the number of possible social interactions also increases.

Although large groups typically have less social solidarity than small ones, they may have more power. However, the relationship between size and power is more complicated than it might initially seem. The power relationship depends on both a group's *absolute* size and its *relative* size (Simmel, 1950/1902–1917; Merton, 1968). The absolute size is the number of members the group actually has; the relative size is the number of potential members. For example, suppose that three hundred people (out of many thousands) who have been the victims of sexual harassment band together to "march on Washington" and demand more stringent enforcement of harassment laws. Although three hundred people is a large number in some contexts, opponents of this group would argue that the low turnout demonstrates that harassment is not as big a problem as some might think. At the same time, the power of a small group to demand change may be based on a "strength in numbers" factor if the group is seen as speaking on behalf of a large number of other people (who are also voters).

Larger groups typically have more formalized leadership structures. Their leaders are expected to perform a variety of roles, some related to the internal workings of the group and others related to external relationships with other groups.

Group Leadership

What role do leaders play in groups? Leaders are responsible for directing plans and activities so that the group completes its task or fulfills its goals. Primary groups generally have informal leadership. For example, most of us do not elect or appoint leaders in our own families. Various family members may assume a leadership role at various times or act as leaders for specific tasks. In traditional families, the father or eldest male is usually the leader. However, in today's more diverse families, leadership and power are frequently in question, and power relationships may be quite different, as discussed later in this text. By comparison, larger groups typically have more formalized leadership structures. Their leaders are expected to perform a variety of roles, some related to the internal workings of the group and others related to external relationships with other groups. For example, leadership in secondary groups (such as colleges, governmental agencies, and corporations) involves a clearly defined chain of command, with written responsibilities assigned to each position in the organizational structure.

Organizations have different leadership styles based on the purpose of the group. How do leadership styles in the military differ from those on college and university campuses?

Leadership Functions Both primary and secondary groups have some type of leadership or positions that enable certain people to be leaders, or at least to wield power over others. From a functionalist perspective, if groups exist to meet the instrumental and expressive needs of their members, then leaders are responsible for helping the group meet those needs. ***Instrumental leadership is goal or task oriented;*** this type of leadership is most appropriate when the group's purpose is to complete a task or reach a particular goal. ***Expressive leadership provides emotional support for members;*** this type of leadership is most appropriate when the group is dealing with emotional issues, and harmony, solidarity, and high morale are needed. Both kinds of leadership are needed for groups to work effectively. Traditionally, instrumental and expressive leadership roles have been limited by gender socialization. Instrumental leadership has been linked with men whereas expressive leadership has been linked with women. Social change in recent years has somewhat blurred the distinction between gender-specific leadership characteristics, but these outdated stereotypes have not completely disappeared (Basow, 1992).

Leadership Styles Three major styles of leadership exist in groups: authoritarian, democratic, and laissez-faire. ***Authoritarian leaders make all major group decisions and assign tasks to members.*** These leaders focus on the instrumental tasks of the group and demand compliance from others. In times of crisis, such as a war or natural disaster, authoritarian leaders may be commended for their decisive actions. In other situations, however, they may be criticized for being dictatorial and for fostering intergroup hostility. By contrast, ***democratic leaders* encourage group discussion and decision making through consensus building.** These leaders may be praised for their expressive, supportive behavior toward group members, but they may also be blamed for being indecisive in times of crisis.

Laissez-faire literally means "to leave alone." ***Laissez-faire leaders* are only minimally involved in decision making and encourage group members to make their own decisions.** On the one hand, laissez-faire leaders may be viewed positively by group members because they do not flaunt their power or position. On the other hand, a group that needs active leadership is not likely to find it with this style of leadership, which does not work vigorously to promote group goals.

Studies of kinds of leadership and decision-making styles have certain inherent limitations. They tend to focus on leadership that is imposed externally on a group (such as bosses or political leaders) rather than leadership that arises within a group. Different decision-making styles may be more effective in one setting than another. For example, imagine attending a college class in which the professor asked the students to determine what should be covered in the course, what the course requirements should be, and how students should be graded. It would be a difficult and cumbersome way to start the semester; students might spend the entire term negotiating these matters and never actually learn anything.

Group Conformity

To what extent do groups exert a powerful influence in our lives? Groups have a significant amount of influence on our values, attitudes, and behavior. In order to gain and then retain our membership in groups, most of us are willing to exhibit a high level of conformity to the wishes of other group members. *Conformity* **is the process of maintaining or changing behavior to comply with the norms established by a society, subculture, or other group.** We often experience powerful pressure from other group members to conform. In some situations, this pressure may be almost overwhelming.

In several studies (which would be impossible to conduct today for ethical reasons), researchers found that the pressure to conform may cause group members to say they see something that is contradictory to what they are actually seeing or to do something that they would otherwise be unwilling to do. As we look at two of these studies, ask yourself what you might have done if you had been involved in this research.

Asch's Research Pressure to conform is especially strong in small groups in which members want to fit in with the group. In a series of experiments conducted by Solomon Asch (1955, 1956), the pressure toward group conformity was so great that participants were willing to contradict their own best judgment if the rest of the group disagreed with them.

One of Asch's experiments involved groups of undergraduate men (seven in each group) who were allegedly recruited for a study of visual perception. All the men were seated in chairs. However, the person in the sixth chair did not know that he was the only actual subject; all the others were assisting the researcher. The participants were first shown a large card with a vertical line on it and then a second card with three vertical lines (see Figure 6.2). Each of the seven participants was asked to indicate which of the three lines on the second card was identical in length to the "standard line" on the first card.

In the first test with each group, all seven men selected the correct matching line. In the second trial, all seven still answered correctly. In the third trial, however, the actual subject became very uncomfortable when all the others selected the incorrect line. The subject could not understand what was happening and became even more confused as the others continued to give incorrect responses on eleven out of the next fifteen trials.

If you had been in the position of the subject, how would you have responded? Would you have continued to give the correct answer, or would you have been swayed by the others? When Asch (1955) averaged the responses of all fifty actual subjects who participated in the study, he found that about 33 percent routinely chose to conform to the group by giving the same (incorrect) responses as Asch's assistants. An-

Figure 6.2 Asch's Cards

Although Line 2 is clearly the same length as the line in the lower card, Solomon Asch's research assistants tried to influence "actual" participants by deliberately picking Line 1 or Line 3 as the correct match. Many of the participants went along rather than risking the opposition of the "group."

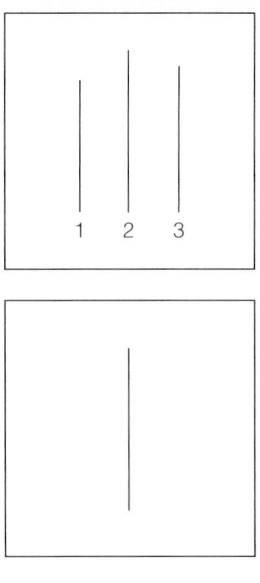

Source: Asch, 1955.

other 40 percent gave incorrect responses in about half of the trials. Although 25 percent always gave correct responses, even they felt very uneasy and "knew that something was wrong." In discussing the experiment afterward, most of the subjects who gave incorrect responses indicated that they had known the answers were wrong but decided to go along with the group in order to avoid ridicule or ostracism.

After conducting additional research, Asch concluded that the size of the group and the degree of social cohesion felt by participants were important influences on the extent to which individuals respond to group pressure. In dyads, for example, the subject was much less likely to conform to an incorrect response from one assistant than in four-member groups. This effect peaked in groups of approximately seven members and then leveled off (see Figure 6.3). Not surprisingly, when groups were not cohesive (when more than one member dissented), group size had less effect. If even a single assistant did not agree with the others, the subject was reassured by hearing someone else question the accuracy of incorrect responses and was much less likely to give a wrong answer himself.

Figure 6.3 Effect of Group Size in the Asch Conformity Studies

As more people are added to the "incorrect" majority, subjects' tendency to conform by giving wrong answers increases—but only up to a point. Adding more than seven people to the incorrect majority does *not* further increase subjects' tendency to conform—perhaps because subjects are suspicious about why so many people agree with one another.

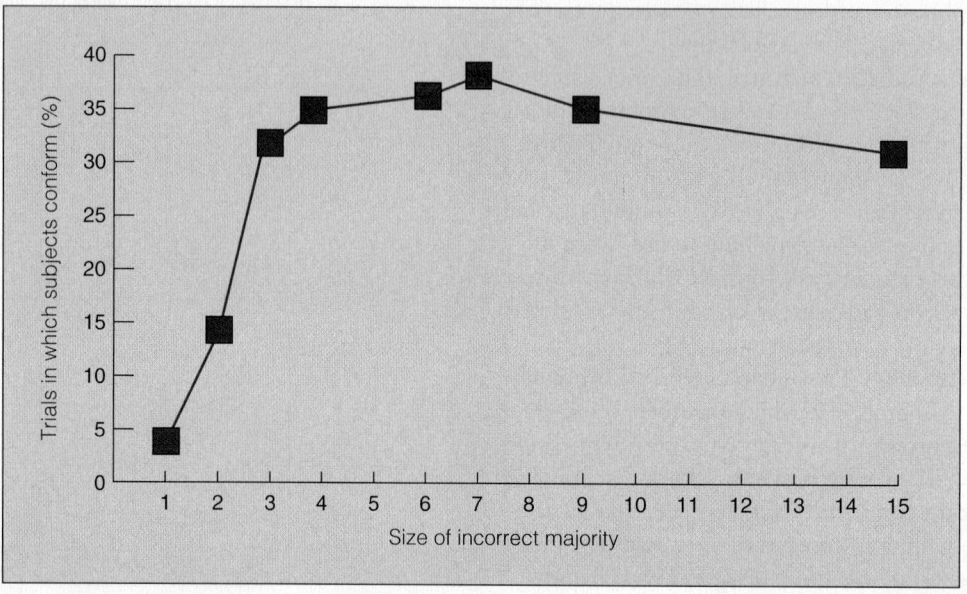

Source: Asch, 1955.

One contribution of Asch's research is the dramatic way in which it calls our attention to the power that groups have to produce a certain type of conformity. *Compliance* is the extent to which people say (or do) things so that they may gain the approval of other people. Certainly, Asch demonstrated that people will bow to social pressure in small-group settings. From a sociological perspective, however, the study was flawed because it involved deception about the purpose of the study and about the role of individual group members. Moreover, the study included only male college students, thus making it impossible for us to generalize its findings to other populations, including women and people who were not undergraduates. Would Asch's conclusions have been the same if women had participated in the study? Would the same conclusions be reached if the study were conducted today? We cannot answer these questions with certainty, but the work of Solomon Asch and his student, Stanley Milgram, have had a lasting impact on social science perceptions about group conformity and obedience to authority.

Milgram's Research How willing are we to do something because someone in a position of authority has told us to do it? How far are we willing to go

in following the demands of that individual? Stanley Milgram (1963, 1974) conducted a series of controversial experiments to find answers to these questions about people's obedience to authority. *Obedience* is a form of compliance in which people follow direct orders from someone in a position of authority.

Milgram's subjects were men who had responded to an advertisement for participants in an experiment. When the first (actual) subject arrived, he was told that the study concerned the effects of punishment on learning. After the second subject (an assistant of Milgram's) arrived, the two men were instructed to draw slips of paper from a hat to get their assignments as either the "teacher" or the "learner." Because the drawing was rigged, the actual subject always became the teacher, and the assistant the learner. Next, the learner was strapped into a chair with protruding electrodes that looked something like an electric chair. The teacher was placed in an adjoining room and given a realistic-looking but nonoperative shock generator. The "generator's" control panel showed levels that went from "Slight Shock" (15 volts) on the left, to "Intense Shock" (255 volts) in the middle, to "DANGER: SEVERE SHOCK" (375 volts), and finally "XXX" (450 volts) on the right.

Figure 6.4   Results of Milgram's Obedience Experiment

Even Milgram was surprised by subjects' willingness to administer what they thought were severely painful and even dangerous shocks to a helpless "learner."

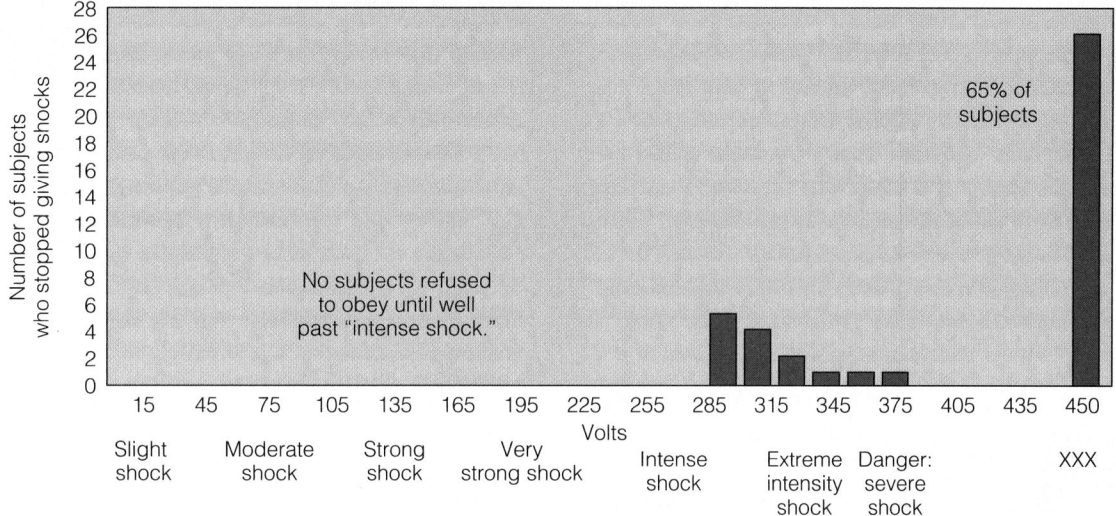

Source: Milgram, 1963.

The teacher was instructed to read aloud a pair of words and then repeat the first of the two words. At that time, the learner was supposed to respond with the second of the two words. If the learner could not provide the second word, the teacher was instructed to press the lever on the shock generator so that the learner would be punished for forgetting the word. Each time the learner gave an incorrect response, the teacher was supposed to increase the shock level by 15 volts. The alleged purpose of the shock was to determine if punishment improves a person's memory.

What was the maximum level of shock that a "teacher" was willing to inflict on a "learner"? The learner had been instructed (in advance) to beat on the wall between him and the teacher as the experiment continued, pretending that he was in intense pain. The teacher was told that the shocks might be "extremely painful" but that they would cause no permanent damage. At about 300 volts, when the learner quit responding at all to questions, the teacher often turned to the experimenter to see what he should do next. When the experimenter indicated that the teacher should give increasingly painful shocks, 65 percent of the teachers administered shocks all the way up to the "XXX" (450-volt) level (see Figure 6.4). By this point in the process, the teachers were frequently sweating, stuttering, or biting on their lip. According to Milgram, the teachers (who were free to leave whenever they wanted to) continued in the experiment because they were being given directions by a person in a position of authority (a university scientist wearing a white coat).

What can we learn from Milgram's study? The study provides evidence that obedience to authority may be more common than most of us would like to believe. None of the "teachers" challenged the process before they had applied 300 volts. Almost two-thirds went all the way to what could have been a deadly jolt of electricity if the shock generator had been real. For many years, Milgram's findings were found to be consistent in a number of different settings and with variations in the research design (Miller, 1986).

This research once again raises some questions originally posed in Chapter 2 concerning research ethics. As was true of Asch's research, Milgram's subjects were deceived about the nature of the study in which they were asked to participate. Many of them found the experiment extremely stressful. Such conditions cannot be ignored by social scientists because subjects may receive lasting emotional scars from such research. It would be virtually impossible today to obtain permission to replicate this experiment in a university setting.

Group Conformity and Sexual Harassment

Let's look at a more contemporary example of how

social science research can help us learn about the ways in which group conformity may contribute to a complex social problem such as sexual harassment, which consists of unwanted sexual advances, requests for sexual favors, or other verbal or physical conduct of a sexual nature.

Psychologist John Pryor (Pryor and McKinney, 1991) has conducted behavioral experiments on college campuses to examine the social dynamics of harassment. In one of his studies, a graduate student (who was actually a member of the research team) led research subjects to believe that they would be training undergraduate women to use a computer. The actual purpose of the experiment was to observe whether the trainers (subjects) would harass the women if given the opportunity and encouraged to do so. By design, the graduate student purposely harassed the women (who were also part of the research team), setting an example for the subjects to follow.

Pryor found that when the "trainers" were led to believe that sexual harassment was condoned and were then left alone with the women, they took full advantage of the situation in 90 percent of the experiments. Shannon Hoffman, one of the women who participated in the research, felt vulnerable because of the permissive environment created by the men in charge:

> It was very uncomfortable for me. I realized that had it been out of the experimental setting that, as a woman, I would have been very nervous with someone that close to me and reaching around me. So it kind of made me feel a little bit powerless as far as that goes because there was nothing I could do about it. But I also realized that in a business setting, if this person really was my boss, that it would be harder for me to send out the negative signals or whatever to try to fend off that type of thing. (PBS, 1992b)

This research suggests a relationship between group conformity and harassment. Sexual harassment is more likely to occur when it is encouraged (or at least not actively discouraged) by others. When people think they can get away with it, they are more likely to engage in such behavior.

Groupthink

As we have seen, individuals often respond differently in a group context than they might if they were alone. Social psychologist Irving Janis (1972, 1989) examined group decision making among political experts and found that major blunders in U.S. history may be attributed to pressure toward group conformity.

To describe this phenomenon, he coined the term *groupthink*—**the process by which members of a cohesive group arrive at a decision that many individual members privately believe is unwise.** Why not speak up at the time? Members usually want to be "team players." They may not want to be the ones who undermine the group's consensus or who challenge the group's leaders. Consequently, members often limit or withhold their opinions and focus on consensus rather than on exploring all of the options and determining the best course of action. Figure 6.5 summarizes the dynamics and results of groupthink.

The tragic 1986 launch of the space shuttle *Challenger*—which exploded 73 seconds into its flight, killing all seven crew members—has been cited as an example of this process. On the day preceding the launch, engineers at the company responsible for designing and manufacturing the shuttle's solid rocket boosters became concerned that freezing temperatures at the launch site would interfere with the proper functioning of the O-ring seals in the boosters, but were overruled by high-level officials at the company and within NASA, where executives were impatient because of earlier delays. A presidential commission that investigated the tragedy concluded that neither the manufacturer nor NASA responded adequately to warnings about the seals (Lippa, 1994).

In February of 2003, the space shuttle *Columbia* also exploded—in this instance, while preparing to land—once again raising concerns about groupthink at NASA. During takeoff, a chunk of insulated foam fell off the bipod ramp of the external fuel tank, striking and damaging the shuttle's left wing. Although some NASA engineers had previously raised concerns that hardened foam popping off the fuel tank could cause damage to the ceramic tiles protecting the shuttle, and although these concerns were again raised following *Columbia*'s liftoff, these concerns were overruled by NASA officials prior to and during the flight (Glanz and Wong, 2003; Schwartz, 2003). One analyst subsequently described the way that NASA dealt with these concerns as an example of "the ways that smart people working collectively can be dumber than the sum of their brains" (Schwartz and Wald, 2003: WK3).

Social Exchange/ Rational Choice Theories

Social exchange/rational choice theories focus on the process by which actors—individuals, groups, corporations, or societies, for example—settle on one optimal outcome out of a range of possible choices. The foundation of this approach is the doctrine of *utilitar-*

Figure 6.5	Janis's Description of Groupthink

In Janis's model, prior conditions such as a highly homogeneous group with committed leadership can lead to potentially disastrous "groupthink," which short-circuits careful and impartial deliberation. Events leading up to the tragic 2003 explosion of the space shuttle *Columbia* have been cited as an example of this process.

Process of Groupthink

PRIOR CONDITIONS

Isolated, cohesive, homogeneous decision-making group

Lack of impartial leadership

High stress

SYMPTOMS OF GROUPTHINK

Closed-mindedness

Rationalization

Squelching of dissent

"Mindguards"

Feelings of righteousness and invulnerability

DEFECTIVE DECISION MAKING

Incomplete examination of alternatives

Failure to examine risks and contingencies

Incomplete search for information

CONSEQUENCES

Poor decisions

Example: *Columbia* Explosion

NASA had previously orchestrated many successful shuttle missions and was under pressure to complete additional space missions that would fulfill agency goals and keep its budget intact.

Although *Columbia*'s left wing had been damaged on takeoff when a chunk of insulated foam from the external fuel tank struck it, NASA did not regard this as a serious problem because it had occurred on previous launches. Some NASA engineers stated that they did not feel free to raise questions about problems.

The debate among engineers regarding whether the shuttle had been damaged to the extent that the wing might burn off on reentry was not passed on to the shuttle crew or to NASA's top officials in a timely manner because either the engineers harbored doubts about their concerns or were unwilling to believe that the mission was truly imperiled.

The shuttle *Columbia* was destroyed during reentry into the Earth's atmosphere, killing all seven crew members and strewing debris across large portions of the United States.

© 2003 AP/Wide World Photos

© 2003 AP/Wide World Photos, Tyler Morning Telegraph, Dr. Scott Lieberman

Sources: Broder, 2003; Glanz and Wong, 2003; Schwartz (with Wald), 2003; Schwartz and Broder, 2003; Schwartz and Wald, 2003.

ianism—a belief that the purpose of all action should be to bring about the greatest happiness to the greatest number of people. An example of this belief is found in the assertion of early economist Adam Smith (1976/1776) that individuals who are allowed to make economic decisions free from the external constraints of government will make the best decisions not only for themselves but also for the entire society.

Social exchange theories are based on the assumption that *self-interest* is the basic motivating factor in people's interactions. According to this approach, people learn to adjust their behavior so that they receive rewards from others rather than negative responses or punishment (Homans, 1974). When people do not *give* and *take* in a manner that is deemed appropriate by other group members, conflict often ensues, and relationships among people may be destabilized (Gouldner, 1960). Consider this example of give and take: A good friend offers you a gift, but you decide to refuse it. What factors contribute to your decision? Self-interest—such as keeping the friendship or acquiring a possession of some worth—might dictate that you accept the gift. However, other factors may also be involved. What if you do not like the gift or do not want to feel obligated to reciprocate in some manner? The offer of the gift forces you to assess your self-interest in the situation: What will you gain or lose by accepting or rejecting the gift? Ultimately, your decision may cause solidarity or conflict; it may stabilize or destabilize your relationship with the other person.

Now, if we think of a similar exchange involving *words* rather than tangible objects, a similar process occurs. In work settings, for example, an exchange might involve conferring a reward (prestige) on someone in return for a valuable contribution (such as expert advice) (see Homans, 1958, 1974; and Blau, 1964, 1975). Based on self-interest, a person may accept or reject the statements or implicit assumptions of another person. In fact, people often compete with one another as they seek to maximize their rewards and minimize their punishments (Blau, 1964, 1975). People do not always gain reciprocal benefits from exchanges with others, particularly in situations where one person in the exchange occupies a dominant power position over another person (Emerson, 1962).

Rational choice theorists have analyzed situations in which the actors have differing amounts of power. *Rational choice theories* are based on the assumption that social life can be explained by means of models of rational individual action (Outhwaite and Bottomore, 1994). Accordingly, rational choice theorists are more concerned with explaining *social outcomes* than in predicting what an individual will do in a particular situation (Hechter and Kanazawa, 1997). Rational choice theory assumes that *actors* are purposeful or intentional in their decisions; however, many theorists acknowledge that not all actions are necessarily rational and that people do not always act rationally (Hechter and Kanazawa, 1997).

What are the key elements of rational choice theories? According to the sociologist James S. Coleman (1990), *actors* and *resources* are important factors in rational choice. Actors may include individuals, groups, corporations, and societies. Resources comprise the things over which actors have control and in which they have some interest. Major constraints on actors' choices are the scarcity of resources and structural restrictions (Marsden, 1983). Actors with fewer resources are less likely to pursue the most highly valued goal or end. They may decide to go for the next-most-attractive goal or end out of fear that they will lose the chance to acquire even the next-most-attractive end if they initially pursue an unrealistic goal or expectation. For example, suppose that Zoe, who has very few economic resources, wants to attend an Ivy League uni-

versity (her most-highly-valued goal). Although her first-choice school sends her a letter of acceptance, it does not offer her a scholarship. Meanwhile, a large state university (her next-most-attractive goal) admits her and offers her a full four-year scholarship. When Zoe weighs her options—attending "Ivy U" by taking out large student loans and getting a job or attending "State U" on a full scholarship that gives her time to study—she may select her next-most-attractive goal, whereas individuals with greater economic resources would probably attend their first-choice institution.

In addition to availability of resources, a second major constraint on actors' choices is social institutions (Friedman and Hechter, 1988). Institutional constraints such as rules, laws, ordinances, corporate policies, and religious doctrines limit actors' available choices. Using a rational choice approach, the sociologist Michael Hechter (1987) studied what happens when people "do their own thing" in an organization. Among other findings, Hechter concluded that organizations with large numbers of employees hire others (including security guards, managers, and inspectors) to control people's behavior so that they will not unduly pursue their propensity to maximize their own gain or pleasure, particularly at the expense of the organization.

FORMAL ORGANIZATIONS IN GLOBAL PERSPECTIVE

Over the past century, the number of formal organizations has increased dramatically in the United States and other industrialized nations. Previously, everyday life was centered in small, informal, primary groups, such as the family and the village. With the advent of industrialization and urbanization (as discussed in Chapter 1), people's lives became increasingly dominated by large, formal secondary organizations. A *formal organization,* you will recall, is a highly structured secondary group formed for the purpose of achieving specific goals in the most efficient manner. Formal organizations (such as corporations, schools, and government agencies) usually keep their basic structure for many years in order to meet their specific goals.

Types of Formal Organizations

We join some organizations voluntarily and others out of necessity. Sociologist Amitai Etzioni (1975) classified formal organizations into three categories—

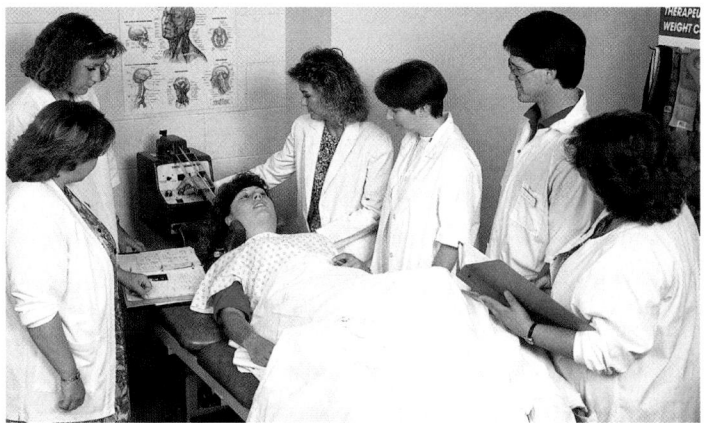

Normative organizations rely on volunteers to fulfill their goals; for example, Red Cross workers in New York City performed many functions in the aftermath of the 2001 terrorist attacks. Coercive organizations rely on involuntary recruitment; these prison inmates in Alabama are being resocialized in a total institution. Utilitarian organizations provide material rewards to participants; in teaching hospitals such as this one, medical students and patients hope that they may benefit from involvement within the organization.

normative, coercive, and utilitarian—based on the nature of membership in each.

Normative Organizations We voluntarily join *normative organizations* when we want to pursue some common interest or gain personal satisfaction or prestige from being a member. Political parties, ecological activist groups, religious organizations, parent–teacher associations, and college sororities and fraternities are examples of normative, or voluntary, associations.

Class, gender, and race are important determinants of a person's participation in a normative association. Class (socioeconomic status based on a person's education, occupation, and income) is the most significant predictor of whether a person will participate in mainstream normative organizations; membership costs may exclude some from joining. Those with higher socioeconomic status are more likely to be not only members but also active participants in these groups. Gender is also an important determinant. Half of the voluntary associations in the United States have all-female memberships; one-fifth are all male. However, all-male organizations usually have higher levels of prestige than all-female ones (Odendahl, 1990).

Throughout history, people of all racial–ethnic categories have participated in voluntary organiza-

tions, but the involvement of women in these groups has largely gone unrecognized. For example, African American women were actively involved in antislavery societies in the nineteenth century and in the civil rights movement in the twentieth century (see Scott, 1990). Other normative organizations focusing on civil rights, self-help, and philanthropic activities in which African American women and men have been involved include the National Association for the Advancement of Colored People (NAACP) and the Urban League. Similarly, Native American women have participated in the American Indian Movement, a group organized to fight problems ranging from police brutality to housing and employment discrimination (Feagin and Feagin, 2003). Mexican American women (as well as men) have held a wide range of leadership positions in La Raza Unida Party and the League of United Latin American Citizens, organizations oriented toward civic activities and protest against injustices (Amott and Matthaei, 1996). The activities of some of these organizations are discussed in more detail in Chapter 10 ("Race and Ethnicity").

One of the central characteristics of normative associations is that membership is voluntary—members function as unpaid workers. However, even when unpaid work has required education and experience

Concept Table 6.A	CHARACTERISTICS OF GROUPS AND ORGANIZATIONS	
Types of Social Groups	primary group	small, less specialized group in which members engage in face-to-face, emotion-based interaction over an extended period of time
	secondary group	larger, more specialized group in which members engage in more impersonal, goal-oriented relationships for a limited period of time
	ingroup	a group to which a person belongs and with which the person feels a sense of identity
	outgroup	a group to which a person does not belong and toward which the person may feel a sense of competitiveness or hostility
	reference group	a group that strongly influences a person's behavior and social attitudes, regardless of whether the person is actually a member
Group Size	dyad	a group composed of two members
	triad	a group composed of three members
	formal organization	a highly structured secondary group formed for the purpose of achieving specific goals
Types of Formal Organizations	normative	organizations we join voluntarily to pursue some common interest or gain personal satisfaction or prestige by joining
	coercive	associations that people are forced to join (total institutions such as boot camps and prisons are examples)
	utilitarian	organizations we join voluntarily when they can provide us with a material reward that we seek

similar to that required by wage labor, this work has been devalued. Conflict theorists argue that the devalued nature of unpaid work derives from the fact that women have historically done most of this kind of work.

Since participation in the organization is voluntary, few formal control mechanisms exist for enforcing norms on members. As a result, people tend to change affiliations rather frequently. They may change groups when personal objectives have been fulfilled or when other groups might better meet their individual needs.

Coercive Organizations Unlike normative organizations, people do not voluntarily become members of *coercive organizations*—associations that people are forced to join. Total institutions, such as boot camps, prisons, and some mental hospitals, are examples of coercive organizations. As discussed in Chapter 4, the assumed goal of total institutions is to resocialize people through incarceration. These environments are characterized by restrictive barriers (such as locks, bars, and security guards) that make it impossible for people to leave freely. When people leave without being officially dismissed, their exit is referred to as an "escape."

Utilitarian Organizations We voluntarily join *utilitarian organizations* when they can provide us with a material reward we seek. To make a living or earn a college degree, we must participate in organizations that can provide us these opportunities. Although we have some choice regarding where we work or attend school, utilitarian organizations are not always completely voluntary. For example, most people must continue to work even if the conditions of their employment are less than ideal. (Concept Table 6.A reviews types of groups, sizes of groups, and types of formal organizations.)

Bureaucracies

The bureaucratic model of organization remains the most universal organizational form in government, business, education, and religion. A *bureaucracy* **is**

an organizational model characterized by a hierarchy of authority, a clear division of labor, explicit rules and procedures, and impersonality in personnel matters.

When we think of a bureaucracy, we may think of "buck-passing," such as occurs when we are directed from one office to the next without receiving an answer to our question or a solution to our problem. We also may view a bureaucracy in terms of "red tape" because of the situations in which there is so much paperwork and so many incomprehensible rules that no one really understands what to do. However, bureaucracy was not originally intended to be this way; it was seen as a way to make organizations *more* productive and efficient.

Sociologist Max Weber (1968/1922) was interested in the historical trend toward bureaucratization that accelerated during the Industrial Revolution. To Weber, the bureaucracy was the most "rational" and efficient means of attaining organizational goals because it contributed to coordination and control. According to Weber, *rationality* **is the process by which traditional methods of social organization, characterized by informality and spontaneity, are gradually replaced by efficiently administered formal rules and procedures.** Bureaucracy can be seen in all aspects of our lives, from small colleges with perhaps a thousand students to multinational corporations employing many thousands of workers worldwide.

In his study of bureaucracies, Weber relied on an ideal-type analysis, which he adapted from the field of economics. An ***ideal type*** **is an abstract model that describes the recurring characteristics of some phenomenon** (such as bureaucracy). To develop this ideal type, Weber abstracted the most characteristic bureaucratic aspects of religious, educational, political, and business organizations. For example, to develop an ideal type for bureaucracy in higher education, you would need to include the relationships among governing bodies (such as boards of regents or trustees), administrators, faculty, staff, and students. You would also have to include the rules and policies that govern the school's activities (such as admissions criteria, grading policies, and graduation requirements). Although no two schools would have exactly the same criteria, the ideal-type constructs would be quite similar. Weber acknowledged that no existing organization would exactly fit his ideal type of bureaucracy (Blau and Meyer, 1987).

Ideal Characteristics of Bureaucracy Weber set forth several ideal-type characteristics of bureaucratic organizations. Although bureaucratic realities often differ from these ideal characteristics, Weber's model highlights the organizational efficiency and productivity that bureaucracies strive for. The model has been criticized for not taking into account informal networks (Roethlisberger and Dickson, 1939; Blau and Meyer, 1987). However, it is not surprising that these patterns are largely ignored in Weber's theory of bureaucracy; he constructed an idealized model that deliberately overlooked imperfections and unintended outcomes (Blau and Meyer, 1987).

Division of Labor Bureaucratic organizations are characterized by specialization, and each member has a specific status with certain assigned tasks to fulfill. This division of labor requires the employment of specialized experts who are responsible for the effective performance of their duties.

Hierarchy of Authority In the sense that Weber described hierarchy of authority, or chain of command, it includes each lower office being under the control and supervision of a higher one. Sociologist Charles Perrow (1986) has noted that all groups with a division of labor are hierarchically structured. Although the chain of command is not always followed, "in a crunch, the chain is there for those higher up to use it." Authority that is distributed hierarchically takes the form of a pyramid; those few individuals at the top have more power and exercise more control than do the many at the lower levels. Hierarchy inevitably influences social interaction. Those who are lower in the hierarchy report to (and often take orders from) those above them in the organizational pyramid. Persons at the upper levels are responsible not only for their own actions but also for those of the individuals they supervise. Hierarchy has been described as a graded system of interpersonal relationships, a society of unequals in which scarce rewards become even more scarce further down the hierarchy (Presthus, 1978).

Rules and Regulations Weber asserted that rules and regulations establish authority within an organization. These rules are typically standardized and provided to members in a written format. In theory, written rules and regulations offer clear-cut standards for determining satisfactory performance. They also provide continuity so that each new member does not have to reinvent the necessary rules and regulations.

Qualification-Based Employment Bureaucracies hire staff members and professional employees based

A ritual of college life is standing in line and waiting one's turn. Students are acutely aware of how academic bureaucracies operate. Although some students are critical of the "red tape" they experience, most realize that bureaucracy is necessary to enable the complex system of the university to operate smoothly.

on specific qualifications. Favoritism, family connections, and other subjective factors not relevant to organizational efficiency are not acceptable criteria for employment. Individual performance is evaluated against specific standards, and promotions are based on merit as spelled out in personnel policies.

Impersonality A detached approach should prevail toward clients so that personal feelings do not interfere with organizational decisions. Officials must interact with subordinates based on their official status, not on the officials' personal feelings.

Contemporary Applications of Weber's Theory

How well do Weber's theory of rationality and his ideal-type characteristics of bureaucracy withstand the test of time? Over 100 years later, many organizational theorists still modify and extend Weber's perspective. For example, the sociologist George Ritzer has applied Weber's theories to an examination of fast-food restaurants such as McDonald's. According to Ritzer, the process of "McDonaldization" has become a global phenomenon as four elements of rationality can be found in fast-food restaurants and other "speedy" or "jiffy" businesses (such as Sir Speedy Printing and Jiffy Lube). Ritzer (2000b: 433) identifies four dimensions of formal rationality—efficiency, predictability, emphasis on quantity rather than qual-ity, and control through nonhuman technologies—that are found in today's fast-food restaurants:

> Efficiency means the search for the best means to the end; in the fast-food restaurant, the drive-through window is a good example of heightening the efficiency of obtaining a meal. Predictability means a world of no surprises; the Big Mac in Los Angeles is indistinguishable from the one in New York; similarly, the one we consume tomorrow or next year will be just like the one we eat today. Rational systems tend to emphasize quantity, usually large quantities, rather than quality. The Big Mac is a good example of this emphasis on quantity rather than quality. Instead of the human qualities of a chef, fast-food restaurants rely on nonhuman technologies like unskilled cooks following detailed directions and assembly-line methods applied to the cooking and serving of food. Finally, such a formally rational system brings with it various irrationalities, most notably the demystification and dehumanization of the dining experience.

Informal Structure in Bureaucracies

When we look at an organizational chart, the official, formal structure of a bureaucracy is readily apparent. In practice, however, a bureaucracy has patterns of activities and interactions that cannot be accounted for by

How do people use the informal "grapevine" to spread information? Is this faster than the organization's official communications channels? Is it more or less accurate than official channels?

its organizational chart. These have been referred to as *bureaucracy's other face* (Page, 1946).

An organization's ***informal structure* is composed of those aspects of participants' day-to-day activities and interactions that ignore, bypass, or do not correspond with the official rules and procedures of the bureaucracy.** An example is an informal "grapevine" that spreads information (with varying degrees of accuracy) much faster than do official channels of communication, which tend to be slow and unresponsive. The informal structure has also been referred to as *work culture* because it includes the ideology and practices of workers on the job. It is the "informal, customary values and rules [that] mediate the formal authority structure of the workplace and distance workers from its impact" (Benson, 1983: 185). Workers create this work culture in order to confront, resist, or adapt to the constraints of their jobs, as well as to guide and interpret social relations on the job (Zavella, 1987). Today, computer networks and e-mail offer additional opportunities for workers to enhance or degrade their work culture. Some organizations have sought to control offensive communications so that workers will not be exposed to a hostile work environment brought about by colleagues, but such control has raised significant privacy issues (see Box 6.3).

Hawthorne Studies and Informal Networks

The existence of informal networks was first established by researchers in the Hawthorne studies. As you will recall from Chapter 2, the Hawthorne effect takes place when research subjects modify their behavior because they know that they are being observed. The Hawthorne studies also made social scientists aware of the effect of informal networks on workers' productivity.

In this particular study, researchers observed fourteen men in the "bank wiring room" who were responsible for making parts of switches for telephone equipment. Although management had offered financial incentives to encourage the men to work harder, the men persisted in working according to their own informal rules and sanctions. For example, they tended to work rapidly in the morning and ease off in the afternoon. They frequently stopped their own work to help another person who had fallen behind. When they got bored, they swapped tasks so that their work was more varied. They played games and made bets on horseraces and on baseball. Two competing cliques formed in the room, each with its own separate games and activities.

Why did these men insist on lagging behind even when they had been offered financial incentives to work harder? Perhaps they feared that the required productivity levels would increase if they showed that they could do more. Some of them may also have feared that they would lose their jobs if the work was finished more rapidly. One fact stood out in the study: The men's productivity level was clearly related to the pressure they received from other members of their informal networks. Those who worked too hard were called "speed kings" and "rate busters"; individuals who worked too slowly were referred to as "chiselers." Those who broke the informal norm against telling a supervisor about someone else's shortcomings were called "squealers." Negative sanctions in the form of "binging" (striking a person on the shoulder) made the workers want to adhere to the informal norms of their clique. Ultimately, the level of productivity was determined by the workers' informal networks, not by the levels set by management (Roethlisberger and Dickson, 1939; Blau and Meyer, 1987).

Positive and Negative Aspects of Informal Structure

Is informal structure good or bad? Should it be controlled or encouraged? Two schools of thought have emerged with regard to these questions. One approach emphasizes control (or eradication) of informal groups; the other suggests that they should be nurtured. Traditional management theories are based on the assumption that people are basically lazy and motivated by greed. Consequently, informal groups must be controlled (or eliminated) in order to ensure greater worker productivity. Proponents of this

Box 6.3 SOCIOLOGY AND SOCIAL POLICY

Computer Privacy in the Workplace

I think I'm getting paranoid. When I'm at work, I think somebody is watching me. When I send an e-mail or search the Web, I wonder who knows about it besides me. Don't get me wrong. I get all my work done first, but when I have some spare time, I may play a computer game or check out some Web site I'm interested in. I know that when I'm at work, my time belongs to the company, but somehow I still feel like it's an invasion of my privacy for some computer to monitor every single thing I do. I mean, I work for a company that makes cardboard boxes, not the CIA!

 —a student in one of the author's classes, expressing her irritation over computer surveillance at work

Do employers really have the right to monitor everything that their employees do on company-owned computers? Generally speaking, the answer is yes, and the practice is widespread. A recent survey found that about one-half of all U.S. companies monitor their employees' e-mail and more than 60 percent of employers monitor Internet connections (American Management Association, 2001).

Employers assert not only that they have the right to engage in such surveillance but also that it may be necessary for them to do so for their own protection. As for their right to do so, they note that they own the computer, pay for the Internet service, and pay the employee to spend his or her time on company business. As for it possibly being necessary for them to monitor their employees' computer usage, employers argue that they may be held legally responsible for harassing or discriminatory e-mail sent on company computers and that surveillance is the only way to protect against such liability. As a result, many employers take the position, according to the Privacy Foundation's Stephen Keating (qtd. in Agonafir, 2002), that "You leave your First Amendment [privacy] rights at the door when you work for a private employer. That's the way it has always been."

In most instances, courts have upheld monitoring even when the employees were not aware of the surveillance, and according to the American Management Association (2001), about one-third of all U.S. employers who engage in computer surveillance do not advise their employees that they are doing so. (At the time of this writing, Connecticut is the only state that requires that employees be notified of monitoring before it can legally occur.) In several high-profile cases, the firing of an employee based on his or her "inappropriate" e-mail has been upheld even when the employer's stated policy was *not* to monitor employee e-mail.

Yet there are valid arguments against computer surveillance as well, and invasion of a worker's privacy is certainly one of them. When an employee makes a personal phone call while at work, and it is a local or toll-free call, the employee usually has a reasonable expectation of privacy—a reasonable belief that neither fellow workers nor his or her employer is eavesdropping on that call. How about "snail mail"? An employee has a reasonable expectation of privacy that the employer will not steam open a personal letter addressed to the employee, read it, and reseal the envelope. Why should an e-mail exchange with friends or relatives not be equally private and protected? If the employer is going to read an employee's e-mail or track the person's Internet activities, shouldn't the employer at least have to make sure that its workers are aware of that policy?

Regarding the argument that personal e-mail wastes the employer's time, privacy advocates state that most employees who exchange personal e-mails or surf the Internet while at work are either doing so on their own time during breaks or at least at times when they have a "free" moment or two. If an employee's productivity falls below expected levels, shouldn't that be obvious to his or her boss without snooping on the employee's e-mail?

Finally, with regard to the employer possibly being held responsible for its employees' actions, Chief Judge Edith H. Jones (qtd. in Gordon, 2001) of the U.S. Fifth Circuit Court has observed that "It seems highly disproportionate to inflict a monitoring program that may invade thousands of people's privacy for the sake of exposing a handful of miscreants." The need to prevent a crime or to protect a company against potential liability must be balanced against each individual's privacy rights.

Ultimately, this balancing must be done by the legislature or the courts. Since the terrorist attacks of September 11, 2001, numerous analysts have called for greater Internet security, whereas other analysts have argued that U.S. citizens are being asked to relinquish too many of their rights in the interest of security. There are no easy answers to this pressing social policy issue, but it should remain a concern for all who live in a democratic society. Where do you stand on this topic? Why?

Women in firefighting are less likely than men to be included in informal networks and more likely to be harassed on the job. What steps could be taken to reduce women's lack of networks in traditionally male-dominated occupations?

© Thomas K. Wanstall/The Image Works

view cite the bank wiring room study as an example of the importance of controlling informal networks.

By contrast, the other school of thought asserts that people are capable of cooperation. Thus, organizations should foster informal groups that permit people to work more efficiently toward organizational goals. Chester Barnard (1938), an early organizational theorist, focused on the functional aspects of informal groups. He suggested that organizations are cooperative systems in which informal groups "oil the wheels" by providing understanding and motivation for participants. In other words, informal networks serve as a means of communication and cohesion among individuals, as well as protect the integrity of the individual (Barnard, 1938; Perrow, 1986).

The *human relations approach,* which is strongly influenced by Barnard's model, views informal networks as a type of adaptive behavior that workers engage in because they experience a lack of congruence between their own needs and the demands of the organization (Argyris, 1960). Organizations typically demand dependent, childlike behavior from their members and strive to thwart the members' ability to grow and achieve "maturity" (Argyris, 1962). At the same time, members have their own needs to grow and mature. Informal networks help workers fill this void. Large organizations would be unable to function without strong informal norms and relations among participants (Blau and Meyer, 1987).

More recent studies have confirmed the importance of informal networks in bureaucracies. Whereas some scholars have argued that women and people of color receive fairer treatment in larger bureaucracies than they do in smaller organizations, others have stressed that they may be categorically excluded from networks that are important for survival and advancement in the organization (Kanter, 1993/1977; South et al., 1982; Benokraitis and Feagin, 1995; Feagin, 1991). A woman firefighter in New York City describes how detrimental, and even hazardous, it is for workers to be excluded from such informal networks because of race/ethnicity, gender, or other attributes:

> I had sort of a "Pollyanna" view of how long it would take before women were really accepted in these nontraditional, very male-dominated jobs [such as being a firefighter]. One always thinks that once I and the other women prove that we can do the job well, people will just accept us and we'll all fit in. . . . [However,] I went to a firehouse where the men refused to eat with me. They would not talk to me. On one occasion my protective gear had been tampered with. It was always a big question as to whether in fact you have anyone there to back you up when you needed them. (PBS, 1992b)

White women and people of color who are employed in positions traditionally held by white men (such as firefighters, police officers, and factory workers) often experience categorical exclusion from the informal structure. Not only do they lack an informal network to "grease the wheels"; they may also be harassed and endangered by their co-workers. In sum, the informal structure is critical for employees—whether or not they are allowed to participate in it.

Figure 6.6   Characteristics and Effects of Bureaucracy

The very characteristics that define Weber's idealized bureaucracy can create or exacerbate the problems that many people associate with this type of organization. Can you apply this model to an organization with which you are familiar?

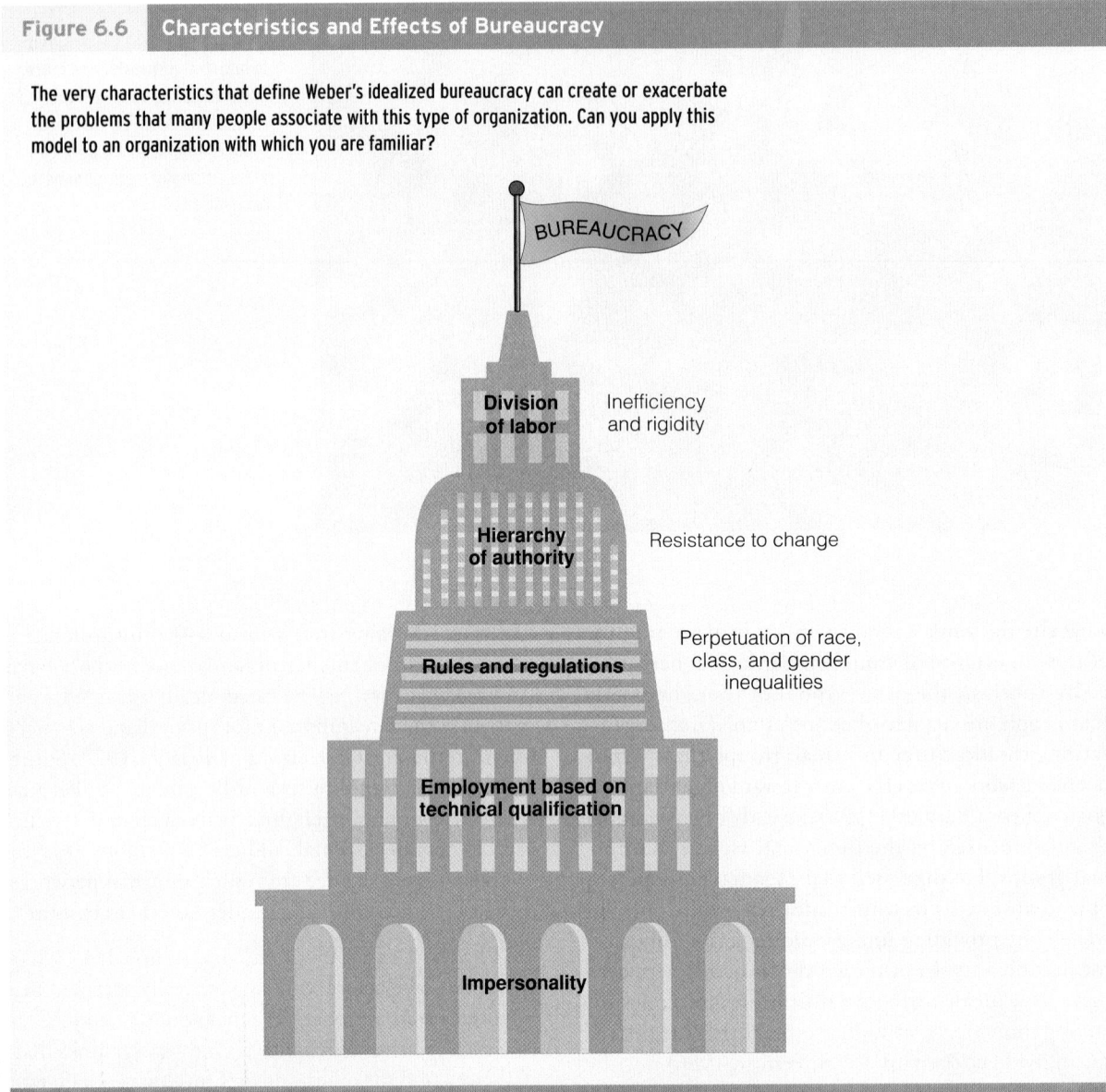

Shortcomings of Bureaucracies

As noted previously, Weber's description of bureaucracy was intentionally an abstract, idealized model of a rationally organized institution. However, the very characteristics that make up this "rational" model have a dark side that has frequently given this type of organization a bad name (see Figure 6.6). Three of the major problems of bureaucracies are (1) inefficiency and rigidity, (2) resistance to change, and (3) perpetuation of race, class, and gender inequalities (see Blau and Meyer, 1987).

Inefficiency and Rigidity   Bureaucracies experience inefficiency and rigidity at both the upper and lower levels of the organization. The self-protective behavior of officials at the top may render the organization inefficient. One type of self-protective behavior is the monopolization of information in order to maintain control over subordinates and outsiders. Information is a valuable commodity in organizations. Budgets and long-range plans are theoretically based on relevant information, and decisions are made based on the best available data. However, those in positions of authority guard information because it is a source of power for them—others cannot "second-guess" their decisions without access to relevant (and often "confidential") information (Blau and Meyer, 1987).

This information blockage is intensified by the hierarchical arrangement of officials and workers. When

Some corporations are resistant to change not only in how they are organized but also in how they expect their employees to dress and act. Is resistance to change always bad in organizations? Why or why not?

those at the top tend to use their power and authority to monopolize information, they also fail to communicate with workers at the lower levels. As a result, they are often unaware of potential problems facing the organization and of high levels of worker frustration. Meanwhile, those at the bottom of the structure hide their mistakes from supervisors, a practice that may ultimately result in disaster for the organization.

Policies and procedures also contribute to inefficiency and rigidity. Sociologists Peter M. Blau and Marshall W. Meyer (1987) have suggested that bureaucratic regulations are similar to bridges and buildings in that they are designed to withstand far greater stresses than they will ever experience. Accordingly, bureaucratic regulations are written in far greater detail than is necessary in order to ensure that almost all conceivable situations are covered. ***Goal displacement* occurs when the rules become an end in themselves rather than a means to an end, and organizational survival becomes more important than achievement of goals** (Merton, 1968). Administrators tend to overconform to the rules because their expertise is knowledge of the regulations, and they are paid to enforce them. Officials are most likely to emphasize rules and procedures when they fear that they may lose their jobs or a "spoils system" that benefits them. They also fear that if they bend the rules for one person, they may be accused of violating the norm of impersonality and engaging in favoritism (Blau and Meyer, 1987).

Inefficiency and rigidity occur at the lower levels of the organization as well. Workers often engage in *ritualism;* that is, they become most concerned with "going through the motions" and "following the rules." According to Robert Merton (1968), the term ***bureaucratic personality* describes those workers who are more concerned with following correct procedures than they are with getting the job done correctly.** Such workers are usually able to handle routine situations effectively but are frequently incapable of handling a unique problem or an emergency. Thorstein Veblen (1967/1899) used the term *trained incapacity* to characterize situations in which workers have become so highly specialized, or have been given such fragmented jobs to do, that they are unable to come up with creative solutions to problems. Workers who have reached this point also tend to experience bureaucratic alienation—they really do not care what is happening around them.

Sociologists have extensively analyzed the effects of bureaucracy on workers. Whereas some may become alienated, others may lose any identity apart from the organization. Sociologist William H. Whyte, Jr. (1957) coined the term *organization man* to identify an individual whose life is controlled by the corporation. C. Wright Mills (1959a) suggested that employees become "cheerful robots" when they are controlled by an organization. Other scholars have argued that most workers do not reach these extremes.

Resistance to Change Once bureaucratic organizations are created, they tend to resist change. Although the formal structure may help an organization survive during periods of crisis, it may also undermine creativity and profitability. In a study of 152 New York architectural firms, for example, the sociologist Judith R. Blau (1984) found that the same characteristics that made these firms the most likely to survive an economic recession also made them the least likely to become highly profitable. Although the large, well-established firms had the corporate clients and the resources to withstand difficult economic times, their greater overhead made them less inclined than smaller firms to take risks. Some of the risky ventures taken by the smaller firms led to greater profits, yet others led to bankruptcy.

Resistance to change occurs in all bureaucratic organizations, including schools, trade unions, businesses, and government agencies. This resistance not only makes bureaucracies virtually impossible to eliminate but also contributes to bureaucratic enlargement. Because of the assumed relationship between size and importance, officials tend to press for larger budgets and more staff and office space. To justify growth, administrators and managers must come up with more tasks for workers to perform. Ultimately, the outcome predicted by "Parkinson's Law" is fulfilled: "Work expands to fill the time available for its completion" (Parkinson, 1957).

Resistance to change may also lead to incompetence. Based on organizational policy, bureaucracies tend to promote people from within the organization. As a consequence, a person who performs satisfactorily in one position is promoted to a higher level in the organization. Eventually, people reach a level that is beyond their knowledge, experience, and capabilities. This process has been referred to as the "Peter Principle": People "rise to the level of their incompetence" (Peter and Hull, 1969: 25). However, neither the Peter Principle nor Parkinson's Law has been systematically tested by sociologists. Although each of these theories may contain some truth, if both were completely accurate, all bureaucracies would be run by incompetents.

Perpetuation of Racial, Class, and Gender Inequalities
Some bureaucracies perpetuate inequalities of race, class, and gender because this form of organizational structure creates a specific type of work or learning environment. This structure was typically created for middle- and upper-middle-class white men, who for many years were the predominant organizational participants.

Racial and Ethnic Inequalities In a study of 209 middle-class African Americans from more than a dozen cities, the sociologist Joe R. Feagin (1991) found that *entry* into dominant white bureaucratic organizations should not be equated with thorough *integration*. Instead, many have experienced an internal conflict between the bureaucratic ideals of equal

opportunity and fairness and the prevailing norms of discrimination and hostility that exist in many organizations. In another study of 125 African American bankers, interviewees repeatedly indicated that they were excluded from informal communications networks and were not "in on things" in their organization. Very few had mentors or anyone else highly placed within their organizations to take an interest in furthering their careers (Irons and Moore, 1985; Cose, 1993). Other research has found that people of color are more adversely affected than dominant-group members by hierarchical bureaucratic structures. These studies have been conducted in a number of organizational settings, ranging from medical schools to canning factories to corporations (see Kendall and Feagin, 1983; Zavella, 1987; and Collins, 1989). We will continue our discussion of the impact of organizations on people of color in Chapter 10 ("Race and Ethnicity").

Social Class Inequalities Like racial inequalities, social class divisions may be perpetuated in bureaucracies (Blau and Meyer, 1987). Sociologists have explored the impact of labor market conditions on the kinds of jobs and wages available to workers. The theory of a "dual labor market" has been developed to explain how social class distinctions are perpetuated through different types of employment. Middle- and upper-middle-class employees are more likely to work in industries characterized by higher wages, more job security, and opportunities for advancement. By contrast, poor and working-class employees work in industries characterized by low wages, lack of job security, and few opportunities for promotion. Even though the "dual economy" is not a perfect model for explaining class-based organizational inequalities, it does illuminate how individuals' employment not only reflects their position in the social class but also perpetuates it. Peter Blau and Marshall Meyer (1987: 160–161) conclude that "over time, then, organizational conditions reinforce social stratification. . . . Bureaucracies create profound differences in the life chances of the people working in them."

Gender Inequalities Gender inequalities are also perpetuated in bureaucracies. Sociologist Rosabeth Moss Kanter (1993/1977) analyzed how the power structure of bureaucratic hierarchies can negatively affect white women and people of color when they are underrepresented within an organization. In such cases, they tend to be more visible ("on display") and feel greater pressure not to make mistakes or stand out too much. They may also find it harder to gain credibility, particularly in management positions. As a result, they are more likely to feel isolated, to be excluded from informal networks, and to have less access to mentors and to power through alliances. By contrast, affluent white men are generally seen as being "one of the group." They find it easier to gain credibility, to join informal networks, and to find sponsorships (Kanter, 1993/1977).

Gender inequality in organizations has additional consequences. People who lack opportunities for integration and advancement tend to be pessimistic and to have lower self-esteem. They seek satisfaction away from work and are less likely to promote change at work. Believing that they have few opportunities, they resign themselves to staying put and surviving at that level. By contrast, those who enjoy full access to organizational opportunities tend to have high aspirations and high self-esteem. They feel loyalty to the organization and typically see their job as a means for mobility and growth.

In addition, women working in occupations and professions traditionally dominated by men have a greater likelihood of becoming the victims of sexual harassment. In recent years, much attention has been given to situations where men in high political office have been sexually involved with women who held subordinate positions in the governmental bureaucracy.

Reports of overt harassment of women have come from virtually all occupational areas, including the armed forces, coal mines, corporate offices, universities, and factories (Lott, 1994). In regard to the military, for example, one U.S. Army report stated that "sexual harassment exists throughout the Army, crossing gender, rank and racial lines" and that the military leadership was to blame (Shenon, 1997: A1). According to the report, most women in the military do not report incidents of sexual harassment because they fear retribution by those doing the harassing. Moreover, the report notes that "Victims are re-victimized by the system. . . . We are firmly convinced that leadership is the fundamental issue. Passive leadership has allowed sexual harassment to persist" (Shenon, 1997: A1).

Covert or subtle harassment is even more difficult to document. To study conductors and train operators for a major rapid transit system, Marian Swerdlow (1989) spent four years working as a conductor. She found that the men (who constituted 96 percent of these employees) did not engage in overt harassment. However, they routinely engaged in subtle harassment, such as sexualizing work relationships, exaggerating women's errors, depicting women's routine

competence as exceptional, and perpetuating a myth that women received special "preference" in hiring.

Frequently, complaints of harassment are disregarded or downplayed by employers and supervisors. When officials fail to pursue grievances, they give the impression of endorsing or condoning the harassment through their inaction (Janofsky, 1993). Although sexual harassment violates equal opportunity employment laws, enforcement is difficult given an employer's economic power to reward or punish a woman employee.

Many women do not report incidents of harassment because they fear that they may lose their job or suffer retaliation from their boss or co-workers. When harassment occurs in the workplace, women may simply quit their jobs. Those who experience harassment in college may drop a class, change majors, or transfer to another institution in order to escape the harasser. Women of color face double jeopardy in that they may experience both racial discrimination and sexual harassment (Benokraitis and Feagin, 1995). Recent studies have brought to light the confluence of gender, race, and class in situations of contrapower sexual harassment—a form of harassment in which the target of harassment has greater formal organizational power than the perpetrator (see Rospenda, Richman, and Nawyn, 1998; D. Smith, 1999). One example of contrapower sexual harassment would be a woman faculty member who is the target of harassment by a male student (D. Smith, 1999). Another example would be a minority woman employed in an administrative staff position at a university library who is continually harassed by a male part-time student worker (Rospenda, Richman, and Nawyn, 1998).

The bottom line is this: Sexual or racial harassment undermines the goal of equality in the workplace and in education. Consequently, organizations must be proactive in establishing guidelines for what is considered acceptable and unacceptable behavior. The elimination of sexual harassment must be viewed as desirable not only for moral, legal, and financial reasons but also for creating and maintaining a positive organizational atmosphere for all participants (see Riggs, Murrell, and Cutting, 1993). Ultimately, gender equality and racial equality in the workplace reflect the "ideal" characteristics of bureaucracy, as well as being guaranteed by law in the United States.

Bureaucracy and Oligarchy

Max Weber believed that bureaucracy was a necessary evil because it achieved coordination and control and thus efficiency in administration (Blau and Meyer,

1987). Sociologist Charles Perrow (1986) has suggested that bureaucracy produces a high standard of living for persons in industrialized countries because of its superiority as a "social tool over other forms of organization." Bureaucratic characteristics (such as a hierarchy of authority, a clear division of labor, explicit rules and procedures, and impersonality in personnel matters) may contribute to organizational efficiency, or they may produce gridlock.

However, Weber was not completely favorable toward bureaucracies. He believed such organizations stifle human initiative and creativity, thus producing an "iron cage." Bureaucracy also places an enormous amount of unregulated and often unperceived social power in the hands of a very few leaders. Such a situation is referred to as an oligarchy—the rule of the many by the few.

Why do a small number of leaders at the top make all the important organizational decisions? According to the German political sociologist Robert Michels (1949/1911), all organizations encounter the **iron law of oligarchy—the tendency to become a bureaucracy ruled by the few.** His central idea was that those who control bureaucracies not only wield power but also have an interest in retaining their power. In his research, Michels studied socialist parties and labor unions in Europe before World War I and concluded that even some of the most radical leaders of these organizations had a vested interest in clinging to their power. In this case, if the leaders lost their power positions, they would become manual laborers once again.

According to Michels, the hierarchical structures of bureaucracies and oligarchies go hand in hand. On the one hand, power may be concentrated in the hands of a few people because rank-and-file members must inevitably delegate a certain amount of decision-making authority to their leaders. Leaders then have access to information that other members do not have. They also have "clout," which they may use to protect their own interests, sometimes at the expense of the interests of others. On the other hand, oligarchy may result when individuals have certain outstanding qualities that make it possible for them to manage, if not control, others. The members choose to look to their leaders for direction; the leaders are strongly motivated to maintain the power and privileges that go with their leadership positions.

Is the iron law of oligarchy correct? Many scholars believe that Michels overstated his case. The leaders in most organizations do not have unlimited power. Divergent groups within a large-scale organization often compete for power, and informal networks can be used to "go behind the backs" of leaders. In addition, members routinely challenge, and sometimes remove, their leaders when they are not pleased with their actions.

ALTERNATIVE FORMS OF ORGANIZATION

Many organizations have sought new and innovative ways to organize work more efficiently than the traditional hierarchical model. In the early 1980s, there was a movement in the United States to *humanize bureaucracy*—to establish an organizational environment that develops rather than impedes human resources. More-humane bureaucracies are characterized by (1) less-rigid hierarchical structures and greater sharing of power and responsibility by all participants, (2) encouragement of participants to share their ideas and try new approaches to problem solving, and (3) efforts to reduce the number of people in dead-end jobs, train people in needed skills and competencies, and help people meet outside family responsibilities while still receiving equal treatment inside the organization (Kanter, 1983, 1985, 1993/ 1977). However, this movement may have been overshadowed by the perceived strengths of the Japanese system of macroplanning, which has included economic planning, teamwork, and job security (Hodson and Sullivan, 2002).

Organizational Structure in Japan

For several decades, the Japanese model of organization has been widely praised for its innovative structure. A number of social scientists and management specialists concluded that guaranteed lifetime employment and a teamwork approach to management were the major reasons that Japanese workers had been so productive since the end of World War II, when Japan's economy was in shambles. When the U.S. manufacturing sector weakened in the 1980s, the Japanese system was widely discussed as an alternative to the prevalent U.S. hierarchical organizational structure. Let's briefly compare the characteristics of large Japanese corporations with their U.S.-based counterparts.

Lifetime Employment Until recently, many large Japanese corporations guaranteed their workers permanent employment after an initial probationary

The Japanese model of organization has viewed the workplace as an extension of the family, and workers have been encouraged to join in shared activities such as the daily exercise program at this shipyard.

© Martin Rogers (Stone)/Getty Images

period. Thus, Japanese employees often remained with the same company for their entire career whereas their American counterparts often changed employers every few years. Likewise, Japanese employers in the past had an obligation not to "downsize" by laying off workers or cutting their wages. Unlike top managers in the United States who gave themselves pay raises, bonuses, and so on even when their companies were financially strapped and laying off workers, Japanese managers took pay cuts. When financial problems occurred, Japanese workers were frequently reassigned or retrained by their company.

According to advocates, the Japanese system encourages worker loyalty and a high level of productivity. Managers move through various parts of the organization and acquire technical knowledge about the workings of many aspects of the corporation, unlike their U.S. counterparts, who tend to become highly specialized (Sengoku, 1985). Clearly, permanent employment has its advantages for workers. They do not have to worry about losing their jobs or having their wages cut. Employers benefit by not having to compete with one another to keep workers.

Quality Circles Small work groups made up of about five to fifteen workers who meet regularly with one or two managers to discuss the group's performance and working conditions are known as *quality circles*. The purpose of this team approach to management is both to improve product quality and to lower product costs. Workers are motivated to save the corporation money because they, in turn, receive bonuses or higher wages for their efforts. Quality circles have

been praised for creating worker satisfaction, helping employees develop their potential, and improving productivity (Ishikawa, 1984). Because quality circles focus on both productivity and worker satisfaction, they (at least ideally) meet the needs of both the corporation and the workers. Although many Japanese corporations have used quality circles to achieve better productivity, many U.S.-based corporations have implemented automation in hopes of controlling the quality of their products.

Limitations Will the Japanese organizational structure continue to work over a period of time? Recently, the notion of lifetime employment has become problematic as an economic recession forced some Japanese factories to close and other companies ran out of subsidiaries willing to take on workers for "reassignment" (Sanger, 1994). Also, the sociologist Robert Coles (1979) has suggested that the Japanese model does not actually provide workers more control over the corporation. This model may give them more control over their own work, but production goals set by managers must still be met.

Although the possibility of implementing the Japanese model in U.S.-based corporations has been widely discussed, its large-scale acceptance is doubtful. Cultural traditions in Japan have focused on the importance of the group rather than the individual. Workers in the United States are not likely to embrace this idea because it directly conflicts with the value of individualism so strongly held by many in this country (Ouchi, 1981). U.S. workers are also unwilling to commit themselves to one corporation for their entire

Box 6.4 YOU CAN MAKE A DIFFERENCE

Creating Small Communities of Our Own Within Large Organizations

We had 47 hours to get [ready for] September 13th, when the bond markets reopened and there was one situation that our technology department had that they spent more time on than anything else. . . . It was getting into the systems, [figuring out] the IDs of the systems because so many people had died and the people that knew how to get into those systems and who knew the backup . . . and the second emergency guys were all gone. The way they got into those systems? They sat around [in a] group, they talked about where they went on vacation, what their kids' names were, what their wives' names were, what their dogs' names were, you know, every imaginable thing about their personal life. And the fact that we knew things about their personal life to break into those IDs and into the systems to be able to get the technology up and running before the bond market opened, I think [that] is probably the number one connection between technology, communication, and sociology.

—an executive describing how, following the September 11, 2001, terrorist attack on the World Trade Center, employees in companies that were devastated by the collapse of the towers and the loss of hundreds of co-workers found themselves needing to quickly reestablish communications networks and to get their operations up and running, not only for the profitability of their own corporations but also for the viability of the nation's economy (qtd. in Kelly and Stark, 2002: 2)

Prior to the attack, sociologist David Stark had been conducting an ethnographic study of the trading room of a major financial firm in the World Financial Center (which was adjacent to the World Trade Center). As a result, he was in a prime position to learn how organizations cope with unimaginable devastation. Like other sociologists who have studied disasters, Stark concluded that it was people, not technology, who saved the situation and made it possible to move forward. Although technology and risk-management plans are important, the microcommunities that people had established through their conversations ultimately saved the day. As the researchers noted, "No one said, 'Our technology saved us,' or, 'Our plan really worked.' To a person, they said, 'It was people'" (qtd. in Schwartz, 2002: WK5).

Perhaps one lesson that we can learn from the experiences of others and from Stark's research is that each of us, through our everyday communications with other people, may establish significant microcommunities within the larger organizations of which we are a part. These microcommunities of informal friendships and smaller organized groups such as campus clubs not only give us a sense of belonging but may also provide vital support systems in difficult times. Each of us may be able to make a difference at our colleges and in the lives of others by trying to create a sense of community among our friends and colleagues, by learning more about them—getting to know not only those people who are similar to us in outlook and world view but also those who may hold very different views on social issues and global concerns.

As we face the future, many analysts believe that lack of meaningful conversation among people from diverse backgrounds will produce a growing sense of "us" versus "them," a feeling of alienation and possibly even antagonism. But how can we improve communications with people we don't really know or understand? One analyst offers these suggestions (Drake, 2002):

- Strive for open and direct communication with the other person, yet make sure that you show respect for that person.
- Focus on the positive aspects of the other person, enhancing that person's self-esteem.
- Focus on the situation that you are discussing—don't let the conversation get personal.
- Share appropriate information with the other person, keeping both of you better informed.
- Look for "win-win" topics, where both of you benefit from the conversation.

Hopefully, none of us will ever face the same situation that the people in the World Financial Center did after the 2001 terrorist attacks—desperately needing to remember key details from our conversations with other people. However, each of us does face the smaller—but still significant—challenge of engaging in effective communication in the groups and organizations to which we belong in order to produce a better school, work, or community environment, and perhaps to enhance the quality of life for ourselves and other people.

work life. Moreover, men typically fare much better than women in Japanese corporations in which, due to patriarchy, many women have found themselves excluded from career-track positions (Brinton, 1989).

In spite of these limitations, many organizations in the United States are turning to a more participatory style of management. The incentives for such changes exist because many of the corporations experience greater worker satisfaction and higher productivity and profits (see Florida and Kenney, 1991).

ORGANIZATIONS IN THE FUTURE

What kind of "help wanted" ad might a company run in the future? One journalist has suggested that a want ad in the new century might read something like this: "WANTED: Bureaucracy basher, willing to challenge convention, assume big risks, and rewrite the accepted rules of industrial order" (Byrne, 1993: 76). Organizational theorists have suggested a *horizontal* model for corporations in which both hierarchy and functional or departmental boundaries would largely be eliminated. Seven key elements of the horizontal corporation have been suggested: (1) work would be organized around "core" processes, not tasks; (2) the hierarchy would be flattened; (3) teams would manage everything and be held accountable for measurable performance goals; (4) performance would be measured by customer satisfaction, not profits; (5) team performance would be rewarded; (6) employees would have regular contact with suppliers and customers; and (7) all employees would be trained in how to use available information effectively to make their own decisions (Byrne, 1993).

In the horizontal structure, a limited number of senior executives would still fill support roles (such as finance and human resources) while everyone else would work in multidisciplinary teams and perform core processes (such as product development or sales generation). Organizations would have fewer layers between company heads and the staffers responsible for any given process. Performance objectives would be related to the needs of customers; people would be rewarded not just for individual performance but for skills development and team performance. If such organizations become a reality, organizational charts of the twenty-first century will more closely resemble a pepperoni pizza, a shamrock, or an inverted pyramid than the traditional pyramid-shaped stack of boxes connected by lines.

Regardless of the organizational structure, in times of crisis we learn how very important *people* are to the functioning of the collectivity. We also become acutely aware of the significance of effective communication with other people (see Box 6.4).

What is the best organizational structure for the future? Of course, this question is difficult to answer because it requires the ability to predict economic, political, and social conditions. Nevertheless, we can make several observations. Ultimately, everyone has a stake in seeing that organizations operate in as humane a fashion as possible and that channels for opportunity are widely available to all people regardless of race, gender, or class. Workers and students alike can benefit from organizational environments that make it possible for people to explore their joint interests without fear of being harassed or being pitted against one another in a competitive struggle for advantage.

CHAPTER REVIEW

■ **How do sociologists distinguish among social groups, aggregates, and categories?**

Sociologists define a social group as a collection of two or more people who interact frequently, share a sense of belonging, and depend on one another. People who happen to be in the same place at the same time are considered an aggregate. Those who share a similar characteristic are considered a category. Neither aggregates nor categories are considered social groups.

■ **How do sociologists classify groups?**

Sociologists distinguish between primary and secondary groups. Primary groups are small and personal, and members engage in emotion-based interactions over an extended period. Secondary groups are larger and more specialized, and members have less personal and more formal, goal-oriented relationships. Sociologists also divide groups into ingroups, outgroups, and reference groups. Ingroups are groups to which we belong and with which we identify. Outgroups are groups we do not belong to or perhaps feel hostile toward. Reference groups are groups that strongly influence people's behavior whether or not they are actually members.

■ What is the significance of group size?

In small groups, all members know one another and interact simultaneously. In groups with more than three members, communication dynamics change and members tend to assume specialized tasks.

■ What are the major styles of leadership?

Leadership may be authoritarian, democratic, or laissez-faire. Authoritarian leaders make major decisions and assign tasks to individual members. Democratic leaders encourage discussion and collaborative decision making. Laissez-faire leaders are minimally involved and encourage members to make their own decisions.

■ What do experiments on conformity show us about the importance of groups?

Groups may have significant influence on members' values, attitudes, and behaviors. In order to maintain ties with a group, many members are willing to conform to norms established and reinforced by group members.

■ What are the strengths and weaknesses of bureaucracies?

A bureaucracy is a formal organization characterized by hierarchical authority, division of labor, explicit procedures, and impersonality. According to Max Weber, bureaucracy supplies a rational means of attaining organizational goals because it contributes to coordination and control. A bureaucracy also has an informal structure, which includes the daily activities and interactions that bypass the official rules and procedures. The informal structure may enhance productivity or may be counterproductive to the organization. A bureaucracy may be inefficient, resistant to change, and a vehicle for perpetuating class, race, and gender inequalities.

■ Are all formal organizations based on oligarchy?

An oligarchy is the rule of the many by the few. In bureaucracies with an oligarchical structure, those in control have not only power but also a great interest in maintaining that power. However, recent trends such as the Japanese model of organization emphasize a teamwork approach to management rather than an oligarchical structure.

KEY TERMS

QUESTIONS FOR CRITICAL THINKING

1. Who might be more likely to conform in a bureaucracy, those with power or those wanting more power?
2. Although there has been much discussion recently concerning what is and what is not sexual harassment, it has been difficult to reach a clear consensus on what behaviors and actions are acceptable. What are some specific ways that both women and men can avoid contributing to an atmosphere of sexual harassment in organizations? Consider team relationships, management and mentor relationships, promotion policies, attitudes, behavior, dress and presentation, and after-work socializing.
3. Do the insights gained from Milgram's research on obedience outweigh the elements of deception and stress that were forced on its subjects?
4. If you were forming a company based on humane organizational principles, would you base the promotional policies on merit and performance or on affirmative action goals?

RESOURCES ON THE INTERNET

Chapter-Related Web Sites

The following Web sites have been selected for their relevance to the topics in this chapter. These sites are among the more stable, but please note that Web site addresses change frequently. For an updated list of chapter-related

Web sites with URL links, please visit the *Sociology in Our Times* Web site (**www.wadsworth.com/KendallSIOT**).

Historical Background of Organizational Behavior

http://web.cba.neu.edu/~ewertheim/introd/history.htm

Created by Edward G. Wertheim of the College of Business Administration at Northeastern University, this site provides a good introduction to the study of organizational behavior and addresses many of the major works in the field.

McDonaldization

http://www.mcdonaldization.com/main.shtml

The objective of this site is to educate the public about the McDonaldization of America. In addition to presenting information on George Ritzer's text *The McDonaldization of Society,* the site features an introduction to the concept of McDonaldization, related topics, articles, news, and interesting links.

ONLINE STUDY AND RESEARCH TOOLS

Accompanying this text are many *free* powerful online study tools that will help you master the material in this chapter, help increase your depth of understanding, and help you make the grade!

SocCoach CD-ROM

Use the SocCoach CD-ROM enclosed with this text to help you formulate a customized study plan for this chapter. After you take the Diagnostic Quiz, SocCoach will generate a customized study plan just for you! It will identify sections of the chapter that you should review and will provide videos, charts, graphs, and excerpts from the text to supplement your studies and enhance your understanding. You'll also find fun, interactive activities such as Virtual Explorations and Map the Stats to apply what you've learned and stretch your sociological imagination.

The Companion Web Site for Sociology in Our Times, *Fifth Edition*

www.wadsworth.com/KendallSIOT

Gain an even better grasp on this chapter by going to the companion Web site to take one of the Tutorial Quizzes, use the Flash Cards to master key terms, or check out the many other study aids you'll find there. You'll also find special features such as GSS Data and Census 2000 information that'll put data and resources at your fingertips to help you with that special project or help you as you do some research on your own.

In this chapter, when you see the icon on the left, it alerts you to a specific exercise found in *Wadsworth's Sociology Online Resources and Writing Companion.* This valuable guide shows you how to use Wadsworth's exclusive online resources—*InfoTrac College Edition,* the *Opposing Viewpoints Resource Center,* and *MicroCase Online*—to assist you in your study of sociology and to build essential research and writing skills.

CHAPTER 7

Deviance and Crime

FELIX (a sociologist): When did you join the Diamonds?

FLACO (a gang member): Four and half years ago. I was a freshman in high school.

FELIX: And what were the reasons for joining?

FLACO: Well, let me see. More or less the reason was because I was already hanging with them since I was small, so I was in the neighborhood. And I had a couple of real good buddies. And I had protection by them.

FELIX: What exactly did your friends say to you?

FLACO: Not much. It wasn't like they were forcing me. They just told me to turn. It was no big deal.

FELIX: And what did you do after this?

FLACO: I turned. I believed them. . . .

FELIX: You mentioned earlier that your friends gave you protection. What's protection?

FLACO: Protection? I had backup. You have to have Folks on your side, 'cause, if you're not in one gang, you're not in another gang, and they always be asking if you got a sign, and if you're on this side, and you get rolled on anyway. And you have nobody on your side, so you're still gonna pick a favorite of your Folks or People. I don't know, kids could stay neutron, but it's very hard, and it's very rarely you see kids that are neutrons. And the ones that are neutrons—they still got a favorite; they like Folks better, or People. And the Folks or the People can't let you see them too much on the other side. Like, if you're a neutron, and you're in favor of Folks, if the Folks see you in People's neighborhoods, they're gonna think something about it. And then that's when you got problems.

FELIX: So one of the reasons why you joined was for protection, for backup?

FLACO: Yeah. I liked chilling out with them, too.

FELIX: What did you like about hanging out with them?

FLACO: Everything's fun. Sometimes we would ditch school or take a day off. We just go, we get a couple of cases of beer, we drink those, and then we'll go up to the top of the roofs and look over. Police come, and they chase us down. We just had a lot of fun, do a lot of kinds of things.

—Sociologist Felix M. Padilla (1993: 80-81) interviewing a gang member as part of Padilla's ethnographic research on why people join gangs

Mark Richards/PhotoEdit

■ Members of the group known as the "Grape Street Gang" typify how gang members use items of clothing such as the bandannas shown here and gang signs made with their hands to assert their identity with the group and solidarity with one another. Some people might view this conduct as deviant behavior, whereas many gang members view it as an act of conformity.

Sociologists and criminologists typically define a *gang* as a group of people, usually young, who band together for purposes generally considered to be deviant or criminal by the larger society. Throughout the past century, gang behavior has been of special interest to sociologists (see Puffer, 1912), who generally agree that youth gangs can be found in many settings and among all racial and ethnic categories. The Office of Juvenile Justice and Delinquency Prevention (2002) estimates that there are over 24,500 gangs with about 772,500 members in the United States.

As unusual as it initially may sound, some important similarities exist between youth gangs and peer cliques, which are typically viewed as conforming to most social norms. At the most basic level, *cliques* are friendship circles, whose members identify one another as mutually connected (Adler and Adler, 1998). However, cliques are much more complex than this definition suggests. According to the sociologists Patricia A. Adler and Peter Adler (1998: 56), cliques "have a hierarchical structure, being dominated by leaders, and are exclusive in nature, so that not all individuals who desire membership are accepted." Moreover, sociologists have found that cliques function as "bodies of power" in schools by "incorporating the most popular individuals, offering the most exciting social lives, and commanding the most interest and attention from classmates" (Adler and Adler, 1998: 56).

In this chapter, we look at the relationship among conformity, deviance, and crime; even in times of national crisis and war, "everyday" deviance and crime occur as usual. People do not stop activities that might be viewed by others—or by law enforcement officials—as violating social norms. An example is gang behavior, which is used in this chapter as an example of deviant behavior. For individuals who find a source of identity, self-worth, and a feeling of protection by virtue of gang membership, no radical change occurs in daily life even as events around them may change. Youth gangs have been present in the United States for many years because they meet perceived needs of members. Some gangs may be thought of as being very similar to youth cliques whereas other gangs engage in activities that constitute crime. Before reading on, take the quiz on peer cliques, youth gangs, and deviance in Box 7.1.

QUESTIONS AND ISSUES

Chapter Focus Question: What do studies of peer cliques and youth gangs tell us about deviance?

What is deviant behavior?

When is deviance considered a crime?

What are the major theoretical perspectives on deviance?

How are crimes classified?

How does the criminal justice system deal with crime?

WHAT IS DEVIANCE?

Deviance **is any behavior, belief, or condition that violates significant social norms in the society or group in which it occurs.** We are most familiar with *behavioral* deviance, based on a person's intentional or inadvertent actions. For example, a person may engage in intentional deviance by drinking too much or robbing a bank, or in inadvertent deviance by losing money in a Las Vegas casino or laughing at a funeral.

Although we usually think of deviance as a type of behavior, people may be regarded as deviant if they express a radical or unusual *belief system*. Members of cults (such as Moonies and satanists) and of far-right-wing or far-left-wing political groups may be considered deviant when their religious or political beliefs become known to people with more-conventional cultural beliefs. However, individuals who are considered to be "deviant" by one category of people may be seen as conformists by another group. For example, adolescents in some peer cliques and youth gangs may shun mainstream cultural beliefs and values but routinely conform to subcultural codes of dress, attitude (such as defiant individualism), and behavior

(Jankowski, 1991). Those who think of themselves as "Goths" may wear black trench coats, paint their fingernails black, and listen to countercultural musicians, as one journalist explains:

> The current manifestation of Gothic culture began with the British punk scene in the early '80s. Bands like Bauhaus . . . created the atmospheric doom-rock sound. A clothing style evolved that was part Johnny Rotten, part Anne Rice and all black. Acolytes sometimes took an interest (purely academic) in subjects such as Satanism and blood drinking, which ensured that this was one rebellion that would never enter the mainstream. In the '90s, shock rockers like [Marilyn] Manson appropriated the image and blurred the lines—until any shaggy-haired, trench-coat-wearing teen could be considered a Goth by his peers. (Taylor, 1999: 45)

As this description suggests, we often interpret other people's belief systems based on how they look. In Littleton, Colorado, many Columbine High School students and teachers had previously identified the two youthful killers and some members of their clique as

Box 7.1 SOCIOLOGY AND EVERYDAY LIFE

How Much Do You Know About Peer Cliques, Youth Gangs, and Deviance?

True	False		
T	F	1.	According to some sociologists, deviance may serve a useful purpose in society.
T	F	2.	Peer cliques on high school campuses have few similarities to youth gangs.
T	F	3.	Most people join gangs to escape from broken homes caused by divorce or the death of a parent.
T	F	4.	Juvenile gangs are an urban problem; few rural areas have problems with gangs.
T	F	5.	Street crime has a much higher economic cost to society than crimes committed in executive suites or by government officials.
T	F	6.	Persons aged fifteen to twenty-four account for almost half of all arrests for property crimes such as burglary, larceny, arson, and vandalism.
T	F	7.	Virtually all gangs are made up of persons from lower-income families.
T	F	8.	Recent studies have shown that peer cliques have become increasingly important to adolescents over the past two decades.
T	F	9.	Many gang members continue their membership into adulthood.
T	F	10.	Gangs are an international problem.

Answers on page 212.

Goths because of their appearance. However, after widespread media coverage of the "Goth link" to the Littleton tragedy, other people in the "Gothic community" stated that they were appalled by the killings, as well as the inference that the murderers had belonged to their culture (Taylor, 1999).

In addition to their behavior and beliefs, individuals may also be regarded as deviant because they possess a specific *condition* or characteristic. A wide range of conditions have been identified as "deviant," including being obese (Degher and Hughes, 1991; Goode, 1996) and having AIDS (Weitz, 1993). For example, research by the sociologist Rose Weitz (1993) has shown that persons with AIDS live with a stigma that affects their relationships with other people, including family members, friends, lovers, colleagues, and health care workers. Chapter 4 defines a *stigma* as any physical or social attribute or sign which so devalues a person's social identity that it disqualifies the person from full social acceptance (Goffman, 1963b). Based on this definition, the stigmatized person has a "spoiled identity" as a result of being negatively evaluated by others (Goffman, 1963b). To avoid or reduce

stigma, many people seek to conceal the characteristic or condition that might lead to stigmatization.

Who Defines Deviance?

Are some behaviors, beliefs, and conditions inherently deviant? In commonsense thinking, deviance is often viewed as inherent in certain kinds of behavior or people. For sociologists, however, deviance is a formal property of social situations and social structure. As the sociologist Kai T. Erikson (1964: 11) explains,

> Deviance is not a property inherent in certain forms of behavior; it is a property conferred upon these forms by the audiences which directly or indirectly witness them. The critical variable in the study of deviance, then, is the social audience rather than the individual actor, since it is the audience which eventually determines whether or not any episode of behavior or any class of episodes is labeled deviant.

Based on this statement, we can conclude that deviance is *relative*—that is, an act becomes deviant

Box 7.1 SOCIOLOGY AND EVERYDAY LIFE

Answers to the Sociology Quiz on Peer Cliques, Youth Gangs, and Deviance

1. **True.** From Durkheim to contemporary functionalists, theorists have regarded some degree of deviance as functional for societies.

2. **False.** Many social scientists believe that there are striking similarities between adolescent cliques and youth gangs, including the demands that are placed on members in each category to conform to group norms pertaining to behavior, appearance, and other people with whom one is allowed to associate.

3. **False.** Recent studies have found that people join gangs for a variety of reasons, including the desire to gain access to money, recreation, and protection.

4. **False.** Gangs are frequently thought of as an urban problem because central-city gangs organized around drug dealing have become prominent in recent years; however, gangs are found in rural areas throughout the country as well.

5. **False.** Although street crime—such as assault and robbery—often has a greater psychological cost, crimes committed by persons in top positions in business (such as accounting and tax fraud) or government (including the Pentagon) have a far greater economic cost, especially for U.S. taxpayers.

6. **True.** This age group accounts for about 46 percent of all arrests for property crimes, the most common crimes committed in the United States.

7. **False.** Many gang members do come from lower-income families; however, some young people from middle- and upper-middle-income backgrounds create or join gangs for status or protection against other gangs.

8. **True.** As more youths grow up in single-parent households or in households where both parents are employed, many adolescents have turned to members of their peer cliques to satisfy their emotional needs and to gain information.

9. **True.** Many gang members do continue their membership into adulthood, and some move into adult criminal activity. However, others cease to participate in gang activities.

10. **True.** Gangs are found in nations around the world. In countries such as Japan, youth gangs are often points of entry for adult crime organizations.

Sources: Based on Adler and Adler, 1998, 2003; Inciardi, Horowitz, and Pottieger, 1993; and Jankowski, 1991.

when it is socially defined as such. Definitions of deviance vary widely from place to place, from time to time, and from group to group. Today, for example, some women wear blue jeans and very short hair to college classes; some men wear an earring and long hair. In the past, such looks violated established dress codes in many schools, and administrators probably would have asked these students to change their appearance or leave school.

Deviant behavior also varies in its *degree of seriousness,* ranging from mild transgressions of folkways, to more serious infringements of mores, to quite serious violations of the law. Have you kept a library book past its due date or cut classes? If so, you have vio-

lated folkways. Others probably view your infraction as relatively minor; at most, you might have to pay a fine or receive a lower grade. Violations of mores—such as falsifying a college application or cheating on an examination—are viewed as more serious infractions and are punishable by stronger sanctions, such as academic probation or expulsion. Some forms of deviant behavior violate the criminal law, which defines the behaviors that society labels as criminal. A **crime is a behavior that violates criminal law and is punishable with fines, jail terms, and/or other negative sanctions.** Crimes range from minor offenses (such as traffic violations) to major offenses (such as murder). A subcategory, ***juvenile delin-***

WHAT IS DEVIANCE? 213

Although most people think of a high school clique as being far different from a gang, patterns of inclusion and exclusion operate similarly in both groups. The three young women talking to one another here are obviously good friends, but their clique does not include the other young woman standing nearby.

Spencer Grant/PhotoEdit

quency, **refers to a violation of law or the commission of a status offense by young people.** Note that the legal concept of juvenile delinquency includes not only crimes but also status offenses, which are illegal only when committed by younger people (such as cutting school or running away from home).

What Is Social Control?

Societies not only have norms and laws that govern acceptable behavior; they also have various mechanisms to control people's behavior. ***Social control* refers to the systematic practices that social groups develop in order to encourage conformity to norms, rules, and laws and to discourage deviance.** Social control mechanisms may be either internal or external. Internal social control takes place through the socialization process: Individuals *internalize* societal norms and values that prescribe how people should behave and then follow those norms and values in their everyday lives. By contrast, external social control involves the use of negative sanctions that proscribe certain behaviors and set forth the punishments for rule breakers and nonconformists. In contemporary societies, the criminal justice system, which includes the police, the courts, and the prisons, is the primary mechanism of external social control.

For some social analysts, maintaining social control is critical for the stability of society. Political scientist James Q. Wilson (1996: xv) uses the image of broken windows to explain how neighborhoods may decay into disorder and crime if no one maintains social control:

If a factory or office window is broken, passersby observing it will conclude that no one cares and no one is in charge. In time, a few will begin throwing rocks to break more windows. Soon all the windows will be broken, and now passersby will think that, not only is no one in charge of the building, no one is in charge of the street on which it faces. Only the young, the criminal, or the foolhardy have any business on an unprotected avenue, and so more and more citizens will abandon the street to those they assume prowl it. Small disorders lead to larger and larger ones, and perhaps even to crime.

But if most actions deemed deviant do little or no direct harm to society or its members, why is social control so important to groups and societies? Why are some actions punished whereas others are not? Why are some people who engage in deviance punished and others not? Why is the same belief or action punished in one group or society and not in another? These questions pose interesting theoretical concerns and research topics for sociologists and criminologists who examine issues pertaining to law, social control, and the criminal justice system. ***Criminology* is the systematic study of crime and the criminal justice system, including the police, courts, and prisons.**

The primary interest of sociologists and criminologists is not questions of how crime and criminals can best be controlled. Instead, as the questions set forth above suggest, sociologists focus on social control as a social product. Likewise, when sociologists study deviance, they do not judge certain kinds of behavior or people as being "good" or "bad." Instead, they attempt

External social control involves the use of negative sanctions by the police and other law enforcement officials to encourage conformity to laws and to discourage deviant behavior. How does this differ from internal social control?

© 2003 AP/Wide World Photos

to determine what types of behavior are defined as deviant, who does the defining, how and why people become deviants, and how society deals with deviants. Although sociologists have developed a number of theories to explain deviance and crime, no one perspective is a comprehensive explanation of all deviance. Each theory provides a different lens through which we can examine aspects of deviant behavior.

FUNCTIONALIST PERSPECTIVES ON DEVIANCE

As we have seen in previous chapters, functionalists focus on societal stability and the ways in which various parts of society contribute to the whole. According to functionalists, a certain amount of deviance contributes to the smooth functioning of society.

What Causes Deviance, and Why Is It Functional for Society?

Sociologist Emile Durkheim believed that deviance is rooted in societal factors such as rapid social change and lack of social integration among people. As you will recall, Durkheim attributed the social upheaval he saw at the end of the nineteenth century to the

shift from mechanical to organic solidarity, which was brought about by rapid industrialization and urbanization. Although many people continued to follow the dominant morals (norms, values, and laws) as best they could, rapid social change contributed to *anomie*—a social condition in which people experience a sense of futility because social norms are weak, absent, or conflicting. According to Durkheim, as social integration (bonding and community involvement) decreased, deviance and crime increased. However, from his perspective, this was not altogether bad because he believed that deviance has positive social functions in terms of its consequences. For Durkheim (1964a/1895), deviance is a natural and inevitable part of all societies. Likewise, contemporary functionalist theorists suggest that deviance is universal because it serves three important functions:

1. *Deviance clarifies rules.* By punishing deviant behavior, society reaffirms its commitment to the rules and clarifies their meaning.
2. *Deviance unites a group.* When deviant behavior is seen as a threat to group solidarity and people unite in opposition to that behavior, their loyalties to society are reinforced.
3. *Deviance promotes social change.* Deviants may violate norms in order to get them changed. For example, acts of *civil disobedience*—including lunch counter sit-ins and bus boycotts—were used to protest and eventually correct injustices such as segregated buses and lunch counters in

Table 7.1 MERTON'S STRAIN THEORY OF DEVIANCE

MODE OF ADAPTATION	METHOD OF ADAPTATION	SEEKS CULTURE'S GOALS	FOLLOWS CULTURE'S APPROVED WAYS
Conformity	Accepts culturally approved goals; pursues them through culturally approved means	Yes	Yes
Innovation	Accepts culturally approved goals; adopts disapproved means of achieving them	Yes	No
Ritualism	Abandons society's goals but continues to conform to approved means	No	Yes
Retreatism	Abandons both approved goals and the approved means to achieve them	No	No
Rebellion	Challenges both the approved goals and the approved means to achieve them	No—seeks to replace	No—seeks to replace

the South. Students periodically stage campus demonstrations to call attention to perceived injustices, such as a tuition increase or the firing of a popular professor.

Functionalists acknowledge that deviance may also be dysfunctional for society. If too many people violate the norms, everyday existence may become unpredictable, chaotic, and even violent. If even a few people commit acts that are so violent that they threaten the survival of a society, then deviant acts move into the realm of the criminal and even the unthinkable. Of course, the examples that stand out in everyone's mind are the terrorist attacks on the United States and the fear that ensued as a result.

Although there are a wide array of contemporary functionalist theories regarding deviance and crime, many of these theories focus on social structure. For this reason, the first theory we will discuss is referred to as a structural functionalist approach. It describes the relationship between the society's economic structure and why people might engage in various forms of deviant behavior.

Strain Theory: Goals and Means to Achieve Them

Modifying Durkheim's (1964a/1895) concept of *anomie,* the sociologist Robert Merton (1938, 1968) developed strain theory. According to **strain theory, people feel strain when they are exposed to cultural goals that they are unable to obtain because they do not have access to culturally approved means of achieving those goals.** The goals may be material possessions and money; the approved means may include an education and jobs. When denied legitimate access to these goals, some people seek access through deviant means. Strain theory is often used to explain deviance by people from lower-income neighborhoods, who are typically depicted as being left out of the economic mainstream, feeling hopeless, and sometimes turning their anger and rage toward other people or things. In this way, the structure of the society and the economic status of the people involved are major factors in why some people commit deviant and/or criminal acts.

Merton identified five ways in which people adapt to cultural goals and approved ways of achieving them: conformity, innovation, ritualism, retreatism, and rebellion (see Table 7.1). According to Merton, *conformity* occurs when people accept culturally approved goals and pursue them through approved means. Persons who want to achieve success through conformity work hard, save their money, and so on. Even people who find that they are blocked from achieving a high level of education or a lucrative career may take a lower-paying job and attend school part time, join the military, or seek alternative (but legal) avenues, such as playing the lottery, to "strike it rich."

Conformity is also crucial for members of middle- and upper-class teen cliques, who often gather in small groups to share activities and confidences. Some

Table 7.2 DEVIANTS OR CONFORMISTS? HIGH SCHOOL "UNIFORMS"	
Jocks, Cheerleaders, and the "In" Crowd	Clothes from Tommy Hilfiger, Abercrombie & Fitch, the Gap, Old Navy, "letter" jackets, white baseball caps. Short hair for boys; long, "frosted," or "streaked" hair for girls.
"Hicks" or "Kickers"	Cowboy boots, big hats, and oversize belt buckles (regional).
"Surfers"	Sun-bleached hair, tropical or other light clothing (regional).
"Skaters" (as in Skateboards)	Grunge look (regional).
"Freaks," "Punks," and "Ravers"	Spiky and/or brightly colored hair (Kool Aid used in some cases), black clothing, extensive body piercing all over face and body, numerous tattoos.
"Outcasts"–Boys	Makeup, including face powder and black eyeliner; dress in feminine ways or in black leather and chains; black T-shirts, trench coats, Doc Martens.
"Outcasts"–Girls	Black nail polish and lipstick; black leather, chains; visible tattoos; Doc Martens or high platform shoes.

Sources: Based on Cohen, 1999; *Newsweek*, 1999.

youths are members of a variety of cliques, and peer approval is of crucial significance to them—being one of the "in" crowd, not an "outcast" or a "loner," is a significant goal for many teenagers. In the aftermath of the recent school shootings, for example, numerous journalists trekked to school campuses to report that athletes ("jocks"), cheerleaders, and other "popular" students enforce the social code at high schools (Adler, 1999; Cohen, 1999). One report suggested that "from who's in which clique to where you sit in the cafeteria, every day [high school] can be a struggle to fit in" (Adler, 1999: 56). A comparison of appearance norms held by some teen peer groups and some juvenile gang members shows that what constitutes conformity within one group may be viewed as deviance within another (see Table 7.2).

Merton classified the remaining four types of adaptation as deviance. *Innovation* occurs when people accept society's goals but adopt disapproved means of achieving them. Innovations for acquiring material possessions or money cover a wide variety of illegal activities, including theft and drug dealing. For example, the journalist Nathan McCall (1994: 6) describes how his innovative behavior took the form of hustling when he was a gang member:

> Hustling seemed like the thing to do. With Shell Shock [a fellow gang member] as my main partner, I tried every nickel-and-dime hustle I came across, focusing mainly on stealing. We stole everything that wasn't nailed down, from schoolbooks, which we sold at half price, to wallets,

which we lifted from guys' rear pockets. We even stole gifts from under the Christmas tree of a girl we visited. . . .

Although Merton primarily focused on deviance committed by persons from lower-income backgrounds, innovation is also used by middle- and upper-income people. For example, affluent adults may cheat on their income taxes or embezzle money from their employer to maintain an expensive lifestyle. Students from middle- and high-income families may cheat on exams in hopes of receiving higher grades and ensuring their admission to a top college.

Merton's third mode of adaptation is *ritualism*, which occurs when people give up on societal goals but still adhere to the socially approved means of achieving them. Ritualism is the opposite of innovation; persons who cannot obtain expensive material possessions or wealth may nevertheless seek to maintain the respect of others by being a "hard worker" or "good citizen." *Retreatism* occurs when people abandon both the approved goals and the approved means of achieving them. Merton included persons such as skid-row alcoholics and drug addicts in this category; however, not all retreatists are destitute. Some may be middle- or upper-income individuals who see themselves as rejecting the conventional trappings of success or the means necessary to acquire them.

The fifth type of adaptation, *rebellion*, occurs when people challenge both the approved goals and the approved means for achieving them, and advocate an alternative set of goals or means. To achieve their

Sociologist Robert Merton identified five ways in which people adapt to cultural goals and approved ways of achieving them. Consider The Scary Guy (now his legal name), who is covered from head to foot with tattoos. Which of Merton's modes of adaptation might best explain The Scary Guy's views on social life?

alternative goals, rebels may use violence (such as vandalism or rioting) or nonviolent tactics (such as civil disobedience).

Opportunity Theory: Access to Illegitimate Opportunities

Expanding on Merton's strain theory, sociologists Richard Cloward and Lloyd Ohlin (1960) suggested that for deviance to occur, people must have access to **illegitimate opportunity structures—circumstances that provide an opportunity for people to acquire through illegitimate activities what they cannot achieve through legitimate channels.** For example, gang members may have insufficient legitimate means to achieve conventional goals of status and wealth but have illegitimate opportunity structures—such as theft, drug dealing, or robbery—through which they can achieve these goals. In his study of the "Diamonds," a Chicago street gang whose members are second-generation Puerto Rican youths, sociologist Felix M. Padilla (1993) found that gang membership was linked to the members' belief that they might reach their aspirations by transforming the gang into a business enterprise. Coco, one of the Diamonds, explains the importance of sticking together in the gang's income-generating business organization:

We are a group, a community, a family—we have to learn to live together. If we separate, we will never have a chance. We need each other even to make sure that we have a spot for selling our supply [of drugs]. You know, there is people around here, like some opposition, that want to take over your *negocio* [business]. And they think that they can do this very easy. So we stick together, and that makes other people think twice about trying to take over what is yours. In our case, the opposition has never tried messing with our hood, and that's because they know it's protected real good by us fellas. (qtd. in Padilla, 1993: 104)

Based on their research, Cloward and Ohlin (1960) identified three basic gang types—criminal, conflict, and retreatist, which emerge on the basis of what type of illegitimate opportunity structure is available in a specific area. The *criminal gang* is devoted to theft, extortion, and other illegal means of securing an income. For young men who grow up in a criminal gang, running drug houses and selling drugs on street corners make it possible for them to support themselves and their families as well as purchase material possessions to impress others. By contrast, *conflict gangs* emerge in communities that do not provide either legitimate or illegitimate opportunities. Members of conflict gangs seek to acquire a "rep" (reputation) by fighting over turf (territory) and adopting a value system of toughness, courage, and similar qualities. Unlike criminal and conflict gangs, members of *retreatist gangs* are unable to gain success through legitimate means and are unwilling to do so through illegal ones. As a result, the consumption of drugs is stressed, and addiction is prevalent.

Sociologist Lewis Yablonsky (1997) has updated Cloward and Ohlin's findings on delinquent gangs. According to Yablonsky, today's gangs are more likely to use and sell drugs, and carry more lethal weapons than gang members did in the past. Today's gangs have become more varied in their activities and are more likely to engage in intraracial conflicts, with "black on black and Chicano on Chicano violence," whereas minority gangs in the past tended to band together to defend their turf from gangs of different racial and ethnic backgrounds (Yablonsky, 1997: 3).

How useful are social structural approaches such as opportunity theory and strain theory in explaining deviant behavior? Although there are weaknesses to these approaches, they focus our attention on one crucial issue: the close association between certain forms of deviance and social class position. According to criminologist Anne Campbell (1984: 267), gangs are a "microcosm of American society, a mirror image in which power, possession, rank, and role . . . are found within a subcultural life of poverty and crime."

However, the social scientists Charles Tittle and Robert Meier (1990) dispute the proposition that class position is the most important factor in explaining why some people commit crimes. According to Tittle and Meier, most people from low-income backgrounds *do not* commit crimes, whereas some people from middle- and upper-income backgrounds *do* commit crimes. Likewise, some activities of gang members from low-income neighborhoods have commonalities with actions taken by nondelinquent youths in suburban cliques. Consider the practice of guarding one's "turf." Both adolescent gang members and high school clique participants often "guard" their favorite location, and a group may be known to others by the place that its members have chosen. Journalists recently described how clique members at Glenbrook, a suburban Chicago high school, jealously guard their turf. Moreover, the cliques are named for their favorite perches: The fashionable "wall people" favor a bench along the wall outside the cafeteria, whereas the punkish "trophy-case" kids sit on the floor under a display of memorabilia (Adler, 1999: 58). The relationship between conventional behavior and deviance is obviously much more complex than either opportunity theory or strain theory might suggest.

SYMBOLIC INTERACTIONIST PERSPECTIVES ON DEVIANCE

As we discussed in Chapter 4, symbolic interactionists focus on *social processes,* such as how people develop a self-concept and learn conforming behavior through socialization. According to this approach, deviance is learned in the same way as conformity—through interaction with others. Although there are a number of symbolic interactionist perspectives on deviance, we will examine four major approaches—differential association and differential reinforcement theories, rational choice theory, control theory, and labeling theory.

Differential Association Theory and Differential Reinforcement Theory

How do people learn deviant behavior through their interactions with others? According to the sociologist

Edwin Sutherland (1939), people learn the necessary techniques and the motives, drives, rationalizations, and attitudes of deviant behavior from people with whom they associate. ***Differential association theory* states that people have a greater tendency to deviate from societal norms when they frequently associate with individuals who are more favorable toward deviance than conformity.** From this approach, criminal behavior is learned within intimate personal groups such as one's family and peer groups. Learning criminal behavior also includes learning the techniques of committing crimes, as former gang member Nathan McCall explains:

> Sometimes I picked up hustling ideas at the 7-Eleven, which was like a criminal union hall: Crapshooters, shoplifters, stickup men, burglars, everybody stopped off at the store from time to time. While hanging up there one day, I ran into Holt. . . . He had a pocketful of cash, even though he had quit school and was unemployed. I asked him, "Yo, man, what you been into?"
>
> "Me and my partner kick in cribs and make a killin'. You oughta come go with us sometimes.". . . I hooked school one day, went with them, and pulled my first B&E [breaking and entering]. Before we went to the house, Hilliard . . . explained his system: "Look, man, we gonna split up and go to each house on the street. Knock on the door. If somebody answers, make up a name and act like you at the wrong crib. If nobody answers, we mark it for a hit.". . .
>
> After I learned the ropes, Shell Shock [another gang member] and I branched out, doing B&Es on our own. We learned to get in and out of houses in no time flat. (McCall, 1994: 93–94)

As McCall's orientation to breaking and entering shows, learning deviance may involve acquisition of certain attitudes and mastery of specialized techniques.

Differential association theory contributes to our knowledge of how deviant behavior reflects the individual's learned techniques, values, attitudes, motives, and rationalizations. It calls attention to the fact that criminal activity is more likely to occur when a person has frequent, intense, and long-lasting interactions with others who violate the law. However, it does not explain why many individuals who have been heavily exposed to people who violate the law still engage in conventional behavior most of the time.

Criminologist Ronald Akers (1998) has combined differential association theory with elements of psychological learning theory to create *differential reinforcement theory,* which suggests that both deviant behavior and conventional behavior are learned through the same social processes. Akers starts with the fact that people learn to evaluate their own behavior through interactions with significant others. If the persons and groups that a particular individual considers most significant in his or her life define deviant behavior as being "right," that individual is more likely to engage in deviant behavior; likewise, if the person's most significant friends and groups define deviant behavior as "wrong," the person is less likely to engage in that behavior. This approach helps explain not only juvenile gang behavior but also how peer cliques on high school campuses have such a powerful influence on people's behavior. Returning to our example of how clique members at Glenbrook High School turfed out territory at the school, notice how one student responded to powerful pressures to conform:

> As an experiment . . . Lauren Barry, a pink-haired trophy-case kid at Glenbrook, switched identities with a well-dressed girl from "the wall." Barry walked around all day in the girl's expensive jeans and Doc Martens, carrying a shopping bag from Abercrombie & Fitch. "People kept saying, 'Oh, you look so pretty,'" she recalls. "I felt really uncomfortable." It was interesting, but the next day, and ever since, she's been back in her regular clothes. (qtd. in Adler, 1999: 58)

Another approach to studying deviance is rational choice theory, which suggests that people weigh the rewards and risks involved in certain types of behavior and then decide which course of action to follow.

Rational Choice Theory

As you may recall from Chapter 6, rational choice theory is based on the assumption that when people are faced with several courses of action, they will usually do what they believe is likely to have the best overall outcome (Elster, 1989). The **rational choice theory of deviance states that deviant behavior occurs when a person weighs the costs and benefits of nonconventional or criminal behavior and determines that the benefits will outweigh the risks involved in such actions.** Rational choice approaches suggest that most people who commit crimes do not engage in random acts of antisocial behavior. Instead, they make careful decisions based on weighing the available information regarding *situational factors,* such as the place of the crime, suitable targets, and the availability of people to deter the behavior, and *personal factors,* such as what rewards they may gain from their criminal behavior (Siegel, 1998).

How useful is rational choice theory in explaining deviance and crime? A major strength of this theory is that it explains why high-risk youths do not constantly engage in delinquent acts: They have learned to balance risk against the potential for criminal gain in each situation. Moreover, rational choice theory is not limited by the underlying assumption of most social structural theories, which is that the primary participants in deviant and criminal behaviors are people in the lower classes. Rational choice theory also has important policy implications regarding crime reduction or prevention, suggesting that people must be taught that the risks of engaging in criminal behavior far outweigh any benefits they may gain from their actions. Thus, people should be taught *not* to engage in crime.

Control Theory: Social Bonding

Control theories focus on another aspect of why some people do not engage in deviant behavior. According to the sociologist Walter Reckless (1967), society produces pushes and pulls that move people toward criminal behavior; however, some people "insulate" themselves from such pressures by having positive self-esteem and good group cohesion. Reckless suggests that many people do not resort to deviance because of *inner containments*—such as self-control, a sense of responsibility, and resistance to diversions—and *outer containments*—such as supportive family and friends, reasonable social expectations, and supervision by others. Those with the strongest containment mechanisms are able to withstand external pressures that might cause them to participate in deviant behavior.

Extending Reckless's containment theory, sociologist Travis Hirschi's (1969) social control theory is based on the assumption that deviant behavior is minimized when people have strong bonds that bind them to families, schools, peers, churches, and other social institutions. **Social bond theory holds that the probability of deviant behavior increases when a person's ties to society are weakened or broken.** According to Hirschi, social bonding consists of (1) *attachment* to other people, (2) *commitment* to conformity, (3) *involvement* in conventional activities, and (4) *belief* in the legitimacy of conventional values and norms. Although Hirschi did not include females in his study, others who have replicated that study with both females and males have found that the theory appears to apply to each (see Naffine, 1987).

What does control theory have to say about delinquency and crime? Control theories suggest that the

Bill Bachmann/PhotoEdit

Michael Newman/PhotoEdit

According to control theory, strong bonds—including close family ties—are a factor in explaining why many people do not engage in deviant behavior. Across class lines, how are early childhood experiences linked to juvenile and adult behavior?

probability of delinquency increases when a person's social bonds are weak and when peers promote antisocial values and violent behavior. However, some critics assert that Hirschi was mistaken in his assumption that a weakened social bond leads to deviant behavior. The chain of events may be just the opposite: People who routinely engage in deviant behavior may find that their bonds to people who would be positive influences are weakened over time (Agnew, 1985; Siegel 1998). Or, as labeling theory suggests, people may engage in deviant and criminal behavior because of destructive social interactions and encounters (Siegel, 1998).

Labeling Theory

Labeling theory **states that deviance is a socially constructed process in which social control agencies designate certain people as deviants, and they, in turn, come to accept the label placed upon them and begin to act accordingly.** Based on the symbolic interaction theory of Charles H. Cooley and George H. Mead (see Chapter 4), labeling theory focuses on the variety of symbolic labels that people are given in their interactions with others. Sociologist Larry J. Siegel (1998: 212) explains the link between labeling and deviance as follows:

Labels imply a variety of behaviors and attitudes; labels thus help define not just one trait but the whole person. For example, people labeled "insane" are also assumed to be dangerous, dishonest, unstable, violent, strange, and otherwise unsound. Valued labels, including "smart," "honest," and "hard worker," which suggest overall competence, can improve self-image and social standing. Research shows that people who are labeled with one positive trait, such as being physically attractive, are assumed to maintain others, such as intelligence and competence. In contrast, negative labels, including "troublemaker," "mentally ill," and "stupid," help stigmatize their targets and reduce their self-image.

How does the process of labeling occur? The act of fixing a person with a negative identity, such as "criminal" or "mentally ill," is directly related to the power and status of those persons who *do* the labeling and those who are *being labeled.* Behavior, then, is not deviant in and of itself; it is defined as such by a social audience (Erikson, 1962). According to the sociologist Howard Becker (1963), *moral entrepreneurs* are often the ones who create the rules about what constitutes deviant or conventional behavior. Becker believes that moral entrepreneurs use their own perspectives on "right" and "wrong" to establish the rules by which they expect other people to live. They also label others as deviant. Often, these rules are enforced on persons with less power than the moral entrepreneurs. Becker (1963: 9) concludes that the deviant is "one to whom the label has successfully been applied; deviant behavior is behavior that people so label."

As the definition of labeling theory suggests, several stages may occur in the labeling process. ***Primary deviance* refers to the initial act of rule breaking** (Lemert, 1951). However, if individuals accept the negative label that has been applied to them as a result of the primary deviance, they are more likely to continue to participate in the type of behavior that the label was initially meant to control. ***Secondary deviance* occurs when a person who has been labeled a deviant accepts that new identity and continues the deviant behavior.** For example, a person may shoplift an item of clothing from a department store but not be apprehended or labeled as a deviant. The person may subsequently decide to forgo such behavior in the future. However, if the person shoplifts the item, is apprehended, is labeled as a "thief," and subsequently accepts that label, then the person may shoplift items from stores on numerous occasions. A few people engage in ***tertiary deviance,* which occurs when a person who has been labeled a deviant seeks to normalize the behavior by relabeling it as non-deviant** (Kitsuse, 1980). An example would be drug users who believe that using marijuana or other illegal drugs is no more deviant than drinking alcoholic beverages and therefore should not be stigmatized.

Can labeling theory be applied to high school peer groups and gangs? In a classic study, the sociologist William Chambliss (1973) documented how the labeling process works in some high schools when he studied two groups of adolescent boys: the "Saints" and the "Roughnecks." Members of both groups were constantly involved in acts of truancy, drinking, wild parties, petty theft, and vandalism. Although the Saints committed more offenses than the Roughnecks, the Roughnecks were the ones who were labeled as "troublemakers" and arrested by law enforcement officials. By contrast, the Saints were described as being the "most likely to succeed," and none of the Saints were ever arrested. According to Chambliss (1973), the Roughnecks were more likely to be labeled as deviants because they came from lower-income families, did poorly in school, and were generally viewed negatively whereas the Saints came from "good families," did well in school, and were generally viewed positively by others. Although both groups engaged in similar behavior, only the Roughnecks were stigmatized by a deviant label.

Another study of juvenile offenders also found that those from lower-income families were more likely to be arrested and indicted than were middle-class juveniles who participated in the same kinds of activities (Sampson, 1986). In determining how to deal with youthful offenders, the criminal justice system frequently takes into account such factors as the offender's family life, educational achievement (or lack thereof), and social class. The individuals most likely to be apprehended, labeled as delinquent, and prosecuted are people of color who are young, male, unemployed, and undereducated, and who reside in urban high-crime areas (Vito and Holmes, 1994). Why might this be true? According to the criminologist Robert J. Sampson (1997), family and neighborhood, more than the individual characteristics of people involved in deviance and crime, are important factors in determining variations in crime rates. For example, parents with the lowest incomes may have the most difficulty with parenting, which may result in young people receiving harsh or erratic discipline and poor supervision (Sampson and Laub, 1993). However, even young people who have been chronically involved in delinquent behavior may reach certain *turning points* in their life, such as marriage or a career, which may cause them to decide against crime (Sampson and Laub, 1993).

How successful is labeling theory in explaining deviance and social control? One contribution of labeling theory is that it calls attention to the way in which social control and personal identity are intertwined: Labeling may contribute to the acceptance of deviant roles and self-images. Critics argue that this does not explain what caused the original acts that constituted primary deviance, nor does it provide insight into why some people accept deviant labels and others do not (Cavender, 1995).

Whereas symbolic interactionist perspectives are concerned with how people learn deviant behavior, identities, and social roles through interaction with others, conflict theorists emphasize the connections among social class, deviance, and social control.

CONFLICT PERSPECTIVES ON DEVIANCE

Who determines what kinds of behavior are deviant or criminal? Different branches of conflict theory offer somewhat divergent answers to this question. One branch emphasizes power as the central factor in defining deviance and crime: People in positions of power maintain their advantage by using the law to

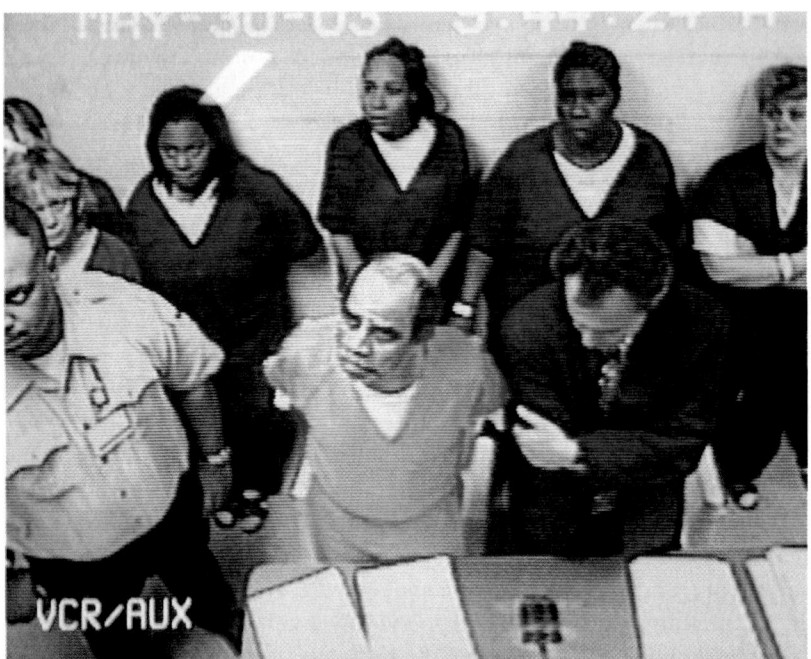

Conflict theorists suggest that criminal law is unequally enforced along class lines. Consider this setting, in which low-income defendants are arraigned by a judge who sees them only on a television monitor. Do you think, as a rule, that these defendants will be as well represented by attorneys as a wealthier defendant might be?

© 2003 AP/Wide World Photos

protect their own interests. Another branch emphasizes the relationship between deviance and capitalism, while a third focuses on feminist perspectives and the confluence of race, class, and gender issues in regard to deviance and crime.

Deviance and Power Relations

Conflict theorists who focus on power relations in society suggest that the lifestyles considered deviant by political and economic elites are often defined as illegal. From this perspective, the law defines and controls two distinct categories of people: (1) *social dynamite*—persons who have been marginalized (including rioters, labor organizers, gang members, and criminals)—and (2) *social junk*—members of stigmatized groups (such as welfare recipients, the homeless, and persons with disabilities) who are costly to society but relatively harmless (Spitzer, 1975). According to this approach, norms and laws are established for the benefit of those in power and do not reflect any absolute standard of right and wrong (Turk, 1969, 1977). As a result, the activities of poor and lower-income individuals are more likely to be defined as criminal than those of persons from middle- and upper-income backgrounds. Moreover, the criminal justice system is more focused on, and is less forgiving of, deviant and criminal behavior engaged in by people in specific categories. For example, research shows that young, single, urban males are more likely to be perceived as members of the *dangerous classes* and receive stricter sentences in criminal courts (Miethe and

Moore, 1987). Power differentials are also evident in how victims of crime are treated. When the victims are wealthy, white, and male, law enforcement officials are more likely to put forth more extensive efforts to apprehend the perpetrator as contrasted with cases in which the victims are poor, black, and female (Smith, Visher, and Davidson, 1984). Recent research generally supports this assertion (Wonders, 1996). This branch of conflict theory shows how power relations in society influence the law and the criminal justice system, often to the detriment of people who are at the bottom of the social structure hierarchy, and it questions functionalist views on conformity and deviance that are based on the assumption that laws reflect a consensus among the majority of people.

Deviance and Capitalism

A second branch of conflict theory—Marxist/critical theory—views deviance and crime as a function of the capitalist economic system. Although the early economist and social thinker Karl Marx wrote very little about deviance and crime, many of his ideas are found in a critical approach that has emerged from earlier Marxist and radical perspectives on criminology. The critical approach is based on the assumption that the laws and the criminal justice system protect the power and privilege of the capitalist class. According to the social scientist Barry Krisberg (1975), *privilege* is the possession of what is most valued by a particular social group in a given historical period. As such, privilege not only includes rights such as life, liberty, and happiness, but also material possessions such as money, luxury items, land, and houses.

As you may recall from Chapter 1, Marx based his critique of capitalism on the inherent conflict that he believed existed between the capitalists (*bourgeoisie*) and the working class (*proletariat*). In a capitalist society, social institutions (such as law, politics, and education, which make up the superstructure) legitimize existing class inequalities and maintain the capitalists' superior position in the class structure. According to Marx, capitalism produces haves and have-nots, who engage in different forms of deviance and crime.

Why do people commit crimes? Some critical theorists believe that members of the capitalist class commit crimes because they are greedy and want more than they have. Corporate or white-collar crimes such as stock market manipulation, land speculation, fraudulent bankruptcies, and crimes committed on behalf of organizations often involve huge sums of money and harm many people. By contrast, street crimes such as robbery and aggravated assault generally involve small sums of money and cause harm to limited numbers of victims. According to these theorists, the poor commit street crimes in order to survive; they find that they cannot afford the essentials, such as food, clothing, shelter, and health care. Thus, some crime represents a rational response by the poor to the unequal distribution of resources in society (Gordon, 1973). Further, living in poverty may lead to violent crime and victimization *of the poor by the poor.* For example, violent gang activity may be a collective response of young people to seemingly hopeless poverty (Quinney, 1979).

According to the sociologist Richard Quinney (2001/1974), people with economic and political power define as criminal any behavior that threatens their own interests. The powerful use law to control those who are without power. For example, drug laws enacted early in the twentieth century were actively enforced in an effort to control immigrant workers, especially the Chinese, who were being exploited by the railroads and other industries (Tracy, 1980). By contrast, antitrust legislation passed at about the same time was seldom enforced against large corporations owned by prominent families such as the Rockefellers, Carnegies, and Mellons. Having antitrust laws on the books merely shored up the government's legitimacy by making it appear responsive to public concerns about big business (Barnett, 1979).

In sum, the Marxist/critical approach argues that criminal law protects the interests of the affluent and powerful. The way that laws are written and enforced benefits the capitalist class by ensuring that individuals at the bottom of the social class structure do not infringe on the property or threaten the safety of those at the top (Reiman, 1998). However, critics assert that critical theorists have not shown that powerful economic and political elites actually manipulate lawmaking and law enforcement for their own benefit. Rather, people of all classes share a consensus about the criminality of certain acts. For example, laws that prohibit murder, rape, and armed robbery protect not only middle- and upper-income people but also low-income people, who are frequently the victims of such violent crimes. To shift the focus from seeing law as always working for the rich and against the poor, structural Marxist theorists such as John Hagan (1989) argue that law can be used to control the members of *any class* who pose a threat to the existence of capitalism. For example, the excessive greed of a few affluent capitalists might "rock the boat" and bring into question the business practices of many others if their behavior is not targeted and sanctioned.

Feminist Approaches

Can theories developed to explain male behavior be used to understand female deviance and crime? According to feminist scholars, the answer is no. A new interest in women and deviance developed in 1975 when two books—Freda Adler's *Sisters in Crime* and Rita James Simons's *Women and Crime*—declared that women's crime rates were going to increase significantly as a result of the women's liberation movement. Although this so-called *emancipation theory* of female crime has been refuted by subsequent analysts, Adler's and Simons's works encouraged feminist scholars (both women and men) to examine more closely the relationship among gender, deviance, and crime. More recently, feminist scholars such as Kathleen Daly and Meda Chesney-Lind (1988) have developed theories and conducted research to fill the void in our knowledge about gender and crime. For example, in a recent study of the female offender, Chesney-Lind (1997) examined the cultural factors in women's lives that may contribute to their involvement in criminal behavior. Although there is no single feminist perspective on deviance and crime, three schools of thought have emerged.

Why do women engage in deviant behavior and commit crimes? According to the *liberal feminist approach,* women's deviance and crime are a rational response to the gender discrimination that women experience in families and the workplace. From this view, lower-income and minority women typically have fewer opportunities not only for education and good jobs but also for "high-end" criminal endeavors. As some feminist theorists have noted, a woman is no more likely to be a big-time drug dealer or an organized crime boss than she is to be a corporate director (Daly and Chesney-Lind, 1988; Simpson, 1989).

By contrast, the *radical feminist approach* views the cause of women's crime as originating in patriarchy (male domination over females). This approach focuses on social forces that shape women's lives and experiences and shows how exploitation may trigger deviant behavior and criminal activities. From this view, arrests and prosecution for crimes such as prostitution reflect our society's sexual double standard whereby it is acceptable for a man to pay for sex but unacceptable for a woman to accept money for such services. Although state laws usually view both the female prostitute and the male customer as violating the law, in most states the woman is far more likely than the man to be arrested, brought to trial, convicted, and sentenced.

The third school of feminist thought, the *Marxist (socialist) feminist approach,* is based on the assumption that women are exploited by both capitalism and patriarchy. From this approach, women's criminal behavior is linked to gender conflict created by the economic and social struggles that often take place in postindustrial societies such as ours. According to the social scientist James Messerschmidt (1986), men control women biologically and economically just as members of the capitalist class control the labor of workers. As a result, women experience "double marginality," which provides women with fewer opportunities to commit certain types of deviance and crime. Because most females have relatively low-wage jobs (if any) and few economic resources, crimes such as prostitution and shoplifting become a means to earn money or acquire consumer goods. However, instead of freeing women from their problems, prostitution institutionalizes women's dependence on men and results in a form of female sexual slavery (Vito and Holmes, 1994). Lower-income women are further victimized by the fact that they are often the targets of violent acts by lower-class males, who perceive themselves as being powerless in the capitalist economic system. Since Western societies value aggressive male behavior, whether in sports or business pursuits, men who feel powerless may "prove" their manliness by *doing gender*—attempting to improve their male self-image through acts of violence or abuse against women or children (Siegel, 1998).

Some feminist scholars have noted that these approaches to explaining deviance and crime neglect the centrality of race and ethnicity and focus on the problems and perspectives of women who are white, middle- and upper-income, and heterosexual without taking into account the views of women of color, lesbians, and women with disabilities (Martin and Jurik, 1996).

Approaches Focusing on Race, Class, and Gender

Some recent studies have focused on the simultaneous effects of race, class, and gender on deviant behavior. In one study, the sociologist Regina Arnold (1990) examined the relationship between women's earlier victimization in their family and their subsequent involvement in the criminal justice system. Arnold interviewed African American women serving criminal sentences and found that adolescent females are often "labeled and processed as deviants—and subsequently as criminals—for refusing to accept or participate in their own victimization." Arnold attributes many of the women's offenses to living in families in which sexual abuse, incest, and other violence left them few choices except to engage in deviance. Economic marginality and racism also contributed to their victimization: "To be young, Black, poor, and female is to be in a high-risk category for victimization and stigmatization on many levels" (Arnold, 1990: 156). According to Arnold, the criminal behavior of the women in her study was linked to class, gender, and racial oppression, which they experienced daily in their families and at school and work.

Feminist sociologists and criminologists believe that research on women as both victims and perpetrators of crime is long overdue. For example, few studies of violent crime—such as robbery and aggravated assault—have included women as subjects or respondents. Some scholars have argued that women are less motivated to commit such crimes, are not as readily exposed to attractive targets, and are more protected from being the victims of such crimes than men are (see Sommers and Baskin, 1993). Other scholars have stressed that research should integrate women into the larger picture of criminology (Simpson, 1989). For example, sociologists Susan Ehrlich Martin and Nancy C. Jurik (1996) examined the role of women in the criminal justice system and found that women continue to experience significant barriers in justice occupations ranging from law enforcement to the legal profession.

POSTMODERNIST PERSPECTIVES ON DEVIANCE

How might postmodernists view deviance and social control? Although the works of social theorist Michel Foucault defy simple categorization, *Discipline and*

Michel Foucault contended that new means of surveillance would make it possible for prison officials to use their knowledge of prisoners' activities as a form of power over the inmates. This prison guard is able to monitor the activities of many prisoners without ever leaving his station.

© 2003 AP/Wide World Photos

Punish (1979) might be considered somewhat post-modernist in its approach to explaining the intertwining nature of power, knowledge, and social control. In his study of prisons from the mid-1800s to the early 1900s, Foucault found that many penal institutions ceased torturing prisoners who disobeyed the rules and began using new surveillance techniques to maintain social control. Although the prisons appeared to be more humane in the post-torture era, Foucault contends that the new means of surveillance impinged more on prisoners and brought greater power to prison officials. To explain, he described the *Panoptican*—a structure that gives prison officials the possibility of complete observation of criminals at all times. For example, the Panoptican might be a tower located in the center of a circular prison from which guards could see all the cells. Although the prisoners know they can be observed at any time, they do not actually know when their behavior is being scrutinized. As a result, prison officials are able to use their knowledge as a form of power over the inmates. Eventually, the guards would not even have to be present all the time because prisoners would believe that they were under constant scrutiny by officials in the observation post. Thus, in this case social control and discipline are based on the use of knowledge, power, and technology. How does Foucault's perspective explain social control in the larger society? According to Foucault, technologies such as the Panoptican make widespread surveillance and disciplinary power possible in many settings, including the state-police network, factories, schools, and hospitals. However, he did not believe that discipline would sweep uniformly through society due to opposing forces that would use their power to oppose such surveillance.

Foucault's view on deviance and social control has influenced other social analysts, including Shoshana Zuboff (1988), who views the computer as a modern Panoptican that gives workplace supervisors virtually unlimited capabilities for surveillance over subordinates.

We have examined functionalist, interactionist, conflict, and postmodernist perspectives on social control, deviance, and crime (see Concept Table 7.A). All of these explanations contribute to our understanding of the causes and consequences of deviant behavior; however, we now turn to the subject of crime itself.

CRIME CLASSIFICATIONS AND STATISTICS

Crime in the United States can be divided into different categories. We will look first at the legal classifications of crime and then at categories typically used by sociologists and criminologists.

How the Law Classifies Crime

Crimes are divided into felonies and misdemeanors. The distinction between the two is based on the seriousness of the crime. A *felony* is a serious crime such as rape, homicide, or aggravated assault, for which punishment typically ranges from more than a year's imprisonment to death. A *misdemeanor* is a minor crime that is typically punished by less than one year in jail. In either event, a fine may be part of the sanction as well. Actions that constitute felonies and misdemeanors are determined by the legislatures in the various states; thus, their definitions vary from jurisdiction to jurisdiction.

Concept Table 7.A THEORETICAL PERSPECTIVE ON DEVIANCE

	Theory	Key Elements
FUNCTIONALIST PERSPECTIVES		
Robert Merton	Strain theory	Deviance occurs when access to the approved means of reaching culturally approved goals is blocked. Innovation, ritualism, retreatism, or rebellion may result.
Richard Cloward/Lloyd Ohlin	Opportunity theory	Lower-class delinquents subscribe to middle-class values but cannot attain them. As a result, they form gangs to gain social status and may achieve their goals through illegitimate means.
SYMBOLIC INTERACTIONIST PERSPECTIVES		
Edwin Sutherland	Differential association	Deviant behavior is learned in interaction with others. A person becomes delinquent when exposure to law-breaking attitudes is more extensive than exposure to law-abiding attitudes.
Travis Hirschi	Social control/ social bonding	Social bonds keep people from becoming criminals. When ties to family, friends, and others become weak, an individual is most likely to engage in criminal behavior.
Howard Becker	Labeling theory	Acts are deviant or criminal because they have been labeled as such. Powerful groups often label less-powerful individuals.
Edwin Lemert	Primary/secondary deviance	Primary deviance is the initial act. Secondary deviance occurs when a person accepts the label of "deviant" and continues to engage in the behavior that initially produced the label.
CONFLICT PERSPECTIVES		
Karl Marx Richard Quinney	Critical approach	The powerful use law and the criminal justice system to protect their own class interests.
Kathleen Daly Meda Chesney-Lind	Feminist approach	Historically, women have been ignored in research on crime. Liberal feminism views women's deviance as arising from gender discrimination, radical feminism focuses on patriarchy, and socialist feminism emphasizes the effects of capitalism and patriarchy on women's deviance.
POSTMODERNIST PERSPECTIVE		
Michel Foucault	Knowledge as power	Power, knowledge, and social control are intertwined. In prisons, for example, new means of surveillance that make prisoners think they are being watched all the time give officials knowledge that inmates do not have. Thus, the officials have a form of power over the inmates.

Figure 7.1 **Distribution of Arrests by Type of Offenses, 2002**

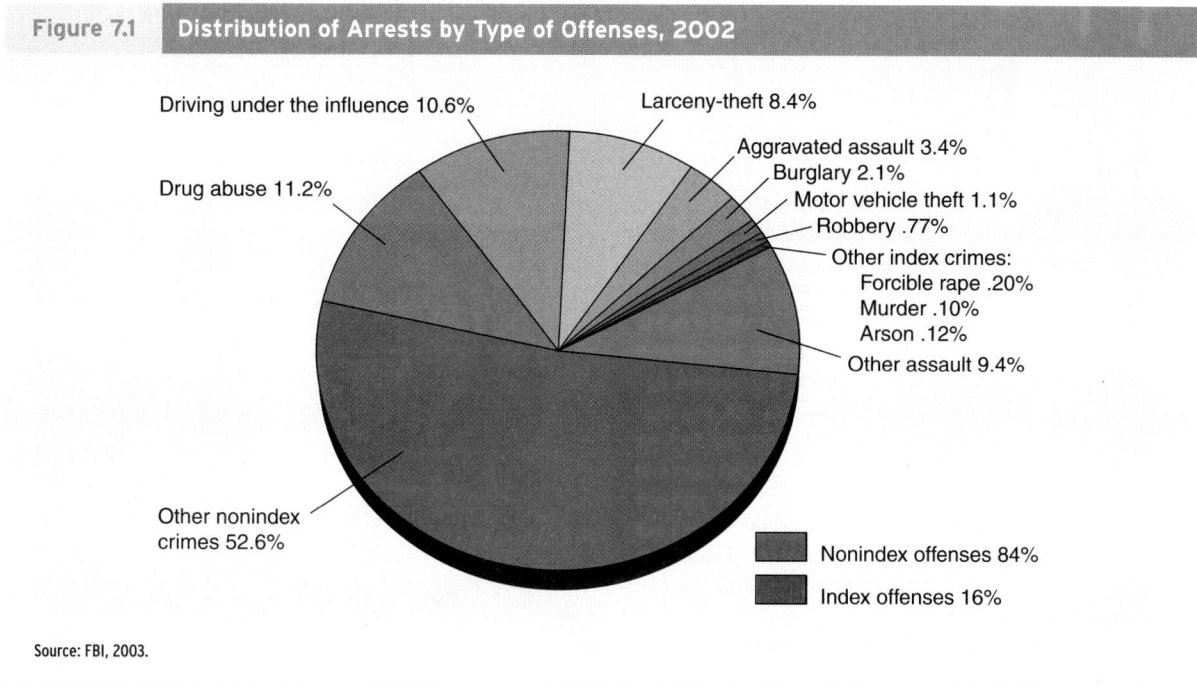

Source: FBI, 2003.

The *Uniform Crime Report* (UCR) is the major source of information on crimes reported in the United States. The UCR has been compiled since 1930 by the Federal Bureau of Investigation based on information filed by law enforcement agencies throughout the country. When we read that the homicide rate in California is higher than the national average, for example, this information is usually based on UCR data. The UCR focuses on eight major crimes, called *index crimes:* murder, rape, robbery, assault, burglary (breaking into private property to commit a serious crime), motor vehicle theft, arson, and larceny (theft of property worth $50 or more). It also contains data on other types of crime. In 2002, about 14 million arrests were made in the United States for all criminal infractions (excluding traffic violations). Of those arrests, about 16 percent were for index crimes, as shown in Figure 7.1. Although the UCR gives some indication of crime in the United States, the figures do not reflect the actual number and kinds of crimes, as will be discussed later.

How Sociologists Classify Crime

Sociologists categorize crimes based on how they are committed and how society views the offenses. We will examine four types: (1) conventional (street) crime, (2) occupational (white-collar) and corporate crime, (3) organized crime, and (4) political crime. As you read about these types of crime, ask yourself

how you feel about them. Should each be a crime? How stiff should the sanctions be against each type? What personal characteristics (such as your race/ethnicity, class, and gender) influence your opinion?

Conventional Crime Most of the UCR's eight index crimes are conventional crimes. ***Conventional (street) crime* is all violent crime, certain property crimes, and certain morals crimes.** Obviously, all street crime does not occur on the street; it frequently occurs in the home, workplace, and other locations.

Violent crime consists of actions involving force or the threat of force against others, including murder, rape, robbery, and aggravated assault. Violent crimes are probably the most anxiety-provoking of all criminal behavior—most of us know someone who has been a victim of violent crime, or we have been so ourselves. Victims are often physically injured or even lose their lives; the psychological trauma may last for years after the event (Parker and Anderson-Facile, 2000). Violent crime receives the most sustained attention from law enforcement officials and the media (see Warr, 2000). Some analysts question whether there might be a link between youth violence and media glorification of physical injury and killing. Violent video games, often enjoyed especially by boys and young men, have particularly been a topic of concern to some analysts in recent years (see Box 7.2).

Nationwide, there is growing concern over juvenile violence. Beginning in 1988, juvenile violent

Box 7.2 CHANGING TIMES: MEDIA AND TECHNOLOGY

Child's Play? Extreme Violence and Video Games

All you die hard *GTA* [*Grand Theft Auto*] fans out there are gonna absolutely fall in love with this work of art [*Grand Theft Auto: Vice City*]! There are loads more guns (RM60, Colt python . . .), more cars, more moves in hand to hand combat and gangs work together to bring you down! S.V. teams absail down from choppers to bag you. As you can now jump from moving vehicles, the police use stinger traps [to get] you. You can smash up cars with [baseball] bats and shoot other people through their own car windows! Although I could go on forever, I'm not gonna. But trust me, you simply, positively have to buy this awesome work of pure genius!!!!

—posted by "Spud" in 2002 (Bizrate.com, 2003)

© Dwayne Newton/PhotoEdit

[*Grand Theft Auto: Vice City*] teaches you it's cool to be a criminal and it promotes violence. I'm surprised it doesn't have the characters smoking dope!!!

—posted by "Online Shopper" in 2002 (Bizrate .com, 2003)

As these postings by "Spud" and "Online Shopper" show, reviews of video games and CDs vary widely on online message boards; however, comments by these game enthusiasts mention only a few of the issues that have been raised about the increasingly violent, and sometimes sexually explicit, nature of some video games.

The game "Spud" and "Online Shopper" are referring to is *Grand Theft Auto: Vice City*, the latest in a series of *Grand Theft Auto* video games and CDs. This series is extremely popular with many video game players who own a PlayStation, a GameCube, or Microsoft's Xbox, or play the game on their home computer. Although *Grand Theft Auto* games are not supposed to be sold to players under the age of seventeen, this series often finds its way into the homes of children and young adolescents.

If you are not familiar with the general themes in this video series, the setting for *GTA: Vice City* is a Miami-like city in the 1980s, and home game players go along with Tommy Vercetti in his quest to find the men who ran off with his drugs. Weapons, sex, and illegal drugs are central features of the game. Here are a few excerpts from a description of *GTA: Vice City* that Rockstar, the game's manufacturer, provides:

Having just made it back onto the streets of Liberty City after a long stretch in maximum secu-

---

crime arrest rates started to rise, a trend that has been linked by some scholars to gang membership (see Inciardi, Horowitz, and Pottieger, 1993; Thornberry et al., 1993). Fear of violence is felt not only by the general public but by gang members themselves, as Charles Campbell commented:

The generation I'm in is going to be lost. Of the circle of friends I grew up in, three are dead, four are in jail, and another is out of school and just does nothing. When he runs out of money he'll sell a couple bags of weed. . . .

I would carry a gun because I am worried about that brother on the fringe. There are some people, there is nothing out there for them. They will blow you away because they have nothing to lose. There are no jobs out there. It's hard to get money to go to school. (qtd. in F. Lee, 1993: 21)

Property crimes include robbery, burglary, larceny, motor vehicle theft, and arson. Some offenses, such as robbery, are both violent crimes and property crimes. In the United States, a property crime occurs, on average, once every three seconds; a violent crime occurs, on average, once every twenty-two seconds (see Figure 7.2). In most property crimes, the primary motive is to obtain money or some other desired valuable.

rity, Tommy Vercetti is sent to Vice City by his old boss, Sonny Forelli. They were understandably nervous about his re-appearance in Liberty City, so a trip down south seemed like a good idea. But all does not go smoothly upon his arrival in the glamorous, hedonistic metropolis of Vice City. He's set up and is left with no money and no merchandise. Sonny wants his money back, but the biker gangs, Cuban gangsters, and corrupt politicians stand in his way. Most of Vice City seems to want Tommy dead. His only answer is to fight back and take over the city himself. . . . (Rockstargames.com, 2003)

Are games like this just good, clean fun? Experts do not agree on the effect that video games have on young people, and studies have only recently tried to determine whether or not such activities might produce antisocial or criminal behavior, particularly in adolescent males. Yet we do know that a recent study of teenagers and young adults who watch a lot of television violence concluded that many of these individuals might be more likely to commit violent crimes and engage in other forms of aggressive behavior later in life (Kolata, 2002). However, researchers in the TV study caution that their findings do not prove that watching excessive amounts of TV violence necessarily *causes* individuals to commit violent acts. Prior studies have also concluded that there is some association (if not necessarily causation) between younger children watching excessive violence on TV and engaging in aggressive behavior. Does this finding potentially hold true for video games as well? Researchers will perhaps be able to reach more definitive conclusions on this matter in the future. However, in the meantime, the question remains about whether the line between fantasy and reality can become blurred for children and adolescents who routinely play violent video games such as the *Grand Theft Auto* series.

Rockstar Games claims that its games are "geared towards mature audiences" and that it "makes every effort to market its games responsibly, targeting advertising and marketing only to adult consumers over the age of 17" (qtd. in Cuomo, 2002). But these games are often played by younger people for whom it may be possible that fantasy and reality can become blurred, particularly as players are bombarded with a continual diet of graphic depictions of physical violence and as they become a part of the process through the interactive component of this and other video games like it. Regardless, there is little likelihood that game developers and manufacturers will voluntarily modify the violent subject matter in their products because, as one journalist stated, "Simulated crime pays—and it pays well. [*GTA: Vice City*] sold 4 million copies before it was even released—and an industry expert says it could eventually sell 10 million copies, bringing in $400 million" (Cuomo, 2002).

WRITING IN SOCIOLOGY ASSIGNMENT

Do video games contribute to deviance and crime in contemporary societies, or are these  games nothing more than harmless entertainment? What do you think?

"Morals" crimes involve an illegal action voluntarily engaged in by the participants, such as prostitution, illegal gambling, the private use of illegal drugs, and illegal pornography. Many people assert that such conduct should not be labeled as a crime; these offenses are often referred to as *victimless crimes* because they involve a willing exchange of illegal goods or services among adults. However, morals crimes can include children and adolescents as well as adults. Young children and adolescents may unwillingly become child pornography "stars" or prostitutes. Members of juvenile gangs often find selling drugs to be a lucrative business in which getting arrested is merely an occupational hazard. Fernando Morales, age six-

teen, is an example of a young man who feels he has nothing to lose by selling drugs: "Sometimes [selling drugs] bothers me. But see, I'm a hustler. I got to look out for myself. I got to be making money. Forget [the customers]. If you put that in your head, you're going to be caught out. You going to be a sucker. You going to be like them" (qtd. in Tierney, 1993: A1, A11).

Occupational and Corporate Crime

Although sociologist Edwin Sutherland (1949) developed the theory of white-collar crime more than fifty years ago, it was not until the 1980s that the public became aware of its nature. **Occupational (white-collar) crime comprises illegal activities committed**

Figure 7.2   The FBI Crime Clock

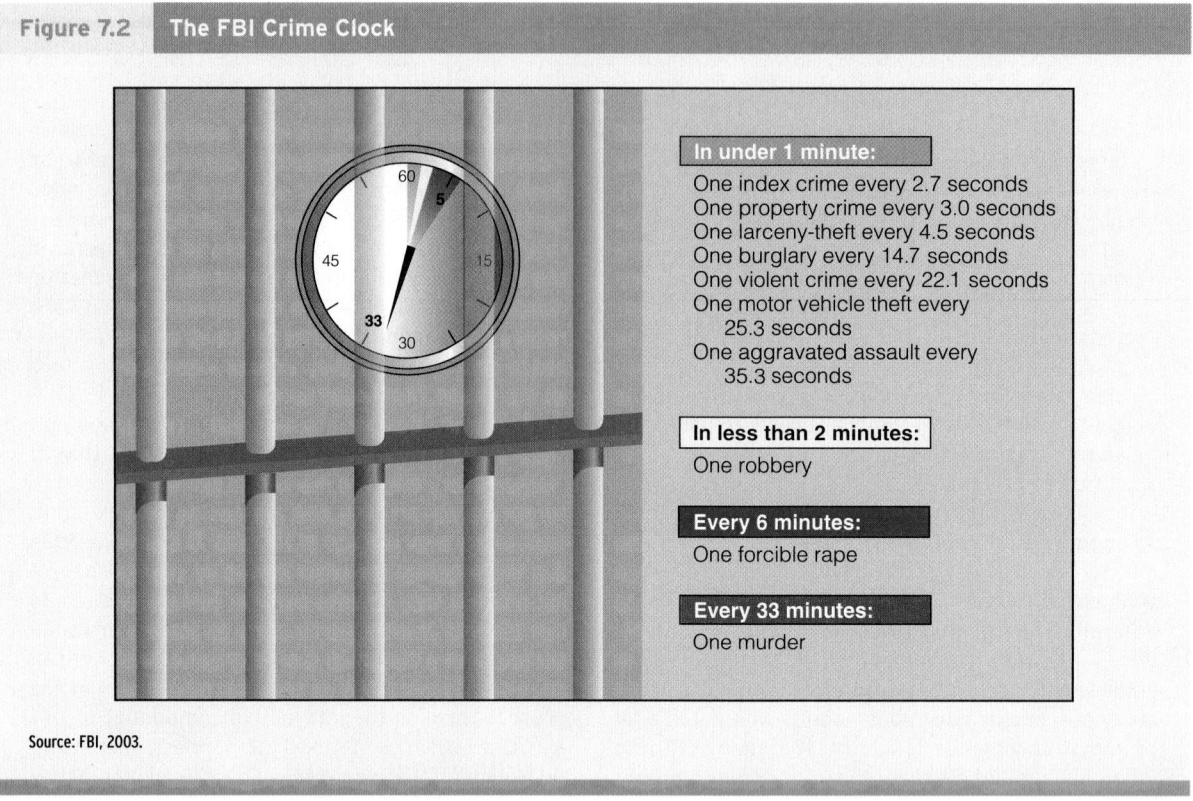

In under 1 minute:
One index crime every 2.7 seconds
One property crime every 3.0 seconds
One larceny-theft every 4.5 seconds
One burglary every 14.7 seconds
One violent crime every 22.1 seconds
One motor vehicle theft every
    25.3 seconds
One aggravated assault every
    35.3 seconds

In less than 2 minutes:
One robbery

Every 6 minutes:
One forcible rape

Every 33 minutes:
One murder

Source: FBI, 2003.

by people in the course of their employment or financial affairs.

In addition to acting for their own financial benefit, some white-collar offenders become involved in criminal conspiracies designed to improve the market share or profitability of their companies. This is known as *corporate crime*—**illegal acts committed by corporate employees on behalf of the corporation and with its support.** Examples include antitrust violations; tax evasion; misrepresentations in advertising; infringements on patents, copyrights, and trademarks; price fixing; and financial fraud (Friedrichs, 1996). These crimes are a result of deliberate decisions made by corporate personnel to enhance resources or profits at the expense of competitors, consumers, and the general public.

Although people who commit occupational and corporate crimes can be arrested, fined, and sent to prison, many people often have not regarded such behavior as "criminal." People who tend to condemn street crime are less sure of how their own (or their friends') financial and corporate behavior should be judged. At most, punishment for such offenses has usually been a fine or a relatively brief prison sentence. The case of Michael Milken illuminates the typical sanctions for white-collar crimes. After his 1990s conviction for illegal insider trading, Milken

was ordered to pay $600 million in fines and restitution and was sentenced to ten years in prison, but he was released from prison after less than two years to do community service (Clines, 1993).

Until recently, public concern and media attention focused primarily on the street crimes disproportionately committed by persons who are poor, powerless, and nonwhite. Beginning in 2002, however, our attention at least temporarily shifted to crimes committed in corporate suites, such as fraud, tax evasion, and insider trading by executives at some large and well-known corporations. High-ranking executives at communications giant WorldCom admitted to overstating its financial situation by almost $4 billion and were indicted for securities and accounting fraud, among other charges. WorldCom, which owned the nation's second-largest long-distance telephone company and once had been a favorite of the stock market because its per-share price doubled annually for a number of years, filed bankruptcy. Accounting fraud charges forced other corporations, including energy trader Enron, into bankruptcy as well; Enron's outside auditor, big-five accounting firm Arthur Andersen, was convicted of obstruction of justice in connection with its handling of the account, and several of Enron's former officers have been indicted for various offenses. Dennis Kozlowski, chief executive of Tyco International, was indicted for evading $1 million in sales tax on art purchases for his own collection. Biopharmaceutical company ImClone's CEO, Sam Waskal, pleaded guilty to charges of insider stock trading, and some of his relatives and friends—including lifestyle expert Martha Stewart—were accused of insider trading in the sale of shares of that company prior to a Food and Drug Administration announcement driving down the price of the company's stock (Peyser, 2002).

Corporate crimes are often more costly in terms of money and lives lost than street crimes. Thousand of jobs and billions of dollars were lost as a result of corporate crime in the year 2002 alone. Deaths resulting from corporate crimes such as polluting the air and water, manufacturing defective products, and selling unsafe foods and drugs far exceed the number of deaths due to homicides each year. Other costs include the effect on the moral climate of society (Clinard and Yeager, 1980; Simon, 1996). The confidence of everyday people throughout the United States in the nation's economy has been shaken badly by the greedy and illegal behavior of corporate insiders.

Organized Crime
Organized crime **is a business operation that supplies illegal goods and services for profit.** Premeditated, continuous illegal activities of organized crime include drug trafficking, prostitution, loan-sharking, money laundering, and large-scale theft such as truck hijackings (Simon, 1996). No single organization controls all organized crime; rather, many groups operate at all levels of society. Organized crime thrives because there is great demand for illegal goods and services. Criminal organizations initially gain control of illegal activities by combining threats and promises. For example, small-time operators running drug or prostitution rings may be threatened with violence if they compete with organized crime or fail to make required payoffs (Cressey, 1969).

Apart from their illegal enterprises, organized crime groups have infiltrated the world of legitimate business. Known linkages between legitimate businesses and organized crime exist in banking, hotels and motels, real estate, garbage collection, vending machines, construction, delivery and long-distance hauling, garment manufacture, insurance, stocks and bonds, vacation resorts, and funeral parlors (National Council on Crime and Delinquency, 1969). In addition, some law enforcement and government officials are corrupted through bribery, campaign contributions, and favors intended to buy them off.

Political Crime
The term *political crime* **refers to illegal or unethical acts involving the usurpa-**

Over the years, there have been many notorious leaders of organized crime syndicates. Shown here is Salvatore ("Sammy the Bull") Gravano, who was sentenced to prison for masterminding a drug ring.

© 2003 AP/Wide World Photos

tion of power by government officials, or illegal/ unethical acts perpetrated against the government by outsiders seeking to make a political statement, undermine the government, or overthrow it. Government officials may use their authority unethically or illegally for the purpose of material gain or political power (Simon, 1996). They may engage in graft (taking advantage of political position to gain money or property) through bribery, kickbacks, or "insider" deals that financially benefit them. For example, in the late 1980s, several top Pentagon officials were found guilty of receiving bribes for passing classified information on to major defense contractors that had garnered many lucrative contracts from the government (Simon, 1996).

Other types of corruption have been costly for taxpayers, including dubious use of public funds and public property, corruption in the regulation of commercial activities (such as food inspection), graft in zoning and land use decisions, and campaign contributions and other favors to legislators that corrupt the legislative process. Whereas some political crimes are for personal material gain, others (such as illegal wiretapping and political "dirty tricks") are aimed at gaining or maintaining political office or influence.

Some acts committed by agents of the government against persons and groups believed to be threats to national security are also classified as political crimes. Four types of political deviance have been attributed to some officials: (1) secrecy and deception designed to manipulate public opinion, (2) abuse of power, (3) prosecution of individuals due to their political activities, and (4) official violence, such as police brutality against people of color or the use of citizens as unwilling guinea pigs in scientific research (Simon, 1996).

Political crimes also include illegal or unethical acts perpetrated against the government by outsiders seeking to make a political statement or to undermine or overthrow the government. Examples include treason, acts of political sabotage, and terrorist attacks on public buildings.

Crime Statistics

How useful are crime statistics as a source of information about crime? As mentioned previously, official crime statistics provide important information on crime; however, the data reflect only those crimes that have been reported to the police. Although the rates have been decreasing slightly during the past few years, the UCR reflects that overall levels of crime increased by nearly two-thirds over the past twenty-five years. However, this increase may reflect (at least partially) an increase in the number of crimes *reported,* not necessarily a change in the number of crimes *committed.* Why are some crimes not reported? People are more likely to report crime when they believe that something can be done about it (apprehension of the perpetrator or retrieval of their property, for example). About half of all assault and robbery victims do not report the crime because they may be embarrassed or fear reprisal by the perpetrator. Thus, the number of crimes reported to police represents only the proverbial "tip of the iceberg" when compared with all offenses actually committed. Official statistics are problematic in social science research because of these limitations.

The National Crime Victimization Survey was developed by the Bureau of Justice Statistics as an alternative means of collecting crime statistics. In this annual survey, the members of 100,000 randomly selected households are interviewed to determine whether they have been the victims of crime, even if the crime was not reported to the police. The most recent victimization survey indicates that 62 percent of all crimes are not reported to the police and are thus not reflected in the UCR (Lightblau, 1999).

Studies based on anonymous self-reports of criminal behavior also reveal much higher rates of crime than those found in official statistics. For example, self-reports tend to indicate that adolescents of all social classes violate criminal laws. However, official statistics show that those who are arrested and placed in juvenile facilities typically have limited financial resources, have repeatedly committed serious offenses,

or both (Steffensmeier and Allan, 2000). Data collected for the Juvenile Court Statistics Program also reflect class and racial bias in criminal justice enforcement. Not all children who commit juvenile offenses are apprehended and referred to court. Children from white, affluent families are more likely to have their cases handled outside the juvenile justice system (for example, a youth may be sent to a private school or hospital rather than to a juvenile correctional facility).

Many crimes committed by persons of higher socioeconomic status in the course of business are handled by administrative or quasi-judicial bodies, such as the Securities and Exchange Commission or the Federal Trade Commission, or by civil courts. As a result, many elite crimes are never classified as "crimes," nor are the businesspeople who commit them labeled as "criminals."

Terrorism and Crime

Since 2001, the United States and other nations have been confronted with a difficult prospect: how to deal with terrorism. As compared with crimes that are committed by the citizens of one country, how are sociologists and criminologists to explain world terrorism, which may have its origins in more than one nation and include diverse "cells" of terrorists who operate in a somewhat gang-like manner but are believed to be following directives from leaders elsewhere? In order to deal with the aftermath of the terrorist attacks on New York City and Washington, D.C., government officials focused on "known enemies" of the United States, including Osama bin Laden and Saddam Hussein, who was the president of Iraq. The anthrax attacks that followed the September 11 terrorism were even more difficult to pinpoint; at one point, the FBI indicated that—despite following up on thousands of leads—it was no closer to knowing the source of these acts of deadly bioterrorism. The nebulous nature of the "enemy" and the limitations on any one government to identify and apprehend the perpetrators of such acts of terrorism have resulted in a "war on terrorism." Social scientists who use a rational choice approach suggest that terrorists are rational actors who constantly calculate the gains and losses of participation in violent—and sometimes suicidal—acts against others. Chapter 14 ("Politics and Government in Global Perspective") further discusses the issue of terrorism.

Street Crimes and Criminals

Given the limitations of official statistics, is it possible to determine who commits crimes? We have much

© 2003 AP/Wide World Photos

Since the terrorist attacks of 2001, more emphasis has been placed on identifying groups and apprehending individuals who may be linked to possible terrorism. Zacarias Moussaoui, shown here, is the only person to stand trial in the United States charged with conspiracy in connection with the 9/11 attacks.

more information available about conventional (street) crime than elite crime; therefore, statistics concerning street crime do not show who commits all types of crime. Gender, age, class, and race are important factors in official statistics pertaining to street crime.

Gender and Crime Before considering differences in crime rates by males and females, three similarities should be noted. First, the three most common arrest categories for both men and women are driving under the influence of alcohol or drugs (DUI), larceny, and minor or criminal mischief types of offenses. These three categories account for about 47 percent of all male arrests and about 49 percent of all female arrests. Second, liquor law violations (such as underage drinking), simple assault, and disorderly conduct are middle-range offenses for both men and women. Third, the rate of arrests for murder, arson, and embezzlement is relatively low for both men and women (Steffensmeier and Allan, 2000).

The most important gender differences in arrest rates are reflected in the proportionately greater involvement of men in major property crimes (such as robbery and larceny-theft) and violent crime, as shown in Figure 7.3. In 2002, men accounted for almost 90 percent of robberies and murders and 63 percent of all larceny-theft arrests in the United States. Of those

types of offenses, males under age 18 accounted for approximately 20 percent of the 2002 arrests. The property crimes for which women most frequently are arrested are nonviolent in nature, including shoplifting, theft of services, passing bad checks, credit card fraud, and employee pilferage. When women are arrested for serious violent and property crimes, they are typically seen as accomplices to men who planned the crime and instigated its commission (Steffensmeier and Allan, 2000). However, one study found that some women play an active role in planning and carrying out robberies (Sommers and Baskin, 1993).

Age and Crime Of all factors associated with crime, the age of the offender is one of the most significant. Arrest rates for index crimes are highest for people between the ages of 13 and 25, with the peak being between ages 16 and 17. In 2002, persons under age 25 accounted for more than 54 percent of all arrests for index crimes (FBI, 2003). Individuals under age 18 accounted for over 23 percent of all arrests for robbery and 29 percent of all arrests for larceny-theft.

Scholars do not agree on the reasons for this age distribution. In one study, the sociologist Mark Warr (1993) found that peer influences (defined as exposure to delinquent peers, time spent with peers, and loyalty to peers) tend to be more significant in explaining delinquent behavior than age itself.

The median age of those arrested for aggravated assault and homicide is somewhat older, generally in the late twenties. Typically, white-collar criminals are even older because it takes time to acquire both a high-ranking position and the skills needed to commit this type of nonindex crime.

Rates of arrest remain higher for males than females at every age and for nearly all offenses. This female-to-male ratio remains fairly constant across all age categories. The most significant gender difference in the age curve is for prostitution (a nonindex crime). In 2002, 58 percent of all women arrested for prostitution were under age 35. For individuals over age 45, many more men than women are arrested for sex-related offenses (including procuring the services of a prostitute). This difference has been attributed to a more stringent enforcement of prostitution statutes when young females are involved (Chesney-Lind, 1997). It has also been suggested that opportunities for prostitution are greater for younger women. This age difference may not have the same impact on males, who continue to purchase sexual services from young females or males (see Steffensmeier and Allan, 2000).

Social Class and Crime Individuals from all social classes commit crimes; they simply commit

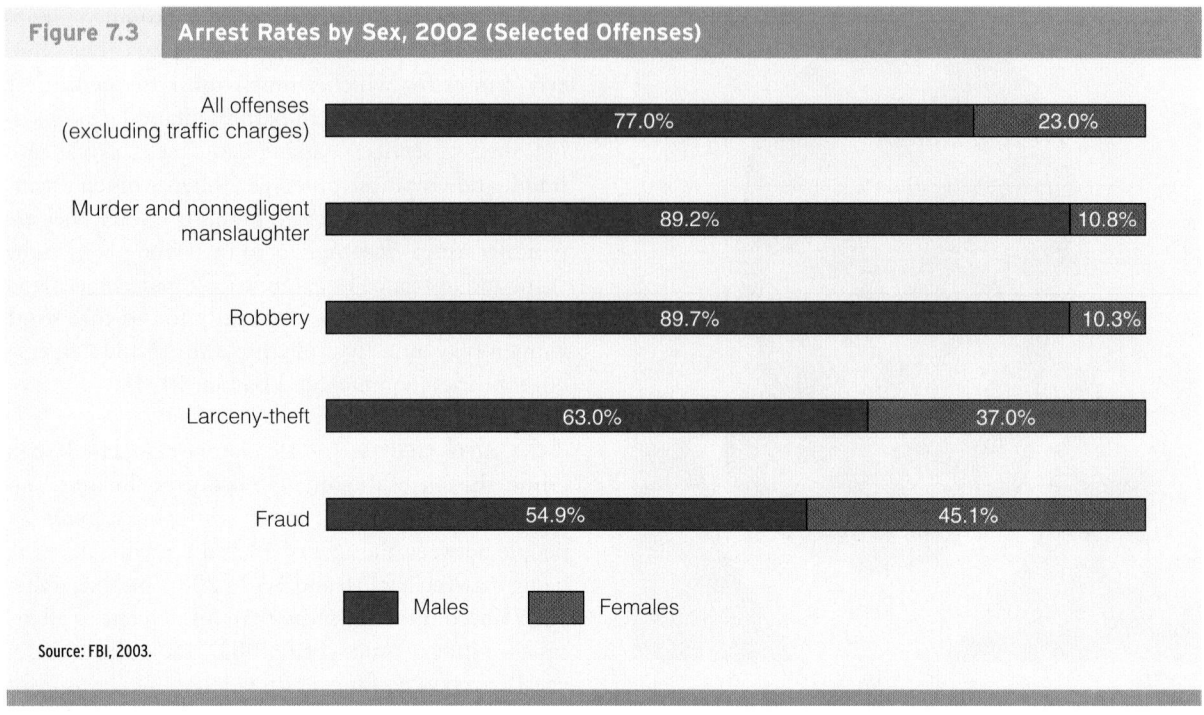

Figure 7.3   Arrest Rates by Sex, 2002 (Selected Offenses)

All offenses (excluding traffic charges): Males 77.0%, Females 23.0%

Murder and nonnegligent manslaughter: Males 89.2%, Females 10.8%

Robbery: Males 89.7%, Females 10.3%

Larceny-theft: Males 63.0%, Females 37.0%

Fraud: Males 54.9%, Females 45.1%

Males / Females

Source: FBI, 2003.

different kinds of crimes. Persons from lower socio-economic backgrounds are more likely to be arrested for violent and property crimes. By contrast, persons from the upper part of the class structure generally commit white-collar or elite crimes, although only a very small proportion of these individuals will ever be arrested or convicted of a crime.

What about social class and recent violence by youths? Between 1994 and 1999, there were 253 violent deaths in U.S. schools (U.S. Department of Education, 2001b). Most of these deaths were not attributed to lower-income, inner-city youths, as popular stereotypes might suggest. Instead, some of these acts of violence were perpetrated by young people who lived in houses that cost anywhere from $75,000 to $5 million or more (Gibbs, 1999).

Similarly, membership in today's youth gangs cannot be identified with just one social class. Increases in gang membership among middle-class suburban youths have been reported (Henneberger, 1993). However, the lower classes are heavily represented in central-city gangs. Today, females are more visible in some previously all-male gangs as well as in female gangs. Studies suggest that both male and female gangs vary along class lines (Taylor, 1993).

In any case, official statistics are not an accurate reflection of the relationship between social class and crime. Self-report data from offenders themselves may be used to gain information on family income,

years of education, and occupational status; however, such reports rely on respondents to report information accurately and truthfully.

Race and Crime In 2002, whites (including Latinos/as) accounted for almost 66 percent of all arrests for index crimes, as shown in Figure 7.4. Compared with African Americans, arrest rates for whites were higher for nonviolent property crimes such as fraud (a nonindex crime) and larceny-theft but were lower for violent crimes such as robbery and murder. In 2002, whites accounted for about 68 percent of all arrests for property crimes and about 60 percent of arrests for violent crimes. African Americans accounted for 38 percent of arrests for violent crimes and 30 percent of arrests for property crimes (FBI, 2003).

Although official arrest records reveal certain trends, these data tell us very little about the actual dynamics of crime by racial–ethnic category. According to official statistics, African Americans are overrepresented in arrest data (Harris, 1991). In 2002, African Americans made up about 12 percent of the U.S. population but accounted for almost 27 percent of all arrests. Latinos/as made up about 12 percent of the U.S. population and accounted for about 13 percent of all arrests. Over two-thirds of their offenses were for nonindex crimes such as alcohol- and drug-related offenses, and disorderly conduct. In 2002, about 1 percent of all arrests were of Asian Americans

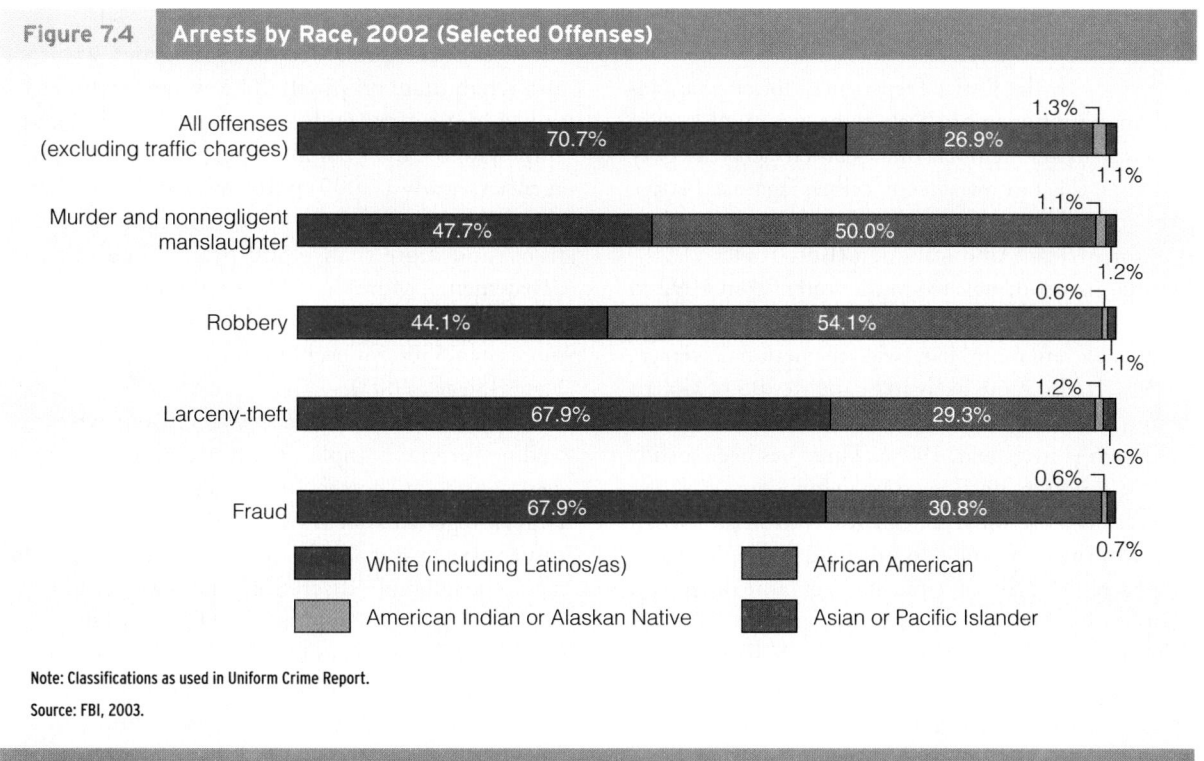

Figure 7.4   Arrests by Race, 2002 (Selected Offenses)

All offenses (excluding traffic charges): White 70.7%, African American 26.9%, American Indian or Alaskan Native 1.3%, Asian or Pacific Islander 1.1%

Murder and nonnegligent manslaughter: White 47.7%, African American 50.0%, American Indian or Alaskan Native 1.1%, Asian or Pacific Islander 1.2%

Robbery: White 44.1%, African American 54.1%, American Indian or Alaskan Native 0.6%, Asian or Pacific Islander 1.1%

Larceny-theft: White 67.9%, African American 29.3%, American Indian or Alaskan Native 1.2%, Asian or Pacific Islander 1.6%

Fraud: White 67.9%, African American 30.8%, American Indian or Alaskan Native 0.6%, Asian or Pacific Islander 0.7%

Legend: White (including Latinos/as); African American; American Indian or Alaskan Native; Asian or Pacific Islander

Note: Classifications as used in Uniform Crime Report.

Source: FBI, 2003.

or Pacific Islanders, and about 1 percent were of Native Americans (designated in the UCR as "American Indian" or "Alaskan Native"). For the general population, the majority of arrests were for larceny-theft, assaults, vandalism, and alcohol- and drug-related violations (FBI, 2003).

Criminologist Coramae Richey Mann (1993) has argued that arrest statistics are not an accurate reflection of the crimes actually committed in our society. Reporting practices differ in accordance with race and social class. Arrest statistics reflect the UCR's focus on index crimes, and especially property crimes, which are committed primarily by low-income people. This emphasis on index crimes draws attention away from the white-collar and elite crimes committed by middle- and upper-income people (Harris and Shaw, 2000). Police may also demonstrate bias and racism in their decisions regarding whom to question, detain, or arrest under certain circumstances (Mann, 1993).

Another reason that statistics may show a disproportionate number of people of color being arrested is because of the focus of law enforcement on certain types of crime and certain neighborhoods in which crime is considered more prevalent. As discussed previously, many poor, young, central-city males turn to forms of criminal activity due to their belief that no opportunities exist for them to earn a living wage

through legitimate employment. Because of the trend of law enforcement efforts to focus on drug-related offenses, arrest rates for young people of color have risen rapidly. These young people are also more likely to live in central-city areas, where there are more police patrols to make arrests.

Finally, arrest should not be equated with guilt: Being arrested does not mean that a person is guilty of the crime with which he or she has been charged. In the United States, individuals accused of crimes are, at least theoretically, "innocent until proven guilty" (Mann, 1993).

Crime Victims

Based on the National Crime Victimization Survey (NCVS), men are more likely to be victimized by crime, although women tend to be more fearful of crime, particularly crimes directed toward them, such as forcible rape (Warr, 2000). Victimization surveys indicate that men are the most frequent victims of most crimes of violence and theft. Among males who are now 12 years old, an estimated 89 percent will be the victims of a violent crime at least once during their lifetime, as compared with 73 percent of females. The elderly also tend to be more fearful of crime but are the least likely to be victimized. Young men of color between the ages of 12 and 24 have the highest criminal victimization rates.

A study by the Justice Department found that Native Americans are more likely to be victims of violent crimes than are members of any other racial category and that the rate of violent crimes against Native American women was nearly 50 percent higher than that for African American men (Butterfield, 1999). During the period covered in the study (from 1992 to 1996), Native Americans were the victims of violent crimes at a rate more than twice the national average. They were also more likely to be the victims of violent crimes committed by members of a race other than their own (Butterfield, 1999). There has been a shift over the past twenty years in which more Native Americans have moved from reservations to urban areas. In the cities they do not tend to live in segregated areas, so they come into contact more often with people of other racial and ethnic groups, whereas African Americans and whites are more likely to live in segregated areas of the city and commit violent crimes against other people in their same racial or ethnic category. According to the survey, the average annual rate at which Native Americans were victims of crime—124 crimes per 1,000 people, ages 12 or older—is about two-and-a-half times the national average of 50 crimes per 1,000 people who are above the age of 12. By comparison, the average annual rate for whites was 49 crimes per 1,000 people, for African Americans, 61 per 1,000, and for Asian Americans, 29 per 1,000 (Butterfield, 1999).

The burden of robbery victimization falls more heavily on some categories of people than others. NCVS data indicate that males are robbed at almost twice the rate of females. African Americans are more than twice as likely to be robbed as whites. Young people have a much greater likelihood of being robbed than middle-aged and older persons. Persons from lower-income families are more likely to be robbed than people from higher-income families (U.S. Bureau of Justice Statistics, 2003b).

THE CRIMINAL JUSTICE SYSTEM

Of all of the agencies of social control (including families, schools, and churches) in contemporary societies, only the criminal justice system has the power to control crime and punish those who are convicted of criminal conduct. The *criminal justice system* refers to the more than 55,000 local, state, and federal agencies that enforce laws, adjudicate crimes, and treat and rehabilitate criminals. The system includes the police, the courts, and corrections facilities, and it employs more than 2 million people in 17,000 police agencies, nearly 17,000 courts, more than 8,000 prosecutorial agencies, about 6,000 correctional institutions, and more than 3,500 probation and parole departments. More than $150 billion is spent annually for civil and criminal justice, which amounts to more than $500 for every person living in the United States (Siegel, 2003).

The term *criminal justice system* is somewhat misleading because it implies that law enforcement agencies, courts, and correctional facilities constitute one large, integrated system when, in reality, the criminal justice system is made up of many bureaucracies that have considerable discretion in how decisions are made. *Discretion* refers to the use of personal judgment by police officers, prosecutors, judges, and other criminal justice system officials regarding whether and how to proceed in a given situation (see Figure 7.5). The police are a prime example of discretionary processes because they have the power to selectively enforce the law and have on many occasions been accused of being too harsh or too lenient on alleged offenders.

The Police

The role of the police in the criminal justice system continues to expand. The police are responsible for crime control and maintenance of order, but local police departments now serve numerous other human service functions, including improving community relations, resolving family disputes, and helping people during emergencies. It should be remembered that not all "police officers" are employed by local police departments; they are employed in more than 25,000 governmental agencies ranging from local jurisdictions to federal levels. However, we will focus primarily on metropolitan police departments because they constitute the vast majority of the law enforcement community.

Metropolitan police departments are made up of a chain of command (similar to the military), with ranks such as officer, sergeant, lieutenant, and captain, and each rank must follow specific rules and procedures. However, individual officers maintain a degree of discretion in the decisions they make as they respond to calls and try to apprehend fleeing or violent offenders. The problem of police discretion is most acute when decisions are made to use force (such as grabbing, pushing, or hitting a suspect) or

| Figure 7.5 | Discretionary Powers in Law Enforcement |

Police
Enforce specific laws
Investigate specific crimes
Search people, vicinities, buildings
Arrest or detain people

Prosecutors
File charges or petitions for judicial decision
Seek indictments
Drop cases
Reduce charges
Recommend sentences

Judges or Magistrates
Set bail or conditions for release
Accept pleas
Determine delinquency
Dismiss charges
Impose sentences
Revoke probation

deadly force (shooting and killing a suspect). Generally, deadly force is allowed only in situations in which a suspect is engaged in a felony, is fleeing the scene of a felony, or is resisting arrest and has endangered someone's life.

Although many police departments have worked to improve their public image in recent years, the practice of *racial profiling*—the use of ethnic or racial background as a means of identifying criminal suspects—remains a highly charged issue. Officers in some police departments have singled out for discriminatory treatment African Americans, Latinos/Latinas, and other people of color, treating them more harshly than white (Euro-American) individuals. However, police department officials typically contend that race is only one factor in determining why individuals are questioned or detained as they go about everyday activities such as driving a car or walking down the street. By contrast, equal-justice advocacy groups argue that differential treatment of minority-group members amounts to a race-based double standard, which they believe exists not only in police work but throughout the criminal justice system (see Cole, 2000).

The belief that differential treatment takes place on the basis of race contributes to a negative image of police among many people of color who believe that they have been hassled by police officers, and this assumption is intensified by the fact that police departments have typically been made up of white male

personnel at all levels. In recent years, this situation has slowly begun to change. Currently, about 22 percent of all *sworn officers*—those who have taken an oath and been given the powers to make arrests and use necessary force in accordance with their duties—are women and minorities (Cole and Smith, 2004). The largest percentage of minority and women police officers are located in cities with a population of 250,000 or more. African Americans make up a larger percentage of the police department in cities with a larger proportion of African American residents (such as Detroit) while Latinos/Latinas constitute a larger percentage in cities such as San Antonio and El Paso, Texas, where Latinos/Latinas make up a larger proportion of the population. Women officers of all races are more likely to be employed in larger departments in cities of more than 250,000 (where they make up 16 percent of all officers) as compared with smaller communities (cities of less than 50,000), where women officers make up only 2 to 5 percent of the force (Cole and Smith, 2004). In the past, women were excluded from police departments and other law enforcement careers largely because of stereotypical beliefs that they were not physically and psychologically strong enough to enforce the law. However, studies have indicated that as more females have entered police work, they receive similar evaluations to male officers from their administrators and that fewer complaints are filed against women officers, which some researchers believe is a function of how female officers more effectively control potentially violent encounters (Brandl, Stroshine, and Frank, 2001).

In the future, the image of police departments may change as greater emphasis is placed on *community-oriented policing*—an approach to law enforcement in which officers maintain a presence in the community, walking up and down the streets or riding bicycles, getting to know people, and holding public service meetings at schools, churches, and other neighborhood settings. Community-oriented policing is often limited by budget constraints and lack of available personnel to conduct this type of "hands-on" community involvement. In many jurisdictions, police officers believe that they have only enough time to keep up with reports of serious crime and life-threatening occurrences and that the level of available personnel and resources does not allow officers to take on a greatly expanded role in the community.

The Courts

Criminal courts determine the guilt or innocence of those persons accused of committing a crime. In theory, justice is determined in an adversarial process in which the prosecutor (an attorney who represents the state) argues that the accused is guilty and the defense attorney asserts that the accused is innocent. In reality, judges wield a great deal of discretion. Working with prosecutors, they decide whom to release and whom to hold for further hearings, and what sentences to impose on those persons who are convicted.

Prosecuting attorneys also have considerable leeway in deciding which cases to prosecute and when to negotiate a plea bargain with a defense attorney. As cases are sorted through the legal machinery, a steady attrition occurs. At each stage, various officials determine what alternatives will be available for those cases still remaining in the system. These discretionary decisions often have a disproportionate impact on youthful offenders who are poor (see Box 7.3).

About 90 percent of criminal cases are never tried in court; instead, they are resolved by plea bargaining, a process in which the prosecution negotiates a reduced sentence for the accused in exchange for a guilty plea (Senna and Siegel, 2002). Defendants (especially those who are poor and cannot afford to pay an attorney) may be urged to plead guilty to a lesser crime in return for not being tried for the more serious crime for which they were arrested. Prison sentences given in plea bargains vary widely from one region to another and even from judge to judge within one state.

Those who advocate the practice of plea bargaining believe that it allows for individualized justice for alleged offenders because judges, prosecutors, and defense attorneys can agree to a plea and punishment that best fits the offense and the offender. They also believe that this process helps reduce the backlog of criminal cases in the court system as well as the lengthy process often involved in a criminal trial. However, those who seek to abolish plea bargaining believe that this practice leads to innocent people pleading guilty to crimes they have not committed or pleading guilty to a crime other than the one they actually committed because they are offered a lesser sentence (Cole and Smith, 2004).

More serious crimes, such as murder, felonious assault, and rape, are more likely to proceed to trial than other forms of criminal conduct; however, many of these cases do not reach the trial stage. For example, one study of 75 of the largest counties in the United States found that only 26 percent of murder cases actually went to trial. By contrast, only about 6 percent of all other cases proceeded to trial (Reaves, 2001).

One of the most important activities of the court system is establishing the sentence of the accused after he or she has been found guilty or has pleaded guilty. Typically, sentencing involves the following kinds of sentences or dispositions: fines, probation, alternative or intermediate sanctions (such as house arrest or electronic monitoring), incarceration, and capital punishment (Siegel, 2003).

Punishment and Corrections

Punishment **is any action designed to deprive a person of things of value (including liberty) because of some offense the person is thought to have committed** (Barlow and Kauzlarich, 2002). Historically, punishment has had four major goals:

1. *Retribution* is punishment that a person receives for infringing on the rights of others (Cole and Smith, 2004). Retribution imposes a penalty on the offender and is based on the premise that the punishment should fit the crime: The greater the degree of social harm, the more the offender should be punished. For example, an individual who murders should be punished more severely than one who shoplifts. This function has received renewed interest over the past three decades as some critics have argued that the concept of rehabilitation is not working to reduce criminal behavior.

2. *General deterrence* seeks to reduce criminal activity by instilling a fear of punishment in the general public. However, we most often focus on *specific deterrence*, which inflicts punishment on

Box 7.3 SOCIOLOGY AND SOCIAL POLICY

Juvenile Offenders and "Equal Justice Under the Law"

When you walk into the U.S. Supreme Court building in Washington, D.C., it is impossible to miss the engraved statement overhead: "Equal Justice Under the Law." Do young people, regardless of race, class, or gender, receive the same treatment under the law?

In courtrooms throughout the nation, judges have a wide range of discretion in their decisions regarding juveniles alleged to have committed some criminal or status offense. Whereas judges in television courtroom dramas are often African Americans, women, or members of other subordinate groups, "real-life" judges typically come from capitalist or managerial and professional backgrounds. Because more than 90 percent are white and most are male, their decisions may reflect a built-in class, racial, and gender bias.

Juvenile courts were established under a different premise than courts for adults. Under the doctrine of *parens patriae* (the state as parent), the stated purpose of juvenile courts has been to care for, rather than punish, youthful offenders. In theory, less weight is given to offenses and more weight to the youth's physical, mental, or social condition. The juvenile court seeks to change or resocialize offenders through treatment or therapy, not to punish them. Consequently, judges in juvenile courts are given relatively wide latitude, or discretion, in the decisions they mete out regarding young offenders.

Unlike adult offenders, juveniles are not always represented by legal counsel. A juvenile hearing is not a trial but rather an informal private hearing before a judge or probation officer with only the young person and a parent or guardian present. No jury is convened, and the juvenile offender does not cross-examine her or his accusers. In addition, the offender is not "sentenced"; rather, the case is "adjudicated" or "dis-

posed of." Finally, the offender is not "punished" but instead may be "remanded to the custody" of a youth authority in order to receive training, treatment, or care.

Because of judicial discretion, courts may treat juveniles differently based on gender. Considerable disparity exists in the disposition of juvenile cases, with much of the variation thought to result from judges' beliefs rather than objective facts in the case. Female offenders are more likely than males to be institutionalized for committing status offenses such as truancy, running away from home, and other offenses that serve as "buffer charges" for suspected sexual misconduct (Chesney-Lind, 1989).

Disparity also exists on the basis of race and class. Judges tend to see youths from white, middle- or upper-class families as being very much like their own children and to believe that the families will take care of the problem on their own. They may view juveniles from lower-income families or other racial–ethnic groups as delinquents in need of attention from authorities. Furthermore, some judges view gang members from impoverished central cities as "guilty by association" because of their companions.

The political climate may have an effect on how judges dispose of juvenile cases. In the process of dealing with the public perception that the juvenile justice system is too lenient, some judges may have inadvertently contributed to other problems. Many more youths have been remanded to overcrowded juvenile detention facilities that are unable to provide necessary educational, health, and social services. Based on a judge's discretion, many juvenile offenders are incarcerated under indeterminate sentences and placed in a detention facility that may serve merely as a school for adult criminality.

Sources: Based on Barlow and Kauzlarich, 2002; Chesney-Lind, 1989; and Inciardi, Horowitz, and Pottieger, 1993.

---

specific criminals to discourage them from committing future crimes. Recently, criminologists have debated whether imprisonment has a deterrent effect, given the fact that high rates of those who are released from prison become recidivists (previous offenders who commit new crimes).

3. *Incapacitation* is based on the assumption that offenders who are detained in prison or are executed will be unable to commit additional crimes. This approach is often expressed as "lock

'em up and throw away the key!" In recent years, more emphasis has been placed on *selective incapacitation*, which means that offenders who repeat certain kinds of crimes are sentenced to long prison terms (Cole and Smith, 2004).

4. *Rehabilitation* seeks to return offenders to the community as law-abiding citizens by providing therapy or vocational or educational training. Based on this approach, offenders are treated, not punished, so that they will not continue their

In recent years, military-style boot camps such as this one have been used as an alternative to prison and long jail terms for nonviolent offenders under age 30. Critics argue that structural solutions—not stopgap measures—are needed to reduce crime.

© Porter Gifford/Getty Images

criminal activity. However, many correctional facilities are seriously understaffed and underfunded in the rehabilitation programs that exist. The job skills (such as agricultural work) that many offenders learn in prison do not transfer to the outside world, nor are offenders given any assistance in finding work that fits their skills once they are released.

Recently, newer approaches have been advocated for dealing with criminal behavior. Key among these is the idea of *restoration,* which is designed to repair the damage done to the victim and the community by an offender's criminal act (Cole and Smith, 2004). This approach is based on the *restorative justice perspective,* which states that the criminal justice system should promote a peaceful and just society; therefore, the system should focus on peacemaking rather than on punishing offenders. Advocates of this approach believe that punishment of offenders actually encourages crime rather than deterring it and are in favor of approaches such as probation with treatment. Opponents of this approach suggest that increased punishment of offenders leads to lower crime rates and that the restorative justice approach amounts to "coddling criminals." However, numerous restorative justice programs are now in operation, and many are associated with community policing programs as they seek to help offenders realize the damage that they have done to their victims and the community and to be reintegrated into society (Senna and Siegel, 2002).

Instead of the term *punishment,* the term *corrections* is often used. Criminologists George F. Cole and Christopher E. Smith (2004: 409) explain corrections as follows:

Corrections refers to the great number of programs, services, facilities, and organizations responsible for the management of people accused or convicted of criminal offenses. In addition to prisons and jails, corrections includes probation, halfway houses, education and work release programs, parole supervision, counseling, and community service. Correctional programs operate in Salvation Army hostels, forest camps, medical clinics, and urban storefronts.

As Cole and Smith (2004) explain, corrections is a major activity in the United States today. Consider the fact that about 6.5 million adults (more than one out of every twenty men and one out of every hundred women) are under some form of correctional control. The rate of African American males under some form of correctional supervision is even greater (one out of every six African American adult men and one out of three African American men in their twenties). Some analysts believe that these figures are a reflection of centuries of underlying racial, ethnic, and class-based inequalities in the United States as well as sentencing disparities that reflect race-based differences in the criminal justice system. However, others argue that newer practices such as determinate or mandatory sentences may help to reduce such disparities over time. A *determinate sentence* sets the term of imprisonment at a fixed period of time (such as three years) for a specific offense. *Mandatory sentencing guidelines* are established by law and require that a person convicted of a specific offense or series of offenses be given a penalty within a fixed range. Although these practices limit judicial discretion in sentencing, many critics are concerned about the effects of these sentencing approaches. Another area of great discord within and outside the criminal justice system is the issue of the death penalty.

The Death Penalty

Historically, removal from the group has been considered one of the ultimate forms of punishment. For many years, capital punishment, or the death penalty, has been used in the United States as an appropriate and justifiable response to very serious crimes. In 2002, 71 inmates were executed and more than 3,500 people awaited execution, having received the death penalty under federal law or the laws of one of the 38 states that have the death penalty (U.S. Bureau of Justice Statistics, 2003a). By far the largest percentage (about two-thirds) of those on death row are in southern states, including Alabama, Florida, North Carolina, Oklahoma, and Texas.

Because of the finality of the death penalty, it has been a subject of much controversy and numerous Supreme Court debates about the decision-making process involved in capital cases. In 1972 the U.S. Supreme Court ruled (in *Furman v. Georgia*) that *arbitrary* application of the death penalty violates the Eighth Amendment to the Constitution but that the death penalty itself is not unconstitutional. In other words, capital punishment is legal if it is fairly imposed. Although there have been a number of cases involving death penalty issues before the Supreme Court since that time, the court typically has upheld the constitutionality of this practice. Yet the fact remains that racial disparities are highly evident in the death row census. African Americans make up about 42 percent of the death row population but less than 13 percent of the U.S. population. The ex-slave states are more likely to execute criminals than are other states (see Figure 7.6). African Americans are eight to ten times more likely to be sentenced to death for homicidal rape than are whites (non-Latinos/as) who have committed the same crime (Marquart, Ekland-Olson, and Sorensen, 1994).

People who have lost relatives and friends as a result of criminal activity often see the death penalty as justified. However, capital punishment raises many doubts for those who fear that innocent individuals may be executed for crimes they did not commit. For still others, the problem of racial discrimination in the sentencing process poses troubling questions. Other questions that remain today involve execution of those who are believed to be insane, of juvenile offenders, and of those defendants who did not have effective legal counsel during their trial. In 2002,

for example, the Supreme Court ruled (in *Atkins v. Virginia*) that executing the mentally retarded is unconstitutional. In another landmark case (*Ring v. Arizona*), the Court ruled that juries, not judges, must decide whether a convicted murderer should receive the death penalty (Cole and Smith, 2004). Although about 250 new death sentences are handed down each year, the issue of the death penalty is far from resolved; the debate, which has taken place for more than two hundred years, no doubt will continue well into the twenty-first century.

DEVIANCE AND CRIME IN THE UNITED STATES IN THE FUTURE

Two pressing questions pertaining to deviance and crime will face us in the future: Is the solution to our "crime problem" more law and order? Is equal justice under the law possible?

Although many people in the United States agree that crime is one of the most important problems in this country, they are divided over what to do about it. Some of the frustration about crime might be based on unfounded fears; studies show that the overall crime rate has been decreasing slightly in recent years.

One thing is clear: The existing criminal justice system cannot solve the "crime problem." If roughly 20 percent of all crimes result in arrest, only half of those lead to a conviction in serious cases, and less than 5 percent of those result in a jail term, the "lock 'em up and throw the key away" approach has little chance of succeeding. Nor does the high rate of recidivism among those who have been incarcerated speak well for the rehabilitative efforts of our existing correctional facilities. Reducing street crime may hinge on finding ways to short-circuit criminal behavior.

One of the greatest challenges is juvenile offenders, who may become the adult criminals of tomorrow. However, instead of military-style boot camps or other stopgap measures, *structural solutions*—such as more and better education and jobs, affordable housing, more equality and less discrimination, and socially productive activities—are needed to reduce street crime. In the past, structural solutions such as these have made it possible for immigrants who initially committed street crimes to leave the streets, get jobs, and lead productive lives. Ultimately, the best

Figure 7.6 Death Row Census, April 2003

Death row inmates are heavily concentrated in certain states. African Americans, who make up about 12 percent of the U.S. population, account for approximately 42 percent of inmates on death row.

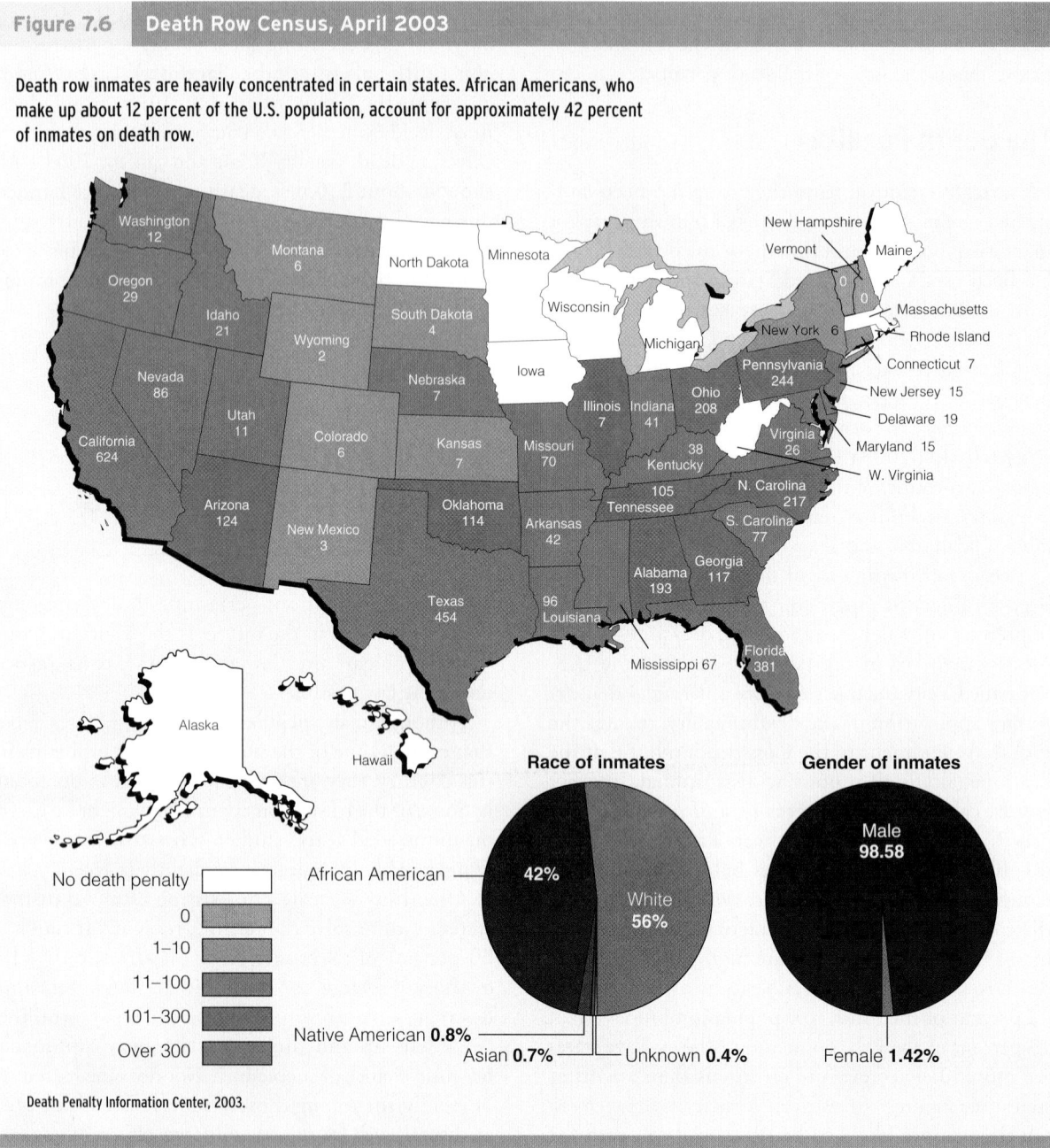

Death Penalty Information Center, 2003.

approach for reducing delinquency and crime would be prevention: to work with young people *before* they become juvenile offenders to help them establish family relationships, build self-esteem, choose a career, and get an education that will help them pursue that career (see Box 7.4). Sociologist Elliott Currie (1998) has proposed that an initial goal in working to prevent delinquency and crime is to pinpoint specifically what kinds of preventive programs work and to establish priorities that make prevention possible.

Among these priorities are preventing child abuse and neglect, enhancing children's intellectual and social development, providing support and guidance to vulnerable adolescents, and working intensively with juvenile offenders (Currie, 1998).

Is equal justice under the law possible? As long as racism, sexism, classism, and ageism exist in our society, people will see deviant and criminal behavior through a selective lens. To solve the problems addressed in this chapter, we must ask ourselves what

Box 7.4 YOU CAN MAKE A DIFFERENCE

Combating Delinquency and Crime at an Early Age

Linda Warsaw of San Bernardino, California, explains that she started Kids Against Crime after being inspired by an eight-year-old girl who insisted that the neighbor who had assaulted her be brought to trial:

> This little girl decided on her own that she wanted this person put away so no other children would have to go through the pain she'd gone through. It was hard for her, but she went through with the trial. After seeing her courage and conviction, I realized kids can make a difference. That's when I went home and drew up the proposal that started Kids Against Crime. (qtd. in Lappé and Du Bois, 1994: 147-148)

When you think about your community, what organization might you start—or volunteer to assist—that could enhance children's lives and help prevent gang violence and delinquency? Consider, for example, these programs related to Kids Against Crime:

- *A peer support hotline.* Peer support hotlines address issues and questions about gangs, drugs, crime, and personal problems.
- *Preventive education programs.* Skits and workshops on topics such as suicide, child abuse, teen pregnancy, and AIDS are presented at shopping malls, schools, and community centers.
- *Improvement projects for neighborhoods.* Children and young people are encouraged to participate in projects to clean up graffiti and improve neighborhoods.
- *Learning public life skills.* Programs include public speaking, planning, active listening, and dealing with the media (see Lappé and Du Bois, 1994).
- *Organizing young people for social change.* Volunteers work with children and young people to organize so that their voices can be heard. For example, here is how an organizer for the Youth

Action Program in New York, which seeks to involve poor teenagers in self-governing training programs in the construction trades to prevent them from becoming involved with gangs and crime, describes this process:

> When we were organizing to appear at the city council, calling for jobs for young people, we practiced every Wednesday night for three months. We practiced walking in an organized way, in single file, filling up every successive seat in a row rather than flowing in to the council in an undisciplined way. We practiced standing in unison and clapping together at the close of every speech of one of our supporters. Nobody wore hats or chewed gum. We knew that this degree of self-discipline, implying no threat but demonstrating internal unity, would have an impact. It did. (qtd. in Lappé and Du Bois, 1994: 62)

For additional information on programs for children and young people, contact one of these sources:

- Kids Against Crime, P.O. Box 22004, San Bernardino, CA 92406. (714) 882-1344.
- KIDS Consortium (Kids Involved Doing Service), P.O. Box 27, East Boothbay, ME 14544. (207) 633-3152.
- Youth Action Program, 1280 Fifth Avenue, New York, NY 10029. (212) 860-8170.
- YouthBuild, 55 Day Street, West Somerville, MA 02144. (617) 623-9900.
- On the Internet, the Center for the Future of Children provides sociological data and other useful information about the well-being of children:

http://futureofchildren.org

- For tips on making crime prevention a family matter, try this book: Susan Hull, *50 Simple Ways to Make Life Safer from Crime* (New York: Pocket Books, 1996).

we can do to ensure the rights of everyone, including the poor, people of color, and women and men alike. Many of us can counter classism, racism, sexism, and ageism where they occur. Perhaps the only way that the United States can have equal justice under the law (and, perhaps, less crime as a result) in the future is to promote social justice for individuals regardless of their race, class, gender, or age.

THE GLOBAL CRIMINAL ECONOMY

Consider this scenario:

> Con men operating out of Amsterdam sell bogus U.S. securities by telephone to Germans; the operation is controlled by an Englishman residing in

Monaco, with his profits in Panama. Which police force should investigate? In which jurisdiction should a prosecution be mounted? There may even be a question about whether a crime has been committed, although if all the actions had taken place in a single country there would be little doubt. (United Nations Development Programme, 1999: 104)

As this example shows, international criminal activity poses new and interesting questions not only for those who are the victims of such actions but also for governmental agencies mandated to control crime.

Global crime—the networking of powerful criminal organizations and their associates in shared activities around the world—is a relatively new phenomenon (Castells, 1998). However, it is an extremely lucrative endeavor as criminal organizations have increasingly set up their operations on a transnational basis, using the latest communication and transportation technologies.

How much money and other resources change hands in the global criminal economy? Although the exact amount of profits and financial flows originating in the global criminal economy is impossible to determine, the 1994 United Nations Conference on Global Organized Crime estimated that about $500 billion (in U.S. currency) a year is accrued in the global trade in drugs alone. Today, profits from all kinds of global criminal activities are estimated to range from $750 billion to over $1.5 trillion a year (United Nations Development Programme, 1999). Some analysts believe that even these figures underestimate the true nature and extent of the global criminal economy (Castells, 1998). The highest income-producing activities of global criminal organizations include trafficking in drugs, weapons, and nuclear material; smuggling of things and people (including many migrants); trafficking in women and children for the sex industry; and trafficking in body parts such as corneas and major organs for the medical industry. Undergirding the entire criminal system is money laundering and various complex financial schemes and international trade networks that make it possible for people to use the resources they obtain through illegal activity for the purposes of consumption and investment in the ("legitimate") formal economy.

Who engages in global criminal activities? According to the sociologist Manuel Castells (1998), the following groups are the major players in the global criminal economy:

The Sicilian *Costa Nostra* (and its associates, *La Camorra, 'Ndrangheta,* and *Sacra Corona Unita*), the American Mafia, the Colombian cartels, the

Mexican cartels, the Nigerian criminal networks, the Japanese *Yakuza,* the Chinese Triads, the constellation of Russian *Mafiyas,* the Turkish heroin traffickers, the Jamaican Posses, and a myriad of regional and local criminal groupings in all countries that come together in a global, diversified network that permeates boundaries and links up ventures of all sorts.

Although these groups have existed for many years in their countries of origin and perhaps surrounding territories, they have expanded rapidly in the era of global communications and rapid transportation networks. In some countries, criminal organizations maintain a quasi-legal existence and are visible in "legitimate" business and political activities. For example, in Japan the *Yakuza* have operated for many years without much scrutiny by the Japanese government. Today, the *Yakuza* have exported their practice of blackmail and extortion of corporations to the United States and other nations, where they send in violent *provocateurs,* known as the *Sokaiya,* to intimidate Japanese executives living abroad (Castells, 1998). Among other global criminal organizations, these Japanese gang members have been able to operate in the United States by investing heavily in real estate and engaging in agreements with the U.S. Mafia and various Russian criminal groups for control over numerous illegal enterprises. According to Castells (1998), networking and strategic alliances between criminal networks have been key factors in the success of the criminal organizations that have sought to expand their criminal activities over the past two decades. It is through such networks that the criminal economy thrives on a global basis, allowing participants to escape police control and live beyond the laws of any one nation.

Can anything be done about global crime? Recent studies have concluded that reducing global crime will require a global response, including the cooperation of law enforcement agencies, prosecutors, and intelligence services across geopolitical boundaries. However, this approach is problematic because countries such as the United States often have difficulty getting the various law enforcement agencies to cooperate within their own nation. Similarly, law enforcement agencies in high-income nations such as the United States and Canada are often suspicious of law enforcement agencies in low-income countries, believing that these officers are corrupt. Regulation by the international community (for example, through the United Nations) would also be necessary to control global criminal activities such as international money laundering and trafficking in people and controlled substances such as drugs and weapons. However, development and enforcement of international agreements on activities such as the smuggling of migrants or trafficking of women and children for the sex industry have been extremely limited thus far. Many analysts acknowledge that economic globalization has provided many opportunities for wealth through global organized crime (Castells, 1998; United Nations Development Programme, 1999).

CHAPTER REVIEW

■ How do sociologists view deviance?

Sociologists are interested in what types of behavior are defined by societies as "deviant," who does that defining, how individuals become deviant, and how those individuals are dealt with by society.

■ What are the main functionalist theories for explaining deviance?

Functionalist perspectives on deviance include strain theory and opportunity theory. Strain theory focuses on the idea that when people are denied legitimate access to cultural goals, such as a good job or a nice home, they may engage in illegal behavior to obtain them. Opportunity theory suggests that for deviance to occur, people must have access to illegitimate means to acquire what they want but cannot obtain through legitimate means.

■ How do symbolic interactionists view deviance?

According to symbolic interactionists, deviance is learned through interaction with others. Differential association theory states that individuals have a greater tendency to deviate from societal norms when they frequently associate with persons who tend toward deviance instead of conformity. According to social control theories, everyone is capable of committing crimes, but social bonding (attachments to family and to other social institutions) keeps many from doing so. According to labeling theory, deviant behavior is that which is labeled deviant. The process of labeling is related to the power and status of those persons who do the labeling and those who are labeled. Generally, those in power label the behavior of others as deviant.

■ How do conflict and feminist perspectives explain deviance?

Conflict perspectives on deviance focus on inequalities in society. According to some conflict theorists, lifestyles considered deviant by those with political and economic power are often defined as illegal. Marxist conflict theorists link deviance and crime to the capitalist society, which divides people into haves and have-nots, leaving crime as the only source of support for those at the bottom of the economic ladder. Feminist approaches to deviance focus on the relationship between gender and deviance. Liberal feminism explains female deviance as a rational response to gender discrimination experienced in work, marriage, and interpersonal relationships. Marxist/radical feminism suggests that patriarchy (male domination of females) contributes to female deviance, especially prostitution. Socialist feminism states that exploitation of women by patriarchy and capitalism is related to women's involvement in criminal acts such as prostitution and shoplifting.

■ What is the postmodernist view on deviance?

Postmodernist views on deviance focus on how control may be maintained through largely invisible forces such as the Panoptican, described by Michel Foucault. According to this approach, everyone—not just persons engaged in deviant or criminal behavior—is likely to be under surveillance by unseen persons. Some postmodern approaches point out the intertwining nature of knowledge, power, and technology while rejecting the grand narratives found in functionalist and conflict approaches.

■ How do sociologists classify crime?

Sociologists identify four main categories of crime: conventional, occupational, organized, and political. Conventional (street) crime includes violent crimes, property crimes, and morals crimes. Occupational (white-collar) crimes are illegal activities committed by people in the course of their employment or financial dealings. Organized crime is a business operation that supplies illegal goods and services for profit. Political crime refers to illegal or unethical acts involving the usurpation of power by government officials, or illegal or unethical acts perpetrated against the government by outsiders seeking to make a political statement, to undermine the government, or to overthrow it.

■ **What are the main sources of crime statistics?**

Official crime statistics are taken from the Uniform Crime Report, which lists crimes reported to the police, and the National Crime Victimization Survey, which interviews households to determine the incidence of crimes, including those not reported to police. Studies show that many more crimes are committed than are officially reported.

■ **How are age and class related to crime statistics?**

Age is the key factor in crime statistics. In 2002, persons under age twenty-five accounted for more than 54 percent of all arrests for index crimes, whereas persons arrested for aggravated assault, homicide, and occupational crimes are usually older. Persons from lower socioeconomic backgrounds are more likely to be arrested for violent and property crimes; white-collar crime is more likely to occur among upper socioeconomic classes.

■ **Who are the most frequent victims of crime?**

Young males of color between ages twelve and twenty-four have the highest criminal victimization rates. The elderly tend to be fearful of crime but are the least likely to be victimized.

■ **How is discretion used in the criminal justice system?**

The criminal justice system, including the police, the courts, and prisons, often has considerable discretion in dealing with offenders. The police often use discretion in deciding whether to act on a situation. Prosecutors and judges use discretion in deciding which cases to pursue and how to handle them.

KEY TERMS

conventional (street) crime 227
corporate crime 230
crime 212
criminology 213
deviance 210
differential association theory 218
illegitimate opportunity structures 216
juvenile delinquency 212
labeling theory 220
occupational (white-collar) crime 229
organized crime 231
political crime 231
primary deviance 221
punishment 238
rational choice theory of deviance 219

secondary deviance 221
social bond theory 219
social control 213
strain theory 215
tertiary deviance 221

QUESTIONS FOR CRITICAL THINKING

1. Does public toleration of deviance lead to increased crime rates? If people were forced to conform to stricter standards of behavior, would there be less crime in the United States?
2. Should so-called victimless crimes, such as prostitution and recreational drug use, be decriminalized? Do these crimes harm society?
3. As a sociologist armed with a sociological imagination, how would you propose to deal with the problem of crime in the United States? What programs would you suggest enhancing? What programs would you reduce?

RESOURCES ON THE INTERNET

Chapter-Related Web Sites

The following Web sites have been selected for their relevance to the topics in this chapter. These sites are among the more stable, but please note that Web site addresses change frequently. For an updated list of chapter-related Web sites with URL links, please visit the *Sociology in Our Times* Web site (**www.wadsworth.com/KendallSIOT**).

Federal Bureau of Investigation (FBI)
http://www.fbi.gov

Visit the FBI's home page to learn more about the agency and to access an array of information and resources on crime. The site features an easy-to-use keyword search engine that allows you to quickly locate information on topics of interest.

Bureau of Justice Statistics (BJS)
http://www.ojp.usdoj.gov/bjs/welcome.html

This Web site provides interesting statistics, reports, and analyses on a number of crime-related topics, including crime victimization, courts and sentencing, firearms and crime, homicide trends, international crime statistics, and crime expenditures.

Court TV.com
http://www.courttv.com

Court TV's Web site provides comprehensive legal coverage and analysis with current events coverage, information on recent criminal and civil cases, legal news about public figures, unusual crime stories, streaming video feeds of trials, legal help links, a searchable crime library, chat rooms, and message boards. Specific topics in the criminal justice system are typically well researched, objectively presented, and very interesting.

ONLINE STUDY AND RESEARCH TOOLS

Accompanying this text are many *free* powerful online study tools that will help you master the material in this chapter, help increase your depth of understanding, and help you make the grade!

SocCoach CD-ROM

Use the SocCoach CD-ROM enclosed with this text to help you formulate a customized study plan for this chapter. After you take the Diagnostic Quiz, SocCoach will generate a customized study plan just for you! It will identify sections of the chapter that you should review and will provide videos, charts, graphs, and excerpts from the text to supplement your studies and enhance your understanding. You'll also find fun, interactive activities such as Virtual Explorations and Map the Stats to apply what you've learned and stretch your sociological imagination.

The Companion Web Site for Sociology in Our Times, *Fifth Edition*

www.wadsworth.com/KendallSIOT

Gain an even better grasp on this chapter by going to the companion Web site to take one of the Tutorial Quizzes, use the Flash Cards to master key terms, or check out the many other study aids you'll find there. You'll also find special features such as GSS Data and Census 2000 information that'll put data and resources at your fingertips to help you with that special project or help you as you do some research on your own.

In this chapter, when you see the icon on the left, it alerts you to a specific exercise found in *Wadsworth's Sociology Online Resources and Writing Companion.* This valuable guide shows you how to use Wadsworth's exclusive online resources—*InfoTrac College Edition,* the *Opposing Viewpoints Resource Center,* and *MicroCase Online*—to assist you in your study of sociology and to build essential research and writing skills.

Global Stratification

When one is poor, she has no say in public, she feels inferior. She has no food, so there is famine in her house; no clothing, and no progress in her family.

—A poor woman in Uganda

A better life for me is to be healthy, peaceful and live in love without hunger. Love is more than anything. Money has no value in the absence of love.

—A poor woman in Ethiopia

If you want to do something and have no power to do it, it is talauchi (poverty).

—A person in Nigeria

For a poor person everything is terrible—illness, humiliation, shame. We are cripples; we are afraid of everything; we depend on everyone. No one needs us. We are like garbage that everyone wants to get rid of.

—A woman in Moldova

Poverty is lack of freedom, enslaved by crushing daily burden, by depression and fear of what the future will bring.

—A person in Georgia (World Bank, 2003b)

In many African and Eastern European nations, and in other low-income nations around the world, poverty is a pressing problem that pervades all aspects of daily life for many people. The wide disparities between high-income and low-income nations were highlighted in 2003, when President George W. Bush traveled to Africa to discuss how the United States might promote economic stability and help fight HIV/AIDS on that continent. At the time of this trip, Jeffrey D. Sachs (2003: A23), director of the Earth Institute at Columbia University, wrote a column in the *New York Times* describing the sharp contrast between the wealth of a few people in the United States and the vast poverty of many on the African continent:

> According to a recent report from the Internal Revenue Service, some 400 superrich Americans had an average income of nearly $174 million each, or a combined income of $69 billion, in 2000. Incredibly, that's more than the combined incomes of the 166 million people living in four of the countries that the president [visited]: Nigeria, Senegal, Uganda and Botswana.

Whether people live in Africa, Eastern Europe, or elsewhere in the world, social and economic inequality are pressing daily concerns. Poverty and inequality know no political boundaries or national borders. In

■ In some areas of the world, running water in households is taken for granted. In nations such as Mali, low-income women continue to carry water to their homes. Will the people of all nations more equally participate in economic prosperity in the twenty-first century?

this chapter, we examine global stratification and inequality, and various perspectives that have been developed to explain the nature and extent of this problem. Before reading on, test your knowledge of global wealth and poverty (see Box 8.1).

QUESTIONS AND ISSUES

Chapter Focus Question: How are global stratification and gender linked?

What is global stratification, and how does it contribute to economic inequality?

How are global poverty and human development related?

What is modernization theory, and what are its stages?

How do conflict theorists explain patterns of global stratification?

WHAT IS SOCIAL STRATIFICATION?

Social stratification **is the hierarchical arrangement of large social groups based on their control over basic resources** (Feagin and Feagin, 2003). Stratification involves patterns of structural inequality that are associated with membership in each of these groups, as well as the ideologies that support inequality. Sociologists examine the social groups that make up the hierarchy in a society and seek to determine how inequalities are structured and persist over time.

Max Weber's term *life chances* **refers to the extent to which individuals have access to important societal resources such as food, clothing, shelter, education, and health care.** According to sociologists, more-affluent people typically have better life chances than the less-affluent because they have greater access to quality education, safe neighborhoods, high-quality nutrition and health care, police and private security protection, and an extensive array of other goods and services. In contrast, persons with low- and poverty-level incomes tend to have limited access to these resources. *Resources* are anything valued in a society, ranging from money and property to medical care and education; they are considered to be scarce because of their unequal distribution among social categories. If we think about the valued resources available in various nations, for example, the differences in life chances are readily apparent. As one analyst suggested, "Poverty narrows and closes life chances. The victims of poverty experience a kind of arteriosclerosis of opportunity. Being poor not only means economic insecurity, it also wreaks havoc on one's mental and physical health" (Ropers, 1991: 25).

Throughout the world, poverty is a major social concern, and it is often related to people's access to scarce resources as well as their nationality, perceived race or ethnicity, gender, age, religion, and many other factors that are used to distinguish between—and sometimes privilege or disadvantage—individuals and groups that are believed to fit into certain categories. For example, all societies distinguish among people by age. Young children typically have less authority and responsibility than older persons. Older persons, especially those without wealth or power, may find themselves at the bottom of the social hierarchy. Similarly, all societies differentiate between females and males: Women are often treated as subordinate to men. From society to society, people are treated differently as a result of their religion, race/ethnicity, appearance, physical strength, disabilities, or other distinguishing characteristics. All of these differentiations result in inequality. However, systems of stratification are also linked to the specific economic and social structure of a society and to a nation's position in the system of global stratification.

GLOBAL SYSTEMS OF STRATIFICATION

Around the globe, one of the most important characteristics of systems of stratification is their degree of flexibility. Sociologists distinguish among such systems based on the extent to which they are open or closed. In an *open system,* the boundaries between levels in the hierarchies are more flexible and may be influenced (positively or negatively) by people's achieved statuses. Open systems are assumed to have some degree of social mobility. *Social mobility* **is the movement of individuals or groups from one level in a stratification system to another** (Rothman, 2001). This movement can be either upward or downward. *Intergenerational mobility* **is the social movement experienced by family members from one generation to the next.** For example, Sarah's father is a carpenter who makes good

Box 8.1 SOCIOLOGY AND EVERYDAY LIFE

How Much Do You Know About Global Wealth and Poverty?

True	False	
T	F	1. The world's ten richest people are U.S. citizens.
T	F	2. Although the percentage of the world's people living in absolute poverty has declined over the past decade, the total number of people living in poverty has increased.
T	F	3. The richest fifth of the world's population receives about 50 percent of the total world income.
T	F	4. The political role of governments in policing the activities of transnational corporations has expanded as companies' operations have become more globalized.
T	F	5. Most analysts agree that the World Bank was created to serve the poor of the world and their borrowing governments.
T	F	6. In low-income countries, the problem of poverty is unequally shared between men and women.
T	F	7. The assets of the 200 richest people are more than the combined income of over 40 percent of the world's population.
T	F	8. In recent years, poverty levels have declined somewhat in East Asia, the Middle East, and North Africa.
T	F	9. The majority of people with incomes below the poverty line live in rural areas of the world.
T	F	10. Poor people in low-income countries meet most of their energy needs by burning wood, dung, and agricultural wastes, which increases health hazards and environmental degradation.

Answers on page 253.

wages in good economic times but is often unemployed when the construction industry slows to a standstill. Sarah becomes a neurologist, earning $350,000 a year, and moves from the working class to the upper-middle class. Between her father's generation and her own, Sarah has experienced upward social mobility.

By contrast, *intragenerational mobility* **is the social movement of individuals within their own lifetime.** Consider, for example, RaShandra, who began her career as a high-tech factory worker and through increased experience and taking specialized courses in her field became an entrepreneur, starting her own highly successful "dot.com" business. RaShandra's advancement is an example of upward intragenerational social mobility. However, both intragenerational mobility and intergenerational mobility may be downward as well as upward.

In a *closed system,* the boundaries between levels in the hierarchies of social stratification are rigid, and people's positions are set by ascribed status. Open and closed systems are ideal-type constructs; no actual stratification system is completely open or closed. The

systems of stratification that we will examine—slavery, caste, and class—are characterized by different hierarchical structures and varying degrees of mobility. Let's examine these three systems of stratification to determine how people acquire their positions in each and what potential for social movement they have.

Slavery

Slavery **is an extreme form of stratification in which some people are owned by others.** It is a closed system in which people designated as "slaves" are treated as property and have little or no control over their lives. According to some social analysts, throughout recorded history only five societies have been slave societies—those in which the social and economic impact of slavery was extensive: ancient Greece, the Roman Empire, the United States, the Caribbean, and Brazil (Finley, 1980). Others suggest that slavery also existed in the Americas prior to European settlement, and throughout Africa and Asia (Engerman, 1995).

People's life chances are enhanced by access to important societal resources such as education. How will the life chances of students who have the opportunity to pursue a college degree differ from those of young people who do not have the chance to go to college?

Those of us living in the United States are most aware of the legacy of slavery in our own country. Beginning in the 1600s, slaves were forcibly imported to the United States as a source of cheap labor. Slavery was defined in law and custom by the 1750s, making it possible for one person to own another person (Healey, 2002). In fact, early U.S. presidents including George Washington, James Madison, and Thomas Jefferson owned slaves. As practiced in the United States, slavery had four primary characteristics: (1) it was for life and was inherited (children of slaves were considered to be slaves); (2) slaves were considered property, not human beings; (3) slaves were denied rights; and (4) coercion was used to keep slaves "in their place" (Noel, 1972). Although most slaves were powerless to bring about change, some were able to challenge slavery—or at least their position in the system—by engaging in activities such as sabotage, intentional carelessness, work slowdowns, or running away from owners and working for the abolition of slavery (Healey, 2002). Despite the fact that slavery officially ended many years ago in this country, sociologists such as Patricia Hill Collins (1990) believe that its legacy is deeply embedded in current patterns of prejudice and discrimination against African Americans.

Slavery is not simply an unfortunate historical legacy. Although slavery is illegal everywhere, and there is no more *legal* ownership of people, the economist Stanley L. Engerman (1995: 175) believes that the world will not be completely free of slavery as long as there are "debt bondage, child labor, contract labor, and other varieties of coerced work for limited periods of time, with limited opportunities for mo-

bility, and with limited political and economic power." As the scholar Kevin Bales (1999) states in his recent book, *Disposable People: New Slavery in the Global Economy,* "When people buy slaves today they don't ask for a receipt or ownership papers, but they do gain *control*—and they use violence to maintain this control. Slaveholders have all the benefits of ownership without the legalities."

Slavery in the United States was a topic of concern for Harriet Martineau, who, as discussed in Chapter 1, was one of the first women social scientists. In *Society in America* (1962/1837), Martineau described her visit to Montgomery, Alabama, in 1835, when she visited certain plantations and observed the lives of the slaves. Seeing through the prejudice of her day, Martineau discredited negative stereotypes about the African Americans who were held in slavery and applauded the few slave owners whom she thought treated their slaves decently:

> I spent some days at a plantation a few miles from Montgomery, and heard there of an old lady who treats her slaves in a way very unusual, but quite safe, as far as appears. She gives them knowledge, which is against the law; but the law leaves her in peace and quiet. She also commits to them the entire management of her estate, requiring only that they should make her comfortable, and letting them take the rest. There is an obligation by law to keep an overseer; to obviate insurrection. How she manages about this, I omitted to inquire: but all goes on well; the cultivation of the estate is creditable, and all parties are contented. This is only a temporary ease and contentment. The old lady must die; and her slaves will either be sold to a new owner, whose temper will be an accident; or, if freed, must leave the State: but the story is satisfactory in as far as it gives evidence of the trustworthiness of the negroes.

Martineau, an Englishwoman, was a strong—but largely unheard—voice advocating the end of slavery in the United States.

How many people are held in slavery today? Bales (1999) estimates that the number of slaves worldwide stands at about 27 million people, most of whom are held in *bonded labor* in India, Pakistan, Bangladesh, and Nepal. *Bonded labor* or *debt bondage* refers to a situation in which people give themselves into slavery as security against a loan or when they inherit a debt from a relative (Bales, 1999). For the most part, slavery is concentrated in Southeast Asia, northern and western Africa, and parts of South America. However, Bales argues that there are slaves in almost every

Box 8.1 SOCIOLOGY AND EVERYDAY LIFE

Answers to the Sociology Quiz on Global Wealth and Poverty

1. **False.** In 2003, eight of the ten richest people were U.S. citizens. These included Bill Gates, Warren Buffett, and four members of the Walton family (which owns 38 percent of the Wal-Mart chain) (*Forbes*, 2003).

2. **True.** Data from the World Bank indicate that the percentage of the world's people living in absolute poverty has declined since the mid-1980s, particularly in Asia. However, other regions have not reduced the incidence of poverty to the same degree, and the total number of people living in poverty rose to approximately 1.4 billion in the mid-1990s (World Bank, 2003c).

3. **False.** According to the Human Development Report published by the United Nations Development Programme (2003), the richest fifth receives more than 80 percent of total world income. This ratio doubled between the 1950s and the 1990s.

4. **False.** As companies have globalized their operations, governmental restrictions have become less effective in controlling their activities. Transnational corporations have very little difficulty sidestepping governmental restrictions based on old assumptions about national economies and foreign policy. For example, Honda is able to circumvent import restrictions that the governments of Taiwan, South Korea, and Israel have placed on its vehicles by shipping vehicles made in Ohio to those locations (Korten, 1996).

5. **False.** Some analysts point out the linkages between the World Bank and the transnational corporate sector on both the borrowing and lending ends of its operation. Although the bank is supposedly owned by its members' governments and lends money only to governments, many of its projects involve vast financial dealings with transnational construction companies, consulting firms, and procurement contractors (see Korten, 1996).

6. **True.** In almost all low-income countries (as well as middle- and high-income countries), poverty is a more chronic problem for women due to sexual discrimination, resulting in a lack of educational and employment opportunities (Hauchler and Kennedy, 1994).

7. **True.** Assets of the 200 richest people are more than the combined income of 41 percent of the world's population (United Nations Development Programme, 2003).

8. **True.** These have been the primary regions in which poverty has decreased somewhat and infant mortality rates have fallen. Factors such as economic growth, oil production, foreign investment, and overall development have been credited for the decrease in poverty in East Asia, the Middle East, and North Africa (United Nations DPCSD, 1997).

9. **True.** The majority of people with incomes below the poverty line live in rural areas of the world; however, the number of poor people residing in urban areas is growing rapidly. In fact, most people living in poverty in Latin America are urban dwellers (United Nations DPCSD, 1997).

10. **True.** Although these fuels are inefficient and harmful to health, many low-income people cannot afford appliances, connection charges, and so forth. In some areas, electric hookups are not available (United Nations DPCSD, 1997).

country of the world, where they work in simple, nontechnical, traditional labor such as brickmaking, cloth and carpet making, domestic service, and prostitution. Although some products made with slave labor are sold at the local level, goods made by people in slavery or debt bondage are eventually sold to people around the globe. According to Bales (1999), *what* slaves make is not as important as the sheer *volume of work* that they can be forced to produce and the number of hours that they can be required to work. Bales believes that poverty is a more important indicator of who will be today's slaves rather than the race, ethnicity, or religion of the people involved.

As modernization and globalization have affected many previously less developed regions of the world, traditional economies and family life have been changed drastically. Subsistence farming can no longer support many people because of the shift to cash-crop agriculture, the loss of common land, and government policies that focus on the production of cheap food for cities rather than subsistence for farmers and agricultural laborers (Bales, 1999).

Various forms of slavery include child prostitution in Thailand, enslaved brickmakers in Pakistan, and domestic slaves in France; however, we will briefly examine modern-day slavery in Brazil as an example of the experiences of many who are enslaved today. Ronald, who was interviewed in a charcoal camp in Mato Grosso, is one example:

> My parents lived in a very dry area and when I got older there was no work, no work at all there . . . but one day a *gato* [labor recruiter] came and began to recruit people to work out here in Mato Grosso. The *gato* said that we would be given good food every day, and we would have good wages besides. He promised that every month his truck would bring people back [home] so that they could visit their families and bring them their pay. He even gave money to some men to give to their families before they left and to buy food to bring with them on the trip. . . .
>
> When we got to Mato Grosso we kept driving further and further into the country. This camp is almost fifty miles from anything; it is just raw *cerrado* for fifty miles before you get to even a ranch, and there is just the one road. When we reached the camp we could see it was terrible: the conditions were not good enough for animals. Standing around the camp were men with guns. And then the *gato* said, "You each owe me a lot of money: there is the cost of the trip, and all that food you ate, and the money I gave you for your families—

so don't even think about leaving." (qtd. in Bales, 1999: 127)

Like many others who are enslaved in Brazil, Ronald was trapped by poverty, circumstances, and the threat of physical force by armed men. The *gatos* even take away the workers' state identity cards and labor cards that are the key to legal employment; consequently, workers cannot find other employment even if they are able to flee the oppressive conditions.

Today, slavery flourishes in rural and other isolated areas of Brazil—on sugarcane plantations and ranches, in gold mines, and in the charcoal industries of the Amazon. The sugarcane fields are harvested by farm workers who toil from sunup to sundown and sleep in hammocks strung in cow stalls. Employers often do not pay wages for the work: Workers instead receive script, which they can redeem for food. Because landowners need to ensure that they will have a readily available supply of cheap labor, they bind laborers by encouraging them to run up unpayable debts at company-owned stores or canteens. According to an international labor organization report, "Workers who try to escape are pursued by gunmen and returned to the estate, where they can be beaten, whipped or subjected to mutilation or sexual abuse" (Brooke, 1993b: 3).

What happens to people who are enslaved? Like people in all nations, enslaved people have hopes and dreams for their children and grandchildren. Doralice Moreira de Souza, a forty-seven-year-old woman whose hands are gnarled from years of cutting five tons of sugarcane daily in Conceicao de Macabu, Brazil, has a dream for Alan, her ten-year-old grandson: "I would like him to study, so that when he grows up, he won't end up a slave like me" (qtd. in Brooke, 1993a: Y3). However, Alan's life chances are already seriously limited because of economic conditions and labor exploitation in his country. Why does such exploitation occur? Social scientists have developed various theories to explain the persistence of extreme global inequality.

The Caste System

Like slavery, caste is a closed system of social stratification. A **caste system is a system of social inequality in which people's status is permanently determined at birth based on their parents' ascribed characteristics.** Vestiges of caste systems exist in contemporary India and South Africa.

In India, caste is based in part on occupation; thus, families typically perform the same type of work from generation to generation. By contrast, the caste sys-

Systems of stratification include slavery, caste, and class. As shown above, the life chances of people living in each of these systems differ widely.

tem of South Africa was based on racial classifications and the belief of white South Africans (Afrikaners) that they were morally superior to the black majority. Until the 1990s, the Afrikaners controlled the government, the police, and the military by enforcing *apartheid*—the separation of the races. Blacks were denied full citizenship and restricted to segregated hospitals, schools, residential neighborhoods, and other facilities. Whites held almost all of the desirable jobs; blacks worked as manual laborers and servants.

In a caste system, marriage is endogamous, meaning that people are allowed to marry only within their own group. In India, parents have traditionally selected marriage partners for their children. In South Africa, interracial marriage was illegal until 1985.

Cultural beliefs and values sustain caste systems. Hinduism, the primary religion of India, reinforced the caste system by teaching that people should accept their fate in life and work hard as a moral duty. Caste systems grow weaker as societies industrialize; during that process, the values reinforcing the system break down, and people start to focus on the types of skills needed for industrialization. However, vestiges of caste systems often remain for hundreds of years beyond the period in which they are "officially" abolished. For example, some social scientists believe that past racial segregation in the United States has contributed to a caste-like condition of racial and ethnic inequality that remains in the present (Cox, 1948; Frankenberg, 1993). We will return to this issue in Chapter 10 ("Race and Ethnicity").

As we have seen, in closed systems of stratification, group membership is hereditary, and moving up within the structure is almost impossible. Custom and law frequently perpetuate privilege and ensure that higher-level positions are reserved for the children of the advantaged (Rothman, 2001).

The Class System

The *class system* is a type of stratification based on the ownership and control of resources and on the type of work people do (Rothman, 2001). At least theoretically, a class system is more open than a caste system because the boundaries between classes are less distinct than the boundaries between castes. In a class system, status comes at least partly through achievement rather than entirely by ascription.

In class systems, people may become members of a class other than that of their parents through both intergenerational mobility and intragenerational mobility, either upward or downward. Horizontal mobility occurs when people experience a gain or loss in position and/or income that does not produce a change in their place in the class structure. For example, a person may get a pay increase and a more prestigious title but still not move from one class to another. By contrast, movement up or down the class structure is *vertical mobility*. Martin, a commercial artist who owns his own firm in the United States, is an example of vertical, intergenerational mobility:

> My family came out of a lot of poverty and were eager to escape it. . . . My [mother's parents] worked in a sweatshop. My grandfather to the day he died never earned more than $14 a week. My grandmother worked in knitting mills while she had five children. . . . My father quit school when he was in eighth grade and supported his mother and his two sisters when he was twelve years old. My grandfather died when my father was four and he basically raised his sisters. He got a man's job when he was twelve and took care of the three of them. (qtd. in Newman, 1993: 65)

Martin's situation reflects upward mobility; however, as previously stated, people may also experience downward mobility, caused by any number of reasons, including a lack of jobs, low wages and employment instability, marriage to someone with fewer resources and less power than oneself, and changing social conditions (Ehrenreich, 1989; Newman, 1988, 1993).

Social classes involve much more than the upward or downward mobility of individuals: The class structure of any society is shaped by the historical, economic, political, and social context in which it is embedded. For example, changes in technology contribute to the changing composition of classes and to how people are distributed within those classes. In many nations, government policies influence the economy and how economic rewards are distributed among citizens. In other words, social classes do not exist apart from the class system of which they are a part. According to many social analysts, class is a relationship that is not confined to economic divisions alone: Class relations shape all human relationships, including how parents bring up their children and how employers and employees interact with each other; thus, the life chances of individuals and their families are tied to their position in the class structure. We will examine these class structures in greater detail in Chapter 9, when we look at the U.S. class system.

WEALTH AND POVERTY IN GLOBAL PERSPECTIVE

What do we mean by global stratification? *Global stratification* refers to the unequal distribution of wealth, power, and prestige on a global basis, resulting in people having vastly different lifestyles and life chances both within and among the nations of the world. Just as the United States is divided into classes, the world is divided into unequal segments characterized by extreme differences in wealth and poverty. For example, the income gap between the richest and the poorest 20 percent of the world population continues to widen (see Figure 8.1). However, when we com-

Figure 8.1   **Income Gap Between the World's Richest and Poorest People**

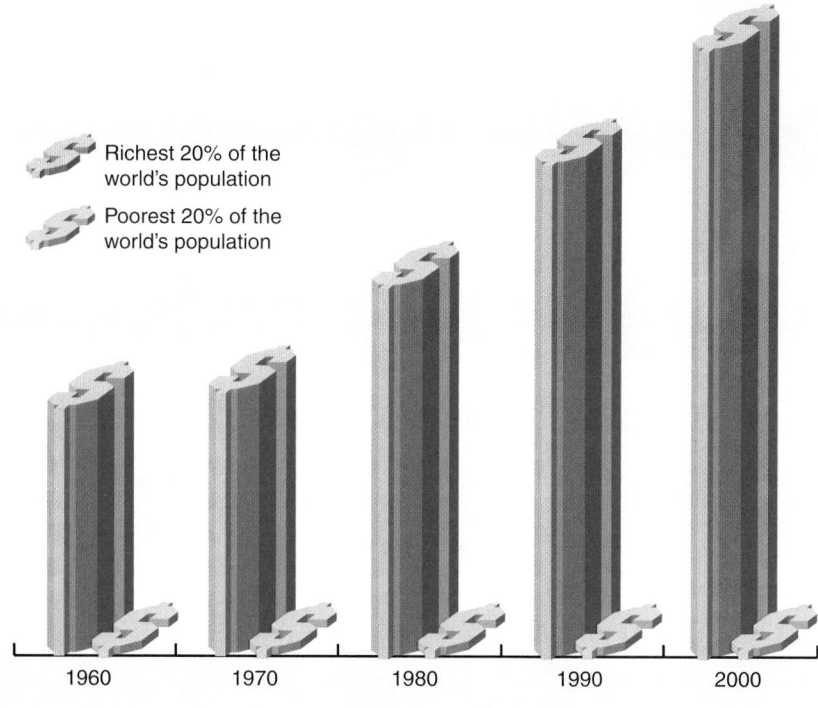

Richest 20% of the world's population

Poorest 20% of the world's population

The income gap between the richest and poorest people in the world continued to grow between 1960 and 2000. As this figure shows, in 1960 the highest-income 20 percent of the world's population received $30 for each dollar received by the lowest-income 20 percent. By 2000, the disparity had increased: $74 to $1.

1960   1970   1980   1990   2000

Source: International Monetary Fund, 1992; United Nations Development Programme, 2003.

pare social and economic inequality within other nations, we find gaps that are more pronounced than they are in the United States. As previously defined, *high-income countries* are nations characterized by highly industrialized economies; technologically advanced industrial, administrative, and service occupations; and relatively high levels of national and per capita (per person) income. In contrast, *middle-income countries* are nations with industrializing economies, particularly in urban areas, and moderate levels of national and personal income. *Low-income countries* are primarily agrarian nations with little industrialization and low levels of national and personal income. Within some nations, the poorest one-fifth of the population has an income that is only a slight fraction of the overall average per capita income for that country. For example, in Brazil, Bolivia, and Honduras, less than 3 percent of total national income accrues to the poorest one-fifth of the population (World Bank, 2003c).

Just as the differences between the richest and poorest people in the world have increased, the gap in global income differences between rich and poor countries has continued to widen over the past fifty years. In 1960, the wealthiest 20 percent of the world population had more than thirty times the income of

the poorest 20 percent. By 2000, the wealthiest 20 percent of the world population had almost eighty times the income of the poorest 20 percent (Hauchler and Kennedy, 1994; United Nations Development Programme, 2003). Income disparities *within* countries were even more pronounced.

However, in examining the income gap, it is important to note that economic inequality is not the only dimension of global stratification. For example, in an earlier study on world poverty, the Swedish economist Gunnar Myrdal (1970: 56) distinguished between social and economic inequality:

Social inequality is clearly related to status and can perhaps best be defined as an extreme lack of social mobility and a severely hampered possibility of competing freely. . . . Economic inequality . . . is related to differences in wealth and income. . . . But there is a close relation between the two, since *social inequality stands as a main cause of economic inequality, while, at the same time, economic inequality supports social inequality.*

Social inequality, which may result from factors such as discrimination based on race, ethnicity, gender, or religion, exacerbates problems of economic inequality. According to Myrdal, social inequality and

© Erik Eckholm/The New York Times Pictures

Global inequality is most striking in nations that are sometimes referred to as "Third World" or "underdeveloped." In China, for example, workers in a small shoemaking factory experience poverty and low standards of living, and they often have shorter life expectancies and higher rates of mortality because of the toxic conditions in which they work.

economic inequality are main causes of the poverty of a nation; therefore, a society must have greater social equality among its citizens as a precondition for the entire country getting out of poverty. Myrdal (1970) believed that the poorest people living in extremely poor nations experience much greater economic hardship than do the poor who live in more-affluent countries of the world.

Many people have sought to address the issue of world poverty and to determine ways in which resources can be used to meet the urgent challenge of poverty. However, not much progress has been made on this front (Lummis, 1992) despite a great deal of talk and billions of dollars in "foreign aid" flowing from high-income to low-income nations. The idea of "development" has become the primary means used in attempts to reduce social and economic inequalities and alleviate the worst effects of poverty in the less industrialized nations of the world. Often, the nations that have not been able to reduce or eliminate poverty are chastised for not making the necessary social and economic reforms to make change possible (Myrdal, 1970). Or, as another social analyst has suggested,

> The *problem* of inequality lies not in poverty, but in excess. "The problem of the world's poor," defined more accurately, turns out to be "the problem of the world's rich." This means that the solution to the problem is not a massive change in the culture of poverty so as to place it on the path of development, but a massive change in the culture of superfluity in order to place it on the path of counterdevelopment. It does not call for a new value system forcing the world's majority to feel shame at their traditionally moderate consumption habits, but for a new value system forcing the world's rich to see the shame and vulgarity of their overconsumption habits, and the double vulgarity of standing on other people's shoulders to achieve those consumption habits. (Lummis, 1992: 50)

As this statement suggests, the increasing interdependence of all the world's nations was largely overlooked or ignored until increasing emphasis was placed on the global marketplace and the global economy. In addition, there are a number of problems inherent in studying global stratification, one of which is what terminology should be used to describe various nations.

PROBLEMS IN STUDYING GLOBAL INEQUALITY

One of the primary problems encountered by social scientists studying global stratification and social and economic inequality is what terminology should be used to refer to the distribution of resources in various nations. During the past fifty years, major changes have occurred in the way that inequality is addressed by organizations such as the United Nations and the World Bank. Most definitions of inequality are based on comparisons of levels of income or economic development, whereby countries are identified in terms of the "three worlds" or upon their levels of economic development.

The "Three Worlds" Approach

After World War II, the terms "First World," "Second World," and "Third World" were introduced by social analysts to distinguish among nations on the basis of their levels of economic development and the standard of living of their citizens. *First World* nations were said to consist of the rich, industrialized nations that primarily had capitalist economic systems and democratic political systems. The most frequently noted First World nations were the United States,

Canada, Japan, Great Britain, Australia, and New Zealand. *Second World* nations were said to be countries with at least a moderate level of economic development and a moderate standard of living. These nations included China, North Korea, Vietnam, Cuba, and portions of the former Soviet Union. According to social analysts, although the quality of life in Second World nations was not comparable to that of life in the First World, it was far greater than that of people living in the *Third World*—the poorest countries, with little or no industrialization and the lowest standards of living, shortest life expectancies, and highest rates of mortality.

Is the "three worlds" approach an accurate representation of global inequality? According to the sociologist Manuel Castells (1998), the Second World has disintegrated with the advent of the Information Age, a period in which industrial capitalism is being superseded by global informational capitalism. Likewise, the Third World is no longer clearly identifiable in regard to specific geographic regions or political regimes as nations have become more diversified in their economic and social development (Castells, 1998). Instead, Castells (1998: 164) adopts the term *Fourth World* to describe the "multiple black holes of social exclusion" that he believes exist throughout the planet in areas ranging from much of sub-Saharan Africa and the impoverished rural areas of Latin America and Asia to Spanish enclaves of mass youth unemployment and U.S. inner-city ghettos. A term first used in the 1970s, the Fourth World originally designated indigenous people who are descended from a country's aboriginal population (such as Native Americans in the United States or First Canadians in Canada) and who are completely or partly deprived of the right to their own territory and its resources. Castells (1998: 165) describes the contemporary Fourth World as follows:

> [The Fourth World] is populated by millions of homeless, incarcerated, prostituted, criminalized, brutalized, stigmatized, sick, and illiterate persons. They are the majority in some areas, the minority in others, and a tiny minority in a few privileged contexts. But, everywhere, they are growing in number and increasing in visibility, as the selective triage of information capitalism, and the political breakdown of the welfare state, intensify social exclusion. In the current historical context, the rise of the Fourth World is inseparable from the rise of informational, global capitalism.

Thus, whether we embrace the concept of the "three worlds" approach or shift to the idea of the Fourth World, the central idea is the extent to which social exclusion occurs. According to Castells, **social exclusion is the process by which certain individuals and groups are systematically barred from access to positions that would enable them to have an autonomous livelihood in keeping with the social standards and values of a given social context.** For Castells, social exclusion is a process, not a condition. As informational capitalism grows around the world, the lack of regular work as a source of income is increasingly a key mechanism in social exclusion. Among persons with a disability or those who are too ill to work, social exclusion means not having health coverage or other benefits that incorporate them into the larger community. At the bottom line, Castells sees global financial markets and their networks of management as the central players in the twenty-first century because they control capital flows on a global scale. According to Castells (1998), rather than the "three worlds" approach, the future will be based on a global economy governed by a set of multilateral institutions, networked among themselves, and at the core of this network will be the International Monetary Fund, the World Bank, and the G-7 nations such as the United States, as described in "High-Income Countries" later in this chapter.

The Levels of Development Approach

Among the most controversial terminology used for describing world poverty and global stratification has been the language of development. Terminology based on levels of development includes concepts such as developed nations, developing nations, less-developed nations, and underdevelopment. Let's look first at the contemporary origins of the idea of "underdevelopment" and "underdeveloped nations."

Following World War II, the concepts of *underdevelopment* and *underdeveloped nations* emerged out of the Marshall Plan (named after U.S. Secretary of State George C. Marshall), which provided massive sums of money in direct aid and loans to rebuild the European economic base destroyed during World War II. Given the Marshall Plan's success in rebuilding much of Europe, U.S. political leaders decided that the Southern Hemisphere nations that had recently been released from European colonialism could also benefit from a massive financial infusion and rapid economic development. Leaders of the developed nations argued that urgent problems such as poverty, disease, and famine could be reduced through the transfer of finance, technology, and experience from the developed nations to lesser-developed countries. From this

According to the levels of development approach, economic development is the primary way to solve the poverty problem in low-income nations. Does the employment of workers such as these reduce the poverty of the poorest people in a country? What other factors are also important?

viewpoint, economic development is the primary way to solve the poverty problem: Hadn't economic growth brought the developed nations to their own high standard of living? Moreover, "self-sustained development" in a nation would require that people in the lesser-developed nations accept the beliefs and values of people in the developed nations.

Ideas regarding *underdevelopment* were popularized by President Harry S Truman in his 1949 inaugural address. According to Truman, the nations in the Southern Hemisphere were "underdeveloped areas" because of their low gross national product, which today is referred to as *gross national income* (GNI)—a term that refers to all the goods and services produced in a country in a given year, plus the net income earned outside the country by individuals or corporations. If nations could increase their GNI, then social and economic inequality among the citizens within the country could also be reduced. Accordingly, Truman believed that it was necessary to assist the people of economically underdeveloped areas to raise their *standard of living,* by which he meant material well-being that can be measured by the quality of goods and services that may be purchased by the per capita national income (Latouche, 1992). Thus, an increase in the standard of living meant that a nation was moving toward economic development, which typically included the improved exploitation of natural resources by industrial development. According to the social scientist Serge Latouche (1992: 250–251), measuring social and economic conditions on the basis of standard of living ultimately denigrates the culture and way of life in some societies:

> While the hope of a satisfactory life is a very human concern, the obsession with this sort of "standard of living" is very recent. Interest in salary levels on the part of wage earners and as a general social preoccupation dates from the industrial era. . . . In fact, looking at the world in terms of "standard of living" is like looking through dark glasses; they make the rich variety of colours disappear, turning all differences into shades of the same colour.

What has happened to the issue of development since the post–World War II era? After several decades of economic development fostered by organizations such as the United Nations and the World Bank, it became apparent by the 1970s that improving a country's GNI did not tend to reduce the poverty of the poorest people in that country. In fact, global poverty and inequality were increasing, and the initial optimism of a speedy end to underdevelopment faded. Although many developing countries had achieved economic growth, it was not shared by everyone in the nation. For example, the poorest 20 percent of the Brazilian population receives less than 3 percent of the total national income, whereas the richest 20 percent of the population receives 63 percent (World Bank, 2003c).

Why did inequality increase even with greater economic development? Some analysts in the developed nations began to link growing social and economic inequality on a global basis to relatively high rates of population growth taking place in the underdeveloped nations. Organizations such as the United Nations and the World Health Organization stepped up their efforts to provide family planning services to the populations so that they could control their own fertility. More recently, however, population researchers have become aware that issues such as population growth, economic development, and environmental problems must be seen as interdependent concerns. This changing perception culminated in the U.N. Conference on Environment and Development in Rio de Janeiro, Brazil (the "Earth Summit"), in 1992; as a result, terms such as *underdevelopment* have largely been dropped in favor of measurements such as sustainable development, and economies are now classified by their levels of income.

CLASSIFICATION OF ECONOMIES BY INCOME

Today, the World Bank (2003c) focuses on three development themes: people, the environment, and the economy. Since the World Bank's primary business is providing loans and policy advice to low- and

Map 8.1 High-, Middle-, and Low-Income Economies in Global Perspective

How does the United States compare with other nations with regard to income?

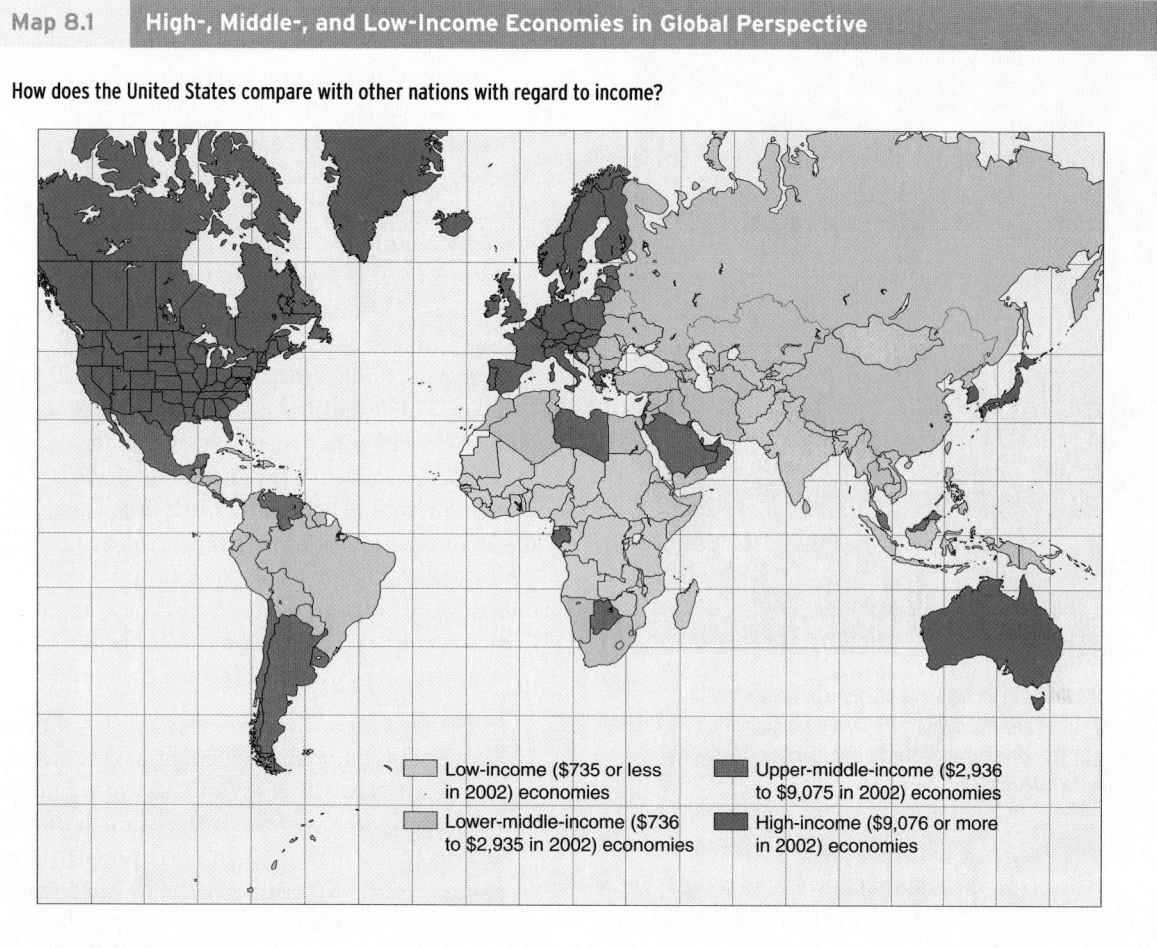

Low-income ($735 or less in 2002) economies

Lower-middle-income ($736 to $2,935 in 2002) economies

Upper-middle-income ($2,936 to $9,075 in 2002) economies

High-income ($9,076 or more in 2002) economies

Source: World Bank, 2003c.

middle-income member countries, it classifies nations into three economic categories: *low-income economies* (a GNI per capita of $735 or less in 2002), *middle-income economies* (a GNI per capita between $736 and $9,075 in 2002), and *high-income economies* (a GNI per capita of more than $9,076 in 2002).

Low-Income Economies

About half the world's population lives in the sixty-four low-income economies, where most people engage in agricultural pursuits, reside in nonurban areas, and are impoverished (World Bank, 2003c). As shown in Map 8.1, low-income economies are primarily found in countries in Asia and Africa, where half of the world's population resides. Included are nations such as Rwanda, Mozambique, Ethiopia, Nigeria, Cambodia, Vietnam, Afghanistan, and Bangladesh. Eastern European countries such as Moldova and Georgia are among the low-income economies of Eastern Europe. In some of these nations, civil war is

both a cause and a product of the poverty in which much of the population lives (see Box 8.2).

Among those most affected by poverty in low-income economies are women and children. Mayra Buvinić, who has served as chief of the women in development program unit at the Inter-American Development Bank, describes the plight of one Nigerian woman as an example:

> On the outskirts of Ibadan, Nigeria, Ade cultivates a small, sparsely planted plot with a baby on her back and other visibly undernourished children nearby. Her efforts to grow an improved soybean variety, which could have improved her children's diet, failed because she lacked the extra time to tend the new crop, did not have a spouse who would help her, and could not afford hired labor. (Buvinić, 1997: 38)

According to Buvinić, Ade's life is typical of many women worldwide who face obstacles to increasing their economic power because they do not have the

John Maier, Jr./The Image Works

Vast inequalities in income and lifestyle are shown in this photo of slums and nearby upper-class housing in Rio de Janeiro, Brazil. Do similar patterns of economic inequality exist in other nations?

time to invest in the additional work that could bring in more income. Think about the daily schedule of rural women in Mali, a schedule that demonstrates how women's time is spent in low-income countries (Ballara, 1991: 26):

4:30 A.M.	Rise
4:30–6 A.M.	Fetch water; prepare breakfast; attend to children
6–10 A.M.	Wash dishes; pound flour; collect vegetables or leaves for mid-day meal; wash clothes; visit market
10 A.M.–3 P.M.	Carry a meal to the fields; farm or help a husband farm the fields
3–6 P.M.	Gather wood to cook the evening meal; collect wild fruit and nuts
6–8 P.M.	Fetch water; pound flour; clean the residence; prepare the evening meal
8–9 P.M.	Card and spin cotton

Many poor women worldwide also do not have access to commercial credit and have been trained only in traditionally female skills that produce low wages. These factors have contributed to the *global feminization of poverty,* whereby women around the world tend to be more impoverished than men. Despite the fact that women have made some gains in terms of

well-being, the income gap between men and women continues to grow wider in the low-income, developing nations as well as in the high-income, developed nations such as the United States (Buvinić, 1997).

In some low-income nations, children are not only affected by poverty but also by child labor practices. In 1997 the United States banned imports of goods made by children in bondage, including handmade imported rugs made by child labor in rural-based economies. Although accurate numbers do not exist for the number of children who work in servitude making rugs alone, estimates range from 300,000 to 1,000,000 children (Iovine, 1997). Since production is often spread out in homes, it is very difficult to count the children involved or monitor labor practices. Moreover, some people believe that it is unfair to evaluate the labor practices of other cultures in terms of Western values and labor standards and to take away the one means of earning income that these people may have. According to a museum curator who specializes in handmade rugs from low-income economies,

> In many cultures the economies are very different from one's own, many are family-based and the set-ups are very different from documentable child abuse. Rug weaving is one of the most perfect examples of a sustainable economy in developing countries. In Turkey virtually every living room will have a loom in it. There is pride and delight felt by the whole family with their rugs. To ban all that could have a devastating effect. (qtd. in Iovine, 1997: B9)

Middle-Income Economies

About one-third of the world's population resides in the ninety-two nations with middle-income economies (a GNI per capita between $736 and $9,075 in 2002). The World Bank divides middle-income economies into lower-middle-income ($736 to $2,935) and upper-middle-income ($2,936 to $9,075). Countries classified as lower-middle-income include the Latin American nations of Bolivia, Columbia, Guatemala, and El Salvador. However, even though these countries are referred to as "middle-income," more than half of the people residing in countries such as Bolivia, Guatemala, and Honduras live in poverty, defined as below $1 per day in purchasing power (World Bank, 2003c).

Other lower-middle-income economies include Russia, Romania, and Kazakhstan. These nations had centrally planned (i.e., socialist) economies until dramatic political and economic changes occurred in the

Box 8.2 SOCIOLOGY IN GLOBAL PERSPECTIVE

Poverty and the Curse of Civil War in Africa

Civil war is development in reverse.
—World Bank (qtd. in *The Economist,* 2003: 11)

When we saw them, everyone ran. I did not stay to see what happened. I was scared. I thought I was going to die. . . . They cut people, shot them with arrows. It is so terrible I can't think about it.
—John Bosco Lossa, a teacher who is an ethnic Hema living in the village of Blukwa in the Democratic Republic of Congo, describing what happened when a group of men from the Lendu (a rival tribe) attacked his village as he was teaching French to a class of five-year-olds (qtd. in Robinson, 2000)

Attacks such as the one described by Lossa have left more than 5,000 people dead and over 150,000 unaccounted for or displaced as a result of fighting between the Hema and the Lendu in the Democratic Republic of Congo (DRC), a nation of about 55 million people in Central Africa. Although these ethnic clashes, which have been going on intermittently since 1975, are not directly linked to the ongoing civil war in the DRC, analysts believe that the ethnic strife and violence are aggravated both by that war and by grinding poverty in the region (Robinson, 2000). A civil war is defined as an internal conflict with at least 1,000 battlefield deaths (Collier, 2000); the civil war in the DRC is Africa's most widespread civil war and one of its longest-lasting internal conflicts. It is estimated that since the current civil war began in 1998, as many as 2 million people have been displaced and 2.5 million people have died from various causes, including malnutrition and preventable diseases (Oxfam, 2002).

Ironically, the Democratic Republic of Congo could be a wealthy nation because it has vast natural resources. However, the resources may be part of what the conflict among the warring parties is all about: Divergent factions (and other nations) want to have access to, and control over, these resources (Oxfam, 2002). As a result of the war and the instability of the government, foreign businesses have curtailed operations in the country, and per capita annual income is about $590.

The relationship between poverty and civil war is not limited to the DRC. According to reports by the World Bank, the poorest nations in the world are disproportionately likely to be at war—especially those nations in which 25 percent of the national income comes from the export of natural resources (Collier, 2000; *The Economist,* 2003). The continual presence of grinding poverty and a stagnant economy combine to produce conflict, which in turn aggravates existing poverty and the problems linked to it.

Some people believe that little can be done about poverty and warfare, and that nations simply have to "grow out of it." Others believe that economic growth and diversification are the basis for reducing both poverty-related problems and civil war. Analysts who advocate a "carrot and stick" approach suggest that those countries that end their internal conflicts should be given foreign aid; critics of this approach argue that such aid often comes in large initial amounts when governments are unable to use it effectively.

Although the problems in the Democratic Republic of Congo seem to be thousands of miles away and perhaps irrelevant to our daily lives, problems of poverty and war ultimately are everyone's concern. What steps do you think we should take to try to help other nations end problems such as the DRC is encountering? Why?

late 1980s and early 1990s. Since then, these nations have been going through a transition to a market economy. According to the World Bank (2003c), some nations have been more successful than others in implementing key elements of change and bringing about a higher standard of living for their citizens. Among other factors, high rates of inflation, the growing gap between the rich and the poor, low life-expectancy rates, and homeless children have been visible signs of problems in the transition toward a free market economy in countries such as Russia.

As compared with lower-middle-income economies, nations having upper-middle-income economies typically have a somewhat higher standard of living and export diverse goods and services, ranging from manufactured goods to raw materials and fuels. Nations with upper-middle-income economies include Argentina, Chile, Hungary, Mexico, and Saudi Arabia. Although these nations are referred to as middle-income economies, many of them have extremely high levels of indebtedness, leaving them with few resources for fighting poverty. According to one analyst, "A major cause of debt accumulation was investment in ill-considered, ill-conceived projects, many involving bloated capital costs and healthy doses of graft" (George, 1993: 88). Additional debts also accrued through military spending, even though a sizable portion of some populations is living in hunger and misery (George, 1993). The World Bank sought to set up funds to reduce the debts of some nations that have debts in excess of 200 to 250 percent of their annual export earnings. Among the nations in this position are Mozambique, Nicaragua, Congo, Bolivia, Ethiopia, and Tanzania. Some high-income nations believed that such an action on the part of the World Bank might set a dangerous precedent, so the debt-reduction process was limited in scope (Lewis, 1996).

As middle-income nations have been required to make payments on their debts, the requisite structural adjustments have necessitated that the countries make spending cuts in areas that formerly helped some of the poor, including subsidized food, education, and health care. These structural adjustments have been required in agreements with lending agencies such as the World Bank and the International Monetary Fund. Thus, there is a high rate of poverty in many countries that are classified as middle-income economies. The United Nations children's relief agency, UNICEF, decries the extent to which paying off debt to these international lending agencies affects the young in Latin America and Africa:

> Debt, in particular, still shackles many developing nations, claiming a large proportion of the resources which might otherwise have been available for human progress. With falling family incomes, and cuts in public spending on services such as health and education, many African and Latin American children are still paying for their opportunity for normal growth, their opportunity to be educated, and often their lives. With no less urgency than at any time in the last five years, UNICEF must again say that it is the antithesis of civilization that so many millions of children should be continuing to pay such a price. (UNICEF, 1991; qtd. in Hauchler and Kennedy, 1994: 62)

High-Income Economies

High-income economies are found in fifty-two nations, including the United States, Canada, Japan, Australia, Portugal, Ireland, Israel, Italy, Norway, and Germany. According to the World Bank, people in high-income economies typically have a higher standard of living than those in low- and middle-income economies. Nations with high-income economies continue to dominate the world economy, despite the fact that shifts in the global marketplace have affected some workers who have found themselves without work due to *capital flight*—the movement of jobs and economic resources from one nation to another—and *deindustrialization*—closing plants and factories because of their obsolescence or the fact that workers in other nations are being hired to do the work more cheaply (see Chapter 13, "The Economy and Work in Global Perspective").

Some of the nations that have been in the high-income category do not have as high a rate of annual economic growth as the newly industrializing nations, particularly in East Asia and Latin America, that are still in the process of development. The only significant group of middle- and lower-income economies to close the gap with the high-income, industrialized economies over the past few decades has been the nations of East Asia. China has experienced a 270-percent increase in per capita income over the past twenty years, but some analysts wonder whether the East Asian "miracle" is over (Crossette, 1997). Differences in the rate of growth between so-called G-7 countries (United States, Canada, Britain, France, Germany, Italy, and Japan) and the nations in East Asia, Pacific, and South Asia are shown in Figure 8.2.

Despite economic growth, the East Asian region remains home to approximately 350 million poor people (see Figure 8.3). And from all signs, it appears that nations such as India will continue to have an extremely large population and rapid annual growth rates.

MEASURING GLOBAL WEALTH AND POVERTY

On a global basis, measuring wealth and poverty is a difficult task because of conceptual problems and problems in acquiring comparable data from various nations. Although gross national income continues to be one of the most widely used measures of national income, in recent years the United Nations and the World Bank have begun to use the gross domestic product (GDP). The *gross domestic product* is all the goods and services produced *within* a country's economy during a given year. Unlike the GNI, the GDP does not include any income earned by individuals or corporations if the revenue comes from sources out-

Figure 8.2   **Economic Growth**

Substantial gains in economic growth were experienced by some East Asian nations over the past decade, as compared with the slower rate of growth among the more-developed G-7 nations. However, some analysts wonder if the East Asian "miracle" is over.

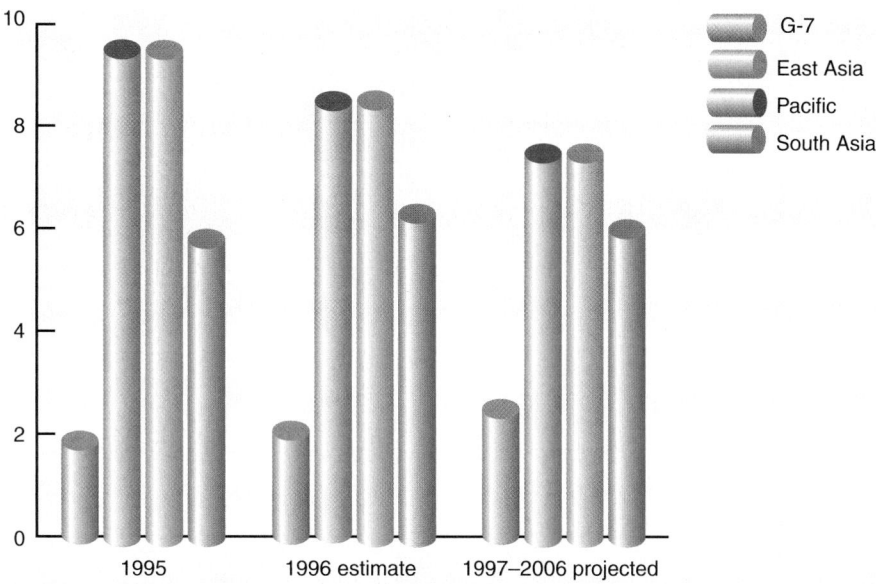

The G-7 countries are Great Britain, Canada, France, Germany, Italy, Japan, and the United States.

Source: World Bank, 2001.

---

side of the country. For example, using GDP as a measure of economic growth, a World Bank report concluded that nations such as China, India, and Indonesia were well on their way to becoming "economic powerhouses" in the next twenty-five years (Stevenson, 1997). Despite setbacks in 2002, the economy in East Asia is expected to grow faster than the economy in any other region of the world (World Bank, 2003a). The World Bank also believes that sub-Saharan Africa, one of the poorest regions of the world, will experience a surge of development that will help alleviate its poverty (Stevenson, 1997). However, some scholars criticize the assumption that economic development always benefits low-income nations and that poor nations *can* and *should* play "catch-up" with the wealthy, industrialized nations (Lummis, 1992).

Absolute, Relative, and Subjective Poverty

How is poverty defined on a global basis? Isn't it more a matter of comparison than an absolute standard? According to social scientists, defining poverty in-

volves more than comparisons of personal or household income: It also involves social judgments made by researchers. From this point of view, *absolute poverty*—previously defined as a condition in which people do not have the means to secure the most basic necessities of life—would be measured by comparing personal or household income or expenses with the cost of buying a given quantity of goods and services. The World Bank has defined absolute poverty as living on less than a dollar a day. Similarly, *relative poverty*—which exists when people may be able to afford basic necessities but are still unable to maintain an average standard of living—would be measured by comparing one person's income with the incomes of others. Finally, *subjective poverty* would be measured by comparing the actual income against the income earner's expectations and perceptions (World Bank, 2003c). However, for low-income nations in a state of economic transition, data on income and levels of consumption are typically difficult to obtain and are often ambiguous when they are available. Defining levels of poverty involves several dimensions: (1) how many people are poor, (2) how far below the poverty line people's incomes fall, and (3) how long they have

Figure 8.3 Percentage of Population Below the Poverty Line in East Asia

Despite economic growth, the East Asian region still has large numbers of people who live below the poverty line.

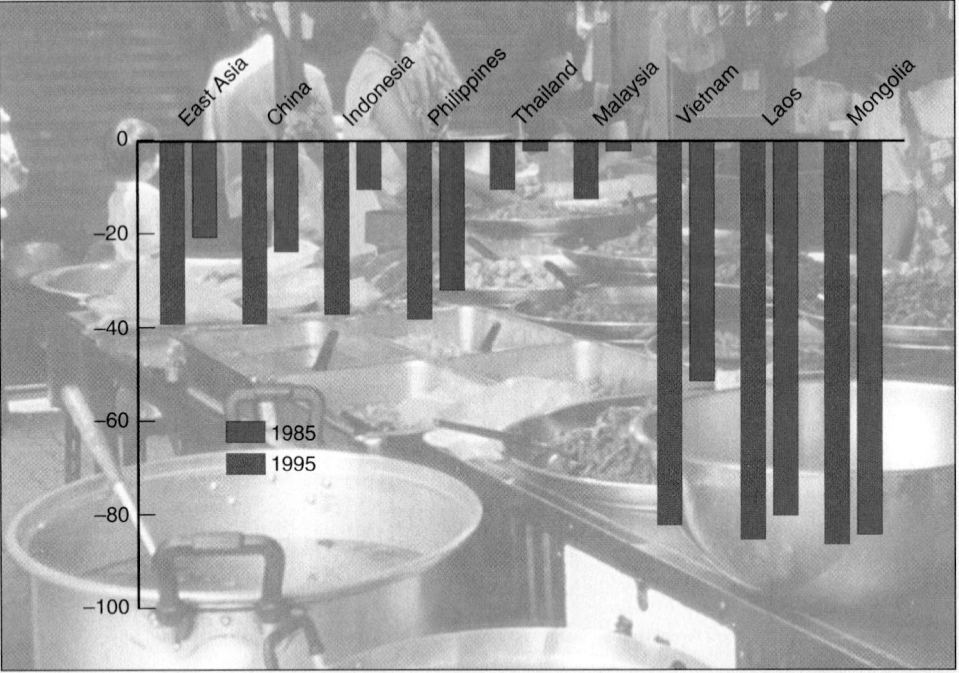

Source: World Bank, 2001.

been poor (is the poverty temporary or long term?) (World Bank, 2003c).

The Gini Coefficient and Global Quality of Life Issues

The World Bank uses as its measure of income inequality what is known as the *Gini coefficient,* which ranges from zero (meaning that everyone has the same income) to 100 (one person receives all the income). Using this measure, the World Bank (2003c) has concluded that inequality has increased in nations such as Bulgaria, the Baltic countries, and the Slavic countries of the former Soviet Union to levels similar to those in the less-equal industrial market economies, such as the United States. In fact, inequality in Russia rose sharply in the 1990s: By 1998, the top 20 percent of the population received a much larger percentage of the total income than it had in 1993. Stark contrasts also exist in countries such as India, where abject poverty still exists side by side with lavish opu-

lence in Calcutta. Note the sharp contrast in lifestyle on one Calcutta street:

> On one side is the Tollygunge Club, 40 hectares of landscaped serenity with an 18-hole golf course, a driving range, riding stables, tennis courts, and two covered swimming pools. . . . On the other side stands the M.R. Bangur Hospital, a sooty building with a morgue [that sometimes takes] the corpses of paupers who die in Calcutta's streets. (Watson, 1997: F1)

This street is symbolic of the sharp chasm that divides the 11 million people who live in Calcutta. In fact, some analysts believe that the scale of poverty in south and east Asia is most visible in the heavily populated states of India and China. It is estimated that 652 million of Asia's people are poor and that 448 million of them live in India alone, where there are more than twice as many poor people as in sub-Saharan Africa (Watson, 1997).

But similar disparities between the rich and the poor can be seen in nations throughout the world. For example, in Montrouis, Haiti, Club Med runs a resort featuring pristine beaches, an Olympic-size swimming pool, and all the amenities that affluent tourists expect from a luxury resort. However, just a short walk away from Club Med, open markets have raw meat crawling with flies, and homeless, malnourished people sleep nearby on the ground. But Club Med staff members and tourists at the resort seldom see the other side of Montrouis: Most never leave the compound except when they are going to and from the airport (Emling, 1997b). In other regions of Haiti as well, starvation and disease are a way of life for most inhabitants. As the poorest nation in the Western Hemisphere, Haiti's ability to feed people has been further reduced by recent droughts. It is estimated that 40 percent of Haitian children are chronically malnourished, and an estimated 80 percent of all Haitians eat fewer than 2,200 calories a day (Emling, 1997a).

GLOBAL POVERTY AND HUMAN DEVELOPMENT ISSUES

As the example of Haiti shows, income disparities are not the only factor that defines poverty and its effect on people. Although the average income per person in lower-income countries has doubled in the past thirty years and for many years economic growth has been seen as the primary way to achieve development in low-income economies, the United Nations since the 1970s has more actively focused on human development as a crucial factor in fighting poverty. In 1990 the United Nations Development Program introduced the Human Development Index (HDI), establishing three new criteria—in addition to GDP—for measuring the level of development in a country: life expectancy, education, and living standards. According to the United Nations, human development is the process of "expanding choices that people have in life, to lead a life to its full potential and in dignity, through expanding capabilities and through people taking action themselves to improve their lives" (United Nations Development Programme, 2002). Figure 8.4 compares indicators such as gross domestic product, life expectancy, and adult literacy in regions around the world. The World Bank calculates that 37 percent of the population in low- and middle-income countries (1.6 billion people) lack the essentials of well-being, whereas only 21 percent (900 million people) are "income poor," as defined by the bank's poverty line.

Women account for most of the "extra" 700 million poor people (Buvinić, 1997: 40).

Life Expectancy

Although some advances have been made in middle- and low-income countries regarding life expectancy, major problems still exist. On the plus side, average life expectancy has increased by about a third in the past three decades and is now more than 70 years in 87 countries (United Nations Development Programme, 2003). Although no country has reached a life expectancy for men of 80 years, 19 countries have now reached an expectancy of 80 years or more for women. On a less positive note, the average life expectancy at birth of people in middle-income countries remains about 12 years less than that of people in high-income countries. Moreover, the life expectancy of people in low-income nations is as much as 23 years less than that of people in high-income nations. Especially striking are the differences in life expectancies in high-income economies and low-income economies such as sub-Saharan Africa, where estimated life expectancy has dropped significantly in Uganda (from 48 to 43 for women and from 46 to 43 for men) over the past two decades.

One major cause of shorter life expectancy in low-income nations is the high rate of infant mortality. The infant mortality rate (deaths per thousand live births) is more than eight times higher in low-income countries than in high-income countries (World Bank, 2003c). Low-income countries typically have higher rates of illness and disease, and they do not have adequate health care facilities. Malnutrition is a common problem among children, many of whom are underweight, stunted, and have anemia—a nutritional deficiency with serious consequences for child mortality (see Chapter 19, "Population and Urbanization"). Consider this journalist's description of a child she saw in Haiti:

> Like any baby, Wisly Dorvil is easy to love. Unlike others, this 13-month-old is hard to hold.
>
> That's because his 10-pound frame is so fragile that even the most minimal of movements can dislocate his shoulders.
>
> As lifeless as a rag doll, Dorvil is starving. He has large, brown eyes and a feeble smile, but a stomach so tender that he suffers from ongoing bouts of vomiting and diarrhea.
>
> Fortunately, though, Dorvil recently came to the attention of U.S. aid workers. With round-the-clock feeding, he is expected to survive.
>
> Others are not so lucky. (Emling, 1997a: A17)

Figure 8.4   Indicators of Human Development

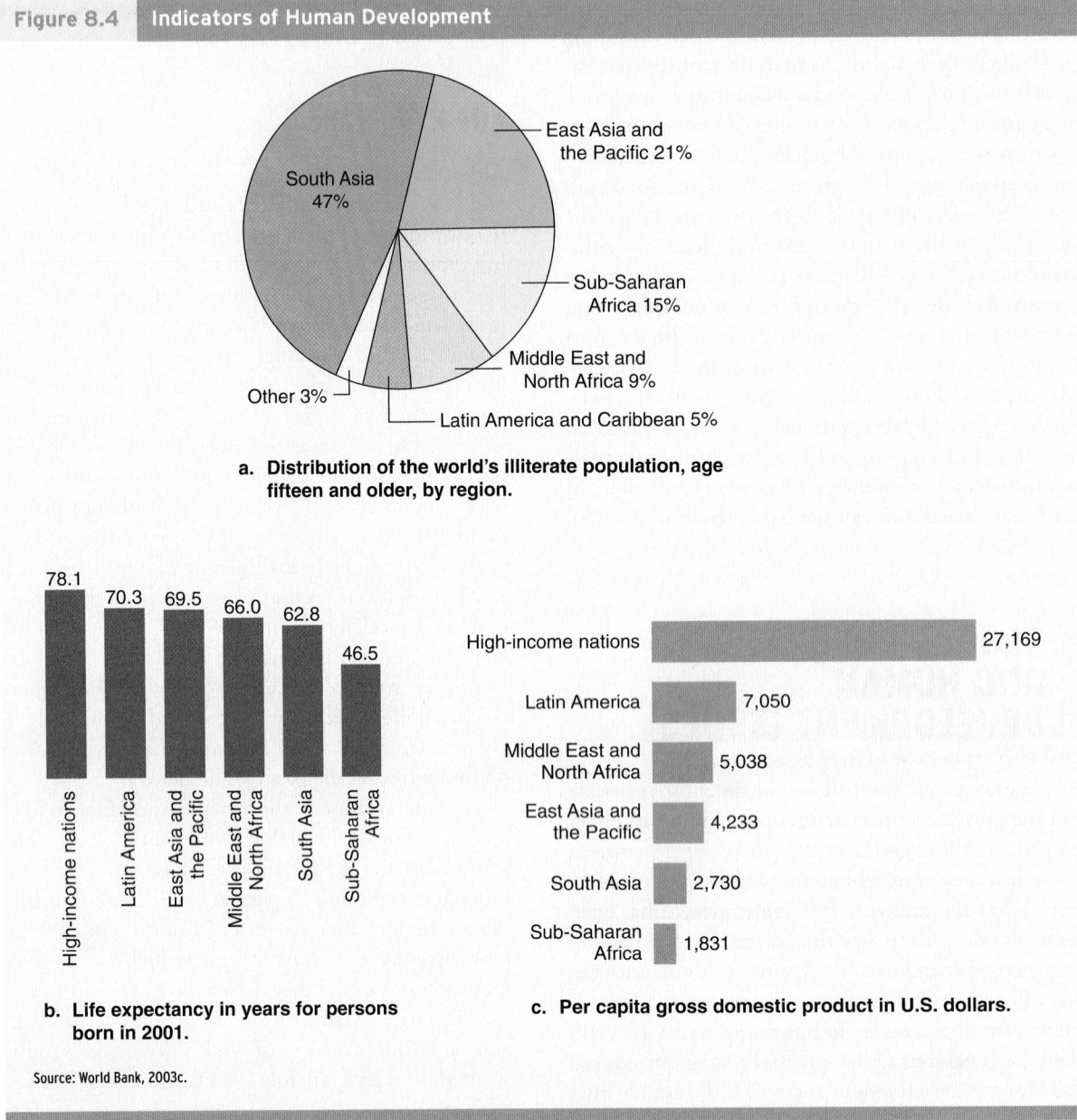

a. Distribution of the world's illiterate population, age fifteen and older, by region.

b. Life expectancy in years for persons born in 2001.

c. Per capita gross domestic product in U.S. dollars.

Source: World Bank, 2003c.

Among adults and children alike, life expectancies are strongly affected by hunger and malnutrition. It is estimated that people in the United States spend more than $5 billion each year on diet products to lower their calorie consumption, whereas the world's poorest 600 million people suffer from chronic malnutrition, and over 40 million people die each year from hunger-related diseases (Kidron and Segal, 1995). To put this figure in perspective, the number of people worldwide dying from hunger-related diseases is the equivalent of more than 300 jumbo-jet crashes per day with no survivors, and half the passengers children (Kidron and Segal, 1995).

Health

Health is defined in the Constitution of the World Health Organization as "a state of complete physical, mental and social well-being and not merely the absence of disease or infirmity." Many people in low-income nations are far from having physical, mental, and social well-being. In fact, about 17 million people die each year from diarrhea, malaria, tuberculosis, and other infectious and parasitic illnesses (Hauchler and Kennedy, 1994). According to the World Health Organization, infectious diseases are far from under control in many nations due to such factors as unsanitary

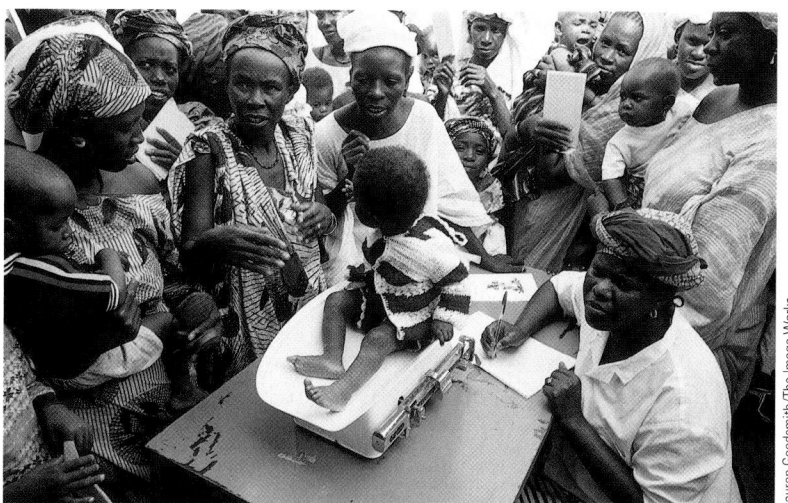

Malnutrition is a widespread health problem in many low-income nations. What kind of health problems are closely linked to middle-income and high-income countries?

Lauren Goodsmith/The Image Works

or overcrowded living conditions. Despite the possible eradication of diseases such as poliomyelitis, leprosy, guinea-worm disease, and neonatal tetanus in the near future, at least thirty new diseases—for which there is no treatment or vaccine—have recently emerged.

Some middle-income countries are experiencing rapid growth in degenerative diseases such as cancer and coronary heart disease, and many more deaths are expected from smoking-related diseases. Despite the decrease in tobacco smoking in high-income countries, there has been an increase in per capita consumption of tobacco in low- and middle-income countries, many of which have been targeted for free samples and promotional advertising by U.S. tobacco companies (United Nations Development Programme, 1997).

Education and Literacy

According to the Human Development Report (United Nations Development Programme, 2003), education is fundamental to reducing both individual and national poverty. As a result, school enrollment is used as one measure of human development. Although school enrollment has increased at the primary and secondary levels in about two-thirds of the lower-income regions, enrollment is not always a good measure of educational achievement because many students drop out of school during their elementary-school years.

What is literacy, and why is it important for human development? The United Nations Educational, Scientific and Cultural Organization (UNESCO) defines a literate person as "someone who can, with understanding, both read and write a short, simple statement on their everyday life" (United Nations, 1997:

89). Based on this definition, people who can write only with figures, their name, or a memorized phrase are not considered literate. The adult literacy rate in the low-income countries is about half that of the high-income countries, and for women the rate is even lower (United Nations, 1997). Women constitute about two-thirds of those who are illiterate: There are approximately 74 literate women for every 100 literate men (United Nations, 1997). Literacy is crucial for women because it has been closely linked to decreases in fertility, improved child health, and increased earnings potential (Hauchler and Kennedy, 1994).

Persistent Gaps in Human Development

Some middle- and lower-income countries have made progress in certain indicators of human development. The gap between some richer and middle- or lower-income nations has narrowed significantly for life expectancy, adult literacy, and daily calorie supply; however, the overall picture for the world's poorest people remains dismal. The gap between the poorest nations and the middle-income nations has continued to widen. Poverty, food shortages, hunger, and rapidly growing populations are pressing problems for at least 1.3 billion people, most of them women and children living in a state of absolute poverty. Although more women around the globe have paid employment than in the past, more and more women are still finding themselves in poverty because of increases in single-person and single-parent households headed by women and the fact that low-wage work is often the only source of livelihood available to them.

In an effort to reduce poverty, some nations have developed adult literacy programs so that people can gain an education that will help lift them out of poverty. In regions such as Iraqi Kurdistan, women's literacy is a particularly crucial issue.

According to an analyst for the Inter-American Development Bank, women experience sexual discrimination not only in terms of employment but also in wages:

> In Honduras, for example, coffee and tobacco farmers prefer to hire girls and women as laborers because they are willing to accept low wages and are more reliable workers. Especially in poor countries, female labor is primarily sought for low-paid positions in services, agriculture, small-scale commerce, and in the growing, unregulated manufacturing and agribusiness industries, which pay their workers individual rather than family wages, offer seasonal or part-time employment, and carry few or no benefits. Hence, this explains the seemingly contradictory trends of women's increased economic participation alongside their growing impoverishment. (Buvinić, 1997: 47)

▌THEORIES OF GLOBAL INEQUALITY

Why is the majority of the world's population growing richer while the poorest 20 percent—over 1 billion people—are so poor that they are effectively excluded from even a moderate standard of living? Social scientists have developed a variety of theories that view the causes and consequences of global inequality somewhat differently. We will examine the development approach and modernization theory, dependency theory, and world systems theory.

Development and Modernization Theory

According to some social scientists, global wealth and poverty are linked to the level of industrialization and economic development in a given society. Although the process by which a nation industrializes may vary somewhat, industrialization almost inevitably brings with it a higher standard of living in a nation and some degree of social mobility for individual participants in the society. Specifically, the traditional caste system becomes obsolete as industrialization progresses. Family status, race/ethnicity, and gender are said to become less significant in industrialized nations than in agrarian-based societies. As societies industrialize, they also urbanize as workers locate their residences near factories, offices, and other places of work. Consequently, urban values and folkways overshadow the beliefs and practices of the rural areas. Analysts using a development framework typically view industrialization and economic development as essential steps that nations must go through in order to reduce poverty and increase life chances for their citizens.

Earlier in the chapter, we discussed the post–World War II Marshall Plan, under which massive financial aid was provided to the European nations to help rebuild infrastructure lost in the war. Based on the success of this infusion of cash in bringing about modernization, President Truman and many other politicians and leaders in the business community believed that it should be possible to help so-called underdeveloped nations modernize in the same manner.

The most widely known development theory is **modernization theory—a perspective that links global inequality to different levels of economic development and suggests that low-income economies can move to middle- and high-income economies by achieving self-sustained economic growth.** According to modernization theory, the low-income, less-developed nations can improve their standard of living only with a period of intensive economic growth and accompanying changes in people's beliefs, values, and attitudes toward work. As a result of modernization, the values of people in developing countries supposedly become more similar to those of people in high-income nations. The number of hours that people work at their jobs each week is one measure of the extent to which individuals subscribe to the *work ethic,* a core value widely believed to be of great significance in the modernization process.

Perhaps the best-known modernization theory is that of Walt W. Rostow (1971, 1978), who, as an economic advisor to U.S. President John F. Kennedy, was

Poverty and war continue to devastate low-income countries such as Afghanistan. Many low-income countries receive aid from industrialized nations through initiatives such as the World Food Program. Modernization theory links global inequality to levels of economic development, but factors such as war and internal conflict also greatly contribute to patterns of global inequality.

© Chris Hondros/Getty Images

highly instrumental in shaping U.S. foreign policy toward Latin America in the 1960s. To Rostow, one of the largest barriers to development in low-income nations was the traditional cultural values held by people, particularly beliefs that are fatalistic, such as viewing extreme hardship and economic deprivation as inevitable and unavoidable facts of life. In cases of fatalism, people do not see any need to work in order to improve their lot in life: If it is predetermined for them, why bother? Based on modernization theory, poverty can be attributed to people's cultural failings, which are further reinforced by governmental policies interfering with the smooth operation of the economy.

Rostow suggested that all countries go through four stages of economic development, with identical content, regardless of when these nations started the process of industrialization. He compared the stages of economic development to an airplane ride. The first stage is the *traditional stage,* in which very little social change takes place, and people do not think much about changing their current circumstances. According to Rostow, societies in this stage are slow to change because the people hold a fatalistic value system, do not subscribe to the work ethic, and save very little money. The second stage is the *take-off stage*—a period of economic growth accompanied by a growing belief in individualism, competition, and achievement. During this stage, people start to look toward the future, to save and invest money, and to discard traditional values. According to Rostow's modernization theory, the development of capitalism is essential for the transformation from a traditional,

simple society to a modern, complex one. With the financial help and advice of the high-income countries, low-income countries will eventually be able to "fly" and enter the third stage of economic development. (Box 8.3 discusses some of the problems that low-income nations may encounter in completing this step today.) In the third stage, the country moves toward *technological maturity.* At this point, the country will improve its technology, reinvest in new industries, and embrace the beliefs, values, and social institutions of the high-income, developed nations. In the fourth and final stage, the country reaches the phase of *high mass consumption* and a correspondingly high standard of living.

Modernization theory has had both its advocates and its critics. According to proponents of this approach, studies have supported the assertion that economic development occurs more rapidly in a capitalist economy. In fact, the countries that have been most successful in moving from low- to middle-income status typically have been those that are most centrally involved in the global capitalist economy. For example, the nations of East Asia have successfully made the transition from low-income to higher-income economies through factors such as a high rate of savings, an aggressive work ethic among employers and employees, and the fostering of a market economy.

Critics of modernization theory point out that it tends to be Eurocentric in its analysis of low-income countries, which it implicitly labels as backward (see Evans and Stephens, 1988). In particular, modernization theory does not take into account the possibility

Box 8.3 CHANGING TIMES: MEDIA AND TECHNOLOGY

A Wired World?

- Sometimes I get angry that we had to wait so long. But now the Internet is all that my friends and I talk about. There are so many things in this world. —Hisham A. Turkistani, age twenty-five, who lives in Riyadh, Saudi Arabia (qtd. in Jehl, 1999: A4)

- A forty page document can be sent from Madagascar to Côte d'Ivoire by five-day courier for $75, by 30-minute fax for $45 or by two-minute email for less than 20 cents—and the email can go to hundreds of people at no extra cost. The choice is easy, if the choice is there. (United Nations Development Programme, 1999: 58)

The Internet and the creation of the World Wide Web have rapidly transformed communications for many people in some areas of the world. The number of Internet hosts—computers with a direct connection to the Internet—rose from under 100,000 in 1988 to over 40 million in 2000. However, as some analysts have suggested, "The collapse of space, time and borders may be creating a global village, but not everyone can be a citizen. The global, professional elite now faces low borders, but billions of others find borders as high as ever" (United Nations Development Programme, 1999: 3). At a time when many global businesses and political leaders are emphasizing the importance of global Internet connections, some have forgotten one basic fact: Many people in lower-income, less-industrialized nations do not have toilets, telephones, radios, or television sets, much less computers, modems, and other equipment necessary for becoming "wired." Nor can they afford it.

Why is rapid global communication important? According to United Nations experts, people who live in lower-income, developing nations need greater access to all kinds of information, particularly information related to health care issues. In lower-income nations, many people experience the world's most virulent and infectious diseases, yet they often have the least access to information for preventing or combating illnesses. Likewise, educational opportunities are extremely limited, and access to distance learning might provide people with needed information even though they live in an "information-poor"

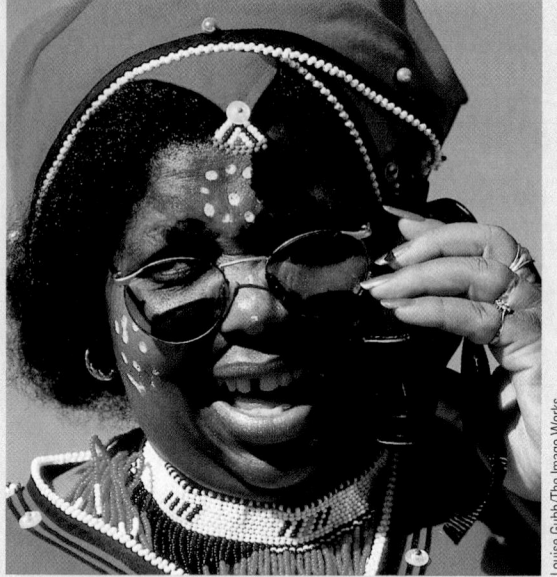

Louise Gubb/The Image Works

nation. Communications technology also opens up new opportunities for people to become part of the global economy and the political arena (United Nations Development Programme, 1999).

Who currently has access to the Internet? Although there has been a significant increase in Internet use worldwide, only about 10 percent of the world's inhabitants have such access—but that figure is up from about 3 percent in 1999. Access varies widely by region, educational level, gender, income, age, and rural versus urban location. Some of the most underrepresented areas are in Africa, where less than 1 percent of the population in nations such as Cameroon, Chad, Ethiopia, Ghana, and Nigeria have Internet access.

Is it possible to make global communications truly global? According to United Nations studies, global communications will not be possible for everyone unless a conscious effort is made to increase connectivity. From this viewpoint, access to the global communications network should be free from commercialization and profitability. Instead, the focus should be community and collaboration on the "information superhighway."

WRITING IN SOCIOLOGY ASSIGNMENT

Do you believe that more people living in the most densely populated regions of the world will gain access to the Internet? What ideas can you  suggest for enhancing global communications for a more diverse group of people?

Figure 8.5   Approaches to Studying Global Inequality

What causes global inequality? Social scientists have developed a variety of explanations, including the four theories shown here.

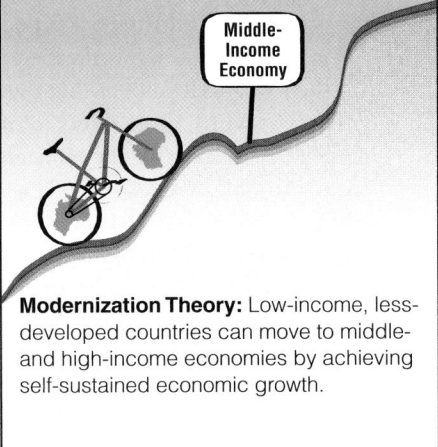

Modernization Theory: Low-income, less-developed countries can move to middle- and high-income economies by achieving self-sustained economic growth.

Dependency Theory: Global poverty can at least partially be attributed to the fact that low-income countries have been exploited by high-income economies; the poor nations are trapped in a cycle of dependency on richer nations.

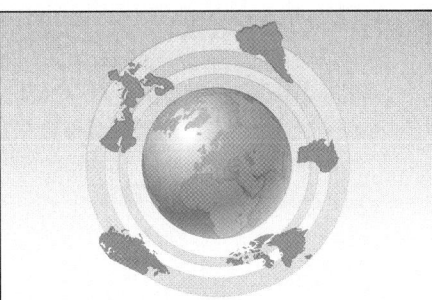

World Systems Theory: How a country is incorporated into the global capitalist economy (e.g., a core, semiperipheral, or peripheral nation) is the key feature in determining how economic development takes place in that nation.

The New International Division of Labor Theory: Commodity production is split into fragments, each of which can be moved (e.g., by a transnational corporation) to whichever part of the world can provide the best combination of capital and labor.

that all nations do not industrialize in the same manner. In contrast, some analysts have suggested that modernization of low-income nations today will require novel policies, sequences, and ideologies that are not accounted for by Rostow's approach (see Gerschenkron, 1962).

Which sociological perspective is most closely associated with the development approach? Modernization theory is based on a market-oriented perspective which assumes that "pure" capitalism is good and that the best economic outcomes occur when governments follow the policy of laissez-faire (or hands-off) business, giving capitalists the opportunity to make the "best" economic decisions, unfettered by government restraints or cumbersome rules and regulations (see

Chapter 13, "The Economy and Work in Global Perspective"). In today's global economy, however, many analysts believe that national governments are no longer central corporate decision makers and that transnational corporations determine global economic expansion and contraction. Therefore, corporate decisions to relocate manufacturing processes around the world make the rules and regulations of any one nation irrelevant and national boundaries obsolete (Gereffi, 1994). Just as modernization theory most closely approximates a functionalist approach to explaining inequality, dependency theory, world systems theory, and the new international division of labor theory are perspectives rooted in the conflict approach. All four of these approaches are depicted in Figure 8.5.

Dependency Theory

Dependency theory states that global poverty can at least partially be attributed to the fact that the low-income countries have been exploited by the high-income countries. Analyzing events as part of a particular historical process—the expansion of global capitalism—dependency theorists see the greed of the rich countries as a source of increasing impoverishment of the poorer nations and their people. Dependency theory disputes the notion of the development approach, and modernization theory specifically, that economic growth is the key to meeting important human needs in societies. In contrast, the poorer nations are trapped in a cycle of structural dependency on the richer nations due to their need for infusions of foreign capital and external markets for their raw materials, making it impossible for the poorer nations to pursue their own economic and human development agendas. For this reason, dependency theorists believe that countries such as Brazil, Nigeria, India, and Kenya cannot reach the sustained economic growth patterns of the more-advanced capitalist economies.

Dependency theory has been most often applied to the newly industrializing countries (NICs) of Latin America, whereas scholars examining the NICs of East Asia found that dependency theory had little or no relevance to economic growth and development in that part of the world. Therefore, dependency theory had to be expanded to encompass transnational economic linkages that affect developing countries, including foreign aid, foreign trade, foreign direct investment, and foreign loans. On the one hand, in Latin America and sub-Saharan Africa, transnational linkages such as foreign aid, investments by transnational corporations, foreign debt, and export trade have been significant impediments to development within a country. On the other hand, East Asian countries such as Taiwan, South Korea, and Singapore have historically also had high rates of dependency on foreign aid, foreign trade, and interdependence with transnational corporations but have still experienced high rates of economic growth despite dependency. According to the sociologist Gary Gereffi (1994), differences in outcome are probably associated with differences in the timing and sequencing of a nation's relationship with external entities such as foreign governments and transnational corporations. However, in her study of the satellite factory system in Taiwan, the sociologist Ping-Chun Hsiung found that managers in the plants use the norms and behavior patterns of Chinese society to derive high profits for the capitalists. As one manager explained to Hsiung,

> We managers are the mediators between the boss and the workers. We have to communicate [the workers'] opinion to the boss, while at the same time watching out for the boss's pocket. . . . It isn't really a bad thing if the workers are not happy with their wages. It implies that they have the potential to be worth more. How to manipulate them all depends on us, the managers. . . . Chinese are humble. We seldom talk about how good we are, not to mention boast. When things come down to wage conflicts, I always turn the issue around by asking the workers to give me a figure. That is, I ask them to tell me exactly how much more they believe their labor is worth. If they can't come out with a concrete price, then, they have to listen to me. I may decide to give them a raise. I may not. It's all up to me. Even if they do give me a price, the chances are it will always be lower than the real value of their labor. For example, if what they really want is a one-hundred-dollar raise, as Chinese they will only say that their work is worth eighty dollars more, at the most. When this happens, I can really cut it to fifty dollars. . . . By handling them this way, their productivity will go up because I do show them that I did recognize their unrest [and give them a raise]. . . . It is a win–win battle for the company when things come down to wage conflict, you know. (qtd. in Hsiung, 1996: 62)

Dependency theory makes a positive contribution to our understanding of global poverty by noting that "underdevelopment" is not necessarily the cause of inequality. Rather, it points out that exploitation not only of one country by another but of countries by transnational corporations may limit or retard economic growth and human development in some nations. However, what remains unexplained is how East Asia and India had successful "dependency management" whereas many Latin American countries did not (Gereffi, 1994). In fact, over the past decade, the annual economic growth in East Asia (excluding Japan) has averaged 8.5 percent, which is four times the rate of the West. Currently, the World Bank forecasts a 6-percent annual compound growth rate in the region during the next decade (World Bank, 2003c). If we compare this rate with the Industrial Revolution, we find that it took Britain nearly sixty years to double its per capita output and the United States nearly fifty years to do the same. In contrast, East Asia doubles its per capita out-

A variety of factors—such as foreign investment and the presence of transnational corporations—have contributed to the economic growth of nations such as Singapore.

James Marshall/The Image Works

come every decade. Other conflict explanations have sought to explain global inequality by using the framework of world systems theory and the new international division of labor.

World Systems Theory

World systems theory suggests that what exists under capitalism is a truly global system that is held together by economic ties. From this approach, global inequality does not emerge solely as a result of the exploitation of one country by another. Instead, economic domination involves a complex world system in which the industrialized, high-income nations benefit from other nations and exploit their citizens. This theory is most closely associated with the sociologist Immanuel Wallerstein (1979, 1984), who believed that a country's mode of incorporation into the capitalist work economy is the key feature in determining how economic development takes place in that nation. According to *world systems theory,* the capitalist world economy is a global system divided into a hierarchy of three major types of nations—core, semiperipheral, and peripheral—in which upward or downward mobility is conditioned by the resources and obstacles that characterize the international system. *Core nations* **are dominant capitalist centers characterized by high levels of industrialization and urbanization.** Core nations such as the United States, Japan, and Germany possess most of the world's capital and technology. Even more importantly for their position of domination, they exert massive control over world trade and economic agreements across national boundaries. Some cities in core nations are referred to as *global cities* because they

serve as international centers for political, economic, and cultural concerns. New York, Tokyo, and London are the largest global cities, and they are often referred to as the "command posts" of the world economy.

Semiperipheral nations **are more developed than peripheral nations but less developed than core nations.** Nations in this category typically provide labor and raw materials to core nations within the world system. These nations constitute a midpoint between the core and peripheral nations that promotes the stability and legitimacy of the three-tiered world economy. These nations include South Korea and Taiwan in East Asia, Mexico and Brazil in Latin America, India in South Asia, and Nigeria and South Africa in Africa. Only two global cities are located in semiperipheral nations: São Paulo, Brazil, which is the center of the Brazilian economy, and Singapore, which is the economic center of a multi-country region in Southeast Asia. According to Wallerstein, semiperipheral nations exploit peripheral nations, just as the core nations exploit both the semiperipheral and the peripheral.

Most low-income countries in Africa, South America, and the Caribbean are *peripheral nations—* **nations that are dependent on core nations for capital, have little or no industrialization (other than what may be brought in by core nations), and have uneven patterns of urbanization.** According to Wallerstein (1979, 1984), the wealthy in peripheral nations benefit from the labor of poor workers and from their own economic relations with core nation capitalists, whom they uphold in order to maintain their own wealth and position. At a global level, uneven economic growth results from capital investment by core nations; disparity between the rich and the

Figure 8.6   Maquiladora Plants

Here is the process by which transnational corporations establish plants in Mexico so that profits can be increased by using low-wage workers there to assemble products that are then brought into the United States for sale.

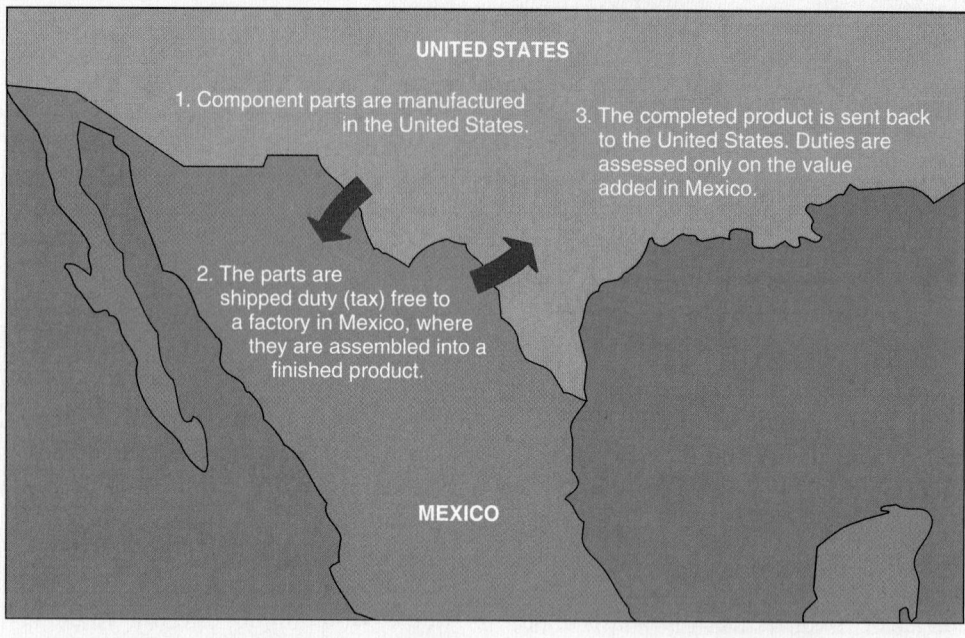

UNITED STATES

1. Component parts are manufactured in the United States.

3. The completed product is sent back to the United States. Duties are assessed only on the value added in Mexico.

2. The parts are shipped duty (tax) free to a factory in Mexico, where they are assembled into a finished product.

MEXICO

poor within the major cities in these nations is increased in the process. The U.S./Mexican border is an example of disparity and urban growth: Transnational corporations have built *maquiladora* plants so that goods can be assembled by low-wage workers to keep production costs down. Figure 8.6 describes this process. In 2001, there were almost 3,800 maquiladora plants in Mexico, employing 1.3 million people (*Austin American-Statesman,* 2001). Because of a demand for a large supply of low-wage workers, thousands of people moved from the rural regions of Mexico to urban areas along the border in hope of earning a higher wage. This influx pushed already overcrowded cities far beyond their capacity. Many people live on the edge of the city in *shantytowns* made from discarded materials or in low-cost rental housing in central-city slums because their wages are low and affordable housing is nonexistent (Flanagan, 2002). In fact, housing shortages are among the most pressing problems in many peripheral nations.

Not all social analysts agree with Wallerstein's perspective on the hierarchical position of nations in the global economy. However, most scholars acknowledge that nations throughout the world are influenced by a relatively small number of cities and

transnational corporations that have prompted a shift from an international to a more global economy (see Knox and Taylor, 1995; Wilson, 1997). Even Wallerstein (1991) acknowledges that world systems theory is an "incomplete, unfinished critique" of long-term, large-scale social change that influences global inequality.

The New International Division of Labor Theory

Although the term *world trade* has long implied that there is a division of labor between societies, the nature and extent of this division have recently been reassessed based on the changing nature of the world economy. According to the *new international division of labor theory,* commodity production is being split into fragments that can be assigned to whichever part of the world can provide the most profitable combination of capital and labor. Consequently, the new international division of labor has changed the pattern of geographic specialization between countries, whereby high-income countries have now become dependent on low-income countries for labor. The low-

The global assembly line includes many women in low-income nations who work diligently to earn wages to support themselves and their families. The women shown here are assembling blue jeans that will be sold in high-income countries.

income countries provide transnational corporations with a situation in which they can pay lower wages and taxes and face fewer regulations regarding workplace conditions and environmental protection (Waters, 1995). Overall, a global manufacturing system has emerged in which transnational corporations establish labor-intensive, assembly-oriented export production, ranging from textiles and clothing to technologically sophisticated exports such as computers, in middle- and lower-income nations (Gereffi, 1994). At the same time, manufacturing technologies are shifting from the large-scale, mass-production assembly lines of the past toward a more flexible production process involving microelectronic technologies. Even service industries—such as processing insurance claims forms—that were formerly thought to be less mobile have become exportable through electronic transmission and the Internet. The global nature of these activities has been referred to as *global commodity chains,* a complex pattern of international labor and production processes that results in a finished commodity ready for sale in the marketplace.

Some commodity chains are producer-driven, whereas others are buyer-driven. *Producer-driven commodity chains* is the term used to describe industries in which transnational corporations play a central part in controlling the production process. Industries that produce automobiles, computers, and other capital- and technology-intensive products are typically producer-driven. In contrast, *buyer-driven commodity chains* is the term used to refer to industries in which large retailers, brand-name merchandisers, and trading companies set up decentralized production networks in various middle- and low-income countries. This type of chain is most common in labor-intensive, consumer-goods industries such as toys, garments, and

footwear (Gereffi, 1994). Athletic footwear companies such as Nike and Reebok and clothing companies like The Gap and Liz Claiborne are examples of the buyer-driven model. Since these products tend to be labor intensive at the manufacturing stage, the typical factory system is very competitive and globally decentralized. Workers in buyer-driven commodity chains are often exploited by low wages, long hours, and poor working conditions. In fact, most workers cannot afford the products they make. Tini Heyun Alwi, who works on the assembly line of the shoe factory in Indonesia that makes Reebok sneakers, is an example: "I think maybe I could work for a month and still not be able to buy one pair" (qtd. in Goodman, 1996: F1). Since Tini earns only 2,600 Indonesian rupiah ($1.28) per day working a ten-hour shift six days a week, her monthly income would fall short of the retail price of the athletic shoes (Goodman, 1996). Sociologist Gary Gereffi (1994: 225) explains the problem with studying the new global patterns as follows:

> The difficulty may lie in the fact that today we face a situation where (1) the political unit is *national,* (2) industrial production is *regional,* and (3) capital movements are *international.* The rise of Japan and the East Asian [newly industrializing countries] in the 1960s and 1970s is the flip side of the "deindustrialization" that occurred in the United States and much of Europe. Declining industries in North America have been the growth industries in East Asia.

As other analysts suggest, these changes have been a mixed bag for people residing in these countries. For example, Indonesia has been able to woo foreign business into the country, but workers have experienced poverty despite working full time in factories making such consumer goods as Nike tennis shoes (Gargan, 1996). As employers feel pressure from workers to raise wages, clashes erupt between the workers and managers or owners. Similarly, the governments in these countries fear that rising wages and labor strife will drive away the businesses, sometimes leaving behind workers who have no other hopes for employment and become more impoverished than they previously were. Moreover, in situations where government officials were benefiting from the presence of the companies, they also become losers if the workers rebel against their pay or working conditions.

Is this theory useful for examining global inequality today? Recent research on the blue-jeans industry in Mexico suggests that it is (see Bair and Gereffi 2001; Gereffi, Spener, and Bair, 2002). Consider the fact that the blue-jeans manufacturing industry in Mexico is primarily an export-oriented business in that jeans made in cities such as Torreon, Mexico, are produced specifically to be exported to high-income nations such as the United States, where the jeans are sold in large retail chains such as Wal-Mart and Target or specialty retailers such as Gap and Limited. When there is an economic slowdown in the United States, fewer jeans orders are placed. Further complicating matters, jeans produced specifically for U.S. corporations cannot be sold directly in Mexico, so the market for already-produced jeans is further limited. As a result, when U.S. retailers cut their orders for products produced in Mexico, this has a detrimental effect on workers in the Mexican cities where they are made and on the local economy in those cities. This is only one example of the problems that emerge as a result of the complex interdependence of high-income nations and low-income nations in regard to labor, production, and consumption of goods. No doubt, issues such as these will be crucial in determining the nature and extent of global inequality in the future.

GLOBAL INEQUALITY IN THE FUTURE

As we have seen, social inequality is vast both within and among the countries of the world. Even in high-income nations where wealth is highly concentrated, many poor people coexist with the affluent. In middle- and low-income countries, there are small pockets of wealth in the midst of poverty and despair.

What are the future prospects for greater equality across and within nations? Not all social scientists agree on the answer to this question. Depending on the theoretical framework they apply in studying global inequality, social analysts may describe either an optimistic or a pessimistic scenario for the future. Moreover, some analysts highlight the human rights issues embedded in global inequality, whereas others focus primarily on an economic framework.

In some regions, persistent and growing poverty continues to undermine human development and future possibilities for socioeconomic change. Gross inequality has high financial and quality-of-life costs to people, even among those who are not the poorest of the poor. In the future, continued population growth, urbanization, environmental degradation, and violent conflict threaten even the meager living conditions of those residing in low-income nations. From this approach, the future looks dim not only for people in low-income and middle-income countries but also for those in high-income countries, who will see their quality of life diminish as natural resources are depleted, the environment is polluted, and high rates of immigration and global political unrest threaten the high standard of living that many people have previously enjoyed. According to some social analysts, transnational corporations and financial institutions such as the World Bank and the International Monetary Fund will further solidify and control a globalized economy, which will transfer the power to make significant choices to these organizations and away from the people and their governments. As a result, further loss of resources and means of livelihood will affect people and countries around the globe.

As a result of global corporate domination, there could be a leveling out of average income around the world, with wages falling in the high-income countries and wages increasing significantly in low- and middle-income countries. If this pessimistic scenario occurs, there is likely to be greater polarization of the rich and the poor and more potential for ethnic and national conflicts over such issues as worsening environmental degradation and who has the right to natural resources. For example, pulp-and-paper companies in Indonesia, along with palm oil plantation owners, have continued clearing land for crops by burning off vast tracts of jungle, producing high levels of smog and pollution across seven Southeast Asian nations and creating havoc for millions of people (Mydans, 1997b). Whose rights should prevail in situations such as this? The smoke from the Indonesian fires has affected agriculture; food shortages and rising prices are inevitable. Reduced sunlight is slowing the growth of crops, and people in record numbers are suffering from respiratory illnesses. The possibility

Box 8.4 YOU CAN MAKE A DIFFERENCE

Global Networking to Reduce World Hunger and Poverty

We, the people of the world, will mobilize the forces of transnational civil society behind a widely shared agenda that binds our many social movements in pursuit of just, sustainable, and participatory human societies. In so doing we are forging our own instruments and processes for redefining the nature and meaning of human progress and for transforming those institutions that no longer respond to our needs. We welcome to our cause all people who share our commitment to peaceful and democratic change in the interest of our living planet and the human societies it sustains.

> —International NGO Forum, United Nations Conference on Environment and Development, Rio de Janeiro, Brazil, June 12, 1992 (qtd. in Korten, 1996: 333)

If everyone lit just one little candle, what a bright world this would be.

> —line from the 1950s theme song for Bishop Fulton J. Sheen's television series *Life Is Worth Living* (Sheen, 1995: 245)

When many of us think about problems such as world poverty, we tend to see ourselves as powerless to bring about change in so vast an issue. However, a recurring message from social activists and religious leaders is that each person can contribute something to the betterment of other people and sometimes the entire world.

An initial way for each of us to become involved is to become more informed about global issues and to learn how we can contribute time and resources to organizations seeking to address social issues such as illiteracy and hunger. We can also find out about meetings and activities of organizations and participate in online discussion forums where we can express our opinions, ask questions, share information, and interact with other people interested in topics such as international relief and development. At first, it may not feel like you are doing much to address global problems; however, information and education are the first steps to promoting greater understanding of social problems and of the world's people, whether they reside in high-, middle-, or low-income countries and regardless of their individual socioeconomic position. Likewise, it is important to help our own nation's children understand that they can make a difference in ending hunger in the United States and other nations.

Would you like to function as a catalyst for change? You can learn how to proceed by gathering information from organizations that seek to reduce problems

■ Willie Colon, a Puerto Rican salsa star and spokesperson for CARE, visited this Bolivian classroom as part of that international relief organization's project to reduce women's poverty.

such as poverty and to provide forums for interacting with other people. Here are a few starting points for your information search:

- CARE International is a confederation of ten national members in North America, Europe, Japan, and Australia. CARE assists the world's poor in their efforts to achieve social and economic well-being. Its work reaches 25 million people in fifty-three nations in Africa, Asia, Latin America, and Eastern Europe. Programs include emergency relief, education, health and population, children's health, reproductive health, water and sanitation, small economic activity development, agriculture, community development, and environment. Contact CARE at 151 Ellis Street, NE, Atlanta, GA 30303-2439. On the Internet:

http://www.care.org

Other organizations fighting world hunger and health problems include the following:

- World Hunger Year ("WHY"):

http://www.worldhungeryear.org

- "Kids Can Make a Difference," an innovative guide developed by WHY:

http://www.kids.maine.org/cando.htm

- World Health Organization:

http://www.who.org

remains that severe, long-term illnesses may also develop over time, especially among the young, the old, and people with respiratory problems (Mydans, 1997b). This could be a future scenario if "business as usual" continues to take place around the world.

On the other hand, a more optimistic scenario is also possible. With modern technology and worldwide economic growth, it might be possible to reduce absolute poverty and to increase people's opportunities. Among the trends cited by the Human Development Report (United Nations Development Programme, 2003) that have the potential to bring about more sustainable patterns of development are the socioeconomic progress made in many low- and middle-income countries over the past thirty years as technological, social, and environmental improvements have occurred. For example, technological innovation continues to improve living standards for some people. Fertility rates are declining in some regions (but remain high in others, where there remains grave cause for concern about the availability of adequate natural resources for the future). Finally, health and education may continue to improve in lower-income countries. According to the Human Development Report (United Nations Development Programme, 2003), healthy, educated populations are crucial for the future in order to reduce global poverty. The education of women is of primary importance in the future if global inequality is to be reduced. As one analyst stated, "If you educate a boy, you educate a human being. If you educate a girl, you educate generations" (Buvinić, 1997: 49). All aspects of schooling and training are crucial for the future, including agricultural extension services in rural areas to help women farmers in regions such as western Kenya produce more crops to feed their families. As we saw at the beginning of the chapter, easier access to water can make a crucial difference in people's lives. Mayra Buvinić, of the Inter-American Development Bank, puts global poverty in perspective for people living in high-income countries by pointing out that their problems are our problems. She provides the following example:

> Reina is a former guerilla fighter in El Salvador who is being taught how to bake bread under a post–civil war reconstruction program. But as she says, "the only thing I have is this training and I don't want to be a baker. I have other dreams for my life."
>
> Once upon a time, women like Reina . . . only migrated [to the United States] to follow or find a husband. This is no longer the case. It is likely that Reina, with few opportunities in her own country, will sooner or later join the rising number of female migrants who leave families and children behind to seek better paying work in the United States and other industrial countries. Wisely spent foreign aid can give Reina the chance to realize her dreams in her *own* country. (Buvinić, 1997: 38, 52)

From this viewpoint, we can enjoy prosperity only by ensuring that other people have the opportunity to survive and thrive in their own surroundings (see Box 8.4 on page 279). The problems associated with global poverty are therefore of interest to a wide-ranging set of countries and people.

We will continue to focus in subsequent chapters on issues pertaining to inequality as we examine social class in the United States, including topics such as why inequalities persist and how the distribution of income and wealth in our nation produces differential access to goods and services, health and nutrition, and educational opportunities.

CHAPTER REVIEW

■ What is stratification, and how does it affect our daily life?

Stratification is the hierarchical arrangement of large social groups based on their control over basic resources. People are treated differently based on where they are positioned within the social hierarchies of class, race, gender, and age.

■ What are the major systems of stratification?

Stratification systems include slavery, caste, and class. Slavery, an extreme form of stratification in which people are owned by others, is a closed system. The caste system is also a closed one in which people's status is determined at birth, based on their parents' position in society. The class system, which exists in the United States, is a type of stratification based on ownership of resources and on the type of work that people do. Class systems are characterized by unequal distribution of resources and by movement up and down the class structure through social mobility.

■ What is global stratification, and how does it contribute to economic inequality?

Global stratification refers to the unequal distribution of wealth, power, and prestige on a global basis, which results in people having vastly different lifestyles and life chances both within and among the nations of the world. Today, the income gap between the richest and the poorest 20 percent of the world population continues to widen, and within some nations the poorest one-fifth of the population has an income that is only a slight fraction of the overall average per capita income for that country.

■ Why is it difficult to study global inequality?

Terminology is a major problem in studying global inequality. Most definitions of inequality are based on comparisons of levels of income or economic development, whereby countries are identified in terms of the "three worlds" or upon their levels of economic development. Today, many sociologists use the World Bank's classification of nations into three economic categories: low-income economies, middle-income economies, and high-income economies. A second problem lies in acquiring comparable data from various nations; although gross national product continues to be one of the most widely used measures of national income, in recent years the United Nations and the World Bank have begun to use the gross domestic product—all of the goods and services produced within a country's economy during a given year.

■ How are global poverty and human development related?

Income disparities are not the only factor that defines poverty and its effect on people. The United Nations' Human Development Index measures the level of development in a country through indicators such as life expectancy, infant mortality rate, proportion of underweight children under age five (a measure of nourishment and health), and adult literacy rate for low-income, middle-income, and high-income countries.

■ What is modernization theory, and what stages did Rostow believe all societies go through?

Modernization theory is a perspective that links global inequality to different levels of economic development and suggests that low-income economies can move to middle- and high-income economies by achieving self-sustained economic growth. According to Rostow, all countries go through four stages of economic development: (1) the traditional stage, in which very little social change takes place; (2) the take-off stage, a period of economic growth accompanied by a growing belief in individualism, competition, and achievement; (3) technological maturity, a period of improving technology, reinvesting in new industries, and embracing the beliefs, values, and social institutions of the high-income, developed nations; and (4) the phase of high mass consumption, which is accompanied by a high standard of living.

■ How does dependency theory differ from modernization theory?

Dependency theory states that global poverty can at least partially be attributed to the fact that the low-income countries have been exploited by the high-income countries. Whereas modernization theory focuses on how societies can reduce inequality through industrialization and economic development, dependency theorists see the greed of the rich countries as a source of increasing impoverishment of the poorer nations and their people.

■ What is world systems theory, and how does it view the global economy?

According to world systems theory, the capitalist world economy is a global system divided into a hierarchy of three major types of nations: Core nations are dominant capitalist centers characterized by high levels of industrialization and urbanization, semiperipheral nations are more developed than peripheral nations but less developed than core nations, and peripheral nations are those countries that are dependent on core nations for capital, have little or no industrialization (other than what may be brought in by core nations), and have uneven patterns of urbanization.

■ What is the new international division of labor theory?

The new international division of labor theory is based on the assumption that commodity production is split into fragments that can be assigned to whichever part of the world can provide the most profitable combination of capital and labor. This division of labor has changed the pattern of geographic specialization between countries, whereby high-income countries have become dependent on low-income countries for labor. The low-income countries provide transnational corporations with a situation in which they can pay lower wages and taxes, and face fewer regulations regarding workplace conditions and environmental protection.

KEY TERMS

QUESTIONS FOR CRITICAL THINKING

1. You have decided to study global wealth and poverty. How would you approach your study? What research methods would provide the best data for analysis? What might you find if you compared your research data with popular presentations—such as films and advertising—of everyday life in low- and middle-income countries?
2. How would you compare the lives of poor people living in the low-income nations of the world with those in central cities and rural areas of the United States? In what ways are their lives similar? In what ways are they different?
3. Should U.S. foreign policy include provisions for reducing poverty in other nations of the world? Should U.S. domestic policy include provisions for reducing poverty in the United States? How are these issues similar? How are they different?
4. Using the theories discussed in this chapter, devise a plan to alleviate global poverty. Assume that you have the necessary wealth, political power, and other resources necessary to reduce the problem. Share your plan with others in your class, and create a consolidated plan that represents the best ideas and suggestions presented.

RESOURCES ON THE INTERNET

Chapter-Related Web Sites

The following Web sites have been selected for their relevance to the topics in this chapter. These sites are among the more stable, but please note that Web site addresses change frequently. For an updated list of chapter-related Web sites with URL links, please visit the *Sociology in Our Times* Web site (**www.wadsworth.com/KendallSIOT**).

National Labor Committee
http://www.nlcnet.org

The mission of the National Labor Committee is to educate the public on human and labor rights abuses by corporations. Its Web site features numerous reports on contemporary slavery around the world, a photo gallery, information on how to become involved, and numerous links to sites displaying human labor rights violations.

The United Nations
http://www.un.org

The United Nations Web site contains a wealth of information regarding social, economic, and political conditions around the world. The UN's publications on economic and social development are a good source of data about global stratification.

The World Bank Group
http://www.worldbank.org

The World Bank Group helps the poorest people and countries around the world by providing development assistance in areas such as education, health, debt relief, and the environment. Click on "Development Topics" and scroll down to "Poverty" to access publications and other information related to global poverty.

ONLINE STUDY AND RESEARCH TOOLS

Accompanying this text are many *free* powerful online study tools that will help you master the material in this chapter, help increase your depth of understanding, and help you make the grade!

SocCoach CD-ROM

Use the SocCoach CD-ROM enclosed with this text to help you formulate a customized study plan for this chapter. After you take the Diagnostic Quiz, SocCoach will generate a customized study plan just for you! It will identify sections of the chapter that you should review and will provide videos, charts, graphs, and excerpts from the text to supplement your studies and enhance your understanding. You'll also find fun, interactive activities such as Virtual Explorations and Map the Stats to apply what you've learned and stretch your sociological imagination.

The Companion Web Site for Sociology in Our Times, *Fifth Edition*

www.wadsworth.com/KendallSIOT

Gain an even better grasp on this chapter by going to the companion Web site to take one of the Tutorial Quizzes, use the Flash Cards to master key terms, or check out the many other study aids you'll find there. You'll also find special features such as GSS Data and Census 2000 information that'll put data and resources at your fingertips to help you with that special project or help you as you do some research on your own.

In this chapter, when you see the icon on the left, it alerts you to a specific exercise found in *Wadsworth's Sociology Online Resources and Writing Companion.* This valuable guide shows you how to use Wadsworth's exclusive online resources—*InfoTrac College Edition,* the *Opposing Viewpoints Resource Center,* and *MicroCase Online*—to assist you in your study of sociology and to build essential research and writing skills.

CHAPTER 9

Social Class in the United States

We treat them in hospitals every day.

They are young brothers, often drug dealers, gang members, or small-time criminals, who show up shot, stabbed, or beaten after a hustle gone bad. To some of our medical colleagues, they are just nameless thugs, perpetuating crime and death in neighborhoods that have seen far too much of these things. But when we look into their faces, we see ourselves as teenagers, we see our friends, we see what we easily could have become as young adults. And we're reminded of the thin line that separates us—three twenty-nine-year-old doctors (an emergency-room physician, an internist, and a dentist)—from those patients whose lives are filled with danger and desperation.

We grew up in poor, broken homes in New Jersey neighborhoods riddled with crime, drugs, and death, and came of age in the 1980s at the height of a crack epidemic that ravaged communities like ours throughout the nation. . . . Two of us landed in juvenile-detention centers before our eighteenth birthdays. But inspired early by caring and imaginative role models, one of us in childhood latched on to a dream of becoming a dentist, steered clear of trouble, and in his senior year of high school persuaded his two best friends to apply to a college program for minority students interested in becoming doctors. We knew we'd never survive if we went after it alone. And so we made a pact:

we'd help one another through, no matter what.

–Drs. Sampson Davis, George Jenkins, and Rameck Hunt (2003: 1-2), describing their path from the streets of Newark to being named among the forty most influential African Americans by *Essence* magazine and thus, in the eyes of many people, achieving the *American Dream*

The remarkable success of Sampson Davis, George Jenkins, and Rameck Hunt as they stuck together and worked diligently to get out of graffiti-covered New Jersey public-housing projects and to ultimately complete their education in medical and dental schools might be described as a contemporary version of the American Dream. What is the American Dream? Simply stated, the American Dream is the belief that if people work hard and play by the rules, they will have a chance to get ahead (see Hochschild, 1995). Moreover, each generation will be able to have a higher standard of living than that of its parents (Danziger and Gottschalk, 1995). The American Dream is based on the assumption that people in the United States have equality of opportunity regardless of their race, creed, color, national origin, gender, or religion.

For middle- and upper-income people, the American Dream typically means that each subsequent generation will be able to acquire more material possessions and wealth than people in the preceding generations. To some people, achieving the American Dream means having a secure job, owning a home, and getting a good education for their children. To others, it is the promise that anyone may rise from poverty to wealth (from "rags to riches") if he or she works hard enough. In this chapter, we examine the U.S. class structure to see how people's opportunities are affected by their position in that structure. However, before we explore class and stratification in the United States, test your knowledge of wealth, poverty, and the American Dream by taking the quiz in Box 9.1.

QUESTIONS AND ISSUES

Chapter Focus Question: How is the American Dream influenced by social stratification?

How do prestige, power, and wealth determine social class?

What role does occupational structure play in a functionalist perspective on class structure?

What role does ownership of resources play in a conflict perspective on class structure?

How are social stratification and poverty linked?

INCOME AND WEALTH DIFFERENCES IN THE UNITED STATES

Throughout human history, people have argued about the distribution of scarce resources in society. Disagreements often concentrate on whether the share we get is a fair reward for our effort and hard work. Recently, social analysts have pointed out that (except during temporary economic downturns) the old maxim "the rich get richer" continues to be valid in the United States. To understand how this happens, we must take a closer look at income and wealth inequality in this country.

Comparing Income and Wealth

When many people discuss financial well-being—and, thus, a person's place in the U.S. class structure—they are usually referring to income. However, it is important to distinguish between income and wealth. ***Income* is the economic gain derived from wages, salaries, income transfers (governmental aid), and ownership of property** (Beeghley, 2000). Or, to put it another way, "income refers to money, wages, and payments that periodically are received as returns from an occupation or investment" (Kerbo 2000: 19). For most of us living in a class system, our major sources of income are a wage or salary, although some people live entirely off the earnings on their investments. But income is only one aspect of wealth. ***Wealth* is the value of all of a person's or family's economic assets, including income, personal property, and income-producing property.** As this definition suggests, wealth refers to accumulated assets in the form of various types of valued goods, including property such as buildings, land, farms, houses, factories, and cars, as well as other assets such as bank accounts, corporate stocks, bonds, and insurance policies. Wealth can be used to generate income that is used to purchase necessities (such as food, clothing, and shelter) or luxuries (such as a diamond ring or a yacht). It has been said that "wealth generates more wealth" (Keister, 2000: 7), meaning that wealthy people may acquire even more wealth as a result of their investments. Wealth makes it possible for people to "buy leisure," in that the owner of wealth can decide whether he or she will work or not. Similarly, wealth can be used to help people gain an advantage that they otherwise would not have. Such advantages include, but are not limited to, gaining high social prestige, political influence, improved opportunities and greater safety for oneself and one's family, high-quality health care, and enhanced life chances (Keister, 2000).

The idea of the American Dream is linked to both wealth and income, and this dream is therefore threatened in periods of economic problems and political unrest, particularly for those persons who must rely on income alone for their family's economic survival. Even in "good" economic times, however, wealth is much more unequally distributed than is income, and we now turn to that topic.

Distribution of Income and Wealth

Money—in the form of both income and wealth—is very unevenly distributed in the United States. Money is essential for acquiring goods and services. People without money cannot purchase food, shelter, clothing, medical care, legal aid, education, and the other things they need or desire. Median household income varies widely from one state to another, for example (see Map 9.1). Among the prosperous nations, the United States is number one in inequality of income distribution (Rothchild, 1995).

Box 9.1 SOCIOLOGY AND EVERYDAY LIFE

How Much Do You Know About Wealth, Poverty, and the American Dream?

True	False	
T	F	1. People no longer believe in the American Dream.
T	F	2. Individuals over age sixty-five have the highest rate of poverty.
T	F	3. Men account for two out of three impoverished adults in the United States.
T	F	4. About one in twenty U.S. people lives in a household whose members sometimes do not get enough to eat.
T	F	5. With the economic slowdown that began in the late 1990s, the twenty-five wealthiest people in this country lost about 50 percent of their fortunes between 1999 and 2002.
T	F	6. Income is more unevenly distributed than wealth.
T	F	7. People who are poor usually have personal attributes that contribute to their impoverishment.
T	F	8. A number of people living below the official poverty line have full-time jobs.
T	F	9. Federal welfare law (Temporary Assistance for Needy Families) does not require recipients with young children to find work.
T	F	10. One in three U.S. children will be poor at some point in their childhood.

Answers on page 288.

Map 9.1   Median Income by State

What factors contribute to the uneven distribution of income in the United States?

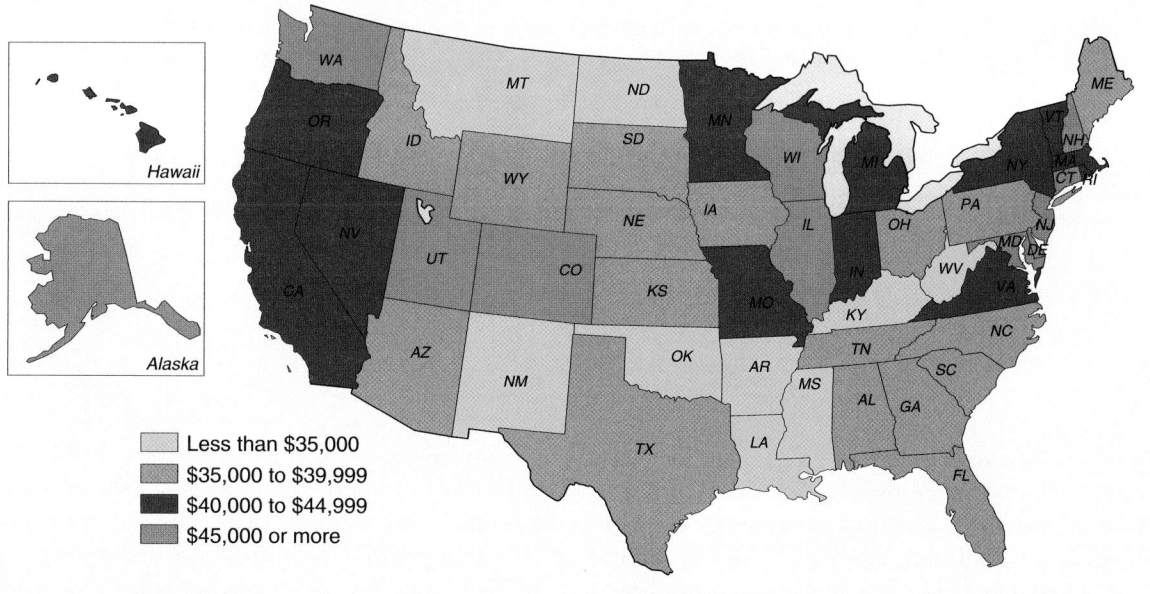

Less than $35,000
$35,000 to $39,999
$40,000 to $44,999
$45,000 or more

Source: U.S. Census Bureau, 2002.

Box 9.1 SOCIOLOGY AND EVERYDAY LIFE

Answers to the Sociology Quiz on Wealth, Poverty, and the American Dream

1. **False.** The American Dream appears to be alive and well. U.S. culture places a strong emphasis on the goal of monetary success, and many people use legal or illegal means to attempt to achieve that goal.

2. **False.** As a group, children have a higher rate of poverty than the elderly. Government programs such as Social Security have been indexed for inflation whereas many of the programs for the young have been scaled back or eliminated. However, many elderly individuals still live in poverty.

3. **False.** Women, not men, account for two out of three impoverished adults in the United States. Reasons include the lack of job opportunities for women, lower pay than men for comparable jobs, lack of affordable day care for children, sexism in the workplace, and a number of other factors.

4. **True.** It is estimated that about 5 percent of the U.S. population (one in twenty people) resides in household units where members do not get enough to eat.

5. **False.** Based on data calculated by *Forbes* magazine, which keeps detailed accounts of the wealthiest people's assets, the twenty-five wealthiest United States citizens lost about 18 percent of their fortunes. Some lost more than others; for example, Bill Gates, the wealthiest person in the nation, saw his net worth decline from $85 billion in 1999 to $43 billion in 2002.

6. **False.** Wealth is more unevenly distributed among the U.S. population than is income. However, both wealth and income are concentrated in very few hands compared with the size of the overall population.

7. **False.** According to one widely held stereotype, the poor are lazy and do not want to work. Rather than looking at structural characteristics of the society, people cite the alleged personal attributes of the poor as the reason for their plight.

8. **True.** Many of those who fall below the official poverty line are referred to as the "working poor" because they work full time but earn such low wages that they are still considered to be impoverished.

9. **False.** The latest federal welfare guidelines require recipients with young children to find work. This change has resulted in an increased demand for both childcare services and jobs for many women with little education or job training. In some states, welfare recipients have been employed by child-care centers that offer them on-site training in child care.

10. **True.** According to recent data from the Children's Defense Fund, one in three U.S. children will live in a family that is below the official poverty line at some point in their childhood. For some of the children, poverty will be a persistent problem throughout their childhood and youth.

Sources: Based on Children's Defense Fund, 2001; *Forbes*, 2002; Gilbert, 2003; and U.S. Census Bureau, 2002.

Income Inequality In regard to income inequality in the United States, the economist Paul Samuelson stated that "If we made an income pyramid out of a child's blocks, with each layer portraying $500 of income, the peak would be far higher than Mount Everest, but most people would be within a few feet of the ground" (Samuelson and Nordhaus, 1989: 644).

Similarly, sociologist Dennis Gilbert (2003) compares the distribution of income to a national pie that has been sliced into portions, ranging in size from stingy to generous, for distribution among segments of the population. As shown in Figure 9.1, in 2001 the wealthiest 20 percent of households received almost 50 percent of the total income "pie" while the

Figure 9.1   Distribution of Pretax Income in the United States

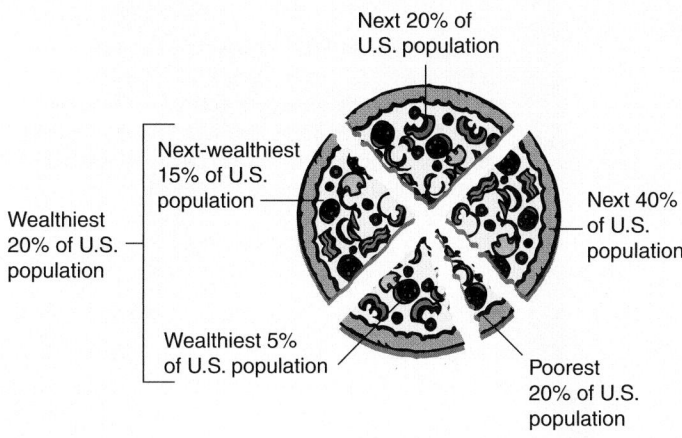

Next 20% of U.S. population

Next-wealthiest 15% of U.S. population

Wealthiest 20% of U.S. population

Next 40% of U.S. population

Wealthiest 5% of U.S. population

Poorest 20% of U.S. population

Thinking of personal income in the United States (before taxes) as a large pizza helps us to see which segments of the population receive the largest and smallest portions. What part do taxes play in redistributing parts of the pizza?

Source: U.S. Census Bureau, 2002.

poorest 20 percent of households received less than 4 percent of all income. The top 5 percent *alone* received more than 22 percent of all income—an amount greater than that received by the bottom 40 percent of all households (DeNavas-Walt and Cleveland, 2002).

In the last two decades of the twentieth century, the gulf between the rich and the poor widened in the United States. Since the early 1990s, the poor have been more likely to stay poor, and the affluent have been more likely to stay affluent. Between 1991 and 2001, the income of the top one-fifth of U.S. families increased by 31 percent; during that same period of time, the income of the bottom one-fifth of families increased by only 10 percent (DeNavas-Walt and Cleveland, 2002) (see Figure 9.2 on page 290).

Income distribution varies by race/ethnicity as well as class. Figure 9.3 on page 291 compares median income by race/ethnicity, showing not only the disparity among groups, but also the consistency of that disparity. Over half of African American (57.2 percent) and Latino/a (51.8 percent) households fall within the lowest two income categories; slightly over one-third (38.0 percent) of whites are in these categories (DeNavas-Walt and Cleveland, 2002). In all categories, differences in the median income of married couples and female-headed households are striking.

Wealth Inequality   To see how wealth inequality has increased in recent decades, let's compare two studies. A 1986 study by the Joint Economic Committee of Congress divided the population into four categories: (1) the super-rich (0.5 percent of house-

holds), who own 35 percent of the nation's wealth, with net assets averaging almost $9 million; (2) the very rich (the next 0.5 percent of households), who own about 7 percent of the nation's wealth, with net assets ranging from $1.4 million to $2.5 million; (3) the rich (9 percent of households), who own 30 percent of the wealth, with net assets of a little over $400,000; and (4) everybody else (the bottom 90 percent), who own about 28 percent of the nation's wealth. However, by 1995, another study indicated that the holdings of super-rich households had risen from 35 percent to almost 40 percent of all assets in the nation (stocks, bonds, cash, life insurance policies, paintings, jewelry, and other tangible assets) (Rothchild, 1995).

For the upper class, wealth often comes from interest, dividends, and inheritance (Haseler, 2000). One analysis of the Forbes 400 list of the wealthiest U.S. citizens found that 42 percent of the people on that list had inherited sufficient wealth to put them on that list (Gilbert, 2003). Inheritors are often three or four generations removed from individuals who amassed the original wealth (Odendahl, 1990). After inheriting a fortune, John D. Rockefeller, Jr., stated that "I was born into [wealth] and there was nothing I could do about it. It was there, like air or food or any other element. The only question with wealth is what to do with it" (qtd. in Glastris, 1990: 26).

Disparities in wealth are more pronounced when compared across racial and ethnic categories. According to the Census Bureau, the net worth of the average white household in 2000 was more than ten times that of the average African American household and

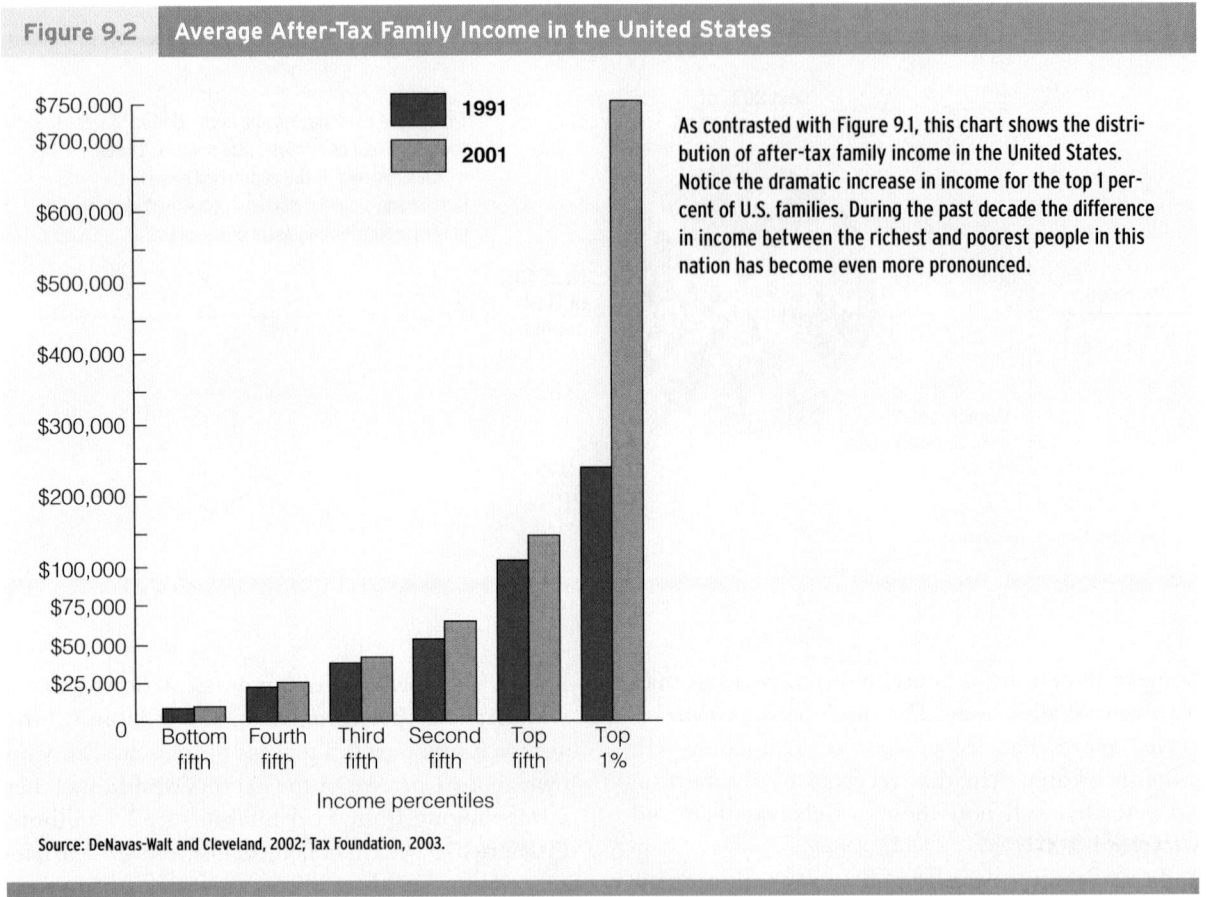

Figure 9.2   Average After-Tax Family Income in the United States

As contrasted with Figure 9.1, this chart shows the distribution of after-tax family income in the United States. Notice the dramatic increase in income for the top 1 percent of U.S. families. During the past decade the difference in income between the richest and poorest people in this nation has become even more pronounced.

Income percentiles

Source: DeNavas-Walt and Cleveland, 2002; Tax Foundation, 2003.

more than eight times that of the average Latina/o household (Orzechowski and Sepielli, 2001). Married couples have a higher net worth than the unmarried, and households headed by people age fifty-five and older are wealthier than those headed by younger persons (Orzechowski and Sepielli, 2001).

CLASSICAL PERSPECTIVES ON SOCIAL CLASS

The issue of economic inequality is not a new concern. For many years, social scientists have developed theories to explain social class differences and how these differences are related to people's resources and opportunities. Early social thinkers such as Karl Marx and Max Weber identified class as an important determinant of social inequality and social change, and their works have had a profound influence on how we view the U.S. class system today.

Karl Marx: Relationship to the Means of Production

According to Karl Marx, class position and the extent of our income and wealth are determined by our work situation, or our relationship to the means of production. As we have previously seen, Marx stated that capitalistic societies consist of two classes—the capitalists and the workers. The **capitalist class (bourgeoisie) consists of those who own the means of production**—the land and capital necessary for factories and mines, for example. The **working class (proletariat) consists of those who must sell their labor to the owners in order to earn enough money to survive** (see Figure 9.4).

According to Marx, class relationships involve inequality and exploitation. The workers are exploited as capitalists maximize their profits by paying workers less than the resale value of what they produce but do not own. Marx believed that a deep level of antagonism exists between capitalists and workers because of extreme differences in the *material interests* of the people in these two classes. According to the sociolo-

Figure 9.3   Median Household Income by Race/Ethnicity in the United States

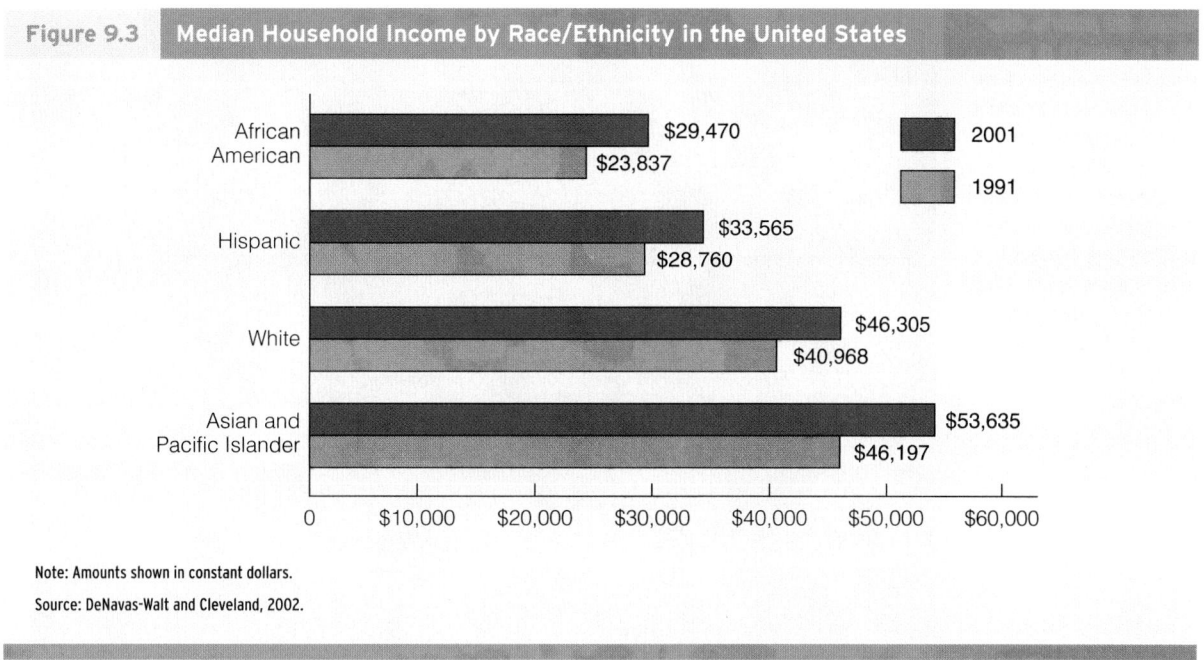

Note: Amounts shown in constant dollars.

Source: DeNavas-Walt and Cleveland, 2002.

gist Erik O. Wright (1997: 5), material interests are "the interests people have in their material standard of living, understood as the package of toil, consumption and leisure. Material interests are thus not interests of maximizing consumption *per se,* but rather interests in the trade-off between toil, leisure and consumption." Wright suggests that *exploitation* is the key concept for understanding Marx's assertion that *interests* are generated by class relations: "In an exploitative relation, the exploiter *needs* the exploited since the exploiter depends upon the effort of the exploited." In other words, the capitalists *need* the workers to derive profits; therefore, capitalists benefit when workers do not have adequate resources to provide for themselves and hence must sell their labor power to the capitalist class. As Marx suggests, exploitation involves ongoing interactions between the two antagonistic classes, which are structured by a set of social relations that binds the exploiter and the exploited together (E. Wright, 1997).

Continual exploitation results in workers' **alienation—a feeling of powerlessness and estrangement from other people and from oneself.** In Marx's view, alienation develops as workers manufacture goods that embody their creative talents but the goods do not belong to them. Workers are also alienated from the work itself because they are forced to perform it in order to live. Because the workers' activities are not their own, they feel self-estrangement. Moreover, the workers are separated from others in

the factory because they individually sell their labor power to the capitalists as a commodity.

According to Marx, the capitalist class maintains its position at the top of the class structure by control of the society's *superstructure,* which is composed of the government, schools, churches, and other social

Figure 9.4   Marx's View of Stratification

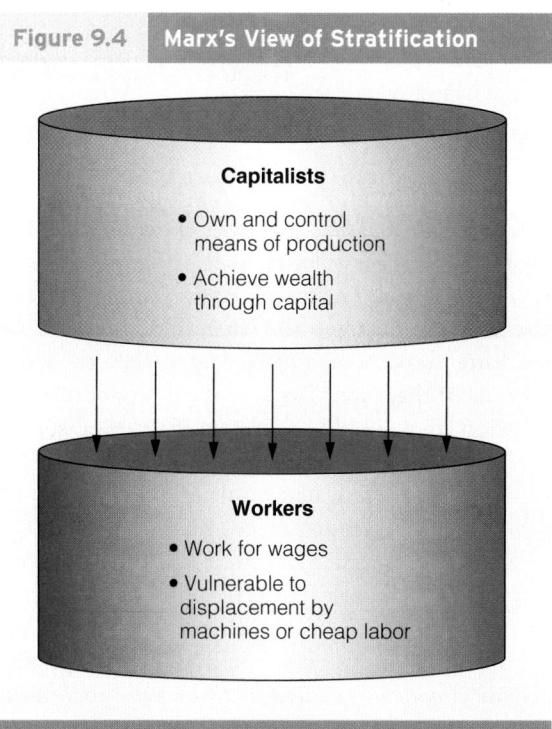

Wealth and poverty influence all aspects of our daily lives, from where we live to how we think about opportunities for social mobility. Children in wealthy families typically have many opportunities that children in poverty-level families do not. What are the long-term consequences of such inequality for individuals and for societies?

William Strode/Woodfin Camp & Associates

Tim Carlson/Stock Boston

institutions that produce and disseminate ideas perpetuating the existing system of exploitation. Marx predicted that the exploitation of workers by the capitalist class would ultimately lead to ***class conflict*—the struggle between the capitalist class and the working class.** According to Marx, when the workers realized that capitalists were the source of their oppression, they would overthrow the capitalists and their agents of social control, leading to the end of capitalism. The workers would then take over the government and create a more egalitarian society.

Why has no workers' revolution occurred? According to the sociologist Ralf Dahrendorf (1959), capitalism may have persisted because it has changed significantly since Marx's time. Individual capitalists no longer own and control factories and other means

of production; today, ownership and control have largely been separated. For example, contemporary transnational corporations are owned by a multitude of stockholders but run by paid officers and managers. Similarly, many (but by no means all) workers have experienced a rising standard of living, which may have contributed to a feeling of complacency. Moreover, as discussed in Chapter 1, many people have become so engrossed in the process of consumption—including acquiring more material possessions and going on outings to shopping malls, movie theaters, and amusement parks such as Disney World—that they are less likely to engage in workers' rebellions against the system that has brought them a relatively high standard of living (Gottdiener, 1997). During the twentieth century, workers pressed for salary in-

This recent strike by service employees reflects the activism of workers throughout the years as they have sought to gain better wages and working conditions. How would Karl Marx explain the problems experienced by employees such as these?

creases and improvements in the workplace through their activism and labor union membership. They also gained more legal protection in the form of workers' rights and benefits such as workers' compensation insurance for job-related injuries and disabilities (Dahrendorf, 1959). For these reasons, and because of a myriad of other complex factors, the workers' revolution predicted by Marx never came to pass. However, the failure of his prediction does not mean that his analysis of capitalism and his theoretical contributions to sociology are without validity.

Marx had a number of important insights into capitalist societies. First, he recognized the economic basis of class systems (Gilbert, 2003). Second, he noted the relationship between people's social location in the class structure and their values, beliefs, and behavior. Finally, he acknowledged that classes may have opposing (rather than complementary) interests. For example, capitalists' best interests are served by a decrease in labor costs and other expenses and a corresponding increase in profits; workers' best interests are served by well-paid jobs, safe working conditions, and job security.

Max Weber: Wealth, Prestige, and Power

Max Weber's analysis of class builds upon earlier theories of capitalism (particularly those by Marx) and of money (particularly those by Simmel, as discussed in Chapter 1). Living in the late nineteenth and early twentieth centuries, Weber was in a unique position to

see the transformation that occurred as individual, competitive, entrepreneurial capitalism went through the process of shifting to bureaucratic, industrial, corporate capitalism. As a result, Weber had more opportunity than Marx to see how capitalism changed over time.

Weber agreed with Marx's assertion that economic factors are important in understanding individual and group behavior. However, Weber emphasized that no single factor (such as economic divisions between capitalists and workers) was sufficient for defining the location of categories of people within the class structure. According to Weber, the access that people have to important societal resources (such as economic, social, and political power) is crucial in determining people's life chances. To highlight the importance of life chances for categories of people, Weber developed a multidimensional approach to social stratification that reflects the interplay among wealth, prestige, and power. In his analysis of these dimensions of class structure, Weber viewed the concept of "class" as an *ideal type* (that can be used to compare and contrast various societies) rather than as a specific social category of "real" people (Bourdieu, 1984).

As previously defined, *wealth* is the value of all of a person's or family's economic assets, including income, personal property, and income-producing property. Weber placed categories of people who have a similar level of wealth and income in the same class. For example, he identified a privileged commercial class of *entrepreneurs*—wealthy bankers, ship owners, professionals, and merchants who possess similar financial resources. He also described a class of *rentiers*—wealthy

| Figure 9.5 | Weber's Multidimensional Approach to Social Stratification |

According to Max Weber, wealth, power, and prestige are separate continuums. Individuals may rank high in one dimension and low in another, or they may rank high or low in more than one dimension. Also, individuals may use their high rank in one dimension to achieve a comparable rank in another. How does Weber's model compare with Marx's approach as shown in Figure 9.4?

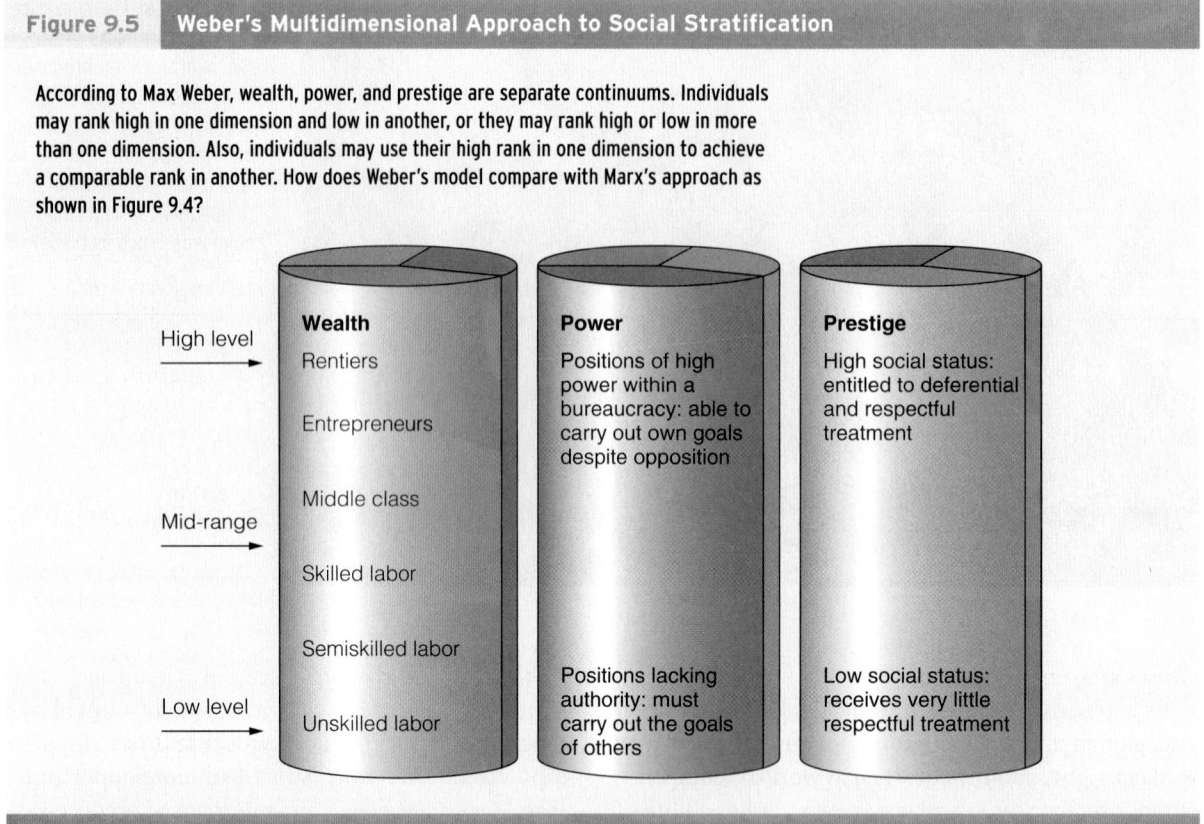

individuals who live off their investments and do not have to work. According to Weber, entrepreneurs and rentiers have much in common. Both are able to purchase expensive consumer goods, control other people's opportunities to acquire wealth and property, and monopolize costly status privileges (such as education) that provide contacts and skills for their children.

Weber divided those who work for wages into two classes: the middle class and the working class. The middle class consists of white-collar workers, public officials, managers, and professionals. The working class consists of skilled, semiskilled, and unskilled workers.

The second dimension of Weber's system of stratification is *prestige*—**the respect or regard with which a person or status position is regarded by others.** Fame, respect, honor, and esteem are the most common forms of prestige. A person who has a high level of prestige is assumed to receive deferential and respectful treatment from others. Weber suggested that individuals who share a common level of social prestige belong to the same status group regardless of their level of wealth. They tend to socialize with one another, marry within their own group of social equals, spend their leisure time together, and safeguard their status by restricting outsiders' opportuni-

ties to join their ranks (Beeghley, 2000). Style of life, formal education, and occupation are often significant factors in establishing and maintaining prestige in industrial and postindustrial societies.

The other dimension of Weber's system is *power*—**the ability of people or groups to achieve their goals despite opposition from others.** The powerful can shape society in accordance with their own interests and direct the actions of others (Tumin, 1953). According to Weber, social power in modern societies is held by bureaucracies; individual power depends on a person's position within the bureaucracy. Weber suggested that the power of modern bureaucracies was so strong that even a workers' revolution (as predicted by Marx) would not lessen social inequality (Hurst, 1998).

Weber stated that wealth, prestige, and power are separate continuums on which people can be ranked from high to low, as shown in Figure 9.5. Individuals may be high on one dimension while being low on another. For example, people may be very wealthy but have little political power (for example, a recluse who has inherited a large sum of money). They may also have prestige but not wealth (for instance, a college professor who receives teaching excellence awards but lives on a relatively low income). In Weber's

Table 9.1 PRESTIGE RATINGS FOR SELECTED OCCUPATIONS IN THE UNITED STATES, 1996 AND 1963

Respondents were asked to evaluate a list of occupations according to their prestige; the individual rankings were averaged and then converted into scores, with 1 the lowest possible score and 99 the highest possible score (Gilbert and Kahl, 1998).

	SCORE				SCORE	
Occupation	1996	1963	Occupation	1996	1963
Physician	86	93	Police officer	60	72
Attorney	75	89	Electrician	51	76
College professor	74	90	Mail carrier	47	66
Dentist	72	88	Garbage collector	28	39
Accountant	65	81	Janitor	22	48
Grade school teacher	64	82	Shoe shiner	9	34

Sources: Hodge, Siegel, and Rossi, 1964; National Opinion Research Center, 1996.

multidimensional approach, people are ranked on all three dimensions. Sociologists often use the term *socioeconomic status (SES)* **to refer to a combined measure that attempts to classify individuals, families, or households in terms of factors such as income, occupation, and education to determine class location.**

What important insights does Weber provide in regard to social stratification and class? Weber's analysis of social stratification contributes to our understanding by emphasizing that people behave according to both their economic interests and their values. He also added to Marx's insights by developing a multidimensional explanation of the class structure and by identifying additional classes.

A substantial advantage of Weber's theory is that it has made empirical investigation of the U.S. class structure possible (Blau and Duncan, 1967). Through his distinctions among wealth, power, and prestige, Weber makes it possible for researchers to examine the different dimensions of social stratification. Weber's enlarged conceptual formulation of stratification is the theoretical foundation for mobility research by sociologists such as Peter Blau and Otis Duncan. Blau and Duncan (1967) measure the three dimensions from Weber's theory through a study of the occupational positions that individuals hold. According to Blau and Duncan, a person's occupational position is not identical to either economic class or prestige, but is closely related to both. As you might expect, different occupations have significantly different levels of status or prestige (see Table 9.1). For the past fifty years, occupational ratings by prestige have been remarkably consistent in the United States

(Gilbert, 2003). These rankings have become the foundation for status attainment research, which uses sophisticated statistical measurements to assess the influence of family background and education on people's occupational mobility and success (see Blau and Duncan, 1967; Duncan, 1968). *Status attainment research* focuses on the process by which people ultimately reach their position in the class structure. Based largely on studies of men, this research uses the father's occupation and the son's education and first job as primary determinants of the eventual class position of the son. Obviously, family background is the central factor in this process because the son's education and first job are linked to the family's economic status. In addition, the family's location in the class system is related to the availability of social ties that may open occupational doors for the son.

Although they have been widely employed in some prestigious sociological research, status attainment models have several serious limitations. One is the focus of this research on the occupational prestige of traditionally male jobs and the exclusion of women's work, which has often been unpaid. In regard to African Americans, Patricia Hill Collins (1990: 45) noted that "the higher rates of Black male unemployment, the racial discrimination that has crowded all African-Americans into a narrow set of occupations, and the existence of household arrangements other than two-parent nuclear families . . . have all combined to make status attainment models less suitable for explaining Black social class dynamics." Moreover, the status attainment model is unable to take into account power differentials rooted in inequalities based on race, ethnicity, or gender.

A significant limitation of occupational prestige rankings is that the level of prestige accorded to a position may not actually be based on the importance of the position to society. The highest ratings may be given to professionals—such as physicians and lawyers—because they have many years of training in their fields and some control their own work, not because these positions contribute the most to society.

CONTEMPORARY SOCIOLOGICAL MODELS OF THE U.S. CLASS STRUCTURE

How many social classes exist in the United States today? What criteria are used for determining class membership? No broad consensus exists about how to characterize the class structure in this country. In fact, many people deny that class distinctions exist (see Eisler, 1983; Parenti, 1994). Most people like to think of themselves as middle class; it puts them in a comfortable middle position—neither rich nor poor. Sociologists have developed several models of the class structure: One is broadly based on a Weberian approach, the second on a Marxian approach, and the third on a study of society today. We will examine all three models briefly.

The Weberian Model of the U.S. Class Structure

Expanding on Weber's analysis of the class structure, the sociologists Dennis Gilbert (2003) and Joseph A. Kahl developed a widely used model of social classes based on three elements: (1) education, (2) occupation of family head, and (3) family income (see Figure 9.6).

The Upper (Capitalist) Class The upper class is the wealthiest and most powerful class in the United States. About 1 percent of the population is included in this class, whose members own substantial income-producing assets and operate on both the national and international levels. According to Gilbert (2003), people in this class have an influence on the economy and society far beyond their numbers.

Some models further divide the upper class into upper-upper ("old money") and lower-upper ("new money") categories (Warner and Lunt, 1941; Coleman and Rainwater, 1978; Kendall, 2002). Members

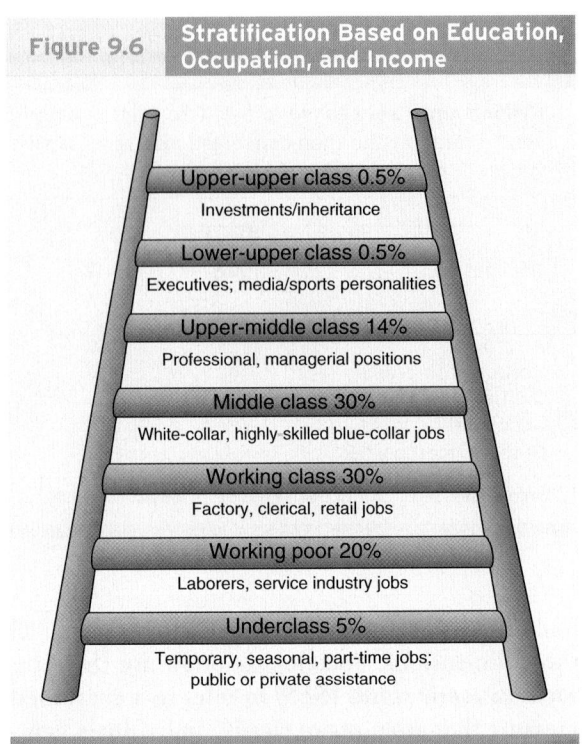

Figure 9.6 — **Stratification Based on Education, Occupation, and Income**

Upper-upper class 0.5%
Investments/inheritance

Lower-upper class 0.5%
Executives; media/sports personalities

Upper-middle class 14%
Professional, managerial positions

Middle class 30%
White-collar, highly-skilled blue-collar jobs

Working class 30%
Factory, clerical, retail jobs

Working poor 20%
Laborers, service industry jobs

Underclass 5%
Temporary, seasonal, part-time jobs; public or private assistance

of the upper-upper class come from prominent families which possess great wealth that they have held for several generations. Family names—such as Rockefeller, Mellon, Du Pont, and Kennedy—are well-known and often held in high esteem. Persons in the upper-upper class tend to have strong feelings of in-group solidarity. They belong to the same exclusive clubs and support high culture (such as the opera, symphony orchestras, ballet, and art museums). Children are educated in prestigious private schools and Ivy League universities; many acquire strong feelings of privilege from birth, as upper-class author Lewis H. Lapham (1988: 14) states:

> Together with my classmates and peers, I was given to understand that it was sufficient accomplishment merely to have been born. Not that anybody ever said precisely that in so many words, but the assumption was plain enough, and I could confirm it by observing the mechanics of the local society. A man might become a drunkard, a concert pianist or an owner of companies, but none of these occupations would have an important bearing on his social rank.

Children of the upper class are socialized to view themselves as different from others; they also learn that they are expected to marry within their own class (Warner and Lunt, 1941; Mills, 1959a; Domhoff, 1983; Kendall, 2002).

Members of the lower-upper class may be extremely wealthy but not have attained as much prestige as members of the upper-upper class. The "new rich" have earned most of their money in their own lifetime as entrepreneurs, presidents of major corporations, sports or entertainment celebrities, or top-level professionals. For some members of the lower-upper class, the American Dream has become a reality. Others still desire the respect of members of the upper-upper class.

The Upper-Middle Class

Persons in the upper-middle class are often highly educated professionals who have built careers as physicians, attorneys, stockbrokers, or corporate managers. Others derive their income from family-owned businesses. According to Gilbert (2003), about 14 percent of the U.S. population is in this category. A combination of three factors qualifies people for the upper-middle class: university degrees, authority and independence on the job, and high income. Of all the class categories, the upper-middle class is the one that is most shaped by formal education. Over the past fifty years, Asian Americans, Latinos/as, and African Americans have placed great importance on education as a means of attaining the American Dream. Many people of color have moved into the upper-middle class by acquiring higher levels of education. James P. Comer, an African American child psychiatrist, describes how his mother, who grew up in abject poverty in the rural South, encouraged her children to get a good education:

> My mother, Maggie, believed that education was the way to achieve her American Dream. When she was denied the opportunity herself, she declared that all her children would be educated. Mom and Dad together gave all five of us the support needed to acquire thirteen college degrees. (Comer, 1988: xxii)

Across racial–ethnic and class lines, children are encouraged to acquire an education suitable for upper-middle-class occupations, which typically have high prestige in the community. However, many social analysts believe that racism still diminishes the life chances for people of color even when they achieve a high income and a prestigious career (see Cose, 1993; Takaki, 1993; Feagin and Sikes, 1994).

The Middle Class

In past decades, a high school diploma was necessary to qualify for most middle-class jobs. Today, two-year or four-year college degrees have replaced the high school diploma as an entry-level requirement for employment in many

© Tom Rosenthal/SuperStock

People's life chances are enhanced by access to important societal resources such as education. How will the life chances of students who have the opportunity to pursue a college degree differ from those of young people who do not have the chance to go to college?

middle-class occupations, including medical technicians, nurses, legal and medical assistants, lower-level managers, semiprofessionals, and nonretail salesworkers. An estimated 30 percent of the U.S. population is in the middle class (Gilbert, 2003).

Traditionally, most middle-class occupations have been relatively secure and have provided more opportunities for advancement (especially with increasing levels of education and experience) than have working-class positions. Recently, however, four factors have eroded the American Dream for this class: (1) escalating housing prices, (2) occupational insecurity, (3) blocked mobility on the job, and (4) the cost-of-living squeeze that has penalized younger workers, even when they have more education and better jobs than their parents (Newman, 1993). Consider, for example, Brenda and Amancio Irizarry of New York, who are struggling to get by on a combined pretax income of $38,000 per year. Although they make more money than their parents did, the Irizarrys cannot afford to buy a home, have no savings, and have never taken a vacation together. They are working hard to send their daughter, Michelle, to college, as Mrs. Irizarry explains:

> We didn't have a college fund for Michelle. You see commercials on television saying to start [saving] when the baby is born. But because of the unforeseen things that happen in life, it's impossible to save money. (qtd. in Jones, 1995: 9)

Michelle appreciates her parents' efforts but eventually wants much more for herself:

> I don't want to live like this the rest of my life. . . . I mean, they have morals and are trying to make it. But unless they hit the Lotto, this is how it's going to be the rest of their lives. And that's sad, to think that's it. This is as good as it's going to get. (qtd. in Jones, 1995: 9)

But Michelle has not given up on the American Dream; she is determined to become a social worker and have a nice place to live (Jones, 1995). As this example suggests, class distinctions between the middle and working classes are sometimes blurred due to overlapping characteristics (Gilbert, 2003).

The Working Class

An estimated 30 percent of the U.S. population is in the working class. The core of this class is made up of semiskilled machine operators who work in factories and elsewhere. Members of the working class also include some workers in the service sector, as well as clerks and salespeople whose job responsibilities involve routine, mechanized tasks requiring little skill beyond basic literacy and a brief period of on-the-job training (Gilbert, 2003). Some people in the working class are employed in ***pink-collar occupations*—relatively low-paying, non-manual, semiskilled positions primarily held by women,** such as day-care workers, checkout clerks, cashiers, and restaurant servers.

How does life in the working-class family compare with that of individuals in middle-class families? According to sociologists, working-class families not only earn less than middle-class families, but they also have less financial security, particularly because of high rates of layoffs and plant closings in some regions of the country. Few people in the working class have more than a high school diploma, and many have less, which makes job opportunities scarce for them in a "high-tech" society (Gilbert, 2003). Others find themselves in low-paying jobs in the service sector of the economy, particularly fast-food restaurants,

a condition that often places them among the working poor.

The Working Poor

The working poor account for about 20 percent of the U.S. population. Members of the working-poor class live from just above to just below the poverty line; they typically hold unskilled jobs, seasonal migrant jobs in agriculture, lower-paid factory jobs, and service jobs (such as counter help at restaurants). Employed single mothers often belong to this class; consequently, children are overrepresented in this category. African Americans and other people of color are also overrepresented among the working poor. To cite only one example, in the United States today, there are two white hospital orderlies to every one white physician, whereas there are twenty-five African American orderlies to every one African American physician (Gilbert, 2003). For the working poor, living from paycheck to paycheck makes it impossible to save money for emergencies such as periodic or seasonal unemployment, which is a constant threat to any economic stability they may have.

Social critic and journalist Barbara Ehrenreich (2001) left her upper-middle-class lifestyle for a period of time to see if it was possible for the working poor to live on the wages that they were being paid as restaurant servers, sales clerks at discount department stores, aides at nursing homes, house cleaners for franchise maid services, or similar jobs. She conducted her research by actually holding those jobs for periods of time and seeing if she could live on the wages that she received. Through her research, Ehrenreich persuasively demonstrated that people who work full time, year-round, for poverty-level wages must develop survival strategies that include such things as help from relatives or constantly moving from one residence to another in order to have a place to live. Like many other researchers, Ehrenreich found that minimum-wage jobs cannot cover the full cost of living, such as rent, food, and the rest of an adult's monthly needs, even without taking into consideration the needs of children or other family members (see also Newman, 1999). According to Ehrenreich (2001: 221),

> The "working poor," as they are approvingly termed, are in fact the major philanthropists of our society. They neglect their own children so that the children of others will be cared for; they live in substandard housing so that other homes will be shiny and perfect; they endure privation so that inflation will be low and stock prices high. To be a member of the working poor is to be an

Box 9.2 CHANGING TIMES: MEDIA AND TECHNOLOGY

Media Stars and the American Dream

I'll never forget finding this Tonka truck. It was missing a wheel, but it was the best toy I ever had.

–Marc Anthony, about the success of his pop album in English and a salsa album in Spanish, recalling the times as a child when he had gone through "rich people's trash" looking for things that his parents–Puerto Rican émigrés in East Harlem, New York–could not afford to purchase for him (qtd. in Buia, 2001)

Today, Marc Anthony is the prototype of the American Dream; his success in the music industry and as an actor has brought him wealth, a 10,000-square-foot mansion, and a 73-foot yacht, among other possessions (Buia, 2001).

Anthony's achievement of the American Dream is an example of how increasing globalization of the music industry has brought new markets as listeners around the world enjoy the music of artists from other cultures and nations. For some, the globalization of culture is a hotly contested issue; for others, it is a means of gaining wealth and worldwide recognition. Although the term *crossover* is frequently used for entertainers such as Ricky Martin and Jennifer Lopez, who have widespread appeal in multiple cultures, Marc Anthony does not believe that his success should be described in that manner. According to Anthony, "What did I cross over from? I'm as

■ Some people believe that singer Marc Anthony reflects the American Dream because he has achieved great wealth after a childhood of poverty.

American as anybody. I was born in your backyard" (qtd. in Buia, 2001: 12).

Although most young people will not become a household name and earn large sums of money the way Marc Anthony has, many individuals hope that they, too, will hit it big in the land of opportunity–the United States–and move into the upper class of our society. What are the actual chances that most people will strike it rich through music, entertainment, sports, a great entrepreneurial idea, or a winning lottery ticket? What produces most of the social mobility in our society? Why?

anonymous donor, a nameless benefactor, to everyone else.

When the children of the working poor become successful, often through entertainment or sports careers or gaining high levels of education, their successes may receive a great deal of media attention (see Box 9.2).

The Underclass According to Gilbert (2003), people in the underclass are poor, seldom employed, and caught in long-term deprivation that results from low levels of education and income and high rates of unemployment. Some are unable to work because of age or disability; others experience discrimination based on race/ethnicity. Single mothers are overrepresented in this class because of the lack of jobs, the lack of affordable child care, and many other impediments to the mother's future and that of her children. People without a "living wage" often must rely on public

or private assistance programs for their survival. About 3 to 5 percent of the U.S. population is in this category, and the chances of their children moving out of poverty are about fifty-fifty (Gilbert, 2003).

Whereas some young people in the underclass are hopeful that they can escape poverty, others view their futures with pessimism and uncertainty. They may have leveled aspirations, as the author Jay MacLeod (1988) learned when he asked members of two male teenage peer groups from underclass families what their lives would be like in twenty years. Three of them replied (in separate interviews):

STONEY: Hard to say, I could be dead tomorrow. Around here, you gotta take life day by day.

BOO-BOO: I dunno. I don't want to think about it. I'll think about it when it comes.

SHORTY: Twenty years? I'm gonna be in jail. (qtd. in MacLeod, 1988: 61)

Those who discussed work saw it solely as a means to an end—money (MacLeod, 1988).

Studies by various social scientists have found that meaningful employment opportunities are the critical missing link for people on the lowest rungs of the class ladder. According to these analysts, job creation is essential in order for people to have the opportunity to earn a decent wage; have medical coverage; live meaningful, productive lives; and raise their children in a safe environment (see Fine and Weis, 1998; Nelson and Smith, 1999; Newman, 1999; W. Wilson, 1996). These issues are closely tied to the American Dream we have been discussing in this chapter.

The Marxian Model of the U.S. Class Structure

The earliest Marxian model of class structure identified ownership or nonownership of the means of production as the distinguishing feature of classes. From this perspective, classes are social groups organized around property ownership, and social stratification is created and maintained by one group in order to protect and enhance its own economic interests. Moreover, societies are organized around classes in conflict over scarce resources. Inequality results from the more powerful exploiting the less powerful.

Contemporary Marxian (or conflict) models examine class in terms of people's relationship with others in the production process. For example, conflict theorists attempt to determine what degree of control that workers have over the decision-making process and the extent to which they are able to plan and implement their own work. They also analyze the type of supervisory authority, if any, that a worker has over other workers. According to this approach, most employees are a part of the working class because they do not control either their own labor or that of others.

Erik Olin Wright (1979, 1985, 1997), one of the leading stratification theorists to examine social class from a Marxian perspective, has concluded that Marx's definition of "workers" does not fit the occupations found in advanced capitalist societies. For example, many top executives, managers, and supervisors who do not own the means of production (and thus would be "workers" in Marx's model) act like capitalists in their zeal to control workers and maximize profits. Likewise, some experts hold positions in which they have control over money and the use of their own time even though they are not owners. Wright views Marx's category of "capitalist" as being too broad as well. For instance, small-business owners might be viewed as capitalists because they own their own tools and have a few people working for them, but they have little in common with large-scale capitalists and do not share the interests of factory workers. Figure 9.7 compares Marx's model and Wright's model.

Wright (1979) argues that classes in modern capitalism cannot be defined simply in terms of different levels of wealth, power, and prestige, as in the Weberian model. Consequently, he outlines four criteria for placement in the class structure: (1) ownership of the means of production, (2) purchase of the labor of others (employing others), (3) control of the labor of others (supervising others on the job), and (4) sale of one's own labor (being employed by someone else). Wright (1978) assumes that these criteria can be used to determine the class placement of all workers, regardless of race/ethnicity, in a capitalist society. Let's take a brief look at Wright's (1979, 1985) four classes—(1) the capitalist class, (2) the managerial class, (3) the small-business class, and (4) the working class—so that you can compare them to those found in the Weberian model.

The Capitalist Class According to Wright, this class holds most of the wealth and power in society through ownership of capital—for example, banks, corporations, factories, mines, news and entertainment industries, and agribusiness firms. The "ruling elites," or "ruling class," within the capitalist class hold political power and are often elected or appointed to influential political and regulatory positions (Parenti, 1994).

The capitalist class is composed of individuals who have inherited fortunes, own major corporations, or are top corporate executives with extensive stock holdings or control of company investments. Even though many top executives have only limited *legal ownership* of their corporations, they have substantial economic ownership and exert extensive control over investments, distribution of profits, and management of resources. The major sources of income for the capitalist class are profits, interest, and very high salaries. Members of this class make important decisions about the workplace, including which products and services to make available to consumers and how many workers to hire or fire (Feagin and Feagin, 2003).

According to *Forbes* magazine's 2003 list of the richest people in the world, Bill Gates (co-founder of Microsoft Corporation, the world's largest microcomputer software company) was the wealthiest capitalist, with a net worth of $40.7 billion, down about one-third from his $63-billion figure in 2000 as a re-

Figure 9.7 Comparison of Marx's and Wright's Models of Class Structure

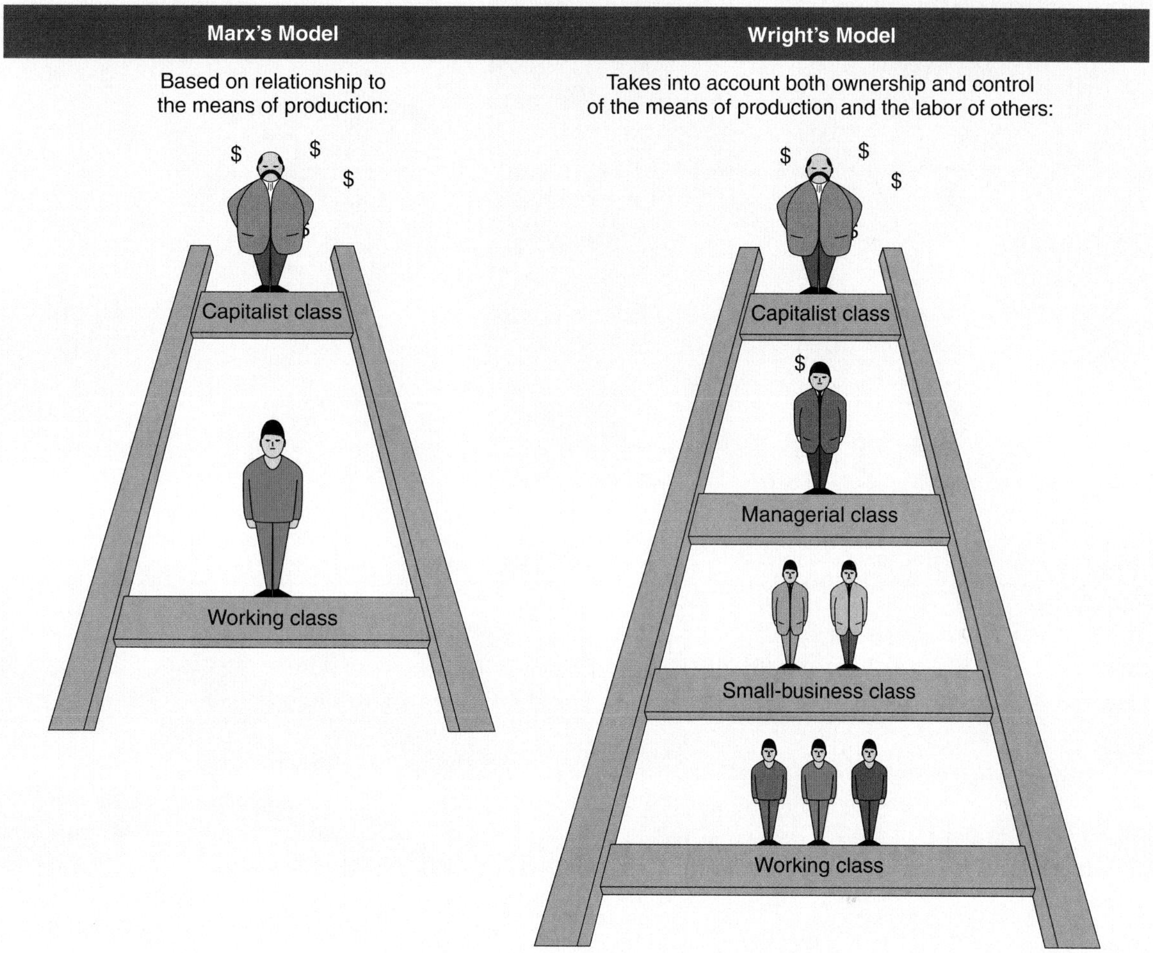

Marx's Model

Based on relationship to the means of production:

- Capitalist class
- Working class

Wright's Model

Takes into account both ownership and control of the means of production and the labor of others:

- Capitalist class
- Managerial class
- Small-business class
- Working class

sult of charitable gifts and steep decreases in the stock market (*Forbes,* 2003). Investor Warren Buffet came in second with $30.5 billion. However, the number of billion-dollar fortunes in the United States rose from 129 in 1995 to 274 in 2000 before falling to 223 in 2003 (*Forbes,* 1996, 2003). Although some of the men who made the *Forbes* list of the wealthiest people have gained their fortunes through entrepreneurship or being CEOs of large corporations, women who made the list typically have acquired their wealth through inheritance, marriage, or both. In 2002, only six women were heads of Fortune 500 companies (Catalyst, 2002).

The Managerial Class People in the managerial class have substantial control over the means of production and over workers. However, these upper-level managers, supervisors, and professionals typically do

not participate in key corporate decisions such as how to invest profits. Lower-level managers may have limited control over employment practices, including the hiring and firing of some workers.

Top professionals such as physicians, attorneys, accountants, and engineers may control the structure of their own work; however, they typically do not own the means of production and may not have supervisory authority over more than a few people. Even so, they may influence the organization of work and the treatment of other workers. Members of the capitalist class often depend on these professionals for their specialized knowledge.

As previously discussed, members of the managerial class occupy a contradictory class location between the capitalist and working classes (Wright, 1979). Like members of the working class, persons in the managerial class do not own the means of production, and

Erik Olin Wright's conflict model of the U.S. class system emphasizes the differing interests of (1) the capitalist class, exemplified by Bill Gates, the co-founder and head of Microsoft Corporation; (2) the managerial class; (3) the small-business class; and (4) the working class.

they usually earn a regular salary. However, they typically have control over the work of others and may exercise considerable authority over the organization of production. They also may invest in corporate stock and gain significant unearned income. As a result, they tend to align themselves with the basic interests of the capitalist class. For people of color and white women in the managerial class, additional contradictions exist based on race and/or gender.

The Small-Business Class This class consists of small-business owners and craftspeople who may hire a small number of employees but largely do their own work. Some members own businesses such as "mom and pop" grocery stores, retail clothing stores, and jewelry stores. Others are doctors and lawyers who receive relatively high incomes from selling their own services. Some of these professionals now share attributes with members of the capitalist class because they

have formed corporations that hire and control the employees who produce profits for the professionals.

It is in the small-business class that we find many people's hopes of achieving the American Dream. Recent economic trends, including corporate downsizing, telecommuting, and the movement of jobs to other countries, have encouraged more people to think about starting their own business. As a result, more people today are self-employed or own a small business than at any time in the past (U.S. Department of Labor, 2003). More women of all races and people of color are in the small-business class than was true previously. According to recent statistics, for example, women own 34 percent of all small businesses. However, gaps in revenues persist between businesses owned by women and those owned by men, with women-owned businesses earning on average about 40 percent less than businesses owned by men (U.S. Department of Labor, 2003).

Throughout U.S. history, immigrants and people of color have owned small businesses (Butler, 1991), seeing such enterprises as a way to achieve the American Dream. Over the past decade, the number of businesses owned by subordinate-group members has increased dramatically, but the share of such businesses owned by people of color is still not proportionate to their numbers in the overall population. African Americans make up more than 12 percent of the U.S. population but own less than 4 percent of businesses; Latinos/as make up more than 13 percent of the population yet own slightly more than 4 percent of all businesses. Asian Americans are closest to being proportional in business ownership: They constitute about 3.6 percent of the population and own about 3.5 percent of all U.S. businesses (U.S. Department of Labor, 2003).

In recent years, small-business ownership has become more prevalent among persons with disabilities, who frequently encounter more prejudices and other problems in seeking traditional employment. Although there are many pitfalls—including low revenues and bankruptcy—to being in the small-business class, many people continue to pursue this avenue to the American Dream.

The Working Class

The working class is made up of a number of subgroups, one of which is blue-collar workers, some of whom are highly skilled and well paid and others of whom are unskilled and poorly paid. Skilled blue-collar workers include electricians, plumbers, and carpenters; unskilled blue-collar workers include janitors and gardeners.

White-collar workers are another subgroup of the working class. Referred to by some as a "new middle class," these workers are actually members of the working class because they do not own the means of production, do not control the work of others, and are relatively powerless in the workplace. Secretaries, other clerical workers, and sales workers are members of the white-collar faction of the working class. They take orders from others and tend to work under constant supervision. Thus, these workers are at the bottom of the class structure in terms of domination and control in the workplace. The working class contains about half of all employees in the United States.

The New Class Society?

According to the sociologists Robert Perrucci and Earl Wysong (1999), the beginning of the twenty-first century is characterized by a new class society in which transnational corporations, high technology, and disposable workers encounter new and increasingly polarized class lines. Perrucci and Wysong believe that class membership is based on access to a new mix of critical resources, including income, investment capital, credentialed skills verified by elite schools, and social connections to organizational leaders. Like Marx, Perrucci and Wysong view the two largest classes—the privileged class and the working class—as possessing fundamentally different and opposed interests; consequently, when the situation of one class improves, the other class loses. For example, significantly improving the income, benefits, or job security of the working class would cut into the capital that members of the privileged class use for consumption and investment.

To picture Perrucci and Wysong's model of the U.S. class structure, think of a figure that looks like a double diamond, with a small diamond on the top and a much larger diamond on the bottom. The small, top diamond is made up of the *privileged class* (about 20 percent of the U.S. population), and the much larger, bottom diamond is made up of the *new working class* (about 80 percent of the population).

The Privileged Class

The top part of the privileged class is the *superclass* (about 1 to 2 percent of the population), which comprises wealthy owners and employers who make a living from investments or business ownership. Six- and seven-figure incomes ($100,000 and up) are customary, and many make far more, thereby providing them with large sums of capital for consumption and investment. Beneath the superclass is the *credentialed class,* which consists of managers (13 to 15 percent) in mid- and upper-level management positions and the CEOs of corporations and public organizations. This class also includes professionals (4 to 5 percent of the population), who use their professional degrees and organizational ties to advance their interests even as they solidify the position of those in the superclass.

The New Working Class

Perrucci and Wysong divide the new working class into the *comfort class* (10 percent), which is made up of nurses, teachers, civil servants, and skilled workers such as carpenters and electricians, and the *contingent class*. Wage earners (such as people in clerical and sales jobs and personal services or crafts jobs) are located in the contingent class, along with those who are self-employed and hire no employees. At the bottom of the new working class is the *excluded class*—people who are in and out of the labor force and whose only source of employment is a variety of unskilled, low-wage, temporary jobs.

Is the new class society an accurate model of the U.S. class structure? What about the American Dream? These are difficult questions to answer; however, if Perrucci and Wysong are correct, the United States currently does not have a middle class: Many people do not have stable, secure resources over time such as a steady job that provides an adequate income, health insurance, and a pension. According to this approach, the growing division in power and resources between the privileged class and the new working class gives the class structure an intergenerational permanence, and the balance of class power continues to shift toward the privileged class, which is able to "devise, disguise, and legitimate corporate practices and government policies that serve its interests but are destructive to working-class interests" (Perrucci and Wysong, 1999: 244). Moreover, institutional classism serves to perpetuate privileged-class dominance. **Classism refers to the belief that persons in the upper or privileged class are superior to those in the lower or working class, particularly in regard to values, behavior, and lifestyles.** However, some people resist classism by standing up against policies they believe will contribute to economic, political, and cultural inequalities. Perrucci and Wysong describe a "structural democracy" approach that seeks to revitalize participation in politics and government as well as encourage people to place demands on corporate structures to enhance democratic governance so that the interests of the working class are not neglected. As Perrucci and Wysong (1999: 273) state, it is necessary to reduce class-based inequalities between the privileged and working classes so that democracy in the United States does not topple to a "permanent resting spot at the bottom of history."

Although Marxian and Weberian models of the U.S. class structure show differences in people's occupations and access to valued resources, neither fully reflects the nature and extent of inequality in the United States. In the next section, we will take a closer look at the unequal distribution of income and wealth in the United States and the effects of inequality on people's opportunities and life chances.

CONSEQUENCES OF INEQUALITY

Income and wealth are not simply statistics; they are intricately related to the American Dream and our individual life chances. Persons with a high income or substantial wealth have more control over their own lives. They have greater access to goods and services; they can afford better housing, more education, and a wider range of medical services. Persons with less income, especially those living in poverty, must spend their limited resources to acquire the basic necessities of life.

Physical and Mental Health and Nutrition

People who are wealthy and well educated and who have high-paying jobs are much more likely to be healthy than are poor people. As people's economic status increases, so does their health status. The poor have shorter life expectancies and are at greater risk for chronic illnesses such as diabetes, heart disease, and cancer, as well as infectious diseases such as tuberculosis.

Children born into poor families are at much greater risk of dying during their first year of life. Some die from disease, accidents, or violence. Others are unable to survive because they are born with low birth weight, a condition linked to birth defects and increased probability of infant mortality (Rogers, 1986). Low birth weight in infants is attributed, at least in part, to the inadequate nutrition received by many low-income pregnant women. Most of the poor do not receive preventive medical and dental checkups; many do not receive adequate medical care after they experience illness or injury.

Many high-poverty areas lack an adequate supply of doctors and medical facilities. Even in areas where such services are available, the inability to pay often prevents people from seeking medical care when it is needed. Some "charity" clinics and hospitals may provide indigent patients (those who cannot pay) with minimal emergency care but make them feel stigmatized in the process. For many of the working poor, medical insurance is out of the question. More than 41 million people in the United States are without health insurance coverage (U.S. Census Bureau, 2002). Many people rely on their employers for health coverage; however, some employers are cutting back on health coverage, particularly for employees' family members. Despite passage of the 1996 Kassenbaum–Kennedy bill by Congress, which makes insurance more readily available for millions of people who change their jobs or lose them, many unemployed workers and their families remain without medical coverage. However, the uninsured are a changing group in the United States—not everyone who becomes uninsured for a month or more remains uninsured throughout a given year. Of all age groups, persons aged eighteen to twenty-four are the most likely to be uninsured; Medicare and other benefit programs provide medical care to most persons sixty-five and over (U.S. Census Bureau, 2002). As shown in Figure 9.8, a high percentage of poor persons do not have health insurance.

Many lower-paying jobs are often the most dangerous and have the greatest health hazards. Black lung disease, cancer caused by asbestos, and other environmental hazards found in the workplace are more likely to affect manual laborers and low-income workers, as are job-related accidents.

Although the precise relationship between class and health is not known, analysts suggest that people with higher income and wealth tend to smoke less, exercise more, maintain a healthy body weight, and eat nutritious meals. As a category, more-affluent persons tend to be less depressed and face less psychological stress, conditions that tend to be directly proportional to income, education, and job status (*Mental Medicine,* 1994).

Good health is basic to good life chances; in turn, adequate nutrition is essential for good health. Hunger is related to class position and income inequality. Recent surveys estimate that 13 percent of children under age twelve are hungry or at risk of being hungry. Among the working poor, almost 75 percent of the children are thought to be in this category. After spending 60 percent of their income on housing, low-income families are unable to provide adequate food for their children. Between one-third and one-half of all children living in poverty consume significantly less than the federally recommended guidelines for caloric and nutritional intake (Children's Defense Fund, 2002). Lack of adequate nutrition has been linked to children's problems in school. Moreover, recent changes in welfare laws have adversely affected millions of U.S. children (see Box 9.3).

Housing

As discussed in Chapter 5 ("Society, Social Structure, and Interaction"), homelessness is a major problem in the United States. The lack of affordable housing is a pressing concern for many low-income individuals and families. With the economic prosperity of the 1990s, low-cost housing units in many cities were replaced with expensive condominiums and luxury single-family residences for affluent people. As unemployment rose dramatically beginning in 2001, partly due to terrorism and a faltering economy, housing costs remained high compared to many families'

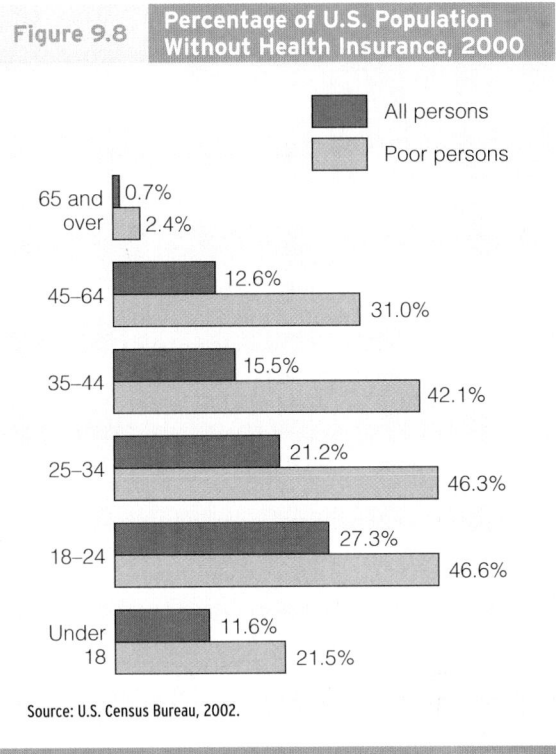

Figure 9.8 **Percentage of U.S. Population Without Health Insurance, 2000**

- All persons
- Poor persons

Age	All persons	Poor persons
65 and over	0.7%	2.4%
45–64	12.6%	31.0%
35–44	15.5%	42.1%
25–34	21.2%	46.3%
18–24	27.3%	46.6%
Under 18	11.6%	21.5%

Source: U.S. Census Bureau, 2002.

ability to pay for food, shelter, clothing, and other necessities.

Lack of *affordable* housing is one central problem brought about by economic inequality. Another concern is *substandard* housing, which refers to facilities that have inadequate heating, air conditioning, plumbing, electricity, or structural durability. Structural problems—due to faulty construction or lack of adequate maintenance—exacerbate the potential for other problems such as damage from fire, falling objects, or floors and stairways collapsing.

Housing is a problem that is directly related to inequality in both the urban and rural areas of the United States. Due to the importance of adequate housing for even a subsistence level of existence, the Census Bureau includes questions about housing in the questions that it asks during the decennial census (see "Census Profiles: Housing Characteristics as a Social Concern"). In some areas of the United States, people live in substandard housing that cannot withstand local weather conditions. An example is the problem of living in trailers and manufactured housing in areas where tornadoes and hurricanes are a possibility. Another example is the housing for migrant farm-worker families in rural areas of Arizona, California, New Mexico, Texas, and other states, where the housing that is provided may be without safe electricity, running water, and indoor toilets.

Box 9.3 SOCIOLOGY AND SOCIAL POLICY

Welfare Reform and Its Aftermath

I'm a hard worker. I'm friendly. I'm dependable. I'm a fast learner. When I was working in the schools, I was there every day on time. And whenever somebody needed me, I was there and I stayed late.

–Linda Bailey, a thirty-three-year-old mother of two young sons, had been on welfare for eight years when she began participating in a job-readiness program in New York City (qtd. in Swarns, 1997: 1)

With her training, Linda found a part-time job in a supermarket. Upon receiving her first paycheck, she stated that "Now I don't feel like a failure anymore" (qtd. in Swarns, 1997: 1). But how far can Linda and other women like her go in supporting themselves and their children on the earnings they receive from jobs that are part time or pay only the federal minimum wage, which has remained at $5.15 per hour since 1997?

When dramatic changes to this nation's welfare laws were enacted by Congress and signed into law by President Clinton in 1996, the stated purposes of that new law included not only getting people off the welfare rolls by requiring welfare recipients to find paid employment but also bringing families together by encouraging parental responsibility. When the law was enacted, more than 14 million U.S. children—one in five—lived in poverty. Today, although children younger than age six remain the poorest age group in the nation, the percentage of U.S. children living in poverty has been reduced (U.S. Department of Health and Human Services, 2000).

How well has welfare reform worked? According to a variety of social analysts, the results are mixed.

There has been a dramatic reduction in the number of people on welfare, which was one of the goals of welfare reform. About 60 percent of the mothers who have left the welfare rolls have found jobs, which was another of the goals of the legislation. However, the Children's Defense Fund (2002) argues that getting people off the welfare rolls and into low-wage work alone is not enough to help these people overcome hardship and poverty. It cites, among others, these examples:

- More than half of those persons who left welfare since 1996 for work no longer have a job.
- For those parents who had left a job, lack of child care was the reason most often reported.
- More than half of the employed parents had been unable to pay the rent, buy food, afford medical care, or pay for their telephone or electric service.

What will be the future of women and children who formerly depended, at least temporarily, on the government for assistance? We do not have a definitive answer for this pressing question. However, some social analysts believe that people typically bring their own political and economic agendas with them when they analyze such concerns as what effects welfare reform will have on families with young children. For some, cutting the welfare rolls is the goal, and this goal is being met. For others, the effects of these welfare changes can be measured only in terms of whether or not they put more women and their children at risk. At the bottom line, eliminating child poverty is crucial not only on the grounds of humanity and compassion, but also as an investment in this nation's future (Children's Defense Fund, 2002).

WRITING IN SOCIOLOGY ASSIGNMENT

Why is welfare a controversial issue in the United States? Is welfare a controversial issue in all nations? What reasons might exist for any differences among nations with regard to whether or not welfare is controversial?

Education

Educational opportunities and life chances are directly linked. Some functionalist theorists view education as the "elevator" to social mobility. Improvements in the educational achievement levels (measured in number of years of schooling completed) of the poor, people of color, and white women have been cited as evidence that students' abilities are now more important than their class, race, or gender. From this perspective, inequality in education is declining, and students have an opportunity to achieve upward mobility through achievements at school. Functionalists generally see the education system as flexible, allowing most students the opportunity to attend college if they apply themselves (Ballantine, 2001).

In contrast, most conflict theorists stress that schools are agencies for reproducing the capitalist class system and perpetuating inequality in society. From this perspective, education perpetuates poverty. Parents with limited income are not able to provide the same educational opportunities for their children as are families with greater financial resources.

Today, great disparities exist in the distribution of educational resources. Because funding for education primarily comes from local property taxes, school districts in wealthy suburban areas generally pay higher teachers' salaries, have newer buildings, and provide state-of-the-art equipment. By contrast, schools in poorer areas have a limited funding base. Students in central-city schools and poverty-stricken rural areas often attend dilapidated schools that lack essential equipment and teaching materials. Author Jonathan Kozol (1991, qtd. in Feagin and Feagin, 1994: 191) documented the effect of a two-tiered system on students:

> Kindergartners are so full of hope, cheerfulness, high expectations. By the time they get into fourth grade, many begin to lose heart. They see the score, understanding they're not getting what others are getting. . . . They see suburban schools on television. . . . They begin to get the point that they are not valued much in our society. By the time they are in junior high, they understand it. "We have eyes and we can see; we have hearts and we can feel. . . . We know the difference."

Poverty extracts such a toll that many young people will not have the opportunity to finish high school, much less enter college.

Crime and Lack of Safety

Along with diminished access to quality health care, nutrition, and housing, and unequal educational opportunities, crime and lack of safety are other consequences of inequality. As discussed in Chapter 7 ("Deviance and Crime"), although people from all classes commit crimes, they commit different kinds of crime. Capitalism and the rise of the consumer society may be factors in the criminal behavior of some upper-middle-class and upper-class people, who may be motivated by greed or the competitive desire to stay ahead of others in their reference group. By contrast, crimes committed by people in the lower classes may be motivated by feelings of anger, frustration, and hopelessness.

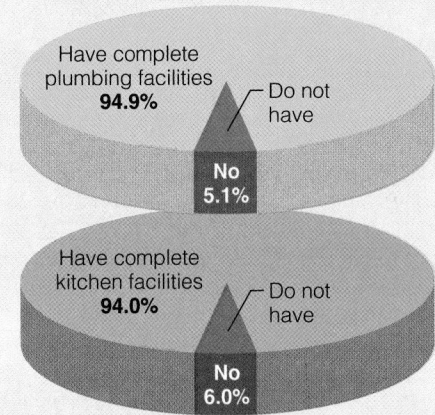

According to Marxist criminologists, capitalism produces social inequalities that contribute to criminality among people, particularly those who are outside the economic mainstream. Poverty and violence are also linked. In his recent ethnographic study of inner-city life in Philadelphia, the sociologist Elijah Anderson (1999: 33) suggests that what some people refer to as "random, senseless street violence" is often not random at all, but instead a response to profound social inequalities in the inner city:

> The inclination to violence springs from the circumstances of life among the ghetto poor—the lack of jobs that pay a living wage, limited basic public services (police response in emergencies, building maintenance, trash pickup, lighting, and other services that middle-class neighborhoods take for granted), the stigma of race, the fallout from rampant drug use and drug trafficking, and the resulting alienation and absence of hope for the future. Simply living in such an environment places young people at special risk of falling victim to aggressive behavior. . . . [T]his environment means that even youngsters whose home lives reflect mainstream values—and most of the homes in the community do—must be able to handle themselves in a street-oriented environment. (Anderson, 1999: 32–33)

As Anderson states, consequences of inequality include both crime and lack of safety on the streets, particularly for people who feel a profound sense of alienation from mainstream society and its institutions. Those who are able to take care of themselves and protect their loved ones against aggression are accorded deference and regard by others. However, Anderson believes that it is wrong to place blame solely on individuals for the problems that exist in urban ghettos; instead, he asserts that the focus should be on the socioeconomic structure and public policy that have threatened the well-being of people who live in poverty.

POVERTY IN THE UNITED STATES

When many people think about poverty, they think of people who are unemployed or on welfare. However, many hardworking people with full-time jobs live in poverty. The U.S. Social Security Administration has established an *official poverty line,* which is based on what is considered to be the minimum amount of money required for living at a subsistence level. The poverty level is computed by determining the cost of a minimally nutritious diet (a low-cost food budget on which a family could survive nutritionally on a short-term, emergency basis) and multiplying this figure by three to allow for nonfood costs. In 2001, almost 33 million people lived below the official government poverty level of $18,104 for a family of four, an increase of slightly more than 1 million people in poverty since 2000 (Proctor and Dalaker, 2002).

When sociologists define poverty, they distinguish between absolute and relative poverty. ***Absolute poverty* exists when people do not have the means to secure the most basic necessities of life.** This definition comes closest to that used by the federal government. Absolute poverty often has life-threatening consequences, such as when a homeless person freezes to death on a park bench. By comparison, ***relative poverty* exists when people may be able to afford basic necessities but are still unable to maintain an average standard of living** (Ropers, 1991). A family must have income substantially above the official poverty line in order to afford the basic necessities, even when these are purchased at the lowest possible cost. At about 155 percent of the official poverty line, families could live on an economy budget. What is it like to live on the economy budget? John Schwarz and Thomas Volgy (1992: 43) offer the following distressing description:

> Members of families existing on the economy budget never go out to eat, for it is not included in the food budget; they never go out to a movie, concert, or ball game or indeed to any public or private establishment that charges admission, for there is no entertainment budget; they have no cable television, for the same reason; they never purchase alcohol or cigarettes; never take a vacation or holiday that involves any motel or hotel or, again, any meals out; never hire a baby-sitter or have any other paid child care; never give an allowance or other spending money to the children; never purchase any lessons or home-learning tools for the children; never buy books or records for the adults or children, or any toys, except in the small amounts available for birthday or Christmas presents ($50 per person over the year); never pay for a haircut; never buy a magazine; have no money for the feeding or veterinary care of any pets; and, never spend any money for preschool for the children, or educational trips for them away from home, or any summer camp or other activity with a fee.

Table 9.2 PERCENTAGE DISTRIBUTION OF POVERTY IN THE UNITED STATES				
	ALL RACES[a]	WHITE	AFRICAN AMERICAN	HISPANIC[b]
By Age				
Under 18 years	16.3	9.5	30.2	28.0
18-24 years	16.3	12.3	26.8	21.0
25-44	9.8	6.7	17.5	17.9
45-64	8.7	6.9	17.2	15.8
65 and above	10.1	8.1	21.9	21.8
By Education				
No high school diploma	22.3	18.0	32.9	25.0
4 years of high school	9.6	7.4	19.6	14.7
Some college (no degree)	6.6	5.5	11.0	9.7
College degree or more	3.3	2.6	5.3	6.5

[a]Includes other races/ethnicities not shown separately.
[b]Includes Hispanic persons of any race.

Source: Proctor and Dalaker, 2002.

Who Are the Poor?

Poverty in the United States is not randomly distributed, but rather is highly concentrated according to age and race/ethnicity, as indicated in Table 9.2, as well as gender.

Age Today, children are at a much greater risk of living in poverty than are older persons. A generation ago, persons over age 65 were at the greatest risk of being poor; however, government programs such as Social Security and pension plans have been indexed for inflation and thus provide for something closer to an adequate standard of living than do other social welfare programs. Even so, older women are twice as likely to be poor as older men; older African Americans and Latinos/as are much more likely to live below the poverty line than are non-Latino/a whites.

The age category most vulnerable to poverty today is the very young. One out of every three persons below the poverty line is under 18 years of age, and a large number of children hover just above the official poverty line. The precarious position of African American and Latino/a children is even more striking. In 2001, about 30 percent of all African Americans under age 18 lived in poverty; 28 percent of Latino/a children were also poor, as compared with less than 10 percent of non-Latino/a white children (Proctor and Dalaker, 2002).

What do such statistics indicate about the future of our society? Children as a group are poorer now than they were at the beginning of the 1980s, whether they live in one- or two-parent families. The majority live in two-parent families in which one or both parents are employed. However, children in single-parent households headed by women have a much greater likelihood of living in poverty. In 2001, approximately 29 percent of white (non-Latino/a) children under age 18 in female-headed households lived below the poverty line, as sharply contrasted with about 50 percent of Latina/o and 49 percent of African American children in the same category (Proctor and Dalaker, 2002). Nor does the future look bright: Many governmental programs established to alleviate childhood poverty and malnutrition have been seriously cut back or eliminated altogether.

Gender About two-thirds of all adults living in poverty are women. In 2001, single-parent families headed by women had a 35-percent poverty rate as compared with a 10-percent rate for two-parent families. Sociologist Diana Pearce (1978) coined a term to describe this problem: The ***feminization of poverty refers to the trend in which women are disproportionately represented among individuals living in poverty.*** According to Pearce (1978), women have a higher risk of being poor because they bear the major economic and emotional burdens of raising children when they are single heads of households but earn between 70 and 80 cents for every dollar a male worker earns. More women than men are unable to obtain regular, full-time, year-round employment, and lack of adequate, affordable day care exacerbates this problem.

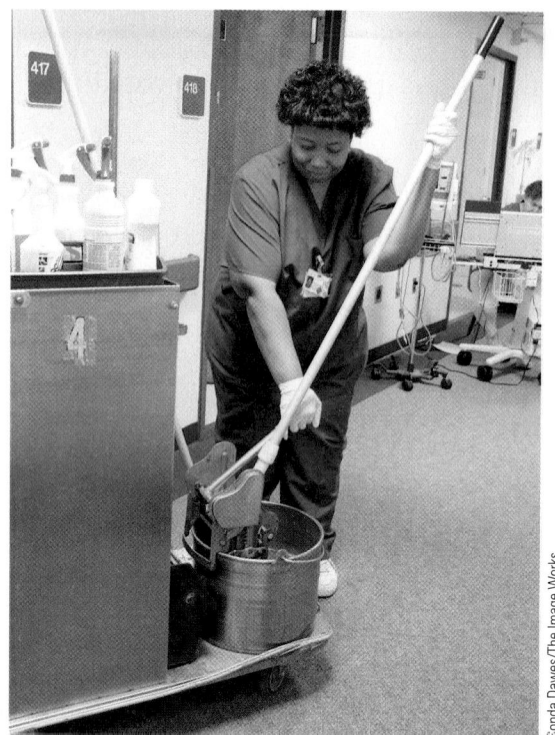

Sonda Dawes/The Image Works

Many women are among the "working poor," who, although employed full time, have jobs in service occupations that are typically lower paying and less secure than jobs in other sectors of the labor market. Does the nature of women's work contribute to the feminization of poverty in the United States?

Does the feminization of poverty explain poverty in the United States today? Is poverty primarily a women's issue? On the one hand, this thesis highlights a genuine problem—the link between gender and poverty. On the other hand, several major problems exist with this argument. First, women's poverty is not a new phenomenon. Women have always been more vulnerable to poverty (see Katz, 1989). Second, all women are not equally vulnerable to poverty. Many in the upper and upper-middle classes have the financial resources, education, and skills to support themselves regardless of the presence of a man in the household. Third, event-driven poverty does not explain the realities of poverty for many women of color, who instead may experience "reshuffled poverty"—a condition of deprivation that follows them regardless of their marital status or the type of family in which they live. Research by Mary Jo Bane (Bane, 1986; Bane and Ellwood, 1994) demonstrates that two out of three African American families headed by a woman were poor before the family event that made the woman a single mother. In addition, the poverty risk for a two-parent African American family is more than twice that for a white two-parent family.

Finally, poverty is everyone's problem, not just women's. When women are impoverished, so are their children. Moreover, many of the poor in our society are men, especially the chronically unemployed, older persons, the homeless, persons with disabilities, and men of color who have spent their adult lives without hope of finding work (Sidel, 1986).

Race/Ethnicity According to some stereotypes, most of the poor and virtually all welfare recipients are people of color. However, this stereotype is false; white Americans (non-Latinos/as) account for approximately two-thirds of those below the official poverty line. However, such stereotypes are perpetuated because a disproportionate percentage of the impoverished in the United States is made up of African Americans, Latinos/as, and Native Americans. About 22 percent of African Americans and Latinas/os were among the officially poor in 2001, as compared with about 8 percent of non-Latino/a whites (Proctor and Dalaker, 2002).

Native Americans are among the most severely disadvantaged persons in the United States. About one-third are below the poverty line, and some of these individuals live in conditions of extreme poverty. For example, a congressional study found that 70 percent of the Navajo households in Arizona, New Mexico, and Utah lived in houses without running water, sewer facilities, or electricity (Benokraitis, 1999). Some Native Americans receive governmental assistance; many others do not fall within the officially defined criteria for social welfare assistance.

Economic and Structural Sources of Poverty

Social inequality and poverty have both economic and structural sources. The low wages paid for many jobs is the major cause: Half of all families living in poverty are headed by someone who is employed, and one-third of those family heads work full time. A person with full-time employment in a minimum-wage job cannot keep a family of four from sinking below the official poverty line.

Structural problems contribute to both unemployment and underemployment. Corporations have been disinvesting in the United States, displacing millions of people from their jobs. Economists refer to this displacement as the *deindustrialization* of America (Bluestone and Harrison, 1982). Even as they have closed their U.S. factories and plants, many corporations have opened new facilities in other countries where "cheap labor" exists because people will, of necessity, work for lower wages. *Job deskilling—***a reduction in the proficiency needed to perform a specific job that leads to a corresponding reduction in the wages for that job**—has resulted from the introduction of computers and other technology (Hodson and Parker, 1988). The shift from manufacturing to service occupations has resulted in the loss of higher-paying positions and their replacement with lower-paying and less-secure positions that do not offer the wages, job stability, or advancement potential of the disappearing manufacturing jobs. Many of the new jobs are located in the suburbs, thus making them inaccessible to central-city residents. The problems of unemployment, underemployment, and poverty-level wages are even greater for people of color and young people in declining central cities. The unemployment rate for African Americans is almost double that of whites (U.S. Bureau of Labor Statistics, 2003b). African Americans have also experienced gender differences in employment that may produce different types of economic vulnerability. African American men who find employment typically earn more than African American women; however, the men's employment is often less secure. In the past decade, African American men have been more likely to lose their jobs because of declining employment in the manufacturing sector (Bane, 1986; Collins, 1990; Bane and Ellwood, 1994).

Solving the Poverty Problem

The United States has attempted to solve the poverty problem in several ways. One of the most enduring is referred to as social welfare. When most people think of "welfare," they think of food stamps and programs such as Temporary Assistance for Needy Families (TANF) or the earlier program it replaced, Aid to Families with Dependent Children (AFDC). However, the primary beneficiaries of social welfare programs are not poor. Some analysts estimate that approximately 80 percent of all social welfare benefits are paid to people who do not qualify as "poor." For example, many recipients of Social Security are older people in middle- and upper-income categories.

When older persons, including members of Congress, accept Social Security payments, they are not stigmatized. Similarly, veterans who receive benefits from the Veterans' Administration are not viewed as "slackers," and farmers who profit because of price supports are not considered to be lazy and unwilling to work. Unemployed workers who receive unemployment compensation are viewed with sympathy because of the financial plight of their families. By contrast, poor women and children who receive minimal benefits from welfare programs tend to be stigmatized and sometimes humiliated.

For the past two decades, "welfare reform" has been debated by presidential administrations, members of Congress, sociologists, and journalists. Several years ago, a law was passed to "end welfare as we know it" and to establish state-level "workfare" programs and mandatory time limits on welfare benefits (see Box 9.3 on page 306).

Recent thinking regarding overcoming poverty suggests that people with training and education are in the best position to "work their way out of poverty."

SOCIOLOGICAL EXPLANATIONS OF SOCIAL INEQUALITY IN THE UNITED STATES

Obviously, some people are disadvantaged as a result of social inequality. Therefore, is inequality always harmful to society?

Functionalist Perspectives

According to the sociologists Kingsley Davis and Wilbert Moore (1945), inequality is not only inevitable but also necessary for the smooth functioning of society. The Davis–Moore thesis, which has become the definitive functionalist explanation for social inequality, can be summarized as follows:

1. All societies have important tasks that must be accomplished and certain positions that must be filled.
2. Some positions are more important for the survival of society than others.
3. The most important positions must be filled by the most qualified people.
4. The positions that are the most important for society and that require scarce talent, extensive training, or both, must be the most highly rewarded.
5. The most highly rewarded positions should be those that are functionally unique (no other

According to conflict theorists, the factory workers shown here and millions of others like them around the globe produce the profits that maintain the capitalist class.

position can perform the same function) and on which other positions rely for expertise, direction, or financing.

Davis and Moore use the physician as an example of a functionally unique position. Doctors are very important to society and require extensive training, but individuals would not be motivated to go through years of costly and stressful medical training without incentives to do so. The Davis–Moore thesis assumes that social stratification results in **meritocracy—a hierarchy in which all positions are rewarded based on people's ability and credentials.**

Critics have suggested that the Davis–Moore thesis ignores inequalities based on inherited wealth and intergenerational family status (Rossides, 1986). The thesis assumes that economic rewards and prestige are the only effective motivators for people and fails to take into account other intrinsic aspects of work, such as self-fulfillment (Tumin, 1953). It also does not adequately explain how such a reward system guarantees that the most qualified people will gain access to the most highly rewarded positions.

Conflict Perspectives

From a conflict perspective, people with economic and political power are able to shape and distribute the rewards, resources, privileges, and opportunities in society for their own benefit. Conflict theorists do not believe that inequality serves as a motivating force

for people; they argue that powerful individuals and groups use ideology to maintain their favored positions at the expense of others. Core values in the United States emphasize the importance of material possessions, hard work, and individual initiative to get ahead, and behavior that supports the existing social structure. According to the conflict perspective, these same values support the prevailing resource distribution system and contribute to social inequality.

Are wealthy people smarter than others? According to conflict theorists, certain stereotypes suggest that this is the case; however, the wealthy may actually be "smarter" than others only in the sense of having "chosen" to be born to wealthy parents from whom they could inherit assets. Conflict theorists also note that laws and informal social norms support inequality in the United States. For the first half of the twentieth century, both legalized and institutionalized segregation and discrimination reinforced employment discrimination and produced higher levels of economic inequality. Although laws have been passed to make these overt acts of discrimination illegal, many forms of discrimination still exist in educational and employment opportunities.

Why is inequality growing in the United States? According to critical conflict (Marxist) theorists, the answer to this question partially lies in the concept of *surplus value*—the value produced, or the profit created, when the *cost of labor* is less than the *cost of the goods or services* that are produced by the workers. Thus, surplus value is created by the workers' labor power, which is bought and employed by members of the capitalist and managerial classes, who work on behalf of the capitalist class. When profits are made, they are either reinvested in the business or used for the enrichment of members of the capitalist class, which includes wealthy shareholders who are not involved in the day-to-day operations of a particular corporation but nevertheless control vast quantities of stock or other interests in that corporation. However, since capitalists deem it necessary to continually increase the rate of surplus value, owners and managers must continue to find new ways to make more profits. These methods are typically at the expense of workers (employees) and their families. In the contemporary workplace, employees are the first to feel the brunt of corporate reorganizations, layoffs, downsizing activities, and production demands that require them to work longer hours, sometimes without additional compensation. In certain situations, employees are replaced by labor-saving machines or technology such as computers and robots. Even in a "booming economy," social inequality persists because not everyone benefits from economic growth and development. The wealth of the upper classes, particularly those at the top of the ladder, continues to increase, whereas the rank-and-file employee has limited opportunities, and the size of the marginal working class and the poor population swells. Other workplace factors that contribute to social inequality, such as minimum-wage jobs and part-time or "temp" work, are discussed in Chapter 13 ("The Economy and Work in Global Perspective"). How the class structure is reproduced generation after generation through social institutions such as education is described in Chapter 16 ("Education").

Symbolic Interactionist Perspectives

Symbolic interactionists focus on microlevel concerns and usually do not analyze larger structural factors that contribute to inequality and poverty. However, many significant insights on the effects of wealth and poverty on people's lives and social interactions can be derived from applying a symbolic interactionist approach. Using qualitative research methods and influenced by a symbolic interactionist approach, researchers have collected the personal narratives of people across all social classes, ranging from the wealthiest to the poorest people in the United States.

Microlevel studies of the wealthy have examined issues such as the social and psychological factors that influence members of the upper class to contribute vast sums of money to charitable and arts organizations (Odendahl, 1990; Ostrower, 1997). Other studies have focused on the social interactions involved in "male bonding" rituals of elite men's organizations such as the Bohemian Club, which has annual two-week retreats at the "Bohemian Grove" in Northern California. Several decades ago, the sociologist G. William Domhoff (1974) engaged in participant observation research (serving as a waiter) at the Bohemian Grove to examine how the interactions of wealthy business executives and powerful political leaders partaking in games, rituals, and other activities helped bind these male members of the upper class into a socially cohesive group. Domhoff (1974) concluded that one of the ways the upper class perpetuates its position is through social cohesion, which is furthered by a wide variety of small groups that encourage face-to-face interaction and ensure status and security for members. Similarly, other social scientists have looked at the social and psychological aspects of life in the middle class (Newman, 1988, 1993; Rubin, 1994) or the working and lower classes (Anderson, 1999; Fine and Weis, 1998; Nelson and Smith, 1999; Newman, 1999).

A few studies provide rare insights into the social interactions between people from vastly divergent class locations. Sociologist Judith Rollins's (1985) study of the relationship between household workers and their employers is one example. Based on in-depth interviews and participant observation, Rollins examined rituals of deference that were often demanded by elite white women of their domestic workers, who were frequently women of color. According to the sociologist Erving Goffman (1967), *deference* is a type of ceremonial activity that functions as a symbolic means whereby appreciation is regularly conveyed to a recipient. In fact, deferential behavior between nonequals (such as employers and employees) confirms the inequality of the relationship and each party's position in the relationship relative to the other. Rollins identified three types of linguistic deference between domestic workers and their employers: use of the first names of the workers, contrasted with titles and last names (Mrs. Adams) of the employers; use of the term *girls* to refer to female household workers regardless of their age; and deferential references to employers, such as "Yes, ma'am." Spatial demeanor, including touching and how close one person stands to another, is an additional factor in deference rituals across class lines. Rollins (1985: 232) concludes that

> The employer, in her more powerful position, sets the essential tone of the relationship; and that tone . . . is one that functions to reinforce the inequality of the relationship, to strengthen

Box 9.4 YOU CAN MAKE A DIFFERENCE

Feeding the Hungry

The great fear among us all is that we are going to have to feed even more people [after changes in the welfare law]. . . . It's not enough to just hand food out anymore.

—Robert Egger, director of the nonprofit Central Kitchen in Washington, D.C. (qtd. in Clines, 1996)

Egger is one of the people responsible for an innovative chef's training program that feeds hope as well as hunger. At the Central Kitchen, located in the nation's capital, staff and guest chefs annually train around forty-eight homeless persons in three-month-long kitchen-arts courses. While the trainees are learning about food preparation, which will help them get starting jobs in the restaurant industry, they are helping feed about 3,000 homeless persons each day. Much of the food is prepared using donated goods such as turkeys that people have received as gifts at office parties and given to the kitchen, and leftover food from grocery stores including 7-Eleven stores, restaurants such as Pizza Hut, hotel food services, and college cafeterias. Central Kitchen got its start using leftovers from President George Bush's inaugural banquet in the late 1980s (Clines, 1996). Recently, donated food has gotten a boost from the

Good Samaritan law passed in 1996, which exempts nonprofit organizations and *gleaners*—volunteers who collect what is left in the field after harvesting—from liability for problems with food that they contribute in good faith (see Burros, 1996).

Can you think of ways that leftover food could be recovered from places where you eat so the food could be redistributed to persons in need? Have you thought about suggesting that members of an organization to which you belong might donate their time to help the Salvation Army, Red Cross, or other voluntary organization to collect, prepare, and serve food to others? If you would like to know more,

- "A Citizens Guide to Food Recovery" is available to help individuals participate in food recovery. Call 800-GLEAN-IT, or call the Salvation Army in your community. On the Internet:

http://www.salvationarmy.org

- World Hunger Year has projects such as Reinvesting in America that try to end hunger:

http://www.worldhungeryear.org

- American Red Cross:

http://www.redcross.org

---

the employer's belief in the rightness of her advantaged class and racial position, and to provide her with justification for the inegalitarian social system.

Many concepts introduced by the sociologist Erving Goffman (1959, 1967) could be used as springboards for examining microlevel relationships between inequality and people's everyday interactions. What could you learn about class-based inequality in the United States by using a symbolic interactionist approach to examine a setting with which you are familiar?

U.S. STRATIFICATION IN THE FUTURE

Will social inequality in the United States increase, decrease, or remain the same in the future? Many social scientists believe that existing trends point to an

increase. First, the purchasing power of the dollar has stagnated or declined since the early 1970s. As families started to lose ground financially, more family members (especially women) entered the labor force in an attempt to support themselves and their families (Gilbert, 2003). Economist and former Secretary of Labor Robert Reich (1993) has noted that in recent years the employed have been traveling on two escalators—one going up and the other going down. The gap between the earnings of workers and the income of managers and top executives has widened.

Second, wealth continues to become more concentrated at the top of the U.S. class structure. As the rich have grown richer, more people have found themselves among the ranks of the poor. Third, federal tax laws in recent years have benefited corporations and wealthy families at the expense of middle- and lower-income families. Finally, structural sources of upward mobility are shrinking whereas the rate of downward mobility has increased.

Are we sabotaging our future if we do not work constructively to eliminate poverty? It has been said that a chain is no stronger than its weakest link. If we apply this idea to the problem of poverty, then it is to our advantage to see that those who cannot find work or do not have a job that provides a living wage receive adequate training and employment. Innovative programs can combine job training with producing something useful to meet the immediate needs of people living in poverty. Children of today—the adults of tomorrow—need nutrition, education, health care, and safety as they grow up (see Box 9.4).

Some social analysts believe that the United States will become a better nation if it attempts to regain the American Dream by attacking poverty. According to the sociologist Michael Harrington (1985: 13), if we join in solidarity with the poor, we will "rediscover our own best selves . . . we will regain the vision of America."

CHAPTER REVIEW

■ How does income differ from wealth?

Whereas income is the economic gain derived from wages, salaries, income transfers (governmental aid), and ownership of property, wealth is the value of all of a person's or family's economic assets, including income, personal property, and income-producing property.

■ How did classical sociologists such as Karl Marx and Max Weber view social class?

Karl Marx and Max Weber acknowledged social class as a key determinant of social inequality and social change. For Marx, people's relationship to the means of production determines their class position. Weber developed a multidimensional concept of stratification that focuses on the interplay of wealth, prestige, and power.

■ What are some of the consequences of inequality in the United States?

The stratification of society into different social groups results in wide discrepancies in income and wealth and in variable access to available goods and services. People with high income or wealth have greater opportunity to control their own lives. People with less income have fewer life chances and must spend their limited resources to acquire basic necessities.

■ How do sociologists view poverty?

Sociologists distinguish between absolute poverty and relative poverty. Absolute poverty exists when people do not have the means to secure the basic necessities of life. Relative poverty exists when people may be able to afford basic necessities but are still unable to maintain an average standard of living.

■ Who are the poor?

Age, gender, and race tend to be factors in poverty. Children have a greater risk of being poor than do the elderly; women have a higher rate of poverty than do men. Although whites account for approximately two-thirds of those below the poverty line, people of color account for a disproportionate share of the impoverished in the United States. As the gap between rich and poor and between employed and unemployed widens, social inequality will clearly continue to increase if we do nothing.

■ What is the functionalist view on class?

Functionalist perspectives view classes as broad groupings of people who share similar levels of privilege on the basis of their roles in the occupational structure. According to the Davis–Moore thesis, stratification exists in all societies, and some inequality is not only inevitable but also necessary for the ongoing functioning of society. The positions that are most important within society and that require the most talent and training must be highly rewarded.

■ What is the conflict view on class?

Conflict perspectives on class are based on the assumption that social stratification is created and maintained by one group in order to enhance and protect its own economic interests. Conflict theorists measure class according to people's relationships with others in the production process. Erik Olin Wright identified four criteria for placement in a capitalist class structure: (1) ownership of the means of production, (2) purchase of the labor of others, (3) control of the labor of others, and (4) sale of one's own labor.

KEY TERMS

QUESTIONS FOR CRITICAL THINKING

1. Based on the Weberian and Marxian models of class structure, what is the class location of each of your ten closest friends or acquaintances? What is their location relative to yours? To one another's? What does their location tell you about friendship and social class?
2. Should employment be based on merit, need, or affirmative action policies?
3. What might happen in the United States if the gap between rich and poor continues to widen?

RESOURCES ON THE INTERNET

Chapter-Related Web Sites

The following Web sites have been selected for their relevance to the topics in this chapter. These sites are among the more stable, but please note that Web site addresses change frequently. For an updated list of chapter-related Web sites with URL links, please visit the *Sociology in Our Times* Web site (**www.wadsworth.com/KendallSIOT**).

U.S. Census Bureau
http://www.census.gov

The U.S. Census Bureau publishes numerous reports that provide comprehensive national data on income, poverty, and other measures that depict the class structure of the United States. In addition, the much-used reference *Statistical Abstract of the United States* is available online at this site.

Inequality.org
http://inequality.org

Inequality.org is a nonprofit organization whose goal is to disseminate information and ideas that are not widely publicized in the media. The site provides a wealth of information on inequality in the United States with regard to health care, technology, economics, education, and other contemporary issues.

Institute for Research on Poverty (IRP)
http://www.ssc.wisc.edu/irp

Established in 1966 at the University of Wisconsin–Madison by the U.S. Office of Economic Opportunity, IRP is a national nonprofit and nonpartisan organization dedicated to researching the causes and consequences of poverty and social inequality in the United States. The site is an excellent source for statistics, links, news, and research on national issues such as health, education, welfare reform, low-wage workers, and child support.

ONLINE STUDY AND RESEARCH TOOLS

Accompanying this text are many *free* powerful online study tools that will help you master the material in this chapter, help increase your depth of understanding, and help you make the grade!

SocCoach CD-ROM

Use the SocCoach CD-ROM enclosed with this text to help you formulate a customized study plan for this chapter. After you take the Diagnostic Quiz, SocCoach will generate a customized study plan just for you! It will identify sections of the chapter that you should review and will provide videos, charts, graphs, and excerpts from the text to supplement your studies and enhance your understanding. You'll also find fun, interactive activities such as Virtual Explorations and Map the Stats to apply what you've learned and stretch your sociological imagination.

The Companion Web Site for Sociology in Our Times, *Fifth Edition*
www.wadsworth.com/KendallSIOT

Gain an even better grasp on this chapter by going to the companion Web site to take one of the Tutorial Quizzes, use the Flash Cards to master key terms, or check out the many other study aids you'll find there.

You'll also find special features such as GSS Data and Census 2000 information that'll put data and resources at your fingertips to help you with that special project or help you as you do some research on your own.

 In this chapter, when you see the icon on the left, it alerts you to a specific exercise found in *Wadsworth's Sociology Online Resources and Writing Companion.* This valuable guide shows you how to use Wadsworth's exclusive online resources—*InfoTrac*

College Edition, the *Opposing Viewpoints Resource Center,* and *MicroCase Online*—to assist you in your study of sociology and to build essential research and writing skills.

Race and Ethnicity

REPORTER: Mr. Ashe, I guess [having AIDS] must be the heaviest burden you have ever had to bear, isn't it?

ARTHUR ASHE: No, it isn't. It's a burden, all right. But AIDS isn't the heaviest burden I have had to bear. . . . Being black is the greatest burden I've had to bear. . . . No question about it. Race has always been my biggest burden. Having to live as a minority in America. Even now it continues to feel like an extra weight tied around me. . . .

—Tennis star Arthur Ashe, a Wimbledon champion in the 1970s and one of the early African American players to break the "color line," comparing racial discrimination with having AIDS (which he had contracted through a blood transfusion), a few months before his death (Ashe and Rampersad, 1994: 139–141)

Would tennis stars such as the African American sisters Serena and Venus Williams—the 2003 Wimbledon women's singles champion and runner-up, respectively—have always been welcomed on tennis courts and in matches around the globe? As strange as it may seem, some of the athletes who today have invigorated sports such as tennis and golf would not have been permitted to play the game a few decades ago because of their race. Historically, sports played in "private club" settings (such as tennis, golf, swimming, sailing, and polo) have been less accessible for people of color because of the exclusionary policies of some of the most prestigious clubs where matches, tournaments, and regattas are likely to be held (Symonds, 1990). In this chapter, sports is used as an example of the effects of race and ethnicity on people's lives—and to examine whether those effects are changing. Sports is used as an example because it shows how, for more than a century, people who have been singled out for negative treatment on the basis of their perceived race or ethnicity have sought to overcome prejudice and discrimination through their determination in endeavors that can be judged based on objective standards ("What's the final score?") rather than subjective standards ("Does this person look the way we want sales associates to look at Abercrombie & Fitch?"). Before reading on, test your knowledge about race, ethnicity, and sports by taking the quiz in Box 10.1.

■ Were it not for early trailblazers such as African American tennis champion Arthur Ashe, top athletes today—including Venus and Serena Williams—might not have the opportunities they do to compete in tournaments and other sporting matches.

QUESTIONS AND ISSUES

Chapter Focus Question: How is sports a reflection of racial and ethnic relations in the United States and other nations?

How do race and ethnicity differ?

How do discrimination and prejudice differ?

How are racial and ethnic relations analyzed according to sociological perspectives?

What are the unique experiences of racial and ethnic groups in the United States?

RACE AND ETHNICITY

What is race? Some people think it refers to skin color (the Caucasian "race"); others use it to refer to a religion (the Jewish "race"), nationality (the British "race"), or the entire human species (the human "race") (Marger, 2003). Popular usages of race have been based on the assumption that a race is a grouping or classification based on *genetic* variations in physical appearance, particularly skin color. However, social scientists and biologists dispute the idea that biological race is a meaningful concept (Johnson, 2000). In fact, the idea of race has little meaning in a biological sense because of the enormous amount of interbreeding that has taken place within the human population. For these reasons, sociologists sometimes place "race" in quotation marks to show that categorizing individuals and population groups on biological characteristics is neither accurate nor based on valid distinctions between the genetic make-up of differently identified "races" (Marshall, 1998). Today, sociologists emphasize that race is a *socially constructed reality,* not a biological one (Feagin and Feagin, 2003). From this approach, the social significance that people accord to race is more significant than any biological differences that might exist among people who are placed in arbitrary categories.

A *race* **is a category of people who have been singled out as inferior or superior, often on the basis of real or alleged physical characteristics such as skin color, hair texture, eye shape, or other subjectively selected attributes** (Feagin and Feagin, 2003). Categories of people frequently thought of as racial groups include Native Americans, Mexican Americans, African Americans, and Asian Americans. This classification is rooted in nineteenth-century distinctions made by some biologists, who divided the world's population into three racial categories: *Caucasian*—people characterized as having relatively light skin and fine hair; *Negroid*—people with darker skin and coarser, curlier hair; and *Mongoloid*—people with yellow or brown skin and distinctively shaped eyelids.

However, racial categorization based on phenotypical differences (such as facial characteristics or skin color) does not correlate with genotypical differences (differences in genetic makeup). As anthropologists and biologists have acknowledged for some time, most of us have more in common genetically with individuals from another "race" than we have with the genetic average of people from our own "race." Moreover, throughout human history, extensive interbreeding has made such classifications unduly simplistic.

As compared with race, *ethnicity* defines individuals who are believed to share common characteristics that differentiate them from the other collectivities in a society. An ***ethnic group* is a collection of people distinguished, by others or by themselves, primarily on the basis of cultural or nationality characteristics** (Feagin and Feagin, 2003). Examples of ethnic groups include Jewish Americans, Irish Americans, and Italian Americans. Ethnic groups share five main characteristics: (1) *unique cultural traits,* such as language, clothing, holidays, or religious practices; (2) *a sense of community;* (3) *a feeling of ethnocentrism;* (4) *ascribed membership from birth;* and (5) *territoriality,* or the tendency to occupy a distinct geographic area (such as Little Italy or Little Havana) by choice and/or for self-protection. Although some people do not identify with any ethnic group, others participate in social interaction with individuals in their ethnic group and feel a sense of common identity based on cultural characteristics such as language, religion, or politics. However, ethnic groups are not only influenced by their own past history but also by patterns of ethnic domination and subordination in societies (Marshall, 1998).

The Social Significance of Race and Ethnicity

How important are race and ethnicity? Sociologist Manuel Castells believes that ethnicity has been a fundamental source of meaning and recognition throughout human history and remains so today:

Box 10.1 SOCIOLOGY AND EVERYDAY LIFE

How Much Do You Know About Race, Ethnicity, and Sports?

True	False	
T	F	1. Because sports are competitive and fans, coaches, and players want to win, the color of the players has not been a factor, only their performance.
T	F	2. In the late 1800s and early 1900s, boxing provided social mobility for some Irish, Jewish, and Italian immigrants.
T	F	3. African Americans who competed in boxing matches in the late 1800s often had to agree to lose before they could obtain a match.
T	F	4. Racially linked genetic traits explain many of the differences among athletes.
T	F	5. All racial and ethnic groups have viewed sports as a means to become a part of the mainstream.
T	F	6. Until recently, the positions of quarterback and kicker in the National Football League have been held almost exclusively by white players.
T	F	7. In recent years, players of color have moved into coaching, management, and ownership positions in professional sports.
T	F	8. The odds are good that many outstanding high school and college athletes will make the pros if they do not get injured.
T	F	9. Racism and sexism appear to be on the decline in sports in the United States.

Answers on page 322.

Ethnicity is a founding structure of social differentiation, and social recognition, as well as discrimination, in many contemporary societies, from the United States to Sub-Saharan Africa. It has been, and it is, the basis for uprisings in search of social justice, as for Mexican Indians in Chiapas . . . as well as the irrational rationale for ethnic cleansing, as practiced by Bosnian Serbs. . . . And it is, to a large extent, the cultural basis that induces networking and trust-based transactions in the new business world, from Chinese business networks . . . to the ethnic "tribes" that determine success in the new global economy. (Castells, 1997: 53)

Race and ethnicity take on great social significance because how people act in regard to these terms drastically affects other people's lives, including what opportunities they have, how they are treated, and even how long they live. According to the sociologists Michael Omi and Howard Winant (1994: 158), race "permeates every institution, every relationship, and every individual" in the United States:

As we . . . compare real estate prices in different neighborhoods, select a radio channel to enjoy while we drive to work, size up a potential client, customer, neighbor, or teacher, stand in line at the unemployment office, or carry out a thousand other normal tasks, we are compelled to think racially, to use the racial categories and meaning systems into which we have been socialized. (Omi and Winant, 1994: 158)

Historically, stratification based on race and ethnicity has pervaded all aspects of political, economic, and social life. Consider sports as an example. Throughout the early history of the game of baseball, many African Americans had outstanding skills as players but were categorically excluded from Major League teams because of their skin color. Even after Jackie Robinson broke the "color line" in 1947 to become the first African American in the Major Leagues, his experience was marred by racial slurs, hate letters, death threats against his infant son, and assaults on his wife (Ashe, 1988; Peterson, 1992). With some professional athletes from diverse racial–ethnic categories having

Box 10.1 SOCIOLOGY AND EVERYDAY LIFE

Answers to the Sociology Quiz on Race, Ethnicity, and Sports

1. **False.** Discrimination has been pervasive throughout the history of sports in the United States. For example, African American athletes, regardless of their abilities, were excluded from white teams for many years.

2. **True.** Irish Americans were the first to dominate boxing, followed by Jewish Americans and then Italian Americans. Boxing, like other sports, was a source of social mobility for some immigrants.

3. **True.** Promoters, who often set up boxing matches that pitted fighters by race, assumed that white fans were more likely to buy tickets if the white fighters frequently won.

4. **False.** Although some scholars and journalists have used biological or genetic factors to explain the achievements of athletes, sociologists view these explanations as being based on the inherently racist assumption that people have "natural" abilities (or disabilities) because of their race or ethnicity.

5. **False.** Some racial and ethnic groups—including Chinese Americans and Japanese Americans—have not viewed sports as a means of social mobility.

6. **True.** As late as the 1990s, whites accounted for about 90 percent of the quarterbacks and kickers on NFL teams. However, this changed early in the twenty-first century, and today there are some African Americans playing virtually every position on all professional football teams.

7. **False.** Although more African American players are employed by these teams (especially in basketball), their numbers have not increased significantly in coaching and management positions. Only one professional sports team is currently owned by an African American.

8. **False.** The odds of becoming a professional athlete are very low. The percentage of high school football players who make it to the pros is estimated at about .14 percent; for basketball, about .06 percent; and for baseball, about .18 percent. The percentages of college athletes who make it to the pros are a little higher: 3.5 percent for football, 2.6 percent for basketball, and 4.3 percent for baseball.

9. **False.** Even as people of color and white women have made gains on collegiate and professional teams, scholars have documented the continuing significance of racial and gender discrimination in sports.

Sources: Based on Coakley, 2004; Hight, 1994; Messner, Duncan, and Jensen, 1993; and Nelson, 1994.

multi-million-dollar contracts and lucrative endorsement deals, it is easy to assume that racism in sports is a thing of the past. However, this *commercialization* of sports does not mean that racial prejudice and discrimination no longer exist (Coakley, 2004).

Racial Classifications and the Meaning of Race

If we examine racial classifications throughout history, we find that in ancient Greece and Rome a person's race was the group to which she or he belonged, associated with an ancestral place and culture (Zack, 1998). From the Middle Ages until about the eighteenth century, a person's race was based on family and ancestral ties, in the sense of a *line,* or to a national group. During the eighteenth century, physical differences such as the darker skin hues of Africans became associated with race, but racial divisions were typically based on differences in religion and cultural tradition rather than on human biology. With the intense (though misguided) efforts that surrounded the attempt to justify black slavery and white dominance in all areas of life during the second half of the nineteenth century, *races* came to be defined as distinct

Miami's Little Havana is an ethnic enclave where people participate in social interaction with other individuals in their ethnic group and feel a sense of shared identity. Ethnic enclaves provide economic and psychological support for recent immigrants as well as for those who were born in the United States.

Jeff Greenberg/PhotoEdit

biological categories of people who were not all members of the same family but who shared inherited physical and cultural traits that were alleged to be different from those traits shared by people in other races. Hierarchies of races were established, placing the "white race" at the top, the "black race" at the bottom, and others in between.

However, racial classifications in the United States have changed over the past century. If we look at U.S. Census Bureau classifications, for example, we can see how the meaning of race continues to change. First, race is defined by perceived skin color: white or nonwhite. Whereas one category exists for "whites" (who vary considerably in actual skin color and physical appearance), all of the remaining categories are considered "nonwhite."

Second, racial purity is assumed to exist. Prior to the 2000 census, for example, the true diversity of the U.S. population was not revealed in census data, since multiracial individuals were forced to either select a single race as being their "race" or to select the vague category of "other." Professional golfer Tiger Woods is an example of how people often have a mixed racial heritage. Woods describes himself as one-half Asian American (one-fourth Thai and one-fourth Chinese), one-eighth white, one-eighth Native American, and one-fourth African American (White, 1997). Census 2000 made it possible—for the first time—for individuals to classify themselves as being of more than one race (see "Census Profile: Percentage Distribution of Persons Reporting Two or More Races").

Third, categories of official racial classifications may (over time) create a sense of group membership or "consciousness of kind" for people within a somewhat arbitrary classification. When people of European descent were classified as "white," some began to see themselves as different from "nonwhite." Consequently, Jewish, Italian, and Irish immigrants may have felt more a part of the Northern European white mainstream in the late eighteenth and early nineteenth centuries. Whether Chinese Americans, Japanese Americans, Korean Americans, and Filipino Americans

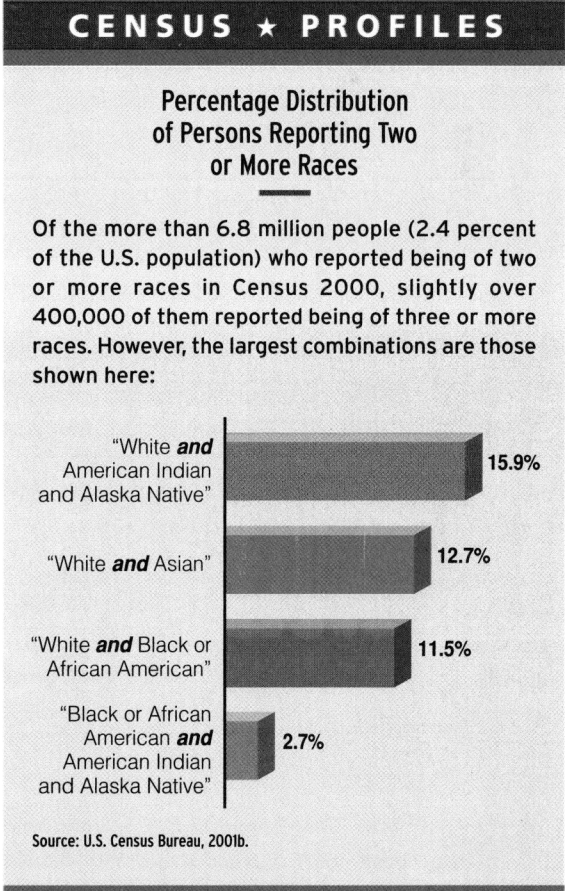

CENSUS ★ PROFILES

Percentage Distribution of Persons Reporting Two or More Races

Of the more than 6.8 million people (2.4 percent of the U.S. population) who reported being of two or more races in Census 2000, slightly over 400,000 of them reported being of three or more races. However, the largest combinations are those shown here:

"White *and* American Indian and Alaska Native" **15.9%**

"White *and* Asian" **12.7%**

"White *and* Black or African American" **11.5%**

"Black or African American *and* American Indian and Alaska Native" **2.7%**

Source: U.S. Census Bureau, 2001b.

come to think of themselves collectively as "Asian Americans" because of official classifications remains to be seen (S. Lee, 1993).

In the future, increasing numbers of children in the United States will be likely to have a mixed racial or ethnic heritage such as this student describes:

> I am part French, part Cherokee Indian, part Filipino, and part black. Our family taught us to be aware of all these groups, and just to be ourselves. But I have never known what I am. People have asked if I am a Gypsy, or a Portuguese, or a Mexican, or lots of other things. It seems to make people curious, uneasy, and sometimes belligerent. Students I don't even know stop me on campus and ask, "What are you anyway?" (qtd. in Davis, 1991: 133)

The way people are classified remains important because such classifications affect their access to employment, housing, social services, federal aid, and many other "publicly or privately valued goods" (Omi and Winant, 1994: 3). However, recent publicity about celebrities of mixed-race heritage, such as Tiger Woods, together with the change in the way that the Census Bureau asks about race, may lead more people to identify themselves as having more than one racial heritage.

Dominant and Subordinate Groups

The terms *majority group* and *minority group* are widely used, but their meanings are less clear as the composition of the U.S. population continues to change. Accordingly, many sociologists prefer the terms *dominant* and *subordinate* to identify power relationships that are based on perceived racial, ethnic, or other attributes and identities. To sociologists, a **dominant group is one that is advantaged and has superior resources and rights in a society** (Feagin and Feagin, 2003). In the United States, whites with Northern European ancestry (often referred to as Euro-Americans, white Anglo-Saxon Protestants, or WASPs) are considered to be a dominant group. A **subordinate group is one whose members, because of physical or cultural characteristics, are disadvantaged and subjected to unequal treatment by the dominant group and who regard themselves as objects of collective discrimination.** Persons of color and white women are considered to be subordinate-group members in the United States, particularly when these individuals are from lower-income categories.

In the sociological sense, *group* as used in these two terms is misleading because people who merely share ascribed racial or ethnic characteristics do not constitute a group. However, the terms *dominant group* and *subordinate group* do give us a way to describe relationships of advantage/disadvantage and power/exploitation that exist in contemporary nations.

PREJUDICE

Although there are various meanings of the word *prejudice*, sociologists typically use it in a specialized sense. From a sociological perspective, **prejudice is a negative attitude based on faulty generalizations about members of selected racial and ethnic groups** (Allport, 1958; Feagin and Feagin, 2003). The term *prejudice* is from the Latin words *prae* ("before") and *judicium* ("judgment"), which means that people may be biased either for or against members of other groups even before they have had any contact with them (Cashmore, 1996). Although prejudice can be either *positive* (bias in favor of a group—often our own) or *negative* (bias against a group—one we deem less worthy than our own), it most often refers to the negative attitudes that people may have about members of other racial or ethnic groups. To some extent, all people have prejudiced feelings against members of other groups. In this case, prejudice is not restricted to racial or ethnic groups but may be applied to virtually any category of people, including an entire nation or continent, to which negative generalizations are applied (Cashmore, 1996). During World War II, for example, Japan was an "enemy" of the United States, and some people in this country referred to everyone in Japan (or originating from that country) as "Japs" (Feagin and Feagin, 2003).

Stereotypes

Prejudice is rooted in ethnocentrism and stereotypes. When used in the context of racial and ethnic relations, *ethnocentrism* refers to the tendency to regard one's own culture and group as the standard—and thus superior—whereas all other groups are seen as inferior. Ethnocentrism is maintained and perpetuated by **stereotypes—overgeneralizations about the appearance, behavior, or other characteristics of members of particular categories.** The term *stereotype* comes from the Greek word *stereos* ("solid") and refers to a fixed mental impression (Cashmore, 1996). Although stereotypes can be either positive or negative, examples of negative stereotyping abound in sports. Think about the Native American names, images, and mascots used by sports teams such as the

As a person from a mixed racial background, professional golfer Tiger Woods has sought to reduce prejudice and discrimination.

Atlanta Braves, Cleveland Indians, Golden State Warriors, Washington Redskins, and Kansas City Chiefs. Members of Native American groups have been actively working to eliminate the use of stereotypic mascots (with feathers, buckskins, beads, spears, and "warpaint"), "Indian chants," and gestures (like the "tomahawk chop"), which they claim trivialize and exploit Native American culture (Churchill, 1994). College and university sports teams with Native American names and logos also remain the subject of controversy in the twenty-first century (see Spindel, 2000). According to sociologist Jay Coakley (2004), the use of stereotypes and words such as *redskin* symbolize a lack of understanding of the culture and heritage of native peoples and are offensive to many Native Americans. Although some people see the use of these names and activities as "innocent fun," others view them as a form of racism.

Racism

What is racism? **Racism is a set of attitudes, beliefs, and practices that is used to justify the superior treatment of one racial or ethnic group and the inferior treatment of another racial or ethnic group.** As discussed later, some forms of racism are based on

biological arguments about the innate inferiority of members of certain racial and ethnic groups; however, more recent forms of racism have attempted to justify the unequal treatment of the same groups of people on other grounds (Cashmore, 1996). The world has seen a long history of racism: It can be traced from the earliest civilizations. At various times throughout U.S. history, various categories of people, including the Irish Americans, Italian Americans, Jewish Americans, African Americans, and Latinos/as, have been the objects of racist ideology. However, not everyone is equally racist. Recent studies have shown that the underlying reasoning behind racism differs according to factors such as gender, age, class, and geography (see Cashmore, 1996; Feagin and Vera, 1995).

Racism may be overt or subtle. Overt racism is more blatant and may take the form of public statements about the "inferiority" of members of a racial or ethnic group. In sports, for example, calling a player of color a derogatory name, participating in racist chanting during a sporting event, and writing racist graffiti in a team's locker room are all forms of overt racism. These racist actions are blatant, but subtle forms of racism are often hidden from sight and more difficult to prove. Examples of subtle racism in sports include those descriptions of African American athletes which suggest that they have "natural" abilities and are better suited for team positions requiring speed and agility. By contrast, whites are described as having the intelligence, dependability, and leadership and decision-making skills needed in positions requiring higher levels of responsibility and control.

Racism tends to intensify in times of economic uncertainty and high rates of immigration. Recently, relatively high rates of immigration in nations such as the United States, Canada, England, France, and Germany have been accompanied by an upsurge in racism and racial conflict (see Box 10.2). Sometimes, intergroup racism and conflicts further exacerbate strained relationships between dominant-group members and subordinate racial and ethnic group members. For example, when animosities have run very high among African American and Salvadoran groups in Long Island, New York, some white Americans have pointed to those hostilities as evidence that both groups are inferior and not deserving of assistance from the U.S. government or from charitable organizations such as the Catholic church (Mahler, 1995).

Theories of Prejudice

Are some people more prejudiced than others? To answer this question, some theories focus on how individuals may transfer their internal psychological

Box 10.2 SOCIOLOGY IN GLOBAL PERSPECTIVE

Racism and Anti-Racism in European Football

The problem of racism is a real issue for the European football family and for the image of football in general. Of course racism at our games is a sad reflection of society in general, but, because of the high-profile nature of football on our continent, we have a particular responsibility to take steps to stamp it out and prevent it occurring in the future.

—a joint statement by Lennart Johansson, president, and Gerhard Aigner, chief executive, of the UEFA (the Union of European Football Associations) (uefa.com, 2002)

Soccer is Europe's equivalent of football in the United States, and European fans are at least equally as passionate about sports as are fans in the United States. However, racism is an issue in European soccer just as it is an issue in U.S. sports. The UEFA, which is European soccer's ruling body, receives numerous reports of racist behavior at soccer matches throughout Europe, and officials have decided that the UEFA must work to reduce racism. Accusations of racist behavior against players of other nationalities or ethnic groups have been lodged against sports fans and some team members at various athletic events throughout Europe. At some soccer matches, fans shout racist comments during the game; throw bottles, cans, and other missiles at players; and pass out derogatory literature about other ethnic groups or nationalities outside the stadium. For example, Slovakian fans were accused of racially abusing English players in a qualifying match, and the Slovakian team was subsequently fined by the UEFA and ordered to play its next home international game in an empty stadium as punishment for its fans' inappropriate behavior (Barclay, 2003). Ironically, the English fans were later charged with similar racist behavior against the Turkish team (globeandmail.com, 2003).

Acknowledging that racism is a problem that "is created and stimulated outside of [soccer], but is often given expression and public focus through the game," UEFA officials recently issued a call for unity and developed a plan to reduce racist behavior at athletic competitions (sportsnetwork.com, 2003).

According to the call for unity issued by the UEFA, "Racism is a problem for all of us, which must be faced. We hope that all parts of European football can come together to unite against racism and do all we can to eradicate it from our game both on and off the field" (uefa.com, 2002). Among the things that the plan calls for are these actions on the part of soccer clubs:

- Soccer clubs should issue a statement in printed programs and on the grounds where the competition takes place that the club will not tolerate racism and should set forth specific actions that the club will take against people who engage in racist chanting.
- Regular public address announcements should be made condemning racist chanting at matches.
- To become a season-ticket holder, people should agree not to take part in any racist abuse.
- Clubs should take action against those who attempt to sell racist literature inside and outside the athletic grounds.
- All racist graffiti should be removed quickly from the grounds.
- Soccer clubs should work with other organizations, including schools, youth clubs, and local businesses, to raise public awareness of the importance of eliminating racial abuse and discrimination. (adapted from uefa.com, 2002)

Will these steps help to reduce racism in European sporting events? Analysts believe that these actions are at least a step in the right direction. If the UEFA enforces the rules that it has established, fans and players will be more aware of the negative sanctions that might occur if they do not "clean up their act." UEFA officials and others have acknowledged that a problem does exists and that the problem is not limited to a small handful of fans and players. But, in the final analysis, as the UEFA officials stated, "Of course no one organization can solve this problem. Everyone involved, including the clubs, fans, players, police and those responsible for stewarding, has a responsibility here" (uefa.com, 2002).

WRITING IN SOCIOLOGY ASSIGNMENT

Is the problem of racism in sports unique to European nations? Are overt and covert forms of racism practiced at U.S. sporting events? How about behind the scenes, such as in people's  everyday school conversations, in locker-room banter, or behind the backs of some athletes? What plan would you develop to help eliminate these forms of inappropriate behavior?

According to the frustration–aggression hypothesis, members of white supremacy groups such as the Ku Klux Klan often use members of subordinate racial and ethnic groups as scapegoats for societal problems over which they have no control.

© Michael Greenlar/The Image Works

problem onto an external object or person. Others look at factors such as social learning and personality types.

The frustration–aggression hypothesis states that people who are frustrated in their efforts to achieve a highly desired goal will respond with a pattern of aggression toward others (Dollard et al., 1939). The object of their aggression becomes the **scapegoat—a person or group that is incapable of offering resistance to the hostility or aggression of others** (Marger, 2003). Scapegoats are often used as substitutes for the actual source of the frustration. For example, members of subordinate racial and ethnic groups are often blamed for societal problems (such as unemployment or an economic recession) over which they have no control.

According to some symbolic interactionists, prejudice results from social learning; in other words, it is learned from observing and imitating significant others, such as parents and peers. Initially, children do not have a frame of reference from which to question the prejudices of their relatives and friends. When they are rewarded with smiles or laughs for telling derogatory jokes or making negative comments about outgroup members, children's prejudiced attitudes may be reinforced.

Psychologist Theodor W. Adorno and his colleagues (1950) concluded that highly prejudiced individuals tend to have an **authoritarian personality, which is characterized by excessive conformity, submissiveness to authority, intolerance, insecurity, a high level of superstition, and rigid, stereotypic thinking** (Adorno et al., 1950). It is most likely to develop in a family environment in which dominating parents who are anxious about status use physical discipline but show very little love in raising their children (Adorno et al., 1950). Other scholars have linked prejudiced attitudes to traits such as submissiveness to authority, extreme anger toward outgroups, and conservative religious and political beliefs (Altemeyer, 1981, 1988; Weigel and Howes, 1985).

Measuring Prejudice

To measure levels of prejudice, some sociologists use the concept of **social distance—the extent to which people are willing to interact and establish relationships with members of racial and ethnic groups other than their own** (Park and Burgess, 1921). Sociologist Emory Bogardus (1925, 1968) developed a scale to measure social distance in specific situations, ranging from minimal contact to marriage. He concluded that some groups were consistently ranked as more desirable than others for close interpersonal contacts. More recently, analysts have found that whites who accept racial stereotypes desire greater social distance from people of color than do whites who reject negative stereotypes (Krysan and Farley, 1993). But can prejudice really be measured? Existing research does not provide us with a conclusive answer to this question. Most social distance research has examined the perceptions of whites; few studies have measured the perceptions of people of color about their interactions with members of the dominant group.

■ DISCRIMINATION

Whereas prejudice is an attitude, **discrimination involves actions or practices of dominant-group members (or their representatives) that have a**

harmful impact on members of a subordinate group (Feagin and Feagin, 2003). Prejudiced attitudes do not always lead to discriminatory behavior. As shown in Figure 10.1, the sociologist Robert Merton (1949) identified four combinations of attitudes and responses. Unprejudiced nondiscriminators are not personally prejudiced and do not discriminate against others. For example, two players on a professional sports team may be best friends although they are of different races. Unprejudiced discriminators may have no personal prejudice but still engage in discriminatory behavior because of peer group pressure or economic, political, or social interests. For example, in some sports, a coach might feel no prejudice toward African American players but believe that white fans will accept only a certain percentage of people of color on the team. Prejudiced nondiscriminators hold personal prejudices but do not discriminate due to peer pressure, legal demands, or a desire for profits. For example, a coach with prejudiced beliefs may hire an African American player to enhance the team's ability to win (Coakley, 2004). Finally, prejudiced discriminators hold personal prejudices and actively discriminate against others. For example, an umpire who is personally prejudiced against African Americans may intentionally call a play incorrectly based on that prejudice.

Discriminatory actions vary in severity from the use of derogatory labels to violence against individuals and groups. The ultimate form of discrimination occurs when people are considered to be unworthy to live because of their race or ethnicity. **Genocide is the deliberate, systematic killing of an entire people or nation.** Examples of genocide include the killing of thousands of Native Americans by white settlers in North America and the extermination of 6 million European Jews by Nazi Germany. More recently, the term *ethnic cleansing* has been used to define a policy of "cleansing" geographic areas (such as in Yugoslavia) by forcing persons of other races or religions to flee— or die.

Discrimination also varies in how it is carried out. Individuals may act on their own, or they may operate within the context of large-scale organizations and institutions, such as schools, churches, corporations, and governmental agencies. How does individual discrimination differ from institutional discrimination? **Individual discrimination consists of one-on-one acts by members of the dominant group that harm members of the subordinate group or their property** (Carmichael and Hamilton, 1967; Feagin and Feagin, 2003). For example, a person may decide not to rent an apartment to someone of a different race. By contrast, **institutional discrimination consists**

Figure 10.1   Merton's Typology of Prejudice and Discrimination

Merton's typology shows that some people may be prejudiced but not discriminate against others. Do you think that it is possible for a person to discriminate against some people without holding a prejudiced attitude toward them? Why or why not?

	Prejudiced attitude?	Discriminatory behavior?
Unprejudiced nondiscriminator	No	No
Unprejudiced discriminator	No	Yes
Prejudiced nondiscriminator	Yes	No
Prejudiced discriminator	Yes	Yes

of the day-to-day practices of organizations and institutions that have a harmful impact on members of subordinate groups (Feagin and Feagin, 2003). For example, a bank might consistently deny loans to people of a certain race. Institutional discrimination is carried out by the individuals who implement the policies and procedures of organizations.

Sociologist Joe R. Feagin has identified four major types of discrimination:

1. *Isolate discrimination* is harmful action intentionally taken by a dominant-group member against a member of a subordinate group. This type of discrimination occurs without the support of other members of the dominant group in the immediate social or community context. For example, a prejudiced judge may give harsher sentences to all African American defendants but may not be supported by the judicial system in that action.

2. *Small-group discrimination* is harmful action intentionally taken by a limited number of dominant-group members against members of subordinate groups. This type of discrimination is not supported by existing norms or other dominant-group members in the immediate social or community context. For example, a small group of white students may deface a professor's

office with racist epithets without the support of other students or faculty members.

3. *Direct institutionalized discrimination* is organizationally prescribed or community-prescribed action that intentionally has a differential and negative impact on members of subordinate groups. These actions are routinely carried out by a number of dominant-group members based on the norms of the immediate organization or community (Feagin and Feagin, 2003). Intentional exclusion of people of color from public accommodations is an example of this type of discrimination.

4. *Indirect institutionalized discrimination* refers to practices that have a harmful impact on subordinate-group members even though the organizationally or community-prescribed norms or regulations guiding these actions were initially established with no intent to harm. For example, special education classes were originally intended to provide extra educational opportunities for children with various types of disabilities. However, critics claim that these programs have amounted to racial segregation in many school districts. Over a period of time, special education classes may have contributed to racial stereotyping that tells young people of color that "they will not make it in life, so why even try" (Richardson, 1994: B8).

SOCIOLOGICAL PERSPECTIVES ON RACE AND ETHNIC RELATIONS

Symbolic interactionist, functionalist, and conflict analysts examine race and ethnic relations in different ways. Functionalists focus on the macrolevel intergroup processes that occur between members of majority and minority groups in society. Conflict theorists analyze power and economic differentials between the dominant group and subordinate groups. Symbolic interactionists examine how microlevel contacts between people may produce either greater racial tolerance or increased levels of hostility.

Symbolic Interactionist Perspectives

What happens when people from different racial and ethnic groups come into contact with one another?

In the *contact hypothesis,* symbolic interactionists point out that contact between people from divergent groups should lead to favorable attitudes and behavior when certain factors are present. Members of each group must (1) have equal status, (2) pursue the same goals, (3) cooperate with one another to achieve their goals, and (4) receive positive feedback when they interact with one another in positive, nondiscriminatory ways (Allport, 1958; Coakley, 2004).

Does participation in interracial sports teams, for example, promote intergroup cohesion and reduce prejudice? Scholars have found that increased contact may have little or no effect on existing prejudices and, in some circumstances, can even lead to an increase in prejudice and conflict. If racial tension between players on teams is high enough, violence may erupt (Edwards, 1973; Lapchick, 1991). Why might sports contribute to prejudice? Sports events are highly competitive, and prejudice may be aggravated by the actions of players or spectators.

What happens when individuals meet someone who does not conform to their existing stereotype? Frequently, they ignore anything that contradicts the stereotype, or they interpret the situation to support their prejudices (Coakley, 2004). For example, a person who does not fit the stereotype may be seen as an exception: "You're not like other [persons of a particular race]."

When a person is seen as conforming to a stereotype, he or she may be treated simply as one of "you people." Former Los Angeles Lakers basketball star Earvin "Magic" Johnson (1992: 31–32) described how he was categorized along with all other African Americans when he was bused to a predominantly white school:

> On the first day of [basketball] practice, my teammates froze me out. Time after time I was wide open, but nobody threw me the ball. At first I thought they just didn't see me. But I woke up after a kid named Danny Parks looked right at me and then took a long jumper. Which he missed.
>
> I was furious, but I didn't say a word. Shortly after that, I grabbed a defensive rebound and took the ball all the way down for a basket. I did it again and a third time, too.
>
> Finally Parks got angry and said, "Hey, pass the [bleeping] ball."
>
> That did it. I slammed down the ball and glared at him. Then I exploded. "I *knew* this would happen!" I said. "That's why I didn't want to come to this [bleeping] school in the first place!"
>
> "Oh, yeah? Well, you people are all the same," he said. "You think you're gonna come in here

Are sports a source of upward mobility for recent immigrants and ethnic minorities, as was true for some in previous generations? Early-twentieth-century Jewish American and Italian American boxers, for example, not only produced intragroup ethnic pride but also earned a livelihood in such sporting events.

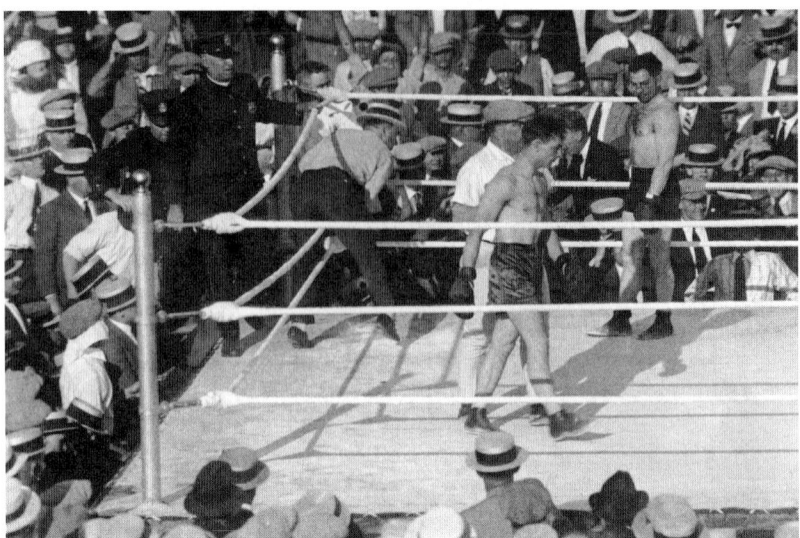

Bettmann/Corbis

and do whatever you want? Look, hotshot, your job is to get the rebound. Let us do the shooting."

The interaction between Johnson and Parks demonstrates that when people from different racial and ethnic groups come into contact with one another, they may treat one another as stereotypes, not as individuals. Eventually, Johnson and Parks were able to work out most of their differences. "There's nothing like winning to help people get along," Johnson explained (1992: 32).

Symbolic interactionist perspectives make us aware of the importance of intergroup contact and the fact that it may either intensify or reduce racial and ethnic stereotyping and prejudice.

Functionalist Perspectives

How do members of subordinate racial and ethnic groups become a part of the dominant group? To answer this question, early functionalists studied immigration and patterns of majority- and minority-group interactions.

Assimilation *Assimilation* **is a process by which members of subordinate racial and ethnic groups become absorbed into the dominant culture.** To some analysts, assimilation is functional because it contributes to the stability of society by minimizing group differences that might otherwise result in hostility and violence (Gordon, 1964).

Assimilation occurs at several distinct levels, including the cultural, structural, biological, and psychological stages. *Cultural assimilation,* or *acculturation,* occurs when members of an ethnic group adopt dominant-group traits, such as language, dress, values, religion,

and food preferences. Cultural assimilation in this country initially followed an "Anglo conformity" model; members of subordinate ethnic groups were expected to conform to the culture of the dominant white Anglo-Saxon population (Gordon, 1964). However, members of some groups refused to be assimilated and sought to maintain their unique cultural identity.

Structural assimilation, or *integration,* occurs when members of subordinate racial or ethnic groups gain acceptance in everyday social interaction with members of the dominant group. This type of assimilation typically starts in large, impersonal settings such as schools and workplaces, and only later (if at all) results in close friendships and intermarriage. *Biological assimilation,* or *amalgamation,* occurs when members of one group marry those of other social or ethnic groups. Biological assimilation has been more complete in some other countries, such as Mexico and Brazil, than in the United States.

Psychological assimilation involves a change in racial or ethnic self-identification on the part of an individual. Rejection by the dominant group may prevent psychological assimilation by members of some subordinate racial and ethnic groups, especially those with visible characteristics such as skin color or facial features that differ from those of the dominant group.

Can sports teach people the thoughts, feelings, and actions that are necessary for assimilation? An important function of sports for some groups has been tension management (Coakley, 2004). For example, when Italian immigrants arrived in New York City in the early 1900s, sports represented "a way to defuse the exuberant adolescent brawling that sometimes threatened to explode into ethnic violence" (Mangione and Morreale, 1992: 373). Boxing was a means to channel energy into positive avenues and to teach values such as competition, patriotism, and physical fitness (Mangione and Morreale, 1992). Newly arrived Jewish immigrants worked to excel in baseball to gain acceptance and to challenge existing stereotypes that they were physically weak and inferior (see Levine, 1992). In turn, Latinos found opportunities in baseball for assimilation and financial gains that they could not find through other avenues. However, few women engaged in these activities (see Nelson, 1994).

Ethnic Pluralism   Instead of complete assimilation, many groups share elements of the mainstream culture while remaining culturally distinct from both the dominant group and other social and ethnic groups. ***Ethnic pluralism is the coexistence of a variety of distinct racial and ethnic groups within one society.***

Equalitarian pluralism, or *accommodation,* is a situation in which ethnic groups coexist in equality with one another. Switzerland has been described as a model of equalitarian pluralism; over 6 million people with French, German, and Italian cultural heritages peacefully coexist there (Simpson and Yinger, 1972). *Inequalitarian pluralism,* or *segregation,* exists when specific ethnic groups are set apart from the dominant group and have unequal access to power and privilege (Marger, 2003). ***Segregation is the spatial and social separation of categories of people by race, ethnicity, class, gender, and/or religion.*** Segregation may be enforced by law. *De jure segregation* refers to laws that systematically enforced the physical and social separation of African Americans in all areas of public life. An example of de jure segregation was the Jim Crow laws, which legalized the separation of the races in public accommodations (such as hotels, restaurants, transportation, hospitals, jails, schools, churches, and cemeteries) in the southern United States after the Civil War (Feagin and Feagin, 2003). Segregation denied African Americans access to opportunities in many areas, including education, jobs, health care, and politics. In sports, Jim Crow laws barred many great African American athletes from playing in "white" leagues. In 1895, John W. (Bud) Fowler, an outstanding baseball player, described how "Jim Crow" affected his life: "I didn't pick up a living; I just existed. . . . My skin is against me. If I had not been quite so black, I might have caught on as a Spaniard or something of that kind. The race prejudice is so strong that my black skin barred me" (qtd. in Peterson, 1992/1970: 40).

Segregation may also be enforced by custom. *De facto segregation*—racial separation and inequality enforced by custom—is more difficult to document than de jure segregation. For example, residential segregation is still prevalent in many U.S. cities; owners, land-

lords, real estate agents, and apartment managers often use informal mechanisms to maintain their properties for "whites only." Even middle-class African Americans find that racial polarization is fundamental to the residential layout of many cities. Most whites who discriminate know that they will not be sanctioned, since fair-housing laws are weak and the agencies responsible for their enforcement are handicapped by a lack of funds (see Feagin and Sikes, 1994).

Although functionalist explanations provide a description of how some early white ethnic immigrants assimilated into the cultural mainstream, they do not adequately account for the persistent racial segregation and economic inequality experienced by people of color.

Conflict Perspectives

Conflict theorists focus on economic stratification and access to power in their analyses of race and ethnic relations. Some emphasize the castelike nature of racial stratification, others analyze class-based discrimination, and still others examine internal colonialism and gendered racism.

The Caste Perspective   The caste perspective views racial and ethnic inequality as a permanent feature of U.S. society. According to this approach, the African American experience must be viewed as different from that of other racial or ethnic groups. African Americans were the only group to be subjected to slavery; when slavery was abolished, a caste system was instituted to maintain economic and social inequality between whites and African Americans (Dollard, 1957/1937).

The caste system was strengthened by *antimiscegenation laws,* which prohibited sexual intercourse or marriage between persons of different races. Most states had such laws, which were later expanded to include relationships between whites and Chinese, Japanese, and Filipinos. These laws were not declared unconstitutional until 1967 (Frankenberg, 1993).

Although the caste perspective points out that racial stratification may be permanent because of structural elements such as the law, it has been criticized for not examining the role of class in perpetuating racial inequality (Cox, 1948).

Class Perspectives   Class perspectives emphasize the role of the capitalist class in racial exploitation. Based on early theories of race relations by the African American scholar W. E. B. Du Bois, the sociologist Oliver C. Cox (1948) suggested that African Americans were enslaved because they were the cheapest and best workers the owners could find for heavy labor in the mines and on plantations. Thus, the profit motive of capitalists, not skin color or racial prejudice, accounts for slavery.

More recently, sociologists have debated the relative importance of class and race in explaining the unequal life chances of African Americans. Sociologist William Julius Wilson (1996) has suggested that race, cultural factors, social psychological variables, and social class must all be taken into account in examining the life chances of "inner-city residents." His analysis focuses on how class-based economic determinants of social inequality, such as deindustrialization and the decline of the central (inner) city, have affected many African Americans, especially in the Northeast. African Americans were among the most severely affected by the loss of factory jobs because work in the manufacturing sector had previously made upward mobility possible. Wilson (1996) is not suggesting that prejudice and discrimination have been eradicated; rather, he is arguing that they may be less important than class in explaining the current status of African Americans.

How do conflict theorists view the relationship among race, class, and sports? Simply stated, sports reflects the interests of the wealthy and powerful. At all levels, sports exploits athletes (even highly paid ones) in order to gain high levels of profit and prestige for coaches, managers, and owners. African American athletes and central-city youths in particular are exploited by the message of rampant consumerism. Many are given the unrealistic expectation that sports can be a ticket out of the ghetto or barrio. If they try hard enough (and wear the right sneakers), they too can become wealthy and famous. However, even for those with virtually no chance of becoming professional athletes, consumerism is encouraged by media blitzes featuring the tennis shoes, clothes, and other gear worn by their favorite college and professional sports "heroes."

Today, athletic gear companies have signed contracts with major universities so that the sponsor's brand becomes the "official" shoe of the university. The university, in turn, agrees to license the use of its name and logo on the T-shirts, sweatpants, and jogging suits that the sponsor sells to the public (Weistart, 1993). These practices support the assertion of conflict theorists that sports is a capitalistic venture in which profit is more important than the health and well-being of the athletes or the enjoyment of the general public.

Internal Colonialism

Why do some racial and ethnic groups continue to experience subjugation after many years? According to the sociologist Robert Blauner (1972), groups that have been subjected to internal colonialism remain in subordinate positions longer than groups that voluntarily migrated to the United States. *Internal colonialism occurs when members of a racial or ethnic group are conquered or colonized and forcibly placed under the economic and political control of the dominant group.*

In the United States, indigenous populations (including groups today known as Native Americans and Mexican Americans) were colonized by Euro-Americans and others who invaded their lands and conquered them. In the process, indigenous groups lost property, political rights, aspects of their culture, and often their lives (Blauner, 1972). The capitalist class acquired cheap labor and land through this government-sanctioned racial exploitation (Blauner, 1972). The effects of past internal colonialism are reflected today in the number of Native Americans who live on government reservations and in the poverty of Mexican Americans who lost their land and had no right to vote.

The experiences of internally colonized groups are unique in three ways: (1) these groups have been forced to exist in a society other than their own; (2) they have been kept out of the economic and political mainstream, so it is difficult for them to compete with dominant-group members; and (3) they have been subjected to severe attacks on their culture, which may lead to its extinction (Blauner, 1972).

The internal colonialism perspective is rooted in the historical foundations of racial and ethnic inequality in the United States. However, it tends to view all voluntary immigrants as having many more opportunities than do members of colonized groups. Thus, this model does not explain the continued exploitation of some immigrant groups, such as the Chinese, Filipinos, Cubans, Vietnamese, and Haitians, and the greater acceptance of others, primarily those from Northern Europe (Cashmore, 1996).

The Split-Labor-Market Theory

Who benefits from the exploitation of people of color? Dual- or split-labor-market theory states that white workers and members of the capitalist class both benefit from the exploitation of people of color. *Split labor market refers to the division of the economy into two areas of employment, a primary sector or upper tier, composed of higher-paid (usually dominant-group) workers in more secure jobs, and a secondary sector or lower tier, composed of lower-paid (often subordinate-group) workers in jobs with little security and hazardous working conditions* (Bonacich, 1972, 1976). According to this perspective, white workers in the upper tier may use racial discrimination

Grinding poverty is a pressing problem for families living along the border between the United States and Mexico. Economic development has been limited in areas where *colonias* such as this one are located, and the wealthy have derived far more benefit than others from recent changes in the global economy.

© Shelly Katz/Getty Images

against nonwhites to protect their positions. These actions most often occur when upper-tier workers feel threatened by lower-tier workers hired by capitalists to reduce labor costs and maximize corporate profits. In the past, immigrants were a source of cheap labor that employers could use to break strikes and keep wages down. Throughout U.S. history, higher-paid workers have responded with racial hostility and joined movements to curtail immigration and thus do away with the source of cheap labor (Marger, 2003).

Proponents of the split-labor-market theory suggest that white workers benefit from racial and ethnic antagonisms. However, these analysts typically do not examine the interactive effects of race, class, and gender in the workplace.

Recent Perspectives on Race and Gender

The term *gendered racism* refers to the interactive effect of racism and sexism on the exploitation of women of color. According to the social psychologist Philomena Essed (1991), women's particular position must be explored within each racial or ethnic group, because their experiences will not have been the same as men's in each grouping.

All workers are not equally exploited by capitalists. Gender and race or ethnicity are important in this exploitation. Historically, the high-paying primary labor market has been monopolized by white men. People of color and most white women more often hold lower-tier jobs. Below that tier is the underground sector of the economy, characterized by illegal or quasi-legal activities such as drug trafficking, prostitution, and working in sweatshops that do not meet minimum wage and safety standards. Many undocumented workers and some white women and people of color attempt to earn a living in this sector (Amott and Matthaei, 1996).

The *theory of racial formation* states that actions of the government substantially define racial and ethnic relations in the United States. Government actions range from race-related legislation to imprisonment of members of groups believed to be a threat to society. Sociologists Michael Omi and Howard Winant (1994) suggest that the U.S. government has shaped the politics of race through actions and policies that cause people to be treated differently because of their race. For example, immigration legislation reflects racial biases. The Naturalization Law of 1790 permitted only white immigrants to qualify for naturalization; the Immigration Act of 1924 favored Northern Europeans and excluded Asians and Southern and Eastern Europeans.

The government's definition of racial realities is periodically challenged by social protest movements of various racial and ethnic groups. When this social rearticulation occurs, people's understanding about race may be restructured somewhat. For example, the African American protest movements of the 1950s and 1960s helped redefine the rights of people of color in the United States.

An Alternative Perspective: Critical Race Theory

Emerging out of scholarly law studies on racial and ethnic inequality, critical race theory derives its foundation from the U.S. civil rights tradition and the writing of persons such as Martin Luther King, Jr., W.E.B. Du Bois, Malcolm X, and César Chávez. Critical race theory has several major premises, including the belief that racism is such an ingrained feature of U.S. society that it appears to be ordinary and natural to many people (Delgado, 1995). As a result, civil rights legislation and affirmative action laws (formal equality) may remedy some of the more overt, blatant forms of racial injustice but have little effect on subtle, business-as-usual forms of racism that people of color experience as they go about their everyday lives. According to this approach, the best way to document racism and ongoing inequality in society is to listen to the lived experiences of people who have experienced such discrimination. In this way, we can learn what actually happens in regard to racial oppression and the many effects it has on people, including alienation, depression, and certain physical illnesses. Central to this argument is the belief that *interest convergence* is a crucial factor in bringing about social change. According to the legal scholar Derrick Bell, white elites tolerate or encourage racial advances for people of color *only* if the dominant-group members believe that their own self-interest will be served in so doing (cited in Delgado, 1995). From this approach, civil rights laws have typically benefited white Americans as much (or more) as people of color because these laws have been used as mechanisms to ensure that "racial progress occurs at just the right pace: change that is too rapid would be unsettling to society at large; change that is too slow could prove destabilizing" (Delgado, 1995: xiv).

Critical race theory is similar to postmodernist approaches in that it calls our attention to the fact that things are not always as they seem. Formal equality under the law does not necessarily equate to actual equality in society. This theory also makes us aware of the ironies and contradictions in civil rights law, which some see as self-serving laws. However, mainstream critics argue that critical race theory is unduly pessimistic because it largely ignores the strides that have been made toward racial equality in the United States. Concept Table 10.A outlines the key aspects of each sociological perspective.

❚ RACIAL AND ETHNIC GROUPS IN THE UNITED STATES

How do racial and ethnic groups come into contact with one another? How do they adjust to one another and to the dominant group over time? Sociologists have explored these questions extensively; however, a detailed historical account of the unique experiences of each group is beyond the scope of this chapter. In-

stead, we will look briefly at intergroup contacts. In the process, sports will be used as an example of how members of some groups have attempted to gain upward mobility and become integrated into society.

Native Americans

Native Americans are believed to have migrated to North America from Asia thousands of years ago, as shown on the time line in Figure 10.2. One of the most widely accepted beliefs about this migration is that the first groups of Mongolians made their way across a natural bridge of land called Beringia into present-day Alaska. From there, they moved to what is now Canada and the northern United States, eventually making their way as far south as the tip of South America (Cashmore, 1996).

As schoolchildren are taught, Spanish explorer Christopher Columbus first encountered the native inhabitants in 1492 and referred to them as "Indians." When European settlers (or invaders) arrived on this continent, the native inhabitants' way of life was changed forever. Experts estimate that approximately 2 million native inhabitants lived in North America at that time (Cashmore, 1996); however, their numbers had been reduced to less than 240,000 by 1900 (Churchill, 1994).

Genocide, Forced Migration, and Forced Assimilation Native Americans have been the victims of genocide and forced migration. Although the United States never had an official policy that set in motion a pattern of deliberate extermination, many Native Americans were either massacred or died from European diseases (such as typhoid, smallpox, and measles) and starvation (Wagner and Stearn, 1945; Cook, 1973). In battle, Native Americans were often no match for the Europeans, who had "modern" weaponry (Amott and Matthaei, 1996). Europeans justified their aggression by stereotyping the Native Americans as "savages" and "heathens" (Takaki, 1993). Early atrocities perpetrated against the Native Americans were described by one social historian:

> The degree of violence that was woven into the texture of early frontier life fairly boggles the mind of our, in some ways, far more delicate age. In the 1650s, Dutch colonists brought back eighty decapitated Indian heads from a massacre and used them as kickballs in the streets of New Amsterdam. . . . An English traveler in the northern colonies casually recorded in his diary, in 1760, that "some people have an Indian's Skin for a Tobacco Pouch," while a Revolutionary War soldier

Concept Table 10.A SOCIOLOGICAL PERSPECTIVES ON RACE AND ETHNIC RELATIONS

	Focus	Theory/Hypothesis
SYMBOLIC INTERACTIONIST	Microlevel contacts between individuals	Contact hypothesis

FUNCTIONALIST	Macrolevel intergroup processes	1. Assimilation
		   a. cultural
		   b. biological
		   c. structural
		   d. psychological
		2. Ethnic pluralism
		   a. equalitarian pluralism
		   b. inequalitarian pluralism (segregation)

CONFLICT	Power/economic differentials between dominant and subordinate groups	1. Caste perspective 2. Class perspective 3. Internal colonialism 4. Split labor market 5. Gendered racism 6. Racial formation 7. Social rearticulation
CRITICAL RACE THEORY	Racism as an ingrained feature of society that affects everyone's daily life	Laws may remedy overt discrimination but have little effect on subtle racism. Interest convergence is required for social change.

Esbin-Anderson/The Image Works

© Roger Allyn/SuperStock

© Richard Heinzen/SuperStock

AP/Wide World Photos

campaigning against the Iroquois could note with equal dispassion that he had been given a pair of boot-tops made from the freshly skinned legs of two enemy braves. (Bordewich, 1996: 35–36)

After the Revolutionary War, the federal government offered treaties to the Native Americans so that more of their land could be acquired for the growing white population. Scholars note that the government broke treaty after treaty as it engaged in a policy of wholesale removal of indigenous nations in order to clear the land for settlement by Anglo-Saxon "pioneers" (Green, 1977; Churchill, 1994). Entire nations were forced to move in order to accommodate the white settlers. The "Trail of Tears" was one of the most disastrous of the forced migrations. In the coldest part of the winter of 1832, over half of the Cherokee Nation died during or as a result of their forced

relocation from the southeastern United States to the Indian Territory in Oklahoma (Thornton, 1984).

Native Americans were made wards of the government (meaning they had a legal status similar to that of minors and incompetents) and were subjected to forced assimilation on the reservations after 1871 (Takaki, 1993). Native American children were placed in boarding schools operated by the Bureau of Indian Affairs to hasten their assimilation into the dominant culture. About 98 percent of native lands had been expropriated by 1920 (see McDonnell, 1991; Churchill, 1994). Even after Native Americans received full citizenship and the right to vote in 1924, the Supreme Court continued to hold that they were wards of the government.

Native Americans Today
Currently, about 2 million Native Americans live in the United States, including Aleuts, Inuit (Eskimos), Cherokee, Navajo,

Figure 10.2 Time Line of Racial and Ethnic Groups in the United States

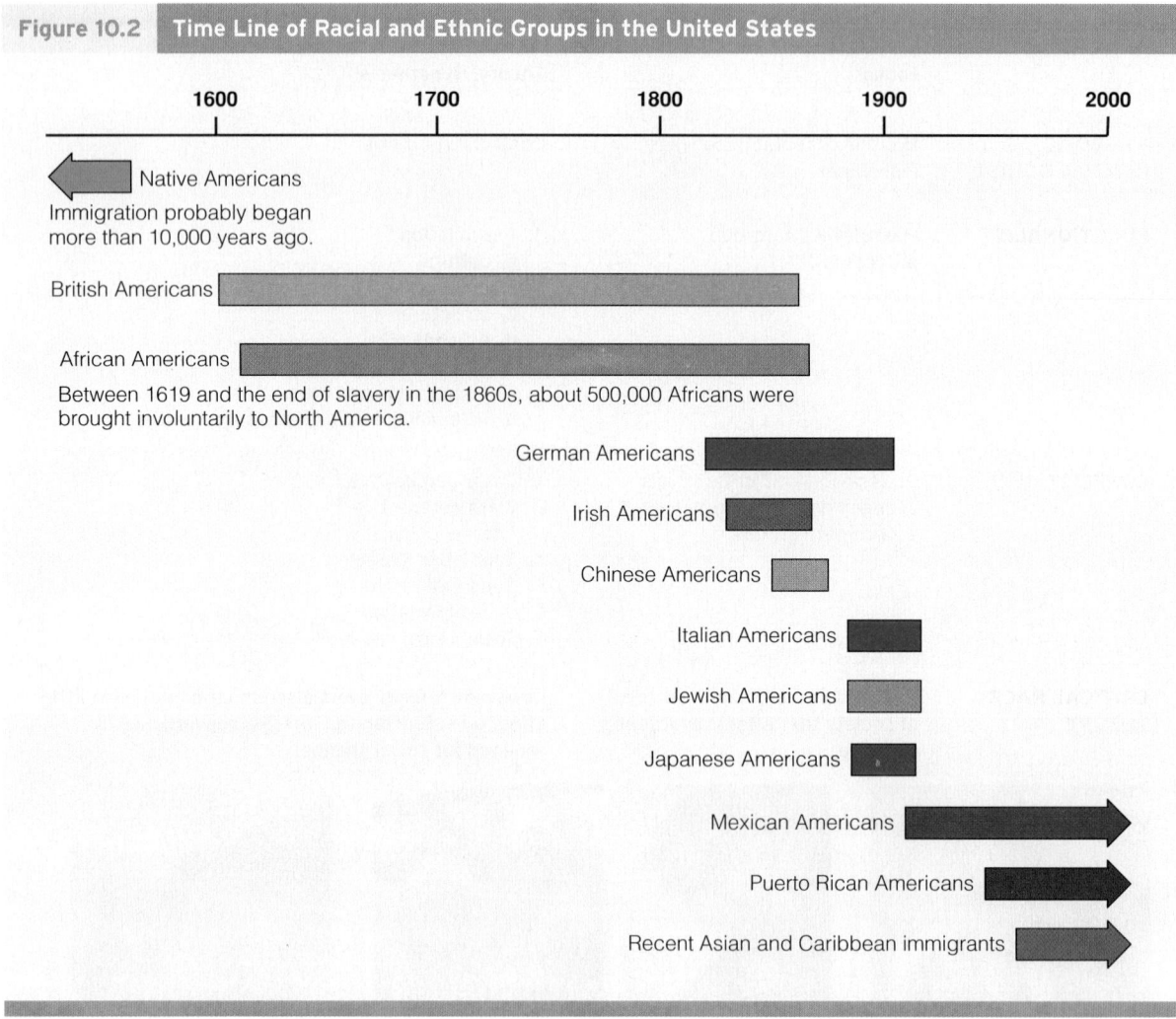

Chippewa, Sioux, and over five hundred other nations of varying sizes and different locales. Most are concentrated in the Southwest, and about one-third live on reservations. Native Americans are the most disadvantaged racial or ethnic group in the United States in terms of income, employment, housing, nutrition, and health. The life chances of Native Americans who live on reservations are especially limited. They have the highest rates of infant mortality and death by exposure and malnutrition. They also have high rates of tuberculosis, alcoholism, and suicide (Bachman, 1992). Reservation-based Native American men have an average life expectancy of less than forty-five years; for women, it is less than forty-eight years (see Churchill, 1994). Native Americans have had very limited educational opportunities and have a very high rate of unemployment. In recent years, however, a network of tribal colleges has been successful in providing some Native Americans with the education they need to move into the ranks of the skilled working class and beyond (Bordewich, 1996).

In spite of the odds against them, many Native Americans resist oppression. The American Indian Movement, Women of All Red Nations, and other groups have demanded the recovery of Native American lands and reparation for past losses. Native American women have publicized the harmful conditions (including radiation sickness and the forced sterilization of women) that exist on reservations. The American Indian Anti-Defamation Council advocates doing away with the word *tribe* because it demeans Native Americans by equating their level of cultural attainment with "primitivism" or "barbarism" (see Churchill, 1994). Moreover, reinterpretation of federal law in the 1990s has made it possible for Native American nations to open lucrative cigarette shops, bingo halls, and casino gambling operations on reservations (Cashmore, 1996).

Native Americans and Sports Early in the twentieth century, Native Americans such as Jim Thorpe gained national visibility as athletes in football, baseball, and track and field. Teams at boarding

Life chances are extremely limited for Native Americans who live on reservations. Although a few Native Americans early in the twentieth century were well-known athletes, Native Americans today have little opportunity to compete in sports at the college, professional, or Olympic level.

schools such as the Carlisle Indian Industrial School in Pennsylvania and the Haskell Institute in Kansas were well-known. However, after the first three decades of the twentieth century, Native Americans became much less prominent in sports. Although some Navajo athletes have been very successful in basketball and some Choctaws have excelled in baseball, Native Americans have seldom been able to compete at the college, professional, or Olympic level (Blanchard, 1980; Oxendine, 2003). Native American scholar Joseph B. Oxendine (2003) attributes the lack of athletic participation to these factors: (1) a reduction in opportunities for developing sports skills, (2) restricted opportunities for participation, and (3) a lessening of Native Americans' interest in competing with and against non-Native Americans. For example, Native Americans have fewer opportunities to participate in college athletics, which serves as the springboard to professional playing careers in football and basketball. Similarly, Native American young people who live on or near reservations do not have access to the highly specialized sports clubs that are often the training grounds for sports such as golf, tennis, ice skating, or fencing. These limited opportunities make it difficult for them to seek upward social mobility and economic success through sports (Coakley, 2004; Oxendine, 2003).

White Anglo-Saxon Protestants (British Americans)

Whereas Native Americans have been among the most disadvantaged peoples, white Anglo-Saxon Protestants (WASPs) have been the most privileged group in this country. Although many English set-

tlers initially came to North America as indentured servants or as prisoners, they quickly emerged as the dominant group, creating a core culture (including language, laws, and holidays) to which all other groups were expected to adapt. Most of the WASP immigrants arriving from northern Europe were advantaged over later immigrants because they were highly skilled and did not experience high levels of prejudice and discrimination.

Class, Gender, and WASPs Like members of other racial and ethnic groups, not all WASPs are alike. Social class and gender affect their life chances and opportunities. For example, members of the working class and the poor do not have political and economic power; men in the capitalist class do. WASPs constitute the majority of the upper class and maintain cohesion through listings such as the *Social Register* and interactions with one another in elite settings such as private schools and country clubs (Higley, 1997; Kendall, 2002). However, WASP women do not always have the same rights as the men of their group. Although WASP women have the privilege of a dominant racial position, they do not have the gender-related privileges of men (Amott and Matthaei, 1996). However, most WASPs do not think of themselves as having race or ethnicity. As Helen Standish explains,

> [Being a white Anglo-Saxon Protestant] didn't seem like a culture because everyone else was the same. . . . ["Others"] are different. . . . They speak Italian, but everybody else in the U.S. speaks English. They eat strange, different food, but I eat the same kind of food as everybody else in the U.S. . . . The way I was brought up was to think that everybody who was the same as me were "Americans," and the other people were of "such and such descent." (qtd. in Frankenberg, 1993: 198)

These beliefs are shared by some WASPs but denied by other individuals because they do not embrace the notion of "whiteness." According to race scholars, *whiteness* primarily signifies superiority and privilege, and it is used as a mechanism to devalue so-called nonwhite skin color (Cashmore, 1996).

WASPs and Sports Family background, social class, and gender play an important role in the sports participation of WASPs. Contemporary North American football was invented at the Ivy League colleges and was dominated by young, affluent WASPs who had the time and money to attend college and participate in sports activities. As Table 10.1 shows, whites are more likely than any other racial or ethnic group

Frank Siteman/Stock Boston

Sociological studies show that many non-Latino/a whites do not think of themselves as having race or ethnicity because they continue to benefit from being members of a dominant societal group that has experienced little prejudice or discrimination from other groups.

to become professional athletes in all sports except football and basketball.

Affluent WASP women participated in intercollegiate women's basketball in the late 1800s, and various other sporting events were used as a means to break free of restrictive codes of femininity (Nelson, 1994). Until recently, however, most women have had little chance for any involvement in college and professional sports.

African Americans

The African American (black) experience has been one uniquely marked by slavery, segregation, and persistent discrimination. There is a lack of consensus about whether *African American* or *black* is the most appropriate term to refer to the 36 million Americans of African descent who live in the United States today. Those who prefer the term *black* point out that it incorporates many African-descent groups living in

this country that do not use *African American* as a racial or ethnic self-description. For example, people who trace their origins to Haiti, Puerto Rico, or Jamaica typically identify themselves as "black" but not as "African American" (Cashmore, 1996).

Although the earliest African Americans probably arrived in North America with the Spanish conquerors in the fifteenth century, most historians trace their arrival to about 1619, when the first groups of indentured servants were brought to the colony of Virginia. However, by the 1660s, indentured servanthood had turned into full-fledged slavery with the enactment of laws in states such as Virginia that sanctioned the enslavement of African Americans. Although the initial status of persons of African descent in this country may not have been too different from that of the English indentured servants, all of that changed with the passage of laws turning human beings into property and making slavery a status from which neither individuals nor their children could escape (Franklin, 1980). Slavery laws dehumanized not only the people who were labeled as slaves but also the slave owners and political leaders who created and maintained this so-called "peculiar institution" (Feagin and Feagin, 2003).

Between 1619 and the 1860s, about 500,000 Africans were forcibly brought to North America, primarily to work on southern plantations, and these actions were justified by the devaluation and stereotyping of African Americans. Some analysts believe that the central factor associated with the development of slavery in this country was the plantation system, which was heavily dependent on cheap and dependable manual labor. Slavery was primarily beneficial to the wealthy southern plantation owners, but many of the stereotypes used to justify slavery were eventually institutionalized in southern custom and practice (Wilson, 1978). However, some slaves and whites engaged in active resistance against slavery and its barbaric practices, eventually resulting in slavery being outlawed in the northern states by the late 1700s. Slavery continued in the South until 1863, when it was abolished by the Emancipation Proclamation (Takaki, 1993).

Segregation and Lynching Gaining freedom did not give African Americans equality with whites. African Americans were subjected to many indignities because of race. Through informal practices in the North and *Jim Crow laws* in the South, African Americans experienced segregation in housing, employment, education, and all public accommodations. (These laws were referred as Jim Crow laws after a derogatory song about a black man.) African

| Table 10.1 | ODDS OF BECOMING A PROFESSIONAL ATHLETE BY RACE/ETHNICITY AND SPORT | | | |

Many young people dream of becoming a professional athlete; however, as this table shows, the odds of actually becoming one are very small.

| | RACE/ETHNICITY | | | |
	White	African American	Latino/a	Asian American
Football	1 in 62,500	1 in 47,600	1 in 2,500,000	1 in 5,000,000
Baseball	1 in 83,300	1 in 333,300	1 in 500,000	1 in 50,000,000
Basketball	1 in 357,100	1 in 153,800	1 in 33,300,000	–
Hockey	1 in 66,700	–	–	–
Golf				
Men's	1 in 312,500	1 in 12,500,000	1 in 33,300,000	1 in 20,000,000
Women's	1 in 526,300	–	1 in 33,300,000	1 in 3,300,000
Tennis				
Men's	1 in 285,700	1 in 2,000,000	1 in 3,300,000	–
Women's	1 in 434,800	1 in 20,000,000	1 in 20,000,000	–

Note: The odds of Native Americans participating in professional football are 1 in 12,500,000; they are not represented in other professional sports.

Source: Leonard and Reyman, 1988: 162–169.

Americans who did not stay in their "place" were often the victims of violent attacks and lynch mobs (Franklin, 1980). *Lynching* is a killing carried out by a group of vigilantes seeking revenge for an actual or imagined crime by the victim. Lynchings were used by whites to intimidate African Americans into staying "in their place." It is estimated that as many as 6,000 lynchings occurred between 1892 and 1921 (Feagin and Feagin, 2003). In spite of all odds, many African American women and men resisted oppression and did not give up in their struggle for equality (Amott and Matthaei, 1996).

Discrimination In the twentieth century, the lives of many African Americans were changed by industrialization and two world wars. When factories were built in the northern United States, many African American families left the rural South in hopes of finding jobs and a better life. As previously discussed, African Americans in the South experienced *de jure segregation,* which was enforced by law, whereas African Americans in the North encountered *de facto segregation,* which was enforced by custom. Many African American women left the South not only to seek employment but also to escape the sexual exploitation that was still being perpetrated by some white men (Amott and Matthaei, 1996).

During World Wars I and II, African Americans were a vital source of labor in war production indus-

tries; however, racial discrimination continued both on and off the job. In World War II, many African Americans fought for their country in segregated units in the military; after the war, they sought—and were denied—equal opportunities in the country for which they had risked their lives.

African Americans began to demand sweeping societal changes in the 1950s. Initially, the Reverend Dr. Martin Luther King, Jr., and the civil rights movement used *civil disobedience*—nonviolent action seeking to change a policy or law by refusing to comply with it—to call attention to racial inequality and to demand greater inclusion of African Americans in all areas of public life. Subsequently, leaders of the Black Power movement, including Malcolm X and Marcus Garvey, advocated black pride and racial awareness among African Americans. Elena Albert, an African American "amateur historian," describes how she was influenced by Black Power advocates:

When I was a young woman, I spent so much money and time to straighten my hair. My mother had hoped that I would speak in a refined low voice—that if I would become as much like the ideal of white womanhood, perhaps I might be accepted. Well, that's not necessary. And when Marcus Garvey said that we were black and beautiful, and when he said, "Up ye mighty race, you could do what ye will" (*sung exultantly*), that

Box 10.3 SOCIOLOGY AND SOCIAL POLICY

The U.S. Supreme Court and the Continuing Debate Over Affirmative Action

WASHINGTON, June 23, 2003—The Supreme Court ruled today that race can be a factor for universities shaping their admissions programs, saying a broad social value may be gained from diversity in the classroom.

But race cannot be an overriding factor for schools' admissions programs, the court ruled, saying that such plans can lead to unconstitutional policies. (CNN.com, 2003b)

When the U.S. Supreme Court rules on complex legal issues such as it recently did in two University of Michigan cases involving that school's affirmative action policies, many people are left wondering exactly what the court's decision really means and what effect, if any, it might have on them. The Court's rulings in *Grutter v. Bollinger* (upholding the admissions policies of the University of Michigan's law school) and *Gratz v. Bollinger* (declaring the undergraduate admissions policies of the same university unconstitutional), both of which were decided in 2003, are an example. What exactly was it that the court held with regard to affirmative action in higher education, and what effect could those two rulings have in other areas such as employment policies?

Both decisions dealt with the concepts of affirmative action and reverse discrimination. *Affirmative action* is a term that describes policies or procedures that are intended to promote equal opportunity for categories of people deemed to have been previously excluded from equality in education, employment, and other fields on the basis of characteristics such as race or ethnicity. Critics of affirmative action often assert that these policies amount to *reverse discrimination,* a term that describes a situation in which a person who is better qualified is denied en-

© 2003 AP/Wide World Photos

rollment in an educational program or employment in a specific position as a result of another person receiving preferential treatment as a result of affirmative action. Both decisions dealt with these concepts as they relate to higher education.

Education was one of the earliest targets of social policy pertaining to civil rights in the United States. Increased educational opportunity has been a goal of many subordinate-group members because of the widely held belief that education is the key to economic and social advancement. In the 1970s, some schools began to establish affirmative action pro-

made us look at ourself. And we are handsome, we are pretty, we are beautiful, we are lovely. (qtd. in Blauner, 1989: 312)

Gradually, racial segregation was outlawed by the courts and the federal government. For example, the Civil Rights Acts of 1964 and 1965 sought to do away with discrimination in education, housing, employment, and health care. Affirmative action programs were instituted in both public-sector and private-sector organizations in an effort to bring about greater opportunities for African Americans and other previ-

ously excluded groups. However, as discussed in Box 10.3, issues such as affirmative action remain highly controversial in the twenty-first century.

African Americans Today African Americans make up about 13 percent of the U.S. population. Some are descendants of families that have been in this country for many generations; others are recent immigrants from Africa and the Caribbean. Black Haitians make up the largest group of recent Caribbean immigrants; others come from Jamaica, Trinidad, and Tobago. Recent African immigrants are

grams, but one such program was challenged by the plaintiff (the person who files a lawsuit) in *Bakke v. The University of California at Davis* on the basis that it resulted in unconstitutional reverse discrimination since the university received public funding. Bakke, a white male, had been denied admission to the University of California at Davis medical school even though his grade point average and Medical College Admissions Test score were higher than those of some students of color who gained admission under a special program. The Supreme Court ruled that Bakke should be admitted to the medical school—that it was unconstitutional to reserve a fixed number of seats for members of subordinate groups—but that the Constitution did not prohibit *any* consideration of race or ethnicity in college admissions policies.

Based on this analysis of permissible race-conscious policies, most public and private colleges developed guidelines for admissions, financial aid, scholarships, and faculty hiring that took race, ethnicity, and gender into account. However, in 1996 the U.S. Fifth Circuit Court of Appeals (in *Hopwood v. State of Texas*) ruled that the University of Texas law school's admissions policies, under which some African Americans and other students of color were admitted despite the fact that they had lower scores on the Law School Admissions Test than did Hopwood and the other plaintiffs, discriminated against whites and were unconstitutional. The Eleventh Circuit Court of Appeals also struck down race-conscious college admissions criteria. As a result, colleges and universities in the states governed by those rulings largely dismantled existing affirmative action programs, and a dramatic drop in minority enrollment took place in many schools until some of them reassessed admissions criteria, such as whether a student was in the top ten percent of his or her high school class or had other unique attributes or skills that enhanced the overall quality of the student population.

In *Grutter v. Bollinger*, however, the Supreme Court upheld a law school admissions policy that took into account (but did not guarantee admission on the basis of) undergraduate grade point average and Law School Admissions Test score but also required admissions officials to look to other criteria, including "diversity which has the potential to enrich everyone's educa-tion. . . ." Under the policy, race and ethnicity could be taken into account in creating that diversity. A majority of the Court held that the university's "narrowly tailored use of race in admissions decisions" was permissible but also cautioned that the Court expected the use of racial preferences to no longer be necessary twenty-five years from now. In an opinion issued the same day, a majority of the Supreme Court held (in *Gratz v. Bollinger*) that an admissions policy under which an applicant from an "underrepresented racial or ethnic minority group" was entitled to about one-fifth of the total points necessary for admission based solely on subordinate-group status was unconstitutional.

Although *Grutter* has been hailed as a victory for affirmative action, it should be noted that the holding applies only to higher education, only to cases in which a college has an affirmative action policy, only to states that do not have a law prohibiting affirmative action in public higher education, and only to narrowly defined criteria for a limited period of time in the future.

Supreme Court decisions, such as those involving affirmative action, are one of the many ways in which social policy is established. However, thousands of other people also take part in the policy-making realm regarding affirmative action in college admissions, including the members of the state legislatures that pass laws on the subject, college administrators who draft guidelines to govern their individual institutions, and college admissions officials who establish and carry out the specific policies on their campuses. Should policy making regarding affirmative action be this decentralized, or should there be a single, national policy on this issue? What do you think?

---

primarily from Nigeria, Ethiopia, Ghana, and Kenya. They have been simultaneously "pushed" out of their countries of origin by severe economic and political turmoil and "pulled" by perceived opportunities for a better life in the United States. Recent immigrants are often victimized by the same racism that has plagued African Americans as a people for centuries.

Since the 1960s, many African Americans have made significant gains in politics, education, employment, and income. Between 1964 and 1999, the number of African Americans elected to political office increased from about 100 to almost 9,000 nationwide (Joint Center for Political and Economic Studies, 2000). African Americans won mayoral elections in many major cities that have large African American populations, such as Atlanta, Houston, New Orleans, Philadelphia, and Washington, D.C. Despite these political gains, African Americans still represent less than 3 percent of all elected officials in the United States.

Some African Americans have made impressive occupational gains and joined the ranks of professionals in the upper middle class. Others have achieved great wealth and fame as entertainers, professional athletes,

As more African Americans have made gains in education and employment, many of them have also made a conscious effort to increase awareness of African culture and to develop a sense of unity, cooperation, and self-determination. The seven-day celebration of Kwanzaa in late December and early January exemplifies this desire to maintain a distinct cultural identity.

Lawrence Migdale/Stock Boston

and entrepreneurs. However, even those who make millions of dollars a year and live in affluent neighborhoods are not always exempt from racial prejudice and discrimination. Although some African Americans have made substantial occupational and educational gains, many more have not. The African American unemployment rate remains twice as high as that of whites.

African Americans and Sports

In recent decades, many African Americans have seen sports as a possible source of upward mobility because other means have been unavailable. However, their achievements in sports have often been attributed to "natural ability" and not determination and hard work. Sociologists have rejected such biological explanations for African Americans' success in sports and have focused instead on explanations rooted in the social structure of society.

During the slavery era, a few African Americans gained better treatment and, occasionally, freedom by winning boxing matches on which their owners had bet large sums of money (McPherson, Curtis, and Loy, 1989). After emancipation, some African Americans found jobs in horse racing and baseball. For example, fourteen of the fifteen jockeys in the first Kentucky Derby (in 1875) were African Americans. A number of African Americans played on baseball teams; a few played in the Major Leagues until the Jim Crow laws forced them out. Then they formed their own "Negro" baseball and basketball leagues (Peterson, 1992/1970). After African Americans were forced out of horse racing in the late 1800s, boxing remained the only sport in which they were allowed to compete, and even there, they often had to agree

to lose before they could obtain a match (McPherson, Curtis, and Loy, 1989).

Since Jackie Robinson broke baseball's "color line" in 1947, many African American athletes have played collegiate and professional sports. Even now, however, winning games and earning profits are more important to many dominant-group coaches and owners than is providing equal opportunity for African Americans. Persistent class inequalities between whites and African Americans are reflected in the fact that, until recently, African Americans have primarily excelled in sports (such as basketball or football) that do not require much expensive equipment and specialized facilities in order to develop athletic skills (Coakley, 2004). According to one sports analyst, African Americans typically participate in certain sports and not others because of the *sports opportunity structure*—the availability of facilities, coaching, and competition in the schools and community recreation programs in their area (Phillips, 1993). Regardless of the sport in which they participate, African American men athletes continue to experience inequalities in assignment of playing positions, rewards and authority structures, and management and ownership opportunities in professional sports (Eitzen and Sage, 1997). For example, in 2002 only 2 of the 32 National Football League head coaches and only 4 of the 115 Division I-A head coaches in college football were African Americans (SportsLine.com, 2002).

Throughout the 1990s, representation of African American men in professional sports leagues remained at over 75 percent in the National Basketball Association, 57 percent in the National Football League, and 21 percent in Major League Baseball (Coakley, 2004). Depictions of African American men in the U.S.

media most frequently show black athletes in these sports and focus on the physical attributes that contribute to their success. Today, African Americans remain significantly underrepresented in other sports, including hockey, skiing, figure skating, golf, volleyball, softball, swimming, gymnastics, sailing, soccer, bowling, cycling, and tennis (Coakley, 2004).

Although African American women athletes tend to experience dual discrimination based on both race and gender, they have excelled in college and professional sports and in international competitions such as the Olympic Games.

White Ethnic Americans

The American Dream initially brought many white ethnics to the United States. The term *white ethnic Americans* is applied to a wide diversity of immigrants who trace their origins to Ireland and to Eastern and Southern European countries such as Poland, Italy, Greece, Germany, Yugoslavia, and Russia and other former Soviet republics. Unlike the WASPs, who immigrated primarily from Northern Europe and assumed a dominant cultural position in society, white ethnic Americans arrived late in the nineteenth century and early in the twentieth century to find relatively high levels of prejudice and discrimination directed at them by nativist organizations that hoped to curb the entry of non-WASP European immigrants. Because many of the people in white ethnic American categories were not Protestant, they experienced discrimination because they were Catholic, Jewish, or members of other religious bodies, such as the Eastern Orthodox churches (Farley, 1995).

Discrimination Against White Ethnics   Many white ethnic immigrants entered the United States between 1830 and 1924. Irish Catholics were among the first to arrive, with over 4 million Irish fleeing the potato famine and economic crisis in Ireland and seeking jobs in the United States (Feagin and Feagin, 2003). When they arrived, they found that British Americans controlled the major institutions of society. The next arrivals were Italians who had been recruited for low-wage industrial and construction jobs. British Americans viewed Irish and Italian immigrants as "foreigners": The Irish were stereotyped as ape-like, filthy, bad-tempered, and heavy drinkers, and the Italians were depicted as lawless, knife-wielding thugs looking for a fight—"dagos" and "wops" (short for "without papers") (Feagin and Feagin, 2003: 79–81, 92–94).

Both Irish Americans and Italian Americans were subjected to institutionalized discrimination in employment. Employment ads read "Help Wanted—No Irish Need Apply" and listed daily wages at $1.30–$1.50 for "whites" and $1.15–$1.25 for "Italians" (Gambino, 1975: 77). In spite of discrimination, white ethnics worked hard to establish themselves in the United States, often establishing mutual self-help organizations and becoming politically active (Mangione and Morreale, 1992).

Between 1880 and 1920, over 2 million Jewish immigrants arrived in the United States and settled in the Northeast. Jewish Americans differ from other white ethnic groups in that some focus their identity primarily on their religion whereas others define their Jewishness in terms of ethnic group membership (Feagin and Feagin, 2003). In any case, Jews continued to be the victims of *anti-Semitism*—prejudice, hostile attitudes, and discriminatory behavior targeted at Jews. For example, signs in hotels read "No Jews Allowed," and some "help wanted" ads stated "Christians Only" (Levine, 1992: 55). In spite of persistent discrimination, Jewish Americans achieved substantial success in many areas, including business, education, the arts and sciences, law, and medicine.

White Ethnics and Sports   Sports provided a pathway to assimilation for many white ethnics. The earliest collegiate football players who were not white Anglo-Saxon Protestants were of Irish, Italian, and Jewish ancestry. Sports participation provided educational opportunities that some white ethnics would not have had otherwise.

Boxing became a way to make a living for white ethnics who did not participate in collegiate sports. Boxing promoters encouraged ethnic rivalries to increase their profits, pitting Italians against Irish or Jews, and whites against African Americans (Levine, 1992; Mangione and Morreale, 1992). Eventually, Italian Americans graduated from boxing into baseball and football. Jewish Americans found that sports lessened the shock of assimilation and gave them an opportunity to refute stereotypes about their physical weaknesses and counter anti-Semitic charges that they were "unfit to become Americans" (Levine, 1992: 272).

Today, assimilation is so complete that little attention is paid to the origins of white ethnic athletes. As former Pittsburgh Steeler running back Franco Harris stated, "I didn't know I was part Italian until I became famous" (qtd. in Mangione and Morreale, 1992: 384).

Asian Americans

The U.S. Census Bureau uses the term *Asian Americans* to designate the many diverse groups with roots

Historically, Chinatowns in major U.S. cities have provided a safe haven and an economic enclave for many Asian immigrants. Some contemporary Chinese Americans reside in these neighborhoods, whereas others visit to celebrate cultural diversity and ethnic pride.

AP/Wide World Photos

in Asia. Chinese and Japanese immigrants were among the earliest Asian Americans. Many Filipinos, Asian Indians, Koreans, Vietnamese, Cambodians, Pakistani, and Indonesians have arrived more recently. Today, Asian Americans belong to the fastest-growing ethnic minority group in the United States.

Chinese Americans The initial wave of Chinese immigration occurred between 1850 and 1880, when over 200,000 Chinese men were "pushed" from China by harsh economic conditions and "pulled" to the United States by the promise of gold in California and employment opportunities in the construction of transcontinental railroads. Far fewer Chinese women immigrated; however, many of them were brought to the United States against their will and forced into prostitution, where they were treated like slaves (Takaki, 1993).

Chinese Americans were subjected to extreme prejudice and stereotyped as "coolies," "heathens," and "Chinks." Historian Ronald Takaki (1989: 13) describes the economic context in which this discrimination occurred:

> Unlike the Irish and other groups from Europe, Asian immigrants could not become "mere individuals, indistinguishable in the cosmopolitan mass of the population." Regardless of their personal merits, they sadly discovered, they could not gain acceptance in the larger society. They were judged not by the content of their character but by their complexion. . . . "Color" in America operated within an economic context. Asian immigrants came here to meet demands for labor—plantation workers, railroad crews, miners, factory operatives, cannery workers, and farm laborers. Employers developed a dual-wage system to pay Asian labor-

ers less than white workers and pitted the groups against each other in order to depress wages for both. Ethnic antagonism . . . led white laborers to demand the restriction of Asian workers already here in a segregated labor market of low-wage jobs and the exclusion of future Asian immigrants. Thus the class interests of white capital as well as white labor needed Asians as "strangers."

As Takaki notes, prejudice and discrimination ran high against Asian immigrants, leading to Congress's passage of the Chinese Exclusion Act of 1882, which brought Chinese immigration to a halt. The Exclusion Act was not repealed until World War II, when Chinese Americans who were contributing to the war effort by working in defense plants pushed for its repeal (Takaki, 1993). After immigration laws were further relaxed in the 1960s, the second and largest wave of Chinese immigration occurred, with immigrants coming primarily from Hong Kong and Taiwan. These recent immigrants have had more education and workplace skills than earlier arrivals, and brought families and capital with them to pursue the American Dream (Chen, 1992).

Today, many Chinese Americans live in large urban enclaves in California, New York, Hawaii, Illinois, and Texas. As a group, they have enjoyed considerable upward mobility. Some own laundries, restaurants, and other businesses; others have professional careers (Chen, 1992). However, many Chinese Americans, particularly recent immigrants, remain in the lower tier of the working class—providing low-wage labor in garment and knitting factories and Chinese restaurants.

Japanese Americans Most of the early Japanese immigrants were men who worked on sugar plantations in the Hawaiian Islands in the 1860s. Like Chinese immigrants, the Japanese American workers were viewed as a threat by white workers, and immigration of Japanese men was curbed in 1908. However, Japanese women were permitted to enter the United States for several years thereafter because of the shortage of women on the West Coast. Although some Japanese women married white men, this practice was stopped by laws prohibiting interracial marriage.

With the exception of the enslavement of African Americans, Japanese Americans experienced one of the most vicious forms of discrimination ever sanctioned by U.S. laws. During World War II, when the United States was at war with Japan, nearly 120,000 Japanese Americans were placed in internment camps, where they remained for more than two years despite the total lack of evidence that they posed a security threat to this country (Takaki, 1993). This action was a direct violation of the citizenship rights of many *Nisei* (second-generation Japanese Americans) who were born in the United States (see Daniels, 1993). Ironically, only Japanese Americans were singled out for such harsh treatment; German Americans avoided this fate even though the United States was also at war with Germany. Four decades later, the U.S. government issued an apology for its actions and eventually paid $20,000 each to some of those who had been placed in internment camps (Daniels, 1993; Takaki, 1993).

Since World War II, many Japanese Americans have been very successful. The median income of Japanese Americans is more than 30 percent above the national average. However, most Japanese Americans (and other Asian Americans) live in states that not only have higher incomes but also higher costs of living than the national average. In addition, many Asian American families have more persons in the paid labor force than do other families (Takaki, 1993).

Korean Americans
The first wave of Korean immigrants were male workers who arrived in Hawaii between 1903 and 1910. The second wave came to the U.S. mainland following the Korean War in 1954 and was made up primarily of the wives of servicemen and Korean children who had lost their parents in the war. The third wave arrived after the Immigration Act of 1965 permitted well-educated professionals to migrate to the United States. Korean Americans have helped one another open small businesses by pooling money through the *kye*—an association that grants members money on a rotating basis to gain access to more capital. According to Takaki (1989), Korean Americans were a hidden minority before 1965 because so few lived in the United States. After that

time, however, Korean Americans have become a very visible group in this country.

Today, many Korean Americans live in California and New York, where there is a concentration of Korean-owned grocery stores, businesses, and churches. Unlike earlier Korean immigrants, more recent arrivals have come as settlers and have brought their families with them. However, their experiences with other subordinate racial and ethnic groups have not always been harmonious. Ongoing discord has existed between African Americans and Korean Americans in New York and among African Americans, Latinos, and Korean Americans in California.

Filipino Americans
Today, Filipino Americans constitute the second largest category of Asian Americans, with over a million population in the United States. To understand the status of Filipino Americans, it is important to look at the complex relationship between the Philippine Islands and the United States government. After Spain lost the Spanish-American War, the United States established colonial rule over the islands, a rule that lasted from 1898 to 1946 (Feagin and Feagin, 2003). Despite control by the United States, Filipinos were not granted U.S. citizenship. But, like the Chinese and the Japanese, male Filipinos were allowed to migrate to Hawaii and the U.S. mainland to work in agriculture and in fish canneries in Seattle and Alaska. Like other Asian Americans, Filipino Americans were accused of taking jobs away from white workers and suppressing wages, and Congress restricted Filipino immigration to fifty people per year between the Great Depression and the aftermath of World War II.

The second wave of Filipino immigrants came following the Immigration Act of 1965, when large numbers of physicians, nurses, technical workers, and other professionals moved to the U.S. mainland. Most Filipinos have not had the start-up capital necessary to open their own businesses, and many have been employed in the low-wage sector of the service economy. However, the average household income of Filipino American families is relatively high because about 75 percent of Filipino American women are employed, and nearly half have a four-year college degree (Espiritu, 1995).

Indochinese Americans
Indochinese Americans include people from Vietnam, Cambodia, Thailand, and Laos, most of whom have come to the United States in the past three decades. Vietnamese refugees who had the resources to flee at the beginning of the Vietnam War were the first to arrive. Next came Cambodians and lowland Laotians, referred to

Until recently, few Asian Americans played on college or professional sports teams. After playing college football at Texas A&M University, linebacker Dat Nguyen became the first Vietnamese American signed by an NFL team, the Dallas Cowboys.

AP/Wide World Photos

as "boat people" by the media. Many who tried to immigrate did not survive at sea; others were turned back when they reached this country or were kept in refugee camps for long periods of time. When they arrived in the United States, inflation was high, the country was in a recession, and many native-born citizens feared that they would lose their jobs to these new refugees, who were willing to work very hard for low wages. The frustrations of many Indochinese American immigrants were expressed by a Hmong refugee from Laos:

> In our old country, whatever we had was made or brought in by our own hands; we never had any doubts that we would not have enough for our mouths. But from now on to the future, that time is over. We are so afraid and worried that there will be one day when we will not have anything for eating or paying the rent, and these days these things are always in our minds. . . . Don't know how to read or write, don't know how to speak the language. My life is only to live day by day until the last day I live, and maybe that is the time when my problems will be solved. (qtd. in Portes and Rumbaut, 1996: 155)

Today, many Indochinese Americans are foreign born; about half live in the western states, especially California. Even though most Indochinese immigrants spoke no English when they arrived in this country, some of their children have done very well in school and have been stereotyped as "brains."

Asian Americans and Sports   Until recently, Asian Americans received little recognition in sports. However, as women's athletic events, including ice

skating and gymnastics, have garnered more media coverage, names of persons such as Kristi Yamaguchi, Michelle Kwan, and Amy Chow have become widely known and often idolized by fans. As one sports analyst stated, "[These athletes] are of Asian descent, but more importantly they are Asian Americans whose actions reflect upon the United States. . . . As role models, particularly for the Asian-American community, they exemplify success, integrity, discipline and a dedicated work ethic" (Shum, 1997).

One Asian American athlete who has captured extensive media attention is football player Dat Nguyen (pronounced Win), who won numerous awards at Texas A&M and joined the Dallas Cowboys football team. Nguyen is the son of Vietnamese American parents who fled a war-torn Vietnam with five children and a sixth child (Dat) on the way and entered the United States as refugees in 1975. When asked by a sports reporter whether it was harder to overcome his lack of "football size" (he is 5'11" and weighed about 230 pounds at the time) or the fact that he was an Asian American in "big-time college football," Dat Nguyen replied: "That's hard to answer. It's probably been overcoming a lot of adversity and not just on the football field. People weren't expecting me to be where I am now and what I've accomplished" (qtd. in Silversten, 1998: 1). In the future, there will, no doubt, be many more well-known Asian American men and women participating in college and professional sporting events.

Latinos/as (Hispanic Americans)

The terms *Latino* (for males), *Latina* (for females), and *Hispanic* are used interchangeably to refer to people who trace their origins to Spanish-speaking Latin America and the Iberian peninsula (Cashmore, 1996). However, as racial–ethnic scholars have pointed out, the label *Hispanic* was first used by the U.S. government to designate people of Latin American and Spanish descent living in the United States (Oboler, 1995), and it has not been fully accepted as a source of identity by the more than 37 million Latinos/as who live in the United States today (Oboler, 1995; Romero, 1997). Instead, many of the people who trace their roots to Spanish-speaking countries think of themselves as Mexican Americans, Chicanos/as, Puerto Ricans, Cuban Americans, Salvadorans, Guatemalans, Nicaraguans, Costa Ricans, Argentines, Hondurans, Dominicans, or members of other categories. Many also think of themselves as having a combination of Spanish, African, and Native American ancestry.

Mexican Americans or Chicanos/as

Mexican Americans—including both native- and foreign-born people of Mexican origin—are the largest segment (approximately two-thirds) of the Latino population in the United States. Most Mexican Americans live in the southwestern region of the United States, although more have recently moved to the Midwest and the Washington, D.C., metropolitan area (Cashmore, 1996).

Immigration from Mexico is the primary vehicle by which the Mexican American population grew in this country. Initially, Mexican-origin workers came to work in agriculture, where they were viewed as a readily available cheap and seasonal labor force. Many initially entered the United States as undocumented workers ("illegal aliens"); however, they were more vulnerable to deportation than other illegal immigrants because of their visibility and the proximity of their country of origin. For more than a century, there has been a "revolving door" between the United States and Mexico that has been open when workers were needed and closed during periods of economic recession and high rates of U.S. unemployment.

The early experiences of Mexican Americans in this country were not always positive. In fact, many have experienced disproportionate poverty as a result of internal colonialism. Following the Mexican-American War, the United States seized land that had previously belonged to Mexico, and many formerly wealthy Mexican ranchers became impoverished farmhands. As sociologist Mary Romero (1997: 12) explains,

> The structure of opportunity for Chicanos . . . is rooted in the history of westward expansion, the geographical proximity and poverty of Mexico

that facilitate continued immigration, and the historical labor functions of Mexican workers in the U.S. economy. Capitalist penetration of the Southwest dispossessed Chicanos of their land, created a cheap labor force and brought about the eventual destruction or transformation of the indigenous social system governing the lives of the Mexican residents.

Mexican Americans have long been seen as a source of cheap labor, while—ironically—at the same time, they have been stereotyped as lazy and unwilling to work. As has been true of other groups, when white workers viewed Mexican Americans as a threat to their jobs, they demanded that the "illegal aliens" be sent back to Mexico. Consequently, U.S. citizens who happen to be Mexican American have been asked for proof of their citizenship, especially when anti-immigration sentiments are running high. Many Mexican American families have lived in the United States for four or five generations—they have fought in wars, made educational and political gains, and consider themselves to be solid U.S. citizens. Thus, it is a great source of frustration for them to be viewed as illegal immigrants or to be asked "How long have you been in this country?"

Puerto Ricans

When Puerto Rico became a possession of the United States in 1917, Puerto Ricans acquired U.S. citizenship and the right to move freely to and from the mainland. In the 1950s, many migrated to the mainland when the Puerto Rican sugar industry collapsed, settling primarily in New York and New Jersey. Although living conditions have improved substantially for some Puerto Ricans, life has been difficult for the many living in poverty in Spanish Harlem and other barrios. Nevertheless, in recent years Puerto Ricans have made dramatic advances in education, the arts, and politics. Increasing numbers have become lawyers, physicians, and college professors (see Rodriguez, 1989).

Cuban Americans

Cuban Americans live primarily in the Southeast, especially Florida. As a group, they have fared somewhat better than other Latinos/as because many Cuban immigrants were affluent professionals and businesspeople who fled Cuba after Fidel Castro's 1959 Marxist revolution. This early wave of Cuban immigrants has median incomes well above those of other Latinos/as; however, this group is still below the national average. The second wave of Cuban Americans, arriving in the 1970s, has fared worse. Many had been released from prisons and mental hospitals in Cuba, and their arrival fueled an

Pedro Martinez of the Boston Red Sox is one of the most visible Latino athletes. Latino and Latina sports figures have gained prominence in a wide variety of sports, including boxing, baseball, golf, and tennis.

© 2003 AP/Wide World Photos

upsurge in prejudice against all Cuban Americans. The more recent arrivals have developed their own ethnic and economic enclaves in Miami's Little Havana, and many of the earlier immigrants have become mainstream professionals and entrepreneurs.

Latinos/as and Sports For most of the twentieth century, Latinos have played Major League Baseball. Originally, Cubans, Puerto Ricans, and Venezuelans were selected for their light skin as well as for their skill as players (Hoose, 1989). Today, Latinos represent more than 20 percent of all major leaguers. If not for a 1974 U.S. Labor Department quota limiting how many foreign-born players can play professional baseball, this number might be even larger (Hoose, 1989).

Recently, Latinos in sports have gained more recognition as books and Web sites have been created to describe their accomplishments. For example, the Web site "Latino Legends in Sports" was created in 1999 to inform people about the contributions of Latino and Latina athletes (see **http://www.latinosportslegends .com**). Recently, the Web site has featured interviews with boxer Oscar De La Hoya, baseball pitcher Pedro Martinez, and Olympic speed skater Derek Parra (Latino Legends in Sports, 2003).

Education is a crucial issue for Latinos/as. Because of past discrimination and unequal educational opportunities, many Latinos/as currently have low levels of educational attainment. Many are unable to attend college or participate in collegiate sports, which is essential for being drafted in professional

sports other than baseball. Consequently, the overall number of Latinas/os in college and professional sports is low compared to the rest of the U.S. population who are in this age bracket.

Middle Eastern Americans

Since 1970, many immigrants have arrived in the United States from countries located in the "Middle East," which is the geographic region from Afghanistan to Libya and includes Arabia, Cyprus, and Asiatic Turkey. Placing people in the "Middle Eastern" American category is somewhat like placing wide diversities of people in the categories of Asian American or Latino/a; some U.S. residents trace their origins to countries such as Bahrain, Egypt, Iran, Iraq, Kuwait, Lebanon, Oman, Qatar, Saudi Arabia, Syria, UAE (United Arab Emirates), and Yemen. Middle Eastern Americans speak a variety of languages and have diverse religious backgrounds: Some are Muslim, some are Coptic Christian, and others are Melkite Catholic. Although some are from working-class families, Lebanese Americans, Syrian Americans, Iranian Americans, and Kuwaiti Americans primarily come from middle- and upper-income family backgrounds. For example, Iranian Americans are scientists, professionals, and entrepreneurs.

In cities across the United States, Muslims have established social, economic, and ethnic enclaves. On the Internet, they have created web sites that provide information about Islamic centers, schools, and lists of businesses and services available from those who adhere to Islam, one of the fastest growing religions in this country (see Chapter 17, "Religion"). In cities such as Seattle, incorporation into the economic mainstream has been relatively easy for Palestinian immigrants who left their homeland in the 1980s. Some have found well-paid employment with corporations such as Microsoft because they bring educational skills and talents to the information-based economy, including the ability to translate software into Arabic for Middle Eastern markets (M. Ramirez, 1999). In the United States, Islamic schools and centers often bring together people from a diversity of countries such as Egypt and Pakistan. Many Muslim leaders and parents focus on how to raise children to be good Muslims and good U.S. citizens. However, recent immigrants continue to be torn between establishing roots in the United States and the continuing divisions and strife that exist in their homelands. Some Middle Eastern Americans experience prejudice and discrimination based on their speech patterns, appearance (such as the *hijabs,* or "head-to-toe covering" that leaves only the face exposed, which many girls and women wear), or the assumption that "all Middle Easterners" are somehow associated with terrorism.

Following the September 11, 2001, attacks on the United States by terrorists whose origins were traced to the Middle East, hate crimes and other forms of discrimination against people who were assumed to be Arabs, Arab Americans, or Muslims escalated in this country. With the passage of the U.S. Patriot Act—a law giving the federal government greater authority to engage in searches and surveillance with less judicial review than previously—in the aftermath of the terrorist attacks, many Arab Americans have expressed concern that this law is being used to target people who appear to be of Middle Eastern origins. To counter this potential oppression and loss of rights, some Arab Americans have engaged in social activism to highlight their concerns (Feagin and Feagin, 2003).

Middle Eastern Americans and Sports   Although more Islamic schools are beginning to focus on sports, particularly for teenage boys, there has been less emphasis on competitive athletics among many Middle Eastern Americans. Based on popular sporting events in their countries of origin, some Middle Eastern Americans play golf or soccer. As well, some Iranian Americans follow the soccer careers of professional players from Iran, who now play for German, Austrian, Belgian, and Greek clubs. Keeping up with global sporting events is easy with all-sports television cable channels and Web sites that provide up-to-the-

minute information about players and competitions. Over time, there will probably be greater participation by Middle Eastern American males in competitions such as soccer and golf; however, girls and women in Muslim families are typically not allowed to engage in athletic activities. Although little research has been done on this issue in the United States, one study of Islamic countries in the Middle East found that female athletes face strong cultural opposition to their sports participation (Dupre and Gains, 1997).

GLOBAL RACIAL AND ETHNIC INEQUALITY IN THE FUTURE

Throughout the world, many racial and ethnic groups seek *self-determination*—the right to choose their own way of life. As many nations are currently structured, however, self-determination is impossible.

Worldwide Racial and Ethnic Struggles

The cost of self-determination is the loss of life and property in ethnic warfare. In recent years, the Cold War has given way to dozens of smaller wars over ethnic dominance. In Europe, for example, ethnic violence has persisted in Yugoslavia, Spain, Britain (between the Protestant majority and the Catholic minority in Northern Ireland), Romania, Russia, Moldova, and Georgia. Ethnic violence continues in the Middle East, Africa, Asia, and Latin America. Hundreds of thousands have died from warfare, disease (such as the cholera epidemic in war-torn Rwanda), and refugee migration.

Ethnic wars have a high price even for survivors, whose life chances can become bleaker even after the violence subsides. In ethnic conflict between Abkhazians and Georgians in the former Soviet Union, for example, as many as two thousand people have been killed and over eighty thousand displaced. More recently, ethnic hatred has devastated the province of Kosovo, which is located in Serbia—Yugoslavia's dominant republic—and brought about the deaths of thousands of ethnic Albanians (Bennahum, 1999).

In the twenty-first century, the struggle between the Israeli government and various Palestinian factions over the future and borders of Palestine continues to

Box 10.4 YOU CAN MAKE A DIFFERENCE

Working for Racial Harmony

Suppose that you are talking with several friends about a series of racist incidents at your college. Having studied the sociological imagination, you decide to start an organization similar to No Time to Hate, which was started at Emory University several years ago to reduce racism on campus. In analyzing racism, your group identifies factors contributing to the problem: (1) divisiveness between different cultural and ethnic communities, (2) persistent lack of trust, (3) the fact that many people never really communicate with one another, (4) the need to bring different voices into the curriculum and college life generally, and (5) the need to learn respect for people from different backgrounds (Loeb, 1994). Your group also develops a set of questions to be answered regarding racism on campus:

- *Encouraging inclusion and acceptance.* Do members of our group reflect the college's racial and ethnic diversity? How much do I know about other people's history and culture? How can I become more tolerant—or accepting—of people who are different from me?
- *Raising consciousness.* What is racism? What causes it? Can people participate in racist language and behavior without realizing what they are doing? What is our college or university doing to reduce racism?
- *Becoming more self-aware.* How much do I know about my own family roots and ethnic background? How do the families and communities in which we grow up affect our perceptions of racial and ethnic relations?
- *Using available resources.* What resources are available for learning more about working to reduce racism? Here are some agencies to contact:

- ACLU (American Civil Liberties Union), 132 West 43rd Street, New York, NY 10036. (212) 944-9800. Online:

http://www.aclu.org

- ADL (Anti-Defamation League of B'nai B'rith), 823 United Nations Plaza, New York, NY 10017. (212) 490-2525. Online:

http://www.adl.org

- NAACP (National Association for the Advancement of Colored People), 4805 Mt. Hope Drive, Baltimore, MD 21215. (410) 358-8900. Online:

http://www.naacp.org

- National Council of La Raza, 1119 19th, NW, Suite 1000, Washington, DC 20036. Online:

http://www.nclr.org

Also available online:

- The American Studies Web/Race and Ethnicity provides links to related Web sites on race and ethnicity in the United States. It also includes information on African Americans, Native Americans, Latinos/as and Chicanas/os, and Asian Americans:

http://www.georgetown.edu/crossroads/asw

What additional items would you add to the list of problem areas on your campus? How might your group's objective be reached? Over time, many colleges and universities have been changed as a result of involvement by students like you!

make headlines. Discord in this region has heightened tensions among people not only in Israel and Palestine but also in the United States and around the world as deadly clashes continue and political leaders are apparently unable to reach a lasting solution to the decades-long strife.

Growing Racial and Ethnic Diversity in the United States

Racial and ethnic diversity is increasing in the United States. African Americans, Latinos/as, Asian Americans,

and Native Americans constitute one-fourth of the U.S. population, whereas whites are a shrinking percentage of the population. In the year 2000, white Americans made up 70 percent of the population, in contrast to 80 percent in 1980. It is predicted that by 2056, the roots of the average U.S. resident will be in Africa, Asia, Hispanic countries, the Pacific islands, and the Middle East—not white Europe (Henry, 1990).

What effect will these changes have on racial and ethnic relations? Several possibilities exist. On the one hand, conflicts may become more overt and confrontational as people continue to use *sincere fictions*—personal beliefs that reflect larger societal mythologies, such as "I am not a racist" or "I have never discriminated against anyone"—even when these are inaccurate perceptions (Feagin and Vera, 1995). Interethnic tensions may increase as competition for education, jobs, and other resources continues to grow.

On the other hand, there is reason for cautious optimism. Throughout U.S. history, members of diverse racial and ethnic groups have struggled to gain the freedom and rights that were previously withheld from them. Today, minority grass-roots organizations are pressing for affordable housing, job training, and educational opportunities (Feagin and Feagin, 2003). As discussed in Box 10.4, movements composed of both whites and people of color continue to oppose racism in everyday life, to seek to heal divisions among racial groups, and to teach children about racial tolerance (Rutstein, 1993). Many groups hope not only to af-

fect their own microcosm but also to contribute to worldwide efforts to end racism (Ford, 1994).

To eliminate racial discrimination, it will be necessary to equalize opportunities in schools and workplaces. As Michael Omi and Howard Winant (1994: 158) have emphasized,

> Today more than ever, opposing racism requires that we notice race, not ignore it, that we afford it the recognition it deserves and the subtlety it embodies. By noticing race we can begin to challenge racism, with its ever-more-absurd reduction of human experience to an essence attributed to all without regard for historical or social context. . . . By noticing race we can develop the political insight and mobilization necessary to make the U.S. a more racially just and egalitarian society.

CHAPTER REVIEW

■ How do race and ethnicity differ?

A race is a category of people who have been singled out as inferior or superior, often on the basis of physical characteristics such as skin color, hair texture, or eye shape. An ethnic group is a collection of people distinguished primarily by cultural or national characteristics, including unique cultural traits, a sense of community, a feeling of ethnocentrism, ascribed membership, and territoriality.

■ What are dominant and subordinate groups?

A dominant group is an advantaged group that has superior resources and rights in society. A subordinate group is a disadvantaged group whose members are subjected to unequal treatment by the majority group. Use of the terms *dominant* and *subordinate* reflects the importance of power in relationships.

■ How is prejudice related to discrimination?

Prejudice is a negative attitude often based on stereotypes, which are overgeneralizations about the appearance, behavior, or other characteristics of all members of a group. Discrimination involves actions or practices of dominant-group members that have a harmful impact on members of a subordinate group. Whereas prejudice involves attitudes, discrimination involves actions. Discriminatory actions range from name-calling to violent actions; genocide is the ultimate form of discrimination.

■ What are the major psychological explanations of prejudice?

According to the frustration–aggression hypothesis of prejudice, people frustrated in their efforts to achieve a highly desired goal may respond with aggression toward others, who then become scapegoats. Another theory of prejudice focuses on the authoritarian personality, marked by excessive conformity, submissiveness to authority, intolerance, insecurity, superstition, and rigid thinking.

■ How do individual discrimination and institutional discrimination differ?

Individual discrimination involves actions by individual members of the dominant group that harm members of subordinate groups or their property. Institutional discrimination involves day-to-day practices of organizations and institutions that have a harmful impact on members of subordinate groups.

■ How do sociologists view racial and ethnic group relations?

Symbolic interactionists suggest that increased contact between people from divergent groups should lead to favorable attitudes and behavior when members of each group (1) have equal status, (2) pursue the same goals, (3) cooperate with one another to achieve goals, and (4) receive positive feedback when they interact with one another. Functionalists stress that members of subordinate groups become a part of the mainstream through assimilation, the process by which members of subordinate groups become absorbed into the dominant culture. Conflict theorists focus on economic stratification and access to power in race and ethnic relations. The caste perspective views inequality as a permanent feature of society, whereas class perspectives focus on the link between capitalism and racial exploitation. According to racial formation theory, the actions of the U.S. government substantially define racial and ethnic relations.

■ **How have the experiences of various racial-ethnic groups differed in the United States?**

Native Americans suffered greatly from the actions of European settlers, who seized their lands and made them victims of forced migration and genocide. Today, they lead lives characterized by poverty and lack of opportunity. White Anglo-Saxon Protestants are the most privileged group in the United States, although social class and gender affect their life chances. White ethnic Americans, whose ancestors migrated from Southern and Eastern European countries, have gradually made their way into the mainstream of U.S. society. Following the abolishment of slavery through the Emancipation Proclamation in 1863, African Americans were still subjected to segregation, discrimination, and lynchings. More recently, despite civil rights legislation and economic and political gains by many African Americans, racial prejudice and discrimination still exist. Asian American immigrants as a group have enjoyed considerable upward mobility in U.S. society in recent decades, but many Asian Americans still struggle to survive by working at low-paying jobs and living in urban ethnic enclaves. Although some Latinos/as have made substantial political, economic, and professional gains in U.S. society, as a group they still are subjected to anti-immigration sentiments. Middle Eastern immigrants to the United States speak a variety of languages and have diverse religious backgrounds. Because they generally come from middle-class backgrounds, they have made inroads into mainstream U.S. society.

KEY TERMS

assimilation 330
authoritarian personality 327
discrimination 327
dominant group 324
ethnic group 320
ethnic pluralism 331
genocide 328
individual discrimination 328
institutional discrimination 328
internal colonialism 332

prejudice 324
race 320
racism 325
scapegoat 327
segregation 331
social distance 327
split labor market 332
stereotypes 324
subordinate group 324

QUESTIONS FOR CRITICAL THINKING

1. Do you consider yourself defined more strongly by your race or by your ethnicity? How so?
2. Given that subordinate groups have some common experiences, why is there such deep conflict between some of these groups?
3. What would need to happen in the United States, both individually and institutionally, for a positive form of ethnic pluralism to flourish in the twenty-first century?

RESOURCES ON THE INTERNET

Chapter-Related Web Sites

The following Web sites have been selected for their relevance to the topics in this chapter. These sites are among the more stable, but please note that Web site addresses change frequently. For an updated list of chapter-related Web sites with URL links, please visit the *Sociology in Our Times* Web site (**www.wadsworth.com/KendallSIOT**).

The Race and Ethnicity Collection

http://eserver.org/race/resources.html

This site consists of reference materials, articles, literary works, and additional resources that address issues of race and ethnicity in the United States. The site also includes a useful search engine to help you locate specific material of interest.

American Studies Crossroads Project

http://www.georgetown.edu/crossroads

Maintained by Crossroads, a project of the American Studies Association (sponsored by Georgetown University), this site houses the largest bibliography of Web-based resources in the field of American studies. On the opening page, click on "New American Studies Web," scroll down the page, and select "Race and Ethnicity." This link will connect you to a well-organized selection of resources related to the study of race and ethnicity in the United States.

ONLINE STUDY AND RESEARCH TOOLS

Accompanying this text are many *free* powerful online study tools that will help you master the material in this chapter, help increase your depth of understanding, and help you make the grade!

SocCoach CD-ROM

Use the SocCoach CD-ROM enclosed with this text to help you formulate a customized study plan for this chapter. After you take the Diagnostic Quiz, SocCoach will generate a customized study plan just for you! It will identify sections of the chapter that you should review and will provide videos, charts, graphs, and excerpts from the text to supplement your studies and enhance your understanding. You'll also find fun, interactive activities such as Virtual Explorations and Map the Stats to apply what you've learned and stretch your sociological imagination.

The Companion Web Site for Sociology in Our Times, *Fifth Edition*
www.wadsworth.com/KendallSIOT

Gain an even better grasp on this chapter by going to the companion Web site to take one of the Tutorial Quizzes, use the Flash Cards to master key terms, or check out the many other study aids you'll find there. You'll also find special features such as GSS Data and Census 2000 information that'll put data and resources at your fingertips to help you with that special project or help you as you do some research on your own.

In this chapter, when you see the icon on the left, it alerts you to a specific exercise found in *Wadsworth's Sociology Online Resources and Writing Companion.* This valuable guide shows you how to use Wadsworth's exclusive online resources—*InfoTrac College Edition,* the *Opposing Viewpoints Resource Center,* and *MicroCase Online*—to assist you in your study of sociology and to build essential research and writing skills.

Sex and Gender

Today was my first appointment with Dr. Gold [a psychiatrist]. . . . Dr. Gold asked if I thought the girls at school who diet are overweight. It was such a stupid question. . . . "*Of course* they aren't overweight, didn't I already say that they were *popular*?" . . . Dr. Gold . . . gave me [some] paper and asked me to draw my "ideal" of what I wanted to look like. . . . So I picked up the pencil and drew a girl I want to look like. She was tall and skinny, but she had my face and hair. When Dr. Gold took the drawing back, he didn't nod. "This is a stick figure," he said, like I didn't understand the assignment the first time. "Try to draw a *realistic* picture of how you'd like to look. Don't worry if you aren't very good at art." He must have thought I was terrible at art. I tried explaining how that was *exactly* the way I wanted to look, but Dr. Gold said I wouldn't be alive if I looked like that drawing. . . .

[During another appointment, Dr. Gold decided that I had to be hospitalized so that I would gain weight. At the hospital, I had to follow a bunch of rules, like getting weighed every morning, being watched by a nurse during meals, not being allowed to exercise, and not being allowed to look in a full-length mirror. After I began eating more, I started being allowed more privileges, however, one of which was being allowed out of the hospital to attend my brother's graduation.]

After the ceremony, we all went out to dinner at this fancy prime rib place, [and] my stomach started hurting because I wasn't used to eating so much at the hospital. I had to go the bathroom pretty bad. . . . The first thing I saw when I walked into the bathroom was a really skinny girl who looked just like the pictures in Dr. Katz's book. Whoever decorated the ladies' bathroom was madly in love with mirrors, so I kept seeing the skinny girl everywhere. *She* was a stick figure, . . . but when I turned around to look, I was the only person in the bathroom. Except the girl couldn't be me because she was so skinny. . . . So I walked closer to the sink to get a better look, but the skinny legs in the mirror kept moving at the same time as mine. Then I smiled, and so did the mouth in the mirror. . . . Finally I turned on the hot water and washed all [the] makeup off my face. That's when I knew for sure it was me. I couldn't believe it! I looked disgusting. I used to want to be a stick figure, but now I'm not so sure.

—Lori Gottlieb, who weighed only fifty-five pounds at age eleven, describing her experience with anorexia nervosa (Gottlieb, 2000: 112-114, 151-152, 204-208)

© Duncan Smith/Photodisc Green-Getty Images

I s Lori's situation an isolated case of an eating disorder? Her comments are similar to those found in numerous sociological studies, including research by the sociologist Shar-lene Hesse-Biber (1996), who found that many college women were extremely critical of their weight and shape. As Lori and many other women are aware, people who deviate significantly from existing weight and appearance norms are often devalued and objectified by others. *Objectification* is the process of treating people as if they were objects or things, not human beings. We objectify people when we judge them on the basis of their physical appearance rather than on the basis of their individual qualities or actions (Schur, 1983). Although men may be objectified, objectification of women is especially common in the United States and other nations (see Table 11.1).

According to Hesse-Biber, college women—more than men—express dissatisfaction with their body weight and shape. She discovered several distinct differences between women and men in their perceptions of body weight:

- Women overestimated their weights and thought their bodies were heavier than the medically desirable weights, whereas men judged their weights more accurately.
- When Hesse-Biber asked her sample of students how much they wanted to gain or lose, the vast majority of women (95 percent) wanted to lose weight whereas the men were almost evenly split between wanting to lose and wanting to gain weight.
- Over three-quarters of the females in her sample, but less than one-third of the males, ever dieted.
- When Hesse-Biber asked the question "When you look in the mirror, which best describes your feelings: proud, content, neutral, anxious, depressed, or repulsed?" she found that 50 percent of the men in her sample were at least content with their body image compared with only 37 percent of the women.
- Many women expressed anxiety about their bodies (28 percent of the women compared to 6 percent of the men). About 7 percent of the women felt depressed and repulsed by their bodies compared with 4 percent of the men.

Other studies have found that both men and women may have negative perceptions about their body size, weight, and appearance (Heywood and Dworkin, 2003). Many men compare themselves unfavorably to muscular bodybuilders and believe that they need to gain weight or muscularity, which for some is associated with masculinity and power (Basow, 1992; Klein, 1993). For women, however, body image is an even greater concern. Women may compare themselves unfavorably to slender stars of film and television, and believe that they need to lose weight. Men are less likely to let concerns about appearance affect how they feel about their own competence, worth, and abilities; among women, dislike of their bodies may affect self-esteem and feelings of self-worth.

Why do women and men feel differently about their bodies? Cultural differences in appearance norms may explain women's greater concern; they tend to be judged more harshly, and they know it (Schur, 1983). Throughout life, men and women receive different cultural messages about body image, food, and eating. Men are encouraged to eat, whereas women are made to feel guilty about eating (Basow, 1992). Hesse-Biber (1996: 11) refers to this phenomenon as the *cult of thinness,* in which people worship the "perfect" body and engage in rituals such as dieting and exercising with "obsessive attention to monitoring progress—weighing the body at least once a day and constantly checking calories." The main criterion for joining the cult of thinness is being female (Hesse-Biber, 1996). Similarly, the authors of a recent study describing "the confident, empowered female athlete" as a new cultural icon acknowledge that eating disorders remain an "occupational hazard of appearance-based sports like figure skating, gymnastics, long-distance running, or even swimming . . ." (Heywood and Dworkin, 2003: 48).

Body image is only one example of the many socially constructed differences between men and women—differences that relate to gender (a social concept) rather than to a person's biological makeup, or sex. In this chapter, we examine the issue of gender: what it is and how it affects us. Before reading on, test your knowledge about body image and gender by taking the quiz in Box 11.1.

QUESTIONS AND ISSUES

Chapter Focus Question: How do expectations about female and male appearance, and especially weight, reflect gender inequality?

How do a society's resources and economic structure influence gender stratification?

What are the primary agents of gender socialization?

How does the contemporary workplace reflect gender stratification?

How do functionalist, conflict, and feminist perspectives on gender stratification differ?

Table 11.1   THE OBJECTIFICATION OF WOMEN

GENERAL ASPECTS OF OBJECTIFICATION	OBJECTIFICATION BASED ON CULTURAL PREOCCUPATION WITH "LOOKS"
Women are responded to primarily as "females," whereas their personal qualities and accomplishments are of secondary importance.	Women are often seen as the objects of sexual attraction, not full human beings—for example, when they are stared at.
Women are seen as "all alike."	Women are seen by some as depersonalized body parts—for example, "a piece of ass."
Women are seen as being subordinate and passive, so things can easily be "done to a woman"—for example, discrimination, harassment, and violence.	Depersonalized female sexuality is used for cultural and economic purposes—such as in the media, advertising, fashion and cosmetics industries, and pornography.
Women are seen as easily ignored or trivialized.	Women are seen as being "decorative" and status-conferring objects to be bought (sometimes collected) and displayed by men and sometimes by other women.
	Women are evaluated according to prevailing, narrow "beauty" standards and often feel pressure to conform to appearance norms.

Source: Schur, 1983.

No wonder many women are extremely concerned about body image; even billboards communicate cultural messages about women's appearance.

SEX: THE BIOLOGICAL DIMENSION

Whereas the word *gender* is often used to refer to the distinctive qualities of men and women (masculinity and femininity) that are culturally created, **sex refers to the biological and anatomical differences between females and males.** At the core of these differences is the chromosomal information transmitted at the moment a child is conceived. The mother contributes an X chromosome and the father either an X (which produces a female embryo) or a Y chromosome (which produces a male embryo). At birth, male and female infants are distinguished by **primary sex characteristics: the genitalia used in the reproductive process.** At puberty, an increased production of hormones results in the development of **secondary sex characteristics: the physical traits (other than reproductive organs) that identify an individual's sex.** For women, these include larger breasts, wider hips, and narrower shoulders; a layer of fatty tissue throughout the body; and menstruation. For men, they include development of enlarged genitals, a deeper voice, greater height, a more muscular build, and more body and facial hair.

Hermaphrodites/Transsexuals

Sex is not always clear-cut. Occasionally, a hormone imbalance before birth produces a **hermaphrodite—a person in whom sexual differentiation is ambiguous or incomplete.** Hermaphrodites tend to have some combination of male and female genitalia. In one case, for example, a chromosomally normal (XY) male was born with a penis just one centimeter long and a urinary opening similar to that of a female. Some people may be genetically of one sex but have a gender identity of the other. That is true for a **transsexual, a**

Box 11.1 SOCIOLOGY AND EVERYDAY LIFE

How Much Do You Know About Body Image and Gender?

True	False	
T	F	1. Most people have an accurate perception of their own physical appearance.
T	F	2. Recent studies show that up to 95 percent of men express dissatisfaction with some aspect of their bodies.
T	F	3. Many young girls and women believe that being even slightly "overweight" makes them less "feminine."
T	F	4. Physical attractiveness is a more central part of self-concept for women than for men.
T	F	5. Virtually no men have eating problems such as anorexia and bulimia.
T	F	6. Thinness has always been the "ideal" body image for women.
T	F	7. Women bodybuilders have gained full acceptance in society.
T	F	8. In school, boys are more likely than girls to ridicule people about their appearance.
T	F	9. Most states have laws prohibiting employment discrimination on the basis of weight.
T	F	10. Young girls and women very rarely die as a result of anorexia or bulimia.

Answers on page 360.

person in whom the sex-related structures of the brain that define gender identity are opposite from the physical sex organs of the person's body. Consequently, transsexuals often feel that they are the opposite sex from that of their sex organs. Transsexuals may become aware of this conflict between gender identity and physical sex as early as the preschool years. Some transsexuals take hormone treatments or have a sex change operation to alter their genitalia in order to achieve a body congruent with their sense of sexual identity (Basow, 1992). Many transsexuals who receive hormone treatments or undergo surgical procedures go on to lead lives that they view as being compatible with their true sexual identity.

Western societies acknowledge the existence of only two sexes; some other societies recognize three—men, women, and *berdaches* (or *hijras* or *xaniths*), biological males who behave, dress, work, and are treated in most respects as women. The closest approximation of a third sex in Western societies is a **transvestite, a male who lives as a woman or a female who lives as a man but does not alter the genitalia.** Although transvestites are not treated as a third sex, they often "pass" for members of that sex because their appearance and mannerisms fall within the range of what is expected from members of the other sex (Lorber, 1994).

Transsexuality may occur in conjunction with homosexuality, but this is frequently not the case. Some researchers believe that both transsexuality and homosexuality have a common prenatal cause such as a critically timed hormonal release due to stress in the mother or the presence of certain hormone-mimicking chemicals during critical steps of fetal development. Researchers continue to examine this issue and debate the origins of transsexuality and homosexuality.

Sexual Orientation

Sexual orientation **refers to an individual's preference for emotional–sexual relationships with members of the opposite sex (heterosexuality), the same sex (homosexuality), or both (bisexuality)** (Lips, 2001). Some scholars believe that sexual orientation is rooted in biological factors that are present at birth; others believe that sexuality has both biological and social components and is not preordained at birth.

The terms *homosexual* and *gay* are most often used in association with males who prefer same-sex relationships; the term *lesbian* is used in association with females who prefer same-sex relationships. Heterosexual individuals, who prefer opposite-sex relationships, are sometimes referred to as *straight*. However, it is important to note that heterosexual people are much less likely to be labeled by their sexual orientation than are people who are gay, lesbian, or bisexual.

What criteria do social scientists use to classify individuals as gay, lesbian, or homosexual? In a definitive study of sexuality in the mid-1990s, researchers at the University of Chicago established three criteria for identifying people as homosexual or bisexual: (1) *sexual attraction* to persons of one's own gender, (2) *sexual involvement* with one or more persons of one's own gender, and (3) *self-identification* as a gay, lesbian, or bisexual (Michael et al., 1994). According to these criteria, then, having engaged in a homosexual act does not necessarily classify a person as homosexual. In fact, many respondents in the University of Chicago study indicated that although they had at least one homosexual encounter when they were younger, they were no longer involved in homosexual conduct and never identified themselves as gay, lesbian, or bisexual.

Recent studies have examined how sexual orientation is linked to identity. Sociologist Kristin G. Esterberg (1997) interviewed lesbian and bisexual women to determine how they "perform" lesbian or bisexual identity through daily activities such as choice of clothing and hairstyles, as well as how they use body language and talk. According to Esterberg (1997), some of the women viewed themselves as being "lesbian from birth" whereas others had experienced shifts in their identities, depending on social surroundings, age, and political conditions at specific periods in their lives. Another study looked at gay and bisexual men. Human development scholar Ritch C. Savin-Williams (2004) found that gay/bisexual youths often believe from an early age that they are different from other boys:

> The pattern that most characterized the youths' awareness, interpretation, and affective responses to childhood attractions consisted of an overwhelming desire to be in the company of men. They wanted to touch, smell, see, and hear masculinity. This awareness originated from earliest childhood memories; in this sense, they "always felt gay."

However, most of the boys and young men realized that these feelings were not typical of other males and were uncomfortable when others attempted to make them conform to the established cultural definitions of masculinity, such as showing a great interest in team sports, competition, and aggressive pursuits.

Recently, the term *transgender* was created to describe individuals whose appearance, behavior, or self-identification does not conform to common social rules of gender expression. Transgenderism is sometimes used to refer to those who cross-dress, to transsexuals, and to others outside mainstream categories. Although some gay and lesbian advocacy groups oppose the concept of transgender as being somewhat meaningless, others applaud the term as one that might help unify diverse categories of people based on sexual identity. Various organizations of gays, lesbians, and transgendered persons have been unified in their desire to reduce hate crimes and other forms of **homophobia—extreme prejudice directed at gays, lesbians, bisexuals, and others who are perceived as not being heterosexual.**

GENDER: THE CULTURAL DIMENSION

Gender **refers to the culturally and socially constructed differences between females and males found in the meanings, beliefs, and practices associated with "femininity" and "masculinity."** Although biological differences between women and men are very important, in reality most "sex differences" are socially constructed "gender differences." According to sociologists, social and cultural processes, not biological "givens," are the most important factors in defining what females and males are, what they should do, and what sorts of relations do or should exist between them. Sociologist Judith Lorber (1994: 6) summarizes the importance of gender:

> Gender is a human invention, like language, kinship, religion, and technology; like them, gender organizes human social life in culturally patterned ways. Gender organizes social relations in everyday life as well as in the major social structures, such as social class and the hierarchies of bureaucratic organizations.

Virtually everything social in our lives is *gendered:* People continually distinguish between males and females and evaluate them differentially. Gender is an integral part of the daily experiences of both women and men (Kimmel and Messner, 2004).

A microlevel analysis of gender focuses on how individuals learn gender roles and acquire a gender identity. *Gender role* **refers to the attitudes, behavior, and activities that are socially defined as appropriate for each sex and are learned through the**

Box 11.1 SOCIOLOGY AND EVERYDAY LIFE

Answers to the Sociology Quiz on Body Image and Gender

1. **False.** Many people do not have a very accurate perception of their own bodies. For example, many young girls and women think of themselves as "fat" when they are not. Some young boys and men tend to believe that they need a well-developed chest and arm muscles, broad shoulders, and a narrow waist.

2. **True.** In recent studies, up to 95 percent of men believed they needed to improve some aspect of their bodies.

3. **True.** More than half of all adult women in the United States are currently dieting, and over three-fourths of normal-weight women think they are "too fat." Recently, very young girls have developed similar concerns. For example, 80 percent of fourth-grade girls in one study were watching their weight.

4. **True.** Women have been socialized to believe that being physically attractive is very important. Studies have found that weight and body shape are the central determinants of women's perception of their physical attractiveness.

5. **False.** Some men do have eating problems such as anorexia and bulimia. These problems have been found especially among gay men and male fashion models and dancers.

6. **False.** The "ideal" body image for women has changed a number of times. A positive view of body fat has prevailed for most of human history; however, in the twentieth century in the United States, this view gave way to "fat aversion."

7. **False.** Although bodybuilding among women has gained some degree of acceptance, women bodybuilders are still expected to be very "feminine" and not to overdevelop themselves.

8. **True.** Boys are especially likely to ridicule girls whom they perceive to be "unattractive" or overweight.

9. **False.** Very few states have laws prohibiting employment discrimination on the basis of weight.

10. **False.** Although the exact number is not known, many young girls and women do die as a result of starvation, malnutrition, and other problems associated with anorexia and bulimia. These are considered to be life-threatening behaviors by many in the medical profession.

Sources: Based on Fallon, Katzman, and Wooley, 1994; Kilbourne, 1994, 1999; Lips, 2001; and Seid, 1994.

socialization process (Lips, 2001). For example, in U.S. society, males are traditionally expected to demonstrate aggressiveness and toughness whereas females are expected to be passive and nurturing. **Gender identity is a person's perception of the self as female or male.** Typically established between eighteen months and three years of age, gender identity is a powerful aspect of our self-concept (Lips, 2001). Although this identity is an individual perception, it is developed through interaction with others. As a result, most people form a gender identity that matches their biological sex: Most biological females think of themselves as female, and most biological males think

of themselves as male. Body consciousness is a part of gender identity. **Body consciousness is how a person perceives and feels about his or her body;** it also includes an awareness of social conditions in society that contribute to this self-knowledge (Thompson, 1994). Consider, for example, these comments by Steve Michalik, a former Mr. Universe:

I was small and weak, and my brother Anthony was big and graceful, and my old man made no bones about loving him and hating me. . . . The minute I walked in from school, it was, "You worthless little s--t, what are you doing home so

In nations around the world, women are often the objects of the male gaze. How might the behavior of others influence our own perceptions of body consciousness?

early?" His favorite way to torture me was to tell me he was going to put me in a home. We'd be driving along in Brooklyn somewhere, and we'd pass a building with iron bars on the windows, and he'd stop the car and say to me, "Get out. This is the home we're putting you in." I'd be standing there sobbing on the curb—I was maybe eight or nine at the time. (qtd. in Klein, 1993: 273)

As we grow up, we become aware, as Michalik did, that the physical shape of our bodies subjects us to the approval or disapproval of others. Being small and weak may be considered positive attributes for women, but they are considered negative characteristics for "true men."

A macrolevel analysis of gender examines structural features, external to the individual, that perpetuate gender inequality. These structures have been referred to as *gendered institutions,* meaning that gender is one of the major ways by which social life is organized in all sectors of society. Gender is embedded in the images, ideas, and language of a society and is used as a means to divide up work, allocate resources, and distribute power. For example, every society uses gender to assign certain tasks—ranging from child rearing to warfare—to females and to males, and differentially rewards those who perform these duties.

These institutions are reinforced by a *gender belief system,* which includes all the ideas regarding masculine and feminine attributes that are held to be valid in a society. This belief system is legitimated by religion, science, law, and other societal values (Lorber, 1994, 2001). For example, gendered belief systems may change over time as gender roles change. Many fathers take care of young children today, and there is a much greater acceptance of this change in roles. However, popular stereotypes about men and women,

as well as cultural norms about gender-appropriate appearance and behavior, serve to reinforce gendered institutions in society.

The Social Significance of Gender

Gender is a social construction with important consequences in everyday life. Just as stereotypes regarding race/ethnicity have built-in notions of superiority and inferiority, gender stereotypes hold that men and women are inherently different in attributes, behavior, and aspirations. Stereotypes define men as strong, rational, dominant, independent, and less concerned with their appearance. Women are stereotyped as weak, emotional, nurturing, dependent, and anxious about their appearance.

The social significance of gender stereotypes is illustrated by eating problems. The three most common eating problems are anorexia, bulimia, and obesity. With *anorexia,* a person has lost at least 25 percent of body weight due to a compulsive fear of becoming fat (Lott, 1994). With *bulimia,* a person binges by consuming large quantities of food and then purges the food by induced vomiting, excessive exercise, laxatives, or fasting. With *obesity,* individuals are 20 percent or more above their desirable weight, as established by the medical profession. For a 5-foot-4-inch woman, that is about twenty-five pounds; for a 5-foot-10-inch man, it is about thirty pounds (Burros, 1994: 1).

Sociologist Becky W. Thompson argues that, based on stereotypes, the primary victims of eating problems are presumed to be white, middle-class, heterosexual women. However, such problems also exist among women of color, working-class women, lesbians, and some men. According to Thompson, explanations regarding the relationship between gender and eating problems must take into account a complex array of social factors, including gender socialization and women's responses to problems such as racism and emotional, physical, and sexual abuse (Thompson, 1994; see also Wooley, 1994).

Bodybuilding is another gendered experience. *Bodybuilding* is the process of deliberately cultivating an increase in the mass and strength of the skeletal muscles by means of lifting and pushing weights (Mansfield and McGinn, 1993). In the past, bodybuilding was predominantly a male activity; musculature connoted power, domination, and virility (Klein, 1993). Today, an increasing number of women engage in this activity. As gendered experiences, eating problems and bodybuilding have more in common than we might think. Women's studies scholar Susan

Bodybuilding may be a source of empowerment for women, or it may be a way in which the body becomes objectified. Competitions such as the one shown here have grown in popularity in recent years.

© Tony Duffy/Getty Images

Bordo (2004) has noted that the anorexic body and the muscled body are not opposites but instead are both united against the common enemy of soft, flabby flesh. In other words, the *body* may be objectified through both compulsive dieting and compulsive bodybuilding.

Sexism

Sexism **is the subordination of one sex, usually female, based on the assumed superiority of the other sex.** Sexism directed at women has three components: (1) negative attitudes toward women; (2) stereotypical beliefs that reinforce, complement, or justify the prejudice; and (3) discrimination—acts that exclude, distance, or keep women separate (Lott, 1994).

Can men be victims of sexism? Although women are more often the target of sexist remarks and practices, men can be victims of sexist assumptions. According to the social psychologist Hilary M. Lips (2001), an example of sexism directed against men is the mistaken idea that it is more harmful for female soldiers to be killed in battle than male soldiers.

Like racism, sexism is used to justify discriminatory treatment. When women participate in what is considered gender-inappropriate endeavors in the workplace, at home, or in leisure activities, they often find that they are the targets of prejudice and discrimination. Obvious manifestations of sexism are found in the undervaluing of women's work and in hiring and promotion practices that effectively exclude women from an organization or confine them to the bottom of the organizational hierarchy. Even today, some women who enter nontraditional occupations (such as firefighting and welding) or professions (such as dentistry and architecture) encounter hurdles that men do not face.

Sexism is interwoven with *patriarchy*—**a hierarchical system of social organization in which cultural, political, and economic structures are controlled by men.** By contrast, *matriarchy* **is a hierarchical system of social organization in which cultural, political, and economic structures are controlled by women;** however, few (if any) societies have been organized in this manner. Patriarchy is reflected in the way men may think of their position as men as a given whereas women may deliberate on what their position in society should be. As the sociologist Virginia Cyrus (1993: 6) explains, "Under patriarchy, men are seen as 'natural' heads of households, Presidential candidates, corporate executives, college presidents, etc. Women, on the other hand, are men's subordinates, playing such supportive roles as housewife, mother, nurse, and secretary." Gender inequality and a division of labor based on male dominance are nearly universal, as we will see in the following discussion on the origins of gender-based stratification.

GENDER STRATIFICATION IN HISTORICAL AND CONTEMPORARY PERSPECTIVE

How do tasks in a society come to be defined as "men's work" or "women's work"? Three factors are important in determining the gendered division of labor in a society: (1) the type of subsistence base, (2) the supply of and demand for labor, and (3) the extent to which women's child-rearing activities are compatible with certain types of work. As defined in Chapter 5, *subsistence* refers to the means by which a society gains the basic necessities of life, including food, shelter, and clothing. You may recall that societies are classified, based on subsistence, as hunting and gathering societies, horticultural and pastoral societies, agrarian societies, industrial societies, and postindustrial societies.

Hunting and Gathering Societies

The earliest known division of labor between women and men is in hunting and gathering societies. While the men hunt for wild game, women gather roots and berries. A relatively equitable relationship exists because neither sex has the ability to provide all the food necessary for survival. When wild game is nearby, both men and women may hunt. When it is far away, hunting becomes incompatible with child rearing (which women tend to do because they breast-feed their young), and women are placed at a disadvantage in terms of contributing to the food supply (Lorber, 1994). In most hunting and gathering societies, women are full economic partners with men; relations between them tend to be cooperative and relatively egalitarian (Chafetz, 1984; Bonvillain, 2001). Little social stratification of any kind is found because people do not acquire a food surplus. A few hunting and gathering societies remain, including the Bushmen of Africa, the aborigines of Australia, the Kaska Indians of Canada, and the Yanomami of South America.

Horticultural and Pastoral Societies

In horticultural societies, which first developed ten to twelve thousand years ago, a steady source of food becomes available. People are able to grow their own food because of hand tools, such as the digging stick and the hoe. Women make an important contribution to food production because hoe cultivation is compatible with child care. A fairly high degree of gender equality exists because neither sex controls the food supply (Basow, 1992).

When inadequate moisture in an area makes planting crops impossible, *pastoralism*—the domestication of large animals to provide food—develops. Herding is primarily done by men, and women contribute relatively little to subsistence production in such societies. In some herding societies, women have relatively low status; their primary value is their ability to produce male offspring so that the family lineage can be preserved and enough males will exist to protect the group against attack (Nielsen, 1990).

Social practices contribute to gender inequality in horticultural and pastoral societies. Male dominance is promoted by practices such as menstrual taboos, bridewealth, and polygyny. *Polygyny*—the marriage of one man to multiple wives—contributes to power differences between women and men. A man with multiple wives can produce many children, who will enhance his resources, take care of him in his "old age," and become heirs to his property (Nielsen, 1990). *Menstrual taboos* place women in a subordinate position by segregating them into menstrual huts for the duration of their monthly flow. Even when women are not officially segregated, they are defined as "unclean." *Bridewealth*—the payment of a price by a man for a wife—turns women into property that can be bought and sold. The man gives the bride's family material goods or services in exchange for their daughter's exclusive sexual services and his sole claim to their offspring.

In contemporary horticultural societies, women do most of the farming while men hunt game, clear land, work with arts and crafts, make tools, participate in religious and ceremonial activities, and engage in war. A combination of horticultural and pastoral activities is found in some contemporary societies in Asia, Africa, the Middle East, and South America. These societies are characterized by more gender inequality than in hunting and gathering societies but less than in agrarian societies (Bonvillain, 2001).

Agrarian Societies

In agrarian societies, which first developed about eight to ten thousand years ago, gender inequality and male dominance become institutionalized. The most extreme form of gender inequality developed about five thousand years ago in societies in the fertile crescent around the Mediterranean Sea (Lorber, 1994). Agrarian societies rely on agriculture—farming done by animal-drawn or mechanically powered plows and equipment. Because agrarian tasks require more labor and greater physical strength than horticultural ones, men become more involved in food production. It has been suggested that women are excluded from these tasks because they are viewed as too weak for the work and because child-care responsibilities are considered incompatible with the full-time labor that the tasks require (Nielsen, 1990).

Why does gender inequality increase in agrarian societies? Scholars cannot agree on an answer; however, some suggest that it results from private ownership of property. When people no longer have to move continually in search of food, they can acquire a surplus. Men gain control over the disposition of the surplus and the kinship system, and this control serves men's interests (Lorber, 1994). The importance of producing "legitimate" heirs to inherit the surplus increases significantly, and women's lives become more secluded and restricted as men attempt to ensure the legitimacy of their children. Premarital virginity and marital fidelity are required; indiscretions are punished (Nielsen, 1990). However, some scholars argue that male dominance existed before the private ownership of property (Firestone, 1970; Lerner, 1986).

Four practices in agrarian societies contribute to the subordination of women. *Purdah,* found primarily among Hindus and Muslims, requires the seclusion of women, extreme modesty in apparel, and the visible subordination of women to men. Women must show deference to men by walking behind them, speaking only when spoken to, and eating only after the men have finished a meal (Nielsen, 1990).

Footbinding is the thwarting of the growth of a female's feet. Footbinding was first practiced in China beginning around A.D. 1000, and the custom continued into the early twentieth century. The toes of young girls were bent under and continually bound tighter to the soles of their feet. As a result, women might experience extreme pain as their toenails grew into their feet or develop serious infections due to lack of blood circulation (Dworkin, 1974).

Suttee (most common in parts of India) is the sacrificial killing of a widow upon the death of her husband. Although some women allegedly choose to make this sacrifice, others are tied to their husband's funeral pyre. The practice is justified on the basis that the widow's sins in a former life are responsible for her husband's death. However, the actual purpose is to ensure that the husband's male relatives, rather than the widow, inherit his property (Nielsen, 1990).

Genital mutilation is a surgical procedure performed on young girls as a method of sexual control. The mutilation involves cutting off all or part of a girl's clitoris and labia, and in some cases stitching her vagina closed until marriage (Simons, 1993). Often justified on the erroneous belief that the Qur' an commands it, these procedures are supposed to ensure that women are chaste before marriage and have no extramarital affairs after marriage. Genital mutilation has resulted in the maiming of many females, some of whom died as a result of hemorrhage, infection, or other complications. It is still practiced in more than twenty-five countries.

In sum, male dominance is very strong in agrarian societies. Women are secluded, subordinated, and mutilated as a means of regulating their sexuality and protecting paternity. Most of the world's population currently lives in agrarian societies in various stages of industrialization. Issues concerning the rights of

Historical Picture Archive/Corbis

Although suttee—the sacrificial killing of a widow upon her husband's death—was banned in India in 1829, it still occurs in rare circumstances in rural India.

women in some of these societies became the subject of media attention after the 2001 terrorist attacks on the United States were linked to an organization with ties to the repressive Taliban regime in Afghanistan (see Box 11.2).

Industrial Societies

An *industrial society* is one in which factory or mechanized production has replaced agriculture as the major form of economic activity. As societies industrialize, the status of women tends to decline further. Industrialization in the United States created a gap between the nonpaid work performed by women at home and the paid work that increasingly was performed by men and unmarried girls (Amott and Matthaei, 1996). When families needed extra money, their daughters worked in the textile mills until they married. In 1900, for example, 22 percent of single white women who had been born in the United States were in the paid labor force. Because factory work was not compatible with child-care responsibilities, only 3 percent of married women were so employed, and only when they were extremely poor (Amott and Matthaei, 1996). As it became more difficult to make a living by farming, many men found work in the factories, where their primary responsibility was often supervising the work of women and children. Men began to press for a clear division between "men's work" and "women's work," as well as corresponding pay differentials (higher for men, lower for women).

In the United States, the division of labor between men and women in the middle and upper classes be-

Box 11.2 SOCIOLOGY IN GLOBAL PERSPECTIVE

Oppression, Resistance, and the Women of Afghanistan

It was a very emotional moment. After years the women of Afghanistan came out in the open. Under the Taliban we all wore burkas and did not know each other. Now we all know each other's faces. . . . I just want to tell the world that women should be able to speak out about their own problems.

> —Soraya Parlika, a prominent Afghan activist, describing how she felt when many women removed their veils after the overthrow of the Taliban regime (qtd. in McCarthy, 2001: 46)

Although the plight of women and girls in Afghanistan was already a topic of concern for some international women's activist groups prior to the September 11, 2001, terrorist attacks, it was not until after this date that U.S. political leaders and the media highlighted the problems that many Afghan women had experienced for nearly a half decade under the Taliban. After the Taliban's takeover of that country in 1996, Afghan women were prohibited from most employment, walking alone on the streets, and teaching or studying in the schools. Residents were required to paint their windows black so that passersby could not see the face of any woman in the home (Lacayo, 2001). During the Taliban era, women were virtually imprisoned in their homes.

However, women had not always been so oppressed in Afghanistan. Prior to the Taliban takeover, Afghan women comprised 60 percent of the teachers and 50 percent of students at Kabul University; they had also been 40 percent of the doctors and 70 percent of the schoolteachers. After the takeover, many schools were closed because of a shortage of teachers and students, and women and girls were forced to wear a head-to-toe covering called a "burqa" (or "burka"), which has only a small mesh opening through which the person can see and breathe (NOW, 2002).

Although women lost virtually all of their rights under the repressive Taliban regime, journalists and other social analysts have documented how, even under these adverse conditions, some Afghan women subtly fought against the Taliban's oppression of women through acts of resistance. Some women ran secret home schools so that girls could continue their education, while other women operated clandestine businesses such as beauty salons and laundries in an effort to obtain enough money to feed their children (Waldman, 2001).

What rights will emerge for the women of Afghanistan? It has been suggested by some analysts that people who live in the United States are not able to fully understand the many factors that are involved in the struggle for women's rights in other countries. According to these observers, U.S. beliefs about women's rights are linked to Western thinking and the ideas of feminist movements in Western Europe and the United States, which are not applicable to people in other nations and cultures. According to Rina Amiri (2001, A21), senior associate for research with the Women Waging Peace Initiative at Harvard's Kennedy School of Government,

> It has come to be assumed in much of the Muslim world that to be a proponent of women's rights is to be pro-Western. This enmeshing of gender and geopolitics has robbed Muslim women of their ability to develop a discourse on their rights independent of a cultural debate between the Western and Muslim worlds.

Many social scientists who specialize in international relations hope that such a discourse—free of cultural bias—will develop so that girls and women will not continue to experience the high levels of subordination that exist in many areas of the world today. Although not as oppressive as the rules established by the Taliban, these patterns of subordination may limit the human potential of more than half of the population of these regions.

---

came much more distinct with industrialization. The men were responsible for being "breadwinners"; the women were seen as "homemakers." In this new "cult of domesticity" (also referred to as the "cult of true womanhood"), the home became a private, personal sphere in which women created a haven for the family. Those who supported the cult of domesticity argued that women were the natural keepers of the domestic sphere and that children were the mother's responsibility. Meanwhile, the "breadwinner" role

placed enormous pressures on men to support their families—being a good provider was considered to be a sign of manhood (Amott and Matthaei, 1996). However, this gendered division of labor increased the economic and political subordination of women. As a result, many white women focused their efforts on acquiring a husband who was capable of bringing home a good wage. Single women and widows and their children tended to live a bleak existence, crowded into run-down areas of cities, where they were often unable to support themselves on their meager wages.

The cult of true womanhood not only increased white women's dependence on men but also became a source of discrimination against women of color, based on both their race and the fact that many of them had to work in order to survive. Employed, working-class white women were similarly stereotyped at the same time that they became more economically dependent on their husbands because their wages were so much lower.

Although industrialization was a source of upward mobility for many whites, most people of color were left behind. For example, the cult of domesticity was distinctly white and middle or upper class. White families with sufficient prosperity hired domestic servants to do much of the household work. In the late 1800s and early 1900s, many African American women were employed as household servants.

As people moved from a rural, agricultural lifestyle to an urban existence, body consciousness increased. People who worked in offices often became sedentary and exhibited physical deterioration from their lack of activity. As gymnasiums were built to fight this lack of physical fitness, a new image of masculinity developed. Whereas the "burly farmer" or "robust workman" had previously been the idealized image of masculinity, now the middle-class man who exercised and lifted weights came to embody this ideal (Klein, 1993).

In the late nineteenth century, middle-class women started to become preoccupied with body fitness. As industrialization progressed and food became more plentiful, the social symbolism of body weight and size changed. Previously, it had been considered a sign of high status to be somewhat overweight, but now a slender body reflected an enhanced social status. To the status-seeking middle-class man, a slender wife became a symbol of the husband's success. At the same time, excess body weight was seen as a reflection of moral or personal inadequacy, or a lack of willpower (Bordo, 2004).

In sum, from hunting and gathering societies to contemporary societies, patriarchy and male dominance have remained pervasive in many aspects of social life, including how the physical appearances of women and men are evaluated by others (Bonvillain, 2001).

Postindustrial Societies

Chapter 5 defines *postindustrial societies* as ones in which technology supports a service- and information-based economy. In such societies, the division of labor in paid employment is increasingly based on whether people provide or apply information or are employed in service jobs such as fast-food restaurant counter help or health care workers. For both women and men in the labor force, formal education is increasingly crucial for economic and social success. However, as some women have moved into entrepreneurial, managerial, and professional occupations, many others have remained in the low-paying service sector, which affords few opportunities for upward advancement.

Will technology change the gendered division of labor in postindustrial societies? Scholars do not agree on the effects of computers, the Internet, the World Wide Web, cellular phones, and many newer forms of communications technology on the role of women in society. For example, some feminist writers had a pessimistic view of the impact of computers and monitors on women's health and safety, predicting that women in secretarial and administrative roles would experience an increase in eyestrain, headaches, and problems such as carpel tunnel syndrome. However, some medical experts now believe that such problems extend to both men and women, as computers have become omnipresent in more people's lives. The term "24–7" has come to mean that a person is available "twenty-four hours a day, seven days a week" via cell phones, pocket pagers, fax machines, e-mail, and other means of communication, whether the individual is at the office or four thousand miles away on "vacation."

How do new technologies influence gender relations in the workplace? Although some analysts presumed that technological developments would reduce the boundaries between women's and men's work, researchers have found that the gender stereotyping associated with specific jobs has remained remarkably stable even when the nature of work and the skills required to perform it have been radically transformed. Today, men and women continue to be segregated into different occupations, and this segregation is particularly visible within individual workplaces (as discussed later in the chapter).

How does the division of labor change in families in postindustrial societies? For a variety of reasons, more households are headed by women with no adult male present. As shown in the Census Profiles feature, the percentage of U.S. households headed by a single mother with children under eighteen has increased. Chapter 15 ("Families and Intimate Relationships") discusses a number of reasons why the current division of labor in household chores in some families is between a woman and her children rather than between women and men. Consider, for example, that almost one-

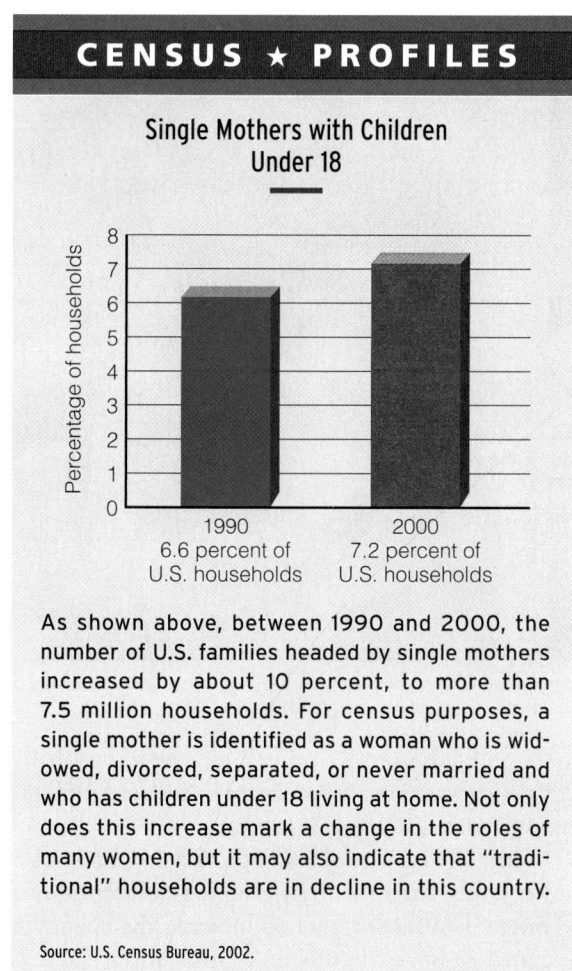

CENSUS ★ PROFILES

Single Mothers with Children Under 18

As shown above, between 1990 and 2000, the number of U.S. families headed by single mothers increased by about 10 percent, to more than 7.5 million households. For census purposes, a single mother is identified as a woman who is widowed, divorced, separated, or never married and who has children under 18 living at home. Not only does this increase mark a change in the roles of many women, but it may also indicate that "traditional" households are in decline in this country.

Source: U.S. Census Bureau, 2002.

the labor force, meaning that finding time to care for children, help aging parents, and meet the demands of the workplace will continue to place a heavy burden on women, despite living in an information- and service-oriented economy.

How people accept new technologies and the effect these technologies have on gender stratification are related to how people are socialized into gender roles. However, gender-based stratification remains rooted in the larger social structures of society, which individuals have little ability to control.

GENDER AND SOCIALIZATION

We learn gender-appropriate behavior through the socialization process. Our parents, teachers, friends, and the media all serve as gendered institutions that communicate to us our earliest, and often most lasting, beliefs about the social meanings of being male or female and about thinking and behaving in masculine or feminine ways. Some gender roles have changed dramatically in recent years; others remain largely unchanged over time.

Many parents prefer boys to girls because of stereotypical ideas about the relative importance of males and females to the future of the family and society (Basow, 1992). Although some parents prefer boys to girls because these parents believe old myths about the biological inferiority of females, research suggests that social expectations also play a major role in this preference. We are socialized to believe that it is important to have a son, especially for a first or only child. For many years, it was assumed that a male child could support his parents in their later years and carry on the family name.

Across cultures, boys are preferred to girls, especially when the number of children that parents can have is limited by law or economic conditions. For example, in China, which strictly regulates the allowable number of children to one per family, a disproportionate number of female fetuses are aborted (Basow, 1992). In India, the practice of aborting female fetuses is widespread, and female infanticide occurs frequently (Burns, 1994). As a result, both India and China have a growing surplus of young men who will face a shortage of women their own age (Shenon, 1994).

In the United States, some sex selection no doubt takes place through abortion. However, most women seek abortions because of socioeconomic factors,

fourth (23 percent) of all U.S. children live with their mother only (as contrasted with just 5 percent who reside with their father only); among African American children, 48 percent live with their mother only (U.S. Census Bureau, 2002). This means that women in these households truly have a double burden, both from family responsibilities and from the necessity of holding gainful employment in the labor force.

Even in single-person or two-parent households, programming "labor-saving" devices (if they can be afforded) often means that a person must have some leisure time to learn how to do the programming. According to analysts, leisure is deeply divided along gender lines, and women have less time to "play in the house" than do men and boys. Some Web sites seek to appeal to women who have economic resources but are short on time, making it possible for them to shop, gather information, "telebank," and communicate with others at all hours of the day and night.

In postindustrial societies such as the United States, more than 60 percent of adult women are in

Are children's toys a reflection of their own preferences and choices? How do toys reflect gender social-ization by parents and other adults?

problematic relationships with partners, health-related concerns, and lack of readiness or ability to care for a child (or another child) (Lott, 1994).

Gender Socialization by Parents

From birth, parents act toward children on the basis of the child's sex. Baby boys are perceived to be less fragile than girls and tend to be treated more roughly by their parents. Girl babies are thought to be "cute, sweet, and cuddly" and receive more gentle treatment (MacDonald and Parke, 1986). Parents strongly in-fluence the gender-role development of children by passing on—both overtly and covertly—their own beliefs about gender (Witt, 1997). When girl babies cry, parents respond to them more quickly, and par-ents are more prone to talk and sing to girl babies (Basow, 1992). However, one early study ("The Fa-vored Infants," 1976) found that African American mothers "rubbed, patted, rocked, touched, kissed and talked to" boy babies more than girl babies.

Children's toys reflect their parents' gender expec-tations. Gender-appropriate toys for boys include computer games, trucks and other vehicles, sports equipment, and war toys such as guns and soldiers. Girls' toys include "Barbie" dolls, play makeup, and homemaking items. Parents' choices of toys for their children are not likely to change in the near future (see Box 11.3). A group of college students in one study were shown slides of toys and asked to decide which ones they would buy for girls and boys. Most said they would buy guns, soldiers, jeeps, carpenter

tools, and red bicycles for boys; girls would get baby dolls, dishes, sewing kits, jewelry boxes, and pink bi-cycles (Fisher-Thompson, 1990).

When children are old enough to help with house-hold chores, they are often assigned different tasks. Maintenance chores (such as mowing the lawn) are assigned to boys whereas domestic chores (such as shopping, cooking, and clearing the table) are as-signed to girls. Chores may also become linked with future occupational choices and personal characteris-tics. Girls who are responsible for domestic chores such as caring for younger brothers and sisters may learn nurturing behaviors that later translate into em-ployment as a nurse or schoolteacher. Boys may learn about computers and other types of technology that lead to different career options.

Because most studies of gender socialization by par-ents focus on white, middle-class families, it is probably inappropriate to assume that the patterns of differential parental treatment of girls and boys hold true in all re-spects for other class and racial groups (Lips, 2001). For example, children from middle- and upper-income families are less likely to be assigned gender-linked chores than those from lower-income backgrounds. In addition, analysts have found that gender-linked chore assignments occur less frequently in African American families, where both sons and daughters tend to be so-cialized toward independence, employment, and child care (Bardwell, Cochran, and Walker, 1986; Hale-Benson, 1986). Sociologist Patricia Hill Collins (1991) suggests that African American mothers are less likely to socialize their daughters into roles as subordinates;

Box 11.3 CHANGING TIMES: MEDIA AND TECHNOLOGY

Red Jack: Revenge of the Brethren, Barbie Super Sports, and Gendered Play

To play *Red Jack: Revenge of the Brethren:*

You take the role of Nicholas Dove, a dreamer who lives in the doomed island of Lizard Point. As you walk into town on a stormy night, you come across a bunch of pirates who have docked to stock up on ale. As you converse with the various crew members, you realize that this is your way off this lowly island. You decide to go with the pirates, who are trying to find out who has been killing the Brethren (a renegade band of privateers, formerly led by Red Jack, who was killed because of an unknown traitor). You join the pirates and the two remaining Brethren to find the traitor. (IGN Entertainment, 2003)

To play *Barbie Super Sports:*

You decide whether you want to be Barbie or one of her friends—Christie, Kira or Teresa. Then, after choosing between in-line skating and snowboarding, you'll go shopping for a coordinating outfit and the appropriate sports-specific footgear (skates or boards). You then go to the sport's practice area and listen to Barbie as she tells you the requirements for the sport. For example, you may be required to race against the clock, collect pink balloons, jump over fountains in an oriental garden, or smash into snowmen on a slippery slope. At the end of each race, you are awarded tickets that can be traded for faster skates and snowboards.

–based on a review of *Barbie Super Sports* by Jim Fleming (2000)

Do newer forms of media, including computer games, bring equal rewards to girls and boys in regard to gender socialization? One analyst answers this question by noting that male control of computers is supported by the vast libraries of games software with characteristics conventionally described as masculine: an emphasis on competition and violence (Grint and Woolgar, 1997). In this case, the *technology* is physically flexible and can easily be used by both girls and boys, but the *social construction* of gender is reproduced by software programs such as *Barbie Super Sports* and *Red Jack: Revenge of the Brethren.*

Through the magic of computer graphics, girls can experience the "perfect body" myth as they are exposed to Barbie's unrealistic measurements and her beautiful clothing even when skating or snowboarding. By contrast, boys playing *Red Jack* are engaged in a computer-animated tale of "villainous betrayal, supernatural curses, and buried treasure" (Herz, 1998a: D4). Children's and young people's encounters with Barbie or Red Jack are examples of *simulacra,* the term that postmodernist theorist Jean Baudrillard uses to describe a situation in which simulation replaces a reality that never existed. For example, instead of playing with a real person, or even a plastic Barbie doll with genuine nylon hair, girls in cyberspace amuse themselves with a virtual doll on a screen. Likewise, boys playing *Red Jack* seek "to uncover the traitor, revenge the great fallen pirate, dig up gold, get the girl and sail off into the Caribbean sunset" (Herz, 1998a: D4).

What influence will virtual Barbie and Red Jack have on young people? According to a Mattel spokesperson, girls who play with virtual Barbie will spend more time on the computer and gain similar skills and equal footing with boys in the job market (Herz, 1998b). Likewise, boys playing with Red Jack may learn "masculine" behavior and problem-solving skills through their analysis of what the virtual actors should do next in their quest to gain the upper hand.

However, to further examine the issue of gendered play, think about these questions: What happens if a boy wants to play with Barbie computerized games? Is it different if a girl wants to play with a "male-oriented" game such as *Red Jack*? What do you think?

---

instead, they are likely to teach them a critical posture that allows them to cope with contradictions.

Just as appropriate "masculine" or "feminine" behavior is learned through interaction with parents and other caregivers, inappropriate behavior such as eating problems can be learned from parents. Nicole Annesi tells how she learned about binging and purging from her mother:

I was seven years old the first time I was exposed to my mother's bulimia. It was after dinner one evening. After Mom and I cleared the table . . . she quickly disappeared into the bathroom. . . . What was unusual about these visits was that they became consistent. After each meal Mom would visit the bathroom and come out a few minutes later looking pale, yet refreshed. . . . So after the dishes were cleared away that evening, I disappeared into the bathroom. I hid in the tub, behind the navy blue, opaque curtain . . . like clockwork, in she came. I peered between the curtains to find my mother bent over the toilet, like she was going to get sick or something. And then I watched her. . . . She placed a popsicle stick down her throat and made herself sick. How weird, I thought to myself. Mom comes into the bathroom every night to stick one of these doctor sticks down her throat! . . .

I began to do some thinking myself. . . . I knew when I got sick, afterwards my stomach would flatten out. I felt lighter. So one day after school I ate a bag of Pecan Sandies. I stuffed myself until I couldn't swallow. Afterwards I made my way to the upstairs bathroom and locked the door behind me. I turned on the faucet so that nothing could be heard. . . . That day I became a seven-year-old bulimic. (Annesi, 1993: 91–93)

Anorexia, bulimia, and compulsive overeating may be rooted in the childhood experiences of people from all racial and ethnic groups and classes. When Joselyn, an African American woman, was a child in Queens, New York, having chubby, healthy children was considered a sign that the family was doing well financially. However, as her family moved up the social ladder, Joselyn's father began to insist that she lose weight:

When my father's business began to boom and my father was interacting more with white businessmen and seeing how they did business, suddenly thin became important. If you were a truly well-to-do family, then your family was slim and elegant. (qtd. in Thompson, 1994: 32)

While Joselyn was being pressured to diet, her father was still bringing home treats for her and serving her large portions of food at meals. When she was put on diet pills and diets before she reached puberty, she began a cycle of dieting, compulsive eating, and bulimia that followed her into adulthood (Thompson, 1994). In families such as Nicole's and Joselyn's, gender socialization by parents becomes intertwined with appearance norms that produce dysfunctional behavior, in this case eating problems (Thompson, 1994).

Many parents are aware of the effect that gender socialization has on their children and make a conscientious effort to provide nonsexist experiences for them. For example, one study found that mothers with nontraditional views encourage their daughters to be independent (Brooks-Gunn, 1986). Many fathers also take an active role in socializing their sons to be thoughtful and caring individuals who do not live by traditional gender stereotypes. However, peers often make nontraditional gender socialization much more difficult for parents and children (see Rabinowitz and Cochran, 1994).

Peers and Gender Socialization

Peers help children learn prevailing gender-role stereotypes, as well as gender-appropriate and gender-inappropriate behavior (Hibbard and Buhrmester, 1998). During the preschool years, same-sex peers have a powerful effect on how children see their gender roles (Maccoby and Jacklin, 1987); children are more socially acceptable to their peers when they conform to implicit societal norms governing the "appropriate" ways that girls and boys should act in social situations and what prohibitions exist in such cases (Martin, 1989).

Male peer groups place more pressure on boys to do "masculine" things than female peer groups place on girls to do "feminine" things (Fagot, 1984). For example, girls wear jeans and other "boy" clothes, play soccer and softball, and engage in other activities traditionally associated with males. By contrast, if a boy wears a dress, plays hopscotch with girls, and engages in other activities associated with being female, he will be ridiculed by his peers. This distinction between the relative value of boys' and girls' behaviors strengthens the cultural message that masculine activities and behavior are more important and more acceptable (Wood, 1999).

During adolescence, peers are often stronger and more effective agents of gender socialization than adults (Hibbard and Buhrmester, 1998). Peers are thought to be especially important in boys' development of gender identity (Maccoby and Jacklin, 1987). Male bonding that occurs during adolescence is believed to reinforce masculine identity (Gaylin, 1992) and to encourage gender-stereotypical attitudes and behavior (Huston, 1985; Martin, 1989). For example, male peers have a tendency to ridicule and bully others about their appearance, size, and weight. Aleta Walker painfully recalls walking down the halls at school when boys would flatten themselves against the lockers and cry, "Wide load!" At lunchtime, the boys made a production of watching her eat lunch and frequently made sounds like grunts or moos (Kolata, 1993). Because peer acceptance is extremely important for both males and females during their first two decades, such actions can have very harmful consequences for the victims.

As young adults, men and women still receive many gender-related messages from peers. Among college students, for example, peer groups are organized largely around gender relations and play an im-

Parents, peers, and the larger society all influence our perceptions about gender-appropriate behavior.

portant role in career choices and the establishment of long-term, intimate relationships (Holland and Eisenhart, 1990). In a study of women college students at two universities (one primarily white, the other predominantly African American), anthropologists Dorothy C. Holland and Margaret A. Eisenhart (1990) found that the peer system propelled women into a world of romance in which their attractiveness to men counted most. Although peers did not initially influence the women's choices of majors and careers, they did influence whether the women continued to pursue their original goals, changed their course of action, or were "derailed" (Holland and Eisenhart, 1981, 1990).

If Holland and Eisenhart's research can be generalized to other colleges and universities, peer pressure is often at its strongest in relation to appearance norms. As other researchers have shown, peer pressure can strongly influence a person's body consciousness. Women in college often face pressure to be very thin, as Karen explains:

> "Do you diet?" asked a friend my freshman year in college, as I was stuffing a third homemade chocolate chip cookie in my mouth. "Do you know how many calories there are in that one cookie?"
>
> Stopping to think for a moment as she and two other friends stared at me, probably wanting to

ask me the same question, I realized that I really didn't even know what a calorie was. . . .

> From that moment, I'd taken on a new enemy, one more powerful and destructive than any human can be. One that nearly fought me to the death—my death. . . .
>
> I just couldn't eat food anymore. I was so obsessed with it that I thought about it every second. . . . In two months, I'd lost thirty pounds. . . . Everyone kept telling me I looked great. . . .
>
> I really didn't realize that anything was wrong with me. . . . There were physical things occurring in my body other than not having my period anymore. My hair was falling out and was getting thinner. . . . I would constantly get head rushes every time I stood up. . . . When my friends would all go out to dinner or to a party I stayed home quite often, afraid that I might have to eat something, and afraid that my friends would find out that I didn't eat. (Twenhofel, 1993: 198)

Feminist scholars have concluded that eating problems are not always psychological "disorders" (as they are referred to by members of the medical profession). Instead, eating (or not eating) may be a strategy for coping with problems such as unrealistic social pressures about slenderness (see Orbach, 1978; Chernin, 1981; Hesse-Biber, 1989) and/or social injustices brought about by racism, sexism, and classism in society (Thompson, 1994).

Teachers, Schools, and Gender Socialization

From kindergarten through college, schools operate as a gendered institution. Teachers provide important messages about gender through both the formal content of classroom assignments and informal interactions with students. Sometimes, gender-related messages from teachers and other students reinforce gender roles that have been taught at home; however, teachers may also contradict parental socialization. During the early years of a child's schooling, teachers' influence is very powerful; many children spend more hours per day with their teachers than they do with their own parents.

According to some researchers, the quantity and quality of teacher–student interactions often vary between the education of girls and that of boys (Wellhousen and Yin, 1997). One of the messages that teachers may communicate to students is that boys are more important than girls. Research spanning the past thirty years shows that unintentional gender bias occurs in virtually all educational settings. ***Gender***

Teachers often use competition between boys and girls because they hope to make a learning activity more interesting. Here, a middle-school girl leads other girls against boys in a Spanish translation contest. What are the advantages and disadvantages of gender-based competition in classroom settings?

Mary Kate Denny/PhotoEdit

bias **consists of showing favoritism toward one gender over the other.** Researchers consistently find that teachers devote more time, effort, and attention to boys than to girls (Sadker and Sadker, 1994). Males receive more praise for their contributions and are called on more frequently in class, even when they do not volunteer. Very often, boys receive attention because they call out in class, demand help, and sometimes engage in disruptive behavior (Sadker and Sadker, 1994). Teachers who do not negatively sanction such behavior may unintentionally encourage it. Boys learn that when they yell out an answer without being called on, their answer will be accepted by the teacher; girls learn that they will be praised when they are compliant and wait for the teacher to call on them. If they call out an answer, they may be corrected with comments such as "Please raise your hand if you want to speak" (Sadker and Sadker, 1994).

The content of teacher–student interactions is very important. In a multiple-year study of more than one hundred fourth-, sixth-, and eighth-graders, education professors Myra and David Sadker (1984) identified four types of teacher comments: praise, acceptance, remediation, and criticism. They found that boys typically received more of all four types of teacher comments than did girls. Teachers also gave more precise and penetrating replies to boys; by contrast, teachers used vague and superficial terms such as "OK" when responding to girls. Because boys receive more specific and intense interaction from teachers, they may gain more insights than girls into the strengths and weaknesses of their answers and thus learn how to improve their responses.

Teacher–student interactions influence not only students' learning but also their self-esteem (Sadker and Sadker, 1985, 1986, 1994). A comprehensive study of gender bias in schools suggested that girls' self-esteem is undermined in school through such experiences as

(1) a relative lack of attention from teachers; (2) sexual harassment by male peers; (3) the stereotyping and invisibility of females in textbooks, especially in science and math texts; and (4) test bias based on assumptions about the relative importance of quantitative and visual–spatial ability, as compared with verbal ability, where girls typically excel. White males may have better self-esteem because they receive more teacher attention than all other students (Sadker and Sadker, 1994). Studies in the 1980s showed, that, although African American girls started out with very positive behaviors, they received little teacher feedback or academic encouragement (Irvine, 1986). Of all groups, African American males received the most unfavorable treatment, as well as the most frequent teacher referrals to special education programs (McIntyre and Pernell, 1985). Whether or not this has changed in the ensuing two decades is the topic of widespread debate.

Teachers also influence how students treat one another during school hours. Many teachers use sex segregation as a way to organize students, resulting in unnecessary competition between females and males. In one study, for example, a teacher divided her class into the "Beastly Boys" and the "Gossipy Girls" for a math game and allowed her students to do the "give me five" hand-slapping ritual when one group outscored the other (Thorne, 1995). Competition based on gender often reinforces existing misconceptions about the skills and attributes of boys and girls and may contribute to overt and subtle discrimination in the classroom and beyond.

The effect of gender bias is particularly problematic if teachers take a "boys will be boys" attitude when boys and young men make derogatory remarks or demonstrate aggressive behavior against girls and young women. When girls complain of *sexual harassment—* **unwanted sexual advances, requests for sexual favors, or other verbal or physical conduct of a sexual** nature—their concerns are sometimes overlooked or downplayed by teachers and school administrators. Sexual harassment is prohibited by law, and teachers and administrators are obligated to investigate such incidents. However, girls are not alone in experiencing sexual harassment at school. A recent report based on a national survey of more than 2,000 public school students in grades eight through eleven found that 83 percent of girls and 79 percent of boys reported having experienced sexual harassment at least once and that one in four of the students surveyed had experienced such harassment often (AAUW, 2001).

Most problems that exist in pre-kindergarten through high school are also found in colleges and universities. Some college instructors pay more attention to men than women in their classes (Hall and Sandler, 1982, 1984). In these cases, women are called on less frequently, receive less encouragement, and are often interrupted, ignored, or devalued. Researchers have found that women college professors sometimes encourage a more participatory classroom environment and do a better job of including both women and men in their interactions (Statham, Richardson, and Cook, 1991).

Over time, some of these obstacles are diminishing, particularly as women now constitute a larger proportion of all college students. However, in some academic fields of study, women remain a distinct minority and may be more marginalized in classrooms and interpersonal interactions with professors and other students. Men still constitute the majority of majors in architecture, engineering, computer technology, and the physical sciences. Despite such obstacles, women are more likely than men to earn an undergraduate college degree, but at the graduate level, the number of women degree recipients declines dramatically.

Sports and Gender Socialization

Children spend more than half of their nonschool time in play and games, but the type of games played sometimes differs with the child's sex. Studies indicate that boys are socialized to participate in highly competitive, rule-oriented games with a larger number of participants than games played by girls. Girls typically are socialized to play with other girls, in groups of two or three, in activities such as hopscotch and jump rope that involve a minimum of competitiveness. Other research shows that boys express more favorable attitudes toward games and sports that involve physical exertion and competition than girls do. Some analysts believe this difference in attitude is linked to ideas about what is gender-appropriate behavior for boys and girls (Brustad, 1996).

In recent years, women have expanded their involvement in professional sports through organizations such as the WNBA; however, most sports remain rigidly divided into female events and male events. Do you think that media coverage of women's and men's college and professional sporting events differs?

For males, competitive sport becomes a means of "constructing a masculine identity, a legitimated outlet for violence and aggression, and an avenue for upward mobility" (Lorber, 1994: 43). For females, being an athlete and a woman may constitute contradictory statuses. One study during the 1980s found that women college basketball players dealt with this contradiction by dividing their lives into segments. On the basketball court, the women "did athlete"; they pushed, shoved, fouled, ran hard, sweated, and cursed. Off the court, they "did woman"; after the game, they showered, dressed, applied makeup, and styled their hair, even if they were only getting on a van for a long ride home (Watson, 1987). However, researchers in a more recent study concluded that perceptions about women athletes may be changing. Specifically, ideas about what constitutes the ideal body image for girls and women are changing as more females become involved in physical fitness activities and athletic competitions. Young women and men in one poll rated the athletic female body higher than that of the anorexic model (Heywood and Dworkin, 2003).

Since passage in 1972 of Title IX, which mandates equal opportunities in academic and athletic programs for females, girls' and young women's participation in athletics has increased substantially. More girls play soccer and softball and participate in other sports formerly regarded as "male" activities. However, even with these changes over the past three decades, only about 42 percent of high school and college athletes are female. According to the sociologist Michael A. Messner

(2002), girls and women have been empowered by their entry into sports; however, sex segregation of female and male athletes, as well as coaches, persists.

Most sports are rigidly divided into female and male events, and funding of athletic programs is often unevenly divided between men's and women's programs. Assumptions about male and female physiology and athletic capabilities influence the types of sports in which members of each sex are encouraged to participate. For example, women who engage in activities that are assumed to be "masculine" (such as bodybuilding) may either ignore their critics or attempt to redefine the activity or its result as "feminine" or womanly (Klein, 1993; Lowe, 1998). Some women bodybuilders do not want their bodies to get "overbuilt." They have learned that they are more likely to win women's bodybuilding competitions if they look and pose "more or less along the lines of fashion models" (Klein, 1993: 179). In her study of more than 100 people connected with women's bodybuilding, the sociologist Maria R. Lowe (1998) found that "women of steel" (the female bodybuilders) live in a world where size and strength must regularly be balanced with a nod toward grace and femininity.

Cautious optimism is possible regarding the changing nature of sports and gender socialization based on several recent studies of women in sports (Heywood and Dworkin, 2003; Messner, 2002). Clearly, changes have occurred that might positively influence the gender socialization of both girls and boys; however, it appears that much remains to be done to bring about greater gender equity in the area of sport. One such area is how the media report on women's and men's sporting events and the attributes (such as physical attractiveness) that they highlight regarding female competitors while they emphasize the athletic skills of male competitors.

The Media and Gender Socialization

The media, including newspapers, magazines, television, and movies, are powerful sources of gender stereotyping. Although some critics argue that the media simply reflect existing gender roles in society, others point out that the media have a unique ability to shape ideas. Think of the impact that television might have on children if they spend one-third of their waking time watching it, as has been estimated. African American children and adolescents spend even more time watching television than their white peers (Brown et al., 1990). Similarly, children from working-class families spend significantly more time in front of the television than those from the middle class (Basow, 1992).

From children's cartoons to adult shows, television programs offer more male than female characters. Furthermore, the male characters act in a strikingly different manner from female ones. Male characters in both children's programs and adult programs are typically aggressive, constructive, and direct, while some female characters defer to others or manipulate them by acting helpless, seductive, or deceitful (Basow, 1992). Because advertisers hope to appeal to boys, who constitute a significant proportion of the viewing audience for some shows, many programs feature lively adventure and lots of loud noise and violence. Until recently, even educational programs such as *Sesame Street* featured more characters with male names and masculine voices, and the audience was typically encouraged to participate in "boys' activities."

Although attempts have been made by media "watchdogs" and some members of the media to eliminate sexism in children's programming, adult daytime and prime-time programs (which children frequently watch) have received less scrutiny. Soap operas are a classic example of gender stereotyping. As media scholar Deborah D. Rogers (1995) has noted, even though more contemporary soaps feature career women, the cumulative effect of these programs is to reconcile women to traditional feminine roles and relationships. Whether the scene takes place at home or in the workplace, female characters typically gossip about romances or personal problems or compete for a man. Meanwhile, male characters are shown as superior beings who give orders and advice to others and do almost anything. In a hospital soap, for example, while female nurses gossip about hospital romances and their personal lives, the same male doctor who delivers babies also handles AIDS patients and treats trauma victims (Rogers, 1995). Prime-time television has provided better portrayals of women in medicine and in leadership positions in hospitals, but this change to more realistic depictions of women is not reflected throughout television programming.

How do the media serve to "construct" gender and our ideas about what is appropriate for women and men? Television and films influence our thinking about the appropriate behavior of women and men in the roles they play in everyday life. Because it is often necessary for an actor to overplay a comic role to gain laughs from the viewing audience or to exaggerate a dramatic role to make a quick impression on the audience, the portrayal of girls and women may (either intentionally or inadvertently) reinforce old stereotypes or create new ones. Consider, for example, the growing number of women who are depicted as having careers or professions (such as law or medicine) in which they may earn more income and have more power than female characters had in the past. Some of these women may be portrayed as well-adjusted individuals; however, many women characters are still shown as overly emotional and unable to resolve their personal problems even as they are able to carry out their professional duties (the title character in *Judging Amy* is an example). Women characters who are supervisors or bosses are sometimes portrayed as loud, bossy, and domineering individuals. Mothers in situation comedies frequently boss their children around and may have a Homer Simpson-type husband who is lazy and incompetent or who engages in aggressive verbal combat with the woman throughout most of the program (*My Wife and Kids, Eight Simple Rules for Dating My Teenage Daughter, The George Lopez Show,* and reruns of *Roseanne,* for example).

The competence of women to engage in certain kinds of activities, such as purchasing a vehicle, is often questioned and serves as the subject of laugh lines in popular situation comedies. One example is an episode of the ABC comedy *According to Jim.* In a Season 2 episode, "Car and Chicks," Jim (the lead male character, played by Jim Belushi) claims that women are not capable of getting a good deal on a car. Consequently, when Cheryl (his wife, played by Courtney Thorne-Smith) helps Dana (her unmarried sister, played by Kimberly Williams) select a new car to purchase, the vehicle turns out to be a lemon, and Dana begs Jim to "use his macho influence at the dealership to help her, but instead he allows sexy sales manager Gretchen (played by Cindy Crawford) to talk him into trading the family van in for a sports car—without consulting Cheryl," according to ABC's (2003) description.

Both married and single women are stereotyped in the media. The models for single women include the "sexy sales manager" in *According to Jim* and the sex-addicted Samantha in *Sex and the City*—for a number of years one of the most popular cable television programs. Programs such as this do not portray single career women as being self-supporting and emotionally balanced. Rather, they depict the single woman as incomplete without "a man," or they suggest that physical attractiveness and sexuality are commodities that women frequently use to their own advantage. As Betsy Israel, author of a recent book on single women, stated, "Single women have always been portrayed and depicted in the mass culture in a negative and nasty way that influenced the lives of many women, and at the same time was completely untrue, and these images, some of them 150 years old, are still being played out and the ideas are just being recycled" (qtd. in Hoban, 2002: A19).

Advertising—whether on television and billboards or in magazines and newspapers—further reinforces ideas about women and physical attractiveness. The intended message is clear to many people: If they embrace femininity and heightened sexuality, their personal and social success is assured; if they purchase the right products and services, they can enhance their appearance and gain power over other people. For example, one study of television commercials reflects that men's roles are portrayed differently from women's roles (Kaufman, 1999). Men are more likely to be away from home or are more likely to be shown working or playing outside the house rather than inside, where women are more likely to be located. There is also a far greater likelihood that women will be doing domestic tasks such as cooking, cleaning, or shopping, whereas men are more likely to be taking care of cars or yards or playing games. As such, television commercials may act as agents of socialization, showing children and others what women's and men's designated activities are (Kaufman, 1999).

A study by the sociologist Anthony J. Cortese (1999) found that women—regardless of what they were doing in a particular ad—were frequently shown in advertising as being young, beautiful, and seductive. Although such depictions may sell products, they may also have the effect of influencing how we perceive ourselves and others with regard to issues of power and subordination. Women are bombarded with media images of ideal beauty and physical appearance (Cortese, 1999), and eating problems such as anorexia and bulimia are a major concern associated with many media depictions of the ideal body image for women. In both forms of eating disorders, distorted body image plays an important part, and this distorted image may be perpetuated by media depictions and advertisements. Clearly, the dramatic increase in eating problems in recent years cannot be attributed solely to advertising and the mass media; however, their potential impact on the gender socialization of young girls and women in establishing role models with whom to identify is extremely important (Kilbourne, 1994, 1999). Although advertisements and other media depictions of gender may sell products, they may also have the effect of influencing how we perceive ourselves and others with regard to issues of power and subordination.

Adult Gender Socialization

Gender socialization continues as women and men complete their training or education and join the work force. Men and women are taught the "appropriate" type of conduct for persons of their sex in a particular job or occupation—both by their employers and by co-workers. However, men's socialization usually does not include a measure of whether their work can be successfully combined with having a family; it is often assumed that men can and will do both. Even today, however, the reason given for women not entering some careers and professions is that this kind of work is not suitable for women because of their assumed child-care responsibilities.

Different gender socialization may occur as people reach their forties and enter "middle age." As discussed in Chapter 12 ("Aging and Inequality Based on Age"), a double standard of aging exists that affects women more than men. Often, men are considered to be at the height of their success as their hair turns gray and their face gains a few wrinkles. By contrast, not only do other people in society make middle-aged women feel as if they are "over the hill," but multi-million-dollar advertising campaigns continually call attention to women's every weakness, every pound gained, and every bit of flabby flesh, wrinkle, or gray hair.

A knowledge of how we develop a gender-related self-concept and learn to feel, think, and act in feminine or masculine ways is important for an understanding of ourselves. Examining gender socialization makes us aware of the impact of our parents, siblings, teachers, friends, and the media on our perspectives about gender. However, the gender socialization perspective has been criticized on several accounts. Childhood gender-role socialization may not affect people as much as some analysts have suggested. For example, the types of jobs that people take as adults may have less to do with how they were socialized in childhood than it does with how they are treated in the workplace. From this perspective, women and men will act in ways that bring them the most rewards and produce the fewest punishments (Reskin and Padavic, 2002). Also, gender socialization theories can be used to blame women for their own subordination. For example, if we assume that women's problems can all be blamed on women themselves, existing social structures that perpetuate gender inequality will be overlooked. We will now examine a few of those structural forces.

CONTEMPORARY GENDER INEQUALITY

According to feminist scholars, women experience gender inequality as a result of economic, political, and educational discrimination. Women's position in the U.S. work force reflects their overall subordination in society.

Gendered Division of Paid Work

Where people are located in the occupational structure of the labor market has a major effect on their earnings (Kemp, 1994). The workplace is another example of a gendered institution. In industrialized countries, most jobs are segregated by gender and by race/ethnicity. Sociologist Judith Lorber (1994: 194) gives this example:

> In a workplace in New York City—for instance, a handbag factory—a walk through the various departments might reveal that the owners and managers are white men; their secretaries and bookkeepers are white and Asian women; the order takers and data processors are African American women; the factory hands are [Latinos] cutting pieces and [Latinas] sewing them together; African American men are packing and loading the finished product; and non-English-speaking Eastern European women are cleaning up after everyone. The workplace as a whole seems integrated by race, ethnic group, and gender, but the individual jobs are markedly segregated according to social characteristics.

Lorber notes that in most workplaces, employees are either gender segregated or all of the same gender. *Gender-segregated work* refers to the concentration of women and men in different occupations, jobs, and places of work (Reskin and Padavic, 2002). Gender-segregated work is most visible in occupations that remain more than 90 percent female (for example, secretary, registered nurse, and bookkeeper/auditing clerk) or more than 90 percent male (for example, carpenter, construction worker, mechanic, truck driver, and electrical engineer) (U.S. Department of Labor, 2002). In 2001, for example, 98 percent of all secretaries in the United States were women; 90 percent of all engineers were men (U.S. Census Bureau, 2002). To eliminate gender-segregated jobs in the United States, more than half of all men or all women workers would have to change occupations. Moreover, women are severely underrepresented at the top of U.S. corporations. Only about 10 percent of the executive jobs at Fortune 500 companies are held by women, and only six women are the CEO of such a company (*Fortune,* 2001).

Although the degree of gender segregation in the professional labor market (including physicians, dentists, lawyers, accountants, and managers) has declined since the 1970s, racial–ethnic segregation has remained deeply embedded in the social structure. As the sociologist Elizabeth Higginbotham (1994) points out, African American professional women

Table 11.2   PERCENTAGE OF WORK FORCE REPRESENTED BY WOMEN, AFRICAN AMERICANS, AND HISPANICS IN SELECTED OCCUPATIONS

The U.S. Census Bureau accumulates data that show what percentage of the total work force is made up of women, African Americans, and Hispanics. As used in this table, *women* refers to females in all racial-ethnic categories, whereas *African American* and *Hispanic* refer to both women and men. Based on this table, in which occupations are white men most likely to be employed? In which occupations are they least likely to be employed?

	WOMEN	AFRICAN AMERICANS	HISPANICS
All occupations	46.6	11.3	10.9
Managerial and professional specialty (all)	50.0	8.3	5.1
Executive, administrative, and managerial	46.0	7.9	5.6
Professional specialty	53.7	8.6	4.7
Technical, sales, and administrative support (all)	63.7	11.4	9.1
Technicians and related support	53.4	10.3	7.5
Sales occupations	49.4	9.1	8.7
Service occupations (all)	60.4	17.9	16.3
Private household workers	96.2	12.1	32.8
Cleaning and building service workers	46.0	20.7	23.8
Health service assistants, aides, nurses, attendants	89.2	29.4	11.5
Operators, fabricators, and laborers	23.3	15.6	17.7

Source: U.S. Census Bureau, 2002.

find themselves limited to employment in certain sectors of the labor market. Most are concentrated in public sector employment (as public schoolteachers, welfare workers, librarians, public defenders, and faculty members at public colleges, for example) rather than in the private sector (for example, in large corporations, major law firms, private educational institutions, or private hospitals). Across all categories of occupations, white women and all people of color are not evenly represented, as shown in Table 11.2.

Labor market segmentation—the division of jobs into categories with distinct working conditions—results in women having separate and unequal jobs (Amott and Matthaei, 1996; Lorber, 1994). The pay gap between men and women is the best-documented consequence of gender-segregated work (Reskin and Padavic, 2002). Most women work in lower-paying, less-prestigious jobs, with little opportunity for advancement. Because many employers assume that men are the breadwinners, men are expected to make more money than women in order to support their families. For many years, women have been viewed as supplemental wage earners in a male-headed household, regardless of the women's marital status. Consequently, women have not been seen as legitimate workers but mainly as wives and mothers (Lorber, 1994, 2001).

Gender-segregated work affects both men and women. Men are often kept out of certain types of jobs. Those who enter female-dominated occupations often have to justify themselves and prove that they are "real men." They have to fight stereotypes (gay, "wimpy," and passive) about why they are interested in such work (Williams, 2004). Even if these assumptions do not push men out of female-dominated occupations, they affect how the men manage their gender identity at work. For example, men in occupations such as nursing emphasize their masculinity, attempt to distance themselves from female colleagues, and try to move quickly into management and supervisory positions (Williams, 2004).

Occupational gender segregation contributes to stratification in society. Job segregation is structural; it does not occur simply because individual workers have different abilities, motivations, and material needs. As a result of gender and racial segregation, employers are able to pay many men of color and all women less money, promote them less often, and provide fewer benefits. If they demand better working conditions or wages, workers are often reminded of

Figure 11.1   The Wage Gap

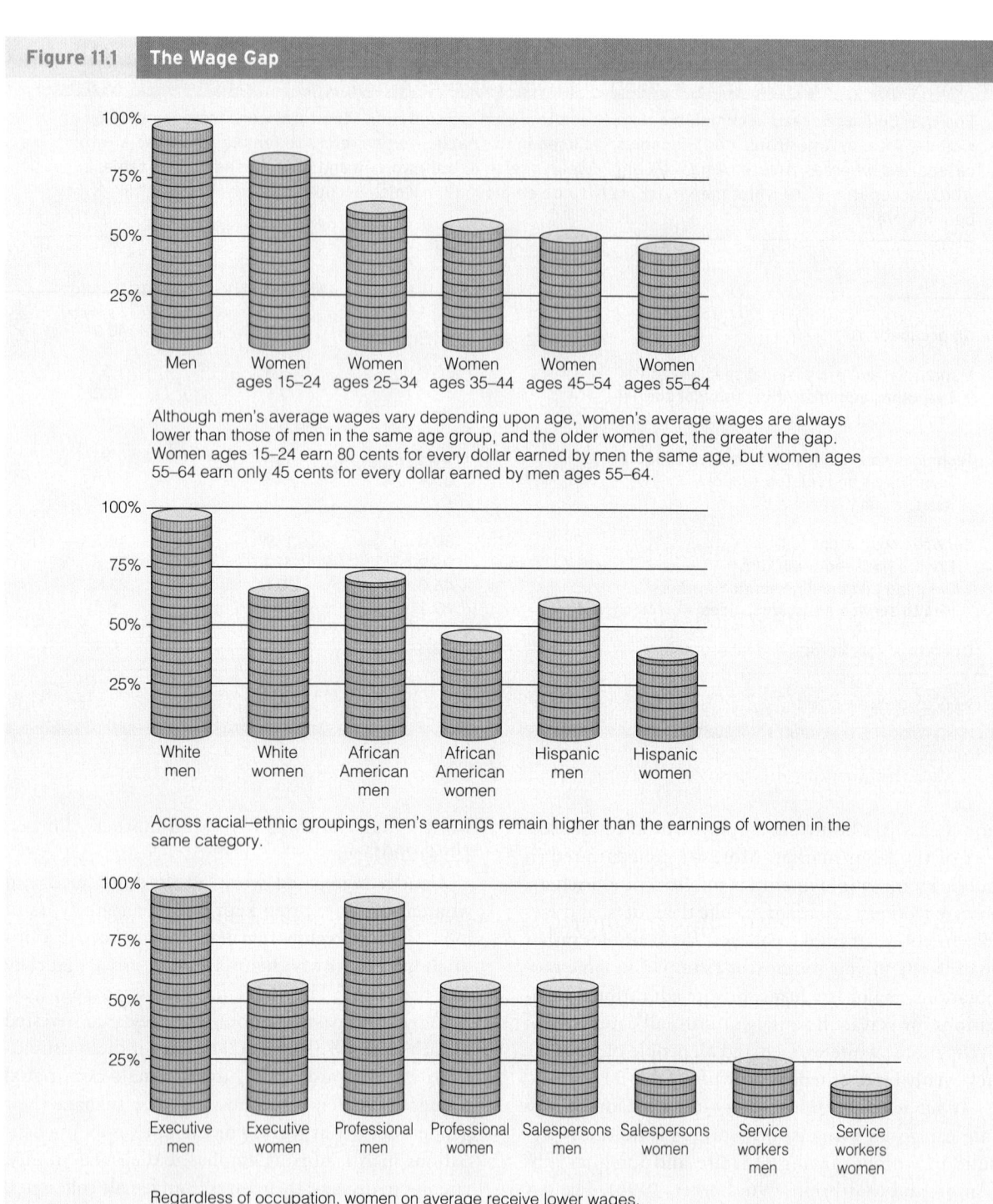

Although men's average wages vary depending upon age, women's average wages are always lower than those of men in the same age group, and the older women get, the greater the gap. Women ages 15–24 earn 80 cents for every dollar earned by men the same age, but women ages 55–64 earn only 45 cents for every dollar earned by men ages 55–64.

Across racial–ethnic groupings, men's earnings remain higher than the earnings of women in the same category.

Regardless of occupation, women on average receive lower wages.

Source: U.S. Census Bureau, 2002.

the number of individuals (members of Marx's "reserve army") who would like to have their jobs.

Pay Equity (Comparable Worth)

Occupational segregation contributes to a ***wage gap***— **the disparity between women's and men's earnings.**

It is calculated by dividing women's earnings by men's to yield a percentage, also known as the *earnings ratio* (Reskin and Padavic, 2002). Overall, women make almost 78 cents for every $1 earned by men. However, as Figure 11.1 shows, women at all levels of educational attainment receive less pay than men with the same levels of education. Across different categories of

What stereotypes are associated with men in female-oriented positions? With women in male-oriented occupations? Do you think such stereotypes will change in the near future?

employment, data continue to demonstrate this disparity. Moreover, the pay gap is even greater for women of color. Although white women in 2000 earned 73 percent as much as white men, African American women earned only 48 percent and Latinas 39 percent of what white male workers earned (U.S. Census Bureau, 2002). The gap increases as a person's salary grows: Whereas 48 percent of people earning $20,000 to $25,000 per year are women, women constitute less than 10 percent of people earning over $200,000 per year (Johnston, 2002). Likewise, among older workers, the pay gap between men and women is larger. Women's earnings tend to rise more slowly than men's when they are younger and then drop off when women reach their late thirties and early forties, whereas the wages of men tend to increase as they age (Reskin and Padavic, 2002).

Pay equity or **comparable worth is the belief that wages ought to reflect the worth of a job, not the gender or race of the worker.** How can the comparable worth of different kinds of jobs be determined? One way is to compare the actual work of women's and men's jobs and see if there is a disparity in the salaries paid for each. To do this, analysts break a job into components—such as the education, training, and skills required, the extent of responsibility for others' work, and the working conditions—and then allocate points for each (Lorber, 1994). For pay equity to exist, men and women in occupations that receive the same number of points should be paid the same. However, pay equity exists for very few jobs, and little change has occurred over the past two decades in many occupations. For example, for every dollar earned by men, women

in the same occupation earn 96 cents as secretaries, stenographers, and typists; 80 cents as restaurant servers; 75 cents as lawyers; and 67 cents as financial managers (U.S. Bureau of Labor Statistics, 2002).

Comparable worth is important for all workers, not just women. Both men and women suffer an economic penalty when they work in jobs that employ mostly women. Compared to what they could make in male-dominated jobs, men suffer a loss of wages in female-dominated jobs such as nurse, secretary, and elementary schoolteacher. Perhaps this helps explain why men are less likely to seek employment in female-dominated fields than women are to seek employment in male-dominated fields (Jacobs, 1993). In their study of men in female-dominated jobs, the sociologists Paula England and Melissa Herbert (1993) found that the differences in pay are due to the cultural devaluation of women. If women were compensated fairly, an employer could not undercut men's wages by hiring women at a cheaper rate (Kessler-Harris, 1990).

Gender-related appearance norms have become an issue in hiring, wages, and retention of workers. For example, age- and weight-based bias has been found to keep employees who do not meet these norms from being paid as much as others. Other issues pertaining to the gendered nature of the workplace are addressed in Chapter 13 ("The Economy and Work in Global Perspective").

Paid Work and Family Work

As previously discussed, the first big change in the relationship between family and work occurred with the

Industrial Revolution and the rise of capitalism. The cult of domesticity kept many middle- and upper-class women out of the work force during this period. Primarily, working-class and poor women were the ones who had to deal with the work/family conflict. Today, however, the issue spans the entire economic spectrum (Reskin and Padavic, 2002). The typical married woman in the United States combines paid work in the labor force with family work as a homemaker. Although this change has occurred at the societal level, individual women bear the brunt of the problem.

Even with dramatic changes in women's workforce participation, the sexual division of labor in the family remains essentially unchanged. Most married women now share responsibility for the breadwinner role, yet many men do not accept their share of domestic responsibilities (Reskin and Padavic, 2002). Consequently, many women have a "double day" or "second shift" because of their dual responsibilities for paid and unpaid work (Hochschild, 1989, 2003). Working women have less time to spend on housework; if husbands do not participate in routine domestic chores, some chores simply do not get done or get done less often. Although the income that many women earn is essential for the economic survival of their families, they still must spend part of their earnings on family maintenance, such as day-care centers, fast-food restaurants, and laundries, in an attempt to keep up with their obligations.

Especially in families with young children, domestic responsibilities consume a great deal of time and energy. Although some kinds of housework can be put off, the needs of children often cannot be ignored or delayed. When children are ill or school events cannot be scheduled around work, parents (especially mothers) may experience stressful role conflicts ("Shall I be a good employee or a good mother?"). Many working women care not only for themselves, their husbands, and their children but also for elderly parents or in-laws. Some analysts refer to these women as "the sandwich generation"—caught between the needs of their young children and of their elderly relatives.

Although both men and women profess that working couples should share household responsibilities, researchers find that family demands remain mostly women's responsibility, even among women who hold full-time paid employment. Many women try to solve their time crunch by forgoing leisure time and sleep. When Arlie Hochschild interviewed working mothers, she found that they talked about sleep "the way a hungry person talks about food" (1989: 9). Perhaps this is one reason that, in more recent research, Hochschild (1997) learned that some married women

with children found more fulfillment at work and that they worked longer hours because they liked work better than facing the pressures of home. Situations in which women are the sole earners in their families while their husbands (labeled by the media as "Mr. Moms" or "Stay-at-Home Dads") take care of the household and children are discussed in Chapter 15, "Families and Intimate Relationships."

PERSPECTIVES ON GENDER STRATIFICATION

Sociological perspectives on gender stratification vary in their approach to examining gender roles and power relationships in society. Some focus on the roles of women and men in the domestic sphere; others note the inequalities arising from a gendered division of labor in the workplace. Still others attempt to integrate both the public and private spheres into their analyses.

Functionalist and Neoclassical Economic Perspectives

As seen earlier, functionalist theory views men and women as having distinct roles that are important for the survival of the family and society. The most basic division of labor is biological: Men are physically stronger, and women are the only ones able to bear and nurse children. Gendered belief systems foster assumptions about appropriate behavior for men and women and may have an effect on the types of work that women and men perform.

The Importance of Traditional Gender Roles According to functional analysts such as Talcott Parsons (1955), women's roles as nurturers and caregivers are even more pronounced in contemporary industrialized societies. While the husband performs the *instrumental* tasks of providing economic support and making decisions, the wife assumes the *expressive* tasks of providing affection and emotional support for the family. This division of family labor ensures that important societal tasks will be fulfilled; it also provides stability for family members.

This view has been adopted by a number of politically conservative analysts who assert that relationships between men and women are damaged when changes in gender roles occur, and family life suffers

According to the human capital model, women may earn less in the labor market because of their childrearing responsibilities. What other sociological explanations are offered for the lower wages that women receive?

as a consequence. From this perspective, the traditional division of labor between men and women is the natural order of the universe (Kemp, 1994).

The Human Capital Model

Functionalist explanations of occupational gender segregation are similar to neoclassical economic perspectives, such as the human capital model (Horan, 1978; Kemp, 1994). According to this model, individuals vary widely in the amount of human capital they bring to the labor market. *Human capital* is acquired by education and job training; it is the source of a person's productivity and can be measured in terms of the return on the investment (wages) and the cost (schooling or training) (Stevenson, 1988; Kemp, 1994).

From this perspective, what individuals earn is the result of their own choices (the kinds of training, education, and experience they accumulate, for example) and of the labor market need (demand) for and availability (supply) of certain kinds of workers at specific points in time. For example, human capital analysts argue that women diminish their human capital when they leave the labor force to engage in child-bearing and child-care activities. While women are out of the labor force, their human capital deteriorates from nonuse. When they return to work, women earn lower wages than men because they have fewer years of work experience and have "atrophied human capital" because their education and training may have become obsolete (Kemp, 1994: 70).

Other neoclassical economic models attribute the wage gap to such factors as (1) the different amounts of energy that men and women expend on their work (women who spend substantial energy on their family and household have less to put into their work), (2) the occupational choices women make (choosing female-dominated occupations so that they can spend more time with their families), and (3) the crowding of too many women into some occupations (suppressing wages because the supply of workers exceeds demand) (Kemp, 1994).

Evaluation of Functionalist and Neoclassical Economic Perspectives

Although Parsons and other functionalists did not specifically endorse the gendered division of labor, their analysis suggests that it is natural and perhaps inevitable. However, critics argue that problems inherent in traditional gender roles, including the personal role strains of men and women and the social costs to society, are minimized by this approach. For example, men are assumed to be "money machines" for their families when they might prefer to spend more time in child-rearing activities. Also, the woman's place is assumed to be in the home, an assumption that ignores the fact that many women hold jobs due to economic necessity.

In addition, the functionalist approach does not take a critical look at the structure of society (especially the economic inequalities) that make educational and occupational opportunities more available to some than to others. Furthermore, it fails to examine the underlying power relations between men and women or to consider the fact that the tasks assigned to women and to men are unequally valued by society (Kemp, 1994). Similarly, the human capital model is rooted in the premise that individuals are evaluated based on their human capital in an open, competitive market where education, training, and other job-enhancing characteristics are taken into account. From this perspective, those who make less money (often men of color and all women) have no one to blame but themselves.

Critics note that instead of blaming people for their choices, we must acknowledge other realities. Wage discrimination occurs in two ways: (1) the wages are higher in male-dominated jobs, occupations, and segments of the labor market, regardless of whether women take time for family duties, and (2) in any job, women and people of color will be paid less (Lorber, 1994).

Conflict Perspectives

According to many conflict analysts, the gendered division of labor in families and in the workplace results from male control of and dominance over women and resources. Differentials between men and women may exist in terms of economic, political, physical, and/or interpersonal power. The importance of a male monopoly in any of these arenas depends on the significance of that type of power in a society (Richardson,

1993). In hunting and gathering and horticultural societies, male dominance over women is limited because all members of the society must work in order to survive (Collins, 1971; Nielsen, 1990). In agrarian societies, however, male sexual dominance is at its peak. Male heads of household gain a monopoly not only on physical power but also on economic power, and women become sexual property.

Although men's ability to use physical power to control women diminishes in industrial societies, men still remain the head of household and control the property. In addition, men gain more power through their predominance in the most highly paid and prestigious occupations and the highest elected offices. By contrast, women have the ability to trade their sexual resources, companionship, and emotional support in the marriage market for men's financial support and social status; as a result, however, women as a group remain subordinate to men (Collins, 1971; Nielsen, 1990).

All men are not equally privileged; some analysts argue that women and men in the upper classes are more privileged, because of their economic power, than men in lower-class positions and all people of color (Lorber, 1994). In industrialized societies, persons who occupy elite positions in corporations, universities, the mass media, and government or who have great wealth have the most power (Richardson, 1993). Most of these are men, however.

Conflict theorists in the Marxist tradition assert that gender stratification results from private ownership of the means of production; some men not only gain control over property and the distribution of goods but also gain power over women. According to Friedrich Engels and Karl Marx, marriage serves to enforce male dominance. Men of the capitalist class instituted monogamous marriage (a gendered institution) so that they could be certain of the paternity of their offspring, especially sons, whom they wanted to inherit their wealth. Feminist analysts have examined this theory, among others, as they have sought to explain male domination and gender stratification.

Feminist Perspectives

Feminism—**the belief that women and men are equal and that they should be valued equally and have equal rights**—is embraced by many men as well as women. It holds in common with men's studies the view that gender is a socially constructed concept that has important consequences in the lives of all people (Craig, 1992). According to the sociologist Ben Agger (1993), men can be feminists and propose feminist theories; both women and men have much in common as they seek to gain a better understanding of the causes and consequences of gender inequality. Over the past three decades, many different organizations have been formed to advocate causes uniquely affecting women or men and to help people gain a better understanding of gender inequality (see Box 11.4).

Feminist theory seeks to identify ways in which norms, roles, institutions, and internalized expectations limit women's behavior. It also seeks to demonstrate how women's personal control operates even within the constraints of relative lack of power (Stewart, 1994).

Liberal Feminism In liberal feminism, gender equality is equated with equality of opportunity. The roots of women's oppression lie in women's lack of equal civil rights and educational opportunities. Only when these constraints on women's participation are removed will women have the same chance for success as men. This approach notes the importance of gender-role socialization and suggests that changes need to be made in what children learn from their families, teachers, and the media about appropriate masculine and feminine attitudes and behavior. Liberal feminists fight for better child-care options, a woman's right to choose an abortion, and elimination of sex discrimination in the workplace.

Radical Feminism According to radical feminists, male domination causes all forms of human oppression, including racism and classism (Tong, 1989). Radical feminists often trace the roots of patriarchy to women's childbearing and child-rearing responsibilities, which make them dependent on men (Firestone, 1970; Chafetz, 1984). In the radical feminist view, men's oppression of women is deliberate, and ideological justification for this subordination is provided by other institutions such as the media and religion. For women's condition to improve, radical feminists claim, patriarchy must be abolished. If institutions are currently gendered, alternative institutions—such as women's organizations seeking better health care, day care, and shelters for victims of domestic violence and rape—should be developed to meet women's needs.

Socialist Feminism Socialist feminists suggest that women's oppression results from their dual roles as paid *and* unpaid workers in a capitalist economy. In the workplace, women are exploited by capitalism; at home, they are exploited by patriarchy (Kemp, 1994). Women are easily exploited in both sectors; they are paid low wages and have few economic resources. Gendered job segregation is "the primary mechanism in capitalist society that maintains the superiority of men over women, because it enforces lower wages for

women in the labor market" (Hartmann, 1976: 139). As a result, women must do domestic labor either to gain a better-paid man's economic support or to stretch their own wages (Lorber, 1994). According to socialist feminists, the only way to achieve gender equality is to eliminate capitalism and develop a socialist economy that would bring equal pay and rights to women.

Multicultural Feminism Recently, academics and activists have been rethinking the experiences of women of color from a feminist perspective. The experiences of African American women and Latinas/Chicanas have been of particular interest to some social analysts. Building on the civil rights and feminist movements of the late 1960s and early 1970s, some contemporary black feminists have focused on the cultural experiences of African American women. A central assumption of this analysis is that race, class, and gender are forces that simultaneously oppress African American women (Hull, Bell-Scott, and Smith, 1982). The effects of these three statuses cannot be adequately explained as "double" or "triple" jeopardy (race + class + gender = a poor African American woman) because these ascribed characteristics are not simply added to one another. Instead, they are multiplicative in nature (race × class × gender); different characteristics may be more significant in one situation than another. For example, a well-to-do white woman (class) may be in a position of privilege when compared to people of color (race) and men from lower socioeconomic positions (class), yet be in a subordinate position as compared with a white man (gender) from the capitalist class (Andersen and Collins, 1998). In order to analyze the complex relationship among these characteristics, the lived experiences of African American women and other previously "silenced people" must be heard and examined within the context of particular historical and social conditions.

Another example of multicultural feminist studies is the work of the psychologist Aida Hurtada (1996), who explored the cultural identification of Latina/ Chicana women. According to Hurtada, distinct differences exist between the world views of the white (non-Latina) women who participate in the women's movement and many Chicanas, who have a strong sense of identity with their own communities. Hurtada (1996) suggests that women of color do not possess the "relational privilege" that white women have because of their proximity to white patriarchy through husbands, fathers, sons, and others. Like other multicultural feminists, Hurtada calls for a "politics of inclusion," creating social structures that lead to positive behavior and bring more people into a dialogue about how to improve social life and reduce inequalities.

Feminists who analyze race, class, and gender suggest that equality will occur only when all women, regardless of race/ethnicity, class, age, religion, sexual orientation, or ability (or disability), are treated more equitably (Andersen and Collins, 1998).

Feminist Perspectives on Eating Problems

Feminist analysts suggest that eating problems are not just individual "disorders" but relate also to the issue of subordination (see Orbach, 1978; Fallon, Katzman, and Wooley, 1994). This analysis focuses on the relationship between eating problems and patriarchy (male dominance) in the labor force and family. Eating problems cannot be viewed solely as psychologi-

The disparity between women's and men's earnings is even greater for women of color. Feminists who analyze race, class, and gender suggest that equality will occur only when all women are treated more equitably.

cal "disorders"; they are symbolic of women's personal and cultural oppression. Anorexia and bulimia reflect women's (and sometimes men's) denial of other problems, disconnection from other people, and disempowerment in society (Peters and Fallon, 1994).

Feminist scholars have begun to look at ways in which race/ethnicity may be linked to eating problems (Root, 1990). By contrast, most early research focused on the problems of white, middle- to upper-class females; women of color were, at most, mentioned in a footnote (for example, see Brumberg, 1988). In a study of women with eating problems, the sociologist Becky Thompson (1994) found that more than half the women of color had been victims of sexual abuse, racism, anti-Semitism, and/or homophobia. However, she suggests that it is impossible to determine a single ethic about socialization and eating problems among women of color. For example, thinness was not highly valued by Latinas with family roots in the Dominican Republic, whereas upper-class women from Argentinean families had been taught that a woman's weight was the primary criterion of her worth (Thompson, 1994: 29). Similarly, African American women from the rural South had been socialized to believe that weight did not determine beauty to the extent that skin color and hair did (Thompson, 1994). African American women who, as girls, attended schools where there were few African Americans were more likely to view thinness as a desirable characteristic of women (Thompson, 1994).

Eating problems may also be associated with social class and sexual orientation. For example, some lower-class women may view binging as a momentary reprieve from poverty and other worries. Some lesbians may develop eating problems in rebellion against cultural expectations that attempt to force heterosexuality or at least heterosexual values on them, in sharp contradiction to their own sexual identities (Thompson, 1994). Two feminist studies comparing lesbian and heterosexual women found that both groups are influenced by cultural pressures to be thin but that lesbians tend to be more satisfied with their bodies and to desire a somewhat higher ideal weight (Brand, Rothblum, and Solomon, 1992; Herzog et al., 1992). Gay men, on the other hand, may be more prone to eating problems because of the importance that some place on low body weight and/or physical attractiveness (see Shisslak and Crago, 1992).

Evaluation of Conflict and Feminist Perspectives Conflict and feminist perspectives provide insights into the structural aspects of gender inequality in society. These approaches emphasize factors external to individuals that contribute to the oppression of white women and people of color; however, they have been criticized for emphasizing the differences between men and women without taking into account the commonalities they share. Feminist approaches have also been criticized for their emphasis on male dominance without a corresponding analysis of the ways in which some men may also be oppressed by patriarchy and capitalism.

Recently, the debate has continued about whether the feminist movement has diminished the well-being of boys and men. Sociologist William J. Goode (1982) suggests that some men have felt "under attack" by women's demands for equality because the men do not see themselves as responsible for societal conditions such as patriarchy but attribute their own achievements to hard work and intelligence, not to built-in societal patterns of male domination and female subordination.

In *Stiffed: The Betrayal of the American Man,* the author and journalist Susan Faludi (1999) argues that men are not as concerned about the possibility of the feminist movement diminishing their own importance as they are experiencing an identity crisis brought about by the current societal emphasis on wealth, power, fame, and looks (often to the exclusion of significant social values). According to Faludi, men are increasingly aware of their body image and are spending ever-increasing sums of money on men's cosmetics, health and fitness gear and classes, and

cigar bars and "gentlemen's clubs." The popularity of her book, among others, suggests that issues of gender inequality and of men's and women's roles in society are far from resolved.

GENDER ISSUES IN THE FUTURE

During the twentieth century, women made significant progress in the labor force (Reskin and Padavic, 2002). Laws were passed to prohibit sexual discrimination in the workplace and school. Affirmative action programs helped make women more visible in education, government, and the professional world. More women entered the political arena as candidates instead of as volunteers in the campaign offices of male candidates (Lott, 1994).

Many men joined movements to raise their consciousness, realizing that what is harmful to women may also be harmful to men. For example, women's lower wages in the labor force suppress men's wages as well; in a two-paycheck family, women who are paid less contribute less to the family's finances, thus placing a greater burden on men to earn more money.

In the midst of these changes, however, many gender issues remain unresolved. In the labor force, gender segregation and the wage gap are still problems. Although women continue to narrow the pay gap, women earn about 78 cents for every dollar compared with men, and analysts believe that the narrowing can partly be attributed to the economic boom (for some) of the 1990s. Employers have had to look for employees based on merit (rather than race, class, and gender) in order to have the number and types of employees they need to meet the demands of global competition (Barakat, 2000).

Although some employers have implemented family-leave policies, these do not relieve women's domestic burden in the family. Analysts believe that the burden of the "double day" or "second shift" will probably preserve women's inequality at home and in the workplace for another generation (Reskin and Padavic, 2002).

In regard to specific problems such as the eating disorders discussed in this chapter, we have a long way to go before all people, particularly women, are valued as human beings rather than as possessions or sex objects. What do you think might be a first step toward reducing gender inequality in the United States and the world?

CHAPTER REVIEW

■ How do sex and gender differ?

Sex refers to the biological categories and manifestations of femaleness and maleness; *gender* refers to the socially constructed differences between females and males. In short, sex is what we (generally) are born with; gender is what we acquire through socialization.

■ How do gender roles and gender identity differ from gendered institutions?

Gender role encompasses the attitudes, behaviors, and activities that are socially assigned to each sex and that are learned through socialization. Gender identity is an individual's perception of self as either female or male. By contrast, gendered institutions are those structural features that perpetuate gender inequality.

■ How does the nature of work affect gender equity in societies?

In most hunting and gathering societies, fairly equitable relationships exist between women and men because neither sex has the ability to provide all of the food necessary for survival. In horticultural societies, a fair degree of gender equality exists because neither sex controls the food supply. In agrarian societies, male dominance is very apparent; agrarian tasks require more labor and physical strength, and females are often excluded from these tasks because they are viewed as too weak or too tied to child-rearing activities. In industrialized societies, a gap exists between nonpaid work performed by women at home and paid work performed by men and women. A wage gap also exists between women and men in the marketplace.

■ What are the key agents of gender socialization?

Parents, peers, teachers and schools, sports, and the media are agents of socialization that tend to reinforce stereotypes of appropriate gender behavior.

■ What causes gender inequality in the United States?

Gender inequality results from economic, political, and educational discrimination against women. In most workplaces, jobs are either gender segregated or the majority of employees are of the same gender. Although the degree of gender segregation in the professional workplace has declined since the 1970s, racial and ethnic segregation remains deeply embedded.

■ How is occupational segregation related to the wage gap?

Many women work in lower-paying, less-prestigious jobs than men. This occupational segregation leads to a disparity, or wage gap, between women's and men's earnings. Even when women are employed in the same job as men, on average they do not receive the same, or comparable, pay.

■ How do functionalists and conflict theorists differ in their view of division of labor by gender?

According to functionalist analysts, women's roles as caregivers in contemporary industrialized societies are crucial in ensuring that key societal tasks are fulfilled. The husband performs the instrumental tasks of economic support and decision making; the wife assumes the expressive tasks of providing affection and emotional support for the family. According to conflict analysts, the gendered division of labor within families and the workplace—particularly in agrarian and industrial societies—results from male control and dominance over women and resources.

KEY TERMS

body consciousness 360
comparable worth 379
feminism 382
gender 359
gender bias 371
gender identity 360
gender role 359
hermaphrodite 357
homophobia 359
matriarchy 362
patriarchy 362
primary sex characteristics 357
secondary sex characteristics 357
sex 357
sexism 362
sexual harassment 372
sexual orientation 358
transsexual 357
transvestite 358
wage gap 378

QUESTIONS FOR CRITICAL THINKING

1. Do the media reflect societal attitudes on gender, or do the media determine and teach gender behavior?

(As a related activity, watch television for several hours and list the roles for women and men depicted in programs and those represented in advertising.)

2. Review the concept of cultural relativism in Chapter 3. Should the U.S. State Department and human rights groups such as Amnesty International protest genital mutilation in other nations and withhold any funding or aid for those nations until they cease the practice?

3. Examine the various academic departments at your college. What is the gender breakdown of the faculty in selected departments? What is the gender breakdown of undergraduates and graduate students in those departments? Are there major differences among social science, science, and humanities departments? What hypothesis can you come up with to explain your observations?

RESOURCES ON THE INTERNET

Chapter-Related Web Sites

The following Web sites have been selected for their relevance to the topics in this chapter. These sites are among the more stable, but please note that Web site addresses change frequently. For an updated list of chapter-related Web sites with URL links, please visit the *Sociology in Our Times* Web site (**www.wadsworth.com/KendallSIOT**).

Women Studies Database: Gender Issues
http://www.inform.umd.edu/EdRes/Topic/ WomenStudies/GenderIssues

This University of Maryland Web site is a gateway to numerous online resources concerning gender issues. Information is readily available on many topics, including women in the work force, sex discrimination, women in sports, and the glass ceiling.

American Men's Studies Association (AMSA)
http://mensstudies.org

This site is maintained by the American Men's Studies Association, a nonprofit professional organization of scholars, therapists, and other individuals interested in the examination of masculinity in modern society. The site houses information about the AMSA, access to the *Journal of Men's Studies,* and interesting Web links.

Women Watch
http://www.un.org/womenwatch

Visit the United Nation's "Internet gateway on the Advancement and Empowerment of Women" to access information on global women's issues. The site features links, reports, analyses, databases, and online forums that address the most current information on women's rights and issues around the world.

ONLINE STUDY AND RESEARCH TOOLS

Accompanying this text are many *free* powerful online study tools that will help you master the material in this chapter, help increase your depth of understanding, and help you make the grade!

SocCoach CD-ROM

Use the SocCoach CD-ROM enclosed with this text to help you formulate a customized study plan for this chapter. After you take the Diagnostic Quiz, SocCoach will generate a customized study plan just for you! It will identify sections of the chapter that you should review and will provide videos, charts, graphs, and excerpts from the text to supplement your studies and enhance your understanding. You'll also find fun, interactive activities such as Virtual Explorations and Map the Stats to apply what you've learned and stretch your sociological imagination.

The Companion Web Site for Sociology in Our Times, *Fifth Edition*
www.wadsworth.com/KendallSIOT

Gain an even better grasp on this chapter by going to the companion Web site to take one of the Tutorial Quizzes, use the Flash Cards to master key terms, or check out the many other study aids you'll find there. You'll also find special features such as GSS Data and Census 2000 information that'll put data and resources at your fingertips to help you with that special project or help you as you do some research on your own.

In this chapter, when you see the icon on the left, it alerts you to a specific exercise found in *Wadsworth's Sociology Online Resources and Writing Companion.* This valuable guide shows you how to use Wadsworth's exclusive online resources—*InfoTrac College Edition,* the *Opposing Viewpoints Resource Center,* and *MicroCase Online*—to assist you in your study of sociology and to build essential research and writing skills.

Aging and Inequality Based on Age

We're a new generation. When we grew up, anybody fifty or sixty was considered old. I remember as a young boy, thirteen, fourteen, attending the twenty-fifth wedding anniversary of my mother and father. Everybody was dancing and singing and having a wonderful time. I remember saying to myself: "What are they so happy about? They're on the verge of dying." They were maybe fifty-five.

There's a new breed now. I'm going to be seventy-nine [soon,] and I don't for a second consider myself old. I still play a good game of golf, and I exercise and swim and am active in business. There are many corporations out there that feel you're old and should be out of it, no matter how you look or feel. . . . I feel I have more vitality than those who call me an old man. You turn around and want to know who . . . they're talking about.

—Jack Culberg, age 78, a former CEO of several large corporations, describing his views on growing older (qtd. in Terkel, 1996: 9)

I f we apply our sociological imagination to Jack Culberg's experience, we see that views on aging and the problems that people experience in growing older are not just personal problems but also public issues of concern to everyone. Eventually, all of us will be affected by aging. ***Aging* is the physical, psychological, and social processes associated with growing older** (Atchley and Barusch, 2004). In the United States and some other high-income countries, older people are the targets of prejudice and discrimination based on myths about aging. For example, older persons may be viewed as incompetent solely on the basis of their age. Although some older people may need assistance from others and support from society, many others are physically, socially, and financially independent.

Almost 90 percent of the people living in the high-income countries today are expected to survive to age 65. In the United States, about 85 percent of the women and 77 percent of the men born between 1995 and 2000 are expected to survive to that age (United Nations Development Programme, 2003). By sharp contrast, in low-income countries such as Zambia, Uganda, and Rwanda, nearly half of the total population in each nation is not expected to survive to *age 40* (United Nations Development Programme, 2003). In this chapter, we examine the sociological aspects of aging. We will also examine how older people seek dignity, autonomy, and empowerment in societies such as the United States that may devalue those who do not fit the ideal norms of youth, beauty, physical fitness, and self-sufficiency. Before reading on, test your knowledge about aging and age-based discrimination in the United States by taking the quiz in Box 12.1.

QUESTIONS AND ISSUES

Chapter Focus Question: Given the fact that aging is an inevitable consequence of living—unless a person dies young—why do many people in the United States devalue older persons?

How does functional age differ from chronological age?

How does age determine a person's roles and statuses in society?

What actions can be taken to bring about a more equitable society for older people?

THE SOCIAL SIGNIFICANCE OF AGE

"How old are you?" This is one of the most frequently asked questions in the United States. Beyond indicating how old or young a person is, age is socially significant because it defines what is appropriate for or expected of people at various stages. For example, child development specialists have identified stages of cognitive development based on children's ages:

> [W]e do not expect our preschool children, much less our infants, to have adultlike memories or to be completely logical. We are seldom surprised when a 4-year-old is misled by appearances; we express little dismay when our 2-year-old calls a duck a chicken. . . . But we would be surprised if our 7-year-olds continued to think segmented routes were shorter than other identical routes or if they continued to insist on calling all reasonably shaggy-looking pigs "doggy." We expect some intellectual (or cognitive) differences between preschoolers and older children. (Lefrançois, 1996: 196)

At the other end of the age continuum, a 75-year-old grandmother who travels through her neighborhood on in-line skates will probably raise eyebrows and perhaps garner media coverage about her actions because she is defying norms regarding age-appropriate behavior.

When people say "Act your age," they are referring to *chronological age*—**a person's age based on date of birth** (Atchley and Barusch, 2004). However, most of us actually estimate a person's age on the basis of *functional age*—**observable individual attributes such as physical appearance, mobility, strength, coordination, and mental capacity that are used to**

Bob Daemmrich/Stock Boston

Childhood is one of the most significant periods in the life course. For some children, these are carefree years; for others, these are times of powerlessness and vulnerability.

assign people to age categories (Atchley and Barusch, 2004). Because we typically do not have access to other people's birth certificates to learn their chronological age, visible characteristics—such as youthful appearance or gray hair and wrinkled skin—may become our criteria for determining whether someone is "young" or "old." According to the historian Lois W. Banner (1993: 15), "Appearance, more than any other factor, has occasioned the objectification of aging. We define someone as old because he or she looks old." In fact, feminist scholars believe that functional age is so subjective that it is evaluated differently for women and men—as men age, they are believed to become more distinguished or powerful, whereas when women grow older, they are thought to be "over-the-hill" or grandmotherly (Banner, 1993).

Box 12.1 SOCIOLOGY AND EVERYDAY LIFE

How Much Do You Know About Aging and Age-Based Discrimination?

True	False	
T	F	1. Most older persons have serious physical or mental disabilities.
T	F	2. Women in the United States have a longer life expectancy than do men.
T	F	3. Scientific studies have documented the fact that women age faster than men do.
T	F	4. Most older persons are economically secure today as a result of Social Security, Medicare, and pensions.
T	F	5. Studies show that advertising no longer stereotypes older persons.
T	F	6. After women reach menopause, they may enjoy sexual activity more than when they were younger.
T	F	7. Organizations representing older individuals have demanded the same rights and privileges as those accorded to younger persons.
T	F	8. The rate of elder abuse in the United States has been greatly exaggerated by the media.

Answers on page 392.

Trends in Aging

Over the past two decades, the U.S. population has been aging. The median age (the age at which half the people are younger and half are older) has increased by 6 years—from 30 in 1980 to 36 in 2000 (U.S. Census Bureau, 2002). This change was partly a result of the Baby Boomers (people born between 1946 and 1964) moving into middle age and partly a result of more people living longer. As shown in Figure 12.1a, the number of older persons—age 65 and above—increased significantly between 1980 and 2000. The population over age 85 has been growing especially fast.

Referred to as the *graying of America,* the aging of the U.S. population resulted from an increase in life expectancy combined with a decrease in birth rates (Atchley and Barusch, 2004). **Life expectancy is the average number of years that a group of people born in the same year could expect to live.** Based on the death rates in the year of birth, life expectancy shows the average length of life of a *cohort*—**a group of people born within a specified period of time.** Cohorts may be established on the basis of one-, five-, or ten-year intervals; they may also be defined by events taking place at the time of their birth, such as

Depression-era babies or Baby Boomers (Moody, 2002). For the cohort born in 2000, as an example, life expectancy at birth was 77.1 years for the overall U.S. population—74.2 for males and 79.9 for females. However, as Figure 12.1b shows, there are significant racial–ethnic and sex differences in life expectancy. Although the life expectancy of people of color has improved over the past 50 years, higher rates of illness and disability—attributed to poverty, inadequate health care, and greater exposure to environmental risk factors—still persist.

Today, a much larger percentage of the U.S. population is over age 65 than in the past. In 1900, about 4 percent of the U.S. population was over age 65; by 1980, that number had risen to approximately 11 percent. In 2000, approximately 14 percent of the population was age 65 or over (U.S. Census Bureau, 2002). According to Census Bureau projections, about 20 percent of the population will be at least age 65 by 2050. One of the fastest growing segments of the population is made up of people age 85 and above. This cohort is expected to almost double in size between 2000 and 2025 (U.S. Census Bureau, 2002); by the year 2050, the Census Bureau predicts that the number of persons age 85 and over will have increased to about 18 million (almost 5 percent of the popula-

Box 12.1 SOCIOLOGY AND EVERYDAY LIFE

Answers to the Sociology Quiz on Aging and Age-Based Discrimination

1. **False.** Only about 14 percent of older people have severe functional limitations; at age 85 or over, 31 percent have severe disabilities.

2. **True.** In 2000, female life expectancy (at birth) was 80 years, as compared with 74 for males. These figures vary by race and ethnicity.

3. **False.** No studies have documented that women actually age faster than men. However, some scholars have noted a "double standard" of aging that places older women at a disadvantage with respect to older men because women's worth in the United States is often defined in terms of physical appearance.

4. **False.** Although some older persons are economically secure, persons who rely solely on Social Security, Medicare, and/or pensions tend to live on low, fixed incomes that do not adequately meet their needs. A number live below the official poverty line.

5. **False.** Studies have shown that advertisements frequently depict older persons negatively—for example, as sickly and silly.

6. **True.** Women may enjoy sexual activity more after menopause because their sexual enjoyment is now separated from the possibility of pregnancy.

7. **True.** Organizations such as the Gray Panthers and AARP have been instrumental in the enactment of legislation beneficial to older persons.

8. **False.** Although cases of abuse and neglect of older persons are highly dramatized in the media, most coverage pertains to problems in hospitals, nursing homes, or other long-term care facilities. We know very little about the nature and extent of abuse that occurs in private homes.

Sources: Based on Atchley and Barusch, 2004; Belsky, 1999; and U.S. Census Bureau, 2002.

---

tion). Even more astonishing is the fact that the number of centenarians (persons 100 years of age and above) in this country will increase more than 12 times, from 66,000 in 1999 to about 834,000 in 2050 (L. Ramirez, 1999).

The current distribution of the U.S. population is depicted in the "age pyramid" in Figure 12.1d. If, every year, the same number of people are born as in the previous year and a certain number die in each age group, the rendering of the population distribution should be pyramid shaped. As you will note, however, Figure 12.1d is not a perfect pyramid, but instead reflects declining birth rates among post-Baby Boomers.

As a result of changing population trends, research on aging has grown dramatically in the past 50 years. **Gerontology is the study of aging and older people.** A subfield of gerontology, *social gerontology,* is the study of the social (nonphysical) aspects of aging, including such topics as the societal consequences of an aging population and the personal experience of

aging (Atchley and Barusch, 2004). According to gerontologists, age is viewed differently from society to society, and its perception changes over time.

Age in Historical and Contemporary Perspectives

People are assigned to different roles and positions based on the age structure and role structure in a particular society. *Age structure* is the number of people of each age level within the society; *role structure* is the number and type of positions available to them (Riley and Riley, 1994). Over the years, the age continuum has been chopped up into finer and finer points. Two hundred years ago, people divided the age spectrum into "babyhood," a *very* short childhood, and then adulthood. What we would consider "childhood" today was quite different two hundred years ago, when agricultural societies needed a large number of strong

Figure 12.1 Trends in Aging and Life Expectancy

a. U.S. Population Growth, 1980–2000
The percentage of persons 65 years of age and above increased dramatically between 1980 and 2000.

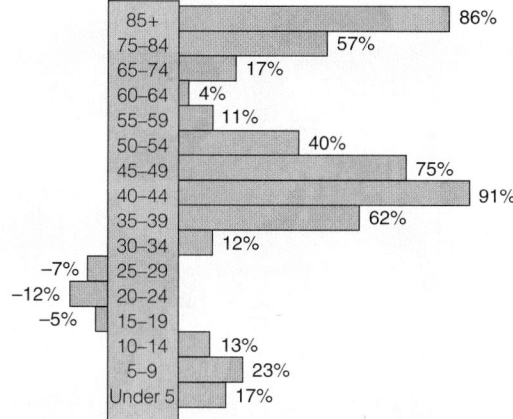

b. Selected Life Expectancies by Race, Ethnicity, and Sex, 2000
There are significant racial–ethnic and sex differences in life expectancy.

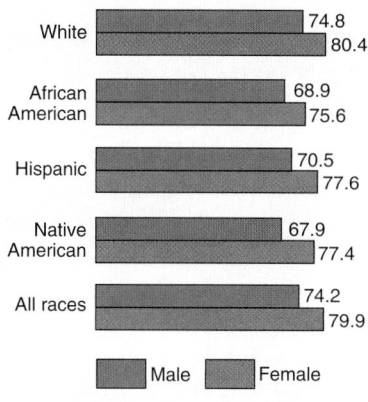

c. Percentage Distribution of U.S. Population by Age, 2000–2050 (projected)
Projections indicate that an increasing percentage of the U.S. population will be over age 65; one of the fastest growing categories is persons age 85 and over.

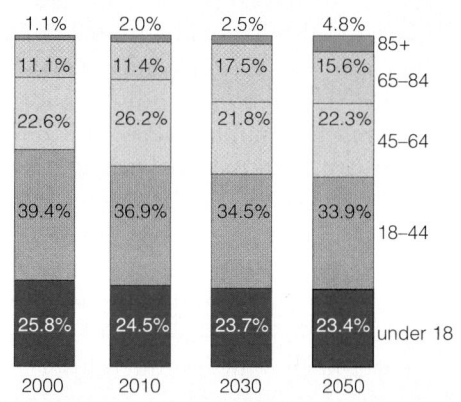

d. U.S. Age Pyramid by Age and Sex, 1997 (in millions)
Declining birth rates and increasing aging of the population are apparent in this age pyramid.

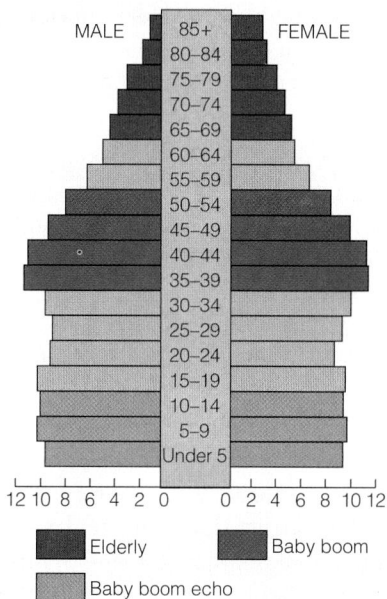

Source: U.S. Census Bureau, 2002.

arms and backs to work on the land to ensure survival. When 95 percent of the population had to be involved in food production, categories such as toddlers, pre-schoolers, preteens, teenagers, young adults, the middle-aged, and older persons did not exist.

If the physical labor of young people is necessary for society's survival, then young people are considered "little adults" and are expected to act like adults and do adult work. Older people are also expected to continue to be productive for the benefit of the society as long as they are physically able. In preindustrial societies, people of all ages help with the work, and little training is necessary for the roles that they fill. During the seventeenth and eighteenth centuries in the United States, for example, older individuals helped with the work and were respected because they were needed—and because few people lived that long (Gratton, 1986).

AGE IN GLOBAL PERSPECTIVE

Physical and sociocultural environments have different effects on how people experience aging and old age. In fact, concepts such as *young* and *old* may vary considerably from culture to culture. Unlike the sophisticated data-gathering techniques used to determine the number of older people in high-income and middle-income nations, we know less about the life expectancies and aging populations in hunting and gathering, horticultural, pastoral, and agrarian societies. However, individuals in every age category in hunting and gathering societies have a greater likelihood of dying. Reaching the age of 30 or 40 is less likely for people in low-income (less-developed) nations than reaching the age of 70 or 80 in many high-income (developed) countries. The proportion of older persons in a society also influences the individual's experience of old age.

Hunting and Gathering Societies

People in hunting and gathering societies typically have shorter life expectancies than people in high-income nations such as the United States. As Chapter 5 discusses, people in hunting and gathering societies are not able to accumulate a food surplus and must spend much of their time seeking food. They do not have permanent housing that protects them from the environment. Lack of food surpluses and lack of housing that protects one from the environment are factors associated with higher death rates and a smaller proportion of the total population living beyond 30 or 40 years of age. In hunting and gathering societies, younger people are a valuable asset in the nomadic lifestyle involved in hunting wild game. Although it is not universally true, older people in some hunting and gathering societies may be viewed as a liability because they typically move more slowly, are less agile, and may be perceived as being less productive.

Horticultural, Pastoral, and Agrarian Societies

With the domestication of plants (gardening) and animals (herding), and the availability of the plow and

© Jacob Halaska/Index Stock Imagery

In lower-income nations, older people may still perform useful economic functions such as gathering food or farming.

other agricultural technologies, death rates begin to decline. Although more people reach older ages in horticultural, pastoral, and agrarian societies, life is still very hard for most people. Since it is possible to accumulate a surplus, older individuals, particularly men, are often the most privileged in a society because they have the most wealth, power, and prestige. In agrarian societies, farming makes it possible for more people to live to adulthood and to more-advanced years. In these societies, the proportion of older people living with other family members is extremely high, with few elderly living alone. Studies of people in traditional West African towns, such as the cocoa farmers in Ghana, have found that even if people leave home for a time, they want to spend their later years in their hometown so that they can participate in kin networks and religious ceremonies and be buried there (Stucki, 1992).

In recent years, a growing number of people are reaching age 60 and above in some middle- and lower-income nations. Consider that India, for example, has about a billion people in its population. If only 5 percent of the population reaches age 60 or above, there will still be a significant increase in the number of older people in that country. Because so much of the world's population resides in India and other low-income nations, the proportion of older people in these countries will increase dramatically during the twenty-first century.

Industrial and Postindustrial Societies

In industrial societies, living standards improve and advances in medicine contribute to greater longevity for more people. Although it is often believed that less-industrialized countries accord greater honor, prestige, and respect to older people, some studies have found that the stereotypical belief that people in such nations will be taken care of by their relatives, particularly daughters and sons, is not necessarily true today (Martin and Kinsella, 1994).

In postindustrial societies, information technologies are extremely important, and a large proportion of the working population is employed in the service sector of the economy. The largest employers are in the fields of education and health care, both of which may benefit older people. "Lifelong learning courses" are offered in many colleges and universities to provide educational and leisure-time opportunities for older members of society. Some more-affluent older people may move away from family and friends upon retirement in pursuit of recreational facilities or a better climate (such as the popular move from the northeastern United States to the southern "sunbelt" states). Others may relocate to be closer to children or other relatives. Innovative forms of transportation and massive interstate highway systems have also contributed to a number of either temporary or permanent migratory patterns among older people, including an activity known as "RVing" (see Box 12.2).

In the United States, the trend is toward multigenerational families in which two or more generations are in the age range that has been traditionally labeled as "old." The shift from a society that was primarily young to a society that is older will bring about major changes in societal patterns and in the needs of the population. Issues that must be addressed include the health care system, the Social Security system, transportation, housing, and recreation.

A Case Study: Aging, Gender, and Japanese Society

The "graying" of the Japanese population has taken only 25 years, whereas it took almost a century in North America and Europe. If the present trend continues in Japan, by 2025 people age 65 and over will make up about 25 percent of the total population,

and more than half of the older population will be over 75 years of age. It is widely assumed that older people are respected and revered in Japan; however, several recent studies suggest that sociocultural changes and population shifts may be bringing about a gradual change in the social importance of the elderly in that nation. Until recently, most of the focus on the aging population in Japan has been on men and the workplace. Currently, feminist activists in that country are questioning why there has been so little attention given to the health and aging of women who are age 40 and above. When women have been discussed in regard to the aging population, the subject has primarily been caregiving for elderly relatives. For example, the "homebody" or "professional housewife" who cares not only for her husband and children but for other relatives, particularly the ill and the aged, has been used as the "ideal" model of the Japanese woman, against whom all others are measured. However, with over 60 percent of Japanese women in the labor force, greater pressure is being placed on Japanese policy makers to consider how the government can play a larger role in the care of the aging population, rather than placing the burden completely on families, particularly women. As one analyst explained,

> Because financially secure middle-class women are assumed to represent Japanese women as a whole, the situation of the majority who must give up work to look after their relatives, often at great cost to the well-being of the entire family, is usually erased from national consciousness. Moreover, many live to be well over ninety years of age, and daughters-in-law in their seventies find themselves . . . caring for one or more incontinent, immobile, and sometimes senile relatives. Furthermore, since stroke is the usual cause of disability among the elderly population in Japan, intensive nursing is often required, but men assist very rarely with this onerous duty. It is, therefore, the debate about home nursing and living together as a three-generation family that takes up most of the energy of activist women in Japan today. . . . (Lock, 1999: 61)

Challenges such as these are presented by the aging of the population not only in Japan but also in the United States and other nations. Social policy issues about older people reflect the intertwining nature of class, gender, and age as individuals and nations adapt to the "graying" of their populations. Those persons who are growing older, whether in Japan or any other nation, seek to have both greater longevity and a high quality of life during their later years.

Box 12.2 CHANGING TIMES: MEDIA AND TECHNOLOGY

"On the Road Again": Older People and the Irrationality of Stereotypes

We have all seen them: a grey-haired couple in a motor home or pulling a trailer with their pickup truck. We see them heading south in November about the time when the furnace starts to come on at night. Their return, like the robins', is one of the sure signs of spring. Some of us have made uncomplimentary comments about their driving skills. "Come on Grandpa. Drive it or park it!" They aren't in a hurry, even if we are. They just amble along, taking their time. Like they have all day. Or all week. And they can be almost impossible to pass, especially on a mountain road or a busy highway. . . . These are the folks with bumper stickers: *HOME IS WHERE I PARK IT.* You may have seen them.

> –Anthropologists Dorothy Ayers Counts and David R. Counts (2001: 15) discussing how many people view older, often retired people who choose to travel the countryside in recreational vehicles (RVs). After the Countses retired, they too joined this community of RVers.

When we think of technology and how it transforms social life, many of us do not think about the many forms of transportation that are available to us. However, from the days of the covered wagon, many people have migrated, either temporarily or permanently, from one place to another, living for extended periods of time under the roof that travels everywhere with them. Today, recreational vehicles range from camper tops that can be placed on a pickup truck to three- and four-bedroom, multiple-bath, luxury units that sell for more than $1 million. Along with the introduction of RVs and the development of massive interstate highway systems that made it possible for people to travel thousands of miles in a matter of hours or days, RV parks were established to give people a place to park their vehicles and participate in social activities with people from other areas of the country.

As the Countses' statement suggests, RVing is often associated with older people who are assumed to have the leisure and the money to engage in such nomadic adventures. However, in their ethnographic study of the world of RVing seniors, the Countses found that this world is not always rosy for older people. Like other RVers, seniors are often viewed as "gypsy tourists" who are somehow unstable because they move around. They are also viewed as violating the work ethic because of their "conspicuous pursuit of leisure." But, most significant of all, older RVers are seen as violating the stereotype of old age. According to Counts and Counts (2001: 89), "Retired RVers are doubly stigmatized because they occupy two categories of marginalized, devalued people: they are both elderly and RVers." Counts and Counts (2001: 90) believe that retired RVers often attempt to challenge the stereotypes of the elderly as "rocking-chair bound, cookie-baking Golden Agers and Perfect Grandparents."

The ultimate irony in regard to the stereotyping of older people, and particularly those who travel in recreational vehicles, is that here are independent people going about their everyday lives and having a good time, and they still are the objects of stereotyping and scorn even though they defy other, lasting stereotypes of older individuals as being tight with their money, being bland and never having a good time, and being dependent and unable to take care of themselves.

The larger sociological issue here is not just the technology that makes it possible for older people to travel from place to place. Rather, it is how people in devalued statuses (such as "the retired" or "old people"), even when they defy traditional stereotypes about themselves, are frequently subjected to yet another set of stereotypes that, at best, carry a kernel of truth to them and that, at worst, reinforce old prejudices and discrimination across generational lines.

Can you think of other examples of how older people, even when defying the traditional stereotypes often associated with aging, are still the objects of stereotyping and prejudice? What about clothing? Music? Other lifestyle choices?

AGE AND THE LIFE COURSE IN CONTEMPORARY SOCIETY

During the twentieth century, life expectancy steadily increased as industrialized nations developed better water and sewage systems, improved nutrition, and made tremendous advances in medical science. However, children today are often viewed as an economic liability; they cannot contribute to the family's financial well-being and must be supported. In industrialized and postindustrial societies, the skills necessary for many roles are more complex and the number of unskilled positions is more limited. Consequently, children are expected to attend school and learn the necessary skills for future employment rather than perform unskilled labor. Further, older people are typically expected to retire so that younger people can assume their economic, political, and social roles. However, when older people have fewer productive roles to fill, age-based inequality tends to increase. For example, the trend in recent years to "downsize" the work force has contributed to pressure on some older workers to retire early, thereby saving employers money and preserving jobs for younger workers. Such "early retirement" is not always voluntary and may pose significant economic risks for individuals who find that they cannot live on their pensions or that their employer-sponsored health benefits have been reduced (Moody, 2002). However, even for the financially secure, retirement often presents problems, as Jack Culberg explains:

> When you suddenly leave [the corporate jungle], life is pretty empty. I was sixty-five, the age people are supposed to retire. I started to miss it quite a bit. The phone stops ringing. The king is dead. You start wanting to have lunch with old friends. At the beginning, they're nice to you, but then you realize that they're busy, they're working. They've got a job to do and just don't have the time to talk to anybody where it doesn't involve their business. I could be nasty and say, "Unless they can make a buck out of it"—but I won't. . . . You hesitate to call them. (qtd. in Terkel, 1996: 9–10)

As Culberg's statement indicates, people tend to think of age in narrowly defined categories, and reaching "retirement age" places many of them out of the mainstream.

In the United States, age differentiation is typically based on categories such as infancy, childhood, adolescence, young adulthood, middle adulthood, and later adulthood. In Chapter 4 ("Socialization"), we examined the socialization process that occurs when people are in various stages of the life course. However, these narrowly defined age categories have had a profound effect on our perceptions of people's capabilities, responsibilities, and entitlement. In this chapter, we will look at what is considered appropriate for or expected of people at various ages. These expectations are somewhat arbitrarily determined and produce *age stratification*—**inequalities, differences, segregation, or conflict between age groups** (Atchley and Barusch, 2004). We will now examine some of those strata and the unique problems associated with each one.

Infancy and Childhood

Infancy (birth to age 2) and childhood (ages 3 to 12) are typically thought of as carefree years; however, children are among the most powerless and vulner-

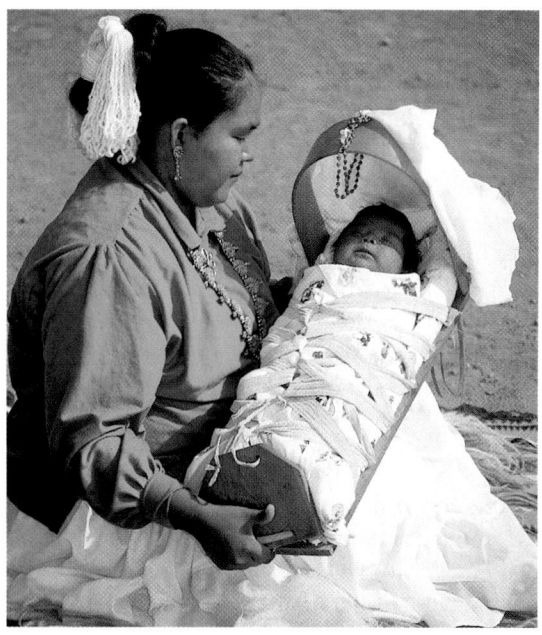

Branson Reynolds/Index Stock Imagery

People in all cultures take seriously their responsibility to socialize the next generation for full participation in social groups. This Navajo mother and infant show that even the physical arrangements associated with infancy, including the types of baby carriers used, vary from culture to culture.

able people in society. Historically, children were seen as the property of their parents, who could do with them as they pleased (Tower, 1996). In fact, whether an infant survives the first year of life depends on a wide variety of parental factors, as a community health scholar explains:

> All infants are not created equal. Those born to teenage mothers or to mothers who smoke cigarettes, drink alcohol, or take drugs are at higher risk for death in their first year. Those born in very rural areas or in inner cities are more likely to die as infants. Those born to black women are at twice the risk as those born to white women. Older mothers carry a high risk for conceiving an infant with Down's syndrome, and Native American women carry a high risk for having a baby with a serious birth defect. Add to the mix a mother's education; her economic, marital, and nutritional status; and whether she had adequate prenatal care, which all play into whether her infant will make it through the first year of life.
>
> But surviving the first year is only one piece of the equation. Quality of life is another. Infants who survive the first year can have lives so compromised that their future is seriously limited. . . . We cannot always predict which infants will survive, and we certainly cannot predict who will be happy. (Schneider, 1995: 26)

Moreover, early socialization plays a significant part in children's experiences and their quality of life. Many children are confronted with an array of problems in their families because of marital instability, an increase in the number of single-parent households, and the percentage of families in which both parents are employed full time. These factors have heightened the need for high-quality, affordable child care for infants and young children. However, many parents have few options regarding who will take care of their children while they work. These statistics from the Children's Defense Fund (2002) point out the perils of infancy and childhood:

> Every day in the United States, 1 out of every 5 infants are born into poverty; over 3,000 children die from gunshot wounds every year, 7 million children are at home alone on a regular basis without adult supervision, and every 11 seconds a child is reported abused or neglected.

As these statistics show, the risk of death or permanent disability from accidents is a major concern in childhood. In fact, two-thirds of all childhood deaths are caused by injuries. (The other third are caused by cancers, birth defects, heart disease, pneumonia, and HIV/AIDS.) Although many previous childhood killers such as polio, measles, and diphtheria are now controlled through immunizations and antibiotics, motor vehicle accidents have become a major source of injury and death for infants and children. (As compared with all other racial–ethnic categories, Native American children have the highest rate of motor vehicle deaths.) The childhood motor vehicle fatality rate is higher in the South and Southwest, where more parents own pickup trucks and allow children to ride in the truck bed (Schneider, 1995). Despite laws and protective measures implemented to protect infants and children, far too many lose their lives at an early age due to the abuse, neglect, or negligence of adults.

Adolescence

In contemporary industrialized countries, adolescence roughly spans the teenage years, although some analysts place the lower and upper ages at 15 and 24. Before the twentieth century, adolescence did not exist as an age category. Today, it is a period in which young people are expected to continue their education and perhaps hold a part-time job.

What inequalities based on age are experienced by adolescents in our society? Adolescents are not granted full status as adults in most societies, but they are not allowed to act "childish" either. Early teens are considered too young to do "adult" things, such as stay out late at night, vote, drive, use tobacco, or consume alcoholic beverages. Many adolescents also face conflicting demands to attend school and to make money. Most states have compulsory school attendance laws requiring young people between the ages of 6 and 16 or 18 to attend school regularly; however, students who see no benefit from school or believe that the money they make working is more important may find themselves labeled as juvenile offenders for missing school. Moreover, juvenile laws define behavior such as truancy or running away from home as forms of delinquency—which would not be offenses if they were committed by an adult. Despite child labor laws implemented to control working conditions for young employees, many adolescents of today are employed in settings with hazardous working conditions, low wages, no benefits, and long work hours.

A variety of reports have labeled contemporary U.S. teenagers as a "generation at risk" because of the many problems that social analysts believe to be profound among today's adolescents (see Zill and Nord, 1994; Carnegie Council on Adolescent Development, 1995). Among the most pressing adolescent problems identified were crime and violence, teen pregnancy, suicide, drug abuse, and excessive peer pressure. However, other social analysts dispute these claims and suggest that teenagers have become the "scapegoat generation" and are widely viewed as being a problem for society. Although defining the "youth problem" in this manner may be harmful for all adolescents, it could be especially harmful for young people of color from low-income families. Without educational and economic opportunities, they are the most likely to constitute the majority of young people in jails, prisons, and detention facilities, where it is believed that they can do less harm to other people or to the nation as a whole (Males, 1996). One analyst finds this trend especially troublesome:

> American adults have regarded adolescents with hope and foreboding [for many decades]. What is transpiring is new and ominous. A particular danger attends older generations indulging "they-deserve-it" myths to justify enriching ourselves at the expense of younger ones. The message . . . adults have spent two decades sending to youths is: *You are not our kids. We don't care about you.* (Males, 1996: 43)

In Males's opinion, the primary way to save the adolescent generation of today is to reduce poverty among children, teenagers, and young families, as well as to move away from the large number of age-based laws that restrict adolescents' opportunities for employment and freedom (Males, 1996).

Young Adulthood

Young adulthood, which follows adolescence and lasts to about age 39, is socially significant because during this time people are expected to get married, have children, and get a job. People who do not fulfill these activities during young adulthood tend to be viewed negatively. Individuals who do not get married by age 39 are often quizzed by relatives and friends about their intentions and sometimes their sexual orientation. Those who do not have a first child by the time they reach the end of their thirties are likely to experience questions from others about whether they plan to become parents or not. Even more significantly, those who are unable to find steady employment tend to become suspect because they have not "settled down" and "taken life seriously" or are viewed as being "lazy and unwilling to work." However, for some young adults, finding a job may be more difficult than for others. As previously discussed, race/ethnicity and gender strongly influence people's opportunities. For example, the sociologist William J. Wilson examined employment opportunities available to young adults living in Chicago's South Side and found that many who wanted to work could find no source of employment. According to one 32-year-old woman in his study,

> There's not enough jobs. . . . There's not enough factories, there's not enough work. Most all the good jobs are in the suburbs. Sometimes it's hard for the people in the city to get to the suburbs, because everybody don't own a car. Everybody don't drive. (qtd. in W. Wilson, 1996: 39)

People who are unable to earn income and pay into Social Security or other retirement plans in early adulthood typically find themselves disadvantaged as they enter middle and late adulthood.

Middle Adulthood

Prior to the twentieth century, life expectancy in the United States was only about 47 years, so the concept of middle adulthood—people between the ages of 40 and 65—did not exist until fairly recently. Normal changes in appearance occur during these years, and although these changes have little relationship to a person's health or physical functioning, they are socially significant to many people (Lefrançois, 1999).

As people progress through middle adulthood, they experience *senescence* (primary aging) in the form of molecular and cellular changes in the body. Wrinkles and gray hair are visible signs of senescence. Less-visible signs include arthritis and a gradual dulling of

What can people across generations learn by taking time to talk to each other? Can the learning process flow in both directions?

the senses of taste, smell, touch, and vision. Typically, reflexes begin to slow down, but the actual nature and extent of change vary widely from person to person.

People also experience a change of life in this stage. Women undergo *menopause*—the cessation of the menstrual cycle caused by a gradual decline in the body's production of the "female" hormones estrogen and progesterone. Menopause typically occurs between the mid-forties and the early fifties and signals the end of a woman's childbearing capabilities. Some women may experience irregular menstrual cycles for several years, followed by hot flashes, retention of body fluids, swollen breasts, and other aches and pains. Other women may have few or no noticeable physical symptoms. The psychological aspects of menopause are often as important as any physical effects. In one study, Anne Fausto-Sterling (1985) concluded that many women respond negatively to menopause because of negative stereotypes associated with menopausal and postmenopausal women. These stereotypes make the natural process of aging in women appear abnormal when compared with the aging process of men. Actually, many women experience a new interest in sexual activity because they no longer have to worry about the possibility of becoming pregnant. On the other hand, a few women have recently chosen to produce children using new medical technologies long after they have undergone menopause.

Men undergo a *climacteric,* in which the production of the "male" hormone testosterone decreases. Some have argued that this change in hormone levels produces nervousness and depression in men; however, it is not clear whether these emotional changes are due to biological changes or to a more general "midlife crisis," in which men assess what they have accomplished (Benokraitis, 2002). Ironically, even as such biological changes may have a liberating effect on some people, they may also reinforce societal stereotypes of older people, especially women, as "sexless."

Along with primary aging, people in middle adulthood also experience *secondary aging,* which occurs as a result of environmental factors and lifestyle choices. For example, smoking, drinking heavily, and engaging in little or no physical activity are factors that affect the aging process. People who live in regions with high levels of environmental degradation and other forms of pollution are also at greater risk of aging more rapidly and having chronic illnesses and diseases associated with these external factors.

On the positive side, middle adulthood for some people represents the time during which (1) they have the highest levels of income and prestige, (2) they leave the problems of child rearing behind them and are content with their spouse of many years, and (3) they may have grandchildren, who give them another tie to the future. Census 2000 asked several questions regarding the status of grandparents and whether or not any of their grandchildren were living with them. The data show that more than 5.6 million grandparents in the United States have one or more grandchildren under age 18 living in their home. Of that number (see the Census Profiles feature), about half of the grandparents are responsible for most of the basic needs of those grandchildren. Even so, persons during middle adulthood know that, given society's current structure, their status may begin to change significantly when they reach the end of this period of their lives. As previously discussed, those who had few opportunities available earlier in life tend to become increasingly disadvantaged as they grow older.

Late Adulthood

Late adulthood is generally considered to begin at age 65—the "normal" retirement age. *Retirement* is the institutionalized separation of an individual from an occupational position, with continuation of income through a retirement pension based on prior years of service (Atchley and Barusch, 2004). Retirement means the end of a status that has long been a source of income and a means of personal identity. Perhaps the loss of a valued status explains why many retired persons introduce themselves by saying "I'm retired now, but I was a (banker, lawyer, plumber, supervisor, and so on) for forty years." As shown by Map 12.1, the percentage of the population age 65 and above varies from state to state—from a low of 5.5 percent in Alaska to a high of 18.3 percent in Florida (U.S. Census Bureau, 2002).

Some gerontologists subdivide late adulthood into three categories: (1) the "young-old" (ages 65 to 74), (2) the "old-old" (ages 75 to 85), and (3) the "oldest-old" (over age 85) (see Moody, 2002). Although these are somewhat arbitrary divisions, the "young-old" are less likely to suffer from disabling illnesses, whereas some of the "old-old" are more likely to suffer such illnesses (Belsky, 1999). However, a recent study found that the prevalence of disability among those 85 and over decreased during the 1980s due to better health care. In fact, it was reported that Jeanne Calment of Paris, France, who died in 1997 at age 122, rode a bicycle until she was 100 (Whitney, 1997).

Map 12.1	Percentage of Resident Population Age 65 and Older by State

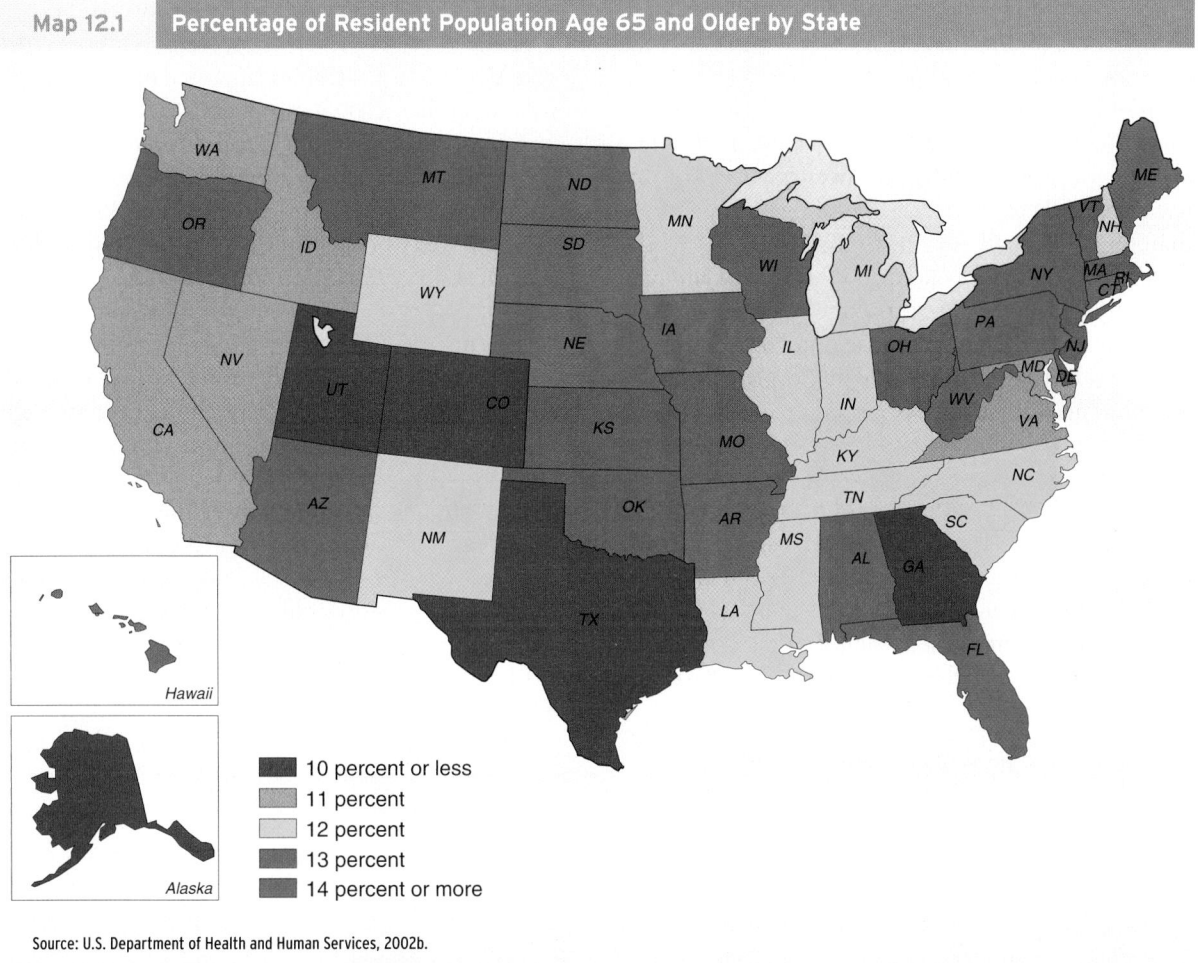

Legend:
- 10 percent or less
- 11 percent
- 12 percent
- 13 percent
- 14 percent or more

Source: U.S. Department of Health and Human Services, 2002b.

The rate of biological and psychological changes in older persons may be as important as their chronological age in determining how they are perceived by themselves and others. As adults grow older, they actually become shorter, partly because bones that have become more porous with age develop curvature. A loss of three inches in height is not uncommon. As bones become more porous, they also become more brittle; simply falling may result in broken bones that take longer to heal. With age, arthritis increases, and connective tissue stiffens joints. Wrinkled skin, "age spots," gray (or white) hair, and midriff bulge appear; however, people sometimes use Oil of Olay, Clairol, or Buster's Magic Tummy Tightener in the hope of avoiding looking older (Atchley and Barusch, 2004).

Older persons also have increased chances of heart attacks, strokes, and cancer, and some diseases affect virtually only persons in late adulthood. Alzheimer's disease (a progressive and irreversible deterioration of brain tissue) is an example; about 55 percent of all organic mental disorders in the older population are caused by Alzheimer's (Atchley and Barusch, 2004). Persons with this disease have an impaired ability to function in everyday social roles; eventually, they cease to be able to recognize people they have always known and lose all sense of their own identity. Finally, they may revert to a speechless, infantile state such that others must feed them, dress them, sit them on the toilet, and lead them around. The disease can last up to 20 years; currently, there is no cure. Alzheimer's affects an estimated 2 percent of people over 65 and as many as 5 percent over age 75 (Lefrançois, 1999).

Fortunately, most older people do not suffer from Alzheimer's and are not incapacitated by their physical condition. Only about 5 percent of older people live in nursing homes, about 10 percent have trouble walking, and about 30 percent have hearing problems. Although most older people experience some decline in strength, flexibility, stamina, and other physical capabilities, much of that decline does not result simply from the aging process and is avoidable; with proper exercise, some of it is even reversible (Lefrançois, 1999).

With the physical changes come changes in the roles that older adults are expected (or even allowed) to perform. For example, people may lose some of the abilities necessary for driving a car safely, such as vision or reflexes. Although it is not true of all older persons, the average individual over age 65 does not react as rapidly as the average person who is younger than 65 (Lefrançois, 1999). In 2003 the issue of elderly drivers was widely debated in the media and political arenas after an 86-year-old man lost control of his car in Santa Monica, California, and raced through a crowded street market, killing 10 people and injuring more than 50 others (see Box 12.3).

The physical and psychological changes that come with increasing age can cause stress. According to Erik Erikson (1963), older people must resolve a tension of "integrity versus despair." They must accept that the life cycle is inevitable, that their lives are nearing an end, and that achieving inner harmony requires accepting both one's past accomplishments and past disappointments. As Hardy Howard, Sr., an 86-year-old African American man, explains,

> I feel good to be as old as I am, and when I think back I never had any problem walking or going up stairs, no false teeth or hearing aids, no nothing like that. . . . You know, what you get out of life is what you put in. Measure unto others as I would have them measure unto me. Anybody I can help, I don't care if they're white, black, green, or gray—anything you want, if I can do you a favor, I'll do it. And this has been my logic since I was a little boy, and I haven't changed from it even today. (qtd. in Mucciolo, 1992: 91–92)

Like many older people, Howard has worked to maintain his dignity and autonomy. In fact, a lot of older people are able to maintain their activities for many years beyond when younger people think it might be possible.

A survey conducted for the AARP (formerly known as the American Association of Retired Persons) found that more than half of U.S. people age 45 or over believed that they had satisfactory sex lives but that sexual activity and contentment diminished with age (Toner, 1999). The most frequent reasons for lack of sexual fulfillment were declining health and lack of a sexual partner. The partner gap was widest for women age 75 or older. Among women and men age 75 and up who did have partners, more than a fourth reported that they had sexual intercourse once a week (Toner, 1999). Other recent surveys of people age 100 and over found that most centenarians have a sense of having "enjoyed the journey to 100" (L. Ramirez, 1999: A4). For example,

Ben Levinson, age 100, who regularly worked out at a gym and set a world record for the shot put in the Nike World Masters Games, appeared on *The Tonight Show* and other television programs talking about his longevity. *Masters games* and *masters athletes* are terms referring to the growing number of adults who are past the average peak performance age of 35 in sporting events but who continue to train with high intensity and compete at various levels (Morgan and Kunkel, 1998).

Will the life stages as we currently understand them accurately reflect aging in the future? Research continues to show that there are limited commonalities between those who are age 65 and those who are centenarians; however, many people tend to place everyone from 65 upward in categories such as "old," "elderly," or "senior citizen." In the future, we will probably see such categorizations revised or deleted as growing numbers of older people reject such labels as forms of "ageism." Some analysts believe that the existing life-course and life-stages models will be modified to reflect a sense of "old age" beginning at age 75 or 80 and that new stages will be added for those who reach age 90 and 100 (Morgan and Kunkel, 1998).

INEQUALITIES RELATED TO AGING

In previous chapters, we have seen how prejudice and discrimination may be directed toward individuals based on ascribed characteristics—such as race/ethnicity or gender—over which they have no control. The same holds true for age.

Ageism

Stereotypes regarding older persons reinforce ***ageism—prejudice and discrimination against people on the basis of age, particularly against older persons.*** Ageism against older persons is rooted in the assumption that people become unattractive, unintelligent, asexual, unemployable, and mentally incompetent as they grow older.

Ageism is reinforced by stereotypes, whereby people have narrow, fixed images of certain groups. One-sided and exaggerated images of older people are used repeatedly in everyday life. Older persons are often stereotyped as thinking and moving slowly; as being bound to themselves and their past, unable to change and grow; as being unable to move forward and often

Box 12.3 SOCIOLOGY AND SOCIAL POLICY

Driving While Elderly: Policies Pertaining to Age and Driving

There are no words to express the feelings my family and I have for those who suffered pain as a result of Wednesday's devastating accident. I am so distraught and my heart is broken over the extent of the tragedy.
—Russell Weller, 86 years of age, issued this statement after his car sped down the length of the crowded farmers' market in Santa Monica, California, killing 10 and injuring 50 others in July 2003 (qtd. in Associated Press, 2003).

Set up as a street fair, the farmers' market in Santa Monica is usually a pleasant, easy-going event where locals and a few tourists select fresh produce and other items to purchase. But on July 16, 2003, that peaceful tranquility was shattered. According to one eyewitness, "I heard the sounds of metal on metal and people screaming and I looked out the window to see a car traveling by. I ran outside with one of my bosses, and to our horror saw a man on the hood of a vehicle and an elderly gentleman being pulled from a car" (CNN.com, 2003a).

Shortly after this tragedy occurred, newspapers and cable TV channels began reporting what had happened and what the possible implications were for older drivers. Should there be a mandatory age at which people must quit driving? Should there be more frequent testing of older drivers? What could be done to prevent future occurrences of this nature? All of these questions buzzed around, but no answers were readily apparent. Advocates for older people emphasized that the Santa Monica tragedy was an isolated incident and that all older drivers should not be penalized for the actions (described as unintentional and perhaps confused) of this one elderly man.

Most of us recall the time when we anxiously awaited our next birthday so that we would be old enough to get our first driver's license. All states in the United States have a minimum age for getting a learner's permit or a first driver's license, but far fewer states have policies regarding drivers over the age of 65. However, this is likely to change in the near future with the aging of the U.S. population. By 2030, one in four drivers in this country will be over age 65. Some people will voluntarily give up their car keys when they believe that they can no longer drive safely. The National Institute on Aging estimates that about 600,000 people age 70 or over give up their driving privileges each year (Neergaard, 2003). But the question remains: What, if anything, should social policy makers, such as state legislators, do about those who either do not know that it is time to quit driving or who simply refuse to give up this privilege, which they often equate to giving up their freedom?

Clearly, there are competing rights and concerns involved in social policies pertaining to older drivers. On the one hand, older drivers point out that most cities lack adequate public transportation and that not being able to drive makes them a burden on other people, if other people are even available to drive them to doctors' appointments, for groceries and other errands, and for recreational activities. On the other hand, pedestrians, other drivers, and the general public have a vested interest in not having individuals of *any* age drive who are unsafe and who might, for whatever reason, constitute a threat to others.

What is happening at this time? The American Medical Association has developed a set of guidelines to help doctors tell when older patients' driving is questionable. The U.S. government has earmarked $1.6 million to start a National Older Drivers Research Center to train certified driving-rehabilitation specialists who can instruct older drivers on additional safety measures they can take (such as avoiding unprotected left turns, unfamiliar roads, and night driving). However, at the bottom line, many decisions will be made at the state level. These may include whether to require more frequent license renewals for older drivers, whether drivers over a certain age should be retested before their license is renewed, which tests (if any) should be performed for license renewal, and what should be done if a person is unable to pass a vision test and show driving competency.

As we think about social policy pertaining to older people, particularly in regard to the right to drive, it is important to have empathy for people who will be quickly affected by new rules requiring them to prove their competence, but it is also important to remember that if we are fortunate to live long enough, each of us will also face such questions as we reach various stages in the course of our lives!

WRITING IN SOCIOLOGY ASSIGNMENT

Surveys have found that factors in addition to the age of the driver and/or malfunctioning of the vehicle contribute to automobile accidents. Some of these factors relate to conduct on the  part of the driver. What kinds of behavior are most closely linked to hazardous driving and automobile accidents?

For many years, advertisers have bombarded women with messages about the importance of a youthful appearance. Increasingly, men, too, are being targeted for advertising campaigns that play on fears about the "ravages" of aging.

moving backward (Belsky, 1999). They are viewed as cranky, sickly, and lacking in social value (Atchley and Barusch, 2004); as egocentric and demanding; as shallow, enfeebled, aimless, and absentminded (Belsky, 1999).

The media contribute to negative images of older persons, many of whom are portrayed as doddering, feebleminded, wrinkled, and laughable men and women, literally standing on their last legs (Lefrançois, 1999). This is especially true with regard to advertising. In one survey, 40 percent of respondents over age 65 agreed that advertising portrays older people as unattractive and incompetent (Pomice, 1990). According to the advertising director of one magazine, "Advertising shows young people at their best and most beautiful, but it shows older people at their worst" (qtd. in Pomice, 1990: 42). Of older persons who do appear on television, most are male; only about one in ten characters appearing to be age 65 or older is a woman, conveying a subtle message that older women are especially unimportant (Pomice, 1990).

Stereotypes also contribute to the view that women are "old" 10 or 15 years sooner than men. The multibillion-dollar cosmetics industry helps perpetuate the myth that age reduces the "sexual value" of women but increases it for men. Men's sexual value is defined more in terms of personality, intelligence, and earning power than physical appearance. For women, however, sexual attractiveness is based on a slender, youthful appearance. By idealizing this "youthful" image of women and playing up the fear of growing older, sponsors sell thousands of products that claim to prevent the "ravages" of aging.

Fortunately, in recent years there has been a growing effort by the media to draw attention to the contributions, talents, and stamina of older persons rather than showing only stereotypical and negative portrayals. For example, the *New York Times,* CNN, and other news sources highlighted more than a dozen runners over age 80 in the New York City Marathon. The media pointed out that some runners were in their nineties and that records are maintained of such accomplishments as being the fastest 85-year-old ever to complete the race (Barron, 1997).

Despite some changes in media coverage of older people, many younger individuals still hold negative stereotypes of "the elderly." In one study, William C. Levin (1988) showed photographs of the same man (disguised to appear as ages 25, 52, and 73 in various photos) to a group of college students and asked them to evaluate these (apparently different) men for employment purposes. Based purely on the photographs, the "73-year-old" was viewed by many of the students as being less competent, less intelligent, and less reliable than the "25-year-old" and the "52-year-old."

Although not all people act on appearances alone, Patricia Moore, an industrial designer, found that many do. At age 27, Moore disguised herself as an 85-year-old woman by donning age-appropriate clothing and placing baby oil in her eyes to create the appearance of cataracts. With the help of a makeup artist, Moore supplemented the "aging process" with latex wrinkles, stained teeth, and a gray wig. For three years, "Old Pat Moore" went to various locations, including a grocery store, to see how people responded to her:

> When I did my grocery shopping while in character, I learned quickly that the Old Pat Moore behaved—and was treated—differently from the Young Pat Moore. When I was 85, people were more likely to jockey ahead of me in the checkout line. And even more interesting, I found that

when it happened, I didn't say anything to the offender, as I certainly would at age 27. It seemed somehow, even to me, that it was okay for them to do this to the Old Pat Moore, since they were undoubtedly busier than I was anyway. And further, they apparently thought it was okay, too! After all, little old ladies have plenty of time, don't they? And then when I did get to the checkout counter, the clerk might start yelling, assuming I was deaf, or becoming immediately testy, assuming I would take a long time to get my money out, or would ask to have the price repeated, or somehow become confused about the transaction. What it all added up to was that people feared I would be trouble, so they tried to have as little to do with me as possible. And the amazing thing is that I began almost to believe it myself. . . . I think perhaps the worst thing about aging may be the overwhelming sense that everything around you is letting you know that you are not terribly important any more. (Moore with Conn, 1985: 75–76)

If we apply our sociological imagination to Moore's study, we find that "Old Pat Moore's" experiences reflect what many older persons already know—it is other people's *reactions* to their age, not their age itself, that places them at a disadvantage.

Many older people buffer themselves against ageism by continuing to view themselves as being in middle adulthood long after their actual chronological age would suggest otherwise. In one study of people aged 60 and over, 75 percent of the respondents stated that they thought of themselves as middle-aged and only 10 percent viewed themselves as being old. When the same people were interviewed again 10 years later, one-third still considered themselves to be middle-aged. Even at age 80, one out of four men and one out of five women said that the word *old* did not apply to them; this lack of willingness to acknowledge having reached an older age is a consequence of ageism in society (Belsky, 1999).

Wealth, Poverty, and Aging

Many of the positive images of aging and suggestions on how to avoid the most negative aspects of ageism are based on an assumption of class privilege, meaning that people can afford plastic surgery, exercise classes, and social activities such as ballroom dancing or golf, and that they have available time and facilities to engage in pursuits that will "keep them young." However, many older people with meager incomes, little savings, and poor health, as well as those who are isolated in rural areas or high-crime sections of central cities, do not have the same opportunities to follow popular recommendations about "successful aging" (Stoller and Gibson, 1997: 76). In fact, for many older people of color, aging is not so much a matter of seeking to defy one's age but rather of attempting to survive in a society that devalues both old age and minority status.

If we compare wealth (all economic resources of value, whether they produce cash or not) with income (available money or its equivalent in purchasing power), we find that older people tend to have more wealth but less income than younger people. Older people are more likely to own a home that has increased substantially in market value; however, some may not have the available cash to pay property taxes, to buy insurance, and to maintain the property (Moody, 2002). Among older persons, a wider range of assets and income is seen than in other age categories. For example, many of the wealthiest people on the *Forbes* list of the richest people in this country (see Chapter 9) are over 65 years of age. On the other hand, more than 10 percent of all people over 65 live in poverty, as shown in Figure 12.2.

Age, Gender, and Inequality

Age, gender, and poverty are intertwined. Although middle-aged and older women make up an increasing portion of the work force, they are paid substantially less than men their age, receive raises at a slower pace, and still work largely in gender-segregated jobs (see Chapter 11). As a result, women do not garner economic security for their retirement years at the same rate that many men do. These factors have contributed to the economic marginality of the current cohort of older women (Gonyea, 1994).

In one study, the gerontologists Melissa A. Hardy and Lawrence E. Hazelrigg (1993) found that gender was more directly related to poverty in older persons than was race/ethnicity, educational background, or occupational status. Hardy and Hazelrigg (1993) suggested that many women who are now age 65 and over spent their early adult lives as financial dependents of husbands or as working nonmarried women trying to support themselves in a culture that did not see women as the heads of households or as sole providers of family income. Because they were not viewed as being responsible for a family's financial security, women were paid less; therefore, older women may have to rely on inadequate income replacement programs originally designed to treat them as dependents. Women also have a greater risk of poverty in their later years; statistically, women tend to marry

Figure 12.2   **Percentage of Persons Age 65+ Below Poverty Level**

Although many of the wealthiest people are over 65 years of age, more than 10 percent of all people over 65 live in poverty. Note the disparities by racial-ethnic group.

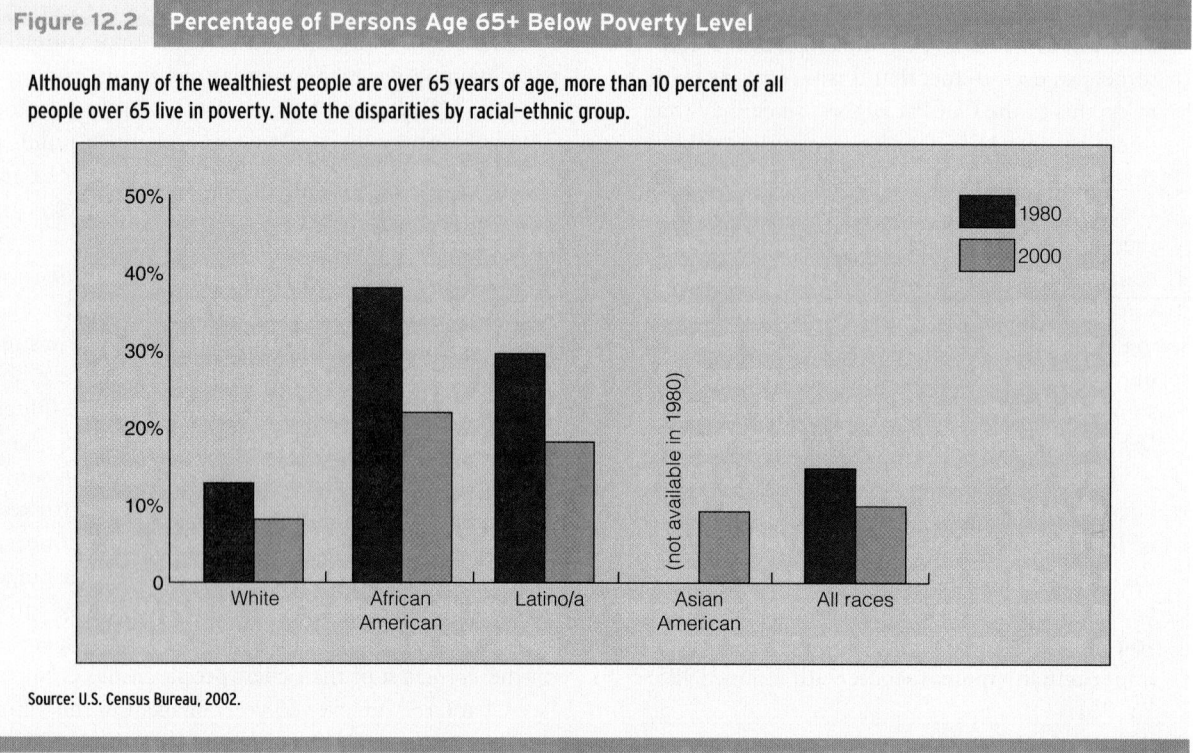

Source: U.S. Census Bureau, 2002.

men who are older than themselves, and women live longer than men. Consequently, nearly half of all women over age 65 are widowed and living alone on fixed incomes (Smith, 2003).

As shown in Figure 12.2, the percentage of persons aged 65 and older living below the poverty line decreased significantly between 1980 and 2000. This largely resulted from increasing benefit levels of ***entitlements—certain benefit payments paid by the government,*** including Social Security, Supplemental Social Income (SSI), Medicare, Medicaid, and civil service pensions, which are the primary source of income for many persons over age 65 (Moody, 2002). Ninety percent of all retired people in the United States draw Social Security benefits (Williamson, Duffy Rinehart, and Blank, 1992). Social Security and SSI provide virtually all of the financial support available to 25 percent of people over age 65.

Analysts note that Social Security keeps more white men out of poverty than it does white women and people of color (Browne and Broderick, 1994). The effectiveness of Social Security as a means of escaping poverty declines as age increases for all groups except white men (Axinn, 1989). Even for white men, Social Security does not provide the financial security in their later years that many of them thought it would when they were younger. For example, when Harold Milton was asked if he had ever worried about how he would manage after he retired, he responded,

"No, it never entered my mind. You see, I was counting on my Social Security." When he retired, Milton found that he could not live on his Social Security benefits of $250 and SSI of $86 per month, so he took a part-time job as a handyman to earn another $100 per month (Margolis, 1990).

Medicare, the other of the two largest entitlement programs, is a nationwide health care program for persons aged 65 and older who are covered by Social Security or who are eligible to "buy into" the program by paying a monthly premium (Atchley and Barusch, 2004). Although Medicare's primary purpose is to provide long-term support, it only partially does so. The reasons for this apparent contradiction are (1) skilled nursing care other than in a hospital is covered only if it immediately follows at least three days in a hospital (and many people are not sick enough to qualify for hospital care), and (2) Medicare reimburses only the first 90 days of hospital care or the first 100 days of skilled nursing care (Atchley and Barusch, 2004). According to many social analysts, Social Security is in need of change and might fail outright before many of today's young people reach retirement age.

Age and Race/Ethnicity

Age, race/ethnicity, and economic inequality are closely intertwined. Inequalities that exist later in life originate in individuals' early participation in the labor force and are amplified in late adulthood. For example, older African Americans continue to feel the impact of segregated schools and overt patterns of job discrimination that were present during their early years. Although African Americans constitute only about 8 percent of the population age 65 and over, they account for 26 percent of the low-income older population. Among persons age 65 and over, 9 percent of whites reported poverty-level incomes in 2001, as compared with 22 percent of African Americans and Latinos/as (U.S. Department of Health and Human Services, 2002b). As you will recall from Chapter 9, the poverty line is determined by estimating how much a low-budget family must spend annually for groceries and then multiplying that amount by three. Based on the assumption that persons in late adulthood eat less than younger people, the poverty line is placed at a lower dollar amount ($8,628 for a single person in 2002) for persons 65 and older than for people below that age ($9,359 in 2002). In other words, at age 64 a person with an income of $8,700 is considered to be below the poverty line (and thus entitled to assistance), but one year later, on the same income, the person is no longer "poor." Although caloric demand does decrease for older people, their nutritional needs remain the same, and no data indicate that their food costs are less (Margolis, 1990).

As previously noted, the primary reason for the lower income status of many older African Americans can be traced to a pattern of limited employment opportunities and periods of unemployment throughout their lives, combined with their concentration in secondary-sector jobs, which pay lower wages, are sporadic, have few benefits, and were not covered by Social Security prior to the 1950s (Hooyman and Kiyak, 2002). Moreover, health problems may force some African Americans out of the labor force earlier than other workers due to a higher rate of chronic diseases such as hypertension, diabetes, and kidney failure (Hooyman and Kiyak, 2002).

Similarly, older Latinas/os have higher rates of poverty than whites (Anglos) because of lack of educational and employment opportunities. Some older Latinos/as entered the country illegally and have had limited opportunities for education and employment, leaving them with little or no Social Security or other benefits in their old age. High rates of poverty among older Latinas/os are associated with poor health conditions, lack of regular care by a physician, and fewer trips to the hospital for medical treatment of illness, disease, or injury.

Older Native Americans are among the most disadvantaged of all categories. Older Native Americans

AP Photo/Gail Oskin

Older people have actively protested changes in Medicare and other programs for senior citizens that they view as detrimental to their well-being. How will the "graying of America" affect political activism in the future?

are more likely to live in high-poverty, rural areas than are other minority older populations. Some studies have found that about 50 percent of all older Native Americans live in poverty, having incomes that are between 40 and 60 percent less than those of older white Americans. In addition to experiencing educational and employment discrimination similar to that of African Americans and Latinos/as, older Native Americans were also the objects of historical oppression and federal policies toward the native nations that exacerbated patterns of economic impoverishment. Consequently, some older Native Americans have the worst living conditions and poorest health of all older people in this country. Research findings regarding the mental health of older Native Americans also indicate a high rate of depression, alcoholism and other drug abuse, and suicide (see Chapter 18, "Health, Health Care, and Disability").

Among older Asian Americans, many who arrived in the United States prior to 1930 have fared less well in their old age than those who were native-born or were more recent arrivals. As discussed in Chapter 10, many older Asian Americans from Japan and China received less education and experienced more economic deprivation than did later cohorts of Japanese Americans and Chinese Americans. Today, many older Asian Americans remain in Chinatown, Japantown, or Koreatown, where others speak their language and provide goods and services that help them maintain their culture, and where mutual-aid and benevolent societies and recreational clubs provide them with social contacts and delivery of services. As a result, many older Asian Americans who qualify for various forms of entitlements, such as Supplemental Security Income, do not apply for it. Moreover, cultural values, including traditional healing practices, may help explain why many Asian Americans, particularly Chinese American elders, do not use physical and mental health services that are available to them.

For many years, studies in sociology and gerontology primarily focused on the attributes and needs of older white Americans from middle- and low-income backgrounds. Overall, more research is needed on the unique needs of older people from diverse racial and ethnic categories.

Older People in Rural Areas

The lives of many older people differ based on whether they reside in urban or rural areas. Despite the stereotypical image of the rural elderly living in a pleasant home located in an idyllic country setting, rural elders, as compared with older urban residents, typically have lower incomes, are more likely to be poor, and have fewer years of schooling. The rural elderly also tend to be in poorer health, and many of them are less likely to receive needed health care because many rural areas lack adequate medical and long-term care facilities (Coburn and Bolda, 1999). With the "graying of America," the population of older adults (age 65 and over) has continued to grow in rural areas; however, this growth varies from region to region, with the Midwest and South having a greater concentration of older persons in rural areas than the West and the Northeast have.

Why are older people in rural areas more likely to have lower incomes and tend to be classified as "poor"? Some social analysts attribute lower income among the rural elderly to factors such as lower Social Security payments based on lower lifetime earnings, limited savings, and fewer opportunities for part-time work (Coburn and Bolda, 1999). Whether or not they are poor, the rural elderly receive a higher proportion of their income from Social Security payments than do the urban elderly. Housing also differs among rural and urban elderly, with older rural residents being more likely to own their own homes than older residents in urban settings. However, the homes of the elderly in rural areas are more likely to have a lower value in the real estate market and to be in greater need of repair than are those owned by urban elderly residents (Coburn and Bolda, 1999).

Elder Abuse

Abuse and neglect of older persons have received increasing public attention in recent years, due both to the increasing number of older people and to the establishment of more-vocal groups to represent their concerns. ***Elder abuse refers to physical abuse, psychological abuse, financial exploitation, and medical abuse or neglect of people age 65 or older*** (Hooyman and Kiyak, 2002). Physical abuse includes malnutrition or injuries such as bruises, welts, sprains, and dislocations. In contrast, psychological abuse is made up of verbal assaults, threats, fear, and social isolation. Financial exploitation typically involves theft or misuse of the older person's money or property by another person. Medical abuse occurs when a person withholds, or improperly administers, medications or health aids such as dentures, glasses, or hearing aids. Neglect is not providing care sufficient for the older person to maintain physical and mental health (Hooyman and Kiyak, 2002).

According to the National Center on Elder Abuse (2003), as many as 1.6 million older people in the United States are the victims of physical or mental abuse each year. Just as with violence against children or women, it is difficult to determine how much abuse against older people occurs. Many victims are understandably reluctant to talk about it. One study indicates that slightly over 2 percent of all older people experience physical abuse (Pillemer and Finkelhor, 1988; Atchley and Barusch, 2004). Although this may appear to be a small percentage, it represents a large number of people. Studies have shown that the victims of elder abuse tend to be concentrated among those over age 75 (Steinmetz, 1987). Most of the victims are white, middle- to lower-middle-class Protestant women, aged 75 to 85, who suffer some form of impairment (Garbarino, 1989; Benokraitis, 2002). Sons, followed by daughters, were the most frequent abusers of older persons (Friedan, 1993).

There has been a widespread belief that elder abuse occurs because the older person is *dependent on other people* (Brandl and Cook-Daniels, 2002). However, studies have found almost no evidence to support that conclusion. According to the sociologist Karl Pillemer (1985), no scientific support exists for the common assumption that dependency on the part of the older person leads to abuse. To the contrary, abusers are very likely to be *dependent on the older person* for housing and financial assistance (Brandl and Cook-Daniels, 2002).

Today, almost every state has enacted mandatory reporting laws regarding elder abuse or has provided some type of governmental protection for older persons. Cases of abuse and neglect of older people are highly dramatized in the media because they offend very central values in the United States—respect for and consideration of older persons (Atchley and Barusch, 2004).

LIVING ARRANGEMENTS FOR OLDER ADULTS

Many frail, older people live alone or in a family setting where care is provided informally by family or friends. Relatives (especially women) provide most of the care. Many women caregivers are employed outside the home; some are still raising a family. Recently, the responsibilities of informal caregivers have become more complex. For some frail, older persons, family members are often involved in nursing regimes—such as chemotherapy and tube feeding—that were previously performed in hospitals (Glazer, 1990). Only about 5 percent of frail, older persons are currently in nursing homes (Benokraitis, 2002).

Support Services, Homemaker Services, and Day Care

Support services help older individuals cope with the problems in their day-to-day care. These services are very expensive even when they are provided through state or federally funded programs, hospitals, or community organizations.

For older persons, homemaker services perform basic chores (such as light housecleaning and laundry); other services (such as Meals on Wheels) deliver meals to homes. Some programs provide balanced meals at set locations, such as churches, synagogues, or senior centers. However, some of these programs have been criticized for their failure to provide meals that take into account the diverse ethnic backgrounds of the people they serve (Gelfand, 2003).

Day-care centers have also been developed to help older persons maintain as much dignity and autonomy as possible. These centers typically provide transportation, activities, some medical personnel (such as a licensed practical nurse) on staff, and nutritious meals. Most centers are very expensive—in some cases, $700 or more per month for a five-day week (Benokraitis, 2002). In some states, Medicaid covers the fees of older persons living below the poverty line; however, only one-third of older persons who are poor receive Medicaid (Neuschler, 1987).

The onset of aging does not have to mean the end of an active life. Many older people are able to enjoy social and recreational activities.

Housing Alternatives

Some older adults remain in the residence where they have lived for many years—a process that gerontologists refer to as *aging in place* (Atchley and Barusch, 2004). Remaining in a person's customary residence is a symbol that he or she is able to maintain independence and preserve ties to his or her neighborhood and surrounding community. However, either by choice or necessity, some older adults move to smaller housing units or apartments.

This type of relocation frequently occurs when older people are living in a residence that does not meet their current needs. Moving to another location may be desirable or necessary when their family has grown smaller, the cost and time necessary to maintain the existing residence become a strain on them, or individuals experience illness or disabilities that make it difficult for them to remain in the same location.

More housing alternatives are available to middle-income and upper-income older people than to low-income individuals. These include *retirement communities* such as Sun City, where residents must be age 55 or above. (Sun City is the name of several privately owned retirement communities located in such states as Arizona and Texas.) Residents of retirement communities purchase their housing units and in some instances pay additional fees for the upkeep of shared areas and amenities such as a golf course, swimming pool, or other recreational facilities. Typically, residents of planned retirement communities are similar to one another with respect to race and ethnic background and social class. Most retirement communities do not provide support services such as health care or transportation.

People needing assistance with daily activities or desiring the regular companionship of other people may move to an *assisted-living* facility. Some of these facilities offer fully independent apartments and provide residents with support services such as bathing, help with dressing, food preparation, and taking medication. Some centers provide residents with transportation to medical appointments, beauty and barber shops, and social events in the community. However, cost is a compelling factor. Assisted-living centers with more-luxurious amenities are primarily available to retired professionals and other upper-middle to upper-income people. A number of assisted-living facilities provide residents with the opportunity to age in place by also having long-term care facilities available if and when the need arises. Some of these facilities are well managed and carefully maintained; however, others are a major source of concern to residents, their families, and state regulators who are responsible for inspecting these facilities and responding to complaints filed against them.

Some lower-income older people live in planned housing projects that are funded by federal, state, or local government agencies or by religious groups. Other low-income individuals receive care and assistance from relatives, neighbors, or members of religious congregations.

Nursing Homes

The most restrictive environment for older persons is the nursing home setting. Many nursing home residents have major physical and/or cognitive problems that prevent them from living in any other setting, or they do not have available caregivers in their family. Women are more likely to enter nursing homes because of their greater life expectancy, higher rates of chronic illness, and higher rates of widowhood. Sociologist Madonna Harrington Meyer (1994: 9) describes the problems inherent in paying for nursing home care:

> Financing of long-term care is a problem for most older persons. Medicare . . . excludes most long-term care and devotes less than 4 percent of its budget to long-term care. Most private insurance policies, including Medigap policies, also exclude long-term care. Just one in six older Americans can afford private policies that do cover long-term care. Nursing home costs average $35,000 a year, yet 60 percent of the elderly have annual incomes of less than one-half that amount. Only Medicaid, the poverty-based health care program for pro-

foundly poor families of all ages and the permanently blind and disabled, includes long-term care coverage. As a result, only the poor elderly receive assistance through the state with long-term care. By default, then, the United States has a poverty-based long-term care system.

Harrington Meyer notes that the current system is based on and perpetuates gender, class, and racial inequalities in society. Because Medicare does not provide broad-based long-term care, people from already disadvantaged groups are particularly unprotected in old age. This system also shifts an even greater burden of informal caregiving onto women, thus magnifying the tension between their paid and unpaid work (Harrington Meyer, 1994).

Some nursing homes have been criticized for a lack of consideration for residents' ethnic backgrounds by staff members performing their daily tasks (Gelfand, 2003). For example, members of different ethnic groups have specific ways in which they prefer to be addressed. When someone violates these rules of etiquette, the older person may view the behavior as a deliberate insult. Likewise, if members of the nursing home staff do not respect the older person's desire for privacy, particularly in regard to dressing, toileting, and bathing, the perception of disrespect grows even stronger (Gelfand, 2003). Language barriers often exist between nursing home assistants and the older nursing home residents; many of the assistants are recent immigrants and do not speak the same language as the residents (Diamond, 1992).

Nursing homes may not be the solution to anyone's problem. Cases of neglect, excessive use of physical restraints, overmedication of patients, and other complaints have been rampant in many of the facilities. Consider Bessie Seday's description of the nursing home in which she resided as a bed-bound 84-year-old:

> I couldn't get anybody's attention, starting on the fourth day. You'd have your call light on for hours, but nobody came. . . . It was like a dungeon. I really would have liked to see the sunshine, but they never put us outside. . . . The screaming is what got to me the worst, the screaming when the lights went out. I couldn't fall asleep until 1 or 2 in the morning with all that screaming going on. (qtd. in Thompson, 1997: 34)

In addition to these problems, Mrs. Seday's bed was not changed, and her room was not properly maintained. In fact, journalists learned that when the nursing home's special washing machine for cleaning dirty bedpans broke down, some of the staff began washing bedpans in the whirlpool, where patients with bedsores were then placed (Thompson, 1997). Mrs. Seday eventually moved in with her daughter because the problems at the home became so bad. Recent investigations have shown that this was not an isolated situation. Across the United States, tens of thousands of complaints have been filed against nursing homes for neglect and failure to medicate patients properly. However, the problem is larger than that, as one journalist indicated:

> Neglectful caregivers are preying not only on elderly residents but also on American taxpayers. More than $45 billion in government funds, mostly from Medicare and Medicaid, is pumped into nursing homes annually, an amount that comes to nearly 60 percent of the national tab for such eldercare. In order to pocket a larger slice of the federal stipend, many nursing homes—largely for-profit enterprises—provide a minimal level of care, if that. (Thompson, 1997: 35)

In a best-case scenario, nursing homes are not without weaknesses and problems. According to author Betty Friedan (1993: 516), even the best-run nursing homes "deny the personhood of age [because they] reify the image of age as inevitable decline and deterioration." What will you do if someone you know must move into a nursing home? A few suggestions are provided in Box 12.4.

SOCIOLOGICAL PERSPECTIVES ON AGING

Sociologists and social gerontologists have developed a number of explanations regarding the social effects of aging. Some of the early theories were based on a microlevel analysis of how individuals adapt to changing social roles. More-recent theories have used a macrolevel approach to examine the inequalities produced by age stratification at the societal level.

Functionalist Perspectives on Aging

Functionalist explanations of aging focus on how older persons adjust to their changing roles in society. According to the sociologist Talcott Parsons (1960), the roles of older persons need to be redefined by society. He suggested that devaluing the contributions of older persons is dysfunctional for society; older persons often have knowledge and wisdom to share with younger people.

How does society cope with the disruptions resulting from its members growing older and dying? According to *disengagement theory,* **older persons make a normal and healthy adjustment to aging when they detach themselves from their social roles and prepare for their eventual death** (Cumming and Henry, 1961). Gerontologists Elaine C. Cumming and William E. Henry (1961) noted that disengagement can be functional for both the individual and society. For example, the withdrawal of older persons from the work force provides employment opportunities for younger people. Disengagement also aids a gradual and orderly transfer of statuses and roles from one generation to the next; an abrupt change would result in chaos. Retirement, then, can be thought of as recognition for years of service and the acknowledgment that the person no longer fits into the world of paid work (Williamson, Duffy Rinehart, and Blank, 1992). The younger workers who move into the vacated positions have received more up-to-date training—for example, the computer skills that are taught to most younger people today.

Critics of this perspective object to the assumption that all older persons want to disengage while they are still productive and still gain satisfaction from their work. Disengagement may be functional for organizations but not for individuals. A corporation that has compulsory retirement may be able to replace higher-paid, older workers with lower-paid, younger workers, but retirement may not be beneficial for some older workers. Contrary to disengagement theory, a number of studies have found that activity in society is *more* important with increasing age.

Symbolic Interactionist Perspectives on Aging

Symbolic interactionist perspectives examine the connection between personal satisfaction in a person's later years and a high level of activity. ***Activity theory* states that people tend to shift gears in late middle age and find substitutes for previous statuses, roles, and activities** (Havighurst, Neugarten, and Tobin, 1968). From this perspective, older people have the same social and psychological needs as middle-aged people and thus do not want to withdraw unless restricted by poor health or disability.

Whether they invest their energies in grandchildren, traveling, hobbies, or new work roles, social activity among retired persons is directly related to longevity, happiness, and health (Palmore, 1981). Psychologist and newspaper columnist Eda LeShan observed a difference in the perceptions of people who do and do not remain active:

Finding a Nursing Home for an Older Relative

It is too late to turn back. I have set in motion events that are tumbling me along like a broken branch in flood-tide. I have run out of choices. I look at my 94-year-old mother. Her thin, hunched body is dwarfed amid a welter of crates, cartons of books, piles of possessions to go to the Salvation Army, and other stacks of worthless items too precious to leave behind. Her expression matches the disheveled room. She doesn't know it yet, but by this time next week, she will be in a nursing home in Colorado, a thousand miles from here, whatever life remains to her drastically changed. . . . My God, what am I doing to her? (Hartney, 1990: 191)

Most of us believe that circumstances will never make it necessary for us to place an older family member in a nursing home. Among more-affluent older people, options such as retirement and elder care centers are often available that provide a mid-point between remaining at home and moving into a nursing home. However, a number of social and demographic trends are affecting many people's ability to care for frail older persons at home. As the number of older people continues to increase, no corresponding increase is occurring in the number of available younger caregivers in many families. Similarly, many people entering the middle years of their lives have become the "sandwich generation" because they have simultaneous responsibilities for both their aging parents and their own children under age 18. According to recent statistics, for example, 43 percent of the daughters and wives and 69 percent of the sons and husbands of older people with disabilities are employed full time (Hooyman and Kiyak, 2002). What happens when the physical and psychological needs of the older parent or other relative are beyond the scope of what younger family members can meet?

Under these circumstances, many people believe that they have no alternative but to turn to nursing homes for elder care. When you or someone you know is in the position of needing to find a nursing home for a loved one, experts suggest that you can make a difference in the kind of care the person receives by following these basic steps:

- Examine your state's record of nursing home inspections. About once a year, each state conducts nursing home inspections and issues a report on deficiencies that affect the residents' health. Beware of homes that are unwilling or unable to produce their most recent survey. Each state has an ombudsperson with information on all nursing homes. For more information, call the National Citizens' Coalition for Nursing Home Reform in Washington, D.C., at (202) 332-2275.

- Visit prospective nursing homes without giving them notice that you are coming. By visiting when personnel have not had time to tidy up or remove smells or waste, you will be able to see how the home really operates on a daily basis.

- Find out from current residents and staff if they feel comfortable there. Look to see if both groups appear to be clean and generally well groomed.

- Carefully observe patients and staff. Watch to see if staff members respond to call lights from patients. Help that arrives in more than five minutes is not adequate. Find out what the menu is, and see if people are being fed nutritious food. Be sure to determine whether large numbers of residents are being detained in restraints.

- If your loved one moves into a nursing home, be sure to visit frequently and to make staff members aware of your presence. Gerontologists often find that patients with attentive families get better care than those whose families and friends never visit. Make holidays and special occasions cheery by taking others with you to sing to residents or to visit with those who would like to have companionship.

Sources: Based on Atchley and Barusch, 2004; and Hooyman and Kiyak, 2002.

The Richardsons came for lunch: friends we hadn't seen for twenty years. . . . Helen and Martin had owned and worked together in a very fine women's clothing shop. . . . Having some mistaken notion they were getting too old and should retire and "enjoy themselves," they sold the business ten years ago.

During lunch, Larry and I realized we were dealing with two seriously depressed people, in excellent health but with no place to go. When Larry asked Helen what she'd been doing, she replied bitterly, "Who has anything to do?" Martin said sadly he was sorry he gave up tennis ten years ago; if he'd kept it up he could still play. . . .

We were embarrassed to indicate we were still so busy that we couldn't see straight. They seemed genuinely shocked that we had no plans to retire at seventy-one and seventy-four. (LeShan, 1994: 221–222)

According to the symbolic interactionist perspective, older people who invest their time and energy in volunteer work or in other enjoyable activities tend to be healthier, be happier, and live longer than those who disengage from society.

Studies have confirmed LeShan's suggestion that healthy people who remain active have a higher level of life satisfaction than do those who are inactive or in ill health (Havighurst, Neugarten, and Tobin, 1968). Among those whose mental capacities decline later in life, deterioration is most rapid in people who withdraw from social relationships and activities.

A variation on activity theory is the concept of *continuity*—that people are constantly attempting to maintain their self-esteem and lifelong principles and practices and that they simply adjust to the feedback from and needs of others as they grow older (Williamson, Duffy Rinehart, and Blank, 1992). From this perspective, aging is a continuation of earlier life stages rather than a separate and unique period. Thus, values and behaviors that have previously been important to an individual will continue to be so as the person ages. People may also turn to their ethnic culture to help them deal with physical changes, role changes, and bereavement issues in their later years. For example, studies have found that the church serves an important function in reducing loneliness, providing support systems, and enhancing self-image in older African Americans (Gelfand, 2003).

Other symbolic interactionist perspectives focus on role and exchange theories. Role theory poses this question: What roles are available for older people? Some theorists have noted that industrialized, urbanized societies typically do not have roles for older people (Cowgill, 1986). Analysts examining the relationship between race/ethnicity and aging have found that many older persons are able to find active roles within their own ethnic group. Although their experiences may not be valued in the larger society, they are esteemed within their ethnic subculture because they provide a rich source of knowledge of ethnic lore and history. For example, Mildred Cleghorn, an 80-year-old Native American woman, passes on information to younger people by the use of dolls:

> I decided . . . to show that we were all not the same, by making dolls that said we were just as different as our clothes are different. I made four dolls in Kansas, representing the four tribes there—then seven more here in Oklahoma for the tribes living here. Now, over the years, I have a collection of forty-one fabric dolls, all different tribes. The trouble is there are thirty-two more to go! (qtd. in Mucciolo, 1992: 23)

Cleghorn's unique knowledge about the various Native American nations has been a valuable source of information for young people who otherwise might be unaware of the great diversity found among Native Americans. According to the gerontologist Donald E. Gelfand (2003), older people can "exchange" their knowledge for deference and respect from younger people.

Conflict Perspectives on Aging

Conflict theorists view aging as especially problematic in contemporary capitalistic societies. As people grow older, their power tends to diminish unless they are able to maintain wealth. Consequently, those who have been disadvantaged in their younger years become even more so in late adulthood. Women age 75 and over are among the most disadvantaged because they often must rely solely on Social Security, having outlived their spouses and sometimes their children (Harrington Meyer, 1990).

Underlying the capitalist system is an ideology which assumes that all people have equal access to the means of gaining wealth and that poverty results from individual weakness. When older people are in need, they may be viewed as not having worked hard enough or planned adequately for their retirement. The family and the private sector are seen as the "proper" agents to respond to their needs. To minimize the demand for governmental assistance, these services are made punitive and stigmatizing to those who need them (Atchley and Barusch, 2004). Class-based theories of inequality assert that government programs for older persons stratify society on the basis of class. Feminist approaches claim that these programs perpetuate inequalities on the basis of gender and race in addition to class (Harrington Meyer, 1994).

Conflict analysis draws attention to the diversity in the older population. Differences in social class, gender, and race/ethnicity divide older people just as they

According to the conflict perspective, differences in class, gender, and race/ethnicity increasingly divide older people in the United States into the "haves" and the "have-nots." The woman pictured here, for instance, lives in a house without indoor plumbing. Older African Americans are especially affected by poverty because of overt discrimination in the past and more subtle forms of contemporary prejudice.

do everyone else. Wealth cannot forestall aging indefinitely, but it can soften the economic hardships faced in later years. The conflict perspective adds to our understanding of aging by focusing on how capitalism devalues older people, especially women. However, critics assert that this approach ignores the fact that industrialization and capitalism have greatly enhanced the longevity and quality of life for many older persons.

■ DEATH AND DYING

Historically, death has been a common occurrence at all stages of the life course. Until the twentieth century, the chances that a newborn child would live to adulthood were very small. Poor nutrition, infectious diseases, accidents, and natural disasters took their toll on men and women of all ages. But in contemporary, industrial societies, death is looked on as unnatural because it has been largely removed from everyday life. Most deaths now occur among older persons and in institutional settings. The association of death with the aging process has contributed to ageism in our society; if people can deny aging, they feel that they can deny death (Atchley and Barusch, 2004).

In the past, explanations for death and dying were rooted in custom or religious beliefs. Today, people who have religious beliefs regarding living an afterlife typically have less anxiety about death, which they may view as the beginning of a better life (Feifel and Nagy, 1981). Research has shown that those who are most fearful of death are people who are confused or uncertain about their religious beliefs, not those who have confirmed their lack of religious belief (Downey, 1984; Hooyman and Kiyak, 2002).

In contemporary, high-technology societies, however, custom and religious beliefs tend to be displaced by medical and legal explanations and definitions of death. In some instances, advanced life-support systems may make it unclear whether death has actually occurred. For example, Nancy Cruzan, age 26, was in an irreversible coma as a result of an automobile accident. After doctors told her parents that she would never recover, Cruzan's parents requested removal of a feeding tube implanted into her stomach. After the hospital refused to do so, the Cruzans took their case to the U.S. Supreme Court, which ruled that states can establish the conditions that apply to the removal of life-support technology. Almost seven years after her accident, a Missouri state court authorized the removal of Nancy Cruzan's feeding tube, and she died twelve days later (Corr, Nabe, and Corr, 2003).

There is no national standard for determining when life-support measures should be ended. Thousands of people are in some kind of permanent vegetative state today, and many thousands more are faced with terminal illnesses. As a result, many people have chosen to have a say in how their lives might end by signing a *living will*—a document stating their wishes about the medical circumstances under which their life should be terminated. Most states recognize living wills. However, many issues pertaining to the quality of life and to death with dignity remain unresolved.

How do people cope with dying? There are three widely known frameworks for explaining how people cope with the process of dying: the *stage-based approach*, the *dying trajectory*, and the *task-based approach*. The *stage-based approach* was popularized by psychiatrist Elisabeth Kübler-Ross (1969), who proposed five stages in the dying process: (1) denial and isolation ("Not me!"), (2) anger and resentment ("Why me?"), (3) bargaining and an attempt to postpone ("Yes me, but . . ."—negotiating for divine intervention), (4) depression and sense of loss, and (5) acceptance. She pointed out that these stages are not the same for all people; some of the stages may exist at the same time. Kübler-Ross (1969: 138) also stated that "the one thing that usually persists through all these stages is hope."

Kübler-Ross's stages were attractive to the general public and the media because they provided common responses to a difficult situation. On the other hand, her stage-based model also generated a great deal of criticism. Some have pointed out that these stages have never been conclusively demonstrated or comprehensively explained.

Second is the *dying trajectory,* which focuses on the perceived course of dying and the expected time of death. For example, a dying trajectory may be sudden, as in the case of a heart attack, or it may be slow, as in the case of lung cancer. According to the dying-trajectory approach, the process of dying involves three phases: the acute phase, characterized by the expression of maximum anxiety or fear; the chronic phase, characterized by a decline in anxiety as the person confronts reality; and the terminal phase, characterized by the dying person's withdrawal from others (Glaser and Strauss, 1968).

Finally, the *task-based approach* is based on the assumption that the dying person can and should go about daily activities and fulfill tasks that make the process of dying easier on family members and friends, as well as on the dying person. Physical tasks can be performed to satisfy bodily needs, whereas psychological tasks can be done to maximize psychological security, autonomy, and richness of experience. Social tasks sustain and enhance interpersonal attachments and address the social implications of dying. Spiritual tasks help people to identify, develop, or reaffirm sources of spiritual energy and to foster hope (Corr, Nabe, and Corr, 2003). In the final analysis, however, how a person dies is shaped by many social and cultural factors. According to the gerontologists Nancy Hooyman and H. Asuman Kiyak (2002), the dying process is not the same for everyone. It is influenced by an individual's personality and philosophy of life, as well as the social context (such as at home or in a hospital) in which the end of life occurs.

In recent years, the process of dying has become an increasingly acceptable topic for public discussion. Such discussions helped further the hospice movement in the 1970s (Weitz, 2004). A **hospice is an organization that provides a homelike facility or home-based care (or both) for people who are terminally ill.** The hospice philosophy asserts that people should participate in their own care and have control over as many decisions pertaining to their life as possible. Pain and suffering should be minimized, but artificial measures should not be used to sustain life. This approach is family based and provides support for family members and friends, as well as for the person who is dying (see Corr, Nabe, and Corr, 2003). Although the hospice movement has been very successful (with some 1,800 facilities), critics claim that the movement has exchanged much of its initial philosophy and goals for social acceptance and financial support (Finn Paradis and Cummings, 1986;

Weitz, 2004). Over time, hospice care has moved toward hospital standards because of the need for hospices to work with the federal government and the American Hospital Association to gain accreditation. Medicare funding and the resultant federal regulations have further changed hospices (see Weitz, 2004).

■ AGING IN THE FUTURE

The size of the older population in the United States will increase dramatically in the early decades of the twenty-first century. By the year 2050, there will be an estimated 80 million people age 65 and older, as compared with 35 million in 2000. Thus, combined with decreasing birth rates, most of the population growth will occur in the older age cohorts during the next 50 years. More people will survive to age 85, and more will even reach the 95-and-over cohort (Atchley and Barusch, 2004). These estimates point out the importance of developing better and more comprehensive ways of assisting people to live full and productive lives as they grow older.

Issues such as governmental assistance for in-home care services; availability of medical services for preventive care, chronic illness, and disability; and housing for older persons will be the focus of many political debates. However, a report issued in 1994 by a bipartisan commission on entitlement and tax reform warned that entitlement benefits are growing so rapidly that (when combined with interest on the national debt) they will consume nearly all federal tax revenues by the year 2012, leaving the government with no money for anything else. As former senator John C. Danforth, then a member of the commission, warned, "There will be no money for national defense, for law enforcement, for the environment, [or] for highways" (qtd. in Rosenblatt, 1994: 1).

Who will assist people with needs they cannot meet themselves? Family members in the future may be less willing or able to serve as caregivers. Women, the primary caregivers in the past, are faced with not just double but *triple* workdays if they attempt to combine working full time with caring for their children and assisting older relatives. Even "superwomen" have a breaking point—no one has unlimited time, energy, and willpower to engage in such demanding activities for extended periods.

As biomedical research on aging continues, new discoveries in genetics may eliminate life-threatening diseases and make early identification of other diseases possible. Technological advances in the diagnosis, prevention, and treatment of Alzheimer's disease may revolutionize people's feelings about growing older (Atchley and Barusch, 2004). Advances in medical technology may lead to a more positive outlook on aging.

If these advances occur, will they help everyone or just some segments of the population? This is a very important question for the future. As we have seen, many of the benefits and opportunities of living in a highly technological, affluent society are not available to all people. Classism, racism, sexism, and ageism all serve to restrict individuals' access to education, medical care, housing, employment, and other valued goods and services in society.

For older persons, the issues discussed in this chapter are not merely sociological abstractions; they are an integral part of their everyday lives. Older people have resisted ageism through organizations such as the Gray Panthers, AARP, and the Older Women's League.

CHAPTER REVIEW

■ **What is aging, and what is the study of aging called?**

Aging refers to the physical, psychological, and social processes associated with growing older. Gerontology is the study of aging and older people. Social gerontology is the study of the social (nonphysical) aspects of aging, including the consequences of an aging population and the personal experience of aging.

■ **How do views of aging differ in preindustrial and industrialized societies?**

In preindustrial societies, people of all ages are expected to share the work, and the contributions of older people are valued. In industrialized societies, however, older people are often expected to retire so that younger people may take their place.

■ **What are ageism and elder abuse, and how are these perpetrated in the United States?**

Ageism is prejudice and discrimination against people on the basis of age, particularly against older persons. Ageism is reinforced by stereotypes of older people. Elder abuse includes physical abuse, psychological abuse, financial exploitation, and medical abuse or neglect of people age 65 or older. Passive neglect is the most common form of abuse.

■ **How do functionalist and symbolic interactionist explanations of aging differ?**

Functionalist explanations of aging focus on how older persons adjust to their changing roles in society; the gradual transfer of statuses and roles from one generation to the next is necessary for the functioning of society. Activity theory, a part of the symbolic interactionist perspective, states that people change in late middle age and find substitutes for previous statuses, roles, and activities. This theory asserts that people do not want to withdraw unless restricted by poor health or disability.

■ **What is the conflict perspective on aging and age-based inequality?**

Conflict theorists link the loss of status and power experienced by many older persons to their lack of ability to produce and maintain wealth in a capitalist economy.

■ **What are the most common living arrangements for older people?**

Many older persons live alone or in an informal family setting. Support services and day care help older individuals who are frail or disabled cope with their day-to-day needs, although many older people do not have the financial means to pay for these services. Nursing homes are the most restrictive environment for older persons. Many nursing home residents have major physical and/or cognitive problems that prevent them from living in any other setting, or they do not have available caregivers in their family.

■ **How is death typically viewed in industrialized societies?**

In industrialized societies, death has been removed from everyday life and is often regarded as unnatural.

■ **What stages in coping with dying were identified by Elisabeth Kübler-Ross?**

Kübler-Ross proposed five stages of coping with dying: denial, anger, bargaining, depression, and acceptance.

KEY TERMS

activity theory 411
age stratification 397
ageism 402
aging 389
chronological age 390
cohort 391
disengagement theory 411

elder abuse 408
entitlements 406
functional age 390
gerontology 392
hospice 415
life expectancy 391

QUESTIONS FOR CRITICAL THINKING

1. Why does activity theory contain more positive assumptions about older persons than disengagement theory does? Analyze your grandparents (or other older persons whom you know well or even yourself if you are older) in terms of disengagement theory and activity theory. Which theory seems to provide the most insights? Why?
2. How are race, class, gender, and aging related?
3. Is it necessary to have a mandatory retirement age? Why or why not?
4. How will the size of the older population in the United States affect society and programs such as Social Security in the future?

RESOURCES ON THE INTERNET

Chapter-Related Web Sites

The following Web sites have been selected for their relevance to the topics in this chapter. These sites are among the more stable, but please note that Web site addresses change frequently. For an updated list of chapter-related Web sites with URL links, please visit the *Sociology in Our Times* Web site (**www.wadsworth.com/KendallSIOT**).

National Institute on Aging (NIA)
http://www.nih.gov/nia

The NIA is a government agency whose mission is to provide leadership in aging research, training, health information dissemination, and other programs relevant to older individuals. The NIA Web site contains the latest research on aging, useful links, information on research funding, and a search feature to locate data on particular topics.

Administration on Aging (AOA)
http://www.aoa.dhhs.gov

The AOA is a government organization under the U.S. Department of Health and Human Services. Click on "Statistics About Older People" to view the most current national and state data on the economic well-being, demographic characteristics, social conditions, and health of the elderly. The site also provides resources for the elderly and their families, information on various programs, a search feature, and helpful links.

American Association of Retired Persons (AARP)
http://www.aarp.org/index.html

The AARP Web site features articles, programs, and resources on issues that affect senior citizens. Issues such as age discrimination, scams, legislative news, health and wellness, and current research are thoroughly explored.

ONLINE STUDY AND RESEARCH TOOLS

Accompanying this text are many *free* powerful online study tools that will help you master the material in this chapter, help increase your depth of understanding, and help you make the grade!

SocCoach CD-ROM

Use the SocCoach CD-ROM enclosed with this text to help you formulate a customized study plan for this chapter. After you take the Diagnostic Quiz, SocCoach will generate a customized study plan just for you! It will identify sections of the chapter that you should review and will provide videos, charts, graphs, and excerpts from the text to supplement your studies and enhance your understanding. You'll also find fun, interactive activities such as Virtual Explorations and Map the Stats to apply what you've learned and stretch your sociological imagination.

The Companion Web Site for Sociology in Our Times, *Fifth Edition*
www.wadsworth.com/KendallSIOT

Gain an even better grasp on this chapter by going to the companion Web site to take one of the Tutorial Quizzes, use the Flash Cards to master key terms, or check out the many other study aids you'll find there. You'll also find special features such as GSS Data and Census 2000 information that'll put data and resources at your fingertips to help you with that special project or help you as you do some research on your own.

In this chapter, when you see the icon on the left, it alerts you to a specific exercise found in *Wadsworth's Sociology Online Resources and Writing Companion.* This valuable guide shows you how to use Wadsworth's exclusive online resources—*InfoTrac College Edition,* the *Opposing Viewpoints Resource Center,* and *MicroCase Online*—to assist you in your study of sociology and to build essential research and writing skills.

The Economy and Work in Global Perspective

I vividly remember the moment it hit me that things were getting Bad. It was in November of 2001, a rare sunny day in San Francisco; I rolled into work at my usual 9:50 A.M. to find that my company had stopped stocking our kitchen with free cereal! "What the . . .?! I have to pay for my own breakfast?! But, I work on the Internet! Cereal Nazis!"

A couple of weeks later I was just another [unemployed] guy sitting in Peet's Coffee at two in the afternoon, looking around at everyone else . . . all of us staring blankly at each other and wondering, "What . . . happened?"

—Greg Penhaligon (2003), a former senior front-end developer for a dotcom company, recalling what it was like when the bubble burst on the information technology (IT) industry and many remaining jobs were relocated to other nations

The irony is that I was a single mother, and I raised five kids by myself and put myself through school. I bought my first house [in Mesa, Arizona,] in 1999—that was a very big deal for me—and now I have to sell it, only because they won't hire Americans. It's devastating.

—Donna Bradley, another IT specialist, describing how she cannot get a job, even for significantly lower wages, because of the practice of "off-shoring" (qtd. in Gongloff, 2003)

Our competitors are doing it [off-shoring jobs,] and we have to do it.

—Tom Lynch, IBM's director for global employee relations (qtd. in Greenhouse, 2003: C1)

It's not about one shore or another shore. It's about investing around the world, including the United States, to build capability and deliver value as defined by our customers.

—Kentra R. Collins, an IBM spokesperson, explaining her company's belief that off-shoring is not the problem that many people believe it to be (qtd. in Greenhouse, 2003: C2)

Information technology professionals and U.S. white-collar employees in many other fields have lost their jobs in recent years. It is estimated, for example, that as many as 2.5 million jobs have been lost in the past few years and that most of them are never coming back to this country (Gongloff, 2003). Many of those who are unemployed blame their current problems not only on economic conditions in the United States but also on corporate labor practices, such as off-shoring, near-shoring, and the hiring of temporary immigrant workers in specialty occupations such as engineering, computer programming, accounting, and law. Let's look briefly at these terms to see how they might have a negative impact on the economy and work in the United States. *Off-shoring* refers to the practice of U.S. companies moving certain of their operations outside of this country. In the past, positions in these operations were primarily jobs in the blue-collar, manufacturing sector. Today, however, more white-collar jobs are being relocated to other nations, including those of customer service representatives (such as specialists who work in call centers for computer "help" lines); people who process insurance claims, payrolls, and income tax forms; and IT specialists who create and maintain Web sites. The term *near-shoring* is used to describe a similar process except that companies send their work to countries such as Canada, Mexico, or Guatemala, which are closer to the United States. Intertwined with both of these practices is the use of temporary immigrant foreign workers in U.S. companies—workers who are admitted to the United States for a specified period of time (usually up to six or seven years) to fulfill certain jobs but not to become permanent residents (TORAW, 2003). Theoretically, nonimmigrant foreign workers fill positions that U.S. employers are unable to fill with American workers; however, this is an assumption that is disputed by many unemployed workers in this country, and in 2003, more than 9.3 million people in the United States were unemployed. Wage and salary growth slowed, and fear was rampant that these factors would seriously harm consumer spending, which accounts for more than two-thirds of the U.S. economy (Gongloff, 2003).

What are the connections between the problems described here and the larger economic structure of our society and the global economy? Among those who are employed, why are some people satisfied with their jobs but others experience alienation? In this chapter, we will discuss the economy as a social institution and how it is linked to the world of work—how people feel about their work, how the nature of work is changing, and what impact these changes may have on your future. Before reading on, test your knowledge about the economy and work by taking the quiz in Box 13.1.

QUESTIONS AND ISSUES

Chapter Focus Question: How are our beliefs about work influenced by changes in the economy and in the world in which we live?

What are the key assumptions of capitalism and socialism?

What contributes to job satisfaction and to worker alienation?

What is the individual's role in the work force?

Why does unemployment occur?

How do workers attempt to gain control over their work situation?

COMPARING THE SOCIOLOGY OF ECONOMIC LIFE WITH ECONOMICS

Perhaps you are wondering how a sociological perspective on the economy differs from the study of economics. Although aspects of the two disciplines overlap, each provides a unique perspective on economic institutions. Economists attempt to explain how the limited resources and efforts of a society are allocated among competing ends. To economists, an imbalance exists between people's wants and society's ability to meet those wants. To illustrate, think about college registration. Would you like to have the "perfect" schedule—with the classes you want, at the times you want, and with "preferred" professors? How many students at your school actually manage to arrange such a schedule? What organizational constraints make it impossible for all people to have what they need or want? Some economists believe that the most important fact of economics is the law of scarcity, which means that there will never be enough resources to meet everyone's wants (Ruffin and Greg-

ory, 2000). Colleges do not have the financial or human resources to provide everything that students (or faculty) want.

Economists focus on the complex workings of economic systems (such as monetary policy, inflation, and the national debt), whereas sociologists focus on interconnections among the economy, other social institutions, and the social organization of work. At the macrolevel, sociologists may study the impact of transnational corporations on industrialized and low-income nations. At the microlevel, sociologists might study people's satisfaction with their jobs. To better understand these issues, we will examine how economic systems came into existence and how they have changed over time.

ECONOMIC SYSTEMS IN GLOBAL PERSPECTIVE

The **economy** is the social institution that ensures the maintenance of society through the production, distribution, and consumption of goods and

Box 13.1 SOCIOLOGY AND EVERYDAY LIFE

How Much Do You Know About the Economy and the World of Work?

True	False	
T	F	1. Terrorism and war have an almost instantaneous effect on an economy.
T	F	2. Professions are largely indistinguishable from other occupations.
T	F	3. Workers' skills are usually upgraded when new technology is introduced in the workplace.
T	F	4. Many of the new jobs being created in the service sector pay poorly and offer little job security.
T	F	5. In the United States and other nations, most assembly line workers are white men.
T	F	6. Labor unions will probably cease to exist sometime during this century.
T	F	7. Few workers resist work conditions that they consider to be oppressive.
T	F	8. Transnational corporations such as McDonald's are seen by many people in other nations as a form of cultural imperialism even if these businesses have created jobs for workers there.
T	F	9. Around the world, positions with the most job security are located in large, transnational corporations.
T	F	10. Assembly lines are rapidly disappearing from all sectors of the U.S. economy.

Answers on page 422.

services. *Goods* are tangible objects that are necessary (such as food, clothing, and shelter) or desired (such as DVDs and electric toothbrushes). *Services* are intangible activities for which people are willing to pay (such as dry cleaning, a movie, or medical care). In high-income nations today, many of the goods and services we consume are information goods. Examples include databases and surveys ("intermediate products") and the mass media, computer software, and the Internet ("information goods").

Some goods and services are produced by human labor (the plumber who unstops your sink, for example); others are primarily produced by capital (such as Internet access available through an Internet service provider). *Labor* refers to the group of people who contribute their physical and intellectual services to the production process in return for wages that they are paid by firms (Boyes and Melvin, 2002). *Capital* is wealth (money or property) owned or used in business by a person or corporation. Obviously, money, or financial capital, is needed to invest in the physical capital (such as machinery, equipment, buildings, warehouses, and factories) used in production. For example, a person who owns a thousand shares of Dell Inc. stock owns financial capital, but these shares also represent an ownership interest in Dell's physical capital.

Preindustrial Economies

Hunting and gathering, horticultural and pastoral, and agrarian societies are all preindustrial economic structures. Most workers in these societies engage in *primary sector production*—**the extraction of raw materials and natural resources from the environment.** These materials and resources are typically consumed or used without much processing. For example, portions of contemporary sub-Saharan Africa have a relatively high rate of exports in primary commodities, and foreign direct investment is concentrated in mineral extraction. Consequently, most of sub-Saharan Africa is highly dependent on primary sector production, which, in turn, is highly vulnerable to the whims of the primary commodity markets (United Nations Development Programme, 2003). In recent years, many people have lost economic ground in sub-Saharan Africa, where per-capita incomes are lower than they were in 1970 (United Nations Development Programme, 2003).

Box 13.1 SOCIOLOGY AND EVERYDAY LIFE

Answers to the Sociology Quiz on the Economy and the World of Work

1. **True.** As terrorism in the United States and the war in its aftermath have shown, the U.S. economy and the economies of many other nations have been almost immediately affected by both terrorism and war.

2. **False.** Professions have four characteristics that distinguish them from other occupations: (1) abstract, specialized knowledge; (2) autonomy; (3) authority over clients and subordinate occupational groups; and (4) a degree of altruism.

3. **False.** Jobs are often deskilled when new technology (such as bar code scanners or computerized cash registers) is installed in the workplace. Some of the workers' skills are no longer needed because a "smart machine" now provides the answers (such as how much something costs or how much change a customer should receive).

4. **True.** Many of the new jobs being created in the service sector, such as nurse's aide, child-care worker, hotel maid, and fast-food server, offer little job security and low pay.

5. **False.** Today, most assembly line workers are young girls and women in developing nations.

6. **False.** Sociologists who have examined organized labor generally predict that unions will continue to exist; however, their strength may wane in the global economy.

7. **False.** Many workers resist work conditions that they believe are unjust or oppressive. Some engage in sabotage; others join unions or participate in other types of pro-worker organizations.

8. **True.** Local facilities of McDonald's and other corporations closely linked to U.S. interests have been attacked and sometimes destroyed by people who oppose global capitalism and the policies of the U.S. government.

9. **False.** Although it is difficult to determine which types of jobs are the most secure, many positions in transnational corporations have been lost through downsizing and plant relocations and closings.

10. **False.** According to some scholars, assembly lines will remain a fact of life for businesses ranging from fast-food restaurants to high-tech semiconductor plants.

Sources: Based on Barboza, 2001; Fox, 2001; Hodson and Sullivan, 2002; and Zuboff, 1988.

Throughout history, the production units in hunting and gathering societies have been small; most goods are produced by family members. The division of labor is by age and gender (Hodson and Sullivan, 2002). The potential for producing surplus goods increases as people learn to domesticate animals and grow their own food. In horticultural and pastoral societies, the economy becomes distinct from family life. The distribution process becomes more complex, with the accumulation of a *surplus* such that some people can engage in activities other than food production. In agrarian societies, production is primarily related to producing food. However, workers have a greater variety of specialized tasks, such as warlord or priest; for example, warriors are necessary to protect the surplus goods from plunder by outsiders (Hodson and Sullivan, 2002). Once a surplus is accumulated, more people can also engage in trade. Initially, the surplus goods are distributed through a system of *barter*—the direct exchange of goods or services considered of equal value by the traders. However, bartering is limited as a method of distribution; equivalencies are difficult to determine (how many fish equal one rabbit?) because there is no way to assign a set value to the items being traded. As a result, *money*, a medium of exchange with a relatively fixed

Even as the United States and other high-income countries have increasingly relied on high-tech economies, other forms of work still exist in the agricultural and industrial sectors of the economy. Here, workers hand-pick a strawberry crop and work on an automobile assembly line.

value, came into use in order to aid the distribution of goods and services in society.

What was the U.S. economy like in the preindustrial era? As previous chapters have stated, the agricultural revolution brought about dramatic changes in the nature of work, including the growing division between work and home. In the preindustrial economy of the colonial period (from the 1600s to the early 1700s), white men earned a livelihood through agricultural work or as small-business owners who ran establishments such as inns, taverns, and shops. During this period, white women worked primarily in their homes, doing such tasks as cooking, cleaning, and child care. Some also developed *cottage industries*—producing goods in their homes that could be sold to nonfamily members. However, a number of white women also worked outside their households as midwives, physicians, nurses, teachers, innkeepers, and shopkeepers (Hess-Biber and Carter, 2000). By contrast, the experiences of people of color were quite different in preindustrial America. As the sociologists Sharlene Hess-Biber and Gregg Lee Carter (2000) explain, the institution of slavery, which came about largely as a result of the demand for cheap agricultural labor, was a major force in the exploitation of many people of color, but it was particularly the women of color who suffered a double burden of oppression in the form of both sexism and racism. For example,

African American women were exploited as workers, as breeders of slaves, and sometimes as sex objects for white men (Hess-Biber and Carter, 2000). By contrast, Native American women in some agricultural communities held greater power because they were able to maintain control over land, tools, and surplus food (Hess-Biber and Carter, 2000).

Do preindustrial forms of work still exist in contemporary high-income nations? In short, yes. Even in high-income nations such as the United States, entire families work in the agricultural sector of the economy, performing tasks such as picking ripened cherries. Here is one journalist's description of "home life" among some seasonal cherry pickers in the state of Washington:

> At the height of cherry-picking season, 12 men were making a patch of woods . . . into a home. Plastic grocery bags hung from nails in the Ponderosa pine bark: the pantry. A shard of mirror was wedged into the trunk of another pine: the bathroom. The kitchen was a Sunbeam propane grill, the bushes served as toilets, and six cheap tents formed the bedroom. . . . Cherry picking requires special timing and skills; the fruit must be picked just as soon as it is ripe, and workers must be careful not to separate the stems from the fruit or damage the branches where next year's

buds will form. . . . Working from 4 A.M. until noon, when heat can damage the fruit, pickers can make $4 for filling a 40-pound box, or up to $80 a day. Many travel for weeks at a time following the ripening from California to Oregon, then from Washington to Montana. (Kelley, 1999: A11)

As this example shows, some parts of the agricultural sector of the U.S. economy have not been changed by industrialization or postindustrialization. The cherry pickers described above are employed in the same region as many high-tech information employees who work for Microsoft or other computer manufacturers or software designers.

Industrial Economies

Industrialization brings sweeping changes to the system of production and distribution of goods and services. Drawing on new forms of energy (such as steam, gasoline, and electricity) and machine technology, factories proliferate as the primary means of producing goods. Most workers engage in **secondary sector production—the processing of raw materials (from the primary sector) into finished goods.** For example, steelworkers process metal ore; autoworkers then convert the ore into automobiles, trucks, and buses. In industrial economies, work becomes specialized and repetitive, activities become bureaucratically organized, and workers primarily work with machines instead of with one another.

Mass production results in larger surpluses that benefit some people and organizations but not others. Goods and services become more unequally distributed because some people can afford anything they want and others can afford very little. Nations engaging primarily in secondary sector production also have some primary sector production, but they rely on less-industrialized nations for the raw materials from which to make many products. In sum, the typical characteristics of industrial economies include the following:

1. *New forms of energy, mechanization, and the growth of the factory system.* With the introduction of the steam engine and steam-powered machines, work becomes centered in factories, which are viewed as separate and distinct spheres from the home and the earlier cottage industries that had been located in the home.
2. *Increased division of labor and specialization among workers.* With industrialization, people carry out a wider diversity of jobs which have different but integrated activities that contribute to the production of specific goods.
3. *Universal application of scientific methods to problem solving and profit making.* Scientific knowledge makes the assembly line and mass production (discussed later) possible. New technologies allow the transformation of raw materials into goods that can be sold as commodities at a far greater cost than the value of the raw materials.
4. *Introduction of wage labor, time discipline, and workers' deferred gratification.* In industrial economies, workers are employees. Managers, supervisors, or owners determine what days people will work and how many hours per day they will be required to be at the work site. Workers typically do not have control over what time they are supposed to be at work or what time they leave at the end of the workday. As a result, workers are expected to engage in deferred gratification, meaning that they are to toil diligently at work and wait until they are on "their own time" before pursuing personal pleasures or recreational activities.
5. *Strengthening of bureaucratic organizational structure.* To develop the most efficient workplace, it is necessary to implement and enforce rules, policies, and procedures, thus requiring the creation of a well-defined, bureaucratic organizational structure (see Chapter 6, "Groups and Organizations"). Ultimately, the goal is to make the workplace more efficient and more profitable for owners and upper-level managers.

All of these characteristics contributed to the development of industrial economies, greater productivity in the workplace, and a dramatic increase in consumption. Although the standard of living improved for many people, the sociologist Thorstein Veblen (1857–1929) was a leading critic of U.S. industrialism. According to Veblen (1967/1899), those who got wealthy from industrialization often engaged in *conspicuous consumption* and *conspicuous leisure.* In *The Theory of the Leisure Class* (1967/1899), Veblen stated that the idle rich, who made their vast sums of money through ownership of the factories and from the toil and sweat of workers, represented a conspicuously consuming, parasitic leisure class. For Veblen, conspicuous consumption is the ostentatious display of symbols of wealth, such as owning numerous mansions and expensive works of art, wearing extravagant jewelry and clothing, or otherwise flaunting the trappings of great wealth. Veblen believed that conspicuous consumption also occurs when rich people participate in wasteful and highly visible leisure activities such as casino gambling or sporting events that require costly gear or excessive travel expenses (such as going on a safari in Africa). If Veblen were alive today, do you think he

CENSUS ★ PROFILES

Civilian Occupations in 2000

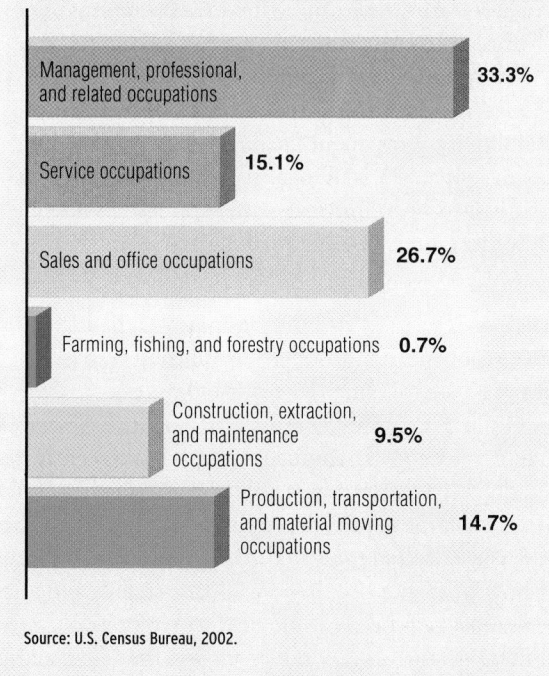

28. Occupation

a. What kind of work was this person doing?
(For example: registered nurse, personnel manager, supervisor of order department, auto mechanic, accountant)

b. What were this person's most important activities or duties? *(For example: patient care, directing hiring policies, supervising order clerks, repairing automobiles, reconciling financial records)*

Census 2000 asked about the occupations of people who are age 16 or over. The majority of people who are not full-time military employees responded that they held jobs in management, professional, service, and sales occupations, as contrasted with primary sector and secondary sector employment:

Occupation	Percent
Management, professional, and related occupations	33.3%
Service occupations	15.1%
Sales and office occupations	26.7%
Farming, fishing, and forestry occupations	0.7%
Construction, extraction, and maintenance occupations	9.5%
Production, transportation, and material moving occupations	14.7%

Source: U.S. Census Bureau, 2002.

might modify his theoretical perspective on conspicuous consumption to incorporate the spending habits of some high-tech entrepreneurs, whose wealth is linked to the postindustrial economy?

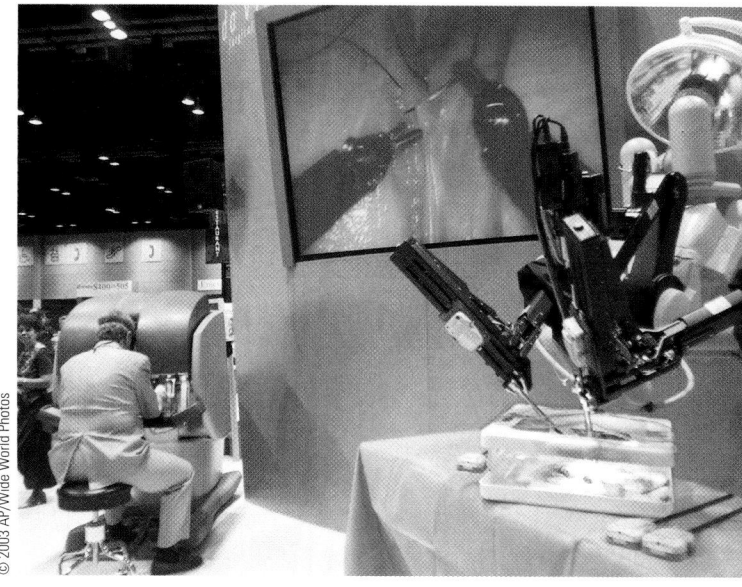

The nature of work has changed dramatically over the past century. Not all social scientists agree on whether robots will eventually replace many humans in the work force. What do you think work will be like in the future?

Postindustrial Economies

A postindustrial economy is based on ***tertiary sector production—the provision of services rather than goods***—as a primary source of livelihood for workers and profit for owners and corporate shareholders. Tertiary sector production includes a wide range of activities, such as fast-food service, transportation, communication, education, real estate, advertising, sports, and entertainment. As shown in the Census Profiles feature, a majority of U.S. jobs are in tertiary sector employment, as contrasted with primary or secondary sector employment.

Several characteristics are central to the postindustrial economy:

1. *Information displaces property as the central preoccupation in the economy.* Postindustrial economies are characterized by ideas, and computer software may eventually become the infrastructure of the future.
2. *Workplace culture shifts away from factories and toward increased diversification of work settings, the workday, the employee, and the manager.* Although many people continue to be employed in traditional workplaces with set workdays, increasing numbers are being affected by layoffs and outsourcing. When other employees are removed, those workers who remain are often expected to do a variety of tasks in addition to the work they were previously assigned.

Figure 13.1   Fastest Growing Occupations, 2000–2010

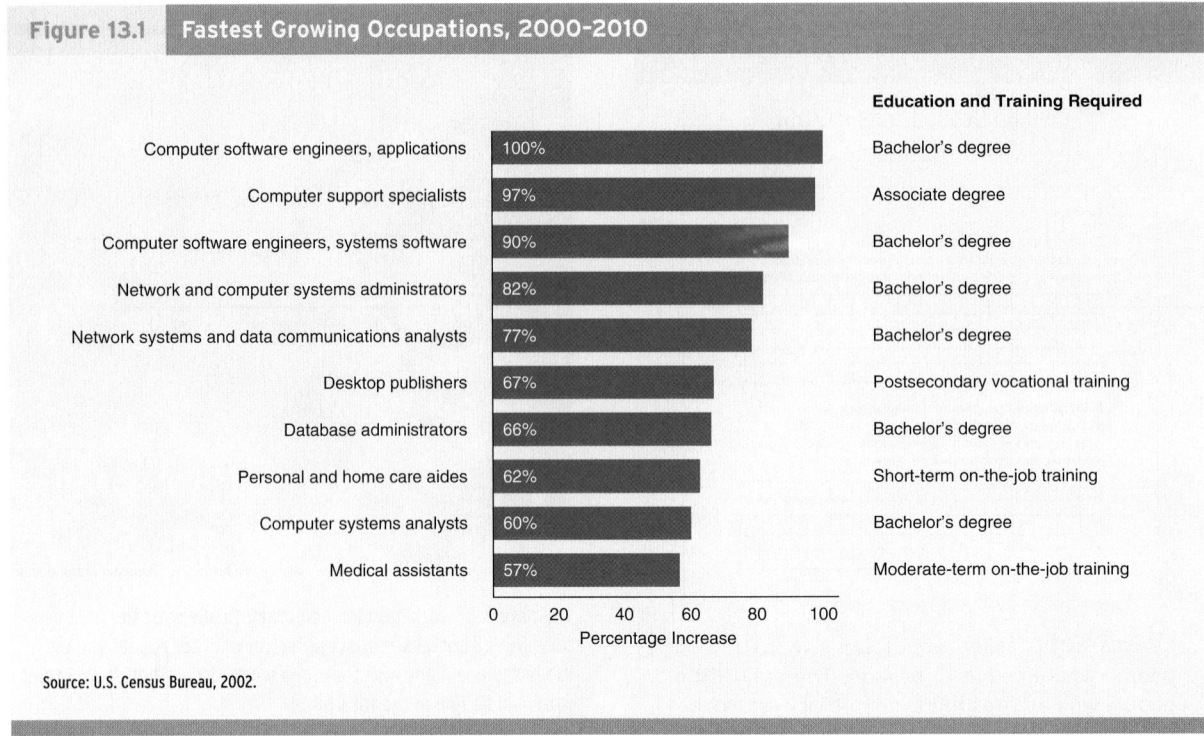

Occupation	Percentage Increase	Education and Training Required
Computer software engineers, applications	100%	Bachelor's degree
Computer support specialists	97%	Associate degree
Computer software engineers, systems software	90%	Bachelor's degree
Network and computer systems administrators	82%	Bachelor's degree
Network systems and data communications analysts	77%	Bachelor's degree
Desktop publishers	67%	Postsecondary vocational training
Database administrators	66%	Bachelor's degree
Personal and home care aides	62%	Short-term on-the-job training
Computer systems analysts	60%	Bachelor's degree
Medical assistants	57%	Moderate-term on-the-job training

Source: U.S. Census Bureau, 2002.

3. *The conventional boundaries between work and home (public life and private life) are breached.* Cell phones, pagers, fax machines, laptop computers, and similar products make it possible for people to work around the clock from locations around the globe. Increasing amounts of remote work activity continue to blur the distinction between one's public life and one's private life.

How far has the United States moved into postindustrialization? Although this question is difficult to answer, projections by the Census Bureau indicate that the fastest-growing occupations of the twenty-first century are in the service sector, as shown in Figure 13.1. Moreover, in his study of the "McDonaldization" of society, the sociologist George Ritzer (2000a) suggested that the number of lower-paying, second-tier service sector positions would continue to increase not only in the United States but worldwide as well. Of course, recent acts of terrorism and war may change the likelihood of this occurring, at least in the short term (see Box 13.2).

Many jobs in the service sector emphasize productivity, often at the expense of workers. Fast-food restaurants are a case in point, as the manager of a McDonald's explains:

> As a manager I am judged by the statistical reports which come off the computer. Which basically means my crew labor productivity. What else can I really distinguish myself by? . . . O.K., it's true,

you can over spend your [maintenance and repair] budget, you can have a low fry yield; you can run a dirty store; you can be fired for bothering high school girls. But basically, every Coke spigot is monitored. Every ketchup squirt is measured. My costs for every item are set. So my crew labor productivity is my main flexibility. . . . Look, you can't squeeze a McDonald's hamburger any flatter. If you want to improve your productivity there is nothing for a manager to squeeze but the crew. (qtd. in Garson, 1989: 33–35)

"McDonaldization" is built on many of the ideas and systems of industrial society, including bureaucracy and the assembly line (Ritzer, 2000a).

Also, class conflict and poverty may increase rather than decrease in postindustrial societies. Recently, researchers have found that employment in the service sector remains gender segregated and, in some types of work, racially segregated. For example, about half of all U.S. service occupations (50.2 percent) and sales positions (43.4 percent) are held by women. But even in these occupations, women occupy the lower-status and lower-paying jobs, where there is little chance for advancement to better-paid positions (Hesse-Biber and Carter, 2000). Domestic service workers are another example. They have always been predominantly female and often minority members. Today, many hotel maids and household domestic service workers are first-generation Latinas (Romero, 1992). According to the sociologist Pierrette Hondagneu-Sotelo

Box 13.2 SOCIOLOGY IN GLOBAL PERSPECTIVE

McDonald's Golden Arches as a Lightning Rod

Hours after the United States launched its first bombing raids in Afghanistan [in the aftermath of the September 11, 2001, terrorist attacks] the backlash began. In Pakistan, crowds vandalized McDonald's outlets in Islamabad and Karachi. In Indonesia, demonstrators burned an American flag outside a McDonald's restaurant in the resort town of Makassar and then stormed it, and across the country in Yogyakarta, other protesters blockaded yet another McDonald's.

> —journalist David Barboza (2001) describing the violence perpetrated against buildings and other structures that reflected the presence of U.S. multinational businesses in other global regions

In the name of Allah, the merciful and gracious, McDonald's Indonesia is owned by an indigenous Muslim.

> —Bambang Rachmadi's six-foot-high green banner outside one of his McDonald's restaurants in Jakarta, Indonesia (qtd. in Solomon, 2001: A1)

© AFP/Cobis

Even before the terrorist attacks on New York City and Washington, D.C., and the war in Afghanistan and Iraq, McDonald's and other transnational businesses received extensive criticism in various nations around the world. When the first McDonald's restaurant opened in 1955 in Southern California, few people predicted that the kingdom of Ray Kroc, the company's founder, would grow to more than 29,000 restaurants serving 50 million customers in 120 nations and employing more than a million people (Barboza, 2001). In the twenty-first century, the success of McDonald's and other corporations with headquarters in the United States has become a topic of hot controversy: McDonald's Golden Arches, as seen in varying forms that supposedly "fit into" different cultural landscapes, have become symbolic of much more than a fast-food restaurant. For some, McDonald's is a quick way to order "standardized" hamburgers; for others, the restaurant chain represents an unacceptable form of globalization. Terms such as "McWorld" have been used to deride a "standardized global village" in which all the people exist on Big Macs and fries (Barboza, 2001). Some analysts refer to this as U.S. cultural imperialism, defined in Chapter 3 as the extensive infusion of one nation's culture into another nation's.

For many years, transnational businesses—even when local franchises are owned by individuals such as Bambang Rachmadi of Indonesia—have been the target of complaints and sometimes of vandalism or destruction because of issues relating to capitalism and cultural erosion. Today, these issues remain volatile but have been overshadowed by political and religious differences among nations. As U.S. companies tighten security to protect their workers and facilities in other nations and as high levels of uncertainty remain regarding the economies of the United States and other high-income countries, many business analysts believe that globalization is going to be in a holding pattern for a while (Fox, 2001; Keefe, 2001).

Will McDonald's, Pizza Hut, Nike, and other transnational giants greatly reduce their economic operations in other nations? Such a prospect is not likely. At the time of this writing, the U.S. economy was struggling in the post-September 11 climate due to massive layoffs in the high-tech industries, the airline and other travel-related businesses, and other sectors of the economy. U.S. consumers had high average household debts—with the average household owing more than $8,500 just on credit-card debt—while retail sales were simultaneously dropping precipitously (Fox, 2001). Add to all of that the fact that many corporations register a large volume of their sales outside the United States; for example, more than half of sales by the McDonald's chain in 2000 took place beyond the borders of the United States (Barboza, 2001). If these conditions continue, executives at large corporations such as McDonald's may believe that they have little choice but to continue to expand in the global marketplace and serve as a lightning rod for worldwide reactions to U.S. government policies and to political and religious sentiment directed at the United States.

(2001: 48), women who are undocumented workers (people who come from other countries but do not have proper papers to be legally employed in the United States) often believe that they can find safety by quickly taking positions as live-in domestic workers: "For newly arrived immigrant women without papers, a live-in job in a private home may feel safer [than other types of work, such as a hotel maid], as private homes in middle- and upper-middle-class neighborhoods are rarely, if ever, threatened by Immigration and Naturalization Service raids." As more middle- and upper-income women have gained full-time employment in better-paid occupations and professional careers, it has often been women of color, particularly recent immigrants, who have taken up the slack in domestic work in their households (Hess-Biber and Carter, 2000; Romero, 1992). As the sociologist Mary Romero (1992: 29) explains, "The cheap domestic labor of women of color is one means by which white middle-class women escape oppressive aspects of their domestic roles."

Although many people in the U.S. work force will continue to be employed in low-paying service jobs in the twenty-first century, others will benefit by continuing changes in technology and global economic conditions. Some analysts believe that the falling prices for manufactured goods and stiff international competition will continue to take a toll on the U.S. manufacturing sector and that "nontraded" services will boom. Nontraded service providers—such as hospitals, schools, nursing homes, and overnight courier services—must remain at the point where the service is consumed and thus cannot be off-shored or near-shored. Neither clients nor providers in settings such as these can be easily relocated. For example, a person who wants a package delivered or needs emergency surgery or care in a nursing home is highly unlikely to relocate to a low-income nation where the workers might be paid lower wages for providing such services.

CONTEMPORARY WORLD ECONOMIC SYSTEMS

Throughout the twentieth century, capitalism and socialism were the principal economic models in industrialized countries. Sociologists often use two criteria—property ownership and market control—to distinguish between types of economies. However, keep in mind that no society has a purely capitalist or socialist economy.

Capitalism

***Capitalism* is an economic system characterized by private ownership of the means of production, from which personal profits can be derived through market competition and without government intervention.** Most of us think of ourselves as "owners" of private property because we own a car, a television, or other possessions. However, most of us are not capitalists; we *spend money* on the things we own rather than *make money* from them. Only a relatively few people own income-producing property from which a profit can be realized by producing and distributing goods and services. Everyone else is a consumer. "Ideal" capitalism has four distinctive features: (1) private ownership of the means of production, (2) pursuit of personal profit, (3) competition, and (4) lack of government intervention.

Private Ownership of the Means of Production
Capitalist economies are based on the right of individuals to own income-producing property, such as land, water, mines, and factories, and the right to "buy" people's labor.

In the early stages of industrial capitalism (1850–1890), virtually all the capital for investment in the United States was individually owned—prior to the Civil War, an estimated two hundred families controlled all major trade and financial organizations. By the 1890s, individual capitalists, including Andrew Carnegie, Cornelius Vanderbilt, and John D. Rockefeller, controlled most of the capital in commerce, agriculture, and industry (Feagin and Feagin, 1997).

As workers grew tired of toiling for the benefit of capitalists instead of for themselves, some of them banded together to form the first national labor union, the Knights of Labor, in 1869. A ***labor union* is a group of employees who join together to bargain with an employer or a group of employers over wages, benefits, and working conditions.** The Knights of Labor included both skilled and unskilled laborers, but the American Federation of Labor (AFL), founded in 1886, targeted groups of skilled workers such as plumbers and carpenters; each of these craft unions maintained autonomy under the "umbrella" of the AFL.

Under early monopoly capitalism (1890–1940), most ownership rapidly shifted from individuals to huge ***corporations*—large-scale organizations that have legal powers, such as the ability to enter into contracts and buy and sell property, separate from their individual owners.** During this period, major industries, including oil, sugar, and grain, came under the control of a few corporations owned by share-

Table 13.1 THE LARGEST NONFINANCIAL U.S. TRANSNATIONAL CORPORATIONS (RANKED BY FOREIGN REVENUES)

CORPORATION	PRODUCT	FOREIGN REVENUES ($ BILLIONS)	TOTAL REVENUES ($ BILLIONS)	ASSETS ($ BILLIONS)
Exxon/Mobil	Oil refining	143.0	206.0	149.0
Chevron Texaco	Oil refining	65.0	117.0	77.6
Ford	Motor vehicles	51.7	170.0	283.4
IBM	Office equipment	51.1	88.4	88.3
General Electric	Electronics	49.5	129.8	437.0
General Motors	Motor vehicles	48.2	184.6	303.1
Wal-Mart Stores	Retail sales	32.1	191.3	78.1
Philip Morris	Tobacco products	32.1	63.3	79.0
Hewlett-Packard	Office equipment	27.5	48.9	34.0
Motorola	Telecommunications	21.8	37.6	42.3

United Nations Conference on Trade and Development, 2002.

holders. As a general rule, individual shareholders—persons who hold or own shares of stock in a corporation—cannot be blamed for the actions of the corporation. Thus, shareholders are able to protect their personal assets while limiting their liability for the corporation's actions or debts.

As automobile and steel plants shifted to mass production, workers once again perceived that their needs were not being met. In 1935 the Congress of Industrial Organizations (CIO) was established to represent both skilled and unskilled workers in entire industries such as automobile manufacturing. According to John Sargent, a steelworker,

I got in the mills in 1936, and I [was] fortunate to be caught up in a great movement of the people in this country. . . . [A] movement of the kind that we had . . . in the CIO was a movement that moved millions of people, literally, and changed not only the course of the working man in this country, but also the nature of the relationship between the working man and the government and between the working man and the boss. . . . (qtd. in Watkins, 1993: 274)

In 1937, GM workers held their first *sit-down strike:* The workers occupied the plants but refused to work, thus paralyzing production (McEachern, 2003). The sit-down strike represented a new approach that came to dominate U.S. labor activism (Watkins, 1993).

In advanced monopoly capitalism (1940–present), ownership and control of major industrial and business sectors have become increasingly concentrated. Economic concentration is the degree to which a rel-

atively small number of corporations control a disproportionately large share of a nation's economic resources. Today, many corporations are global in scope. *Transnational corporations* **are large corporations that are headquartered in one country but sell and produce goods and services in many countries.** These corporations play a major role in the economies and governments of many nations. Ten of the largest U.S. transnational corporations are listed in Table 13.1.

Transnational corporations are not dependent on the labor, capital, or technology of any one country and may move their operations to countries where wages and taxes are lower and the potential profits are higher. Corporate considerations of this kind help explain why many jobs formerly located in the United States have shifted to lower-income nations where few employment opportunities exist and workers can be paid significantly less than their U.S. counterparts. This appears to be a fact of life, whether workers are producing parts for automobiles and computers or cooking hamburgers in a fast-food restaurant owned by a transnational corporation such as McDonald's, which prides itself on placing "Golden Arches" throughout the world.

Pursuit of Personal Profit A tenet of capitalism is the belief that people are free to maximize their individual gain through personal profit; in the process, the entire society will benefit from their activities (Smith, 1976/1776). Economic development is assumed to benefit both capitalists and workers, and the general public also benefits from public expenditures

Capitalism is built on the pursuit of personal profit, and the opening-bell ceremony at the New York Stock Exchange is a symbol of that pursuit. Although corporations and shareholders may reap large economic gain from the stock market, most U.S. families do not own any type of stock.

(such as for roads, schools, and parks) made possible through an increase in business tax revenues.

During the period of industrial capitalism, however, specific individuals and families (not the general public) were the primary recipients of profits. For many generations, descendants of some of the early industrial capitalists have benefited from the economic deeds (and misdeeds) of their ancestors. In early monopoly capitalism, some stockholders derived massive profits from companies that held near-monopolies on specific goods and services. Examples included American Tobacco Company, Westinghouse Electric Corporation, and the former F. W. Woolworth Corporation (Hodson and Sullivan, 2002).

In advanced (late) monopoly capitalism, profits have become even more concentrated. Although many people own *some* stock, they do not own *control;* in other words, they are unable to participate in establishing the policies that determine the size of the profit or the rate of return on investments (which affects the profits they derive). In booming economic times, however, shareholders of lucrative stocks are less concerned about the fact that they do not own control of corporations such as Nike or Dell Computer.

Competition   In theory, competition acts as a balance to excessive profits. When producers vie with one another for customers, they must be able to offer innovative goods and services at competitive prices. However, from the time of early industrial capitalism, the trend has been toward less, rather than more, competition among companies. In early monopoly capitalism, competition was diminished by increasing concentration *within* a particular industry, a classic case being the virtual monopoly on oil held by John D. Rockefeller's Standard Oil Company (Tarbell, 1925/1904; Lundberg, 1988). Today, Microsoft Corporation so dominates certain areas of the computer software industry that it has virtually no competitors in those areas.

What appears to be competition among producers *within* an industry may actually be "competition" among products, all of which are produced and distributed by relatively few corporations. An *oligopoly* **exists when several companies overwhelmingly control an entire industry.** An example is the music industry, in which a few giant companies are behind many of the labels and artists known to consumers (see Table 13.2). More specifically, a *shared monopoly* **exists when four or fewer companies supply 50 percent or more of a particular market.** Examples include U.S. automobile manufacturers (referred to as the "Big Three") and cereal companies (three of which control 77 percent of the market). In 1995 some members of Congress called for an investigation of cereal companies for antitrust violations that had allegedly resulted in a 90-percent increase in the cost of cold cereal over a ten-year period (*Austin American-Statesman,* 1995).

In advanced monopoly capitalism, mergers also occur *across* industries: Corporations gain near-monopoly control over all aspects of the production and distribution of a product by acquiring both the companies that supply the raw materials and the companies that are the outlets for their products. For example, an oil company may hold leases on the land where the oil is pumped out of the ground, own the plants that convert the oil into gasoline, and own

Table 13.2 THE MUSIC INDUSTRY'S BIG FIVE

COMPANY	COUNTRY	LEADING ARTISTS
BMG Entertainment (RCA, Arista)	Germany	Avril Lavigne Foo Fighters Alan Jackson
Universal Music Group (MCA, Polygram)	France	50 Cent Eminem Shania Twain
Sony Music Entertainment	Japan	Dixie Chicks John Mayer Bow Wow
EMI Group (Capitol, EMI, Virgin)	United Kingdom	Norah Jones Radiohead Goo Goo Dolls
Warner Music Group (Atlantic, Elektra)	United States	Faith Hill Linkin Park The Donnas

the individual gasoline stations that sell the product to the public.

Corporations with control both within and across industries are often formed by a series of mergers and acquisitions across industries. These corporations are referred to as *conglomerates*—**combinations of businesses in different commercial areas, all of which are owned by one holding company.** Media ownership is a case in point; companies such as Time Warner and Viacom have extensive holdings in radio and television stations, cable television companies, book publishing firms, and film production and distribution companies, to name only a few.

Competition is reduced over the long run by *interlocking corporate directorates*—**members of the board of directors of one corporation who also sit on the board(s) of other corporations.** Although the Clayton Antitrust Act of 1914 made it illegal for a person to sit simultaneously on the boards of directors of two corporations that are in *direct* competition with each other, a person may serve simultaneously on the board of a financial institution (a bank, for example) and the board of a commercial corporation (a computer manufacturing company or a furniture store chain, for example) that borrows money from the bank. Directors of competing corporations may also serve together on the board of a third corporation that is not in direct competition with the other two. In 1996 former Defense Secretary Frank C. Carlucci sat on the corporate boards of fourteen Fortune 1,000 companies; former Labor Secretary Ann D. McLaughlin sat on eleven; and former Health, Education, and Welfare Secretary Joseph A. Califano, Jr., sat on nine (Dobrzynski, 1996). An example of interlocking directorates is depicted in Figure 13.2. Compensation for members of the boards of top corporations can be as high (taking into account stock, stock options, and pensions) as $400,000 to $500,000 per board position per year (Dobrzynski, 1996).

Interlocking directorates diminish competition by producing interdependence. Individuals who serve on multiple boards are often able to forge cooperative arrangements that benefit their corporations but not necessarily the general public. When several corporations are controlled by the same financial interests, they are more likely to cooperate with one another than to compete (Mintz and Schwartz, 1985).

Lack of Government Intervention Ideally, capitalism works best without government intervention in the marketplace. The policy of *laissez-faire* (les-ay-FARE, which means "leave alone") was advocated by economist Adam Smith in his 1776 treatise *An Inquiry into the Nature and Causes of the Wealth of Nations.* Smith argued that when people pursue their own selfish interests, they are guided "as if by an invisible hand" to promote the best interests of society (see Smith, 1976/1776). Today, terms such as *market economy* and *free enterprise* are often used, but the underlying assumption is the same: that free market competition, not the government, should regulate prices and wages. However, the "ideal" of unregulated markets benefiting all citizens has seldom been realized. Individuals and companies in pursuit of higher

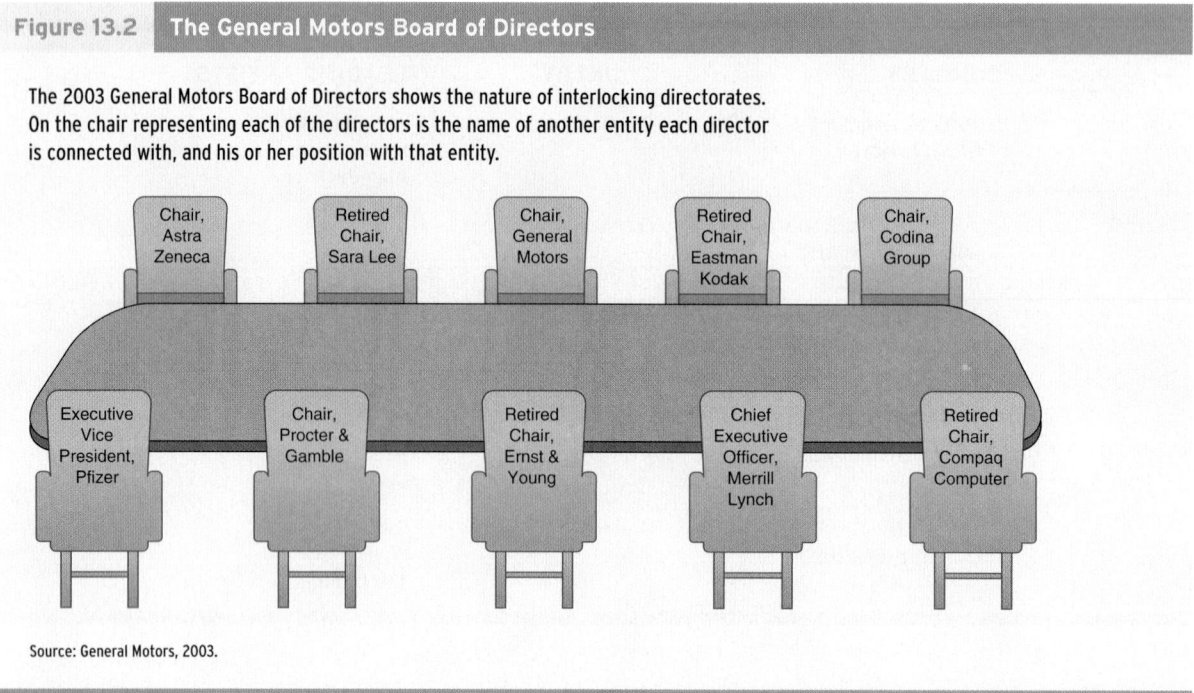

Figure 13.2   The General Motors Board of Directors

The 2003 General Motors Board of Directors shows the nature of interlocking directorates. On the chair representing each of the directors is the name of another entity each director is connected with, and his or her position with that entity.

Source: General Motors, 2003.

profits have run roughshod over weaker competitors, and small businesses have grown into large, monopolistic corporations. Accordingly, government regulations were implemented in an effort to curb the excesses of the marketplace brought about by laissez-faire policies.

However, much of what is referred to as government intervention has been in the form of aid to business. Between 1850 and 1900, corporations received government assistance in the form of public subsidies and protection from competition by tariffs, patents, and trademarks. The federal government also gave large tracts of land to the privately owned railroads to encourage their expansion across the nation. Antitrust laws originally intended to break up monopolies were used instead against labor unions that supported workers' interests (Parenti, 1996).

Government intervention in the past two decades included billions of dollars in subsidies to farmers, tax credits for corporations, and large subsidies or loan guarantees to automakers, aircraft companies, railroads, and others. Overall, most corporations have gained much more than they have lost as a result of government involvement in the economy.

Socialism

Socialism is an economic system characterized by public ownership of the means of production, the pursuit of collective goals, and centralized decision making. Like "pure" capitalism, "pure" socialism does not exist. Karl Marx described socialism as a temporary stage en route to an ideal communist society. Although the terms *socialism* and *communism* are associated with Marx and are often used interchangeably, they are not identical. Marx defined communism as an economic system characterized by common ownership of all economic resources (Marshall, 1998). In the *Communist Manifesto* and *Das Kapital,* he predicted that the working class would become increasingly impoverished and alienated under capitalism. As a result, the workers would become aware of their own class interests, revolt against the capitalists, and overthrow the entire system (see Turner, Beeghley, and Powers, 2002). After the revolution, private property would be abolished and capital would be controlled by collectives of workers who would own the means of production. The government (previously used to further the interests of the capitalists) would no longer be necessary. People would contribute according to their abilities and receive according to their needs (Marx and Engels, 1967/1848; Marx, 1967/1867). Over the years, state control was added as an organizing principle for communist societies. This structure is referred to as a system of "state socialism," and the former Soviet Union (the "Union of Soviet Socialist Republics," what today is Russia and certain other nearby countries) became the best example.

Why did state socialism in the Soviet Union not evolve into a communist economic system? For one thing, Marx's ideas about a truly communistic society were based on his perceptions of *industrialized* nations. At the time that an attempt was made to implement communism in Russia, however, it was largely a feudal society that had not experienced the type of conditions Marx envisioned would be necessary for the development of communism (Rosengarten, 1995). The system of state socialism failed for other reasons as well, as discussed later in this chapter.

Public Ownership of the Means of Production

In a truly socialist economy, the means of production are owned and controlled by a collectivity or the state, not by private individuals or corporations. Prior to the early 1990s, the state owned all the natural resources and almost all the capital in the Soviet Union. In the 1980s, for example, state-owned enterprises produced more than 88 percent of agricultural output and 98 percent of retail trade, and owned 75 percent of the urban housing space (Boyes and Melvin, 2002). At least in theory, goods were produced to meet the needs of the people. Access to housing and medical care was considered to be a right.

Leaders of the Soviet Union and some Eastern European nations decided to abandon government ownership and control of the means of production because the system was unresponsive to the needs of the marketplace and offered no incentive for increased efficiency (Boyes and Melvin, 2002). Shortages and widespread unrest led to a reform movement by then-President Mikhail Gorbachev in the late 1980s.

Since the 1990s, Russia and other states in the former Soviet Union have attempted to privatize ownership of production. China—previously the world's other major communist economy—announced in 1997 that it would privatize most state industries (Serrill, 1997). In *privatization,* resources are converted from state ownership to private ownership; the government takes an active role in developing, recognizing, and protecting private property rights (Boyes and Melvin, 2002). However, the emergent economies of Russia and China saw the value of their stock markets and currencies plunge in the economic crisis that started in Southeast Asia and spread to Russia and South America between 1997 and 1999. In the late 1990s, Russian capital flight averaged as much as $1 billion per month, moving to places like Great Britain and Switzerland, which affluent, powerful Russians believed they could trust, and the economy collapsed (Banerjee, 1999; Kristof and WuDunn, 1999).

Pursuit of Collective Goals

Ideal socialism is based on the pursuit of collective goals rather than on personal profits. Equality in decision making replaces hierarchical relationships (such as between owners and workers or political leaders and citizens). Everyone shares in the goods and services of society, especially necessities such as food, clothing, shelter, and medical care, based on need, not on ability to pay. In reality, however, few societies can or do pursue purely collective goals.

Centralized Decision Making

Another tenet of socialism is centralized decision making. In theory, economic decisions are based on the needs of society; the government is responsible for aiding the production and distribution of goods and services. Central planners set wages and prices to ensure that the production process works. When problems such as shortages and unemployment arise, they can be dealt with quickly and effectively by the central government (Boyes and Melvin, 2002).

Centralized decision making is hierarchical. In the former Soviet Union, for example, broad economic policy decisions were made by the highest authorities of the Communist Party, who also held political power. The production units (the enterprises and farms) at the bottom of the structure had little voice in the decision-making process. Wages and prices were based on political priorities and eventually came to be completely unrelated to actual supply and demand.

The collapse of state socialism in the former Soviet Union was due partly to the declining ability of the Communist Party to act as an effective agent of society and partly to the growing incompatibility of central planning with the requirements of a modern economy (see Misztal, 1993). A decade after centralized decision making was abolished in Russia, people are still faced with poverty, soaring unemployment, and high crime rates. Organized criminal groups have muscled their way into business and trade; many workers feel that their future is very dim. According to a thirty-year-old machinist in Moscow, "I was raised in a country that cared about its workers. . . . Life used to be simple, but we had a deal. We worked hard, and the company took care of the rest" (qtd. in Specter, 1994: A16). Ideas such as this remain strong in Russia, where, as one economic analyst noted,

> Russia has always had this megalomania. The people who run things are from the old system, and they think if they develop five or six big companies, that will drive growth. What could happen with more entrepreneurs entering manufacturing is that it would create a middle class, which would have a huge influence on the social, political and economic life of Russia. (qtd. in Banerjee, 1999: C23)

Mixed Economies

As we have seen, no economy is truly capitalist or socialist; most economies are mixtures of both. A *mixed economy* combines elements of a market economy (capitalism) with elements of a command economy (socialism). Sweden, Great Britain, and France have mixed economies, sometimes referred to as *democratic socialism*—an economic and political system that combines private ownership of some of the means of production, governmental distribution of some essential goods and services, and free elections. For example, government ownership in Sweden is limited primarily to railroads, mineral resources, a public bank, and liquor and tobacco operations (Feagin and Feagin, 1997). Compared with capitalist economies, however, the government in a mixed economy plays a larger role in setting rules, policies, and objectives.

The government is also heavily involved in providing services such as medical care, child care, and transportation. In Sweden, for example, all residents have health insurance, housing subsidies, child allowances, paid parental leave, and day-care subsidies. National insurance pays medical bills associated with work-related injuries, and workplaces are specially adapted for persons with disabilities. College tuition is free, and public funds help subsidize cultural institutions such as theaters and orchestras ("General Facts on Sweden," 1988; Kelman, 1991). Recently, some analysts have suggested that the United States has assumed many of the characteristics of a *welfare state* (a state that uses extensive government action to provide support and services to its citizens) as it has attempted to meet the basic needs of older persons, young children, unemployed people, and persons with a disability (Esping-Andersen, 1990).

PERSPECTIVES ON ECONOMY AND WORK IN THE UNITED STATES

Functionalists, conflict theorists, and symbolic interactionists view the economy and the nature of work from a variety of perspectives. We first examine functionalist and conflict views of the economy; then we focus on the symbolic interactionist perspective on job satisfaction and alienation.

Functionalist Perspective

Functionalists view the economy as a vital social institution because it is the means by which needed goods and services are produced and distributed. When the economy runs smoothly, other parts of society function more effectively. However, if the system becomes unbalanced, such as when demand does not keep up with production, a maladjustment occurs (in this case, a surplus). Some problems can be easily remedied in the marketplace (through "free enterprise") or through government intervention (such as paying farmers *not* to plant when there is an oversupply of a crop). However, other problems, such as periodic *peaks* (high points) and *troughs* (low points) in the business cycle, are more difficult to resolve. The *business cycle* is the rise and fall of economic activity relative to long-term growth in the economy (McEachern, 2003).

From this perspective, peaks occur when "business" has confidence in the country's economic future. During a peak, or *expansion period,* the economy thrives: Plants are built, raw materials are ordered, workers are hired, and production increases. In addition, upward social mobility for workers and their families becomes possible. As a result, workers may hope their children will not have to follow their footsteps into the factory. Ben Hamper (1992: 13) describes how GM workers felt:

> Being a factory worker in Flint, Michigan, wasn't something purposely passed on from generation to generation. To grow up believing that you were brought into this world to follow in your daddy's footsteps, just another chip-off-the-old-shoprat, was to engage in the lowest possible form of negativism. Working the line for GM was something fathers did so that their offspring wouldn't have to.

The American Dream of upward mobility is linked to peaks in the business cycle. Once the peak is reached, however, the economy turns down because too large a surplus of goods has been produced. In part, this is due to inflation—a sustained and continuous increase in prices (McEachern, 2003). Inflation erodes the value of people's money, and they are no longer able to purchase as high a percentage of the goods that have been produced. Because of this lack of demand, fewer goods are produced, workers are laid off, credit becomes difficult to obtain, and people cut back on their purchases even more, fearing unemployment. Eventually, this produces a distrust of the economy, resulting in a *recession*—a decline in an economy's total production that lasts six months or longer. To combat a recession, the government lowers interest rates (to make borrowing easier and to get more money back into circulation) in an attempt to spur the beginning of the next expansion period.

Conflict Perspective

Conflict theorists view business cycles and the economic system differently. From a conflict perspective, business cycles are the result of capitalist greed. In order to maximize profits, capitalists suppress the wages of workers. As the prices of products increase, workers are not able to purchase them in the quantities that have been produced. The resulting surpluses cause capitalists to reduce production, close factories, and lay off workers, thus contributing to the growth of the reserve army of the unemployed, whose presence helps reduce the wages of the remaining workers. In some situations, workers are replaced with machines or nonunionized workers. In 1994, for example, some trucking companies claimed that they could not afford to pay union truck drivers over $17 an hour, so they hired nonunion freight carriers who made about $9 an hour. Drivers such as Tim Hart claimed that pay was the only difference between his work and that of a union driver: "I bust my tail, doing the exact same thing the union drivers do. And they make double what I do. It bothers me. I mean, if those companies can afford to pay that much, why can't mine?" (qtd. in D. Johnson, 1994: 10).

Karl Marx referred to the propensity of capitalists to maximize profits by reducing wages as the *falling rate of profit,* which he believed to be one of the inherent contradictions of capitalism that would produce its eventual downfall. According to the political sociologist Michael Parenti, business *is* the economic system. Parenti believes that political leaders treat the health of the capitalist economy as a necessary condition for the health of the nation and that the goals of big business (rapid growth, high profits, and secure markets at home and abroad) become the goals of government (Parenti, 1996). In sum, to some conflict theorists, capitalism is the problem; to some functionalist theorists, capitalism is the solution to society's problems.

Symbolic Interactionist Perspective

Sociologists who focus on microlevel analyses are interested in how the economic system and the social organization of work affect people's attitudes and behavior. Symbolic interactionists, in particular, have examined the factors that contribute to job satisfaction.

According to symbolic interactionists, work is an important source of self-identity for many people; it can help people feel positive about themselves, or it can cause them to feel alienated. *Job satisfaction* refers to people's attitudes toward their work, based on (1) their job responsibilities, (2) the organizational structure in which they work, and (3) their individual needs and values (Hodson and Sullivan, 2002). Studies have found that worker satisfaction is highest when employees have some degree of control over their work, when they are part of the decision-making process, when they are not too closely supervised, and when they feel that they play an important part in the outcome (Kohn et al., 1990). Persons in management and supervisory positions may feel more of a sense of ownership of the products or services produced by their organization. For example, Deborah Kent, the first African American woman to head a vehicle assembly plant at the Ford Motor Company, feels such a sense of ownership that she sometimes asks strangers on the street how they like their vehicle if it was produced in her plant:

> I feel this sense of ownership, responsibility and pride about every Econoline, Villager, and Quest on the road. Particularly since I know we're the only manufacturing plant producing those vehicles. And if something's wrong, I want to know about it. (qtd. in Williams, 1995: F7)

Job satisfaction is often related to both intrinsic and extrinsic factors. Intrinsic factors pertain to the nature of the work itself, whereas extrinsic factors include such things as vacation and holiday policies, parking privileges, on-site day-care centers, and other amenities that contribute to workers' overall perception that their employer cares about them. In one study, the sociologist Ruth Milkman (1997) found that autoworkers were ambivalent about their jobs. Although they hated the chronic stress and humiliation of factory work, they liked the high pay and good benefits.

Alienation occurs when workers' needs for self-identity and meaning are not met and when work is done strictly for material gain, not a sense of personal satisfaction. According to Marx, workers are resistant to having very little power and no opportunities to make workplace decisions. This lack of control contributes to an ongoing struggle between workers and employers. Job segmentation, isolation of workers, and the discouragement of any type of pro-worker organizations (such as unions) further contribute to feelings of helplessness and frustration. Some occupations may be more closely associated with high levels of alienation than others. Sometimes, factors such as the amount of time and mental energy that it takes to get to work affect people's perceptions about their

Workplace anger such as exhibited by this man has become an increasing concern. Conflict theorists might argue that such stress is more pronounced among alienated employees who believe that they are nothing more than cogs in the wheel of the capitalist economy.

employment. As discussed in Box 13.3, when the economy was booming in the Silicon Valley, more people were sitting in their cars on a traffic-clogged freeway, trying to get to work, but gridlock has been reduced with a downturn in the economy and a higher rate of unemployment, particularly in the high-tech sector of the economy.

THE SOCIAL ORGANIZATION OF WORK

How do societies organize work? As societies grow larger in population and more complex in their division of labor, different categories of work are identified by government bureaucracies such as the U.S. Department of Labor and the U.S. Census Bureau. These classifications are often somewhat arbitrary and may not reflect the actual job descriptions and tasks associated with the work that many people do. For example, sociologists often find that Census Bureau categories such as "executive, administrators, and managerial" or "technical and related support" encompass such a wide range of jobs and differences in compensation that it is difficult to make generalizations from statistical data about workers in these employment categories.

Occupations

Occupations **are categories of jobs that involve similar activities at different work sites** (Reskin and Padavic, 2002). Over 500 different occupations and 21,000 occupational specialties, ranging from motion picture cartoonist to drop-hammer operator, are currently listed by the U.S. Department of Labor (1993). Historically, occupations were classified as blue collar and white collar. Blue-collar workers were primarily factory and craft workers who did manual labor; white-collar workers were office workers and professionals. However, contemporary workers in the service sector do not easily fit into either of these categories; neither do the so-called pink-collar workers, primarily women, who are employed in occupations such as preschool teacher, dental assistant, secretary, and clerk (Hodson and Sullivan, 2002).

Sociologists establish broad occupational categories by distinguishing between employment in the primary labor market and in the secondary labor market. The *primary labor market* **consists of high-paying jobs with good benefits that have some degree of security and the possibility of future advancement.** By contrast, the *secondary labor market* **consists of low-paying jobs with few benefits and very little job security or possibility for future advancement** (Bonacich, 1972). Some jobs in the secondary labor market are also characterized by hazardous working conditions and unscrupulous employers. Jane H. Lii, a Chinese-speaking journalist on an undercover assignment, described such problems when she briefly worked in a garment factory in Brooklyn, New York:

> Seven days later, after 84 hours of work, I got my reward, in the form of a promise that in three weeks I would be paid $54.24, or 65 cents an hour [much less than the minimum wage]. I also walked away from the lint-filled factory with aching shoulders, a stiff back, a dry cough, and a burning sore throat. (Lii, 1995: 1)

Like many workers in the lower tier of the secondary labor market, garment factory workers have relatively little formal education, may be working in violation of child labor laws, and may be recent immigrants who feel powerless to call attention to their plight. By contrast, professionals in the upper tier of the primary labor market are characterized by extensive education or training and some degree of control over their work.

Professions

What occupations are professions? Athletes who are paid for playing sports are referred to as "professional athletes." Dog groomers, pest exterminators, auto-

© Paul Simcock (Image Bank)/Getty Images

Box 13.3 CHANGING TIMES: MEDIA AND TECHNOLOGY

As the Economy Goes, So Goes the Traffic

> I had never dreamed that I would be able to drive [California Highway 17] at 50 mph pretty much the whole time, but that's what happens.
>
> —Jeff Hill, a marketing manager, discussing his commute from Scotts Valley to Los Gatos in the aftermath of a major economic downturn in California's Silicon Valley (qtd. in Edwards, 2003: D3)

Why would a comment like this be newsworthy in the high-tech corridor known as the Silicon Valley? Because it is a reflection of how much the economy has changed for the worse in that area in recent years. Several years ago, when this box was originally written, it was titled "On Top of the World, Stuck in Traffic," and the issue discussed was how some people had great jobs in high-tech industries but were fighting traffic jams and giving up their valuable time waiting to get from one place to another. As one person wrote,

> There will be an impact when a person only has time to work, drive and sleep. [People] don't have time to get involved in their communities. They can't get home in time for Little League games.
>
> —Steve Tedesco, president of the San Jose Silicon Valley Chamber of Commerce (qtd. in Holson, 1999: A16)

Although traffic congestion and even gridlock persist at certain times in the greater San Jose area, the overall decrease in the amount of traffic on the streets and highways is more than just an indicator of how many cars there are in the area: It is also an indicator of the economy and the unemployment rate. And, sadly, the news is not good. The high-tech boom of the 1980s and 1990s that brought thousands of new jobs to Silicon Valley and lined the pockets of some people with billions of dollars turned into a technology bust and an economic nightmare for many people, leaving them unemployed and with few prospects of being immediately hired elsewhere.

Whether the terms used were "laid off," "downsized," or "terminated," the effect was the same. Tens of thousands of high-tech workers lost salaries, benefits, and sometimes shares of stock when their jobs (and sometimes the companies they worked for) disappeared. From a sociological perspective, there is an association here: The more the economy in Silicon Valley falters, the easier it is for vehicles to flow through that area's highway arteries (Edwards, 2003). In the past, commuter traffic had been backed up for three hours or more on California Highway 101 (another highway in that area), but the heaviest traffic there now lasts for less than an hour. According to one highway official, "If the economy starts moving and companies start hiring all those people they've laid off, traffic will pick up" (Edwards, 2003: D3).

Many of us have never thought of the relationship between traffic and employment, but it does exist in some cases. Economic booms bring more people to an area; money flows more freely, and people are more willing to purchase vehicles and buy houses that are a greater distance from their job site. Economic busts result in some people leaving one metropolitan area to seek work elsewhere, and some temporarily move in with relatives. Such downturns make it much more difficult for individuals to purchase vehicles and houses, and those who have savings are less likely to spend money on things they previously might have purchased without much thought. During the first half of 2003, the jobless rate in Silicon Valley (Santa Clara County) remained at or above 8 percent, and prospects did not look good for a significant improvement to occur soon (*Silicon Valley/San Jose Business Journal*, 2003). However, like other economic booms and busts, when things do turn around again in that area—as inevitably they will—two things are likely to be true: Traffic will pick up again, perhaps to the level of gridlock, and some people will, once again, be left behind in the next boom as they were in the previous one.

---

mobile mechanics, and nail technicians (manicurists) also refer to themselves as professionals. Although sociologists do not always agree on exactly which occupations are professions, they do agree that the number of people categorized as "professionals" has grown dramatically since World War II. According to the sociologist Steven Brint (1994), the contemporary professional middle class includes most doctors, natural scientists, engineers, computer scientists, certified

public accountants, economists, social scientists, psychotherapists, lawyers, policy experts of various sorts, professors, at least some journalists and editors, some clergy, and some artists and writers.

Characteristics of Professions Professions are high-status, knowledge-based occupations that have five major characteristics (Freidson, 1970, 1986; Larson, 1977):

Professionals such as the lawyer shown here are expected to have specialized knowledge of their field based on formal education and prior experience.

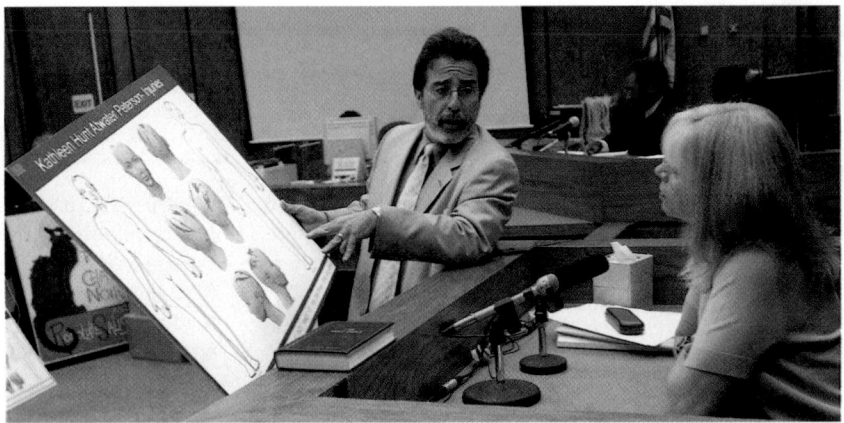

© 2003 AP/Wide World Photos

1. *Abstract, specialized knowledge.* Professionals have abstract, specialized knowledge of their field based on formal education and interaction with colleagues. Education provides the credentials, skills, and training that allow professionals to have job opportunities and assume positions of authority within organizations (Brint, 1994).

2. *Autonomy.* Professionals are autonomous in that they can rely on their own judgment in selecting the relevant knowledge or the appropriate technique for dealing with a problem. Consequently, they expect patients, clients, or students to respect that autonomy.

3. *Self-regulation.* In exchange for autonomy, professionals are theoretically self-regulating. All professions have licensing, accreditation, and regulatory associations that set professional standards and that require members to adhere to a code of ethics as a form of public accountability. Realistically, however, professionals are often constrained by organizational power structures. Many work within large-scale bureaucracies that have rules, policies, and procedures to which professionals, like everyone else, must adhere. For example, physicians who are salaried employees in for-profit hospital corporations must follow bureaucratic rules, regulations, and controls (Ritzer, 2000a).

4. *Authority.* Because of their authority, professionals expect compliance with their directions and advice. Their authority is based on mastery of the body of specialized knowledge and on their profession's autonomy: Professionals do not expect the client to argue about the professional advice rendered. Professionals also have authority over persons in subordinate occupations; for example, attorneys may delegate routine legal work, such as the completion of standard forms or docu-

ments, to paralegals or legal secretaries (Hodson and Sullivan, 2002).

5. *Altruism.* Ideally, professionals have concern for others, not just their own self-interest. The term *altruism* implies some degree of self-sacrifice whereby professionals go beyond their self-interest or personal comfort so that they can help a patient or client (Hodson and Sullivan, 2002). Professionals also have a responsibility to protect and enhance their knowledge and to use it for the public interest.

In the past, job satisfaction among professionals has generally been very high because of relatively high levels of income, autonomy, and authority. In the future, professionals may either become the backbone of a postindustrial society or suffer from "intellectual obsolescence" if they cannot keep up with the knowledge explosion (Leventman, 1981).

Reproduction of Professionals Although higher education is one of the primary qualifications for a profession, the emphasis on education gives children whose parents are professionals a disproportionate advantage early in life (Brint, 1994). There is a direct linkage between parental education and income and children's scores on college admissions tests such as the SAT, as shown in Figure 13.3. In turn, test scores are directly related to students' ability to gain admission to colleges and universities, which serve as springboards to most professions.

Race and gender are also factors in access to the professions. Historically, people of color have been underrepresented. Today, as more persons from underrepresented groups have gained access to professions such as law and medicine, their children have also gained the educational and mentoring opportunities necessary for professional careers. Between 1980 and

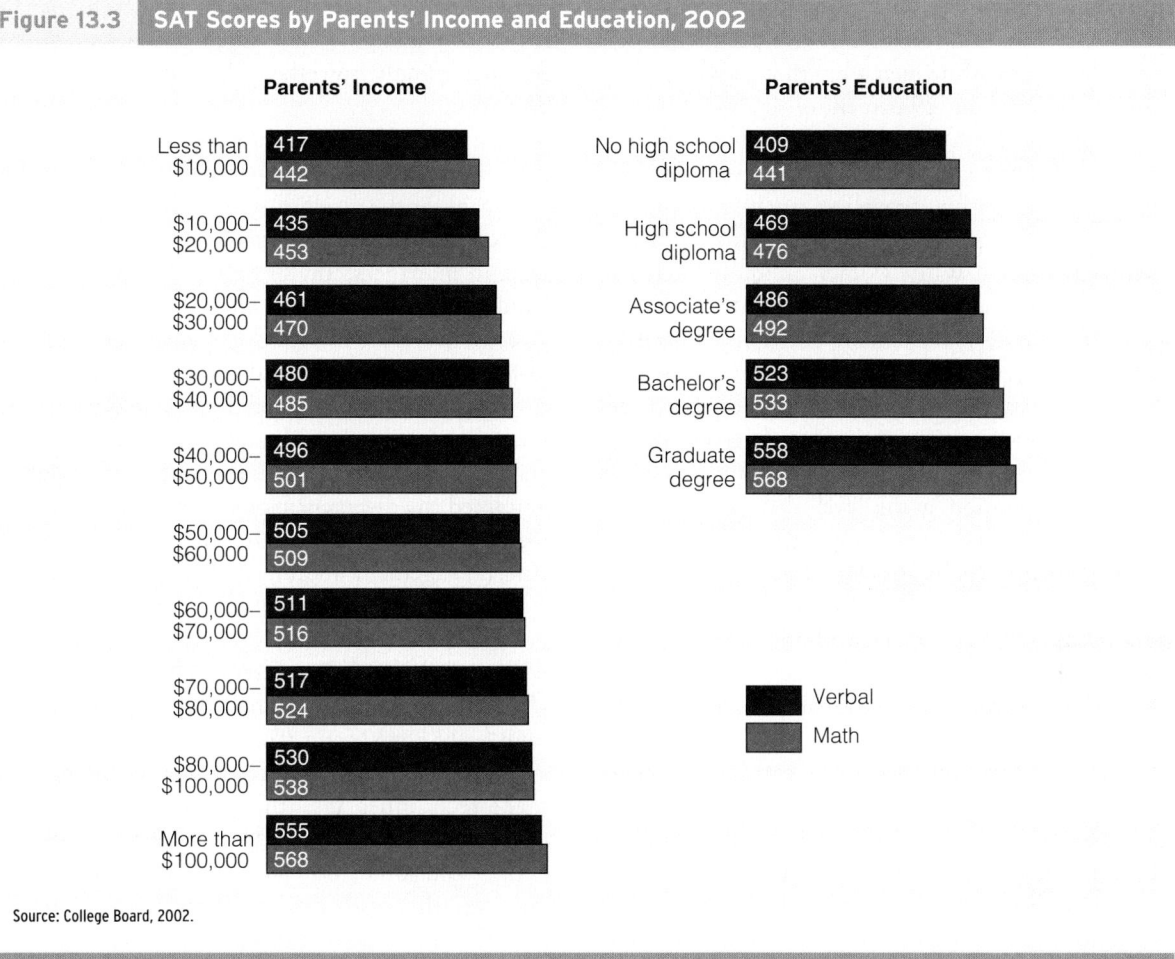

Figure 13.3   SAT Scores by Parents' Income and Education, 2002

Parents' Income

	Verbal	Math
Less than $10,000	417	442
$10,000–$20,000	435	453
$20,000–$30,000	461	470
$30,000–$40,000	480	485
$40,000–$50,000	496	501
$50,000–$60,000	505	509
$60,000–$70,000	511	516
$70,000–$80,000	517	524
$80,000–$100,000	530	538
More than $100,000	555	568

Parents' Education

	Verbal	Math
No high school diploma	409	441
High school diploma	469	476
Associate's degree	486	492
Bachelor's degree	523	533
Graduate degree	558	568

Source: College Board, 2002.

1997, there was a 63-percent increase in the number of African Americans, age 25 and above, with at least 4 years of college education (U.S. Census Bureau, 2002). However, by 2002, only 17 percent of African Americans and 11 percent of Latinos/as age 25 and above had graduated from college, as compared with 29 percent of whites and 47 percent of Asian Americans (U.S. Census Bureau, 2003b).

Deprofessionalization   Certain professions are undergoing a process of *deprofessionalization* in which some of the characteristics of a profession are eliminated (Hodson and Sullivan, 2002). Occupations such as pharmacist have already been *deskilled,* as Nino Guidici explains:

> In the old days [people] took druggists as doctors. . . . All we do [now] is count pills. Count out twelve on the counter, put 'em in here, count out twelve more. . . . Doctors used to write out their own formulas and we made most of these things. Most of the work is now done in the laboratory.

The real druggist is found in the manufacturing firms. They're the factory workers and they're the pharmacists. We just get the name of the drugs and the number and the directions. It's a lot easier. (qtd. in Terkel, 1990/1972)

However, colleges of pharmacy in many universities have fought against deprofessionalization by upgrading the degrees awarded to pharmacy graduates from the traditional B.S. in pharmacy to a Ph.D. This upgrading of degrees has also occurred over the past three decades in law schools, where the Bachelor of Laws (LL.B.) has been changed to the Juris Doctor (J.D.) degree.

Managers and the Managed

A wide variety of occupations are classified as "management" positions. The generic term *manager* is often used to refer to executives, managers, and administrators (Hodson and Sullivan, 2002). At the upper level of a workplace bureaucracy are *executives* (Hodson and

Sullivan, 2002). Today, 95 of every 100 senior management positions at the level of vice president and above are held by white men, even though white men constitute only about 29 percent of the overall work force (Kilborn, 1995). *Administrators* often work for governmental bureaucracies or organizations dealing with health, education, or welfare (such as hospitals, colleges and universities, and nursing homes) and are usually appointed. White women are more likely to gain access to management positions at this level, especially middle management positions (Kilborn, 1995). *Managers* typically have responsibility for workers, physical plants, equipment, and the financial aspects of a bureaucratic organization. In recent years, more women and men of color have gained access to management positions at this level.

Management in Bureaucracies

Managers are essential in contemporary bureaucracies, in which work is highly specialized and authority structures are hierarchical (see Chapter 6). Managers often control workers by applying organizational rules. Workers at each level of the hierarchy take orders from their immediate superiors and perhaps give orders to a few subordinates. Upper-level managers are typically responsible for coordination of activities and control of workers. The *span of control,* or the number of workers a manager supervises, is affected by the organizational structure and by technology. Some analysts believe that hierarchical organization is necessary to coordinate the activities of a large number of people; others suggest that it produces apathy and alienation among workers (Blauner, 1964). Lack of worker control over the labor process was built into the earliest factory systems through techniques known as scientific management (Taylorism) and mass production (Fordism).

Scientific Management (Taylorism)

At the beginning of the twentieth century, industrial engineer Frederick Winslow Taylor revolutionized management with a system he called *scientific management.* In an effort to increase productivity in factories, Taylor did numerous *time-and-motion* studies of workers whom he considered to be reasonably efficient. From these studies, he broke down each task into its smallest components to determine the "one best way" of doing each of them. Workers were then taught to perform the tasks in a concise series of steps. Skilled workers became less essential since unskilled workers could be trained by management to follow routinized procedures. The process of breaking up work into specialized tasks and extremely small operations contributed to the *deskilling* of work and shifted much

of the control of knowledge from workers to management (Braverman, 1974). As this occurred, workers felt increasingly powerless (Westrum, 1991; Ritzer, 2000a).

The differential piece-rate system, a central component of scientific management in which workers were paid for the number of units they produced, further contributed to the estrangement between workers and managers. Taylor believed that this system would reduce the antagonism that workers felt about direct supervision and control by managers. However, just the opposite often tended to occur: Workers felt distrustful and overworked because managers often increased the number of pieces required when workers met their quotas. Overall, scientific management amplified the divergent interests of management and workers rather than lessening them (Zuboff, 1988). And management became even more removed from workers with the advent of mass production.

Mass Production Through Automation (Fordism)

Fordism, named for Henry Ford, the founder of Ford Motor Company, incorporated hierarchical authority structures and scientific management techniques into the manufacturing process (Collier and Horowitz, 1987). Assembly lines, machines, and robots became a means of *technical control* over the work process (Edwards, 1979). The *assembly line,* a system in which workers perform a specialized operation on an unfinished product as it is moved by conveyor past their work station, increased efficiency and productivity. On Ford's assembly line, for example, a Model T automobile could be assembled in one-eighth the time formerly required. The assembly line also allowed managers to control the pace of work by speeding up the line when they wanted to increase productivity. As productivity increased, however, workers began to grow increasingly alienated as they saw themselves becoming robot laborers (Collier and Horowitz, 1987).

The role of contemporary managers was strongly influenced by the development of the assembly line and mass production techniques, which made it possible to use interchangeable parts in a variety of products. According to the sociologist George Ritzer (2000a), the assembly line and machine technology have come to dominate even work settings such as fast-food restaurants. For example, Burger King uses a conveyor belt to cook hamburgers, and food is produced in assembly line-fashion. McDonald's has a soft-drink dispenser with a sensor that shuts off when the glass is full so that the employee does not have to make this decision (Ritzer, 2000a). What do managers do in such a highly rationalized, technically controlled setting? According to Jon DeAngelo, a McDonald's manager,

> You have to trust the computer to do a lot of the job. These computers . . . calculate the payrolls, because they're hooked into the time clocks. My payroll is paid out of a bank in Chicago. . . . It's so fast that the manager hasn't got time to think about it. He has to follow the procedures like the crew. And if he follows the procedures everything is going to come out more or less as it's supposed to. So basically the computer manages the store. (qtd. in Garson, 1989: 36)

Managers also hire workers, settle disputes, and take care of other tasks, but in many work settings automation has dramatically deskilled their jobs (Garson, 1989; Zuboff, 1988; Ritzer, 2000a).

Automation has also contributed to increased surveillance of employees and concern about the "public image" of the corporation or organization. With linked or integrated computer networks, managers can measure productivity without employees even realizing it. With the Internet, managers can also read what employees post online, and they are more likely to monitor online postings if employees do their writing on the company's computer and on company time. Blogging—creating and maintaining an online journal or Web log (*blog,* for short)—is a popular pastime for many employees. On these Web pages, authors post diary-like entries about their lives, opinions, and sometimes how they feel about their work and their bosses. However, some bloggers have learned that their coworkers and bosses may read, and sometimes take exception to, what they write. Negative comments about the company or "insider" information is especially frowned upon by many executives and supervisors, so some bloggers choose to remain anonymous. However, this does not always work within the job setting, as one lawyer advised the associates in his firm: "I say to my associates, 'Watch what you say in public because it could be on the front page tomorrow.' It's gotten to the point so that you need to exercise caution so you don't do something in public that might embarrass you" (qtd. in Seper, 2003). Whereas early forms of management involved direct supervision of workers, technology now makes it possible for a limited number of managers to control many more workers (see Zuboff, 1988).

The Lower Tier of the Service Sector and Marginal Jobs

Positions in the lower tier of the service sector are part of the secondary labor market, characterized by low wages, little job security, few chances for advancement,

Occupational segregation by race and gender is clearly visible in personal service industries, such as restaurants and fast-food chains. Women and people of color are disproportionately represented in marginal jobs such as waitperson or fast-food server—jobs that typically do not meet societal norms for minimum pay, benefits, or job security.

higher unemployment rates, and very limited (if any) unemployment benefits. Typical lower-tier positions include janitor, waitress, messenger, sales clerk, typist, file clerk, migrant laborer, and textile worker. Large numbers of young people, people of color, recent immigrants, and white women are employed in this sector (Callaghan and Hartmann, 1991). In a study of Mexican American women, for example, the sociologist Denise Segura (1994) concluded that people are often channeled into lower-tier jobs based on their race/ethnicity, class, and gender. In addition, schooling and employment training programs rarely provide Mexican American women with a sense that they have employment options outside those traditionally ascribed to Latinas (Segura, 1994).

According to the employment norms of this country, a job should (1) be legal; (2) be covered by government work regulations, such as minimum standards of pay, working conditions, and safety standards; (3) be relatively permanent; and (4) provide adequate pay with sufficient hours of work each week to make a living (Hodson and Sullivan, 2002). However, many lower-tier service sector jobs do not meet these norms and are therefore marginal. ***Marginal jobs* differ from the employment norms of the society in which they are located;** examples in the U.S. labor market include personal service workers and private household workers.

Personal Service Workers More than 11 million workers are employed in personal service industries. Almost 6 million of them are employed in eating and drinking places, about 2 million in hotels, and another 3 million in laundries, beauty shops, and household service, primarily maid service (Hodson and Sullivan, 2002).

Service workers are often viewed by customers as subordinates or personal servants. Frequently, they are required to wear a uniform that reflects their lowly status as a waitperson, fast-food server, maid, or porter. Occupational segregation by race and gender (see chapters 10 and 11) is clearly visible in personal service industries. In 2000, for example, 23.8 percent of all employed Latinas age 16 and over were employed in service occupations, as compared with 16.4 percent of white (non-Latina) females and 9.1 percent of white (non-Latino) males (U.S. Bureau of Labor Statistics, 2000).

Private Household Workers

Household service work has shifted from domestics who work full time with one employer to part-time workers who may work several hours a week in the homes of several different employers. Private household workers include launderers, cooks, maids, housekeepers, gardeners, babysitters, nannies, chauffeurs, and butlers. These workers often must travel long distances to get to work, and having to rely on public transportation can add hours to their workday.

Household work is marginal: It lacks regularity, stability, and adequacy. The jobs are excluded from most labor legislation, the workers are not unionized, and employers often feel that they can flaunt any rules or regulations. Many employers pay cash in order to avoid payroll taxes and Social Security; the jobs typically have no unemployment, health insurance, or retirement benefits (Hodson and Sullivan, 2002). In many cities, an increasing number of household workers are recent immigrants, some of whom are undocumented and could easily be deported by immigration authorities if their whereabouts were known. Consequently, these workers are among the most powerless. They cannot rely on any resources, including local law enforcement, to protect them from abuse—whether financial, physical, or sexual—by their employers. Even employers who are not overtly abusive often seek to control the labor of domestic workers more than those workers believe to be necessary or humane. In her ethnographic research on immigrant domestic workers, Pierrette Hondagneu-Sotelo (2001: 158) recorded this statement from Erlinda Castro (a pseudonym), a Guatemalan woman employed as a housecleaner in Los Angeles:

The bad [employer] for us is the one who's there *craneando* [using her cranium], as we say, trying to

Workers on the global assembly line—often women or young girls—make or assemble a number of products sold in the United States and other high-income countries. How do jobs such as these benefit workers in low-income countries? How might work such as this be to their detriment?

© Charles Gupton (Stone)/Getty Images

devise new work to make the exact eight hours. "Clean the patio around the pool! Wash the patio chairs! Clean the garage! Go sweep outside!" That's the bad one, the one who wants a full eight hours. It's inconsiderate, and among ourselves we say, "They are wringing us out!" They are wringing us out for forty or fifty dollars.

Internationalization of Marginal Jobs

Not only in domestic work but also in other types of marginal employment, many workers feel that they are being "wrung out" by their employers. Some manufacturing jobs fall into this marginal category. Marginal jobs are more likely to be found in peripheral industries than in core, or essential, industries. Peripheral industries tend to be in highly competitive industries (such as garment or microelectronics manufacturing) that operate in markets where prices are subject to sudden, intense fluctuations and where labor is a significant part of the cost of the goods sold (Hodson and Sullivan, 2002).

Although "high-tech" occupations (such as engineering and computer technology) have been viewed by many as the wave of the future, many jobs in this field are marginal and, as discussed at the beginning of the chapter, are now more likely to be off-shored than in the past. Along with an increasing number of professional jobs in information processing and other high-tech industries, many assembly jobs have been exported to other nations. The use of workers in other countries to manufacture or assemble goods to be sold in higher-income nations has been referred to as the *global assembly line*—a situation in which corporations hire workers, often girls and young women, in low-income nations at very low wages to work, sometimes under undesirable and hazardous conditions.

To gain the same benefits of "cheap labor," some U.S. companies hire people who have recently migrated to this country. This hiring practice makes it possible for management to pay workers less and to divide and control them by race and nationality (Hossfeld, 1992, 1994). In her study of immigrant women workers in Silicon Valley, the sociologist Karen J. Hossfeld (1994: 69) noted that "capitalist economic development utilizes and thrives on racism as a method of labor division and control." Gender segregation is also readily apparent in much of this type of work. Some women may become home-workers. Home-work, like piece-work in previous centuries, moves the workplace into women's homes and often pays low wages and provides no employee benefits.

Concerns about marginal jobs in many fields have only grown worse in recent years, as businesses have sought new ways to cut overhead and reduce the number of workers they have on their payrolls. Consequently, contingent work has become increasingly popular as an economic strategy of employers.

Contingent Work

***Contingent work* is part-time work, temporary work, or subcontracted work** that offers advantages to employers but that can be detrimental to the welfare of workers. Contingent work is found in every segment of the work force, including colleges and universities, where tenure-track positions are fewer in number than in the past and a series of one-year, non-tenure-track appointments at the lecturer or instructor level has become a means of livelihood for many professionals (Mydans, 1995). The federal government is part of this trend, as is private enterprise. In addition, the health care field continues to undergo significant change, as physicians, nurses, and other health care workers are increasingly employed through temporary agencies (Rosengarten, 1995).

Employers benefit by hiring workers on a part-time or temporary basis; they are able to cut costs, maximize profits, and have workers available only

Hiring contingent workers can increase the profitability of many corporations. Other companies make their profits by furnishing these contingent workers to the corporations. What message does this ad convey to corporate employers?

when they need them. As some companies have cut their work force, or downsized, they have replaced "staffing table" employees who had higher salaries and full benefit packages with part-time and hourly employees who receive lower wages and no benefits. Although some people voluntarily work part time (such as through job sharing and work sharing), many people are forced to do so because they lack opportunities for full-time employment. About 6 million persons, or 29 percent of the 21 million part-time workers, hold part-time jobs involuntarily. Most part-time workers are women, teenagers, and older persons; women constitute 75 percent of all part-time workers (Applebaum and Schettkat, 1989).

Temporary workers are the fastest-growing segment of the contingent work force, and agencies that "place" them have increased dramatically in number in the last decade. The agencies provide workers on a contract basis to employers for an hourly fee; workers are paid a portion of this fee. Temporary workers

When the economy is booming, fewer people are unemployed. However, long lines at the unemployment offices are a recurring sight throughout the United States in periods when workers are laid off from their jobs. As the sociological imagination reminds us, when one person is unemployed, it may be a personal problem, but when large numbers of people are laid off, the concern becomes a pressing social issue.

Rob Crandall/The Image Works

usually have lower wages and fewer benefits than permanent, full-time employees in the same field. Among manual laborers who earn the minimum wage or slightly more, very little is left after the temporary agency takes its fee.

Subcontracted work is another form of contingent work that often cuts employers' costs at the expense of workers. Instead of employing a large work force, many companies have significantly reduced the size of their payrolls and benefit plans by **subcontracting— an agreement in which a corporation contracts with other (usually smaller) firms to provide specialized components, products, or services to the larger corporation.** Companies ranging from athletic shoe manufacturers to high-tech industries have found that it is more cost efficient (and profitable) to subcontract some of their work to small contractors. For example, a construction company may enter into a contract to build an airport or commercial building and then subcontract components of the work (such as electrical, plumbing, and carpentry) to other firms. Hiring and paying workers on the project becomes the responsibility of the subcontractor, not the general contractor. The employment practices of some subcontractors are unethical, if not illegal. In the garment industry in El Paso, Texas, for example, subcontractors' plants are referred to as *talleres de hambre* (workshops of hunger) (Myerson, 1994).

Many contingent workers experience *underemployment,* as do other workers when their job is permanent but provides fewer hours of work than they desire or when their qualifications are far greater than those required for the job. Age, race, gender, and dis-

ability are factors frequently associated with underemployment and unemployment.

UNEMPLOYMENT

Even when national or regional unemployment rates are low, there are still people looking for a job. As companies restructure, downsize, or go out of business, tens of thousands are at least temporarily jobless. Disputes with employers cause some workers to quit a job and look for another one, and other workers may leave a job because they believe that they need time to find a job that pays more money than they are currently making (Hight, 1999).

There are three major types of unemployment— cyclical, seasonal, and structural. *Cyclical unemployment* occurs as a result of lower rates of production during recessions in the business cycle; although massive layoffs initially occur, some of the workers will eventually be rehired, largely depending on the length and severity of the recession. *Seasonal unemployment* results from shifts in the demand for workers based on conditions such as the weather (in agriculture, the construction industry, and tourism) or the season (holidays and summer vacations). Both of these types of unemployment tend to be relatively temporary in nature.

By contrast, structural unemployment may be permanent. *Structural unemployment* arises because the skills demanded by employers do not match the skills of the unemployed or because the unemployed do not live where jobs are located (McEachern, 2003). This type of unemployment often occurs when a number of plants in the same industry are closed or when new technology makes certain jobs obsolete. For example, workers previously employed in the Appalachian coal industry or in the midwestern steel industry found that their job skills did not transfer to other types of industries when the plants closed. Workers who lose their jobs when a factory closes down or moves or when their positions or shifts are abolished experience worker displacement. Between 1979 and 1983, over 5 million workers were displaced, and another 4 million lost their jobs between 1985 and 1989 (Amott, 1993). Early in the 2000s, many other workers became unemployed due to stagnation in the economy. Those who were employed in secondary sector factory jobs (toy and microcomputer assembly lines, for example) were the most likely to become unemployed, primarily because they lacked union protection and typically did not have the education and skills necessary to find better jobs. Structural unemployment often results from capital flight—the investment of capital in foreign facilities, as previously discussed. Today, many workers fear losing their jobs, exhausting their unemployment benefits (if any), and still not being able to find another job.

The *unemployment rate* **is the percentage of unemployed persons in the labor force actively seeking jobs.** In 2000 the U.S. unemployment rate was 4.0 percent, and workers who had historically suffered from high unemployment rates made some gains. For example, unemployment among African Americans, Latinas/os, women, and workers with high school educations or less were at near-historic lows (Nasar, 1999). In 2001, after massive layoffs at many companies, the unemployment rate had increased to about 5 percent. However, like other types of "official" statistics, unemployment rates may be misleading. For example, Bureau of Labor Statistics data do not include the approximately 3.2 million people who are working part time because full-time work is unavailable or the 0.3 million who are "discouraged" workers (Nasar, 1999: B3). Individuals who become discouraged in their attempt to find work and who are no longer actively seeking employment are not counted as unemployed.

Unemployment compensation provides unemployed workers with short-term income while they look for other jobs. Since this concept was first introduced in the 1930s, federal laws have set the guidelines and minimum standards for the program, which is administered by the states. Although some states exceed the minimum standards, many provide the bare minimum of benefits required by law. For a person to

Seeking to improve economic and social opportunities for farm workers, the late César Chávez held rallies and engaged in other protest activities in an effort to better the workers' lives.

acquire benefits, his or her eligibility must be established. In many states, for example, a person must have worked at least twenty weeks to be eligible. Workers who quit their jobs without "good cause" or who are fired for misconduct or are engaged in a work stoppage due to a labor dispute are usually not eligible for benefits. Fear of losing unemployment benefits, coupled with high rates of underemployment and unemployment, appears to have limited worker resistance and activism in recent years.

WORKER RESISTANCE AND ACTIVISM

In their individual and collective struggles to improve their work environment and gain some measure of control over their work-related activities, workers have used a number of methods to resist workplace alienation. Many have also joined labor unions to gain strength through collective actions.

Labor Unions

U.S. labor unions came into being in the mid-nineteenth century, as previously discussed. Unions have been credited with gaining an eight-hour workday, a five-day workweek, health and retirement benefits, sick leave and unemployment insurance, and workplace health and safety standards for many employees. Most of these gains have occurred through *collective bargaining*—negotiations between employers and labor union leaders on behalf of workers. In some cases, union leaders have called strikes to force

| Figure 13.4 | Major Work Stoppages in the United States, 1960-2002 |

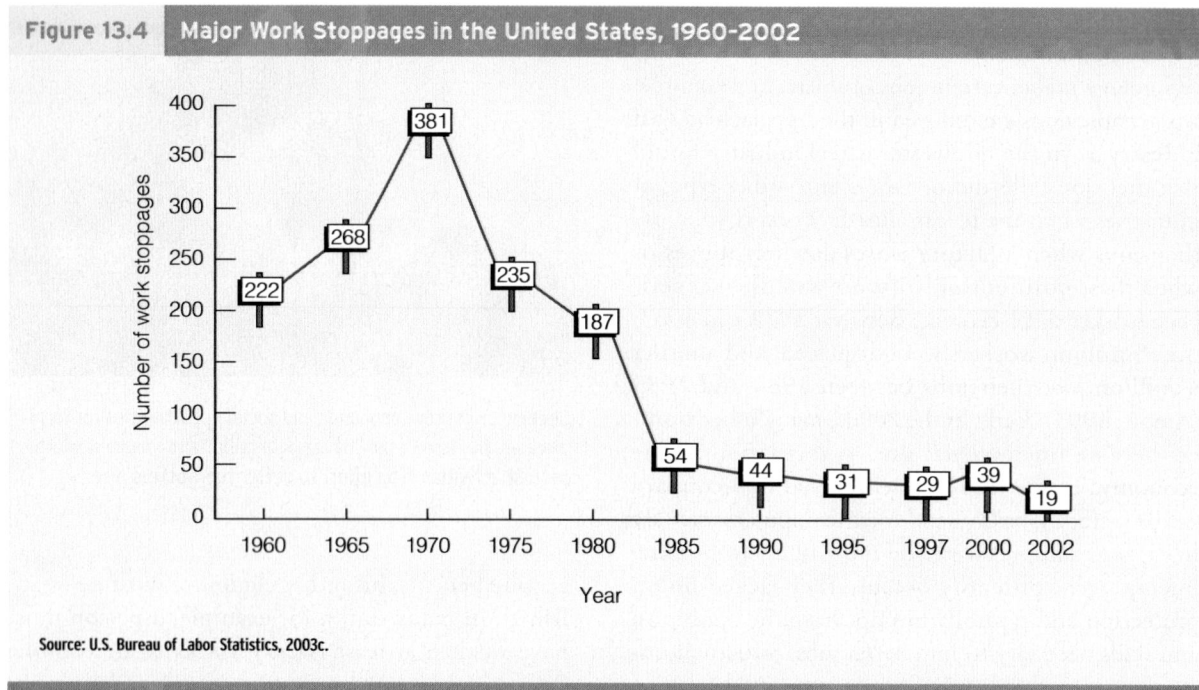

Source: U.S. Bureau of Labor Statistics, 2003c.

employers to accept the union's position on wages and benefits. While on strike, workers may picket in front of the workplace to gain media attention, to fend off "scabs" (nonunion workers) who might take over their jobs, and in some cases to discourage customers from purchasing products made or sold by their employer. In recent years, strike activity has diminished significantly as many workers have feared losing their jobs. As Figure 13.4 shows, in 2002 only 19 strikes, or work stoppages, involving more than 1,000 workers were reported, as compared with significantly higher numbers in the 1960s and 1970s—especially when compared with 1970, when there were twenty times as many strikes as in 2002. The number of workers involved in the actions declined from a peak of more than 2.5 million in 1971 to 339,000 in 1997 (U.S. Census Bureau, 2002). However, in 1997 a strike by the 185,000 members of the Teamsters Union against United Parcel Service (UPS), the nation's largest package-shipping company, virtually closed UPS down until a settlement was reached under which the union gained many of the benefits it had sought from the company.

Although the overall *number* of union members in the United States has increased since the 1960s, primarily as a result of the growth of public employee unions (such as the American Federation of Teachers), the *proportion* of all employees who are union members has declined. Today, about 17 percent of all U.S. employees belong to unions or employee associations. Part of the decline in union participation has been

attributed to the structural shift away from manufacturing and manual work and toward the service sector, where employees have been less able to unionize.

Historically, labor unions contributed to the employment gains of people of color. In 2001, for example, the highest rates of union affiliation were among African American men (19 percent) and women (15 percent), as compared with white men (15 percent) and women (11 percent), and Latinos (12 percent) and Latinas (11 percent) (U.S. Census Bureau, 2002). African Americans have higher rates of union affiliation because of extensive public sector employment, where workers are more likely to be unionized (Costello and Stone, 1994). In the past, many African Americans moved into the middle class through unionized jobs in heavy industry. More recently, the loss of many of these jobs has been especially damaging for people of color (Higginbotham, 1994).

A "spillover" effect exists between union and nonunion firms in the area of wages and benefits; nonunion firms must compete for labor with union firms. Some nonunion firms pay higher wages in order to dissuade union membership. However, the spillover effect may work in the opposite direction as well: If unionized workers lose wages and benefits, nonunion workers may also see a lowering of their wages and benefits (Amott, 1993). Usually, union members earn higher wages than nonunion workers in comparable jobs. In 2001 union workers averaged over $7,000 a year more in wages than did nonunion workers (U.S. Census Bureau, 2002). Women in unions earn an average weekly wage that is 1.3 times that of nonunionized women.

Although labor union membership has declined in most industrialized countries, including the United States, unions are still highly influential in labor negotiations involving wages and working conditions in Scandinavia, Germany, the Netherlands, Ireland, and much of Eastern Europe (Greenhouse, 1997). In most industrialized countries, collective bargaining by unions has been dominated by men. However, in countries such as Sweden, Germany, Austria, and Great Britain, women workers have gained some important concessions as a result of labor union participation.

Absenteeism, Sabotage, and Resistance

Absenteeism is one means by which workers resist working conditions that they consider to be oppressive. Ben Hamper (1992: 47) describes how employees were chastised in a meeting for their absenteeism at the Flint, Michigan, GM plant:

> The Plant Manager introduced the man in charge of overseeing worker attendance. . . . He didn't seem happy at all. The attendance man unveiled a large chart illustrating the trends in absenteeism. . . . He pointed to Monday. . . . Monday was an unpopular day attendance-wise. He moved the pointer over to Tuesday and Wednesday which showed a significant gain in attendance. The chart peaked way up high on Thursday. Thursday was pay night. Everybody showed up on Thursday.
>
> "Then we arrive at Friday," the attendance man announced. A guilty wave of laughter spread through the workers. None of the bossmen appeared at all amused. Friday was an unspoken Sabbath for many of the workers. Paychecks in their pockets, the leash was temporarily loosened. To get a jump on the weekend was often a temptation too difficult to resist.

Other workers use sabotage to bring about informal work stoppages. The phrase "throwing a monkey wrench in the gears" originated with the practice of workers "losing" a tool in assembly line machinery. Sabotage of this sort effectively brought the assembly line to a halt until the now-defective piece of machinery could be repaired. Industrial sabotage has been described as an expression of workers' deep-seated frustrations about poor work environments (Feagin and Feagin, 1997).

Although most workers do not sabotage machinery, a significant number do resist what they perceive to be oppression from supervisors and employers. In a study of Asian American women who work in low-status occupations such as hotel housekeepers, the sociologist Esther Ngan-Ling Chow (1994) found that many of the women felt they had little to lose if they chose to resist or rebel. Onyoung, one of the Korean American women interviewed by Chow, described how she felt:

> Many of us depend on this job to support our families. When I am blamed for the things that I do not do and have no part of it, I will yell my guts out to protest. At most I will lose this job. I want my dignity, to be an honorable person, which my parents taught me. (qtd. in Chow, 1994: 214)

Resistance helps people in lower-tier service jobs survive at work (Dill, 1988; Chow, 1994). In interactions with their employers, domestic workers attempt to maintain their personal dignity and to gain mastery of situations in which they are defined as objects, not as human beings. Scholars have documented resistance by African American domestic workers (Rollins, 1985; Dill, 1988), Chicana private housekeepers (Romero, 1992; Hondagneu-Sotelo, 2001), and Japanese domestic helpers (Glenn, 1986). According to Chow (1994: 218), many Asian American women become "active agents and goal-oriented actors, capable of taking charge of their own lives" in their interactions with supervisors and co-workers. Workers who feel that they can take charge of their own lives may be less likely to experience high levels of alienation than workers who feel powerless over their plight.

Studies of worker resistance are very important to our understanding of how people deal with the social organization of work. Previously, workers (especially women) have been portrayed as passive "victims" of their work environment. As we have seen, this is not always the case for either women or men; many workers resist problematic situations and demand better wages and work conditions.

EMPLOYMENT OPPORTUNITIES FOR PERSONS WITH A DISABILITY

An estimated 48 million persons in the United States have one or more physical or mental disabilities that differentially affect their opportunities for employment. This number continues to increase for several

© Mark Newman/PhotoEdit

Workers with a disability are able to engage in a wide variety of occupations when they are offered the opportunity to do so.

reasons. First, with advances in medical technology, many people who formerly would have died from an accident or illness now survive, although with an impairment. Second, as more people live longer, they are more likely to experience diseases that may have disabling consequences. Third, persons born with serious disabilities are more likely to survive infancy because of medical technology. However, less than 15 percent of persons with a disability today were born with it; accidents, disease, and war account for most disabilities in this country.

In 1990 the United States became the first nation to formally address the issue of equality for persons with a disability. When Congress passed the Americans with Disabilities Act (ADA), this law established "a clear and comprehensive prohibition of discrimination on the basis of disability." Combined with previous disability rights laws (such as those that provide for the elimination of architectural barriers from new federally funded buildings and for the maximum integration of schoolchildren with disabilities), the ADA is a legal mandate for the full equality of people with disabilities.

Despite this law, about two-thirds of working-age persons with a disability are unemployed today. Most persons with a disability believe they could work if they were offered the opportunity. However, even when persons with a disability are able to find jobs, they earn less than persons without a disability (Yelin, 1992). On the average, workers with a disability make 85 percent (for men) and 70 percent (for women) of what their co-workers without disabilities earn, and the gap is growing (Shapiro, 1993). Studies have shown a thirty-year overall decline in the economic condition of persons with disabilities (Yelin, 1992; Burkhauser, Haveman, and Wolfe, 1993). The problem has been particularly severe for African Americans and Latinos/as with disabilities (Kirkpatrick, 1994). Among Latinos/as with disabilities, only 10 percent are employed full time; those who work full time earn 73 percent of what white (non-Latino) persons with disabilities earn.

What does it cost to "mainstream" persons with disabilities? A survey of personnel directors and other executives responsible for making hiring decisions for their companies found that the average cost of workplace modifications to accommodate employees with a disability was less than $500 (Heldrich Center for Workforce Development, 2003). Other accommodations cost about $1,000, with the largest single cost being $14,500 each for special Braille computer displays for visually impaired employees (Noble, 1995). As employment opportunities change in the global economy, it is important that employers use common sense and a creative approach to thinking about how persons with disabilities can fit into the marketplace (see Box 13.4).

THE GLOBAL ECONOMY IN THE FUTURE

How will the U.S. economy look in the future? What about the global economy? Although sociologists do not have a crystal ball with which to predict the future, some general trends can be suggested.

The U.S. Economy

Many of the trends we examined in this chapter will produce dramatic changes in the organization of the economy and work in the twenty-first century. U.S. workers may find themselves fighting for a larger piece of an ever-shrinking economic pie that includes a trade deficit and a national debt of over $6 trillion. In July 2003, our national debt was $6,729,454,960,736.46, making each citizen's share $23,080.13! Even as the gender and racial–ethnic composition of the labor

Box 13.4 YOU CAN MAKE A DIFFERENCE

Creating Access to Information Technologies for Workers with Disabilities

People are shocked I can do this. They have never imagined a blind person could find a use for a computer, much less make a business out of it.
   –Cathy Murtha, a partner in Magical Mist Creations, a company that designs and evaluates sites on the World Wide Web (qtd. in Sreenivasan, 1996: C7)

If you are a person with a visual impairment or a person who plans to supervise workers—some of whom may be visually impaired—you can make a difference in the workplace by learning about new technologies that make jobs more accessible for workers with disabilities. For example, new computer hardware and software programs and the Internet are having a profound effect on workers who are visually impaired. According to Larry Scadden, technological consultant for persons with disabilities at the National Science Foundation, "The Internet has changed forever the lives of blind people, mainly because it provides independent access to information" (qtd. in Sreenivasan, 1996: C7). The goal of making the Internet and the World Wide Web accessible to visually impaired persons requires several things:

- Text must be in such a form that it can be read aloud by a speech synthesizer with text-to-speech software or be output in a Braille format.
- The user must be able to navigate the screen with a keyboard, not a mouse.
- Good descriptions must be given of photos and graphics that the user will not be able to see.

To find out more about creating greater access to information for persons with visual impairment and other disabilities, contact the following agencies:

- American Council of the Blind, 1155 15th St., NW, Suite 720, Washington, DC 20005. (202) 393-3666 or (800) 424-8666.
- American Foundation for the Blind, 15 W. 16th St., New York, NY 10011. (212) 620-2000 or (800) 232-5463.

On the Internet:

- The Americans with Disabilities Act page provides information on the Act and links to disability-related sites, organized by subject:

http://www.usdoj.gov/ada/adahom1.htm

- Blind- and Deaf-Specific Resources includes resources targeted to vision- and hearing-impaired persons and links to general disability Web sites such as webABLE, Handsnet, and Disability Net, as well as assistive technology sites such as Apple's Disability Solutions Online:

http://198.234.201.48/spedpointers.html

- Electronic Resources Regarding Persons with Disabilities is a list of sites pertaining to people with disabilities. Computer information and disability information are available:

http://www.gallaudet.edu/~cadsweb /disability.html

- International Disability Information Resources is a site connecting to other links that feature organizations, guides, and directories to assist persons with disabilities. Among them are the Alliance for Technology Access and the Handicap Information Resources:

http://www.contact.org/disabled.html

To reach Web sites that are "user friendly" for persons with visual impairments, contact the following sites:

- Caldwell College:

http://www.caldwell.edu

- On Island Communications:

http://www.onisland.com

- Cathy's Newstand:

http://www.esrin.esa.it:8080/handy

- Young Opportunities Inc.:

http://www.youngopp.com

- Ann Morris Enterprises:

http://www.annmorris.com

WRITING IN SOCIOLOGY ASSIGNMENT

Discuss how colleges and universities have made higher education more accessible for stu-  dents with disabilities. What key areas still need improvements in such access on your campus?

The fate of high-income countries is increasingly linked to the globalization of markets. Dramatic changes in stock markets in cities such as Hong Kong and New York City create reverberations that are felt in financial markets around the world.

© Paul A. Souders/Corbis

force continues to diversify, many workers will remain in race- and/or gender-segregated occupations and industries. At the same time, workers may increasingly be fragmented into two major labor market divisions: (1) those who work in the innovative, primary sector and (2) those whose jobs are located in the growing secondary, marginal sector. In the innovative sector, increased productivity will be the watchword as corporations respond to heightened international competition. In the marginal sector, alienation will grow as temporary workers, sometimes professionals, look for avenues of upward mobility or at least a chance to make their work life more tolerable. Labor unions will be unable to help workers unless the unions embark on innovative programs to recruit new members, improve their image, and recover their former political clout (Hodson and Sullivan, 2002). However, in spite of these problems, most analysts predict that the United States will remain a major player in the world economy.

Global Economic Interdependence and Competition

Most social analysts predict that transnational corporations will become even more significant in the global economy of this century. As they continue to compete for world market share, these corporations will become even less aligned with the values of any one nation. However, those who advocate increased globalization typically focus on its potential impact on industrialized nations, not the effect it may have on the 80 percent of the world's population that resides in low-income and middle-income nations. Persons in low-income nations may become increasingly baffled and resentful when they are bombarded with media images of Western affluence and consumption; billions of "have-nots" may feel angry at the "haves"—including the engineers and managers of transnational companies living and working in their midst.

In recent years, the average worker in the United States and other high-income countries has benefited more from global economic growth than have workers in lower-income countries. For example, the average citizen of Switzerland has an income several hundred times that of a resident of Ethiopia. More than a billion of the world's people live in abject poverty; for many, this means attempting to survive on less than $400 a year.

A global workplace is emerging in which telecommunications networks will link workers in distant locations. In the developed world, the skills of some professionals will transcend the borders of their countries. For example, there is a demand for the services of international law specialists, engineers, and software designers across countries. Even as nations become more dependent on one another, they will also become more competitive in the economic sphere. Although the prospects for greater economic equality do not appear hopeful, we should not be unduly pessimistic. According to the sociologist G. William Domhoff (1990: 285), "No one foresaw the New Deal [during the Depression], and no one expected a massive civil rights movement [in the 1950s and 1960s]. If history teaches us anything, it is that no one can predict the future."

CHAPTER REVIEW

■ What is the primary function of the economy?

The economy is the social institution that ensures the maintenance of society through the production, distribution, and consumption of goods and services.

■ What are the three sectors of economic production?

Economic production is divided into primary, secondary, and tertiary sectors. In primary sector production, workers extract raw materials and natural resources from the environment and use them without much processing. Industrial societies engage in secondary sector production, which is based on the processing of raw materials (from the primary sector) into finished goods. Postindustrial societies engage in tertiary sector production by providing services rather than goods.

■ How do the major contemporary economic systems differ?

Throughout the twentieth century, capitalism, socialism, and mixed economies were the main systems in industrialized countries. Capitalism is characterized by ownership of the means of production, pursuit of personal profit, competition, and limited government intervention. Socialism is characterized by public ownership of the means of production, the pursuit of collective goals, and centralized decision making. In mixed economies, elements of a capitalist, market economy are combined with elements of a command, socialist economy. These mixed economies are often referred to as democratic socialism.

■ What are the functionalist, conflict, and symbolic interactionist perspectives on the economy and work?

According to functionalists, the economy is a vital social institution because it is the means by which needed goods and services are produced and distributed. Conflict theorists suggest that the capitalist economy is based on greed. In order to maximize profits, capitalists suppress the wages of workers, who, in turn, cannot purchase products, making it necessary for capitalists to reduce production, close factories, lay off workers, and adopt other remedies that are detrimental to workers and society. Symbolic interactionists focus on the microlevel of the economic system, particularly on the social organization of work and its effects on workers' attitudes and behavior. Many workers experience job satisfaction when they like their job responsibilities and the organizational structure in which they work and when their individual needs and values are met. Alienation occurs when workers do not gain a sense of self-identity from their jobs and when their work is done completely for material gain and not for personal satisfaction.

■ What are occupations, and how do they differ in the primary and secondary labor markets?

Occupations are categories of jobs that involve similar activities at different work sites. The primary labor market consists of well-paying jobs with good benefits that have some degree of security and the possibility for future advancement. The secondary labor market consists of low-paying jobs with few benefits and very little job security or possibility for future advancement.

■ What are the characteristics of professions?

Professions are high-status, knowledge-based occupations characterized by abstract, specialized knowledge, autonomy, authority over clients and subordinate occupational groups, and a degree of altruism.

■ What are marginal jobs?

Marginal jobs differ in some manner from mainstream employment norms: Jobs should be legal, be covered by government regulations, be relatively permanent, and provide adequate hours and pay in order to make a living. Marginal jobs fall below some or all of these norms.

■ What is contingent work?

Contingent work is part-time work, temporary work, or subcontracted work that offers advantages to employers but may be detrimental to workers. Through the use of contingent workers, employers are able to cut costs and maximize profits, but workers have little or no job security.

KEY TERMS

capitalism 428
conglomerates 431
contingent work 443
corporations 428
democratic socialism 434
economy 420
interlocking corporate directorates 431
labor union 428

QUESTIONS FOR CRITICAL THINKING

1. If you were the manager of a computer software division, how might you encourage innovation among your technical employees? How might you encourage efficiency? If you were the manager of a fast-food restaurant, how might you increase job satisfaction and decrease job alienation among your employees?

2. Using Chapter 2 as a guide, design a study to determine the degree of altruism in certain professions. What might be your hypothesis? What variables would you study? What research methods would provide the best data for analysis?

3. What types of occupations will have the highest prestige and income in 2020? The lowest prestige and income? What, if anything, does your answer reflect about the future of the U.S. economy?

RESOURCES ON THE INTERNET

Chapter-Related Web Sites

The following Web sites have been selected for their relevance to the topics in this chapter. These sites are among the more stable, but please note that Web site addresses change frequently. For an updated list of chapter-related Web sites with URL links, please visit the *Sociology in Our Times* Web site (**www.wadsworth.com/KendallSIOT**).

National Institute for Occupational Safety and Health (NIOSH)
http://www.cdc.gov/niosh/homepage.html

NIOSH is the federal agency responsible for conducting research and making recommendations for the prevention of work-related disease and injury. The institute's Web site features extensive information, publications, and databases on policies and laws regarding worker safety and health, as well as alerts about occupational illnesses, injuries, and deaths.

National Bureau of Economic Research, Inc. (NBER)
http://www.nber.org

This organization is a private, nonprofit, nonpartisan research organization that strives to promote a greater understanding of how the economy works. In addition to interesting links, publications, and research, this Web site contains a detailed data section with information and statistics from the macro to the micro perspective.

U.S. Department of Labor
http://www.dol.gov

This site is continually updated with new initiatives and reports, and it contains a vast amount of information to inform the work force on topics such as wages, health and safety, job training, job discrimination, and unemployment.

ONLINE STUDY AND RESEARCH TOOLS

Accompanying this text are many *free* powerful online study tools that will help you master the material in this chapter, help increase your depth of understanding, and help you make the grade!

SocCoach CD-ROM

Use the SocCoach CD-ROM enclosed with this text to help you formulate a customized study plan for this chapter. After you take the Diagnostic Quiz, SocCoach will generate a customized study plan just for you! It will identify sections of the chapter that you should review and will provide videos, charts, graphs, and excerpts from the text to supplement your studies and enhance your understanding. You'll also find fun, interactive activities such as Virtual Explorations and Map the Stats to apply what you've learned and stretch your sociological imagination.

The Companion Web Site for Sociology in Our Times, *Fifth Edition*
www.wadsworth.com/KendallSIOT

Gain an even better grasp on this chapter by going to the companion Web site to take one of the Tutorial Quizzes, use the Flash Cards to master key terms, or check out the many other study aids you'll find there. You'll also find special features such as GSS Data and Census 2000 information that'll put data and resources at your fingertips to help you with that special project or help you as you do some research on your own.

 In this chapter, when you see the icon on the left, it alerts you to a specific exercise found in *Wadsworth's Sociology Online Resources and Writing Companion.* This valuable guide shows you how to use Wadsworth's exclusive online resources—*InfoTrac College Edition,* the *Opposing Viewpoints Resource Center,* and *MicroCase Online*—to assist you in your study of sociology and to build essential research and writing skills.

Politics and Government in Global Perspective

Aboard *USS Lincoln,* March 14, 2003.—The media who have been "embedded" with military units have been promised unprecedented access to cover the build-up to a possible conflict with Iraq and a war itself [which subsequently took place]. But less than a week into the embedding, there is already a lot of bristling among some of the journalists aboard the *USS Abraham Lincoln,* an aircraft carrier deployed to the Gulf region. Some of them—us—began muttering when we were told we had to be accompanied by our escorts—members of the ship's crew including some public affairs officers—everywhere we went except the restroom. The escorts also monitor our interviews with ship's personnel.

A few days into the embed, [we] were told to keep a record of who we speak to and what we talk about. We were told it was so they would have a log of how many people were spoken to and for what stories. Reporters are famously resistant to anyone monitoring them working. . . .

Cynics had predicted that the military and journalists would never mix. The cynics are no doubt chuckling to themselves.

—Ron Claiborne (2003), a reporter for ABC, expressing his frustration over the limited access that he believed journalists had to military personnel when the U.S. government allowed media personnel to "embed" with troops prior to and during the 2003 war in Iraq

> We had problems with [the] media following along so closely they were in fact interfering with some of the military actions. We had to watch and make sure we didn't accidentally target them because they were getting in the line of fire. I would definitely recommend more training for the media when dealing with the military in the future.
>
> —Sergeant Ralph Hensley of the Third Brigade, telling a reporter how he felt about having the media embedded with troops, although some others in his group thought that it was a "real good experience" (qtd. in WTVM, 2003)

■ Unlike most previous wars, journalists were embedded with some troops in Iraq, making it possible for them to see firsthand what life was like in a war zone. What effect might this practice have on the accuracy of media accounts of war?

Along with Ron Claiborne, about six hundred journalists were embedded with the U.S. military on aircraft carriers in the Persian Gulf, at Marine and Army posts in Kuwait and Iraq, and in other settings during the U.S. military operations in Iraq in 2003. For the first time in U.S. history, embedding was designed to allow members of the media to become "living and breathing members of units, from front-line rifle companies to support outfits" (Nolte, 2003). Since the U.S. military had been accused of limiting journalists' and photographers' access to war zones during the Vietnam War, Pentagon officials decided that, at least in theory, today's journalists should be allowed to go to the front lines with the troops if they so desired. Consequently, large media entities such as CNN (owned by Time Warner) rented entire hotels in locations such as Kuwait City and delivered fleets of Hummer SUVs, equipped with satellite uplinks and other gear, to the war zone so that journalists could produce "insiders' views" of the war for audiences worldwide (Nolte, 2003).

Although the results of embedding members of the media with U.S. troops are apparently mixed, the extent to which members of the media become "insiders" in government operations in a democratic society remains an issue. For example, questions have previously arisen about the extent to which journalists should be allowed to "ride along" with police to document the activities of law enforcement officials, a practice that has been greatly reduced since the U.S. Supreme Court ruled that the police violated the Fourth Amendment prohibition on unreasonable searches and seizures when they permitted the media to accompany them to a home to photograph the apprehension of alleged suspects who turned out not to be the individuals actually sought by the police.

Activities such as embedding journalists with the military and media personnel taking part in ride-alongs with police officers raise interesting, and often complicated, questions about the relationship between a free press in a democratic society and the best interests of the military and law enforcement personnel as they fulfill their responsibilities. Questions are also raised about our right to know what's going on. Should journalists have unlimited access to the inner workings of politics and government? Do we have a right to know about everything that takes place? Concerns such as these are pressing issues that must be dealt with by officials in the legislative, executive, and judicial bodies of our nation. Likewise, members of the media must balance the best interests of the United States and the private rights of individuals against the need to increase corporate revenues by attracting more readers, viewers, and advertisers. In this chapter, we discuss the intertwining effect of politics, the economy, and the media as we grapple with issues such as these. Before reading on, test your knowledge of the media by taking the quiz in Box 14.1.

QUESTIONS AND ISSUES

Chapter Focus Question: What effect does the intertwining of politics and the media have on the United States and other nations?

What are the major political systems around the world?

How does the center of power differ in the pluralist and the elite models of the U.S. power structure?

How is government shaped by political parties and political attitudes?

Why are government bureaucracies so powerful?

What is the place of democracy in the future?

POLITICS, POWER, AND AUTHORITY

Politics **is the social institution through which power is acquired and exercised by some people and groups.** In contemporary societies, the government is the primary political system. *Government* **is the formal organization that has the legal and political authority to regulate the relationships among members of a society and between the society and those outside its borders.** Some social analysts refer to the government as the *state*—**the political entity that possesses a legitimate monopoly over the use of force within its territory to achieve its goals.** The components of the state include an elaborate, extremely bureaucratized political institution that consists of executive, central, and local administrations; the legislature; the courts; and the armed forces and police.

How does a sociological perspective on politics and government differ from that of political science? Although some areas of political science overlap with political sociology, the focuses of the two are somewhat different.

Political Science and Political Sociology

For most of the twentieth century, *political science* primarily focused on power and its distribution in different types of political systems. Today, some political scientists focus on the machinery of the federal government (such as its legislative, executive, and judicial branches) and the operative political processes (including political parties, public opinion, elections, and political participation). Other political scientists compare and contrast political systems in different countries.

Box 14.1 SOCIOLOGY AND EVERYDAY LIFE

How Much Do You Know About the Media?

True	False		
T	F	1.	U.S. mass media companies earn about $220 billion a year.
T	F	2.	In the United States, no media are publicly owned.
T	F	3.	The top ten U.S. newspaper chains own 20 percent of the daily newspapers.
T	F	4.	The Federal Communications Commission (FCC) has regulations preventing vertical integration of media companies.
T	F	5.	Television's first live marathon broadcast coverage of a political scandal occurred during the Clinton administration in the 1990s.
T	F	6.	Thirty-minute nightly news programs on the major television networks (NBC, CBS, and ABC) typically contain 26 minutes of news and 4 minutes of commercials.
T	F	7.	Almost all movies in the United States are distributed by six large studios.
T	F	8.	Both major U.S. political parties have been accused of purchasing television commercials with questionable campaign contributions.
T	F	9.	Most journalists identify themselves as Republicans.
T	F	10.	Some analysts believe that "reality" TV shows such as *Cops* and *America's Most Wanted* blur the distinction between "news" and "entertainment."

Answers on page 458.

Still others analyze the interrelationships among national governments, multinational corporations, and international organizations such as the United Nations.

By contrast, ***political sociology*** **is the area of sociology that examines the nature and consequences of power within or between societies, as well as the social and political conflicts that lead to changes in the allocation of power** (Orum, 2001). In other words, political sociology primarily focuses on the *social circumstances* of politics and explores how the political arena and its actors are intertwined with social institutions such as the economy, religion, education, and the media. According to the political analyst Michael Parenti (1998: 7–8), many people underrate the significance of politics in daily life:

> Politics is something more than what politicians do when they run for office. Politics is concerned with the struggles that shape social relations within societies and affairs between nations. The taxes and prices we pay and the jobs available to us, the chances that we will live in peace or perish in war, the costs of education and the availability of scholarships, the safety of the airliner or highway we travel on, the quality of the food we eat and the air we breathe, the availability of affordable housing and medical care, the legal protections against racial and sexual discrimination—all the things that directly affect the quality of our lives are influenced in some measure by politics. . . . To say you are not interested in politics, then, is like saying you are not interested in your own well-being.

Using a power–conflict framework for his analysis, Parenti (1998) suggests that the media often distort—either intentionally or unintentionally—the information they provide to citizens. According to Parenti, the media have the power to influence public opinion in a way that favors management over labor, corporations over their critics, affluent whites over minority-group members in central cities, political officials over protestors, and free-market capitalism over public-sector development. Parenti's assertion raises an interesting issue about the distribution of power in the United States and other industrialized nations: Do the media distort information to suit their own interests?

Power and Authority

Power **is the ability of persons or groups to achieve their goals despite opposition from others** (Weber, 1968/1922). Through the use of persuasion, authority,

Box 14.1 SOCIOLOGY AND EVERYDAY LIFE

Answers to the Sociology Quiz on the Media

1. **True.** Mass media companies in the United States earn about $220 billion annually. Newspapers earn about $55 billion, television earns $49 billion, movies earn $35 billion, and book publishing, magazines, recordings, and radio account for the rest.

2. **False.** Although most media outlets are privately owned, Public Broadcasting Service (PBS) and National Public Ratio (NPR) are funded by government support, grants from nonprofit foundations, and donations from viewers and listeners. However, corporate underwriters now play an increasing role in their funding.

3. **True.** The fact that the top ten newspaper chains own 20 percent (one-fifth) of the nation's daily newspapers is an example of *concentration of ownership.*

4. **False.** When the FCC began deregulating the broadcast media in the 1980s, a corresponding increase occurred in *vertical integration* (when a company attempts to control several related aspects of a business at once). For example, Time Warner owns the AOL Internet service and Warner movie studios, is a major book and magazine publisher, and owns and operates many cable TV systems and channels, such as Home Box Office (HBO) and CNN.

5. **False.** Previous marathon broadcast coverage occurred during the Johnson administration's involvement in the Vietnam War, the Nixon administration's involvement in the Watergate scandal, and the Reagan administration's involvement in Iran-Contra, a congressional investigation of how weapons were illegally supplied to Nicaraguan guerrillas.

6. **False.** News programs such as *NBC Nightly News* typically have 21 minutes of news content, 8 minutes of commercials, and 1 minute of "self-promotions" for other NBC programs and properties.

7. **True.** Six major studios—Columbia, Paramount, 20th Century-Fox, MCA/Universal, Time Warner, and Walt Disney—distribute not only the films they produce but also most of the films made by independent producers.

8. **True.** "Soft money" contributions, which are made outside the limits imposed by federal election laws, have allegedly been used by both parties for campaign-style ads, but leaders of the national political parties have claimed that their own ads were about social issues, not candidates.

9. **False.** Forty-four percent of journalists identify themselves as Democrats, 16.3 percent as Republicans, and 34 percent as independents.

10. **True.** Since TV "reality shows" imitate news stories by reenacting events and interviewing crime victims, it is difficult for some viewers to determine whether they are watching "real" news.

Sources: Based on Bagby, 1998; Biagi, 1998; and *Brill's Content*, 1998.

---

or force, some people are able to get others to acquiesce to their demands. Consequently, power is a *social relationship* that involves both leaders and followers. Power is also a dimension in the structure of social stratification. Persons in positions of power control valuable resources of society—including wealth, status, comfort, and safety—and are able to direct the actions of others while protecting and enhancing the

privileged social position of their class (Domhoff, 2002). For example, the sociologist G. William Domhoff (2002) argues that the media tend to reflect "the biases of those with access to them—corporate leaders, government officials, and policy experts." However, although Domhoff believes the media can amplify the message of powerful people and marginalize the concerns of others, he does not think the media are as important as government officials and corporate leaders are in the U.S. power equation.

What about power on a global basis? Although the most basic form of power is physical violence or force, most political leaders do not want to base their power on force alone. The sociologist Max Weber suggested that force is not the most effective long-term means of gaining compliance because those who are being ruled do not accept as legitimate those who are doing the ruling. Consequently, most leaders do not want to base their power on force alone; they seek to legitimize their power by turning it into **authority—power that people accept as legitimate rather than coercive.**

Ideal Types of Authority

Who is most likely to accept authority as legitimate and adhere to it? People have a greater tendency to accept authority as legitimate if they are economically or politically dependent on those who hold power. They may also accept authority more readily if it reflects their own beliefs and values (Turner, Beeghley, and Powers, 2002). Weber's outline of three *ideal types* of authority—traditional, charismatic, and rational–legal—shows how different bases of legitimacy are tied to a society's economy.

Traditional Authority According to Weber, ***traditional authority* is power that is legitimized on the basis of long-standing custom.** In preindustrial societies, the authority of traditional leaders, such as kings, queens, pharaohs, emperors, and religious dignitaries, is usually grounded in religious beliefs and custom. For example, British kings and queens historically traced their authority from God. Members of subordinate classes obey a traditional leader's edicts out of economic and political dependency and sometimes personal loyalty. However, as societies industrialize, traditional authority is challenged by a more complex division of labor and by the wider diversity of people who now inhabit the area as a result of high immigration rates. In industrialized societies, people do not share the same viewpoint on many issues and tend to openly question traditional authority. As the division of labor in a society becomes more complex, political and economic institutions become increasingly interdependent (Durkheim, 1933/1893).

Gender, race, and class relations are closely intertwined with traditional authority. Weber noted that traditional authority is often based on a system of patriarchy in which men are assumed to have authority in the household and in other small groups. Political

scientist Zillah R. Eisenstein (1994) suggests that *racialized patriarchy*—the continual interplay of race and gender—reinforces traditional structures of power in contemporary societies. According to Eisenstein (1994: 2), "Patriarchy differentiates women from men while privileging men. Racism simultaneously differentiates people of color from whites and privileges whiteness. These processes are distinct but intertwined." Although racialized patriarchy has been increasingly challenged, many believe that it remains a reality in both preindustrial and industrialized nations. Class relations may also be linked to traditional authority in industrial nations such as the United States. In some upper-class families, for example, holding political office is considered to be a birthright and a family tradition. In families such as the Kennedys, Rockefellers, and Du Ponts, capitalism and some degree of traditional authority in politics appear to have been mutually reinforcing (see Baltzell, 1958; Domhoff, 1983).

By contrast, Weber predicted that traditional authority would inhibit the development of capitalism. He stressed that capitalism cannot fully develop when rules are not logically established, when officials follow rules arbitrarily, and when leaders are not technically trained (Weber, 1968/1922; Turner, Beeghley, and Powers, 2002). Weber believed that capitalism thrives best in systems of rational–legal authority. As societies industrialize, traditional authority is challenged by a more complex division of labor and by the wider diversity of people who now inhabit the area as a result of high immigration rates.

Charismatic Authority ***Charismatic authority* is power legitimized on the basis of a leader's exceptional personal qualities** or the demonstration of extraordinary insight and accomplishment that inspire loyalty and obedience from followers. According to Weber, charismatic individuals are able to identify themselves with the central facts of people's lives and—through the force of their personalities—communicate their inspirations to others and lead them in new directions (Turner, Beeghley, and Powers, 2002). Charismatic leaders may be politicians, soldiers, and entertainers, among others (Shils, 1965; Bendix, 1971).

From Weber's perspective, a charismatic leader may be either a tyrant or a hero. Thus, charismatic authority has been attributed to such diverse historical figures as Jesus Christ, Napoleon, Julius Caesar, Adolf Hitler, Winston Churchill, Franklin Roosevelt, Martin Luther King, Jr., and César Chávez. Few women other than Joan of Arc, Mother Teresa, Indira Gandhi of India, Eva Perón of Argentina, and Margaret Thatcher of the United Kingdom have had the opportunity to become charismatic leaders due to the predominantly patriarchal social structures in recorded history. Since women are seldom permitted to assume positions of leadership, they are much less likely to become charismatic leaders.

Charismatic authority tends to be temporary and relatively unstable; it derives primarily from individual leaders (who may change their minds, leave, or die) and from an administrative structure usually limited to a small number of faithful followers. For this reason, charismatic authority often becomes routinized. The ***routinization of charisma* occurs when charismatic authority is succeeded by a bureaucracy controlled by a rationally established authority or by a combination of traditional and bureaucratic authority** (Turner, Beeghley, and Powers, 2002). According to Weber (1968/1922: 1148), "It is the fate of charisma to recede . . . after it has entered the permanent structures of social action."

Rational–Legal Authority According to Weber, ***rational–legal authority* is power legitimized by law or written rules and regulations.** Rational–legal authority—also known as *bureaucratic authority*—is based on an organizational structure that includes a clearly defined division of labor, hierarchy of authority, formal rules, and impersonality. Power is legitimized by procedures; if leaders obtain their positions in a procedurally correct manner (such as by election or appointment), they have the right to act.

Rational–legal authority is held by elected or appointed government officials and by officers in a formal organization. However, authority is invested in the *office,* not in the *person* who holds the office. For example, although the U.S. Constitution grants rational–legal authority to the office of the presidency, a president who fails to uphold the public trust may be removed from office. In contemporary society, the media may play an important role in bringing to light allegations about presidents or other elected officials. Examples include the media blitzes surrounding the Watergate investigation of the 1970s that led to the resignation of President Richard M. Nixon and the late-1990s political firestorm over campaign fund-raising and the sex scandal involving President Bill Clinton.

In a rational–legal system, the governmental bureaucracy is the apparatus responsible for creating and enforcing rules in the public interest. Weber believed that rational–legal authority was the only means to attain efficient, flexible, and competent regulation under a rule of law (Turner, Beeghley, and Powers,

2002). Weber's three types of authority are summarized in Concept Table 14.A.

POLITICAL SYSTEMS IN GLOBAL PERSPECTIVE

Political systems as we know them today have evolved slowly. In the earliest societies, politics was not an entity separate from other aspects of life. As we will see, however, all groups have some means of legitimizing power.

Hunting and gathering societies do not have political institutions as such because they have very little division of labor or social inequality. Leadership and authority are centered in the family and clan. Individuals acquire leadership roles due to personal attributes such as great physical strength, exceptional skills, and charisma (Nolan and Lenski, 1999).

Political institutions first emerged in agrarian societies as they acquired surpluses and developed greater social inequality. Elites took control of politics and used custom or traditional authority to justify their position. When cities developed circa 3500–3000 B.C.E., the *city-state*—a city whose power extended to adjacent areas—became the center of political power. Both the Roman and Persian empires were composed of a number of city-states, each of which had its own monarchy. Thus, in these societies, political authority was decentralized. After each of these empires fell, the individual city-states lived on.

Nation-states as we know them began to develop in Europe between the twelfth and fifteenth centuries (see Tilly, 1975). A *nation-state* is a unit of political organization that has recognizable national boundaries and whose citizens possess specific legal rights and obligations. Nation-states emerge as countries develop specific geographic territories and acquire greater ability to defend their borders. Improvements in communication and transportation make it possible for people in a larger geographic area to share a common language and culture. As charismatic and traditional authority are superseded by rational–legal authority, legal standards come to prevail in all areas of life, and the nation-state claims a monopoly over the legitimate use of force (P. Kennedy, 1993).

Approximately 190 nation-states currently exist throughout the world; today, everyone is born, lives, and dies under the auspices of a nation-state (see Skocpol and Amenta, 1986). Four main types of political systems are found in nation-states: monarchy, authoritarianism, totalitarianism, and democracy.

Concept Table 14.A WEBER'S THREE TYPES OF AUTHORITY

Max Weber's three types of authority are shown here in global perspective. Sultan Ali Mirah of Ethiopia is an example of traditional authority sanctioned by custom. Charismatic authority is exemplified by Dr. Martin Luther King, Jr., of the United States, whose leadership was based on personal qualities. Australian Supreme Court justices represent rational-legal authority, which depends upon established rules and procedures.

	Description	Examples	
TRADITIONAL	Legitimized by long-standing custom Subject to erosion as traditions weaken	Patrimony (authority resides in traditional leader supported by larger social structures, as in old British monarchy) Patriarchy (rule by men occupying traditional positions of authority, as in the family)	
CHARISMATIC	Based on leader's personal qualities Temporary and unstable	Napoleon Adolf Hitler Martin Luther King, Jr. César Chávez Mother Teresa	
RATIONAL-LEGAL	Legitimized by rationally established rules and procedures Authority resides in the office, not the person	Modern British Parliament U.S. presidency, Congress, federal bureaucracy	

Betty Press/Woodfin Camp & Associates

SuperStock

Cary Wolinsky/Stock Boston

Monarchy

Monarchy is a political system in which power resides in one person or family and is passed from generation to generation through lines of inheritance. Monarchies are most common in agrarian societies and are associated with traditional authority patterns. However, the relative power of monarchs has varied across nations, depending on religious, political, and economic conditions.

Absolute monarchs claim a hereditary right to rule (based on membership in a noble family) or a divine right to rule (a God-given right to rule that legitimizes the exercise of power). In limited monarchies, rulers depend on powerful members of the nobility to retain their thrones. Unlike absolute monarchs, *limited monarchs* are not considered to be above the law. In *constitutional monarchies,* the royalty serve as symbolic rulers or heads of state while actual authority is held by elected officials in national parliaments. In present-day monarchies such as the United Kingdom, Sweden, Spain, and the Netherlands, members of royal families primarily perform ceremonial functions. In the United Kingdom, for example, the media often focus large

Box 14.2 SOCIOLOGY IN GLOBAL PERSPECTIVE

The European Union: Transcending National Borders and Governments

What do people living in the following nations have in common?

Austria	Germany	The Netherlands
Belgium	Greece	Portugal
Denmark	Ireland	Spain
Finland	Italy	Sweden
France	Luxembourg	United Kingdom

If you answered that all of these countries are part of the European Union (EU), you are correct! What is the EU? The European Union is a treaty-based, institutional framework that defines and manages economic and political cooperation among its member nations. Based on a series of treaties beginning after World War II, the EU became a sustained effort to create communities that share sovereignty in matters of coal and steel production, trade, and nuclear energy. Perhaps more than anything else, the EU was created to make sure that another war in Europe would be unthinkable. According to information from the EU organization, its fundamental goal is the creation of an ever-closer union among the peoples of Europe.

Although the EU has produced cooperation among the many governmental bodies in these nations, two forms of cooperation are most visible to the outsider who travels in EU countries. First, the euro has been established as the official currency in all of the

■ This dramatic sculpture representing the Euro—the European currency—is located near the European Centralbank in Frankfurt, Germany, and is symbolic of the spirit of cooperation emerging among a number of European countries.

nations except Denmark (where the krone is used), Sweden (where the krona is used), and the United Kingdom (where the British pound is used), which makes international commerce and tourism easier. The euro banknotes (similar to dollar bills) are identical in all of the other countries, but each country produces its own coins, with one common side and one

---

Through its many ups and downs, the British royal family has remained a symbol of that nation's monarchy. Monarchies typically pass power from generation to generation in one family. Shown here are the likely successors to the British throne in the twenty-first century.

amounts of time and attention on the royal family, especially the personal lives of its members. Recently, the European Union (of which the United Kingdom, Spain, Sweden, and the Netherlands are all members) has also received media attention as a form of governmental cooperation across national boundaries but not one that weakens the powers of the present-day monarchies (see Box 14.2, "Sociology in Global Perspective").

Authoritarianism

Authoritarianism **is a political system controlled by rulers who deny popular participation in government.** A few authoritarian regimes have been absolute monarchies whose rulers claimed a hereditary right to their position. Today, Saudi Arabia and Kuwait are examples of authoritarian absolute monarchies. In *dictatorships,* power is gained and held by a single individual. Pure dictatorships are rare; all rulers

national side. However, both the notes and coins can be used anywhere in the euro zone. Second, travel within the EU countries is easier and faster because a citizen of any of the EU nations does not have to present a visa or other official document each time he or she passes across a national border. "The right to move freely and to stay in the territory of Member States" is one of the rights guaranteed to citizens in the EU. Each person's citizenship is still linked to one of the member countries, but now she or he is also a member of the larger European community.

Sociologically speaking, it will be interesting to see whether the EU can bring about changes in cultural and media practices in the various nations in this union. Recently, an effort has been made to enact legislation to reduce all forms of discrimination beyond the usual boundaries of work, trade, and labor, which the European Commission (a governing body within the EU) has previously addressed. The 1997 Treaty of Amsterdam gave officials in the European Union the mandate to combat sexual discrimination, as well as discrimination based on age, race, religion, and disability, among other things. The current European commissioner for employment and social affairs

would like for the various nations to agree to eliminate sexism in all areas of public life, including television advertisements, schoolbooks, and insurance rates. Many analysts believe that these efforts will ultimately be unsuccessful, but the discussion of them raises an interesting question: How far will the individual nations within the EU be willing to go in letting a unified body mandate what will take place in each country in regard to issues of culture, media, and long-term problems of discrimination?

Overall, analysts believe that the EU has produced more than fifty years of stability, peace, and economic prosperity, and that this strong alliance will be a model for international governmental cooperation. At this time, a number of nations in Eastern Europe (including Bulgaria, Czech Republic, Hungary, Poland, and Turkey) are considered to be "candidate countries" for inclusion in the EU in the future. Will the inclusion of thirteen or fourteen additional nations further strengthen the EU? Some people believe that such a European Union would produce an even stronger, more unified voice for Europe; however, others believe that greater inclusion might make cooperation across borders more difficult.

Sources: Based on Alvarez, 2003; and Europa, 2003.

WRITING IN SOCIOLOGY ASSIGNMENT

Are there lessons that we in the United States might learn from the EU experience? Do you think that there will be more or less cooperation  among the nations of the world in the future? Explain your answer.

need the support of the military and the backing of business elites to maintain their position. *Military juntas* result when military officers seize power from the government, as has happened in recent decades in Argentina, Chile, and Haiti. Today, authoritarian regimes exist in Fidel Castro's Cuba and in the People's Republic of China. Authoritarian regimes seek to control the media and to suppress coverage of any topics or information that does not reflect upon the regime in a favorable light.

Totalitarianism

***Totalitarianism* is a political system in which the state seeks to regulate all aspects of people's public and private lives.** Totalitarianism relies on modern technology to monitor and control people; mass propaganda and electronic surveillance are widely used to influence people's thinking and control their ac-

tions. One example of a totalitarian regime was the National Socialist (Nazi) Party in Germany during World War II; military leaders there sought to control all aspects of national life, not just government operations. Other examples include the former Soviet Union and contemporary Iraq before the fall of Saddam Hussein's regime.

To keep people from rebelling, totalitarian governments enforce conformity: People are denied the right to assemble for political purposes, access to information is strictly controlled, and secret police enforce compliance, creating an environment of constant fear and suspicion.

Many nations do not recognize totalitarian regimes as being the legitimate government for a particular country. Afghanistan in the year 2001 was an example. As the war on terrorism began in the aftermath of the September 11 terrorist attacks on the United States, many people developed a heightened

© AP/Wide World Photos

What do you know about the "war on terrorism"? Most of what we know is based on media coverage. As shown here, the start of a professional football game in Philadelphia was delayed so that fans could watch a large-screen television while President Bush announced military action against the Taliban in Afghanistan.

awareness of the Taliban regime, which ruled most of Afghanistan and was engaged in fierce fighting to capture the rest of the country. The Taliban regime maintained absolute control over the Afghan people in most of that country. For example, it required that all Muslims take part in prayer five times each day and that men attend prayer at mosques, where women were forbidden (Marquis, 2001). Taliban leaders claimed that their actions were based on Muslim law and espoused a belief in never-ending *jihad*—a struggle against one's perceived enemies. Although the totalitarian nature of the Taliban regime was difficult for many people, it was particularly oppressive for women, who were viewed by this group as being "biologically, religiously and prophetically" inferior to men (McGeary, 2001: 41). Consequently, this regime made the veil obligatory and banned women from public life. U.S. government officials believed that the Taliban regime was protecting Osama bin Laden, the man thought to have been the mastermind behind numerous terrorist attacks on U.S. citizens and facilities, both on the mainland and abroad. As a to-

talitarian regime, the Taliban leadership was recognized by only three other governments, despite controlling most of Afghanistan.

Once the military action commenced in Afghanistan, most of what U.S. residents knew about the Taliban and about the war on terrorism was based on media accounts and "expert opinions" that were voiced on television. According to the political analyst Michael Parenti (1998), the media play a significant role in framing the information we receive about the political systems of other countries. *Framing* refers to how news is packaged, including the amount of exposure given to a story, its placement, the positive or negative tone of the story, the headlines and photographs, and the accompanying visual and auditory effects if the story is being broadcast. The war in Afghanistan, like other wars in the past, was typically framed as a fight to save freedom and democracy.

Democracy

***Democracy* is a political system in which the people hold the ruling power either directly or through elected representatives.** The literal meaning of *democracy* is "rule by the people" (from the Greek words *demos,* meaning "the people," and *kratein,* meaning "to rule"). In an ideal-type democracy, people would actively and directly rule themselves. *Direct participatory democracy* requires that citizens be able to meet together regularly to debate and decide the issues of the day. However, if all 292 million people in the United States came together in one place for a meeting, they would occupy an area of more than seventy square miles, and a single round of five-minute speeches would require more than five thousand years (based on Schattschneider, 1969).

In countries such as the United States, Canada, Australia, and the United Kingdom, people have a voice in the government through *representative democracy,* whereby citizens elect representatives to serve as bridges between themselves and the government. The U.S. Constitution requires that each state have two senators and a minimum of one member in the House of Representatives. The current size of the House (435 seats) has not changed since the apportionment following the 1910 census. Therefore, based on Census 2000, those 435 seats were reapportioned based on the increase or decrease in a state's population between 1990 and 2000 (see "Census Profiles: Political Representation and Shifts in the U.S. Population").

In a representative democracy, elected representatives are supposed to convey the concerns and interests of those they represent, and the government is expected to be responsive to the wishes of the people.

CENSUS ★ PROFILES

Political Representation and Shifts in the U.S. Population

The increase or decrease in the number of members of the House of Representatives elected by the voters in each state is a reflection of how the U.S. population has shifted over the last decade. Increases in population occurred in a number of states located in the Southeast, Southwest, and West, as contrasted with a relative decline in the population of other states, primarily in the Midwest and Northeast. What sociological factors do you think may have contributed to this shift in the U.S. population?

States with an Increase in House Seats

Arizona (2 more seats)
California (1 more seat)
Colorado (1 more seat)
Florida (2 more seats)
Georgia (2 more seats)
Nevada (1 more seat)
North Carolina (1 more seat)
Texas (2 more seats)

States with a Decrease in House Seats

Connecticut (1 fewer seat)
Illinois (1 fewer seat)
Indiana (1 fewer seat)
Michigan (1 fewer seat)
Mississippi (1 fewer seat)
New York (2 fewer seats)
Ohio (1 fewer seat)
Oklahoma (1 fewer seat)
Pennsylvania (2 fewer seats)
Wisconsin (1 fewer seat)

Source: U.S. Census Bureau, 2002.

Elected officials are held accountable to the people through elections. However, representative democracy is not always equally accessible to all people in a nation. Throughout U.S. history, members of subordinate racial–ethnic groups have been denied full participation in the democratic process. Gender and social class have also limited some people's democratic participation. For example, women have not always had the same rights as men. Full voting rights were not gained by women until the ratification of the Nineteenth Amendment in 1920. However, women were divided by class in their perceptions about the necessity of suffrage.

Even representative democracies are not all alike. As compared to the winner-takes-all elections in the United States, which are usually decided by who wins the most votes, the majority of European elections are based on a system of proportional representation, meaning that each party is represented in the national legislature according to the proportion of votes that party received. For example, a party that won 40 percent of the vote would receive 40 seats in a 100-seat legislative body, and a party receiving 20 percent of the votes would receive 20 seats.

PERSPECTIVES ON POWER AND POLITICAL SYSTEMS

Is political power in the United States concentrated in the hands of the few or distributed among the many? Sociologists and political scientists have suggested many different answers to this question; however, two prevalent models of power have emerged: pluralist and elite.

Functionalist Perspectives: The Pluralist Model

The pluralist model is rooted in a functionalist perspective which assumes that people share a consensus on central concerns, such as freedom and protection from harm, and that the government serves important functions no other institution can fulfill. According to Emile Durkheim (1933/1893), the purpose of government is to socialize people to be good citizens, to regulate the economy so that it operates effectively, and to provide necessary services for citizens. Contemporary functionalists state the four main functions as follows: (1) maintaining law and order, (2) planning and directing society, (3) meeting social needs, and (4) handling international relations, including warfare.

But what happens when people do not agree on specific issues or concerns? Functionalists suggest that divergent viewpoints lead to a system of political pluralism in which the government functions as an arbiter between competing interests and viewpoints. According to the *pluralist model,* **power in political systems is widely dispersed throughout many competing interest groups** (Dahl, 1961).

Key Elements Political scientists Thomas R. Dye and Harmon Zeigler (2003) have summarized the key elements of pluralism as follows:

- The diverse needs of women and men, people of all religions and racial–ethnic backgrounds,

Special interest groups help people advocate their own interests and further their causes. Advocates may run for public office and gain a wider voice in the political process. An example is U.S. Senator Ben Nighthorse Campbell of Colorado (left), a Cheyenne chief, who is shown here as he recently participated in a ground-breaking ceremony for the National Museum of the American Indian in Washington, D.C.

© Paul Hosefros/The New York Times

and the wealthy, middle class, and poor are met by political leaders who engage in a process of bargaining, accommodation, and compromise.

- Competition among leadership groups in government, business, labor, education, law, medicine, and consumer organizations, among others, helps prevent abuse of power by any one group. These groups often operate as *veto groups* that attempt to protect their own interests by keeping others from taking actions that would threaten those interests.

- Power is widely dispersed in society. Leadership groups that wield influence on some decisions are not the same groups that may be influential in other decisions.

- Public policy is not always based on majority preference; rather, it reflects a balance among competing interest groups.

- Everyday people can influence public policy by voting in elections, participating in existing special interest groups, or forming new ones to gain access to the political system.

Special Interest Groups *Special interest groups are political coalitions made up of individuals or groups that share a specific interest they wish to protect or advance with the help of the political system* (Greenberg and Page, 2002). Examples of special interest groups include the AFL-CIO (representing the majority of labor unions) and public interest or citizens groups such as the American Conservative Union and Zero Population Growth.

What purpose do special interest groups serve in the political process? According to some analysts, spe-

cial interest groups help people advocate their own interests and further their causes. Broad categories of special interest groups include banking, business, education, energy, the environment, health, labor, persons with a disability, religious groups, retired persons, women, and those espousing a specific ideological viewpoint; obviously, many groups overlap in interests and membership.

Special interest groups are also referred to as *pressure groups* (because they put pressure on political leaders) or *lobbies*. Lobbies are often referred to in terms of the organization they represent or the single issue on which they focus—for example, the "gun lobby" or the "dairy lobby." The people who are paid to influence legislation on behalf of specific clients are referred to as *lobbyists*.

Lobbying can be conducted as either an inside game or an outside game. The *inside game,* or "old-boy network," refers to situations in which interest group representatives are in direct contact with officials in the government and try to build influence on the basis of personal relationships (Berry, 1989). In the *outside game,* interest groups attempt to apply pressure on officials by mobilizing their constituents to telephone or write letters demanding a specific course of action, by stirring up public opinion, and by getting their membership involved in activities to make their wishes known (Greenberg and Page, 2002).

Over the past four decades, special interest groups have become more involved in "single-issue politics," in which political candidates are often supported or rejected solely on the basis of their views on a specific issue—such as abortion, gun control, gay and lesbian rights, or the environment. Single-issue groups derive

Many special interest groups focus on a single issue, such as tax cuts, gun control, or the environment. At this tax day rally, some demonstrators supported tax cuts while others opposed such measures. According to the pluralist model, what function does the government serve in resolving competing interests and viewpoints?

their strength from the intensity of their beliefs; leaders have little room to compromise on issues.

Political Action Committees The funding of lobbying efforts has become more complex in recent years. Reforms in campaign finance laws in the 1970s set limits on direct contributions to political candidates and led to the creation of *political action committees* (PACs)—**organizations of special interest groups that solicit contributions from donors and fund campaigns to help elect (or defeat) candidates based on their stances on specific issues.**

As the cost of running for political office has skyrocketed, candidates have relied more on PACs for financial assistance. PACs contributed more than $245,000,000 to candidates for the U.S. House and Senate during 1999 and 2000, an increase of 19 percent over the preceding two-year period (Federal Election Commission, 2001). Advertising, staff, direct-mail operations, telephone banks, computers, consultants, travel expenses, office rentals, and other expenses incurred in political campaigns make PAC money vital to candidates.

Some PACs represent the "public interest" and ideological interest groups such as gay rights or the National Rifle Association. Other PACs represent the capitalistic interests of large corporations. Realistically, members of the least-privileged sectors of society are not represented by PACs. As one senator pointed out, "There aren't any Poor PACs or Food Stamp PACs or Nutrition PACs or Medicare PACs" (qtd. in Greenberg and Page, 1993: 240). Critics of pluralism argue that "Big Business" wields such disproportionate power in U.S. politics that it undermines the democratic process (see Lindblom, 1977; Domhoff, 1978).

As an outgrowth of record-setting campaign spending in the 1996 national election, campaign financing abuses were alleged by both Republicans and Democrats in Washington. At the center of the controversy was the issue of "soft money" contributions, which are made outside the limits imposed by federal election laws. In 2002, Congress passed the McCain-Feingold campaign finance law, prohibiting soft money contributions in federal elections, and the U.S. Supreme Court upheld the soft money provisions of that law in 2003. However, the McCain-Feingold law applies only to federal elections and does not bar soft money contributions in state and local elections.

Conflict Perspectives: Elite Models

Although conflict theorists acknowledge that the government serves a number of important purposes in society, they assert that government exists for the benefit of wealthy or politically powerful elites who use the government to impose their will on the masses. According to the *elite model,* **power in political systems is concentrated in the hands of a small group of elites, and the masses are relatively powerless.** Early Italian sociologist Vilfredo Pareto (1848–1923) first used the term *elite* to refer to "the few who rule the many" (Marshall, 1998). Similarly, Karl Marx claimed that under capitalism, the government serves the interests of the ruling (or capitalist) class that controls the means of production.

Key Elements Contemporary elite models are based on the following assumptions (Dye and Zeigler, 2003):

- Decisions are made by the elite, which possesses greater wealth, education, status, and other resources than do the "masses" it governs.
- Consensus exists among the elite on the basic values and goals of society; however, consensus

Running for political office takes large sums of money. Actor Arnold Schwarzenegger, shown here with his wife, Maria Shriver, contributed large sums of money to his successful campaign for governor of California and also received large contributions from supporters because of his name recognition. Although some candidates for state and national office raise money from individuals, funding for major campaigns increasingly comes from special interest groups and political action committees.

does not exist among most people in society on these important social concerns.

- Power is highly concentrated at the top of a pyramid-shaped social hierarchy; those at the top of the power structure come together to set policy for everyone.
- Public policy reflects the values and preferences of the elite, not the preferences of the people.

The pluralist and elite models are compared in Figure 14.1.

C. Wright Mills and the Power Elite
Who makes up the U.S. power elite? According to the sociologist C. Wright Mills (1959a), the **power elite is made up of leaders at the top of business, the executive branch of the federal government, and the military.** Of these three, Mills speculated that the "corporate rich" (the highest-paid officers of the biggest corporations) were the most powerful because of their unique ability to parlay the vast economic resources at their disposal into political power. At the middle level of the pyramid, Mills placed the legislative branch of government, special interest groups, and local opinion leaders. The bottom (and widest layer) of the pyramid is occupied by the unorganized masses, who are relatively powerless and are vulnerable to economic and political exploitation.

Mills emphasized that individuals who make up the power elite have similar class backgrounds and interests; many of them also interact on a regular basis. In

addition, many frequently shift back and forth among the business, government, and military sectors. For example, it is not unusual for corporate executives to assume positions in a president's cabinet and then return to the business world. Similarly, a "revolving door" exists between the military and the executive branch, as well as between the military and corporations. Members of the power elite are able to influence many important decisions, including federal spending.

G. William Domhoff and the Ruling Class
Sociologist G. William Domhoff (2002) asserts that this nation in fact has a *ruling class*—the corporate rich, who constitute less than 1 percent of the U.S. population. Domhoff uses the term *ruling class* to signify a relatively fixed group of privileged people who wield power sufficient to constrain political processes and serve underlying capitalist interests. Although the power elite controls the everyday operation of the political system, who *governs* is less important than who *rules*.

Like Mills, Domhoff asserts that individuals in the upper echelon are members of a business class based on the ownership and control of large corporations. The intertwining of the upper class and the corporate community produces economic and social cohesion. Economic interdependence among members of the ruling class is rooted in common stock ownership and is visible in interlocking corporate directorates that serve as a communications network (Domhoff, 2002). Members of the ruling class are also socially linked through exclusive clubs, expensive private schools and debutante parties for their children, and listings in the Social Register (an address book for upper-class families in major metropolitan areas). In a recent study, the sociologists Richard L. Zweigenhaft and G. William Domhoff (1998) argue that more women and people of color have entered the power elite in recent years; however, due to prior discrimination and shared class interests among members of the white power elite, few subordinate-group members have brought about significant changes in how elites conduct their business in organizations and the larger society.

Domhoff (2002) suggests that the corporate rich influence the political process in three ways. First, they affect the candidate selection process by helping to finance campaigns and providing favors to political candidates. Second, through participation in the special interest process, the corporate rich are able to obtain favors, tax breaks, and favorable regulatory rulings. Finally, the corporate rich gain access to the policy-making process through their appointments to governmental advisory committees, presidential commissions, and other governmental positions. Today,

© 2003 AP/Wide World Photos

Figure 14.1   Pluralist and Elite Models

PLURALIST MODEL

ELITE MODEL

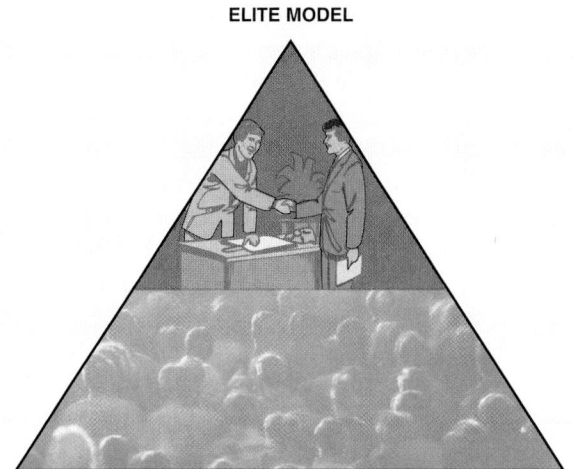

- Decisions are made on behalf of the people by leaders who engage in bargaining, accommodation, and compromise.

- Competition among leadership groups makes abuse of power by any one group difficult.

- Power is widely dispersed, and people can influence public policy by voting.

- Public policy reflects a balance among competing interest groups.

- Decisions are made by a small group of elite people.

- Consensus exists among the elite on the basic values and goals of society.

- Power is highly concentrated at the top of a pyramid-shaped social hierachy.

- Public policy reflects the values and preferences of the elite.

---

some members of the ruling class influence international politics through their involvement in banking, business services, and law firms that have a strong interest in overseas sales, investments, or raw materials extraction (Domhoff, 1990).

Clearly, owners of some media conglomerates would be classified in Domhoff's power elite. Consider, for example, journalist Ken Aulette's discussion with Rupert Murdock, head of News Corporation—the conglomerate that owns over a hundred newspapers worldwide, major movie studios, publishing interests, the Fox TV network, and numerous cable channels—about the enormous growth of Murdock's media empire:

> Does [he] tire of the constant competition to get bigger, to win each war? When does Murdock say *enough*? I asked him this question at the end of a long night that started with a drink and a stroll over his six acre property in Beverly Hills. . . .
>
> After dinner in the dining room of his comfortable Spanish-style home, Murdock sipped a glass of California Chardonnay and treated the

question as something so alien as to be incomprehensible. "You go on," he responded, opaquely. "There is a global village in some sense. You are competing everywhere. . . . And I just enjoy it." (Aulette, 1998: 287)

On another occasion, when Murdock was questioned regarding mergers, joint ventures, and cross-ownership of media firms, he stated that "We [media and telecommunications firms] can join forces now, or we can kill each other and then join forces" (McChesney, 1998: 15). Murdock and other media CEOs believe that two or three companies will eventually dominate the media and telecommunications industry; however, in their view this is not necessarily bad. As Murdock commented, "Monopoly is a terrible thing until you have it" (McChesney, 1998: 15). But media critics are concerned that with fewer companies, powerful executives such as Murdock and Sumner Redstone, chairman of Viacom, will have more ability to exert their influence on news and entertainment programming. Such programming may also be more vulnerable to pressures from interest groups (Bates and Eller, 1999).

Class Conflict Perspectives Most contemporary elite models are based on the work of Karl Marx; however, there are divergent viewpoints about the role of the state within the Marxist (or class conflict) perspective. *Instrumental Marxists* argue that the state invariably acts to perpetuate the capitalist class. From this perspective, capitalists control the government through special interest groups, lobbying, campaign financing, and other types of "influence peddling" to get legislatures and the courts to make decisions favorable to their class (Miliband, 1969; Domhoff, 2002). In other words, the state exists only to support the interests of the dominant class (Marger, 1987).

By contrast, *structural Marxists* contend that the state is not simply a passive instrument of the capitalist class. Because the state must simultaneously preserve order and maintain a positive climate for the accumulation of capital, not all decisions can favor the immediate wishes of the dominant class (Quadagno, 1984; Marger, 1987). For example, at various points in the history of this country, the state has had to institute social welfare programs, regulate business, and enact policies that favor unions in order to placate people and maintain "law and order." Ultimately, however, such actions serve the long-range interests of the capitalists by keeping members of subordinate groups from rebelling against the dominant group (O'Connor, 1973).

Critique of Pluralist and Elite Models

Pluralist and elite models each make a unique contribution to our understanding of power. The pluralist model emphasizes that many different groups, not just elected officials, compete for power and advantage in society. This model also shows how coalitions may shift over time and how elected officials may be highly responsive to public opinion on some occasions. However, critics counter that extensive research evidence contradicts the notion that the U.S. political system is pluralistic. They note instead that our system has the "veneer" of pluralism when, in fact, it is remarkably elitist for a society that claims to value ordinary people's input (Domhoff, 2002). For example, a wide disparity exists between the resources and political clout of "Big Business" when compared with those of interest groups that represent infants and children or persons with a disability. According to critics, consensus is difficult, if not impossible, in populations composed of people from different classes, religions, and racial–ethnic and age groups.

Mills's power-elite model highlights the interrelationships of the economic, political, and military sectors of society and makes us aware that the elite may

be a relatively cohesive group. Similarly, Domhoff's ruling-class model emphasizes the role of elites in setting and implementing policies that benefit the capitalist class. Power-elite models call our attention to a central concern in contemporary U.S. society: the ability of democracy and its ideals to survive in the context of the increasingly concentrated power held by capitalist oligarchies (see Chapter 13).

THE U.S. POLITICAL SYSTEM

The U.S. political system is made up of formal elements, such as the legislative process and the duties of the president, and informal elements, such as the role of political parties in the election process. We now turn to an examination of these informal elements, including political parties, political socialization, and voter participation.

Political Parties and Elections

A ***political party*** **is an organization whose purpose is to gain and hold legitimate control of government;** it is usually composed of people with similar attitudes, interests, and socioeconomic status. A political party (1) develops and articulates policy positions, (2) educates voters about issues and simplifies the choices for them, and (3) recruits candidates who agree with those policies, helps those candidates win office, and holds the candidates responsible for implementing the party's policy positions. In carrying out these functions, a party may try to modify the demands of special interests, build a consensus that could win majority support, and provide simple and identifiable choices for the voters on election day. Political parties create a *platform,* a formal statement of the party's political positions on various social and economic issues.

Since the Civil War, the Democratic and Republican parties have dominated the U.S. political system. Although one party may control the presidency for several terms, at some point the voters elect the other party's nominee, and control shifts. See Figure 14.2 for a look at the major political parties in U.S. history.

Ideal Type Versus Reality How well do the parties measure up to the ideal-type characteristics of a political party as described here? Although both parties have been quite successful at getting their candidates elected at various times, they generally do not

The cost of mounting a campaign for president of the United States has grown dramatically with each passing decade. Prior to celebrating at the presidential gala of 2001, President George W. Bush raised a record $44 million for Republicans at presidential dinners.

© Manny Ceneta/AFP/Corbis

meet the ideal characteristics of a political party. Thomas Dye and Harmon Zeigler (2003) suggest several reasons for that failure. First, the two parties do not offer voters clear policy alternatives. Most voters view themselves as being close to the center of the political spectrum (extremely liberal being the far left of that spectrum and extremely conservative being the far right). Although the definitions of *liberal* and *conservative* vary over time, *liberals* tend to focus on equality of opportunity and the need for government regulation and social safety nets. By contrast, *conservatives* are more likely to emphasize economic liberty and freedom from government interference (Greenberg and Page, 2002). Because most voters consider themselves to be moderates, neither party has much incentive to veer very far from the middle.

Second, the two parties are oligarchies, dominated by active elites. The active party elites hold views that are further from the center of the political spectrum (Democrats to the left, Republicans to the right, usually) than are those of a majority of members of their party. Thus, platforms state the policy goals of the oligarchies, whereas the goal of electing candidates necessitates nominating more-centrist candidates.

Third, primary elections (in which the nominees of political parties for most offices other than president and vice president are selected) determine nominees. Thus, voters in the primaries may select nominees whose views are closer to the center of the political spectrum and further away from the party's own platform. The nominee may establish a personal organization (committee) to work for the nominee's election *outside* the party organization. That committee may contain both Democratic and Republican members, thus being more centrist than the party platform or organization.

Fourth, party loyalties are declining. Many people today vote in one party's primary but then cast their ballot in general elections without total loyalty to that party. They may cast a "split-ticket" ballot (voting for one party's candidate in one race and another party's candidate in another). As a result, it is harder for the party to meet the "ideal" of holding the candidate accountable for implementing the party's platform.

Finally, the media have replaced the party as a means of political communication. Often, the candidate who wins does so as a result of media presentation, not the political party's platform. Candidates no longer need political parties to carry their message to the people.

Despite the fact that the Republican and Democratic parties do not have all of the ideal characteristics of political parties, they both flourish. On the local and state levels, other political parties—or even candidates who identify themselves as "independents"—may win office, but the Republican and Democratic parties have basically controlled the U.S. political scene for almost 150 years. In general, voters tend to select candidates and political parties based on social and economic issues that they consider to be important in their lives.

Social Issues Social issues are those relating to moral judgments or civil rights, ranging from abortion rights to gun control. Other social issues include the rights of people of color and persons with a disability, school prayer, and the environment. For example, persons with a liberal perspective on social issues tend to believe that women have the right to an abortion (at least under certain circumstances), that criminals should be rehabilitated instead of

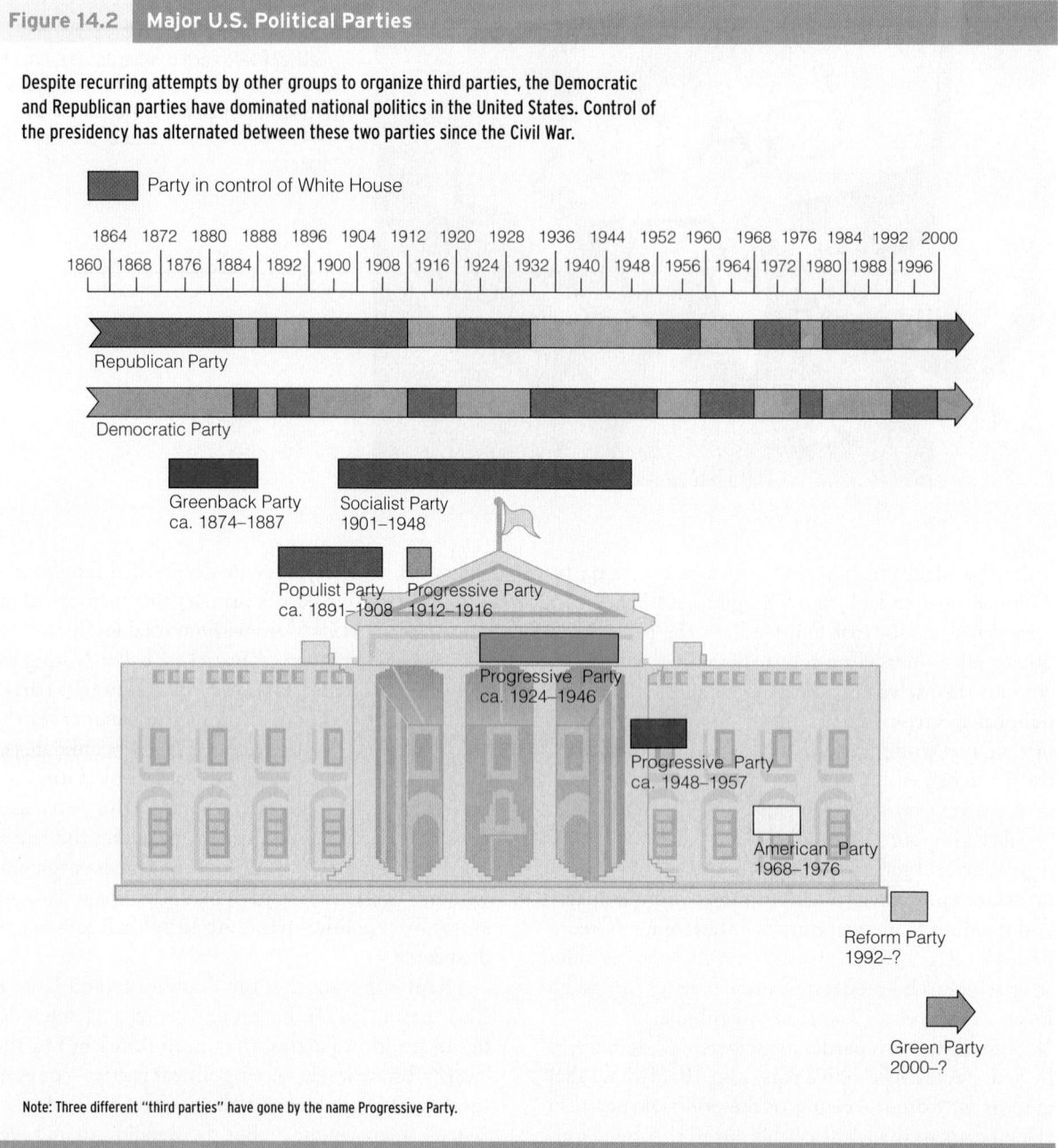

Figure 14.2   Major U.S. Political Parties

Despite recurring attempts by other groups to organize third parties, the Democratic and Republican parties have dominated national politics in the United States. Control of the presidency has alternated between these two parties since the Civil War.

Party in control of White House

1860 1864 1868 1872 1876 1880 1884 1888 1892 1896 1900 1904 1908 1912 1916 1920 1924 1928 1932 1936 1940 1944 1948 1952 1956 1960 1964 1968 1972 1976 1980 1984 1988 1992 1996 2000

Republican Party

Democratic Party

Greenback Party
ca. 1874–1887

Socialist Party
1901–1948

Populist Party
ca. 1891–1908

Progressive Party
1912–1916

Progressive Party
ca. 1924–1946

Progressive Party
ca. 1948–1957

American Party
1968–1976

Reform Party
1992–?

Green Party
2000–?

Note: Three different "third parties" have gone by the name Progressive Party.

punished, and that the government has an obligation to protect the rights of subordinate groups. Conservatives tend to believe in limiting the expansion of individual rights on social issues and tend to oppose social programs that they see as promoting individuals on the basis of minority status rather than merit. Based on these distinctions, Democrats are more likely to seek passage of social programs that make the government a more active participant in society, promoting social welfare and equality. By contrast, Republicans are more likely to believe that government should not be responsible for financial equality.

Economic Issues Economic issues fall into two broad categories: (1) the amount that should be spent on government programs and (2) the extent to which these programs should encourage a redistribution of income and assets. Liberals believe that without governmental intervention, income and assets would become concentrated in the hands of even fewer people and that the government must act to redistribute wealth, thus ensuring that everyone gets a "fair slice" of the economic "pie." In order to accomplish this, liberals envision that larger sums of money must be raised and spent by the government on such pro-

When people move from one nation to another, they learn new political attitudes, values, and behavior. Immigrants to the United States are required to undergo a process of political socialization before being "naturalized." Ironically, newcomers who have experienced poverty and repression in other lands are often stalwart in their devotion to democratic ideals.

grams. Conservatives contend that such programs are not only unnecessary but also counterproductive. That is, programs financed by tax increases lower people's incentive to work and to be innovative, and make people dependent upon the government. However, conservatives do believe in taxes that help maintain the status quo—funds for education, for the criminal justice system to maintain law and order, and for a strong military establishment.

Political Participation and Voter Apathy

Why do some people vote whereas others do not? How do people come to think of themselves as being conservative, moderate, or liberal? Key factors include individuals' political socialization, attitudes, and participation.

Political Socialization *Political socialization* **is the process by which people learn political attitudes, values, and behavior.** For young children, the family is the primary agent of political socialization, and children tend to learn and hold many of the same opinions as their parents. By the time children reach school age, they typically identify with the political party (if any) of their parents (Burnham, 1983). As children grow older, other agents of socialization begin to affect their political beliefs, including peers, teachers, and the media. Over time, these other agents may cause people's political attitudes and values to change, and individuals may cease to identify with the political party of their parents. Even for adults, political socialization continues through the media, friends, neighbors, and colleagues in the workplace.

Political Attitudes In addition to the socialization process, people's socioeconomic status affects their political attitudes, values, and beliefs. For example, individuals who are very poor or are unable to find employment tend to believe that society has failed them and therefore tend to be indifferent toward the political system (Zipp, 1985; Pinderhughes, 1986). Believing that casting a ballot would make no difference to their own circumstances, they do not vote.

People in the upper classes tend to be more conservative on economic issues and more liberal on social issues. Based on a philosophy of *noblesse oblige,* which asserts that those who are well-off have a responsibility for the welfare of the poor and disadvantaged, upper-class conservatives generally favor equality of opportunity but do not want their own income and assets taxed heavily to abolish poverty or other societal problems that they believe some people bring upon themselves. By contrast, people in the lower classes tend to be conservative on social issues, such as school prayer or abortion rights, but liberal on economic issues, such as increasing the minimum wage.

Political Participation Democracy in the United States has been defined as a government "of the people, by the people, and for the people." Accordingly, it would stand to reason that "the people" would actively participate in their government at any or all of four levels: (1) voting, (2) attending and taking part in political meetings, (3) actively participating in political campaigns, and (4) running for and/or holding political office. At most, about 10 percent of the voting-age population in this country participates at a level higher than simply voting, and over the past forty years, less than half of the voting-age population has voted in nonpresidential elections. Even in

Box 14.3 SOCIOLOGY AND SOCIAL POLICY

The Electoral College: Is School Over for This Body?

Sociologists who study political power are interested in the social and political conflicts that arise over the allocation of power in a society. Such a conflict occurred in the 2000 election for president of the United States, resulting in some social policy advocates calling for a change in how this important office is filled.

On Tuesday, November 7, 2000, more than 100 million U.S. citizens cast their votes in the presidential election. As is customary, the news media watched and reported: First, they reported that Texas governor George W. Bush had won the election, then that U.S. vice president Albert Gore had overcome Bush's lead in various "key states," so the outcome was not certain.

In electing the U.S. president, why do "key states" matter? Why not just be bound by the total vote? If the result of the election were determined by who received the most votes, Gore—who received about a half million more votes than did Bush—would have been elected president, yet Bush was sworn in as the next president. Why? The answer is that the framers of the Constitution wanted the president of the United States to be elected by the *states*, not by a plurality of the nation's voters. Accordingly, the U.S. Constitution provides that the president is elected by vote of "electors" selected by the various states and that the number of electors allocated to each state is equal to that state's number of U.S. senators and representatives. In 2000, Bush won more states with more electors than did Gore, although even that was not certain until December 12, 2000, when the U.S. Supreme Court ruled (in a seven-to-two decision in the case of *George W. Bush v. Albert Gore*) that a Florida Supreme Court order requiring a recount of certain inaccurately marked ballots to determine each voter's intent had "constitutional problems" and therefore had to be set aside. As a result, Bush received Florida's electoral votes. It was the first time since 1888 that the person receiving the highest number of votes nationwide was not the winner.

Numerous political analysts and social policy advocates began to call for a constitutional amendment that would eliminate the Electoral College and elect as president the candidate receiving the most votes. Obviously, many of these were people who had supported Gore, and their reaction was viewed by others as simply that of a poor loser. However, before either agreeing or disagreeing with them, it is important to look at the social policy reasons for having—or doing away with—the current system of electing the nation's president.

The delegates to the Constitutional Convention in 1787 realized that they needed to create a national government that was strong enough to provide national security, promote domestic tranquility, and regulate interstate commerce—factors that remain vitally important to us today. But the founders disagreed about how strong that central government should be. Delegates from the wealthiest and most populous states—Massachusetts, Pennsylvania, and Virginia—wanted a strong government and a legislature whose seats would be apportioned based on population. Delegates from the smaller states believed that such a government would be controlled by those populous states, so they wanted a legislature in which each state would have one seat. The Constitution represented a compromise between those two points of view: Each state would have equal representation in one house of the legislature, whereas seats in the other house would be apportioned on the basis of population (Greenberg and Page, 2002). With presidential electors apportioned on the combined representation of each state in the legislature, neither the populous states nor the larger number of states with smaller populations would be able to elect a president without support from some of the others.

As a result of this provision, a presidential candidate has to have broad appeal to voters in many states in order to be elected. Critics of the current method of electing presidents argue that, with the rapid flow of information and people across state lines, state boundaries and interests are no longer relevant in electing the nation's chief executive. Supporters of the electoral system counter with the argument that state boundaries and interests remain important and that the current system is necessary in order to stop the election of a candidate who represents only the interests of some particular region of the nation.

What are the sociological implications of social policy questions such as this? How are the lives of everyday people affected by political processes such as the presidential election? What do you think?

---

presidential elections, voter turnout often is relatively low. Slightly over 51 percent of the voting-age population (age eighteen and older) voted in the 2000 presidential election—an election so close that its outcome wasn't certain until more than a month after election day (see Box 14.3).

The United States has one of the lowest percentages of voter turnout of all Western nations. In many of the other Western nations, the average turnout is between 80 and 90 percent of all eligible voters. Why is it that so many eligible voters in this country stay away from the polls? During any election, millions of

Table 14.1   VOTER PREFERENCES IN THE 2000 PRESIDENTIAL ELECTION, BY SELECTED CHARACTERISTICS

		REPUBLICAN	DEMOCRAT
Gender	Men	52%	43%
	Women	43	54
Race/Ethnicity	Whites	53	42
	African Americans	8	90
	Latinos/as	32	64
	Asian Americans	38	57
Age	18-29	45	48
	30-44	48	48
	45-59	49	47
	60 and Older	51	23
Sexual Orientation	Gay, lesbian, bisexual	26	67
Education	Did not graduate from high school	39	59
	High school graduate	49	48
	Some college	50	45
	College graduate	49	46
Region	Eastern U.S.	39	56
	Midwest	49	47
	Southern U.S.	52	45
	Western U.S.	45	48
Family Income	Under $15,000	37	57
	$15,000-$29,999	41	53
	$30,000-$49,999	46	49
	Over $50,000	51	46

Note: Due to the presence of other candidates on the ballot, columns do not total 100 percent.

Source: Lester, 2000.

voting-age persons do not go to the polls due to illness, disability, lack of transportation, nonregistration, or absenteeism. However, these explanations do not account for why many other people do not vote. According to some conservative analysts, people may not vote because they are satisfied with the status quo or because they are apathetic and uninformed—they lack an understanding of both public issues and the basic processes of government.

By contrast, liberals argue that people stay away from the polls because they feel alienated from politics at all levels of government—federal, state, and local—due to political corruption and influence peddling by special interests and large corporations. Participation in politics is influenced by gender, age, race/ethnicity, and, especially, socioeconomic status (SES). The rate of participation increases as a person's SES increases. (Table 14.1 shows voting preferences by gender, race/ethnicity, age, marital status, sexual orientation, education, region, and income in the 2000 presidential election.) One explanation for the higher rates of political participation at higher SES levels is that advanced levels of education may give

people a better understanding of government processes, a belief that they have more at stake in the political process, and greater economic resources to contribute to the process. Some studies suggest that during their college years, many people develop assumptions about political participation that continue throughout their lives.

GOVERNMENTAL BUREAUCRACY

When most people think of political power, they overlook one of its major sources—the governmental bureaucracy. As previously discussed, Weber's rational–legal authority finds its contemporary embodiment in bureaucratic organizations. Negative feelings about bureaucracy are perhaps strongest when people are describing the "faceless bureaucrats" and "red tape" with which they must deal in government. But who are these "faceless bureaucrats," and what do they do?

Massive government buildings filled with "faceless bureaucrats" is a negative image that many people have of the U.S. government.

© James Pickerell/The Image Works

Characteristics of the Federal Bureaucracy

Bureaucratic power tends to take on a life of its own. During the nineteenth century, the government had a relatively limited role in everyday life. In the 1930s, however, the scope of government was extended greatly during the Great Depression to deal with labor–management relations, public welfare, and the regulation of the securities markets. With dramatic increases in technology and increasing demands from the public that the government "do something" about problems facing society, the government has grown still more in recent decades. Today, even with slight reductions in size, the federal bureaucracy employs more than 2 million people in civilian positions.

Much of the actual functioning of the government is carried on by its bureaucracy. As strange as it may seem, even the president, the White House staff, and cabinet officials have difficulty establishing control over the bureaucracy (Dye and Zeigler, 2003). Many employees in the federal bureaucracy have seen a number of presidents "come and go." For example, when President Clinton promised to make his administration "look like America," analysts watched to see who would be appointed to his cabinet but did not watch for changes in the *permanent government* in Washington, made up of top-tier civil service bureaucrats who have built a major power base. As shown in Figure 14.3, roughly 75 percent of the top-echelon positions in this elite continue to be held by white men.

Because rising to the top of the bureaucracy may take as much as twenty years, some analysts argue that white women and all people of color simply have not held positions in the federal government long enough

to reach the top positions. Others argue that the entrenched rules of this bureaucracy, one of the most powerful in the world, weigh against the promotion of those individuals who are not white and male.

The governmental bureaucracy has been able to perpetuate itself and expand because many of its employees have highly specialized knowledge and skills and cannot be replaced easily by "outsiders." In addition, as the United States has grown in size and complexity, public policy is increasingly made by bureaucrats rather than by elected officials. For example, offices and agencies have been established to create rules, policies, and procedures for dealing with complex issues such as nuclear power, environmental protection, and drug safety; bureaucracies announce an estimated twenty rules or regulations for every one law passed by Congress (Dye and Zeigler, 2003). Typically, these bureaucracies receive little, if any, direction from Congress or the president. The actions of these agencies are subject to challenge in the courts, but most agencies still operate in a highly autonomous manner.

The executive branch is also highly bureaucratized, as shown in Figure 14.4. Cabinet-level secretaries, who are appointed by the president and approved by the Senate, head fifteen departments that carry out governmental functions. Like all other areas of governmental bureaucracy, these departments have grown in number and size over the years. Adding to the layers of bureaucracy are the bureaus and agencies that are subdivisions within cabinet departments, as well as government corporations—agencies organized like private companies and operating in a market setting. For example, the U.S. Postal Service competes with United Parcel Service, FedEx, Airborne Express, and other delivery services (Greenberg and Page, 2002).

Figure 14.3   **Percentage of All Federal Civilian Jobs Held by White Men, 2000**

White men hold the vast majority of civilian jobs in the highest-paid levels of the federal bureaucracy. Note, for example, that 75.6 percent of all senior-level positions with annual salaries of at least $125,400 are held by white men.

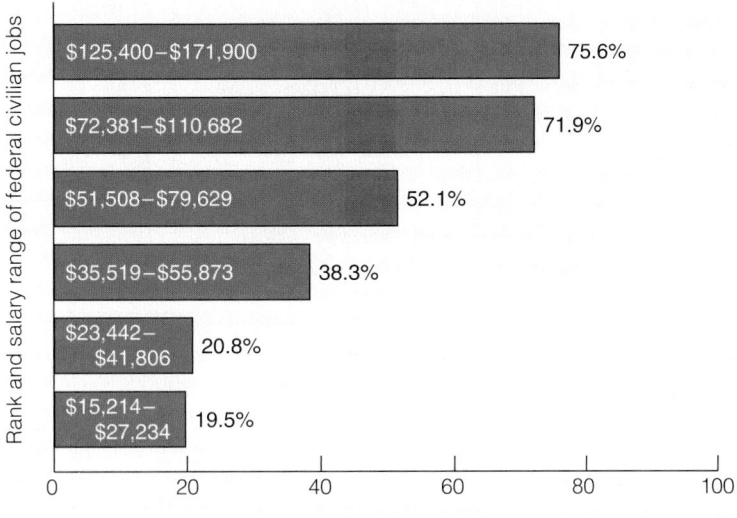

Source: Computed by author based on U.S. Office of Personnel Management, 2001; U.S. Office of Personnel Management, 2003.

The federal budget is the central ingredient in the bureaucracy. Preparing the annual federal budget is a major undertaking for the president and the Office of Management and Budget, one of the most important agencies in Washington. Getting the budget approved by Congress is an even more monumental task; however, as Dye and Zeigler (2003) point out, even with the highly publicized wrangling over the budget by the president and Congress, the final congressional appropriations are usually within 2–3 percent of the budget originally proposed by the president.

What role do special interest groups play in influencing the federal bureaucracy? Interest groups have an effect on the budgets received by various agencies and departments. Even though the president has budgetary authority over the bureaucracy, any agency that feels it did not get its "fair share" can raise a public outcry by contacting friendly interest groups and congressional subcommittees. This outcry may force the president to restore funding to the agency or prod Congress to appropriate money not requested by the president, who then may cooperate to avoid a confrontation.

Another way in which special interest groups exert a powerful influence on the bureaucracy is the *iron triangle of power*—a three-way arrangement in which a private interest group (usually a corporation), a con-

gressional committee or subcommittee, and a bureaucratic agency make the final decision on a political issue that is to be decided by that agency. Figure 14.5 illustrates the alliance among the Defense Department (Pentagon), private military (or defense) contractors, and members of Congress. We will now examine this relationship more closely.

The Iron Triangle and the Military–Industrial Complex

What exactly is the iron triangle, and how does it work? According to the sociologist Joe Feagin,

> The Iron Triangle has a revolving door of money, influence, and jobs among these three sets of actors, involving trillions of dollars. Military contractors who receive contracts from the Defense Department serve on the advisory committees that recommend what weapons they believe are needed. Many people move around the triangle from job to job, serving in the military, then in the Defense Department, then in military industries. (Feagin and Feagin, 1994: 405)

The long-term relationships found in this arrangement are part of what is referred to as the ***military–industrial complex***—the mutual interdependence

Figure 14.4 Organization of the U.S. Government

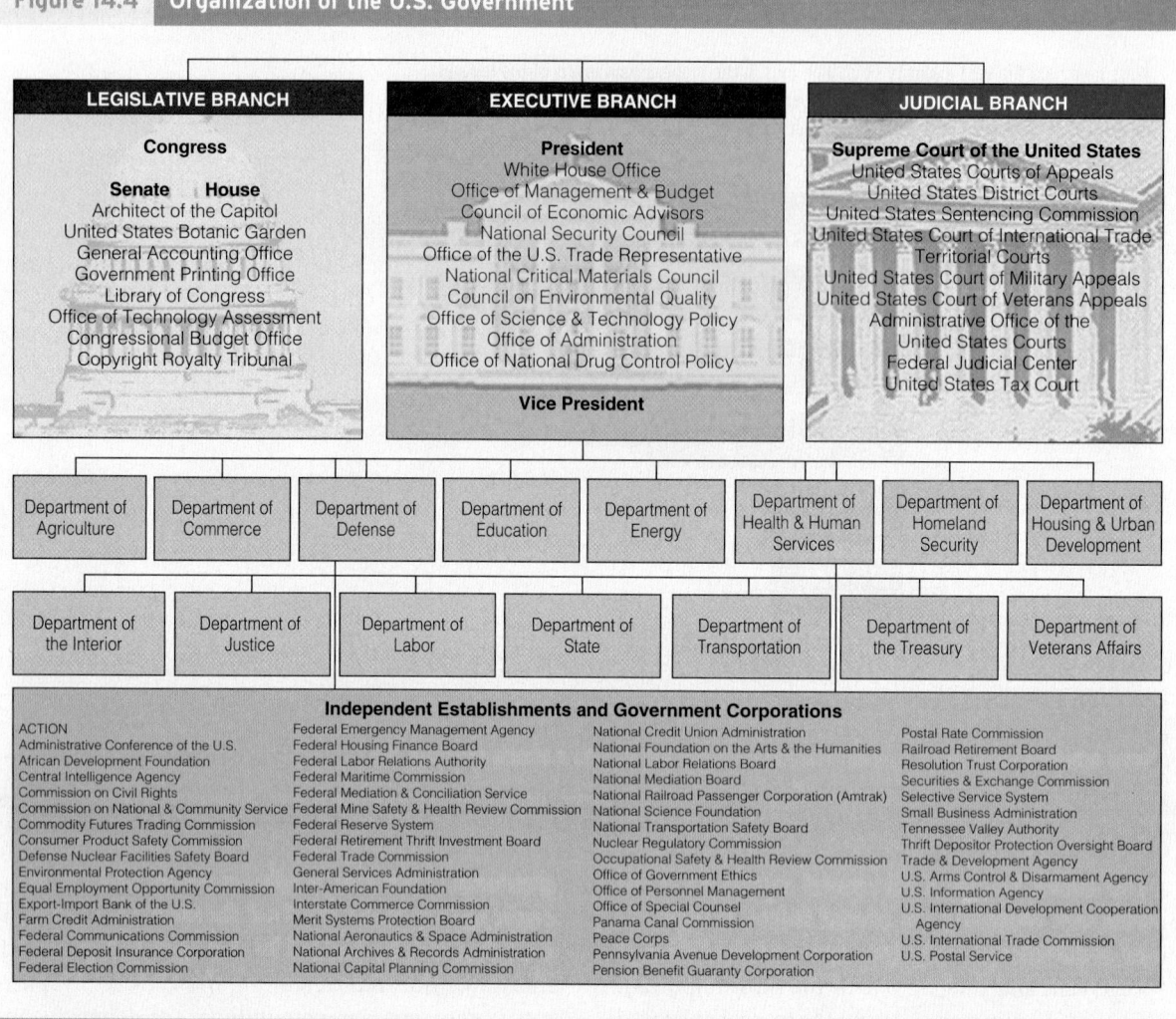

of the military establishment and private military contractors. The term was used by President Dwight D. Eisenhower, a retired five-star general, in his presidential farewell address in 1961, when he warned against the potential power of a huge military establishment working with a large arms industry:

> The conjecture of an immense military establishment and a huge arms industry is new in the American experience. The total influence—economic, political, and even spiritual—is felt in every city, every state house, and every office of the federal government. . . . In the councils of government, we must guard against the acquisition of unwarranted influence, whether sought or unsought, by the military–industrial complex. (Eisenhower, 1961, qtd. in Hartung, 1999)

Sociologists such as C. Wright Mills (1976) have stated that an alliance of economic, military, and political power could result in a "permanent war economy" or "military economy." However, the economist John Kenneth Galbraith (1985) argued that war and the threat of war can benefit the economy. If the nation is seen as having dangerous enemies, government money will be spent on weapons; in turn, these expenditures will stimulate the private sector of the economy, create jobs, and encourage spending. For example, at the beginning of the war on terrorism, the Pentagon awarded what was expected to become the largest military contract in U.S. history to Lockheed Martin to build a new generation of supersonic stealth fighter jets. The contract was expected to be worth more than $200 billion, creating thousands of new jobs in an economy that was struggling at the time of the announcement (Dao, 2000).

Background Until World War I, most U.S. military actions took place within this country (such as

Figure 14.5 Example of the Iron Triangle of Power

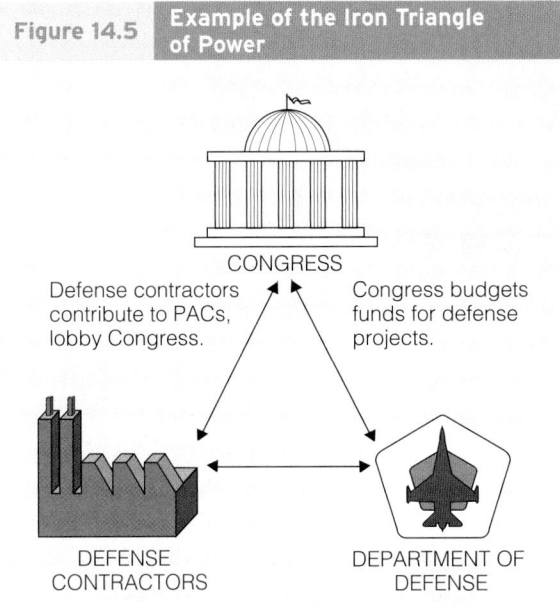

CONGRESS

Defense contractors contribute to PACs, lobby Congress.

Congress budgets funds for defense projects.

DEFENSE CONTRACTORS

DEPARTMENT OF DEFENSE

Defense Department awards contracts to contractors; personnel move back and forth between employment by Defense Department and by defense contractors.

the conquest of Native Americans and the Civil War) or close to home (as in the war with Mexico). After proving its military might at the international level in World War I, the United States took on the role of leader of the free world at the end of World War II.

During World War II, a massive expansion of the industrial infrastructure of the economy occurred as factories were built to produce uniforms, tanks, airplanes, ships, and so on. After the war, a major issue was what to do with the large military establishment that had helped the nation become a superpower. If the country wanted to remain the world leader, many argued, then it had to maintain this military establishment and make defense spending a national priority. The nation did just that. Starting in the 1950s, the United States and the Soviet Union engaged in a *cold war*—a conflict between nations based on the threat of war rather than on actual armed conflict. Because the Soviet Union wanted to be a superpower, as well as maintain control over the nations on its periphery in Eastern and Southern Europe, it entered into an arms race with the United States (Kennedy, 1987).

Defense Contracts and Arms Sales from 1945 to 2000

Between 1945 (the end of World War II) and the early 1990s, the U.S. government spent an estimated $10.2 trillion on national defense (in constant 1987 dollars). Later in the 1990s and continuing until immediately prior to the war on terror-

ism, the United States spent more than $270 billion per year on the military budget, a figure that is dwarfed by the costs of waging a war. The military budgets for 2003 and 2004 were almost $400 billion per year. Most defense spending on weapons was directed toward the "Big Three" weapons manufacturers—Lockheed Martin, Boeing, and Raytheon—which were created in a rash of military-industry mergers (Hartung, 1999). For example, Lockheed Martin was created by the merger of Lockheed with Martin Marietta, Loral Defense, the General Dynamics combat aircraft division, and many other military companies. The result is a $35-billion megacorporation that receives $18 billion annually in Pentagon contracts (Hartung, 1999).

Some of the largest defense contractors have a virtual monopoly over defense contracts because they function in a market with only one buyer (the U.S. government) and few (if any) competitors who can meet Pentagon specifications for a particular item. This monopoly is especially strong when *grease men* (independent consultants who act as a liaison between the Pentagon and defense contractors) provide "insider" information to the corporations. Thus, defense contractors have nothing to lose; they do not have to compete on costs, and their profits are guaranteed. Frequently, these corporations do not have to prove the efficacy of their products because the Pentagon is traditionally anxious to publicize, and occasionally deploy, new weapons systems before rivals can do so.

Some members of Congress have been willing to support measures that expand the military–industrial complex because doing so provides a unique opportunity for them to assist their local constituencies by authorizing funding for defense-related industries, military bases, and space centers in their home districts. These activities constitute a long-standing political practice known as *pork* or *pork barrel*—projects designed to bring jobs and public monies to the home state of members of Congress, for which they can then take credit (Greenberg and Page, 2002). Another reason for congressional support of the defense industry is that the politically powerful military industry employs millions and spends extravagantly on contributions to candidates and political parties.

Moreover, terrorist attacks on the United States have resulted in large expenditures on national security and on domestic security. As a result of the sudden, unexpected attack on the United States on September 11, 2001, many political leaders believed that the ability of the government to quickly respond with troops and sophisticated weapons systems justified the expenditures that had been made on the military–industrial complex over the years.

U.S. government expenditures for weapons and fighter jets such as the one shown here have contributed to the expansion of the military–industrial complex. What are the costs and benefits of advanced military technologies?

© Reuters NewMedia Inc./Corbis

THE MILITARY AND MILITARISM

Militarism **is a societal focus on military ideals and an aggressive preparedness for war.** Core U.S. values such as patriotism, courage, reverence, loyalty, obedience, and faith in authority help to support militarism, as the sociologist Cynthia H. Enloe (1987: 542–543) notes:

> Military expenditures, militaristic values, and military authority now influence the flow of foreign trade and determine which countries will or will not receive agricultural assistance. They shape the design and marketing of children's toys and games and of adult fashions and entertainment. Military definitions of progress and security dominate the economic fate of entire geographic regions. The military's ways of doing business open or shut access to information and technology for entire social groups. Finally, military mythologies of valor and safety influence the sense of self-esteem and well-being of millions of people.

However, note that *militarism* is not synonymous with *war;* it involves an ideology that perpetuates societal activities directed toward military preparedness and enormous spending on weapons production and "defense" as compared with other national priorities (Cock, 1994).

Explanations for Militarism

Sociologists have proposed several reasons for militarism. One is the economy. According to Enloe (1987: 527), people who see capitalism as the moving force behind the military's influence "believe that government officials enhance the status, resources, and authority of the military in order to protect the interests of private enterprises at home and overseas." From this perspective, the origin of militarism is found in the boardroom, not the war room. Because government and military funding is also a prime source of support for research in the physical, biological, and social sciences, administrators and faculty in some institutions of higher learning feel the need to support war-related spending to obtain grants for research. In addition, workers come to rely on military spending for jobs; labor unions have supported defense spending because it has provided well-paid, stable employment for union members.

A second explanation focuses on the role of the nation and its inclination toward coercion in response to perceived threats. From this perspective, nations will inevitably use force to ensure compliance within their society and to protect themselves from outside attacks. A third explanation is based primarily on patriarchy and the relationship between militarism and masculinity. Across cultures and over time, the military has been a male institution, and the "meanings attached to masculinity appear to be so firmly linked to compliance with military roles that it is often impossible to disentangle the two" (Enloe, 1987: 531). Symbolic interactionists might view this perspective as the *social construction of masculinity.* In other words, certain assumptions, teachings, and expectations that serve as the standard for appropriate male behavior—in this case, values of dominance, power, aggression, and violence—are created and re-created. Presumably, such qualities may be learned through gender socialization, including that which is received in military training. Historically, the development of manhood and male superiority has been linked to militarism and combat—the ultimate test of a man's masculinity (Enloe, 1987; Cock, 1994).

Gender, Race, and the Military

In recent years, some significant changes have occurred in the policies and composition of the U.S. armed forces. Especially with the introduction of the

Recent studies have shown that a weak U.S. economy and stepped-up spending on recruitment efforts have resulted in more recruits for the U.S. military. Although the number of women (of all races) and African American recruits has remained relatively unchanged, the share of Latinos entering the military has increased. What are the sociological implications of trends such as this?

all-volunteer force and the end of the draft in the early 1970s, the focus of the military shifted from training "good citizen-soldiers" to recruiting individuals who would enlist in the military in the same way that another person might take a job in the private sector. Now, in return for joining the military, a person is rewarded with a reasonable salary, medical and dental care, the chance to travel, educational opportunities, and other benefits (Stiehm, 1989).

Since the introduction of the volunteer forces, considerable pressure has been placed on the military to recruit women. In 2002, for example, women made up approximately 15 percent of the total uniformed force. As a percentage of total personnel in each branch of the military, women ranged from a high of about 19 percent of Air Force personnel to a low of less than 6 percent of Marine Corps personnel (Women's Research & Education Institute, 2002).

In spite of changes in the gender composition of the military, it remains essentially a male-dominated world. Only since 1991 have women gained the right to be trained for combat duty, and until recently they have been largely restricted from front-line positions. As Enloe (1987) notes, most of women's roles in the military do not disturb the existing world of male militarism.

In examining militarism and the volunteer military, it is important to note how and why militarism uses and affects women differently based on their race/ethnicity. In 2001, for example, African American women made up 47 percent of all enlisted (nonofficer) women in the U.S. Army (Women's Research & Education Institute, 2002), a percentage *four times* their proportion in the population. Thus, it can be concluded that "the U.S. military's appeal to

or claim on women is far from race neutral" (Enloe, 1987: 533). Enloe suggests that militarism may adversely affect women who are in the most precarious position in the economic system; they become willing to risk their lives in war so that they can acquire shelter, health benefits for their children, and a steady income from the military. In other words, people of color and poor people from all racial–ethnic groups may enlist in the military not because of their militaristic tendencies but because of the more limited options available to them in society at large.

■ TERRORISM AND WAR

In times of peace in the United States, people typically view the military as another occupational choice. However, in times of crisis—such as a terrorist attack and the resulting war against those deemed to be the enemy—public opinion on war and the military may shift dramatically. We turn now to an examination of political issues related to terrorism and war.

Types of Terrorism

Terrorism **is the use of calculated, unlawful physical force or threats of violence against a government, organization, or individual to gain some political, religious, economic, or social objective.** Terrorist tactics include bombing, kidnapping, hostage taking, hijacking, assassination, and extortion (Vetter and Perlstein, 1991). Although terrorists sometimes attack government officials and members of the military, they more often target civilians as a way of pressuring the government. Until 2001, most international acts of terrorism that targeted U.S. interests took place outside of this country. The events of September 2001 were a frightening and shocking experience for people in the United States, who realized that the terrorists—whose objectives were not clearly known and whose linkage to any one government or political regime was vague—had wreaked immense destruction within our nation's borders, leaving thousands dead and the entire nation fearful for its long-term safety.

Around the world, acts of terrorism extract a massive toll by producing rampant fear, widespread loss of human life, and extensive destruction of property. One form of terrorism—political terrorism—is actually considered a form of unconventional warfare. *Political terrorism* uses intimidation, coercion, threats of harm, and other violence that attempts to bring about a significant change in or overthrow an existing government. There are three types of political terrorism: revolutionary terrorism, state-sponsored terrorism, and repressive terrorism.

Revolutionary terrorism refers to acts of violence against civilians that are carried out by enemies of the government who want to bring about political change. Some groups believe that if they perpetrate enough random acts of terrorism, they will achieve a political goal. Members of radical religious or revolutionary movements may engage in terrorist activities such as bombings, kidnappings, and assassinations of leading officials. The goal of the terrorists is to traumatize a civilian population so that people will put their government under such pressure that political officers and departments of the government can no longer work effectively and can be brought down (Jacquard, 2001). Although exact linkages may be difficult to pinpoint, some revolutionary terrorists may receive assistance from governments that support their objectives (Vetter and Perlstein, 1991).

Unlike revolutionary terrorism, *state-sponsored terrorism* occurs when a government provides financial resources, weapons, and training for terrorists who conduct their activities in other nations. In Libya, for example, Colonel Muammar Qaddafi has provided money and training for terrorist groups such as the Arab National Youth Organization, which was responsible for skyjacking a Lufthansa airplane over Turkey and forcing the German government to free the surviving members of Black September. Black September was the terrorist group responsible for killing Israeli Olympic athletes in the 1970s (Parry, 1976).

The third type of political terrorism, *repressive terrorism,* is conducted by a government against its own citizens for the purpose of protecting an existing political order. Repressive terrorism has taken place in many countries around the world, including Haiti, the People's Republic of China, and Cambodia, where the Pol Pot regime killed more than a million people in the five years between 1975 and 1979 (Mydans, 1997a).

Terrorism in the United States

The years 1995 and 2001 stand out in recent U.S. history as the periods when this nation suffered from the two worst terrorist attacks that had ever occurred in the continental United States. The April 1995 bombing of the federal building in Oklahoma City took the lives of 168 adults and children, and injured 850 others. Many other people were left with psychological scars. The Oklahoma City bombing was described as an act of domestic terrorism because it was attributed to two men who had no apparent connection to external enemies of the United States.

By contrast, external enemies (who had planted perpetrators in the United States) were believed to be responsible for the attacks on New York's World Trade Center, first in 1993 in the form of a bombing that killed six people and injured more than a thousand, and then in the 2001 destruction of the Trade Center's twin towers (and damage to the Pentagon, in Washington, D.C., that same morning) by suicide hijackers using passenger airplanes as lethal weapons, resulting in the deaths of almost three thousand people.

Prior to the events of September 11, 2001, terrorism in the form of biological and chemical attacks was considered by most analysts to be a remote possibility in the United States. However, with the deaths of postal workers and private citizens from the delivery of anthrax, suspicions deepened that, for the first time, terrorists—whether home-grown or international—might use germ warfare to achieve their subversive political goals. In the twenty-first century, people in the United States are confronted with threats of terrorism and problems associated with war that have been facts of life in some other countries for many years.

War

Generally speaking, **war is organized, armed conflict between nations or distinct political factions.** Social scientists define war to include both *declared wars* between nations or parties and *undeclared wars:* civil and guerrilla wars, covert operations, and some forms of terrorism. War is an institution that involves *violence*—behavior intended to bring pain, physical injury, and/or psychological stress to people or to harm or destroy property (Sullivan, 2000). As such, war is a form of *collective violence* by people who are seeking to promote their cause or resist social policies or practices that they consider oppressive (Sullivan, 2000).

As previously discussed, early U.S. military action took place at home or close to home, but the location and the nature of U.S. military action changed dramatically in the twentieth century. Wars were now fought on foreign soil, and larger numbers of U.S. military personnel were involved. In World War I, for example, about 5 million people served in the U.S. armed forces; more than 16 million men and women served in World War II (Ehrenreich, 1997). It remains to be seen what types of military response will be required in the war against terrorism in the twenty-first century.

The direct effects of war are loss of human life and the serious physical and psychological effects on some survivors. Although it is impossible to determine how many human lives have been lost in wars throughout human history, social analyst Ruth Sivard (1991, 1993) estimated that 589 wars have been fought by 142 countries since 1500 and that approximately 142 million lives have been lost. However, according to Sivard, more lives were lost in wars during the twentieth century than in all of previous history. World War I took the lives of approximately 8 million combatants and 1 million civilians. In World War II, more than 50 million people (17 million combatants and 35 million civilians) lost their lives. During World War II, U.S. casualties alone totaled almost 300,000, and more than 600,000 were wounded. The number of casualties in conventional warfare pales when compared to the potential loss of human life if all-out nuclear, biological, or chemical warfare were to occur. The devastation would be beyond description.

Recent wars in Afghanistan and Iraq, although not involving nuclear, biological, or chemical warfare, have still taken an extreme toll on both civilians and military personnel. For example, some estimates place the number of civilian casualties in Afghanistan at between 3,000 and 5,000 during the primary time of fighting (between October 7, 2001, and April 2002) against the Taliban regime there. Some analysts believe that many of these civilian deaths can be attributed to the U.S. military firing missiles into and dropping bombs on heavily populated areas of Afghanistan (Herold, 2002). Others have disputed the accuracy of these numbers, citing problems with how the data were collected and analyzed (see Isaac, 2002).

In the war in Iraq, estimates of the number of civilian casualties begin at about 800 dead and reach upward into the thousands, depending on how counts are determined (Herold, 2003). These estimates also do not include the many people who were injured or the military personnel who lost their lives as a result of these wars. News sources estimate that 500 U.S. military personnel had lost their lives in the fighting in Iraq as of January 18, 2004.

Media watch groups have asked why the major news networks in the United States have made few efforts to accurately tally the number of deaths in recent wars. Some cite the fact that they believe the media are more likely to report the spin control of elected and appointed federal officials in the White House and the Pentagon rather than engage in investigative reporting on the true nature and extent of damage and deaths in these war-torn regions (see FAIR, 2003, for example).

While the conflicts in Afghanistan and Iraq continue (as of this writing in 2003), the United States is also entering Liberia in an attempt to stop a civil war within that nation. No doubt, civilian and military casualties will continue to increase in various regions around the world as we have moved from a brief era of peacetime in the United States into what appears to be a protracted era of wartime in various regions of the world.

POLITICS AND GOVERNMENT IN THE FUTURE

Thinking about U.S. politics and government in the future is very much like the old story about optimists and pessimists. According to the story, an eight-ounce cup containing exactly four ounces of water is placed on a table. The optimist comes in, sees the cup, and says, "The cup is half full." The pessimist comes in, sees the cup, and says, "The cup is half empty." Clearly, both the pessimist and the optimist are looking at the same cup containing the same amount of water, but their perspective on what they see is quite different. For some analysts, looking at the future of the U.S. government and its political structure is very much like this.

One group of analysts is very distressed about the future of our country. Kevin Phillips (2003: 422), a well-known author and political and economics commentator, is an example of this viewpoint:

Box 14.4 YOU CAN MAKE A DIFFERENCE

Keeping an Eye on the Media

Do we get all of the news that we should about how our government operates and about the pressing social problems of our nation? Consider this list (shown by number) of the twenty-five top news stories that some media analysts believe were not adequately covered by the U.S. media in recent years:

1. FCC Moves to Privatize Airwaves
2. New Trade Treaty Seeks to Privatize Global Social Services
3. United States' Policies in Columbia Support Mass Murder
24. Wal-Mart Takes Union Busting to the State Level
25. Federal Government Bails Out Failing Private Prisons (Project Censored, 2003)

According to the group that compiles this list, Project Censored, which is managed through the Department of Sociology at Sonoma State University, many important stories are either missing from the news altogether or do not receive the attention they deserve. Through the work of more than 200 people, Project Censored keeps an eye on the news media and publishes an annual list of significant stories that its researchers believe were neglected. (To view the entire list, visit Project Censored's Web site at **http://www.projectcensored.org**.) According to a statement by Project Censored, "Corporate media seem to have abdicated their First Amendment responsibility to keep the public informed. The traditional journalist values of supporting democracy by maintaining an educated electorate now take second place to profits and ratings" (Phillips and Project Censored, 2002).

What should be the role of the media in keeping us informed? The media are often referred to as the "Fourth Estate" or the "Fourth Branch of the Government" because they are supposed to provide people relevant information on important topics regarding how the government operates in a democratic society. This information can then be used by citizens to decide how they will vote on candidates and issues presented for their approval or disapproval on the election ballot. However, some analysts believe that much media coverage, especially the political talk program genre that has proliferated on network and cable TV channels, may actually *reduce* public understanding of issues rather than add to it (Fallows, 1997). Why? Because the media "hammer" home the message that "issues" do not matter except as items for politicians to squabble about (Fallows, 1997). In other words, people are socialized to view pressing social concerns primarily as things for the elected leaders to debate and the media pundits to argue loudly about, not as issues that affect their everyday lives and might have lasting consequences for this nation and many others.

The first step in keeping an eye on the news is to become more analytical about the "news" that we do receive. How can we evaluate the information we receive from the media? In *How to Watch TV News*, the media analysts Neil Postman and Steve Powers (1992: 160–168) suggest the following:

1. When watching a news show, we should keep in mind a firm idea of what is important.
2. We should also keep in mind that television news shows are called "shows" for a reason. They are not a public service or a public utility.
3. We should never underestimate the power of commercials, which tell us much about our society.
4. We should learn about the economic and political interests of those who run television stations or own a controlling interest in a media conglomerate.
5. We should pay attention to the *language* of newscasts, not just the visual imagery. For example, a *question* may reveal as much about the *questioner* as the person answering the question.

Becoming aware of the media's role in influencing people's opinions about how our government is run is the first step toward becoming an informed participant in the democratic political process.

The second step in keeping an eye on the news is becoming aware of national and international events that should receive more coverage than they do or that might not be reported in a fair and unbiased manner. To form your own opinion about media coverage of the news, you may wish to also visit these sites, which examine media practices:

- FAIR (Fairness and Accuracy in Reporting):

http://www.fair.org

- MediaChannel.org, which states, "As the media watch the world, we watch the media":

http://www.mediachannel.org

Of course, this is only one side of the media debate. By contrast, other analysts would argue that the media provide a wide array of information to keep people informed and that to know about every single thing that happens in government and in the world would be overwhelming for audiences and readers. What do you think? Do you believe that the media accurately and fairly report all of the news that we need to know?

As the twenty-first century gets underway, the imbalance of wealth and democracy in the United States is unsustainable, at least by traditional yardsticks. Market theology [based on the virtues of capitalism] and unelected leadership have been displacing politics and elections. Either democracy must be renewed, with politics brought back to life, or wealth is likely to cement a new and less democratic regime—plutocracy by some other name. Over the coming decades, American exceptionalism [the belief that a nation is exceptional or unique] may face its greatest test simply in convincing the American people to continue to believe in its comfort and reassurance.

As Philips states, one view of U.S. politics and government is that the future may be dim if major changes do not take place soon. According to this perspective, the staggering wealth and great power of a relatively small number of individuals and corporations have worked together to perpetuate privilege at the expense of the national interest and those who are in the middle and lower classes. Consequently, our nation has become less democratic, and the democratic process is continually subverted by the privileged few who are able to do as they so choose.

Other views of the future of politics and government focus more narrowly on specific questions of concern to the United States:

- Are political parties in a state of decline? Is it possible for one party to so dominate the political landscape that many other voices are never heard and their issues never considered?
- Are global corporate interests and the concerns of the wealthy overshadowing the needs and interests of everyday people?

- Is it possible to prevent future "9/11" type attacks on the United States (and other nations as well) through tightening organizational intelligence and reorganization of some governmental bureaucracies?
- Will we be able to balance the need for national security with the individual's right to privacy and freedom of movement within this country?
- How will elected politicians and appointed government officials handle the challenges regarding the changing demographics of the United States? For example, how will the aging population influence their decisions? Will increasing racial, ethnic, and religious diversity influence their decisions?
- Will immigration and employment policies be based on the best interests of the largest number of people, or will these policies be based on the best interests of a small elite who are major contributors to political campaigns?
- Do the media accurately report what is going on at all levels of government? To what extent can individuals and grassroots organizations influence the media and the political process? (See Box 14.4, "You Can Make a Difference.")

In the first decade of the twenty-first century, these are a few of the many questions regarding politics and the government that face us and people in other nations. How we (and our elected officials) answer these (and related) questions will in large measure determine the future of politics and government in the United States. Our answers will also have a profound influence on people and governments in other countries—whether they be high-, middle-, or low-income nations—around the world.

CHAPTER REVIEW

■ What is power?

Power is the ability of persons or groups to carry out their will even when opposed by others.

■ What is the relationship between power and politics?

Politics is the social institution through which power is acquired and exercised by some people or groups. A strong relationship between politics and power exists in

all countries, and the military is closely tied to the political system.

■ What are the three types of authority?

Max Weber identified these types of authority: traditional, charismatic, and rational–legal. Traditional authority is based on long-standing custom. Charismatic authority is power based on a leader's personal qualities. Rational–legal authority is based on law or written rules and regulations, as found in contemporary bureaucracies.

■ **What are the main types of political systems?**

The main types of political systems are monarchies, authoritarian systems, totalitarian systems, and democratic systems. In a monarchy, one person is the hereditary ruler of the nation. In authoritarian systems, rulers tolerate little or no public opposition and generally cannot be removed from office by legal means. In totalitarian systems, the state seeks to regulate all aspects of society and to monopolize all societal resources in order to exert complete control over both public and private life. In democratic systems, the powers of government are derived from the consent of all the people.

■ **How do pluralist and power elite perspectives view power in the United States?**

According to the pluralist (functionalist) model, power is widely dispersed throughout many competing interest groups. People influence policy by voting, joining special interest groups and political action campaigns, and forming new groups. According to the elite (conflict) model, power is concentrated in a small group of elites, whereas the masses are relatively powerless.

■ **Who makes up the power elite, and why are they important?**

According to C. Wright Mills, the power elite is composed of influential business leaders, key government leaders, and the military. The elites possess greater resources than the masses, and public policy reflects their preferences.

■ **What is the military–industrial complex?**

The military–industrial complex refers to a mutual interdependence of the government military establishment and private military contractors.

■ **What is terrorism?**

Terrorism is the use of calculated, unlawful physical force or threats of violence against a government, organization, or individual to gain some political, religious, economic, or social objective.

■ **What are the major types of political terrorism?**

There are three types of political terrorism: revolutionary terrorism, state-sponsored terrorism, and repressive terrorism. Revolutionary terrorism refers to acts of violence against civilians that are carried out by enemies of the government who want to bring about political change. State-sponsored terrorism occurs when a government provides financial resources, weapons, and training for terrorists who conduct their activities in other nations. Repressive terrorism is conducted by a government against its own citizens for the purpose of protecting an existing political order.

KEY TERMS

authoritarianism 462
authority 459
charismatic authority 459
democracy 464
elite model 467
government 456
militarism 480
military–industrial complex 477
monarchy 461
pluralist model 465
political action committee 467
political party 470
political socialization 473
political sociology 457
politics 456
power 457
power elite 468
rational–legal authority 460
routinization of charisma 460
special interest group 466
state 456
terrorism 481
totalitarianism 463
traditional authority 459
war 482

QUESTIONS FOR CRITICAL THINKING

1. Who is ultimately responsible for decisions and policies that are made in a democracy such as the United States—the people or their elected representatives?
2. How would you design a research project to study the relationship between campaign contributions to elected representatives and their subsequent voting records? What would be your hypothesis? What kinds of data would you need to gather? How would you gather accurate data?
3. How does your school (or workplace) reflect a pluralist or elite model of power and decision making?
4. Can democracy survive in a context of rising concentrated power in the capitalist oligarchies? Why or why not?

RESOURCES ON THE INTERNET

Chapter-Related Web Sites

The following Web sites have been selected for their relevance to the topics in this chapter. These sites are among the more stable, but please note that Web site addresses change frequently. For an updated list of chapter-related

Web sites with URL links, please visit the *Sociology in Our Times* Web site (**www.wadsworth.com/KendallSIOT**).

Roll Call
http://www.rollcall.com

Roll Call is an organization that tracks political news and provides commentary on current political events. In addition to reporting on current events, the site features useful links, information on policy briefings, editorial cartoons, and editorial opinions.

Federal Election Commission (FEC)
http://www.fec.gov

An independent regulatory agency, the FEC was created in 1975 to administer and enforce the Federal Election Campaign Act (FECA), which governs the financing of federal elections. The FEC's Web site includes current news releases, information on elections and campaign finance, and various guides to local, state, and federal election systems.

Open Secrets
http://www.opensecrets.org

This comprehensive site provides information on campaign financing and profiles such current topics as nuclear energy, electricity deregulation, and managed care. The site allows you to view the contribution trends in specific industries, access specific campaign finance information for candidates, and conduct searches for information pertinent to your locality.

ONLINE STUDY AND RESEARCH TOOLS

Accompanying this text are many *free* powerful online study tools that will help you master the material in this chapter, help increase your depth of understanding, and help you make the grade!

SocCoach CD-ROM

Use the SocCoach CD-ROM enclosed with this text to help you formulate a customized study plan for this chapter. After you take the Diagnostic Quiz, SocCoach will generate a customized study plan just for you! It will identify sections of the chapter that you should review and will provide videos, charts, graphs, and excerpts from the text to supplement your studies and enhance your understanding. You'll also find fun, interactive activities such as Virtual Explorations and Map the Stats to apply what you've learned and stretch your sociological imagination.

The Companion Web Site for Sociology in Our Times, *Fifth Edition*
www.wadsworth.com/KendallSIOT

Gain an even better grasp on this chapter by going to the companion Web site to take one of the Tutorial Quizzes, use the Flash Cards to master key terms, or check out the many other study aids you'll find there. You'll also find special features such as GSS Data and Census 2000 information that'll put data and resources at your fingertips to help you with that special project or help you as you do some research on your own.

In this chapter, when you see the icon on the left, it alerts you to a specific exercise found in *Wadsworth's Sociology Online Resources and Writing Companion*. This valuable guide shows you how to use Wadsworth's exclusive online resources—*InfoTrac College Edition,* the *Opposing Viewpoints Resource Center,* and *MicroCase Online*—to assist you in your study of sociology and to build essential research and writing skills.

Families and Intimate Relationships

My son began commuting between his two homes at age 4. He traveled with a Hello Kitty suitcase with a pretend lock and key until it embarrassed him. . . . He graduated to a canvas backpack filled with a revolving arsenal of essential stuff: books and journals, plastic vampire teeth, "Star Trek" Micro Machines, a Walkman, CD's, a teddy bear.

The commuter flights between San Francisco and Los Angeles were the only times a parent wasn't lording over him, so he was able to order Coca-Cola, verboten at home. . . . But such benefits were insignificant when contrasted with his preflight nightmares about plane crashes. . . .

Like so many divorcing couples, we divided the china and art and our young son. . . . First he was ferried back and forth between our homes across town, and then, when his mother moved to Los Angeles, across the state. For the eight years since, he has been one of the thousands of American children with two homes, two beds, two sets of clothes and toys and two toothbrushes.

—several years ago, David Sheff (1995: 64) explaining how divorce and joint custody had affected his son, Nick Sheff

Four years later, Nick Sheff (1999: 16) stated his own view on long-distance joint custody:

I am 16 now and I still travel back and forth, but it's mostly up to me to decide when. I've chosen to spend more time with my friends at the expense of visits with my mom. When I do go to L.A., it's like my stepdad put it: I have a cameo role in their lives. I say my lines and I'm off. It's painful.

What's the toll of this arrangement? I'm always missing somebody. When I'm in northern California, I miss my mom and stepdad. But when I'm in L.A., I miss hanging out with my friends, my other set of parents and little brothers and sisters. After all these back-and-forth flights, I've learned not to get too emotionally attached. I have to protect myself. . . . No child should be subjected to the hardship of long-distance joint custody. To prevent it, maybe there should be an addition to the marriage vows: Do you promise [that] if you ever have children and wind up divorced, [you will] stay within the same geographical area as your kids? . . . Or how about some common sense? If you move away from your children, *you* have to do the traveling to see them. (Sheff, 1999: 16)

■ Changing family patterns and relocations for employment opportunities have made sights such as this increasingly familiar, as children travel alone to various destinations to be reunited with a parent or other family member.

Nick's family is one of millions around the globe experiencing major change as a result of divorce, death, or some other significant event. Although in the past, children in the United States who did not grow up with both biological parents were often considered to be the "ex-

ceptions," we cannot continue to ignore such children, whose numbers continue to increase (Coontz, 1997).

Extensive sociological research has been conducted to examine the causes of divorce and to identify the possible effects it may have on adults and children. In one study, the sociologists Frank F. Furstenberg, Jr., and Andrew J. Cherlin (1991) found that although children describe divorce and parental separation as an unsettling experience, most children eventually make a successful adaptation if two factors are present: (1) the mother does reasonably well financially and psychologically after the divorce, and (2) the divorced parents have a low level of conflict between them. Another study by the sociologist Demie Kurz (1995) focused on how divorce affects women and children. According to Kurz (1995), most of the women in her study were happy to see their marriages end because they found their marriages to be very difficult and in some cases abusive. By con-

trast, divorce offered them freedom from domination, violence, and destructive emotional circumstances.

In this chapter, we examine the increasing complexity and diversity of contemporary families in the United States and other nations. Pressing social issues such as divorce, child care, and new reproductive technologies will be used as examples of how families and intimate relationships continue to change. Before reading on, test your knowledge about some contemporary trends in U.S. family life by taking the quiz in Box 15.1.

QUESTIONS AND ISSUES

Chapter Focus Question: Why are many family-related concerns—such as divorce and child care—viewed primarily as personal problems rather than as social concerns requiring macrolevel solutions?

Why is it difficult to define family?

How do marriage patterns vary around the world?

What are the key assumptions of functionalist, conflict/feminist, and symbolic interactionist perspectives on families?

What significant trends affect many U.S. families today?

FAMILIES IN GLOBAL PERSPECTIVE

As the nature of family life and work has changed in high-, middle-, and low-income nations, the issue of what constitutes a "family" has been widely debated. For many years, the standard sociological definition of *family* has been a group of people who are related to one another by bonds of blood, marriage, or adoption and who live together, form an economic unit, and bear and raise children (Benokraitis, 2002). Many people believe that this definition should not be expanded—that social approval should not be extended to other relationships simply because the persons in those relationships wish to consider themselves a family. However, others challenge this definition because it simply does not match the reality of family life in contemporary society (Lamanna and Riedmann, 2003). Today's families include many types of living arrangements and relationships, including single-parent households, unmarried couples, lesbian and gay couples, and multiple generations (such as grandparent, parent, and child) living in the same house-

hold. To accurately reflect these changes in family life, some sociologists believe that we need a more encompassing definition of what constitutes a family. Accordingly, we will define *families* **as relationships in which people live together with commitment, form an economic unit and care for any young, and consider their identity to be significantly attached to the group.** Sexual expression and parent–child relationships are a part of most, but not all, family relationships (based on Benokraitis, 2002; Lamanna and Riedmann, 2003).

Although families differ widely around the world, they also share certain common concerns in their everyday lives. For example, women and men of all racial–ethnic categories, nationalities, and income levels face problems associated with child care. Various nongovernmental agencies of the United Nations have established international priorities to help families. Suggestions have included the development of "social infrastructures for the care and education of children of working parents in order to reduce their . . . burden" and implementation of flexible working hours so that parents will have more opportunities to spend time with their children (Pietilä and Vickers, 1994: 115).

Box 15.1 SOCIOLOGY AND EVERYDAY LIFE

How Much Do You Know About Contemporary Trends in U.S. Family Life?

True	False		
T	F	1.	Today, people in the United States are more inclined to get married than at any other time in history.
T	F	2.	Most U.S. family households are composed of a married couple with one or more children under age 18.
T	F	3.	One in two U.S. preschoolers has a mother in the paid labor force.
T	F	4.	Recent studies have found that sexual activity is more satisfying to people who are in sustained relationships such as marriage.
T	F	5.	People under age 45 are more likely to cohabit than are people over age 45.
T	F	6.	With the strong U.S. economy at the end of the 1990s, the number of children living in extreme poverty went down significantly.
T	F	7.	Recent studies have found that adult children of divorced parents are more likely to dissolve their own marriages than they were two decades ago.
T	F	8.	Teenage pregnancies have increased dramatically over the past 30 years.
T	F	9.	Couples who already have children at the beginning of their marriage are less likely to divorce.
T	F	10.	The number of children living with grandparents and with no parent in the home increased by more than 50 percent between 1990 and 2000.

Answers on page 492.

How do sociologists approach the study of families? In our study of families, we will use our sociological imagination to see how our personal experiences are related to the larger happenings in society. At the microlevel, each of us has a "biography," based on our experience within our family; at the macrolevel, our families are embedded in a specific social context that has a major effect on them (Aulette, 1994). We will examine the institution of the family at both of these levels, starting with family structure and characteristics.

Family Structure and Characteristics

In preindustrial societies, the primary form of social organization is through kinship ties. **Kinship refers to a social network of people based on common ancestry, marriage, or adoption.** Through kinship networks, people cooperate so that they can acquire the basic necessities of life, including food and shelter. Kinship systems can also serve as a means by which property is transferred, goods are produced and distributed, and power is allocated.

In industrialized societies, other social institutions fulfill some of the functions previously taken care of by the kinship network. For example, political systems provide structures of social control and authority, and economic systems are responsible for the production and distribution of goods and services. Consequently, families in industrialized societies serve fewer and more-specialized purposes than do families in preindustrial societies. Contemporary families are responsible primarily for regulating sexual activity, socializing children, and providing affection and companionship for family members.

Families of Orientation and Procreation

During our lifetime, many of us will be members of two different types of families—a family of orientation and a family of procreation. The ***family of orientation* is the family into which a person is born and in which early socialization usually takes place.** Although most people are related to members of their family of orientation by blood ties, those who are adopted have a legal tie that is patterned after a blood relationship (Aulette, 1994). The ***family of***

Box 15.1 SOCIOLOGY AND EVERYDAY LIFE

Answers to the Sociology Quiz on Contemporary Trends in U.S. Family Life

1. **False.** Census data show that the marriage rate has gone down by about one-third since 1960. In 1960, there were about 73 marriages per 1,000 unmarried women age 15 and up, whereas today the rate is about 49 per 1,000.

2. **False.** Less than 25 percent of all family households are composed of married couples with one or more children under age 18.

3. **True.** One in two preschoolers in the United States has a mother in the paid labor force, which makes affordable, high-quality child care an important issue for many families.

4. **True.** Recent research indicates that people who are married or are in a sustained relationship with their sexual partner express greater satisfaction with their sexual activity than do those who are not.

5. **True.** People under age 45 are more likely to cohabit than are older individuals.

6. **False.** Despite the strong economy prior to 2001, the number of children living in extreme poverty—with incomes below one-half of the poverty line—crept upward by almost 400,000 in recent years.

7. **False.** Adult children of divorced parents are less likely to dissolve their own marriages than they were two decades ago.

8. **False.** Teenage pregnancies have actually decreased over the past 30 years. However, the percentage of *unmarried* teenage pregnancies has increased dramatically.

9. **False.** Couples who already have children at the beginning of their marriage are *more* likely to divorce than those who do not.

10. **True.** The number of children living with grandparents and with no parent in the home grew by 51.5 percent between 1990 and 2000.

Sources: Based on Children's Defense Fund, 2002; Coontz, 1992, 1997; *Fox News,* 1999; and Popenoe and Whitehead, 1999.

---

procreation **is the family that a person forms by having or adopting children** (Benokraitis, 2002). Both legal and blood ties are found in most families of procreation. The relationship between a husband and wife is based on legal ties; however, the relationship between a parent and child may be based on either blood ties or legal ties, depending on whether the child has been adopted (Aulette, 1994).

In the United States, although many young people leave their family of orientation as they reach adulthood, finish school, and/or get married, recent studies have found that many people maintain family ties across generations, particularly as older persons remain actively involved in relationships with their adult children. Author Betty Friedan explains how an intergenerational pattern of "extended family" has emerged in her own family:

> With my own children . . . I have noticed an interesting development. In their twenties and thirties, they started coming back home again—

Thanksgiving, Passover, summers—bringing wives, husbands, their own children now, to my house in Sag Harbor. I relished the renewed sense of family which they clearly wanted again and enjoyed. I realized that they were confident enough now, in their grown-up selves, in their busy careers and the rich texture of their marriages . . . that they did not need to defend themselves anymore against their childish need for mother. (Friedan, 1993: 297)

Extended and Nuclear Families

Sociologists distinguish between extended and nuclear families based on the number of generations that live within a household. An ***extended family*** **is a family unit composed of relatives in addition to parents and children who live in the same household.** These families often include grandparents, uncles, aunts, or other relatives who live close to the parents and children, making it possible for family members to share

Bob Daemmrich/The Image Works

Steven Rubin/The Image Works

While the relationship between a husband and wife is based on legal ties, relationships between parents and children may be established either by blood ties or by legal ties.

resources. In horticultural and agricultural societies, extended families are extremely important; having a large number of family members participate in food production may be essential for survival. Today, extended family patterns are found in Latin America, Africa, Asia, and some parts of Eastern and Southern Europe (Busch, 1990). With the advent of industrialization and urbanization, maintaining the extended family pattern becomes more difficult in societies. Increasingly, young people move from rural to urban areas in search of employment in the industrializing sector of the economy. At that time, the nuclear family typically becomes the predominant family form in the society.

A *nuclear family* is a family composed of one or two parents and their dependent children, all of whom live apart from other relatives. A traditional definition specifies that a nuclear family is made up of a "couple" and their dependent children; however, this definition became outdated when a significant shift occurred in the family structure. A comparison of Census Bureau data from 1970 and 2000 shows that there has been a significant decline in the percentage of U.S. households comprising a married couple with their own children under eighteen years of age (see "Census Profiles: Household Composition"). Conversely, there has been an increase in the percentage of households in which either a woman or a man lives alone.

Marriage Patterns

Across cultures, families are characterized by different forms of marriage. *Marriage* is a legally recognized and/or socially approved arrangement between two or more individuals that carries certain rights and obligations and usually involves sexual activity. In most societies, marriage involves a mutual commitment by each partner, and linkages between two individuals and families are publicly demonstrated.

In the United States, the only legally sanctioned form of marriage is *monogamy—a marriage between two partners, usually a woman and a man.* For some people, marriage is a lifelong commitment that ends only with the death of a partner. For others, marriage is a commitment of indefinite duration. Through a pattern of marriage, divorce, and remarriage, some people practice *serial monogamy—a succession of marriages in which a person has several spouses over a lifetime but is legally married to only one person at a time.

Polygamy is the concurrent marriage of a person of one sex with two or more members of the opposite sex (Marshall, 1998). The most prevalent form of polygamy is *polygyny—the concurrent marriage of one man with two or more women.* Polygyny has been practiced in a number of Islamic societies, including some regions of contemporary Africa and southern Russia. For example, government officials in Africa estimate that 20 percent of Zambian marriages today are polygynous (Chipungu, 1999). How many wives and children might a polygynist have at one time? According to one report, Rodger Chilala of southern Zambia claimed to have 14 wives and more than 40 children; he stated that he previously had 24 wives but found that he could not afford the expenses associated with that many spouses (Chipungu, 1999). Although the cost of providing for multiple wives and many children makes the practice of polygyny implausible for all but the wealthiest men, such marriages still exist in both urban and rural areas of Africa and among people of

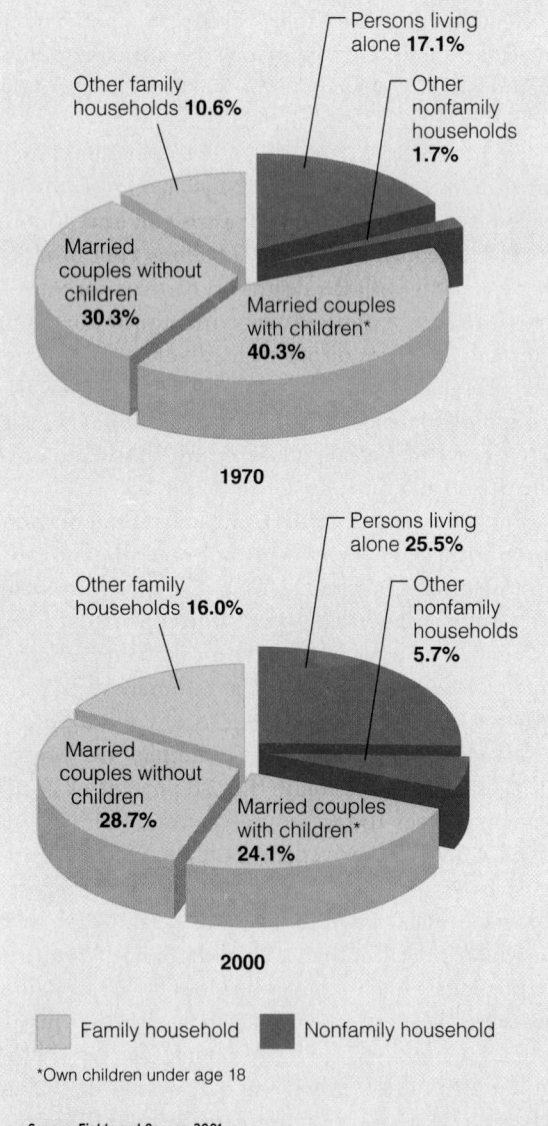

CENSUS ★ PROFILES

Household Composition, 1970 and 2000

Census 2000 quizzed respondents about their marital status and asked them to fill out an additional page of data for each other individual residing in the household. Based on this data, the Census Bureau reported that the percentage distribution of nonfamily and family households had changed during the past 30 years. The most noticeable trend is the decline in the number of married-couple households with their own children living with them, which decreased from about 40 percent of all households in 1970 to about 24 percent in 2000. The distribution shown below reflects this significant change in family structure.

1970

- Persons living alone **17.1%**
- Other nonfamily households **1.7%**
- Married couples with children* **40.3%**
- Married couples without children **30.3%**
- Other family households **10.6%**

2000

- Persons living alone **25.5%**
- Other nonfamily households **5.7%**
- Married couples with children* **24.1%**
- Married couples without children **28.7%**
- Other family households **16.0%**

☐ Family household   ■ Nonfamily household

*Own children under age 18

Source: Fields and Casper, 2001.

divergent social status and age groups (Chipungu, 1999). Polygyny is also allowed in southern Russia, where efforts are under way to revive Islamic traditions (ITAR/TASS, 1999b). However, after legislation was passed in the Republic of Ingushetia that allowed men to have up to four wives, only three men took advantage of the new law (ITAR/TASS, 1999a). Some analysts believe that the practice of polygamy contributes to the likelihood that families will live in poverty (Chipungu, 1999).

The second type of polygamy is *polyandry*—**the concurrent marriage of one woman with two or more men.** Polyandry is very rare; when it does occur, it is typically found in societies where men greatly outnumber women because of high rates of female infanticide or where marriages are arranged between two brothers and one woman ("fraternal polyandry"). According to recent research, polyandry is never the only form of marriage in a society: Whenever polyandry occurs, polygyny co-occurs (Trevithick, 1997). Although Tibetans are the most frequently studied population where polyandry exists, anthropologists have also identified the Sherpas, Paharis, Sinhalese, and various African groups as sometimes practicing polyandry (Trevithick, 1997). A recent anthropological study of Nyinba, an ethnically Tibetan population living in northwestern Nepal, found that fraternal polyandry (two brothers sharing the same wife) is the normative form of marriage and that the practice continues to be highly valued culturally (Levine and Silk, 1997). Despite the fact that some polyandrous marriages fail, societies such as this pass on the practice from one generation to the next (Levine and Silk, 1997).

Patterns of Descent and Inheritance

Even though a variety of marital patterns exist across cultures, virtually all forms of marriage establish a system of descent so that kinship can be determined and inheritance rights established. In preindustrial societies, kinship is usually traced through one parent (unilineally). The most common pattern of unilineal descent is *patrilineal descent*—**a system of tracing descent through the father's side of the family.** Patrilineal systems are set up in such a manner that a legitimate son inherits his father's property and sometimes his position upon the father's death. In nations such as India, where boys are seen as permanent patrilineal family members but girls are seen as only temporary family members, girls tend to be considered more expendable than boys (O'Connell, 1994). Recently, some scholars have concluded that cultural and racial nationalism in China is linked to the idea

Tony Howarth/Woodfin Camp & Associates

Polygamy is the concurrent marriage of a person of one sex with two or more persons of the opposite sex. Although most people in Iran do not practice this pattern of marriage, some men are married to more than one wife.

of patrilineal descent being crucial to the modern Chinese national identity (Dikotter, 1996).

Even with the less common pattern of *matrilineal descent*—**a system of tracing descent through the mother's side of the family**—women may not control property. However, inheritance of property and position is usually traced from the maternal uncle (mother's brother) to his nephew (mother's son). In some cases, mothers may pass on their property to daughters.

By contrast, kinship in industrial societies is usually traced through both parents (bilineally). The most common form is *bilateral descent*—**a system of tracing descent through both the mother's and father's sides of the family.** This pattern is used in the United States for the purpose of determining kinship and inheritance rights; however, children typically take the father's last name.

Power and Authority in Families

Descent and inheritance rights are intricately linked with patterns of power and authority in families. The most prevalent forms of familial power and authority are patriarchy, matriarchy, and egalitarianism. A *patriarchal family* **is a family structure in which authority is held by the eldest male (usually the father).** The male authority figure acts as head of the

household and holds power and authority over the women and children, as well as over other males. A *matriarchal family* **is a family structure in which authority is held by the eldest female (usually the mother).** In this case, the female authority figure acts as head of the household. Although there has been a great deal of discussion about matriarchal families, scholars have found no historical evidence to indicate that true matriarchies ever existed.

The most prevalent pattern of power and authority in families is patriarchy. Across cultures, men are the primary (and often sole) decision makers regarding domestic, economic, and social concerns facing the family. The existence of patriarchy may give men a sense of power over their own lives, but it also can create an atmosphere in which some men feel greater freedom to abuse women and children (O'Connell, 1994). Moreover, some economists believe that the patriarchal family structure (along with prevailing market conditions and public policy) limits people's choices in employment in the United States (Heath, Ciscel, and Sharp, 1998). According to this view, the patriarchal family structure has remained largely unchanged in this country, even as familial responsibilities in the paid labor market have undergone dramatic transformation. In the postindustrial age, for example, gender-specific roles may have been reduced; however, women's choices remain limited by the patriarchal tradition, in which women do most of the unpaid labor, particularly in the family. Despite dramatic increases in the number of women in the paid work force, there has been a lack of movement toward gender equity, which would equalize women's opportunities (Heath, Ciscel, and Sharp, 1998). This issue is not limited to the United States: The patriarchal family has placed a heavy burden on women around the globe. However, in nations such as Japan, more women are changing their outlook on family life and the roles that they play within the family (see Box 15.2).

An *egalitarian family* **is a family structure in which both partners share power and authority equally.** Recently, a trend toward more-egalitarian relationships has been evident in a number of countries as women have sought changes in their legal status and increased educational and employment opportunities. Some degree of economic independence makes it possible for women to delay marriage or to terminate a problematic marriage (O'Connell, 1994). However, one study of the effects of egalitarian values on the allocation and performance of domestic tasks in the family found that changes were relatively slow in coming. According to the study, fathers were more likely to share domestic tasks in nonconventional families where members held more-egalitarian values. Similarly,

Box 15.2 SOCIOLOGY IN GLOBAL PERSPECTIVE

Fukugan Shufu: Changing Terminology and Family Life in Japan

I work until 8 in the evening, but there are plenty of times when I work much later. That's just the social reality in Japan. There are some other women in my milieu, but most of them have just one child and don't plan for more.

—Haruko Takachi, a postal manager in Tokyo, explaining why she is glad her child was accepted into a government-run nursery school (qtd. in French, 2003: A3)

Haruko Takachi is an example of the working woman in Japan who faces dual responsibilities in the workplace and in her family. Even with a significant increase in the number of employed women, a study by the Japanese prime minister's office found that most people still viewed household chores, such as cleaning, washing, cooking, and cleaning up after meals, as the wife's responsibility (White, 2002). However, according to the anthropologist Merry Isaacs White (2002), who has extensively studied family life in Japan, women's roles are becoming much more complex, and families in Japan are as diverse and sometimes as divisive as families elsewhere.

Although Japanese women make up 41 percent of all workers in Japan, the increase in their numbers in the paid work force is only one of many changes that have affected families in that country. Today, more Japanese women who marry are keeping their maiden names. However, this practice—known as *bessei,* or "separate names"—may create additional problems for many of them because they may be hassled by employers, refused passport documents, or denied access to their husband's insurance benefits (White, 2002). Along with the practice of separate names, more couples are choosing to cohabit or to not register their marriage with the government.

New terminology reflects some of the changes in women's roles within the Japanese family. *Fukugan shufu* (meaning many-faced housewife) refers to a

"housewife who does not stay home but instead has created a busy collage of a life of activities" (White, 2002: 93–94). If the traditional Japanese wife was the "indoor housewife" (*oku-sama*) who devoted her time to domestic tasks and child care, the "outdoor housewife" (*soto-sama*) is now busy pursuing leisure activities such as playing tennis, attending lectures, or taking pottery classes. The term *tenuki okusan*—meaning the "no-hands" housewife—is used to describe women who typically are not employed but choose to pick up prepared meals or foods from local shops to serve for dinner (White, 2002: 94).

While these new terms reflect changes in the role of more-affluent married women in Japan, another phrase describes what many young women think of as an ideal husband—*baba nuki, kaa tsuki* ("without grandma, with car"). As White (2002: 83) explains, "Mothers-in-law represented the oppression and immobility of the old-style family; cars represented the free-wheeling youthful style of the modern couple." This thinking represents a significant shift from the past, when it was assumed that the dutiful daughter-in-law would look after elderly parents (Ikegami, 1998; Kakuchi, 1998). As younger women have sought greater freedom and felt less commitment to traditional marriage and family patterns, some analysts in Japan have expressed concern about the future of the family and who will take care of the aging population in that nation. Perhaps more than anything else, people in Japan and other nations are forced to admit that the previously held idealized views of the traditional Japanese family as a paragon of homogeneity, stability, and dutiful compliance are not an accurate depiction of what family life is actually like in Japan today.

Can you see similarities in discussions about "family values" in the United States and the controversy over changing family values in Japan? Why or why not?

children's gender-role typing was more closely linked to their parents' egalitarian values and nonconventional lifestyles than to the domestic tasks they were assigned (Weisner, Garnier, and Loucky, 1994).

Residential Patterns

Residential patterns are interrelated with the authority structure and method of tracing descent in fami-

lies. **Patrilocal residence refers to the custom of a married couple living in the same household (or community) as the husband's family.** Across cultures, patrilocal residency is most common. One example of contemporary patrilocal residency can be found in al-Barba, a lower-middle-class neighborhood in the Jordanian city of Irbid (McCann, 1997). According to researchers, the high cost of renting an apartment or building a new home has resulted in many sons building their own living quarters onto their parents' home, resulting in multi-family households consisting of an older married couple, their unmarried children, their married sons, and their sons' wives and children. The family units often have separate dwellings within the confines of the home. By sharing resources, all members of the extended family can maintain a higher standard of living than if each family unit lived independently. Thus, socioeconomic factors can be an important factor in the establishment of contemporary patrilocal residences (McCann, 1997).

Few societies have residential patterns known as *matrilocal residence*—**the custom of a married couple living in the same household (or community) as the wife's parents.** In industrialized nations such as the United States, most couples hope to live in a *neolocal residence*—**the custom of a married couple living in their own residence apart from both the husband's and the wife's parents.** For many couples, however, economic conditions, availability of housing, and other considerations may make neolocal residency impossible, at least initially.

Up to this point, we have examined a variety of marriage and family patterns found around the world. Even with the diversity of these patterns, most people's behavior is shaped by cultural rules pertaining to endogamy and exogamy. *Endogamy* **is the practice of marrying within one's own group.** In the United States, for example, most people practice endogamy: They marry people who come from the same social class, racial–ethnic group, religious affiliation, and other categories considered important within their own social group. Social scientists refer to this practice as *homogamy*—**the pattern of individuals marrying those who have similar characteristics, such as race/ethnicity, religious background, age, education, and/or social class.** Homogamy is similar to endogamy; however, issues pertaining to *why* people marry within their own group (endogamy) are somewhat different from the issues associated with why people choose to marry persons with similar characteristics and social status (homogamy). Various reasons have been given to explain why endogamy is so prevalent. One reason may be the proximity of other individuals in one's own group as contrasted with those who are geographically separated from it. Another reason may be that a person's marriage choice is often influenced by the opinions of parents, friends, and other people with whom the person associates (Kalmijn, 1998).

Although endogamy is the strongest marital pattern in the United States, more people now marry outside their own group. *Exogamy* **is the practice of marrying outside one's own social group or category.** Depending on the circumstances, exogamy may not be noticed at all, or it may result in a person being ridiculed or ostracized by other members of the "in" group. The three most important sources of positive or negative sanctions for intermarriage are the family, the church, and the state. Participants in these social institutions may look unfavorably on the marriage of an in-group member to an "outsider" because of the belief that it diminishes social cohesion in the group (Kalmijn, 1998). However, educational attainment is also a strong indicator of marital choices. Higher education emphasizes individual achievement, and college-educated people may be less likely than others to identify themselves with their social or cultural roots and thus more willing to marry outside their own social group or category if their potential partner shares a similar level of educational attainment (Hwang, Saenz, and Aguirre, 1995; Kalmijn, 1998).

THEORETICAL PERSPECTIVES ON FAMILIES

The *sociology of family* **is the subdiscipline of sociology that attempts to describe and explain patterns of family life and variations in family structure.** Functionalist perspectives emphasize the functions that families perform at the macrolevel of society whereas conflict and feminist perspectives focus on families as a primary source of social inequality. Symbolic interactionists examine microlevel interactions that are integral to the roles of different family members.

Functionalist Perspectives

Functionalists emphasize the importance of the family in maintaining the stability of society and the well-being of individuals. According to Emile Durkheim, marriage is a microcosmic replica of the larger society; both marriage and society involve a mental and moral fusion of physically distinct individuals (Lehmann, 1994). Durkheim also believed that a division of labor contributes to greater efficiency in all areas of life—including marriages and families—even though he acknowledged that this division imposes significant limitations on some people.

In the United States, Talcott Parsons was a key figure in developing a functionalist model of the family. According to Parsons (1955), the husband/father

Functionalist theorists believe that families serve a variety of functions that no other social institution can adequately fulfill. In contrast, conflict and feminist analysts believe that families may be a source of conflict over values, goals, and access to resources and power. Children in upper-class families have many advantages and opportunities that are not available to other children, as shown in this photo of an affluent family in front of its horse barn.

Alan Carey/The Image Works

fulfills the *instrumental role* (meeting the family's economic needs, making important decisions, and providing leadership), whereas the wife/mother fulfills the *expressive role* (running the household, caring for children, and meeting the emotional needs of family members).

Contemporary functionalist perspectives on families derive their foundation from Durkheim. Division of labor makes it possible for families to fulfill a number of functions that no other institution can perform as effectively. In advanced industrial societies, families serve four key functions:

1. *Sexual regulation.* Families are expected to regulate the sexual activity of their members and thus control reproduction so that it occurs within specific boundaries. At the macrolevel, incest taboos prohibit sexual contact or marriage between certain relatives. For example, virtually all societies prohibit sexual relations between parents and their children and between brothers and sisters.

However, some societies exclude remotely related individuals such as second and third cousins from such prohibitions. Sexual regulation of family members by the family is supposed to protect the *principle of legitimacy*—the belief that all children should have a socially and legally recognized father (Malinowski, 1964/1929).

2. *Socialization.* Parents and other relatives are responsible for teaching children the necessary knowledge and skills to survive. The smallness and intimacy of families make them best suited for providing children with the initial learning experiences they need.

3. *Economic and psychological support.* Families are responsible for providing economic and psychological support for members. In preindustrial societies, families are economic production units; in industrial societies, the economic security of families is tied to the workplace and to macrolevel economic systems. In recent years, psychological support and emotional security have been increasingly important functions of the family.

4. *Provision of social status.* Families confer social status and reputation on their members. These statuses include the ascribed statuses with which individuals are born, such as race/ethnicity, nationality, social class, and sometimes religious affiliation. One of the most significant and compelling forms of social placement is the family's class position and the opportunities (or lack thereof) resulting from that position. Examples of class-related opportunities include access to quality health care, higher education, and a safe place to live.

Functionalist explanations of family problems examine the relationship between family troubles and a decline in other social institutions. Changes in the economy, in religion, in the educational system, and in the law or government programs can all contribute to family problems. For example, when Tony Bardolino was laid off from his job due to a recession, he was unable to find work for about a year; his wife, Marianne, describes the consequences:

So he took this job as a dishwasher in this restaurant. It's one of those new kind of places with an open kitchen, so there he was, standing there washing dishes in front of everybody. I mean, we used to go there to eat sometimes, and now he's washing the dishes and the whole town sees him doing it. He felt so ashamed, like it was such a comedown, that he'd come home even worse than when he wasn't working. (qtd. in Rubin, 1994: 219)

Functionalists assert that erosion of family values may occur when the institution of religion becomes less important in everyday life. Likewise, changes in law (such as recognition of "no fault" divorce) contribute to high rates of divorce and dramatic increases in single-parent households. According to some functionalists, children are the most affected by these trends because they receive less nurturance and guidance from their parents (see Popenoe, 1993).

Conflict and Feminist Perspectives

Conflict and feminist analysts view functionalist perspectives on the role of the family in society as idealized and inadequate. Rather than operating harmoniously and for the benefit of all members, families are sources of social inequality and conflict over values, goals, and access to resources and power (Benokraitis, 2002).

According to some conflict theorists, families in capitalist economies are similar to workers in a factory. Women are dominated by men in the home in the same manner that workers are dominated by capitalists and managers in factories (Engels, 1970/1884). Although childbearing and care for family members in the home contribute to capitalism, these activities also reinforce the subordination of women through unpaid (and often devalued) labor. Other conflict analysts are concerned with the effect that class conflict has on the family. The exploitation of the lower classes by the upper classes contributes to family problems such as high rates of divorce and overall family instability.

Some feminist perspectives on inequality in families focus on patriarchy rather than class. From this viewpoint, men's domination over women existed long before capitalism and private ownership of property (Mann, 1994). Women's subordination is rooted in patriarchy and men's control over women's labor power (Hartmann, 1981). According to one scholar, "Male power in our society is expressed in economic terms even if it does not originate in property relations; women's activities in the home have been undervalued at the same time as their labor has been controlled by men" (Mann, 1994: 42). In addition, men have benefited from the privileges they derive from their status as family breadwinners (see Firestone, 1970; Daly, 1978; Goode, 1982).

Many women resist male domination. Women can control their reproductive capabilities through contraception and other means, and they can take control of their labor power by working for wages outside the home (Mann, 1994). However, the sociologist Jane Riblett Wilkie (1993) found that men are often reluctant to relinquish their status as family breadwinner. Although only 15 percent of families in the United States are supported solely by a male breadwinner, some men continue to construct their ideal of masculinity around this cultural value. Wilkie found that male respondents did not object to women's economic role as long as it was defined simply as earning money. When women also seemed to be challenging the ideal roles of male breadwinner/female homemaker, the men were much less positive. Wilkie (1993: 276) concluded that the persistence of this traditional ideology reflects men's "resistance to the loss of breadwinner status and the privileges in the home to which being the breadwinner has entitled them" (see also Firestone, 1970; Daly, 1978; Goode, 1982).

Conflict and feminist perspectives on families focus primarily on the problems inherent in relationships of dominance and subordination. Specifically, feminist theorists have developed explanations that take into account the unequal political relationship between women and men in families and outside of families (Aulette, 1994). Some feminist analysts explain family violence as a conscious strategy used by men to control women and perpetuate gender inequality (Kurz, 1989). To remedy problems such as wife battering, the subordination of women in all areas of society would have to be eliminated. A starting point might be elimination of factors fostering violence, such as media images that "glorify and romanticize assaults and other violent behavior" (Aulette, 1994: 329).

Symbolic Interactionist Perspectives

Early symbolic interactionists such as Charles Horton Cooley and George Herbert Mead provided key insights on the roles we play as family members and how we modify or adapt our roles to the expectations of others—especially significant others such as parents, grandparents, siblings, and other relatives. How does the family influence the individual's self-concept and identity? Contemporary symbolic interactionist perspectives examine the roles of husbands, wives, and children as they act out their own part and react to the actions of others. From such a perspective, what people think, as well as what they say and do, is very important in understanding family dynamics.

According to the sociologists Peter Berger and Hansfried Kellner (1964), interaction between marital partners contributes to a shared reality. Although newlyweds bring separate identities to a marriage, over time they construct a shared reality as a couple.

Violence in families not only has a detrimental effect on the individuals who are the objects of the violence but also on children and other relatives. What part should the larger society play in seeking to remedy problems such as wife battering?

In the process, the partners redefine their past identities to be consistent with new realities. Development of a shared reality is a continuous process, taking place not only in the family but in any group in which the couple participates together. Divorce is the reverse of this process; couples may start with a shared reality and, in the process of uncoupling, gradually develop separate realities (Vaughan, 1985).

Symbolic interactionists explain family relationships in terms of the subjective meanings and everyday interpretations that people give to their lives. As the sociologist Jessie Bernard (1982/1973) pointed out, women and men experience marriage differently. Although the husband may see *his* marriage very positively, the wife may feel less positive about *her* marriage, and vice versa. Researchers have found that husbands and wives may give very different accounts of the same event and that their "two realities" frequently do not coincide (Safilios-Rothschild, 1969).

How do symbolic interactionists view problems within the family? Some focus on the terminology used to describe these problems, examining the extent to which words convey assumptions or "realities" about the nature of the problem. For example, vio-

lence between men and women in the home is often referred to as "spouse abuse" or "domestic violence." However, these terms imply that women and men play equal roles in perpetrating violence in families, overlooking the more active part that men usually play in such aggression. In addition, the term *domestic violence* suggests that this is the "kind of violence that women volunteer for, or inspire, or provoke" (Jacobs, 1994: 56). Some scholars and activists use terms such as *wife battering* or *wife abuse* to highlight the gendered nature of such behavior (see Bograd, 1988). However, others argue that *battered woman* suggests a "woman who is more or less permanently black and blue and helpless" (Jacobs, 1994: 56).

Analysts using a social constructionist approach note that definitions concerning family violence are not only socially constructed but also have an effect on how people are treated. In one study of a shelter for battered women, analysts found that workers made decisions about whom to assist and how to assist them based on their own understanding of what constitutes a "real" battered woman (Loseke, 1992). (Family violence is discussed later in this chapter.)

Other symbolic interactionists have examined ways in which individuals communicate with one another and interpret these interactions. According to Lenore Walker (1979), females are socialized to be passive and males are socialized to be aggressive long before they take on the adult roles of battered and batterer. However, even women who have not been socialized by their parents to be helpless and passive may be socialized into this behavior by abusive husbands. Three factors contribute to the acceptance of the roles of batterer and battered: (1) low self-esteem on the part of both people involved, (2) a limited range of behaviors (he only knows how to be jealous and possessive/she only knows how to be dependent and anxious to make everyone happy), and (3) a belief by both in stereotypic gender roles (she should be feminine and pampered/he should be aggressive and dominant). Other analysts suggest that this pattern is changing as more women are gaining paid employment and becoming less dependent on their husbands or male companions for economic support.

Postmodernist Perspectives

Although postmodern theorists disparage the idea that a universal theory can be developed to explain social life, a postmodernist perspective might provide insights on questions such as this: How is family life different in the "information age"? Social scientist David Elkind (1995) describes the postmodern family as *permeable*—capable of being diffused or invaded in such a manner that an entity's original purpose is modified or changed. According to Elkind (1995), if the nuclear family is a reflection of the age of modernity, the permeable family reflects the postmodern assumptions of difference, particularity, and irregularity. Difference is evident in the fact that the nuclear family is now only one of many family forms. Similarly, the idea of romantic love under modernity has given way to the idea of consensual love: Individuals agree to have sexual relations with others whom they have no intention of marrying or, if they marry, do not necessarily see the marriage as having permanence. Maternal love has also been transformed into shared parenting, which includes not only mothers and fathers but also caregivers who may either be relatives or nonrelatives (Elkind, 1995).

Today, many people value the autonomy of the individual family member more highly than the family unit. As Elkind (1995: 13) states, "If the nuclear home was a haven, the permeable home is more like a busy railway station with people coming in for rest and sustenance before moving out on another track."

Urbanity is another characteristic of the postmodern family. The boundaries between the public sphere (the workplace) and the private sphere (the home) are becoming much more open and flexible. In fact, family life may be negatively affected by the decreasing distinction between what is work time and what is family time. As more people are becoming connected "24/7" (twenty-four hours a day, seven days a week), the boss who would not call at 11:30 P.M. or when an employee is on vacation may send an e-mail asking for an immediate response to some question that has arisen while the person is away with family members (Leonard, 1999). According to some postmodern analysts, this is an example of the "power of the new communications technologies to integrate and control labour despite extensive dispersion and decentralization" (Haraway, 1994: 439).

Social theorist Jean Baudrillard's idea that the simulation of reality may come to be viewed by some people as "reality" can be applied to family interactions in the "Information Age." Does the ability to contact someone anywhere and any time of the day or night provide greater happiness and stability in families? Or is "reach out and touch someone" merely an ideology promulgated by the consumer society? Journalists have written about the experience of watching a family gathering at an amusement park, restaurant, mall, or other location only to see family members pick up their cell phones to receive or make calls to individuals not present, rather than spending

"face time" with those family members who are present. Similarly, the Internet provides many people with the opportunity to send e-mail messages, birthday cards, flowers, or other presents to friends and family around the world. However, e-mail and chat groups also provide new avenues for miscommunication, whereby people may be offended by the comments, particularly sharp retorts, made by others. As one journalist stated, "E-mail allows us to act before we can think—the perfect tool for a culture of hyperstimulation" (Leonard, 1999: 60). This journalist acknowledges that he has used e-mail to do everything from coping with the pressures of new fatherhood to arguing with his mother:

> [E-mail] saved me time and money without ever requiring me to leave the house; it salvaged my social life, allowed me to conduct interviews as a reporter and kept a lifeline open to my far-flung extended family. Indeed, I finally knew for sure that the digital world was viscerally potent when I found myself in the middle of a bitter fight with my mother—on e-mail. Again, new medium, old story. (Leonard, 1999: 59)

Some analysts paint a bleak future for families in the age of the "integrated circuit." Social scientist Donna Haraway (1994: 443) provides one scenario of the postmodern family:

> Women-headed households, serial monogamy, flight of men, old women alone, technology of domestic work, paid homework, reemergence of home sweat-shops, home-based businesses and telecommuting, electronic cottage, urban homelessness, migration, module architecture, reinforced (simulated) nuclear family, intense domestic violence.

However, Haraway believes that we should not demonize technology but should instead embrace the difficult task of "reconstructing the boundaries of daily life, in partial connections with others, in communication with all our parts" (Haraway, 1994: 452).

Even as postmodern perspectives call our attention to cyberspace, consumerism, and the hyperreal, it is important to recall that there is a growing "digital divide" and a "new kind of cyber class warfare," as some journalists refer to it, going on in the United States and around the world (Alter, 1999). New economic trends that are making the richest 2.7 million Americans even richer are making the poorest fifth of Americans even poorer. Although in 2000 more than 50 percent of all U.S. households owned computers and 41 percent of all households had Internet access, many families were

Concept Table 15.A THEORETICAL PERSPECTIVES ON FAMILIES			
	Focus	Key Points	Perspective on Family Problems
FUNCTIONALIST	Role of families in maintaining stability of society and individuals' well-being.	In modern societies, families serve the functions of sexual regulation, socialization, economic and psychological support, and provision of social status.	Family problems are related to changes in social institutions such as the economy, religion, education and law/government.
CONFLICT/FEMINIST	Families as sources of conflict and social inequality.	Families both mirror and help perpetuate social inequalities based on class and gender.	Family problems reflect social patterns of dominance and subordination.
SYMBOLIC INTERACTIONIST	Family dynamics, including communication patterns and the subjective meanings that people assign to events.	Interactions within families create a shared reality.	How family problems are perceived and defined depends on patterns of communication, the meanings that people give to roles and events, and individuals' interpretations of family interactions.
POSTMODERNIST	Permeability of families.	In postmodern societies, families are diverse and fragmented. Boundaries between workplace and home are blurred.	Family problems are related to cyberspace, consumerism, and the hyperreal in an age increasingly characterized by high-tech "haves" and "have-nots."

left out as the gap in Internet access between those at the highest and the lowest income levels increased.

Concept Table 15.A summarizes sociological perspectives on the family. Taken together, these perspectives on the social institution of families reflect various ways in which familial relationships may be viewed in contemporary societies. Now we shift our focus to love, marriage, intimate relationships, and family issues in the United States.

DEVELOPING INTIMATE RELATIONSHIPS AND ESTABLISHING FAMILIES

The United States has been described as a "nation of lovers"; it has been said that we are "in love with love." Why is this so? Perhaps the answer lies in the fact that our ideal culture emphasizes *romantic love,* which refers to a deep emotion, the satisfaction of significant needs, a caring for and acceptance of the person we love, and involvement in an intimate relationship (Lamanna and Riedmann, 2003).

Love and Intimacy

In the late nineteenth century, during the Industrial Revolution, people came to view work and home as separate spheres in which different feelings and emotions were appropriate (Coontz, 1992). The public sphere of work—men's sphere—emphasized self-reliance and independence. By contrast, the private sphere of the home—women's sphere—emphasized the giving of services, the exchange of gifts, and love. Accordingly, love and emotions became the domain of women, and work and rationality became the domain of men (Lamanna and Riedmann, 2003). Although the roles of women and men changed dramatically in the twentieth century, women and men still may not

Mary Kate Denny/PhotoEdit

In the United States, the notion of romantic love is deeply inter-twined with our beliefs about how and why people develop inti-mate relationships and establish families. Not all societies share this concern with romantic love.

share the same perceptions about romantic love today. According to the sociologist Francesca Cancian (1990), women tend to express their feelings verbally whereas men tend to express their love through nonverbal actions, such as running an errand for someone or repairing a child's broken toy.

Love and intimacy are closely intertwined. Intimacy may be psychic ("the sharing of minds"), sexual, or both. Although sexuality is an integral part of many intimate relationships, perceptions about sexual activities vary from one culture to the next and from one time period to another. For example, kissing is found primarily in Western cultures; many African and Asian cultures view kissing negatively (Reinisch, 1990).

For over forty years, the work of the biologist Alfred C. Kinsey was considered to be the definitive research on human sexuality, even though some of his methodology had serious limitations. Recently, the work of Kinsey and his associates has been superseded by the National Health and Social Life Survey conducted by the National Opinion Research Center at the University of Chicago (see Laumann et al., 1994; Michael et al., 1994). Based on interviews with over 3,400 men and women aged 18 to 59, this random survey tended to reaffirm the significance of the dominant sexual ideologies. Most respondents reported

that they engaged in heterosexual relationships, although 9 percent of the men said they had at least one homosexual encounter resulting in orgasm. Although 6.2 percent of men and 4.4 percent of women said that they were at least somewhat attracted to others of the same gender, only 2.8 percent of men and 1.4 percent of women identified themselves as gay or lesbian. According to the study, persons who engaged in extra-marital sex found their activities to be more thrilling than those with a marital partner, but they also felt more guilt. Persons in sustained relationships such as marriage or cohabitation found sexual activity to be the most satisfying emotionally and physically.

Cohabitation and Domestic Partnerships

Attitudes about cohabitation have changed in the past three decades. Although cohabitation is still defined by the Census Bureau as the sharing of a household by one man and one woman who are not related to each other by kinship or marriage, many sociologists now use a more inclusive definition. For our purposes, we will define **cohabitation as referring to a couple who live together without being legally married.** It is not known how many people actually cohabit because the Census Bureau does not ask about emotional or sexual involvement between unmarried individuals sharing living quarters or between gay and lesbian couples.

Based on Census Bureau data, the people who are most likely to cohabit are under age 45, have been married before, or are older individuals who do not want to lose financial benefits (such as retirement benefits) that are contingent upon not remarrying. Among younger people, employed couples are more likely to cohabit than college students.

For couples who plan to eventually get married, cohabitation somewhat follows the *two-stage marriage* pattern set out by the anthropologist Margaret Mead, who argued that dating patterns in the United States are not adequate preparation for marriage and parenting responsibilities. Instead, Mead suggested that marriage should occur in two stages, each with its own ceremony and responsibilities. In the first stage, the *individual marriage,* two people would make a serious commitment to each other but agree not to have children during this stage. In the second stage, the *parental marriage,* the couple would decide to have children and to share responsibility for the children's upbringing (Lamanna and Riedmann, 2003).

Today, some people view cohabitation as a form of "trial marriage." Some people who have cohabited do eventually marry the person with whom they have been living whereas others do not. A recent study of 11,000 women found that there was a 70-percent marriage rate for women who remained in a cohabiting relationship for at least 5 years. However, of the women in that study who cohabited and then married their partner, 40 percent became divorced within a 10-year period (Bramlett and Mosher, 2001). Whether these findings will be supported by subsequent research remains to be seen. But we do know that studies over the past decade have supported the proposition that couples who cohabit before marriage do not necessarily have a stable relationship following marriage (Bumpass, Sweet, and Cherlin, 1991; London, 1991).

In the United States, many lesbian and gay couples cohabit because they cannot enter into a legally recognized marital relationship except in the state of Vermont. Recently, some gay and lesbian activists have sought recognition of ***domestic partnerships***—**household partnerships in which an unmarried couple lives together in a committed, sexually intimate relationship and is granted the same rights and benefits as those accorded to married heterosexual couples** (Aulette, 1994; Gerstel and Gross, 1995). Benefits such as health and life insurance coverage are extremely important to *all* couples, as Gayle, a lesbian, points out: "It makes me angry that [heterosexuals] get insurance benefits and all the privileges, and Frances [her partner] and I take a beating financially. We both pay our insurance policies, but we don't get the discounts that other people get and that's not fair" (qtd. in Sherman, 1992: 197).

Despite laws to the contrary, many lesbian and gay couples consider themselves married and in a lifelong commitment. To make their commitment public, some couples exchange rings and vows under the auspices of churches such as the Metropolitan Community Church (the national gay church). Others, like Gayle, do not feel the need to be married:

> Within my home I feel married to Frances, but I don't consider us "married." The marriage part is still very heterosexual to me. One of the reasons I don't like to associate with marriage is because heterosexual marriage seems to be in trouble. It's like booking passage on the *Titanic*. Frances is my life partner; that's how I'm accustomed to thinking of her. (qtd. in Sherman, 1992: 189–190)

Marriage

Why do people get married? Couples get married for a variety of reasons. Some do so because they are "in love," desire companionship and sex, want to have children, feel social pressure, are attempting to escape from a bad situation in their parents' home, or believe that they will have more money or other resources if they get married. These factors notwithstanding, the selection of a marital partner is actually fairly predictable. As previously discussed, most people in the United States tend to choose marriage partners who are similar to themselves. As previously defined, *homogamy* refers to the pattern of individuals marrying those who have similar characteristics, such as race/ethnicity, religious background, age, education, or social class. However, homogamy provides only the general framework within which people select their partners; people are also influenced by other factors. For example, some researchers claim that people want partners whose personalities match their own in significant ways. Thus, people who are outgoing and friendly may be attracted to other people with those same traits. However, other researchers claim that people look for partners whose personality traits differ from but complement their own.

Regardless of the individual traits of marriage partners, research indicates that communication and emotional support are crucial to the success of marriages. Common marital problems include a lack of emotional intimacy, poor communication, and lack of companionship. One study concluded that for many middle- and upper-income couples, women's paid work was critical to the success of their marriages. People who have a strong commitment to their work have two distinct sources of pleasure—work and family. For members of the working class, however, work may not be a source of pleasure. For all women and men, balancing work and family life is a challenge.

Housework and Child-Care Responsibilities

Today, over 50 percent of all marriages in the United States are ***dual-earner marriages***—**marriages in which both spouses are in the labor force.** Over half of all employed women hold full-time, year-round jobs. Even when their children are very young, most working mothers work full time. For example, in 2001 more than 76 percent of employed mothers with children under age 6 worked full time (U.S. Census Bureau, 2002). Moreover, as Chapter 11 points out, many married women leave their paid employment at the end of the day and then go home to perform hours of housework and child care. Sociologist Arlie Hochschild (1989, 2003) refers to this as the ***second shift***—**the domestic work that employed women perform at home after they complete their**

Juggling housework, child care, and a job in the paid work force are all part of the average day for many women. Why does sociologist Arlie Hochschild believe that many women work a "second shift"?

© Jonathan Nourok (The Image Bank)/Getty Images

workday on the job. Thus, many married women today contribute to the economic well-being of their families and also meet many, if not all, of the domestic needs of family members by cooking, cleaning, shopping, taking care of children, and managing household routines. According to Hochschild, the unpaid housework that women do on the second shift amounts to an extra month of work each year. In households with small children or many children, the amount of housework increases. Across race and class, numerous studies have confirmed that domestic work remains primarily women's work (Gerstel and Gross, 1995). Hochschild (2003: 28) states that continuing problems regarding the second shift in many families are a sign that the gender revolution has stalled:

> The move of masses of women into the paid workforce has constituted a revolution. But the slower shift in ideas of "manhood," the resistance of sharing work at home, the rigid schedules at work make for a "stall" in this gender revolution. It is a stall in the change of institutional arrangement of which men are the principal keepers.

As Hochschild points out, the second shift remains a problem for many women in dual-earner marriages.

In recent years, more husbands have attempted to share some of the household and child-care responsibilities, especially in families in which the wife's earnings are essential to family finances. Overall, when husbands share some of the household responsibilities, they typically spend much less time in these activities than do their wives. Women and men perform different household tasks, and the deadlines for their work vary widely. Recurring tasks that have specific times for completion (such as bathing a child or cooking a meal) tend to be the women's responsibility; by contrast, men are more likely to do the periodic tasks that have no highly structured schedule (such as mowing the lawn or changing the oil in the car) (Hochschild, 1989). Men are also more reluctant to perform undesirable tasks such as scrubbing the toilet or diapering a baby, or to give up leisure pursuits.

Couples with more-egalitarian ideas about women's and men's roles tend to share more equally in food preparation, housework, and child care (Wright et al., 1992). For some men, the shift to a more-egalitarian household occurs gradually, as Wesley, whose wife works full time, explains:

> It was me taking the initiative, and also Connie pushing, saying, "Gee, there's so much that has to be done." At first I said, "But I'm supposed to be the breadwinner," not realizing she's also the breadwinner. I was being a little blind to what was going on, but I got tired of waiting for my wife to come home to start cooking, so one day I surprised the hell out [of] her and myself and the kids, and I had supper waiting on the table for her. (qtd. in Gerson, 1993: 170)

In the United States, millions of parents rely on child care so that they can work and so that their young children can benefit from early educational experiences that will help in their future school endeavors. For millions more parents, after-school care for school-age children is an urgent concern. Nearly 5 million children are home alone after school each week in this country. The children need productive and safe activities to engage in while their parents are working (Children's Defense Fund, 2002). Although child care is often unavailable or unaffordable for many parents, those children who are in day care for extended hours often come to think of child-care workers and other caregivers as members of their extended families because they may spend nearly as many hours with them as they do with their own parents. For children of divorced parents and other young people living in single-parent households, the issue of child care is often a

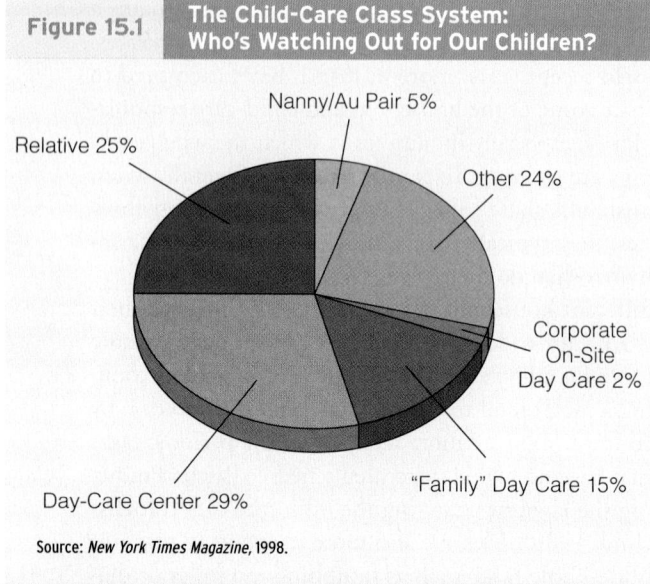

Figure 15.1 The Child-Care Class System: Who's Watching Out for Our Children?

- Nanny/Au Pair 5%
- Relative 25%
- Other 24%
- Corporate On-Site Day Care 2%
- "Family" Day Care 15%
- Day-Care Center 29%

Source: *New York Times Magazine*, 1998.

pressing concern because of the limited number of available adults and lack of financial resources.

In the United States, child-care arrangements are a reflection of the class system (see Figure 15.1). For example, in the upper-income categories, preschool-age children of working mothers are more likely to be cared for in their own homes by nannies or governesses. The "Six-Figure Nanny" is the best paid and most prestigious home caregiver because this individual may earn up to $2,500 per week, plus benefits. The Six-Figure Nanny is college educated and speaks several languages, whereas the typical nanny's language skills and level of experience vary widely. Typical nannies are paid between $250 and $600 per week; an au pair (a young person from another country) makes about $150 per week, plus room and board, taking care of children for about 45 hours a week. Middle-income families with working mothers typically send preschoolers to day-care centers or family day care, which takes place in the home of the caregiver. In low-income families, most preschoolers with working mothers are taken care of by nonparental relatives.

CHILD-RELATED FAMILY ISSUES AND PARENTING

Not all couples become parents. Those who decide not to have children often consider themselves to be "child-free," whereas those who do not produce children through no choice of their own may consider themselves "childless."

Deciding to Have Children

Cultural attitudes about having children and about the ideal family size began to change in the United States in the late 1950s. Women, on average, are now having two children each. However, rates of fertility differ across racial and ethnic categories. In 2000, for example, Latinas (Hispanic women) had a total fertility rate of 2.6, which was 50 percent above that of white (non-Hispanic) women. Among Latinas, the highest rate of fertility was found among Mexican American women, whereas Puerto Rican and Cuban American women had relatively lower rates. Similarly, African American women, on average, had a higher rate of fertility—2.0 as compared to 1.8—than white (non-Hispanic) women in the mid-1990s (Bachu and O'Connell, 2001).

Advances in birth control techniques over the past four decades—including the birth control pill and contraceptive patches and shots—now make it possible for people to decide whether or not they want to have children, how many they wish to have, and to determine (at least somewhat) the spacing of their births. However, sociologists suggest that fertility is linked not only to reproductive technologies but also to women's beliefs that they do or do not have other opportunities in society that are viable alternatives to childbearing (Lamanna and Riedmann, 2003).

Today, the concept of reproductive freedom includes both the desire *to have* or *not to have* one or more children. According to the sociologists Leslie King and Madonna Harrington Meyer (1997), many U.S. women spend up to one-half of their life attempting to control their reproductivity. Other analysts have found that women, more often than men, are the first to choose a child-free lifestyle (Seccombe, 1991). However, the desire not to have children often comes in conflict with our society's *pronatalist bias,* which assumes that having children is the norm and can be taken for granted, whereas those who choose not to have children believe they must justify their decision to others (Lamanna and Riedmann, 2003).

However, some couples experience the condition of *involuntary infertility,* whereby they want to have a child but find that they are physically unable to do so. *Infertility* is defined as an inability to conceive after a year of unprotected sexual relations. Today, infertility affects nearly 5 million U.S. couples, or 1 in 12 couples in which the wife is between the ages of 15 and 44 (Gabriel, 1996). Research suggests that fertility problems originate in females in approximately 30–40 percent of the cases and with males in about 40 percent of the cases; in the other 20 percent of the cases, the cause is impossible to determine (Gabriel, 1996). A leading

Box 15.3 CHANGING TIMES: MEDIA AND TECHNOLOGY

The Reproductive Revolution: Who Are My Parents?

Consider the possibilities of in vitro fertilization, which involves fertilizing a woman's eggs in a Petri dish and then transferring them to the woman's womb. The eggs must come from the woman herself or from a female donor; the sperm can come from the woman's partner or another male donor. An ovum transfer involves moving a five-day-old embryo from one female to another, who then carries the embryo to term. When these are combined, it means that a child can have at least five parents, not counting any later changes with remarriage: a donor mother, a birth mother, a social mother (the one who raises the child), a donor father, and a social father. (Coontz, 1997: 93)

According to the sociologist Stephanie Coontz (1997), medical and technological change has created many new issues in family life. For example, in recent years biological mothers have been pitted against biological fathers in surrogate custody disputes. Ex-spouses have fought bitter court battles over custody of frozen embryos stored while the couple was still married (in the event that they had decided to use in vitro fertilization).

Many infertile couples have had their hopes raised that they might be able to have a child through various forms of assisted reproductive technology. The main forms of this technology have been outlined by Gabriel (1996):

- *In vitro fertilization.* Eggs that were produced as a result of administering fertility drugs are removed from the woman's body and fertilized by sperm in a laboratory dish. The embryos that result from the process are transferred to the woman's uterus.
- *Micromanipulation.* A specialist, looking through a microscope, manipulates egg and sperm in a laboratory dish to improve the chances of a pregnancy.
- *Cryopreserved embryo transfer.* Embryos that were frozen after a previous assisted reproductive technology procedure are thawed and then transferred to the uterus.
- *Egg donation.* Eggs are removed from a donor's uterus, fertilized in a laboratory dish, and transferred to an infertile woman's uterus.
- *Surrogacy.* An embryo is implanted in the uterus of a woman who is paid to carry the fetus until birth. The egg may come from either the legal or the surrogate mother, and the sperm may come from either the legal father or a donor.

How successful are these procedures in cases of infertility? The cost of these procedures prohibits many women from trying one of them. However, among those couples who receive assisted reproductive technology, it is estimated that only one in five will become parents (Gabriel, 1996).

WRITING IN SOCIOLOGY ASSIGNMENT

What influence will new reproductive technology have on families and society in the future? Are questions arising from the reproductive  revolution a matter of *cultural lag,* or are they linked to larger social and ethical issues in our society?

cause of infertility is sexually transmitted diseases, especially those cases that develop into pelvic inflammatory disease (Gold and Richards, 1994). It is estimated that about half of infertile couples who seek treatments such as fertility drugs, artificial insemination, and surgery to unblock fallopian tubes can be helped; however, some are unable to conceive despite expensive treatments such as *in vitro fertilization,* which costs as much as $11,000 per attempt (Gabriel, 1996). (See Box 15.3 for a discussion of issues arising from reproductive technology.) According to the sociologist Charlene Miall (1986), women who are involuntarily childless engage in "information management" to combat the social stigma associated with childlessness. Their tactics range from avoiding people who make them uncomfortable to revealing their infertility so that others will not think of them as "selfish" for being childless. Some people who are involuntarily childless may choose to become parents by adopting a child.

Adoption

Adoption is a legal process through which the rights and duties of parenting are transferred from a child's biological and/or legal parents to new legal parents. This procedure gives the adopted child all the rights

Increasingly, people who adopt children create households in which individuals from diverse racial and ethnic groups learn to live together.

© Dan Habib/Concord Monitor photo/Corbis-SABA

of a biological child. In most adoptions, a new birth certificate is issued and the child has no future contact with the biological parents; however, some states have "right-to-know" laws under which adoptive parents must grant the biological parents visitation rights.

Matching children who are available for adoption with prospective adoptive parents can be difficult. The available children have specific needs, and the prospective parents often set specifications on the type of child they want to adopt. Some adoptions are by relatives of the child; others are by infertile couples (although many fertile couples also adopt) who cannot produce a child of their own. Although thousands of children are available for adoption each year in the United States, many prospective parents seek out children in developing nations such as Romania, South Korea, and India. The primary reason is that the available children in the United States are thought to be "unsuitable." They may have disabilities, or they may be sick, nonwhite (most of the prospective parents are white), or too old (Zelizer, 1985). In addition, fewer infants are available for adoption today than in the past because better means of contraception exist, abortion is more readily available, and more unmarried teenage parents decide to keep their babies.

Ironically, although many couples who would like to have a child are unable to do so, other couples conceive a child without conscious intent. Consider the fact that for the approximately 6.4 million women who become pregnant each year in the United States, about 2.8 million (44 percent) pregnancies are intended whereas about 3.6 million (56 percent) are unintended (Gold and Richards, 1994). As with women and motherhood, some men feel that their fatherhood was planned; others feel that it was thrust upon them

(Gerson, 1993). Unplanned pregnancies usually result from failure to use contraceptives or from using contraceptives that do not work. Even with "planned pregnancies," it is difficult to plan exactly when conception will occur and a child will be born.

Teenage Pregnancies

Teenage pregnancies are a popular topic in the media and political discourse, and the United States has the highest rate of teen pregnancy in the Western industrialized world (National Campaign to Prevent Teen Pregnancy, 1997). However, in 2000 the total number of live births per 1,000 women age 15 to 19 was 59.7, down from 62.1 in 1991 (Bachu and O'Connell, 2001). Perhaps teenage pregnancy is seen as a major problem because teen birth rates have not declined as rapidly as rates for older married women.

Teen pregnancies have also been of concern to analysts who suggest that teenage mothers may be less skilled at parenting, are less likely to complete high school than their counterparts without children, and possess few economic and social supports other than their relatives (Maynard, 1996; Moore, Driscoll, and Lindberg, 1998). In addition, the increase in births among unmarried teenagers may have negative long-term consequences for mothers and their children, who have severely limited educational and employment opportunities and a high likelihood of living in poverty. Moreover, the Children's Defense Fund estimates that among those who first gave birth between the ages of 15 and 19, 43 percent will have a second child within 3 years (Benokraitis, 2002).

Teenage fathers have largely been left out of the picture. According to the sociologist Brian Robinson

(1988), a number of myths exist regarding teenage fathers: (1) they are worldly wise "superstuds" who engage in sexual activity early and often, (2) they are "Don Juans" who sexually exploit unsuspecting females, (3) they have "macho" tendencies because they are psychologically inadequate and need to prove their masculinity, (4) they have few emotional feelings for the women they impregnate, and (5) they are "phantom fathers" who are rarely involved in caring for and rearing their children. However, these myths overlook the fact that some teenage males try to be good fathers; for example, Adan Chamul of Northridge, California, decided not to be like his own father, who left Adan's mother when Adan was 4 months old:

> [My father] didn't want to have anything to do with me until I was able to work and earn money for him. I said "No." [When my girlfriend became pregnant,] my friends were real upset. They told me, "Kick her out of the house. Dump her." But I didn't want my baby to grow up like that, without a dad. It's not the baby's fault. It was our mistake. . . . I regret having her at such a young age, but I love her and I'll never regret that she is here. We're going to grow up. All three of us. (qtd. in Gleick, Reed, and Schindehette, 1994: 55)

Many discussions of teenage pregnancy focus on the negative consequences of behavior rather than on the human faces of the young people involved.

Single-Parent Households

In recent years, there has been a significant increase in single- or one-parent households due to divorce and to births outside of marriage. Even for a person with a stable income and a network of friends and family to help with child care, raising a child alone can be an emotional and financial burden. Single-parent households headed by women have been stereotyped by some journalists, politicians, and analysts as being problematic for children. About 42 percent of all white children and 86 percent of all African American children will spend part of their childhood living in a household headed by a single mother who is divorced, separated, never married, or widowed (Garfinkel and McLanahan, 1986). According to sociologists Sara McLanahan and Karen Booth (1991), children from mother-only families are more likely than children in two-parent families to have poor academic achievement, higher school absentee and dropout rates, early marriage and parenthood, higher rates of divorce, and more drug and alcohol abuse. Does living in a one-parent family *cause* all of this?

Bob Daemmrich/Stock Boston

Mothers and fathers in single-parent households are confronted with the necessity of meeting most of their children's daily needs without help from others. However, even in two-parent households, children are not guaranteed a happy childhood simply because both parents live in the same household.

Certainly not! Many other factors—including poverty, discrimination, unsafe neighborhoods, and high crime rates—contribute to these problems.

Lesbian mothers and gay fathers are counted in some studies as single parents; however, they often share parenting responsibilities with a same-sex partner. Due to homophobia (hatred and fear of homosexuals and lesbians), lesbian mothers and gay fathers are more likely to lose custody to a heterosexual parent in divorce disputes (Falk, 1989; Robson, 1992). In any case, between 1 and 3 million gay men in the United States and Canada are fathers. Some gay men are married natural fathers, others are single gay men, and still others are part of gay couples that have adopted children. Very little research exists on gay fathers; what research does exist tends to show that noncustodial gay fathers try to maintain good relationships with their children (Bozett, 1988).

Single fathers who do not have custody of their children may play a relatively limited role in the lives of those children. Although many remain actively involved in their children's lives, others may become

"Disneyland daddies" who take their children to recreational activities and buy them presents for special occasions but have a very small part in their children's day-to-day life. Sometimes, this limited role is by choice, but more often it is caused by workplace demands on time and energy, location of the ex-wife's residence, and limitations placed on visitation by custody arrangements.

Currently, men head about 17 percent of white one-parent families and 7 percent of African American one-parent families; among many of the men, a pattern of "involved fatherhood" has emerged (Gerson, 1993). For example, in a study of men who became single fathers because of their wife's death, desertion, or relinquishment of custody, the sociologist Barbara Risman (1987) found that the men had very strong relationships with their children.

Two-Parent Households

Parenthood in the United States is idealized, especially for women. According to the sociologist Alice Rossi (1992), maternity is the mark of adulthood for women, whether or not they are employed. By contrast, men secure their status as adults by their employment and other activities outside the family (Hoffnung, 1995).

For families in which a couple truly shares parenting, children have two primary caregivers. Some parents share parenting responsibilities by choice; others share out of necessity because both hold full-time jobs. Some studies have found that men's taking an active part in raising the children is beneficial not only for mothers (who then have a little more time for other activities) but also for the men and the children. The men benefit through increased access to children and greater opportunity to be nurturing parents (Coltrane, 1989). However, other researchers have found that men and women do not experience child rearing in the same way: Child rearing tends to place women and men in separate social worlds. For example, during a child's preschool years, women's social worlds tend to be restricted by a reduction in the number of people with whom women interact and a reduction in the time that women spend interacting with those people (Munch, McPherson, and Smith-Lovin, 1997).

Children in two-parent families are not guaranteed a happy childhood simply because both parents reside in the same household. Children whose parents argue constantly, are alcoholics, or abuse them may be worse off than children in a single-parent family in which the environment is more conducive to their well-being.

TRANSITIONS AND PROBLEMS IN FAMILIES

Families go through many transitions and experience a wide variety of problems, ranging from high rates of divorce and teen pregnancy to domestic abuse and family violence. These all-too-common experiences highlight two important facts about families: (1) for good or ill, families are central to our existence, and (2) the reality of family life is far more complicated than the idealized image of families found in the media and in many political discussions. Although some families provide their members with love, warmth, and satisfying emotional experiences, other families may be hazardous to the individual's physical and mental well-being. Because of this dichotomy in family life, sociologists have described families as both a "haven in a heartless world" (Lasch, 1977) and a "cradle of violence" (Gelles and Straus, 1988).

Family Violence

Violence between men and women in the home is often called spouse abuse or domestic violence. *Spouse abuse* refers to any intentional act or series of acts—whether physical, emotional, or sexual—that causes injury to a female or male spouse (Wallace, 2002). According to sociologists, the term *spouse abuse* refers not only to people who are married but also to those who are cohabiting or involved in a serious relationship, as well as those individuals who are separated or living apart from their former spouse (Wallace, 2002).

How much do we know about family violence? Women, as compared with men, are more likely to be the victim of violence perpetrated by intimate partners. Recent statistics indicate that women are five times more likely than men to experience such violence and that many of these women live in households with children younger than twelve. However, we cannot know the true extent of family violence because much of it is not reported to police. For example, it is estimated that only about one-half of the intimate partner violence against women was reported to police in the 1990s (U.S. Department of Justice, 2000b). African American women were more likely than other women to report such violence, which may further skew data about who is most likely to be victimized by a domestic partner (U.S. Department of Justice, 2000b).

Although everyone in a household where family violence occurs is harmed psychologically, whether or not they are the victims of violence, children are especially affected by household violence. It is estimated that between three and ten million children witness some form of domestic violence in their homes each year, and there is evidence to suggest that domestic violence and child maltreatment often take place in the same household (Children's Defense Fund, 2002). According to some experts, domestic violence is an important indicator that child abuse and neglect are also taking place in the household.

In some situations, family violence can be reduced or eliminated through counseling, the removal of one parent from the household, or other steps that are taken either by the family or by social service or law enforcement officials. However, children who witness violence in the home may display certain emotional and behavioral problems that adversely affect their school life and communication with other people. In some families, the problems of family violence are great enough that the children are removed from the household and placed in foster care.

Children in Foster Care

Not all of the children in foster care have come from violent homes, but many foster children have been in dysfunctional homes where parents or other relatives lacked the ability to meet the children's daily needs. *Foster care* refers to institutional settings or residences where adults other than a child's own parents or biological relatives serve as caregivers. States provide financial aid to foster parents, and the intent of such programs is that the children will either return to their own families or be adopted by other families. However, this is often not the case for "difficult to place" children, particularly those who are over ten years of age, have illnesses or disabilities, or are perceived as suffering from "behavioral problems." More than 568,000 children are in foster care at any given time (Barovick, 2001). About 60 percent of children in foster care are children of color, with about 42 percent of them being African American—almost three times the percentage of African American children in the total U.S. child population (Children's Defense Fund, 2002). Even when the number of children entering foster care for the first time remains relatively stable, the total number of children in foster care continues to increase because fewer children are leaving foster care and being adopted or placed in permanent homes (Children's Defense Fund, 2002). Many children in foster care have limited prospects for finding a permanent home; however, a few innovative programs offer hope for children who previously had

been moved from one foster care setting to another (see Box 15.4).

Problems in the family contribute to the large numbers of children who are in foster care. Such factors include parents' illness, unemployment, or death; violence or abuse in the family; and high rates of divorce.

Divorce

Divorce is the legal process of dissolving a marriage that allows former spouses to remarry if they so choose. Most divorces today are granted on the grounds of *irreconcilable differences,* meaning that there has been a breakdown of the marital relationship for which neither partner is specifically blamed. Prior to the passage of more-lenient divorce laws, many states required that the partner seeking the divorce prove misconduct on the part of the other spouse. Under *no-fault divorce laws,* however, proof of "blameworthiness" is generally no longer necessary.

Over the past 100 years, the U.S. divorce rate (number of divorces per 1,000 population) has varied from a low of 0.7 in 1900 to an all-time high of 5.3 in 1981; by 1995, it had stabilized at 4.4 (U.S. Census Bureau, 2002). Although many people believe that marriage should last for a lifetime, others believe that marriage is a commitment that may change over time.

Recent studies have shown that 43 percent of first marriages end in separation or divorce within fifteen years (National Centers for Disease Control, 2001). Many first marriages do not last even fifteen years: One in three first marriages ends within ten years while one in five ends within five years. The likelihood of divorce goes up with each subsequent marriage in the serial monogamy pattern. When divorces that terminate second or subsequent marriages are taken into account, there are about half as many divorces each year in the United States as there are marriages. These data concern researchers in agencies such as the Centers for Disease Control because separation and divorce often have adverse effects on the health and well-being of both children and adults (National Centers for Disease Control, 2001).

Causes of Divorce Why do divorces occur? As you will recall from Chapter 2, sociologists look for correlations (relationships between two variables) in attempting to answer questions such as this. Existing research has identified a number of factors at both the macrolevel and microlevel that make some couples more or less likely to divorce. At the macrolevel, societal factors contributing to higher rates of divorce include changes in social institutions, such as religion

Box 15.4 YOU CAN MAKE A DIFFERENCE

Providing Hope and Help for Children

I take it personally when I see kids mistreated. I just think they need an advocate to fight for them. . . . For me, it's very simple: The kids' needs come first. That's the bottom line at Hope Meadows. We make decisions as if these are our own children, and when you think that way, your decisions are different than if you are just trying to work within a bureaucratic system.

> —sociologist Brenda Eheart, describing why she founded Hope Meadows (qtd. in Smith, 2001: 22)

After five years of research into the adoptions of older children, the sociologist Brenda Eheart realized that foster families faced many problems when they tried to help children who had been moved from home to home. Thinking that she might be able to make a difference, Eheart developed the plan for Hope Meadows, a community established in 1994 on an abandoned Air Force base in Illinois. Hope Meadows is made up of a three-block-long series of ranch houses that provide multigenerational and multiracial housing for foster children, their temporary families, and older adults who live and work with the children. Older adults who interact with the children receive reduced rent in exchange for at least six hours per week of volunteer work with the children. Foster families that reside at Hope Meadows gain a feeling of community as they work together to help children who have experienced severe abuse or neglect, have been exposed to drugs and numerous foster homes, and often have physical, emotional, and behavioral problems.

Since its commencement, Hope Meadows has been largely successful in helping children in getting adopted. However, children are not the only beneficiaries of this community: Older residents gain the benefit of interacting with children and feeling that they can *make a positive contribution* to the lives of others (Barovick, 2001). Debbie Calhoun, a foster parent at Hope Meadows, has suggested things that chil-

dren need the most when they come there, and we can make a difference by providing the children in our lives with these same things (based on Smith, 2001):

- *Understanding.* We need to gain an awareness of how children feel and why they say and do certain things.
- *Trust.* We need to help children to see us as someone they can rely on and believe in as people they can trust.
- *Love.* We must show children that they are loved and that they will still be loved even when they make mistakes.
- *Compassion.* We must show children compassion because they must experience compassion in order to be able to show it to others.
- *Time.* We must give children time to be a part of our lives, and we must also give them time to adjust and to start over when they need to do so.
- *Security.* We must help children to feel secure in their surroundings and to believe that there is stability or permanency in their living arrangements.
- *Praise.* We must tell children when they are doing well and not always be critical of them.
- *Discipline.* We must let children know what behavior is acceptable and what behavior is not, all the while showing them that we love them, even when discipline is necessary.
- *Self-Esteem.* We must help children feel good about themselves.
- *Pride.* We should provide opportunities for children to learn to take pride in their accomplishments and in themselves.

If these suggestions are beneficial for children in foster care settings, then they are certainly useful ideas for each of us to implement in our own families and communities, as well. What other ideas would you add to the list? Why?

---

and family. Some religions have taken a more lenient attitude toward divorce, and the social stigma associated with divorce has lessened. Further, as we have seen in this chapter, the family institution has undergone a major change that has resulted in less economic and emotional dependency among family members—and thus reduced a barrier to divorce.

At the microlevel, a number of factors contribute to a couple's "statistical" likelihood of divorcing. Here are some of the primary social characteristics of those most likely to get divorced:

- Marriage at an early age (59 percent of marriages to brides under eighteen end in separation or divorce within fifteen years) (National Centers for Disease Control, 2001)
- A short acquaintanceship before marriage
- Disapproval of the marriage by relatives and friends
- Limited economic resources and low wages
- A high school education or less (although deferring marriage to attend college may be more of a factor than education per se)

- Parents who are divorced or have unhappy marriages
- The presence of children (depending on their gender and age) at the beginning of the marriage

The interrelationship of these and other factors is complicated. For example, the effect of age is intertwined with economic resources; persons from families at the low end of the income scale tend to marry earlier than those at more affluent income levels. Thus, the question becomes whether age itself is a factor or whether economic resources are more closely associated with divorce.

The relationship between divorce and factors such as race, class, and religion is another complex issue. Although African Americans are more likely than whites of European ancestry to get a divorce, other factors—such as income level and discrimination in society—must also be taken into account. Latinos/as share some of the problems faced by African Americans, but their divorce rate is only slightly higher than that of whites of European ancestry. As the sociologist Demie Kurz (1995: 21) notes,

> A person's socioeconomic position has a strong influence on the likelihood that they will divorce. Despite the stereotype that divorce is a middle-class phenomenon, and despite increases in the divorce rate in all socio-economic levels, divorce rates have always been higher among lower-income people, and among those with less education. These patterns are related to higher rates of unemployment and job insecurity among workers from lower socio-economic backgrounds.

Religion may affect the divorce rate of some people, including many Latinos/as who are Roman Catholic. However, despite the Catholic doctrine that discourages divorce, the rate of Catholic divorces is now approximately equal to that of Protestant divorces.

Consequences of Divorce Divorce may have a dramatic economic and emotional impact on family members. An estimated 60 percent of divorcing couples have one or more children. By age sixteen, about one out of every three white children and two out of every three African American children will experience divorce within their families. As a result, most of them will remain with their mothers and live in a single-parent household for a period of time (Thornton and Freedman, 1983). In recent years, there has been a debate over whether children who live with their same-sex parent after divorce are better off than their peers who live with an opposite-sex parent. However, sociologists have found virtually no evidence to support the belief that children are better off living with a same-sex parent (Powell and Downey, 1997).

One of the most dramatic consequences of divorce affects children, many of whom divide their time between the separate households of their mother and father. Although some divorced fathers spend time and money on activities with their children, the burden of child rearing in single-parent households often falls most heavily on women.

Although divorce decrees provide for parental joint custody of approximately 100,000–200,000 children annually, this arrangement may create unique problems for some children, as the personal narratives of David and Nick Sheff at the beginning of this chapter demonstrate. Furthermore, some children experience more than one divorce during their childhood because one or both of their parents may remarry and subsequently divorce again.

Divorce changes relationships not only for the couple involved but also for other relatives. In some divorces, grandparents feel that they are the big losers. To see their grandchildren, the grandparents have to keep in touch with the parent who has custody. In-laws are less likely to be welcomed and may be seen as being on the "other side" simply because they are the parents of the ex-spouse. Recently, some grandparents have sued for custody of minor grandchildren. Generally, they have not been successful except in cases where questions existed about the emotional stability of the biological parents or the suitability of a foster care arrangement.

But divorce does not have to be always negative. For some people, divorce may be an opportunity to terminate destructive relationships (Lund, 1990; Kurz, 1995). For others, it may represent a means to

Remarriage and blended families create new opportunities and challenges for parents and children alike.

© Michelle D. Bridwell/PhotoEdit

achieve personal growth by managing their lives and social relationships and establishing their own social identity (see Reissman, 1991). Still others choose to remarry one or more times.

Remarriage

Most people who divorce get remarried. In recent years, more than 40 percent of all marriages were between previously married brides and/or grooms. Among individuals who divorce before age thirty-five, about half will remarry within three years of their first divorce (Bramlett and Mosher, 2001). Most divorced people remarry others who have been divorced. However, remarriage rates vary by gender and age. At all ages, a greater proportion of men than women remarry, often relatively soon after the divorce. Among women, the older a woman is at the time of divorce, the lower her likelihood of remarrying. Women who have not graduated from high school and who have young children tend to remarry relatively quickly; by contrast, women with a college degree and without children are less likely to remarry.

As a result of divorce and remarriage, complex family relationships are often created. Some people become part of stepfamilies or *blended families,* which consist of a husband and wife, children from previous marriages, and children (if any) from the new marriage. At least initially, levels of family stress may be fairly high because of rivalry among the children and hostilities directed toward stepparents or babies born into the family. In spite of these problems, however, many blended families succeed. The family that results from divorce and remarriage is typically a com-

plex, binuclear family in which children may have a biological parent and a stepparent, biological siblings and stepsiblings, and an array of other relatives, including aunts, uncles, and cousins.

According to the sociologist Andrew Cherlin (1992), the norms governing divorce and remarriage are ambiguous. Because there are no clear-cut guidelines, people must make decisions about family life (such as whom to invite for a birthday celebration or wedding) based on their beliefs and feelings about the people involved.

DIVERSITY IN FAMILIES

Although marriage at increasingly younger ages was the trend in the United States during the first half of the twentieth century, by the 1960s the trend had reversed, and many more adults were remaining single. Currently, almost 80 million single adults reside in the United States (U.S. Census Bureau, 2002). Some people remain single by choice, whereas others are single because they are not in the age brackets most likely to be married.

Diversity Among Singles

Never-married singles who remain single by choice may be more interested in opportunities for a career (especially for women), find readily available sexual partners without marriage, believe that the single

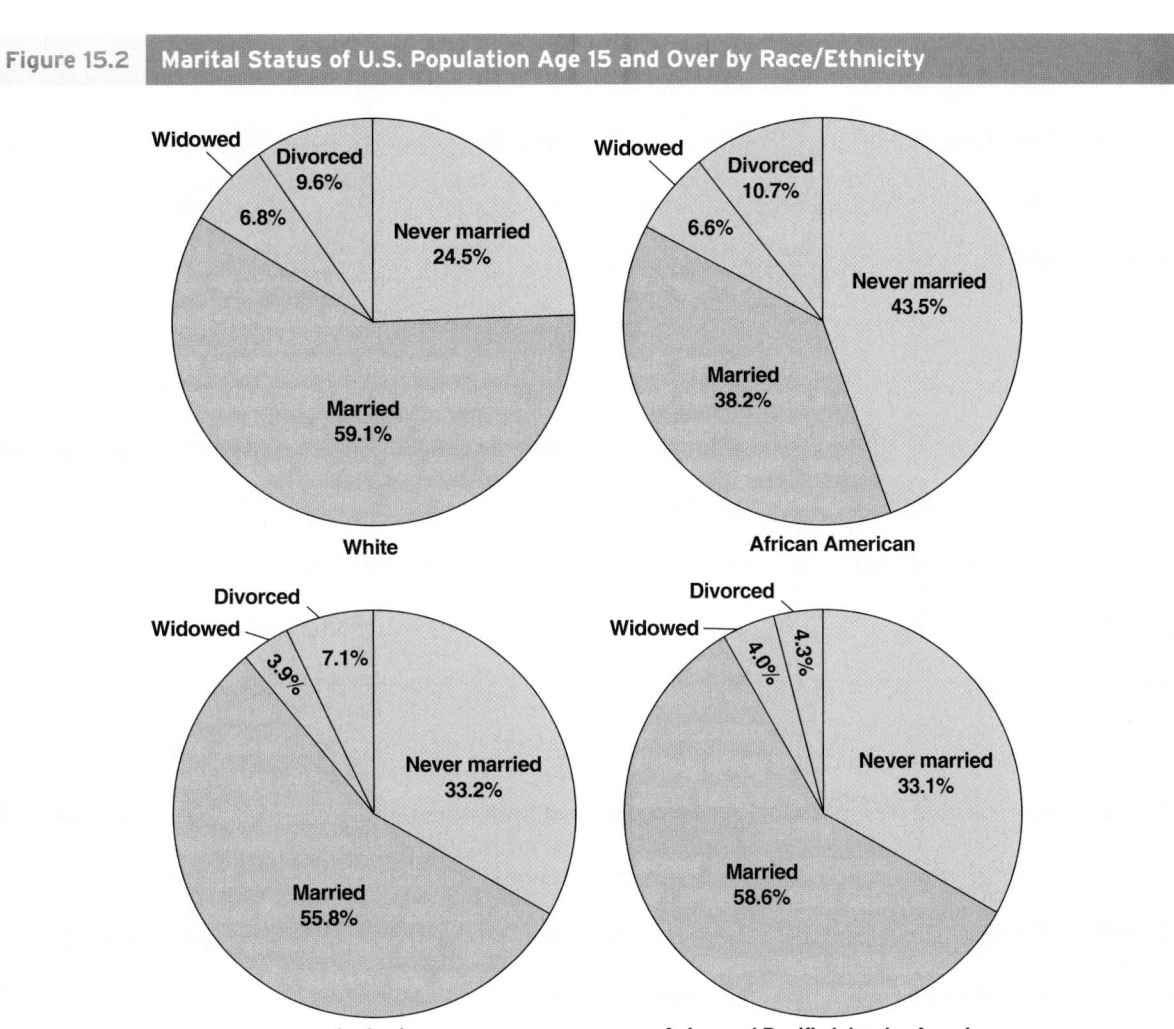

Figure 15.2 Marital Status of U.S. Population Age 15 and Over by Race/Ethnicity

White
- Widowed 6.8%
- Divorced 9.6%
- Never married 24.5%
- Married 59.1%

African American
- Widowed 6.6%
- Divorced 10.7%
- Never married 43.5%
- Married 38.2%

Latino/a
- Widowed 3.9%
- Divorced 7.1%
- Never married 33.2%
- Married 55.8%

Asian and Pacific Islander American
- Widowed 4.0%
- Divorced 4.3%
- Never married 33.1%
- Married 58.6%

Source: U.S. Census Bureau, 2002.

lifestyle is full of excitement, or have a desire for self-sufficiency and freedom to change and experiment (Stein, 1976, 1981). According to some marriage and family analysts, individuals who prefer to remain single typically hold more-individualistic values and are less family oriented than those who choose to marry. They also tend to value friends and personal growth more highly than getting married and having children (Cargan and Melko, 1982; Alwin, Converse, and Martin, 1985).

Other never-married singles remain single out of necessity. For some people, being single is an economic necessity: They simply cannot afford to marry and set up their own household. Structural changes in the economy limit the options of young people from working-class and low-income family backgrounds. Even some college graduates have found that they cannot earn enough money to set up a separate household

away from their parents. Consequently, a growing proportion of young adults are living with one or both parents. Approximately 13 percent of young adults between the ages of 25 and 34 reside with their parents; 16 percent of men live with their parents, as compared with 9 percent of women. According to one young man who lived at home,

> The rent is low and utilities are free. There is hot food on the table and clean socks in the drawer. Mom nags a little and dad scowls a lot, but mostly they don't get in the way. And there's money left at the end of the month for a car payment. (qtd. in Gross, 1991: A1)

The proportion of singles varies significantly by racial and ethnic group, as shown in Figure 15.2. Among persons age 15 and over, 43.5 percent of African Americans have never married, compared with 33.2 percent of Latinos/as, 33.1 percent of Asian and Pacific Islander Americans, and 24.5 percent of whites. Among women age 20 and over, the difference is even more pronounced; almost twice as many African American women in this age category have never married, compared with U.S. women of the same age in general (U.S. Census Bureau, 2002).

Five factors contribute to the lower marriage rate of African American women:

1. There are more African American women than men in the United States as a result of the high rate of mortality among young African American men (Staples, 1994). In addition, college-educated African American women significantly outnumber college-educated African American men and tend to earn more money than the men (Lichter, LeClere, and McLaughlin, 1991; Roberts, 1994; Staples, 1994).

2. African American males who have been subjected to discriminatory practices and limited opportunities may perceive that their only economic options are to serve in the military or participate in criminal activity, thus making them poor candidates for marriage (Staples, 1994).

3. A higher rate of homosexuality exists among African American men than among women (Staples, 1994).

4. More African American men than women marry members of other racial–ethnic groups (Staples, 1994).

5. Working-class African American families often stress education for their children and encourage their daughters to choose education over marriage (Higginbotham, 1991; Higginbotham and Weber, 1995).

Although a number of studies have examined why African Americans remain single, few studies have focused on Latinos/as. Some analysts cite the diversity of experiences among Mexican Americans, Cuban Americans, and Puerto Ricans as the reason for this lack of research. Existing research attributes increased rates of singlehood among Latinas/os to several factors, including the youthful age of the Latino/a population and the economic conditions experienced by many young Latinas/os (see Mindel, Habenstein, and Wright, 1988).

African American Families

As with other racial–ethnic groups, there is no such thing as *the* African American family (McAdoo, 1990). Although many African Americans live in nuclear families, a higher proportion of African Ameri-

cans than whites live in extended family households (Hofferth, 1984). The extended family often provides emotional and financial support not otherwise available. When an emergency arises, three or more generations may work together to support and care for one another (Taylor, Chatters, and Mays, 1988). Intergenerational care and concern by family members contribute to the well-being of adults and children alike, as Dawn March acknowledges:

> They have been there more so during my adult years than a lot of other families that I know about. My mother kept all of my children until they were old enough to go to day care. And she not only kept them, she'd give them a bath for me during the daytime and feed them before I got home from work. Very, very supportive people. So, I really would say I owe them for that. (qtd. in Higginbotham and Weber, 1995: 141)

As in March's case, mutually beneficial relationships often exist between older African American women and their grandchildren, nieces, and nephews. However, among middle- and upper-middle-class African American families, nuclear family patterns are more prevalent than extended family ties. In these nuclear families, visits with relatives outside the immediate family are limited to special occasions such as births, weddings, and funerals even when family members live close to one another (Willie, 1991).

Latina/o Families

Family support systems are also found in many Latina/o families. Referred to as *la familia,* this network spans a wide array of relatives, including a married couple and their children, parents, aunts and uncles, cousins, brothers and sisters, and their children. *La familia* may also include *compadres* (co-parents), *padrinos* (godfathers), and *madrinas* (godmothers) who are not biologically related to the family but become members of the extended family. *Compadres* participate in the major religious ceremonies in the children's lives, including baptism, confirmation, first communion, and marriage (see Griswold del Castillo, 1984). Poet Cherríe Moraga (1994: 41) describes *la familia* as "cross-generational bonding, deep emotional ties. . . . It is finding familia among friends where blood ties are formed through suffering and celebration shared." However, sociologists question the extent to which extended families and *familialism* (a strong sense of the importance of family held by all its members) exist among Latinos/as across social classes. According to the sociologist Norma Williams (1990), extended family networks have been disap-

Eating together is a family practice that cuts across all racial and ethnic groups and cultures. Family members may gather not only for nourishment but also to share their thoughts—and sometimes their disagreements—with one another.

pearing, especially among economically advantaged Mexican Americans in urban centers. Similar patterns have been observed among Puerto Rican families, a trend that is not always pleasing to older family members such as Gloria Santos:

> [My son and daughter] both have good salaries but call me only once or twice a month. I hardly know my grandchildren. . . . They only visit me once a year and only for one or two days. I've told my daughter that instead of sending me money she could call me more often. I was a good mother and worked hard in order for them to get a good education and have everything. All I expected from them was to show me they care, that they love me. (qtd. in Sanchez-Ayendez, 1995: 265)

Asian American Families

Many Asian Americans live in nuclear families; however, others (especially those residing in Chinatowns) have extended family networks. The family of Tony Hom, a twenty-five-year-old Chinese American, is an example of what is known as a *semi-extended family*

because other relatives live close by but not necessarily in the same household:

> My grandparents were always close by. When we lived in an apartment, they were always a few floors up. Now, my parents have a two family house, and my grandparents live next door. So this was always pointed out to me. Maybe they are fearful that one day we would become too Americanized and not take care of them. You always hear them talking about how the "lo-fans" (referring to Caucasians) don't take care of the elders, while the Chinese community prides itself on how it takes care of its old family members. (qtd. in Lee, 1992: 153)

Unlike Tony's family, extended family networks of some Vietnamese Americans are limited because some family members died in the war in Vietnam and others did not emigrate to the United States (Tran, 1988).

Native American Families

Family ties are also strong among many Native Americans, as Mary Crow Dog states:

Our people have always been known for their strong family ties, for people within one family group caring for each other, for the "helpless ones," the old folks and especially the children, the coming generation. . . . At the center of the old Sioux society was the tiyospaye, the extended family group, the basic hunting band, which included grandparents, uncles, aunts, in-laws, and cousins. The tiyospaye was like a warm womb cradling all within it. Kids were never alone, always fussed over by not one but several mothers, watched and taught by several fathers. (Crow Dog and Erdoes, 1991: 12–13)

Today, extended family patterns are common among lower-income Native Americans living on reservations; most others live in nuclear families (Sandefur and Sakamoto, 1988). Many Native Americans still share a sense of their extended family ties even when they do not live close to one another.

Biracial Families

Since the 1970s, there has been a 300-percent increase in marriages between people of different races. When these couples produce offspring, their children are considered to be biracial (Kalish, 1992). Interracial marriage—which is most often thought of as marriage between whites and blacks—was illegal in sixteen states until the U.S. Supreme Court overturned miscegenation laws in 1967.

Today, the term *interracial marriage,* or racially mixed marriage, has taken on a much broader interpretation as people from nations around the globe, representing a wide diversity of racial–ethnic and cultural backgrounds, form couples, marry, and produce children of mixed racial parentage. Skin color and place of birth have ceased to be reliable indicators of a person's identity or origin.

What effect does interracial parenting have on children and families? "Penny Yang," a Hmong woman whose family immigrated to the United States from Laos, describes how her mixed racial–ethnic background affects her family life:

My mom is Japanese and my dad is Hmong and my stepmom is American. It feels kind of different because I never met anyone who is Hmong and Japanese and American before, but I'm proud of it. I kind of get background from all sides, exposure from all different cultures. I wish, though, that I knew how to speak Hmong or Japanese really, really well. . . . When I'm with my dad's relatives, I wish I could understand more of what's

going on. I wish I could communicate with them because it's kind of put a barrier between me and them that I don't speak Hmong. Even though I love them and know that we're a family and everything, it's holding me back because I don't know the language. So in a lot of ways I feel left out, and I know I would feel different if I could speak Hmong. . . . If my Hmong or Japanese relatives don't understand me, then I really can't do anything about it. That's just the way I am. I'm an American with a mixed background, just like a lot of Americans. (qtd. in Faderman, 1998: 238–239)

Many biracial and bicultural children find through the lens of their personal experiences that they have an enhanced ability to understand the meaning of race and culture as these factors influence daily life. They also have opportunities to map a new ethnic terrain for the United States and other nations that transcends traditional racial and cultural divisions (O'Hearn, 1998). Journalist Lise Funderburg (1994), who identifies herself as biracial, sums up the significance of the increasing number of biracial marriages and biracial and bicultural children in this way:

To some extent, as the number of biracial people increases in this country, the choices they make about how to raise their children and how to influence their children's attitudes toward race will help determine not just how future generations of biracial children will be welcomed or shunned by society at large but also how all Americans will view and value race. . . . (Funderburg, 1994: 348)

FAMILY ISSUES IN THE FUTURE

As we have seen, families and intimate relationships changed dramatically during the twentieth century. Some people believe that the family as we know it is doomed. Others believe that a return to traditional family values will save this important social institution and create greater stability in society. However, the sociologist Lillian Rubin (1986: 89) suggests that clinging to a traditional image of families is hypocritical in light of our society's failure to support families: "We are after all, the only advanced industrial nation that has no public policy of support for the family whether with family allowances or decent publicly-sponsored childcare facilities." Some laws even have the effect of hurting children whose families do not fit the traditional model. For example, cutting back on government programs that provide food and medical care for pregnant women and infants will result in seriously ill children rather than model families (Aulette, 1994).

According to the psychologist Bernice Lott (1994: 155), people's perceptions about what constitutes a family will continue to change in the future:

> Persons on whom one can depend for emotional support, who are available in crises and emergencies, or who provide continuing affections, concern, and companionship can be said to make up a family. Members of such a group may live together in the same household or in separate households, alone or with others. They may be related by birth, marriage, or a chosen commitment to one another that has not been legally formalized.

Some of these changes are already becoming evident. For example, many men are attempting to take an active role in raising their children and helping with household chores. Many couples terminate abusive relationships and marriages.

Regardless of problems facing families today, many people still demonstrate their faith in the future by choosing to have children. As members of the Baby Boom generation (born between 1946 and 1964) entered their thirties and forties, many decided that they wanted children. New waves of immigrants have also become parents. Thus, the "boomlet" generation (born since 1988) is far more racially and ethnically diverse than the Baby Boom generation. According to futurist Watts Wacker, the boomlet generation knows "how to take care of themselves and be their own boss, because their parents have always worked. They have an unbelievable sense of being self-sufficient and confident. . . . They feel very comfortable about handling what life's throwing at them" (qtd. in Gabriel, 1995: 15). It will be interesting to see what the families created by the boomlet generation will be like. What will your family be like?

CHAPTER REVIEW

■ What is the family?

Today, families may be defined as relationships in which people live together with commitment, form an economic unit and care for any young, and consider their identity to be significantly attached to the group.

■ How does the family of orientation differ from the family of procreation?

The family of orientation is the family into which a person is born; the family of procreation is the family a person forms by having or adopting children.

■ What pattern of marriage is legally sanctioned in the United States?

In the United States, monogamy is the only form of marriage sanctioned by law. Monogamy is a marriage between two partners, usually a woman and a man.

■ What are the functionalist, conflict, and symbolic interactionist perspectives on families?

Functionalists emphasize the importance of the family in maintaining the stability of society and the well-being of individuals. Functions of the family include sexual regulation, socialization, economic and psychological support, and provision of social status. Conflict and feminist perspectives view the family as a source of social inequality and an arena for conflict. Symbolic interactionists explain family relationships in terms of the subjective meanings and everyday interpretations that people give to their lives. Reflecting the individualism, particularity, and irregularity of social life in the Information Age, postmodern analysts view families as being permeable, capable of being diffused or invaded so that their original purpose is modified.

■ How are families in the United States changing?

Families are changing dramatically in the United States. Cohabitation has increased significantly in the past three decades. With the increase in dual-earner marriages, women have increasingly been burdened by the second shift—the domestic work that employed women perform at home after they complete their workday on the job. Many single-parent families also exist today.

■ What is divorce, and what are some of its causes?

Divorce is the legal process of dissolving a marriage. At the macrolevel, changes in social institutions may contribute to an increase in divorce rates; at the microlevel, factors contributing to divorce include age at marriage, length of acquaintanceship, economic resources, education level, and parental marital happiness.

KEY TERMS

QUESTIONS FOR CRITICAL THINKING

1. In your opinion, what constitutes an ideal family? How might functionalist, conflict, feminist, and symbolic interactionist perspectives describe the ideal family?

2. Suppose that you wanted to find out about women's and men's perceptions about love and marriage. What specific issues might you examine? What would be the best way to conduct your research?

3. You have been appointed to a presidential commission on child-care problems in the United States. How to provide high-quality child care at affordable prices is a key issue for the first meeting. What kinds of suggestions would you take to the meeting? How do you think your suggestions should be funded? How does the future look for children in high-, middle-, and low-income families in the United States?

RESOURCES ON THE INTERNET

Chapter-Related Web Sites

The following Web sites have been selected for their relevance to the topics in this chapter. These sites are among the more stable, but please note that Web site addresses change frequently. For an updated list of chapter-related Web sites with URL links, please visit the *Sociology in Our Times* Web site (**www.wadsworth.com/KendallSIOT**).

American Association for Marriage and Family Therapy (AAMFT)
http://www.aamft.org

AAMFT is a professional association of marriage and family therapists that provides resources for therapists and clients. Its Web site offers information and resources on a variety of topics, including bereavement and loss, infertility, adolescent behavior problems, alcohol, ADHD, children and divorce, and post-traumatic stress disorder.

Family Violence Prevention Fund (FVPF)
http://endabuse.org

The FVPF is a national, nonprofit organization that focuses on policy and education to prevent and to eventually end domestic violence. The Web site includes the latest research, news reports, state and national statistics, celebrity watch, personal stories, advice on how to get help, and additional resources.

National Council on Family Relations (NCFR)
http://www.ncfr.org

In addition to promoting family well-being through the formulation of family-friendly policies, the NCFR provides the opportunity for family researchers, educators, and practitioners to share in the development and distribution of information about families and family relationships. The Web site offers family policy information, research, links, news, and a searchable database of information provided by NCFR members on topics such as couple relationships, parenting, divorce, and families and technology.

ONLINE STUDY AND RESEARCH TOOLS

Accompanying this text are many *free* powerful online study tools that will help you master the material in this chapter, help increase your depth of understanding, and help you make the grade!

SocCoach CD-ROM

Use the SocCoach CD-ROM enclosed with this text to help you formulate a customized study plan for this chapter. After you take the Diagnostic Quiz, SocCoach will generate a customized study plan just for you! It will identify sections of the chapter that you should review and will provide videos, charts, graphs, and excerpts from the text to supplement your studies and enhance your understanding. You'll also find fun, interactive activities such as Virtual Explorations and Map the Stats to apply what you've learned and stretch your sociological imagination.

The Companion Web Site for
Sociology in Our Times, *Fifth Edition*

www.wadsworth.com/KendallSIOT

Gain an even better grasp on this chapter by going to the companion web site to take one of the Tutorial Quizzes, use the Flash Cards to master key terms, or check out the many other study aids you'll find there. You'll also find special features such as GSS Data and Census 2000 information that'll put data and resources at your fingertips to help you with that special project or help you as you do some research on your own.

In this chapter, when you see the icon on the left, it alerts you to a specific exercise found in *Wadsworth's Sociology Online Resources and Writing Companion.* This valuable guide shows you how to use Wadsworth's exclusive online resources—*InfoTrac College Edition,* the *Opposing Viewpoints Resource Center,* and *MicroCase Online*—to assist you in your study of sociology and to build essential research and writing skills.

Education

During my first semester at Santa Barbara City College . . . my classes included aerobics, typing and remedial math. Who would have guessed that after spending three years there, I would be attending Harvard University?

Without the opportunity to study at my local community college, I probably wouldn't have gone to college at all. My high school grades would have sufficed to get me into a decent university, but I didn't consider myself college material. After all, none of my six brothers and sisters had attended college; most didn't even finish high school. My parents have the equivalent of a second-grade education. Mexican immigrants who do not speak English, my mother worked as a maid and my father as a dishwasher. Considering my resources, I thought the convenience and limited cost of a community college made it my most viable option.

But enrolling in a community college was also one of the smartest decisions I ever made. Despite my slow start, I learned the skills I needed to move ahead academically. . . . I also began to explore educational alternatives. And transferring to a reputable four-year university became my most important goal. I specifically wanted to transfer to an Ivy League university because I thought that would open more doors for me.

—Cynthia G. Inda (1997: 31) describing how a community college gave her the opportunity to pursue her educational and career goals

C ynthia is one of hundreds of thousands of people who have attended community colleges to gain knowledge and skills that will benefit them and enhance their opportunities in the future. In fact, a substantial proportion of college students start their higher education at community colleges.

From pre-kindergarten through post-graduate studies, education is one of the most significant social institutions in the United States and other high-income nations. Although most social scientists agree that schools are supposed to be places where people acquire knowledge and skills, not all of them agree on how a large number of factors—including class, race, gender, age, religion, and family background—affect individuals' access to educational opportunities or to the differential rewards that accrue at various levels of academic achievement. Consider this question, for example: Is higher education (at the college and university level) stratified by social class? According to the sociologist Dennis Gilbert (2003: 168), the answer is both yes and no:

(1) The American system of higher education is so big and so open that it provides major opportunities for talented youths from lower and middle-class families to prepare themselves for successful careers that raise them above the level of their parents, and (2) the American system of higher education is sufficiently stratified that its

■ Community colleges, especially those located on reservations, have increased access to higher education for some Native Americans. However, Native Americans as a category remain far below median levels of educational attainment in the United States.

main function is to reproduce for each generation of children the status positions held by their parents. Does the American system of higher education promote mobility or reproduction? The answer appears to be both.

In this chapter, we will look further into this issue as we discuss education as a key social institution and analyze some of the problems that affect contemporary elementary, secondary, and higher education. Before reading further, take the quiz in Box 16.1 to see what you know about education.

QUESTIONS AND ISSUES

Chapter Focus Question: How do race, class, and gender affect people's access to and opportunities in education?

How do educational goals differ in various nations?

What are the key assumptions of functionalist, conflict, and symbolic interactionist perspectives on education?

What major problems are being faced by U.S. schools today?

How are the issues in higher education linked to the problems of the larger society?

AN OVERVIEW OF EDUCATION

Education **is the social institution responsible for the systematic transmission of knowledge, skills, and cultural values within a formally organized structure.** Education is a powerful and influential force in contemporary societies. As a social institution, education imparts values, beliefs, and knowledge considered essential to the social reproduction of individual personalities and entire cultures (Bourdieu and Passeron, 1990). Education grapples with issues of societal stability and social change, reflecting society even as it attempts to shape it. Early socialization is primarily informal and takes place within our families and friendship networks; socialization then passes to the schools and other, more-formalized organizations created for the specific purpose of educating people. Today, education is such a significant social institution that an entire subfield of sociology—the *sociology of education*—is devoted to its study.

How did education emerge as such an important social institution in contemporary industrialized nations? To answer this question, we begin with a brief examination of education in historical–global perspective.

EDUCATION IN HISTORICAL-GLOBAL PERSPECTIVE

Education serves an important purpose in all societies. At the microlevel, people must acquire the basic knowledge and skills they need to survive in society. At the macrolevel, the social institution of education is an essential component in maintaining and perpetuating the culture of a society across generations. *Cultural transmission*—**the process by which children** and recent immigrants become acquainted with the dominant cultural beliefs, values, norms, and accumulated knowledge of a society—occurs through informal and formal education. However, the process of cultural transmission differs in preliterate, preindustrial, and industrial nations.

Informal Education in Preliterate Societies

Preliterate societies existed before the invention of reading and writing. These societies have no written language and are characterized by very basic technology and a simple division of labor. Daily activity often centers around the struggle to survive against natural forces, and the earliest forms of education are survival oriented. People in these societies acquire knowledge and skills through *informal education*—**learning that occurs in a spontaneous, unplanned way.** Through direct informal education, parents and other members of the group provide information about how to gather food, find shelter, make weapons and tools, and get along with others. For example, a boy might learn skills such as hunting, gathering, fishing, and farming from his father, whereas a girl might learn from her mother how to plant, gather, and prepare food or how to take care of her younger sisters and brothers. Such informal education often occurs through storytelling or ritual ceremonies that convey cultural messages and provide behavioral norms. Over time, the knowledge shared through information education may become the moral code of the group.

Formal Education in Preindustrial, Industrial, and Postindustrial Societies

Although *preindustrial societies* have a written language, few people know how to read and write, and

Box 16.1 SOCIOLOGY AND EVERYDAY LIFE

How Much Do You Know About U.S. Education?

True	False	
T	F	1. The United States has the highest ratio of university students per 100,000 population of any developed nation.
T	F	2. Women and men earn about the same number of doctoral degrees in the United States each year.
T	F	3. Public education in the United States dates back more than 150 years.
T	F	4. Equality of opportunity is a vital belief in the U.S. educational system.
T	F	5. The United States has never established nationwide educational goals.
T	F	6. Gender differences in science achievement are decreasing in U.S. public schools.
T	F	7. Core classes such as history and mathematics were never conducted in languages other than English in public schools in the United States before the 1960s civil rights movement.
T	F	8. The federal government has limited control over how funds are spent by school districts because most of the money comes from the state and local levels.

Answers on page 527.

formal education is often reserved for the privileged. Education becomes more formalized in preindustrial and industrial societies. *Formal education* **is learning that takes place within an academic setting such as a school, which has a planned instructional process and teachers who convey specific knowledge, skills, and thinking processes to students.** Perhaps the earliest formal education occurred in ancient Greece and Rome, where philosophers such as Socrates, Plato, and Aristotle taught elite males the skills required to become thinkers and orators who could engage in the art of persuasion (Ballantine, 2001). During the Middle Ages, the first colleges and universities were developed under the auspices of the Catholic church.

The Renaissance and the Industrial Revolution had a profound effect on education. During the Renaissance, the focus of education shifted to the importance of developing well-rounded and liberally educated people. With the rapid growth of industrial capitalism and factories during the Industrial Revolution, it became necessary for workers to have basic skills in reading, writing, and arithmetic. However, from the Middle Ages until the end of World War I, only the sons of the privileged classes were able to attend European universities. Agriculture was the eco-

nomic base of society, and literacy for people in the lower classes was not deemed important.

As societies industrialize, the need for formal education of the masses increases significantly. In the United States, the free public school movement was started in 1848 by Horace Mann, who stated that education should be the "great equalizer." By the mid-1850s, the process of mass education had begun in the United States as all states established free, tax-supported elementary schools that were readily available to children throughout the country. *Mass education* **refers to providing free, public schooling for wide segments of a nation's population.**

As industrialization and bureaucratization intensified, managers and business owners demanded that schools educate students beyond the third or fourth grade so that well-qualified workers would be available for rapidly emerging "white-collar" jobs in management and clerical work (Bailyn, 1960). By the middle of the nineteenth century, the idea of universal free public schooling was fairly well established in the United States. In addition to educating the next generation of children for the workplace, public schools were also supposed to serve as the primary agents of socialization for millions of European immigrants arriving in the United States seeking economic opportunities

Some early forms of mass education took place in one-room schoolhouses such as the one shown here, where children in various grades were all taught by the same teacher. How do changes in the larger society bring about changes in education?

and a better life. By the 1920s, educators had introduced the "core" curriculum: courses such as mathematics, social sciences, natural sciences, and English. This core is reflected in the contemporary "back to basics" movement, which calls for teaching the "three R's" (reading, 'riting, and 'rithmetic) and enforcing stricter discipline in schools.

Contemporary U.S. education attempts to meet the needs of the industrial and postindustrial society by teaching a wide diversity of students a myriad of topics, including history and science, computer skills, how to balance a checkbook, and how to avoid contracting AIDS. According to sociologists, many functions performed by other social institutions in the past are now under the auspices of the public schools. For example, full-day kindergartens and extended-day programs for school-age children are provided by many U.S. school districts because of the growing number of working parents who need high-quality, affordable care for their children. Within the regular classroom setting, many teachers feel that their job description encompasses too many divergent tasks. According to Ruth Prale, an elementary reading specialist,

> A teacher today is a social worker, surrogate parent, a bit of disciplinarian, a counselor, and someone who has to see to it that they eat. Many teachers are mandated to teach sex education, drug awareness, gang awareness. And we're supposed to be benevolent. And we haven't come to teaching yet! (qtd. in Collins and Frantz, 1993: 83)

At all levels of U.S. education, from kindergarten through graduate school, controversy exists over *what* should be taught, *how* it should be taught, and *who* should teach it. Do other nations have similar questions regarding their educational systems? Let's take a brief look at how educational systems are organized in Japan and Bosnia.

Contemporary Education in Other Nations

In this section, we will examine schools in two nations that frequently show up in the U.S. media. Because the United States and Japan are often compared to each other in the global marketplace, social analysts frequently compare the educational systems of these high-income countries. Similarly, since Bosnia is known for racial–ethnic factions and problems with racism, social analysts often examine how Bosnian schools handle ethnic and religious discord.

Education in Japan   Like other countries, Japan did not make public education mandatory for children until the country underwent industrialization. During the Meiji period (1868–1912), feudalism was eliminated, and Japan embarked on a new focus on youth and "bureaucratic" universal education (White, 1994). In hopes of catching up with the West, Japanese officials created an educational system and national educational goals. Education was viewed as a form of economic and national development and as a means of identifying talent for a new technological elite (White, 1994).

Today, Japanese educators, parents, students, and employers all view education as a crucial link in Japan's economic success. Japanese schools not only emphasize conformity and nationalism, but also highlight the importance of obligation to one's family

Box 16.1 SOCIOLOGY AND EVERYDAY LIFE

Answers to the Sociology Quiz on U.S. Education

1. **False.** Canada has the most university students (7,197) per 100,000 population. The United States is second, with 5,653 students per 100,000 population, followed by New Zealand (4,232) and South Korea (4,208) (Ash, 1997).

2. **False.** Men earn 58 percent of the doctorates, whereas women earn 42 percent. In the physical sciences, however, men earn 76 percent of the degrees, as compared to 24 percent for women (*Chronicle of Higher Education,* 2001).

3. **True.** As far back as 1848, free public education was believed to be important in the United States because of the high rates of immigration and the demand for literacy so that the country would have an informed citizenry that could function in a democracy (J. Wright, 1997).

4. **True.** Despite the fact that equality of educational opportunity has not been achieved, many people still subscribe to the belief that this country's educational system provides equality of opportunity to the masses, who can make of it what they will (J. Wright, 1997).

5. **False.** In 1994, Congress passed the Goals 2000: Educate America Act, which established eight major goals for education. Among these goals were a high school graduation rate of 90 percent and having U.S. students become first in the world in science and mathematics achievement (AAUW, 1995).

6. **False.** Gender differences in science achievement are not decreasing and may actually be increasing. According to the National Assessment of Educational Progress, gender differences in science achievement for nine- and thirteen-year-olds increased from the late 1970s to the 1990s due to a significant increase in the performance of male students and a lag in performance for female students (AAUW, 1995).

7. **False.** Late in the nineteenth and early in the twentieth century, some classes were conducted in the language of recent immigrants, such as the Italians, the Polish, and the Germans.

8. **True.** Most funding for public education comes from state and local property taxes, and similar sources of revenue.

and of learning the skills necessary for employment. Beginning at about three years of age, many Japanese toddlers are sent to cram schools (*jukus*) to help them qualify for good preschools.

In both cram schools and public schools, students learn discipline and thinking skills, along with physical activities such as karate and gymnastics to improve agility. By the time children reach elementary school, they are expected to engage in cooperative activities with their classmates. In some schools, children are responsible for preparing, serving, and cleaning up after the midday meal. At the end of the day, children may be seen cleaning the chalkboards and even mopping the floors, all as a part of the spirit of cooperation they are being taught.

In Japan, the middle-school years are especially crucial: Students' futures are based on the academic status track on which they are placed while in middle school. Moreover, the kind of jobs they will hold in the future depends on the university or vocational–technical school they attend, which in turn depends on the high school they attend. By the third year of middle school, most students are acutely aware of their academic future and of how their lives and friendships change as they go through the educational process, as one former student recalled:

> Middle school was fun; I was with my friends and I could throw myself into sports and I didn't have to study very hard. I couldn't be sure which of my friends would be with me in high school. Now that I am in high school, I know these will be my friends for life—they've been with me through very tough times. (qtd. in White, 1994: 77)

Up until high school, the student population of a school typically reflects the neighborhoods in which

Alan Oddie/PhotoEdit

Schools in Japan emphasize conformity even at an early age. Many Japanese people believe that high-quality education is a crucial factor in their country's economic success.

children live. However, at the high school level, entrance to a particular school is based on ability: Some Japanese students enter vocational schools that teach them skills for the workplace; others enter schools that are exclusively for the college-bound. Many analysts have noted the extent to which most Japanese vocational schools provide state-of-the-art equipment and instruction in fields with wide employment opportunities. Consequently, graduates of most vocational high schools do not have problems finding relatively well-paid employment upon graduation.

Typically, the instruction in Japanese high schools for the college-bound is highly structured, and all students are expected to respond in unison to questions posed by the teacher. Students are expected to be fluent in more than one language; many Japanese high schools teach English, particularly vocabulary and sentence construction. Science and math courses are challenging, and Japanese students often take courses such as algebra and calculus several years before their U.S. counterparts. Students must be prepared for a variety of college entrance examinations because each college and university gives its own test, and all tests occur within a few weeks of one another.

Girls and young women in the United States would likely feel stifled by the lack of educational opportunities experienced by their counterparts in Japan. Although there have been some changes, many parents and educators still believe that a good junior college education is all that young women need in order to be employable and marriageable (White, 1994). At the college and university level, the absence of women as students and professors is especially profound. Although a woman recently became the first-ever female president of a state-run university (Nara Women's University) in Japan, women account for

fewer than 5 percent of all presidents of colleges and universities (Findlay-Kaneko, 1997). Moreover, the lack of child-care facilities within the universities remains a pressing problem for women students and faculty in Japanese higher education.

Young men also experience extreme pressures in the Japanese system. In fact, high rates of school truancy occur as tens of thousands of students balk at going to school, and still others experience school-related health problems such as stomach ulcers, allergy disorders, and high blood pressure (White, 1994).

Education in Bosnia To understand education in contemporary Bosnia, it is important to consider the country's history. Located in eastern Europe and frequently subject to ethnic strife, the nation of Bosnia-Herzegovina has sought to maintain schools throughout violent clashes among its three principal ethnic groups: Muslims, Eastern Orthodox Serbs, and Roman Catholic Croats. Despite the existence of a formal peace agreement in 1995 between these groups, animosities remain strong.

Journalists and social scientists have found that each of the principal ethnic groups has taken control over how its children are educated and how cultural ideas are presented to them by teachers and in textbooks. Many schools identify and educate students by their ethnic background. For example, in history classes, students are segregated into ethnically distinct classrooms and taught different versions of history, language, and art, depending on their ethnic and religious identity (Hedges, 1997). The fact that schools with integrated classrooms have seen an increase in conflicts between students from different ethnic groups undoubtedly contributes to this situation. According to one Serbian sociologist, his teenage daughter came home in tears because of comments made in her class about the Serbian "aggressors": "She would not return to school for 10 days. Many of her friends now cross the line at the edge of the city to take classes in schools run by the Bosnian Serbs to avoid ridicule and harassment" (qtd. in Hedges, 1997: A4).

What will happen in the future is unknown. There is some cause for optimism that educational opportunities for children in Bosnia and Herzegovina will improve as organizations such as the United Nations Children's Fund (UNICEF) and some universities in the United States seek to help Bosnia with educational reform (University of Utah, 2003). Even with efforts at greater inclusion in Bosnian schools, many problems of discrimination and exclusion remain unresolved. According to recent reports, "Apartheid is a word still used to describe education in Bosnia" (Bosnia Action Coalition, 2001). Three issues con-

Despite continuing political unrest and violence in Bosnia-Herzegovina, some schoolchildren find solace in their classroom. It is difficult for educators in highly diverse nations to maintain a safe and positive learning environment for all children when the world outside their classrooms is full of strife.

Durand Patrick/Corbis Sygma

tinue to be central points of contention in Bosnian education: language, religion, and control of the schools (McCreight, 2001). The case of Bosnian education illustrates how difficult it is for teaching and learning to take place in an environment where there is little consensus about a shared sense of values, history, and what is important for the future. How much consensus exists in the United States regarding education? Let's examine that question, using the three main theoretical perspectives.

SOCIOLOGICAL PERSPECTIVES ON EDUCATION

Sociologists have divergent perspectives on the purpose of education in contemporary society. Functionalists believe that education contributes to the maintenance of society and provides people with an opportunity for self-enhancement and upward social mobility. Conflict theorists argue that education perpetuates social inequality and benefits the dominant class at the expense of all others. Symbolic interactionists focus on classroom dynamics and the effect of self-concept on grades and aspirations. Each of the three perspectives can provide valuable insights.

Functionalist Perspectives

Functionalists view education as one of the most important components of society. According to Emile Durkheim, education is crucial for promoting social solidarity and stability in society: Education is the "influence exercised by adult generations on those that are not yet ready for social life" (Durkheim, 1956: 28) and helps young people travel the great distance that it has taken people many centuries to cover. In other words, we can learn from what others have already experienced. Durkheim also asserted that *moral education* is very important because it conveys moral values—the foundation of a cohesive social order. He believed that schools are responsible for teaching a commitment to the common morality.

From this perspective, students must be taught to put the group's needs ahead of their individual desires and aspirations. Contemporary functionalists suggest that education is responsible for teaching U.S. values. According to the sociologist Amitai Etzioni (1994: 258–259),

> We ought to teach those values Americans share, for example, that the dignity of all persons ought to be respected, that tolerance is a virtue and discrimination abhorrent, that peaceful resolution of conflicts is superior to violence, that . . . truth telling is morally superior to lying, that democratic government is morally superior to totalitarianism and authoritarianism, that one ought to give a day's work for a day's pay, that saving for one's own and one's country's future is better than squandering one's income and relying on others to attend to one's future needs.

Etzioni suggests that "shared" values should be transmitted by schools from kindergarten through college. However, not all analysts agree on what those shared values should be or what functions education should serve in contemporary societies. In analyzing the values and functions of education, sociologists using a functionalist framework distinguish between manifest

What values are these schoolchildren being taught? Is there a consensus about what today's schools should teach? Why or why not?

functions and latent functions, which are compared in Figure 16.1.

Manifest Functions of Education Some functions of education are *manifest functions*—previously defined as open, stated, and intended goals or consequences of activities within an organization or institution. Education serves five major manifest functions in society:

1. *Socialization.* From kindergarten through college, schools teach students the student role, specific academic subjects, and political socialization. In kindergarten, children learn the appropriate attitudes and behavior for the student role (Ballantine, 2001). In primary and secondary schools, students are taught specific subject matter appropriate to their age, skill level, and previous educational experience. At the college level, students focus on more detailed knowledge of subjects they have previously studied and are exposed to new areas of study and research. Throughout their schooling, students receive political socialization in the form of history and civics lessons.
2. *Transmission of culture.* Schools transmit cultural norms and values to each new generation and play an active part in the process of assimilation, whereby recent immigrants learn dominant cultural values, attitudes, and behavior so that they can be productive members of society. However, questions remain as to *whose* culture is being transmitted. Because of the great diversity in the United States today, it is virtually impossible to define a single culture.
3. *Social control.* Schools are responsible for teaching values such as discipline, respect, obedience, punctuality, and perseverance. Schools teach

conformity by encouraging young people to be good students, conscientious future workers, and law-abiding citizens. The teaching of conformity rests primarily with classroom teachers.

4. *Social placement.* Schools are responsible for identifying the most-qualified people to fill the positions available in society. As a result, students are channeled into programs based on individual ability and academic achievement. Graduates receive the appropriate credentials to enter the paid labor force.
5. *Change and innovation.* Schools are a source of change and innovation. As student populations change over time, new programs are introduced to meet societal needs; for example, sex education, drug education, and multicultural studies have been implemented in some schools to help students learn about pressing social issues. Innovation in the form of new knowledge is required in colleges and universities. Faculty members are encouraged—and sometimes required—to engage in research and to share the results with students, colleagues, and others. In medical schools, for example, innovative technologies (such as new drugs) and new techniques (such as gene splicing) are developed and tested.

Latent Functions of Education All social institutions, including education, have *latent functions*—previously defined as hidden, unstated, and sometimes unintended consequences of activities within an organization or institution. Education serves at least three latent functions:

1. *Restricting some activities.* Early in the twentieth century, all states passed *mandatory education laws* that require children to attend school until

Figure 16.1 Manifest and Latent Functions of Education

Manifest functions—open, stated, and intended goals or consequences of activities within an organization or institution. In education, these are:

- socialization
- transmission of culture
- social control
- social placement
- change and innovation

Latent functions—hidden, unstated, and sometimes unintended consequences of activities within an organization. In education, these include:

- matchmaking and production of social networks
- restricting some activities
- creation of a generation gap

they reach a specified age (usually age sixteen) or complete a minimum level of formal education (generally the eighth grade). The assumption was that an educated citizenry and work force are necessary for the smooth functioning of democracy and capitalism. Out of these laws grew one latent function of education, which is to keep students off the streets and out of the full-time job market for a number of years, thus helping keep unemployment within reasonable bounds (Braverman, 1974).

2. *Matchmaking and production of social networks.* Because schools bring together people of similar ages, social class, and race/ethnicity, young people often meet future marriage partners and develop social networks that may last for many years.

3. *Creation of a generation gap.* Students may learn information in school that contradicts beliefs held by their parents or their religion. Debates over the content of textbooks and library books typically center on information that parents deem unacceptable for their children. When education conflicts with parental attitudes and beliefs, a generation gap is created if students embrace the newly acquired perspective.

Dysfunctions of Education Functionalists acknowledge that education has certain dysfunctions. Some analysts argue that U.S. education is not promoting the high-level skills in reading, writing, science, and mathematics that are needed in the workplace and the global economy. For example, mathematics and science education in the United States does not compare favorably with that found in many other industrialized countries (see Table 16.1). In the latest available (1999) Trends in International Mathematics and Science Study (TIMSS), which compares the mathematics and science performance of U.S. students with that of their peers in thirty-eight nations, U.S. students scored lower than did students in seventeen other nations, including Singapore, the Republic of Korea, and Japan (National Center for Education Statistics, 2003). More discouraging is the fact that a cohort of U.S. students who had been tested when they were in the fourth grade scored lower in comparison with their peers from other nations when retested in the eighth grade during 1999 than they had in the earlier round of tests (National Center for Education Statistics, 2003). Results from the 2003 study will be released late in 2004, and it will be interesting to see if the consistent

Table 16.1 TRENDS IN INTERNATIONAL MATHEMATICS AND SCIENCE STUDY (TIMSS)

Results of the 1999 TIMSS (selected nations)

COUNTRY	MATH SCORE	SCIENCE SCORE
Singapore	604	568
Korea, Republic of	587	549
Chinese Taipei	585	569
Hong Kong SAR	582	530
Japan	579	550
Netherlands	540	545
Hungary	532	552
Canada	531	533
Slovenia	530	533
Czech Republic	520	539
Australia	525	540
United States	**502**	**515**
England	496	538
Israel	466	468
Jordan	428	450
Chile	392	420
South Africa	275	243

Source: National Center for Education Statistics, 2003.

patterns of previous studies are maintained in the first decade of the new century.

Are U.S. schools dysfunctional as a result of scores such as these? Analysts do not agree on what these score differences mean. For example, Japanese students may outperform their U.S. counterparts because their schools are more structured and teachers focus on drill and practice (Celis, 1994). Educators David C. Berliner and Bruce J. Biddle (1995) believe that data on cross-cultural differences in educational attainment actually involve comparisons of "apples" and "oranges": They found that studies such as this often compare the achievement of eighth-grade Japanese students who have already taken algebra with the achievement of U.S. students who typically take such courses a year or two later. Among U.S. students who had already completed an algebra course, most did at least as well as their Japanese counterparts on the mathematics exam (Berliner and Biddle, 1995).

Clearly, test scores are subject to a variety of interpretations; however, for most functionalist analysts, lagging test scores are a sign that dysfunctions exist in the nation's educational system. According to this approach, improvements will occur only when more stringent academic requirements are implemented for students. Dysfunctions may also be reduced by more thorough teacher training and consistent testing of instructors. Overall, functionalists typically advocate the importance of establishing a more rigorous academic environment in which students are required to learn the basics that will make them competitive in school and job markets. An underlying assumption of functionalist thinking is that education can be an effective agent in the fight to reduce social inequality in society.

Conflict Perspectives

Conflict theorists do not believe that public schools reduce social inequality in society; rather, they believe that schools often perpetuate class, racial–ethnic, and gender inequalities as some groups seek to maintain their privileged position at the expense of others (Ballantine, 2001).

Cultural Capital and Class Reproduction

Although many factors—including intelligence, family income, motivation, and previous achievement—are important in determining how much education a person will attain, conflict theorists argue that access to quality education is closely related to social class. From this approach, education is a vehicle for reproducing existing class relationships. According to the French sociologist Pierre Bourdieu, the school legitimates and reinforces the social elites by engaging in specific practices that uphold the patterns of behavior and the attitudes of the dominant class. Bourdieu asserts that students from diverse class backgrounds come to school with differing amounts of ***cultural capital—social assets that include values, beliefs, attitudes, and competencies in language and culture*** (Bourdieu and Passeron, 1990). Cultural capital involves "proper" attitudes toward education, socially approved dress and manners, and knowledge about books, art, music, and other forms of high and popular culture. Middle- and upper-income parents endow their children with more cultural capital than do working-class and poverty-level parents. Because cultural capital is essential for acquiring an education, children with less cultural capital have fewer opportunities to succeed in school. For example, standardized tests that are used to group students by ability and to assign them to classes often measure students' cultural capital rather than their "natural" intelligence or aptitude. Thus, a circular effect occurs: Students with dominant cultural values are more highly rewarded by the educational system. In turn, the educational system teaches and reinforces those values that sustain the elite's position in society.

In her study of working-class women who returned to school after dropping out, the sociologist Wendy Luttrell (1997: 113–114) concluded that schools play a critical role in class reproduction and determining people's self-concept and identity:

> School denied but at the same time protected certain students' unearned advantages related to class, gender, and skin color in ways that made the women doubt their own value, voice, and abilities. . . . [When] the women were degraded by teachers and school officials for their speech, styles of dress, deportment, physical appearance, skin color, and forms of knowledge, they learned to recognize as "intelligent" or "valuable" only the styles, traits, and knowledge possessed by the economically advantaged students[:] . . . white, middle-class . . . behaviors and [appearance] . . . and urban or suburban mannerisms and styles of speech. Most important, [it was assumed that] those who possessed such cultural capital [were] entitled to their superior positions.

Tracking and Social Inequality Closely linked to the issue of cultural capital is how tracking in schools is related to social inequality. ***Tracking refers to the practice of assigning students to specific curriculum groups and courses on the basis of their test scores, previous grades, or other criteria.*** Conflict theorists believe that tracking seriously affects many students' educational performance and

their overall academic accomplishments. In elementary schools, tracking is often referred to *ability grouping* and is based on the assumption that it is easier to teach a group of students who have similar abilities. However, class-based factors also affect which children are most likely to be placed in "high," "middle," or "low" groups, often referred to by such innocuous terms as "Blue Birds," "Red Birds," and "Yellow Birds." This practice is described by Ruben Navarrette, Jr. (1997: 274–275), who tells us about his own experience with tracking:

> One fateful day, in the second grade, my teacher decided to teach her class more efficiently by dividing it into six groups of five students each. Each group was assigned a geometric symbol to differentiate it from the others. There were the Circles. There were the Squares. There were the Triangles and Rectangles.
>
> I remember being a Hexagon. . . . I remember something else, an odd coincidence. The Hexagons were the smartest kids in the class. These distinctions are not lost on a child of seven. . . . Even in the second grade, my classmates and I knew who was smarter than whom. And on the day on which we were assigned our respective shapes, we knew that our teacher knew, too. As Hexagons, we would wait for her to call on us, then answer by hurrying to her with books and pencils in hand. We sat around a table in our "reading group," chattering excitedly to one another and basking in the intoxication of positive learning. We did not notice, did not care to notice, over our shoulders, the frustrated looks on the faces of Circles and Squares and Triangles who sat quietly at their desks, doodling on scratch paper or mumbling to one another. We knew also that, along with our geometric shapes, our books were different and that each group had different amounts of work to do. . . . The Circles had the easiest books and were assigned to read only a few pages at a time. . . . Not surprisingly, the Hexagons had the most difficult books of all, those with the biggest words and the fewest pictures, and we were expected to read the most pages.
>
> The result of all of this education by separation was exactly what the teacher had imagined that it would be: Students could, and did, learn at their own pace without being encumbered by one another. Some learned faster than others. Some, I realized only [later], did not learn at all.

As Navarette suggests, tracking does make it possible for students to work together based on their perceived abilities and at their own pace; however, it also extracts a serious toll for students who are labeled as "underachievers" or "slow learners." Race, class, language, gender, and many other social categories may determine the placement of children in elementary tracking systems as much or more than their actual academic abilities and interests.

The practice of tracking continues in middle school/junior high and high school. Although schools in some communities bring together students from diverse economic and racial/ethnic backgrounds, the students do not necessarily take the same courses or move on the same academic career paths (Gilbert, 2003). Numerous studies over the past three decades have found that ability grouping and tracking affect students' academic achievements and career choices (Oakes, 1985; Welner and Oakes, 2000). Education scholar Jeannie Oakes found that tracking affects students' perceptions of classroom goals and achievements, as the following statements from high- and low-track students suggest:

- I want to be a lawyer and debate has taught me to dig for answers and get involved. I can express myself. (High Track English)
- To understand concepts and ideas and experiment with them. Also to work independently. (High Track Science)
- To behave in class. (Low Track English)
- To be a better listener in class. (Low Track English)
- I have learned that I should do my questions for the book when he asks me to. (Low Track Science) (Oakes, 1985: 86–89)

Perceptions of the students on the "low tracks" reflect the impact that years of tracking and lowered expectations can have on people's educational and career aspirations. Often, the educational track—vocational or college-bound—on which high school students are placed has a significant influence on their future educational and employment opportunities. Although the stated purpose of tracking systems is to permit students to study subjects that are suitable to their skills and interests, most research reveals that this purpose has not been achieved (Oakes, 1985; Welner and Oakes, 2000). Moreover, some social scientists believe that tracking is one of the most obvious mechanisms through which students of color and those from low-income families receive a diluted academic program, making it much more likely that they will fall even further behind their white, middle-class counterparts (see Miller, 1995). For example, a recent study of Latinas concluded that school practices such as tracking impose low expectations that create self-fulfilling prophecies for many of these young women (Ginorio

and Huston, 2000). Instead of enhancing school performance, tracking systems may result in students dropping out of school or ending up in "dead-end" situations because they have not taken the courses required to go to college if they choose to do so.

The Hidden Curriculum According to conflict theorists, the ***hidden curriculum*** **is the transmission of cultural values and attitudes, such as conformity and obedience to authority, through implied demands found in the rules, routines, and regulations of schools** (Snyder, 1971). As compared to the published curriculum of schools, which includes information such as what courses a student is required to take, the hidden curriculum includes "the often unarticulated and unacknowledged things that students are taught in school" (Johnson, 2000: 143). For students from dominant groups in society, the way they are treated and what they learn in school tend to enhance their self-esteem and expectations that they will attain success. By contrast, students of color, students from lower-income families or families that have recently migrated to this country, along with many young women, may be treated in such a manner that they acquire inferior self-images and hold less optimistic views of their education options and future prospects. Let's look more closely at this argument.

Social Class and the Hidden Curriculum Although students from all social classes are subjected to the hidden curriculum, working-class and poverty-level students may be affected the most adversely (Polakow, 1993; Ballantine, 2001; Oakes and Lipton, 2003). When teachers from middle- and upper-middle-class backgrounds instruct students from working- and lower-income families, the teachers often have a more structured classroom and a more controlling environment for students. These teachers may also have lower expectations for students' academic achievements. For example, one study of five elementary schools in different communities found significant differences in how knowledge was transmitted to students even though the general curriculum of the school was organized similarly (Anyon, 1980, 1997). Schools for working-class students emphasize procedures and rote memorization without much decision making, choice, or explanation of why something is done a particular way. Schools for middle-class students stress the processes (such as figuring and decision making) involved in getting the right answer. Schools for affluent students focus on creative activities in which students express their own ideas and apply them to the subject under consideration. Schools for students from elite families work to develop students' analytical powers

Signs in this elementary classroom list the rules, rewards, and consequences of different types of student behavior. According to conflict theorists, schools impose rules on working-class and poverty-level students so that they will learn to follow orders and to be good employees in the workplace. How would functionalists and symbolic interactionists interpret these same signs?

and critical thinking skills, applying abstract principles to problem solving.

Through the hidden curriculum, schools make working-class and poverty-level students aware that they will be expected to take orders from others, arrive at work punctually, follow bureaucratic rules, and experience high levels of boredom without complaining (Ballantine, 2001). Over time, these students may be disqualified from higher education and barred from obtaining the credentials necessary for well-paid occupations and professions (Bowles and Gintis, 1976). Educational credentials are extremely important in societies that emphasize *credentialism*—**a process of social selection in which class advantage and social status are linked to the possession of academic qualifications** (Collins, 1979; Marshall, 1998). Credentialism is closely related to *meritocracy*—previously defined as a social system in which status is assumed to be acquired through individual ability and effort (Young, 1994/1958). Persons who acquire the appropriate credentials for a job are assumed to have gained the position through what they know, not who they are or who they know. According to conflict theorists, the hidden curriculum determines in advance that the most valued credentials will primarily stay in the hands of the elites, so the United States is not actually as meritocratic as some might claim.

Gender Bias and the Hidden Curriculum According to conflict theorists, gender bias is embedded in both the formal and the hidden curricula of schools. Although most girls and young women in the United States have a greater opportunity for education than

those living in developing nations, their educational opportunities are not equal to those of boys and young men in their social class (see AAUW, 1995; Orenstein, 1995). For many years, reading materials, classroom activities, and treatment by teachers and peers contributed to a feeling among many girls and young women that they were less important than male students. Over time, this kind of differential treatment undermined females' self-esteem and discouraged them from taking certain courses, such as math and science, which were usually dominated by male teachers and students (Raffalli, 1994).

In recent years, some improvements have taken place in girls' education, as more females have enrolled in advanced placement or honors courses and in academic areas, such as math and science, where they had previously lagged (AAUW, 1998). However, girls are still not enrolled in higher-level science (such as physics) and computer sciences courses in the same numbers as boys. Girls and young women continue to make up only a small percentage of students in computer science and computer design classes, and the gender gap grows even wider from grade eight to eleven. Whereas male students are more likely to enroll in courses where they learn how to develop computer programs, female students are more likely to enroll in clerical and data-entry classes (AAUW, 1998). Some researchers find that the hidden curriculum works against young women in that some educators do not provide females with as much information about economic trends and the relationship among curriculum, course-taking choices, and career options as they provide to male students from middle- and upper-income families. Latinas are particularly affected by the hidden curriculum as it relates to gender, race/ethnicity, and language (AAUW, 1998).

Ethnicity, Language, and the Hidden Curriculum

How are ethnicity and language related to the hidden curriculum? Teachers who are responsible for English as a Second Language (ESL) courses frequently find that they are not only teaching children English but also teaching them social skills, such as how to talk to other people in a polite manner that is acceptable to people in the prevalent culture. One ESL instructor described how the hidden curriculum works in her classes:

> I believe that teaching our ESL students what is considered good manners in the United States is very important. I spend a lot of time teaching students how to give and receive compliments, to thank someone for something, to answer the telephone, to ask directions, to say "thank you" and to make small talk. . . . Many educators believe that second language learners will acquire socially appropriate behavior and language simply by being with English-speaking natives. I don't think this should be left to chance. (Haynes, 1999)

Through the process of teaching a new language to students, some teachers also instill certain values and behaviors that they believe will be beneficial for the children as they adapt to their new cultural environment. However, this socialization process is not always viewed in the same manner by their parents. ESL teacher Judie Haynes (1999) describes what happened when a third-grade ESL student uttered a swear word in the classroom:

> An excited ESL student in my school recently used an "x-rated" expression in his third grade classroom. The teacher was understandably distressed and required that the student write an apology for homework. Even more upsetting to the teacher was that the student's parents did not take the infraction of school rules seriously. I explained to her that swearing does not have the same "shock" value in a person's second language as it does in their first. So the parents were not "shocked" by their child's use of this language. What is considered "shocking" or inappropriate language for the classroom must sometimes be directly taught.

As in other manifestations of the hidden curriculum, this example shows that students learn far more—both positively and negatively—than just the subject matter that is being taught in classrooms. They are exposed to a wide range of beliefs, values, attitudes, and behavioral expectations that are not directly related to subjects such as English, algebra, or history. Conflict theorists use these examples to show how inequality is structurally produced and reproduced by formal and informal socialization processes in schools and other educational settings.

Symbolic Interactionist Perspectives

Unlike functionalist analysts, who focus on the functions and dysfunctions of education, and conflict theorists, who focus on the relationship between education and inequality, symbolic interactionists focus on classroom communication patterns and educational practices, such as labeling, that affect students' self-concept and aspirations.

Labeling and the Self-Fulfilling Prophecy

Chapter 7 explains that *labeling* is the process whereby a person is identified by others as possessing a specific characteristic or exhibiting a certain pattern of behavior (such as being deviant). According to symbolic interactionists, the process of labeling is directly related to the power and status of those persons who do the labeling and those who are being labeled. In schools, teachers and administrators are empowered to label children in various ways, including grades, written comments on deportment (classroom behavior), and placement in classes. For example, based on standardized test scores or classroom performance, educators label some children as "special ed" or low achievers, whereas others are labeled as average or "gifted and talented." For some students, labeling amounts to a *self-fulfilling prophecy*—previously defined as an unsubstantiated belief or prediction resulting in behavior that makes the originally false belief come true (Merton, 1968). A classic form of labeling and the self-fulfilling prophecy occurs through the use of IQ (intelligence quotient) tests, which claim to measure a person's inherent intelligence, apart from any family or school influences on the individual. In many school systems, IQ tests are used as one criterion in determining student placement in classes and ability groups.

Using Labeling Theory to Examine the IQ Debate
The relationship between IQ testing and labeling theory has been of special interest to sociologists. In the 1960s, two social scientists conducted an experiment in an elementary school during which they intentionally misinformed teachers about the intelligence test scores of students in their classes

(Rosenthal and Jacobson, 1968). Despite the fact that the students were randomly selected for the study and had no measurable differences in intelligence, the researchers informed the teachers that some of the students had extremely high IQ test scores, whereas others had average to below-average scores. As the researchers observed, the teachers began to teach "exceptional" students in a different manner from other students. In turn, the "exceptional" students began to outperform their "average" peers and to excel in their classwork. This study called attention to the labeling effect of IQ scores.

However, experiments such as this also raise other important issues: What if a teacher (as a result of stereotypes based on the relationship between IQ and race) believes that some students of color are less capable of learning? Will that teacher (often without realizing it) treat such students as if they are incapable of learning? In their controversial book *The Bell Curve: Intelligence and Class Structure in American Life,* Richard J. Herrnstein and Charles Murray (1994) argue that intelligence is genetically inherited and that people cannot be "smarter" than they are born to be, regardless of their environment or education. According to Herrnstein and Murray, certain racial–ethnic groups differ in average IQ and are likely to differ in "intelligence genes" as well. For example, they point out that on average, people living in Asia score higher on IQ tests than white Americans and that African Americans score 15 points lower on average than white Americans. Based on an all-white sample, the authors also concluded that low intelligence leads to social pathology, such as high rates of crime, dropping out of school, and winding up poor. In contrast, high intelligence typically leads to success, and family background plays only a secondary role.

Many scholars disagree with Herrnstein and Murray's research methods and conclusions. Two major flaws found in their approach were as follows: (1) the authors used biased statistics that underestimate the impact of hard-to-measure factors such as family background, and (2) they used scores from the Armed Forces Qualification Test, an exam that depends on the amount of schooling that people have completed. Thus, what the authors claim is immutable intelligence is actually acquired skills (Weinstein, 1997). Despite this refutation, the idea of inherited mental inferiority tends to acquire a life of its own when people want to believe that such differences exist (Duster, 1995; Hauser, 1995; Taylor, 1995). According to researchers, many African American and Mexican American children are placed in special education classes on the basis of IQ scores when the students are not fluent in English and thus cannot understand the directions given for the test. Moreover, when children are labeled as "special ed" students or as being *learning disabled,* these terms are social constructions that may lead to stigmatization and become a self-fulfilling prophecy (Carrier, 1986; Coles, 1987).

Labeling students based on IQ scores has been an issue for many decades. Immigrants from southern and eastern Europe—particularly from Italy, Poland, and Russia—who arrived in this country at the beginning of the twentieth century had lower IQ scores on average than did northern European immigrants who had arrived earlier from nations such as Great Britain. For many of the white ethnic students, IQ testing became a self-fulfilling prophecy: Teachers did not expect them to do as well as children from a northern European (WASP) family background and thus did not encourage them or give them an opportunity to overcome language barriers or other educational obstacles. Although many students persisted and achieved an education, the possibility that differences in IQ scores could be attributed to linguistic, cultural, and educational biases in the tests was largely ignored (Feagin and Feagin, 2003). Debates over the possible intellectual inferiority of white ethnic groups are unthinkable today, but arguments pertaining to African Americans and IQ continue to surface.

A self-fulfilling prophecy can also result from labeling students as gifted. Gifted students are considered to be those with above-average intellectual ability, academic aptitude, creative or productive thinking, or leadership skills (Ballantine, 2001). When some students are labeled as better than others, they may achieve at a higher level because of the label. Ironically, such labeling may also result in discrimination against these students. For example, according to law professor Margaret Chon (1995: 238), the "myth of the superhuman Asian" creates a self-fulfilling prophecy for some Asian American students:

When I was in college, I applied to the Air Force ROTC program. . . . I was given the most complete physical of my life [, and] I took an intelligence test. When I reported back to the ROTC staff, they looked glum. What is it? I thought. Did the physical turn up some life-threatening defect? . . . It turned out I had gotten the highest test score ever at my school. . . . Rather than feeling pleased and flattered, I felt like a sideshow freak. The recruiters were not happy either. I think our reactions had a lot to do with the fact that I did not resemble a typical recruit. I am a woman of East Asian, specifically Korean, descent.

. . . They did not want me in ROTC no matter how "intelligent" I was.

According to Chon, painting Asian Americans as superintelligent makes it possible for others to pretend that they do not exist: "Governments ignore us because we've already made it. Schools won't recruit us because we do so well on the SATs. . . . Asian Americans seem almost invisible, except when there is a grocery store boycott—or when we're touted as the model minority" (Chon, 1995: 239–240).

Labeling and the self-fulfilling prophecy are not unique to U.S. schools. Around the globe, students are labeled by batteries of tests and teachers' evaluations of their attitudes, academic performance, and classroom behavior. Other problems in education, ranging from illiteracy and school discipline to unequal school financing and educational opportunities for students with disabilities, are concerns in elementary and secondary education in many nations.

Postmodernist Perspectives

As discussed in Chapter 15, postmodern theories often highlight *difference* and *irregularity* in society. From this perspective, education—like the family—is a social institution characterized by its permeability. In contemporary schools, a wide diversity of family kinship systems is recognized, and educators attempt to be substitute parents and promulgators of self-esteem in students. Urbanity is reflected in multicultural and anti-bias curricula that are initially introduced in early-childhood education. Similarly, autonomy is evidenced in policies such as voucher systems, under which parents have a choice about which schools their children will attend. Since the values of individual achievement and competition have so permeated contemporary home and school life, social adjustment (how to deal with others) has become of little importance to many people (Elkind, 1995).

How might a postmodern approach describe higher education? Postmodern views of higher education might incorporate the ideas of the sociologist George Ritzer (1998), who believes that "McUniversity" can be thought of as a means of educational consumption that allows students to consume educational services and eventually obtain "goods" such as degrees and credentials:

> Students (and often, more importantly, their parents) are increasingly approaching the university as consumers; the university is fast becoming little more than another component of the consumer society. . . . Parents are, if anything, likely to be

Courtesy of Baylor University

Many colleges and universities now offer students a wide array of new amenities and recreational facilities, such as this 52-foot climbing wall located in the student life building at Baylor University.

even more adept as consumers than their children and because of the burgeoning cost of higher education more apt to bring a consumerist mentality to it. (Ritzer, 1998: 151–152)

Savvy college and university administrators are aware of the permeability of higher education and the "students-as-consumers" model:

> [Students] want education to be nearby and to operate during convenient hours—preferably around the clock. They want to avoid traffic jams, to have easy, accessible and low cost parking, short lines, and polite and efficient personnel and services. They also want high-quality products but are eager for low costs. They are willing to shop—placing a premium on time and money. (Levine, 1993: 4)

To attract new students and enhance current students' opportunities for consumption, many campuses have student centers equipped with amenities such as food courts, ATMs, video games, Olympic-sized swimming pools, and massive rock-climbing walls. "High-tech" or "wired" campuses are also a major attraction for student consumers, and virtual classrooms make it possible for some students to earn college credit without having to look for a parking place at the traditional brick-and-mortar campus.

The permeability of contemporary universities may be so great that eventually it will be impossible to distinguish higher education from other means of consumption. For example, Ritzer (1998) believes that officials of "McUniversity" will start to emphasize the same kinds of production values as CNN or MTV, resulting in a simulated world of education somewhat like postmodernist views of Disneyland. Based on Baudrillard's fractal stage, where everything interpenetrates, Ritzer (1998: 160) predicts that we may enter a "transeducational" era: "Since education will be everywhere, since everything will be educational, in a sense nothing will be educational." Based on a postmodern approach, what do you believe will be the dominant means by which future students will consume educational services and goods at your college or university?

INEQUALITIES AMONG ELEMENTARY AND SECONDARY SCHOOLS

As in colleges and universities, education in kindergarten through high school is a microcosm of many of the issues and problems facing the United States and other nations. Today, there are almost 15,000 school districts in the United States in what is probably the most decentralized system of public education in any high-income, developed nation of the world. Countries such as France have an education ministry that is officially responsible for every elementary school in the nation. Countries such as Japan have a centrally controlled curriculum; in nations such as England, national achievement tests are administered by the government, and students must pass them in order to advance to the next level of education (Lemann, 1997).

Inequality in Public Schools Versus Private Schools

Often, there is a perceived competition between public schools and private schools for students and financial resources. However, far more students and their parents are dependent on public schools than private ones for providing a high-quality education. Enrollment in U.S. elementary and secondary education (grades K through 12) totals 53 million, and these numbers are projected to increase to over 55 million by 2020 (Children's Defense Fund, 2002). Almost 90 percent of U.S. elementary and secondary students are educated in public schools. About 9.5 percent of all students are educated in low-tuition private schools, primarily Catholic (parochial) schools; only 1.5 percent of all students attend private schools with tuition of more than $5,000 a year (Lemann, 1997).

Private secondary boarding schools tend to be reserved for students from high-income families and for a few lower-income or subordinate-group students who are able to acquire academic or athletic scholarships that cover their tuition, room and board, and other expenses. The cost for seven-day tuition and room and board at secondary boarding schools is between $16,000 and $25,000 per year (American Universities Admission Program, 2003).

Among parents whose children attend private secondary schools, an important factor for a majority (51 percent) of the parents is the emphasis on academics that they believe exists in private (as opposed to public) schools. The most important factor for 17 percent of the parents is the moral and ethical standards that they believe private secondary schools instill in students. Overall, many families believe that private schools are a better choice for their children because they feel that these schools are more academically demanding, motivate students to learn, provide more-stringent discipline, and do not have many of the inadequacies found in public schools. However, according to some social analysts, there is little evidence to substantiate the claim that private schools (other than elite academies attended by the children of the wealthiest and most-influential families) are inherently better than public schools (Berliner and Biddle, 1995).

Unequal Funding of Public Schools

One of the biggest problems in public education today results from unequal funding. Why does unequal funding exist? Most educational funds come from state legislative appropriations and local property taxes. State and local governments contribute about 47 percent *each* toward educational expenses, and the federal government pays the remaining 6 percent, largely for special programs for students who are disadvantaged (e.g., the Head Start program) or have disabilities (Bagby, 1997). Although public school spending increased by 28 percent between 1991 and

"Rich" schools and "poor" schools are readily identifiable by their buildings and equipment. What are the long-term social consequences of unequal funding for schools?

1996, there was an 8.4-percent increase in enrollment during that same period, and special education programs received an increasing share of school budgets (*New York Times,* 1997). According to one economist, "The largest share of the new money has gone to special programs, not to regular education. We have shifted much of the . . . responsibility for children with disabilities from private charities and institutions to the public schools" (qtd. in *New York Times,* 1997: A15).

Per-capita spending on public and secondary education varies widely from state to state (see Map 16.1). In part, this is because the local property-tax base has been eroding in central cities as major industries have relocated or gone out of business. Many middle- and upper-income families have moved to suburban areas with their own property-tax base so their children can attend relatively new schools equipped with the latest textbooks and state-of-the-art computers—advantages that schools in central cities and poverty-ridden rural areas lack (see Kozol, 1991; Ballantine, 2001).

Unequal funding, along with underfunding, of public schools has opened a door to the commercialization of education as corporations have provided media equipment, computers, money, and other resources for schools. However, there is a potential downside to the largesse of corporations with regard to public education (see Box 16.2).

Inequality Within Public School Systems

Another major problem in public school systems is the inequality that exists *within* local public school systems. Although the problem described in this section could be found in any major school district in the United States, we will focus on a case study conducted by education scholar John Devine (1996) regarding the New York City public schools. According to Devine, public high schools in New York City are stratified into higher-, middle-, and lower-tiered schools. Although the New York school system does not officially designate high schools on this basis, the higher-tier schools are at the pinnacle of the educational structure. These schools provide students with a completely different educational environment and learning experience than the lower-tier schools. Higher-tier schools are highly specialized; admission is limited to students who pass a highly competitive entrance examination, and dropout rates are low. These schools turn out winners of academic achievement awards, and a large number of graduates attend prestigious universities. At the next level are middle-stratum schools that are academically comprehensive and located in fairly well-integrated neighborhoods. Referred to as "ed op" (educational option) schools, these high schools offer training for specific careers, such as health or computer science, and college prep courses. Some "ed op" schools serve as *magnet schools,* which offer a specialized curriculum that focuses on areas such as science, music, or art, and enroll high-achieving students from other neighborhoods.

According to Devine (1996: 23), in New York City it is not only the private schools of Manhattan and the Catholic schools that "siphon the best and the brightest—and the wealthiest—away from the public system . . . rather, the system itself does the most thorough job of drawing 'the best' to the top." In sharp contrast, students who attend schools in the lower tier tend to have low test scores and poor attendance.

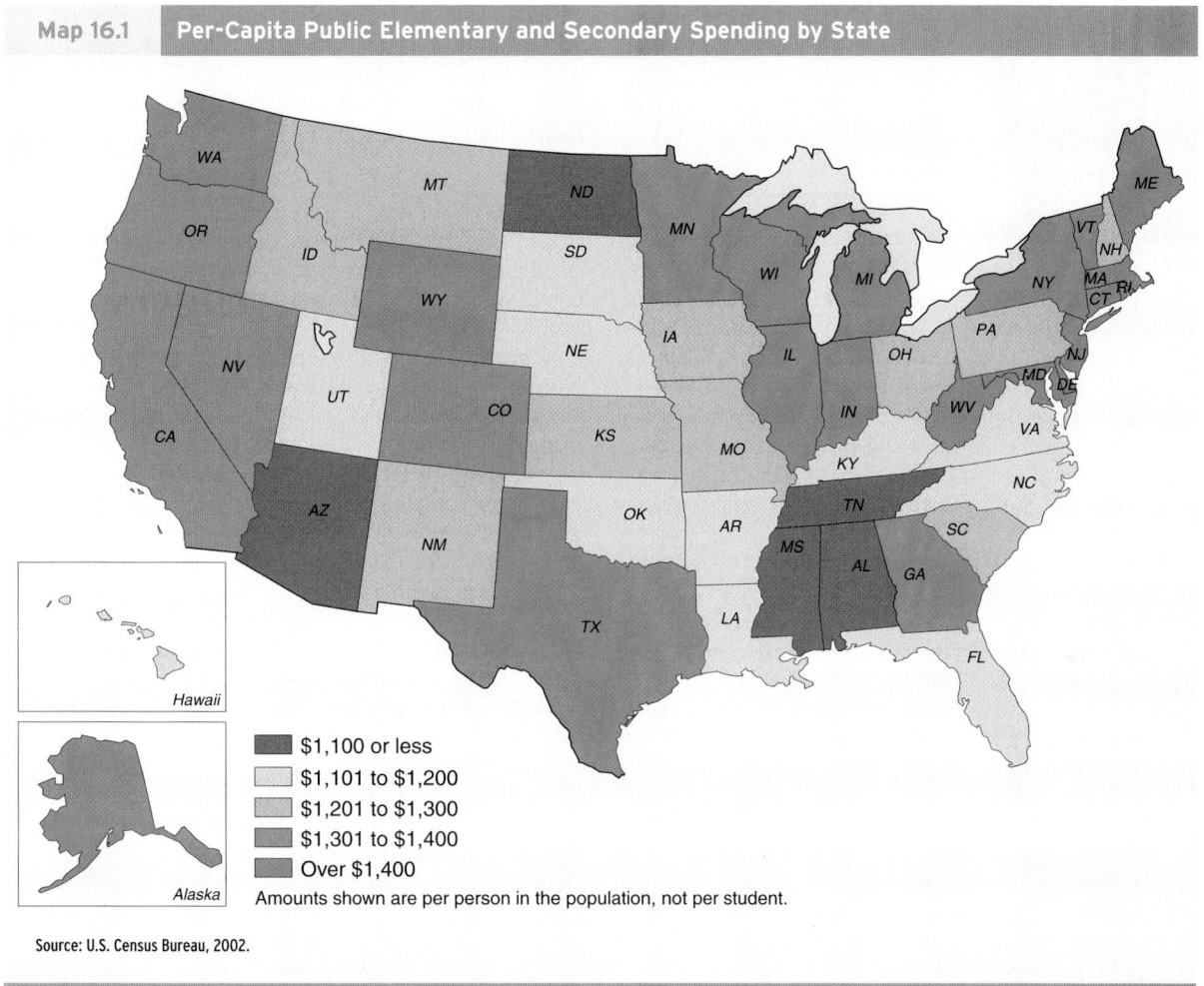

| Map 16.1 | Per-Capita Public Elementary and Secondary Spending by State |

Legend:
- $1,100 or less
- $1,101 to $1,200
- $1,201 to $1,300
- $1,301 to $1,400
- Over $1,400

Amounts shown are per person in the population, not per student.

Source: U.S. Census Bureau, 2002.

Most lower-tier schools are very large, overcrowded neighborhood schools located in highly segregated, deteriorating, and violent neighborhoods. Among New York City public schools, the lower-tier high schools have the highest dropout rates, the lowest graduation rates, the worst scores on standardized tests, the poorest attendance patterns, and the worst statistics on assault and possession of weapons.

Recent studies have found that the official dropout rate in New York City schools is about 20 percent, but if students who are "pushed out" are included, that number could be as high as 25 to 30 percent (Lewin and Medina, 2003). So that some of the schools will not have low pass rates on stringent Regents exams, which a student must pass before receiving a high school diploma, some schools have allegedly been pushing out low-achieving students, getting them to leave school of their own accord so that the schools' statistics will not be tarnished if these students fail to graduate on time (Lewin and Medina, 2003). (The term *push out* was also used by Latino and Latina students in the 1960s to describe the prac-

tice in the East Los Angeles schools of ignoring or discouraging Hispanic students in hopes that they would seek alternative schooling, take the GED high school equivalency exam, or simply drop out of their own accord so that they would not cause "problems.") Push-out rates and other related problems exist in school districts across the nation because, as the early sociologist Emile Durkheim suggested, schools are a reflection of the larger political, economic, and social environment in which they exist. As such, we might expect that schools also reflect the racial and ethnic divisions within the larger society.

Racial Segregation and Resegregation

In many areas of the United States, schools remain racially segregated or have become resegregated after earlier attempts at integration failed. In 1954 the U.S. Supreme Court ruled (in *Brown v. The Board of Education of Topeka, Kansas*) that "separate but equal"

Box 16.2 CHANGING TIMES: MEDIA AND TECHNOLOGY

"But It's Just a TV!" Is Commercialization in Schools a Problem?

Public schools are largely financed by funds that are allocated by state legislatures and local school boards. As a result of shortages of tax dollars and other public funds, public schools have increasingly come under the influence of corporations that willingly supply money and other resources for classrooms and athletic programs (Giroux, 2000). Education professor Henry A. Giroux (2000) has pointed out that corporations derive major benefits from these "gifts" to public education. According to Giroux (2000), by placing various forms of advertising in schools, corporations reach an estimated 43 million school-age children who spend over $108 billion each year and also influence how their parents spend money. However, some school trustees and administrators have shown little concern for the effects that commercialization may have on young people, including how students are educated to think of themselves first and foremost as consumers of corporate products. An example is Channel One, a for-profit television network that each day broadcasts into the classroom a ten-minute program of news and current events along with two minutes of commercials, typically for candy, soft drinks, athletic shoes, video games, and similar products.

Why do school officials agree to arrangements such as this? One factor may be that they believe that it is important to bring news and public affairs programming to students who otherwise might not keep up with current events through newspapers or television broadcasts. However, an equally compelling factor in their decisions may be the availability of "free" electronic equipment from a corporation. Channel One, for example, provides schools with as much as $25,000 in equipment such as VCRs, television monitors, and satellite dishes if the schools agree to let students watch the twelve-minute sequence of news and commercials during the school day (Channel One Network, 2001).

Channel One is not alone in its corporate contributions to public schools and, correspondingly, increased commercialization in classrooms. Other examples include a McDonald's curriculum package for elementary schools that teaches children how a McDonald's restaurant is built, how the restaurant operates, and how the students can eventually apply for a job there (*Business Week,* 1997), as well as other corporate sponsors, such as Pizza Hut, that give away textbook covers promoting their products.

Some legislators and school board members justify such advertising by describing it as a learning experience for students, not as an economic aid for cash-strapped school districts. However, other analysts believe that the process of commercialization will undermine what public education is supposed to be all about, namely, as Giroux (2000: 83) states, "providing students with the critical capacities, knowledge, and values that enable them to become active citizens striving to build a stronger democratic society." These analysts believe that schooling should be adequately financed with tax dollars and that concerned citizens should challenge the encroachment of corporate power in schools. In some states, for example, these concerned citizens have created organizations to lobby for legislation that would ban brand-name advertising and corporate logos on school property.

Will school boards and legislatures become more wary of commercialization of public schooling, or will this trend become even more prevalent in the future? What do you think will happen with regard to the commercialization of education in the future?

---

segregated schools are unconstitutional because they are inherently unequal. However, four decades later, racial segregation remains a fact of life in education.

Efforts to bring about *desegregation*—the abolition of legally sanctioned racial–ethnic segregation—or *integration*—the implementation of specific action to change the racial–ethnic and/or class composition of the student body—have failed in many districts throughout the country. Some school districts have bused students across town to achieve racial integration. Others have changed school attendance boundaries or introduced magnet schools with specialized

programs such as science or the fine arts to change the racial–ethnic composition of schools. But school segregation does not exist in isolation. Racially segregated housing patterns are associated with the high rate of school segregation experienced by African American and Latina/o students. Although most racial segregation has affected African Americans, Latinas/os in the Southwest are even more likely than African American or white (non-Latina/o) students to attend segregated schools (Brooks and South, 1995).

Racial segregation and class segregation are apparent in many urban school districts where students of color constitute the vast majority of the student body. In contrast, private urban schools or upscale public schools in the suburbs are attended primarily by middle-class or upper-middle-class white students. In some school districts, more than 90 percent of the students are African American or Latina/o, and many live in low-income homes where English is not spoken (Judson, 1995).

Why are racially segregated schools a problem? Most research shows that such schools have serious negative consequences for minority students. According to recent studies, predominantly minority schools have lower student-retention rates, have higher teacher–student ratios, and employ less-qualified teachers who typically have lower expectations for their students (Feagin and Feagin, 2003).

Even in more-integrated schools, resegregation often occurs at the classroom level (Mickelson and Smith, 1995). Because of past racial discrimination and current socioeconomic inequalities, many children of color are placed in lower-level courses and special education classes. At the same time, non-Latina/o white and Asian American students are more likely to be enrolled in high-achievement courses and programs for the gifted and talented. However, racial segregation occurs even in required courses such as physical education (McLarin, 1994). Thus, the achievement gap between African American and white first-graders is much less than the gap that exists between African American and white twelfth-graders. According to researchers, schools tend to reinforce—rather than eliminate—the disadvantages of race and class during the years of a child's educational experience (Mickelson and Smith, 1995).

PROBLEMS WITHIN ELEMENTARY AND SECONDARY SCHOOLS

The inequalities in educational opportunities we have previously discussed focus primarily on macrolevel issues in grades K through 12 and frequently involve decisions made by state legislatures, school boards, and school administrators. The problems we examine in this section, including school discipline, bullying and sexual harassment, and school violence typically take place within the microcosm of individual schools, although the effect of these problems is much more widespread than at the original locations where they take place.

School Discipline and Teaching Styles

Is it possible for teachers and school administrators to handle the wide array of discipline problems that occur in many contemporary schools? Over the past three decades, this question has been asked by sociologists and education scholars; however, a lack of consensus exists regarding whether it is possible to make changes within the schools without reducing or eliminating larger structural problems in society that are intertwined with school problems. For example, one study conducted by the California Department of Education compared school discipline problems over the past sixty years. In the 1940s, some of the leading discipline problems included getting out of place in line, talking without permission during class, or not putting paper in the trash can (Collins and Frantz, 1993). By contrast, in the 1990s and into the twenty-first century, the leading discipline problems include violence, drug abuse, suicide, robbery, assault, and other forms of aggressive behavior.

Although schools in the past used corporal punishment (usually in the form of spankings or "paddling" of students) to discipline errant students, most states now prohibit the use of corporal punishment. Some school districts also prohibit disciplinary actions such as keeping children in during recess or expelling them from class. Moreover, many teachers and school administrators do not believe that corporal punishment or any other form of negative sanctions will deter the kinds of behavior that are occurring in classrooms and on the school grounds, as Evelyn Campbell, who has been teaching for over thirty years, explains:

> Not in this day and time, I don't think corporal punishment will work. We've seen violence among students escalate in this district—kids carrying knives and weapons, kids with Uzis in their cars. A teacher dare not take that risk to hit a kid. For that reason alone, we need to stay away from corporal punishment. It could endanger a teacher's life. When they gave kids rights and parents rights, it meant hands off. We still have detention, expulsion. That's all we do. (qtd. in Collins and Frantz, 1993: 143)

Maintaining classroom order and discipline is often a more difficult task for those who teach students whose family background and life experiences are vastly different from their own. For example, one study found that students whose earlier experiences had been shaped by traditional cultural and educational systems—including Caribbean, Chinese, African, and Indian systems—were shocked when they entered the lower-tier New York City high schools by how most of the students behaved, but they did not find support from teachers for acting otherwise. Over time, the newer arrivals tended to act much more like the indigenous student body (Devine, 1996).

Teaching and disciplinary styles may differ across racial and ethnic lines as well. Most studies on this issue have focused on the styles of white and African American teachers. According to one study, many white teachers try to ignore racial and ethnic differences in their students. As one teacher stated, "I don't see color, I only see children" (qtd. in Delpit, 1995: 177). However, other analysts believe that this approach suggests that there is something wrong with being black or brown and that color *should* be noticed. According to the education scholar Lisa Delpit (1995: 177), "If one does not see color, then one does not really see children. Children made 'invisible' in this manner become hard-pressed to see themselves as worthy of notice."

Can educational reforms reduce or eliminate school discipline problems? According to the education scholar Jean Anyon (1997), educational reform has a long way to go in bringing about better discipline—and other needed changes—because of structural inequalities in the larger society. Thus, when "blackness and whiteness, extreme poverty and relative affluence, cultural marginalization and social legitimacy come together—and conflict—within a school," the prospects of better discipline and other educational reform, including a reduction in problems such as bullying, teasing, sexual harassment, and school violence, are greatly diminished (based on Anyon, 1997: 36–37).

Bullying, Teasing, and Sexual Harassment

Although some people think of harassment as a "women's issue," in schools throughout the United States, both boys and girls report that they have experienced harassment, bullying, and teasing from their classmates. According to a recent study of 2,064 public school students in eighth through eleventh grades conducted by Harris Interactive (a worldwide research firm that conducts the Harris poll) on behalf of the American Association of University Women (AAUW, 2001),

- Eighty-three percent of girls and 79 percent of boys report having ever experienced harassment, with many students reporting that sexual harass-

ment was an ongoing experience that occurred "often."
- Seventy-six percent of students have experienced nonphysical harassment (such as taunting, rumors, graffiti, jokes, or gestures) while 58 percent have experienced physical harassment.
- Although both boys and girls report that they have been harassed, girls are more likely to report being negatively affected by it in ways such as feeling "self conscious," "embarrassed," or "less confident."
- Substantial numbers of students fear that they will be sexually harassed or hurt in school.

Researchers in this study provided students with the definition of sexual harassment as "unwanted and unwelcome sexual behavior that interferes with your life. Sexual harassment is not behaviors that you like or want (for example, wanted kissing, touching, or flirting)." Based on this definition, many of the students believed that their academic experience had been diminished by the actions of other students, even as early as elementary school. And, as a spokesperson for the National Education Agency commented, "For children who are constantly picked on, ridiculed or harassed, school becomes torture" (AAUW, 2001). In fact, many analysts believe that there is a link between some persistent forms of harassment and school violence. When violence occurs in a school, some officials are quick to point out that the perpetrator (as in the case of a school shooting) was an "outsider" or someone who had been ridiculed by other students. Schools no doubt have a long way to go in educating students and training teachers and administrators to deal appropriately with the complex problems of bullying, teasing, and sexual harassment, as well as how to diffuse situations that may result in violent behavior.

School Violence

Violence and fear of violence continue to be problems in schools throughout the United States. In the 1990s, violent acts resulted in numerous deaths in schools across the nation. Schools in communities such as Pearl, Mississippi; West Paducah, Kentucky; Jonesboro, Arkansas; Springfield, Oregon; and Littleton, Colorado, witnessed a series of killings in schools by students that shocked people across the world.

Throughout the nation, there has been a call for greater school security. As a result, more students are in an academic environment that is similar to a prison. For example, students and faculty must present identity cards upon entering the school grounds

Each incident of school violence leaves an aftermath of grief, fear, and concern about the future of schools, students, teachers, and administrators. What can be done to reduce acts of violence that have occurred in schools such as Columbine High School in Littleton, Colorado?

© Gary Caskey/Reuters/Landov

or buildings. In some cases, they must wear photo IDs around their necks while on school premises. Frequently, students, faculty, and staff are required to pass through a weapons-scanning metal detector and have their backpacks and other items inspected or sent through an X-ray machine. Many schools are equipped with magnetic door locks and have a large staff of security guards who carry walkie-talkies and weapons. Faculty and security guards often use criminal justice vernacular such as "scanning," "holding areas," and "corridor sweeps" to describe their daily activities at the schools (Devine, 1996).

However, many education analysts believe that technology and increased law enforcement should not be the primary tools for achieving school discipline and ending violence and crime on school campuses. Education scholar John Devine (1996) emphasizes that schools must be understood as part of a larger, tumultuous society: Violence in schools will not end until guns are eliminated from U.S. society. Given the fact that some schools do not motivate students to learn and are arranged much like a prison, is it any wonder that school dropout rates are high?

Dropping Out

Although there has been a decrease in the overall school dropout rate over the past two decades, about 10 percent of people between the ages of fourteen and twenty-four have left school before earning a high school diploma. However, ethnic and class differences are significant in dropout rates. For example, Latinos/as (Hispanics) have the highest dropout rate (34.4 percent), followed by African Americans (17.1 percent) and non-Hispanic whites (13.7 percent) (U.S. Census Bureau, 2000c). The dropout rate also varies by region. In New York City, for example, African American, Puerto Rican, and Italian American youths have the highest dropout rates (Pinderhughes, 1997).

Why are Latinos/as more likely to drop out of high school than are other racial and ethnic groups? First, the category of "Hispanic" or "Latino/a" incorporates a wide diversity of young people—including those who trace their origins to Mexico, Puerto Rico, Haiti, and countries in Central and South America—who may leave school for a variety of reasons. Second, some students may drop out of school partly because their teachers have labeled them as "troublemakers." Some students have been repeatedly expelled from school before they actually become "dropouts." In a study by the sociologist Howard Pinderhughes, one of the respondents—Rocco—discusses how he was expelled several times before he finally dropped out:

> I never liked school. The teachers all hated my guts because I didn't just accept what they had to say. I was kind of a wiseass, and that got me in trouble. But that's not why they kept kicking me out of school. Either I didn't come to school or I got in fights when I did come. It got so I got a reputation that people heard about and there was always some kid challengin' me. (qtd. in Pinderhughes, 1997: 58)

Some students may view school as a waste of time. Another respondent in Pinderhughes's study—Ronnie—explains why he dropped out:

> I thought [school] was a big waste of time. I knew older guys who had dropped out of the tenth grade and within a couple of years were makin' cash money as a mechanic or a carpenter. My brothers did the same thing. I didn't think I was learnin' no skill that I could get a job with in school. I wasn't very good in school. It was pretty damn boring. So I left. (qtd. in Pinderhughes, 1997: 58)

Students who drop out of school may be skeptical about the value of school even while they are still attending because they believe that school will not increase their job opportunities. Upon leaving school, many dropouts have high hopes of making some money and enjoying their newfound freedom. However, these feelings often turn to disappointment when they find that few jobs are available and that they do not meet the minimum educational requirements for any "good" jobs that exist (Pinderhughes, 1997).

In a study of at-risk Latino/a high school students, the sociologists Harriett Romo and Toni Falbo (1996) found that a growing number of students were taking the GED (General Equivalency Diploma) exam rather than graduating from high school. The most common reasons given for dropping out of school included lack of interest in school, serious personal problems, serious family problems, poor grades, and alcohol and/or drug problems (Romo and Falbo, 1996). Despite stereotypes to the contrary, Romo and Falbo (1996) found that becoming a teen mother was not necessarily linked to dropping out of high school. According to these researchers, many girls who dropped out were neither pregnant nor mothers at the time they left school. In fact, some schools had better programs for teen mothers to help them stay in school until graduation than they had for other students.

Other studies continue to confirm that Latina/o students are disadvantaged in school districts that tend to view bilingualism and some cultural values as a liability rather than an asset. A study conducted by the AAUW found that the high school graduation rate for Latinas was lower than for girls in any other racial or ethnic group, although it was still higher than the graduation rate for Latinos (Ginorio and Huston, 2000). Compared to their white or Asian counterparts, Latinas were also less likely to take the SAT exam and were the least likely of any group of women to complete a bachelor's degree. However, Latinas and Latinos are only one of many categories of students that face significant challenges in grades K through 12 and in their pursuit of a college degree. We will now look at some of the key problems in higher education.

PROBLEMS IN COLLEGES AND UNIVERSITIES

Who attends college? What sort of college or university do they attend? For students who complete high school, access to colleges and universities is deter-

Soaring costs of both public and private institutions of higher education are a pressing problem for today's college students and their parents. What factors have contributed to the higher overall costs of obtaining a college degree?

mined not only by prior academic record but also by the ability to pay.

The Soaring Cost of a College Education

What does a college education cost? Recent studies by the College Board (a nonprofit organization that provides tests and many other educational services for students, schools, and colleges) have found that a college education is quite expensive and that increases in average yearly tuition for four-year colleges are higher than the overall rate of inflation (Bagby, 1997). Although public institutions such as community colleges and state colleges and universities typically have lower tuition and overall costs—because they are funded primarily by tax dollars—than private colleges have, the cost of attending public institutions has increased dramatically over the past decade. A substantial proportion of college students start at the lower cost level of junior or community colleges. However, the overall enrollment of low-income students has dropped as a result of declining scholarship funds since the 1980s and also because many students must work full time or part time to pay for their education.

As a result, many students drop out before they have completed a two-year degree. In contrast, students from affluent families are more likely to attend prestigious public universities or private colleges, where annual costs range from $12,000 to more than $30,000 annually (Famighetti, 1997).

According to some social analysts, a college education is a bargain—even at about $90 a day for private schools or $35 for public schools—because for their money students receive instruction, room, board, and other amenities such as athletic facilities and job placement services. However, other analysts believe that the high cost of a college education reproduces the existing class system: Students who lack money may be denied access to higher education, and those who are able to attend college tend to receive different types of education based on their ability to pay. For example, a community college student who receives an associate's degree or completes a certificate program may be prepared for a position in the middle of the occupational status range, such as a dental assistant, computer programmer, or auto mechanic (Gilbert, 2003). In contrast, university graduates with four-year degrees are more likely to find initial employment with firms where they stand a chance of being promoted to high-level management and executive positions. Although higher education may be a source of upward mobility for talented young people from poor families, the U.S. system of higher education is sufficiently stratified that it may also reproduce the existing class structure (Gilbert, 2003).

Racial and Ethnic Differences in Enrollment

How does college enrollment differ by race and ethnicity? People of color (who are more likely than the average white student to be from lower-income families) are underrepresented in higher education. However, some increases in minority enrollment have occurred over the past three decades. Latina/o enrollment as a percentage of total college enrollment increased from about 3.5 percent to 8.9 percent between 1976 and 2000. However, Latinos/as are underrepresented in higher education: Today, Latinos/as account for 13.4 percent of the total U.S. population (*Chronicle of Higher Education,* 2002; U.S. Census Bureau, 2002).

Although African American enrollment increased somewhat between 1970 and 2000, it remains steady today at about 11 percent. Gender differences are evident in African American enrollment: Women accounted for more than 63 percent of all African

American college students in 1999 (*Chronicle of Higher Education,* 2002). Native American enrollment rates have remained stagnant at about 0.9 percent from the 1970s to the 2000s; however, two-year community colleges—known as tribal colleges—on reservations have experienced an increase in Native American student enrollment. Founded to overcome racism experienced by Native American students in traditional four-year colleges and to shrink the high dropout rate among Native American college students, there are now 27 colleges chartered and run by the Native American nations (Holmes, 1997). Unlike other community colleges, however, the tribal colleges receive no funding from state and local governments and, as a result, are chronically short of funds to fulfill their academic mission. But for most Native Americans residing on reservations, these schools provide the best hope for attaining a higher education (Holmes, 1997).

The proportionately low number of people of color enrolled in colleges and universities is reflected in the educational achievement of people age 25 and over, as shown in the "Census Profiles" feature. If we focus on persons who receive doctorate degrees, the underrepresentation of persons of color is even more striking. Minority-group members accounted for about 18 percent of the doctorates awarded in 1999–2000: African Americans earned 6.6 percent, whereas Asian Americans earned 7.1 percent, Latinos/as 2.8 percent, and Native Americans 0.5 percent (*Chronicle of Higher Education,* 2002).

Despite some recent gains in higher education, students of color continue to experience problems of prejudice and discrimination at predominantly white colleges and universities. According to the sociologists Joe R. Feagin, Hernán Vera, and Nikitah Imani (1996), racial discrimination presents major barriers for African American students in many major colleges and universities. The barriers are erected by white students, professors, administrators, police officers, and other staff members. The authors concluded the following:

[T]he black students and parents we interviewed call attention to African American students not being recognized as full members of the campus community. From university publications to the daily rhythm of life on campus, many symbols, comments, and actions suggest to black students that they really do not belong. The strength of their perceptions and feelings on these matters underscores the need for making majority-white educational institutions more welcoming to students whose cultures are not those of the dominant group. (Feagin, Vera, and Imani, 1996: 173)

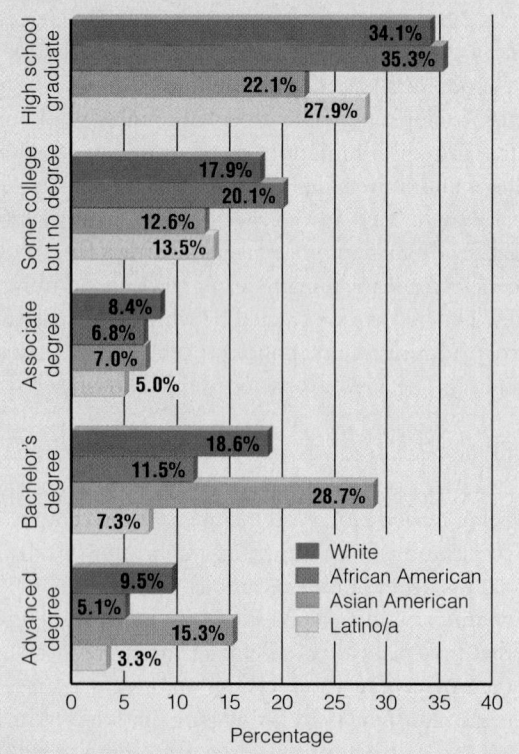
Similarly, the sociologist Felix M. Padilla (1997) found that Latino/a students and professors in a major university sought to overcome isolation and discrimination by forging their own alliances and developing specialized curricula. According to Padilla (1997), Latino/a professors must speak out against injustices against Latino/a students, but this is difficult on campuses where Latina/o scholars are underrepresented in tenured faculty positions.

The Lack of Faculty Diversity

Despite the widely held assumption that there has been a significant increase in the number of minority professors, a recent nationwide survey found that African Americans, Latinas/os, Asian Americans, and other people of color accounted for about 15 percent of all full-time faculty members (*Chronicle of Higher Education,* 2002). Although there has been a slight increase in their overall representation, minority scholars are in the lowest tiers of the academic profession. According to one study, 35 percent of the white faculty members surveyed were full professors (with tenure), compared with only 22 percent of the African American faculty members and 25 percent of the Latino faculty members surveyed (*Chronicle of Higher Education,* 2002). As a category, Asian American scholars have fared the best: Thirty-one percent of those in the study were full professors. Figure 16.2 shows the academic rank and tenure status of faculty members by racial and ethnic group and sex. Gender is also a factor in faculty diversity: Across racial and ethnic categories, women are underrepresented at the level of full professor and overrepresented at the assistant professor and instructor levels. Although assistant professors may be on a "tenure track," neither of these last two ranks provides the security of tenured positions.

Minority faculty members are more likely than their white counterparts to be employed at two-year colleges. For example, 36 percent of Native American faculty members are employed at tribal colleges located on the reservations, such as Little Big Horn College on the Crow Reservation in south-central Montana, or other community colleges. Likewise, 39 percent of Latino/a faculty members are at two-year colleges. Only 23 percent of white (non-Latina/o) faculty members are employed at two-year institutions (Schneider, 1997). According to social analysts, Native Americans and Latinas/os are underrepresented on the faculty at four-year colleges and universities because fewer minority-group members hold earned doctorates or the highest technical degree in their teaching field. The underrepresentation of Native Americans and Latinos/as as college students, es-

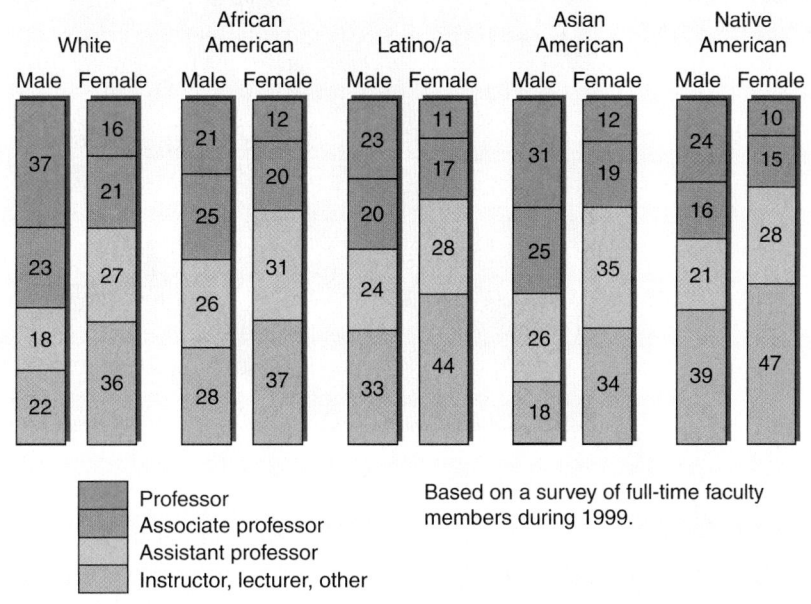

Figure 16.2 Academic Rank and Tenure Status by Race/Ethnicity and Sex

Professor
Associate professor
Assistant professor
Instructor, lecturer, other

Based on a survey of full-time faculty members during 1999.

Source: *Chronicle of Higher Education*, 2002.

pecially in Ph.D. programs, perpetuates the problem of lack of diversity among faculty members. Asian American scholars are the exception to this problem: They are more likely to have a terminal degree in their field and to be employed at a research university (Schneider, 1997).

According to a recent nationwide study, "Race and Ethnicity in the American Professorate," many minority scholars believe that subtle discrimination routinely takes place at their institution. Some scholars of color reported that colleagues devalued their teaching and research because it focused on so-called minority issues. Despite a decrease in some reported forms of discrimination, nearly half of all minority faculty respondents reported that they experienced stress due to subtle discrimination (Schneider, 1997). Based on the findings of such studies, many analysts believe that affirmative action programs must remain in higher education; however, others adamantly disagree, stating that affirmative action has fulfilled its purpose and should now become a thing of the past.

The Debate Over Affirmative Action

In Chapter 10 ("Race and Ethnicity"), we examined the effects of recent U.S. Supreme Court decisions on affirmative action (see Box 10.3). As you will recall, in 2003 the Court ruled in *Grutter v. Bollinger* (involv-

White women and people of color who have been historically underrepresented in law schools and medical schools have gained greater access to these institutions through affirmative action. Although some analysts believe that affirmative action has fulfilled its purpose and is no longer necessary, other analysts believe that the playing field remains uneven for many white women and people of color.

ing admissions policies of the University of Michigan's law school) and *Gratz v. Bollinger* (involving the undergraduate admissions policies of the same university) that race can be a factor for universities in shaping their admissions programs, but only within carefully defined limits. Some analysts believe that these cases

Obtaining a quality education is a problem across all age categories. Today, many adults are enrolled in classes that will help them learn to read, write, and speak English.

© Don Smetzer/PhotoEdit

were a victory for affirmative action; other analysts believe that the effect of these decisions will be limited in scope. One thing remains clear in either event: Our discussions regarding access to higher education—particularly regarding the way that access is influenced by income, race/ethnicity, gender, nationality, and other characteristics and attributes—are far from over as our country grows increasingly diverse in its population. Access is only one of the many educational issues that we face, both in colleges and universities and in grades K through 12, in the twenty-first century.

FUTURE ISSUES AND TRENDS IN EDUCATION

Many books have been written about the issues facing education in the United States and other nations throughout the world. In this section, we will examine only a few of the concerns that are currently being discussed.

Academic Standards and Functional Illiteracy

Some social analysts believe that academic standards are not high enough in U.S. schools. As part of their evidence for this concern, they cite the rate of functional illiteracy in this country. *Functional illiteracy* **is the inability to read and/or write at the skill level necessary for carrying out everyday tasks.** It is estimated that 56 percent of adult Latinas/os are function-

ally illiterate in English, compared with 44 percent of adult African Americans and 16 percent of adult (non-Latino/a) whites. How can the United States have such a high rate of functional illiteracy when we have a relatively high rate of high school graduation? According to one study conducted by the U.S. Department of Education, many people who are functionally illiterate have graduated from high school. In fact, half of the people who scored in the lowest 20 percent on a literacy test were high school graduates. Today, 15 to 20 percent of people who have graduated from high school cannot read at the sixth-grade level (Kaplan, 1993).

However, we must distinguish between *functional illiteracy* and illiteracy. The National Literacy Act of 1991 defines *literacy* as "an individual's ability to read, write, and speak English and compute and solve problems at levels of proficiency necessary to function on the job and in society, to achieve one's goals, and to develop one's knowledge and potential" (U.S. Department of Education, 1993: 3). Based on this definition, the U.S. Department of Education concluded in its research that nearly half (95 million people) of the adult U.S. population was illiterate when it comes to tasks such as understanding a bus schedule, filling out a bank deposit slip, or computing the cost of having some work done (Kaplan, 1993). However, other social analysts note that this report classified people as illiterate on the basis of *one* reading comprehension test and did not take into account that English is not the native language of many respondents or that some respondents were visually impaired or had physical or mental disabilities (Berliner and Biddle, 1995).

Most suggestions for reducing both illiteracy and functional illiteracy have focused on school reforms

Table 16.2 NATIONS WITH THE LOWEST LITERACY RATES

COUNTRY	LITERACY RATE
Niger	15%
Burkina Faso	22%
Mali	31%
Sierra Leone	31%
Afghanistan	31%
Ethiopia	36%
Guinea	36%
Senegal	38%

Source: VirtualSources.com, 2003.

such as more testing of students and teachers, increasing the requirements for high school graduation, and increasing the number of school days per year. Although some critics believe that the illiteracy problem is largely tied to high rates of immigration, the author Jonathan Kozol (1986), who has extensively studied the problem, argues that illiteracy is a homegrown problem that is closely linked to problems of poverty and segregated schools, both of which exist even without immigration. However, illiteracy is a global problem as well as a national one (see Table 16.2).

The Debate Over Bilingual Education

As the United States becomes more racially, ethnically, and culturally diverse, schools become the focus of debate on how to best educate children who do not speak English as their primary language. The issue of bilingual and bicultural education in the twenty-first century is not a new one. During the large waves of immigration to the United States in the late nineteenth and early twentieth centuries, children often received classroom instruction from teachers who spoke the children's native language. For example, Italian American students studied Italian in school, and some students were instructed in the core subjects in their native language until they achieved proficiency in English. During World War I, however, members of white ethnic groups such as German Americans believed that they should show their loyalty to the United States and—as a consequence—that their children should be taught in English. As a result, there was little bilingual education until the years following World War II, when a large influx of Mexican and Asian immigrants arrived in the United States, and many schools enrolled large numbers of students who did not speak English.

The issue of bilingual education resurfaced in the 1960s with the civil rights movement. In 1968, Congress mandated that public schools must provide non-English-speaking students with *bilingual education*—instruction in both a non-English language and in English. However, few school districts had the necessary money or zeal to hire teachers who could provide instruction in two or more languages. Only after a U.S. Supreme Court decision did bilingual education become a reality in this nation. In *Lau v. Nichols* (1974), the Court affirmed the responsibility of the states and local districts for providing appropriate education for minority-language students. As officials in school districts implemented bilingual programs, however, they were confronted by opposition from some members of the general public and from some political leaders, even as the number of non-English-speaking people living in the United States increased dramatically.

How successful are bilingual education programs? Not all social analysts and education scholars agree on the answer to this question. Critics point out that many children remain in the programs far too long. One report criticized schools in New York for allowing tens of thousands of students to remain in transitional bilingual education (TBE) programs for up to six years (Ravitch and Viteritti, 1997). These critics believe that TBE programs are much less efficient in bringing students into the English-speaking mainstream than are classes conducted primarily in English (Ravitch and Viteritti, 1997). But other social analysts disagree because they believe that TBE programs are highly effective in helping children build competency both in their native language and in English. Moreover, advocates of TBE believe that all children in U.S. public schools should have exposure to more than one language (Berliner and Biddle, 1995).

One analyst has suggested that we need more positive models and more information on schools that have been successful in educating children from diverse cultural backgrounds rather than the typical media coverage focusing on "failures" in bilingual education (Delpit, 1995).

Equalizing Opportunities for Students with Disabilities

Another recent concern in education has been how to provide better educational opportunities for students with disabilities. As discussed in Chapter 18, the term *disability* has a wide range of definitions. For purposes of this chapter, disability is regarded as any physical and/or mental condition that limits students' access to, or full involvement in, school life. Along with other provisions, the Americans with Disabilities Act of 1990 requires schools and government agencies to make their facilities, services, activities, and programs accessible to people with disabilities. The law that specifically covers the treatment of children with disabilities is the Individuals with Disabilities Education Act (IDEA), which mandates that students with disabilities must receive free and appropriate education. Many schools have attempted to *mainstream* children with disabilities by providing *inclusion programs,* under which the special education curriculum is integrated with the regular education program and each child receives an *individualized education plan* that provides annual educational goals (Weinhouse and Weinhouse, 1994). *Inclusion* means that children with disabilities work with a wide variety of people; over the course of a day, children may interact with their regular education teacher, the special education teacher, a speech therapist, an occupational therapist, a physical therapist, and a resource teacher, depending on the child's individual needs.

Although much remains to be done, recent measures to enhance education for children with disabilities has increased the inclusion of many young people who were formerly excluded or marginalized in the educational system. But the problem of equal educational opportunities does not end at the elementary and high school levels for students with disabilities. If these students complete high school and continue on to college, they find new sets of physical and academic barriers that limit their access to higher education. However, many colleges and universities have provided relatively inexpensive accommodations to make facilities more accessible to students with disabilities. For example, special computers and wheelchair-accessible showers can make a major difference in whether or not the students can complete a college education. Author Connie Panzarino (1994: 219),

Computers make it possible for students with a disability to gain educational opportunities previously unavailable to them. Without a special reading screen, this student with a severe vision problem would be unable to read, write, or complete his homework.

who was born with the rare disease Spinal Muscular Atrophy Type III, recalls how difficult it was to attend college without being able to take a shower:

> A few weeks later I received a letter from the Disabled Students Office informing me that the administration had to indefinitely put off installing wheelchair-accessible showers. "How the Hell am I supposed to stay healthy if I can't stay clean?" I thought to myself. . . . Several of the disabled students had already fallen while trying to take showers or baths in the undersized, nonregulation tubs in our bathrooms. My Hoyer bathtub lift would not fit into these tubs.

But, as other students with disabilities have learned, a little activism sometimes resolves the problem. After Panzarino called a "shower strike" in which students with and without disabilities refused to take a bath until something was done about the showers, the university constructed accessible showers for each wheelchair-accessible dorm room (Panzarino, 1994). Meeting the specialized needs of students with disabilities is one of many problems confronting U.S. colleges and universities today.

Despite the difficulties associated with U.S. public schools, many people have not lost hope that schools can be improved. Some individuals have begun their own initiatives to make a difference in education (see

Breaking the Rules to Change a School!

Why should that be unique? There's nothing [the principal] is doing that couldn't be done everywhere else. Why do we have to depend on a single charismatic individual?

—Robert Slavin, education researcher at Johns Hopkins University, describing the recent work of Craig Spilman, principal of Canton Middle School in Baltimore (qtd. in Larson, 1997: 92)

Like many schools in a working-class neighborhood overlooking an industrial area of a city, Canton Middle School, located on Baltimore's Inner Harbor, looks like a fortress. There are grates over the windows of the austere brick building; the entrance has steel doors. Of the 750 students at Canton, 55 percent are white ethnics who trace their ancestry to Poland, Italy, or Greece. African Americans make up 35 percent of the student body; of the remainder, 5 percent are Latinos/as, and 5 percent are Native Americans. Most of the students come from poverty-level family backgrounds; 86 percent qualify for the federal government's free lunch program. Is this school another urban jungle? Not according to one journalist who recently visited Canton Middle School:

With classes in session, the halls are quiet. Bright. The floors gleam. More important: the school has become an academic showcase, with test scores advancing year to year—all this within a broader school system described by its own interim chief executive officer as "academically bankrupt." (Larson, 1997: 92)

Why is Canton different from other schools? Analysts attribute the success of this school to the difference that one person can make. Craig Spilman, the principal, took over at a time when absenteeism was at an all-time high and test scores were at an all-time low. Today, this situation has been reversed. To bring about change, Spilman hired teachers who were not currently employed in the Baltimore school system. He also abandoned traditional classroom models and adopted an "Expeditionary Learning" protocol in which children work in teams conducting field research aimed at producing a final product such as an exhibition or a presentation. Thus far, presentations have included the application of mathematics, physics, and historical research techniques to interior design plans for an old Baltimore cannery that is being converted to office space. Although Spilman is a school principal, other analysts suggest that many people could accomplish similar results—if they believe change is possible and are willing to work to bring it about (Larson, 1997).

For example, firefighters at a fire station on Chicago's South Side have served as friends and mentors to the schoolchildren who live in a nearby housing project. When children began skipping school to hang out at the firehouse, the firefighters started rewarding the children for perfect school attendance or good grades with prizes and small presents (Fedarko, 1997). Each child adopts a firefighter who helps with homework and imparts lessons on how to behave—for example, the children are not allowed to swear, and their faces and hands must be clean (Fedarko, 1997).

Today, many college students also tutor low-income children or serve as mentors or informal "big brothers" or "big sisters" for the children. Some do their volunteer work through sororities, fraternities, and campus service organizations; others link up with church groups, schools, or community organizations such as those that teach literacy to adults.

What can *you* do to enhance someone's educational opportunities?

Box 16.3). Others have sought alternatives for their children's education, including school vouchers, charter schools, and home schooling.

School Vouchers

Some elected officials, business leaders, and educators have shifted their focus to other ways of improving education. Some are advocating school voucher programs, which give parents the choice of what school their child will attend. Although vouchers have been widely discussed for many years, only a few school systems currently have programs in existence under which public funds (tax dollars) are provided to students so that they can pay tuition to attend either public or private schools of their choice. In about thirty public school systems, private donations provide funding for the vouchers (Lacayo, 1997). According to researchers who assessed the voucher program in Cleveland, Ohio (one of the districts in which such a program exists, although vouchers are provided only to low-income students there),

Box 16.4   SOCIOLOGY AND SOCIAL POLICY

The Ongoing Debate Over School Vouchers

As discussed in Chapter 3 ("Culture"), one of the core American values is equality—at least equality of *opportunity*, an assumed equal opportunity to achieve success. However, with problems in public education such as are discussed in this chapter, does every child really have that equal opportunity? If it is at least possible that a child could get a better education at a private school than at an inadequately funded inner-city school, should the government provide that child's family with the resources to send the child to a private school in order to have the same opportunity as a child from a wealthier family? Should the government provide that funding even if the private school teaches specific religious beliefs as part of the academic curriculum? Questions such as these have been of concern to social policy makers, educators, parents, and others in the United States for decades, and it appears that the controversy is far from being resolved. To understand the issues involved in the school voucher controversy, let's briefly look at the intended purpose of vouchers and at the arguments for and against voucher programs.

The original idea, first proposed in 1955, was to provide a voucher (equal in value to the average amount spent per student by the local school district) to the family of any student who left the public schools to attend a private school (Henderson, 1997). The private school could exchange the voucher for that amount of money and apply it to the student's tuition at the private school; the school district would save an equivalent amount of money as a result of the student not enrolling in the public schools. Over the years, variations on the original idea have been proposed, such as allowing vouchers to be used for transfers between public schools in the same school district, transfers from one public school district to another, transfers only to schools with no religious connections, and vouchers for use only by students from low-income families.

Although the legislatures in at least twenty-six states have rejected school voucher programs (*New York Times*, 2002a), school districts in a number of cities have adopted various types of school choice programs. Some of those plans authorize only transfers between public schools in the same district; other plans authorize the use of vouchers for transfers to private schools. For example, the Cleveland, Ohio, school district allows students to attend participating public or private schools of their parents' choosing—even public schools in adjacent districts—and provides tuition aid to the parents according to financial need. The tuition aid goes to the public or private school in which the student is enrolled. In 1999-2000, 96 percent of the students participating in this program were enrolled in religiously affiliated schools. The program was challenged in court on the basis that it violated the constitutional separation of church and state.

In *Zelman v. Simmons-Harris*, a divided U.S. Supreme Court upheld the Cleveland program against

---

two-thirds of the parents whose children received vouchers were "very satisfied" with the academic quality of the private or parochial school their child attended, whereas less than 30 percent of the parents who applied for vouchers but whose children remained in the public schools were satisfied with their children's education (Lewis, 1997). Critics of the school voucher system believe that such programs undermine public education and might cause the public school system to collapse. Supporters of the voucher system believe that fairness is a central issue: The poor should have the same chance for a quality education as the more affluent (Lacayo, 1997). According to this approach, vouchers make it possible for poor children to leave behind the problems of inner-city public schools and find better educational opportunities

elsewhere. The ongoing debate over school vouchers is further discussed in Box 16.4.

Charter Schools and "For Profit" Schools

As compared to the voucher system, the charter school movement creates public schools that are free from many of the bureaucratic rules that often limit classroom performance. These schools operate under a charter contract negotiated by the school's organizers (often parents or teachers) and a sponsor (usually a local school board, a state board of education, or a university) that oversees the provisions of the contract. A charter school is freed from the day-to-day bureaucracy

the constitutional challenge, with the majority's opinion holding that the plan was adopted for the purpose of providing educational assistance to poor children and was neutral toward religion. Would the same result have been reached if the plan was for *all* children instead of only those from low-income families?

According to some analysts, the Court's decision shifted the battle regarding voucher programs from a legal argument based on a constitutional question to a political debate involving social policy and politics (Nagourney, 2002). Advocates of the Cleveland program argue that the Court's decision was good because children from lower-income families can now have a wider range of educational choices, a range similar to that currently available to children from middle-income families. They assert that competition for students' vouchers will improve public education, forcing school administrators and teachers to perform at a higher level and thereby producing greater competencies in students. By contrast, opponents of school vouchers argue that the Court's decision could be the beginning of the end of public schools, that public education may be unable to withstand the loss of funds and the transfer of gifted students to private schools.

What are the major social policy issues surrounding voucher systems? Many people still strongly believe that using public tax dollars (even though they are technically given to the students and their parents rather than the schools) for funding private, religiously based schools violates the separation of church and state. Despite the Court's decision, they will continue to make that argument, whether on constitutional grounds or on the basis of public policy. Ultimately, from a social policy standpoint, the future of school vouchers may depend on the willingness (or unwillingness) of political leaders to authorize vouchers and the willingness of voters to pay for them.

How the social issues are framed may be significant in shaping the future of vouchers. If vouchers are issued only to the families of low-income children, will they be thought of as a welfare program, and thus be stigmatized? If they are available to all students, would they in effect be a tax subsidy for private schools, becoming more costly than the public school systems that currently exist? What effect would vouchers have on children's educational opportunities and the already struggling public schools? What are your thoughts on this important social policy issue?

WRITING IN SOCIOLOGY ASSIGNMENT

Discuss the pros and cons of school vouchers. If you were in a position to develop a voucher sys-  tem (for instance, if you were a state legislator), would you do so? Why or why not?

of a larger school district and may provide more autonomy for individual students and teachers. Critics of the charter school movement argue that it takes money away from conventional public schools. But according to some education scholars, charter schools serve an important function in education: A large number of minority students receive a higher-quality education than they would in the public schools (Bierlein, 1997).

Other recent approaches for improving education include "contracting out"—hiring for-profit companies on a contract basis to operate public schools. Most contracting is done by three private companies: Whittle Communications' Edison Project, Sabis International Education Alternatives Incorporated, and Alternative Public Schools (Hill, 1997). For example, the Edison Project has managed schools in Boston;

Mount Clemens, Michigan; Sherman, Texas; and Wichita, Kansas. Although many have applauded the innovative nature of this approach, critics are concerned about the privatization of education, which they believe undermines the public school system.

Home Schooling

A final alternative, home schooling, has been chosen by some parents who hope to avoid the problems of public schools while providing a quality education for their children. In the 2000s, estimates range from 850,000 to 1.7 million children in grades K through 12 who are educated in home-school programs (National Home Education Research Institute, 2002; U.S. Department of Education, 2001a). According to

Home schooling has grown in popularity in recent decades as parents have sought to have more control over their children's education. Although some home school settings may resemble a regular classroom, other children learn in more informal settings such as the family's kitchen.

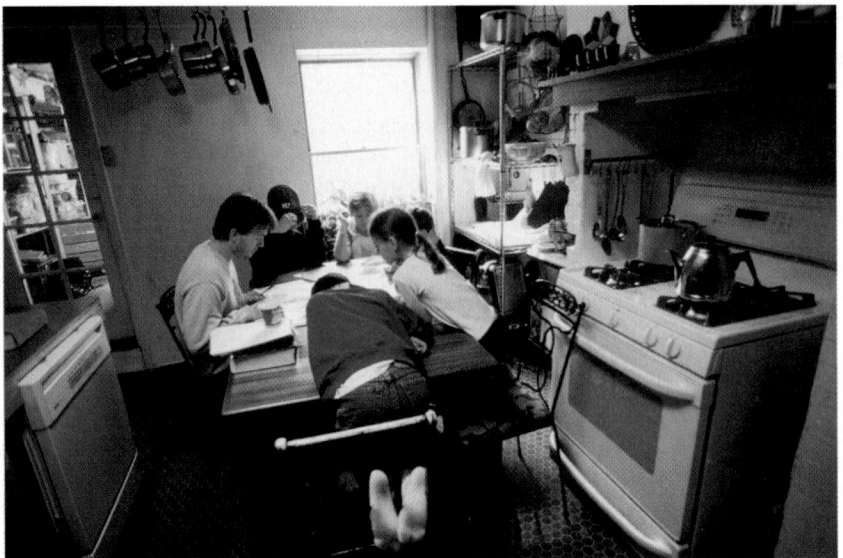

© Bridget Besaw Gorman/Aurora Photos

a recent study by the U.S. Department of Education, home-schoolers are more likely than other students to live with two or more siblings in a two-parent family, with only one parent working outside the home. Typically, home-schoolers' parents are better educated, on average, than other parents; however, their income is about the same. Researchers have found that boys and girls are equally likely to be home-schooled.

Parents who educate their children at home believe that their children are receiving a better education at home because instruction can be individualized to the needs and interests of their children. Some parents also indicate religious reasons for their decision to home-school their children. An association of home-schoolers now provides communication links for parents and children, and technological advances in computers and the Internet have made it possible for home-schoolers to gain information and communicate with one another. According to home-schooling advocates, home-school students typically have high academic achievements and high rates of employment.

Critics of home schooling question the knowledge and competence of the typical parents to educate their own children at home, particularly in rapidly changing academic subjects such as science and computer technology. Some states have passed laws mandating accountability that must be met by parents who teach their children at home. For example, Florida requires that parents register their children with their school districts or with umbrella schools that maintain and submit records to the state. Parents must maintain portfolios of their children's work and submit annual evaluations, including grades on standardized exams, to school officials (Miller, 2003).

Many people believe that the home-schooling movement will grow rapidly if problems in public education are not alleviated in the near future.

Concluding Thoughts

What will education be like in the future? How will it accommodate the increasing diversity of the U.S. population? According to Gary D. Fenstermacher (1994), the education we need for the future is different from that of the past:

> For this education is not bent on assimilation, to the melting of different cultures and languages into some common American pot, or to merely readying today's children for tomorrow's workforce. In contrast, the education that must engage us today and in the future is how to form common space and common speech and common commitment while respecting and preserving our differences in heritage, race, language, culture, gender, sexual orientation, spiritual values, and political ideologies. It is a new challenge for America.

Will we be able to meet this challenge? President George W. Bush has signed into law the No Child Left Behind Act of 2001. This law requires that schools must be accountable for students' learning and sets forth steps toward producing a more "accountable" education system:

- States will create a set of standards—beginning with math and reading—for what all children should know at the end of each grade in school; other standards will be developed over a period of time.

- States must test every student's progress toward meeting those standards.
- States, individual school districts, and each school will be expected to make yearly progress toward meeting the standards.
- School districts must report results regarding the progress of individual schools toward meeting these standards.
- Schools and districts that do not make adequate progress will be held accountable and could lose funding and pupils; in some cases, parents will be allowed to move their children from low-performing schools to schools that are meeting the standards.

Will these changes make a difference? It will be some time before we can assess how effective these changes are in bringing about higher-quality education for students and better graduation rates across class and racial/ethnic categories. Of course, other pressing social problems, such as harassment, guns, and violence, will remain major issues that affect the future of U.S. education. Ultimately, education is an important social institution that must be maintained and enhanced in a variety of ways because we have learned that simply spending larger sums of money on education does not guarantee that many of the problems facing this social institution will be resolved.

CHAPTER REVIEW

■ What is education?

Education is the social institution responsible for the systematic transmission of knowledge, skills, and cultural values within a formally organized structure.

■ What is the functionalist perspective on education?

According to functionalists, education has both manifest functions (socialization, transmission of culture, social control, social placement, and change and innovation) and latent functions (keeping young people off the streets and out of the job market, matchmaking and producing social networks, and creating a generation gap).

■ What is the conflict perspective on education?

From a conflict perspective, education is used to perpetuate class, racial–ethnic, and gender inequalities through tracking, ability grouping, and a hidden curriculum that teaches subordinate groups conformity and obedience.

■ What is the symbolic interactionist perspective on education?

Symbolic interactionists examine classroom dynamics and study ways in which practices such as labeling may become a self-fulfilling prophecy for some students, such that these students come to perform up—or down—to the expectations held for them by teachers.

■ How are U.S. public schools funded?

Most educational funds come from state legislative appropriations and local property taxes. State and local governments each contribute about 47 percent of total educational expenses, and the federal government pays the remaining 6 percent.

■ What is the purpose of "magnet schools"?

Magnet schools offer a specialized curriculum that focuses on a certain area of study, such as art, music, or science. These schools are often created as a means of giving students more-specialized educational opportunities while at the same time increasing racial and social class integration in a school district.

■ What are some of the leading discipline problems in today's schools?

Schools have been plagued by the same array of problems that are occurring in the larger society, including violence, drug abuse, suicide, robbery, assault, and other forms of aggressive behavior.

■ How is the increasing racial and ethnic diversity of the United States affecting public education?

Nearly 40 percent of U.S. schoolchildren are African American, Latino/a, Asian American, or Native American, whereas most teachers are middle-class white females.

■ What is functional illiteracy, and how does it differ from illiteracy?

Functional illiteracy is the inability to read and/or write at the skill level necessary for carrying out everyday tasks. In contrast, illiteracy is the inability to read and/or write at the most basic level. Both forms of illiteracy are a problem in the United States.

■ **What controversies persist in education?**

Racial segregation and integration, bilingual education, and how to equalize educational opportunities for students with disabilities are among the pressing issues facing U.S. public education today.

■ **What are the major problems in higher education?**

Among the most pressing problems are the high cost of a college education, the underrepresentation of minorities as students and faculty in many schools and degree programs, and the continuing debate over affirmative action.

KEY TERMS

credentialism 535
cultural capital 533
cultural transmission 524
education 524
formal education 525
functional illiteracy 550
hidden curriculum 534
informal education 524
mass education 525
tracking 533

QUESTIONS FOR CRITICAL THINKING

1. What are the major functions of education for individuals and for societies?
2. Why do some theorists believe that education is a vehicle for decreasing social inequality whereas others believe that education reproduces existing class relationships?
3. Why does so much controversy exist over what should be taught in U.S. public schools?
4. How are the values and attitudes that you learned from your family reflected in your beliefs about education?

RESOURCES ON THE INTERNET

Chapter-Related Web Sites

The following Web sites have been selected for their relevance to the topics in this chapter. These sites are among

the more stable, but please note that Web site addresses change frequently. For an updated list of chapter-related Web sites with URL links, please visit the *Sociology in Our Times* Web site (**www.wadsworth.com/KendallSIOT**).

National Center for Education Statistics (NCES)
http://www.ed.gov/NCES/index.html

The NCES is the federal organization primarily responsible for the collection and distribution of data on topics related to education in the United States and around the world. The NCES Web site features information on literacy, education costs, selecting colleges, student assessment, students with disabilities, and much more.

American Association of University Women (AAUW)
http://www.aauw.org

The AAUW is a national organization that advocates gender equity in education and promotes awareness of issues related to gender and education. The AAUW Web site features research reports, public policy initiatives, news updates, informative Web links, and information on how to take action against gender inequities in education.

International Bureau of Education (IBE)
http://www.ibe.unesco.org

The IBE is a private, nongovernmental organization that was created in 1925 to centralize information related to public and private education and to promote scientific research in the education field. In addition to profiles of national education systems and reports on the development of education, the IBE Web site provides access to extensive data banks, including the World Data on Education collection.

ONLINE STUDY AND RESEARCH TOOLS

Accompanying this text are many *free* powerful online study tools that will help you master the material in this chapter, help increase your depth of understanding, and help you make the grade!

SocCoach CD-ROM

Use the SocCoach CD-ROM enclosed with this text to help you formulate a customized study plan for this chapter. After you take the Diagnostic Quiz, SocCoach will generate a customized study plan just for you! It will identify sections of the chapter that you should review and will provide videos, charts, graphs, and excerpts from the text to supplement your studies and enhance your understanding. You'll also find fun, interactive activities such as

Virtual Explorations and Map the Stats to apply what you've learned and stretch your sociological imagination.

The Companion Web Site for Sociology in Our Times, *Fifth Edition*

www.wadsworth.com/KendallSIOT

Gain an even better grasp on this chapter by going to the companion web site to take one of the Tutorial Quizzes, use the Flash Cards to master key terms, or check out the many other study aids you'll find there. You'll also find special features such as GSS Data and Census 2000 information that'll put data and resources at your fingertips to help you with that special project or help you as you do some research on your own.

In this chapter, when you see the icon on the left, it alerts you to a specific exercise found in *Wadsworth's Sociology Online Resources and Writing Companion.* This valuable guide shows you how to use Wadsworth's exclusive online resources—*InfoTrac College Edition,* the *Opposing Viewpoints Resource Center,* and *MicroCase Online*—to assist you in your study of sociology and to build essential research and writing skills.

CHAPTER 17

Religion

I am a fundamentalist in the sense that I believe that the Bible is the inerrant word of God. . . . I am a student of the Bible as well as I am a student of biology, and my responsibility is to teach the truth in the classroom, and I believe that there is a major deception occurring in our public education system today, and that is the myth that evolutionism is a scientific theory that can be substantiated with scientific evidence. I have found that it is not. . . .

I also believe there is a conspiracy to cover up what I am attempting to expose, and that is we do have a state-sponsored religion, it's called the religion of humanism with evolutionism being the primary tenet of it, and it's being promoted in the public classroom despite the Constitution that prohibits the establishment of a particular religious view. Now of course, it comes out as if I am attempting to promote my religious view, but I believe that's a smoke screen. I believe that it is an attempt to divert the public from the real issue. And the real issue is: Is evolutionism science or is it a religious belief system?

—John Peloza, a biology teacher at a Mission Viejo, California, public high school, explaining why he believes that he has the right to teach the theory of creationism (a belief in the divine origins of the universe and human beings based on a literal interpretation of the Bible) in addition to the theory of evolution (PBS, 1992a)

What's at stake here, it seems to me, is what kind of society are we going to have. Are we going to let religious fundamentalists of any ilk, I don't care if they are Muslim or Christian or Jewish or what they are, are we going to let religious fundamentalists of any ilk make public policy based on their religion?

—Jim Corbett, a journalism advisor at the same high school, stating why he disagrees with Peloza (PBS, 1992a)

UPI/Corbis

This argument is only one in a lengthy history of debates about the appropriate relationship between religion and other social institutions. In education, for example, controversies have arisen over topics such as the teaching of creationism versus evolutionism, moral education, sex education, school prayer, and the subject matter of textbooks and library books. More than seventy years ago, evolutionism versus creationism was hotly debated in the famous "Scopes Monkey Trial," so named because of Charles Darwin's assertion that human beings had evolved from lower primates. In this case, John Thomas Scopes, a substitute high school biology teacher in Tennessee, was found guilty of teaching evolution, which denied the "divine creation of man as taught in the Bible." Although an appeals court later overturned Scopes's conviction and $100 fine (on the grounds that the fine was excessive), teaching evolution in Tennessee's public schools remained illegal until

■ The Scopes trial—in which attorneys Clarence Darrow and William Jennings Bryan (shown here) debated whether or not a public schoolteacher had a right to teach evolution—was perhaps the beginning of the evolution versus creation battle in the United States.

1967 (Chalfant, Beckley, and Palmer, 1994). By contrast, recent U.S. Supreme Court rulings have looked unfavorably on the teaching of creationism in public schools, based on a provision in the Constitution that requires a "wall of separation" between church (religion) and state (government) (Albanese, 1999). Initially, this wall of separation was erected to protect religion from the state, not the state from religion (Carter, 1994).

As these examples show, religion can be a highly controversial topic. One group's deeply held beliefs or cherished religious practices may be a source of irritation to another. Today, religion is a source of both stability and conflict not only in the United States but throughout the world (Kurtz, 1995). In this chapter, we examine how religion influences life in the United States and in other areas of the world. Before reading on, test your knowledge about how religion affects public education in this country by taking the quiz in Box 17.1.

QUESTIONS AND ISSUES

Chapter Focus Question: What is the relationship between society and religion, and what role does religion play in people's everyday lives?

What are the key components of religion?

How do functionalist, conflict, and symbolic interactionist perspectives on religion differ?

What are the central beliefs of the world's religions?

How do religious bodies differ in organizational structure?

What is the future of religion in the United States?

THE SOCIOLOGICAL STUDY OF RELIGION

What is religion? *Religion* **is a system of beliefs and practices (rituals)—based on some sacred or supernatural realm—that guides human behavior, gives meaning to life, and unites believers into a single moral community** (Durkheim, 1995/1912). Religion is one of the most significant social institutions in society. As such, it consists of a variety of elements, including beliefs about the sacred or supernatural, rituals, and a social organization of believers drawn together by their common religious tradition (Kurtz, 1995). This system of beliefs seeks to bridge the gap between the known and the unknown, the seen and the unseen, and the sacred ("holy, set apart, or forbidden") and the secular (things of this world). Most religions attempt to answer fundamental questions such as those regarding the meaning of life and how the world was created. Most religions also provide comfort to persons facing emotional traumas such as illness, suffering, grief, and death. According to the sociologist Lester Kurtz (1995: 9), religious beliefs are typically woven into a series of narratives, including stories about how ancestors and other significant figures had meaningful experiences with supernatural powers. Moreover, religious beliefs are linked to practices that bind people together and to rites of passage such as birth, marriage, and death. People with similar religious beliefs and practices often gather themselves together in a moral community (such as a church, mosque, temple, or synagogue) where they can engage in religious beliefs and practices with similarly minded people (Kurtz, 1995).

Given the diversity and complexity of religion, how is it possible for sociologists to study this social institution? Most sociologists studying religion are committed to the pursuit of "disinterested scholarship," meaning that they do not seek to make value judgments about religious beliefs or to determine whether particular religious bodies are "right" or "wrong." However, many acknowledge that it is impossible to completely rid themselves of those values and beliefs into which they were socialized (Bruce, 1996). Therefore, for the most part, sociologists study religion by using sociological methods such as historical analysis, experimentation, participant observation, survey research, and content analysis that can be verified and replicated (Roberts, 2004). As a result, most studies in the sociology of religion focus on tangible elements that can be *seen,* such as written texts, patterns of behavior, or individuals' opinions about religious matters, and that can be studied using standard sociological research tools. According to the sociologist Keith A. Roberts (2004: 28), beliefs constitute only a small part of a sociological examination of religion:

The sociological approach focuses on religious groups and institutions (their formation, maintenance, and demise), on the behavior of individuals within these groups (e.g., social processes that affect conversion, ritual behavior, or decision to

Box 17.1 SOCIOLOGY AND EVERYDAY LIFE

How Much Do You Know About the Effect of Religion on U.S. Education?

True	False	
T	F	1. The Constitution of the United States originally specified that religion should be taught in the public schools.
T	F	2. Virtually all sociologists have advocated the separation of moral teaching from academic subject matter.
T	F	3. Parochial schools have decreased in enrollment as interest in religion has waned in the United States.
T	F	4. Conservative religious groups have targeted school board seats as a way of gaining control of public education in some cities and counties.
T	F	5. The number of children from religious backgrounds other than Christianity and Judaism has grown steadily in public schools over the past three decades.
T	F	6. Debates over the content of textbooks focus only on elementary education because of the vulnerability of young children.
T	F	7. Most members of the Baby Boom generation have no religion today because they received no religious instruction in school.
T	F	8. Prayer or a moment of silence in public schools will cease to be a political issue in the United States during this century.

Answers on page 565.

---

defect to another group), and on conflicts between religious groups (such as Catholic vs. Protestant, Christian vs. [Muslim], mainline denomination vs. cult). For the sociologist, beliefs are only one small part of religion.

Recently, more U.S. scholars have started examining religion from a global perspective to determine "ways in which religious ideas are performed on the world stage" (Kurtz, 1995: 16). As Kurtz (1995: 211) points out, conflicts in the global village are often deeply intertwined with religious differences: "In the twentieth century, the twin crises of modernism and multiculturalism . . . added a religious dimension to many ethnic, economic, and political battles, providing cosmic justifications for the most violent struggles." Of course, conflict is not always inherently bad. It can be the source of constructive change in communities and societies.

How does the sociological study of religion differ from the theological approach? Unlike the sociological approach, which primarily focuses on the visible aspects of religion, *theologians* study specific religious doctrines or belief systems, including answers to questions such as what is the nature of God or the gods and what is the relationship among supernatural

power, human beings, and the universe. Many theologians primarily study the religious beliefs of a specific religion (such as Christianity, Judaism, Buddhism, or Hinduism), denomination (such as the Baptists, Catholics, Methodists, or Episcopalians), or religious leader (such as the Reverend Sun Myung Moon or L. Ron Hubbard, founder of Scientology) so that they can interpret this information for laypersons who seek answers for seemingly unanswerable questions about the meaning of life and death.

Religion and the Meaning of Life

Religion seeks to answer important questions such as why we exist, why people suffer and die, and what happens when we die. Sociologist Peter Berger (1967) referred to religion as a *sacred canopy*—a sheltering fabric hanging over people that gives them security and provides answers for the questions of life. However, this sacred canopy requires that people have **faith—unquestioning belief that does not require proof or scientific evidence.** Science and medicine typically rely on existing scientific evidence to respond to questions of suffering, death, and injustice, whereas religion seeks to explain such phenomena by referring

© Tom Carter/PhotoEdit

College students—like people in all walks of life—may turn to religion for answers to important questions for which there are no easy answers. Rituals help individuals outwardly express their beliefs and provide a sense of cohesion and belonging.

to the sacred. According to Emile Durkheim (1995/1912), *sacred* **refers to those aspects of life that are extraordinary or supernatural**—in other words, those things that are set apart as "holy." People feel a sense of awe, reverence, deep respect, or fear for that which is considered sacred. Across cultures and in different eras, many things have been considered sacred, including invisible gods, spirits, specific animals or trees, altars, crosses, holy books, and special words or songs that only the initiated could speak or sing (Collins, 1982). Those things that people do not set apart as sacred are referred to as *profane*—**the everyday, secular ("worldly") aspects of life** (Collins, 1982). Sacred beliefs are rooted in the holy or supernatural, whereas secular beliefs have their foundation in scientific knowledge or everyday explanations. In the educational debate over creationism and evolutionism, for example, advocates of creationism view their belief as founded in sacred (Biblical) teachings, but advocates of evolutionism argue that their beliefs are based on provable scientific facts.

In addition to beliefs, religion also comprises symbols and rituals. According to the anthropologist Clifford Geertz (1966), religion is a set of cultural symbols that establishes powerful and pervasive moods and motivations to help people interpret the meaning of life and establish a direction for their behavior. People often act out their religious beliefs in the form of *rituals*—**regularly repeated and carefully prescribed forms of behaviors that symbolize a cherished value or belief** (Kurtz, 1995). Rituals range from songs and prayers to offerings and sacrifices that worship or praise a supernatural being, an ideal, or a set of supernatural principles. For example, Muslims bow toward Mecca, the holy city of Islam,

five times a day at fixed times to pray to God, whereas Christians participate in the celebration of communion (or the "Lord's Supper") to commemorate the life, death, and resurrection of Jesus. Rituals differ from everyday actions in that they involve very strictly determined behavior. The rituals involved in praying or in observing communion are carefully orchestrated and must be followed with precision. According to the sociologist Randall Collins (1982: 34), "In rituals, it is the forms that count. Saying prayers, singing a hymn, performing a primitive sacrifice or a dance, marching in a procession, kneeling before an idol or making the sign of the cross—in these, the action must be done the right way."

Not all sociologists believe that the "sacred canopy" metaphor suggested by Berger accurately describes contemporary religion. Some analysts believe that a more accurate metaphor for religion in the global village is that of the *religious marketplace,* in which religious institutions and traditions compete for adherents, and worshipers shop for a religion in much the same way that consumers decide what goods and services they will purchase in the marketplace (Moore, 1995). However, other analysts do not believe that moral and ethical beliefs are bought and sold like groceries, shoes, or other commodities, and they note that many of the world's religions have persisted from earlier eras to advanced technological societies. But this poses another question: When did the earliest religions begin?

Although it is difficult to establish exactly when religious rituals first began, anthropologists have concluded that all known groups over the past hundred thousand years have had some form of religion (Haviland, 1999). Religions have been classified into four main categories based on their dominant belief: simple supernaturalism, animism, theism, and transcendent idealism (McGee, 1975). In very simple preindustrial societies, religion often takes the form of *simple supernaturalism*—**the belief that supernatural forces affect people's lives either positively or negatively.** This type of religion does not acknowledge specific gods or supernatural spirits but focuses instead on impersonal forces that may exist in people or natural objects. For example, simple supernaturalism has been used to explain mystifying events of nature, such as sunrises and thunderstorms, and ways that some objects may bring a person good or bad luck. By contrast, *animism* **is the belief that plants, animals, or other elements of the natural world are endowed with spirits or life forces that have an impact on events in society.** Animism is associated with early hunting and gathering societies and with many Native American societies, in which everyday life is

Box 17.1 SOCIOLOGY AND EVERYDAY LIFE

Answers to the Sociology Quiz on Religion and U.S. Education

1. False. Due to the diversity of religious backgrounds of the early settlers, no mention of religion was made in the original Constitution. Even the sole provision that currently exists (the establishment clause of the First Amendment) does not speak directly of the issue of religious learning in public education.

2. False. Obviously, contemporary sociologists hold strong beliefs and opinions on many subjects; however, most of them do not think that it is their role to advocate specific stances on a topic. Early sociologists were less inclined to believe that they had to be "value-free." For example, Durkheim strongly advocated that education should have a moral component and that schools had a responsibility to perpetuate society by teaching a commitment to the common morality.

3. False. Just the opposite has happened. As parents have felt that their children were not receiving the type of education they desired in public schools, parochial schools have flourished. Christian schools have grown to over five thousand; Jewish parochial schools have also grown rapidly over the past decade.

4. True. In a number of cities and counties, "new right" fundamentalist groups have supported candidates for school boards who are sympathetic to their views. In San Diego County, California, for example, the Christian Coalition won a number of seats in 1993 and moved on to Los Angeles County in an attempt to do the same. Throughout the country, politicians who do not vote according to the beliefs of these groups are targeted for replacement in the next election.

5. True. Although about 86 percent of those age eighteen and over in the forty-eight contiguous states of the United States describe their religion as some Christian denomination, there has still been a significant increase in those who adhere either to no religion (7.5 percent) or who are Jewish, Muslim/Islamic, Unitarian-Universalist, Buddhist, or Hindu.

6. False. Attempts to remove textbooks occur at all levels of schooling. A recent case involved the removal of Chaucer's "The Miller's Tale" and Aristophanes's *Lysistrata* from a high school curriculum.

7. False. Some Baby Boomers received religious instruction, either in private schools or in public schools where prayer and Bible reading took place. Many have returned to religion today, believing "something was missing" from their lives.

8. False. Regardless of whether the current proposed school prayer amendment becomes part of the Constitution, school prayer will remain an issue due to the diversity of the population in this country.

Sources: Based on Ballantine, 2001; Gibbs, 1994; Greenberg and Page, 2002; C. Johnson, 1994; Kosmin and Lachman, 1993; and Roof, 1993.

not separated from the elements of the natural world (Albanese, 1999).

The third category of religion is *theism*—**a belief in a god or gods.** Horticultural societies were among the first to practice *monotheism*—**a belief in a single, supreme being or god who is responsible for significant events such as the creation of the world.** Three of the major world religions—Christianity, Judaism, and Islam—are monotheistic. By contrast, Hinduism, Shinto, and a number of the indigenous religions of

Africa are forms of *polytheism*—**a belief in more than one god.** The fourth category of religion, transcendent idealism, is a *nontheistic religion*—**a religion based on a belief in divine spiritual forces such as sacred principles of thought and conduct, rather than a god or gods.** Transcendent idealism focuses on principles such as truth, justice, affirmation of life, and tolerance for others, and its adherents seek an elevated state of consciousness in which they can fulfill their true potential.

Religion and Scientific Explanations

During the Industrial Revolution, scientific explanations began to compete with religious views of life. Rapid growth in scientific and technological knowledge gave rise to the idea that science would ultimately answer questions that had previously been in the realm of religion. Many scholars believed that increases in scientific knowledge would result in ***secularization***—**the process by which religious beliefs, practices, and institutions lose their significance in sectors of society and culture** (Berger, 1967). Secularization involves a decline of religion in everyday life and a corresponding increase in organizations that are highly bureaucratized, fragmented, and impersonal (Chalfant, Beckley, and Palmer, 1994).

As previously discussed, some people argue that science and technology have overshadowed religion in the United States, but others point to the resurgence of religious beliefs in recent years and an unprecedented development of alternative religions (Kosmin and Lachman, 1993; Roof, 1993; Singer and Lalich, 1995). For some people, secularization has contributed to a decline in morality and traditional family values. This approach is seen in the conflict between education and religion in public schools, as described by a woman from Ohio:

> It bothers me what they are teaching kids in school. . . . They are changing history around. This country was founded on God. The people that came and founded this country were Godly people, and they have totally taken that out of history. They are trying to get rid of everything that ever says anything about God to please someone who is offended by it. That bothers me. . . . (qtd. in Roof, 1993: 98)

However, not all parents would agree with this mother's view on the relationship between religion and education; many parents believe that only religious organizations should be responsible for teaching religious beliefs.

SOCIOLOGICAL PERSPECTIVES ON RELIGION

Religion as a social institution is a powerful, deeply felt, and influential force in human society. Sociologists study the social institution of religion because of the importance that religion holds for many people;

they also want to know more about the influence of religion on society, and vice versa. For example, some people believe that the introduction of prayer or religious instruction in public schools would have a positive effect on the teaching of values such as honesty, compassion, courage, and tolerance because these values could be given a moral foundation. However, society has strongly influenced the practice of religion in the United States as a result of court rulings and laws that have limited religious activities in public settings, including schools.

The major sociological perspectives have different outlooks on the relationship between religion and society. Functionalists typically emphasize the ways in which religious beliefs and rituals can bind people together. Conflict explanations suggest that religion can be a source of false consciousness in society. Symbolic interactionists focus on the meanings that people give to religion in their everyday lives.

Functionalist Perspectives on Religion

Emile Durkheim was one of the first sociologists to emphasize that religion is essential to the maintenance of society. He suggested that religion is a cultural universal found in all societies because it meets basic human needs and serves important societal functions.

Durkheim on Religion In *The Elementary Forms of Religious Life* (1995/1912: 44), Durkheim defined *religion* as "a unified system of beliefs and practices relative to sacred things, that is to say, things set apart and forbidden—beliefs and practices which unite into one single moral community . . . all those who adhere to them." According to Durkheim, all religions share three elements: (1) beliefs held by adherents, (2) practices (rituals) engaged in collectively by believers, and (3) a moral community based on the group's shared beliefs and practices pertaining to the sacred.

For Durkheim, the central feature of all religions is the presence of sacred beliefs and rituals that bind people together in a collectivity. In his studies of the religion of the Australian aborigines, for example, Durkheim found that each clan had established its own sacred totem, which included kangaroos, trees, rivers, rock formations, and other animals or natural creations. To clan members, their totem was sacred; it symbolized some unique quality of their clan. People developed a feeling of unity by performing ritual dances around their totem, causing them to abandon individual self-interest. Durkheim suggested that the correct performance of the ritual gives rise to religious conviction. Religious beliefs and rituals are *collective representations*—group-held meanings that express something important about the group itself (McGuire, 2002). Because of the intertwining of group consciousness and society, functionalists suggest that religion is functional because it meets basic human needs.

Functions of Religion

From a functionalist perspective, religion has three important functions in any society: (1) providing meaning and purpose to life, (2) promoting social cohesion and a sense of belonging, and (3) providing social control and support for the government.

Meaning and Purpose Religion offers meaning for the human experience. Some events create a profound sense of loss on both an individual basis (such as injustice, suffering, and the death of a loved one) and a group basis (such as famine, earthquake, economic depression, and subjugation by an enemy). Inequality may cause people to wonder why their own situation is no better than it is. Most religions offer explanations for these concerns. Explanations may differ from one religion to another, yet each tells the individual or group that life is part of a larger system of order in the universe (McGuire, 2002). Some (but not all) religions even offer hope of an afterlife for persons who follow the religion's tenets of morality in this life. Such beliefs help make injustices easier to endure.

In a study of religious beliefs among Baby Boomers (persons born between 1946 and 1964) in the United States, religion and society scholar Wade Clark Roof (1993) found that a number of people had returned to organized religion as part of a personal quest for meaning. Roof noted that they were looking "for something to believe in, for answers to questions about life," as reflected in these comments by a woman in North Carolina:

> Something was missing. You turn around and you go, is this it? I have a nice husband, I have a nice house; I was just about to finish graduate school. I knew I was going to have a very marketable degree. I wanted to do it. And you turn and you go, here I am. This is it. And there were just things that were missing. I just didn't have stimulation. I didn't have the motivation. And I guess when you mentioned faith, I guess that's what was gone. (qtd. in Roof, 1993: 158)

Social Cohesion and a Sense of Belonging By emphasizing shared symbolism, religious teachings and practices help promote social cohesion. An example is the Christian ritual of communion, which not only commemorates a historical event but also allows followers to participate in the unity ("communion") of themselves with other believers (McGuire, 2002). All religions have some form of shared experiences that rekindles the group's consciousness of its own unity.

Religion has played an important part in helping members of subordinate groups develop a sense of social cohesion and belonging even when they are the objects of prejudice and discrimination by dominant-group members. For example, some scholars suggest that African Americans initially brought into the United States as slaves found cohesion and stability in religion:

> Common religious beliefs and practices provided a new form of social cohesion in place of the old forms that had been destroyed when the slaves were seized in Africa and forcibly brought to America. Black churches brought about a distinctive culture and worldview that paralleled rather than replicated the culture of the land in which blacks resided involuntarily. The terms of their faith—salvation, freedom, and the Kingdom of God—were rooted in the black experience and expressed themselves in joy and jubilation tinged with mournfulness. (Kosmin and Lachman, 1993: 31)

Religion has also been important to those who voluntarily migrated to the United States. For example, Irish Catholic and Italian Catholic churches helped Irish Americans and Italian Americans preserve a sense of identity and belonging (Greeley, 1972; Roberts, 2004). Since the 1960s, Korean Americans have found religious and ethnic fellowship in more than two thousand Korean American churches, mostly Presbyterian, Southern Baptist, and United Methodist (Kosmin and Lachman, 1993). Since the late 1980s and early 1990s, Russian Jewish immigrants have found a sense of belonging in some congregations, and even though they did not initially speak the language of their new country, they still shared established religious rituals and a sense of history. Shared experiences such as these strengthen not only the group but also the individual's commitment to the group's expectations and goals (McGuire, 2002).

Social Control and Support for the Government How does religion help bind society together and maintain social control? All societies attempt to maintain social control through systems of rewards and punishments. Sacred symbols and beliefs establish powerful, pervasive, long-lasting motivations based on the concept of a general order of existence. In

Box 17.2 SOCIOLOGY AND SOCIAL POLICY

Should Prayer Be Allowed in Public Schools? Issues of Separation of Church and State

What we want is actual prayer [as opposed to a moment of silence for contemplation]. It happened to have been around on Sept. 11. The next day at some . . . schools, there was open prayer all through the schools. Even the [U.S.] president is asking for prayer. But [in] the very institutions that we need to have prayer the most, it has been outlawed. So why not [have it] where it is needed the most and where it can have a lasting effect?
  —Ronald J. Waters, a city alderman in Harvey, Illinois, explaining why he believes students should be able to pray in school, particularly in the aftermath of a national disaster such as the 9/11 terrorist attacks (qtd. in Fountain, 2001: A18)

In U.S. public schools, including Thornton Township High School in Harvey (a suburb south of Chicago), some students meet outside of regular school hours in classrooms or at other school facilities to have prayer and Bible study, although such activities are believed by many critics to violate the separation of church and state mandated by law. Known as Prayer Warriors for Christ, the group of Harvey students and some city officials in their community are waging a war on the wall that separates church and state. They want group prayer during the school day (Fountain, 2001). Many people in school districts across the land are waging that same war. For example, Arkansas governor Mike Huckabee proclaimed October 2001 "Student Religious Liberty Month" and sent a letter urging school districts to allow students to pray (Morse, 2001).

Why is the issue of prayer and other religious observances in public schools such a concern? As we

■ Should prayer be permitted in the classroom? On the school grounds? At school athletic events? Given the diversity of beliefs that U.S. people hold, arguments and court cases over activities such as prayer around the school flagpole will no doubt continue in the future.

think about this question, it is necessary to understand how social policy has been historically used to establish a division between church and state. When the U.S. Constitution was ratified in 1789, the colonists who made up the majority of the population of the original states were of many different faiths. Due to this diversity, there was no mention of religion in

other words, if individuals consider themselves to be part of a larger order that holds the ultimate meaning in life, they will feel bound to one another (and to past and future generations) in a way that might not be possible otherwise (McGuire, 2002).

Religion also helps maintain social control in society by conferring supernatural legitimacy on the norms and laws of a society. In some societies, social control occurs as a result of direct collusion between the dominant classes and the dominant religious organizations. Niccolo Machiavelli, an influential sixteenth-century statesman and author, wrote that it was "the duty of princes and heads of republics to uphold the foundations of religion in their countries,

for then it is easy to keep their people religious, and consequently well conducted and united" (qtd. in McGuire, 2002: 242). And, as discussed in Chapter 14, absolute monarchs have often claimed a divine right to rule.

In the United States, the separation of church and state reduces religious legitimation of political power (see Box 17.2). Nevertheless, political leaders often use religion to justify their decisions, stating that they have prayed for guidance in deciding what to do. This informal relationship between religion and the state has been referred to as *civil religion*—**the set of beliefs, rituals, and symbols that makes sacred the values of the society and places the nation in the**

the original Constitution. However, in 1791 the First Amendment added the following provision (to this day, the only reference to religion in that document): "Congress shall make no law respecting an establishment of religion, or prohibiting the free exercise thereof." The ban was binding only on the federal government; however, in 1947 the U.S. Supreme Court held that the Fourteenth Amendment ("No State shall make or enforce any law which shall abridge the privileges or immunities of citizens") had the effect of making the First Amendment's separation of church and state applicable to state governments as well.

Historically, the Supreme Court has been called on to define the boundary between permissible and impermissible governmental action with regard to religion. In 1947 the Court held that the ban on establishing a religion made it unconstitutional for a state to use tax revenues to support an institution that taught religion. In 1962 and 1963, the Court expanded this ruling to include many types of religious activities at schools or in connection with school activities, including group prayer, invocations at sporting events, and distribution of religious materials at school. However, in 1990 the Court ruled that religious groups (such as the Prayer Warriors) could meet on school property if certain conditions were met, including that attendance must be voluntary, the meetings must be organized and run by students, and the activities must occur outside of regular class hours.

Compromises regarding prayer in schools have been attempted in some states as legislators passed laws that either permit or mandate a daily moment of silence in public schools; however, the effect of such compromises has often been that neither side in the debate is appeased. For example, those who favor school prayer have proposed constitutional amendments such as "Nothing in this Constitution shall be construed to prohibit individual or group prayer in public schools or other public institutions. No person shall be required by the United States or by any state to participate in prayer. Neither the United States nor any state shall compose the words of any prayer to be said in public schools."

The church–state division continues to be a topic of extensive debate among school officials and political leaders. Advocates of allowing more religious activities in public education during school hours as well as after school believe that the constitutional dictate prohibiting "establishment of religion" was not intended to keep religion out of the public schools and that students need greater access to religious and moral training. Opponents believe that any entry of religious training and religious observances into public education and taxpayer-supported school facilities or events violates the Constitution and might be used by some people to promote their religion over others, or over a person's right to have no religion at all. Where should the line be drawn? What do you think?

Sources: Based on Albanese, 1999; Fountain, 2001; Greenberg and Page, 2002; and Morse, 2001.

WRITING IN SOCIOLOGY ASSIGNMENT

Why is school prayer a controversial issue in a diverse society? What major arguments can be made *for* and *against* religious activities in public schools?

---

context of the ultimate system of meaning. Civil religion is not tied to any one denomination or religious group; it has an identity all its own. For example, many civil ceremonies in the United States have a marked religious quality. National values are celebrated on "high holy days" such as Memorial Day and the Fourth of July. Political inaugurations and courtroom trials often require people to place their hand on a Bible while swearing to do their duty or tell the truth, as the case may be. The U.S. flag is the primary sacred object of our civil religion, and the Pledge of Allegiance includes the phrase "one nation under God." U.S. currency bears the inscription "In God We Trust."

Some critics have attempted to eliminate all vestiges of civil religion from public life. However, the sociologist Robert Bellah (1967), who has studied civil religion extensively, argues that civil religion is not the same thing as Christianity; rather, it is limited to affirmations that members of any denomination can accept. As the sociologist Meredith McGuire (2002: 203) explains,

Civil religion is appropriate to actions in the official public sphere, and Christianity and other religions are granted full liberty in the sphere of personal piety and voluntary social action. This division of spheres of relevance is particularly

Separation of church and state is a constitutional requirement that is often contested by people who believe that religion should be a part of public life. These workers are complying with a federal court order to remove a monument bearing the Ten Commandments from the rotunda of the Alabama State Judicial Building.

© 2003 AP/Wide World Photos

important for countries such as the United States, where religious pluralism is both a valued feature of sociopolitical life and a barrier to achieving a unified perspective for decision making.

However, this assertion may not resolve the problem for those who do not believe in the existence of God or for those who believe that *true* religion is trivialized by civil religion.

Civil religion may serve either a *priestly function* by celebrating the greatness of the United States or a *prophetic function* by pointing out discrepancies between the nation's ideals and the realities of its actions. For example, in his famous "I Have a Dream" speech, Martin Luther King, Jr., appealed to people of all religions to end racial discrimination in the United States based on patriotic values (Roberts, 2004). However, civil religion was also used to justify two opposing stances on U.S. involvement in the war in Vietnam; some bumper stickers read "America—Love It or Leave It!" while others demanded "America—Change It or Lose It!" (McGuire, 2002).

Conflict Perspectives on Religion

Although many functionalists view religion, including civil religion, as serving positive functions in society, some conflict theorists view religion negatively.

Karl Marx on Religion   For Marx, *ideologies*—systematic views of the way the world ought to be—are embodied in religious doctrines and political values (Turner, Beeghley, and Powers, 2002). These ideologies

serve to justify the status quo and retard social change. The capitalist class uses religious ideology as a tool of domination to mislead the workers about their true interests. For this reason, Marx wrote his now-famous statement that religion is the "opiate of the masses." People become complacent because they have been taught to believe in an afterlife in which they will be rewarded for their suffering and misery in this life. Although these religious teachings soothe the masses' distress, any relief is illusory. Religion unites people in a "false consciousness" that they share common interests with members of the dominant class (Roberts, 2004).

From a conflict perspective, religion tends to promote strife between groups and societies. For example, the new religious right in the United States has incorporated both the priestly and prophetic functions into its agenda. While calling for moral reform, it also calls the nation back to a covenant with God (Roberts, 2004). According to McGuire (2002), Weber's distinction between people's "class situation" (stratification based on economic factors) and "status situation" (stratification based on lifestyle, honor, and prestige) is useful in understanding how the new religious right can press for change while at the same time demanding a "return" to traditional family values and prayer in public schools. McGuire (2002: 241) suggests that members of new-right religious organizations who may feel that their status (prestige or honor) is "threatened by changing cultural norms assert their values politically in order to re-establish the ideological basis of their status."

According to conflict theorists, conflict may be *between* religious groups (for example, anti-Semitism), *within* a religious group (for example, when a splinter group leaves an existing denomination), or between a religious group and the *larger society* (for example, the conflict over religion in the classroom). Conflict theorists assert that in attempting to provide meaning and purpose in life while at the same time promoting the status quo, religion is used by the dominant classes to impose their own control over society and its resources. Many feminist scholars object to the patriarchal nature of most religions, some advocate a break from traditional religions, and others seek to reform religious language, symbols, and rituals in order to eliminate the elements of patriarchy.

Max Weber's Response to Marx Whereas Marx believed that religion retards social change, Weber argued just the opposite. For Weber, religion could be a catalyst to produce social change. In *The Protestant Ethic and the Spirit of Capitalism* (1976/ 1904–1905), Weber asserted that the religious teachings of John Calvin are directly related to the rise of capitalism. Calvin emphasized the doctrine of *predestination*—the belief that even before they are born, all people are divided into two groups, the saved and the damned, and only God knows who will go to heaven (the elect) and who will go to hell. Because people cannot know whether they will be saved, they tend to look for earthly signs that they are among the elect. According to the Protestant ethic, those who have faith, perform good works, and achieve economic success are more likely to be among the chosen of God. As a result, people work hard, save their money, and do not spend it on worldly frivolity; instead, they reinvest it in their land, equipment, and labor (Chalfant, Beckley, and Palmer, 1994).

The spirit of capitalism grew in the fertile soil of the Protestant ethic. Even as people worked ever harder to prove their religious piety, structural conditions in Europe led to the Industrial Revolution, free markets, and the commercialization of the economy—developments that worked hand in hand with Calvinist religious teachings. From this viewpoint, wealth is an unintended consequence of religious piety and hard work. However, Weber (1976/1904–1905: 182) recognized that for some people "the pursuit of wealth, stripped of its religious and ethical meaning, tends to become associated with purely mundane passions, which actually give it the character of sport."

With the contemporary secularizing influence of wealth, people often think of wealth and material possessions as the major (or only) reason to work. Although the "Protestant ethic" is rarely invoked today, many people still refer to the "work ethic" in somewhat the same manner that Weber did. For example, political and business leaders in the United States often claim that "the work ethic is dead."

Like Marx, Weber was acutely aware that religion could reinforce existing social arrangements, especially the stratification system. The wealthy can use religion to justify their power and privilege: It is a sign of God's approval of their hard work and morality. As for the poor, if they work hard and live a moral life, they will be richly rewarded in another life. The Hindu belief in reincarnation is an example of religion reinforcing the stratification system. Because a person's social position in the current life is the result of behavior in a former life, the privileges of the upper class must be protected so that each person may enjoy those privileges in another incarnation.

Does Weber's thesis about the relationship between religion and the economy withstand the test of time? Recently, the sociologist Randall Collins reexamined Weber's assertion that the capitalist breakthrough occurred just in Christian Europe and concluded that this belief is only partially accurate. According to Collins, Weber was correct that religious institutions are among the most likely places within agrarian societies for capitalism to begin. However, Collins believes that the foundations for capitalism in Asia, particularly in Japan, were laid in the Buddhist monastic economy of late medieval Japan. Collins (1997: 855) states that "The temples were the first entrepreneurial organizations in Japan: the first to combine control of the factors of labor, capital, and land so as to allocate them for enhancing production." Due to an ethic of self-discipline and restraint on consumption, high levels of accumulation and investment took place in medieval Japanese Buddhism. Gradually, secular capitalism emerged from temple capitalism as new guilds arose that were independent of the temples, and the gap between the clergy and everyday people narrowed through property transformation brought about by uprisings of the common people and wars with outside entities. Moreover, the capitalist dynamic in the monasteries was eventually transferred to the secular economy, opening the way to the Industrial Revolution in Japan (Collins, 1997). From the works of Weber and Collins, we can conclude that the emergence of capitalism through a religious economy happened in several parts of the world, not just one, and that it occurred in both Christian and Buddhist forms (Collins, 1997).

Symbolic Interactionist Perspectives on Religion

Thus far, we have been looking at religion primarily from a macrolevel perspective. Symbolic interactionists focus their attention on a microlevel analysis that examines the meanings people give to religion in their everyday lives.

Religion as a Reference Group

For many people, religion serves as a reference group to help them define themselves. For example, religious symbols have meaning for large bodies of people. The Star of David holds special significance for Jews, just as the crescent moon and star do for Muslims and the cross does for Christians. For individuals as well, a symbol may have a certain meaning beyond that shared by the group. For instance, a symbolic gift given to a child may have special meaning when he or she grows up and faces war or other crises. It may not only remind the adult of a religious belief but also create a feeling of closeness with a relative who is now deceased. It has been said that the symbolism of religion is so very powerful because it "expresses the essential facts of our human existence" (Collins, 1982: 37).

Her Religion and His Religion

Not all people interpret religion in the same way. In virtually all religions, women have much less influence in establishing social definitions of appropriate gender roles both within the religious community and in the larger community. Therefore, women and men may belong to the same religious group, but their individual religion will not necessarily be a carbon copy of the group's entire system of beliefs. According to McGuire (2002), women's versions of a certain religion probably differ markedly from men's versions. For example, although an Orthodox Jewish man may focus on his public ritual roles and his discussion of sacred texts, women have few ritual duties and are more likely to focus on their responsibilities in the home. Consequently, the meaning of being Jewish may be different for women than for men.

Religious symbolism and language typically create a social definition of the roles of men and women. For example, religious symbolism may depict the higher deities as male and the lower deities as female. Females are sometimes depicted as negative or evil spiritual forces. For example, the Hindu goddess Kali represents men's eternal battle against the evils of materialism (Daly, 1973). Historically, language has defined women as being nonexistent in the world's major religions. Phrases such as *for all men* in Catholic and Episcopal services have gradually been changed to *for all;* however, some churches retain the traditional liturgy. And although there has been resistance, especially by women, to some traditional terms, inclusive language is less common, overall, than older male terms for God (Briggs, 1987).

Orthodox Jews at the Western Wall in Jerusalem—a wall that holds special significance for all Jews—express their faith in God and in the traditions of their ancestors.

Many women resist the subordination that they have experienced in organized religion. They have worked to change the existing rules that have excluded them or placed them in a clearly subordinate position.

■ WORLD RELIGIONS

Although there are many localized religions throughout the world, those religions classified as *world religions* cover vast expanses of the Earth and have millions of followers. Six world religions—Hinduism, Buddhism, Confucianism, Judaism, Islam, and Christianity—have more than 4 billion adherents, almost 75 percent of the world's population. These six religions are compared in Table 17.1.

Hinduism

We begin with Hinduism because it is believed to be one of the world's oldest current religions, having originated along the banks of the Indus River in Pakistan between 3,500 and 4,500 years ago. Since Hinduism began before written records were kept,

Table 17.1   MAJOR WORLD RELIGIONS

		CURRENT FOLLOWERS	FOUNDER/DATE	BELIEFS
	Christianity	1.7 billion	Jesus; 1st century C.E.	Jesus is the Son of God. Through good moral and religious behavior (and/or God's grace), people achieve eternal life with God.
	Islam	1 billion	Muhammad; ca. 600 C.E.	Muhammad received the Qur' an (scriptures) from God. On Judgment Day, believers who have submitted to God's will, as revealed in the Qur' an, will go to an eternal Garden of Eden.
	Hinduism	719 million	No specific founder; ca. 1500 B.C.E.	Brahma (creator), Vishnu (preserver), and Shiva (destroyer) are divine. Union with ultimate reality and escape from eternal reincarnation are achieved through yoga, adherence to scripture, and devotion.
	Buddhism	309 million	Siddhartha Gautama; 500 to 600 B.C.E.	Through meditation and adherence to the Eightfold Path (correct thought and behavior), people can free themselves from desire and suffering, escape the cycle of eternal rebirth, and achieve nirvana (enlightenment).
	Judaism	18 million	Abraham, Isaac, and Jacob; ca. 2000 B.C.E.	God's nature and will are revealed in the Torah (Hebrew scripture) and in His intervention in history. God has established a covenant with the people of Israel, who are called to a life of holiness, justice, mercy, and fidelity to God's law.
	Confucianism	5.9 million	K'ung Fu-Tzu (Confucius); circa 500 B.C.E.	The sayings of Confucius (collected in the *Analects*) stress the role of virtue and order in the relationships among individuals, their families, and society.

modern scholars have only limited information about its earliest leaders and their teachings (Kurtz, 1995). Hindu beliefs and practices have been preserved through an oral tradition and expressed in texts and hymns known as the *Vedas* (meaning "knowledge" or "wisdom"); however, this religion does not have a "sacred" book, such as the Judeo-Christian Bible or the Islamic Qur' an (Sharma, 1995). Consequently, Hindu beliefs and practices emerged over the centuries across the subcontinent of India in a variety of forms, reflecting the influence of the various regional cultures (Kurtz, 1995). Since Hinduism has no scriptures that are thought to be inspired by a god or gods and is not based on the teachings of any one person, religion scholars refer it as to as an *ethical religion*—a system of beliefs that calls upon adherents to follow an ideal way of life. For most Hindus, this is partly achieved by adhering to the expectations of the caste system (see Chapter 8).

Central to Hindu teachings is the belief that individual souls (*jivas*) enter the world and roam the universe until they break free into the limitless atmosphere of illumination (*moska*) by discovering their own *dharma*—duties or responsibilities. According to Hinduism, individual *jivas* pass through a sequence of bodies over time as they undergo a process known as reincarnation (*samsara*)—an endless passage through cycles of life, death, and rebirth until the soul earns liberation (Kurtz, 1995). The soul's acquisition of each new body is tied to the law of *karma* (deed or act), which is a doctrine of the moral law of cause and effect. The present condition of the soul—how happy or unhappy it is, for example—is directly related to what it has done in the past, and its present thoughts and decisions are the ultimate determinants of what its future will be. The ultimate goal of Hindu existence is entering the state of *nirvana*—becoming liberated from the world by uniting the individual soul with the universal soul (*Brahma*).

Hinduism has been devoid of some of the social conflict experienced by other religions. Since Hinduism is based on the assumption that there are many paths to the "Truth" and that the world's religions are alternate paths to that goal, Hindus typically have not engaged in religious debates or "holy wars" with those holding differing beliefs. One of the best-known

According to Marx and Weber, religion serves to reinforce social stratification in a society. For example, according to Hindu belief, a person's social position in his or her current life is a result of behavior in a former life.

Mark O'Neill/Canada Wide

Hindu leaders of modern times was Mohandas ("Mahatma") Gandhi, the champion of India's independence movement, who was devoted to the Hindu ideals of nonviolence, honesty, and courage (Sharma, 1995). However, some social analysts note that Hinduism has been closely associated with the perpetuation of the caste system in India. Although people in the lower castes are taught to live out their lives with dignity, even in the face of poverty and despair, they may also come to believe that their lowly position is the acceptable and appropriate place for them to be—which allows the upper castes to exploit them (Kurtz, 1995).

The Hindu religion is almost as diverse as the wide array of people who adhere to its teachings. It is estimated that there are more than 700 million Hindus in the world today, with 95 percent of them residing in India, over 80 percent of whose population is Hindu (Sharma, 1995).

Since changes in U.S. immigration laws in the 1960s brought a wave of immigrants from India, the number of Hindus in this country has increased significantly (Albanese, 1999). Most Asian Indian immigrants to the United States have been well-educated professionals who have joined the ranks of the U.S. middle- and upper-middle classes and have been active in supporting the more than forty Hindu temples in the nation. For most people of Asian Indian descent in the United States, these temples are sites of worship and gathering places where they can maintain a sense of community. They are also ritual centers where language, arts, and practices from their ethnic past can be preserved (Albanese, 1999).

How have Hindus fared in the United States? Intolerance has been an ongoing problem for many members of the Asian Indian community in this country. In view of numerous violent hate crimes perpetrated against "dot heads" (so-called by adversaries because some Asian Indian women wear a dot in the middle of their forehead), many have bound together to fight intolerance across lines of race, class, and religion.

Despite discrimination and initial confrontations with members of other racial–ethnic groups, the influence of Hindus in the United States is likely to be profound as their number increases. It is estimated that there are about 800,000 adherents in the United States today—a number eight times as large as it was twenty years ago (Shorto, 1997).

Buddhism

When Buddhism first emerged in India some twenty-five hundred years ago, it was thought of as a "new religious movement," arising as it did around the sixth century B.C.E. ("before the common era"), after many earlier religions had become virtually defunct. Buddhism's founder, Siddhartha Gautama of the Sakyas (also known as Gautama Buddha), was born about 563 B.C.E. into the privileged caste. His father was King Suddhodana (who was more like a feudal lord than a king because many kingdoms existed on the Indian subcontinent during that era). According to historians, the king attempted to keep his young son in the palace at all times so that he would neither see how poor people lived nor experience the suffering present in the outside world. As a result, Siddhartha was oblivious to social inequality until he began to make forays beyond the palace walls and into the "real" world, where he observed how other people lived. On one excursion, he saw a monk with a shaven head and became aware that some people withdraw from the secular world and live a life of strict asceticism. Later, Siddhartha engaged in intense meditation underneath a bodhi tree in what is now Nepal, eventually declaring that he had obtained Enlightenment—an awakening to the true nature of reality (Kurtz, 1995). From

Worshippers at this Buddhist temple in Los Angeles celebrate the Thai New Year.

Gary Conner/PhotoEdit

that day forward, Siddhartha was referred to as *Buddha,* meaning "the Enlightened One" or the "Awakened One," and spent his life teaching others how to reach nirvana (Smith, 1991).

Because of the efforts of a series of invaders, Buddhism had ceased to exist in India by the thirteenth century but had already expanded into other nations in various forms. *Theravadin Buddhism,* which focuses on the life of the Buddha and seeks to follow his teachings, gained its strongest toehold in Southeast Asia. *Mahayana Buddhism* is centered in Japan, China, and Korea and primarily focuses on meditation and the Four Noble Truths:

1. Life is *dukkha*—physical and mental suffering, pain, or anguish that pervades all human existence.
2. The cause of life's suffering is rooted in *tanha*—grasping, craving, and coveting.
3. One can overcome *tanha* and be released into Ultimate Freedom in Perfect Existence (nirvana).
4. Overcoming desire can be accomplished through the Eightfold Path to Nirvana. This path is a way of living that avoids extremes of indulgence and suggests that a person can live in the world but not be worldly. The path's eight steps are *right view* (proper belief), *right intent* (renouncing attachment to the world), *right speech* (not lying, slandering, or using abusive talk), *right action* (avoiding sexual indulgence), *right livelihood* (avoiding occupations that do not enhance spiritual advancement), *right effort* (preventing potential evil from arising), *right mindfulness,* and *right concentration* (overcoming sensuous appetites and evil desires). (Smith, 1991; Matthews, 2004)

The third major branch of Buddhism—*Vajrayana*—incorporates the first two branches along with some aspects of Hinduism; it emerged in Tibet in the seventh century (Albanese, 1999). Like Hinduism, the teachings of this type of Buddhism—and specifically those of the Dalai Lama, the Tibetan Buddhist leader—emphasize the doctrine of *ahimsa,* or nonharmfulness, and discourage violence and warfare. It is believed that Buddhism may have suppressed caste-related tensions resulting from the vast economic inequality found in early India (Kurtz, 1995).

When did Buddhism first arrive in this country? According to most scholars, some branches appeared as early as the 1840s, when Chinese immigrants arrived on the West Coast. Shortly thereafter, temples were erected in San Francisco's Chinatown; however, ethnic Buddhism in the United States expanded after the Civil War, when Japanese immigrants arrived first in Hawaii (then a U.S. possession) and a decade later in California. The branch of Pure Land Buddhism brought by the Japanese probably had the greatest chance of succeeding in the United States because it most closely resembled Christianity. Pure Land Buddhism included teachings about God and His son Jesus, Who brought human beings to an eternal life in paradise (Albanese, 1999). Buddhism went through the process of Americanization, which is reflected in the contemporary use of terms such as *church, bishop,* and *Sunday school*—terms previously unknown in this religion.

Today, Buddhism is one of the fastest-growing Eastern religions in the United States. Zen and Tibetan Buddhism are extremely popular forms. Zen sprang up among white, middle-class young people, many of whom were well-educated and lived in California or New York. Buddhist monks fleeing the takeover of Tibet by the Communist Chinese in the 1960s brought Tibetan Buddhism to this country. Recently, Tibetan Buddhism has received extensive media coverage at least partly due to the conversion of some U.S. celebrities, including Richard Gere, Steven Seagal, and Adam Yauch (Van Biema, 1997).

Confucianism

Confucianism—which means the "family of scholars"—started as a school of thought or a tradition of learning before its eventual leader, Confucius, was born (Wei-ming, 1995). Confucius (the Latinized form of K'ung Fu-tzu) lived in China between 551 and 479 B.C.E. and emerged as a teacher at about the

PhotoEdit

Confucianism is based on the ethical teachings formulated by Confucius, shown here in a portrait created by a Manchu prince in 1735.

same time that the Buddha became a significant figure in India.

Confucius—whose sayings are collected in the *Analects*—taught that people must learn the importance of *order* in human relationships and must follow a strict code of moral conduct, including respect for others, benevolence, and reciprocity (Kurtz, 1995). A central teaching of Confucius was that humans are by nature good and that they learn best by having an example or a role model. As a result, he created a *junzi* ("chun-tzu"), or model person, who has such attributes as being upright regardless of outward circumstances, being magnanimous by expressing forgiveness toward others, being directed by internal principles rather than external laws, being sincere in speech and action, and being earnest and benevolent. Confucius wanted to demonstrate these traits, and he believed that he should be a role model for his students. The *junzi*'s behavior is to be based on the Confucian principle of *Li,* meaning righteousness or propriety, which refers both to ritual and to correct conduct in public. One of the central attributes exhibited by the *junzi* is *ren (jen),* which means having deep empathy or compassion for other humans (Matthews, 2004).

Confucius established the foundation for social hierarchy—and potential conflict—when he set forth his Five Constant Relationships: *ruler–subject, husband–wife, elder brother–younger brother, elder friend–junior friend,* and *father–son.* In each of these pairs, one person is unequal to the other, but each is expected to carry out specific responsibilities to the other (Matthews, 2004). Confucius taught that due authority is not automatic; it must be earned. The subject does not owe loyalty to the ruler or authority figure if that individual does not fulfill his or her end of the bargain.

Confucianism is based on the belief that Heaven and Earth are not separate places but rather a continuum in which both realms are constantly in touch with each other. According to this approach, those who inhabit Heaven are ruled over by a supreme ancestor and are the ancestors of those persons who are on Earth. These forefathers are eventually joined by those who are currently on Earth; therefore, death is nothing more than the promotion to a more honorable estate.

Until the Communist takeover and the establishment of the People's Republic, Confucianism was the official religion of China. After the takeover, political leaders sought to replace Confucianism, Taoism, and Buddhism with Maoism—belief in the teachings of the Chinese Communist leader—but the Confucian influence remains in East Asian countries such as Japan, Korea, Taiwan, and Singapore.

How prominent is Confucianism in the United States? It is difficult to provide an accurate answer to this question because some people view Confucianism as a set of ethical teachings rather than as a religion (Smith, 1991). However, many recent immigrants from China and Southeast Asia adhere to the teachings of Confucius, although perhaps mixed with those of other great Eastern religious philosophers and teachers. Neo-Confucianism, which has emerged on the West Coast, is heavily influenced by Buddhism and Taoism (Matthews, 2004).

We now move from the Eastern religions (Hinduism, Buddhism, and Confucianism), which tend to be based on ethics or values more than on a deity or Supreme Being, to the Western religions (Judaism, Christianity, and Islam), which are founded on the Abrahamic tradition and place an emphasis on God and a relationship between human beings and a Supreme Being. The original locations of all six of these religions are shown in Map 17.1.

Judaism

Although Judaism has fewer adherents worldwide than some other major religions, its influence is deeply felt in Western culture. Today, there are an estimated 18 million Jews residing in about 134 countries worldwide; however, the majority reside in the United States or Israel (J. Wright, 1997).

Map 17.1 **Original Locations of the World's Major Religions**

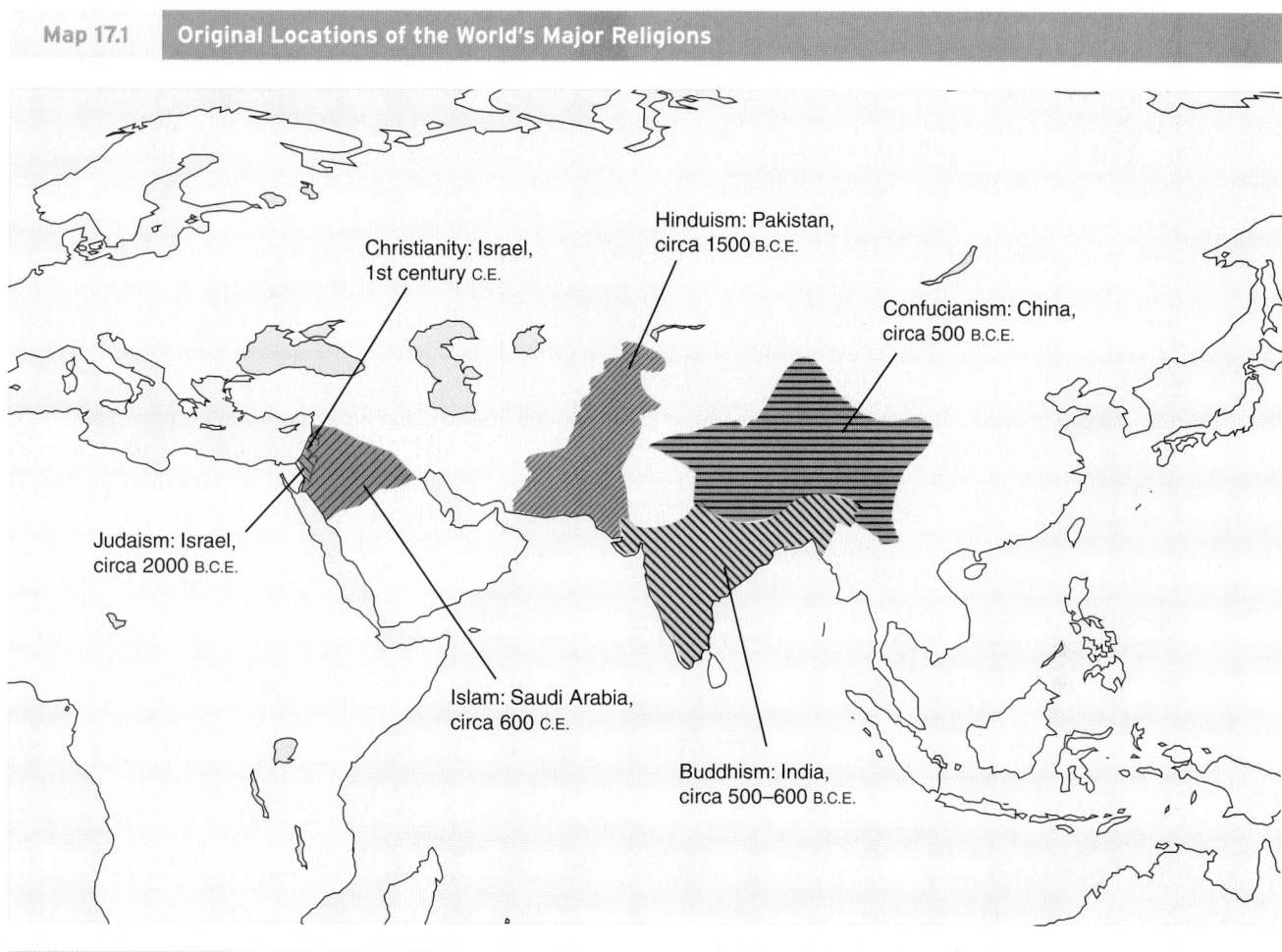

Central to contemporary Jewish belief is monotheism, the idea of a single god, called Yahweh, the God of Abraham, Isaac, and Jacob. The Hebrew tradition emerged out of the relationship of Abraham and Sarah, a husband and wife, with Yahweh. According to Jewish tradition, the God of the Jews made a covenant with Abraham and Sarah—His chosen people—that He would protect and provide for them if they swore Him love and obedience. When God appeared to Abraham in about the eighteenth century B.C.E., He encouraged Abraham to emigrate to the area near the Sea of Galilee and the Dead Sea (what is now Israel), leaving behind the ancient fertile crescent of the Middle East (present-day Iraq).

The descendants of Abraham and Sarah migrated to Egypt, where they became slaves of the Egyptians. In a vision, God's chosen leader, Moses, was instructed to liberate His chosen people from the bondage and slavery imposed upon them by the pharaoh. After experiencing a series of ten plagues, the pharaoh decided to free the slaves. The tenth and final plague had involved killing all of the firstborn in

the land of Egypt—human beings and lower animals, as well—except the firstborn children of the Hebrews, who had put the blood of a lamb on the doorposts of their houses so that they would be passed over. This practice inspired the Jewish holiday Passover, which commemorates God's deliverance of the Hebrews from slavery in Egypt during the time of Moses (Matthews, 2004). It is believed that the first Passover took place in Egypt in about 1300–1200 B.C.E. (Matthews, 2004).

Wandering in the desert after their release, the Hebrews established a covenant with God, Who promised that if they would serve Him exclusively, He would give the Hebrews a promised land and make them a great nation. Known as the Ten Commandments (or *Decalogue*), the covenant between God and human beings was given to Moses on top of Mount Sinai. The Ten Commandments and discussions of moral, ceremonial, and cultural laws are contained in four books of the Torah: Exodus, Leviticus, Numbers, and Deuteronomy. The Torah, also known as the Pentateuch, is the sacred book of contemporary Judaism.

The Jewish people believe that they have a unique relationship with God, affirmed on the one hand by His covenant and on the other by His law. Judaism has three key components: God (the deity), Torah (God's teachings), and Israel (the community or holy nation). Although God guides human destiny, people are responsible for making their own ethical choices in keeping with His law; when they fail to act according to the law, they have committed a sin. Also fundamental to Judaism is the belief that one day the Messiah will come to Earth, ushering in an age of peace and justice for all.

Today, Jews worship in synagogues in congregations led by a *rabbi*—a teacher or ordained interpreter and leader of Judaism. The Sabbath is observed from sunset Friday to sunset Saturday, based on the story of Creation in Genesis, especially the belief that God rested on the seventh day after He had created the world. Worship services consist of readings from scripture, prayer, and singing. Jews celebrate a set of holidays distinct from U.S. dominant cultural religious celebrations. The most important holidays in the Jewish calendar are Rosh Hashana (New Year), Yom Kippur (Day of Atonement), Hanukkah (Festival of Lights), and Pesach (Passover).

Throughout their history, Jews have been the object of prejudice and discrimination. The Holocaust, which took place in Nazi Germany (and several other nations that the Germans occupied) between 1933 and 1945, remains one of the saddest eras in history. After the rise of Hitler in Germany in 1933 and the Nazi invasion of Poland, Jews were singled out with special registrations, passports, and clothing. Many of their families were separated by force, and some family members were sent to slave labor camps while others were sent to "resettlement." Eventually, many Jews were imprisoned in death camps, where six million lost their lives.

Anti-Semitism has been a continuing problem in the United States since the late nineteenth century. Like other forms of prejudice and discrimination, anti-Semitism has extracted a heavy toll on multiple generations of Jewish Americans. In the 1980s and 1990s, for example, there was an increase in inter-ethnic violence between African Americans and Jews in large urban centers. These confrontations are often triggered by a situation such as when a Hasidic Jew driving a motor vehicle ran over and killed an African American youth, and members of the black community believed that the vehicle's driver did not receive the legal penalty he deserved for his actions. However, problems between Jews and members of other racial–ethnic and religious groups can be traced back to the earliest encounters between immigrants arriving in this country from nations throughout the world.

Today, Judaism has three main branches—Orthodox, Reform, and Conservative. Orthodox Judaism follows the traditional practices and teachings, including eating only kosher foods prepared in a designated way, observing the traditional Sabbath, segregating women and men in religious services, and wearing traditional clothing.

Reform Judaism, which began in Germany in the nineteenth century, is based on the belief that the Torah is binding only in its moral teachings and that adherents should no longer be required to follow all of the Talmud, the compilation of Jewish law setting forth the strict rabbinic teachings on practices such as food preparation, rituals, and dress. In some Reform congregations, gender-segregated seating is no longer required. In the United States, services are conducted almost entirely in English, a Sunday Sabbath is observed, and less emphasis is placed on traditional Jewish holidays (Albanese, 1999).

Conservative Judaism emerged between 1880 and 1914 with the arrival of many Jewish immigrants in the United States from countries such as Russia, Poland, Rumania, and Austria. Seeking freedom and an escape from persecution, these new arrivals settled in major cities such as New York and Chicago, where they primarily became factory workers, artisans, and small shopkeepers. Conservative Judaism, which became a middle ground between Orthodox and Reform Judaism, teaches that the Torah and Talmud must be followed and that *Zionism*—the movement to establish and maintain a Jewish homeland in Israel—is crucial to the future of Judaism. In Conservative synagogues, worship services are typically performed in Hebrew. Men are expected to wear head coverings, and women have roles of leadership in the congregation; some may become ordained rabbis (Matthews, 2004). Despite centuries of religious hatred and discrimination, Judaism persists as one of the world's most influential religions.

Islam

Like Judaism, Islam is a religion in the Abrahamic tradition; both religions arise through sons of Abraham—Judaism through Isaac and Islam through Ishmael. Islam, whose followers are known as Muslims, is based on the teachings of its founder, Muhammad, who was born in Mecca (now in Saudi Arabia) in about 570 C.E. According to Muhammad, followers must adhere to the five Pillars of Islam: (1) believing that there is no god but Allah, (2) participating in five periods of prayer each day, (3) paying taxes to help support the needy, (4) fasting during the daylight hours in the month of Ramadan, and (5) making at

The Muslim women shown here pray at a mosque courtyard in Bangladesh during the fasting month of Ramadan. According to Muslim teaching, Ramadan marks God's revelation of the Qur' an to the Prophet Mohammad.

least one pilgrimage to the Sacred House of Allah in Mecca (Matthews, 2004).

The Islamic faith is based on the Qur' an—the holy book of the Muslims—as revealed to the Prophet Muhammad through the Angel Gabriel at the command of God. According to the Qur' an, it is up to God, not humans, to determine which individuals are deserving of punishment and what kinds of violence are justified under various conditions.

The Islamic notion of *jihad*—meaning "struggle"—is a core belief. The Greater Jihad is believed to be the internal struggle against sin within a person's heart, whereas the Lesser Jihad is the external struggle that takes place in the world, including violence and war (Ferguson, 1977; Kurtz, 1995). The term *jihad* is typically associated with religious fundamentalism. Despite the fact that fundamentalism is found in most of the world's religions, some social analysts believe that Islamic fundamentalism is uniquely linked to the armed struggles of groups such as Hamas, an alleged terrorist organization, and the militant Islamic Jihad, which is believed to engage in continual conflict (see Barber, 1996).

Today, more than 19 percent of the world's population considers itself to be Muslim. Most of the more than one billion adherents of this religion reside in the Middle East, but the majority of people residing in northern Africa and western Asia also consider themselves to be Muslim. Other large populations of Muslims are located in Pakistan, India, Bangladesh, Indonesia, and the southern regions of the former Soviet Union.

Islam is one of the fastest-growing religions in the United States. Driven by recent waves of migration and a relatively high rate of conversion, there has been a significant increase in the number of Muslims in this country. At about six million, they outnumber members of several major Protestant denominations, including the Presbyterians (estimated at 3.5 million) and Episcopalians (estimated at 2.3 million). Recent Muslim arrivals in the United States typically have come from countries such as Pakistan, Iran, and Saudi Arabia. Most have settled in the Midwest or on the East Coast. Approximately 30 percent of Muslims are from southern Asia, 25 percent are African American, and about 20 percent are Arab Americans (Shorto, 1997). The oldest U.S. Islamic group is the Federation of Islamic Associations, and many university campuses now have Muslim student associations. The largest "umbrella" organization of U.S. Muslims is the Islamic Society of North America (Matthews, 2004).

Muslims in the United States have not always felt welcome. During the Carter administration, the Iranian hostage crisis triggered negative stereotypes, including movies and television programs depicting Muslims as terrorists. The bombing of the World Trade Center in New York in 1993 and the terrorist attacks on the United States in 2001, which claimed the lives of about 3,000 people, intensified stereotypes regarding people appearing to be Muslim or from countries associated with Islam. According to some analysts, media distortions of "Islam" and "Muslims" intensify prejudice against Arab Americans (see Box 17.3).

Despite these negative stereotypes and some acts of violence against Muslims, the more than six hundred mosques and Islamic centers in this country today continue to serve as centers of worship and as social and educational centers, where people can maintain cultural ties and hold ceremonial rites of passage such as weddings and funerals (Matthews, 2004).

Box 17.3 CHANGING TIMES: MEDIA AND TECHNOLOGY

Islamophobia and the Media

New York Times, May 28, 2003:

I'M AN AMERICAN *AND* I'M A MUSLIM

MY NAME IS DR. J. AISHA SIMON. I attended the Medical College of Virginia, completed my residency at Georgetown University and I'll be attending Harvard University to earn a master's degree in public health. I'm a family physician, a wife, and mother. I'm also involved in international relief work, traveling to places like Bosnia and Africa, and coordinating medical volunteers to serve in Guatemala....

I'm an American Muslim woman and I believe in the importance of charity and service to my community. The values I learned from my family and my religion while growing up in America have led me to a life of service. Islam calls upon us to strive with one another in hastening to good deeds, and to care for the less fortunate as we care for ourselves. The Prophet Muhammad taught us that when we serve our brother or sister, we are serving God. **I'M AN AMERICAN MUSLIM.**

> –Dr. J. Aisha Simon, one of the people featured in "Islam in America," a year-long ad campaign designed to foster greater understanding of Islam and to counter a rising tide of anti-Muslim rhetoric (CAIR, 2003)

Prior to the terrorist attacks of September 11, 2001, Arabs and Arab Americans had already experienced biased media coverage and were often the objects of negative stereotyping, as one analyst stated:

> Nowadays one ethnic group in the United States ... consistently bears the brunt of media discrimination: Arab Americans. It seems that everyone can make pejorative remarks on TV and in newspapers against them and get away with it: They are now a focal point of, and a substitute for, prejudice and stereotyping that used to be directed against a number of other ethnic and ethnoreligious groups.... (Khleif, 1998: 290)

According to media analyst Bud B. Khleif (1998: 290), "Superficial knowledge of Arab culture and deep ignorance about the Middle East and its ethnic groups fuel Arab-bashing on American television, in American congressional attitudes, and in editorials and cartoons appearing in magazines and newspapers."

Negativism after September 11 has not only focused on Arabs and Arab Americans but has also placed the spotlight on Muslims, particularly as repeated terror alerts and 24/7 media coverage of bombings in countries such as Israel and Indonesia have kept many people fearful of terrorist attacks in the United States and worldwide. According to some

analysts, negativity toward Muslims is so significant that the term *Islamophobia*–dread or hatred of Islam, resulting in fear or dislike of all or most Muslims–has been coined to describe prejudice and discrimination directed toward people because of their perceived religion or nationality of origin (Runnymede Trust, 1997).

Positive ad campaigns such as the one involving Dr. Simon are an effort to counteract this rising tide of negative sentiment against Muslims. The problem is not limited to the United States: Media in Britain and Canada (along with many others) have been accused of perpetuating distorted and discriminatory images of Muslims and Muslim life (see Farag, 2003; Younge, 2002). British journalist Gary Younge (2002) has asked what we might believe if all we knew about Muslims was what we have seen recently on television:

> You would "know" them as mugshots who have or might commit acts of terror, have been accused of committing acts of terror and demonstrators waving holy books. You would be "familiar" with bearded older men standing outside mosques, young men on the rampage and women in headscarves. You would "know" that they come from the adjectives "fundamentalist," "terrorist" and "moderate." If you are Muslim ... you may recognize many of these characters as part of your community, but wonder why the media have become so obsessed with them that they ignore all the other parts. If you are not Muslim, then you will now "know" Muslims as "other"–a group of people with whom you have little in common. You will know them as a "problem"–a community that appears to hate you and which you would rather avoid. In short, you will know just enough to know that you don't really know them at all.

Although media depictions of Muslims were problematic before September 11, the events of that day helped to exaggerate the trend, exacerbating existing tensions and compounding the exclusion that already existed. As Younge (2002) concludes, "[I]t globalized a conflict which until then had been local."

Can the media and people who do not identify themselves as Muslim gain more accurate insights on the many contributions of millions of Muslims in the United States and other nations? According to some analysts, media coverage of the entire Islamic community, rather than of a few on the radical fringe, would be a good beginning for creating positive images rather than negative ones. However, a pressing question for journalists, as well as for everyone else, remains: How does fear of terrorism continue to fuel negative depictions of Muslims in the media and thereby maintain and perpetuate prejudice and discrimination in our nation and others?

Christianity

Along with Judaism and Islam, Christianity follows the Abrahamic tradition, tracing its roots to Abraham and Sarah. Although Jews and Christians share common scriptures in the portions of the Bible known to Christians as the "Old Testament," they interpret them differently. The Christian teachings in the "New Testament" present a world view in which the old covenant between God and humans, as found in the Old Testament, is obsolete in light of God's offer of a new covenant to the followers of Jesus, whom Christians believe to be God's only son (Matthews, 2004).

As described in the New Testament, Jesus was born to the virgin Mary and her husband, Joseph. After a period of youth in which He prepared himself for the ministry, Jesus appeared in public and went about teaching and preaching, including performing a series of miracles—events believed to be brought about by divine intervention—such as raising people from the dead.

The central themes in the teachings of Jesus are the kingdom of God and standards of personal conduct for adherents of Christianity. Jesus emphasized the importance of righteousness before God and of praying to the Supreme Being for guidance in the daily affairs of life (Matthews, 2004).

One of the central teachings of Christianity is linked to the unique circumstances surrounding the death of Jesus. Just prior to His death, Jesus and His disciplines held a special supper, now referred to as "the last supper," which is commemorated in contemporary Christianity in the sacrament of Holy Communion. Afterward, Jesus was arrested by a group sent by the priests and scribes for claiming to be king of the Jews. After being condemned to death by political leaders, Jesus was executed by crucifixion, which made the cross a central symbol of the Christian religion. According to the New Testament, Jesus died, was placed in a tomb, and on the third day was resurrected—restored to life—establishing that He is the son of God. Jesus then remained on Earth for forty days, after which He ascended into heaven on a cloud. Two thousand years later, many Christian churches teach that one day Jesus will "come again in glory" and that His second coming will mark the end of the world as we know it.

Whereas Judaism is basically an inherited religion and most adherents are born into the community, Christianity has universalistic criteria for membership—meaning that it does not have ethnic or tribal qualifications—based on acceptance of a set of beliefs. Becoming a Christian requires personal belief that Jesus is the son of God; that He died, was buried, and on the third day rose from the dead; and that the

Christians around the world have been drawn to cathedrals such as Notre-Dame de Paris (built between 1163 and 1257) to worship God and celebrate their religious beliefs.

Supreme Being is a sacred *trinity*—the Christian belief of "God the Three in One," comprising God the Father, Jesus the Son, and the Holy Spirit—a presence that lives within those who have accepted Jesus as savior (Barna, 1996). To become members of the religious community, believers must affirm their faith and go through a rite of passage of baptism (Matthews, 2004). According to the teachings of Christianity, those who believe in Jesus as their savior will be resurrected from death and live eternally in the presence of God, whereas those who are wicked will endure an eternity in hell (Matthews, 2004).

For centuries, Christians, Jews, and Muslims have lived together in peace and harmony in some areas of the world; in others, however, they have engaged in strife and fighting. The wounds and animosities between Muslims and Christians have remained since the early Christian era; nevertheless, some religion scholars believe that there is hope for a genuine Christian–Muslim dialogue in the future (Cox, 1995).

Today, almost one-third of the world's population (between 1.5 billion and 2 billion) refer to themselves as Christians. The majority of Christians live in North or South America or in Europe. According to Kurtz (1995), a sociological analysis of Christianity would reveal that it became the dominant religion not necessarily due to its theology but because of its alliance with the power structures of Western civilization, beginning with the fourth-century conversion of Roman Emperor Constantine and following with its expansion into

Western Europe during the Middle Ages. Missionary movements helped spread Christianity outward from Europe to other regions of the world in the nineteenth century, as missionaries also conquered land, cultures, and the economic and political resources of indigenous populations (Kurtz, 1995). As Christianity moved across cultures, it underwent dramatic transformations and became, in actuality, a tremendous variety of "Christianities" rather than just one highly integrated body of religious beliefs and practices (Kurtz, 1995).

Roman Catholics sent the first missionaries to North America, where they established forty missions and converted about 26,000 Native Americans to Christianity before the early Protestant settlers arrived on this continent. From the earliest days of the British colonies in this country, a variety of religions were represented, including Anglicans (the forerunners of the Protestant Episcopal church), Baptists, Quakers, Presbyterians, Methodists, and Lutherans. Freedom of religion provided people with the opportunity to establish other denominations and sects and generally to worship as they pleased.

The African American church was the center of community life first for slaves and freed slaves, and then for generations of blacks who experienced on-going prejudice and discrimination based on their race. The theory of nonviolent protest and civil disobedience used in the civil rights movement in the 1960s, and largely orchestrated by the Rev. Dr. Martin Luther King, Jr., emerged from the African American church in the South. Over the years, members of other minority groups, including Latinos/as and Asian Americans, have benefited from religious and social ties to various Christian denominations, including the Roman Catholic church. Today, about 160 million people in the United States are associated with Christian churches, of whom about 63 million consider themselves to be Catholics.

TYPES OF RELIGIOUS ORGANIZATION

Religious groups vary widely in their organizational structure. Although some groups are large and somewhat bureaucratically organized, others are small and have a relatively informal authority structure. Some require total commitment of their members; others expect members to have only a partial commitment. Sociologists have developed typologies or ideal types of religious organization to enable them to study a wide variety of religious groups. The most common

categorization sets forth four types: ecclesia, church, sect, and cult.

Ecclesia

Some countries have an official or state religion known as the *ecclesia*—**a religious organization that is so integrated into the dominant culture that it claims as its membership all members of a society.** Membership in the ecclesia occurs as a result of being born into the society, rather than by any conscious decision on the part of individual members. The linkages between the social institutions of religion and government are often very strong in such societies. Although no true ecclesia exists in the contemporary world, the Anglican church (the official church of England), the Lutheran church in Sweden and Denmark, the Roman Catholic Church in Italy and Spain, and the Islamic mosques in Iran and Pakistan come fairly close.

The Church-Sect Typology

To help explain the different types of religious organizations found in societies, Ernst Troeltsch (1960/1931) and his teacher, Max Weber (1963/1922), developed a typology that distinguishes between the characteristics of churches and sects (see Table 17.2). Unlike an ecclesia, a church is not considered to be a state religion; however, it may still have a powerful influence on political and economic arrangements in society. A *church* is **a large, bureaucratically organized religious organization that tends to seek accommodation with the larger society in order to maintain some degree of control over it.** Church membership is largely based on birth; typically, children of church members are baptized as infants and become lifelong members of the church. Older children and adults may choose to join the church, but they are required to go through an extensive training program that culminates in a ceremony similar to the one that infants go through. Churches have a bureaucratic structure, and leadership is hierarchically arranged. Usually, the clergy have many years of formal education. Churches have very restrained services that appeal to the intellect rather than the emotions. Religious services are highly ritualized; they are led by clergy who wear robes, enter and exit in a formal processional, administer sacraments, and read services from a prayer book or other standardized liturgical format. The Lutheran church and the Episcopal church are two examples.

Midway between the church and the sect is the *denomination*—**a large organized religion characterized by accommodation to society but frequently lacking in ability or intention to dominate society**

Table 17.2 CHARACTERISTICS OF CHURCHES AND SECTS

CHARACTERISTIC	CHURCH	SECT
Organization	Large, bureaucratic organization, led by a professional clergy	Small, faithful group, with high degree of lay participation
Membership	Open to all; members usually from upper and middle classes	Closely guarded membership, usually from lower classes
Type of Worship	Formal, orderly	Informal, spontaneous
Salvation	Granted by God, as administered by the church	Achieved by moral purity
Attitude Toward Other Institutions and Religions	Tolerant	Intolerant

(Niebuhr, 1929). Denominations have a trained ministry, and although involvement by lay members is encouraged more than in the church, their participation is usually limited to particular activities, such as readings or prayers. Denominations tend to be more tolerant and are less likely than churches to expel or excommunicate members. This form of organization is most likely to thrive in societies characterized by *religious pluralism*—a situation in which many religious groups exist because they have a special appeal to specific segments of the population. Perhaps because of its diversity, the United States has more denominations than any other nation. Table 17.3 shows this diversity in Christian denominations. Today, denominations range from Baptists and members of the Church of Christ to Unitarians and Congregationalists.

A *sect* **is a relatively small religious group that has broken away from another religious organization to renew what it views as the original version of the faith.** Unlike churches, sects offer members a more personal religion and an intimate relationship with a supreme being, depicted as taking an active interest in the individual's everyday life. Sects have informal prayers composed at the time they are given, whereas churches use formalized prayers, often from a prayer book. Typically, religious sects appeal to those who might be characterized as lower class, whereas denominations primarily appeal to the middle and upper-middle classes, and churches focus on the upper classes.

According to the church–sect typology, as members of a sect become more successful economically and socially, they tend to focus more on this world and less on the next. However, sect members who do not achieve financial success often believe that they are being left behind as the other members, and sometimes the minister, shift their priorities to things of this world. Eventually, this process weakens some religious organizations, and the dissatisfied or downwardly mobile split off to create new, less worldly versions of the original group that will be more com-

mitted to "keeping the faith." Those who defect to form a new religious organization may start another sect or form a cult (Stark and Bainbridge, 1981).

Cults

A *cult* **is a loosely organized religious group with practices and teachings outside the dominant cultural and religious traditions of a society.** Although many people view cults negatively, some major religions (including Judaism, Islam, and Christianity) and some denominations (such as the Mormons) started as cults. Cult leadership is based on *charismatic* characteristics (personal magnetism or mystical leadership) of the individual leader, including an unusual ability to communicate and to form attachments with other people. An example is the religious movement started by Reverend Sun Myung Moon, a Korean electrical engineer who believed that God had revealed to him that the Judgment Day was rapidly approaching. Out of this movement, the Unification church, or "Moonies,"

AP/Wide World Photos

This mass wedding ceremony of thousands of brides and grooms brought widespread media attention to the Reverend Sun Myung Moon and the Unification Church, which many people view as a religious cult.

Table 17.3 MAJOR U.S. DENOMINATIONS THAT SELF-IDENTIFY AS CHRISTIAN

RELIGIOUS BODY	MEMBERS	CHURCHES
Roman Catholic Church	63,683,000	19,544
Southern Baptist Convention	15,960,000	41,588
United Methodist Church	8,341,000	35,469
National Baptist Convention, U.S.A.[a]	8,200,000	33,000
Church of God in Christ[a]	5,500,000	15,300
Church of Jesus Christ of Latter Day Saints	5,209,000	11,562
Evangelical Lutheran Church in America	5,126,000	10,816
National Baptist Convention of America	3,500,000	2,500
Presbyterian Church (U.S.A.)	3,485,000	11,178
Assemblies of God	2,578,000	12,084
Lutheran Church–Missouri Synod	2,554,000	6,150
African Methodist Episcopal Church	2,500,000	6,200
National Missionary Baptist Convention of America[a]	2,500,000	(N/A)
Progressive National Baptist Convention[a]	2,500,000	2,000
Episcopal Church	2,311,000	7,359
Churches of Christ[a]	1,500,000	15,000
Greek Orthodox Church	1,500,000	508
Pentecostal Assemblies of the World[a]	1,500,000	1,750
American Baptist Churches in U.S.A.	1,437,000	5,756
United Church of Christ	1,377,000	5,923
African Methodist Episcopal Zion Church	1,297,000	3,218
Baptist Bible Fellowship, International	1,200,000	4,500
Christian Churches and Churches of Christ[a]	1,072,000	5,579
Orthodox Church in America	1,000,000	721
Jehovah's Witnesses	998,000	11,636

[a]Current data not available; prior data used may no longer be comparable.

Source: U.S. Census Bureau, 2002.

grew and flourished, recruiting new members through their personal attachments to present members. Some cult leaders have not fared well, including Jim Jones, whose ill-fated cult ended up committing mass suicide in Guyana in 1978, and more recently Marshall Herff Applewhite ("Do"), who led his thirty-eight Heaven's Gate followers to commit mass suicide in 1997 at their mansion in Rancho Santa Fe, California. Applewhite's followers were convinced that the comet Hale-Bopp, which swung by Earth in late March of that year, would be their celestial chariot taking them to a higher level (*Newsweek,* 1997).

Are all cults short-lived? Over time, some cults disappear; however, others undergo transformation into sects or denominations. For example, cult leader Mary Baker Eddy's Christian Science church has become an established denomination with mainstream methods of outreach, such as a Christian Science Reading Room strategically placed in an office building or shopping mall, where persons who otherwise might know nothing of the organization learn of its beliefs during their routine activities.

TRENDS IN RELIGION IN THE UNITED STATES

As we have seen throughout this chapter, religion in the United States is very diverse. Pluralism and religious freedom are among the cultural values most widely espoused, and no state church or single denomination predominates. As shown in Table 17.4, Protestants constitute the largest religious body in the United States, followed by Roman Catholics, Muslims, Jews, Eastern churches, and others.

Religion and Social Inequality

Social science research continues to identify significant patterns between social class and religious denominations. Historically, those denominations with the most-affluent members and the highest social status have been the mainline liberal churches, such as the Episcopalians, the Presbyterians, and the Congregationalists. The earliest members of some of these

Table 17.4	U.S. RELIGIOUS BODIES MEMBERSHIP
RELIGIOUS BODY	**NUMBER OF MEMBERS**
Protestants	91,500,000
Roman Catholics	63,683,000
Muslims	6,000,000
Jews	5,602,000
Orthodox Christians	5,631,000
Buddhists	1,864,000
Hindus	795,000

Sources: Ash, 2003; J. Wright, 1997.

churches in the United States were immigrants from Britain, who were also instrumental in establishing the U.S. government and the capitalist economy. Since that time, people with high levels of economic and political power have been closely linked to these denominations; however, members of these denominations also come from the middle and working classes and from minority groups. Although the Presbyterian church is typically characterized as middle class and the Baptist church as working class and lower income, both of these denominations also have members from a wide diversity of income levels (Kosmin and Lachman, 1993). According to some analysts, the socioeconomic position of families that adhere to conservative religious beliefs may sometimes be lower because traditional gender roles assign women to the home and men to the workplace, thus limiting the family's income, as contrasted with religions that do not discourage women from working outside the home (Kosmin and Lachman, 1993). A comparison of different denominations yields some information on class and income differences, but a significant difference in race and class is seen in many central-city and suburban churches.

Race and Class in Central-City and Suburban Churches

As more middle- and upper-income individuals and families moved to the suburbs during the twentieth century, some churches followed the members of their congregations to the suburbs. This pattern is known as "upgrading" and results in the church having newer facilities and more members who are in the upper-middle and upper classes. Meanwhile, the church's former building site is often taken over by a minority congregation or by a denomination that

appeals primarily to members of the working class, older members living on a fixed income, or recent immigrant groups such as Vietnamese Americans (Hudnut-Beumler, 1994).

The sociologist Nancy Tatom Ammerman (1997) and her associates examined churches struggling to survive in central cities and concluded that the efforts of many churches that remain in the central city and engage in community outreach have had mixed results. When church leaders seek to include people from the immediate neighborhood in the life of the church, some longtime members find the experience too disruptive and discourage future endeavors to reach out to the surrounding community (Ammerman, 1997). As a result, the church may stagnate as older members die or move away.

On the other hand, racial and cultural minorities who feel overpowered by their lack of economic resources may be drawn to churches that help them establish a sense of dignity and personal integrity that is otherwise missing in their daily social interactions. For example, African American churches have provided members with a sense of personal dignity and worth in the face of persistent racial prejudice and discrimination while serving as a family life and social service center, community organization, and a means of providing hope for the next generation. As the sociologist Andrew Billingsley (1992: 349) explains, "Over the centuries, the church has become the strongest institution in [the African American] community. It is prevalent, independent, and has extensive outreach." Consider how one pastor in a contemporary African American church described his role:

As a pastor . . . I have watched families struggle with an assortment of devastating problems. I have shared the pain of families in which members have been accused or convicted of theft, drug addiction, prostitution, rape, and murder. I have been involved with homeless families who have been so desperate for a place to live that squatting in abandoned houses was their only recourse. I have witnessed elderly persons lose all sense of autonomy because of homelessness, illness, and loneliness. I have heard the cries of children, parents, and the elderly as they faced conditions of hopelessness. Through it all I have witnessed a remarkable fact—for these persons the church has been the central authenticating reality in their lives. When the world has so often been willing to say only "no" to these people, the church has said "yes." For black people the church has been the one place where they have been able to experience unconditional positive regard. (qtd. in Smith, 1985: 14)

Storefront churches such as this seek to win religious converts and offer solace to people in some low-income areas.

Michael Newman/PhotoEdit

As previously noted, members of other subordinate racial and ethnic groups have also found support and hope in their churches.

Secularization and the Rise of Religious Fundamentalism

How strong is the influence of religion in contemporary life? Some analysts believe that *secularization*—the decline in the significance of the sacred in daily life—has occurred; consequently, there has been a resurgence of religious fundamentalism among some groups. ***Fundamentalism is a traditional religious doctrine that is conservative, is typically opposed to modernity, and rejects "worldly pleasures" in favor of other-worldly spirituality.*** In the United States, traditional fundamentalism primarily appealed to people from lower-income, rural, southern backgrounds; however, the "new" fundamentalist movement of the 1980s and 1990s has had a much wider appeal to people from all socioeconomic levels, geographical areas, and occupations. One reason for the rise of fundamentalism has been a reaction against modernization (Shupe and Hadden, 1989). Around the world, those who adhere to fundamentalism—whether they are Muslims, Christians, or followers of one of the other world religions—believe that sacred traditions must be revitalized. In the United States, public education has been the focus of some who follow the tenets of Christian fundamentalism. For example, various religious and political leaders have vowed to bring the Christian religion "back" into the public life of this country. They have been especially critical of educators who teach what they perceive to be *secular humanism*—the belief that human beings can become better through their own efforts rather than through belief in God and a religious conversion. According to Christian fundamentalists, elementary schoolchildren do not receive a fair and balanced picture of the Christian religion, but instead are taught that their parents' religion is inferior and perhaps irrational (Carter, 1994).

But how might students and teachers who come from the diverse religious heritages that we have examined in this chapter feel about Christian religious instruction or organized prayer in public schools? Many social analysts believe that such practices would cause conflict and perhaps discrimination on the basis of religion. For example, Rick Nelson—a teacher in the Fairfax County, Virginia, public school system—describes the potential effect of religious teaching and group prayer on students in his classroom:

> I think it really trivializes religion when you try to take such a serious topic with so many different viewpoints and cover it in the public schools. At my school we have teachers and students who are Hindu. They are really devout, but they are not monotheistic. . . . I am not opposed to individual prayer by students. I expect students to pray when I give them a test. They need to do that for my tests. . . . But when there is group prayer . . . who's going to lead the group? And if I had my Hindu students lead the prayer, I will tell you it will disrupt many of my students and their parents. It will disrupt the mission of my school, unfortunately . . . if my students are caused to participate in a group Hindu prayer. (CNN, 1994)

RELIGION IN THE FUTURE

Debates over religion, particularly issues such as secularization and fundamentalism, will no doubt continue; however, many social scientists believe that religion is alive and well in the United States and in other nations of the world. Not only are we seeing the creation of new religious forms, but we are also seeing a dramatic revitalization of traditional forms of religious life (Kurtz, 1995).

One example of this change is ***liberation theology—the Christian movement that advocates freedom from political subjugation within a traditional perspective and the need for social transformation to benefit the poor and downtrodden*** (Kurtz, 1995). Although liberation theology initially emerged in Latin America as people sought to free themselves from the historical oppression of that area, this perspective has been embraced by a wide variety of people, ranging from African and African American Christians to German theologians and some feminists.

Another example of changes in the nature of theology is found in some feminist movements that have turned to pagan religions and witchcraft as a means of countering what they consider to be the patriar-

Understanding and Tolerating Religious and Cultural Differences

[The small strip of kente cloth that hangs on the marble altar and the flags of a dozen West Indies nations, including Trinidad, Jamaica, and Barbados,] represent all of the people in this parish. It's an opportunity for us to celebrate everyone's culture. You won't find your typical Episcopal church looking like that.

> —Rev. Cannon Peter P. Q. Golden, rector of St. Paul's Episcopal Church in Brooklyn, New York (qtd. in Pierre-Pierre, 1997: A11)

One part of the history of St. Paul's Episcopal Church in Flatbush, Brooklyn, is etched into its huge stained-glass windows, which tell the story of the prominent descendants of English and Dutch settlers who founded the church. However, the church's more recent history is told in the faces of its parishioners—most of whom are Caribbean immigrants from the West Indies who labor as New York City's taxi drivers, factory workers, accountants, and medical professionals (Pierre-Pierre, 1997).

Traveling only a short distance, we find a growing enclave of Muslims in Brooklyn. It is their wish that people will recognize that Islam shares a great deal with Christianity and Judaism, including the fact all three believe in one God, are rooted in the same part of the world, and share some holy sites (Sengupta, 1997). Some Muslim adherents also hope that more people in this country will learn greater tolerance toward those who have a different religion and celebrate different holidays. For example, in some years the Islamic holy season, Ramadan, falls at about the same time as Christmas and Hanukkah, but some Muslim children and adults are disparaged by their neighbors because they do not celebrate the same holidays as others. As one person observed, "As Muslims, we have to respect all religions" (Sengupta, 1997: A12). Left un-

said was the belief that other people should do likewise. How can each of us—regardless of our race, color, creed, or national origin—help to bring about greater tolerance of religious diversity in this country? Here is one response to this question, from Huston Smith (1991: 389-390), a historian of religion:

> Whether religion is, for us, a good word or bad; whether (if on balance it is a good word) we side with a single religious tradition or to some degree open our arms to all: How do we comport ourselves in a pluralistic world that is riven by ideologies, some sacred, some profane? . . . We listen. . . . If one of the [world's religions] claims us, we begin by listening to it. Not uncritically, for new occasions teach new duties and everything finite is flawed in some respects. Still, we listen to it expectantly, knowing that it houses more truth than can be encompassed in a single lifetime.
>
> But we also listen to the faith of others, including the secularists. We listen first because . . . our times require it. The community today can be no single tradition; it is the planet. Daily, the world grows smaller, leaving understanding the only place where peace can find a home. . . . Those who listen work for peace, a peace built not on ecclesiastical or political hegemonies but on understanding and mutual concern.

Perhaps this is how each of us can make a difference—by *learning* more about our own beliefs and about the diverse denominations and world religions represented in the United States and around the globe, and *listening* to what other people have to say about their own beliefs and religious experiences.

Will religious tolerance increase in the United States? In the world? What steps can you take to help make a difference?

---

chal structure and content of the world's religions. For instance, the *Goddess movement* encompasses a variety of countercultural beliefs based on paganism and feminism, and is rooted in acknowledgment of the legitimacy of female power as a "beneficent and independent power" (Christ, 1987: 121).

What significance will religion have in the future? Religion will continue to be important in the lives of many people. Moreover, the influence of religion may be felt even by those who claim no religious beliefs of their own. In many nations, the rise of *religious nationalism* has led to the blending of strongly held religious and political beliefs. The rise of religious nationalism is especially strong in the Middle East,

where Islamic nationalism has spread rapidly and where the daily lives of people, particularly women and children, have been greatly affected (Juergensmeyer, 1993). Similarly, in the United States the influence of religion will be evident in ongoing political battles over social issues such as school prayer, abortion, gay and lesbian rights, and family issues. On the one hand, religion may unify people; on the other, it may result in tensions and confrontations among individuals and groups. However, according to the legal scholar Stephen L. Carter (1994), the tension between religion and other social institutions is not always negative. Carter believes that one of the most cherished freedoms in the United States is religious liberty.

Maintaining an appropriate balance between religion and other aspects of social life will be an important challenge for the future. In contrast, totalitarian states find all conflict threatening and often quickly remove religious liberty when a statist dictator takes control.

As we have seen in this chapter, the debate continues over what religion is, what it should do, and what its relationship to other social institutions such as education should be. It will be up to your generation to understand other religions and to work for greater understanding among the diverse people who make up our nation and the world (see Box 17.4). But there is reason for hope, as one scholar explains:

> [People] know that religion, for all its institutional limitations, holds a vision of life's unity and meaningfulness, and for that reason it will continue to have a place in their narrative. In a very basic sense, religion itself was never the problem, only social forms of religion that stifle the human spirit. The sacred lives on and is real to those who can access it. (Roof, 1993: 261)

CHAPTER REVIEW

■ What is religion, and what purpose does it serve in society?

Religion is a system of beliefs, symbols, and rituals, based on some sacred or supernatural realm, that guides human behavior, gives meaning to life, and unites believers into a community.

■ What is the functionalist perspective on religion?

According to functionalists, religion has three important functions in any society: (1) providing meaning and purpose to life, (2) promoting social cohesion and a sense of belonging, and (3) providing social control and support for the government.

■ What is the conflict perspective on religion?

From a conflict perspective, religion can have negative consequences in that the capitalist class uses religion as a tool of domination to mislead workers about their true interests. However, Max Weber believed that religion could be a catalyst for social change.

■ What is the symbolic interactionist perspective on religion?

Symbolic interactionists focus on a microlevel analysis of religion, examining the meanings that people give to religion and the meanings that they attach to religious symbols in their everyday life.

■ What are the major types of religious organization?

Religious organizations can be categorized as ecclesia, churches, denominations, sects, and cults.

■ What are the major world religions?

The major world religions are Buddhism, Hinduism, Judaism, Islam, and Christianity. More than 75 percent of the world's population is represented in one of these religions.

■ How has modernization contributed to the growth of religious orthodoxy (fundamentalism)?

Fundamentalism has emerged in many religions because people do not like to see social changes taking place that affect their most treasured beliefs and values. In some situations, people have viewed modernity, especially science and new technologies, as a threat to their traditional beliefs and practices, which are important components of self-identity and group cohesion. As a result, fundamentalism opposes religious accommodation to the things of this world and demands higher standards of adherents, including a code of conduct and acceptance of specific beliefs.

■ Will religion continue as a major social institution?

Although a few scholars of religion have "written off" this social institution, most believe that it is not in decline but rather is experiencing changes and some new definitions. One thing appears certain: With the influx of recent immigrants and increasing cultural diversity, religious congregations in the United States in the future will be much more diverse and will hold a wider variety of beliefs than did the traditional mainline denominations of the past.

KEY TERMS

animism 564
church 582
civil religion 568
cult 583
denomination 582
ecclesia 582
faith 563
fundamentalism 586

QUESTIONS FOR CRITICAL THINKING

1. Why do people who believe they have "no religion" subscribe to civil religion? How would you design a research project to study the effects of civil religion on everyday life?
2. How is religion a force for social stability? How is it a force for social change?
3. What is the relationship among race, class, gender, and religious beliefs in contemporary society?
4. If Durkheim, Marx, and Weber were engaged in a discussion about religion, on what topics might they agree? On what topics would they disagree?

RESOURCES ON THE INTERNET

Chapter-Related Web Sites

The following Web sites have been selected for their relevance to the topics in this chapter. These sites are among the more stable, but please note that Web site addresses change frequently. For an updated list of chapter-related Web sites with URL links, please visit the *Sociology in Our Times* Web site (**www.wadsworth.com/KendallSIOT**).

American Religion Data Archive (ARDA)
http://www.arda.tm

Funded by the Lilly Endowment, the ARDA is a project dedicated to collecting quantitative data for the study of American religion. The site offers extensive data on churches, church memberships, religious professionals, and religious groups, as well as maps and useful Web links.

The Religious Movements Homepage
http://religiousmovements.lib.virginia.edu

Developed at the University of Virginia, this Web site seeks to provide an understanding of the processes and changes that religious groups undergo and to promote the tolerance and appreciation of all faiths. In addition to profiles of religious movements, the site provides information on cult controversies, world religions, religious freedom, and many other topics.

The American Religious Experience
http://are.as.wvu.edu

Created by Brian Turley at West Virginia University, the American Religious Experience is a project devoted to scholarship about American religious history. The Web site features a site-specific search engine, a film archive, student projects, articles, and links to information on topics such as women in religion and ethnic groups.

ONLINE STUDY AND RESEARCH TOOLS

Accompanying this text are many *free* powerful online study tools that will help you master the material in this chapter, help increase your depth of understanding, and help you make the grade!

SocCoach CD-ROM

Use the SocCoach CD-ROM enclosed with this text to help you formulate a customized study plan for this chapter. After you take the Diagnostic Quiz, SocCoach will generate a customized study plan just for you! It will identify sections of the chapter that you should review and will provide videos, charts, graphs, and excerpts from the text to supplement your studies and enhance your understanding. You'll also find fun, interactive activities such as Virtual Explorations and Map the Stats to apply what you've learned and stretch your sociological imagination.

The Companion Web Site for Sociology in Our Times, *Fifth Edition*
www.wadsworth.com/KendallSIOT

Gain an even better grasp on this chapter by going to the companion Web site to take one of the Tutorial Quizzes, use the Flash Cards to master key terms, or check out the many other study aids you'll find there. You'll also find special features such as GSS Data and Census 2000 information that'll put data and resources at your fingertips to help you with that special project or help you as you do some research on your own.

In this chapter, when you see the icon on the left, it alerts you to a specific exercise found in *Wadsworth's Sociology Online Resources and Writing Companion*. This valuable guide shows you how to use Wadsworth's exclusive online resources—*InfoTrac College Edition*, the *Opposing Viewpoints Resource Center*, and *MicroCase Online*—to assist you in your study of sociology and to build essential research and writing skills.

CHAPTER 18

Health, Health Care, and Disability

My name is Heather Neumann and I'm an 18 year old freshman at the University of Texas at Austin. When I was nine, I was diagnosed with Juvenile Rheumatoid Arthritis and given all sorts of different medications. I learned that if I didn't go into a remission, my life wouldn't be too swell. Well, I was pretty upset about it all! I played piano, loved all sports, was good at art and was being told that I might not be able to do these things when I got older. That's when I decided to take charge of my life. I took things one day at a time and pushed myself to give 110% at everything.

I continued to take piano until I went to high school, even though it was not always possible to practice when my hands got really bad. And I played every sport offered at my school, however tough it was on my body. Every time someone said, "You can't do that, you're sick" I strived not only to do it, but to do it better than anyone else. . . . It was frustrating for me when my mind wanted to do more than my body would let it. Sometimes my hands would swell and I couldn't wear my class ring. . . . Sometimes I had to ask my boyfriend to carry my books because my shoulder hurt or because my knees decided not to work half-way down the hall. I was often weak. . . .

I am still keeping myself busy in college. I still have [juvenile rheumatoid arthritis,] and I still take two million pills. There are still days when I can't get out of bed, but I accomplish so much in the meantime. I am playing Varsity Field Hockey, I am the service chair for my sorority, a member of student government, an editor of the literary arts journal and an honors student. On days when it feels like I can't cope, I remind myself that there is nothing I cannot do if I put my mind to it.

—Heather Neumann describing her experience with chronic illness (*Band-Aides & Blackboards*, 1999)

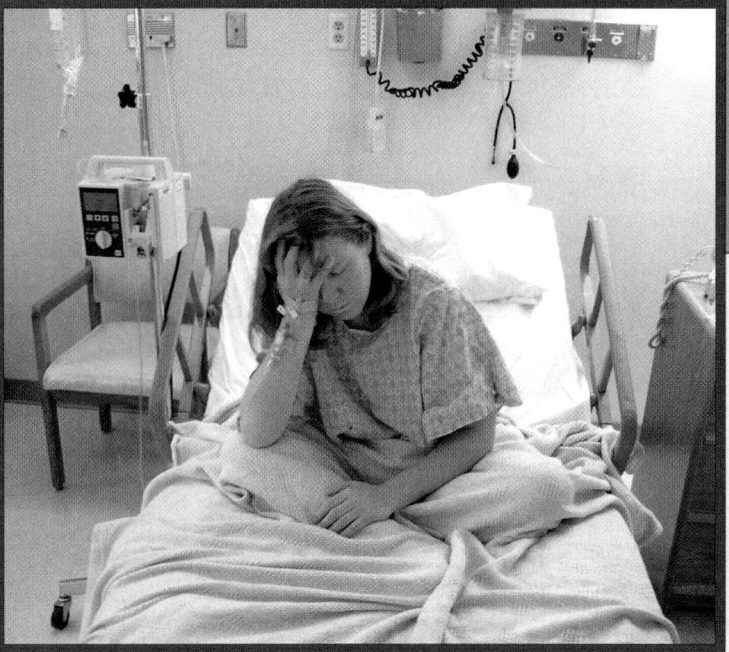

■ Concern over chronic health problems is not limited to older adults. Many young people must also learn how to live with chronic illnesses or disabilities.

Like many other young people, Heather Neumann has learned how to live with chronic medical problems. She has chosen to share her story, "Forever Heather," with other young people, particularly those with chronic illnesses or disabilities, on *Band-Aides and Blackboards*, a Web site maintained by Joan Fleitas, a nursing professor and pediatric clinician. According to Fleitas (1999), the goal of this Web site is to communicate with children and young adults so that they can understand medical problems and increase their acceptance of them. The site is also designed to sensitize others to the harmful effects of everyday practices, such as teasing a child living with a chronic illness or attaching a stigma to illness or disability (Fleitas, 1999). This Web site is only one of the many ways in which health professionals and laypersons are transforming the Internet into a source of information about illness and disability and helping those who seek moral and psychological support when coping with such conditions.

In this chapter, we will explore the dynamics of health, health care, and disability from a sociological perspective, as well as look at issues through the eyes of those who have experienced medical problems. Before reading on, test your knowledge about health, illness, and health care by taking the quiz in Box 18.1.

QUESTIONS AND ISSUES

Chapter Focus Question: Why are health, health care, and disability significant concerns not only for individuals but also for entire societies?

What is the relationship between the social environment and health and illness?

What are the major issues in U.S. health care?

How do functionalist, conflict, and symbolic interactionist approaches differ in their analysis of health and health care?

What is mental illness, and why is it a difficult topic for sociological research?

What are some of the consequences of disability?

What does the concept of health mean to you? At one time, health was considered to be simply the absence of disease. However, the World Health Organization (2003:7) defines *health* as **a state of complete physical, mental, and social well-being.** According to this definition, health involves not only the absence of disease, but also a positive sense of wellness. In other words, health is a multidimensional phenomenon: It includes physical, social, and psychological factors.

What is illness? Illness refers to an interference with health; like health, illness is socially defined and may change over time and between cultures. For example, in the United States and Canada, obesity is viewed as unhealthy, whereas in other times and places, obesity indicated that a person was prosperous and healthy.

What happens when a person is perceived to have an illness or disease? Healing involves both personal and institutional responses to perceived illness and disease. One aspect of institutional healing is health care and the health care delivery system in a society. *Health care* **is any activity intended to improve health.** When people experience illness, they often seek medical attention in hopes of having their health restored. A vital part of health care is *medicine*—**an institutionalized system for the scientific diagnosis, treatment, and prevention of illness.**

HEALTH IN GLOBAL PERSPECTIVE

Studying health and health care issues around the world offers insights on illness and how political and economic forces shape health care in nations. Disparities in health are glaringly apparent between high-

income and low-income nations when we examine factors such as the prevalence of life-threatening diseases, rates of life expectancy and infant mortality, and access to health services. In regard to global health, for example, the number of people infected with HIV/AIDS more than doubled between 1990 and 2000 (from less than 15 million to more than 34 million). AIDS has cut life expectancy by 5 years in Nigeria, 18 years in Kenya, and 33 years in Zimbabwe (U.S. Census Bureau, 2002). *Life expectancy* **refers to an estimate of the average lifetime of people born in a specific year.** AIDS results in higher mortality rates in childhood and young adulthood, stages in the life course when mortality is otherwise low. However, AIDS is not the only disease reducing life expectancy in some nations. Most deaths in low- and middle-income nations are linked to infectious and parasitic diseases that are now rare in high-income, industrialized nations. Among these diseases are tuberculosis, polio, measles, diphtheria, meningitis, hepatitis, malaria, and leprosy. Although it is estimated that only 13 percent of U.S. citizens and 9 percent of Canadians will die prior to age 60, health experts estimate that more than 1.5 billion people around the world will die prior to age 60. This is particularly true in low-income nations such as Zambia, where 80 percent of the people are not expected to see their sixtieth birthday.

The *infant mortality rate* **is the number of deaths of infants under 1 year of age per 1,000 live births in a given year.** The infant mortality rate in some low-income nations is staggering: 149 infants under 1 year of age die per 1,000 live births in Sierra Leone, 196 die in Angola, and 122 die in Malawi (U.S. Census Bureau, 2002). In fact, almost 14 percent of all children born in low-income nations die before they reach their first birthday. The World Health Organization (1999) reports that about 5 million babies born in low-income countries during 1998

Box 18.1 SOCIOLOGY AND EVERYDAY LIFE

How Much Do You Know About Health, Illness, and Health Care?

True	False	
T	F	1. Some social scientists view sickness as a special form of deviant behavior.
T	F	2. The field of epidemiology focuses primarily on how individuals acquire disease and bodily injury.
T	F	3. The primary reason that African Americans have shorter life expectancies than whites is the high rate of violence in central cities and the rural South.
T	F	4. Native Americans have shown dramatic improvement in their overall health level since the 1950s.
T	F	5. Health care in most high-income, developed nations is organized on a fee-for-service basis as it is in the United States.
T	F	6. The medical–industrial complex has operated in the United States with virtually no regulation, and allegations of health care fraud have largely been overlooked by federal and state governments.
T	F	7. Media coverage of chronic depression and other mental conditions focuses almost exclusively on these problems as "women's illnesses."
T	F	8. It is extremely costly for employers to "mainstream" persons with disabilities in the workplace.

Answers on page 595.

died during the *first month* of life. A child born in Latin America or Asia can expect to live between 7 and 13 fewer years, on average, than one born in North America or Western Europe (Epidemiological Network for Latin America and the Caribbean, 2000).

There are many reasons for these differences in life expectancy and infant mortality. Many people in low-income countries have insufficient or contaminated food; lack access to pure, safe water; and do not have adequate sewage and refuse disposal. Added to these hazards is a lack of information about how to maintain good health. Many of these nations also lack qualified physicians and health care facilities with up-to-date equipment and medical procedures.

Nevertheless, tremendous progress has been made in saving the lives of children and adults over the past 15 years. Life expectancy at birth has risen to more than 70 years in 84 countries, up from only 55 countries in 1990. Life expectancy in low-income nations increased on average from 53 to 62 years, and mortality of children under 5 years of age dropped from 149 to 85 per 1,000 live births. Although this increase has been attributed to a number of factors, an especially important advance has been the development of a safe water supply. The percentage of the

world's population with access to safe water nearly doubled between 1990 and 2000 (United Nations Development Programme, 2003).

Will improvements in health around the world continue to occur? Organizations such as the United Nations argue that both public-sector and private-sector initiatives will be required to improve global health conditions. For example, a United Nations report states that in the era of globalization and dominance by transnational corporations, "money talks louder than need" when "cosmetic drugs and slow-ripening tomatoes come higher on the list [of priorities] than a vaccine against malaria or drought-resistant crops for marginal lands" (United Nations Development Programme, 1999: 68).

Recently, pressing questions have arisen about the availability of new technologies and life-saving drugs around the world. An example is the problem of providing access to drugs in countries with high rates of HIV/AIDS. Many people cannot afford to pay for drugs, such as the three-drug combination therapy that prolongs the life of many AIDS patients. Transnational pharmaceutical companies fear that if they provide their name-brand drugs at a lower price in low-income countries, that might undercut their

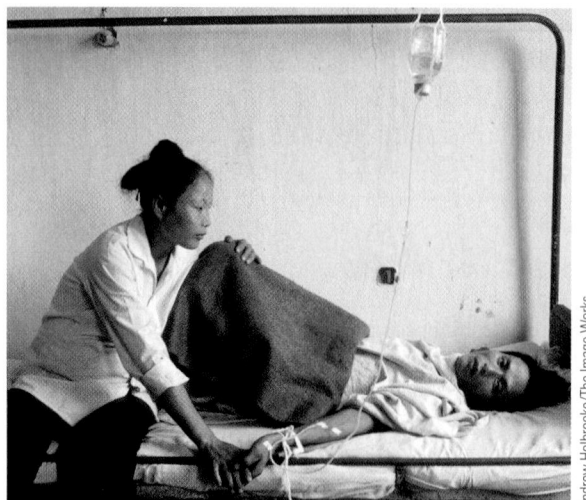
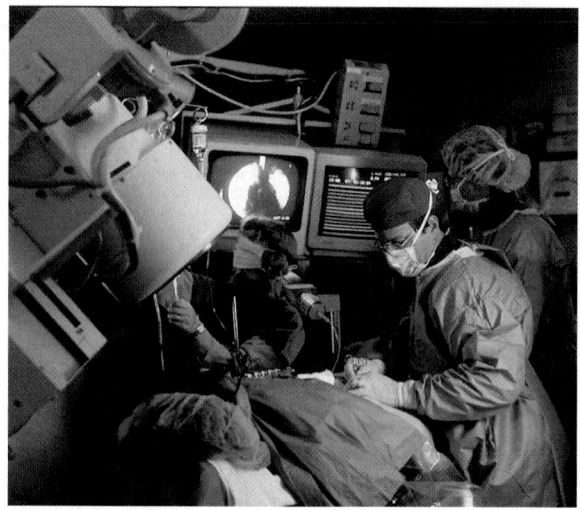

Andrew Holbrooke/The Image Works

© Lew Lause/SuperStock

Access to quality health care is much greater for some people than for others. The factors that are involved vary not only for people within one nation but also across the nations of the world.

major sales base in high-income countries if those drugs become available as generic products (which are less costly and can be made by more than one manufacturer) or are re-imported into the high-income countries at a reduced price. The companies claim that they need the money generated from sales of their name-brand drugs in order to fund research on other products that will reduce suffering and sometimes prolong human life. Pharmaceutical companies that hold the patents on various drugs see their products as something that needs to be protected by law, whereas people in human relief agencies around the world are concerned about the fact that one-third of the world's population does not have access to essential medicines and that—even worse—this figure rises to one-half in the poorest parts of Africa and Asia (United Nations Development Programme, 2003). If we are to see a significant improvement in life expectancy and health among people in all of the nations of the world, improvements are needed in the availability of new medical technologies and life-saving drugs.

How about improvements in health and health care within one nation? Is there a positive relationship between the amount of money that a society spends on health care and the overall physical, mental, and social well-being of its people? Not necessarily. If there were such a relationship, people in the United States would be among the healthiest and most physically fit people in the world. We spend more than one trillion dollars—the equivalent of $3,925 per person—on health care each year, and the amount has almost doubled in the past 10 years (Anell and Willis, 2000). But by comparing health care expenditures in Sweden and the United States with infant mortality rates in these

two countries, we can see that large expenditures for health care do not always produce better health care for individuals. Sweden spends an average of $1,701 per person on health care and has an infant mortality rate of 3.5; by contrast, the United States has an infant mortality rate of 6.8 (Anell and Willis, 2000; U.S. Census Bureau, 2002). Similarly, a child born in Sweden in 2000 had a life expectancy of 79.8 years, whereas a child born in the United States that same year had a life expectancy of 77.1 years (U.S. Census Bureau, 2002).

HEALTH IN THE UNITED STATES

Even if we limit our discussion (for the moment) to people in the United States, why are some of us healthier than others? Is it biology—our genes—that accounts for this difference? Does the environment within which we live have an effect? How about our own individual lifestyle?

Social Epidemiology

The field of social epidemiology attempts to answer questions such as these. **Social epidemiology is the study of the causes and distribution of health, disease, and impairment throughout a population** (Weiss and Lonnquist, 2003). Typically, the target of the investigation is disease agents, the environment,

Box 18.1 SOCIOLOGY AND EVERYDAY LIFE

Answers to the Sociology Quiz on Health, Illness, and Health Care

1. True. Some social scientists view sickness as a special form of deviant behavior. However, it is not equivalent to other forms of deviance such as crime or violent behavior. Unlike many who are defined as criminal, the sick are provided with therapeutic care so that their health will be restored and they can fulfill their roles in society (Weiss and Lonnquist, 2003).

2. False. The primary focus of the epidemiologist is on the health problems of social aggregates or large groups of people, not on individuals as such (Cockerham, 2004).

3. False. The lower life expectancy of African Americans as a category is due to a higher prevalence of life-threatening illnesses, such as cancer, heart disease, hypertension, and AIDS. However, it should be noted that African American males do have the highest death rates from homicide of any racial-ethnic category in the United States (Cockerham, 2004).

4. True. Native Americans (including American Indians and native Alaskans) as a category have had significant improvement in health in recent decades. Some analysts attribute this change to better nutrition and health care services. However, other analysts point out that Native Americans continue to have high rates of mortality from diabetes, alcohol-related illnesses, and suicide (Cockerham, 2004).

5. False. The United States is one of only two high-income, developed nations that do not have some form of universal health coverage. In the United States, health care has traditionally been purchased by the patient. In most other high-income nations, health care is provided or purchased by the government (Cockerham, 2004).

6. False. In the mid-to-late 1990s, government investigation focused on rising health care payments and allegations of fraud in the health care delivery system. Thus far, billing frauds have been found in Medicare and Medicaid payments to physicians, hospitals, nursing homes, home health agencies, medical labs, and medical equipment manufacturers (Findlay, 1997).

7. False. Until recently, chronic depression and other mental conditions were most often depicted as "female" problems. However, in the late 1990s, male celebrities such as Mike Wallace, the news correspondent who is co-editor of *60 Minutes,* have made the general public more aware of male depression. Wallace produced a documentary on chronic depression, and author Terrence Real, a psychotherapist, wrote a book about depression in men: *I Don't Want to Talk About It: Overcoming the Secret Legacy of Male Depression* (Brody, 1997). However, it should be noted that women average higher levels of depression than men do, perhaps because women and men have unequal adult statuses (see Mirowsky, 1996).

8. False. Although disability expenditures nationwide may be costly, individual employers often find that they can accommodate the workplace needs of a worker with a disability for costs ranging from zero to several thousand dollars, thus opening up new opportunities for people previously excluded from certain types of jobs and careers.

and the human host. *Disease agents* include biological agents such as insects, bacteria, and viruses that carry or cause disease; nutrient agents such as fats and carbohydrates; chemical agents such as gases and pollutants in the air; and physical agents such as temperature, humidity, and radiation. The *environment* includes the physical (geography and climate), biological (presence or absence of known disease agents), and social (socioeconomic status, occupation, and location of home) environments. The human *host* takes into account demographic factors (age, sex, and race/ethnicity), physical condition, habits and customs,

To increase awareness of breast cancer, some high schools have introduced short courses to inform students about self-examination for this life-threatening disease. This course uses an artificial breast—held by this young woman—to help educate the students.

and lifestyle (Weiss and Lonnquist, 2003). Let's look briefly at some of these factors.

Age Rates of illness and death are highest among the old and the young. Mortality rates drop shortly after birth and begin to rise significantly during middle age. After age 65, rates of chronic diseases and mortality increase rapidly. *Chronic diseases* **are illnesses that are long term or lifelong and that develop gradually or are present from birth;** in contrast, *acute diseases* **are illnesses that strike suddenly and cause dramatic incapacitation and sometimes death** (Weitz, 2004). Basha, an 89-year-old woman, is an example of how older people cope with conditions associated with aging. Here is how she describes the start of her day:

> Every morning, I wake up in pain. I wiggle my toes. Good. They still obey. I open my eyes. Good. I can see. Everything hurts but I get dressed. I walk down to the ocean. Good. It's still there. Now my day can start. About tomorrow I never know. After all, I'm eighty-nine. I can't live forever. (qtd. in Myerhoff, 1994: 1)

Two of the most common sources of chronic disease and premature death are tobacco use, which increases mortality among both smokers and people who breathe the tobacco smoke of others, and alcohol abuse, both of which are discussed later in this chapter. The fact that rates of chronic diseases increase rapidly after age 65 has obvious implications not only for people reaching that age (and their families) but also for society. The Census Bureau projects that about 20 percent of the U.S. population will be at least age 65 by the year 2050 and that the popula-

tion of persons age 85 and over will have tripled from about 4 million (1.5 percent) in 2000 to about 12 million (5 percent). The cost of caring for many of these people—especially those who must be institutionalized—will increase in at least direct proportion to their numbers.

Sex Prior to the twentieth century, women had lower life expectancies than men because of high mortality rates during pregnancy and childbirth. Preventive measures have greatly reduced this cause of female mortality, and women now live longer than men. For babies born in the United States in 2000, for example, life expectancy at birth was 73.9 years for males and 79.4 years for females. Females have a slight biological advantage over males in this regard from the beginning of life, as can be seen in the fact that they have lower mortality rates both in the prenatal stage and in the first month of life (Weiss and Lonnquist, 2003). However, the sociologist Ingrid Waldron (1994) notes that gender roles and gender socialization also contribute to the difference in life expectancy. Men are more likely to work in dangerous occupations such as commercial fishing, mining, construction, and public safety/firefighting. As a result of gender roles, males may be more likely than females to engage in risky behavior such as drinking alcohol, smoking cigarettes (there is more social pressure on women not to smoke), using drugs, driving dangerously, and engaging in fights. Finally, women are more likely to use the health care system, with the result that health problems are identified and treated earlier (while there is a better chance of a successful outcome), whereas many men are more reluctant to consult doctors.

Because women on average live longer than men, it is easy to jump to the conclusion that they are healthier than men. However, although men at all ages have higher rates of fatal diseases, women have higher rates of chronic illness (Waldron, 1994).

Race/Ethnicity and Social Class

Although race/ethnicity and social class are related to issues of health and mortality, recent research tends to suggest that income and factors such as the neighborhood in which a person lives may be more significant than race or ethnicity with respect to these issues. How is it possible that the neighborhood you live in may significantly affect your risk of dying during the next year? According to a study by the Stanford Center for Research in Disease Prevention (Winkleby and Cubbin, 2003), people have a higher survival rate if they live in better-educated or wealthier neighborhoods than if the neighborhood is low-income and has low levels of education. Among the reasons researchers believe that neighborhoods make a difference are the availability (or lack thereof) of safe areas to exercise, grocery stores with nutritious foods, and access to transportation, education, and good jobs. Many low-income neighborhoods are characterized by fast-food restaurants, liquor stores, and other facilities that do not afford residents healthy options.

As discussed in prior chapters, people of color are more likely to have incomes below the poverty line, and the poorest people typically receive less preventive care and less optimal management of chronic diseases than do other people (Leary, 1996). People living in central cities, where there are high levels of poverty and crime, or in remote rural areas generally have greater difficulty in getting health care because most doctors prefer to locate their practice in a "safe" area, particularly one with a patient base that will produce a high income. Although rural Americans make up 20 percent of the U.S. population, only 9 percent of the nation's physicians practice in rural areas, and fewer specialists such as cardiologists are available in these areas (Ricketts, 1999).

Another factor is occupation. People with lower incomes are more likely to be employed in jobs that expose them to danger and illness—working in the construction industry or around heavy equipment in a factory, for example, or holding a job as a convenience store clerk or other position that exposes a person to the risk of armed robbery. Finally, people of color and poor people are more likely to live in areas that contain environmental hazards.

However, although Latinas/os are more likely than non-Latino/a whites to live below the poverty line, they have lower death rates from heart disease, can-cer, accidents, and suicide, and an overall lower death rate. One explanation may be dietary factors and the strong family life and support networks found in many Latina/o families (Weiss and Lonnquist, 2003). Obviously, more research is needed on this point, for the answer might be beneficial to all people.

Lifestyle Factors

As noted previously, social epidemiologists also examine lifestyle choices as a factor in health, disease, and impairment. We will examine three lifestyle factors as they relate to health: drugs, sexually transmitted diseases, and diet and exercise.

Drug Use and Abuse

What is a drug? There are many different definitions, but for our purposes, a **drug is any substance—other than food and water—that, when taken into the body, alters its functioning in some way.** Drugs are used for either therapeutic or recreational purposes. *Therapeutic* use occurs when a person takes a drug for a specific purpose such as reducing a fever or controlling a cough. In contrast, *recreational* use occurs when a person takes a drug for no purpose other than achieving a pleasurable feeling or psychological state. Alcohol and tobacco are examples of drugs that are primarily used for recreational purposes; their use by people over a fixed age (which varies from time to time and place to place) is lawful. Other drugs—such as some anti-anxiety drugs or tranquilizers—may be used legally only if prescribed by a physician for therapeutic use but are frequently used illegally for recreational purposes.

Alcohol The use of alcohol is considered an accepted part of the dominant culture in the United States. Adults in this country consume an average of 2 gallons of wine, 22 gallons of beer, and 1 gallon of liquor a year (U.S. Census Bureau, 2002). In fact, adults consume more beer on average than milk or coffee. However, these statistics overlook the fact that among people who drink, 10 percent account for roughly half the total alcohol consumption in this country (Levinthal, 2002).

Although the negative short-term effects of alcohol are usually overcome, chronic heavy drinking or alcoholism can cause permanent damage to the brain or other parts of the body (Fishbein and Pease, 1996). For alcoholics, the long-term negative health effects include *nutritional deficiencies* resulting from poor eating habits (chronic heavy drinking contributes to high caloric consumption but low nutritional intake); *cardiovascular problems* such as inflammation and enlargement of the heart muscle, high blood pressure,

Peer groups may contribute to problems associated with drug use and abuse. However, peer groups may also provide the encouragement and support that people need to overcome an existing drug problem.

Mary Kate Denny/PhotoEdit

and stroke; and eventually to *alcoholic cirrhosis*—a progressive development of scar tissue that chokes off blood vessels in the liver and destroys liver cells by interfering with their use of oxygen (Levinthal, 2002). Alcoholic cirrhosis is the ninth most frequent cause of death in the United States. The social consequences of heavy drinking are not always limited to the person doing the drinking. For example, abuse of alcohol and other drugs by a pregnant woman can damage the unborn fetus.

Can alcoholism be overcome? Many people get and stay sober. Some rely on organizations such as Alcoholics Anonymous or church support groups to help them overcome their drinking problem. Others undergo medical treatment and therapy sessions to learn more about the root causes of their problem. Caroline Knapp (1996: 242–243), who started drinking at age 14 and continued throughout college, describes how she felt after she regained sobriety and came to view herself as a recovering alcoholic:

> I still think about drinking, and not drinking, many many times each day, and sometimes I think I always will. We live in an alcohol-saturated world; it's simply impossible to avoid the stuff. When I read the papers now, I find myself scanning the pages for items about drink-related disasters, things that will reinforce my sense that I've made the right choice: what celebrity got pulled over for drunk driving; what college kid got drunk and plunged out of a five-story window; what alcohol-fueled argument between a couple fired up into a case of domestic violence. Evidence of the havoc liquor can wreak is there in black and white, almost every day, but it's not nearly as prevalent as the other messages: the liquor ads, the images of gaiety and romance,

phrases like CHAMPAGNE BRUNCH: $19.95. At times, I've grumbled to friends about longing to return to Prohibition, a gripe that stems from the feeling, familiar among many alcoholics, that if *I* can't drink, no one should. But alcohol occupies a large role in the social world and it's important for me to remember that I have to come to terms with it, that I still have a relationship with liquor, even if the relationship now has the quality of a divorce rather than an active involvement.

Nicotine (Tobacco) The nicotine in tobacco is a toxic, dependency-producing psychoactive drug that is more addictive than heroin. It is classified as a stimulant because it stimulates central nervous system receptors and activates them to release adrenaline, which raises blood pressure, speeds up the heartbeat, and gives the user a temporary sense of alertness. Although the overall proportion of smokers in the general population has declined somewhat since the 1964 Surgeon General warning that smoking is linked to cancer and other serious diseases, tobacco is still responsible for about one in every five deaths in this country (Akers, 1992). Even people who never light up a cigarette are harmed by *environmental tobacco smoke*—the smoke in the air inhaled by nonsmokers as a result of other people's tobacco smoking (Levinthal, 2002). Researchers have found that environmental smoke is especially hazardous for nonsmokers who carpool or work with heavy smokers.

Illegal Drugs Marijuana is the most extensively used illegal drug in the United States. About one-third of all people over age twelve have tried marijuana at least once. Although most marijuana users are between the ages of eighteen and twenty-five, use by teenagers has

Michael Newman/PhotoEdit

Despite a variety of warnings from the U.S. Surgeon General about the potentially harmful effects of smoking, many people continue to light up cigarettes. Even those who do not smoke may be affected by environmental tobacco smoke.

more than doubled during the past decade. High doses of marijuana smoked during pregnancy can disrupt the development of a fetus and result in congenital abnormalities and neurological disturbances (Fishbein and Pease, 1996). Furthermore, some studies have found an increased risk of cancer and other lung problems associated with marijuana because its smokers are believed to inhale more deeply than tobacco users.

Another widely used illegal drug is cocaine: About 23 million people over the age of twelve in the United States report that they have used cocaine at least once, and about one million acknowledge having used it during the past month (Substance Abuse and Mental Health Services Administration, 2000). People who use cocaine over extended periods of time have higher rates of infection, heart problems, internal bleeding, hypertension, stroke, and other neurological and cardiovascular disorders than do nonusers. Intravenous cocaine users who share contaminated needles are also at risk for contracting AIDS.

Each of the drugs discussed in these few paragraphs represents a lifestyle choice that affects health. Whereas age, race/ethnicity, sex, and—at least to some degree—social class are ascribed characteristics, taking drugs is a voluntary action on a person's part.

Sexually Transmitted Diseases

The circumstances under which a person engages in sexual activity is another lifestyle choice with health implications. Although most people find sexual activity enjoyable, it can result in transmission of certain *sexually trans-* *mitted diseases* (STDs), including AIDS, gonorrhea, syphilis, and genital herpes. Prior to 1960, the incidence of STDs in this country had been reduced sharply by barrier-type contraceptives (e.g., condoms) and the use of penicillin as a cure. However, in the 1960s and 1970s the number of cases of STDs increased rapidly with the introduction of the birth control pill, which led to women having more sexual partners and couples being less likely to use barrier contraceptives.

Gonorrhea and Syphilis

Until the 1960s, gonorrhea (today the second-most-common STD) and syphilis (which can be acquired not only by sexual intercourse but also by kissing or coming into intimate bodily contact with an infected person) were the principal STDs in this country. Today, however, they constitute less than 15 percent of all cases of STDs reported in U.S. clinics. Untreated gonorrhea may spread from the sexual organs to other parts of the body, among other things negatively affecting fertility; it can also spread to the brain or heart and cause death. Untreated syphilis can, over time, cause cardiovascular problems, brain damage, or even death. Penicillin can cure most cases of either gonorrhea or syphilis as long as the disease has not spread.

Genital Herpes

This sexually transmitted disease produces a painful rash on the genitals. Genital herpes cannot be cured: Once the virus enters the body, it stays there for the rest of a person's life, regardless of treatment. However, the earlier that treatment is received, the more likely it is that the severity of the symptoms will be reduced. About 40 percent of persons infected with genital herpes have only a first attack of symptoms of the disease; the remaining 60 percent may have attacks four or five times a year for several years.

AIDS

AIDS (acquired immunodeficiency syndrome), which is caused by HIV (human immunodeficiency virus), is among the most significant health problems that this nation—and the world—faces today. Although AIDS almost inevitably ends in death, no one actually dies *of* AIDS. Rather, AIDS reduces the body's ability to fight diseases, making a person vulnerable to many diseases—such as pneumonia—that result in death.

AIDS was first identified in 1981, and the total number of AIDS-related deaths in the United States through 1985 was only 12,493; however, the numbers rose rapidly and precipitously after that. The number of *new* reported cases of AIDS in this country in calendar year 1993 was 103,533; in 2001,

Figure 18.1 Adults and Children Living with HIV/AIDS

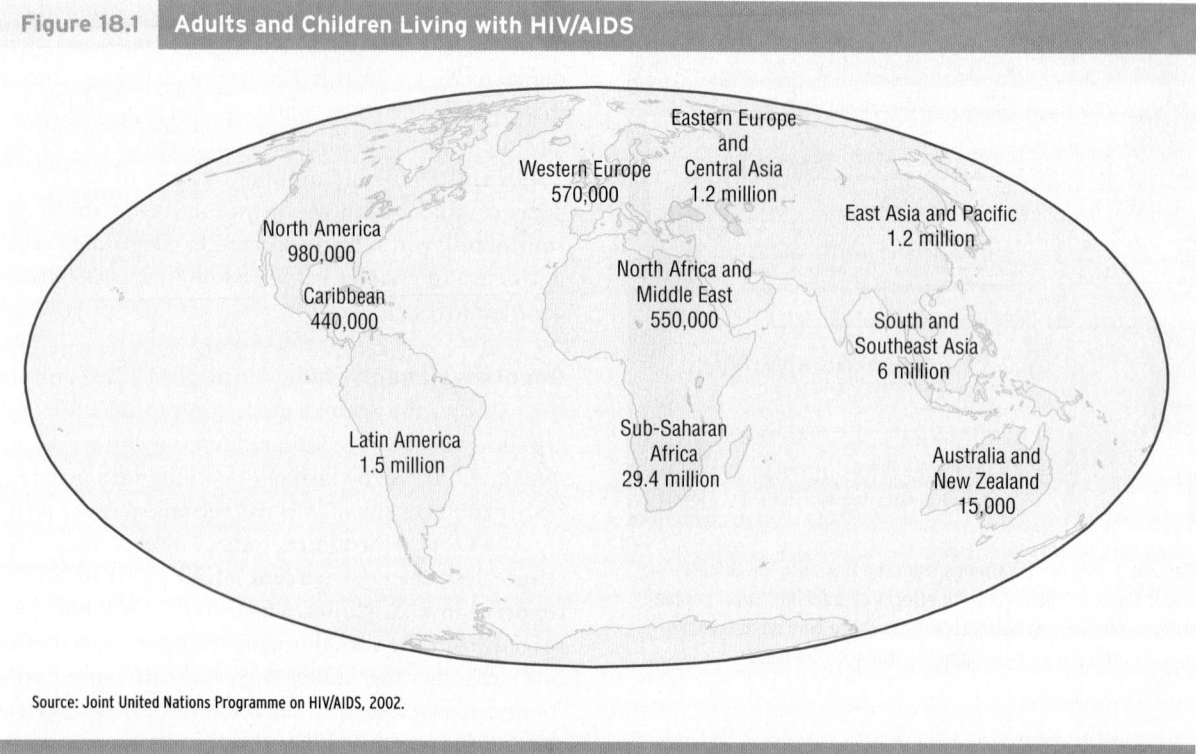

Source: Joint United Nations Programme on HIV/AIDS, 2002.

15,603 people in the United States died of AIDS-related diseases (U.S. Department of Health and Human Services, 2002a). Fortunately, awareness of the syndrome and how it can be acquired has begun to produce some results: The 15,603 deaths in 2001 were less than one-third the number of such deaths in 1995 (U.S. Department of Health and Human Services, 2002a).

Worldwide, however, the number of people with HIV or AIDS is increasing at an alarming rate. The Department of Health and Human Services estimates that in 2002, a total of 42 million people had HIV/AIDS; 14 percent of all new cases worldwide are children, and children under the age of 15 account for about 610,000 AIDS-related deaths each year. More than two-thirds of the people with HIV/AIDS live in sub-Saharan Africa; 14 percent of people with HIV/AIDS live in South and Southeast Asia (see Figure 18.1).

HIV is transmitted through unprotected (or inadequately protected) sexual intercourse with an infected partner (either male or female), by sharing a contaminated hypodermic needle with someone who is infected, by exposure to blood or blood products (usually from a transfusion), and by an infected woman who passes the virus on to her child during pregnancy, childbirth, or breast feeding. It is not transmitted by casual contact such as shaking hands.

Staying Healthy: Diet and Exercise Lifestyle choices also include positive actions such as a healthy diet and good exercise. Over the past several decades, a dramatic improvement in our understanding of food and diet has taken place, and many people in the United States have begun to improve their dietary habits. A significant portion of the population now eats larger amounts of vegetables, fruits, and cereals, and substitutes unsaturated fats and oils for saturated fats. These changes have contributed to a significant decrease in the incidence of heart disease and of some types of cancer.

Exercise is another factor. Regular exercise (at least three times a week) keeps the heart, lungs, muscles, and bones in good health and slows the aging process.

HEALTH CARE IN THE UNITED STATES

Understanding health care as it exists in the United States today requires a brief examination of its history. During the nineteenth century, people became doctors in this country either through apprenticeships, purchasing a mail-order diploma, completing high school and attending a series of lectures, or

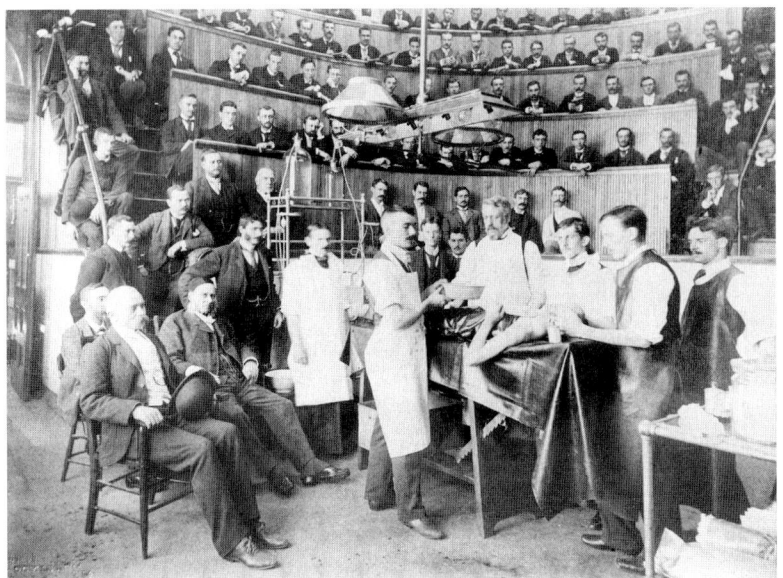

A sight similar to this might have greeted Abraham Flexner as he produced his report on medical education in the United States. In early teaching hospitals, students observed as physicians performed surgical procedures. Although today's operating theaters look quite different, this form of observation is used in many medical schools to inform students about various procedures.

Bettmann/Corbis

obtaining bachelor's and M.D. degrees and studying abroad for a number of years. At that time, medical schools were largely proprietary institutions, and their officials were often more interested in acquiring students than in enforcing standards. The state licensing boards established to improve medical training and stop the proliferation of "irregular" practitioners failed to slow the growth of medical schools, and their number increased from 90 in 1880 to 160 in 1906. Medical school graduates were largely poor and frustrated because of the overabundance of doctors and quasi-medical practitioners, so doctors became highly competitive and anxious to limit the number of new practitioners. The obvious way to accomplish this was to reduce the number of medical schools and set up licensing laws to eliminate unqualified or irregular practitioners (Kendall, 1980).

The Rise of Scientific Medicine and Professionalism

Although medicine had been previously viewed more as an art than as a science, several significant discoveries during the nineteenth century in areas such as bacteriology and anesthesiology began to give medicine increasing credibility as a science (Nuland, 1997). At the same time that these discoveries were occurring, the ideology of science was being advocated in all areas of life, and people came to believe that almost any task could be done better if the appropriate scientific methods were used. To make medicine in the United States more scientific (and more profitable), the Carnegie Foundation (at the request

of the American Medical Association and the forerunner of the Association of American Medical Colleges) commissioned an official study of medical education. The "Flexner report" that resulted from this study has been described as the catalyst of modern medical education but has also been criticized for its lack of objectivity.

The Flexner Report　To conduct his study, Abraham Flexner met with the leading faculty at the Johns Hopkins University School of Medicine to develop a model of how medical education should take place; he next visited each of the 155 medical schools then in existence, comparing them with the model. Included in the model was the belief that a medical school should be a full-time, research-oriented, laboratory facility that devoted all of its energies to teaching and research, not to the practice of medicine (Kendall, 1980). It should employ "laboratory men" to train students in the "science" of medicine, and the students should then apply the principles they had learned in the sciences to the illnesses of patients (Brown, 1979). Only a few of the schools Flexner visited were deemed to be equipped to teach scientific medicine; nonetheless, his model became the standard for the profession (Duffy, 1976).

As a result of the Flexner report (1910), all but two of the African American medical schools then in existence were closed, and only one of the medical schools for women survived. As a result, white women and people of color were largely excluded from medical education for the first half of the twentieth century. Until the civil rights movement and the

Although women today make up about half of the first-year class in U.S. medical schools, they are underrepresented in tenured faculty positions and in specializations such as surgery.

© Flash! Light/Stock Boston

women's movement of the 1960s and 1970s, virtually all physicians were white, male, and upper- or upper-middle class.

The Professionalization of Medicine

Despite its adverse effect on people of color and women who might desire a career in medicine, the Flexner report did help professionalize medicine. When we compare post-Flexner medicine with the characteristics of professions (see Chapter 13), we find that it meets those characteristics:

1. *Abstract, specialized knowledge.* Physicians undergo a rigorous education that results in a theoretical understanding of health, illness, and medicine. This education provides them with the credentials, skills, and training associated with being a professional.
2. *Autonomy.* Physicians are autonomous and (except as discussed subsequently in this chapter) rely on their own judgment in selecting the appropriate technique for dealing with a problem. They expect patients to respect that autonomy.
3. *Self-regulation.* Theoretically, physicians are self-regulating. They have licensing, accreditation, and regulatory boards and associations that set professional standards and require members to adhere to a code of ethics as a form of public accountability.
4. *Authority.* Because of their authority, physicians expect compliance with their directions and advice. They do not expect clients to argue about the advice rendered (or the price to be charged).
5. *Altruism.* Physicians perform a valuable service for society rather than acting solely in their own self-interest. Many physicians go beyond their self-interest or personal comfort so that they can help a patient.

However, with professionalization, licensed medical doctors gained control over the entire medical establishment, a situation that has continued until the present and—despite current efforts at cost control by insurance companies and others—may continue into the future.

Medicine Today

Throughout its history in the United States, medical care has been on a *fee-for-service* basis: Patients are billed individually for each service they receive, including treatment by doctors, laboratory work, hospital visits, prescriptions, and other health-related expenses. Fee-for-service is an expensive way to deliver health care because there are few restrictions on the fees charged by doctors, hospitals, and other medical providers.

There are both good and bad sides to the fee-for-service approach. The good side is that in the "true spirit" of capitalism, coupled with the hard work and scholarship of many people, this approach has resulted in remarkable advances in medicine. Among recent medical innovations are the following:

- *Bloodless surgery*—operations that previously would have required blood transfusions are in some instances now performed without transfusions, using combinations of iron supplements and vitamins, large doses of a blood-building drug (synthetic erythropoietin) that stimulates the bone marrow to produce red blood cells, and intravenous fluids to increase circulation (Langone, 1997).
- *Robotdoc*—a computer-controlled robot used in hip replacement surgery to bore precision holes in a patient's bone, a task that surgeons must

Figure 18.2 Increase in Cost of Health Care, 1970-2000

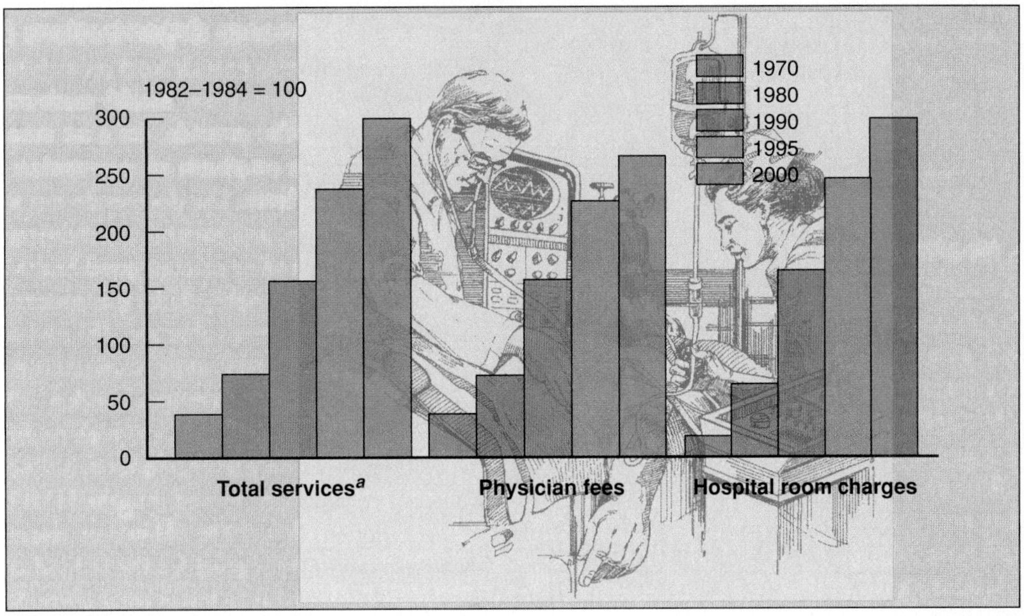

*a*Includes other medical care services.

Source: Computed by author based on U.S. Health Care Financing Administration, 2000.

otherwise do manually (and with much less precision) with a mallet and chisel, is an example of the future use of robots in medical operations (Olmos, 1997).

The bad side of fee-for-service medicine is its inequality of distribution. In effect, the United States has a two-tier system of medical care. Those who can afford it are able to get top-notch medical treatment. And where they receive it may not be much like the hospitals that most of us have visited:

> Every afternoon, between three and five, high above New York's Fifth Avenue, the usual quiet of Eleven West is broken by the soft rustle of white linen cloths and the clink of silver and china as high tea is served room by room. . . . Down the hall, a concierge waits to take your dinner order, provide a video from a list of over 950 titles, arrange for a manicure or massage, or send up that magazine or best-seller that you wanted to read. No, this is not a hitherto unknown outpost of the Four Seasons or the Ritz, but a 19-room wing of the Mt. Sinai Medical Center, one of the nation's leading hospitals. . . .
>
> Here at Mt. Sinai and a few other top hospitals . . . sheets are 250-count cotton, the bathrooms are marble and stocked with toiletries, and there is

ample room to accommodate a nice seating group of leather wing chairs and a brocade sofa. And here no call button is pressed in vain. Hospital personnel not only come when summoned but are eagerly waiting to cater to your every need. . . .

> On these select floors, multi-tiered food service carts and their clattering, plastic trays are gone. "Room service" is in full force. Meals are presented, often course by course, by bow-tied, black-jacketed waiters from rolling, linen-covered tables. (Winik, 1997)

The additional charges for rooms on floors such as those described above may run from $250 to $1,000 per night more than the cost of the standard private room (Winik, 1997). Obviously, this sort of medical care is not within the budget of most of us. However, the cost of health care per person in the United States rose from $141 in 1960 to $3,925 (more than 25 times as much) in 2000 (U.S. Census Bureau, 2002) and is still increasing. Figure 18.2 reflects this cost increase. Keeping in mind the issues that have been raised throughout this text regarding income disparity in the United States, the questions to be considered at this point are "Who pays for medical care, and how?" and "What about the people who simply cannot afford adequate medical care?"

Paying for Medical Care in the United States

The United States and the Union of South Africa are the only developed nations without some form of universal health coverage for all citizens. Before we examine the health care systems of several other nations, however, let's look more closely at our own system.

Private Health Insurance Part of the reason that the cost of fee-for-service health care in the United States escalated rapidly beginning in the 1960s (see Figure 18.2) was the expansion of medical insurance programs at that time. Third-party providers (public and private insurers) began picking up large portions of doctor and hospital bills for insured patients. With third-party fee-for-service payment, patients pay premiums into a fund that in turn pays doctors and hospitals for each treatment the patient receives. According to the medical sociologist Paul Starr (1982), third-party fee-for-service is the main reason for medical inflation because it gives doctors and hospitals an incentive to increase medical services. In other words, the more services they provide, the more fees they charge, and the more money they make. Medical sociologists Gregory L. Weiss and Lynne E. Lonnquist (2003) note that the principle of supply and demand (health care providers, in competition with one another, should be motivated to offer the best possible service at the lowest price in order to attract customers) works ineffectively in health care. Patients have no incentive to limit their visits to doctors or hospitals because they have already paid their premiums and feel entitled to medical care, regardless of the cost. Likewise, many patients simply depend on the advice of their physicians to determine what treatment to have; they do not independently decide on what medical care they need.

Public Health Insurance The United States has two nationwide public health insurance programs, Medicare and Medicaid (see Box 18.2). Medicare is a program for persons age 65 or older who are covered by Social Security or who are eligible and "buy into" the program by paying a monthly premium (Atchley and Barusch, 2004). Medicare pays part of the health care costs of these people. Medicaid, a jointly funded federal–state–local program, was established to make health care more available to the poor. However, both the Medicaid program and the Medicare program are in financial difficulty. Current projections call for Medicaid spending to double and Medicare

spending to triple in the next few years (Weiss and Lonnquist, 2003).

Health Maintenance Organizations (HMOs)

Created in an effort to provide workers with health coverage by keeping costs down, ***health maintenance organizations (HMOs)*** **provide, for a set monthly fee, total care with an emphasis on prevention to avoid costly treatment later.** The doctors do not work on a fee-for-service basis, and patients are encouraged to get regular checkups and to practice good health practices (e.g., exercise and eat right). As long as patients use only the doctors and hospitals that are affiliated with their HMO, they pay no fees, or only small co-payments, beyond their insurance premiums (Anders, 1996). Seeing HMOs as a potential source of high profits because of their emphasis on preventing serious illness, many for-profit corporations moved into the HMO business in the 1980s (Anders, 1996). However, research shows that preventive care is good for the individual's health but does not necessarily lower total costs. Moreover, critics charge that some HMOs require their physicians to withhold vital information from their patients if it could cost the HMO money to provide the needed procedure or hospitalization (Gray, 1996).

Managed Care Another approach to controlling health care costs in the United States is known as ***managed care—any system of cost containment that closely monitors and controls health care providers' decisions about medical procedures, diagnostic tests, and other services that should be provided to patients*** (Weitz, 2004). In most managed care programs, patients choose a primary-care physician from a list of participating doctors. When patients need medical services, they must first contact the primary-care physician; if a specialist is needed for treatment, the primary-care physician refers the patient to a specialist who participates in the program. Doctors must get approval before they perform certain procedures or admit a patient to a hospital; if they fail to obtain such advance approval, the insurance company has the right to refuse to pay for the treatment or hospital stay.

The Uninsured and the Underinsured Despite public and private insurance programs, about one-third of all U.S. citizens are without health insurance or had difficulty getting or paying for medical care at some time in the last year. As shown on Map 18.1, the number of people not covered by health insurance varies from state to state. Of the people not covered by health insurance, 8.5 million are children (Mills, 2002). An estimated 41.2 million people in the

Box 18.2 SOCIOLOGY AND SOCIAL POLICY

Medicare and Medicaid: The Pros and Cons of Government-Funded Medical Care

Should people receive health care even if they cannot afford to pay for it? This question has been a concern in the United States, particularly for older adults who live on a fixed annual income. Consequently, in 1965 Congress enacted the Medicare program. This federal program for people age sixty-five and over (who are covered by Social Security or railroad retirement insurance or who have been permanently and totally disabled for two years or more) is primarily funded through Social Security taxes paid by current workers. We refer to Medicare as an entitlement program because people who receive benefits under the plan must have paid something to be covered. At the time of this writing, Medicare Part A provides coverage for some inpatient hospital expenses (for up to about ninety days) and limited coverage for skilled nursing care at home. By contrast, Part B (which pays some outpatient expenses) requires that the person who is covered must pay an additional premium amounting to about $50 per month.

How successful has Medicare been as a social policy? Medicare has created greater access to health care services for many older people. However, it also has serious limitations, including the fact that it primarily provides for acute, short-term care and is limited to partial payment for ninety days of hospital care. With the aging population, it is problematic that this plan covers only a restricted amount of skilled nursing care and home health services. Patients must pay a deductible and copayments (fees paid by people each time they see a health care provider), and the plan pays only 80 percent of allowable charges, not the actual amount charged by health care providers. Furthermore, Medicare does not include many of the things that some older people need the most, such as dental care, hearing aids, eyeglasses, long-term care in nursing homes, or custodial or nonmedical service, including adult day care or homemaker services. There are such serious gaps

in Medicare coverage that many older people who can afford to do so buy medigap policies, which are supplementary private insurance policies that fill the gaps in the Medicare coverage.

As we have seen, one form of government-funded health care for older people—Medicare—is considered to be an entitlement program. The other one that we will examine—Medicaid—is more often thought of as a welfare program because it primarily serves low-income people. Medicaid is the joint federal–state means-tested welfare program that provides health care insurance for poor people of any age who meet specific eligibility requirements. Although people over age sixty-five constitute about 13 percent of Medicaid recipients, they account for more than 40 percent of the program's total expenditures (Hooyman and Kiyak, 2002). Unlike Medicare recipients, who are seen as "worthy" of their health care benefits, some Medicaid recipients have been stigmatized for participation in this "welfare program." Today, many physicians refuse to take Medicaid patients because the administrative paperwork is burdensome and reimbursements are low—typically less than one-half of what private insurance companies pay for the same services.

When we analyze Medicare and Medicaid as social policies, we find that both programs provide broad coverage for older people. However, both programs have serious flaws, since they are extremely expensive and are highly vulnerable to costly fraud by health care providers and some recipients. Finally, across lines of race/ethnicity and class, some studies have shown that these medical assistance programs typically provide higher-quality health care to older people who are white and more affluent than they do for people of color, very-low-income individuals, people with disabilities, and those who live in rural areas (Pear, 1994). These issues no doubt will remain with us for many years to come.

WRITING IN SOCIOLOGY ASSIGNMENT

How will the aging of the U.S. population affect government-funded health care programs in the future? Are there better solutions than Medicare

and Medicaid for providing better health care for older people at less cost?

Map 18.1 Persons Not Covered by Health Insurance, by State

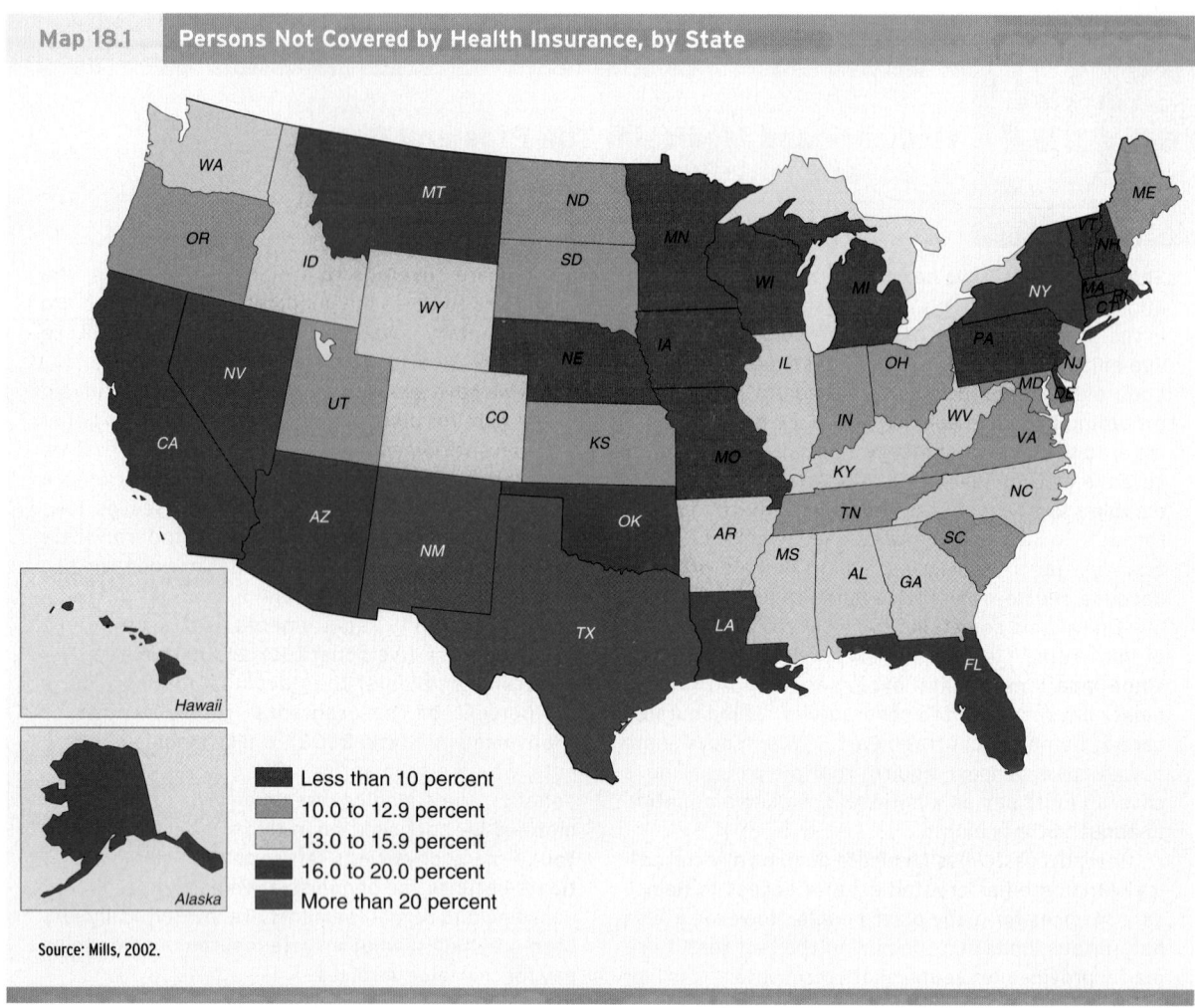

Less than 10 percent
10.0 to 12.9 percent
13.0 to 15.9 percent
16.0 to 20.0 percent
More than 20 percent

Source: Mills, 2002.

United States had no health insurance in 2001—approximately 14.6 percent of the nation's population (Mills, 2002). The working poor constitute a substantial portion of this category, and it is estimated that 18 million of the uninsured hold full-time jobs. They make too little to afford health insurance but too much to qualify for Medicaid, and their employers do not provide health insurance coverage. What happens when they need medical treatment? "They do without," explains Ray Hanley, medical services director for the Arkansas Department of Human Services (qtd. in Kilborn, 1997: A10).

Paying for Medical Care in Other Nations

Other industrialized and industrializing countries do not leave their citizens in the situation in which some people in the United States find themselves. Let's examine how other nations pay for health care.

Canada Prior to the 1960s, Canada's health care system was similar to that of the United States today. However, in 1962 the government of the province of Saskatchewan implemented a health insurance plan despite opposition from doctors, who went on strike to protest the program. The strike was not successful, as the vast majority of citizens supported the government, which maintained health services by importing doctors from Great Britain. The Saskatchewan program proved itself viable in the years following the strike, and by 1972 all Canadian provinces and territories had coverage for medical and hospital services (Kendall, Lothian Murray, and Linden, 2004). As a result, Canada has a ***universal health care*** system—**a health care system in which all citizens receive medical services paid for by tax revenues.** In Canada, these revenues are supplemented by insurance premiums paid by all taxpaying citizens.

One major advantage of the Canadian system over that in the United States is a significant reduction in administrative costs. Whereas more than 20 percent of the U.S. health care dollar represents administrative costs, in Canada the corresponding figure is 10 percent (Weiss and Lonnquist, 2003). However, the system is not without its critics, who claim that it is costly and often wasteful. For example, Canadians are allowed unlimited trips to the doctor, and doctors can increase their income by ordering extensive tests and repeat visits, just as in the United States (Kendall, Lothian Murray, and Linden, 2004).

The Canadian health care system does not constitute what is referred to as **socialized medicine—a health care system in which the government owns the medical care facilities and employs the physicians.** Rather, Canada has maintained the private nature of the medical profession. Although the government pays most health care costs, the physicians are not government employees and have much greater autonomy than do physicians in the health care system in Great Britain.

Great Britain In 1946, Great Britain passed the National Health Service Act, which provided for all health care services to be available at no charge to the entire population. Although physicians work out of offices or clinics—as in the United States or Canada—the government sets health care policies, raises funds and controls the medical care budget, owns health care facilities, and directly employs physicians and other health care personnel (Weiss and Lonnquist, 2003). Unlike the Canadian model, the health care system in Great Britain *does* constitute socialized medicine. Physicians receive capitation payments from the government: a fixed annual fee for each patient in their practice regardless of how many times they see the patient or how many procedures they perform. They also receive supplemental payments for each low-income or elderly patient in their practice, to compensate for the extra time such patients may require; bonus payments if they meet targets for providing preventive services, such as immunizations against disease; and financial incentives if they practice in medically underserved areas (Weitz, 2004). Physicians may accept private patients, but such patients rarely constitute more than a small fraction of a physician's practice; hospitals reserve a small number of beds for private patients (Weiss and Lonnquist, 2003). Why would anyone want to be a private patient who pays for her or his own care or hospital bed? The answer is primarily found in the desire to avoid the long waits ("queues") that the general population encounters and the fact that private patients can enter the hospital for surgery at times convenient to the consumer rather than wait upon the convenience of the system (Gill, 1994).

China After a lengthy civil war, in 1949 the Communist Party won control of mainland China but found itself in charge of a vast nation with a population of one billion people, most of whom lived in poverty and misery. Malnutrition was prevalent, life expectancies were short, and infant and maternal mortality rates were high. In the cities, only the elite could afford medical care; in the rural areas where most of the population resided, Western-style health care barely existed (Weitz, 2004). With a lack of both financial resources and trained health care personnel, China needed to adopt innovative strategies in order to improve the health of its populace. One such policy was developing a large number of *physician extenders* and sending them out into the cities and rural areas to educate the public regarding health and health care and to treat illness and disease. Referred to as *street doctors* in urban areas and *barefoot doctors* in the countryside, these individuals had little formal training and worked under the supervision of trained physicians (Weitz, 2004).

Over the past two decades, medical training has become more rigorous, and supervision has increased. All doctors receive training in both Western and traditional Chinese medicine. Doctors who work in hospitals receive a salary; all other doctors now work on a fee-for-service basis. In urban areas, the cost of health care is paid by employers; however, 85 percent of the population lives in rural areas, where most work on family farms and are expected to pay for their own health care. The cost of health care generally remains low, but the cost of hospital care has risen; accordingly, many Chinese—if they can afford it—purchase health care insurance to cover the cost of hospitalization (Weitz, 2004). As a low-income country, China spends only 2 percent to 3 percent of its gross domestic product on health care, but the health of its citizens is only slightly below that of most industrialized nations.

Regardless of which system of delivering medical care that a nation may have, the health care providers and the general population of the nation are having to face new issues that arise as a result of new technology.

Social Implications of Advanced Medical Technology

Advances in medical technology are occurring at a speed that is almost unbelievable; however, sociologists and other social scientists have identified specific social implications of some of the new technologies (see Weiss and Lonnquist, 2003):

High-tech medicine has extended many people's lives in high-income countries, but it has also significantly increased overall health care costs and raised new questions about the nature and quality of life.

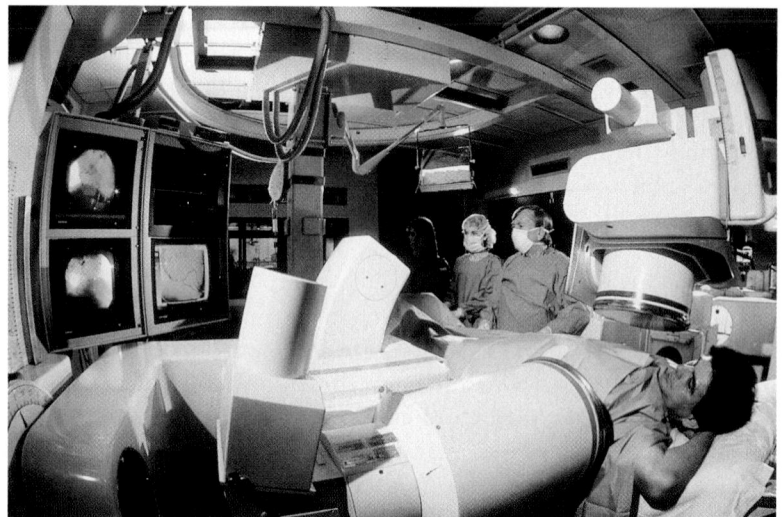

Howard Sochurek/Woodfin Camp & Associates

1. *The new technologies create options for people and for society, but options that alter human relationships.* An example is the ability of medical personnel to sustain a life that in earlier times would have ended as the result of disease or an accident. Although this can be beneficial, technologically advanced equipment that can sustain life after consciousness is lost and there is no likelihood that the person will recover can create a difficult decision for the family of that person if he or she has not left a *living will*—a document stating the person's wishes regarding the medical circumstances under which his or her life should be terminated. Federal law requires all hospitals and other medical facilities to honor the terms of a living will.

2. *The new technologies increase the cost of medical care.* For example, the computerized axial tomography (CT or CAT) scanner—which combines a computer with X-rays that are passed through the body at different angles—produces clear images of the interior of the body that are invaluable in investigating disease. However, the cost of such a scanner is around $1 million. Magnetic resonance imaging (MRI) equipment that allows pictures to be taken of internal organs ranges in cost from $1 million to $2.5 million. Can the United States afford such equipment in every hospital for every patient? The money available for health care is not unlimited, and when it is spent on high-tech equipment and treatment, it is being reallocated from other health care programs that might be of greater assistance to more people.

3. *The new technologies raise provocative questions about the very nature of life.* In Chapter 15, we briefly discuss in vitro fertilization—a form of as-sisted reproductive technology. But during 1997, Dr. Ian Williams and his associates in Scotland took in vitro fertilization a step further: They cloned a lamb (that they named Dolly) from the DNA of an adult sheep. Subsequently, scientists have cloned other animals in the same manner, raising a number of profound questions: If scientists can duplicate mammals from adult DNA, is it possible to clone a perfect (whatever that may be) human being instead of taking a chance on a child that is born to a couple? If it is possible, would it be ethical? For example, if—as discussed earlier in this text—most parents prefer a boy over a girl if they are going to have only one child, would the world suddenly have substantially more boys than girls? Would everyone start to look alike, eliminating diversity? If a child were born other than through cloning and had some sort of biological defect, would the child have the right to sue his or her parents for negligence?

However, at the same time that high-tech medicine is becoming a major part of overall health care, many people are turning to holistic medicine and alternative healing practices.

Holistic Medicine and Alternative Medicine

When examining the subject of medicine, it is easy to think only in terms of conventional (or mainstream) medical treatment. By contrast, **holistic medicine is an approach to health care that focuses on prevention of illness and disease and is aimed at treating the whole person—body and mind—rather than just the part or parts in which symptoms occur.**

The use of herbal therapies is a form of alternative medicine that is increasing in popularity in the United States. Here, a Korean American doctor measures out herbs for one of his patients.

Under this approach, it is important that people not look solely to medicine and doctors for their health, but rather that people engage in health-promoting behavior. Likewise, medical professionals must not only treat illness and disease but also work with the patient to promote a healthy lifestyle and self-image.

Many practitioners of *alternative medicine*—healing practices inconsistent with dominant medical practice—take a holistic approach, and today many people are turning to alternative medicine either in addition to or in lieu of traditional medicine. Jill Neimark tells how she feels about the two types of medicine:

> I'd like to make a confession: I believe. I believe in the brilliance of Western medicine, in the technological wonders of organ transplants, brain scans, and laparoscopic surgery—and I also believe that shamanic healers from indigenous cultures have discovered treatments for illnesses that traditional medicine can't touch, that the mind alone can trigger and then reverse illness, that dietary changes, along with vitamin and herbal supplements, can powerfully impact the course of disease. I believe in merging the best of both worlds—the orthodox and the alternative.
>
> I'm not alone in my belief. This is a season of unprecedented possibility in medicine. And yet in spite of the tremendous changes, there is still some resistance from mainstream medicine, as well as the government and the pharmaceutical industry. One can almost feel the tectonic plates of conventional and holistic medicine crashing up against each other and realigning the landscape of health care. (Neimark, 1997)

Are medical doctors as opposed to alternative medicine as Neimark indicates? In understanding the medical establishment's reaction to alternative medicine, it is important to keep in mind the philosophy of scientific medicine—that medicine is a science, not an art. Thus, to the extent to which alternative medicine is "nonscientific," it must be quackery and therefore something that is undoubtedly worthless and possibly harmful. Undoubtedly, self-interest is also involved in mainstream medicine's reaction to alternative medicine: If the public can be persuaded that scientific medicine is the only legitimate healing practice, fewer health care dollars will be spent on a form of medical treatment that is (at least to some extent) in competition with the medical establishment (Weiss and Lonnquist, 2003). But if all forms of alternative medicine (including chiropractic, massage, and spiritual) are taken into account, people spend more money on unconventional therapies than they do for all hospitalizations (Weiss and Lonnquist, 2003).

SOCIOLOGICAL PERSPECTIVES ON HEALTH AND MEDICINE

Functionalist, conflict, symbolic interactionist, and postmodernist perspectives focus on different aspects of health and medicine; each provides us with significant insights on the problems associated with these pressing social concerns.

A Functionalist Perspective: The Sick Role

According to the functionalist approach, if society is to function as a stable system, it is important for people to be healthy and to contribute to their society. Consequently, sickness is viewed as a form of deviant behavior that must be controlled by society. This view was initially set forth by the sociologist Talcott Parsons (1951) in his concept of the *sick role*—**the set of patterned expectations that defines the norms and values appropriate for individuals who are sick and for those who interact with them.** According to Parsons, the sick role has four primary characteristics:

1. People who are sick are not responsible for their condition. It is assumed that being sick is not a deliberate and knowing choice of the sick person.
2. People who assume the sick role are temporarily exempt from their normal roles and obligations.

According to the functionalist perspective, the sick role exempts the patient from routine activities for a period of time but assumes that the individual will seek appropriate medical attention and get well as soon as possible.

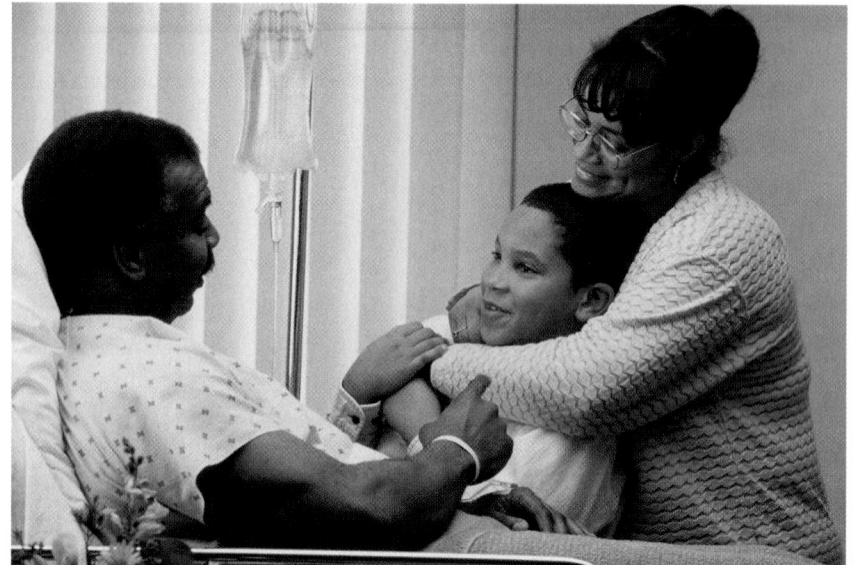

For example, people with illnesses are typically not expected to go to school or work.

3. People who are sick must want to get well. The sick role is considered to be a temporary role that people must relinquish as soon as their condition improves sufficiently. Those who do not return to their regular activities in a timely fashion may be labeled as hypochondriacs or malingerers.

4. People who are sick must seek competent help from a medical professional to hasten their recovery.

As these characteristics show, Parsons believed that illness is dysfunctional for both individuals and the larger society. Those who assume the sick role are unable to fulfill their necessary social roles, such as being parents or employees. Similarly, people who are ill lose days from their productive roles in society, thus weakening the ability of groups and organizations to fulfill their functions. During the winter months, for instance, the media often carry stories about how much illnesses such as flu are "costing" U.S. employers because many workers are at home ill. The sick role affects the bottom line in business because the employee who is not at work contributes to a loss in productivity.

According to Parsons, it is important for the society to maintain social control over people who enter the sick role. Physicians are empowered to determine who may enter this role and when patients are ready to exit it. Because physicians spend many years in training and have specialized knowledge about illness and its treatment, they are certified by the society to be "gatekeepers" of the sick role. When patients seek the advice of a physician, they enter into the patient–physician relationship, which does not contain equal power for both parties. The patient is expected to follow the "doctor's orders" by adhering to a treatment regime, recovering from the malady, and returning to a normal routine as soon as possible.

What are the major strengths and weaknesses of Parsons's model and, more generally, of the functionalist view of health and illness? Parsons's analysis of the sick role was pathbreaking when it was introduced. Some social analysts believe that Parsons made a major contribution to our knowledge of how society explains illness-related behavior and how physicians have attained their gatekeeper status. In contrast, other analysts believe that the sick-role model does not take into account racial–ethnic, class, and gender variations in the ways that people view illness and interpret this role. For example, this model does not take into account the fact that many individuals in the working class may choose not to accept the sick role unless they are seriously ill—because they cannot afford to miss time from work and lose a portion of their earnings. Moreover, people without health insurance may not have the option of assuming the sick role.

A Conflict Perspective: Inequalities in Health and Health Care

Unlike the functionalist approach, conflict theory emphasizes the political, economic, and social forces that affect health and the health care delivery system. Among the issues of concern to conflict theorists are the ability of all people to obtain health care; how race, class, and gender inequalities affect health and health care; power relationships between doctors and other health care workers; the dominance of the medical model of health care; and the role of profit in the health care system.

Who is responsible for problems in the U.S. health care system? According to many conflict theorists, problems in U.S. health care delivery are rooted in the capitalist economy, which views medicine as a commodity that is produced and sold by the medical–industrial complex. The **medical–industrial complex encompasses both local physicians and hospitals as well as global health-related industries such as insurance companies and pharmaceutical and medical supply companies** (Relman, 1992).

The United States is one of the few industrialized nations that rely almost exclusively on the medical–industrial complex for health care delivery and do not have universal health coverage, which provides some level of access to medical treatment for all people. Consequently, access to high-quality medical care is linked to people's ability to pay and to their position within the class structure. Those who are affluent or have good medical insurance may receive high-quality, state-of-the-art care in the medical–industrial complex because of its elaborate technologies and treatments. However, people below the poverty level and those just above it have greater difficulty gaining access to medical care. Referred to as the *medically indigent,* these individuals do not earn enough to afford private medical care but earn just enough money to keep them from qualifying for Medicaid (Weiss and Lonnquist, 2003). In the profit-oriented capitalist economy, these individuals are said to "fall between the cracks" in the health care system.

Who benefits from the existing structure of medicine? According to conflict theorists, physicians—who hold a legal monopoly over medicine—benefit from the existing structure because they can charge inflated fees. Similarly, clinics, pharmacies, laboratories, hospitals, supply manufacturers, insurance companies, and many other corporations derive excessive profits from the existing system of payment in medicine. In recent years, large drug companies and profit-making hospital corporations have come to occupy a larger and larger part of health care delivery. As a result, medical costs have risen rapidly, and the federal government and many insurance companies have placed pressure for cost containment on other players in the medical–industrial complex (Tilly and Tilly, 1998).

Theorists using a radical conflict framework for their analysis believe that the only way to reduce inequalities in the U.S. health care structure is to eliminate capitalism or curb the medical–industrial complex. From this approach, capitalism is implicated in both the rates of illness and how health care is delivered. An example of how rates of illness are related to capitalism is the predominance of workplace hazards and environmental pollution that are dangerous but are often not corrected because corrective measures would be too costly, reducing corporate profits. Capitalistic forms of health care delivery are found in recent medical entrepreneurship on the Internet, whereby companies promote products and medical devices, some of which may not have approval from regulatory agencies such as the Food and Drug Administration (see Box 18.3).

Conflict theorists increase our awareness of inequalities of race, class, and gender as these statuses influence people's access to health care. They also inform us about the problems associated with health care becoming "big business." However, some analysts believe that the conflict approach is unduly pessimistic about the gains that have been made in health status and longevity—gains that are at least partially due to large investments in research and treatment by the medical–industrial complex.

A Symbolic Interactionist Perspective: The Social Construction of Illness

Symbolic interactionists attempt to understand the specific meanings and causes that we attribute to particular events. In studying health, symbolic interactionists focus on the meanings that social actors give their illness or disease and how these affect people's self-concept and relationships with others. According to symbolic interactionists, we socially construct "health" and "illness" and how both should be treated. For example, some people explain disease by blaming it on those who are ill. If we attribute cancer to the acts of a person, we can assume that we will be immune to that disease if we do not engage in the same behavior. Nonsmokers who learn that a lung cancer victim had a two-pack-a-day habit feel comforted that they are unlikely to suffer the same fate. Similarly, victims of AIDS are often blamed for promiscuous sexual conduct or intravenous drug use, regardless of how they contracted HIV. In this case, the social definition of the illness leads to the stigmatization of individuals who suffer from the disease.

Although biological characteristics provide objective criteria for determining medical conditions such as heart disease, tuberculosis, or cancer, there is also a subjective component to how illness is defined. This subjective component is very important when we look at conditions such as childhood hyperactivity, mental illness, alcoholism, drug abuse, cigarette

Box 18.3 CHANGING TIMES: MEDIA AND TECHNOLOGY

Medicine.com—The Wave of the Future?

When nine-year-old Robert Lord snapped his neck in a fall from a rope ladder, a San Diego doctor told Robert's parents that their son might never regain the use of his body below his arms. Unwilling to give up hope, Robert's father asked the advice of a researcher/friend at the Mayo Clinic in Rochester, Minnesota, who suggested that Mr. Lord go online to see what he could learn. By typing the words *spinal cord injury* and searching the Internet all night, Robert's father learned about GM-1 ganglioside, an experimental drug reported to improve chances of recovery from a spinal cord injury when administered within 72 hours of the accident. After Mr. Lord reported his findings, his son's doctor acquired the drug from the government and administered it to his patient 78 hours after the injury. Amazingly, 11 weeks later, Robert Lord left the hospital using only a walker (Hafner, 1998).

Was Robert's improved health status a result of the drug that his father learned about on the Internet, or was it part of the boy's normal recovery? Questions such as this arise more frequently as millions log on to the Internet seeking health information and advice (Hafner, 1998).

What are the advantages of receiving health information online? According to some analysts, people are able to gain insights on an impressive array of health-related topics, ranging from colds and the flu to leprosy, the Ebola virus, and mercury poisoning (Grady, 1998). Online advertisements for medicines, dietary supplements, and medical products, including contraceptives, hearing aids, and contact lenses, offer consumers a wide variety of products, sometimes at reduced prices (Lu, 1998). Some physicians use e-mail to communicate with their patients (Hafner, 1998), and more hospitals are providing personal medical records directly to patients online (Freudenheim, 1998).

What problems are associated with "medicine .com"? Although patients and family members can find large *quantities* of medical information online, the *quality* of the information remains in question (Park, 1999). Health care specialists believe that people should acquire information from reliable sources. For example, the federal Department of Health and Human Services maintains a site that provides links to information from government health agencies, public health and professional groups, universities, support groups, medical journals, and some news sites (**www.healthfinder.gov**). Similarly, Medscape (**www.medscape.com**) offers links to Medline, the medical database at the National Library of Medicine, as well as access to journals and medical news (Grady, 1998).

Other problems also exist. For example, what about unregulated medicine sales on the Web? Although international drug sales have never been simple (e.g., what is identified as a prescription drug in one country may be a nutritional supplement in another), the Internet further complicates the problem and increases the potential for fraud (Lu, 1998). Since no U.S. agency or organization routinely tracks online drug sales, we do not know the extent to which individuals and companies bypass the regulations of the U.S. Food and Drug Administration (FDA). For example, manufactured L-tryptophan, an amino acid removed from the U.S. market in the early 1990s because of potentially fatal side-effects, can be purchased on the Net from companies in other countries. Other products that are prescription drugs in the United States may also be purchased from foreign corporations. Currently, the World Health Organization urges organizations and countries to engage in continuous self-regulation and consumer education to reduce situations that might undermine national drug policies.

In the future, will "medicine at the click of a mouse" become a reality? If so, what effect do you think this might have on people's health in the United States and other nations?

---

smoking, and overeating, all of which have been medicalized. The term ***medicalization*** **refers to the process whereby nonmedical problems become defined and treated as illnesses or disorders.** Medicalization may occur on three levels: (1) the conceptual level (e.g., the use of medical terminology to define the problem), (2) the institutional level (e.g., physicians are supervisors of treatment and gatekeepers to applying for benefits), and (3) the interactional level (e.g., when physicians treat patients' conditions as

medical problems). For example, the sociologists Deborah Findlay and Leslie Miller (1994: 277) explain how gambling has been medicalized:

> Habitual gambling . . . has been regarded by a minority as a sin, and by most as a leisure pursuit—perhaps wasteful but a pastime nevertheless. Lately, however, we have seen gambling described as a psychological illness—"compulsive gambling." It is in the process of being medical-

For most people, occasional gambling is simply a leisure pursuit. However, when gambling is medicalized, it becomes defined as a disease for which there may be treatment.

Jodi Buren/Woodfin Camp & Associates

ized. The consequences of this shift in discourse (that is, in the way of thinking and talking) about gambling are considerable for doctors, who now have in gamblers a new market for their services or "treatment"; perhaps for gambling halls, which may find themselves subject to new regulations, insofar as they are deemed to contribute to the "disease"; and not least, for gamblers themselves, who are no longer treated as sinners or wastrels, but as patients, with claims on our sympathy, and to our medical insurance plans as well.

Sociologists often refer to this form of medicalization as the *medicalization of deviance* because it gives physicians and other medical professionals greater authority to determine what should be considered "normal" and "acceptable" behavior and to establish the appropriate mechanisms for controlling "deviant behaviors."

According to symbolic interactionists, medicalization is a two-way process: Just as conditions can be medicalized, so can they be demedicalized. **Demedicalization refers to the process whereby a problem ceases to be defined as an illness or a disorder.** Examples include the removal of certain behaviors (such as homosexuality) from the list of mental disorders compiled by the American Psychiatric Association and the deinstitutionalization of mental health patients. The process of demedicalization also continues in women's health as advocates seek to redefine childbirth and menopause as natural processes rather than as illnesses (Conrad, 1996).

In addition to how health and illness are defined, symbolic interactionists examine how doctors and patients interact in health care settings. Some physicians may hesitate to communicate certain kinds of medical information to patients, such as why they are pre-

scribing certain medications or what side-effects or drug interactions may occur (Kendall, 2004).

Symbolic interactionist perspectives on health and health care provide us with new insights on the social construction of illness and how health and illness cannot be strictly determined by medical criteria. Symbolic interactionists also make us aware of the importance of communication between physicians and patients, including factors that may reduce effective medical treatment for some individuals. However, these approaches have been criticized for suggesting that few objective medical criteria exist for many illnesses and for overemphasizing microlevel issues without giving adequate recognition to macrolevel issues such as the effects on health care of managed care, health maintenance organizations, and for-profit hospital chains.

A Postmodernist Perspective: The Clinical Gaze

In *The Birth of the Clinic* (1994/1963), postmodern theorist Michel Foucault questioned existing assumptions about medical knowledge and the power that doctors have gained over other medical personnel and everyday people. Foucault asserted that truth in medicine—as in all other areas of life—is a social construction, in this instance one that doctors have created. Foucault believed that doctors gain power through the *clinical* (or "observing") *gaze*, which they use to gather information. Doctors develop the clinical gaze through their observation of patients; as the doctors begin to diagnose and treat medical conditions, they also start to speak "wisely" about everything. As a result, other people start to believe that doctors can "penetrate illusion and see . . . the hidden truth" (Shawver, 1998).

According to Foucault, the prestige of the medical establishment was further enhanced when it became possible to categorize all illnesses within a definitive network of disease classification under which physicians can claim that they know why patients are sick. Moreover, the invention of new tests made it necessary for physicians to gaze upon the naked body, to listen to the human heart with an instrument, and to run tests on the patient's body fluids. Patients who objected were criticized by the doctors for their "false modesty" and "excessive restraint" (Foucault, 1994/1963: 163). As the new rules allowed for the patient to be touched and prodded, the myth of the doctor's diagnostic wisdom was further enhanced, and "medical gestures, words, gazes took on a philosophical density that had formerly belonged only to mathematical thought" (Foucault, 1994/1963: 199). For Foucault, the formation of clinical medicine was merely one of the more-visible ways in which the fundamental structures of human experience have changed throughout history.

Foucault's work provides new insights on medical dominance, but it has been criticized for its lack of attention to alternative viewpoints. Among these is the possibility that medical breakthroughs and new technologies actually help physicians become wiser and more scientific in their endeavors. Another criticism is that Foucault's approach is based upon the false assumption that people are passive individuals who simply comply with doctors' orders—he does not take into account that people (either consciously or unconsciously) may resist the myth of the "wise doctor" and not follow "doctors' orders" (Lupton, 1997). Foucault's analysis (1988/1961) was not limited to doctors who treat bodily illness; he also critiqued psychiatrists and the treatment of insanity.

■ MENTAL ILLNESS

Mental illness affects many people; however, it is a difficult topic for sociological research. Social analysts such as Thomas Szasz (1984) have argued that mental illness is a myth. According to this approach, "mental illnesses" are actually individual traits or behaviors that society deems unacceptable, immoral, or deviant. According to Szasz, labeling individuals as "mentally ill" harms them because they often come to accept the label and are then treated accordingly by others.

Is mental illness a myth? After decades of debate on this issue, social analysts are no closer to reaching a consensus than they were when Szasz originally introduced his ideas. However, many scholars believe that mental illness is a reality that has biological, psychological, environmental, social, and other factors involved. Many medical professionals distinguish between a *mental disorder*—a condition that makes it difficult or impossible for a person to cope with everyday life—and *mental illness*—a condition in which a person has a severe mental disorder requiring extensive treatment with medication, psychotherapy, and sometimes hospitalization.

How many people are affected by mental illness? Some answers to this question have come from a systematic national study known as the National Comorbidity Survey. The term *comorbidity* refers to the physical and mental conditions—such as physical illness and depression—that compound each other and undermine the individual's overall well-being (Angel and Angel, 1993). Researchers in the study found that among respondents between the ages of 15 and 54, nearly 50 percent had been diagnosed with a mental disorder at some time in their lives (Kessler, 1994). However, severe mental illness—such as schizophrenia, bipolar affective disorder (manic-depression), and major depression—typically affects less than 15 percent of U.S. adults at some time in their lives (Bourdon et al., 1992; Kessler, 1994). According to the researchers, the prevalence of mental disorders in the United States is greater than most analysts had previously believed. The most widely accepted classification of mental disorders is the American Psychiatric Association's (1994) *Diagnostic and Statistical Manual of Mental Disorders (DSM),* which is illustrated in Table 18.1. Even among those with a lifetime history of three or more diagnosed mental disorders, the proportion who obtain treatment is less than 50 percent (Kessler, 1994).

The Treatment of Mental Illness

In *Madness and Civilization,* Michel Foucault (1988/1961) examined the "archaeology of madness" from 1500 to 1800 to determine how ideas of mental illness have changed over time and to describe the "birth of the asylum." According to Foucault, early in this period insanity was considered part of everyday life, and people with mental illnesses were free to walk the streets; however, beginning with the Renaissance and continuing into the seventeenth and eighteenth centuries, the mentally ill were viewed as a threat to others. During that time, asylums were built, and a clear distinction was drawn between the "insane" and the rest of humanity. According to Foucault (1988/1961: 252), people came to see "madness" as a minority status that does not have the right to autonomy:

Table 18.1 MENTAL DISORDERS IDENTIFIED BY THE AMERICAN PSYCHIATRIC ASSOCIATION

Anxiety Disorders	Disorders characterized by anxiety that is manifested in phobias, panic attacks, or obsessive-compulsive disorder
Dissociative Disorders	Problems involving a splitting or dissociation of normal consciousness such as amnesia and multiple personality
Disorders First Evident in Infancy, Childhood, or Adolescence	Including mental retardation, attention-deficit hyperactivity, anorexia nervosa, bulimia nervosa, and stuttering
Eating or Sleeping Disorders	Including such problems as anorexia and bulimia, or insomnia and other problems associated with sleep
Impulse Control Disorders	Including the inability to control undesirable impulses such as kleptomania, pyromania, and pathological gambling
Mood Disorders	Emotional disorders such as major depression and bipolar (manic-depressive) disorder
Organic Mental Disorders	Psychological or behavioral disorders associated with dysfunctions of the brain caused by aging, disease, or brain damage
Personality Disorders	Maladaptive personality traits that are generally resistant to treatment, such as paranoid and antisocial personality types
Schizophrenia and Other Psychotic Disorders	Disorders with symptoms such as delusions or hallucinations
Somatoform Disorders	Psychological problems that present themselves as symptoms of physical disease, such as hypochondria
Substance-Related Disorders	Disorders resulting from abuse of alcohol and/or drugs such as barbiturates, cocaine, or amphetamines

Source: Adapted from American Psychiatric Association, 1994.

Madness is childhood. Everything at the [asylum] is organized so that the insane are transformed into minors. They are regarded as children who have an overabundance of strength and make dangerous use of it. They must be given immediate punishments and rewards; whatever is remote has no effect on them.

At the beginning of the twenty-first century, many people with mental disorders do not receive professional treatment. However, mental disorders—particularly substance-related ones (due to alcohol and other drug abuse)—are the leading cause of hospitalization for men between the ages of eighteen and forty-four, and the second leading cause (after childbirth) for women in that age group (Agency for Healthcare Research and Quality, 2000).

Many people seeking psychiatric assistance are treated with medications or psychotherapy—which is believed to help patients understand the underlying reasons for their problem—and sometimes treatment in psychiatric wards of local hospitals or in private psychiatric hospitals. However, the introduction of new psychoactive drugs to treat mental disorders and the deinstitutionalization movement in the 1960s have created dramatic changes in how people with mental disorders are treated. ***Deinstitutionalization refers to the practice of rapidly discharging patients from mental hospitals into the community.*** Originally devised as a *solution* for the problem of "warehousing" mentally ill patients in large, prison-like mental hospitals in the first half of the twentieth century, deinstitutionalization is now viewed as the *problem* by many social scientists. The theory behind this process was that patients' rights were being violated because many patients experienced involuntary commitment (i.e., without their consent) to the hospitals, where they remained for extended periods of time. Instead, some professionals believed that the patients' mental disorders could be controlled with proper medications and treatment from community-based mental health services. Advocates of deinstitutionalization also believed that this practice would relieve the stigma attached to mental illness and hospitalization. However, critics of deinstitutionalization argue that this process exacerbated long-term problems associated with inadequate care for people with mental illness.

Admitting people to mental hospitals on an involuntary basis ("involuntary commitment") has always been controversial; however, it remains the primary method by which police officers, judges, social workers, and other officials deal with people—particularly the homeless—whom they have reason to believe are mentally ill and imminently dangerous to others if not detained (Monahan, 1992). State mental hospitals continue to provide most of the chronic inpatient care for poor people with mental illnesses; these institutions tend to serve as a revolving door to poverty-level board-and-care homes, nursing homes, or homelessness, as contrasted with the situation of patients who pay their bills at private psychiatric facilities through private insurance coverage or Medicare (Brown, 1985).

According to the sociologist Erving Goffman (1961a), mental hospitals are a classic example of a *total institution,* previously defined as a place where people are isolated from the rest of society for a period of time and come under the complete control of the officials who run the institution.

Race, Class, Gender, and Mental Disorders

Most studies examining race/ethnicity and mental disorders have compared African Americans and white Americans and have uncovered no significant differences in diagnosable mental illness. However, researchers have found that African Americans report psychological distress or demoralization more often than white Americans do. In a qualitative study about the effects of racism on the everyday lives of middle-class African Americans, social scientists Joe R. Feagin and Melvin P. Sikes (1994) concluded that repeated personal encounters with racial hostility deeply affect the psychological well-being of most African Americans, regardless of their level of education or social class. In a subsequent study, Feagin and Hernán Vera (1995) found that white Americans also pay a high "psychic cost" for the prevalence of racism because it contradicts deeply held beliefs about the great American dream of equality under the law. In earlier work on the effects of discrimination on mental well-being, the social psychologist Thomas Pettigrew (1981) suggested that about 15 percent of whites have such high levels of racial prejudice that they tend to exhibit symptoms of serious mental illness. According to Pettigrew, racism in all its forms constitutes a mentally unhealthy situation in which people do not achieve their full potential.

A limited number of social science studies have focused on mental disorders among racial–ethnic groups such as Mexican immigrants and Mexican Americans, and some of these studies have yielded contradictory results. Researchers in one study found that strong extended (intergenerational) families—which are emphasized in Mexican culture—provide individuals with social support and sources of self-esteem, even if these individuals possess low levels of education and income (Mirowsky and Ross, 1980). Other analysts have challenged this assumption, noting that retention of Mexican culture, as compared to adopted Anglo culture, is not the primary consideration in whether or not Mexican Americans develop mental disorders (Burnham et al., 1987).

Although cultural factors may be important in determining rates of mental disorders, researchers have consistently agreed that social class has a significant effect on this issue. For example, one study examining the relationship among mental disorders, race, and class (as measured by socioeconomic status) found that rates of mental disorders decrease as social class increases for both white Americans and African Americans. For example, at the lowest socioeconomic level, researchers found few race differences (Williams, Takeuchi, and Adair, 1992). Despite agreement among many researchers about the relationship between class and mental disorders, not all of them agree on whether lower social class status causes mental illness or mental illness causes lower social class position (Weitz, 2004). Analysts using the *social stress framework* to examine schizophrenia—the disorder most consistently linked to class—believe that stresses associated with lower-class life lead to greater mental disorders. In contrast, analysts using the *social drift framework* argue that mental disorders cause people to drift downward in class position. To support their argument, they note that individuals diagnosed with schizophrenia typically hold lower-class jobs than would be expected based on their family backgrounds (Eaton, 1980; Weitz, 2004).

Researchers have also consistently found that the rate of diagnosable depression is about twice as high for women as for men. This gender difference typically emerges in puberty (Cleary, 1987) and increases in adulthood as women and men enter and live out their unequal adult statuses (Mirowsky, 1996). However, researchers have found no consistent relationship between gender and the rate of schizophrenia or other serious mental disorders in which the individual loses touch with reality. Although women have higher rates of minor depression and other disorders that cause psychological distress, men have higher rates of personality disorders such as compulsive gambling and drinking, as well as maladaptive personality traits such as antisocial behavior (Link and Dohrenwend, 1989; Weitz, 2004). Some analysts suggest that differences in mental disorders between women and men are linked to gender-role socialization that instills aggressiveness in men and learned helplessness in women. According to the *learned helplessness theory,* people become depressed when they believe they cannot control their lives (Seligman, 1975). Embedded within this theory is the assumption that women contribute to their own helplessness because of the *subjective* perception that they have no control over their lives. However, feminist analysts argue instead that powerlessness is an *objective* condition that may contribute to depression in many women's lives (Jack, 1993). Support for the latter assumption comes from numerous studies indicating that women in high-income, high-status jobs typically have higher levels of psychological well-being and fewer symptoms of mental disorders, regardless of their marital status (Horowitz, 1982; Angel and Angel, 1993).

Although there have been numerous studies about women with mild depression, women with serious mental illnesses have almost been ignored (Mowbray, Herman, and Hazel, 1992). Similarly, women's mental disorders have often been misdiagnosed, and on many occasions physical illnesses have been confused with psychiatric problems (see Busfield, 1996; Lerman, 1996; Klonoff, 1997). According to some analysts, additional studies focusing on women's diversity across lines of race, class, age, religion, sexual orientation, and other factors will be necessary before we can accurately assess the relationship between gender and mental illness and how women are treated in the mental health care industry (Gatz, 1995).

DISABILITY

What is a disability? There are many different definitions. In business and government, disability is often defined in terms of work—for instance, "an inability to engage in gainful employment." Medical professionals tend to define it in terms of organically based impairments—the problem being entirely within the body (Albrecht, 1992). However, not all disabilities are visible to others or necessarily limit people physically. **Disability refers to a reduced ability to perform tasks one would normally do at a given stage of life and that may result in stigmatization or discrimination against the person with disabilities.** In other words, the notion of disability is based not only on physical conditions but also on social attitudes and the social and physical environments in which people live. In an elevator, for example, the buttons may be beyond the reach of persons using a wheelchair. In

One of the most compelling concerns for many people with a disability is access to buildings and other public accommodations. For persons using a wheelchair, steps such as these are a formidable obstacle in everyday life.

Michael Newman/PhotoEdit

this context, disability derives from the fact that certain things have been made inaccessible to some people (Weitz, 2004). According to disability rights advocates, disability must be thought of in terms of how society causes or contributes to the problem—not in terms of what is "wrong" with the person with a disability.

An estimated 48 million persons in the United States have one or more physical or mental disabilities. This number continues to increase for several reasons. First, with advances in medical technology, many people who once would have died from an accident or illness now survive, although with an impairment. Second, as more people live longer, they are more likely to experience diseases (such as arthritis) that may have disabling consequences (Albrecht, 1992). Third, persons born with serious disabilities are more likely to survive infancy because of medical technology. However, less than 15 percent of persons with a disability today were born with it; accidents, disease, and war account for most disabilities in this country.

Although anyone can become disabled, some people are more likely to be or to become disabled than others. African Americans have higher rates of disability than whites, especially more serious disabilities; persons with lower incomes also have higher rates of disability (Weitz, 2004). However, "disability knows no socioeconomic boundaries. You can become disabled from your mother's poor nutrition or from falling off your polo pony," says Patrisha Wright, a spokesperson for the Disability Rights Education and Defense Fund (qtd. in Shapiro, 1993: 10).

For persons with chronic illness and disability, life expectancy may take on a different meaning. Knowing that they will likely not live out the full life expectancy for persons in their age cohort, they may come to "treasure each moment," as did James Keller, a former baseball coach:

> In December 1992, I found out I have Lou Gehrig's disease—amyotrophic lateral sclerosis, or ALS. I learned that this disease destroys every muscle in the body, that there's no known cure or treatment and that the average life expectancy for people with ALS is two to five years after diagnosis.
>
> Those are hard facts to accept. Even today, nearly two years after my diagnosis, I see myself as 42-year-old career athlete who has always been blessed with excellent health. Though not an hour goes by in which I don't see or hear in my mind that phrase "two to five years," I still can't quite believe it. Maybe my resistance to those words is exactly what gives me the strength to live with them and the will to make the best of every day in every way. (Keller, 1994)

As Keller's comments illustrate, disease and disability are intricately linked.

Environment, lifestyle, and working conditions may all contribute to either temporary or chronic disability. For example, air pollution in automobile-clogged cities leads to a higher incidence of chronic respiratory disease and lung damage, which may result in severe disability for some people. Eating certain types of food and smoking cigarettes increase the risk for coronary and cardiovascular diseases (Albrecht, 1992). In contemporary industrial societies, workers in the second tier of the labor market (primarily recent immigrants, white women, and people of color) are at the greatest risk for certain health hazards and disabilities. Employees in data processing and service-oriented jobs may also be affected by work-related disabilities. The extensive use of computers has been shown to harm some workers' vision; to produce joint problems such as arthritis, low-back pain, and carpal tunnel syndrome; and to place employees under high levels of stress that may result in neuroses and other mental health problems (Albrecht,

1992). As shown in Table 18.2, nearly one out of four people in the United States (23 percent) has a "chronic health condition which, given the physical, attitudinal, and financial barriers built into the social system, makes it difficult to perform one or more activities generally considered appropriate for persons of their age" (Weitz, 2004).

Can a person in a wheelchair have equal access to education, employment, and housing? If public transportation is not accessible to those in wheelchairs, the answer is certainly no. As disability rights activist Mark Johnson put it, "Black people fought for the right to ride in the front of the bus. We're fighting for the right to get on the bus" (qtd. in Shapiro, 1993: 128).

Many disability rights advocates argue that persons with a disability have been kept out of the mainstream of society. They have been denied equal opportunities in education by being consigned to special education classes or special schools. For example, people who grow up deaf are often viewed as disabled; however, many members of the deaf community instead view themselves as a "linguistic minority" that is part of a unique culture (Lane, 1992; Cohen, 1994). They believe that they have been restricted from entry into schools and the work force not due to their own limitations, but by societal barriers.

Living with disabilities is a long-term process. For infants born with certain types of congenital (present at birth) problems, their disability first acquires social significance for their parents and caregivers. In a study of children with disabilities in Israel, the sociologist Meira Weiss (1994) challenged the assumption that parents automatically bond with infants, especially those born with visible disabilities. She found that an infant's appearance may determine how parents will view the child. Parents are more likely to be bothered by external, openly visible disabilities than by internal or disguised ones; some parents are more willing to consent to or even demand the death of an "appearance-impaired" child (Weiss, 1994). According to Weiss, children born with internal (concealed) disabilities are at least initially more acceptable to parents because they do not violate the parents' perceived body images of their children. Weiss's study provides insight into the social significance that people attach to congenital disabilities.

Among persons who acquire disabilities through disease or accidents later in life, the social significance of their disability can be seen in how they initially respond to their symptoms and diagnosis, how they view the immediate situation and their future, and how the illness and disability affect their lives. When confronted with a disability, most people adopt one of two strategies—avoidance or vigilance. Those who

Table 18.2 PERCENTAGE OF U.S. POPULATION WITH DISABILITIES

CHARACTERISTIC	PERCENTAGE[a]
With a disability	23.0
Severe	14.8
Not severe	8.3
Has difficulty or is unable to:	
See words and letters	3.7
Hear normal conversation	3.8
Have speech understood	1.1
Lift or carry ten pounds	7.3
Use stairs	9.5
Walk	9.4
Has difficulty or needs assistance with:	
Getting around inside the house	1.8
Getting in/out of bed or a chair	3.0
Taking a bath or shower	2.4
Dressing	1.7
Eating	0.7
Getting to or using the toilet	1.1
Has difficulty or needs assistance with:	
Going outside the home alone	4.1
Managing money and bills	2.2
Preparing meals	2.3
Doing light housework	3.1
Using the telephone	1.4

[a]Percentage of persons age 15 and older.

Source: Stern, 2001.

use the avoidance strategy deny their condition in order to maintain hopeful images of the future and elude depression; for example, some individuals refuse to participate in rehabilitation following a traumatic injury because they want to pretend that the disability does not exist. By contrast, those using the vigilance strategy actively seek knowledge and treatment so that they can respond appropriately to the changes in their bodies (Weitz, 2004).

Sociological Perspectives on Disability

How do sociologists view disability? Those using the functionalist framework often apply Parsons's sick-role model, which is referred to as the *medical model* of disability. According to the medical model, people with disabilities become, in effect, chronic patients under the supervision of doctors and other medical personnel, subject to a doctor's orders or a program's rules, and not to their own judgment (Shapiro, 1993). From this perspective, disability is deviance.

The deviance framework is also apparent in some symbolic interactionist perspectives. According to symbolic interactionists, people with a disability ex-perience *role ambiguity* because many people equate disability with deviance (Murphy et al., 1988). By labeling individuals with a disability as "deviant," other people can avoid them or treat them as outsiders. Society marginalizes people with a disability because they have lost old roles and statuses and are labeled as "disabled" persons. According to the sociologist Eliot Freidson (1965), how the people are labeled results from three factors: (1) their degree of responsibility for their impairment, (2) the apparent seriousness of their condition, and (3) the perceived legitimacy of the condition. Freidson concluded that the definitions of and expectations for people with a disability are socially constructed factors.

Finally, from a conflict perspective, persons with a disability are members of a subordinate group in conflict with persons in positions of power in the government, in the health care industry, and in the rehabilitation business, all of whom are trying to control their destinies (Albrecht, 1992). Those in positions of power have created policies and artificial barriers that keep people with disabilities in a subservient position (Asch, 1986; Hahn, 1987). Moreover, in a capitalist economy, disabilities are big business. When people with disabilities are defined as a social problem and public funds are spent to purchase goods and services for them, rehabilitation becomes a commodity that can be bought and sold by the medical–industrial complex (Albrecht, 1992). From this perspective, persons with a disability are objectified. They have an economic value as consumers of goods and services that will allegedly make them "better" people. Many persons with a disability endure the same struggle for resources faced by people of color, women, and older persons. Individuals who hold more than one of these ascribed statuses, combined with experiencing disability, are doubly or triply oppressed by capitalism.

Social Inequalities Based on Disability

People with visible disabilities are often the objects of prejudice and discrimination, which interfere with their everyday life. For example, Marylou Breslin, executive director of the Disability Rights Education and Defense Fund, was wearing a businesswoman's suit, sitting at the airport in her battery-powered wheelchair, and drinking a cup of coffee while waiting for a plane. A woman walked by and plunked a coin in the coffee cup that Breslin held in her hand, splashing coffee on Breslin's blouse (Shapiro, 1993). Why did the woman drop the coin in Breslin's cup? The answer to this question is found in stereotypes built on lack of knowledge about or exaggeration of the characteristics of people with a disability. Some stereotypes project the image that persons with disabilities are deformed individuals who may also be horrible deviants. For example, slasher movies such as the *Nightmare on Elm Street* series show a villain who was turned into a hateful, sadistic killer because of disfigurement resulting from a fire. Lighter fare such as the *Batman* movies depict villains as individuals with disabilities: the Joker, disfigured by a fall into a vat of acid, and the Penguin, born with flippers instead of arms (Shapiro, 1993). Other stereotypes show persons with disabilities as pathetic individuals to be pitied. Fund-raising activities by many charitable organizations—such as "poster child" campaigns showing a photograph of a friendly-looking child with a visible disability—sometimes contribute to this perception. Even apparently positive stereotypes become harmful to people with a disability. An example is what some disabled persons refer to as "supercrips"—people with severe disabilities who seem to excel despite the impairment and who receive widespread press coverage in the process. Disability rights advocates note that such stereotypes do not reflect the day-to-day reality of most persons with disabilities, who must struggle constantly with smaller challenges (Shapiro, 1993).

Today, many working-age persons with a disability in the United States are unemployed (see "Census Profiles: Disability and Employment Status"). Most of them believe that they could and would work if offered the opportunity. However, even when persons with a severe disability are able to find jobs, they typically earn less than persons without a disability (Yelin, 1992). On average, workers with a severe disability make 50 percent (for men) and 64 percent (for women) of what their co-workers without disabilities earn, and the gap is growing (U.S. Census Bureau, 2000). The problem has been particularly severe for African Americans and Latinos/as with disabilities. Among Latinos/as with a severe disability, only 26 percent are employed; those who work earn 80 percent of what white (non-Latino) persons with a severe disability earn (U.S. Census Bureau, 2000).

Employment, poverty, and disability are related. On the one hand, people may become economically disadvantaged as a result of chronic illness or disability. On the other hand, poor people are less likely to be educated and more likely to be malnourished and have inadequate access to health care—all of which contribute to risk of chronic illness, physical and mental disability, and the inability to participate in the labor force. In addition, the type of employment available to people with limited resources increases their chances of becoming disabled. They may work in hazardous places such as mines, factory assembly lines, and chemical plants, or in the construction industry, where the chance of becoming seriously disabled is much higher (Albrecht, 1992).

Generalizations about the relationship between disability and income are difficult to make for at least three reasons. First, most research on disability is organized around specific conditions or impairments, making it problematic to reach conclusions about how much income a person loses because of a disability (DeJong, Batavia, and Griss, 1989). Second, generalizations about entire categories of people (such as whites, African Americans, or Latinos/as) tend to be inaccurate because of differences *within* each group. For example, the sociologist Ronald Angel (1984) found significant differences in the effect of disabilities on Latinos depending on whether they were Mexican American, Puerto Rican, or Cuban American. Angel concluded that although both Mexican American and Puerto Rican men had lower rates of full-time employment and lower hourly wages relative to (non-Latino) whites, Puerto Ricans were worse off than Mexican Americans in terms of earnings and number of work hours. Third, there is the problem of determining which factor occurred first. For example, some of the problems for Puerto Ricans with disabilities are

Disability and Employment Status

One of the questions on Census 2000 asked respondents about long-lasting conditions such as a physical, mental, or emotional condition that substantially limited important basic activities. It also asked if they were working at a job or business. The answers allowed the Census Bureau to determine how many people with a disability and how many people without a disability were employed at the time the census was conducted. As you can see from the figure set forth below, for the population aged 21 to 64 years (the period during which people are most likely to be employed), less than 50 percent of persons with a disability are employed, compared with almost 80 percent of persons without a disability who are employed. Is employment among persons with a disability related primarily to their disability status, or do other facts—such as prejudice or lack of willingness to make the necessary accommodations that would allow such persons to hold a job—play a significant part in the high rate of unemployment of persons with a disability?

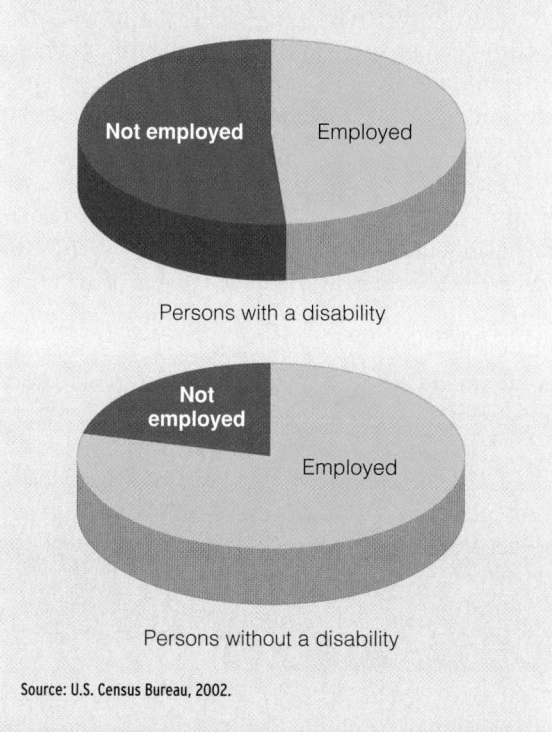

Persons with a disability

Persons without a disability

Source: U.S. Census Bureau, 2002.

tied to their overall position of economic disadvantage (see Chapter 9). Although disability is associated with lower earnings and higher rates of unemployment for both males and females, a disability has a stronger negative effect on women's labor force participation than it does on men's. Compared to men, women with disabilities are overrepresented as clerical and service workers and underrepresented as managers and administrators. Women with disabilities are also much less likely to be covered by pension and health plans than are men (Russo and Jansen, 1988).

What does it cost to "mainstream" persons with disabilities? Disability expenditures for 1986 were estimated to be $169.4 billion for the working-age population, taking into account money spent by both the government and the private sector (Albrecht, 1992). However, between 1978 and 1992, Sears, Roebuck found that the average cost of an accommodation for a worker with a disability was only $126, with some accommodations involving little or no cost (such as flexible schedules, back-support belts, rest periods, and changes in employee work stations). Other accommodations cost about $1,000, with the largest single cost being $14,500 each for special Braille computer displays for visually impaired employees (Noble, 1995). Future opportunities for persons with disabilities are closely related to current expenditures that provide greater access to education and jobs.

HEALTH CARE IN THE FUTURE

A central question regarding the future of health care is the role that advanced technologies will play in medical diagnosis and treatment. Technology is a major stimulus for social change, and the health care systems in high-income nations such as the United States reflect the rapid rate of technological innovation that has occurred in the last few decades. In the future, advanced health care technologies will no doubt provide even more accurate and quicker diagnosis, effective treatment techniques, and increased life expectancy.

However, technology alone cannot solve many of the problems confronting us in health and health care delivery. In fact, some aspects of technological innovation may be dysfunctional for individuals and society. As we have seen, some technological "advances" raise new ethical concerns, such as the moral and legal issues surrounding the cloning of human life. Some "advances" also may fail: A new prescription drug may be found to cause side-effects that are more serious than the

Joining the Fight Against Illness and Disease!

Each day trained volunteers nurture female patients in hospitals and long-term care facilities by administering complimentary makeup and moisturizer. Among the benefits to patients receiving this caring social interaction and gentle touch are higher self-esteem and shorter recovery times. (Fiffer and Fiffer, 1994: 120)

Beverly Barnes did not intend to start the program now known as Patient Pride when she went to the hospital to visit a friend who was recuperating from surgery. However, by the time she left the hospital, having helped her friend put on makeup and freshen her appearance, Barnes realized how much seemingly small acts of kindness can mean to people who are ill or recuperating from surgery. Eventually, Patient Pride, Inc., became a nonprofit corporation that relies on volunteers to visit patients, and new chapters servicing hospitals and long-term health care facilities have opened throughout the country. Recently, the program has sought to include male patients as well as women.

Patient Pride is only one of many examples of how everyday people can make a difference in the lives of people experiencing illness or disability. Here are some Internet sources that you can check out to learn more about volunteering and participating in the fight against diseases such as cancer and cardiovascular disease:

- The American Cancer Society's home page (which has links to various volunteer activities):

http://www.cancer.org

- The American Heart Association's home page:

http://www.amhrt.org

You can also learn more about the health problems discussed in this chapter. To learn more about addictive behavior, go to this site:

http://www.habitsmart.com

---

illness that it was supposed to remedy. Whether advanced technology succeeds or fails in some areas, it will probably continue to increase the cost of health care in the future. As a result, the gap between the rich and the poor in the United States will contribute to inequalities of access to vital medical services. On a global basis, new technologies may lower the death rate in some low-income countries, but it will primarily be the wealthy in those nations who will have access to the level of health care that many people in higher-income countries take for granted.

The organization of U.S. health care will continue to change in the future. Despite recent legal challenges, many managed care companies continue their cost-containment strategies and intervene in medical decisions formerly reserved for physicians and other health care professionals. Consequently, managed care and new technologies will continue to change the traditional doctor–patient relationship, even in how people communicate with one another.

Finally, to a degree, health care in the future will be up to each of us. What measures will we take to safeguard ourselves against illness and disorders? How can we help others who are the victims of acute and chronic diseases or disabilities? Although we cannot change global or national health problems, there are some small (but not insignificant) things we can do to make a difference (see Box 18.4).

CHAPTER REVIEW

■ **What is health, and why are sociologists interested in studying health and medicine?**

According to the World Health Organization, health is a state of complete physical, mental, and social well-being. In other words, health is not only a biological issue but also a social issue. For this reason, sociologists are interested in studying health and medicine. As a so-cial institution, medicine is one of the most important components of quality of life.

■ **Do nations that spend the most on health care have the healthiest citizens?**

There is not always a direct relationship between health care expenditures and people's health, particularly in terms of factors such as infant mortality and life expectancy. For example, on average the United States spends more per person on health care than Sweden, but that country has a lower infant mortality rate and greater life expectancy than the United States.

■ How do acute and chronic diseases differ, and which is most closely linked to spiraling health care costs?

Acute diseases are illnesses that strike suddenly and cause dramatic incapacitation and sometimes death. Chronic diseases, such as arthritis, diabetes, and heart disease, are long-term or lifelong illnesses that develop gradually or are present at birth. Treatment of chronic diseases is typically more costly because of the duration of these problems.

■ What are some lifestyle factors that contribute to illness?

Social epidemiologists identify a number of lifestyle factors that contribute to illness, including abuse of alcohol and other drugs, use of nicotine (tobacco), and abuse of illegal drugs such as marijuana and cocaine. Other lifestyle choices contribute to contracting sexually transmitted diseases, including HIV/AIDS.

■ How is health care paid for in the United States?

Throughout most of the past hundred years, medical care in the United States has been paid for on a fee-for-service basis. The term *fee for service* means that patients are billed individually for each service they receive. This approach to paying for medical services is expensive because few restrictions are placed on the fees that doctors, hospitals, and other medical providers can charge patients. Recently, there have been efforts at cost containment, and HMOs and managed care have produced both positive and negative results in the contemporary practice of medicine. Health maintenance organizations (HMOs) provide, for a set monthly fee, total care with an emphasis on prevention to avoid costly treatment later. Managed care refers to any system of cost containment that closely monitors and controls health care providers' decisions about medical procedures, diagnostic tests, and other services that should be provided to patients.

■ What is holistic medicine, and how does it differ from traditional health care?

Holistic medicine is an approach to health care that focuses on prevention of illness and disease and is aimed at treating the whole person—body and mind—rather than just the part or parts in which symptoms occur. Under this approach, it is important that people not look solely to medicine and doctors for their health, but rather that people engage in health-promoting behavior.

■ What is the functionalist perspective on health and illness?

According to the functionalist approach, if society is to function as a stable system, it is important for people to be healthy and to contribute to their society. Consequently, sickness is viewed as a form of deviant behavior that must be controlled by society. Sociologist Talcott Parsons (1951) described the sick role—the set of patterned expectations that defines the norms and values appropriate for individuals who are sick and for those who interact with them. Although individuals are given permission to not perform their usual activities for a period of time, they are expected to seek medical attention and get well as soon as possible so that they can go about their normal routine.

■ What is the conflict perspective on health and illness?

Conflict theory tends to emphasize the political, economic, and social forces that affect health and the health care delivery system. Among these issues are the ability of all people to obtain health care; how race, class, and gender inequalities affect health and health care; power relations between doctors and other health care workers; the dominance of the medical model of health care; and the role of profit in the health care system.

■ What is the symbolic interactionist perspective on health and illness?

In studying health, symbolic interactionists focus on the fact that the meaning that social actors give their illness or disease will affect their self-concept and their relationships with others. According to symbolic interactionists, we socially construct "health" and "illness" and how both should be treated. Symbolic interactionists also examine medicalization—the process whereby nonmedical problems become defined and treated as illnesses or disorders.

■ What is the postmodernist perspective on health and illness?

Postmodern theorists such as Michel Foucault argue that doctors and the medical establishment have gained control over illness and patients at least partly because of the physicians' clinical gaze, which replaces all other systems of knowledge. The myth of the wise doctor was also supported by the development of disease classification systems and new tests.

■ What is mental illness, and how prevalent is this condition in the United States?

Mental illness refers to a condition in which a person has a severe mental disorder requiring extensive treatment with medication, psychotherapy, and sometimes hospitalization. Researchers in one study found that among respondents between the ages of 15 and 54, nearly 50 percent had been diagnosed with a mental disorder at some time in their lives. However, severe mental illness—such as schizophrenia, bipolar affective disorder (manic-depression), and major depression—typically affects less than 15 percent of U.S. adults at some time in their lives.

■ What is a disability, and how prevalent are disabilities in the United States?

Disability is a physical or health condition that stigmatizes or causes discrimination. An estimated 48 million persons in the United States have one or more physical or mental disabilities. This number continues to increase for several reasons. First, with advances in medical technology, many people who once would have died from an accident or illness now survive, although with an impairment. Second, as more people live longer, they are more likely to experience diseases (such as arthritis) that may have disabling consequences. Third, persons born with serious disabilities are more likely to survive infancy because of medical technology. However, less than 15 percent of persons with a disability today were born with it; accidents, disease, and war account for most disabilities in this country.

KEY TERMS

acute diseases 596
chronic diseases 596
deinstitutionalization 615
demedicalization 613
disability 617
drug 597
health 592
health care 592
health maintenance organization (HMO) 604
holistic medicine 608
infant mortality rate 592
life expectancy 592
managed care 604
medical–industrial complex 611
medicalization 612
medicine 592
sick role 609
social epidemiology 594
socialized medicine 607
universal health care 606

QUESTIONS FOR CRITICAL THINKING

1. Why is it important to explain the social, as well as the biological, aspects of health and illness in societies?

2. In what ways are race, class, and gender intertwined with physical and mental disorders?
3. How would functionalists, conflict theorists, and symbolic interactionists suggest that health care delivery might be improved in the United States?
4. Based on this chapter, how do you think illness and disability will be handled in the United States in the near future? Are there things that we can learn from other nations regarding the delivery of health care? Why or why not?

RESOURCES ON THE INTERNET

Chapter-Related Web Sites

The following Web sites have been selected for their relevance to the topics in this chapter. These sites are among the more stable, but please note that Web site addresses change frequently. For an updated list of chapter-related Web sites with URL links, please visit the *Sociology in Our Times* Web site (**www.wadsworth.com/KendallSIOT**).

World Health Organization (WHO)
http://www.who.int

The Web site of the World Health Organization is an incredibly rich source of information on global health issues, covering diseases, environmental hazards, health systems, and much more.

National Institutes of Health (NIH)
http://www.nih.gov

The NIH is a federal organization dedicated to the discovery of new knowledge aimed at improving the health of individuals. In addition to news and events, scientific resources, and health information, the Web site provides links to all of the NIH institutes' home pages, each of which has its own databases, FAQs, quick facts, and publications.

Office of Minority Health (OMH)
http://www.omhrc.gov

The OMH was created in 1985 by the U.S. Department of Health and Human Services to address and improve the health of minority populations. The Web site offers data and statistics, information on health disparities, a resource center, Web links, and publications.

ONLINE STUDY AND RESEARCH TOOLS

Accompanying this text are many *free* powerful online study tools that will help you master the material in this chapter, help increase your depth of understanding, and help you make the grade!

SocCoach CD-ROM

Use the SocCoach CD-ROM enclosed with this text to help you formulate a customized study plan for this chapter. After you take the Diagnostic Quiz, SocCoach will generate a customized study plan just for you! It will identify sections of the chapter that you should review and will provide videos, charts, graphs, and excerpts from the text to supplement your studies and enhance your understanding. You'll also find fun, interactive activities such as Virtual Explorations and Map the Stats to apply what you've learned and stretch your sociological imagination.

The Companion Web Site for Sociology in Our Times, *Fifth Edition*

www.wadsworth.com/KendallSIOT

Gain an even better grasp on this chapter by going to the companion Web site to take one of the Tutorial Quizzes, use the Flash Cards to master key terms, or check out the many other study aids you'll find there. You'll also find special features such as GSS Data and Census 2000 information that'll put data and resources at your fingertips to help you with that special project or help you as you do some research on your own.

In this chapter, when you see the icon on the left, it alerts you to a specific exercise found in *Wadsworth's Sociology Online Resources and Writing Companion.* This valuable guide shows you how to use Wadsworth's exclusive online resources—*InfoTrac College Edition,* the *Opposing Viewpoints Resource Center,* and *MicroCase Online*—to assist you in your study of sociology and to build essential research and writing skills.

Population and Urbanization

In another half hour of walking, I'll arrive at the highway where I'll catch a bus to take me to Oaxaca City. From there another bus will carry me to Mexico City, then yet another will take me to Nuevo Laredo, on the border. My plan is to go to the United States as a *mojado,* or wetback.

It didn't take a lot of thinking for me to decide to make this trip. It was a matter of following the tradition of the village. . . . For several decades, Macuiltiangus—that's the name of my village—has been an emigrant village, and our people have spread out like the roots of a tree under the earth, looking for sustenance. My people have had to emigrate to survive. First, they went to Oaxaca City, then to Mexico City, and for the past thirty years up to the present, the compass has always pointed to the United States.

I, too, joined the emigrant stream. For a year I worked in Mexico City as a night watchman in a parking garage. I earned the minimum wage and could barely pay living expenses. A lot of the time I had to resort to severe diets and other limitations, just to pay the rent on the apartment where I lived, so that one day I wouldn't come home and find that the owner had put my belongings outside.

After that year, I quit as a night watchman and came back to work at my father's side in the little carpentry shop that supplies the village with simple items of furniture. During the years when I worked at carpentry, I noticed that going to the United States was a routine of village people. People went so often that it was like they were visiting a nearby city. I'd see them leave and come home as changed people. The trips erased for a while the lines that the sun, the wind and the dust put in a peasant's skin. People came home with good haircuts, good clothes, and most of all, they brought dollars in their pockets. . . . Today when somebody says "I'm going to *Los,*" everybody understands that he's referring to Los Angeles, California, the most common destination of us villagers.

But I'm not going to Los Angeles, at least not now. This time, I want to try my luck in the state of Texas, specifically, in Houston, where a friend of mine has been living for several years. He's lent me money for the trip.

–Ramón "Tianguis" Pérez, explaining why he decided to become an undocumented immigrant worker in the United States (Pérez, 1991: 12-14)

© A. Ramey/PhotoEdit

■ Many undocumented workers seek to enter the United States so that they can work and obtain a better quality of life for themselves and their families. However, the aspirations of these workers often run contrary to the mandate of law enforcement officials to patrol the border and keep out individuals who seek to enter the country illegally.

Pérez is one of the many migrants to the United States who come from Mexico. It is estimated that 20 percent of legal migrants and slightly over 50 percent of undocumented migrants to the United States come from—or at least through—Mexico (U.S. Immigration and Naturalization Service, 2000).

When many people think about population growth, they primarily focus on the number of births minus the number of deaths in a specific region or in the world. However, international population flows—such as Pérez describes in talking about his decision to migrate to the United States—have significantly changed the size and composition of cities and rural areas in many nations, particularly those that are identified as high-income, industrialized countries. In recent years, new technologies such as the Internet and cell telephones have provided those who wish to migrate with an expanded network of friends and contacts that makes it possible for them to plan for either temporary or permanent immigration to countries such as the United States. Sociologists are among the scholars who study migration and its effects on human population and on the urban areas of the world.

In this chapter, we explore the dynamics of migration as it affects population growth and urban change in societies. Before reading on, test your knowledge about issues regarding migration by taking the quiz in Box 19.1.

QUESTIONS AND ISSUES

Chapter Focus Question: What effect does migration have on cities and on shifts in the global population?

How are people affected by population changes?

How do ecological/functionalist models and political economy/conflict models differ in their explanations of urban growth?

What is meant by the experience of urban life, and how do sociologists seek to explain this experience?

What are the best-case and worst-case scenarios regarding population and urban growth in the twenty-first century, and how might some of the worst-case scenarios be averted?

DEMOGRAPHY: THE STUDY OF POPULATION

Although population growth has slowed in the United States, the world's population of 6.3 billion in 2003 is increasing by more than 76 million people per year as a result of the larger number of births than deaths worldwide (see Figure 19.1). Between 2000 and 2030, almost all of the world's 1.4-percent annual population growth will occur in low-income countries in Africa, Asia, and Latin America (Population Reference Bureau, 2001).

What causes the population to grow rapidly in some nations? This question is of interest to scholars who specialize in the study of *demography*—**a subfield of sociology that examines population size, composition, and distribution.** Many sociological studies use demographic analysis as a component of the research design because all aspects of social life are affected by demography. For example, an important relationship exists between population size and the availability of food, water, energy, and housing. Population size, composition, and distribution are also connected to issues such as poverty, racial and ethnic diversity, shifts in the age structure of society, and concerns about environmental degradation (Weeks, 2002).

Increases or decreases in population can have a powerful impact on the social, economic, and political structures of societies. As used by demographers, a *population* is a group of people who live in a specified geographic area. Changes in populations occur as a result of three processes: fertility (births), mortality (deaths), and migration.

Fertility

Fertility **is the actual level of childbearing for an individual or a population.** The level of fertility in a society is based on biological and social factors, the primary biological factor being the number of women of childbearing age (usually between ages 15 and 45). Other biological factors affecting fertility include the general health and level of nutrition of women of childbearing age. Social factors influencing the level of fertility include the roles available to women in a society and prevalent viewpoints regarding what constitutes the "ideal" family size.

Based on biological capability alone, most women could produce twenty or more children during their childbearing years. *Fecundity* is the potential number of children who could be born if every woman reproduced at her maximum biological capacity. Fertility rates are not as high as fecundity rates because people's biological capabilities are limited by social factors such as practicing voluntary abstinence and refraining from sexual intercourse until an older age, as well as by contraception, voluntary sterilization, abortion, and infanticide. Additional social factors

Box 19.1 SOCIOLOGY AND EVERYDAY LIFE

How Much Do You Know About Migration?

True	False	
T	F	1. Adults in their thirties and forties are more likely to migrate from one region or country to another than are younger individuals.
T	F	2. Migration adds more people to a population than just the initial number of individuals who move into an area.
T	F	3. The United States is one of only a few nations that have "guest worker" programs or work-related visas that allow individuals to legally enter the country for a period of time to work and then return to their country of origin.
T	F	4. The number of people living outside their native country is lower today than it has been in the past.
T	F	5. Over the past 200 years, the majority of migrants to the United States were from Europe.
T	F	6. With regard to intentional migration, most people move to escape political or religious persecution.
T	F	7. Census 2000, conducted by the U.S. government, made little effort to count those people residing in this country without official documentation.
T	F	8. Trends in migration within the United States have shown a decidedly westward movement over the past fifty years.

Answers on page 631.

affecting fertility include significant changes in the number of available partners for sex and/or marriage (as a result of war, for example), increases in the number of women of childbearing age in the work force, and high rates of unemployment. In some countries, governmental policies also affect the fertility rate. For example, China's two-decades-old policy of allowing only one child per family in order to limit population growth will result in that country's population starting to decline in 2042, according to United Nations projections (Beech, 2001). Over time, China will be faced with a declining fertility rate combined with a rapidly aging population, many of whom will be dependent on younger people for their care and economic contributions.

The most basic measure of fertility is the **crude birth rate—the number of live births per 1,000 people in a population in a given year.** In 2000 the crude birth rate in the United States was 14.2 per 1,000, as compared with an all-time high rate of 27 per 1,000 in 1947 (following World War II). This measure is referred to as a "crude" birth rate because it is based on the entire population and is not "refined" to incorporate significant variables affecting fertility, such as age, marital status, religion, and race/ethnicity.

In most areas of the world, women are having fewer children. Women who have six or seven children tend to live in agricultural regions of the world, where children's labor is essential to the family's economic survival and child mortality rates are very high.

Betty Press/Woodfin Camp & Associates

Women tend to have more children in agricultural regions of the world, such as Kenya, where children's labor is essential to the family's economic survival and child mortality rates are very high.

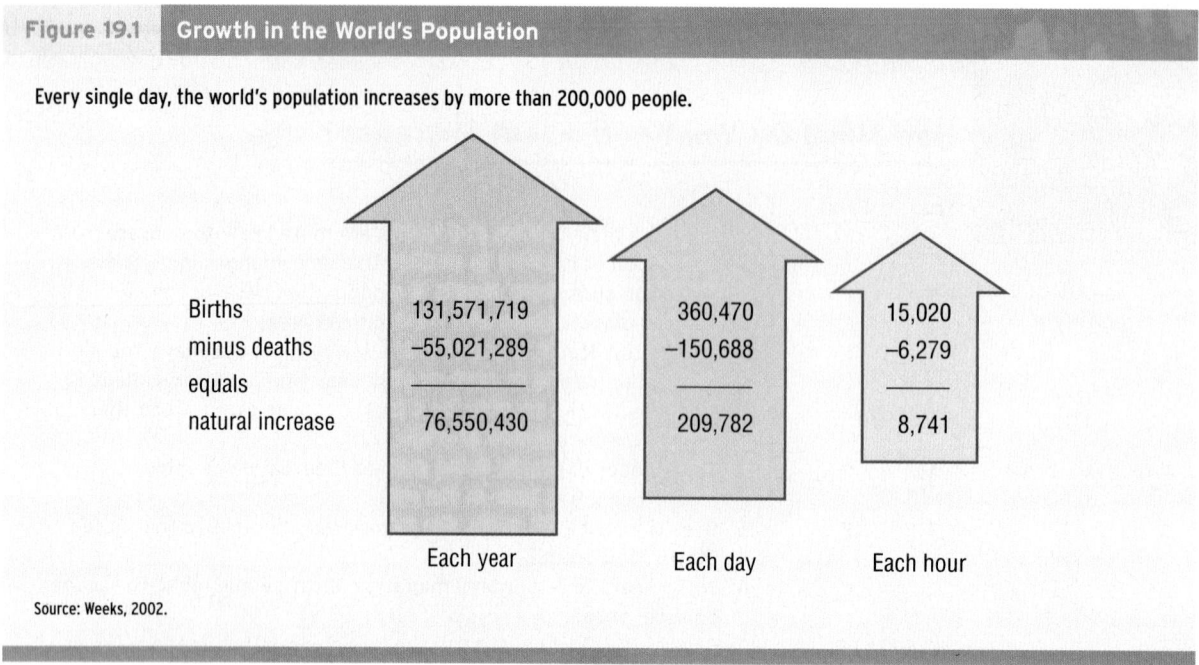

Figure 19.1   Growth in the World's Population

Every single day, the world's population increases by more than 200,000 people.

	Each year	Each day	Each hour
Births	131,571,719	360,470	15,020
minus deaths	−55,021,289	−150,688	−6,279
equals			
natural increase	76,550,430	209,782	8,741

Source: Weeks, 2002.

For example, Uganda has a crude birth rate of 47.5 per 1,000, as compared with 14.2 per 1,000 in the United States (U.S. Census Bureau, 2002). However, in Uganda and some other African nations, families need to have many children in order to ensure that one or two will live to adulthood due to high rates of poverty, malnutrition, and disease.

Mortality

The primary cause of world population growth in recent years has been a decline in **mortality—the incidence of death in a population.** The simplest measure of mortality is the **crude death rate—the number of deaths per 1,000 people in a population in a given year.** In 2001 the U.S. crude death rate was 8.7 per 1,000 (U.S. Census Bureau, 2002). In high-income, developed nations, mortality rates have declined dramatically as diseases such as malaria, polio, cholera, tetanus, typhoid, and measles have been virtually eliminated by vaccinations and improved sanitation and personal hygiene (Weeks, 2002). Just as smallpox appeared to be eradicated, however, HIV/AIDS rapidly rose to surpass the 30-percent fatality rate of smallpox. (The ten leading causes of death are shown in Table 19.1.) In low-income, less-developed nations, infectious diseases remain the leading cause of death; in some areas, mortality rates are increasing rapidly as a result of HIV/AIDS. Children under age fifteen constitute a growing number of those who are infected with HIV.

But many children do not survive long enough to contract communicable diseases. On a global basis, large numbers of newborn infants do not live to see their first birthday. The measure of these deaths is referred to as the *infant mortality rate,* which is defined in Chapter 18 as the number of deaths of infants under 1 year of age per 1,000 live births in a given year. The infant mortality rate is an important reflection of a society's level of preventive (prenatal) medical care, maternal nutrition, childbirth procedures, and neonatal care for infants. Differential levels of access to these services are reflected in the divergent infant mortality rates for African Americans and whites. In 1999, for example, the U.S. mortality rate for white infants was 5.8 per 1,000 live births, as compared with 14.6 per 1,000 live births for African American infants (U.S. Census Bureau, 2002).

As discussed in Chapter 18, *life expectancy* is an estimate of the average lifetime in years of people born in a specific year. For persons born in the United States in 2000, for example, life expectancy at birth was 76.9 years, as compared with 80.2 years in Japan and 50 years or less in the African nations of Congo, Ethiopia, Kenya, and Uganda. Life expectancy varies by sex; for instance, females born in the United States in 2000 could expect to live about 79.5 years as compared with 74.1 years for males. Life expectancy also varies by race; for example, African American men have a life expectancy at birth of about 68.2 years, compared to 74.8 years for white males (U.S. Census Bureau, 2002).

Box 19.1 SOCIOLOGY AND EVERYDAY LIFE

Answers to the Sociology Quiz on Migration

1. **False.** Young adults are more likely to be migrants than are older individuals. The peak ages for migration are between twenty and twenty-nine years of age.

2. **True.** Over time, migration to a specific area contributes more than just the initial number of immigrants who move into the area. Children and grandchildren born to the immigrant population add several times the original number to the population base.

3. **False.** Many nations, including England, France, Germany, and the Netherlands, as well as countries in the Middle East and in Africa, have guest worker programs of varying kinds.

4. **False.** Intentional migration reached an all-time high in terms of absolute numbers at the beginning of the twenty-first century. It is estimated that about 145 million people lived outside their country of origin in the mid-1990s and that this number increases by anywhere from two to four million people annually.

5. **True.** Although most contemporary U.S. immigrants are from Latin America and Asia—with only 10 percent from Europe—the majority of immigrants over the past 200 years came from Europe. In the early decades of the twentieth century, more than 90 percent of immigrants to the United States were from Europe.

6. **False.** Most people move for economic reasons; however, political oppression and religious persecution are also factors in why some people leave certain countries.

7. **False.** Many organized efforts were made by officials conducting the census enumeration to locate people residing in this country, whether or not those people possessed valid visas or citizenship papers.

8. **True.** Migration within the United States has shown a decidedly westward movement over the past few decades as the share of the U.S. population living in the Northeast fell from 26 percent in 1950 to 19 percent in 2000.

Sources: Based on Population Research Bureau, 2001; Schaechter, 2001; and Weeks, 2002.

Table 19.1 THE TEN LEADING CAUSES OF DEATH IN THE UNITED STATES, 1900 AND 2000

CAUSE OF DEATH—1900	RANK	CAUSE OF DEATH—2000
Influenza/pneumonia	1	Heart disease
Tuberculosis	2	Cancer
Stomach/intestinal disease	3	Stroke
Heart disease	4	Chronic lung disease
Cerebral hemorrhage	5	Accidents
Kidney disease	6	Pneumonia and influenza
Accidents	7	Diabetes
Cancer	8	HIV
Diseases in early infancy	9	Suicide
Diptheria	10	Homicide

Sources: Hostetler, 1994; Cockerham, 1995; and U.S. Census Bureau, 2002.

Political unrest, violence, and war are "push" factors that encourage people to leave their country of origin. By contrast, job opportunities—such as agricultural work in the United States—are a major "pull" factor for people from low-income nations.

Migration

Migration **is the movement of people from one geographic area to another for the purpose of changing residency.** Migration affects the size and distribution of the population in a given area. *Distribution* refers to the physical location of people throughout a geographic area. In the United States, people are not evenly distributed throughout the country; many of us live in densely populated areas. *Density* is the number of people living in a specific geographic area. In urbanized areas, density may be measured by the number of people who live per room, per block, or per square mile.

Migration may be either international (movement between two nations) or internal (movement within national boundaries). Internal migration has occurred throughout U.S. history and has significantly changed the distribution of the population over time. In the late nineteenth and early twentieth centuries, a major population shift occurred as thousands of people moved from rural to urban areas. In 1800, about 5 percent of the population resided in urban areas; by 2000, almost 80 percent did so (U.S. Census Bureau, 2002).

Migration involves two types of movement: immigration and emigration. *Immigration* is the movement of people into a geographic area to take up residency. Each year, about 1 million people enter the United States, primarily from Latin America and Asia; however, immigration rates are not an accurate reflection of the actual number of immigrants who enter a country. The U.S. Immigration and Naturalization Service records only legal immigration based on entry visas and change-of-immigration-status forms. Similarly, few records are maintained regarding *emigration*—the movement of people out of a geographic area to take up residency elsewhere. To determine the net migration in a geographic area, the number of people leaving that area to take up permanent or semipermanent residence elsewhere (emigrants) is subtracted from the number of people entering that area to take up residence there (immigrants), unless more people are moving out of the area than into it, in which case the mathematical process is reversed.

People migrate either voluntarily or involuntarily. *Pull* factors at the international level, such as a democratic government, religious freedom, employment opportunities, or a more temperate climate, may draw voluntary immigrants into a nation. Within nations, people from large cities may be pulled to rural areas by lower crime rates, more space, and a lower cost of living. People such as Ramón Pérez, whose decision to migrate to the United States is described at the beginning of this chapter, are drawn by pull factors such as greater economic opportunities at their destination and are pushed by factors such as low wages and few employment opportunities in their previous place of residence. *Push* factors at the international level, such as political unrest, violence, war, famine, plagues, and natural disasters, may encourage people to leave one area and relocate elsewhere. Push factors in regional U.S. migration include unemployment, harsh weather conditions, a high cost of living, inadequate school systems, and high crime rates.

Involuntary, or forced, migration usually occurs as a result of political oppression, such as when Jews fled

Nazi Germany in the 1930s or when Afghans left their country to escape oppression there in the early 2000s. Slavery is the most striking example of involuntary migration; for example, the 10–20 million Africans forcibly transported to the Western Hemisphere prior to 1800 did not come by choice (see Chapter 10).

Population Composition

Changes in fertility, mortality, and migration affect the *population composition*—**the biological and social characteristics of a population,** including age, sex, race, marital status, education, occupation, income, and size of household.

One measure of population composition is the *sex ratio*—**the number of males for every hundred females in a given population.** A sex ratio of 100 indicates an equal number of males and females in the population. If the number is greater than 100, there are more males than females; if it is less than 100, there are more females than males. In the United States, the estimated sex ratio for 2000 was 96.3, which means there were about 96 males per 100 females. Although approximately 124 males are conceived for every 100 females, male fetuses miscarry at a higher rate. From birth to age 14, the sex ratio is 105; in the age 35–44 category, however, the ratio shifts to 98.9, and from this point on, women outnumber men. By age 65, the sex ratio is about 82.3— that is, there are 82 men for every 100 women. As "Census Profiles: Sex Ratios of the U.S. Population Compared by Race and Ethnicity" demonstrates, the ratio of males to females varies among racial and ethnic categories in addition to varying by age.

For demographers, sex and age are significant population characteristics; they are key indicators of fertility and mortality rates. The age distribution of a population has a direct bearing on the demand for schooling, health, employment, housing, and pensions. The current distribution of a population can be depicted in a *population pyramid*—**a graphic representation of the distribution of a population by sex and age.** Population pyramids are a series of bar graphs divided into five-year age cohorts; the left side of the pyramid shows the number or percentage of males in each age bracket; the right side provides the same information for females. The age/sex distribution in the United States and other high-income nations does not have the appearance of a classic pyramid, but rather is more rectangular or barrel-shaped. By contrast, low-income nations, such as Mexico and Iran, which have high fertility and mortality rates, do fit the classic population pyramid.

Sex Ratios of the U.S. Population Compared by Race and Ethnicity

The U.S. Census Bureau asks respondents to indicate the sex of everyone living in their household. Census 2000 found that the U.S. population was made up of 143.4 million females (50.9 percent of the population) and 138.1 million males (49.1 percent of the population). Using this data, the Census Bureau computed the sex ratio (which it refers to as the male-female ratio) for various classifications of people by multiplying the number of males times 100, divided by the number of females.

When the sex ratios for various racial or ethnic categories of people are compared, some pronounced differences are evident, as shown below. As you can see, the categories with the highest male-female ratios (more males than females) are Latino/a and Native Hawaiian or other Pacific Islander. To determine reasons for differences in the sex ratio, it would be important to take into account a number of factors, including the age of the people in the various categories. For example, up to age 24, the sex ratio of the U.S. population is about 105, reflecting the fact that more boys than girls are born every year and that boys continue to outnumber girls until the age 35 to 44 category, when the ratio slips to 98.9. More than 63 percent of the U.S. Latino/a population is younger than 35 years of age, compared to slightly less than 47 percent of the U.S. white population. However, more than 57 percent of the African American population in the United States is younger than age 35; if age were the only factor involved in these differences, the African American sex ratio would be between that for Latinos/as and whites. Since the African American male-female ratio is lower than that for the other categories shown, age is obviously not the only factor involved in these differences.

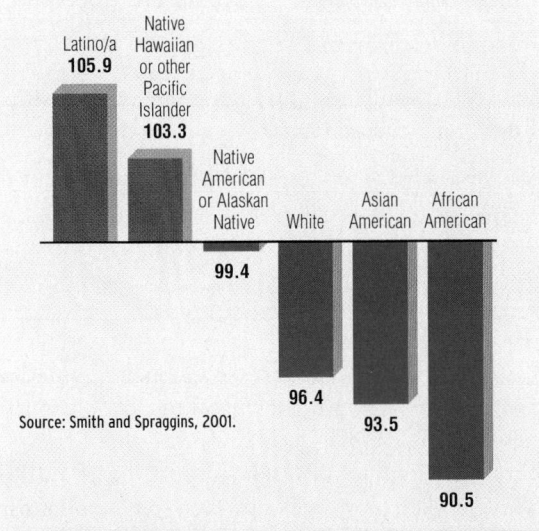

Source: Smith and Spraggins, 2001.

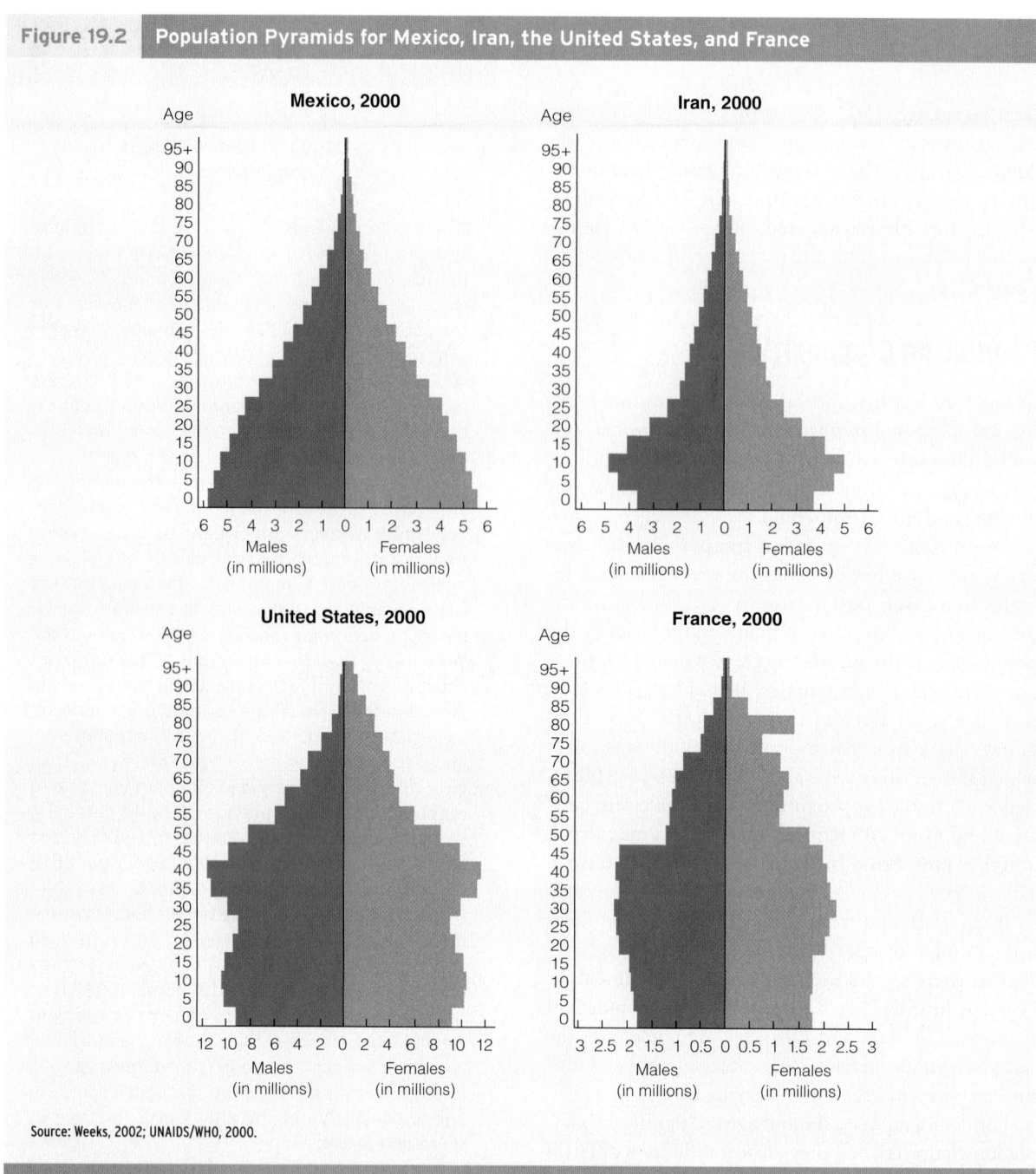

Source: Weeks, 2002; UNAIDS/WHO, 2000.

Figure 19.2 compares the demographic composition of the United States, France, Mexico, and Iran.

POPULATION GROWTH IN GLOBAL CONTEXT

What are the consequences of global population growth? Scholars do not agree on the answer to this question. Some biologists have warned that Earth is a finite ecosystem that cannot support the 10 billion people predicted by 2050; however, some economists have emphasized that free-market capitalism is capa-

ble of developing innovative ways to solve such problems. The debate is not a new one; for several centuries, strong opinions have been voiced about the effects of population growth on human welfare.

The Malthusian Perspective

English clergyman and economist Thomas Robert Malthus (1766–1834) was one of the first scholars to systematically study the effects of population. Displeased with societal changes brought about by the Industrial Revolution in England, Malthus (1965/1798: 7) anonymously published *An Essay on the Principle of Population, As It Affects the Future Improvement of*

Society, in which he argued that "the power of population is infinitely greater than the power of the earth to produce subsistence [food] for man."

According to Malthus, the population, if left unchecked, would exceed the available food supply. He argued that the population would increase in a geometric (exponential) progression (2, 4, 8, 16 . . .) while the food supply would increase only by an arithmetic progression (1, 2, 3, 4 . . .). In other words, a *doubling effect* occurs: Two parents can have four children, sixteen grandchildren, and so on, but food production increases by only one acre at a time. Thus, population growth inevitably surpasses the food supply, and the lack of food ultimately ends population growth and perhaps eliminates the existing population (Weeks, 2002). Even in a best-case scenario, overpopulation results in poverty.

However, Malthus suggested that this disaster might be averted by either positive or preventive checks on population. *Positive checks* are mortality risks such as famine, disease, and war; *preventive checks* are limits to fertility. For Malthus, the only acceptable preventive check was *moral restraint;* people should practice sexual abstinence before marriage and postpone marriage as long as possible in order to have only a few children. Although Malthus later found data disproving his model and wrote essays modifying his earlier statements, his original text is more widely cited (Keyfitz, 1994).

Malthus has had a lasting impact on the field of population studies. Most demographers refer to his dire predictions when they examine the relationship between fertility and subsistence needs (Davis, 1955). Overpopulation is still a daunting problem that capitalism and technological advances thus far have not solved, especially in middle- and low-income nations with rapidly growing populations and very limited resources.

The Marxist Perspective

Among those who attacked the ideas of Malthus were Karl Marx and Frederick Engels. According to Marx and Engels, the food supply is not threatened by overpopulation; technologically, it is possible to produce the food and other goods needed to meet the demands of a growing population. Marx and Engels viewed poverty as a consequence of exploitation of workers by the owners of the means of production. For example, they argued that England had poverty because the capitalists skimmed off some of the workers' wages as profits. The labor of the working classes was used by capitalists to earn profits, which, in turn, were used to purchase machinery that could replace the workers rather than supply food for all.

From this perspective, overpopulation occurs because capitalists desire to have a surplus of workers (an industrial reserve army) so as to suppress wages and force workers concerned about losing their livelihoods to be more productive. Marx believed that overpopulation would contribute to the eventual destruction of capitalism: Unemployment would make the workers dissatisfied, resulting in a class consciousness based on their shared oppression and the eventual overthrow of the system.

According to some contemporary economists, the greatest crisis today facing low-income nations is capital shortage, not food shortage. Through technological advances, agricultural production has reached the level at which it can meet the food needs of the world if food is distributed efficiently. Capital shortage refers to the lack of adequate money or property to maintain a business; it is a problem because the physical capital of the past no longer meets the needs of modern economic development. In the past, self-contained rural economies survived on local labor, using local materials to produce the capital needed for other laborers. For example, in a typical village a carpenter made the loom needed by the weaver to make cloth. Today, in the global economy, the one-to-one exchange between the carpenter and the weaver is lost. With an antiquated, locally made loom, the weaver cannot compete against electronically controlled, mass-produced looms. Therefore, the village must purchase capital from the outside, using its own meager financial resources. In the process, the complementary relationship between labor and capital is lost; modern technology brings with it steep costs and results in village noncompetitiveness and underemployment (see Keyfitz, 1994).

Marx and Engels made a significant contribution to the study of demography by suggesting that poverty, not overpopulation, is the most important issue with regard to food supply in a capitalist economy. Although Marx and Engels offer an interesting counterpoint to Malthus, some scholars argue that the Marxist perspective is self-limiting because it attributes the population problem solely to capitalism. In actuality, nations with socialist economies also have demographic trends similar to those in capitalist societies.

The Neo-Malthusian Perspective

More recently, *neo-Malthusians* (or "new Malthusians") have reemphasized the dangers of overpopulation. To neo-Malthusians, Earth is "a dying planet" with too many people and too little food, compounded by environmental degradation. Overpopulation and rapid population growth result in global

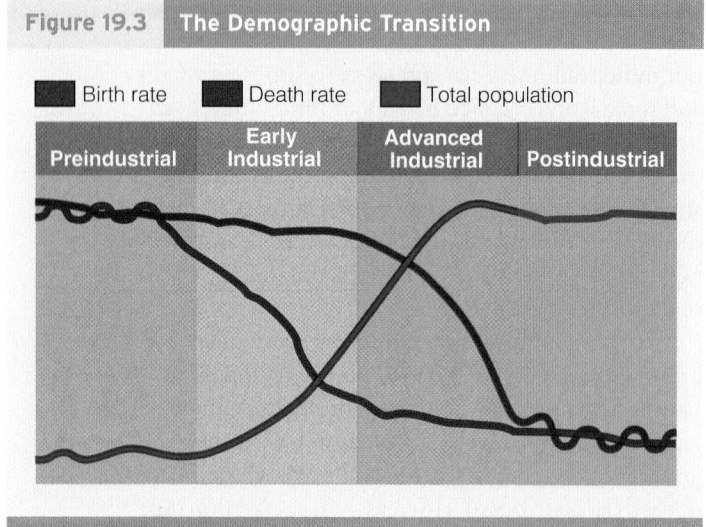

Figure 19.3 The Demographic Transition

■ Birth rate   ■ Death rate   ■ Total population

Preindustrial | Early Industrial | Advanced Industrial | Postindustrial

environmental problems, ranging from global warming and rain-forest destruction to famine and vulnerability to epidemics (Ehrlich, Ehrlich, and Daily, 1995). Unless significant changes are made, including improving the status of women, reducing racism and religious prejudice, reforming the agriculture system, and shrinking the growing gap between rich and poor, the consequences will be dire (Ehrlich, Ehrlich, and Daily, 1995).

Early neo-Malthusians published birth control handbooks, and widespread acceptance of birth control eventually reduced the connection between people's sexual conduct and fertility (Weeks, 2002). Later neo-Malthusians have encouraged people to be part of the solution to the problem of overpopulation by having only one or two children in order to bring about **zero population growth**—**the point at which no population increase occurs from year to year** because the number of births plus immigrants is equal to the number of deaths plus emigrants (Weeks, 2002).

Demographic Transition Theory

Some scholars who disagree with the neo-Malthusian viewpoint suggest that the theory of demographic transition offers a more accurate picture of future population growth. **Demographic transition is the process by which some societies have moved from high birth and death rates to relatively low birth and death rates as a result of technological development.** Demographic transition is linked to four stages of economic development (see Figure 19.3):

- *Stage 1: Preindustrial societies.* Little population growth occurs because high birth rates are offset by high death rates. Food shortages, poor

sanitation, and lack of adequate medical care contribute to high rates of infant and child mortality.
- *Stage 2: Early industrialization.* Significant population growth occurs because birth rates are relatively high whereas death rates decline. Improvements in health, sanitation, and nutrition produce a substantial decline in infant mortality rates. Overpopulation is likely to occur because more people are alive than the society has the ability to support.
- *Stage 3: Advanced industrialization and urbanization.* Very little population growth occurs because both birth rates and death rates are low. The birth rate declines as couples control their fertility through contraceptives and become less likely to adhere to religious directives against their use. Children are not viewed as an economic asset; they consume income rather than produce it. Societies in this stage attain zero population growth, but the actual number of births per year may still rise due to an increased number of women of childbearing age.
- *Stage 4: Postindustrialization.* Birth rates continue to decline as more women gain full-time employment and the cost of raising children continues to increase. The population grows very slowly, if at all, because the decrease in birth rates is coupled with a stable death rate.

Debate continues as to whether this evolutionary model accurately explains the stages of population growth in all societies. Advocates note that demographic transition theory highlights the relationship between technological development and population growth, thus making Malthus's predictions obsolete. Scholars also point out that demographic transitions occur at a faster rate in now-low-income nations than they previously did in the nations that are already developed. For example, nations in the process of development have higher birth rates and death rates than the now-developed societies did when they were going through the transition. The death rates declined in the now-developed nations as a result of internal economic development—not, as is the case today, through improved methods of disease control (Weeks, 2002). Critics suggest that this theory best explains development in Western societies.

Other Perspectives on Population Change

In recent decades, other scholars have continued to develop theories about how and why changes in population growth patterns occur. Some have studied the relationship between economic development and a decline in fertility; others have focused on the process of *secularization*—the decline in the significance of the sacred in daily life—and how a change from believing that otherworldly powers are responsible for one's life to a sense of responsibility for one's own well-being is linked to a decline in fertility. Based on this premise, some analysts argue that the processes of industrialization and economic development are typically accompanied by secularization, but the relationship between these factors is complex when it comes to changes in fertility.

Shifting from the macrolevel to the microlevel, education and social psychological factors also play into the decisions that individuals make about how many children to have. Family planning information is more readily available to people with more years of formal education and may cause them to engage in decision making in accord with *rational choice theory,* which is based on the assumption that people make decisions based on a calculated cost–benefit analysis ("What do I gain and lose from a specific action?"). In low-income countries or other settings in which children are identified as an economic resource for their parents throughout life, fertility rates are higher than in higher-income countries. However, as modernization and urbanization occur in such societies, the positive economic effects of having more children may be offset by the cost of caring for those children and the lowered economic advantage gained from having children in an industrialized nation.

As demographers have reformulated the demographic transition theory, they have highlighted additional factors that are likely to be causes of fertility decline, and they have suggested that demographic transition is not just one process, but rather a set of intertwined transitions. One is the epidemiological transition—the shift from deaths at younger ages due to acute, communicable diseases. Another is the fertility transition—the shift from natural fertility to controlled fertility, resulting in a decrease in the fertility rate. Other transitions include the migration transition, the urban transition, the age transition, and the family and household transition, which occur as a result of lower fertility, longer life, an older age structure, and predominantly urban residence.

As the demographer John R. Weeks (2002) points out, we can best understand demographic events and behavior by studying the context of global change to determine how factors such as political change, economic development, and perhaps the process of "westernization" may influence population growth and patterns of migration.

A BRIEF GLIMPSE AT INTERNATIONAL MIGRATION THEORIES

Why do people relocate from one nation to another? Several major theories have been developed in an attempt to explain international migration. The *neoclassical economic approach* assumes that migration patterns occur based on geographic differences in the supply of and demand for labor. The United States and other high-income countries that have had growing economies and a limited supply of workers for certain types of jobs have paid higher wages than are available in areas with a less-developed economy and a large labor force. As a result, people move to gain higher wages and sometimes better living conditions. They also may take jobs in other countries so that they can send money to their families in their country of origin. For example, it is estimated that Mexican workers in the United States send about half of what they earn to their families across the border, an amount that may total nearly $7 billion per year in good economic times (Ferriss, 2001).

Unlike the neoclassical explanation of migration, which focuses on individual decision making, the *new households economics of migration approach* emphasizes the part that entire families or households play in the migration process. From this approach, the previous example of Mexican workers' temporary migration to the United States would be examined not only from the perspective of the individual worker but also in terms of what the entire family gains from the process of having one or more migrant family members work in another country. By having a diversity of family income (originating from more than one source), the family is cushioned from the economic woes of the nation that most of the family members think of as "home."

Two conflict perspectives on migration add to our knowledge of why people migrate. Split-labor market theory (as previously discussed in Chapter 10) suggests that immigrants from low-income countries are often recruited for secondary labor market positions: dead-end jobs with low wages, unstable employment, and sometimes hazardous working conditions. By contrast, migrants from higher-income countries may migrate for primary-sector employment—jobs in which well-educated workers are paid high wages and receive benefits such as health insurance and a retirement plan. The global migration of some high-tech workers is an example of this process, whereas the migration of farm workers and construction helpers is an example of secondary labor market migration.

Finally, world systems theory (discussed later in this chapter) views migration as linked to the problems

An increasing proportion of the world's population lives in cities. How does this street scene in Hong Kong compare with major U.S. cities?

Paul Conklin/PhotoEdit

caused by capitalist development around the world (Massey et al., 1993). As the natural resources, land, and work force in low-income countries with little or no industrialization have come under the influence of international markets, there has been a corresponding flow of migrants from those nations to the highly industrialized, high-income countries, especially those with which the poorer nations have had the most economic, political, or military contact.

After flows of migration commence, the pattern may continue because potential migrants have personal ties with relatives and friends who now live in the country of destination and can serve as a source of stability when the potential migrants relocate to the new country. Known as *network theory*, this approach suggests that once migration has commenced, it takes on a life of its own and that the migration pattern which ensues may be different from the original *push* or *pull* factors that produced the earlier migration. Another approach, *institutional theory*, suggests that migration may be fostered by groups—such as humanitarian aid organizations relocating refugees or smugglers bringing people into a country illegally—and that the actions of these groups may produce a larger stream of migrants than would otherwise be the case.

As you can see from these diverse approaches to explaining contemporary patterns of migration, the reasons that people migrate are numerous and complex, involving processes occurring at the individual, family, and societal levels.

▌URBANIZATION IN GLOBAL PERSPECTIVE

Urban sociology is a subfield of sociology that examines social relationships and political and economic structures in the city. According to urban sociologists,

a *city* is a relatively dense and permanent settlement of people who secure their livelihood primarily through nonagricultural activities. Although cities have existed for thousands of years, only about 3 percent of the world's population lived in cities 200 years ago, as compared with almost 50 percent today. Current estimates suggest that two out of every three people around the world will live in urban areas by 2050 (United Nations, 2000). Thus, the process of urbanization continues on a global basis.

Emergence and Evolution of the City

Cities are a relatively recent innovation when compared with the length of human existence. The earliest humans are believed to have emerged anywhere from 40,000 to 1,000,000 years ago, and permanent human settlements are believed to have begun first about 8000 B.C.E. However, some scholars date the development of the first city between 3500 and 3100 B.C.E., depending largely on whether a formal writing system is considered as a requisite for city life (Sjoberg, 1965; Weeks, 2002; Flanagan, 2002).

According to the sociologist Gideon Sjoberg (1965), three preconditions must be present in order for a city to develop:

1. *A favorable physical environment,* including climate and soil favorable to the development of plant and animal life and an adequate water supply to sustain both.
2. *An advanced technology* (for that era) that could produce a social surplus in both agricultural and nonagricultural goods.
3. *A well-developed social organization,* including a power structure, in order to provide social stability to the economic system.

Based on these prerequisites, Sjoberg places the first cities in the Middle Eastern region of Mesopotamia or in areas immediately adjacent to it at about 3500

During the industrial era, people not only moved from the countryside into cities, but some people also moved from the cities to the suburbs after transportation became available to make getting from home to work and back again an easier process.

Tony Freeman/PhotoEdit

B.C.E. However, not all scholars concur; some place the earliest city in Jericho (located in present-day Jordan) at about 8000 B.C.E., with a population of about 600 people (see Kenyon, 1957).

The earliest cities were not large by today's standards. The population of the larger Mesopotamian centers was between five and ten thousand (Sjoberg, 1965). The population of ancient Babylon (probably founded around 2200 B.C.E.) may have grown as large as 50,000 people; Athens may have held 80,000 people (Weeks, 2002). Four to five thousand years ago, cities with at least 50,000 people existed in the Middle East (in what today is Iraq and Egypt) and Asia (in what today is Pakistan and China), as well as in Europe. About 3,500 years ago, cities began to reach this size in Central and South America.

Preindustrial Cities

The largest preindustrial city was Rome; by 100 C.E., it may have had a population of 650,000 (Chandler and Fox, 1974). With the fall of the Roman Empire in 476 C.E., the nature of European cities changed. Seeking protection and survival, those persons who lived in urban settings typically did so in walled cities containing no more than 25,000 people. For the next 600 years, the urban population continued to live in walled enclaves, as competing warlords battled for power and territory during the "dark ages." Slowly, as trade increased, cities began to tear down their walls.

Preindustrial cities were limited in size by a number of factors. For one thing, crowded housing conditions and a lack of adequate sewage facilities increased the hazards from plagues and fires, and death rates were high. For another, food supplies were limited. In order to generate food for each city resident, at least fifty farmers had to work in the fields (Davis,

1949), and animal power was the only means of bringing food to the city. Once foodstuffs arrived in the city, there was no effective way to preserve them. Finally, migration to the city was difficult. Many people were in serf, slave, and caste systems whereby they were bound to the land. Those able to escape such restrictions still faced several weeks of travel to reach the city, thus making it physically and financially impossible for many people to become city dwellers.

In spite of these problems, many preindustrial cities had a sense of *community*—a set of social relationships operating within given spatial boundaries or locations that provides people with a sense of identity and a feeling of belonging. The cities were full of people from all walks of life, both rich and poor, and they felt a high degree of social integration. You will recall that Ferdinand Tönnies (1940/1887) described such a community as *Gemeinschaft*—a society in which social relationships are based on personal bonds of friendship and kinship and on intergenerational stability, such that people have a commitment to the entire group and feel a sense of togetherness. By contrast, industrial cities were characterized by Tönnies as *Gesellschaft*—societies exhibiting impersonal and specialized relationships, with little long-term commitment to the group or consensus on values (see Chapter 5). In *Gesellschaft* societies, even neighbors are "strangers" who perceive that they have little in common with one another.

Industrial Cities

The Industrial Revolution changed the nature of the city. Factories sprang up rapidly as production shifted from the primary, agricultural sector to the secondary, manufacturing sector. With the advent of factories came many new employment opportunities not available to people in rural areas. Emergent technology, including new forms of transportation and agricultural production, made it easier for people to leave the countryside and move to the city. Between 1700 and 1900, the population of many European cities mushroomed. For example, the population of London increased from 550,000 to almost 6.5 million. Although the Industrial Revolution did not start in the United States until the mid-nineteenth century, the effect was similar. Between 1870 and 1910, for example, the population of New York City grew by 500 percent. In fact, New York City became the first U.S. *metropolis*—one or more central cities and their surrounding suburbs that dominate the economic and cultural life of a region. Nations, such as Japan and Russia, that became industrialized after England and the United States experienced a delayed pattern of urbanization, but this process moved quickly once it commenced in those countries.

Postindustrial Cities

Since the 1950s, postindustrial cities have emerged in nations such as the United States as their economies have gradually shifted from secondary (manufacturing) production to tertiary (service and information-processing) production. Postindustrial cities increasingly rely on an economic structure that is based on scientific knowledge rather than industrial production, and as a result, a class of professionals and technicians grows in size and influence. Postindustrial cities are dominated by "light" industry, such as software manufacturing; information-processing services, such as airline and hotel reservation services; educational complexes; medical centers; convention and entertainment centers; and retail trade centers and shopping malls. Most families do not live close to a central business district. Technological advances in communication and transportation make it possible for middle- and upper-income individuals and families to have more work options and to live greater distances from the workplace; however, these options are not often available to people of color and those at the lower end of the class structure (see Box 19.2).

On a global basis, cities such as New York, London, and Tokyo appear to fit the model of the postindustrial city (see Sassen, 2001). These cities have experienced a rapid growth in knowledge-based industries such as financial services. London, Tokyo, and New York have—at least until recently—experienced an increase in the number of highly paid professional jobs, and more workers have been in high-income categories. Many people have benefited for a number of years from these high incomes and have created a

Bill Varie/Corbis

Despite an increase in telecommuting and more diverse employment opportunities in the high-tech economy, our highways have grown increasingly congested. Can we implement measures to reduce the problems of urban congestion and environmental pollution, or will these problems grow worse with each passing year?

lifestyle that is based on materialism and the gentrification of urban spaces. Meanwhile, those persons outside the growing professional categories have seen their own quality of life further deteriorate and their job opportunities become increasingly restricted to secondary labor markets in their respective "global" cities.

PERSPECTIVES ON URBANIZATION AND THE GROWTH OF CITIES

Urban sociology follows in the tradition of early European sociological perspectives that compared social life with biological organisms or ecological processes. For example, Auguste Comte pointed out that cities are the "real organs" that make a society function. Emile Durkheim applied natural ecology to his analysis of *mechanical solidarity*, characterized by a simple

Box 19.2 CHANGING TIMES: MEDIA AND TECHNOLOGY

"Transit-Dependent" People in the High-Tech Age

At about 3 on weekday afternoons, Dorothy L. Johnson heads to her job in the suburbs cleaning office buildings. Two buses and two hours later, she reports to work. At 11 P.M., her shift ends and she begins the trip back home. A little before midnight, she steps off a suburban bus just inside the Detroit border, and waits for the city bus that will get her home by 12:30. (Meredith, 1998: A12)

Dorothy Johnson is like 75 percent of the people living in central Detroit in that she cannot afford a car and must rely on the public transit system to get to her job in the suburbs. Nowadays, most jobs that pay anywhere near a living wage are located in the suburban areas outside the central city. However, the greater Detroit bus systems contain vestiges of a dual system in the past whereby city and suburban bus routes were intentionally designed so that they would not match up, thereby making it difficult, if not impossible, for central-city residents (75 percent of whom are African American) to reach the mostly white suburbs. In contrast, white suburbanites commuted by automobile to the city for work and recreation.

Although the bus system in the greater Detroit area is somewhat better today than in the past, most central-city residents find the two-hour bus trips a significant impediment to pursuing employment opportunities in the suburbs. The same story could be told for tens of thousands of people living in central cities throughout the United States (see Bullard and Johnson, 1997).

Throughout U.S. history, the predominant mode of transportation in a specific era was a reflection of the level of technology available in the society at that time. The development of the railroad, trolley, bus, and automobile had a significant influence on how people lived their daily lives. Availability of and access to various modes of transportation are intimately connected to race, class, and politics in the United States.

Today, public transportation is a form of technology that is largely paid for by taxpayers and passengers. In the past, many cities had a two-tiered transit system, divided between private transportation (cars) and public transportation (buses). Historically, the public transit system of Los Angeles showed this division:

While most Angelenos of all races drove cars, the public transportation system ... was understood to be the avenue of last resort for the urban poor, the elderly, the disabled and students; and as the city's urban poor became increasingly Latino, black, and Asian/Pacific Islander, so did the composition of most of the bus ridership. . . . For many years, bus lines to predominantly white suburbs ... had better service, more direct express routes, and newer buses. This was justified with the argument that mass transit in the suburbs had to compete with auto use, whereas for the urban poor, most of whom were minority group members, many of them were "transit-dependent" since they could not afford to purchase a car. (Mann, 1997: 69)

Today, many analysts believe that urban transit systems are often three-tiered: private transportation (cars), public transportation in central cities (buses), and public transportation for more-affluent suburban commuters (high-tech, light-rail commuter trains). In cities such as Atlanta, where newer modes of light-rail transit have been introduced, advocates for lower-income African American residents have worked aggressively to try to narrow transportation gaps between central-city residents and suburbanites. However, according to the technology scholar Sidney Davis (1997: 96), much still remains to be done in regard to the transit-dependent across the nation:

The issue of mobility for individuals is more than just having available physical facilities. Low incomes constrain the array of choices of where people live and work. A superior road network that encourages the dispersal of activity is a disadvantage to those whose travel mode choices are limited to transit. This is especially true for black women, whose incomes typically are lower than those of their male counterparts and who are, to a significant degree, transit-dependent.

In the past, transportation development policies did not emerge in a race- and class-neutral society (Bullard and Johnson, 1997). In the future, do you think race and class will remain barriers to transportation or to other useful technologies that could benefit those people who have previously had fewer opportunities in society?

Figure 19.4   Three Models of the City

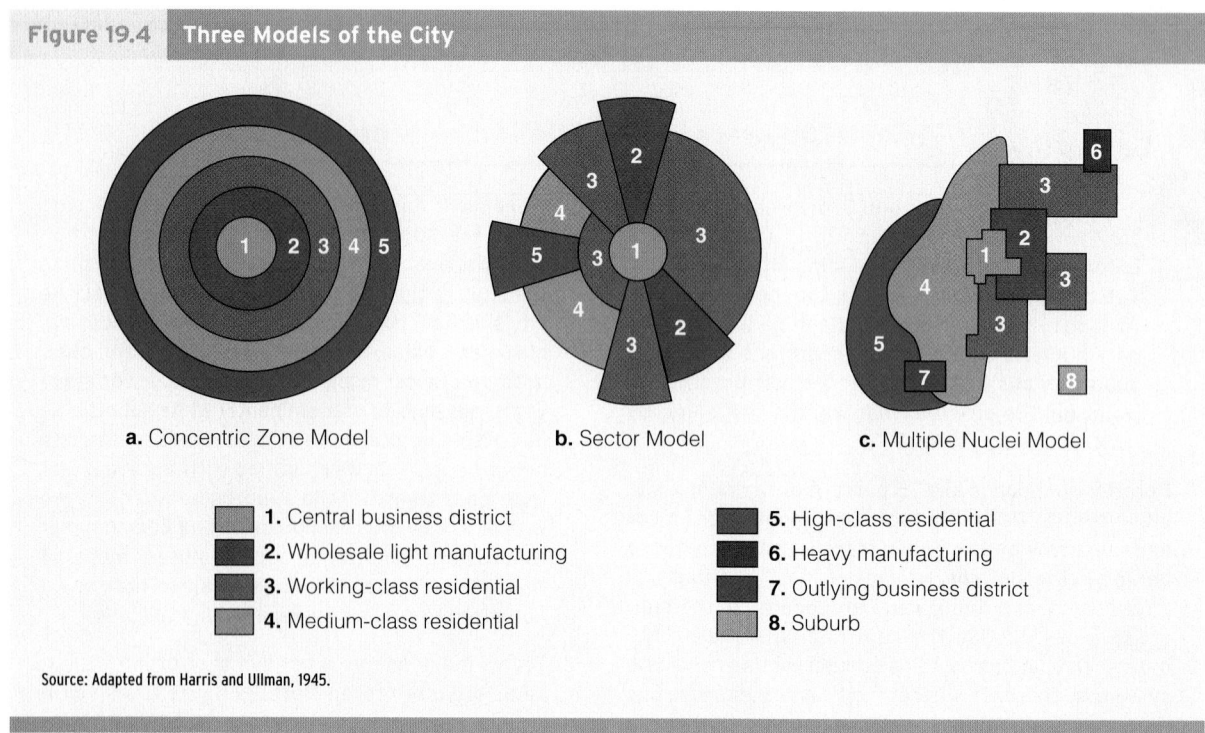

a. Concentric Zone Model   **b.** Sector Model   **c.** Multiple Nuclei Model

1. Central business district   **5.** High-class residential
2. Wholesale light manufacturing   **6.** Heavy manufacturing
3. Working-class residential   **7.** Outlying business district
4. Medium-class residential   **8.** Suburb

Source: Adapted from Harris and Ullman, 1945.

division of labor and shared religious beliefs such as are found in small, agrarian societies, and *organic solidarity,* characterized by interdependence based on the elaborate division of labor found in large, urban societies (see Chapter 5). These early analyses became the foundation for ecological models/functionalist perspectives in urban sociology.

Functionalist Perspectives: Ecological Models

Functionalists examine the interrelations among the parts that make up the whole; therefore, in studying the growth of cities, they emphasize the life cycle of urban growth. Like the social philosophers and sociologists before him, the University of Chicago sociologist Robert Park (1915) based his analysis of the city on *human ecology*—the study of the relationship between people and their physical environment. According to Park (1936), economic competition produces certain regularities in land-use patterns and population distributions. Applying Park's idea to the study of urban land-use patterns, the sociologist Ernest W. Burgess (1925) developed the concentric zone model, an ideal construct that attempted to explain why some cities expand radially from a central business core.

The Concentric Zone Model   Burgess's *concentric zone model* is a description of the process of urban

growth that views the city as a series of circular areas or zones, each characterized by a different type of land use, that developed from a central core (see Figure 19.4a). *Zone 1* is the central business district and cultural center. In *Zone 2,* houses formerly occupied by wealthy families are divided into rooms and rented to recent immigrants and poor persons; this zone also contains light manufacturing and marginal businesses (such as secondhand stores, pawnshops, and taverns). *Zone 3* contains working-class residences and shops and ethnic enclaves. *Zone 4* comprises homes for affluent families, single-family residences of white-collar workers, and shopping centers. *Zone 5* is a ring of small cities and towns populated by persons who commute to the central city to work and by wealthy people living on estates.

Two important ecological processes are involved in the concentric zone theory: invasion and succession. ***Invasion* is the process by which a new category of people or type of land use arrives in an area previously occupied by another group or type of land use** (McKenzie, 1925). For example, Burgess noted that recent immigrants and low-income individuals "invaded" Zone 2, formerly occupied by wealthy families. ***Succession* is the process by which a new category of people or type of land use gradually predominates in an area formerly dominated by another group or activity** (McKenzie, 1925). In Zone 2, for example, when some of the single-family residences were sold and subsequently divided into multiple housing units, the remaining single-family owners moved out because the "old" neighborhood had changed. As a result of their move, the process of invasion was complete and succession had occurred.

Invasion and succession theoretically operate in an outward movement: Those who are unable to "move out" of the inner rings are those without upward social mobility, so the central zone ends up being primarily occupied by the poorest residents—except when gentrification occurs. **Gentrification is the process by which members of the middle and upper-middle classes, especially whites, move into the central-city area and renovate existing properties.** Centrally located, naturally attractive areas are the most likely candidates for gentrification. To urban ecologists, gentrification is the solution to revitalizing the central city. To conflict theorists, however, gentrification creates additional hardships for the poor by depleting the amount of affordable housing available and by "pushing" them out of the area (Flanagan, 2002).

The concentric zone model demonstrates how economic and political forces play an important part in the location of groups and activities, and it shows how a large urban area can have internal differentiation (Gottdiener, 1985). However, the model is most applicable to older cities that experienced high levels of immigration early in the twentieth century and to a few midwestern cities such as St. Louis (Queen and Carpenter, 1953). No city, including Chicago (on which the model is based), entirely conforms to this model.

The Sector Model

In an attempt to examine a wider range of settings, urban ecologist Homer Hoyt (1939) studied the configuration of 142 cities. Hoyt's *sector model* emphasizes the significance of terrain and the importance of transportation routes in the layout of cities. According to Hoyt, residences of a particular type and value tend to grow outward from the center of the city in wedge-shaped sectors, with the more-expensive residential neighborhoods located along the higher ground near lakes and rivers or along certain streets that stretch in one direction or another from the downtown area (see Figure 19.4b). By contrast, industrial areas tend to be located along river valleys and railroad lines. Middle-class residential zones exist on either side of the wealthier neighborhoods. Finally, lower-class residential areas occupy the remaining space, bordering the central business area and the industrial areas. Hoyt (1939) concluded that the sector model applied to cities such as Seattle, Minneapolis, San Francisco, Charleston (South Carolina), and Richmond (Virginia).

The Multiple Nuclei Model

According to the *multiple nuclei model* developed by urban ecologists Chauncey Harris and Edward Ullman (1945), cities do not have one center from which all growth radiates, but rather have numerous centers of development based on specific urban needs or activities (see Figure 19.4c). As cities began to grow rapidly, they annexed formerly outlying and independent townships that had been communities in their own right. In addition to the central business district, other nuclei developed around entities such as an educational institution, a medical complex, or a government center. Residential neighborhoods may exist close to or far away from these nuclei. A wealthy residential enclave may be located near a high-priced shopping center, for instance, whereas less-expensive housing must locate closer to industrial and transitional areas of town. This model may be applicable to cities such as Boston. However, critics suggest that it does not provide insights about the uniformity of land-use patterns among cities and relies on an after-the-fact explanation of why certain entities are located where they are (Flanagan, 2002).

Contemporary Urban Ecology

Urban ecologist Amos Hawley (1950) revitalized the ecological tradition by linking it more closely with functionalism. According to Hawley, urban areas are complex and expanding social systems in which growth patterns are based on advances in transportation and communication. For example, commuter railways and automobiles led to the decentralization of city life and the movement of industry from the central city to the suburbs (Hawley, 1981).

Other urban ecologists have continued to refine the methodology used to study the urban environment. *Social area analysis* examines urban populations in terms of economic status, family status, and ethnic classification (Shevky and Bell, 1966). For example, middle- and upper-middle-class parents with school-aged children tend to cluster together in "social areas" with a "good" school district; young single professionals may prefer to cluster in the central city for entertainment and nightlife.

The influence of human ecology on the field of urban sociology is still very strong today (see Frisbie and Kasarda, 1988). Contemporary research on European and North American urban patterns is often based on the assumption that spatial arrangements in cities conform to a common, most efficient design (Flanagan, 2002). However, some critics have noted that ecological models do not take into account the influence of powerful political and economic elites

According to conflict theorists, exploitation by the capitalist class increasingly impoverishes poor whites and low-income minority-group members. Increasing rates of homelessness have made scenes such as this a recurring sight in many cities.

© Andrew Holbrooke/The Image Works

on the development process in urban areas (Feagin and Parker, 1990).

Conflict Perspectives: Political Economy Models

Conflict theorists argue that cities do not grow or decline by chance. Rather, they are the product of specific decisions made by members of the capitalist class and political elites. These far-reaching decisions regarding land use and urban development benefit the members of some groups at the expense of others (see Castells, 1977/1972). Karl Marx suggested that cities are the arenas in which the intertwined processes of class conflict and capital accumulation take place; class consciousness and worker revolt are more likely to develop when workers are concentrated in urban areas (Flanagan, 2002).

According to the sociologists Joe R. Feagin and Robert Parker (1990), three major themes prevail in political economy models of urban growth. First, both economic *and* political factors affect patterns of urban growth and decline. Economic factors include capitalistic investments in production, workers, workplaces, land, and buildings. Political factors include governmental protection of the right to own and dispose of privately held property as owners see fit and the role of government officials in promoting the interests of business elites and large corporations.

Second, urban space has both an exchange value and a use value. *Exchange value* refers to the profits that industrialists, developers, bankers, and others make from buying, selling, and developing land and buildings. By contrast, *use value* is the utility of space,

land, and buildings for everyday life, family life, and neighborhood life. In other words, land has purposes other than simply for generating profits—for example, for homes, open spaces, and recreational areas. Today, class conflict exists over the use of urban space, as is evident in battles over rental costs, safety, and development of large-scale projects (see Tabb and Sawers, 1984).

Third, both structure and agency are important in understanding how urban development takes place. *Structure* refers to institutions such as state bureaucracies and capital investment circuits that are involved in the urban development process. *Agency* refers to human actors, including developers, business elites, and activists protesting development, who are involved in decisions about land use.

Capitalism and Urban Growth in the United States

According to political economy models, urban growth is influenced by capital investment decisions, power and resource inequality, class and class conflict, and government subsidy programs. Members of the capitalist class choose corporate locations, decide on sites for shopping centers and factories, and spread the population that can afford to purchase homes into sprawling suburbs located exactly where the capitalists think they should be located (Feagin and Parker, 1990).

Today, a few hundred financial institutions and developers finance and construct most major and many smaller urban development projects around the country, including skyscrapers, shopping malls, and suburban housing projects. These decision makers set limits on the individual choices of the ordinary citizen with regard to real estate, just as they do with regard to other choices (Feagin and Parker, 1990). They can make housing more affordable or totally unaffordable for many people. Ultimately, their motivation rests not in benefiting the community, but rather in making a profit; the cities that they produce reflect this mindset.

One of the major results of these urban development practices is *uneven development*—the tendency of some neighborhoods, cities, or regions to grow and prosper whereas others stagnate and decline (Perry and Watkins, 1977). Conflict theorists argue that uneven development reflects inequalities of wealth and power in society. The problem not only affects areas in a state of decline but also produces external costs, even in "boom" areas, that are paid by the entire community. Among these costs are increased pollution, increased traffic congestion, and rising rates of crime and violence. According to the sociologist Mark Gottdiener (1985: 214), these costs are "intrinsic to the very core of capitalism, and those who profit the most from development are not called upon to remedy its side effects."

The Gated Community in the Capitalist Economy

The growth of *gated communities*—subdivisions or neighborhoods surrounded by barriers such as walls, fences, gates, or earth banks covered with bushes and shrubs, along with a secured entrance—is an example to many people of how developers, builders, and municipalities have encouraged an increasing division between public and private property in capitalist societies. Many gated communities are created by developers who hope to increase their profits by offering potential residents a semblance of safety, privacy, and luxury that they might not have in nongated residential areas. Other gated communities have been developed after the fact in established neighborhoods by adding walls, gates, and sometimes security guard stations. For example, a recent study noted situations in which residents of elite residential enclaves, such as the River Oaks area of Houston or the "Old Enfield" area of Austin, Texas, were able to gain approval from the city to close certain streets and create cul de sacs, or to erect other barriers to discourage or prevent outsiders from driving through the neighborhood (Kendall, 2002). Gated communities for upper-middle-class and upper-class residents convey the idea of exclusivity and privilege, whereas such communities for middle- and lower-income residents typically focus on such features as safety for children and the ability to share amenities such as a "community" swimming pool or recreational center with other residents.

Regardless of the social and economic reasons given for the development of gated communities, many analysts agree that these communities reflect a growing divide between public and private space in urban areas. According to a recent qualitative study by the anthropologist Setha Low (2003), gated communities do more than simply restrict access to the residents' homes: They also limit the use of public spaces, making it impossible for others to use the roads, parks, and open space contained within the enclosed community. Low (2003) refers to this phenomenon as the "fortressing of America."

Gender Regimes in Cities

Feminist perspectives have only recently been incorporated in urban studies (Garber and Turner, 1995). From this perspective, urbanization reflects the workings not only of the political economy but also of patriarchy. According to the sociologist Lynn M. Appleton (1995), different kinds of cities have different *gender regimes*—prevailing ideologies of how women and men should think, feel, and act; how access to social positions and control of resources should be managed; and how relationships between men and women should be conducted. The higher density and greater diversity found in central cities such as New York City serve as a challenge to the private patriarchy found in the home and workplace in lower-density, homogeneous areas such as suburbs and rural areas. *Private patriarchy* is based on a strongly gendered division of labor in the home, gender-segregated paid employment, and women's dependence on men's income. At the same time, cities may foster *public patriarchy* in the form of women's increasing dependence on paid work and the state for income and their decreasing emotional interdependence with men. At this point, gender often intersects with class and race as a form of oppression because lower-income women of color often reside in central cities. Public patriarchy may be perpetuated by cities through policies that limit women's access to paid work and public transportation. However, such cities may also be a forum for challenging patriarchy; all residents who differ in marital status, paternity, sexual orientation, class, and/or race/ethnicity tend to live close to one another and may hold a common belief that both public and private patriarchy should be eliminated (Appleton, 1995).

Symbolic Interactionist Perspectives: The Experience of City Life

Symbolic interactionists examine the *experience* of urban life. How does city life affect the people who live in a city? Some analysts answer this question positively; others are cynical about the effects of urban living on the individual.

Simmel's View of City Life According to the German sociologist Georg Simmel (1950/1902–1917), urban life is highly stimulating, and it shapes people's thoughts and actions. Urban residents are influenced by the quick pace of the city and the pervasiveness of economic relations in everyday life. Due to the intensity of urban life, people have no choice but to become somewhat insensitive to events and individuals around them. Many urban residents avoid emotional involvement with one another and try to ignore events taking place around them. Urbanites feel wary toward other people because most interactions in the city are economic rather than social. Simmel suggests that attributes such as punctuality and exactness are rewarded but that friendliness and warmth in interpersonal relations are viewed as personal weaknesses. Some people act reserved to cloak their deeper feelings of distrust or dislike toward others. However, Simmel did not view city life as completely negative; he also pointed out that urban living could have a liberating effect on people because they had opportunities for individualism and autonomy (Flanagan, 2002).

Urbanism as a Way of Life Based on Simmel's observations on social relations in the city, the early Chicago School sociologist Louis Wirth (1938) suggested that urbanism is a "way of life." *Urbanism* refers to the distinctive social and psychological patterns of life typically found in the city. According to Wirth, the size, density, and heterogeneity of urban populations typically result in an elaborate division of labor and in spatial segregation of people by race/ethnicity, social class, religion, and/or lifestyle. In the city, primary-group ties are largely replaced by secondary relationships; social interaction is fragmented, impersonal, and often superficial. Even though people gain some degree of freedom and privacy by living in the city, they pay a price for their autonomy, losing the group support and reassurance that come from primary-group ties.

From Wirth's perspective, people who live in urban areas are alienated, powerless, and lonely. A sense of community is obliterated and replaced by the "mass society"—a large-scale, highly institutionalized society in which individuality is supplanted by mass messages, faceless bureaucrats, and corporate interests.

Gans's Urban Villagers In contrast to Wirth's gloomy assessment of urban life, the sociologist Herbert Gans (1982/1962) suggested that not everyone experiences the city in the same way. Based on research in the west end of Boston in the late 1950s, Gans concluded that many residents develop strong

Robert Brenner/PhotoEdit

Festive occasions such as this New York City street fair provide opportunities for urban villagers to mingle with others, enjoying entertainment and social interaction.

loyalties and a sense of community in central-city areas that outsiders may view negatively. According to Gans, there are five major categories of adaptation among urban dwellers. *Cosmopolites* are students, artists, writers, musicians, entertainers, and professionals who choose to live in the city because they want to be close to its cultural facilities. *Unmarried people and childless couples* live in the city because they want to be close to work and entertainment. *Ethnic villagers* live in ethnically segregated neighborhoods; some are recent immigrants who feel most comfortable within their own group. The *deprived* are poor individuals with dim future prospects; they have very limited education and few, if any, other resources. The *trapped* are urban dwellers who can find no escape from the city; this group includes persons left behind by the process of invasion and succession, downwardly mobile individuals who have lost their former position in society, older persons who have nowhere else to go, and individuals addicted to alcohol or other drugs. Gans concluded that the city is a pleasure and a challenge for some urban dwellers and an urban nightmare for others.

Gender and City Life In their everyday lives, do women and men experience city life differently? According to the scholar Elizabeth Wilson (1991), some men view the city as *sexual space* in which women, based on their sexual desirability and accessibility, are categorized as prostitutes, lesbians, temptresses, or virtuous women in need of protection. Wilson suggests that more-affluent, dominant-group women are more likely to be viewed as virtuous women in need of protection by their own men or police officers. Cities offer a paradox for women: On the one hand, cities offer more freedom than is found in comparatively isolated rural, suburban, and domestic settings; on the other, women may be in greater physical danger in the city. For Wilson, the answer to women's vulnerability in the city is not found in offering protection for them, but rather in changing people's perceptions that they can treat women as sexual objects because of the impersonality of city life (Wilson, 1991).

Cities and Persons with a Disability

Chapter 18 describes how disability rights advocates believe that structural barriers create a "disabling" environment for many people, particularly in large urban settings. Many cities have made their streets and sidewalks more user-friendly for persons in wheelchairs and individuals with visual disability by constructing concrete ramps with slide-proof surfaces at intersections or installing traffic lights with sounds designating when to "Walk." However, both urban and rural areas have a long way to go before many persons with disabilities will have the access to the things they need to become productive members of the community: educational and employment opportunities. Because some persons with disabilities cannot navigate the streets and sidewalks of their communities or face obstacles getting into buildings that marginally, at best, meet the accessibility standards of the Americans with Disabilities Act, many persons with a disability are unemployed.

Political scientist Harlan Hahn (1997: 177–178) traces the problem of lack of access to the beginnings of industrialism:

The rise of industrialism produced extensive changes in the lives of disabled as well as non-disabled people. As factories replaced private dwellings as the primary sites of production, routines and architectural configurations were standardized to suit nondisabled workers. Both the design of worksites and of the products that were manufactured gave virtually no attention to the needs of people with disabilities. As a result, patterns of aversion and avoidance toward disabled persons were embedded in the construction of commodities, landscapes, and buildings that would remain for centuries. . . .

The social and economic changes fostered by industrialization may have been exacerbated by the accompanying process of urbanization. As

workers increasingly moved from farms and rural villages to live near the institutions of mass production, the character of community life appeared to shift perceptibly. Deviant or atypical personal characteristics that may have gradually become familiar in a small community seemed bizarre or disturbing in an urban milieu.

As Hahn's statement suggests, historical patterns in the dynamics of industrial capitalism contributed to discrimination against persons with disabilities, and this legacy remains evident in contemporary cities. Structural barriers are further intensified when other people do not respond favorably toward persons with disabilities. Scholar and disability rights advocate Sally French (1999: 25–26), who is visually disabled, describes her own experience:

I have lived in the same house for 16 years and yet I cannot recognize my neighbors. I know nothing about them at all; which children belong to whom, who has come and gone, who is old or young, ill or well, black or white. . . . On moving to my present house I informed several neighbors that, because of my inability to recognize them, I would doubtless pass them by in the street without greeting them. One neighbor, who had previously seen me striding confidently down the road, refused to believe me, but the others said they understood and would talk to me if our paths crossed. For the first couple of weeks it worked and I was surprised how often we met, but after that their greetings rapidly decreased and then ceased altogether. Why this happened I am not sure, but I suspect that my lack of recognition strained the interaction and limited the social reward they received from the encounter.

Concept Table 19.A examines the multiple perspectives on urban growth and urban living.

PROBLEMS IN GLOBAL CITIES

As we have seen, although people have lived in cities for thousands of years, the time is rapidly approaching when more people worldwide will live in or near a city than live in a rural area. In the middle-income and low-income regions of the world, Latin America is becoming the most urbanized: Four megacities—Mexico City (18 million), Buenos Aires (12 million), Lima (7 million), and Santiago (5 million)—already contain more than half of the region's population and continue

FUNCTIONALIST PERSPECTIVES: ECOLOGICAL MODELS	Concentric zone model	Due to invasion, succession, and gentrification, cities are a series of circular zones, each characterized by a particular land use.
	Sector model	Cities consist of wedge-shaped sectors, based on terrain and transportation routes, with the most-expensive areas occupying the best terrain.
	Multiple nuclei model	Cities have more than one center of development, based on specific needs and activities.
CONFLICT PERSPECTIVES: POLITICAL ECONOMY MODELS	Capitalism and urban growth	Members of the capitalist class choose locations for skyscrapers and housing projects, limiting individual choices by others.
	Gender regimes in cities	Different cities have different prevailing ideologies regarding access to social positions and resources for men and women.
	Global patterns of growth	Capital investment decisions by core nations result in uneven growth in peripheral and semiperipheral nations.
SYMBOLIC INTERACTIONIST PERSPECTIVES: THE EXPERIENCE OF CITY LIFE	Simmel's view of city life	Due to the intensity of city life, people become somewhat insensitive to individuals and events around them.
	Urbanism as a way of life	The size, density, and heterogeneity of urban population result in elaborate division of labor and space.
	Gans's urban villagers	Five categories of adaptation occur among urban dwellers, ranging from cosmopolites to trapped city dwellers.
	Gender and city life	Cities offer women a paradox: more freedom than in more isolated areas, yet greater potential danger.

to grow rapidly. By 2010, Rio de Janeiro and São Paulo are expected to have a combined population of about 40 million people living in a 350-mile-long megalopolis (Petersen, 1994). By 2015, New York City will be the only U.S. city among the world's 10 most populous (Figure 19.5 shows current populations).

Natural increases in population (higher birth rates than death rates) account for two-thirds of new urban growth, and rural-to-urban migration accounts for the rest. Some people move from rural areas to urban areas because they have been displaced from their land. Others move because they are looking for a better life. No matter what the reason, migration has caused rapid growth in cities in sub-Saharan Africa, India, Algeria, and Egypt. At the same time that the population is growing rapidly, the amount of farmland available for growing crops to feed people is decreasing. In Egypt, for example, land that was previously used for growing crops is now used for petroleum refineries, food-processing plants, and other factories (Kaplan, 1996).

Rapid global population growth in Latin America and other regions is producing a wide variety of urban problems, including overcrowding, environmental pollution, and the disappearance of farmland. In fact, many cities in middle- and low-income nations are quickly reaching the point at which food, housing, and basic public services are available to only a limited segment of the population (Crossette, 1996). With urban populations growing at a rate of 170,000 people per day, cities such as Cairo, Lagos, Dhaka, Beijing, and São Paulo are likely to soon have acute water shortages; Mexico City is already experiencing a chronic water shortage (*New York Times,* 1996).

As global urbanization has increased over the past three decades, differences in urban areas based on economic development at the national level have become apparent. Some cities in what Immanuel Wallerstein's (1984) world systems theory describes as core nations (see Chapter 8) are referred to as *global cities*—interconnected urban areas that are centers of political, economic, and cultural activity. New York, Tokyo, and London are generally considered the largest global cities. These cities are the sites of new and innovative product development and marketing, and they are often the "command posts" for the world economy (Sassen, 2001). But economic prosperity is not shared equally by all of the people in the core-nation global cities. Sometimes the living conditions of workers in low-wage service sector jobs or in assembly production jobs more closely resemble the living conditions of workers in semiperipheral nations than they resemble the conditions of middle-class workers in their own country.

Figure 19.5 The World's Ten Largest Metropolises

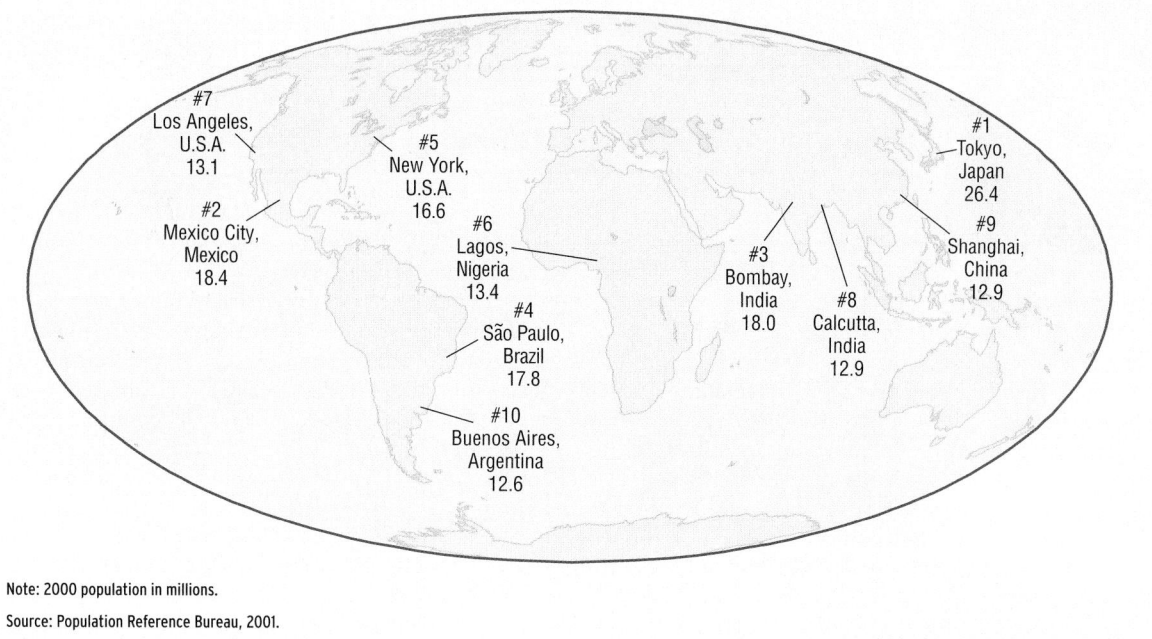

Note: 2000 population in millions.

Source: Population Reference Bureau, 2001.

Most African countries and many countries in South America and the Caribbean are *peripheral* nations, previously defined as nations that depend on core nations for capital, have little or no industrialization (other than what may be brought in by core nations), and have uneven patterns of urbanization. According to Wallerstein (1984), the wealthy in peripheral nations support the exploitation of poor workers by core nation capitalists in return for maintaining their own wealth and position. Poverty is thus perpetuated, and the problems worsen because of the unprecedented population growth in these countries.

In regard to the semiperipheral nations, only two cities are considered to be global cities: São Paulo, Brazil, the center of the Brazilian economy, and Singapore, the economic center of a multicountry region in Southeast Asia (Friedmann, 1995). Like peripheral nations, semiperipheral nations—such as India, Iran, and Mexico—are confronted with unprecedented population growth. In addition, a steady flow of rural migrants to large cities is creating enormous urban problems (see Box 19.3). What is the outlook for cities in the United States?

URBAN PROBLEMS IN THE UNITED STATES

Even the most optimistic of observers tends to agree that cities in the United States have problems brought on by years of neglect and deterioration. As we have seen in previous chapters, poverty, crime, racism, sexism, homelessness, inadequate public school systems, alcoholism and other drug abuse, gangs and guns, and other social problems are most visible and acute in urban settings. Issues of urban growth and development are intertwined with many of these problems.

Divided Interests: Cities, Suburbs, and Beyond

Since World War II, a dramatic population shift has occurred in this country as thousands of families have moved from cities to suburbs. Even though some people lived in suburban areas prior to the twentieth century, it took the involvement of the federal government and large-scale development to spur the dramatic shift that began in the 1950s (Palen, 1995). According to urban historian Kenneth T. Jackson (1985), postwar suburban growth was fueled by aggressive land developers, inexpensive real estate and construction methods, better transportation, abundant energy, government subsidies, and racial stress in the cities. However, the sociologist J. John Palen (1995) suggests that the Baby Boom following World War II and the liberalization of lending policies by federal agencies such as the Veterans Administration (VA) and the Federal Housing Authority (FHA) were significant factors in mass suburbanization.

Regardless of its causes, mass suburbanization has created a territorial division of interests between cities and suburban areas (Flanagan, 2002). Although many suburbanites rely on urban centers for their

Box 19.3 SOCIOLOGY IN GLOBAL PERSPECTIVE

Urban Migration and the "Garbage Problem"

Why do people around the world move from one location to another within their own country? Like people in the United States, individuals and families around the globe are more likely to move from rural to urban areas than vice versa. In some low-income countries, people move to cities primarily because they have been displaced from their land. However, others move in hopes of finding new opportunities and a better quality of life. No matter what the reason, rural-to-urban migration has produced rapid growth in many cities, including ones located in Latin America, sub-Saharan Africa, India, and Egypt. As the population continues to grow rapidly in these regions, the amount of farmland available for growing crops to feed people decreases.

Although rapid global population growth and strong patterns of rural-to-urban migration have produced a wide variety of urban problems—including overcrowding, environmental pollution, and the disappearance of farmland—one pressing problem in many cities around the world is the collection of household garbage, something that people in high-income countries simply take for granted. We put our garbage in the apartment dumpster or place our garbage can at curbside outside our residence, and it is picked up on the appointed day, never to be seen by us again. Obviously, we are not without garbage problems even in the United States and other high-income countries, as our landfills overflow and our city "dumps" become a source of environmental pollution and shame. However, our problems are not as large as those in urban areas in low-income countries, where some large cities have inadequate household-garbage-collection facilities. Consider the fact that less than 10 percent of the population in some cities benefits from the regular collection of household wastes. Estimates show that the proportion of garbage *not* collected by official means in Accra (Ghana) and Kampala (Uganda) is 90 percent and is 65 percent in Dar es Salaam (Tan-

zania) and 50 percent in Bogotá (Columbia) (see Middleton, 1999). Of course, these figures do not mean that garbage is not collected or "recycled" at all.

In the United States, we tend to think of recycling in terms of formal programs that encourage people to separate papers, cans, and bottles from other forms of trash so that these recyclables can be processed and used again. In low-income countries, however, the process is different. In Dar es Salaam, for example, informal scavenging at city dumps is a source of employment for many people. The scavenged resources are used as inputs to small-scale manufacturing industries that produce low-cost buckets, charcoal stoves, and lamps, which are then sold to residents of the city's squatter settlements for a lower cost than if the products had been made from imported raw materials (Middleton, 1999). This informal pattern of recycling and reuse takes place because of individual initiative and necessity rather than urban policies directed toward recycling. However, Dar es Salaam residents are not the only ones who deal with some aspects of the "garbage problem" in this manner. Mexico City and Cairo also have large squatter communities whose residents support themselves by living and working at official or unofficial rubbish dump sites.

Overall, the picture of the global garbage problem is not rosy: Uncollected garbage is a major problem in that it can be a serious fire hazard and a health hazard, attracting pests and becoming a breeding grounds for certain diseases. In Kampala, carnivorous Marabou storks live on the garbage and the rodents that are attracted to it. Today, demographers and other social analysts are concerned—just as analysts have been concerned since the days of Thomas Malthus—about rapid population growth and the patterns of migration to the largest cities of the world, many of which have already far exceeded their capacity to provide a safe urban environment for existing residents.

---

employment, entertainment, and other services, they pay their property taxes to suburban governments and school districts. Some affluent suburbs have state-of-the-art school districts, police and fire departments, libraries, and infrastructures (such as roads, sewers, and water treatment plants). By contrast, central-city services and school districts languish for lack of funds. Affluent families living in "gentrified" properties typically send their children to elite private schools, whereas the children of poor families living in racially

segregated public housing projects attend underfunded (and often substandard) public schools.

Race, Class, and Suburbs The intertwining impact of race and class is visible in the division between central cities and suburbs. About 41 percent of central-city residents are persons of color, although they constitute a substantially smaller portion of the nation's population; just 27 percent of all African Americans live in suburbs. For most African American suburban-

Affluent gated communities and enclaves of million-dollar homes stand in sharp contrast to low-income housing when we see them on the urban landscape. What sociological theories help us describe the disparity of lifestyles and life chances shown in these two settings?

ites, class is more important than race in determining one's neighbors. According to Vincent Lane, chairman of the Chicago Housing Authority, "Suburbanization isn't about race now; it's about class. Nobody wants to be around poor people, because of all the problems that go along with poor people: poor schools, unsafe streets, gangs" (qtd. in De Witt, 1994: A12).

Nationally, most suburbs are predominantly white, and many upper-middle- and upper-class suburbs remain virtually white. For example, only 5 percent of the population in northern Fulton County (adjoining Atlanta, Georgia) is African American. Likewise, in Plano (adjoining Dallas, Texas), nearly nine out of ten students in the public schools are white, whereas the majority of students in the Dallas Independent School District are African American, Latina/o, or Asian American. In the suburbs, people of color (especially African Americans) often become resegregated (see Feagin and Sikes, 1994). An example is Chicago, which remains one of the most-segregated metropolitan areas in the country in spite of its fair-housing ordinance. African Americans who have fled the high crime of Chicago's South Side primarily reside in nearby suburbs such as Country Club Hill and Chicago Heights, whereas suburban Asian Americans are most likely to live in Skokie and Naperville and suburban Latinos/as to reside in Maywood, Hillside, and Bellwood (De Witt, 1994). Similarly, suburban Latinas/os are highly concentrated in eight metropolitan areas in California, Texas, and Florida; by far, the largest such racial–ethnic concentration is found in the Los Angeles–Long Beach metropolitan area, with over 1.7 million Latinas/os. Like other groups, affluent Latinas/os live in affluent suburbs, whereas poorer Latinas/os remain segregated in less-desirable central-city areas (Palen, 1995).

Some analysts argue that the location of one's residence is a matter of personal choice. However, other analysts suggest that residential segregation reflects discriminatory practices by landlords, homeowners, and white realtors and their agents, who engage in *steering* people of color to different neighborhoods than those shown to their white counterparts. Lending practices of banks (including the *redlining* of certain properties so that acquiring a loan is virtually impossible) and the behavior of neighbors further intensify these problems (see Feagin and Sikes, 1994). In a study of suburban property taxes, the sociologist Andrew A. Beveridge found that African American homeowners are taxed more than whites on comparable homes in 58 percent of the suburban regions and 30 percent of the cities (cited in Schemo, 1994). Some analysts suggest that African Americans are more likely to move to suburbs with declining tax bases because they have limited finances, because they are steered there by real estate agents, or because white flight occurs as African American homeowners move in, leaving a heavier tax burden for the newcomers and those who remain behind. Longer-term residents may not see their property reassessed or their taxes go up for some period of time; in some cases, reassessment does not occur until the house is sold (Schemo, 1994).

Beyond the Suburbs: Edge Cities
New urban fringes (referred to as *edge cities*) have been springing up beyond central cities and suburbs in recent years (Garreau, 1991). The Massachusetts Turnpike corridor west of Boston and the Perimeter area north of Atlanta are examples. Edge cities initially develop as residential areas; then retail establishments and office parks move into the area, creating the unincorporated edge city. Commuters from the edge city are able to travel around (rather than in and out of) the metropolitan region's center and can avoid its rush-hour traffic quagmires. Edge cities may not have a governing body or correspond to municipal boundaries; however, they drain taxes from central cities and older suburbs. Many businesses and industries have moved physical plants and tax dollars to these areas; land is cheaper, and utility rates and property taxes are lower.

Lower taxes are a contributing factor to another recent development in the United States—the growth of Sunbelt cities in the southern and western states. In the 1970s, millions of people moved from the north-central and northeastern states to this area. Four reasons are generally given for this population shift: (1) more jobs and higher wages; (2) lower taxes; (3) pork-barrel programs that funneled federal money into projects in the Sunbelt, creating jobs and encouraging industry; and (4) easier transition to new industry, since most industry in the northern states was based on heavy manufacturing rather than high technology.

The Continuing Fiscal Crises of the Cities

The largest cities in the United States have faced periodic fiscal crises for many years. In the twenty-first century, these crises have been intensified by terrorist attacks that have at least temporarily reduced major sources of revenue in many urban areas. However, earlier fiscal crises can be traced to the 1980s, when federal aid to cities was cut drastically. Many cities were faced with cutting services or raising taxes (or both) at a time when the tax base was already shrinking because of suburban flight.

The fiscal crisis of the cities became quite visible in the decaying infrastructure. Old water mains threatened to burst, roads became plagued by potholes, and sections of highways and bridges crumbled. In the economic boom of the 1990s, many cities regained their economic footing as growth took place in many sectors of the marketplace and massive development and gentrification occurred. However, unequal development resulted in even greater disparity between the "good" areas of cities and those areas that were considered "less desirable." Among other things, the increase in high-priced commercial and residential property in central cities produced a corresponding increase in homelessness and the significant concentration of poor people in other areas of the cities (Sassen, 2001).

Shortly after the twenty-first century began, however, the nation's economy headed into troubled waters, a situation that was worsened by the terrorist attacks of September 11, 2001. Consider, for example, that tourism in New York City prior to those attacks was a $25-billion-per-year industry employing more than 280,000 people. The revenue received from tourism was the equivalent of $3,100 for every person living in New York City, although it was obviously not distributed that way (Herbert, 2001). Other cities, including Las Vegas, Nevada; San Francisco, California; and Orlando, Florida—all of which are top tourist destinations—experienced significant slumps in revenue as people chose to remain at home in the aftermath of the attacks. Loss of tax revenues (such as from

hotel and restaurant taxes), loss of jobs, and the need for heightened security to prevent future terrorist attacks have taken a significant toll on many cities that had been able to make some progress during the earlier periods of economic growth. Consider the effect on New York City alone, where 80,000 people became unemployed as a result of the 2001 attacks and another 75,000 people (although still employed) had their work schedules cut back or could no longer rely on tips, which often accounted for half of their income (Greenhouse, 2001). The economic effect of the terrorist attacks on cities around the world remains to be assessed in the years to come, but the immediate picture appears to be somewhat bleak.

RURAL COMMUNITY ISSUES IN THE UNITED STATES

Although most people think of the United States as highly urbanized, about 25 percent of the U.S. population resides in rural areas, identified as communities of 2,500 people or less by the U.S. Census Bureau. In the United States, incorporated towns with more than 2,500 people are considered to be urban, based on criteria established in the 1920s (Weeks, 2002). Sociologists typically identify *rural communities* as small, sparsely settled areas that have a relatively homogeneous population of people who primarily engage in agriculture (Johnson, 2000). However, rural communities today are more diverse than this definition suggests.

Unlike the standard migration patterns from rural to urban places in the past, recently more people have moved from large urban areas and suburbs into rural areas. As a result, some rural counties grew in population between the 1990 and 2000 censuses, reversing a trend that began decades earlier, when people left rural areas seeking education and job opportunities in the cities. Many of those leaving urban areas today want to escape the high cost of living, crime, traffic congestion, and environmental pollution that make daily life difficult. Technological advances make it easier for people to move to outlying rural areas and still be connected to urban centers if they need to be. The proliferation of computers, cell phones, commuter airlines, and highway systems has made previously remote areas seem much more accessible to many people. However, many recent immigrants to rural areas do not face some traditional problems experienced by long-term rural residents, particularly

Rural communities are especially hard hit by economic downturns and by the closing of long-established businesses in the area. Is it possible to revitalize small communities such as the one shown here?

farmers, small-business owners, teachers, doctors, and medical personnel in these rural communities.

For individuals in rural areas who have made their livelihood through farming and other agricultural endeavors, recent decades have been very difficult for many, both financially and emotionally. Rural crises such as droughts, crop failures, and the loss of small businesses in the community have had a negative effect on many adults and their children. Like their urban counterparts, rural families have experienced problems of divorce, alcoholism, abuse, and other crises, but these issues have sometimes been exacerbated by such events as the loss of the family farm or business (Pitzer, 2003). Since home is also the center of work in farming families, the loss of the farm may also mean the loss of family and social life, and the loss of such things dear to children as their 4-H projects—often an animal that a child raises to show and sell (Pitzer, 2003). Some rural children and adolescents are also subject to injuries associated with farm work, such as livestock kicks or crushing, falling out of a tractor or pickup, and operating machinery designed for adults, that are not typically experienced by their urban counterparts (Schutske, 2002).

Economic opportunities are limited in many rural areas, and average salaries are typically lower than in urban areas, based on the assumption that a family can live on less money in rural communities than in cities. An example is rural teachers, who earn substantially less than their urban and suburban counterparts. Recent implementation of the "No Child Left Behind" Act has

brought with it a specific set of guidelines for small and rural school districts because of the unique problems experienced in these areas. Rural schools typically face a severe teacher shortage and often pay salaries that lag far behind those in urban and suburban school districts (Rural School and Community Trust, 2003). Some rural areas have lost many teachers and administrators to higher-paying districts in other cities, and some recruitment takes place across state lines.

Although many of the problems we have examined in this book are intensified in rural areas, one of the most pressing is the availability of health services and doctors. Recently, some medical schools have established clinics and practices in outlying rural regions of the states in which they are located in an effort to increase the number of physicians available to rural residents. Typically, physicians who have just started to practice medicine have chosen to work in large urban centers with accessible high-tech medical facilities. Because of the pressing time constraints of tending to patients with life-threatening problems, such as heart attacks and strokes, the availability of community clinics and hospitals in rural areas may be a life-or-death matter for some residents. Loss of these facilities can have a devastating effect on people's health and life chances.

In addition to the movement of some urban dwellers to rural areas, two other factors have changed the face of rural America in some regions. One is the proliferation of superstores, such as Wal-Mart, PetsMart, Lowes, and Home Depot. In some cases, these superstores have effectively put small businesses such as hardware stores and pet shops out of business because local merchants cannot meet the prices established by these large-volume discount chains. The development of superstores and outlet malls along the rural highways of this country has raised new concerns about environmental issues such as air and water pollution, and has brought about new questions regarding whether these stores benefit the rural communities where they are located.

A second factor that has changed the face of some rural areas (and is sometimes related to the growth of mega-stores and outlet malls) is an increase in tourism in rural America (Brown, 2003). According to a recent study, about 87 million people (nearly two-thirds of all U.S. adults) have taken a trip to a rural destination, usually for leisure purposes, over the past few years. Tourism produces jobs; however, many of the positions are for food servers, retail clerks, and hospitality workers, which are often low-paying, seasonal jobs that have few benefits. Tourism may improve a community's tax base, but this does not occur when the outlet malls, hotels, and fast-food restaurants are

Figure 19.6 Growth of the World's Population

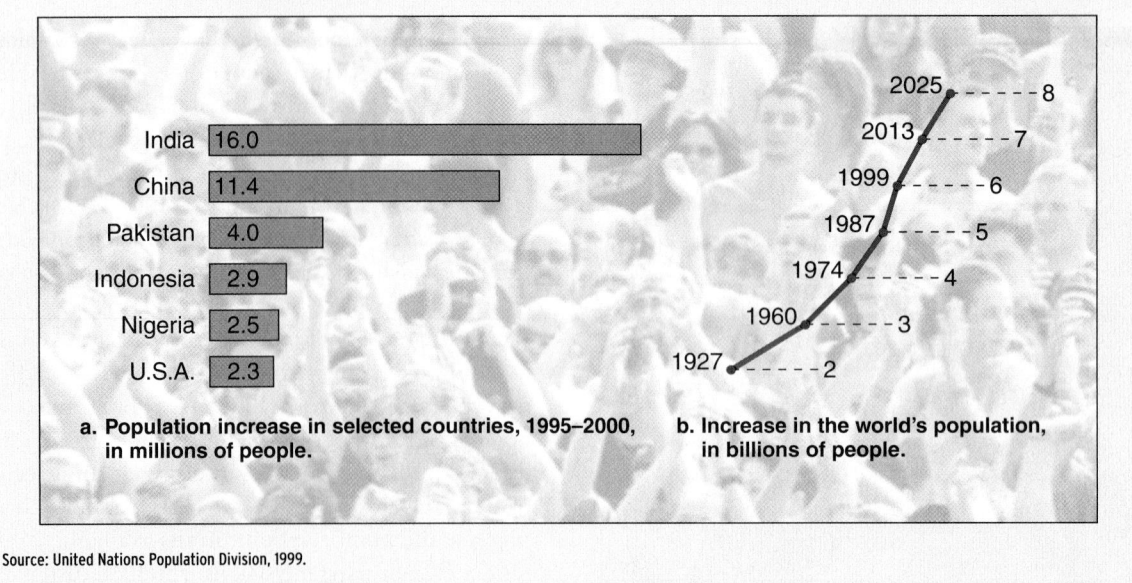

a. Population increase in selected countries, 1995–2000, in millions of people.

India	16.0
China	11.4
Pakistan	4.0
Indonesia	2.9
Nigeria	2.5
U.S.A.	2.3

b. Increase in the world's population, in billions of people.

Source: United Nations Population Division, 1999.

located outside of the rural community's taxing authority, as frequently occurs when developers decide where to locate malls and other tourist amenities.

Although rural areas throughout the world do not share all of the same problems as rural areas in the United States, some commonalities exist across regions and nations. Among these are problems of health and nutrition, concerns about educational opportunities for children, and the issue of how to provide food, housing, and other necessities of life for the diverse populations that call rural regions of the world "home." No doubt these issues will continue to be of concern for people in both rural and urban areas during the twenty-first century.

POPULATION AND URBANIZATION IN THE FUTURE

In the future, rapid global population growth is inevitable. Although death rates have declined in many low-income nations, there has not been a corresponding decrease in birth rates. Between 1985 and 2025, 93 percent of all global population growth will have occurred in Africa, Asia, and Latin America; 83 percent of the world's population will live in those regions by 2025 (Petersen, 1994). Perhaps even more amazing is the fact that in the five-year span between 1995 and 2000, 21 percent of the entire world's population increase occurred in two countries: China and India. Figure 19.6a shows the net annual additions to the populations of six countries during that five-year period. Figure 19.6b shows the growth of the world's

population from 1927 to 1999 and the expected growth to eight billion by the year 2025.

In the future, low-income countries will have an increasing number of poor people. While the world's population will *double,* the urban population will *triple* as people migrate from rural to urban areas in search of food, water, and jobs.

One of the many effects of urbanization is greater exposure of people to the media. Increasing numbers of poor people in less-developed nations will see images from the developed world that are beamed globally by news networks such as CNN. As futurist John L. Petersen (1994: 119) notes, "For the first time in history, the poor are beginning to understand how relatively poor they are compared to the rich nations. They see, in detail, how the rest of the world lives and feel their increasing disenfranchisement."

Infants born today will be teenagers in the year 2020. By then, in a worst-case scenario, central cities and nearby suburbs in the United States will have experienced bankruptcy exacerbated by sporadic race- and class-oriented violence. The infrastructure will be beyond the possibility of repair. Families and businesses with the ability to do so will have long since moved to "new cities," where they will inevitably diminish the quality of life that they originally sought there. Areas that we currently think of as being relatively free from such problems will be characterized by depletion of natural resources and greater air and water pollution (see Ehrlich, Ehrlich, and Daily, 1995). If social and environmental problems become too great in one nation, members of the capitalist class may simply move to another country.

By contrast, in a best-case scenario, the problems brought about by rapid population growth in low-

Box 19.4 YOU CAN MAKE A DIFFERENCE

Working Toward Better Communities

We've definitely touched a nerve. People believe that sprawl and traffic are out of control, and the vast majority want more open space, reliable public transit and neighborhood reinvestment. All the evidence shows that Americans support smarter growth, and our elected officials better start paying attention.

–Don Chen, director of the Smart Growth America coalition (qtd. in Smart Growth America, 2001)

According to a recent study by Smart Growth America, a new nationwide coalition of more than sixty public interest groups, many people in the United States support the concept of "smart growth" to reduce traffic congestion, preserve their communities, and protect the environment and open space. Advocates of smart growth suggest that many citizens are ahead of their public officials in wanting to find a solution to problems such as urban sprawl, and believe that it is possible for everyday people to have a voice in how cities give priority to improving services—such as schools, roads, affordable housing, and public transportation—in existing communities, rather than encouraging new housing, commercial development, and highways in the countryside.

The Smart Growth Network began in 1996, when the U.S. Environmental Protection Agency joined with several nonprofit and government organizations to address the issue of community concerns about boosting the economy, protecting the environment, and enhancing community vitality (Smart Growth Network, 2001). Among the "principles of smart growth" are the importance of providing quality housing for people of all income levels, creating walkable neighborhoods, and encouraging community collaboration on how and where a city should grow. (To see all of the principles, visit the Smart Growth Web site [**http://www.smartgrowth.org**].)

How can you make a difference regarding issues of urban planning and change? In most cities, urban planning is community based in that elected bodies such as the city council or planning commission invite "citizen participation" in their dialogues regarding community development. In addition to such official channels, many people continue to use grassroots efforts to encourage the development of adequate housing, protection of open space and the environment, and good public-health and community services.

As the urban population in the United States continues to grow, each of us has a unique opportunity to learn about the issues that will affect the quality of life and profoundly influence what kind of lives we and our neighbors throughout the entire community will be able to live. There is no single "right" answer or set of answers as to how to achieve smart growth in any particular city; rather, the citizens of each city must find their own best solutions, taking into account the existing structure of their community, the extent of likely population growth, and the resources that are available to the city's government. But how your city will handle future growth and change is something that should be important to you. Today, most large cities—and some smaller cities—have a Web site with links to other sites that discuss these issues regarding that particular city; such information can also be found in local newspapers and on city-sponsored television stations. You can make a difference in your community by learning about these issues and what solutions are being proposed, and by making sure that your opinions regarding these issues get taken into account.

WRITING IN SOCIOLOGY ASSIGNMENT

Learn about the "growth plans" of a city with which you are familiar, and write a paper dis-  cussing what you think are the strengths and weaknesses of these plans for that city.

income nations will be remedied by new technologies that make goods readily available to people. International trade agreements such as NAFTA (North American Free Trade Agreement) and GATT (General Agreement on Trade and Tariffs) are removing trade barriers and making it possible for all nations to fully engage in global trade. People in low-income nations will benefit by gaining jobs and opportunities to purchase goods at lower prices. Of course, the opposite also may occur: People may be exploited as inexpensive labor, and their country's natural resources may be depleted as transnational corporations buy up raw materials without contributing to the long-term economic stability of the nation.

In the United States, a best-case scenario for the future might include improvements in how tax revenues are collected and spent. Some analysts have suggested that regional governments must be developed. Regionalization of government would create wider service areas than existing city and county (or parish or borough) governments. Regional governments would be responsible for water, wastewater (sewage), transportation, schools, parks, hospitals, and other public services over a wider area. Revenues would be shared among central cities, affluent suburbs, and edge cities based on the assumption that everyone will benefit if the quality of life is improved throughout the region. With regard to pollution in urban areas, some futurists predict that environmental activism will increase dramatically as people see irreversible changes in the atmosphere and experience firsthand the effects of environmental hazards and pollution on their health and well-being. Environmental activism

is discussed in Chapter 20 ("Collective Behavior, Social Movements, and Social Change").

At the macrolevel, we may be able to do little about population and urbanization; however, at the microlevel, we may be able to exercise some degree of control over our communities and our lives (see Box 19.4). In both cases, futurists suggest that as we approach the future, we must "leave the old ways and invent new ones" (Petersen, 1994: 340). What aspects of our "old ways" do you think we should discard? Can you help invent new ones?

CHAPTER REVIEW

■ **What are the processes that produce population changes?**

Populations change as the result of fertility (births), mortality (deaths), and migration.

■ **What is the Malthusian perspective?**

Over two hundred years ago, Thomas Malthus warned that overpopulation would result in poverty, starvation, and other major problems that would limit the size of the population. According to Malthus, the population would increase geometrically while the food supply would increase only arithmetically, resulting in a critical food shortage and poverty.

■ **What are the views of Karl Marx and the neo-Malthusians on overpopulation?**

According to Karl Marx, poverty is the result of capitalist greed, not overpopulation. More recently, neo-Malthusians have reemphasized the dangers of overpopulation and encouraged zero population growth—the point at which no population increase occurs from year to year.

■ **What are the stages in demographic transition theory?**

Demographic transition theory links population growth to four stages of economic development: (1) the preindustrial stage, with high birth rates and death rates; (2) early industrialization, with relatively high birth rates and a decline in death rates; (3) advanced industrialization and urbanization, with low birth rates and death rates; and (4) postindustrialization, with additional decreases in the birth rate coupled with a stable death rate.

■ **How do preindustrial cities differ from industrial and postindustrial cities?**

Because of their limited size, preindustrial cities tend to provide a sense of community and a feeling of belonging. The Industrial Revolution changed the size and na-

ture of the city; people began to live close to factories and to one another, resulting in overcrowding and poor sanitation. In postindustrial cities, some people live and work in suburbs or outlying edge cities.

■ **What are the three functionalist models of urban growth?**

Functionalists view urban growth in terms of ecological models. The concentric zone model sees the city as a series of circular areas, each characterized by a different type of land use. The sector model describes urban growth in terms of terrain and transportation routes. The multiple nuclei model views cities as having numerous centers of development from which growth radiates.

■ **What is the political economy model/conflict perspective on urban growth?**

According to political economy models/conflict perspectives, urban growth is influenced by capital investment decisions, power and resource inequality, class and class conflict, and government subsidy programs. At the global level, capitalism also influences the development of cities in core, peripheral, and semiperipheral nations.

■ **How do symbolic interactionists view urban life?**

Symbolic interactionist perspectives focus on how people experience urban life. Some analysts view the urban experience positively; others believe that urban dwellers become insensitive to events and to people around them.

■ **How did the U.S. population change during the second half of the twentieth century?**

During the 1950s, a dramatic population shift occurred in the United States as people moved from cities to suburbs; almost 80 percent of the U.S. population lives in urban areas today. Edge cities have also developed beyond central cities and suburbs, first with residential areas and then with retail establishments and office parks.

KEY TERMS

crude birth rate 629
crude death rate 630
demographic transition 636
demography 628
fertility 628
gentrification 643
invasion 642
migration 632
mortality 630
population composition 633
population pyramid 633
sex ratio 633
succession 642
zero population growth 636

QUESTIONS FOR CRITICAL THINKING

1. What impact might a high rate of immigration have on culture and personal identity in the United States?
2. If you were designing a study of growth patterns for the city where you live (or one you know well), which theoretical model(s) would provide the most useful framework for your analysis?
3. What do you think everyday life in U.S. cities, suburbs, and rural areas will be like in 2020? Where would you prefer to live? What does your answer reflect about the future of U.S. cities?

RESOURCES ON THE INTERNET

Chapter-Related Web Sites

The following Web sites have been selected for their relevance to the topics in this chapter. These sites are among the more stable, but please note that Web site addresses change frequently. For an updated list of chapter-related Web sites with URL links, please visit the *Sociology in Our Times* Web site (**www.wadsworth.com/KendallSIOT**).

Population Reference Bureau (PRB)

http://www.prb.org

The PRB Web site features extensive information on population topics in the United States, including a state-by-state database. You can find reports on world and U.S. population trends, including family and work patterns, migration, urbanization, and gender.

World Resources Institute (WRI)

http://www.wri.org

WRI's Web site provides extensive information on critical environmental issues that require social policy solutions, including sustainable development and preservation of the world's resources. You can access WRI's annual report, *World Resources,* which focuses on a different topic each year.

U.S. Census Bureau: International Demographic Data

http://www.census.gov/ipc/www/idbsum.html

This Census Bureau service has been documenting global population data since 1950. The site features a drop-down menu that allows you to select detailed demographic data and population pyramids for countries around the world.

ONLINE STUDY AND RESEARCH TOOLS

Accompanying this text are many *free* powerful online study tools that will help you master the material in this chapter, help increase your depth of understanding, and help you make the grade!

SocCoach CD-ROM

Use the SocCoach CD-ROM enclosed with this text to help you formulate a customized study plan for this chapter. After you take the Diagnostic Quiz, SocCoach will generate a customized study plan just for you! It will identify sections of the chapter that you should review and will provide videos, charts, graphs, and excerpts from the text to supplement your studies and enhance your understanding. You'll also find fun, interactive activities such as Virtual Explorations and Map the Stats to apply what you've learned and stretch your sociological imagination.

The Companion Web Site for Sociology in Our Times, *Fifth Edition*
www.wadsworth.com/KendallSIOT

Gain an even better grasp on this chapter by going to the companion Web site to take one of the Tutorial Quizzes, use the Flash Cards to master key terms, or check out the many other study aids you'll find there. You'll also find special features such as GSS Data and Census 2000 information that'll put data and resources at your fingertips to help you with that special project or help you as you do some research on your own.

In this chapter, when you see the icon on the left, it alerts you to a specific exercise found in *Wadsworth's Sociology Online Resources and Writing Companion.* This valuable guide shows you how to use Wadsworth's exclusive online resources—*InfoTrac College Edition,* the *Opposing Viewpoints Resource Center,* and *MicroCase Online*—to assist you in your study of sociology and to build essential research and writing skills.

Collective Behavior, Social Movements, and Social Change

I was raised along the [Hudson] river. I was in the first generation that was taught the river was unsafe—not because of tides that might pull you down but because of water quality. As a young adult, I found a legacy I had been kept from inheriting. The lives of my family had swirled around the river; my grandfather was a fisherman; that's where families gathered. I discovered that connection. But then there was a larger connection. It seemed that every community on the river had lost touch with it and with the notion that the river was their home. The greatest single tragedy on the Hudson is that hundreds of years of history are disappearing. It's like burning down a museum or trashing a library. The loss is devastating and profound.

—John Cronin, a former commercial fisherman who was appointed Hudson Riverkeeper, describing how he became an activist in an effort to save a river he loves from environmental degradation (qtd. in Rosenblatt, 1999: 76)

To me [the movement to save the Hudson River] is a struggle of good and evil—between short-term greed and ignorance and a long-term vision of building communities that are dignified and enriching and that meet the obligations to future generations. There are two visions of America. One is that this is just a place where you make a pile for yourself and keep moving. And the other is that you put down roots and build communities that are examples to the rest of humanity.

–Robert F. Kennedy, Jr., chief prosecuting attorney for the Hudson Riverkeepers and senior attorney for the Natural Resources Defense Council, explaining why he believes it is important to crack down on those who pollute the environment (qtd. in Rosenblatt, 1999: 76)

■ The Riverkeepers and similar organizations that seek to protect rivers and lakes from environmental degradation are a good example of how social movements and the hard work of individual volunteers can improve existing conditions and bring about positive social change.

In their best-selling book *The Riverkeepers* (1999), John Cronin and Robert F. Kennedy, Jr., describe how they became environmental activists seeking social change. Sociologists define **social change as the alteration, modification, or transformation of public policy, culture, or social institutions over time;** such change is usually brought about by collective behavior and social movements. Although the early endeavors of Cronin and Kennedy to protect the Hudson River constituted collective behavior, their actions led to an environmental movement that now includes 110 other Waterkeeper groups in the United States and throughout the world. The Riverkeeper movement is an environmental neighborhood watch program, a citizens patrol to protect the nation's waters (Riverkeeper.org, 2003).

In this chapter, we will examine collective behavior, social movements, and social change from a sociological perspective. We will use environmental activism as an example of how people may use social movements as a form of mass mobilization and social transformation (Buechler, 2000). Before reading on, test your knowledge about collective behavior and environmental issues by taking the quiz in Box 20.1.

QUESTIONS AND ISSUES

Chapter Focus Question: Can collective behavior and social movements make people aware of important social issues such as environmental pollution?

What causes people to engage in collective behavior?

What are some common forms of collective behavior?

How can different types of social movements be distinguished from one another?

What draws people into social movements?

What factors contribute to social change?

COLLECTIVE BEHAVIOR

Collective behavior **is voluntary, often spontaneous activity that is engaged in by a large number of people and typically violates dominant-group norms and values.** Unlike the *organizational behavior* found in corporations and voluntary associations (such as labor unions and environmental organizations), collective behavior lacks an official division of labor, hierarchy of authority, and established rules and procedures. Unlike *institutional behavior* (in education, religion, or politics, for example), it lacks institutionalized norms to govern behavior. Collective behavior can take various forms, including crowds, mobs, riots, panics, fads, fashions, and public opinion.

According to the sociologist Steven M. Buechler (2000), early sociologists studied collective behavior because they lived in a world that was responding to the processes of modernization, including urbanization, industrialization, and proletarianization of workers. Contemporary forms of collective behavior, particularly social protests, are variations on the themes that originated during the transition from feudalism to capitalism and the rise of modernity in Europe (Buechler, 2000). Today, some forms of collective behavior and social movements are directed toward public issues such as air pollution, water pollution, and the exploitation of workers in global sweatshops by transnational corporations (see Shaw, 1999). For example, Riverkeeper, the group of environmental activists who protect the Hudson River, mounts campaigns to lobby members of Congress. The members of this organization and its volunteers also focus more widely on a number of environmental concerns on waterways in many regions.

Conditions for Collective Behavior

Collective behavior occurs as a result of some common influence or stimulus that produces a response from a collectivity. A *collectivity* is a number of people who act together and may mutually transcend, bypass, or subvert established institutional patterns and structures. Three major factors contribute to the likelihood that collective behavior will occur: (1) structural factors that increase the chances of people responding in a particular way, (2) timing, and (3) a breakdown in social control mechanisms and a corresponding feeling of normlessness (McPhail, 1991; Turner and Killian, 1993).

A common stimulus is an important factor in collective behavior. For example, the publication of *Silent Spring* (1962) by former Fish and Wildlife Service biologist Rachel Carson is credited with triggering collective behavior directed at demanding a clean environment and questioning how much power large corporations should have. Carson described the dangers of pesticides such as DDT, which was then being promoted by the chemical industry as the miracle that could give the United States the unchallenged position as food supplier to the world (Cronin and Kennedy, 1999). Carson's activism has been described in this way:

> Carson was not a wild-eyed reformer intent on bringing the industrial age to a grinding halt. She wasn't even opposed to pesticides per se. She was a careful scientist and brilliant writer whose painstaking research on pesticides proved that the "miraculous" bursts of agricultural productivity had long-term costs undisclosed in the chemical industry's exaggerated puffery. Americans were losing things—their health, many birds and fishes,

Box 20.1 SOCIOLOGY AND EVERYDAY LIFE

How Much Do You Know About Collective Behavior and Environmental Issues?

True	False	
T	F	1. The environmental movement in the United States started in the 1960s.
T	F	2. People who hold strong attitudes regarding the environment are very likely to be involved in social movements to protect the environment.
T	F	3. Environmental groups may engage in civil disobedience or use symbolic gestures to call attention to their issue.
T	F	4. Most sociologists believe that people act somewhat irrationally when they are in large crowds.
T	F	5. People are most likely to believe rumors when no other information is readily available on a topic.
T	F	6. Influencing public opinion is a very important activity for many social movements.
T	F	7. Social movements are more likely to flourish in democratic societies.
T	F	8. Most social movements in the United States seek to improve society by changing some specific aspect of the social structure.
T	F	9. Sociologists have found that people in a community respond very similarly to natural disasters and to disasters caused by technological failures.
T	F	10. People have the capacity to change the environment for better or for worse.

Answers on page 662.

and the purity of their waterways—that they should value more than modest savings at the grocery store. (Cronin and Kennedy, 1999: 151)

Timing is another significant factor in bringing about collective behavior. For example, in the 1960s smog had started staining the skies in this country; in Europe, birds and fish were dying from environmental pollution; and oil spills from tankers were provoking public outrage worldwide (Cronin and Kennedy, 1999). People in this country were ready to acknowledge that problems existed. By writing *Silent Spring*, Carson made people aware of the hazards of chemicals in their foods and the destruction of wildlife. However, that is not all she produced: As a consequence of her careful research and writing, she also produced anger in people at a time when they were beginning to wonder if they were being deceived by the very industries that they had entrusted with their lives and their resources. Once aroused to action, many people began demanding an honest, compre-

hensive accounting of where pollution was occurring and how it might be endangering public health and environmental resources. Public outcries also led to investigations in courts and legislatures throughout the United States as people began to demand legal recognition of the right to a clean environment (Cronin and Kennedy, 1999).

A breakdown in social control mechanisms has been a powerful force in triggering collective behavior regarding environmental protection and degradation. During the 1970s, people in the "Love Canal" area of Niagara Falls, New York, became aware that their neighborhood and their children's school had been built over a canal where tons of poisonous waste had been dumped by a chemical company between 1930 and 1950. After the company closed the site, covered it with soil, and sold it (for $1) to the city of Niagara Falls, homes and a school were built on the sixteen-acre site. Over the next two decades, an oily black substance began oozing into the homes in the area and killing the trees and grass on the lots; schoolchildren

Box 20.1 SOCIOLOGY AND EVERYDAY LIFE

Answers to the Sociology Quiz on Collective Behavior and Environmental Issues

1. **False.** The environmental movement in the United States is the result of more than 100 years of collective action. The first environmental organization, the American Forestry Association (now American Forests), originated in 1875.

2. **False.** Since the 1980s, public opinion polls have shown that the majority of people in the United States have favorable attitudes regarding protection of the environment and banning nuclear weapons; however, far fewer individuals are actually involved in collective action to further these causes.

3. **True.** Environmental groups have held sit-ins, marches, boycotts, and strikes, which sometimes take the form of civil disobedience. Others have hanged political leaders in effigy or held officials hostage. Still others have dressed as grizzly bears to block traffic in Yellowstone National Park or created a symbolic "crack" (made of plastic) on the Glen Canyon Dam on the Colorado River to denounce development in the area.

4. **False.** Although some early social psychological theories were based on the assumption of "crowd psychology" or "mob behavior," most sociologists believe that individuals act quite rationally when they are a part of a crowd.

5. **True.** Rumors are most likely to emerge and circulate when people have very little information on a topic that is important to them. For example, rumors abound in times of technological disasters, when people are fearful and often willing to believe a worst-case scenario.

6. **True.** Many social movements, including grassroots environmental activism, attempt to influence public opinion so that local decision makers will feel obliged to correct a specific problem through changes in public policy.

7. **True.** Having a democratic process available is important for dissenters. Grassroots movements have used the democratic process to bring about change even when elites have sought to discourage such activism.

8. **True.** Most social movements are reform movements that focus on improving society by changing some specific aspect of the social structure. Examples include environmental movements and the disability rights movement.

9. **False.** Most sociological studies have found that people respond differently to natural disasters, which usually occur very suddenly, and to technological disasters, which may occur gradually. One of the major differences is the communal bonding that tends to occur following natural disasters, as compared with the extreme social conflict that may follow technological disasters.

10. **True.** One of the goals of most environmental movements is to stress the importance of "thinking globally and acting locally" to protect the environment.

Sources: Based on Adams, 1991; Gamson, 1990; Hynes, 1990; Worster, 1985; and Young, 1990.

reported mysterious illnesses and feelings of malaise. Tests indicated that the dump site contained more than two hundred different chemicals, many of which could cause cancer or other serious health problems. Upon learning this information, Lois Gibbs, a mother of one of the schoolchildren, began a grassroots campaign to force government officials to relocate community members injured by seepages from the chemical dump. The collective behavior of neighborhood volunteers was not only successful in eventually bringing about social change but also inspired others to engage in collective behavior regarding environmental problems in their communities. After the Love Canal occurrence, Lois Gibbs founded Citizens' Clearinghouse for Hazardous Waste, which has assisted more than 1,700 community groups in fighting pollution in their neighborhoods (see Cable and Cable, 1995; Cronin and Kennedy, 1999). Similarly, the Hudson Valley Riverkeeper originated in a grass-roots movement that was fueled by "people power and never lost touch with its power base" (Cronin and Kennedy, 1999: 277).

Dynamics of Collective Behavior

To better understand the dynamics of collective behavior, let's briefly examine several questions. First, how do people come to transcend, bypass, or subvert established institutional patterns and structures? Some environmental activists have found that they cannot get their point across unless they go outside established institutional patterns and organizations. For example, Lois Gibbs and other Love Canal residents initially tried to work within established means through the school administration and state health officials to clean up the problem. However, they quickly learned that their problems were not being solved through "official" channels. As the problem appeared to grow worse, organizational responses became more defensive and obscure. Accordingly, some residents began acting outside of established norms by holding protests and strikes (Gibbs, 1982). Some situations are more conducive to collective behavior than others. When people can communicate quickly and easily with one another, spontaneous behavior is more likely (Turner and Killian, 1993). When people are gathered together in one general location (whether lining the streets or assembled in a massive stadium), they are more likely to respond to a common stimulus.

Second, how do people's actions compare with their attitudes? People's attitudes (as expressed in public opinion surveys, for instance) are not always reflected in their political and social behavior. Issues pertaining to the environment are no exception. For example, people may indicate in survey research that they believe the quality of the environment is very important, but the same people may not turn out on election day to support propositions that protect the environment or candidates who promise to focus on environmental issues. Likewise, individuals who indicate on a questionnaire that they are concerned about increases in ground-level ozone—the primary component of urban smog—often drive single-occupant, oversized vehicles that government studies have shown to be "gas guzzlers" that contribute to lowered air quality in urban areas. As a result, smog levels increase, contributing to human respiratory problems and dramatically reduced agricultural crop yields (Voynick, 1999).

Third, why do people act collectively rather than singly? As the sociologists Ralph H. Turner and Lewis M. Killian (1993: 12) note, people believe that there is strength in numbers: "the rhythmic stamping of feet by hundreds of concert-goers in unison is different from isolated, individual cries of 'bravo.'" Likewise, people may act as a collectivity when they believe it is the only way to fight those with greater power and resources. Collective behavior is not just the sum total of a large number of individuals acting at the same time; rather, it reflects people's joint response to some common influence or stimulus.

Distinctions Regarding Collective Behavior

People engaging in collective behavior may be divided into crowds and masses. A **crowd is a relatively large number of people who are in one another's immediate vicinity** (Lofland, 1993). Examples of crowds include the audience in a movie theater or people at a pep rally for a sporting event. By contrast, a **mass is a number of people who share an interest in a specific idea or issue but who are not in one another's immediate vicinity** (Lofland, 1993). An example is the popularity of blogging on the Internet. A *blog,* which is short for Web log, is an online journal maintained by an individual who frequently records entries that are maintained in a chronological order. People who self-publish blogs are widely diverse in their interests. Some may include poetry, diary entries, or discussions of such activities as body piercings. However, others express their beliefs about social issues such as the environment, terrorism, and war, and their concerns about the future. Readers often share a common interest with the blogger on the topics the person is writing about, but these individuals have never met—and probably will never meet—each other in a face-to-face encounter. To distinguish between crowds and masses, think of the difference between a riot and a rumor: People who participate in a riot must be in the same general location; those who spread a rumor may be thousands of miles apart, communicating by telephone or the Internet.

Collective behavior may also be distinguished by the dominant emotion expressed. According to the sociologist John Lofland (1993: 72), the *dominant emotion* refers to the "publicly expressed feeling perceived by participants and observers as the most prominent in an episode of collective behavior." Lofland suggests that fear, hostility, and joy are three fundamental emotions found in collective behavior;

however, grief, disgust, surprise, or shame may also predominate in some forms of collective behavior.

Types of Crowd Behavior

When we think of a crowd, many of us think of *aggregates,* previously defined as a collection of people who happen to be in the same place at the same time but who share little else in common. However, the presence of a relatively large number of people in the same location does not necessarily produce collective behavior. Sociologist Herbert Blumer (1946) developed a typology in which crowds are divided into four categories: casual, conventional, expressive, and acting. Other scholars have added a fifth category, protest crowds.

Casual and Conventional Crowds *Casual crowds* are relatively large gatherings of people who happen to be in the same place at the same time; if they interact at all, it is only briefly. People in a shopping mall or a subway car are examples of casual crowds. Other than sharing a momentary interest, such as a clown's performance or a small child's fall, a casual crowd has nothing in common. The casual crowd plays no active part in the event—such as the child's fall—which likely would have occurred whether or not the crowd was present; the crowd simply observes.

Conventional crowds are made up of people who come together for a scheduled event and thus share a common focus. Examples include religious services, graduation ceremonies, concerts, and college lectures. Each of these events has preestablished schedules and norms. Because these events occur regularly, interaction among participants is much more likely; in turn, the events would not occur without the crowd, which is essential to the event.

Expressive and Acting Crowds *Expressive crowds* provide opportunities for the expression of some strong emotion (such as joy, excitement, or grief). People release their pent-up emotions in conjunction with other persons experiencing similar emotions. Examples include worshippers at religious revival services; mourners lining the streets when a celebrity, public official, or religious leader has died; and revelers assembled at Mardi Gras or on New Year's Eve at Times Square in New York.

Acting crowds are collectivities so intensely focused on a specific purpose or object that they may erupt into violent or destructive behavior. Mobs, riots, and panics are examples of acting crowds, but casual and conventional crowds may become acting crowds under some circumstances. A *mob* is a highly emo-

© AP/Wide World Photos

What kind of crowd is this? Based on the crowd's purpose and the form of interaction that takes place, sociologists would identify these 1,000 shoppers, who are trying to beat the holiday rush, as a casual crowd. How do casual crowds differ from other types of crowds?

tional crowd whose members engage in, or are ready to engage in, violence against a specific target—a person, a category of people, or physical property. Mob behavior in this country has included lynchings, fire bombings, effigy hangings, and hate crimes. Mob violence tends to dissipate relatively quickly once a target has been injured, killed, or destroyed. Sometimes, actions such as an effigy hanging are used symbolically by groups that are not otherwise violent. For example, Lois Gibbs and other Love Canal residents called attention to their problems with the chemical dump site by staging a protest in which they "burned in effigy" the governor and the health commissioner to emphasize their displeasure with the lack of response from these public officials (A. Levine, 1982).

Compared with mob actions, riots may be of somewhat longer duration. A *riot* **is violent crowd behavior that is fueled by deep-seated emotions but not directed at one specific target.** Riots are often triggered by fear, anger, and hostility; however, not all riots are caused by deep-seated hostility and hatred—people may be expressing joy and exuberance when rioting occurs. Examples include celebrations after sports victories such as those that occurred in Montreal, Canada, following a Stanley Cup win and in Vancouver following a playoff victory (Kendall, Lothian Murray, and Linden, 2004).

A *panic* **is a form of crowd behavior that occurs when a large number of people react to a real or perceived threat with strong emotions and self-destructive behavior.** The most common type of panic occurs when people seek to escape from a perceived danger, fearing that few (if any) of them will be able to get away from that danger. Panics can also arise in response to events that people believe are beyond their control—such as a major disruption in the economy. Although panics are relatively rare, they receive massive media coverage because they provoke strong feelings of fear in readers and viewers, and the number of casualties may be large.

Protest movements take place around the world when people believe that an injustice has occurred or that great inequalities exist in the distribution of societal resources. These Zapatista rebels in Mexico are part of a movement seeking political inclusion and greater access to economic opportunities.

Protest Crowds Sociologists Clark McPhail and Ronald T. Wohlstein (1983) added protest crowds to the four types of crowds identified by Blumer. *Protest crowds* engage in activities intended to achieve specific political goals. Examples include sit-ins, marches, boycotts, blockades, and strikes. In 1997 an International Day of Protest was staged against Nike's labor practices in countries such as Vietnam (Shaw, 1999). Some protests take the form of ***civil disobedience***—**nonviolent action that seeks to change a policy or law by refusing to comply with it.** Acts of civil disobedience may become violent, as in a confrontation between protesters and police officers; in this case, a protest crowd becomes an *acting crowd.* In the 1960s, African American students and sympathetic whites used sit-ins to call attention to racial injustice and demand social change (see Morris, 1981; McAdam, 1982). Some of these protests can escalate into violent confrontations even when violence was not the intent of the organizers.

As you may recall, collective action often puts individuals into a situation where they engage in some activity as a group that they might not do on their own. Does this mean that people's actions are produced by some type of "herd mentality"? Some analysts have answered this question affirmatively; however, sociologists typically do not agree with that assessment.

Explanations of Crowd Behavior

What causes people to act collectively? How do they determine what types of action to take? One of the earliest theorists to provide an answer to these questions was Gustave Le Bon, a French scholar who focused on crowd psychology in his contagion theory.

Contagion Theory *Contagion theory* focuses on the social–psychological aspects of collective behavior; it attempts to explain how moods, attitudes, and behavior are communicated rapidly and why they are accepted by others (Turner and Killian, 1993). Le Bon (1841–1931) argued that people are more likely to engage in antisocial behavior in a crowd because they are anonymous and feel invulnerable. Le Bon (1960/1895) suggested that a crowd takes on a life of its own that is larger than the beliefs or actions of any one person. Because of its anonymity, the crowd transforms individuals from rational beings into a single organism with a collective mind. In essence, Le Bon asserted that emotions such as fear and hate are contagious in crowds because people experience a decline in personal responsibility; they will do things as a collectivity that they would never do when acting alone.

Le Bon's theory is still used by many people to explain crowd behavior. However, critics argue that the "collective mind" has not been documented by systematic studies.

Social Unrest and Circular Reaction Sociologist Robert E. Park was the first U.S. sociologist to investigate crowd behavior. Park believed that Le Bon's analysis of collective behavior lacked several important elements. Intrigued that people could break away from the powerful hold of culture and their established routines to develop a new social order, Park added the concepts of social unrest and circular reaction to contagion

Convergence theory is based on the assumption that crowd behavior involves shared emotions, goals, and beliefs, such as the importance of protecting the environment. An example is the Earth Day events that brought together these children carrying this banner to foster environmental causes.

theory. According to Park, social unrest is transmitted by a process of *circular reaction*—the interactive communication between persons such that the discontent of one person is communicated to another, who, in turn, reflects the discontent back to the first person (Park and Burgess, 1921).

Convergence Theory

Convergence theory focuses on the shared emotions, goals, and beliefs that many people may bring to crowd behavior. Because of their individual characteristics, many people have a predisposition to participate in certain types of activities (Turner and Killian, 1993). From this perspective, people with similar attributes find a collectivity of like-minded persons with whom they can express their underlying personal tendencies. Although people may reveal their "true selves" in crowds, their behavior is not irrational; it is highly predictable to those who share similar emotions or beliefs.

Convergence theory has been applied to a wide array of conduct, from lynch mobs to environmental movements. In social psychologist Hadley Cantril's (1941) study of one lynching, he found that the participants shared certain common attributes: They were poor and working-class whites who felt that their status was threatened by the presence of successful African Americans. Consequently, the characteristics of these individuals made them susceptible to joining a lynch mob even if they did not know the target of the lynching.

Convergence theory adds to our understanding of certain types of collective behavior by pointing out how individuals may have certain attributes—such as racial hatred or fear of environmental problems that directly threaten them—that initially bring them together. However, this theory does not explain how the attitudes and characteristics of individuals who take some collective action differ from those who do not.

Emergent Norm Theory

Unlike contagion and convergence theories, *emergent norm theory* emphasizes the importance of social norms in shaping crowd behavior. Drawing on the symbolic interactionist perspective, the sociologists Ralph Turner and Lewis Killian (1993: 12) asserted that crowds develop their own definition of a situation and establish norms for behavior that fit the occasion:

> Some shared redefinition of right and wrong in a situation supplies the justification and coordinates the action in collective behavior. People do what they would not otherwise have done when they panic collectively, when they riot, when they engage in civil disobedience, or when they launch terrorist campaigns, because they find social support for the view that what they are doing is the right thing to do in the situation.

According to Turner and Killian (1993: 13), emergent norms occur when people define a new situation as highly unusual or see a long-standing situation in a new light.

Sociologists using the emergent norm approach seek to determine how individuals in a given collectivity develop an understanding of what is going on, how they construe these activities, and what type of norms are involved. For example, in a study of audience participation, the sociologist Steven E. Clayman (1993) found that members of an audience listening to a speech applaud promptly and independently but wait to coordinate their booing with other people; they do not wish to "boo" alone.

Some emergent norms are permissive—that is, they give people a shared conviction that they may disregard ordinary rules such as waiting in line, taking turns, or treating a speaker courteously. Collective activity such as mass looting may be defined (by participants) as taking what rightfully belongs to them and punishing those who have been exploitative. For example, following the Los Angeles riots of 1992, some analysts argued that Korean Americans were targets of rioters because they were viewed by Latinos/as and African Americans as "callous and greedy invaders" who became wealthy at the expense of members of other racial–ethnic groups (Cho, 1993). Thus, rioters who used this rationalization could view looting and burning as a means of "paying back" Korean Americans or of gaining property (such as TV sets and microwave ovens) from those who had already taken from them. Once a crowd reaches some agreement on the norms, the collectivity is supposed to adhere to them. If crowd members develop a norm that condones looting or vandalizing property, they will proceed to cheer for those who conform and ridicule those who are unwilling to abide by the collectivity's new norms.

Emergent norm theory points out that crowds are not irrational. Rather, new norms are developed in a rational way to fit the immediate situation. However, critics note that proponents of this perspective fail to specify exactly what constitutes a norm, how new ones emerge, and how they are so quickly disseminated and accepted by a wide variety of participants. One variation of this theory suggests that no single dominant norm is accepted by everyone in a crowd; instead, norms are specific to the various categories of actors rather than to the collectivity as a whole (Snow, Zurcher, and Peters, 1981). For example, in a study of football victory celebrations, the sociologists David A. Snow, Louis A. Zurcher, and Robert Peters (1981) found that each week, behavioral patterns were changed in the postgame revelry, with some being modified, some added, and some deleted.

Mass Behavior

Not all collective behavior takes place in face-to-face collectivities. **Mass behavior is collective behavior that takes place when people (who often are geographically separated from one another) respond to the same event in much the same way.** For people to respond in the same way, they typically have common sources of information that provoke their collective behavior. The most frequent types of mass behavior are rumors, gossip, mass hysteria, public opinion, fashions, and fads. Under some circumstances, social movements constitute a form of mass behavior. However, we will examine social movements separately because they differ in some important ways from other types of dispersed collectivities.

Rumors and Gossip *Rumors* **are unsubstantiated reports on an issue or subject** (Rosnow and Fine, 1976). Whereas a rumor may spread through an assembled collectivity, rumors may also be transmitted among people who are dispersed geographically, including people posting messages on the Internet or talking by cell phone. Although they may initially contain a kernel of truth, rumors may be modified as they spread to serve the interests of those repeating them. Rumors thrive when tensions are high and when little authentic information is available on an issue of great concern. For example, when the blackout of August 2003 occurred, leaving 50 million people in eight states and parts of Canada without electricity, the earliest rumors about the power failures reflected the turbulent times in which we live. One of the first rumors that began to spread was that the blackout was an act of terrorism. As one person stated in an e-mail (uncorrected by your author),

> I think its an act of terrorism. They are planning to strike. They need to know how the power system works and what cities and states will be affected so they can plan there strategy. In the meantime, USA is assuming that its natural. Typical of americans. Don't put your guards down. . . . (Cromwell, 2003)

Television broadcasters and public officials quickly tried to deflect this rumor for fear that people might panic and be injured as they sought to leave their workplaces in cities such as New York and make their way home. Other rumors that initially circulated ranged from the humorous to the accusatory: Some attributed the power failure to "UFOs," "the energy companies that want to gouge everyone," "President Bush's buddies at Enron," "homeland security crooks," and a wide variety of other explanations. Nearly a month after the massive blackout, investigators were still attempting to analyze the data to determine specific causes; however, the rumor mill had provided people with instant speculation about why this event had occurred. Fortunately, most people acted responsibly, and the riots and looting that took place during prior blackouts in New York City did not recur (*BBC News*, 2003; CNN.com, 2003c; Gibbs, 2003).

As the example of the blackout of 2003 shows, people are willing to give rumors credence when no opposing information is available. Environmental issues are similar. For example, in the past, when residents of Love Canal waited for information from health department officials about their exposure to the toxic

When unexpected events such as the massive 2003 power outage in the United States and Canada occur, people frequently rely on rumors to help them know what is going on. Getting accurate information out quickly helped prevent people from panicking as tens of thousands of workers in Manhattan sought to get home any way they could while electricity was unavailable in the city.

chemicals and from the government about possible relocation at state expense to another area, new waves of rumors also spread through the community daily. By the time a meeting was called by health department officials to provide homeowners with the results of air-sample tests for hazardous chemicals (such as chloroform and benzene) performed on their homes, already fearful residents were ready to believe the worst, as Lois Gibbs (1982: 25) describes:

> Next to the names [of residents] were some numbers. But the numbers had no meaning. People stood there looking at the numbers, knowing nothing of what they meant but suspecting the worst.
>
> One woman, divorced and with three sick children, looked at the piece of paper with numbers and started crying hysterically: "No wonder my children are sick. Am I going to die? What's going to happen to my children?" No one could answer. . . .
>
> The night was very warm and humid, and the air was stagnant. On a night like that, the smell of Love Canal is hard to describe. It's all around you.

It's as though it were about to envelop you and smother you. By now, we were outside, standing in the parking lot. The woman's panic caught on, starting a chain reaction. Soon, many people there were hysterical.

Once a rumor begins to circulate, it seldom stops unless compelling information comes to the forefront that either proves the rumor false or makes it obsolete.

In industrialized societies with sophisticated technology, rumors come from a wide variety of sources and may be difficult to trace. Print media (newspapers and magazines) and electronic media (radio and television), fax machines, cellular networks, satellite systems, and the Internet aid the rapid movement of rumors around the globe. In addition, modern communications technology makes anonymity much easier. In a split second, messages (both factual and fictitious) can be disseminated to thousands of people through e-mail, computerized bulletin boards, and Internet newsgroups.

Whereas rumors deal with an issue or a subject, *gossip* **refers to rumors about the personal lives of individuals.** Charles Horton Cooley (1963/1909) viewed gossip as something that spread among a small group of individuals who personally knew the person who was the object of the rumor. Today, this is frequently not the case; many people enjoy gossiping about people whom they have never met. Tabloid newspapers and magazines such as the *National Inquirer* and *People,* and television "news" programs that purport to provide "inside" information on the lives of celebrities, are sources of contemporary gossip, much of which has not been checked for authenticity.

Mass Hysteria and Panic

Mass hysteria is a form of dispersed collective behavior that occurs when a large number of people react with strong emotions and self-destructive behavior to a real or perceived threat. Does mass hysteria actually occur? Although the term has been widely used, many sociologists believe that this behavior is best described as a panic with a dispersed audience.

An example of mass hysteria or a panic with a widely dispersed audience was actor Orson Welles's 1938 Halloween eve radio dramatization of H. G. Wells's science fiction classic *The War of the Worlds.* A CBS radio dance music program was interrupted suddenly by a news bulletin informing the audience that Martians had landed in New Jersey and were in the process of conquering Earth. Some listeners became extremely frightened even though an announcer had indicated before, during, and after the performance that the broadcast was a fictitious dramatization. According to some reports, as many as 1 million of the estimated 10 million listeners believed that this astonishing event had occurred. Thousands were reported to have hidden in their storm cellars or to have gotten in their cars so that they could flee from the Martians (see Brown, 1954). In actuality, the program probably did not generate mass hysteria, but rather a panic among gullible listeners. Others switched stations to determine if the same "news" was being broadcast elsewhere. When they discovered that it was not, they merely laughed at the joke being played on listeners by CBS. In 1988, on the fiftieth anniversary of the broadcast, a Portuguese radio station rebroadcast the program; once again, a panic ensued.

Fads and Fashions

As you will recall from Chapter 3, a *fad* is a temporary but widely copied activity enthusiastically followed by large numbers of people. Fads can be embraced by widely dispersed collectivities; news networks such as CNN and Internet Web sites may bring the latest fad to the attention of audiences around the world. Recently, people in a number of countries have participated in a relatively new fad known as the "flash mob" (see Box 20.2).

Unlike fads, fashions tend to be longer lasting. In Chapter 3, *fashion* is defined as a currently valued style of behavior, thinking, or appearance. Fashion also applies to art, music, drama, literature, architecture, interior design, and automobiles, among other things. However, most sociological research on fashion has focused on clothing, especially women's apparel (Davis, 1992).

In preindustrial societies, clothing styles remained relatively unchanged. With the advent of industrialization, items of apparel became readily available at low prices because of mass production. Fashion became more important as people embraced the "modern" way of life and as advertising encouraged "conspicuous consumption."

Georg Simmel, Thorstein Veblen, and Pierre Bourdieu have all viewed fashion as a means of status differentiation among members of different social classes. Simmel (1957/1904) suggested a classic "trickle-down" theory (although he did not use those exact words) to describe the process by which members of the lower classes emulate the fashions of the upper class. As the fashions descend through the status hierarchy, they are watered down and "vulgarized" so that they are no longer recognizable to members of the upper class, who then regard them as unfashionable and in bad taste (Davis, 1992). Veblen (1967/1899) asserted that fashion serves mainly to institutionalize conspicuous consumption among the wealthy. Almost eighty years later, Bourdieu (1984) similarly (but more subtly) suggested that "matters of taste," including fashion sensibility, constitute a large share of the "cultural capital" possessed by members of the dominant class.

Herbert Blumer (1969) disagreed with the trickle-down approach, arguing that "collective selection" best explains fashion. Blumer suggested that people in the middle and lower classes follow fashion because it is *fashion,* not because they desire to emulate members of the elite class. Blumer thus shifts the focus on fashion to collective mood, tastes, and choices: "Tastes are themselves a product of experience. . . . They are formed in the context of social interaction, responding to the definitions and affirmation given by others. People thrown into areas of common interaction and having similar runs of experience develop common tastes" (qtd. in Davis, 1992: 116). Perhaps one of the best refutations of the trickle-down approach is the way in which fashion today often originates among people in the lower social classes and is mimicked by the elites. In the mid-1990s, the so-called grunge look was a prime example of this.

Public Opinion

Public opinion consists of the attitudes and beliefs communicated by ordinary citizens to decision makers (Greenberg and Page, 2002). It is measured through polls and surveys, which use research methods such as interviews and questionnaires, as described in Chapter 2. Many people are not interested in all aspects of public policy but are concerned about issues they believe are relevant to themselves. Even on a single topic, public opinion will vary widely based on race/ethnicity, religion, region, social class, education level, gender, age, and so on.

Scholars who examine public opinion are interested in the extent to which the public's attitudes are communicated to decision makers and the effect (if any) that public opinion has on policy making (Turner and Killian, 1993). Some political scientists argue that public opinion has a substantial effect on decisions at all levels of government (see Greenberg and Page, 2002); others strongly disagree. For example, Thomas Dye and Harmon Zeigler (2000: 447) argue that

Public policy does not reflect demands of "the people," but rather the preferences, interests, and values of the very few who participate in the policy-making process. Changes or innovations in public policy come about when elites redefine their own interests or modify their own values. Policies decided by elites need not be oppressive or exploitative of the masses. Elites may be very public-regarding, and the welfare of the masses may be an important consideration in elite decision making. Yet it is the *elites* that make policy, not the *masses.*

Box 20.2 SOCIOLOGY IN GLOBAL PERSPECTIVE

"Flash Mobs": Collective Behavior in the Information Age

News Item: About 200 people, mostly in their 20s and 30s, crowd into the card section of the Harvard Coop bookstore, pretending to look for a card for "Bill." On cue, they burst into spontaneous applause. (Ray, 2003)

News Item: In Rome, Italy, hundreds of people enter a music and bookshop asking employees for non-existent books and authors. (Shmueli, 2003)

News Item: In Berlin, Germany, about 40 people take out their mobile phones in the middle of a busy street and shout, "Yes, yes!" and then applaud. (Shmueli, 2003)

What do these news items have in common? They are examples of the "flash mobs" that are spreading quickly across the United States and Europe, Australia, and China. A *flash mob* is a large group of people who gather in a usually predetermined location, perform some brief action, and then quickly disperse (Wordspy.com, 2003). The flash mob is related to a *smart mob,* which was introduced by the futurist Howard Rheingold in his book *Smart Mobs: The Next Social Revolution* (2003). Smart mobs are made up of people who use technologies such as e-mail, Web sites, and cell phones to organize an event that is often political in nature. An example is the street demonstrators in the 1999 protests against the World Trade Organization in Seattle, Washington, who organized by using Web sites, cell phones, and other devices that possess both communication and computing capabilities.

Most flash mobs are primarily organized for fun and produce no harmful effects. Usually, the organizers are the only ones who know the details of what is to take place: Participants learn when they arrive at the site. The flash mob at the Grand Hyatt Hotel in New York is an example. Participants received an e-mail telling them to assemble at the food court in Grand Central Station. When they arrived, the organizers (who were each holding a copy of the *New York Review of Books* so they could be identified) presented participants with printed instructions to assemble on the mezzanine level of the Grand Hyatt Hotel next to Grand Central Station. According to Fred Hoysted, a participant,

> I was pretty apprehensive beforehand since all I really knew was what I had read on the Web and thought it sounded fun. The mob itself was slightly bizarre. There were about 200 of us standing at the balcony railings on the mezzanine floor of the Hyatt Hotel, next to Grand Central Station. At the appointed time, we burst into applause for 15 seconds as instructed. The look of joy on people's faces was incredible. And even though I'd felt somewhat detached from the proceedings, I couldn't help but smile and join. I knew this was something I wanted to do again. (qtd. in Shmueli, 2003)

Flash mobs have recently grown in popularity; however, some sociologists might view these mobs as another fad that will pass quickly. According to one designer and Web logger, flash mobs may be more long term because they actually have roots that can be traced to the late 1970s:

> There's a vague, growing interest in grass-roots activities that transcends more traditional institutions. They prove people can still form ad hoc communities and make things happen that are beyond the reach of the gigantic, corrupt corporate and governmental powers that seem to dominate so much of modern life. But maybe I'm reading too much into it. (qtd. in Kahney, 2003)

Even if the flash mobs of today are short lived, the methods of instant communication that make such mobs possible will continue to change how people organize many kinds of events. How long do you think flash mobs will last? Why?

From this perspective, polls may artificially create the appearance of a public opinion; pollsters may ask questions that those being interviewed had not even considered before the survey.

As the masses attempt to influence elites and vice versa, a two-way process occurs with the dissemination of *propaganda*—**information provided by individuals or groups that have a vested interest in furthering their own cause or damaging an opposing one.** Although many of us think of propaganda in negative terms, the information provided can be correct and can have a positive effect on decision making.

In recent decades, grassroots environmental activists (including the Love Canal residents and the

Martin Luther King, Jr., a leader of the civil rights movement in the 1950s and 1960s, advocated nonviolent protests that sometimes took the form of civil disobedience. Here he marches along-side his wife, Coretta Scott King, who took over many of Dr. King's activities after he was assassinated.

Riverkeepers) have attempted to influence public opinion. In a study of public opinion on environmental issues, the sociologist Riley E. Dunlap (1992) found that public awareness of the seriousness of environmental problems and support for environmental protection increased dramatically between the late 1960s and the 1990s. However, it is less clear that public opinion translates into action by either decision makers in government and industry or by individuals (such as a willingness to adopt a more ecologically sound lifestyle).

Initially, most grassroots environmental activists attempt to influence public opinion so that local decision makers will feel the necessity of correcting a specific problem through changes in public policy. Although activists usually do not start out seeking broader social change, they often move in that direction when they become aware of how widespread the problem is in the larger society or on a global basis. One of two types of social movements often develops at this point—one focuses on NIMBY ("not in my backyard"), whereas the other focuses on NIABY ("not in anyone's backyard") (Freudenberg and Steinsapir, 1992).

SOCIAL MOVEMENTS

Although collective behavior is short-lived and relatively unorganized, social movements are longer lasting, are more organized, and have specific goals or purposes. A **social movement is an organized group that acts consciously to promote or resist change through collective action** (Goldberg, 1991). Because social movements have not become institutionalized and are outside the political mainstream, they offer "outsiders" an opportunity to have their voices heard.

Social movements are more likely to develop in industrialized societies than in preindustrial societies, where acceptance of traditional beliefs and practices makes such movements unlikely. Diversity and a lack of consensus (hallmarks of industrialized nations) contribute to demands for social change, and people who participate in social movements typically lack power and other resources to bring about change without engaging in collective action. Social movements are most likely to spring up when people come to see their personal troubles as public issues that cannot be solved without a collective response.

Social movements make democracy more available to excluded groups (see Greenberg and Page, 2002). Historically, people in the United States have worked at the grassroots level to bring about changes even when elites sought to discourage activism (Adams, 1991). For example, the civil rights movement brought into its ranks African Americans in the South who had never been allowed to participate in politics (see Killian, 1984). The women's suffrage movement gave voice to women who had been denied the right to vote (Rosenthal et al., 1985). Similarly, a grassroots environmental movement gave the working-class residents of Love Canal a way to "fight city hall" and Hooker Chemicals (A. Levine, 1982), as Lois Gibbs (1982: 38–40) explains:

> People were pretty upset. They were talking and stirring each other up. I was afraid there would be violence. We had a meeting at my house to try to put everything together [and] decided to form a homeowners' association. We got out the word as best we could and told everyone to come to the

Frontier Fire Hall on 102d Street. . . . The fire-house was packed with people, and more were outside. . . .

I was elected president. . . . I took over the meeting but I was scared to death. It was only the second time in my life I had been in front of a microphone or a crowd. . . . We set four goals right at the beginning—(1) get all the residents within the Love Canal area who wanted to be evacuated, evacuated and relocated, especially during the construction and repair of the canal; (2) do something about propping up property values; (3) get the canal fixed properly; and (4) have air sampling and soil and water testing done throughout the whole area, so we could tell how far the contamination had spread. . . .

Most social movements rely on volunteers like Lois Gibbs to carry out the work. Traditionally, women have been strongly represented in both the membership and the leadership of many grassroots movements (A. Levine, 1982; Freudenberg and Steinsapir, 1992).

The Love Canal activists set the stage for other movements that have grappled with the kind of issues that the sociologist Kai Erikson (1994) refers to as a "new species of trouble." Erikson describes the "new species" as environmental problems that "contaminate rather than merely damage . . . they pollute, befoul, taint, rather than just create wreckage . . . they penetrate human tissue indirectly rather than just wound the surfaces by assaults of a more straightforward kind. . . . And the evidence is growing that they scare human beings in new and special ways, that they elicit an uncanny fear in us" (Erikson, 1991: 15). The chaos that Erikson (1994: 141) describes is the result of technological disasters—"meaning everything that can go wrong when systems fail, humans err, designs prove faulty, engines misfire, and so on." A recent example of such a disaster occurred in Japan, where more than 300,000 residents living within six miles of the nuclear plant at Tokaimura were told to stay indoors in the aftermath of three workers' mishandling of stainless-steel pails full of uranium, causing the worst nuclear accident in Japan's history (Larimer, 1999). Although no lives were immediately lost, workers in the plant soaked up potentially lethal doses of radiation, some of which also leaked from the plant into the community. The fifty-two nuclear power plants in Japan have been plagued by other accidents and radiation leaks, causing concern for many people around the globe (Larimer, 1999).

Social movements provide people who otherwise would not have the resources to enter the game of politics a chance to do so. We are most familiar with those movements that develop around public policy issues considered newsworthy by the media, ranging from abortion and women's rights to gun control and environmental justice. However, a number of other types of social movements exist as well.

Types of Social Movements

Social movements are difficult to classify; however, sociologists distinguish among movements on the basis of their *goals* and the *amount of change* they seek to produce (Aberle, 1966; Blumer, 1974). Some movements seek to change people whereas others seek to change society.

Reform Movements   Grassroots environmental movements are an example of *reform movements,* which seek to improve society by changing some specific aspect of the social structure. Members of reform movements usually work within the existing system to attempt to change existing public policy so that it more adequately reflects their own value system. Examples of reform movements (in addition to the environmental movement) include labor movements, animal rights movements, antinuclear movements, Mothers Against Drunk Driving, and the disability rights movement.

Sociologist Lory Britt (1993) suggested that some movements arise specifically to alter social responses to and definitions of stigmatized attributes. From this perspective, social movements may bring about changes in societal attitudes and practices while at the same time causing changes in participants' social emotions. For example, the civil rights and gay rights movements helped replace shame with pride (Britt, 1993). Such a sense of pride may extend beyond current members of a reform movement. The late César Chávez, organizer of a Mexican American farm workers' movement that developed into the United Farm Workers Union, noted that the "consciousness and pride raised by our union is alive and thriving inside millions of young Hispanics who will never work on a farm!" (qtd. in Ayala, 1993: E4).

Revolutionary Movements   Movements seeking to bring about a total change in society are referred to as *revolutionary movements.* These movements usually do not attempt to work within the existing system; rather, they aim to remake the system by replacing existing institutions with new ones. Revolutionary movements range from utopian groups seeking to establish an ideal society to radical terrorists who use fear tactics to intimidate those with whom they disagree ideologically (see Alexander and Gill, 1984; Berger, 1988; Vetter and Perlstein, 1991).

As the issue of immigration illustrates, clashes are inevitable between members of proactive social movements and their counterparts in resistance movements. Why is immigration a pressing social issue that attracts the attention of some social movement organizers?

Movements based on terrorism often use tactics such as bombings, kidnappings, hostage taking, hijackings, and assassinations (Vetter and Perlstein, 1991). Over the past thirty years, a number of movements in the United States have engaged in terrorist activities or supported a policy of violence. However, the terrorist attacks in New York City and Washington, D.C., on September 11, 2001, and the events that followed those attacks proved to all of us that terrorism within this country can originate from the activities of revolutionary terrorists from outside the country as well.

Religious Movements Social movements that seek to produce radical change in individuals are typically based on spiritual or supernatural belief systems. Also referred to as *expressive movements, religious movements* are concerned with renovating or renewing people through "inner change." Fundamentalist religious groups seeking to convert nonbelievers to their belief system are an example of this type of movement. Some religious movements are *millenarian*—that is, they forecast that "the end is near" and assert that an immediate change in behavior is imperative. Relatively new religious movements in industrialized Western societies have included Hare Krishnas, the Unification Church, Scientology, and the Divine Light Mission, all of which tend to appeal to the psychological and social needs of young people seeking meaning in life that mainstream religions have not provided for them.

Alternative Movements Movements that seek limited change in some aspect of people's behavior are referred to as *alternative movements.* For example, early in the twentieth century the Women's Christian Temperance Union attempted to get people to abstain from drinking alcoholic beverages. Some analysts place "therapeutic social movements" such as Alcoholics Anonymous in this category; however, others do not, due to their belief that people must change their lives completely in order to overcome alcohol abuse (see Blumberg, 1977). More recently, a variety of "New Age" movements have directed people's behavior by emphasizing spiritual consciousness combined with a belief in reincarnation and astrology. Such practices as vegetarianism, meditation, and holistic medicine are often included in the self-improvement category. Beginning in the 1990s, some alternative movements have included the practice of yoga (usually without its traditional background in the Hindu religion) as a means by which the self can be liberated and union can be achieved with the supreme spirit or universal soul.

Resistance Movements Also referred to as *regressive movements, resistance movements* seek to prevent change or to undo change that has already occurred. Virtually all of the proactive social movements previously discussed face resistance from one or more reactive movements that hold opposing viewpoints and want to foster public policies that reflect their own beliefs. Examples of resistance movements are groups organized since the 1950s to oppose school integration, civil rights and affirmative action legislation, and domestic partnership initiatives. However, perhaps the most widely known resistance movement includes many who label themselves as "pro-life" advocates—such as Operation Rescue, which seeks to close abortion clinics and make abortion illegal under all circumstances (Gray, 1993; Van Biema, 1993). Protests by some radical anti-abortion groups have grown violent, resulting in the death of several doctors and clinic workers, and creating fear among health professionals and patients seeking abortions (Belkin, 1994).

Stages in Social Movements

Do all social movements go through similar stages? Not necessarily, but there appear to be identifiable stages in virtually all movements that succeed beyond their initial phase of development.

In the *preliminary* (or *incipiency*) *stage,* widespread unrest is present as people begin to become aware of a problem. At this stage, leaders emerge to agitate others into taking action. In the *coalescence stage,* people begin to organize and to publicize the problem. At this stage, some movements become formally organized at local and regional levels. In the *institutionalization* (or *bureaucratization*) *stage,* an organizational structure develops, and a paid staff (rather than volunteers) begins to lead the group. When the movement reaches this stage, the initial zeal and idealism of members may diminish as administrators take over management of the organization. Early grassroots supporters may become disillusioned and drop out; they may also start another movement to address some as-yet-unsolved aspect of the original problem. For example, some national environmental organizations—such as the Sierra Club, the National Audubon Society, and the National Parks and Conservation Association—that started as grassroots conservation movements are currently viewed by many people as being unresponsive to local environmental problems (Cable and Cable, 1995). As a result, new movements have arisen.

SOCIAL MOVEMENT THEORIES

What conditions are most likely to produce social movements? Why are people drawn to these movements? Sociologists have developed several theories to answer these questions.

Relative Deprivation Theory

According to relative deprivation theory, people who are satisfied with their present condition are less likely to seek social change. Social movements arise as a response to people's perception that they have been deprived of their "fair share" (Rose, 1982). Thus, people who suffer relative deprivation are more likely to feel that change is necessary and to join a social movement in order to bring about that change. *Relative deprivation* refers to the discontent that people may feel when they compare their achievements with those of similarly situated persons and find that they have less than they think they deserve (Orum and Orum, 1968). Karl Marx captured the idea of relative deprivation in this description: "A house may be large or small; as long as the surrounding houses are small it satisfies all social demands for a dwelling. But let a palace arise beside the little house, and it shrinks from a little house to a hut" (qtd. in Ladd, 1966: 24). Movements based on relative deprivation are most likely to occur when an upswing in the standard of living is followed by a period of decline, such that people have *unfulfilled rising expectations*—newly raised hopes of a better lifestyle that are not fulfilled as rapidly as the people expected or are not realized at all.

Although most of us can relate to relative deprivation theory, it does not fully account for why people experience social discontent but fail to join a social movement. Even though discontent and feelings of deprivation may be necessary to produce certain types of social movements, they are not sufficient to bring movements into existence. In fact, the sociologist Anthony Orum (1974) found the best predictor of participation in a social movement to be prior organizational membership and involvement in other political activities.

Value-Added Theory

The value-added theory developed by sociologist Neil Smelser (1963) is based on the assumption that certain conditions are necessary for the development of a social movement. Smelser called his theory the "value-added" approach based on the concept (borrowed from the field of economics) that each step in the production process adds something to the finished product. For example, in the process of converting iron ore into automobiles, each stage "adds value" to the final product (Smelser, 1963). Similarly, Smelser asserted, six conditions are necessary and sufficient to produce social movements when they combine or interact in a particular situation:

Resource mobilization theory focuses on the ability of leaders of a social movement to mobilize people. For many years, college students have played an active part in mobilizing campaigns for the causes they believe to be right and against the injustices they perceive to be wrong.

© 2003 AP/Wide World Photos

1. *Structural conduciveness.* People must become aware of a significant problem and have the opportunity to engage in collective action. According to Smelser, movements are more likely to occur when a person, class, or agency can be singled out as the source of the problem; when channels for expressing grievances either are not available or fail; and when the aggrieved have a chance to communicate among themselves.

2. *Structural strain.* When a society or community is unable to meet people's expectations that something should be done about a problem, strain occurs in the system. The ensuing tension and conflict contribute to the development of a social movement based on people's belief that the problem would not exist if authorities had done what they were supposed to do.

3. *Spread of a generalized belief.* For a movement to develop, there must be a clear statement of the problem and a shared view of its cause, effects, and possible solution.

4. *Precipitating factors.* To reinforce the existing generalized belief, an inciting incident or dramatic event must occur. With regard to technological disasters, some (including Love Canal) gradually emerge from a long-standing environmental threat whereas others (including the Three Mile Island nuclear power plant) involve a suddenly imposed problem.

5. *Mobilization for action.* At this stage, leaders emerge to organize others and give them a sense of direction.

6. *Social control factors.* If there is a high level of social control on the part of law enforcement officials, political leaders, and others, it becomes more difficult to develop a social movement or engage in certain types of collective action.

Value-added theory takes into account the complexity of social movements and makes it possible to test Smelser's assertions regarding the necessary and sufficient conditions that produce such movements. However, critics note that the approach is rooted in the functionalist tradition and views structural strains as disruptive to society. Smelser's theory has been described as a mere variant of convergence theory, which, you will recall, is based on the assumption that people with similar predispositions will be activated by a common event or object (Quarantelli and Hundley, 1993).

Resource Mobilization Theory

Smelser's value-added theory tends to underemphasize the importance of resources in social movements. By contrast, *resource mobilization theory* focuses on the ability of members of a social movement to acquire resources and mobilize people in order to advance their cause (Oberschall, 1973; McCarthy and Zald, 1977). Resources include money, people's time and skills, access to the media, and material goods such as property and equipment. Assistance from outsiders is essential for social movements. For example, reform movements are more likely to succeed when they gain the support of political and economic elites (Oberschall, 1973).

Resource mobilization theory is based on the assumption that participants in social movements are rational people. According to the sociologist Charles Tilly (1973, 1978), movements are formed and dissolved, mobilized and deactivated, based on rational decisions about the goals of the group, available resources, and the cost of mobilization and collective action. Resource mobilization theory also assumes that participants must have some degree of economic and political resources to make the movement a success. In other words, widespread discontent alone cannot produce a social movement; adequate resources and motivated people are essential to any concerted social action (Aminzade, 1973; Gamson, 1990). Based on an analysis of fifty-three U.S. social protest groups ranging from labor unions to peace movements between 1800 and 1945, the sociologist William Gamson (1990) concluded that the organization and tactics of a movement strongly influence its chances of success. However, critics note that this theory fails to account for social changes brought about by groups with limited resources.

Scholars continue to modify resource mobilization theory and to develop new approaches for investigating the diversity of movements at the beginning of the twenty-first century (see Buechler, 2000). For example, emerging perspectives based on resource mobilization theory emphasize the ideology and legitimacy of movements as well as material resources (see Zald and McCarthy, 1987; McAdam, McCarthy, and Zald, 1988).

Social Constructionist Theory: Frame Analysis

Recent theories based on a symbolic interactionist perspective focus on the importance of the symbolic presentation of a problem to both participants and the general public (see Snow et al., 1986; Capek, 1993). Social constructionist theory is based on the assumption that a social movement is an interactive, symbolically defined, and negotiated process that involves participants, opponents, and bystanders (Buechler, 2000).

Research based on this perspective often investigates how problems are framed and what names they are given. This approach reflects the influence of the sociologist Erving Goffman's *Frame Analysis* (1974), in which he suggests that our interpretation of the particulars of events and activities is dependent on the framework from which we perceive them. According to Goffman (1974: 10), the purpose of frame analysis is "to try to isolate some of the basic frameworks of understanding available in our society for making sense out of events and to analyze the special vulnerabilities to which these frames of reference are subject." In other words, various "realities" may be simultaneously occurring among participants engaged

in the same set of activities. Sociologist Steven M. Buechler (2000: 41) explains the relationship between frame analysis and social movement theory:

> Framing means focusing attention on some bounded phenomenon by imparting meaning and significance to elements within the frame and setting them apart from what is outside the frame. In the context of social movements, framing refers to the interactive, collective ways that movement actors assign meanings to their activities in the conduct of social movement activism. The concept of framing is designed for discussing the social construction of grievances as a fluid and variable process of social interaction—and hence a much more important explanatory tool than resource mobilization theory has maintained.

Sociologists have identified at least three ways in which grievances are framed. First, *diagnostic framing* identifies a problem and attributes blame or causality to some group or entity so that the social movement has a target for its actions. Second, *prognostic framing* pinpoints possible solutions or remedies, based on the target previously identified. Third, *motivational framing* provides a vocabulary of motives that compel people to take action (Benford, 1993; Snow and Benford, 1988). When successful framing occurs, the individual's vague dissatisfactions are turned into well-defined grievances, and people are compelled to join the movement in an effort to reduce or eliminate those grievances (Buechler, 2000).

Beyond motivational framing, additional frame alignment processes are necessary in order to supply a continuing sense of urgency to the movement. *Frame alignment* is the linking together of interpretive orientations of individuals and social movement organizations so that there is congruence between individuals' interests, beliefs, and values and the movement's ideologies, goals, and activities (Snow et al., 1986). Four distinct frame alignment processes occur in social movements: (1) *frame bridging* is the process by which movement organizations reach individuals who already share the same world view as the organization, (2) *frame amplification* occurs when movements appeal to deeply held values and beliefs in the general population and link those to movement issues so that people's preexisting value commitments serve as a "hook" that can be used to recruit them, (3) *frame extension* occurs when movements enlarge the boundaries of an initial frame to incorporate other issues that appear to be of importance to potential participants, and (4) *frame transformation* refers to the process whereby the creation and maintenance of new values, beliefs, and meanings induce movement participation by redefining activities and events in such a manner that people believe they must become involved in collective action (Buechler, 2000). Some or all of these frame alignment processes are used by social movements as they seek to define grievances and recruit participants.

Sociologist William Gamson (1995) believes that social movements borrow, modify, or create frames as they seek to advance their goals. For example, social activism regarding nuclear disasters involves differential framing over time. When the Fermi reactor had a serious partial meltdown in 1966, there was no sustained social opposition. However, when Pennsylvania's Three Mile Island nuclear power plant experienced a similar problem in 1979, the situation was defined as a technological disaster. At that time, social movements opposing nuclear power plants grew in number, and each intensified its efforts to reduce or eliminate such plants. This example shows that people react toward something on the basis of the meaning that thing has for them, and such meanings typically emerge from the political culture, including the social activism, of that specific era (Buechler, 2000).

Frame analysis provides new insights on how social movements emerge and grow when people are faced with problems such as technological disasters, about which greater ambiguity typically exists, and when people are attempting to "name" the problems associated with things such as nuclear or chemical contamination. However, frame analysis has been criticized for its "ideational biases" (McAdam, 1996). According to the sociologist Doug McAdam (1996), frame analyses of social movements have looked almost exclusively at ideas and their formal expression, whereas little attention has been paid to other significant factors such as movement tactics, mobilizing structures, and changing political opportunities that influence the signifying work of movements. In this context, "political opportunity" means government structure, public policy, and political conditions that set the boundaries for change and political action. These boundaries are crucial variables in explaining why various social movements have different outcomes (Meyer and Staggenborg, 1996; Gotham, 1999).

New Social Movement Theory

New social movement theory looks at a diverse array of collective actions and the manner in which those actions are based in politics, ideology, and culture. It also incorporates factors of identity, including race, class, gender, and sexuality, as sources of collective action and social movements. Examples of "new social movements" include ecofeminism and environmental

justice movements. Ecofeminism emerged in the late 1970s and early 1980s out of the feminist, peace, and ecology movements. Prompted by the near-meltdown at the Three Mile Island nuclear power plant, ecofeminists established World Women in Defense of the Environment. *Ecofeminism* is based on the belief that patriarchy is a root cause of environmental problems. According to ecofeminists, patriarchy not only results in the domination of women by men but also contributes to a belief that nature is to be possessed and dominated, rather than treated as a partner (see Ortner, 1974; Merchant, 1983, 1992; Mies and Shiva, 1993).

Another "new social movement" focuses on environmental justice and the intersection of race and class in the environmental struggle. Sociologist Stella M. Capek (1993) investigated a contaminated landfill in the Carver Terrace neighborhood of Texarkana, Texas, and found that residents were able to mobilize for change and win a federal buyout and relocation by symbolically linking their issue to a larger *environmental justice* framework. Since the 1980s, the emerging environmental justice movement has focused on the issue of **environmental racism—the belief that a disproportionate number of hazardous facilities (including industries such as waste disposal/treatment and chemical plants) are placed in low-income areas populated primarily by people of color** (Bullard and Wright, 1992). These areas have been

Andrew Lichtenstein/The Image Works

Referred to as "Cancer Alley," this area of Baton Rouge, Louisiana, is home to a predominantly African American population and also to many refineries that heavily pollute the region. Sociologists suggest that environmental racism is a significant problem in the United States and other nations. What do you think?

Concept Table 20.A SOCIAL MOVEMENT THEORIES	
	Key Components
RELATIVE DEPRIVATION	People who are discontent when they compare their achievements with those of others consider themselves relatively deprived and join social movements in order to get what they view as their "fair share," especially when there is an upswing in the economy followed by a decline.
VALUE-ADDED	Certain conditions are necessary for a social movement to develop: (1) structural conduciveness, such that people are aware of a problem and have the opportunity to engage in collective action; (2) structural strain, such that society or the community cannot meet people's expectations for taking care of the problem; (3) growth and spread of a generalized belief as to causes and effects of and possible solutions to the problem; (4) precipitating factors, or events that reinforce the beliefs; (5) mobilization of participants for action; and (6) social control factors, such that society comes to allow the movement to take action.
RESOURCE MOBILIZATION	A variety of resources (money, members, access to media, and material goods such as equipment) are necessary for a social movement; people participate only when they feel the movement has access to these resources.
SOCIAL CONSTRUCTION THEORY: FRAME ANALYSIS	Based on the assumption that social movements are an interactive, symbolically defined, and negotiated process involving participants, opponents, and bystanders, frame analysis is used to determine how people assign meaning to activities and processes in social movements.
NEW SOCIAL MOVEMENT	The focus is on sources of social movements, including politics, ideology, and culture. Race, class, gender, sexuality, and other sources of identity are also factors in movements such as ecofeminism and environmental justice.

left out of most of the environmental cleanup that has taken place in the last two decades (Schneider, 1993). Capek concludes that linking Carver Terrace with environmental justice led to it being designated as a cleanup site. She also views this as an important turning point in new social movements: "Carver Terrace is significant not only as a federal buyout and relocation of a minority community, but also as a marker of the emergence of environmental racism as a major new component of environmental social movements in the United States" (Capek, 1993: 21).

Sociologist Steven M. Buechler (2000) has argued that theories pertaining to twenty-first-century social movements should be oriented toward the structural, macrolevel contexts in which movements arise. These theories should incorporate both political and cultural dimensions of social activism:

Social movements are historical products of the age of modernity. They arose as part of a sweeping social, political, and intellectual change that led a significant number of people to view society as a social construction that was susceptible to social reconstruction through concerted collective effort.

Thus, from their inception, social movements have had a dual focus. Reflecting the political, they have always involved some form of challenge to prevailing forms of authority. Reflecting the cultural, they have always operated as symbolic laboratories in which reflexive actors pose questions of meaning, purpose, identity, and change. (Buechler, 2000: 211)

Concept Table 20.A summarizes the main theories of social movements.

As we have seen, social movements may be an important source of social change. Throughout this text, we have examined a variety of social problems that have been the focus of one or more social movements during the past hundred years. In the process of bringing about change, most movements initially develop innovative ways to get their ideas across to decision makers and the public. Some have been successful in achieving their goals; others have not. As the historian Robert A. Goldberg (1991) has suggested, gains made by social movements may be fragile, acceptance brief, and benefits minimal and easily lost. For this reason, many groups focus on preserving their gains while simultaneously fighting for those that they believe they still desire.

SOCIAL CHANGE IN THE FUTURE

In this chapter, we have focused on collective behavior and social movements as potential forces for social change in contemporary societies. A number of other factors also contribute to social change, including the physical environment, population trends, technological development, and social institutions.

The Physical Environment and Change

Changes in the physical environment often produce changes in the lives of people; in turn, people can make dramatic changes in the physical environment, over which we have only limited control. Throughout history, natural disasters have taken their toll on individuals and societies. Major natural disasters—including hurricanes, floods, and tornados—can devastate an entire population. Even comparatively "small" natural disasters change the lives of many people. As the sociologist Kai Erikson (1976, 1994) has suggested, the trauma that people experience from disasters may outweigh the actual loss of physical property—memories of such events can haunt people for many years.

Some natural disasters are exacerbated by human decisions. For example, floods are viewed as natural disasters, but excessive development may contribute to a flood's severity. As office buildings, shopping malls, industrial plants, residential areas, and highways are developed, less land remains as groundcover to absorb rainfall. When heavier-than-usual rains occur, flooding becomes inevitable; some regions of the United States have remained under water for days or even weeks in recent years. Clearly, humans cannot control the rain, but human decisions can worsen the consequences.

Since flooding is one of the many problems that face the world today, it may seem strange that one of the major concerns in the twenty-first century is the availability of water. Many experts warn that water is a finite resource that is necessary for both human survival and the production of goods. However, water is being wasted and polluted, and the supply of *potable* (drinkable) water is limited. People are causing—or at least contributing to—that problem (see Box 20.3).

People also contribute to changes in the Earth's physical condition. Through soil erosion and other degradation of grazing land, often at the hands of people, an estimated 24 billion tons of topsoil is lost annually. As people clear forests to create farmland and pastures and to acquire lumber and firewood, the Earth's tree cover continues to diminish. As millions of people drive motor vehicles, the amount of carbon dioxide in the environment continues to rise each year, possibly resulting in global warming.

Just as people contribute to change in the physical environment, human activities must also be adapted to changes in the environment. For example, we are being warned to stay out of the sunlight because of increases in ultraviolet rays, a cause of skin cancer, as a result of the accelerating depletion of the ozone layer. If the ozone warnings are accurate, the change in the physical environment will dramatically affect those who work or spend their leisure time outside.

Population and Change

Changes in population size, distribution, and composition affect the culture and social structure of a society and change the relationships among nations. As discussed in Chapter 19, the countries experiencing the most rapid increases in population have a less-developed infrastructure to deal with those changes. How will nations of the world deal with population growth as the global population continues to move toward seven billion? Only time will provide a response to this question.

In the United States, a shift in population distribution from central cities to suburban and exurban areas has produced other dramatic changes. Central cities have experienced a shrinking tax base as middle- and upper-middle-income residents and businesses have moved to suburban and outlying areas. As a result, schools and public services have declined in many areas, leaving those people with the greatest needs with the fewest public resources and essential services. The changing composition of the U.S. population has resulted in children from more diverse cultural backgrounds entering school, producing a demand for new programs and changes in curricula. An increase in the birth rate has created a need for more child care; an increase in the older population has created a need for services such as medical care, placing greater demands on programs such as Social Security.

As we have seen in previous chapters, population growth and the movement of people to urban areas have brought profound changes to many regions and intensified existing social problems. Among other

Box 20.3 SOCIOLOGY AND SOCIAL POLICY

The Fight Over Water Rights

Who controls water rights—the rights to the world's finite supply of potable (drinkable) water? (Figure 1 shows how finite that supply is.) Is the water supply something that individual people or governments should be able to *own*? Answers to that question may vary depending on whether an individual is one of the "haves" or the "have-nots" with regard to water. Court decisions speak of water rights in terms of *sovereignty* with regard to governmental actions and *riparian rights* with

Figure 1   The Earth's Supply of Potable Water

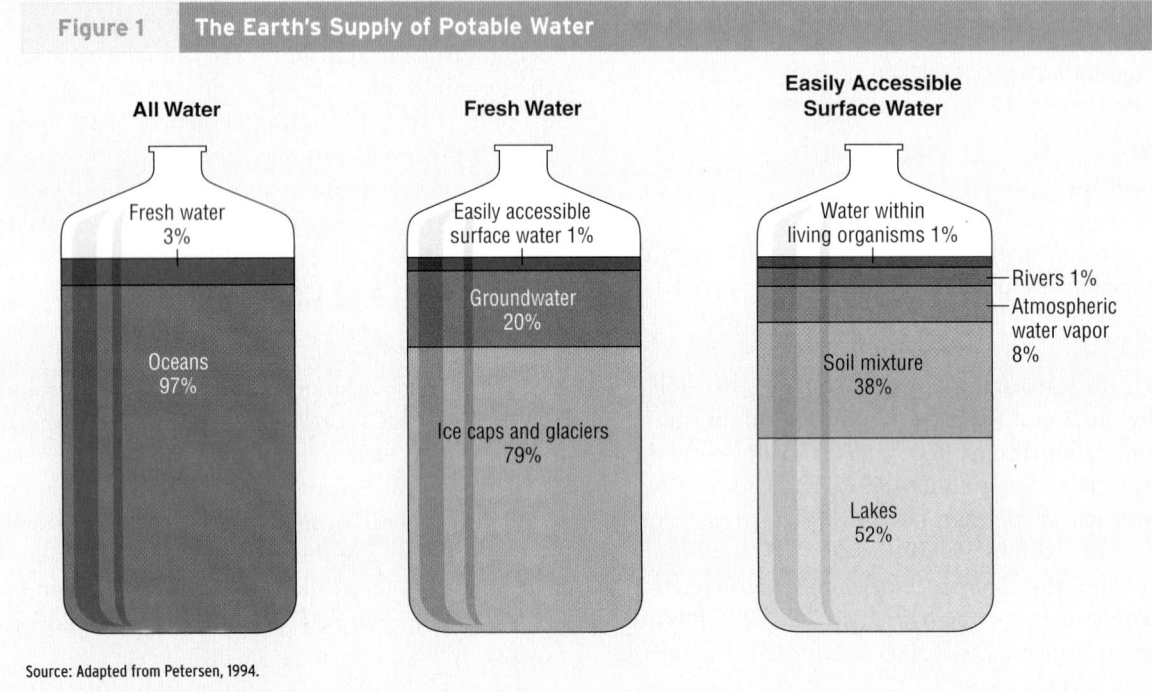

Source: Adapted from Petersen, 1994.

factors, growth in the global population is one of the most significant driving forces behind environmental concerns such as the availability and use of natural resources.

Technology and Change

Technology is an important force for change; in some ways, technological development has made our lives much easier. Advances in communication and transportation have made instantaneous worldwide communication possible but have also brought old belief systems and the status quo into question as never before. Today, we are increasingly moving information instead of people—and doing it almost instantly. Advances in science and medicine have made significant changes in people's lives in high-income countries. The light bulb, the automobile, the airplane, the as-

sembly line, and recent high-tech developments have contributed to dramatic changes in people's lives. Individuals in high-income nations have benefited from the use of the technology; those in low-income nations may have paid a disproportionate share of the cost of some of these inventions and discoveries.

Scientific advances will continue to affect our lives, from the foods we eat to our reproductive capabilities. Genetically engineered plants have been developed and marketed in recent years, and biochemists are creating potatoes, rice, and cassava with the same protein value as meat (Petersen, 1994). Advances in medicine have made it possible for those formerly unable to have children to procreate; women well beyond menopause are now able to become pregnant with the assistance of medical technology. Advances in medicine have also increased the human lifespan, especially for white and middle- or upper-class indi-

regard to individual or group water rights. *Riparian* refers to the rights of a person (or group) to water by virtue of owning or occupying the bank of a river or lake. Historically, if a river passed through your property, you had the right to take and use as much of its water as you wanted or needed, without regard to the effect this had on people farther down the river. Accordingly, those who lived higher up a river could build a dam and divert water into lavish agricultural irrigation projects even when this resulted in water rationing for people farther down the river. Often, untreated wastewater was intentionally discharged back into the river, again without regard to the effect such action had on those farther down the river.

In the United States—as well as in the rest of the world—the assertion of riparian rights is being challenged by those whose water supply is threatened, whether by dwindling supplies or by pollution. A nation, a state, or a city may assert absolute sovereignty over the natural resources within its territory (such as its water supply), whereas nations, states, or cities farther downstream may assert another principle—governmental integrity, or the right to a supply that is adequate (in terms of both quantity and quality) to meet their own survival requirements.

One example of this conflict can be found in the Edwards Aquifer debate in Texas (an *aquifer* is an underground water supply). The Edwards Aquifer reaches from Austin to San Antonio (the nation's tenth-largest city) and beyond. San Antonio relies on the aquifer for its potable water supply, but other—smaller—cities and many rural businesses also depend on the aquifer, including farmers who need the water to grow their crops. When the supply in the aquifer drops, the farmers must compete with the cities for water; attorneys representing both groups often go before the courts and regulatory agencies to argue over how much of the water each should be entitled to receive and use.

Referred to as the "western water wars," water policies have angered many people in Nevada and California. Over the past eighty years in Nevada, billions of tax dollars have been used to finance irrigation projects that siphoned off water from Pyramid Lake and diverted it to alfalfa farms and cattle ranches in the middle of the high desert east of Reno. However, the U.S. government recently started buying back much of that water and giving it to the Pyramid Lake Paiute Indians to restore fish runs and wetlands. The owners of the irrigated farms in the desert are extremely frustrated at this change in policy (Egan, 1997). In California, large volumes of water are being transferred from the big farms in the Central Valley to help depleted fish runs in the San Joaquin and Sacramento rivers, and an area "sucked dry" of its water by Los Angeles has demanded its water back (Brandon, 1997).

What will the future hold in regard to water? Many people agree that water policies are necessary; they just do not agree on what those policies should be. However, most of us believe that taking care of the environment—for today and for the future—is a worthy and necessary goal. Maintaining an adequate and nonpolluted supply of water is an integral part of that task; however, as you can see, social policy in this area often produces a great deal of conflict, and it is likely to produce even greater conflict in the future.

viduals in high-income nations; medical advances have also contributed to the declining death rate in low-income nations, where birth rates have not yet been curbed.

Just as technology has brought about improvements in the quality and length of life for many, it has also created the potential for new disasters, ranging from global warfare to localized technological disasters at toxic waste sites. As the sociologist William Ogburn (1966) suggested, when a change in the material culture occurs in society, a period of *cultural lag* follows in which the nonmaterial (ideological) culture has not caught up with material development. The rate of technological advance at the level of material culture today is mind-boggling. Many of us can never hope to understand technological advances in the areas of artificial intelligence, holography, virtual reality, biotechnology, cold fusion, and robotics.

One of the ironies of twenty-first-century high technology is the increased vulnerability that results from the increasing complexity of such systems. As futurist John L. Petersen (1994: 70) notes, "The more complex a system becomes, the more likely the chance of system failure. There are unknown secondary effects and particularly vulnerable nodes." He also asserts that most of the world's population will not participate in the technological revolution that is occurring in high-income nations (Petersen, 1994).

Technological disasters may result in the deaths of tens of thousands, especially if we think of modern warfare as a technological disaster. Nuclear power, which can provide energy for millions, can also be the source of a nuclear war that could devastate the planet. As a government study on even limited nuclear war concluded,

Natural resources would be destroyed; surviving equipment would be designed to use materials and skills that might no longer exist; and indeed some regions might be almost uninhabitable. Furthermore, pre-war patterns of behavior would surely change, though in unpredictable ways. (U.S. Congress, 1979, qtd. in Howard, 1990: 320)

Even when lives are not lost in technological disasters, families are uprooted and communities cease to exist as people are relocated. In many cases, the problem is not solved; people are moved either temporarily or permanently away from the site.

Social Institutions and Change

Many changes occurred in the family, religion, education, the economy, and the political system during the twentieth century and early in the twenty-first century. As we saw in Chapter 15, the size and composition of families in the United States changed with the dramatic increase in the number of single-person and single-parent households. Changes in families produced changes in the socialization of children, many of whom spend large amounts of time in front of a television set or in child-care facilities outside their own homes. Although some political and religious leaders advocate a return to "traditional" family life, numerous scholars argue that such families never worked quite as well as some might wish to believe.

Public education changed dramatically in the United States during the twentieth century. This country was one of the first to provide "universal" education for students regardless of their ability to pay. As a result, at least until recently, the United States has had one of the most highly educated populations in the world. Today, the United States still has one of the best public education systems in the world for the top 15 percent of the students, but it badly fails the bottom 25 percent. As the nature of the economy changes, schools almost inevitably will have to change, if for no other reason than demands from leaders in business and industry for an educated work force that allows U.S. companies to compete in a global economic environment.

Political systems experienced tremendous change and upheaval in some parts of the world during the twentieth century. The United States participated in two world wars and numerous other "conflicts," the largest and most divisive being in Vietnam and the most recent being the war on terrorism. The cost of these wars and conflicts, and of increased security measures as a result of terrorism, is staggering.

A new concept of world security is emerging, requiring the cooperation of high-income countries in

Jeff Greenberg/PhotoEdit

Pollution of lakes, rivers, and other bodies of water has an adverse effect on food supplies, air quality, and the entire environment. What influence does a "business as usual" approach have on environmental quality in your area?

halting the proliferation of weapons of mass destruction and combating terrorism. That concept also requires the cooperation of *all* nations in reducing the plight of the poorest people in the low-income countries of the world.

Although we have examined changes in the physical environment, population, technology, and social institutions separately, they all operate together in a complex relationship, sometimes producing large, unanticipated consequences. As we move further into the twenty-first century, we need new ways of conceptualizing social life at both the macrolevel and the microlevel. The sociological imagination helps us think about how personal troubles—regardless of our race, class, gender, age, sexual orientation, or physical abilities and disabilities—are intertwined with the public issues of our society and the global community of which we are a part. As one analyst noted regarding Lois Gibbs and Love Canal,

If Love Canal has taught Lois Gibbs—and the rest of us—anything, it is that ordinary people become very smart very quickly when their lives are threatened. They become adept at detecting absurdity, even when it is concealed in bureaucratese and scientific jargon. Lois Gibbs learned that one cannot always rely on government to act in the best interests of ordinary citizens—at least, not without considerable prodding. She determined that she would prod them until her objectives were attained. She led one of the most successful, single-purpose grass roots efforts of our time. (M. Levine, 1982: xv)

Box 20.4 YOU CAN MAKE A DIFFERENCE

Recycling for Tomorrow

Numerous social movements have called our attention to the fact that our use of resources, and hence the production of waste, is a major source of pollution in industrialized nations.

Environmental advocates have suggested that recycling is an important first step that any of us can take to improve our communities. Here are a few questions about the basic facts of recycling (the answers appear below):

1. How many times can a glass bottle be recycled?
   a. 1
   b. 5
   c. 10
   d. an unlimited number of times
2. How much paper (paper products) is currently recycled in the United States?
   a. about 25%
   b. about 50%
   c. about 75%
   d. almost 100%
3. Which of the following is not one of the three Rs of recycling?
   a. reclaim
   b. reduce
   c. recycle
   d. reuse
4. Which type of soft-drink container is the most recycled?
   a. aluminum cans
   b. boxes
   c. glass
   d. plastic bottles

5. What is the most recycled product on Earth (by percentage)?
   a. appliances
   b. automobiles
   c. boats
   d. food

How did you score on this brief quiz? If each of us not only learns about recycling but also implements plans in our household, we can become part of the solution rather than part of the environmental problem.

Environmental sociologists have determined that in many cases, recycling helps diminish waste disposal problems by reducing the amount of waste hauled off to landfills. Paper products (magazines, newspapers, and packaging), glass and plastic bottles, aluminum and steel cans, and appliances (refrigerators, home computers, and televisions) are all recyclable.

Advocates believe that recycling is worth the time and effort it takes because the process reduces the amount of waste and saves energy and natural resources. From a sociological perspective, recycling may also create a sense of community among people as they work toward a cleaner city and, in some cases, use recycling programs as a way to raise funds for schools, youth groups, churches, and other community-based organizations. To learn more about starting a recycling program, visit the following Web site:

http://environment.about.com/library/weekly

Answers: 1–d, 2–b, 3–a, 4–a, 5–b

WRITING IN SOCIOLOGY ASSIGNMENT

Using the sociological imagination that you have gained in this course, what are some positive steps that you believe might be taken in the United States to make our society a better place  for everyone? What types of collective behavior and/or social movements might be required in order to take those steps?

Taking care of the environment is an example of something that government and each of us as individuals can do to help (see Box 20.4). And it is vitally important that we all do everything that we can in order to protect the environment.

A Few Final Thoughts

In this text, we have covered a substantial amount of material, examined different perspectives on a wide variety of social issues, and have suggested different methods by which to deal with them. The purpose of this text is not to encourage you to take any particular point of view; rather, it is to allow you to understand different viewpoints and ways in which they may be helpful to you and to society in dealing with the issues of the twenty-first century. Possessing that understanding, we can hope that the future will be something we can all look forward to—producing a better way of life, not only in this country but worldwide as well.

CHAPTER REVIEW

■ **What is the relationship between social change and collective behavior?**

Social change—the alteration, modification, or transformation of public policy, culture, or social institutions over time—is usually brought about by collective behavior, which is defined as relatively spontaneous, unstructured activity that typically violates established social norms.

■ **When is collective behavior likely to occur?**

Collective behavior occurs when some common influence or stimulus produces a response from a relatively large number of people.

■ **What is a crowd?**

A crowd is a relatively large number of people in one another's immediate presence. Sociologist Howard Blumer divided crowds into four categories: (1) casual crowds, (2) conventional crowds, (3) expressive crowds, and (4) acting crowds (including mobs, riots, and panics). A fifth type of crowd is a protest crowd.

■ **What causes crowd behavior?**

Social scientists have developed several theories to explain crowd behavior. Contagion theory asserts that a crowd takes on a life of its own as people are transformed from rational beings into part of an organism that acts on its own. A variation on this is social unrest and circular reaction—people express their discontent to others, who communicate back similar feelings, resulting in a conscious effort to engage in the crowd's behavior. Convergence theory asserts that people with similar attributes find other like-minded persons with whom they can release underlying personal tendencies. Emergent norm theory asserts that as a crowd develops, it comes up with its own norms that replace more conventional norms of behavior.

■ **What are the primary forms of mass behavior?**

Mass behavior is collective behavior that occurs when people respond to the same event in the same way even if they are not in geographic proximity to one another. Rumors, gossip, mass hysteria, fads and fashions, and public opinion are forms of mass behavior.

■ **What are the major types of social movements, and what are their goals?**

A social movement is an organized group that acts consciously to promote or resist change through collective action. Reform, revolutionary, religious, and alternative movements are the major types identified by sociologists. Reform movements seek to improve society by changing some specific aspect of the social structure.

Revolutionary movements seek to bring about a total change in society—sometimes by the use of terrorism. Religious movements seek to produce radical change in individuals by way of spiritual or supernatural belief systems. Alternative movements seek limited change to some aspect of people's behavior. Resistance movements seek to prevent change or to undo change that has already occurred.

■ **How do social movements develop?**

Social movements typically go through three stages: (1) a preliminary stage (unrest results from a perceived problem), (2) coalescence (people begin to organize), and (3) institutionalization (an organization is developed, and paid staff replaces volunteers in leadership positions).

■ **How do relative deprivation theory, value-added theory, and resource mobilization theory explain social movements?**

Relative deprivation theory asserts that if people are discontented when they compare their accomplishments with those of others similarly situated, they are more likely to join a social movement than are people who are relatively content with their status. According to value-added theory, six conditions are required for a social movement: (1) a perceived problem, (2) a perception that the authorities are not resolving the problem, (3) a spread of the belief to an adequate number of people, (4) a precipitating incident, (5) mobilization of other people by leaders, and (6) a lack of social control. By contrast, resource mobilization theory asserts that successful social movements can occur only when they gain the support of political and economic elites, who provide access to the resources necessary to maintain the movement.

■ **What is the primary focus of research based on frame analysis and new social movement theory?**

Research based on frame analysis often highlights the social construction of grievances through the process of social interaction. Various types of framing occur as problems are identified, remedies are sought, and people feel compelled to take action. Like frame analysis, new social movement theory has been used in research that looks at technological disasters and cases of environmental racism.

KEY TERMS

civil disobedience 665
collective behavior 660
crowd 663
environmental racism 677

QUESTIONS FOR CRITICAL THINKING

1. What types of collective behavior in the United States do you believe are influenced by inequalities based on race/ethnicity, class, gender, age, or disabilities? Why?

2. Which of the four explanations of crowd behavior (contagion theory, social unrest and circular reaction, convergence theory, and emergent norm theory) do you believe best explains crowd behavior? Why?

3. In the text, the Love Canal environmental movement is analyzed in terms of the value-added theory. How would you analyze that movement under the relative deprivation and resource mobilization theories?

RESOURCES ON THE INTERNET

Chapter-Related Web Sites

The following Web sites have been selected for their relevance to the topics in this chapter. These sites are among the more stable, but please note that Web site addresses change frequently. For an updated list of chapter-related Web sites with URL links, please visit the *Sociology in Our Times* Web site (**www.wadsworth.com/KendallSIOT**).

Terrorism Research Center (TRC)

http://www.terrorism.com

The mission of the TRC is to inform the public about terrorism, computer security, law enforcement, national security, and defense policy. The site provides information on current events, profiles of terrorist organizations and counterterrorism groups, original analyses, references, and links to numerous resources.

Envirolink

http://www.envirolink.org

Envirolink is an objective online information clearinghouse providing information about hundreds of environmental action groups. The Web site presents the latest environmental news, education resources, government links, career resources, articles, publications, and much more.

Environmental Protection Agency (EPA)

http://www.epa.gov

The mission of the EPA is to protect the natural environment and human health. Visit the Web site's newsroom to access the latest headlines, locate environmental information about where you live, search a comprehensive list of topics related to the environment, and access links and educational resources.

ONLINE STUDY AND RESEARCH TOOLS

Accompanying this text are many *free* powerful online study tools that will help you master the material in this chapter, help increase your depth of understanding, and help you make the grade!

SocCoach CD-ROM

Use the SocCoach CD-ROM enclosed with this text to help you formulate a customized study plan for this chapter. After you take the Diagnostic Quiz, SocCoach will generate a customized study plan just for you! It will identify sections of the chapter that you should review and will provide videos, charts, graphs, and excerpts from the text to supplement your studies and enhance your understanding. You'll also find fun, interactive activities such as Virtual Explorations and Map the Stats to apply what you've learned and stretch your sociological imagination.

The Companion Web Site for Sociology in Our Times, *Fifth Edition*

www.wadsworth.com/KendallSIOT

Gain an even better grasp on this chapter by going to the companion Web site to take one of the Tutorial Quizzes, use the Flash Cards to master key terms, or check out the many other study aids you'll find there. You'll also find special features such as GSS Data and Census 2000 information that'll put data and resources at your fingertips to help you with that special project or help you as you do some research on your own.

In this chapter, when you see the icon on the left, it alerts you to a specific exercise found in *Wadsworth's Sociology Online Resources and Writing Companion.* This valuable guide shows you how to use Wadsworth's exclusive online resources—*InfoTrac College Edition*, the *Opposing Viewpoints Resource Center*, and *MicroCase Online*—to assist you in your study of sociology and to build essential research and writing skills.

G L O S S A R Y

absolute poverty a level of economic deprivation that exists when people do not have the means to secure the most basic necessities of life.

achieved status a social position that a person assumes voluntarily as a result of personal choice, merit, or direct effort.

activity theory the proposition that people tend to shift gears in late middle age and find substitutes for previous statuses, roles, and activities.

acute diseases illnesses that strike suddenly and cause dramatic incapacitation and sometimes death.

age stratification the inequalities, differences, segregation, or conflict between age groups.

ageism prejudice and discrimination against people on the basis of age, particularly against older people.

agents of socialization the persons, groups, or institutions that teach us what we need to know in order to participate in society.

aggregate a collection of people who happen to be in the same place at the same time but share little else in common.

aging the physical, psychological, and social processes associated with growing older.

agrarian societies societies that use the technology of large-scale farming, including animal-drawn or energy-powered plows and equipment, to produce their food supply.

alienation a feeling of powerlessness and estrangement from other people and from oneself.

animism the belief that plants, animals, or other elements of the natural world are endowed with spirits or life forces having an effect on events in society.

anomie Emile Durkheim's designation for a condition in which social control becomes ineffective as a result of the loss of shared values and of a sense of purpose in society.

anticipatory socialization the process by which knowledge and skills are learned for future roles.

ascribed status a social position conferred at birth or received involuntarily later in life based on attributes over which the individual has little or no control, such as race/ethnicity, age, and gender.

assimilation a process by which members of subordinate racial and ethnic groups become absorbed into the dominant culture.

authoritarian leaders people who make all major group decisions and assign tasks to members.

authoritarian personality a personality type characterized by excessive conformity, submissiveness to authority, intolerance, insecurity, a high level of superstition, and rigid, stereotypic thinking.

authoritarianism a political system controlled by rulers who deny popular participation in government.

authority power that people accept as legitimate rather than coercive.

bilateral descent a system of tracing descent through both the mother's and father's sides of the family.

body consciousness a term that describes how a person perceives and feels about his or her body.

bourgeoisie (or **capitalist class**) Karl Marx's term for the class that consists of those who own the means of production.

bureaucracy an organizational model characterized by a hierarchy of authority, a clear division of labor, explicit rules and procedures, and impersonality in personnel matters.

bureaucratic personality a psychological construct that describes those workers who are more concerned with following correct procedures than they are with getting the job done correctly.

capitalism an economic system characterized by private ownership of the means of production, from which personal profits can be derived through market competition and without government intervention.

capitalist class (or **bourgeoisie**) Karl Marx's term for the class that consists of those who own and control the means of production.

caste system a system of social inequality in which people's status is permanently determined at birth based on their parents' ascribed characteristics.

category a number of people who may never have met one another but share a similar characteristic, such as education level, age, race, or gender.

charismatic authority power legitimized on the basis of a leader's exceptional personal qualities.

chronic diseases illnesses that are long term or lifelong and that develop gradually or are present from birth.

chronological age a person's age based on date of birth.

church a large, bureaucratically organized religious organization that tends to seek accommodation with the larger society in order to maintain some degree of control over it.

civil disobedience nonviolent action that seeks to change a policy or law by refusing to comply with it.

civil religion the set of beliefs, rituals, and symbols that makes sacred the values of the society and places the nation in the context of the ultimate system of meaning.

class conflict Karl Marx's term for the struggle between the capitalist class and the working class.

class system a type of stratification based on the ownership and control of resources and on the type of work that people do.

classism the belief that persons in the upper or privileged class are superior to those in the lower or working class, particularly in regard to values, behavior, and lifestyles.

cohabitation a situation in which two people live together, and think of themselves as a couple, without being legally married.

cohort a group of people born within a specified period in time.

collective behavior voluntary, often spontaneous activity that is engaged in by a large number of people and typically violates dominant-group norms and values.

comparable worth (or **pay equity**) the belief that wages ought to reflect the worth of a job, not the gender or race of the worker.

conflict perspectives the sociological approach that views groups in society as engaged in a continuous power struggle for control of scarce resources.

conformity the process of maintaining or changing behavior to comply with the norms established by a society, subculture, or other group.

conglomerate a combination of businesses in different commercial areas, all of which are owned by one holding company.

content analysis the systematic examination of cultural artifacts or various forms of communication to extract thematic data and draw conclusions about social life.

contingent work part-time work, temporary work, or subcontracted work that offers advantages to employers but that can be detrimental to the welfare of workers.

control group in an experiment, the group containing the subjects who are not exposed to the independent variable.

conventional (street) crime all violent crime, certain property crimes, and certain morals crimes.

core nations according to world systems theory, dominant capitalist centers characterized by high levels of industrialization and urbanization.

corporate crime illegal acts committed by corporate employees on behalf of the corporation and with its support.

corporations large-scale organizations that have legal powers, such as the ability to enter into contracts and buy and sell property, separate from their individual owners.

correlation a relationship that exists when two variables are associated more frequently than could be expected by chance.

counterculture a group that strongly rejects dominant societal values and norms and seeks alternative lifestyles.

credentialism a process of social selection in which class advantage and social status are linked to the possession of academic qualifications.

crime behavior that violates criminal law and is punishable with fines, jail terms, and other sanctions.

criminology the systematic study of crime and the criminal justice system, including the police, courts, and prisons.

crowd a relatively large number of people who are in one another's immediate vicinity.

crude birth rate the number of live births per 1,000 people in a population in a given year.

crude death rate the number of deaths per 1,000 people in a population in a given year.

cult a religious group with practices and teachings outside the dominant cultural and religious traditions of a society.

cultural capital Pierre Bourdieu's term for people's social assets, including values, beliefs, attitudes, and competencies in language and culture.

cultural imperialism the extensive infusion of one nation's culture into other nations.

cultural lag William Ogburn's term for a gap between the technical development of a society (material culture) and its moral and legal institutions (nonmaterial culture).

cultural relativism the belief that the behaviors and customs of any culture must be viewed and analyzed by the culture's own standards.

cultural transmission the process by which children and recent immigrants become acquainted with the dominant cultural beliefs, values, norms, and accumulated knowledge of a society.

cultural universals customs and practices that occur across all societies.

culture the knowledge, language, values, customs, and material objects that are passed from person to person and from one generation to the next in a human group or society.

culture shock the disorientation that people feel when they encounter cultures radically different from their own and believe they cannot depend on their own taken-for-granted assumptions about life.

deinstitutionalization the practice of rapidly discharging patients from mental hospitals into the community.

demedicalization the process whereby a problem ceases to be defined as an illness or a disorder.

democracy a political system in which the people hold the ruling power either directly or through elected representatives.

democratic leaders leaders who encourage group discussion and decision making through consensus building.

democratic socialism an economic and political system that combines private ownership of some of the means of production, governmental distribution of some essential goods and services, and free elections.

demographic transition the process by which some societies have moved from high birth rates and death rates to relatively low birth rates and death rates as a result of technological development.

demography a subfield of sociology that examines population size, composition, and distribution.

denomination a large organized religion characterized by accommodation to society but frequently lacking in ability or intention to dominate society.

dependency theory the belief that global poverty can at least partially be attributed to the fact that the low-income countries have been exploited by the high-income countries.

dependent variable a variable that is assumed to depend on or be caused by one or more other (independent) variables.

deviance any behavior, belief, or condition that violates cultural norms.

differential association theory the proposition that individuals have a greater tendency to deviate from societal norms when they frequently associate with persons who are more favorable toward deviance than conformity.

diffusion the transmission of cultural items or social practices from one group or society to another.

disability a physical or health condition that stigmatizes or causes discrimination.

discovery the process of learning about something previously unknown or unrecognized.

discrimination actions or practices of dominant-group members (or their representatives) that have a harmful effect on members of a subordinate group.

disengagement theory the proposition that older persons make a normal and healthy adjustment to aging when they detach themselves from their social roles and prepare for their eventual death.

domestic partnerships household partnerships in which an unmarried couple lives together in a committed, sexually intimate relationship and is granted the same rights and benefits as those accorded to married heterosexual couples.

dominant group a group that is advantaged and has superior resources and rights in a society.

dramaturgical analysis the study of social interaction that compares everyday life to a theatrical presentation.

drug any substance—other than food and water—that, when taken into the body, alters its functioning in some way.

dual-earner marriages marriages in which both spouses are in the labor force.

dyad a group composed of two members.

ecclesia a religious organization that is so integrated into the dominant culture that it claims as its membership all members of a society.

economy the social institution that ensures the maintenance of society through the production, distribution, and consumption of goods and services.

education the social institution responsible for the systematic transmission of knowledge, skills, and cultural values within a formally organized structure.

egalitarian family a family structure in which both partners share power and authority equally.

ego according to Sigmund Freud, the rational, reality-oriented component of personality that imposes restrictions on the innate pleasure-seeking drives of the id.

elder abuse a term used to describe physical abuse, psychological abuse, financial exploitation, and medical abuse or neglect of people aged sixty-five or older.

elite model a view of society that sees power in political systems as being concentrated in the hands of a small group of elites whereas the masses are relatively powerless.

endogamy cultural norms prescribing that people marry within their social group or category.

entitlements certain benefit payments made by the government (Social Security, for example).

environmental racism the belief that a disproportionate number of hazardous facilities (including industries such as waste disposal/treatment and chemical plants) are placed in low-income areas populated primarily by people of color.

ethnic group a collection of people distinguished, by others or by themselves, primarily on the basis of cultural or nationality characteristics.

ethnic pluralism the coexistence of a variety of distinct racial and ethnic groups within one society.

ethnocentrism the assumption that one's own culture and way of life are superior to all others.

ethnography a detailed study of the life and activities of a group of people by researchers who may live with that group over a period of years.

ethnomethodology the study of the commonsense knowledge that people use to understand the situations in which they find themselves.

exogamy cultural norms prescribing that people marry outside their social group or category.

experiment a research method involving a carefully designed situation in which the researcher studies the impact of certain variables on subjects' attitudes or behavior.

experimental group in an experiment, this group contains the subjects who are exposed to an independent variable (the experimental condition) to study its effect on them.

expressive leadership an approach to leadership that provides emotional support for members.

extended family a family unit composed of relatives in addition to parents and children who live in the same household.

faith an unquestioning belief that does not require proof or scientific evidence.

families relationships in which people live together with commitment, form an economic unit and care for any young, and consider their identity to be significantly attached to the group.

family of orientation the family into which a person is born and in

which early socialization usually takes place.

family of procreation the family that a person forms by having or adopting children.

feminism the belief that all people—both women and men—are equal and that they should be valued equally and have equal rights.

feminization of poverty the trend in which women are disproportionately represented among individuals living in poverty.

fertility the actual level of childbearing for an individual or a population.

field research the study of social life in its natural setting: observing and interviewing people where they live, work, and play.

folkways informal norms or everyday customs that may be violated without serious consequences within a particular culture.

formal education learning that takes place within an academic setting such as a school, which has a planned instructional process and teachers who convey specific knowledge, skills, and thinking processes to students.

formal organization a highly structured group formed for the purpose of completing certain tasks or achieving specific goals.

functional age a term used to describe observable individual attributes such as physical appearance, mobility, strength, coordination, and mental capacity that are used to assign people to age categories.

functional illiteracy the inability to read and/or write at the skill level necessary for carrying out everyday tasks.

functionalist perspectives the sociological approach that views society as a stable, orderly system.

fundamentalism a traditional religious doctrine that is conservative, is typically opposed to modernity, and rejects "worldly pleasures" in favor of otherworldly spirituality.

Gemeinschaft (guh-MINE-shoft) a traditional society in which social relationships are based on personal bonds of friendship and kinship and on intergenerational stability.

gender the culturally and socially constructed differences between females and males found in the meanings, beliefs, and practices associated with "femininity" and "masculinity."

gender bias   behavior that shows favoritism toward one gender over the other.

gender identity   a person's perception of the self as female or male.

gender role   the attitudes, behavior, and activities that are socially defined as appropriate for each sex and are learned through the socialization process.

gender socialization   the aspect of socialization that contains specific messages and practices concerning the nature of being female or male in a specific group or society.

generalized other   George Herbert Mead's term for the child's awareness of the demands and expectations of the society as a whole or of the child's subculture.

genocide   the deliberate, systematic killing of an entire people or nation.

gentrification   the process by which members of the middle and upper-middle classes, especially whites, move into a central-city area and renovate existing properties.

gerontology   the study of aging and older people.

Gesellschaft   (guh-ZELL-shoft) a large, urban society in which social bonds are based on impersonal and specialized relationships, with little long-term commitment to the group or consensus on values.

goal displacement   a process that occurs in organizations when the rules become an end in themselves rather than a means to an end, and organizational survival becomes more important than achievement of goals.

gossip   rumors about the personal lives of individuals.

government   the formal organization that has the legal and political authority to regulate the relationships among members of a society and between the society and those outside its borders.

groupthink   the process by which members of a cohesive group arrive at a decision that many individual members privately believe is unwise.

Hawthorne effect   a phenomenon in which changes in a subject's behavior are caused by the researcher's presence or by the subject's awareness of being studied.

health   a state of complete physical, mental, and social well-being.

health care   any activity intended to improve health.

health maintenance organizations (HMOs)   companies that provide, for a set monthly fee, total care with an emphasis on prevention to avoid costly treatment later.

hermaphrodite   a person in whom sexual differentiation is ambiguous or incomplete.

hidden curriculum   the transmission of cultural values and attitudes, such as conformity and obedience to authority, through implied demands found in rules, routines, and regulations of schools.

high-income countries (sometimes referred to as **industrial countries**) nations with highly industrialized economies; technologically advanced industrial, administrative, and service occupations; and relatively high levels of national and personal income.

holistic medicine   an approach to health care that focuses on prevention of illness and disease and is aimed at treating the whole person—body and mind—rather than just the part or parts in which symptoms occur.

homogamy   the pattern of individuals marrying those who have similar characteristics, such as race/ethnicity, religious background, age, education, or social class.

homophobia   extreme prejudice directed at gays, lesbians, bisexuals, and others who are perceived as not being heterosexual.

horticultural societies   societies based on technology that supports the cultivation of plants to provide food.

hospice   an organization that provides a homelike facility or home-based care (or both) for people who are terminally ill.

hunting and gathering societies   societies that use simple technology for hunting animals and gathering vegetation.

hypothesis   in research studies, a tentative statement of the relationship between two or more concepts.

id   Sigmund Freud's term for the component of personality that includes all of the individual's basic biological drives and needs that demand immediate gratification.

ideal type   an abstract model that describes the recurring characteristics of some phenomenon (such as bureaucracy).

illegitimate opportunity structures   circumstances that provide an opportunity for people to acquire through illegitimate activities what they cannot achieve through legitimate channels.

impression management (presentation of self)   Erving Goffman's term for people's efforts to present themselves to others in ways that are most favorable to their own interests or image.

income   the economic gain derived from wages, salaries, income transfers (governmental aid), and ownership of property.

independent variable   a variable that is presumed to cause or determine a dependent variable.

individual discrimination   behavior consisting of one-on-one acts by members of the dominant group that harm members of the subordinate group or their property.

industrial societies   societies based on technology that mechanizes production.

industrialization   the process by which societies are transformed from dependence on agriculture and hand-made products to an emphasis on manufacturing and related industries.

infant mortality rate   the number of deaths of infants under 1 year of age per 1,000 live births in a given year.

informal education   learning that occurs in a spontaneous, unplanned way.

informal structure   those aspects of participants' day-to-day activities and interactions that ignore, bypass, or do not correspond with the official rules and procedures of the bureaucracy.

ingroup   a group to which a person belongs and with which the person feels a sense of identity.

institutional discrimination   the day-to-day practices of organizations and institutions that have a harmful impact on members of subordinate groups.

instrumental leadership goal- or task-oriented leadership.

intergenerational mobility the social movement (upward or downward) experienced by family members from one generation to the next.

interlocking corporate directorates members of the board of directors of one corporation who also sit on the board(s) of other corporations.

internal colonialism according to conflict theorists, a practice that occurs when members of a racial or ethnic group are conquered or colonized and forcibly placed under the economic and political control of the dominant group.

interview a research method using a data collection encounter in which an interviewer asks the respondent questions and records the answers.

intragenerational mobility the social movement (upward or downward) of individuals within their own lifetime.

invasion the process by which a new category of people or type of land use arrives in an area previously occupied by another group or land use.

invention the process of reshaping existing cultural items into a new form.

iron law of oligarchy according to Robert Michels, the tendency of bureaucracies to be ruled by a few people.

job deskilling a reduction in the proficiency needed to perform a specific job that leads to a corresponding reduction in the wages for that job.

juvenile delinquency a violation of law or the commission of a status offense by young people.

kinship a social network of people based on common ancestry, marriage, or adoption.

labeling theory the proposition that deviants are those people who have been successfully labeled as such by others.

labor union a group of employees who join together to bargain with an employer or a group of employers over wages, benefits, and working conditions.

laissez-faire leaders leaders who are only minimally involved in decision making and who encourage group members to make their own decisions.

language a set of symbols that expresses ideas and enables people to think and communicate with one another.

latent functions unintended functions that are hidden and remain unacknowledged by participants.

laws formal, standardized norms that have been enacted by legislatures and are enforced by formal sanctions.

liberation theology the Christian movement that advocates freedom from political subjugation within a traditional perspective and the need for social transformation to benefit the poor and downtrodden.

life chances Max Weber's term for the extent to which individuals have access to important societal resources such as food, clothing, shelter, education, and health care.

life expectancy an estimate of the average lifetime of people born in a specific year.

looking-glass self Charles Horton Cooley's term for the way in which a person's sense of self is derived from the perceptions of others.

low-income countries (sometimes referred to as **underdeveloped countries**) nations with little industrialization and low levels of national and personal income.

macrolevel analysis an approach that examines whole societies, large-scale social structures, and social systems.

managed care any system of cost containment that closely monitors and controls health care providers' decisions about medical procedures, diagnostic tests, and other services that should be provided to patients.

manifest functions functions that are intended and/or overtly recognized by the participants in a social unit.

marginal jobs jobs that differ from the employment norms of the society in which they are located.

marriage a legally recognized and/or socially approved arrangement between two or more individuals that carries certain rights and obligations and usually involves sexual activity.

mass a number of people who share an interest in a specific idea or issue but who are not in one another's immediate vicinity.

mass behavior collective behavior that takes place when people (who often are geographically separated from one another) respond to the same event in much the same way.

mass education the practice of providing free, public schooling for wide segments of a nation's population.

master status the most important status that a person occupies.

material culture a component of culture that consists of the physical or tangible creations (such as clothing, shelter, and art) that members of a society make, use, and share.

matriarchal family a family structure in which authority is held by the eldest female (usually the mother).

matriarchy a hierarchical system of social organization in which cultural, political, and economic structures are controlled by women.

matrilineal descent a system of tracing descent through the mother's side of the family.

matrilocal residence the custom of a married couple living in the same household (or community) as the wife's parents.

mechanical solidarity Emile Durkheim's term for the social cohesion in preindustrial societies, in which there is minimal division of labor and people feel united by shared values and common social bonds.

medical–industrial complex local physicians, local hospitals, and global health-related industries such as insurance companies and pharmaceutical and medical supply companies that deliver health care today.

medicalization the process whereby nonmedical problems become defined and treated as illnesses or disorders.

medicine an institutionalized system for the scientific diagnosis, treatment, and prevention of illness.

meritocracy a hierarchy in which all positions are rewarded based on people's ability and credentials.

microlevel analysis sociological theory and research that focus on small groups rather than on large-scale social structures.

middle-income countries (sometimes referred to as **developing countries**) nations with industrializing economies and moderate levels of national and personal income.

migration the movement of people from one geographic area to another for the purpose of changing residency.

militarism a term used to describe a societal focus on military ideals and an aggressive preparedness for war.

military–industrial complex the mutual interdependence of the military establishment and private military contractors.

mixed economy an economic system that combines elements of a market economy (capitalism) with elements of a command economy (socialism).

mob a highly emotional crowd whose members engage in, or are ready to engage in, violence against a specific target—a person, a category of people, or physical property.

modernization theory a perspective that links global inequality to different levels of economic development and suggests that low-income economies can move to middle- and high-income economies by achieving self-sustained economic growth.

monarchy a political system in which power resides in one person or family and is passed from generation to generation through lines of inheritance.

monogamy a marriage between two partners, usually a woman and a man.

monotheism belief in a single, supreme being or god who is responsible for significant events such as the creation of the world.

mores strongly held norms with moral and ethical connotations that may not be violated without serious consequences in a particular culture.

mortality the incidence of death in a population.

neolocal residence the custom of a married couple living in their own residence apart from both the husband's and the wife's parents.

network a web of social relationships that links one person with other people and, through them, with other people they know.

nonmaterial culture a component of culture that consists of the abstract or intangible human creations of society (such as attitudes, beliefs, and values) that influence people's behavior.

nontheism a religion based on a belief in divine spiritual forces such as sacred principles of thought and conduct, rather than a god or gods.

nonverbal communication the transfer of information between persons without the use of speech.

norms established rules of behavior or standards of conduct.

nuclear family a family composed of one or two parents and their dependent children, all of whom live apart from other relatives.

occupational (white-collar) crime illegal activities committed by people in the course of their employment or financial affairs.

occupations categories of jobs that involve similar activities at different work sites.

oligopoly a condition existing when several companies overwhelmingly control an entire industry.

organic solidarity Emile Durkheim's term for the social cohesion found in industrial societies, in which people perform very specialized tasks and feel united by their mutual dependence.

organized crime a business operation that supplies illegal goods and services for profit.

outgroup a group to which a person does not belong and toward which the person may feel a sense of competitiveness or hostility.

panic a form of crowd behavior that occurs when a large number of people react to a real or perceived threat with strong emotions and self-destructive behavior.

participant observation a research method in which researchers collect data while being part of the activities of the group being studied.

pastoral societies societies based on technology that supports the domestication of large animals to provide food, typically emerging in mountainous regions and areas with low amounts of annual rainfall.

patriarchal family a family structure in which authority is held by the eldest male (usually the father).

patriarchy a hierarchical system of social organization in which cultural, political, and economic structures are controlled by men.

patrilineal descent a system of tracing descent through the father's side of the family.

patrilocal residence the custom of a married couple living in the same household (or community) as the husband's family.

peer group a group of people who are linked by common interests, equal social position, and (usually) similar age.

peripheral nations according to world systems theory, nations that are dependent on core nations for capital, have little or no industrialization (other than what may be brought in by core nations), and have uneven patterns of urbanization.

personal space the immediate area surrounding a person that the person claims as private.

pink-collar occupations relatively low-paying, nonmanual, semiskilled positions primarily held by women, such as day-care workers, checkout clerks, cashiers, and waitpersons.

pluralist model an analysis of political systems that views power as widely dispersed throughout many competing interest groups.

political action committees (PACs) organizations of special interest groups that solicit contributions from donors and fund campaigns to help elect (or defeat) candidates based on their stances on specific issues.

political crime illegal or unethical acts involving the usurpation of power by government officials, or illegal/unethical acts perpetrated against the government by outsiders seeking to make a political statement, undermine the government, or overthrow it.

political party an organization whose purpose is to gain and hold legitimate control of government.

political socialization the process by which people learn political attitudes, values, and behavior.

political sociology the area of sociology that examines the nature and consequences of power within or between societies as well as the social and political conflicts that lead to changes in the allocation of power.

politics the social institution through which power is acquired and exercised by some people and groups.

polyandry the concurrent marriage of one woman with two or more men.

polygamy the concurrent marriage of a person of one sex with two or more members of the opposite sex.

polygyny the concurrent marriage of one man with two or more women.

polytheism a belief in more than one god.

popular culture the component of culture that consists of activities, products, and services that are assumed to appeal primarily to members of the middle and working classes.

population composition the biological and social characteristics of a population, including age, sex, race, marital status, education, occupation, income, and size of household.

population pyramid a graphic representation of the distribution of a population by sex and age.

positivism a term describing Auguste Comte's belief that the world can best be understood through scientific inquiry.

postindustrial societies societies in which technology supports a service- and information-based economy.

postmodern perspectives the sociological approach that attempts to explain social life in modern societies that are characterized by postindustrialization, consumerism, and global communications.

power according to Max Weber, the ability of people or groups to achieve their goals despite opposition from others.

power elite C. Wright Mills's term for the group made up of leaders at the top of business, the executive branch of the federal government, and the military.

prejudice a negative attitude based on faulty generalizations about members of selected racial and ethnic groups.

presentation of self (impression management) Erving Goffman's term for people's efforts to present themselves to others in ways that are most favorable to their own interests or image.

prestige the respect or regard with which a person or status position is regarded by others.

primary deviance the initial act of rule-breaking.

primary group Charles Horton Cooley's term for a small, less specialized group in which members engage in face-to-face, emotion-based interactions over an extended period of time.

primary labor market the sector of the labor market that consists of high-paying jobs with good benefits that have some degree of security and the possibility of future advancement.

primary sector production the sector of the economy that extracts raw materials and natural resources from the environment.

primary sex characteristics the genitalia used in the reproductive process.

probability sampling choosing participants for a study on the basis of specific characteristics, possibly including such factors as age, sex, race/ethnicity, and educational attainment.

profane the everyday, secular, or "worldly" aspects of life.

professions high-status, knowledge-based occupations.

proletariat (or **working class**) those who must sell their labor to the owners in order to earn enough money to survive.

propaganda information provided by individuals or groups that have a vested interest in furthering their own cause or damaging an opposing one.

public opinion the attitudes and beliefs communicated by ordinary citizens to decision makers.

punishment any action designed to deprive a person of things of value (including liberty) because of some offense the person is thought to have committed.

questionnaire a printed research instrument containing a series of items to which subjects respond.

race a category of people who have been singled out as inferior or superior, often on the basis of physical characteristics such as skin color, hair texture, and eye shape.

racial socialization the aspect of socialization that contains specific messages and practices concerning the nature of one's racial or ethnic status.

racism a set of attitudes, beliefs, and practices that is used to justify the superior treatment of one racial or ethnic group and the inferior treatment of another racial or ethnic group.

random sampling a study approach in which every member of an entire population being studied has the same chance of being selected.

rational choice theory of deviance the belief that deviant behavior occurs when a person weighs the costs and benefits of nonconventional or criminal behavior and determines that the benefits will outweigh the risks involved in such actions.

rational–legal authority power legitimized by law or written rules and procedures. Also referred to as *bureaucratic authority.*

rationality the process by which traditional methods of social organization, characterized by informality and spontaneity, are gradually replaced by efficiently administered formal rules and procedures.

reference group a group that strongly influences a person's behavior and social attitudes, regardless of whether that individual is an actual member.

relative poverty a condition that exists when people may be able to afford basic necessities but are still unable to maintain an average standard of living.

reliability in sociological research, the extent to which a study or research instrument yields consistent results when applied to different individuals at one time or to the same individuals over time.

religion a system of beliefs, symbols, and rituals, based on some sacred or supernatural realm, that guides human behavior, gives meaning to life, and unites believers into a community.

research methods specific strategies or techniques for systematically conducting research.

resocialization the process of learning a new and different set of attitudes, values, and behaviors from those in one's background and experience.

respondents persons who provide data for analysis through interviews or questionnaires.

riot violent crowd behavior that is fueled by deep-seated emotions but is not directed at one specific target.

rituals regularly repeated and carefully prescribed forms of behaviors that symbolize a cherished value or belief.

role a set of behavioral expectations associated with a given status.

role conflict a situation in which incompatible role demands are placed on a person by two or more statuses held at the same time.

role exit a situation in which people disengage from social roles that have been central to their self-identity.

role expectation a group's or society's definition of the way that a specific role *ought* to be played.

role performance how a person *actually* plays a role.

role strain a condition that occurs when incompatible demands are built into a single status that a person occupies.

role-taking the process by which a person mentally assumes the role of another person in order to understand the world from that person's point of view.

routinization of charisma the process by which charismatic authority is succeeded by a bureaucracy controlled by a rationally established authority or by a combination of traditional and bureaucratic authority.

rumor an unsubstantiated report on an issue or subject.

sacred those aspects of life that are extraordinary or supernatural.

sanctions rewards for appropriate behavior or penalties for inappropriate behavior.

Sapir–Whorf hypothesis the proposition that language shapes the view of reality of its speakers.

scapegoat a person or group that is incapable of offering resistance to the hostility or aggression of others.

second shift Arlie Hochschild's term for the domestic work that employed women perform at home after they complete their workday on the job.

secondary analysis a research method in which researchers use existing material and analyze data that were originally collected by others.

secondary deviance the process that occurs when a person who has been labeled a deviant accepts that new identity and continues the deviant behavior.

secondary group a larger, more specialized group in which members engage in more-impersonal, goal-oriented relationships for a limited period of time.

secondary labor market the sector of the labor market that consists of low-paying jobs with few benefits and very little job security or possibility for future advancement.

secondary sector production the sector of the economy that processes raw materials (from the primary sector) into finished goods.

secondary sex characteristics the physical traits (other than reproductive organs) that identify an individual's sex.

sect a relatively small religious group that has broken away from another religious organization to renew what it views as the original version of the faith.

secularization the process by which religious beliefs, practices, and institutions lose their significance in sectors of society and culture.

segregation the spatial and social separation of categories of people by race, ethnicity, class, gender, and/or religion.

self-concept the totality of our beliefs and feelings about ourselves.

self-fulfilling prophecy the situation in which a false belief or prediction produces behavior that makes the originally false belief come true.

semiperipheral nations according to world systems theory, nations that are more developed than peripheral nations but less developed than core nations.

sex the biological and anatomical differences between females and males.

sex ratio a term used by demographers to denote the number of males for every hundred females in a given population.

sexism the subordination of one sex, usually female, based on the assumed superiority of the other sex.

sexual harassment unwanted sexual advances, requests for sexual favors, or other verbal or physical conduct of a sexual nature.

sexual orientation a person's preference for emotional–sexual relationships with members of the opposite sex (heterosexuality), the same sex (homosexuality), or both (bisexuality).

shared monopoly a condition that exists when four or fewer companies supply 50 percent or more of a particular market.

sick role the set of patterned expectations that defines the norms and values appropriate for individuals who are sick and for those who interact with them.

significant others those persons whose care, affection, and approval are especially desired and who are most important in the development of the self.

simple supernaturalism the belief that supernatural forces affect people's lives either positively or negatively.

slavery an extreme form of stratification in which some people are owned by others.

small group a collectivity small enough for all members to be acquainted with one another and to interact simultaneously.

social bond theory the proposition that the probability of deviant behavior increases when a person's ties to society are weakened or broken.

social change the alteration, modification, or transformation of public policy, culture, or social institutions over time.

social construction of reality the process by which our perception of reality is shaped largely by the subjective meaning that we give to an experience.

social control systematic practices developed by social groups to encourage conformity and to discourage deviance.

social Darwinism Herbert Spencer's belief that those species of animals, including human beings, best adapted to their environment survive and prosper, whereas those poorly adapted die out.

social devaluation a situation in which a person or group is considered to have less social value than other individuals or groups.

social distance the extent to which people are willing to interact and establish relationships with members of racial and ethnic groups other than their own.

social epidemiology the study of the causes and distribution of health, disease, and impairment throughout a population.

social exclusion Manuel Castells's term for the process by which certain individuals and groups are systematically barred from access to positions that would enable them to have an autonomous livelihood in keeping with the social standards and values of a given social context.

social facts Emile Durkheim's term for patterned ways of acting, thinking, and feeling that exist *outside* any one individual but that exert social control over each person.

social group a group that consists of two or more people who interact frequently and share a common identity and a feeling of interdependence.

social institution a set of organized beliefs and rules that establishes how a society will attempt to meet its basic social needs.

social interaction the process by which people act toward or respond to other people; the foundation for all relationships and groups in society.

social mobility the movement of individuals or groups from one level in a stratification system to another.

social movement an organized group that acts consciously to promote or resist change through collective action.

social stratification the hierarchical arrangement of large social groups based on their control over basic resources.

social structure the stable pattern of social relationships that exists within a particular group or society.

socialism an economic system characterized by public ownership of the means of production, the pursuit of collective goals, and centralized decision making.

socialization the lifelong process of social interaction through which individuals acquire a self-identity and the physical, mental, and social skills needed for survival in society.

socialized medicine a health care system in which the government owns the medical care facilities and employs the physicians.

society a large social grouping that shares the same geographical territory and is subject to the same political authority and dominant cultural expectations.

sociobiology the systematic study of how biology affects social behavior.

socioeconomic status (SES) a combined measure that, in order to determine class location, attempts to classify individuals, families, or households in terms of factors such as income, occupation, and education.

sociological imagination C. Wright Mills's term for the ability to see the relationship between individual experiences and the larger society.

sociology the systematic study of human society and social interaction.

sociology of family the subdiscipline of sociology that attempts to describe and explain patterns of family life and variations in family structure.

special interest groups political coalitions composed of individuals or groups that share a specific interest that they wish to protect or advance with the help of the political system.

split labor market a term used to describe the division of the economy into two areas of employment, a primary sector or upper tier, composed of higher-paid (usually dominant-group) workers in more-secure jobs, and a secondary sector or lower tier, composed of lower-paid (often subordinate-group) workers in jobs with little security and hazardous working conditions.

state the political entity that possesses a legitimate monopoly over the use of force within its territory to achieve its goals.

status a socially defined position in a group or society characterized by certain expectations, rights, and duties.

status symbol a material sign that informs others of a person's specific status.

stereotypes overgeneralizations about the appearance, behavior, or other characteristics of particular groups.

strain theory the proposition that people feel strain when they are exposed to cultural goals that they are unable to obtain because they do not have access to culturally approved means of achieving those goals.

subcontracting an agreement in which a corporation contracts with other (usually smaller) firms to provide specialized components, products, or services to the larger corporation.

subculture a group of people who share a distinctive set of cultural beliefs and behaviors that differs in some significant way from that of the larger society.

subordinate group a group whose members, because of physical or cultural characteristics, are disadvantaged and subjected to unequal treatment by the dominant group and who regard themselves as objects of collective discrimination.

succession the process by which a new category of people or type of land use gradually predominates in an area formerly dominated by another group or activity.

superego Sigmund Freud's term for the conscience, consisting of the moral and ethical aspects of personality.

survey a poll in which the researcher gathers facts or attempts to determine the relationships between facts.

symbol anything that meaningfully represents something else.

symbolic interactionist perspectives the sociological approach that views society as the sum of the interactions of individuals and groups.

taboos mores so strong that their violation is considered to be extremely offensive and even unmentionable.

technology the knowledge, techniques, and tools that allow people to transform resources into a usable form and the knowledge and skills required to use what is developed.

terrorism the calculated unlawful use of physical force or threats of violence against persons or property in order to intimidate or coerce a government, organization, or individual for the purpose of gaining some political, religious, economic, or social objective.

tertiary deviance deviance that occurs when a person who has been labeled a deviant seeks to normalize the behavior by relabeling it as nondeviant.

tertiary sector production the sector of the economy that is involved in the provision of services rather than goods.

theism a belief in a god or gods.

theory a set of logically interrelated statements that attempts to describe, explain, and (occasionally) predict social events.

total institution Erving Goffman's term for a place where people are isolated from the rest of society for a set period of time and come under the control of the officials who run the institution.

totalitarianism a political system in which the state seeks to regulate all aspects of people's public and private lives.

tracking the assignment of students to specific courses and educational programs based on their test scores, previous grades, or both.

traditional authority power that is legitimized on the basis of long-standing custom.

transnational corporations large corporations that are headquartered in one country but sell and produce goods and services in many countries.

transsexual a person who believes that he or she was born with the body of the wrong sex.

transvestite a male who lives as a woman or a female who lives as a man but does not alter the genitalia.

triad a group composed of three members.

unemployment rate the percentage of unemployed persons in the labor force actively seeking jobs.

universal health care a health care system in which all citizens receive medical services paid for by tax revenues.

unstructured interview an extended, open-ended interaction between an interviewer and an interviewee.

urbanization the process by which an increasing proportion of a population lives in cities rather than in rural areas.

validity in sociological research, the extent to which a study or research instrument accurately measures what it is supposed to measure.

values collective ideas about what is right or wrong, good or bad, and desirable or undesirable in a particular culture.

wage gap a term used to describe the disparity between women's and men's earnings.

war organized, armed conflict between nations or distinct political factions.

wealth the value of all of a person's or family's economic assets, including income, personal property, and income-producing property.

working class (or **proletariat**) those who must sell their labor to the owners in order to earn enough money to survive.

zero population growth the point at which no population increase occurs from year to year.

REFERENCES

AAUW (American Association of University Women). 1995. *How Schools Shortchange Girls/The AAUW Report: A Study of Major Findings on Girls and Education.* New York: Marlowe.

___. 1998. *Gender Gap: Where Schools Still Fail Our Children.* Washington, DC: American Association of University Women. Retrieved Aug. 7, 2003. Online: http://aauw.org/research/girls_education/gg.cfm

___. 2001. *Hostile Hallways: Bullying, Teasing, and Sexual Harassment in School.* Washington, DC: American Association of University Women. Retrieved Aug. 6, 2003. Online: http://www.aauw.org/research/girls_education/hostile.cfm

ABA Banking Journal. 1990. "Easing Borrowers Off the Road to Bankruptcy," p. 42. Quoted in George Ritzer, *Expressing America: A Critique of the Global Credit Card Society.* Thousand Oaks, CA: Pine Forge, 1995, p. 82.

ABC. 2003. "According to Jim." Retrieved Aug. 23, 2003. Online: http://abc.go.com/primetime/accordingtojim/show.html

Aberle, D. F., A. K. Cohen, A. K. Davis, M. J. Leng, Jr., and F. N. Sutton. 1950. "The Functional Prerequisites of Society." *Ethics,* 60 (January): 100–111.

Aberle, David F. 1966. *The Peyote Religion Among the Navaho.* Chicago: Aldine.

Adams, Tom. 1991. *Grass Roots: How Ordinary People Are Changing America.* New York: Citadel.

Adler, Jerry. 1999. "The Truth About High School." *Newsweek* (May 10): 56–58.

Adler, Patricia A., and Peter Adler. 1998. *Peer Power: Preadolescent Culture and Identity.* New Brunswick, NJ: Rutgers University Press.

___. 2003. *Constructions of Deviance: Social Power, Context, and Interaction* (4th ed.). Belmont, CA: Wadsworth.

Adorno, Theodor W., Else Frenkel-Brunswick, Daniel J. Levinson, and R. Nevitt Sanford. 1950. *The Authoritarian Personality.* New York: Harper & Row.

Agency for Healthcare Research and Quality. 2000. "Hospitalization in the United States, 1997." Retrieved Aug. 7, 2003. Online: http://www.ahcpr.gov/data/hcup/factbk1/hcupfbk1.pdf

Agger, Ben. 1993. *Gender, Culture, and Power: Toward a Feminist Postmodern Critical Theory.* Westport, CT: Praeger.

Agnew, Robert. 1985. "Social Control Theory and Delinquency: A Longitudinal Test." *Criminology,* 23: 47–61.

Agonafir, Rebecca. 2002. "Workplace Surveillance of Internet and E-mail Usage." Retrieved July 6, 2002. Online: http://firstclass.wellesley.edu/~ragonafi/cs100.rp1.html

Aiello, John R., and S. E. Jones. 1971. "Field Study of Proxemic Behavior of Young School Children in Three Subcultural Groups." *Journal of Personality and Social Psychology,* 19: 351–356.

Akers, Ronald. 1992. *Drugs, Alcohol, and Society: Social Structure, Process, and Policy.* Belmont, CA: Wadsworth.

Akers, Ronald L. 1998. *Social Learning and Social Structure: A General Theory of Crime and Deviance.* Boston: Northeastern University Press.

Albanese, Catherine L. 1999. *America, Religions and Religion* (3rd ed.). Belmont, CA: Wadsworth.

Albas, Daniel, and Cheryl Albas. 1988. "Aces and Bombers: The Post-Exam Impression Management Strategies of Students." *Symbolic Interaction,* 11 (Fall): 289–302.

Albrecht, Gary L., 1992. *The Disability Business: Rehabilitation in America.* Newbury Park, CA: Sage.

Alexander, Peter, and Roger Gill (Eds.). 1984. *Utopias.* London: Duckworth.

Alireza, Marianne. 1990. "Lifting the Veil of Tradition." *Austin American-Statesman* (Sept. 23): C1, C7.

Allport, Gordon. 1958. *The Nature of Prejudice* (abridged ed.). New York: Doubleday/Anchor.

Altemeyer, Bob. 1981. *Right-Wing Authoritarianism.* Winnipeg: University of Manitoba Press.

___. 1988. *Enemies of Freedom: Understanding Right-Wing Authoritarianism.* San Francisco: Jossey-Bass.

Alter, Jonathan. 1999. "Bridging the Digital Divide." *Newsweek* (Sept. 20): 55.

Alvarez, Lizette. 2003. "A Push to Make la Différence Verboten in the New Europe." *New York Times* (July 27): YT8.

Alwin, Duane, Philip Converse, and Steven Martin. 1985. "Living Arrangements and Social Integration." *Journal of Marriage and the Family,* 47: 319–334.

American Academy of Child and Adolescent Psychiatry. 1997. "Children and Watching TV." Retrieved June 22, 1999. Online: http://www.aacap.org/factsfam/tv.htm

American Anthropological Association. 2001. "What Is Anthropology?" Retrieved July 21, 2001. Online: http://www.aaanet.org/anthbroc.htm

American Association of Suicidology. 2003. "Supplemental Suicide Statistics 2000." Prepared by John L. McIntosh for the American Association of Suicidology. Retrieved June 21, 2003. Online: http://www.iusb.edu/~jmcintos/SuicideStats.html

American Bankruptcy Institute. 2003. "U.S. Bankruptcy Filing Statistics." Retrieved June 14, 2003. Online: http://www.abiworld.org/stats/newstatsfront.html

American Bar Association. 2003. "How Will You Maintain Your Privacy?" ABA Section of Business Law, Committee on Cyberspace Law, Subcommittee on Electronic Commerce. Retrieved June 8, 2003. Online: http://www.safeshopping.org/privacy/main.html

American Management Association. 2001. "Workplace Monitoring and Surveillance Survey, 2001." Retrieved July 8, 2002. Online: http://www.amanet.org/research/pdfs/emsfu_short.pdf

American Psychiatric Association. 1994. *Diagnostic and Statistical Manual of Mental Disorders IV.* Washington, DC: American Psychiatric Association.

American Sociological Association. 1997. *Code of Ethics.* Washington, DC: American Sociological Association (orig. pub. 1971).

American Universities Admission Program. 2003. "Prep School USA." Retrieved Aug. 9, 2003. Online: http://www.auap.com/prepschoolusa.html

Aminzade, Ronald. 1973. "Revolution and Collective Political Violence: The Case of the Working Class of Marseille, France, 1830–1871." Working Paper #86, Center for Research on Social Organization. Ann Arbor: University of Michigan, October 1973.

Amiri, Rina. 2001. "Muslim Women as Symbols—and Pawns." *New York Times* (Nov. 27): A21.

Ammerman, Nancy Tatom. 1997. *Congregation and Community.* New Brunswick, NJ: Rutgers University Press.

Amott, Teresa. 1993. *Caught in the Crisis: Women and the U.S. Economy Today.* New York: Monthly Review Press.

Amott, Teresa, and Julie Matthaei. 1996. *Race, Gender, and Work: A Multicultural Economic History of Women in the United States* (rev. ed.). Boston: South End.

Anders, George. 1996. *Health Against Wealth: HMOs and the Breakdown of Medical Trust.* Boston: Houghton Mifflin.

Andersen, Margaret L., and Patricia Hill Collins (Eds.). 1998. *Race, Class, and Gender: An Anthology* (3rd ed.). Belmont, CA: Wadsworth.

Anderson, David C. 1993. "Ellen Baxter." *New York Times Magazine* (Dec. 19): 36–39.

Anderson, Elijah. 1990. *Streetwise: Race, Class, and Change in an Urban Community.* Chicago: University of Chicago Press.

___. 1999. *Code of the Street: Decency, Violence, and the Moral Life of the Inner City.* New York: Norton.

Anell, Anders, and Michael Willis. 2000. "International Comparison of Health Care Systems Using Resource Profiles." Geneva, Switzerland: World Health Organization. Retrieved Aug. 7, 2003. Online: http://www.who.int/docstore/bulletin/pdf/2000/issue6/bu0585.pdf

Angel, Ronald. 1984. "The Costs of Disability for Hispanic Males." *Social Science Quarterly,* 65: 426–443.

Angel, Ronald J., and Jacqueline L. Angel. 1993. *Painful Inheritance: Health and the New Generation of Fatherless Families.* Madison: University of Wisconsin Press.

Angier, Natalie. 1993. "'Stopit!' She Said. 'Nomore!'" *New York Times Book Review* (Apr. 25): 12.

Annesi, Nicole. 1993. "Like Mother, Like Daughter." In Leslea Newman (Ed.), *Eating Our Hearts Out: Personal Accounts of Women's Relationship to Food.* Freedom, CA: Crossing, pp. 91–95.

Anyon, Jean. 1980. "Social Class and the Hidden Curriculum of Work." *Journal of Education,* 162: 67–92.

___. 1997. *Ghetto Schooling: A Political Economy of Urban Educational Reform.* New York: Teachers College Press.

APA Online 2000. "Psychiatric Effects of Violence." Public Information: APA Fact Sheet Series. Washington, DC: American Psychological Association. Retrieved Apr. 5, 2000. Online: http://www.psych.org/psych/htdocs/public_info/media_violence.html

Applebaum, Eileen R., and Ronald Schettkat. 1989. "Employment and Industrial Restructuring: A Comparison of the U.S. and West Germany." In E. Matzner (Ed.), *No Way to Full Employment.* Research Unit Labor Market and Employment, Discussion Paper FSI 89–16 (July): 394–448.

Appleton, Lynn M. 1995. "The Gender Regimes in American Cities." In Judith A. Garber and Robyne S. Turner (Eds.), *Gender in Urban Research.* Thousand Oaks, CA: Sage, pp. 44–59.

Arendt, Hannah. 1973. *On Revolution.* London: Penguin.

Argyris, Chris. 1960. *Understanding Organizational Behavior.* Homewood, IL: Dorsey.

___. 1962. *Interpersonal Competence and Organizational Effectiveness.* Homewood, IL: Dorsey.

Arnold, Regina A. 1990. "Processes of Victimization and Criminalization of Black Women." *Social Justice,* 17 (3): 153–166.

Asch, Adrienne. 1986. "Will Populism Empower Disabled People?" In Harry G. Boyle and Frank Reissman (Eds.), *The New Populism: The Power of Empowerment.* Philadelphia: Temple University Press, pp. 213–228.

Asch, Solomon E. 1955. "Opinions and Social Pressure." *Scientific American,* 193 (5): 31–35.

___. 1956. "Studies of Independence and Conformity: A Minority of One Against a Unanimous Majority." *Psychological Monographs,* 70 (9) (Whole No. 416).

Ash, Russell. 2003. *The Top 10 of Everything, 2004.* London: DK Publishing.

Ashe, Arthur, and Arnold Rampersad. 1994. *Days of Grace: A Memoir.* New York: Ballantine.

Ashe, Arthur R., Jr. 1988. *A Hard Road to Glory: A History of the African-American Athlete.* New York: Warner.

Associated Press. 1999. "Consumer Groups Accuse Credit Card Industry of Hooking College Students." Retrieved June 1, 1999. Online: http://abcnews.go.com/politics/AP19990609_16.html

___. 2003. "Elderly Driver: My Heart Is Broken." Retrieved July 21, 2003. Online: http://story.news.yahoo.com/news?tmpl=story&cid=519&ncid=519&e=10&u=/ap/20030720/ap_on_re_us/market_accident_67

Atchley, Robert C., and Amanda Barusch. 2004. *Social Forces and Aging: An Introduction to Social Gerontology* (10th ed.). Belmont: CA: Wadsworth.

Aulette, Judy Root. 1994. *Changing Families.* Belmont, CA: Wadsworth.

Aulette, Ken. 1998. *The Highwaymen: Warriors of the Information Superhighway.* San Diego, CA: Harvest/Harcourt.

Austin American-Statesman. 1995. "Lawmakers Ask U.S. to Put Cost of Cereals in Check." (Mar. 8): A1, A7.

___. 2001. "High-Tech Manufacturing in Mexico." (Sept. 24): D1.

Axinn, June. 1989. "Women and Aging: Issues of Adequacy and Equity." In J. D. Barner and Susan O. Mercer (Eds.), *Women As They Age: Challenges, Opportunity and Triumph.* Binghamton, NY: Haworth, pp. 339–362.

Axtell, Roger E. 1991. *Gestures: The Do's and Taboos of Body Language Around the World.* New York: Wiley.

Ayala, Elaine. 1993. "Unfinished Work: Cesar Chavez Is Dead But His Struggle Has Been Reborn, Followers Say." *Austin American-Statesman* (Sept. 6): E1, E4. Quoted from a speech by Cesar Chavez to the Commonwealth Club of San Francisco, Nov. 9, 1984.

Babbie, Earl. 1998. "Plagiarism." Retrieved June 14, 2003. Online: http://www.csubak.edu/ssric/Modules/Other/plagiarism.htm

___. 2001. *The Practice of Social Research* (9th ed.). Belmont, CA: Wadsworth.

Bachman, Ronet. 1992. *Death and Violence on the Reservation: Homicide, Family Violence, and Suicide in American Indian Populations.* New York: Auburn.

Bachu, Amara, and Martin O'Connell. 2001. *Fertility of American Women: June 2000.* Current Population Reports, P20–543RV. Washington, DC: U.S. Census Bureau.

Bagby, Meredith (Ed.). 1997. *Annual Report of the United States of America 1997.* New York: McGraw-Hill.

Bagby, Meredith. 1998. *Annual Report of the United States of America: 1998 Edition.* New York: McGraw-Hill.

Bahr, Howard M., and Theodore Caplow. 1991. "Middletown as an Urban Case Study." In Joe R. Feagin, Anthony M. Orum, and Gideon Sjoberg (Eds.), *A Case for the Case Study.* Chapel Hill: University of North Carolina Press, pp. 80–120.

Bailyn, Bernard. 1960. *Education in the Forming of American Society.* New York: Random House.

Bair, Jennifer, and Gary Gereffi. 2001. "Local Clusters in Global Chains: The Causes and Consequences of Export Dynamism in Torreon's Blue Jeans Industry." *World Development,* 29 (11): 1885–1903.

Baker, Robert. 1993. "'Pricks' and 'Chicks': A Plea for 'Persons.'" In Anne Minas (Ed.),

Gender Basics: Feminist Perspectives on Women and Men. Belmont, CA: Wadsworth, pp. 66–68.
Bales, Kevin. 1999. *Disposable People: New Slavery in the Global Economy.* Berkeley: University of California Press.
Balikci, Asen. 1968. "The Netsilik Eskimos: Adaptive Processes." In Richard B. Lee and Irven DeVore (Eds.), *Man the Hunter.* Chicago: Aldine-Atherton, pp. 78–82.
Ballantine, Jeanne H. 2001. *The Sociology of Education: A Systematic Analysis* (5th ed.). Englewood Cliffs, NJ: Prentice Hall.
Ballara, Marcela. 1991. *Women and Literacy.* Prepared for the UN/NGO Group on Women and Development. Atlantic Highlands, NJ: Zed.
Baltzell, E. Digby. 1958. *Philadelphia Gentlemen: The Making of a National Upper Class.* New York: Free Press.
Band-Aides & Blackboards. 1999. "Forever Heather." Retrieved Oct. 2, 1999. Online: http://funrsc.fairfield.edu/~jfleitas//heather.html
Bane, Mary Jo. 1986. "Household Composition and Poverty: Which Comes First?" In Sheldon H. Danziger and Daniel H. Weinberg (Eds.), *Fighting Poverty: What Works and What Doesn't.* Cambridge, MA: Harvard University Press.
Bane, Mary Jo, and David T. Ellwood. 1994. *Welfare Realities: From Rhetoric to Reform.* Cambridge, MA: Harvard University Press.
Banerjee, Neela. 1999. "Russia's Embryos of Enterprise." *New York Times* (July 20): C1, C23.
Banner, Lois W. 1993. *In Full Flower: Aging Women, Power, and Sexuality.* New York: Vintage.
Barakat, Matthew. 2000. "Survey: Women's Salaries Beat Men's in Some Fields." *Austin American-Statesman* (July 4): D1, D3.
Barber, Benjamin R. 1996. *Jihad vs. McWorld: How Globalism and Tribalism Are Reshaping the World.* New York: Ballantine.
Barboza, David. 2001. "From Golden Arches to Lightning Rod." *International Herald Tribune* (Oct. 15). Retrieved Oct. 22, 2001. Online: http://www.iht.com
Barclay, Bill. 2003. "UEFA Urges Arsenal to Lodge Complaint Over Racist Abuse." *eircom.net.* Retrieved July 17, 2003. Online: http://www.eircom.net
Bardwell, Jill R., Samuel W. Cochran, and Sharon Walker. 1986. "Relationship of Parental Education, Race, and Gender to Sex Role Stereotyping in Five-Year-Old Kindergarteners." *Sex Roles,* 15: 275–281.
Barlow, Hugh D., and David Kauzlarich. 2002. *Introduction to Criminology* (8th ed.). Upper Saddle River, NJ: Prentice Hall.
Barna, George. 1996. *Index of Leading Spiritual Indicators.* Dallas: Word.
Barnard, Chester. 1938. *The Functions of the Executive.* Cambridge, MA: Harvard University Press.
Barnett, Harold. 1979. "Wealth, Crime and Capital Accumulation." *Contemporary Crises,* 3: 171–186.
Baron, Dennis. 1986. *Grammar and Gender.* New Haven, CT: Yale University Press.
Baron, Harold M. 1969. "The Web of Urban Racism." In Louis L. Knowles and Kenneth Prewitt (Eds.), *Institutional Racism in America.* Englewood Cliffs, NJ: Prentice Hall, pp. 134–176.
Barovick, Harriet. 2001. "Hope in the Heartland." *Time* (special edition, July): G1–G3.
Barron, James. 1997. "A Life, Like a Race, May Be Long." *New York Times* (Nov. 1): A13.
Barrymore, Drew, with Todd Gold. 1994. *Little Girl Lost.* New York: Pocket. In Jay David (Ed.), *The Family Secret: An Anthology.* New York: Morrow, pp. 171–199.
Basow, Susan A. 1992. *Gender Stereotypes and Roles* (3rd ed.). Pacific Grove, CA: Brooks/Cole.
Bates, James, and Claudia Eller. 1999. "For Better or Worse, More Mergers Likely." latimes.com (Sept. 8). Retrieved Sept. 10, 1999. Online: http://www.latimes.com
Baudrillard, Jean. 1983. *Simulations.* New York: Semiotext.
___. 1998. *The Consumer Society: Myths and Structures.* London: Sage (orig. pub. 1970).
Baxter, J. 1970. "Interpersonal Spacing in Natural Settings." *Sociology,* 36 (3): 444–456.
BBC. 2002. "Inside China's 'Me' Generation." *BBC News.* Retrieved June 10, 2003. Online: http://news.bbc.co.uk/1/hi/business/2403661.stm
___. 2003. "Citigroup's Deal for Credit Cards in China." *BBC News.* Retrieved June 10, 2003. Online: http://news.bbc.co.uk/1/hi/business/2624459.stm
BBC News. 1999. "World: Asia-Pacific-Japan on Suicide Alert." *BBC Online Network* (July 2). Retrieved Aug. 25, 2001. Online: http://news.bbc.co.uk/hi/english/world/asia-pacific/newsid_383000/383823.stm
___. 2003. "What Caused the Blackouts?" Retrieved Aug. 15, 2003. Online: http://news.bbc.co.uk/1/hi/business/3153237.stm
Becker, Howard S. 1963. *Outsiders: Studies in the Sociology of Deviance.* New York: Free Press.
Beech, Hannah. 2001. "China's Lifestyle Choice." *Time* (Aug. 6): 32.
Beeghley, Leonard. 2000. *The Structure of Social Stratification in the United States* (3rd ed.). Boston: Allyn & Bacon.
Belkin, Lisa. 1994. "Kill for Life?" *New York Times Magazine* (Oct. 30): 47–51, 62–64, 76, 80.
Bell, Daniel. 1973. *The Coming of Post-Industrial Society.* New York: Basic.
___. 1976. *The Cultural Contradictions of Capitalism.* New York: Basic.
Bellah, Robert N. 1967. "Civil Religion." *Daedalus,* 96: 1–21.
Belsky, Janet K. 1999. *The Psychology of Aging: Theory, Research, and Interventions* (3rd ed.). Belmont, CA: Wadsworth.
Bendix, Reinhard. 1971. "Charismatic Leadership." In Reinhard Bendix and Guenther Roth (Eds.), *Scholarship and Partisanship: Essays on Max Weber.* Berkeley: University of California Press, pp. 170–187.
Benford, Robert D. 1993. "'You Could Be the Hundredth Monkey': Collective Action Frames and Vocabularies of Motive Within the Nuclear Disarmament Movement." *Sociological Quarterly,* 34: 195–216.
Benjamin, Lois. 1991. *The Black Elite: Facing the Color Line in the Twilight of the Twentieth Century.* Chicago: Nelson-Hall.
Bennahum, David S. 1999. "For Kosovars, an On-Line Phone Directory of a People in Exile." *New York Times* (July 15): D7.
Benokraitis, Nijole V. 1999. *Marriages and Families: Changes, Choices, and Constraints* (3rd ed.). Upper Saddle River, NJ: Prentice Hall.
___. 2002. *Marriages and Families: Changes, Choices, and Constraints* (4th ed.). Upper Saddle River, NJ: Prentice-Hall.
Benokraitis, Nijole V., and Joe R. Feagin. 1995. *Modern Sexism: Blatant, Subtle, and Covert Discrimination* (2nd ed.). Englewood Cliffs, NJ: Prentice Hall.
Benson, Susan Porter. 1983. "The Customers Ain't God: The Work Culture of Department Store Saleswomen, 1890–1940." In Michael H. Frisch and Daniel J. Walkowitz (Eds.), *Working Class America: Essays on Labor, Community, and American Society.* Urbana: University of Illinois Press, pp. 185–211.
Berg, Bruce L. 1998. *Qualitative Research Methods for the Social Sciences.* Boston: Allyn & Bacon.
Berger, Bennett M. 1988. "Utopia and Its Environment." *Society* (January/February): 37–41.
Berger, Peter. 1963. *Invitation to Sociology: A Humanistic Perspective.* New York: Anchor.
___. 1967. *The Sacred Canopy: Elements of a Sociological Theory of Religion.* New York: Doubleday.
Berger, Peter, and Hansfried Kellner. 1964. "Marriage and the Construction of Reality." *Diogenes,* 46: 1–32.
Berger, Peter, and Thomas Luckmann. 1967. *The Social Construction of Reality: A Treatise in the Sociology of Knowledge.* Garden City, NY: Anchor.
Berliner, David C., and Bruce J. Biddle. 1995. *The Manufactured Crisis: Myths, Fraud, and the Attack on America's Public Schools.* Reading, MA: Addison-Wesley.
Bernard, Jessie. 1982. *The Future of Marriage.* New Haven, CT: Yale University Press (orig. pub. 1973).
Berry, Jeffrey M. 1989. *The Interest Group Society.* Glenview, IL: Scott, Foresman/Little, Brown.
Betschwar, Karl. 2002. "Role Reversal—The Stay-at-Home-Dad's Perspective." Retrieved June 30, 2003. Online: http://www.homedad.org.uk/feature_twins.html
Biagi, Shirley. 1998. *Media/Impact: An Introduction to Mass Media* (3rd ed.). Belmont, CA: Wadsworth.
Biblarz, Arturo, R. Michael Brown, Dolores Noonan Biblarz, Mary Pilgram, and Brent F. Baldree. 1991. "Media Influence on Attitudes Toward Suicide." *Suicide and Life-Threatening Behavior,* 21 (4): 374–385.
Bierlein, Louann A. 1997. "The Charter School Movement." In Diana Ravitch and Joseph P. Viteritti (Eds.), *New Schools for a New Century: The Redesign of Urban Education.* New Haven, CT: Yale University Press, pp. 37–60.
Billingsley, Andrew. 1992. *Climbing Jacob's Ladder: The Enduring Legacy of African-American Families.* New York: Touchstone.
Bizrate.com. 2003. "Grand Theft Auto: Vice City Reviews." Retrieved July 6, 2003. Online: http://www.bizrate.com/marketplace/product_info/review_prod_id—6544713,cat_id
Blanchard, Kendall. 1980. "Sport and Ritual in Choctaw Society: Structure and Perspective." In Helen Schwartzman (Ed.), *Play and Culture.* Champaign, IL: Leisure, pp. 83–91.
Blau, Judith R. 1984. *Architects and Firms: A Sociological Perspective on Architectural Practice.* Cambridge, MA: MIT Press.
Blau, Peter. 1964. *Exchange and Power in Social Life.* New York: Wiley.
___. 1975. *Approaches to the Study of Social Structure.* New York: Free Press.
Blau, Peter M., and Otis Dudley Duncan. 1967. *The American Occupational Structure.* New York: Wiley.
Blau, Peter M., and Marshall W. Meyer. 1987. *Bureaucracy in Modern Society* (3rd ed.). New York: Random House.
Blauner, Bob. 1989. *Black Lives, White Lives.* Berkeley: University of California Press.
Blauner, Robert. 1964. *Alienation and Freedom.* Chicago: University of Chicago Press.
___. 1972. *Racial Oppression in America.* New York: Harper & Row.
Bluestone, Barry, and Bennett Harrison. 1982. *The Deindustrialization of America.* New York: Basic.
Blumberg, Leonard. 1977. "The Ideology of a Therapeutic Social Movement: Alcoholics Anonymous." *Journal of Studies on Alcohol,* 38: 2122–2143.
Blumer, Herbert G. 1946. "Collective Behavior." In Alfred McClung Lee (Ed.), *A New Outline of the Principles of Sociology.* New York: Barnes & Noble, pp. 167–219.
___. 1969. *Symbolic Interactionism: Perspective and Method.* Englewood Cliffs, NJ: Prentice Hall.
___. 1974. "Social Movements." In R. Serge Denisoff (Ed.), *The Sociology of Dissent.* New York: Harcourt, pp. 74–90.
___. 1986. *Symbolic Interactionism: Perspective and Method.* Berkeley: University of California Press (orig. pub. 1969).
Bogardus, Emory S. 1925. "Measuring Social Distance." *Journal of Applied Sociology,* 9: 299–308.
___. 1968. "Comparing Racial Distance in Ethiopia, South Africa, and the United States." *Sociology and Social Research,* 52 (2): 149–156.
Bograd, Michele. 1988. "Feminist Perspectives on Wife Abuse: An Introduction." In Kersti Yllo and Michele Bograd (Eds.), *Feminist Perspectives on Wife Abuse.* Newbury Park, CA: Sage, pp. 11–26.
Bologh, Roslyn Wallach. 1992. "The Promise and Failure of Ethnomethodology from a Feminist Perspective: Comment on Rogers." *Gender & Society,* 6 (2): 199–206.
Bonacich, Edna. 1972. "A Theory of Ethnic Antagonism: The Split Labor Market." *American Sociological Review,* 37: 547–549.
___. 1976. "Advanced Capitalism and Black–White Relations in the United States: A Split Labor Market Interpretation." *American Sociological Review,* 41: 34–51.
Bonvillain, Nancy. 2001. *Women & Men: Cultural Constructs of Gender* (3rd ed.). Upper Saddle River, NJ: Prentice Hall.
Bordewich, Fergus M. 1996. *Killing the White Man's Indian.* New York: Anchor/Doubleday.
Bordo, Susan. 2004. *Unbearable Weight: Feminism, Western Culture, and the Body* (10th anniversary edition). Berkeley: University of California Press.
Bosnia Action Coalition. 2001. "This Week in Bosnia-Hercegovina: Education 'Apartheid.'" Retrieved Aug. 6, 2003. Online: http://www.applicom.com/twibih/twib200112.html
Bourdieu, Pierre. 1984. *Distinction: A Social Critique of the Judgement of Taste.* Trans. Richard Nice. Cambridge, MA: Harvard University Press.
Bourdieu, Pierre, and Jean-Claude Passeron. 1990. *Reproduction in Education, Society and Culture.* Newbury Park, CA: Sage.
Bourdon, Karen H., Donald S. Rae, Ben Z. Locke, William E. Narrow, and Darrel A. Regier. 1992. "Estimating the Prevalence of Mental Disorders in U.S. Adults from the Epidemiological Catchment Area Survey." *Public Health Reports,* 107: 663–668.
Bowles, Samuel, and Herbert Gintis. 1976. *Schooling in Capitalist America: Education and the Contradictions of Economic Life.* New York: Basic.
Boyes, William, and Michael Melvin. 2002. *Economics* (5th ed.). Boston: Houghton Mifflin.
Bozett, Frederick. 1988. "Gay Fatherhood." In Phyllis Bronstein and Carolyn Pape Cowan (Eds.), *Fatherhood Today: Men's Changing Role in the Family.* New York: Wiley, pp. 60–71.
Bramlett, Matthew D., and William D. Mosher. 2001. "First Marriage Dissolution, Divorce, and Remarriage: United States." DHHS publication no. 2001–1250 01–0384 (5/01). Hyattsville, MD: Department of Health and Human Services.
Brand, Pamela A., Esther D. Rothblum, and L. J. Solomon. 1992. "A Comparison of Lesbians, Gay Men, and Heterosexuals on Weight and Restrained Eating." *International Journal of Eating Disorders,* 11: 253–259.
Brandl, Bonnie, and Loree Cook-Daniels. 2002. "Domestic Abuse in Later Life." Retrieved July 26, 2003. Online: http://www.elderabusecenter.org/pdf/research/abusers.pdf
Brandl, Steven, Meghan Stroshine, and James Frank. 2001. "Who Are the Complaint-Prone Officers? An Examination of the Relationship Between Police Officers' Attributes, Arrest Activity, Assignment, and Citizens' Complaints About Excessive Force." *Journal of Criminal Justice,* 29: 521–529.
Brandon, Karen. 1997. "Area Sucked Dry by L.A. Wants Its Water Back." *Austin American-Statesman* (Aug. 10): K3.
Braverman, Harry. 1974. *Labor and Monopoly Capital.* New York: Monthly Review Press.
Breault, K. D. 1986. "Suicide in America: A Test of Durkheim's Theory of Religious and Family Integration, 1933–1980." *American Journal of Sociology,* 92 (3): 628–656.

Bremner, Brian. 2000. "A Japanese Way of Death." *Business Week* (Aug. 22). Retrieved Aug. 25, 2001. Online: http://www.businessweek.com/bwdaily/dnflash/aug2000/nf20000822_176.htm

Briggs, Sheila. 1987. "Women and Religion." In Beth B. Hess and Myra Marx Ferree (Eds.), *Analyzing Gender: A Handbook of Social Science Research*. Newbury Park, CA: Sage, pp. 408–441.

Brill's Content. 1998. "Tale of the Videotapes." (September): 89.

Brint, Steven. 1994. *In an Age of Experts: The Changing Role of Professionals in Politics and Public Life*. Princeton, NJ: Princeton University Press.

Brinton, Mary E. 1989. "Gender Stratification in Contemporary Urban Japan." *American Sociological Review*, 54 (August): 549–564.

Britt, Lory. 1993. "From Shame to Pride: Social Movements and Individual Affect." Paper presented at the 88th annual meeting of the American Sociological Association, Miami, August.

Broder, John M. 2003. "Debris Is Now Leading Suspect in Shuttle Catastrophe." *New York Times* (Feb. 4): A1–A25.

Brody, Jane E. 1997. "Despite the Despair of Depression, Few Men Seek Treatment." *New York Times* (Dec. 30): F7.

Bronfenbrenner, Urie. 1989. "Ecological Systems Theory." In Ross Vasta (Ed.), *Annals of Child Development: A Research Annual* (Vol. 6). Greenwich, CT: JAI.

___. 1990. "Five Critical Processes for Positive Development." From "Discovering What Families Do" in *Rebuilding the Nest: A New Commitment to the American Family*. Retrieved June 29, 1999. Online: http://www.montana.edu/wwwctf/process.html

Brooke, James. 1993a. "Attack on Brazilian Indians Is Worst Since 1910." *New York Times* (Aug. 21): Y3.

___. 1993b. "Slavery on Rise in Brazil, As Debt Chains Workers." *New York Times* (May 23): 3.

Brooks, A. Phillips, and Jeff South. 1995. "School Choice Plans Worry Resegregation Critics." *Austin American-Statesman* (Apr. 9): A1, A18.

Brooks-Gunn, Jeanne. 1986. "The Relationship of Maternal Beliefs About Sex Typing to Maternal and Young Children's Behavior." *Sex Roles*, 14: 21–35.

Brown, Dennis M. 2003. "Rural Tourism: An Annotated Bibliography." United States Department of Agriculture. Retrieved Aug. 9, 2003. Online: http://www.nal.usda.gov/ric/ricpubs/rural_tourism.html#summary

Brown, E. Richard. 1979. *Rockefeller Medicine Men*. Berkeley: University of California Press.

Brown, Jane, Kim Walsh Childers, Karl E. Bauman, and Gary G. Koch. 1990. "The Influence of New Media and Family Structure on Young Adolescents' Television and Radio Use." *Communications Research*, 17: 65–82.

Brown, Phil. 1985. *The Transfer of Care: Psychiatric Deinstitutionalization and Its Aftermath*. Boston: Routledge & Kegan Paul.

Brown, Robert W. 1954. "Mass Phenomena." In Gardner Lindzey (Ed.), *Handbook of Social Psychology* (vol. 2). Reading, MA: Addison-Wesley, pp. 833–873.

Brown, Russell. 2003. "Illusions of Choice: The Selling of *American Idol* and *The Matrix Reloaded*." Retrieved June 21, 2003. Online: http://www.thesimon.com/article_of_week/281

Browne, Colette, and Alice Broderick. 1994. "Asian and Pacific Island Elders: Issues for Social Work Practice and Education." *Social Work*, 39 (3): 252–260.

Bruce, Steve. 1996. *Religion in the Modern World*. New York: Oxford University Press.

Brumberg, Joan Jacobs. 1988. *Fasting Girls: The Emergence of Anorexia as a Modern Disease*. Cambridge, MA: Harvard University Press.

Brustad, Robert J. 1996. "Attraction to Physical Activity in Urban Schoolchildren: Parental Socialization and Gender Influence." *Research Quarterly for Exercise and Sport*, 67: 316–324.

Buckingham, Simon. 1999. "Shift Down to Gear Up." Retrieved June 16, 1999. Online: http://www.unorg.com/a46.htm

Buechler, Steven M. 2000. *Social Movements in Advanced Capitalism: The Political Economy and Cultural Construction of Social Activism*. New York: Oxford University Press.

Buia, Carole. 2001. "The Best of Both Worlds." *Time* (Special Issue: "Music Goes Global," Fall 2001): 10–13.

Bullard, Robert B., and Glenn S. Johnson (Eds.). 1997. *Just Transportation: Dismantling Race and Class Barriers to Mobility*. Gabriola Island, BC: New Society.

Bullard, Robert B., and Beverly H. Wright. 1992. "The Quest for Environmental Equity: Mobilizing the African-American Community for Social Change." In Riley E. Dunlap and Angela G. Mertig (Eds.), *American Environmentalism: The U.S. Environmental Movement, 1970–1990*. New York: Taylor & Francis, pp. 39–49.

Bumpass, Larry, James E. Sweet, and Andrew J. Cherlin. 1991. "The Role of Cohabitation in Declining Rates of Marriage." *Journal of Marriage and the Family*, 53: 913–927.

Burawoy, Michael. 1991. "Introduction." In Michael Burawoy, Alice Burton, Ann Arnett Ferguson and others, *Ethnography Unbounded: Power and Resistance in the Modern Metropolis*. Berkeley: University of California Press, pp. 1–7.

Burciaga, Jose Antonio. 1993. *Drink Cultura*. Santa Barbara, CA: Capra.

Burgess, Ernest W. 1925. "The Growth of the City." In Robert E. Park and Ernest W. Burgess (Eds.), *The City*. Chicago: University of Chicago Press, pp. 47–62.

Burkhauser, Richard V., Robert H. Haveman, and Barbara L. Wolfe. 1993. "How People with Disabilities Fare When Public Policies Change." *Journal of Policy Analysis and Management*, 12: 251–269.

Burnham, M. Audrey, Richard L. Hough, Marvin Karno, Javier I. Escobar, and Cynthia A. Telles. 1987. "Acculturation and Lifetime Prevalence of Psychiatric Disorders Among Mexican Americans in Los Angeles." *Journal of Health and Social Behavior*, 28: 89–102.

Burnham, Walter Dean. 1983. *Democracy in the Making: American Government and Politics*. Englewood Cliffs, NJ: Prentice Hall.

Burns, John F. 1994. "India Fights Abortion of Female Fetuses." *New York Times* (Aug. 27): 5.

Burns, Tom. 1992. *Erving Goffman*. New York: Routledge.

Burros, Marian. 1994. "Despite Awareness of Risks, More in U.S. Are Getting Fat." *New York Times* (July 17): 1, 8.

___. 1996. "Eating Well: A Law to Encourage Sharing in a Land of Plenty." *New York Times* (Dec. 11): B6.

Burt, Martha R. 1992. *Over the Edge: The Growth of Homelessness in the 1980s*. New York: Russell Sage Foundation.

Busch, Ruth C. 1990. *Family Systems: Comparative Study of the Family*. New York: Lang.

Busfield, Joan. 1996. *Men, Women and Madness: Understanding Gender and Mental Disorder*. Houndmills, Basingstoke, Hampshire: MacMillan.

Business Week. 1997. "This Lesson Is Brought to You by." (June 30): 69.

Butler, John S. 1991. *Entrepreneurship and Self-Help Among Black Americans*. New York: SUNY Press.

Butterfield, Fox. 1999. "Indians Are Crime Victims at Rate Above U.S. Average." *New York Times* (Feb. 15): A12.

Buvinić, Mayra. 1997. "Women in Poverty: A New Global Underclass." *Foreign Policy* (Fall): 38–53.

Byrne, John A. 1993. "The Horizontal Corporation: It's About Managing Across, Not Up and Down." *Business Week* (Dec. 20): 76–81.

Cable, Sherry, and Charles Cable. 1995. *Environmental Problems, Grassroots Solutions: The Politics of Grassroots Environmental Conflict*. New York: St. Martin's.

CAIR (Council on American-Islamic Relations). 2003. "Islam in America." Retrieved Aug. 8, 2003. Online: http://www.cair-net.org

Callaghan, Polly, and Heidi Hartmann. 1991. *Contingent Work*. Washington, DC: Economic Policy Institute.

Campbell, Anne. 1984. *The Girls in the Gang* (2nd ed.). Cambridge, MA: Basil Blackwell.

Cancian, Francesca M. 1990. "The Feminization of Love." In C. Carlson (Ed.), *Perspectives on the Family: History, Class, and Feminism*. Belmont, CA: Wadsworth, pp. 171–185.

___. 1992. "Feminist Science: Methodologies That Challenge Inequality." *Gender & Society*, 6 (4): 623–642.

Canetto, Silvia Sara. 1992. "She Died for Love and He for Glory: Gender Myths of Suicidal Behavior." *OMEGA*, 26 (1): 1–17.

Canter, R. J., and S. S. Ageton. 1984. "The Epidemiology of Adolescent Sex-Role Attitudes." *Sex Roles*, 11: 657–676.

Cantor, Muriel G. 1980. *Prime-Time Television: Content and Control*. Newbury Park, CA: Sage.

___. 1987. "Popular Culture and the Portrayal of Women: Content and Control." In Beth B. Hess and Myra Marx Ferree (Eds.), *Analyzing Gender: A Handbook of Social Science Research*. Newbury Park, CA: Sage, pp. 190–214.

Cantril, Hadley. 1941. *The Psychology of Social Movements*. New York: Wiley.

Capek, Stella M. 1993. "The 'Environmental Justice' Frame: A Conceptual Discussion and Application." *Social Problems*, 40 (1): 5–23.

Cargan, Leonard, and Matthew Melko. 1982. *Singles: Myths and Realities*. Newbury Park, CA: Sage.

Carmichael, Stokely, and Charles V. Hamilton. 1967. *Black Power: The Politics of Liberation in American Education*. New York: Random House.

Carnegie Council on Adolescent Development. 1995. *Great Transitions: Preparing Adolescents for a New Century*. New York: Carnegie Foundation.

Carrier, James G. 1986. *Social Class and the Construction of Inequality in American Education*. New York: Greenwood.

Carroll, John B. (Ed.). 1956. *Language, Thought, and Reality: Selected Writings of Benjamin Lee Whorf*. Cambridge, MA: MIT Press.

Carson, Rachel. 1962. *Silent Spring*. Boston: Houghton Mifflin.

Carter, Stephen L. 1994. *The Culture of Disbelief: How American Law and Politics Trivializes Religious Devotion*. New York: Anchor/Doubleday.

Cashmore, E. Ellis. 1996. *Dictionary of Race and Ethnic Relations* (4th ed.). London: Routledge.

Castells, Manuel. 1977. *The Urban Question*. London: Edward Arnold (orig. pub. 1972 as *La Question Urbaine*, Paris).

___. 1997. *The Power of Identity*. Malden, MA: Blackwell.

___. 1998. *End of Millennium*. Malden, MA: Blackwell.

Catalyst. 2002. "Fact Sheet: Women CEOs." Retrieved July 19, 2003. Online: http://www.catalystwomen.org/press_room/factsheets/fact_women_ceos.htm

Cavender, Gray. 1995. "Alternative Theory: Labeling and Critical Perspectives." In Joseph F. Sheley (Ed.), *Criminology: A Contemporary Handbook* (2nd ed.). Belmont, CA: Wadsworth, pp. 349–371.

Cavender, Nick. 2001. "It's a Dad's Life." Retrieved June 30, 2003. Online: http://www.homedad.org.uk/feature_dadslife.html

Celis, William, III. 1994. "Nations Envied for Schools Share Americans' Worries." *New York Times* (July 13): B4.

Chafetz, Janet Saltzman. 1984. *Sex and Advantage: A Comparative, Macro-Structural Theory of Sex Stratification*. Totowa, NJ: Rowman & Allanheld.

Chagnon, Napoleon A. 1992. *Yanomamo: The Last Days of Eden*. New York: Harcourt (rev. from 4th ed., *Yanomamo: The Fierce People*, published by Holt, Rinehart & Winston).

Chalfant, H. Paul, Robert E. Beckley, and C. Eddie Palmer. 1994. *Religion in Contemporary Society* (3rd ed.). Itasca, IL: Peacock.

Chambliss, William J. 1973. "The Saints and the Roughnecks." *Society*, 11: 24–31.

Chandler, Tertius, and Gerald Fox. 1974. *3000 Years of Urban History*. New York: Academic Press.

Channel One Network. 2001. "Channel One Network." Retrieved Nov. 11, 2001. Online: http://www.k-iii.com/html2/education/channel1/channel1.html

Charmaz, K., and V. Olesen. 1997. "Ethnographic Research in Medical Sociology: Its Foci and Distinctive Contributions." *Sociological Methods and Research*, 25: 452–494.

Chen, Hsiang-shui. 1992. *Chinatown No More: Taiwan Immigrants in Contemporary New York*. Ithaca, NY: Cornell University Press.

Cherlin, Andrew J. 1992. *Marriage, Divorce, Remarriage*. Cambridge, MA: Harvard University Press.

Chernin, Kim. 1981. *The Obsession: Reflections on the Tyranny of Slenderness*. New York: Harper & Row.

Chesney-Lind, Meda. 1989. "Girls' Crime and Woman's Place: Toward a Feminist Model of Female Delinquency." *Crime and Delinquency*, 35 (1): 5–29.

___. 1997. *The Female Offender*. Thousand Oaks, CA: Sage.

Children's Defense Fund, 2001. *The State of America's Children: Yearbook 2001*. Boston: Beacon.

___. 2002. *The State of Children in America's Union: A 2002 Action Guide to Leave No Child Behind*. Retrieved June 29, 2003. Online: http://www.childrensdefense.org/pdf/minigreenbook.pdf

Chipungu, Joel. 1999. "Polygamy Is Alive and Well in Zambia." African News Service (July 22). Retrieved Sept. 11, 1999. Online: http://www.comtex.news.com

Cho, Sumi K. 1993. "Korean Americans vs. African Americans: Conflict and Construction." In Robert Gooding-Williams (Ed.), *Reading Rodney King, Reading Urban Uprising*. New York: Routledge, pp. 196–211.

Chon, Margaret. 1995. "The Truth About Asian Americans." In Russell Jacoby and Naomi Glauberman (Eds.), *The Bell Curve Debate: History, Documents, Opinions*. New York: Times Books, pp. 238–240.

Chow, Esther Ngan-Ling. 1994. "Asian American Women at Work." In Maxine Baca Zinn and Bonnie Thornton Dill (Eds.), *Women of Color in U.S. Society*. Philadelphia: Temple University Press, pp. 203–227.

Christ, Carol P. 1987. *Laughter of Aphrodite: Reflections on a Journey to the Goddess*. San Francisco: Harper & Row.

Christians, Clifford G. G., Kim B. Rotzoll, and Mark Fackler. 1987. *Media Ethics*. New York: Longman.

Chronicle of Higher Education. 2001. "Almanac Issue: 2001–2002 (Aug. 30): 7–36.

___. 2002. "Almanac 2002–3" (Aug. 30): 12–38.

Churchill, Ward. 1994. *Indians Are Us? Culture and Genocide in Native North America*. Monroe, ME: Common Courage.

Claiborne, Ron. 2003. "Tight Lips: Embedded Journalists Learn About Restricted Access Early On." *ABCNews.com* (Mar. 14). Retrieved July 26, 2003. Online: http://abcnews.go.com/sections/wnt/World/iraq030314_usslincoln_notebook1.html

Clayman, Steven E. 1993. "Booing: The Anatomy of a Disaffiliative Response." *American Sociological Review*, 58 (1): 110–131.

Cleary, Paul D. 1987. "Gender Differences in Stress-Related Disorders." In Rosalind C. Barnett, Lois Biener, and Grace K. Baruch (Eds.), *Gender and Stress.* New York: Free Press, pp. 39–72.

Clinard, Marshall B., and Peter C. Yeager. 1980. *Corporate Crime.* New York: Free Press.

Clines, Francis X. 1993. "An Unfettered Milken Has Lessons to Teach." *New York Times* (Oct. 16): 1, 9.

___. 1996. "A Chef's Training Program That Feeds Hope as Well as Hunger." *New York Times* (Dec. 11): B1, B6.

Cloward, Richard A., and Lloyd E. Ohlin. 1960. *Delinquency and Opportunity: A Theory of Delinquent Gangs.* New York: Free Press.

Clymer, Adam. 1993. "A Daughter of Slavery Makes the Senate Listen." *New York Times* (July 23): A10.

CNN. 1994. "Both Sides: School Prayer." (Nov. 26).

CNN.com. 2003a. "Crash Witness: Rescue Was 'Collective Effort.'" Retrieved July 20, 2003. Online: http://www.cnn.com/2003/US/West/07/17/cnna.crisman/index.html

___. 2003b. "Narrow Use of Affirmative Action Preserved in College Admissions." Retrieved June 23, 2003. Online: http://www.cnn.com/2003/LAW/06/23/scotus.affirmative.action/index.html

___. 2003c. "U.S. Energy Secretary Says Weeks Needed to Analyze Blackout Data." Retrieved Aug. 28, 2003. Online: http://www.cnn.com/2003/US/Northeast/08/27/blackout.investigation.ap/index.html

Coakley, Jay J. 2004. *Sport in Society: Issues and Controversies* (8th ed.). New York: McGraw-Hill.

Coburn, Andrew F., and Elise J. Bolda. 1999. "The Rural Elderly and Long-Term Care." In Thomas C. Ricketts, III (Ed.), *Rural Health in the United States.* New York: Oxford University Press, pp. 179–189.

Cock, Jacklyn. 1994. "Women and the Military: Implications for Demilitarization in the 1990s in South Africa." *Gender & Society,* 8 (2): 152–169.

Cockerham, William C. 1995. *Medical Sociology* (6th ed.). Englewood Cliffs, NJ: Prentice Hall.

___. 2004. *Medical Sociology* (9th ed.). Upper Saddle River, NJ: Prentice Hall.

Cohen, Adam. 1999. "A Curse of Cliques." *Time* (May 3): 44–45.

Cohen, Leah Hager. 1994. *Train Go Sorry: Inside a Deaf World.* Boston: Houghton Mifflin.

Cole, David. 2000. *No Equal Justice: Race and Class in the American Criminal Justice System.* New York: New Press.

Cole, George F., and Christopher E. Smith. 2004. *The American System of Criminal Justice* (10th ed.). Belmont, CA: Wadsworth.

Coleman, James. 1990. *Foundations of Social Theory.* Cambridge: Belknap Press of Harvard University Press.

Coleman, Richard P., and Lee Rainwater. 1978. *Social Standing in America: New Dimensions of Class.* New York: Basic.

Coles, Gerald. 1987. *The Learning Mystique: A Critical Look at "Learning Disabilities."* New York: Pantheon.

Coles, Robert. 1979. *Work Mobility and Participation: A Comparative Study of American and Japanese Industry.* Berkeley: University of California Press.

College Board. 2002. "2002 College-Bound Seniors: A Profile of SAT Program Test Takers." Retrieved July 26, 2003. Online: http://www.collegeboard.com/prod_downloads/about/news_info/cbsenior/yr2002/pdf/2002_TOTAL_GROUP_REPORT.pdf

Collier, Paul. 2000. "Economic Causes of Civil Conflict and Their Implications for Policy." World Bank. Retrieved July 12, 2003. Online: http://www.globalpolicy.org/security/issues/diamond/wb.htm

Collier, Peter, and David Horowitz. 1987. *The Fords: An American Epic.* New York: Summit.

Collins, Ann Marie. 1991. *How to Live in America: A Guide for the Japanese.* Tokyo: Yohan.

Collins, Catherine, and Douglas Frantz. 1993. *Teachers: Talking Out of School.* Boston: Little, Brown.

Collins, Patricia Hill. 1990. *Black Feminist Thought: Knowledge, Consciousness, and the Politics of Empowerment.* London: HarperCollins Academic.

___. 1991. "The Meaning of Motherhood in Black Culture." In Robert Staples (Ed.), *The Black Family: Essays and Studies.* Belmont, CA: Wadsworth, pp. 169–178. Orig. pub. in *SAGE: A Scholarly Journal on Black Women,* 4 (Fall 1987): 3–10.

___. 1998. *Fighting Words: Black Women and the Search for Justice.* Minneapolis: University of Minnesota Press.

Collins, Randall. 1971. "A Conflict Theory of Sexual Stratification." *Social Problems,* 19 (1): 3–21.

___. 1979. *The Credential Society: An Historical Sociology of Education.* New York: Academic Press.

___. 1982. *Sociological Insight: An Introduction to Non-Obvious Sociology.* New York: Oxford University Press.

___. 1987. "Interaction Ritual Chains, Power, and Property: The Micro–Macro Connection as an Empirically Based Theoretical Problem." In Jeffrey C. Alexander et al. (Eds.), *The Micro–Macro Link.* Berkeley: University of California Press, pp. 193–206.

___. 1994. *Four Sociological Traditions.* New York: Oxford University Press.

___. 1997. "An Asian Route to Capitalism: Religious Economy and the Origins of Self-Transforming Growth in Japan." *American Sociological Review,* 62 (December): 843–865.

Collins, Sharon M. 1989. "The Marginalization of Black Executives." *Social Problems,* 36: 317–331.

Coltrane, Scott. 1989. "Household Labor and the Routine Production of Gender." *Social Problems,* 36: 473–490.

Comarow, Murray. 1993. "Are Sociologists Above the Law?" *Chronicle of Higher Education* (Dec. 15): A44.

Comer, James P. 1988. *Maggie's American Dream: The Life and Times of a Black Family.* New York: New American Library/Penguin.

Condry, Sandra McConnell, John C. Condry, Jr., and Lee Wolfram Pogatshnik. 1983. "Sex Differences: A Study of the Ear of the Beholder." *Sex Roles,* 9: 697–704.

Conrad, Peter. 1996. "Medicalization and Social Control." In Phil Brown (Ed.), *Perspectives in Medical Sociology* (2nd ed.). Prospect Heights, IL: Waveland, pp. 137–162.

Cook, Sherburn F. 1973. "The Significance of Disease in the Extinction of the New England Indians." *Human Biology,* 45: 485–508.

Cookson, Peter W., Jr., and Caroline Hodges Persell. 1985. *Preparing for Power: America's Elite Boarding Schools.* New York: Basic.

Cooley, Charles Horton. 1963. *Social Organization: A Study of the Larger Mind.* New York: Schocken (orig. pub. 1909).

___. 1998. "The Social Self—the Meaning of 'I.'" In Hans-Joachim Schubert (Ed.), *On Self and Social Organization—Charles Horton Cooley.* Chicago: University of Chicago Press, pp. 155–175. Reprinted from Charles Horton Cooley, *Human Nature and the Social Order.* New York: Schocken, 1902.

Coontz, Stephanie. 1992. *The Way We Never Were: American Families and the Nostalgia Trap.* New York: Basic.

___. 1997. *The Way We Really Are: Coming to Terms with America's Changing Families.* New York: Basic.

Corr, Charles A., Clyde M. Nabe, and Donald M. Corr. 2003. *Death and Dying, Life and Living* (4th ed.). Pacific Grove, CA: Brooks/Cole.

Corsaro, William A. 1985. *Friendship and Peer Culture in the Early Years.* Norwood, NJ: Ablex.

___. 1992. "Interpretive Reproduction in Children's Peer Cultures." *Social Psychology Quarterly,* 55 (2): 160–177.

___. 1997. *Sociology of Childhood.* Thousand Oaks, CA: Pine Forge.

Cortese, Anthony J. 1999. *Provocateur: Images of Women and Minorities in Advertising.* Latham, MD: Rowman & Littlefield.

Cose, Ellis. 1993. *The Rage of a Privileged Class.* New York: HarperCollins.

Coser, Lewis A. 1956. *The Functions of Social Conflict.* Glencoe, IL: Free Press.

Costello, Cynthia, and Anne J. Stone (Eds.), for the Women's Research and Education Institute. 1994. *The American Woman, 1994–95.* New York: Norton.

Coughlin, Ellen K. 1993. "Author of a Noted Study on Black Ghetto Life Returns with a Portrait of Homeless Women." *Chronicle of Higher Education* (Mar. 31): A7–A8.

Counts, Dorothy Ayers, and David R. Counts. 2001. *Over the Next Hill: An Ethnography of RVing Seniors in North America.* Orchard Park, NY: Broadview.

Cowgill, Donald O. 1986. *Aging Around the World.* Belmont, CA: Wadsworth.

Cox, Harvey. 1995. "Christianity." In Arvind Sharma (Ed.), *Our Religions.* San Francisco: HarperCollins, pp. 359–423.

Cox, Oliver C. 1948. *Caste, Class, and Race.* Garden City, NY: Doubleday.

Craig, Steve. 1992. "Considering Men and the Media." In Steve Craig (Ed.), *Men, Masculinity, and the Media.* Newbury Park, CA: Sage, pp. 1–7.

Crawford, Elizabeth. 2003. "Campus Castoffs." *Chronicle of Higher Education* (June 20): A6.

Cressey, Donald. 1969. *Theft of the Nation.* New York: Harper & Row.

Creswell, John W. 1998. *Qualitative Inquiry and Research Design: Choosing Among Five Traditions.* Thousand Oaks, CA: Sage.

Cromwell, Larcenia. 2003. "Terrorist." E-mail message posted Aug. 14, 2003.

Cronin, John, and Robert F. Kennedy, Jr. 1999. *The Riverkeepers: Two Activists Fight to Reclaim Our Environment as a Basic Human Right.* New York: Touchstone.

Crossette, Barbara. 1996. "Hope, and Pragmatism, for U.S. Cities Conference." *New York Times* (June 3): A3.

___. 1997. "The 21st Century Belongs to. . . ." *New York Times* (Oct. 19): WK3.

Crow Dog, Mary, and Richard Erdoes. 1991. *Lakota Woman.* New York: HarperPerennial.

Cumming, Elaine C., and William E. Henry. 1961. *Growing Old: The Process of Disengagement.* New York: Basic.

Cuomo, Chris. 2002. "A Delinquent's Dream: The Video Game *Grand Theft Auto* Has Riled Up the Critics." *abcNEWS.com.* Retrieved Nov. 23, 2002. Online: http://www.abcnews.com

Currie, Elliott. 1998. *Crime and Punishment in America.* New York: Metropolitan.

Curtiss, Susan. 1977. *Genie: A Psycholinguistic Study of a Modern Day "Wild Child."* New York: Academic Press.

Cyrus, Virginia. 1993. *Experiencing Race, Class, and Gender in the United States.* Mountain View, CA: Mayfield.

Dahl, Robert A. 1961. *Who Governs?* New Haven, CT: Yale University Press.

Dahrendorf, Ralf. 1959. *Class and Class Conflict in an Industrial Society.* Stanford, CA: Stanford University Press.

Daly, Kathleen, and Meda Chesney-Lind. 1988. "Feminism and Criminology." *Justice Quarterly,* 5: 497–533.

Daly, Mary. 1973. *Beyond God the Father.* Boston: Beacon.

___. 1978. *Gyn/ecology: The Meta-Ethics of Radical Feminism.* Boston: Beacon.

Daniels, Roger. 1993. *Prisoners Without Trial: Japanese-Americans in World War II.* New York: Hill & Wang.

Danziger, Sheldon, and Peter Gottschalk. 1995. *America Unequal.* Cambridge, MA: Harvard University Press.

Dao, James. 2000. "Lockheed Wins $200 Billion Deal for Fighter Jet." *New York Times* (Oct. 27): A1, A9.

Darley, John M., and Thomas R. Shultz. 1990. "Moral Rules: Their Content and Acquisition." *Annual Review of Psychology,* 41: 525–556.

Dart, Bob. 1999. "Kids Get More Screen Time Than School Time." *Austin American-Statesman* (June 28): A1, A5.

Davis, F. James. 1991. *Who Is Black?* University Park: Pennsylvania State University Press.

Davis, Fred. 1992. *Fashion, Culture, and Identity.* Chicago: University of Chicago Press.

Davis, Kingsley. 1940. "Extreme Social Isolation of a Child." *American Journal of Sociology,* 45 (4): 554–565.

___. 1949. *Human Society.* New York: Macmillan.

___. 1955. "Malthus and the Theory of Population." In Paul Lazarsfeld and M. Rosenberg (Eds.), *The Language of Social Research.* New York: Free Press, pp. 540–553.

Davis, Kingsley, and Wilbert Moore. 1945. "Some Principles of Stratification." *American Sociological Review,* 7 (April): 242–249.

Davis, Sampson, George Jenkins, and Rameck Hunt (with Lisa Frazier Page). 2003. *The Pact: Three Young Men Make a Promise and Fulfill a Dream.* New York: Riverhead.

Davis, Sidney. 1997. "Race and the Politics of Transportation in Atlanta." In Robert B. Bullard and Glenn S. Johnson (Eds.), *Just Transportation: Dismantling Race and Class Barriers to Mobility.* Gabriola Island, BC: New Society, pp. 84–96.

Dean, L. M., F. N. Willis, and J. N. la Rocco. 1976. "Invasion of Personal Space as a Function of Age, Sex and Race." *Psychological Reports,* 38 (3) (pt. 1): 959–965.

Death Penalty Information Center. 2003. "Death Row." Retrieved Sept. 6, 2003. Online: http://www.deathpenaltyinfo.org/article.php?did=413&scid=9

Deegan, Mary Jo. 1988. *Jane Addams and the Men of the Chicago School, 1892–1918.* New Brunswick, NJ: Transaction.

Degher, Douglas, and Gerald Hughes. 1991. "The Identity Change Process: A Field Study of Obesity." *Deviant Behavior,* 12: 385–402.

DeJong, Gerben, Andrew I. Batavia, and Robert Griss. 1989. "America's Neglected Health Minority: Working-Age Persons with Disabilities." *Milbank Quarterly,* 67 (suppl. 2): 311–351.

Delgado, Richard. 1995. "Introduction." In Richard Delgado (Ed.), *Critical Race Theory: The Cutting Edge.* Philadelphia: Temple University Press, pp. xiii–xvi.

Delpit, Lisa. 1995. *Other People's Children: Cultural Conflict in the Classroom.* New York: New Press.

DeNavas-Walt, Carmen, and Robert W. Cleveland. 2002. "Money Income in the United States: 2001." U.S. Census Bureau, Current Population Reports, P60–218. Washington, DC: U.S. Government Printing Office.

Denzin, Norman K. 1989. *The Research Act* (3rd ed.). Englewood Cliffs, NJ: Prentice Hall.

Derber, Charles. 1983. *The Pursuit of Attention: Power and Individualism in Everyday Life.* New York: Oxford University Press.

Devine, John. 1996. *Maximum Security: The Culture of Violence in Inner-City Schools.* Chicago: University of Chicago Press.

De Witt, Karen. 1994. "Wave of Suburban Growth Is Being Fed by Minorities." *New York Times* (Aug. 15): A1, A12.

Diamond, Timothy. 1992. *Making Gray Gold: Narratives of Nursing Home Care.* Chicago: University of Chicago Press.

Dikotter, Frank. 1996. "Culture, 'Race' and Nation: The Formation of Identity in Twentieth Century China." *Journal of International Affairs* (Winter): 590–605.

Dill, Bonnie Thornton. 1988. "'Making Your Job Good Yourself': Domestic Service and the Construction of Personal Dignity." In Ann Bookman and Sandra Morgen (Eds.), *Women and the Politics of Empowerment.* Philadelphia: Temple University Press, pp. 33–52.

Dobrzynski, Judith H. 1996. "When Directors Play Musical Chairs." *New York Times* (Nov. 17): F1, F8, F9.

Dohan, Daniel, and Martin Sanchez-Jankowski. 1998. "Using Computers to Analyze Ethnographic Field Data: Theoretical and Practical Considerations." *Annual Review of Sociology,* 24: 477–499.

Dollard, John. 1957. *Caste and Class in a Southern Town.* Garden City, NY: Doubleday (orig. pub. 1937).

Dollard, John, Neal E. Miller, Leonard W. Doob, O. H. Mowrer, and Robert R. Sears. 1939. *Frustration and Aggression.* New Haven, CT: Yale University Press.

Domhoff, G. William. 1974. *The Bohemian Grove and Other Retreats.* New York: Harper and Row.

___. 1978. *The Powers That Be: Processes of Ruling Class Domination in America.* New York: Random House.

___. 1983. *Who Rules America Now? A View for the '80s.* Englewood Cliffs, NJ: Prentice Hall.

___. 1990. *The Power Elite and the State: How Policy Is Made in America.* New York: Aldine De Gruyter.

___. 2002. *Who Rules America? Power and Politics* (4th ed.). New York: McGraw-Hill.

Downey, A. 1984. "Relationship of Religiosity to Death Anxiety of Middle-Aged Males." *Psychological Reports,* 54: 811–822. Cited in N. Hooyman and H. A. Kiyak, 1991.

Drake, Jeffrey. 2002. "Communication Keys for Success." Retrieved July 5, 2002. Online: http://www.achievemax.com/newsletter/00issue/communication-keys.htm

Driskell, Robyn Bateman, and Larry Lyon. 2002. "Are Virtual Communities True Communities? Examining the Environments and Elements of Community." *City & Community* 1 (4): 1–18.

Du Bois, W. E. B. 1967. *The Philadelphia Negro: A Social Study.* New York: Schocken (orig. pub. 1899).

Dubowitz, Howard, Maureen Black, Raymond H. Starr, Jr., and Susan Zuravin. 1993. "A Conceptual Definition of Child Neglect." *Criminal Justice and Behavior,* 20 (1): 8–26.

Duffy, John. 1976. *The Healers.* New York: McGraw-Hill.

Dugger, Celia W. 1997. "In Historical Surge to Be Americans, 5,000 Take Oath." *New York Times* (July 5): 1, 27.

Duncan, Otis Dudley. 1968. "Social Stratification and Mobility: Problems in Measurement of Trend." In E. B. Sheldon and W. E. Moore (Eds.), *Indicators of Social Change.* New York: Russell Sage Foundation.

Dunlap, Riley E. 1992. "Trends in Public Opinion Toward Environmental Issues: 1965–1990." In Riley E. Dunlap and Angela G. Mertig (Eds.), *American Environmentalism: The U.S. Environmental Movement, 1970–1990.* New York: Taylor & Francis, pp. 89–113.

Dupre, Roslyn, and Paul Gains. 1997. "Fundamental Differences." *Women's Sports & Fitness* (October): 63–68.

Durkheim, Emile. 1933. *The Division of Labor in Society.* Trans. George Simpson. New York: Free Press (orig. pub. 1893).

___. 1956. *Education and Sociology.* Trans. Sherwood D. Fox. Glencoe, Il.: Free Press.

___. 1964a. *The Rules of Sociological Method.* Trans. Sarah A. Solovay and John H. Mueller. New York: Free Press (orig. pub. 1895).

___. 1964b. *Suicide.* Trans. John A. Sparkling and George Simpson. New York: Free Press (orig. pub. 1897).

___. 1995. *The Elementary Forms of Religious Life.* Trans. Karen E. Fields. New York: Free Press (orig. pub. 1912).

Duster, Troy. 1995. "Symposium: The Bell Curve." *Contemporary Sociology: A Journal of Reviews,* 24 (2): 158–161.

Dworkin, Andrea. 1974. *Woman Hating.* New York: Dutton.

Dye, Thomas R., and Harmon Zeigler. 2003. *The Irony of Democracy: An Uncommon Introduction to American Politics* (12th ed.). Belmont, CA: Wadsworth.

Dyson, Michael Eric. 1993. "Be Like Mike? Michael Jordan and the Pedagogy of Desire." In Michael Eric Dyson (Ed.), *Reflecting Black: African-American Cultural Criticism.* Minneapolis: University of Minnesota Press, pp. 64–75.

Early, Kevin E. 1992. *Religion and Suicide in the African-American Community.* Westport, CT: Greenwood.

Eaton, William W. 1980. "A Formal Theory of Selection for Schizophrenia." *American Journal of Sociology,* 86: 149–158.

Ebaugh, Helen Rose Fuchs. 1988. *Becoming an EX: The Process of Role Exit.* Chicago: University of Chicago Press.

The Economist. 1999. "Who'd Have Credited It?" Retrieved June 5, 1999. Online: http://www.economist.com/archives

___. 2003. "Civil Wars: The Poor Man's Curse." *The Economist* (May 24–30): 11.

Eder, Donna. 1995. *School Talk: Gender and Adolescent Culture* (with Catherine Colleen Evans and Stephen Parker). New Brunswick, NJ: Rutgers University Press.

Edgerton, Robert B. 1992. *Sick Societies: Challenging the Myth of Primitive Harmony.* New York: Free Press.

Edwards, Gary. 2003. "Traffic Congestion Disappearing Along with Jobs in Silicon Valley." *San Jose Mercury News.* Reprinted in *Austin American-Statesman* (July 21): D3.

Edwards, Harry. 1973. *Sociology of Sport.* Homewood, IL: Dorsey.

Edwards, Richard. 1979. *Contested Terrain.* New York: Basic.

Egan, Timothy. 1997. "Where Water Is Power, the Balance Shifts." *New York Times* (Nov. 30): A1, A16.

Ehrenreich, Barbara. 1989. *Fear of Falling: The Inner Life of the Middle Class.* New York: HarperPerennial.

___. 1997. *Blood Rites: Origins and History of the Passions of War.* New York: Metropolitan.

___. 2001. *Nickel and Dimed: On (Not) Getting by in America.* New York: Metropolitan.

Ehrlich, Paul R., Anne H. Ehrlich, and Gretchen C. Daily. 1995. *The Stork and the Plow: The Equity Answer to the Human Dilemma.* New Haven, CT: Yale University Press.

Eighner, Lars. 1993. *Travels with Lizbeth.* New York: St. Martin's.

Eisenhower, Dwight D. 1961. "Farewell Address to the Nation." Quoted in William D. Hartung, "Military–Industrial Complex Revisited: How Weapons Makers Are Shaping U.S. Foreign and Military Policies." Retrieved Sept. 11, 1999. Online: http://www.foreignpolicy-infocus.org/paper/micr/index.html

Eisenstein, Zillah R. 1994. *The Color of Gender: Reimaging Democracy.* Berkeley: University of California Press.

Eisler, Benita. 1983. *Class Act: America's Last Dirty Secret.* New York: Franklin Watts.

Eitzen, D. Stanley, and George H. Sage. 1997. *The Sociology of North American Sport* (6th ed.). Dubuque, IA: Brown.

Elkin, Frederick, and Gerald Handel. 1989. *The Child and Society: The Process of Socialization* (5th ed.). New York: Random House.

Elkind, David. 1995. "School and Family in the Postmodern World." *Phi Delta Kappan* (September): 8–21.

Ellison, Christopher G., Jeffrey A. Burr, and Patricia L. McCall. 1997. "Religious Homo-geneity and Metropolitan Suicide Rates." *Social Forces,* 76 (September): 273–300.

Elster, Jon. 1989. *Nuts and Bolts for the Social Sciences.* Cambridge, England: Cambridge University Press.

Emerson, Richard M. 1962. "Power–Dependency Relations." *American Sociological Review,* 27: 31–41.

Emling, Shelley. 1997a. "Haiti Held in Grip of Another Drought." *Austin American-Statesman* (Sept. 19): A17, A18.

___. 1997b. "In Haiti, It's Resort vs. Reality." *Austin American-Statesman* (Sept. 27): A17, A19.

Engels, Friedrich. 1970. *The Origins of the Family, Private Property, and the State.* New York: International (orig. pub. 1884).

Engerman, Stanley L. 1995. "The Extent of Slavery and Freedom Throughout the World as a Whole and in Major Subareas." In Julian L. Simon (Ed.), *The State of Humanity.* Cambridge, MA: Blackwell, pp. 171–177.

England, Paula, and Melissa S. Herbert. 1993. "The Pay of Men in 'Female' Occupations: Is Comparable Worth Only for Women?" In Christine L. Williams (Ed.), *Doing "Women's Work": Men in Nontraditional Occupations.* Newbury Park, CA: Sage, pp. 28–48.

Enloe, Cynthia H. 1987. "Feminists Thinking About War, Militarism, and Peace." In Beth H. Hess and Myra Marx Ferree (Eds.), *Analyzing Gender: A Handbook of Social Science Research.* Newbury Park, CA: Sage, pp. 526–547.

Epidemiological Network for Latin America and the Caribbean. 2000. "HIV and AIDS in the Americas: An Epidemic with Many Faces." Retrieved Nov. 23, 2001. Online: http://www.census.gov/ipc/www/hivaidinamerica.pdf

Epstein, Cynthia Fuchs. 1988. *Deceptive Distinctions: Sex, Gender, and the Social Order.* New Haven, CT: Yale University Press.

Erikson, Erik H. 1963. *Childhood and Society.* New York: Norton.

___. 1980. *Identities and the Life Cycle.* New York: Norton (orig. pub. 1959).

Erikson, Kai T. 1962. "Notes on the Sociology of Deviance." *Social Problems,* 9: 307–314.

___. 1964. "Notes on the Sociology of Deviance." In Howard S. Becker (Ed.), *The Other Side: Perspectives on Deviance.* New York: Free Press, pp. 9–21.

___. 1976. *Everything in Its Path: Destruction of Community in the Buffalo Creek Flood.* New York: Simon & Schuster.

___. 1991. "A New Species of Trouble." In Stephen Robert Couch and J. Stephen Kroll-Smith (Eds.), *Communities at Risk: Collective Responses to Technological Hazards.* New York: Land, pp. 11–29.

___. 1994. *A New Species of Trouble: Explorations in Disaster, Trauma, and Community.* New York: Norton.

Esping-Andersen, Gosta. 1990. *The Three Worlds of Welfare Capitalism.* Cambridge, MA: Polity.

Espiritu, Yen Le. 1995. *Filipino American Lives.* Philadelphia: Temple University Press.

Essed, Philomena. 1991. *Understanding Everyday Racism.* Newbury Park, CA: Sage.

Esterberg, Kristin G. 1997. *Lesbian and Bisexual Identities: Constructing Communities, Constructing Self.* Philadelphia: Temple University Press.

Etkind, Marc. 1997. . . . *Or Not to Be: A Collection of Suicide Notes.* New York: Riverhead.

Etzioni, Amitai. 1975. *A Comparative Analysis of Complex Organizations: On Power, Involvement, and Their Correlates* (rev. ed.). New York: Free Press.

___. 1994. *The Spirit of Community: The Reinvention of American Society.* New York: Touchstone.

Europa. 2003. "The European Union at a Glance." Retrieved July 26, 2003. Online: http://www.europa.eu.int

Evans, Glen, and Norman L. Farberow. 1988. *The Encyclopedia of Suicide.* New York: Facts on File.

Evans, Peter B., and John D. Stephens. 1988. "Development and the World Economy." In Neil J. Smelser (Ed.), *Handbook of Sociology.* Newbury Park, CA: Sage, pp. 739–773.

Ezell, William Bruce, Jr. 1993. "Letters to the Editor: The Symbolism of the Confederate Battle Flag." *Chronicle of Higher Education* (Nov. 3): B2.

Faderman, Lillian (with Ghia Ziong). 1998. *I Begin My Life All Over: The Hmong and the American Immigrant Experience.* Boston: Beacon.

Fagot, Beverly I. 1984. "Teacher and Peer Reactions to Boys' and Girls' Play Styles." *Sex Roles,* 11: 691–702.

FAIR (Fairness & Accuracy in Reporting). 2003. "How Many Dead? Major Networks Aren't Counting?" *FAIR* (Dec. 12). Retrieved July 26, 2003. Online: http://www.fair.org/activism/afghanistan-casualties.html

Falk, Patricia. 1989. "Lesbian Mothers: Psychological Assumptions in Family Law." *American Psychologist,* 44: 941–947.

Fallon, Patricia, Melanie A. Katzman, and Susan C. Wooley. 1994. *Feminist Perspectives on Eating Disorders.* New York: Guilford.

Fallows, James. 1997. *Breaking the News: How the Media Undermine American Democracy.* New York: Vintage.

Faludi, Susan. 1999. *Stiffed: The Betrayal of the American Man.* New York: Morrow.

Famighetti, Robert. 1997. *The World Almanac and Book of Facts 1997.* New York: St. Martin's.

Farag, Joseph. 2003. "Racism: Alive and Well in Our Media." *Laurier Journal of Political Affairs.* Waterloo, Ontario: Wilfrid Laurier University.

Farb, Peter. 1973. *Word Play: What Happens When People Talk.* New York: Knopf.

Farley, John E. 1995. *Majority–Minority Relations* (3rd ed.). Englewood Cliffs, NJ: Prentice Hall.

Fausto-Sterling, Anne. 1985. *Myths of Gender: Biological Theories About Women and Men.* New York: Basic.

"The Favored Infants." 1976. *Human Behavior* (June): 49–50.

Feagin, Joe R. 1991. "The Continuing Significance of Race: Antiblack Discrimination in Public Places." *American Sociological Review,* 56 (February): 101–116.

Feagin, Joe R., and Clairece Booher Feagin. 1994. *Social Problems: A Critical Power–Conflict Perspective* (4th ed.). Englewood Cliffs, NJ: Prentice Hall.

___. 1997. *Social Problems: A Critical Power–Conflict Perspective* (5th ed.). Englewood Cliffs, NJ: Prentice Hall.

___. 2003. *Racial and Ethnic Relations* (7th ed.). Upper Saddle River, NJ: Prentice Hall.

Feagin, Joe R., Anthony M. Orum, and Gideon Sjoberg (Eds.). 1991. *A Case for the Case Study.* Chapel Hill: University of North Carolina Press.

Feagin, Joe R., and Robert Parker. 1990. *Building American Cities: The Urban Real Estate Game* (2nd ed.). Englewood Cliffs, NJ: Prentice Hall.

Feagin, Joe R., and Melvin P. Sikes. 1994. *Living with Racism: The Black Middle-Class Experience.* Boston: Beacon.

Feagin, Joe R., and Hernán Vera. 1995. *White Racism: The Basics.* New York: Routledge.

Feagin, Joe R., Hernán Vera, and Nikitah Imani. 1996. *The Agony of Education: Black Students at White Colleges and Universities.* New York: Routledge.

Fedarko, Kevin. 1997. "How a Few Firemen Created a Safe Haven for Some Chicago Kids." *Time* (Nov. 17): 72–73.

Federal Bureau of Investigation (FBI). 2001. *Crime in the United States: 2000.* Washington, DC: U.S. Government Printing Office.

___. 2003. *Crime in the United States: 2002.* Washington, DC: U.S. Government Printing Office.
Federal Election Commission, 2001. "FEC Reports on Congressional Financial Activity for 2000." Retrieved Oct. 24, 2001. Online: http://www.fec.gov/press/051501congfinact/051501congfinact.html
Federal Trade Commission. 2000. "Marketing Violent Entertainment to Children: A Review of Self-Regulation and Industry Practices in the Motion Picture, Music Recording & Electronic Game Industries." Retrieved Sept. 9, 2001. Online: http://www.ftc.gov/opa/2000/09/youthviol.htm
Feifel, H., and W. T. Nagy. 1981. "Another Look at Fear of Death." *Journal of Consulting and Clinical Psychology,* 49: 278–286. Cited in N. Hooyman and H. A. Kiyak, 1991.
Fenstermacher, Gary D. 1994. "The Absence of Democratic and Educational Ideals from Contemporary Educational Reform Initiatives." The Elam Lecture, presented to the Educational Press Association of America, Chicago, June 10.
Ferguson, John. 1977. *War and Peace in the World's Religions.* New York: Oxford University Press.
Ferraro, Gary. 1992. *Cultural Anthropology: An Applied Perspective.* St. Paul, MN: West.
Ferriss, Susan. 2001. "Cold Spell: U.S. Recession Chills Mexico's Economic Hot Spot." *Austin American-Statesman* (Nov. 25): E1, E4 (based on data from the Bank of Mexico).
Fields, Jason, and Lynne M. Casper. 2001. *America's Family and Living Arrangements: March 2000.* Current Population Reports, P20–537. Washington, DC: U.S. Census Bureau.
Fiffer, Steve, and Sharon Sloan Fiffer. 1994. *50 Ways to Help Your Community.* New York: Mainstream/Doubleday.
Findlay, Deborah A., and Leslie J. Miller. 1994. "Through Medical Eyes: The Medicalization of Women's Bodies and Women's Lives." In B. Singh Bolaria and Harley D. Dickinson (Eds.), *Health, Illness, and Health Care in Canada* (2nd ed.). Toronto: Harcourt, pp. 276–306.
Findlay, Steven. 1997. "Health Care Industry Feeling a Little Discomfort." *USA Today* (Aug. 6): 4B.
Findlay-Kaneko, Beverly. 1997. "In a Breakthrough for Japan, a Woman Takes Over at a National University." *Chronicle of Higher Education* (June 20): A41–A42.
Fine, Michelle, and Lois Weis. 1998. *The Unknown City: The Lives of Poor and Working-Class Young People.* Boston: Beacon.
Fink, Arlene. 1995. *How to Sample in Surveys.* Thousand Oaks, CA: Sage.
Finley, M. I. 1980. *Ancient Slavery and Modern Ideology.* New York: Viking.
Finn Paradis, Leonora, and Scott B. Cummings. 1986. "The Evolution of Hospice in America Toward Organizational Homogeneity." *Journal of Health and Social Behavior,* 27: 370–386.
Firestone, Shulamith. 1970. *The Dialectic of Sex.* New York: Morrow.
Fishbein, Diana H., and Susan E. Pease. 1996. *The Dynamics of Drug Abuse.* Boston: Allyn & Bacon.
Fisher-Thompson, Donna. 1990. "Adult Sex-Typing of Children's Toys." *Sex Roles,* 23: 291–303.
Fjellman, Stephen M. 1992. *Vinyl Leaves: Walt Disney World & America.* Boulder, CO: Westview.
Flanagan, William G. 2002. *Urban Sociology: Images and Structures* (4th ed.). Boston: Allyn & Bacon.
Fleitas, Joan. 1999. *Band-Aides & Blackboards.* Retrieved Oct. 2, 1999. Online: http://funrsc.fairfield.edu/~jfleitas.html
Fleming, Jim. 2000. "Barbie Super Sports." Retrieved July 13, 2003. Online: http://www.gradingthemovies.com/html/games/barbie_sports.shtml
Flexner, Abraham. 1910. *Medical Education in the United States and Canada.* New York: Carnegie Foundation.
Florida, Richard, and Martin Kenney. 1991. "Transplanted Organizations: The Transfer of Japanese Industrial Organization to the U.S." *American Sociological Review,* 56 (3): 381–398.
Forbes. 1996. "The 100 Largest U.S. Multinationals" (July 15): 288–290.
___. 2002. "The Forbes 400: The Richest Americans." Retrieved July 19, 2003. Online: http://www.forbes.com/2002/09/13/rich400land.html, and accompanying lists from prior years.
___. 2003. "Special Report: The World's Richest People." Retrieved July 12, 2003. Online: http://www.forbes.com/2003/02/26/billionaireland.html
Ford, Clyde W. 1994. *We Can All Get Along: 50 Steps You Can Take to Help End Racism.* New York: Dell.
Fortune. 2001. "Most Powerful Women in Business." Retrieved July 20, 2002. Online: http://www.fortune.com/lists/women/index.html?_requestid=13939
Foucault, Michel. 1979. *Discipline and Punish: The Birth of the Prison.* New York: Vintage.
___. 1988. *Madness and Civilization: A History of Insanity in the Age of Reason.* New York: Vintage (orig. pub. 1961).
___. 1994. *The Birth of the Clinic: An Archeology of Medical Perception.* New York: Vintage (orig. pub. 1963).
Fountain, John W. 2001. "Prayer Warriors Fight Church–State Division." *New York Times* (Nov. 18): A18.
Fox, Justin. 2001. "War and Recession." *Fortune.com.* Retrieved Oct. 27, 2001. Online: http://www.fortune.com
Fox News. 1999. "Divorce Rates for Children of Broken Families on Decline." Retrieved Sept. 18, 1999. Online: http://www.foxnews.com/js_index.sml?content=/news/national/0812/d_ap_0812_11.sml
Frank, Robert H. 1999. *Luxury Fever: Why Money Fails to Satisfy in an Era of Excess.* New York: Free Press.
Frankenberg, Ruth. 1993. *White Women, Race Matters: The Social Construction of Whiteness.* Minneapolis: University of Minnesota Press.
Franklin, John Hope. 1980. *From Slavery to Freedom: A History of Negro Americans.* New York: Vintage.
Freidson, Eliot. 1965. "Disability as Social Deviance." In Marvin B. Sussman (Ed.), *Sociology and Rehabilitation.* Washington, DC: American Sociology Association, pp. 71–99.
___. 1970. *Profession of Medicine.* New York: Dodd, Mead.
___. 1986. *Professional Powers.* Chicago: University of Chicago Press.
French, Howard W. 2003. "Japan's Neglected Resource: Female Workers." *New York Times* (July 25): A3.
French, Sally. 1999. "The Wind Gets in My Way." In Mairian Corker and Sally French (Eds.), *Disability Discourse.* Buckingham, England: Open University Press, pp. 21–27.
Freud, Sigmund. 1924. *A General Introduction to Psychoanalysis* (2nd ed.). New York: Boni & Liveright.
Freudenberg, Nicholas, and Carl Steinsapir. 1992. "Not in Our Backyards: The Grassroots Environmental Movement." In Riley E. Dunlap and Angela G. Mertig (Eds.), *American Environmentalism: The U.S. Environmental Movement, 1970–1990.* New York: Taylor & Francis, pp. 27–37.
Freudenheim, Milt. 1998. "Medicine at the Click of a Mouse." *New York Times* (Aug. 12): C1, C8.
Friedan, Betty. 1993. *The Fountain of Age.* New York: Simon & Schuster.
Friedman, Debra, and Michael Hechter. 1988. "The Contribution of Rational Choice Theory to Macrosociological Research." *Sociological Theory,* 6: 201–218.
Friedmann, John. 1995. "The World City Hypothesis." In Paul L. Knox and Peter J. Taylor (Eds.), *World Cities in a World-System.* Cambridge, England: Cambridge University Press, pp. 317–331.
Friedrichs, David O. 1996. *Trusted Criminals: White Collar Crime in Contemporary Society.* Belmont, CA: Wadsworth.
Frisbie, W. Parker, and John D. Kasarda. 1988. "Spatial Processes." In Neil Smelser (Ed.), *The Handbook of Sociology.* Newbury Park, CA: Sage, pp. 629–666.
Funderburg, Lise. 1994. *Black, White, Other: Biracial Americans Talk About Race and Identity.* New York: Morrow.
Furstenberg, Frank F., Jr., and Andrew J. Cherlin. 1991. *Divided Families: What Happens to Children When Parents Part.* Cambridge, MA: Harvard University Press.
Gabriel, Trip. 1995. "A Generation's Heritage: After the Boom, a Boomlet." *New York Times* (Feb. 12): 1, 15.
___. 1996. "High-Tech Pregnancies Test Hope's Limits." *New York Times* (Jan. 7): 1, 10–11.
Galbraith, John Kenneth. 1985. *The New Industrial State* (4th ed.). Boston: Houghton Mifflin.
Gambino, Richard. 1975. *Blood of My Blood.* New York: Doubleday/Anchor.
Gamson, William. 1990. *The Strategy of Social Protest* (2nd ed.). Belmont, CA: Wadsworth.
___. 1995. "Constructing Social Protest." In Hank Johnston and Bert Klandermans (Eds.), *Social Movements and Culture.* Minneapolis: University of Minnesota Press, pp. 85–106.
Gandara, Ricardo. 1995. "*Dichos de la Vida:* Homespun Proverbs Link Hispanic Culture's Past with the Present." *Austin American-Statesman* (Jan. 21): E1, E10.
Gans, Herbert. 1974. *Popular Culture and High Culture: An Analysis and Evaluation of Tastes.* New York: Basic.
___. 1982. *The Urban Villagers: Group and Class in the Life of Italian Americans* (updated and expanded ed.; orig. pub. 1962). New York: Free Press.
Garbarino, James. 1989. "The Incidence and Prevalence of Child Maltreatment." In L. Ohlin and M. Tonry (Eds.), *Family Violence.* Chicago: University of Chicago Press, pp. 219–261.
Garber, Judith A., and Robyne S. Turner. 1995. "Introduction." In Judith A. Garber and Robyne S. Turner (Eds.), *Gender in Urban Research.* Thousand Oaks, CA: Sage, pp. x–xxvi.
Garcia Coll, Cynthia T. 1990. "Developmental Outcomes of Minority Infants: A Process-Oriented Look into Our Beginnings." *Child Development,* 61: 270–289.
Gardner, Carol Brooks. 1989. "Analyzing Gender in Public Places: Rethinking Goffman's Vision of Everyday Life." *American Sociologist,* 20 (Spring): 42–56.
Garfinkel, Harold. 1967. *Studies in Ethnomethodology.* Englewood Cliffs, NJ: Prentice Hall.
Garfinkel, Irwin, and Sara S. McLanahan. 1986. *Single Mothers and Their Children: A New American Dilemma.* Washington, DC: Urban Institute Press.
Gargan, Edward A. 1996. "An Indonesian Asset Is Also a Liability." *New York Times* (Mar. 16): 17, 18.
Garreau, Joel. 1991. *Edge City: Life on the New Frontier.* New York: Doubleday.
___. 1993. "GAK Attack." *Austin American-Statesman* (Jan. 9): D1, D6.
Garson, Barbara. 1989. *The Electronic Sweatshop: How Computers Are Transforming the Office of the Future into the Factory of the Past.* New York: Penguin.
Gatz, Margaret (Ed.). 1995. *Emerging Issues in Mental Health and Aging.* Washington, DC: American Psychological Association.
Gaylin, Willard. 1992. *The Male Ego.* New York: Viking/Penguin.
Geertz, Clifford. 1966. "Religion as a Cultural System." In Michael Banton (Ed.), *Anthropological Approaches to the Study of Religion.* London: Tavistock, pp. 1–46.
Gelfand, Donald E. 2003. *Aging and Ethnicity: Knowledge and Services* (2nd ed.). New York: Springer.
Gelles, Richard J., and Murray A. Straus. 1988. *Intimate Violence: The Definitive Study of the Causes and Consequences of Abuse in the American Family.* New York: Simon & Schuster.
"General Facts on Sweden." 1988. *Fact Sheets on Sweden.* Stockholm: Swedish Institute.
General Motors. 2003. *Proxy Statement for Annual Meeting of Shareholders.* Retrieved July 26, 2003. Online: http://www.gm.com/company/investor_information/docs/stockholder_info/gmpxy/pxy03/Gm03ps2.pdf
George, Susan. 1993. "A Fate Worse Than Debt." In William Dan Perdue (Ed.), *Systemic Crisis: Problems in Society, Politics, and World Order.* Fort Worth: Harcourt, pp. 85–96.
Gerbner, George, Larry Gross, Michael Morton, and Nancy Signorielli. 1987. "Charting the Mainstream: Television's Contributions to Political Orientations." In Donald Lazere (Ed.), *American Media and Mass Culture: Left Perspectives.* Berkeley: University of California Press, pp. 441–464.
Gereffi, Gary. 1994. "The International Economy and Economic Development." In Neil J. Smelser and Richard Swedberg (Eds.), *The Handbook of Economic Sociology.* Princeton, NJ: Princeton University Press, pp. 206–233.
Gereffi, Gary, David Spener, and Jennifer Bair (Eds.). 2002. *Free Trade and Uneven Development: The North American Apparel Industry After NAFTA.* Philadelphia: Temple University Press.
Gerschenkron, Alexander. 1962. *Economic Backwardness in Historical Perspective.* Cambridge, MA: Harvard University Press.
Gerson, Kathleen. 1993. *No Man's Land: Men's Changing Commitment to Family and Work.* New York: Basic.
Gerstel, Naomi, and Harriet Engel Gross. 1995. "Gender and Families in the United States: The Reality of Economic Dependence." In Jo Freeman (Ed.), *Women: A Feminist Perspective* (5th ed.). Mountain View, CA: Mayfield, pp. 92–127.
Gibbs, Lois Marie, as told to Murray Levine. 1982. *Love Canal: My Story.* Albany: SUNY Press.
Gibbs, Nancy. 1994. "Home Sweet School." *Time* (Oct. 31): 62–63.
___. 1999. "The Littleton Massacre." *Time* (May 3): 25–36.
___. 2003. "Lights Out." *Time* (Aug. 25): 30–39.
Gilbert, Dennis. 2003. *The American Class Structure in an Age of Growing Inequality* (6th ed.). Belmont, CA: Wadsworth.
Gilbert, Dennis, and Joseph A. Kahl. 1998. *The American Class Structure* (5th ed.). Belmont, CA: Wadsworth.
Gill, Derek. 1994. "A National Health Service: Principles and Practice." In Peter Conrad and Rochelle Kern (Eds.), *The Sociology of Health and Illness* (4th ed.). New York: St. Martin's, pp. 480–494.
Gilligan, Carol. 1982. *In a Different Voice: Psychological Theory and Women's Development.* Cambridge, MA: Harvard University Press.
Gilmore, David D. 1990. *Manhood in the Making: Cultural Concepts of Masculinity.* New Haven, CT: Yale University Press.
Ginorio, Angela, and Michelle Huston. 2000. *¡Sí Puede! Yes, We Can: Latinas in School.* Washington, DC: American Association of University Women.
Giroux, Henry A. 2000. *Stealing Innocence: Youth, Corporate Power, and the Politics of Culture.* New York: St. Martin's.
Glanz, James, and Edward Wong. 2003. "'97 Report Warned of Foam Damaging Tiles." *New York Times* (Feb. 4): A1–A26.
Glaser, Barney, and Anselm Strauss. 1967. *Discovery of Grounded Theory: Strategies for Qualitative Research.* Chicago: Aldine.
___. 1968. *Time for Dying.* Chicago: Aldine.

Glastris, Paul. 1990. "The New Way to Get Rich." *U.S. News & World Report* (May 7): 26–36.

Glazer, Nona. 1990. "The Home as Workshop: Women as Amateur Nurses and Medical Care Providers." *Gender & Society,* 4: 479–499.

Gleick, Elizabeth, Susan Reed, and Susan Schindehette. 1994. "The Baby Trap." *People* (Oct. 24): 39–56.

Glenn, Evelyn Nakano. 1986. *Issei, Nisei, War Bride: Three Generations of Japanese American Women in Domestic Service.* Philadelphia: Temple University Press.

globeandmail.com. 2003. "England Faces Racism Penalty." *globeandmail.com* (Apr. 11). Retrieved July 17, 2003. Online: http://www.theglobeandmail.com

Goffman, Erving. 1956. "The Nature of Deference and Demeanor." *American Anthropologist,* 58: 473–502.

___. 1959. *The Presentation of Self in Everyday Life.* Garden City, NY: Doubleday.

___. 1961a. *Asylums: Essays on the Social Situation of Mental Patients and Other Inmates.* Chicago: Aldine.

___. 1961b. *Encounters: Two Studies in the Sociology of Interaction.* London: Routledge and Kegan Paul.

___. 1963a. *Behavior in Public Places: Notes on the Social Structure of Gatherings.* New York: Free Press.

___. 1963b. *Stigma: Notes on the Management of Spoiled Identity.* Englewood Cliffs, NJ: Prentice Hall.

___. 1967. *Interaction Ritual: Essays on Face to Face Behavior.* Garden City, NY: Anchor.

___. 1974. *Frame Analysis: An Essay on the Organization of Experience.* Boston: Northeastern University Press.

Gold, Rachel Benson, and Cory L. Richards. 1994. "Securing American Women's Reproductive Health." In Cynthia Costello and Anne J. Stone (Eds.), *The American Woman 1994–95.* New York: Norton, pp. 197–222.

Goldberg, Robert A. 1991. *Grassroots Resistance: Social Movements in Twentieth Century America.* Belmont, CA: Wadsworth.

Golden, Stephanie. 1992. *The Women Outside: Meanings and Myths of Homelessness.* Berkeley: University of California Press.

Gongloff, Mark. 2003. "U.S. Jobs Jumping Ship." *CNNMoney.com.* Retrieved July 23, 2003. Online: http://money.cnn.com/2003/07/22/news/economy/jobless_offshore/index.htm

Gonyea, Judith G. 1994. "The Paradox of the Advantaged Elder and the Feminization of Poverty." *Social Work,* 39 (1): 35–42.

Goode, Erich. 1996. "The Stigma of Obesity." In Erich Goode (Ed.), *Social Deviance.* Boston: Allyn & Bacon, pp. 332–340.

Goode, William J. 1960. "A Theory of Role Strain." *American Sociological Review,* 25: 483–496.

___. 1982. "Why Men Resist." In Barrie Thorne with Marilyn Yalom (Eds.), *Rethinking the Family: Some Feminist Questions.* New York: Longman, pp. 131–150.

Goodman, Mary Ellen. 1964. *Race Awareness in Young Children* (rev. ed.). New York: Collier.

Goodman, Peter S. 1996. "The High Cost of Sneakers." *Austin American-Statesman* (July 7): F1, F6.

Gordon, David. 1973. "Capitalism, Class, and Crime in America." *Crime and Delinquency,* 19: 163–186.

Gordon, Milton. 1964. *Assimilation in American Life: The Role of Race, Religion, and National Origins.* New York: Oxford University Press.

Gordon, Philip L. 2001. "Federal Judge's Victory Just the First Shot in the Battle Over Workplace Monitoring." Retrieved July 8, 2002. Online: http://www.privacyfoundation.org/workplace/law/law_show.asp?id=75&action=0

Gotham, Kevin Fox. 1999. "Political Opportunity, Community Identity, and the Emergence of a Local Anti-Expressway Movement." *Social Problems,* 46: 332–354.

Gottdiener, Mark. 1985. *The Social Production of Urban Space.* Austin: University of Texas Press.

___. 1997. *The Theming of America.* Boulder, CO: Westview.

Gottlieb, Lori. 2000. *Stick Figure: A Diary of My Former Self.* New York: Simon & Schuster.

Gouldner, Alvin W. 1960. "The Norm of Reciprocity: A Preliminary Statement." *American Sociological Review,* 25: 161–179.

___. 1970. *The Coming Crisis of Western Sociology.* New York: Basic.

Grady, Denise. 1998. "Sorting Health Facts from Fiction on Line." *New York Times* (July 9): D7.

Gratton, Bruce. 1986. "The New History of the Aged." In David Van Tassel and Paul N. Stearns (Eds.), *Old Age in a Bureaucratic Society.* Westport, CT: Greenwood, pp. 3–29.

Gray, Paul. 1993. "Camp for Crusaders." *Time* (Apr. 19): 40.

___. 1996. "Gagging the Doctors." *Time* (Jan. 8): 50.

Greeley, Andrew M. 1972. *The Denominational Society.* Glenview, IL: Scott, Foresman.

Green, Donald E. 1977. *The Politics of Indian Removal: Creek Government and Society in Crisis.* Lincoln: University of Nebraska Press.

Greenberg, Edward S., and Benjamin I. Page. 1993. *The Struggle for Democracy.* New York: HarperCollins.

___. 2002. *The Struggle for Democracy* (5th ed.). Boston: Allyn & Bacon.

Greenhouse, Steven. 1997. "Union Membership Drops Worldwide, U.N. Reports." *New York Times* (Nov. 4): A8.

___. 2001. "Also Hurt by Sept. 11: A Legion Still Working But Making Much Less." *New York Times* (Nov. 29): A28.

___. 2003. "I.B.M. Explores Shift of Some Jobs Overseas." *New York Times* (July 22): C1–C2.

Grimes, William. 1999. "Eyes on the Fries." *New York Times Magazine* (May 30): 11.

Grint, Keith, and Steve Woolgar. 1997. *The Machine at Work: Technology, Work, and Organization.* Cambridge: Polity.

Griswold del Castillo, R. 1984. *La Familia: Chicano Families in the Urban Southwest, 1848 to the Present.* Notre Dame, IN: University of Notre Dame Press.

Gross, Jane. 1991. "More Young Single Men Clinging to Apron Strings." *New York Times* (June 16): A1.

Hadden, Richard W. 1997. *Sociological Theory: An Introduction to the Classical Tradition.* Peterborough, Ontario: Broadview.

Hafner, Katie. 1998. "Can the Internet Cure the Common Cold?" *New York Times* (July 9): D1, D7.

Hagan, John. 1989. *Structural Criminology.* New Brunswick, NJ: Rutgers University Press.

Hahn, Harlan. 1987. "Civil Rights for Disabled Americans: The Foundation of a Political Agenda." In Alan Gartner and Tom Joe (Eds.), *Images of the Disabled, Disabling Images.* New York: Praeger, pp. 181–203.

___. 1997. "Advertising the Acceptably Employable Image." In Lennard J. Davis (Ed.), *The Disability Studies Reader.* New York: Routledge, pp. 172–186.

Haines, Valerie A. 1997. "Spencer and His Critics." In Charles Camic (Ed.), *Reclaiming the Sociological Classics: The State of the Scholarship.* Malden, MA: Blackwell, pp. 81–111.

Halberstadt, Amy G., and Martha B. Saitta. 1987. "Gender, Nonverbal Behavior, and Perceived Dominance: A Test of the Theory." *Journal of Personality and Social Psychology,* 53: 257–272.

Hale-Benson, Janice E. 1986. *Black Children: Their Roots, Culture and Learning Styles* (rev. ed.). Provo, UT: Brigham Young University Press.

Hall, Edward. 1966. *The Hidden Dimension.* New York: Anchor/Doubleday.

Hall, Roberta M., with Bernice R. Sandler. 1982. *The Classroom Climate: A Chilly One for Women?* Washington, DC: Association of American Colleges, Project on the Status and Education of Women.

___. 1984. *Out of the Classroom: A Chilly Campus Climate for Women.* Washington, DC: Association of American Colleges, Project on the Status and Education of Women.

Halle, David. 1993. *Inside Culture: Art and Class in the American Home.* Chicago: University of Chicago Press.

Hamper, Ben. 1992. *Rivethead: Tales from the Assembly Line.* New York: Warner.

Haraway, Donna. 1994. "A Cyborgo Manifesto: Science, Technology, and Socialist-Feminism in the Late Twentieth Century." In Anne C. Herrmann and Abigail J. Stewart (Eds.), *Theorizing Feminism: Parallel Trends in the Humanities and Social Sciences.* Boulder, CO: Westview, pp. 424–457.

Harding, Sandra. 1986. *The Science Question in Feminism.* Ithaca, NY: Cornell University Press.

Hardy, Melissa A., and Lawrence E. Hazelrigg. 1993. "The Gender of Poverty in an Aging Population." *Research on Aging,* 15 (3): 243–278.

Harlow, Harry F., and Margaret Kuenne Harlow. 1962. "Social Deprivation in Monkeys." *Scientific American,* 207 (5): 137–146.

___. 1977. "Effects of Various Mother–Infant Relationships on Rhesus Monkey Behaviors." In Brian M. Foss (Ed.), *Determinants of Infant Behavior* (vol. 4). London: Methuen, pp. 15–36.

Harrington, Michael. 1985. *The New American Poverty.* New York: Viking/Penguin.

Harrington Meyer, Madonna. 1990. "Family Status and Poverty Among Older Women: The Gendered Distribution of Retirement Income in the United States." *Social Problems,* 37: 551–563.

___. 1994. "Gender, Race, and the Distribution of Social Assistance: Medicaid Use Among the Frail Elderly." *Gender & Society,* 8 (1): 8–28.

Harris, Anthony, and James W. Shaw. 2000. "Looking for Patterns: Race, Class, and Crime." In Joseph F. Sheley (Ed.), *Criminology: A Contemporary Handbook* (3rd ed.). Belmont, CA: Wadsworth, pp. 128–163.

Harris, Anthony R. 1991. "Race, Class, and Crime." In Joseph F. Sheley (Ed.), *Criminology: A Contemporary Handbook.* Belmont, CA: Wadsworth, pp. 94–119.

Harris, Chauncey D., and Edward L. Ullman. 1945. "The Nature of Cities." *Annals of the Academy of Political and Social Sciences* (November): 7–17.

Harris, Marvin. 1974. *Cows, Pigs, Wars, and Witches.* New York: Random House.

___. 1985. *Good to Eat: Riddles of Food and Culture.* New York: Simon & Schuster.

Harrison, Algea O., Melvin N. Wilson, Charles J. Pine, Samuel Q. Chan, and Raymond Buriel. 1990. "Family Ecologies of Ethnic Minority Children." *Child Development,* 61 (2): 347–362.

Harrison, Janine. 2001. "Welfare Reports Document Increasing Homelessness in Australia." *World Socialist Web Site.* Retrieved Sept. 11, 2001. Online: http://wsws.orgarticles/2001/jun2001/home-j07_prn.shtml

Hartmann, Heidi. 1976. "Capitalism, Patriarchy, and Job Segregation by Sex." *Signs: Journal of Women in Culture and Society,* 1 (Spring): 137–169.

___. 1981. "The Unhappy Marriage of Marxism and Feminism." In Lydia Sargent (Ed.), *Women and Revolution.* Boston: South End.

Hartney, Lou L. 1990. "Nursing Homes Are Beneficial." In Karin Swisher (Ed.), *The Elderly: Opposing Viewpoints.* San Diego, CA: Greenhaven, pp. 190–196.

Hartung, William D. 1999. "Military–Industrial Complex Revisited: How Weapons Makers Are Shaping U.S. Foreign and Military Policies." Retrieved Sept. 11, 1999. Online: http://www.foreignpolicy-infocus.org/paper/micr/index.html

Harvard Law Review. 1987. "'Official English': Federal Limits on Efforts to Curtail Bilingual Services in the States." *HLR,* 100 (6): 1345–1362.

Harvard Law School Seminar. 1996. "Cultural Imperialism." Retrieved June 23, 1999. Online: http://roscoe.law.harvard.edu/courses/techseminar96/course/sessions/culturalimperialism/selene2.html

Haseler, Stephen. 2000. *The Super-Rich: The Unjust New World of Global Capitalism.* New York: St. Martin's.

Hastorf, Albert, and H. Cantril. 1954. "They Saw a Game: A Case Study." *Journal of Abnormal and Social Psychology,* 40 (2): 129–134.

Hauchler, Ingomar, and Paul M. Kennedy (Eds.). 1994. *Global Trends: The World Almanac of Development and Peace.* New York: Continuum.

Hauser, Robert M. 1995. "Symposium: The Bell Curve." *Contemporary Sociology: A Journal of Reviews,* 24 (2): 149–153.

Havighurst, Robert J., Bernice L. Neugarten, and Sheldon S. Tobin. 1968. "Patterns of Aging." In Bernice L. Neugarten (Ed.), *Middle Age and Aging.* Chicago: University of Chicago Press, pp. 161–172.

Haviland, William A. 1993. *Cultural Anthropology* (7th ed.). Orlando, FL: Harcourt.

___. 1999. *Cultural Anthropology* (9th ed.). Orlando, FL: Harcourt.

Hawley, Amos. 1950. *Human Ecology.* New York: Ronald.

___. 1981. *Urban Society* (2nd ed.). New York: Wiley.

Hawton, Keith, Sue Simkin, Jonathan J. Deeks, Susan O'Connor, Allison Keen, Douglas G. Altman, Greg Philo, and Christopher Bulstrode. 1999. "Effects of a Drug Overdose in a Television Drama on Presentations to Hospital for Self Poisoning: Time Series and Questionnaire Study." *British Medical Journal,* 318 (Apr. 10): 972–988.

Haynes, Judie. 1999. "English Language Learners and the 'Hidden Curriculum.'" Retrieved Aug. 7, 2003. Online: http://www.everythingesl.net/inservices/goal3.php

Hays, Laurie. 1996. "Banks' Marketing Blitz Yields Rash of Defaults." *Wall Street Journal* (Sept. 25): B1, B6; cited in Robert H. Frank, *Luxury Fever: Why Money Fails to Satisfy in an Era of Excess.* New York: Free Press, 1999, pp. 47–48.

Healey, Joseph F. 2002. *Race, Ethnicity, Gender, and Class: The Sociology of Group Conflict and Change* (3rd ed.). Thousand Oaks, CA: Pine Forge.

Heath, Julia A., David H. Ciscel, and David C. Sharp. 1998. "The Work of Families: The Provision of Market and Household Labor and the Role of Public Policy." *Review of Social Economy* (Winter): 501–520.

Hechter, Michael. 1987. *Principles of Group Solidarity.* Berkeley: University of California Press.

Hechter, Michael, and Satoshi Kanazawa. 1997. "Sociological Rational Choice Theory." *Annual Review of Sociology,* 23: 191–215.

Hedges, Chris. 1997. "In Bosnia's Schools, 3 Ways Never to Learn from History." *New York Times* (Nov. 25): A1, A4.

Heilbron, Johan. 1995. *The Rise of Social Theory.* Trans. Sheila Gogol. Minneapolis: University of Minnesota Press.

Heldrich Center for Workforce Development. 2003. "Work Trends Survey of Employers About People with Disabilities." Retrieved Aug. 23, 2003. Online: http://www.heldrich.rutgers.edu

Henderson, Rick. 1997. "Schools of Thought." *Reasonline* (January). Retrieved July 24, 2002. Online: http://reason.com/9701/fe.rick.shtml

Henley, Nancy. 1977. *Body Politics: Power, Sex, and Nonverbal Communication.* Englewood Cliffs, NJ: Prentice Hall.

Henneberger, Melinda. 1993. "Gang Membership Grows in Middle-Class Suburbs." *New York Times* (July 24): 1, 12.

Henry, William A., III. 1990. "Beyond the Melting Pot." *Time* (Apr. 9): 28–35.

Henslin, James M. 1997. *Sociology: A Down-to-Earth Approach* (3rd ed.). Boston: Allyn & Bacon.
Herbert, Bob. 2001. "The Tourism Crisis." *New York Times* (Nov. 29): A31.
Herek, Gregory M., and Kevin T. Berrill (Eds.). 1992. *Hate Crimes: Confronting Violence Against Lesbians and Gay Men.* Newbury Park, CA: Sage.
Heritage, John. 1984. *Garfinkel and Ethnomethodology.* Cambridge, MA: Polity.
Herold, Marc W. 2002. "A Dossier on Civilian Victims of United States' Aerial Bombing of Afghanistan: A Comprehensive Accounting [revised]." Retrieved July 26, 2003. Online: http://www.cursor.org/stories/civilian_deaths.htm
___. 2003. "When 'Precision' Bombing Really Isn't: The Evil, the Grotesque and the Official Lies." Retrieved July 26, 2003. Online: http://www.cursor.org/stories/archivistan.htm
Herrnstein, Richard J., and Charles Murray. 1994. *The Bell Curve: Intelligence and Class Structure in American Life.* New York: Free Press.
Herz, J. C. 1998a. "New Title on the Cutlass Edge of Software." *New York Times* (Sept. 3): D4.
___. 1998b. "Puzzling Over the Allure of Virtual Barbie." *New York Times* (Mar. 19): D4.
Herzog, David B., K. L. Newman, Christine J. Yeh, and M. Warshaw. 1992. "Body Image Satisfaction in Homosexual and Heterosexual Women." *International Journal of Eating Disorders,* 11: 391–396.
Heshka, Stanley, and Yona Nelson. 1972. "Interpersonal Speaking Distances as a Function of Age, Sex, and Relationship." *Sociometry,* 35 (4): 491–498.
Hesse-Biber, Sharlene. 1989. "Eating Patterns and Disorders in a College Population: Are College Women's Eating Problems a New Phenomenon?" *Sex Roles,* 26: 71–89.
___. 1996. *Am I Thin Enough Yet? The Cult of Thinness and the Commercialization of Identity.* New York: Oxford University Press.
Hesse-Biber, Sharlene, and Gregg Lee Carter. 2000. *Working Women in America: Split Dreams.* New York: Oxford University Press.
Heywood, Leslie, and Shari L. Dworkin. 2003. *Built to Win: The Female Athlete as Cultural Icon.* Minneapolis: University of Minnesota Press.
Hibbard, David R., and Duane Buhrmester. 1998. "The Role of Peers in the Socialization of Gender-Related Social Interaction Styles." *Sex Roles,* 39: 185–203.
Higginbotham, Elizabeth. 1991. "Is Marriage a Priority: Class Differences in Marital Options of Educated Black Women." In Peter Stein (Ed.), *Single Life: Unmarried Adults in Social Context.* New York: St. Martin's.
___. 1994. "Black Professional Women: Job Ceilings and Employment Sectors." In Maxine Baca Zinn and Bonnie Thornton Dill (Eds.), *Women of Color in U.S. Society.* Philadelphia: Temple University Press, pp. 113–131.
Higginbotham, Elizabeth, and Lynn Weber. 1995. "Moving Up with Kin and Community: Upward Social Mobility for Black and White Women." In Margaret L. Andersen and Patricia Hill Collins (Eds.), *Race, Class, and Gender: An Anthology* (2nd ed.). Belmont, CA: Wadsworth, pp. 134–147.
Hight, Bruce. 1994. "A Level Playing Field: Critics Say Minorities Need a Shot Off the Field." *Austin American-Statesman* (June 8): A1, A11.
___. 1999. "Job Hunters: Not Yet Extinct." *Austin American-Statesman* (Sept. 5): J1, J5.
Higley, Stephen Richard. 1997. "Privilege, Power, and Place: The Geography of the American Upper Class." In Diana Kendall (Ed.), *Race, Class, and Gender in a Diverse Society: A Text-Reader.* Boston: Allyn & Bacon, pp. 70–82.
Hill, Paul T. 1997. "Contracting in Public Education." In Diana Ravitch and Joseph P. Viteritti (Eds.), *New Schools for a New Century: The Redesign of Urban Education.* New Haven, CT: Yale University Press, pp. 61–85.
Hilsenrath, Jon. 1996. "In China, a Taste of Buy-Me TV." *New York Times* (Nov. 17): F1, F11.
Hirschi, Travis. 1969. *Causes of Delinquency.* Berkeley: University of California Press.
Hoban, Phoebe. 2002. "Single Girls: Sex but Still No Respect." *New York Times* (Oct. 12): A19, A21.
Hochschild, Arlie Russell. 1983. *The Managed Heart: Commercialization of Human Feeling.* Berkeley: University of California Press.
___. 1997. *The Time Bind: When Work Becomes Home and Home Becomes Work.* New York: Metropolitan.
___. 2003. *The Commercialization of Intimate Life: Notes from Home and Work.* Berkeley: University of California Press.
Hochschild, Arlie Russell, with Ann Machung. 1989. *The Second Shift: Working Parents and the Revolution at Home.* New York: Viking/Penguin.
Hochschild, Jennifer L. 1995. *Facing Up to the American Dream: Race, Class, and the Soul of the Nation.* Princeton, NJ: Princeton University Press.
Hodge, Robert W., Paul Siegel, and Peter Rossi. 1964. "Occupational Prestige in the United States, 1925–63." *American Journal of Sociology,* 70 (November): 286–302.
Hodson, Randy, and Robert E. Parker. 1988. "Work in High Technology Settings: A Review of the Empirical Literature." *Research in the Sociology of Work,* 4: 1–29.
Hodson, Randy, and Teresa A. Sullivan. 2002. *The Social Organization of Work* (3rd ed.). Belmont, CA: Wadsworth.
Hoecker-Drysdale, Susan. 1992. *Harriet Martineau: First Woman Sociologist.* Oxford, England: Berg.
Hofferth, Sandra L. 1984. "Kin Networks, Race, and Family Structure." *Journal of Marriage and the Family,* 46: 791–806.
Hoffnung, Michele. 1995. "Motherhood: Contemporary Conflict for Women." In Jo Freeman (Ed.), *Women: A Feminist Perspective* (5th ed.). Mountain View, CA: Mayfield, pp. 162–181.
Holland, Dorothy C., and Margaret A. Eisenhart. 1981. *Women's Peer Groups and Choice of Career.* Final report for the National Institute of Education. ERIC ED 199 328. Washington, DC.
___. 1990. *Educated in Romance: Women, Achievement, and College Culture.* Chicago: University of Chicago Press.
Holmes, Steven A. 1997. "Bringing Hope and Education to the Reservation." *New York Times Education Life Supplement* (Aug. 3): 28–29, 34–35.
Holson, Laura M. 1999. "On Top of the World, Stuck in Traffic." *New York Times* (Aug. 22): A16.
Homans, George. 1958. "Social Behavior as Exchange." *American Journal of Sociology,* 63: 597–606.
___. 1974. *Social Behavior: Its Elementary Forms* (rev. ed.). New York: Harcourt.
Hondagneu-Sotelo, Pierrette. 2001. *Doméstica: Immigrant Workers Cleaning and Caring in the Shadow of Affluence.* Berkeley: University of California Press.
Hoose, Phillip M. 1989. *Necessities: Racial Barriers in American Sports.* New York: Random House.
Hoover, Eric. 2001. "The Lure of Easy Credit Leaves More Students Struggling with Debt." *Chronicle of Higher Education* (June 15): A35–A36.
Hoover, Kenneth R. 1992. *The Elements of Social Scientific Thinking.* New York: St. Martin's.
Hooyman, Nancy R. R., and H. Asuman Kiyak. 2002. *Social Gerontology: A Multidisciplinary Approach* (6th ed.). Boston: Allyn & Bacon.
Horan, Patrick M. 1978. "Is Status Attainment Research Atheoretical?" *American Sociological Review,* 43: 534–541.
Horowitz, Allan V. 1982. *Social Control of Mental Illness.* New York: Academic.
Horowitz, R. 1997. "Barriers and Bridges to Class Mobility and Formation: Ethnographies of Stratification." *Sociological Methods and Research,* 25: 495–538.
Horsburgh, Susan. 2003. "Daddy Day Care." *People* (June 23): 79–81.
Hossfeld, Karen. 1992. *Small, Foreign, and Female: Immigrant Women Workers in Silicon Valley.* Berkeley: University of California Press.
___. 1994. "Hiring Immigrant Women: Silicon Valley's 'Simple Formula.'" In Maxine Baca Zinn and Bonnie Thornton Dill (Eds.), *Women of Color in U.S. Society.* Philadelphia: Temple University Press, pp. 65–93.
Hostetler, A. J. 1994. "U.S. Death Rate Falls to Lowest Level Ever Despite Rise in AIDS." *Austin American-Statesman* (Dec. 16): A4.
Howard, Michael E. 1990. "On Fighting a Nuclear War." In Francesca M. Cancian and James William Gibson (Eds.), *Making War, Making Peace: The Social Foundations of Violent Conflict.* Belmont, CA: Wadsworth, pp. 314–322.
Hoyt, Homer. 1939. *The Structure and Growth of Residential Neighborhoods in American Cities.* Washington, DC: Federal Housing Administration.
Hsiung, Ping-Chun. 1996. *Living Rooms as Factories: Class, Gender, and the Satellite Factory System in Taiwan.* Philadelphia: Temple University Press.
Huang, Larke Namhe, and Y. Ying. 1989. "Chinese American Children and Adolescents." In Jenel Taylor Gibbs and Larke Namhe Huang (Eds.), *Children of Color: Psychological Interventions with Minority Youth.* San Francisco: Jossey-Bass, pp. 30–66.
Hudnut-Beumler, James. 1994. *Looking for God in the Suburbs: The Religion of the American Dream and Its Critics, 1945–1965.* New Brunswick, NJ: Rutgers University Press.
Hughes, Everett C. 1945. "Dilemmas and Contradictions of Status." *American Journal of Sociology,* 50: 353–359.
Hull, Gloria T., Patricia Bell-Scott, and Barbara Smith. 1982. *All the Women Are White, All the Blacks Are Men, But Some of Us Are Brave.* Old Westbury, NY: Feminist.
Humphreys, Laud. 1970. *Tearoom Trade: Impersonal Sex in Public Places.* Chicago: Aldine.
Hurst, Charles E. 1998. *Social Inequality: Forms, Causes, and Consequences* (3rd ed.). Boston: Allyn & Bacon.
Hurtada, Aida. 1996. *The Color of Privilege: Three Blasphemies on Race and Feminism.* Ann Arbor: University of Michigan Press.
Huston, Aletha C. 1985. "The Development of Sex Typing: Themes from Recent Research." *Developmental Review,* 5: 2–17.
Huyssen, Andreas. 1984. *After the Great Divide.* Bloomington: Indiana University Press.
Hwang, S. S., R. Saenz, and B. E. Aguirre. 1995. "The SES-Selectivity of Interracially Married Asians." *International Migration Review,* 29: 469–491.
Hynes, H. Patricia. 1990. *Earth Right: Every Citizen's Guide.* Rocklin, CA: Prima.
Ibrahim, Youseff M. 1990. "Saudi Tradition: Edicts from Koran Produce Curbs on Women." *New York Times* (Nov. 6): A6.
IGN Entertainment. 2003. "Review of Red Jack: Revenge of the Brethren." Retrieved July 19, 2003. Online: http://pc.ign.com/articles/160/160369p1.html
Ikegami, Naoki. 1998. "Growing Old in Japan." *Age and Ageing* (May): 277–283.
Inciardi, James A., Ruth Horowitz, and Anne E. Pottieger. 1993. *Street Kids, Street Drugs, Street Crime: An Examination of Drug Use and Serious Delinquency in Miami.* Belmont, CA: Wadsworth.
Inda, Cynthia G. 1997. "Why I Took a Chance on Learning: From Community College to Harvard." *New York Times Education Life Supplement* (Aug. 3): 31, 40, 42.
International Monetary Fund. 1992. *World Economic Outlook.* Washington, DC: International Development Fund.
Iovine, Julie V. 1997. "An Industry Monitors Child Labor." *New York Times* (Oct. 16): B1, B9.
Irons, Edward D., and Gilbert W. Moore. 1985. *Black Managers: The Case of the Banking Industry.* New York: Praeger.
Irvine, Jacqueline Jordan. 1986. "Teacher–Student Interactions: Effects of Student Race, Sex, and Grade Level." *Journal of Educational Psychology,* 78 (1): 14–21.
Isacc, Jeffrey C. 2002. "Civilian Casualties in Afghanistan: The Limits of Marc Herold's 'Comprehensive Accounting.'" Retrieved July 26, 2003. Online: http://www.opendemocracy.net/debates/article-2-48-182.jsp
Ishikawa, Kaoru. 1984. "Quality Control in Japan." In Naoto Sasaki and David Hutchins (Eds.), *The Japanese Approach to Product Quality: Its Applicability to the West.* Oxford: Permagon, pp. 1–5.
ITAR/TASS News Agency. 1999a. "Only Three Ingush Men Used Their Right to Polygamy" (Aug. 31). Retrieved Sept. 11, 1999. Online: http://www.comtexnews.com
___. 1999b. "Polygamy Allowed in Southern Russia" (July 21). Retrieved Sept. 11, 1999. Online: http://www.comtexnews.com
Jack, Dana Crowley. 1993. *Silencing the Self: Women and Depression.* New York: Harper-Perennial.
Jackson, Kenneth T. 1985. *Crabgrass Frontier: The Suburbanization of the United States.* New York: Oxford University Press.
Jacobs, Gloria. 1994. "Where Do We Go from Here? An Interview with Ann Jones." *Ms.* (September/October): 56–63.
Jacobs, Jerry A. 1993. "Men in Female-Dominated Fields." In Christine L. Williams (Ed.), *Doing "Women's Work": Men in Nontraditional Occupations.* Newbury Park, CA: Sage, pp. 49–63.
Jacquard, Roland. 2001. "The Guidebook of Jihad." *Time* (Oct. 29): 58.
Jameson, Fredric. 1984. "Postmodernism, or, The Cultural Logic of Late Capitalism." *New Left Review,* 146: 59–92.
Janis, Irving. 1972. *Victims of Groupthink.* Boston: Houghton Mifflin.
___. 1989. *Crucial Decisions: Leadership in Policymaking and Crisis Management.* New York: Free Press.
Jankowski, Martin Sanchez. 1991. *Islands in the Street: Gangs and American Urban Society.* Berkeley: University of California Press.
Janofsky, Michael. 1993. "Race and the American Workplace." *New York Times* (June 20): F1, F6.
Jaramillo, P. T., and Jesse T. Zapata. 1987. "Roles and Alliances Within Mexican-American and Anglo Families." *Journal of Marriage and the Family,* 49 (November): 727–735.
Jary, David, and Julia Jary. 1991. *The Harper Collins Dictionary of Sociology.* New York: HarperPerennial.
Jehl, Douglas. 1999. "The Internet's 'Open Sesame' Is Answered Warily." *New York Times* (Mar. 18): A4.
Jensen, Robert. 1995. "Men's Lives and Feminist Theory." *Race, Gender & Class,* 2 (2): 111–125.
Jewell, K. Sue. 1993. *From Mammy to Miss America and Beyond: Cultural Images and the Shaping of US Social Policy.* New York: Routledge.
Johnson, Allan G. 2000. *The Blackwell Dictionary of Sociology* (2nd ed.). Malden, MA: Blackwell.
Johnson, Claudia. 1994. *Stifled Laughter: One Woman's Story About Fighting Censorship.* Golden, CO: Fulcrum.
Johnson, Dirk. 1994. "Equal Loads, Not Pay for Nonunion Drivers." *New York Times* (Apr. 10): 10.

Johnson, Earvin "Magic," with William Novak. 1992. *My Life*. New York: Fawcett Crest.

Johnston, David Cay. 2002. "As Salary Grows, So Does a Gender Gap." *New York Times* (May 12): BU8.

Joint Center for Political and Economic Studies. 2000. "Joint Center Releases 1999 Count of Black Elected Officials." Retrieved Sept. 25, 2002. Online: http://www.jointcenter.org/pressrel/2000_beo.htm

Joint United Nations Programme on HIV/AIDS. 2002. "AIDS Epidemic Update: December 2002." Retrieved Aug. 7, 2003. Online: http://www.unaids.org/worldaidsday/2002/press/update/epiupdate_en.pdf

Jones, Charisse. 1995. "Family Struggles on Brink of Comfort." *New York Times* (Feb. 18): 1, 9.

Judson, George. 1995. "Connecticut Wins School Bias Suit." *New York Times* (Apr. 13): A1, A11.

Juergensmeyer, Mark. 1993. *The New Cold War? Religious Nationalism Confronts the Secular State*. Berkeley: University of California Press.

Kabagarama, Daisy. 1993. *Breaking the Ice: A Guide to Understanding People from Other Cultures*. Boston: Allyn & Bacon.

Kahney, Leander. 2003. "E-Mail Mobs Materialize All Over." *Wired News*. Retrieved Aug. 11, 2003. Online: http://www.wired.com/news/culture/0,1284,59518,00.html

Kakuchi, Suvendrini. 1998. "Population: Japan Desperate for a Baby Boom." World News: InterPress Service. Retrieved Sept. 21, 1999. Online: http://www.oneworld.org/ips2/nov/japan.html

Kalish, Susan. 1992. "Interracial Baby Boomlet in Progress?" *Population Today*, 20 (December): 1–2, 9.

Kalmijn, Matthijs. 1998. "Intermarriage and Homogamy: Causes, Patterns, Trends." *Annual Review of Sociology*, 24: 395–422.

Kanter, Rosabeth Moss. 1983. *The Change Masters: Innovation and Entrepreneurship in the American Corporation*. New York: Simon & Schuster.

___. 1985. "All That Is Entrepreneural Is Not Gold." *Wall Street Journal* (July 22): 18.

___. 1993. *Men and Women of the Corporation*. New York: Basic (orig. pub. 1977).

Kantrowitz, Barbara. 2003. "Hoping for the Best, Ready for the Worst." *Newsweek* (May 12): 50–51.

Kaplan, David A. 1993. "Dumber Than We Thought." *Newsweek* (Sept. 20): 44–45.

Kaplan, Robert D. 1996. "Cities of Despair." *New York Times* (June 6): A19.

Kaspar, Anne S. 1986. "Consciousness Re-evaluated: Interpretive Theory and Feminist Scholarship." *Sociological Inquiry*, 56 (1): 30–49.

Katz, Michael B. 1989. *The Undeserving Poor: From the War on Poverty to the War on Welfare*. New York: Pantheon.

Katzer, Jeffrey, Kenneth H. Cook, and Wayne W. Crouch. 1991. *Evaluating Information: A Guide for Users of Social Science Research*. New York: McGraw-Hill.

Kaufman, Gayle. 1999. "The Portrayal of Men's Family Roles in Television Commercials." *Sex Roles*, 313: 439–451.

Kaufman, Tracy L. 1996. *Out of Reach: Can America Pay the Rent?* Washington, DC: National Low Income Housing Coalition.

Keefe, Bob. 2001. "Overseas Employees in Harm's Way." *Austin American-Statesman* (Oct. 28): K1–K6.

Keister, Lisa A. 2000. *Wealth in America: Trends in Wealth Inequality*. Cambridge, U.K.: Cambridge University Press.

Keller, James. 1994. "'I Treasure Each Moment.'" *Parade* (Sept. 4): 4–5.

Kelley, Tina. 1999. "For That Bowl of Cherries, A Hard Life." *New York Times* (Aug. 11): A11.

Kelly, John, and David Stark. 2002. "Crisis, Recovery, Innovation: Learning from 9/11." *Working Papers: Center on Organizational Innovation*. New York: Center on Organizational Innovation, Columbia University. Retrieved June 26, 2002. Online: http://www.coi.columbia.edu/pdf/kelly_stark_cri.pdf

Kelman, Steven. 1991. "Sweden Sour? Downsizing the 'Third Way.'" *New Republic* (July 29): 19–23.

Kemp, Alice Abel. 1994. *Women's Work: Degraded and Devalued*. Englewood Cliffs, NJ: Prentice Hall.

Kendall, Diana. 1980. Square Pegs in Round Holes: Non-Traditional Students in Medical Schools. Unpublished doctoral dissertation, Department of Sociology, the University of Texas at Austin.

___. 2002. *The Power of Good Deeds: Privileged Women and the Social Reproduction of the Upper Class*. Lanham, MD: Rowman & Littlefield.

___. 2004. *Social Problems in a Diverse Society* (3rd ed.). Boston: Allyn & Bacon.

Kendall, Diana, and Joe R. Feagin. 1983. "Blatant and Subtle Patterns of Discrimination: Minority Women in Medical Schools." *Journal of Intergroup Relations*, 9 (Summer): 21–27.

Kendall, Diana, Jane Lothian Murray, and Rick Linden. 2004. *Sociology in Our Times* (3rd Canadian edition). Scarborough, Ontario: Nelson Thomson Learning.

Kennedy, Paul. 1987. *The Rise and Fall of the Great Powers*. New York: Random House.

___. 1993. *Preparing for the Twenty-First Century*. New York: Random House.

Kennedy, Randy. 1993. "To Homelessness and Back: A Man's Journey to Respect." *New York Times* (Nov. 28): A15.

Kenyon, Kathleen. 1957. *Digging Up Jericho*. London: Benn.

Kephart, William M., and William W. Zellner. 1994. *Extraordinary Groups: An Examination of Unconventional Life-Styles* (2nd ed.). New York: St. Martin's.

Kerbo, Harold R. 2000. *Social Stratification and Inequality: Class Conflict in Historical, Comparative, and Global Perspective* (4th ed.). New York:McGraw-Hill.

Kessler, Ronald C. 1994. "Lifetime and 12-Month Prevalence of DSM-III-R Psychiatric Disorders in the United States: Results of the National Comorbidity Survey." *JAMA, the Journal of the American Medical Association*, 271 (Mar. 2): 654D.

Kessler-Harris, Alice. 1990. *A Woman's Wage: Historical Meanings and Social Consequences*. Lexington: University Press of Kentucky.

Keyfitz, Nathan. 1994. "The Scientific Debate: Is Population Growth a Problem? An Interview with Nathan Keyfitz." *Harvard International Review* (Fall): 10–11, 74.

Khleif, Bud B. 1998. "Distortions of 'Islam' and 'Muslims' in American Academic Discourse: Some Observations on the Sociology of Vested Enmity." In Yahya R. Kamalipour and Theresa Carilli (Eds.), *Cultural Diversity and the U.S. Media*. Albany: State University of New York Press, pp. 279–291.

Kidron, Michael, and Ronald Segal. 1995. *The State of the World Atlas*. New York: Penguin.

Kilborn, Peter T. 1995. "Women and Minorities Still Face 'Glass Ceiling.'" *New York Times* (Mar. 16): C22.

___. 1997. "Illness Is Turning into Financial Catastrophe for More of the Uninsured." *New York Times* (Aug. 1): A10.

Kilbourne, Jean. 1994. "Still Killing Us Softly: Advertising and the Obsession with Thinness." In Patricia Fallon, Melanie A. Katzman, and Susan C. Wooley (Eds.), *Feminist Perspectives on Eating Disorders*. New York: Guilford, pp. 395–454.

___. 1999. *Deadly Persuasion: The Addictive Power of Advertising*. New York: Simon & Schuster.

Killian, Lewis. 1984. "Organization, Rationality, and Spontaneity in the Civil Rights Movement." *American Sociological Review*, 49: 770–783.

Kimmell, Michael S., and Michael A. Messner. 2004. *Men's Lives* (6th ed.). Boston: Allyn & Bacon.

King, Gary, Robert O. Keohane, and Sidney Verba. 1994. *Designing Social Inquiry: Scientific Inference in Qualitative Research*. Princeton, NJ: Princeton University Press.

King, Leslie, and Madonna Harrington Meyer. 1997. "The Politics of Reproductive Benefits: U.S. Insurance Coverage of Contraceptive and Infertility Treatments." *Gender and Society*, 11 (1): 8–30.

Kirkpatrick, P. 1994. "Triple Jeopardy: Disability, Race and Poverty in America." *Poverty and Race*, 3: 1–8.

Kitsuse, John I. 1980. "Coming Out All Over: Deviance and the Politics of Social Problems." *Social Problems*, 28: 1–13.

Klein, Alan M. 1993. *Little Big Men: Bodybuilding Subculture and Gender Construction*. Albany: SUNY Press.

Klonoff, Elizabeth A. 1997. *Preventing Misdiagnosis of Women: A Guide to Physical Disorders That Have Psychiatric Symptoms*. Thousand Oaks, CA: Sage.

Kluckhohn, Clyde. 1961. "The Study of Values." In Donald N. Barrett (Ed.), *Values in America*. South Bend, IN: University of Notre Dame Press, pp. 17–46.

Knapp, Caroline. 1996. *Drinking: A Love Story*. New York: Dial.

Knox, Paul L., and Peter J. Taylor (Eds.). 1995. *World Cities in a World-System*. Cambridge, England: Cambridge University Press.

Knudsen, Dean D. 1992. *Child Maltreatment: Emerging Perspectives*. Dix Hills, NY: General Hall.

Kohlberg, Lawrence. 1969. "Stage and Sequence: The Cognitive–Developmental Approach to Socialization." In David A. Goslin (Ed.), *Handbook of Socialization Theory and Research*. Chicago: Rand McNally, pp. 347–480.

___. 1981. *The Philosophy of Moral Development: Moral Stages and the Idea of Justice*, vol. 1: *Essays on Moral Development*. San Francisco: Harper & Row.

Kohn, Melvin L. 1977. *Class and Conformity: A Study in Values* (2nd ed.). Homewood, IL: Dorsey.

Kohn, Melvin L., Atsushi Naoi, Carrie Schoenbach, Carmi Schooler, and Kazimierz M. Slomczynski. 1990. "Position in the Class Structure and Psychological Functioning in the United States, Japan, and Poland." *American Journal of Sociology*, 95: 964–1008.

Kolata, Gina. 1993. "Fear of Fatness: Living Large in a Slimfast World." *Austin American-Statesman* (Jan. 3): C1, C6.

___. 2002. "A Study Finds More Links Between TV and Violence." *New York Times* (Mar. 29): A20.

Korsmeyer, Carolyn. 1981. "The Hidden Joke: Generic Uses of Masculine Terminology." In Mary Vetterling-Braggin (Ed.), *Sexist Language: A Modern Philosophical Analysis*. Totowa, NJ: Littlefield, Adams, pp. 116–131.

Korten, David C. 1996. *When Corporations Rule the World*. West Hartford, CT: Kumarian.

Kosmin, Barry A., and Seymour P. Lachman. 1993. *One Nation Under God: Religion in Contemporary American Society*. New York: Crown.

Kovacs, M., and A. T. Beck. 1977. "The Wish to Live and the Wish to Die in Attempted Suicides." *Journal of Clinical Psychology*, 33: 361–365.

Kozol, Jonathan. 1986. *Illiterate America*. New York: Anchor/Doubleday.

___. 1988. *Rachael and Her Children: Homeless Families in America*. New York: Fawcett Columbine.

___. 1991. *Savage Inequalities: Children in America's Schools*. New York: Crown.

Kraft, Scott. 1994. "Agog at Euro Disneyland." *Los Angeles Times* (Jan. 18): H1.

Kramnick, Isaac. (Ed.). 1995. *The Portable Enlightenment Reader*. New York: Penguin.

Krisberg, Barry. 1975. *Crime and Privilege: Toward a New Criminology*. Englewood Cliffs, NJ: Prentice Hall.

Kristof, Nicholas D., and Sheryl WuDunn. 1999. "Of World Markets, None an Island." *New York Times* (Feb. 17): A1, A8–A9.

Kroloff, Charles A. 1993. *54 Ways You Can Help the Homeless*. Southport, CT: Hugh Lauter Levin Associates; and West Orange, NJ: Behrman.

Krysan, Maria, and Reynolds Farley. 1993. "Racial Stereotypes: Are They Alive and Well? Do They Continue to Influence Race Relations?" Paper presented at the annual meeting of the American Sociological Association, Miami Beach, Florida, Aug. 16.

Kübler-Ross, Elisabeth. 1969. *On Death and Dying*. New York: Macmillan.

Kuisel, Richard. 1993. *Seducing the French: The Dilemma of Americanization*. Berkeley: University of California Press.

Kurtz, Lester. 1995. *Gods in the Global Village: The World's Religions in Sociological Perspective*. Thousand Oaks, CA: Sage.

Kurz, Demie. 1989. "Social Science Perspectives on Wife Abuse: Current Debates and Future Directions." *Gender & Society*, 3 (4): 489–505.

___. 1995. *For Richer, for Poorer: Mothers Confront Divorce*. New York: Routledge.

Kvale, Steinar. 1996. *Interviews: An Introduction to Qualitative Research Interviewing*. Thousand Oaks, CA: Sage.

Lacayo, Richard. 1997. "They'll Vouch for That." *Time* (Oct. 27): 72–74.

___. 2001. "About Face: An Inside Look at How Women Fared Under Taliban Oppression and What the Future Holds for Them Now." *Time* (Dec. 3): 36–49.

Ladd, E. C., Jr. 1966. *Negro Political Leadership in the South*. Ithaca, NY: Cornell University Press.

Lamanna, Marianne, and Agnes Riedmann. 2003. *Marriages and Families: Making Choices and Facing Change* (8th ed.). Belmont, CA: Wadsworth.

Lamar, Joe. 2000. "Suicides in Japan Reach a Record High." *British Medical Journal* (Sept. 2). Retrieved Aug. 25, 2001. Online: http://www.findarticles.com/cf_dls/m0999/7260_321/66676910/p1/article.jhtml

Lane, Harlan. 1992. *The Mask of Benevolence: Disabling the Deaf Community*. New York: Vintage.

Langone, John. 1997. "Bloodless Surgery." *Time* (Fall special issue): 74–76.

Lapchick, Richard E. 1991. *Five Minutes to Midnight: Race and Sport in the 1990s*. Lanham, MD: Madison.

Lapham, Lewis H. 1988. *Money and Class in America: Notes and Observations on Our Civil Religion*. New York: Weidenfeld & Nicolson.

Lappé, Frances Moore, and Paul Martin Du Bois. 1994. *The Quickening of America: Rebuilding Our Nation, Remaking Our Lives*. San Francisco: Jossey-Bass.

Lapsley, Daniel K., 1990. "Continuity and Discontinuity in Adolescent Social Cognitive Development." In Raymond Montemayor, Gerald R. Adams, and Thomas P. Gullota (Eds.), *From Childhood to Adolescence: A Transitional Period?* (*Advances in Adolescent Development*, vol. 2). Newbury Park, CA: Sage.

Larimer, Tim. 1999. "The Japan Syndrome." *Time* (Oct. 11): 50–51.

Larson, Erik. 1997. "It's Not the Money, It's the Principal." *Time* (Oct. 27): 92–93.

Larson, Magali Sarfatti. 1977. *The Rise of Professionalism: A Sociological Analysis*. Berkeley: University of California Press.

Lasch, Christopher. 1977. *Haven in a Heartless World*. New York: Basic.

Lash, Scott, and John Urry. 1994. *Economies of Signs and Space*. London: Sage.

"Latino Legends in Sports." 1999. "Sports News." Retrieved Aug. 15, 1999. Online: http://www.latinosportslegends.com/news.htm

___. 2003. "Sports News." Retrieved July 26, 2003. Online: http://www.latinosportslegends.com

Latouche, Serge. 1992. "Standard of Living." In Wolfgang Sachs (Ed.), *The Development Dictionary*. Atlantic Highlands, NJ: Zed, pp. 250–263.

Laumann, Edward O., John H. Gagnon, Robert T. Michael, and Stuart Michaels. 1994. *The Social Organization of Sexuality*. Chicago: University of Chicago Press.

Le Bon, Gustave. 1960. *The Crowd: A Study of the Popular Mind*. New York: Viking (orig. pub. 1895).

Leary, Warren E. 1996. "Even When Covered by Insurance, Black and Poor People Receive Less Health Care." *New York Times* (Sept. 12): A10.

Lee, Felicia R. 1993. "Where Guns and Lives Are Cheap." *New York Times* (Mar. 21): 21.

Lee, Joann Faung Jean. 1992. *Asian Americans: Oral Histories of First to Fourth Generation Americans from China, the Philippines, Japan, India, the Pacific Islands, Vietnam and Cambodia*. New York: New Press.

Lee, Sharon M. 1993. "Racial Classifications in the U.S. Census: 1890–1990." *Ethnic and Racial Studies,* 16 (1): 75–94.

Leenaars, Antoon A. 1988. *Suicide Notes: Predictive Clues and Patterns*. New York: Human Sciences Press.

Leenaars, Antoon A. (Ed.). 1991. *Life Span Perspectives of Suicide: Time-Lines in the Suicide Process*. New York: Plenum.

Lefrançois, Guy R. 1996. *The Lifespan* (5th ed.). Belmont, CA: Wadsworth.

_____. 1999. *The Lifespan* (6th ed.). Belmont, CA: Wadsworth.

Lehmann, Jennifer M. 1994. *Durkheim and Women*. Lincoln: University of Nebraska Press.

Leidner, Robin. 1993. *Fast Food, Fast Talk: Service Work and the Routinization of Everyday Life*. Berkeley: University of California Press.

Lemann, Nicholas. 1997. "Let's Guarantee the Key Ingredients." *Time* (Oct. 27): 96.

Lemert, Charles. 1997. *Postmodernism Is Not What You Think*. Malden, MA: Blackwell.

Lemert, Edwin M. 1951. *Social Pathology*. New York: McGraw-Hill.

Lengermann, Patricia Madoo, and Jill Niebrugge-Brantley. 1998. *The Women Founders: Sociology and Social Theory, 1830–1930*. New York: McGraw-Hill.

Lenzer, Gertrud (Ed.). 1998. *The Essential Writings: Auguste Comte and Positivism*. New Brunswick, NJ: Transaction.

Leonard, Andrew. 1999. "We've Got Mail—Always." *Newsweek* (Sept. 20): 58–61.

Leonard, Wilbert M., and Jonathan E. Reyman. 1988. "The Odds of Attaining Professional Athlete Status: Refining the Computations." *Sociology of Sport Journal:* 162–169.

Lerman, Hannah. 1996. *Pigeonholing Women's Misery: A History and Critical Analysis of the Psychodiagnosis of Women in the Twentieth Century*. New York: Basic.

Lerner, Gerda. 1986. *The Creation of Patriarchy*. New York: Oxford University Press.

LeShan, Eda. 1994. *I Want More of Everything*. New York: New Market.

Lester, David. 1988. *Why Women Kill Themselves*. Springfield, IL: Thomas.

_____. 1992. *Why People Kill Themselves: A 1990s Summary of Research Findings of Suicidal Behavior* (3rd ed.). Springfield, IL: Thomas.

Lester, Will. 2000. "Voters Say What Drove the Choice for a Leader." *Austin American-Statesman* (Nov. 8): A11.

Lev, Michael A. 1998. "Suicide Imbedded in Japan's Culture." *Chicago Tribune* (Feb. 27). Retrieved Aug. 25, 2001. Online: http://seattletimes.nwsource.com/news/nation-world/html98/altjpan_022798.html

Leventman, Paula Goldman. 1981. *Professionals Out of Work*. New York: Free Press.

Levin, Jack, and Jack McDevitt. 1993. *Hate Crimes: The Rising Tide of Bigotry and Bloodshed*. New York: Plenum.

Levin, William C. 1988. "Age Stereotyping: College Student Evaluations." *Research on Aging,* 10 (1): 134–148.

Levine, Adeline Gordon. 1982. *Love Canal: Science, Politics, and People*. Lexington, MA: Lexington.

Levine, Arthur. 1993. "Student Expectations of College." *Change* (September/October): 4.

Levine, Murray. 1982. "Introduction." In Lois Marie Gibbs, *Love Canal: My Story*. Albany: SUNY Press.

Levine, Nancy E., and Joan B. Silk. 1997. "Why Polyandry Fails: Sources of Instability in Polyandrous Marriages." *Current Anthropology* (June): 375–399.

Levine, Peter. 1992. *Ellis Island to Ebbets Field: Sport and the American Jewish Experience*. New York: Oxford University Press.

Levinthal, Charles F. 2002. *Drugs, Behavior, and Modern Society* (3rd ed.). Boston: Allyn & Bacon.

Levy, Janice C., and Eva Y. Deykin. 1989. "Suicidality, Depression, and Substance Abuse in Adolescence." *American Journal of Psychiatry,* 146 (11): 1462–1468.

Lewin, Tamar, and Jennifer Medina. 2003. "To Cut Failure Rates, Schools Shed Students." *New York Times* (July 31): A1, A22.

Lewis, Paul. 1996. "World Bank Moves to Cut Poorest Nations' Debts." *New York Times* (Mar. 16): 17, 18.

_____. 1998. "Marx's Stock Resurges on a 150-Year Tip." *New York Times* (June 27): A17, A19.

Lewis, Tamar. 1997. "School Voucher Program Succeeds in Cleveland." *Austin American-Statesman* (Sept. 21): A31.

Lichter, David T., Felicia B. LeClere, and Diane K. McLaughlin. 1991. "Local Marriage Markets and the Marital Behavior of Black and White Women." *American Journal of Sociology,* 96 (4): 843–867.

Liebow, Elliot. 1967. *Tally's Corner: A Study of Negro Streetcorner Men*. Boston: Little, Brown.

_____. 1993. *Tell Them Who I Am: The Lives of Homeless Women*. New York: Free Press.

Lightblau, Eric. 1999. "U.S. Crime Statistics Drop Again, Hitting 25-Year Low." *Austin American-Statesman* (July 19): A3.

Lii, Jane H. 1995. "Week in Sweatshop Reveals Grim Conspiracy of the Poor." *New York Times* (Mar. 12): 1, 16.

Lindblom, Charles. 1977. *Politics and Markets*. New York: Basic.

Link, Bruce G., and Bruce P. Dohrenwend. 1989. "The Epidemiology of Mental Disorders." In Howard E. Freeman and Sol Levine (Eds.), *Handbook of Medical Sociology* (4th ed.). Englewood Cliffs, NJ: Prentice Hall, pp. 102–127.

Linton, Ralph. 1936. *The Study of Man*. New York: Appleton-Century-Crofts.

Lippa, Richard A. 1994. *Introduction to Social Psychology*. Pacific Grove, CA: Brooks/Cole.

Lips, Hilary M. 2001. *Sex and Gender: An Introduction* (4th ed.). New York: McGraw-Hill.

Lock, Margaret. 1999. "The Politics of Health, Identity, and Culture." In Richard J. Contrada and Richard D. Ashmore (Eds.), *Self, Social Identity, and Physical Health: Interdisciplinary Explorations*. New York: Oxford University Press, pp. 43–68.

Loeb, Paul Rogat. 1994. *Generation at the Crossroads: Apathy and Action on the American Campus*. New Brunswick, NJ: Rutgers University Press.

Lofland, John. 1993. "Collective Behavior: The Elementary Forms." In Russell L. Curtis, Jr., and Benigno E. Aguirre (Eds.), *Collective Behavior and Social Movements*. Boston: Allyn & Bacon, pp. 70–75.

London, Kathryn A. 1991. "Advance Data Number 194: Cohabitation, Marriage, Marital Dissolution, and Remarriage: United States 1988." U.S. Department of Health and Human Services: Vital and Health Statistics of the National Center, January 4.

Lorber, Judith. 1994. *Paradoxes of Gender*. New Haven, CT: Yale University Press.

Lorber, Judith (Ed.). 2001. *Gender Inequality: Feminist Theories and Politics* (2nd ed.). Los Angeles: Roxbury.

Loseke, Donileen. 1992. *The Battered Woman and Shelters: The Social Construction of Wife Abuse*. Albany: SUNY Press.

Lott, Bernice. 1994. *Women's Lives: Themes and Variations in Gender Learning* (2nd ed.). Pacific Grove, CA: Brooks/Cole.

Low, Setha. 2003. *Behind the Gates: Life, Security, and the Pursuit of Happiness in Fortress America*. New York: Routledge.

Lowe, Maria R. 1998. *Women of Steel: Female Bodybuilders and the Struggle for Self-Definition*. New York: New York University Press.

Lu, Stacy. 1998. "World Medical Community Frets Over Unregulated Medicine Sales on Web." *New York Times* (Mar. 23): C3.

Lummis, C. Douglas. 1992. "Equality." In Wolfgang Sachs (Ed.), *The Development Dictionary*. Atlantic Highlands, NJ: Zed, pp. 38–52.

Lund, Kristina. 1990. "A Feminist Perspective on Divorce Therapy for Women." *Journal of Divorce,* 13 (3): 57–67.

Lundberg, Ferdinand. 1988. *The Rich and the Super-Rich: A Study in the Power of Money Today*. Secaucus, NJ: Lyle Stuart.

Lupton, Deborah. 1997. "Foucault and the Medicalisation Critique." In Alan Petersen and Robin Bunton (Eds.), *Foucault: Health and Medicine*. London: Routledge, pp. 94–110.

Lurie, Alison. 1981. *The Language of Clothes*. New York: Random House.

Luttrell, Wendy. 1997. *School-Smart and Mother-Wise: Working-Class Women's Identity and Schooling*. New York: Routledge.

Lynd, Robert S., and Helen M. Lynd. 1929. *Middletown*. New York: Harcourt.

_____. 1937. *Middletown in Transition*. New York: Harcourt.

Maccoby, Eleanor E., and Carol Nagy Jacklin. 1987. "Gender Segregation in Childhood." *Advances in Child Development and Behavior,* 20: 239–287.

MacDonald, Kevin, and Ross D. Parke. 1986. "Parental–Child Physical Play: The Effects of Sex and Age of Children and Parents." *Sex Roles,* 15: 367–378.

Mack, Raymond W., and Calvin P. Bradford. 1979. *Transforming America: Patterns of Social Change* (2nd ed.). New York: Random House.

MacLeod, Jay. 1988. *Ain't No Makin' It: Leveled Aspirations in a Low-Income Neighborhood*. Boulder, CO: Westview.

Maggio, Rosalie. 1988. *The Non-Sexist Word Finder: A Dictionary of Gender-Free Usage*. Boston: Beacon.

Mahler, Sarah J. 1995. *American Dreaming: Immigrant Life on the Margins*. Princeton, NJ: Princeton University Press.

Males, Mike A. 1996. *The Scapegoat Generation: America's War on Adolescents*. Monroe, ME: Common Courage.

Malinowski, Bronislaw. 1922. *Argonauts of the Western Pacific*. New York: Dutton.

_____. 1964. "The Principle of Legitimacy: Parenthood, the Basis of Social Structure." In Rose Laub Coser (Ed.), *The Family: Its Structure and Functions*. New York: St. Martin's (orig. pub. 1929).

Malthus, Thomas R. 1965. *An Essay on Population*. New York: Augustus Kelley (orig. pub. 1798).

Mangione, Jerre, and Ben Morreale. 1992. *La Storia: Five Centuries of the Italian American Experience*. New York: HarperPerennial.

Mann, Coramae Richey. 1993. *Unequal Justice: A Question of Color*. Bloomington: Indiana University Press.

Mann, Eric. 1997. "Confronting Transit Racism in Los Angeles." In Robert B. Bullard and Glenn S. Johnson (Eds.), *Just Transportation: Dismantling Race and Class Barriers to Mobility*. Gabriola Island, BC: New Society, pp. 68–83.

Mann, Patricia S. 1994. *Micro-Politics: Agency in a Postfeminist Era*. Minneapolis: University of Minnesota Press.

Manning, P. K., and B. Cullum-Swan. 1994. "Narrative, Content, and Semiotic Analysis." In Norman K. Denzin and Y. S. Lincoln (Eds.), *Handbook of Qualitative Research*. Thousand Oaks, CA: Sage.

Manning, Robert D. 1999. *Credit Card Nation*. New York: Basic.

Mansfield, Alan, and Barbara McGinn. 1993. "Pumping Irony: The Muscular and the Feminine." In Sue Scott and David Morgan (Eds.), *Body Matters: Essays on the Sociology of the Body*. London: Falmer, pp. 49–58.

Marger, Martin N. 1987. *Elites and Masses: An Introduction to Political Sociology* (2nd ed.). Belmont, CA: Wadsworth.

_____. 1994. *Race and Ethnic Relations: American and Global Perspectives*. Belmont, CA: Wadsworth.

_____. 2003. *Race and Ethnic Relations: American and Global Perspectives* (6th ed.). Belmont, CA: Wadsworth.

Margolis, Richard J. 1990. *Risking Old Age in America*. Boulder, CO: Westview.

Marquart, James W., Sheldon Ekland-Olson, and Jonathan R. Sorensen. 1994. *The Rope, the Chair, and the Needle*. Austin: University of Texas Press.

Marquis, Christopher. 2001. "An American Report Finds the Taliban's Violation of Religious Rights 'Particularly Severe.'" *New York Times* (Oct. 27): B3.

Marsden, Peter V. 1983. "Restricted Access in Networks and Models of Power." *American Journal of Sociology,* 88: 686–717.

Marshall, Gordon. 1998. *A Dictionary of Sociology* (2nd ed.). New York: Oxford University Press.

Martin, Carol L. 1989. "Children's Use of Gender-Related Information in Making Social Judgments." *Developmental Psychology,* 25: 80–88.

Martin, Linda, and Kevin Kinsella. 1994. "Research in the Demography of Aging in Developing Countries." In Linda Martin and Samuel Preston (Eds.), *Demography of Aging*. Washington, DC: National Academic Press.

Martin, Michael T., and Howard Cohen. 1980. "Race and Class Consciousness: A Critique of the Marxist Concept of Race Relations." *Western Journal of Black Studies,* (4) 2: 84–91.

Martin, Susan Ehrlich, and Nancy C. Jurik. 1996. *Doing Justice, Doing Gender*. Thousand Oaks, CA: Sage.

Martineau, Harriet. 1962. *Society in America* (edited, abridged). Garden City, NY: Doubleday (orig. pub. 1837).

_____. 1988. *How to Observe Morals and Manners*. Ed. Michael R. Hill. New Brunswick, NJ: Transaction (orig. pub. 1838).

Marx, Karl. 1967. *Capital: A Critique of Political Economy*. Ed. Friedrich Engels. New York: International (orig. pub. 1867).

Marx, Karl, and Friedrich Engels. 1967. *The Communist Manifesto*. New York: Pantheon (orig. pub. 1848).

_____. 1970. *The German Ideology,* Part 1. Ed. C. J. Arthur. New York: International (orig. pub. 1845–1846).

Massey, Douglas J., G. Hugo Arango, A. Kowasuci, A. Pellegrino, and J. E. Taylor. 1993. "Theories of International Migration: A Review and Appraisal." *Population and Development Review,* 19: 431–466.

Matthews, Warren. 2004. *World Religions* (4th ed.). Belmont, CA: Wadsworth.

Maynard, R. A. 1996. *Kids Having Kids: A Robin Hood Foundation Special Report on the Costs of Adolescent Childbearing*. New York: Robin Hood Foundation.

McAdam, Doug. 1982. *Political Process and the Development of Black Insurgency*. Chicago: University of Chicago Press.

_____. 1996. "Conceptual Origins, Current Problems, Future Directions." In Doug

McAdam, John D. McCarthy, and Meyer N. Zald (Eds.), *Comparative Perspectives on Social Movements.* New York: Cambridge University Press, pp. 23–40.

McAdam, Doug, John D. McCarthy, and Mayer N. Zald. 1988. "Social Movements." In Neil J. Smelser (Ed.), *Handbook of Sociology.* Newbury Park, CA: Sage, pp. 695–737.

McAdoo, Harriet Pipes. 1990. "A Portrait of African American Families in the United States." In S. E. Rix (Ed.), *The American Woman, 1990–91: A Status Report.* New York: Norton.

McCall, George J., and Jerry L. Simmons. 1978. *Identities and Interactions: An Explanation of Human Associations in Everyday Life.* New York: Free Press.

McCall, Nathan. 1994. *Makes Me Wanna Holler: A Young Black Man in America.* New York: Random House.

McCann, Lisa M. 1997. "Patrilocal Co-Residential Units (PCUs) in Al-Barba: Dual Household Structure in a Provincial Town in Jordan." *Journal of Comparative Family Studies* (Summer): 113–136.

McCarthy, John D., and Mayer N. Zald. 1977. "Resource Mobilization and Social Movements: A Partial Theory." *American Journal of Sociology,* 82: 1212–1241.

McCarthy, Terry. 2001. "Stirrings of a Woman's Movement." *Time* (Dec. 3): 46.

McChesney, Robert W. 1998. "The Political Economy of Global Communication." In Robert W. McChesney, Ellen Meiksins Wood, and John Bellamy Foster (Eds.), *Capitalism and the Information Age: The Political Economy of the Global Communication Revolution.* New York: Monthly Review Press, pp. 1–26.

McCreight, Doug. 2001. *Education in Bosnia: Language, Religion, and Control.* Stillwater, OK: New Forums.

McDonald, Patrick Range. 1997. "Financial Disaster 101." New Mass Media, Inc.: Advocate: Back to School. Retrieved May 30, 1999. Online: http://www.newhavenadvocate.com/articles/back2school/back2school/html

McDonnell, Janet A. 1991. *The Dispossession of the American Indian, 1887–1934.* Bloomington: Indiana University Press.

McEachern, William A. 2003. *Economics: A Contemporary Introduction.* Mason, OH: Thomson/South-Western.

McGeary, Johanna. 2001. "The Taliban Troubles." *Time* (Oct. 1): 36–43.

McGee, Reece. 1975. *Points of Departure.* Hinsdale, IL: Dryden.

McGuire, Meredith R. 2002. *Religion: The Social Context* (5th ed.). Belmont, CA: Wadsworth.

McIntyre, L. D., and E. Pernell. 1985. "The Impact of Race on Teacher Recommendations for Special Education Placement." *Journal of Multicultural Counseling and Development,* 13: 112–120.

McKenzie, Roderick D. 1925. "The Ecological Approach to the Study of the Human Community." In Robert Park, Ernest Burgess, and Roderick D. McKenzie, *The City.* Chicago: University of Chicago Press.

McLanahan, Sara, and Karen Booth. 1991. "Mother-Only Families." In Alan Booth (Ed.), *Contemporary Families: Looking Forward, Looking Backward.* Minneapolis: National Council on Family Relations, pp. 405–428.

McLarin, Kimberly J. 1994. "A New Jersey Town Is Troubled by Racial Imbalance Between Classrooms: Would End to Tracking Harm Quality?" *New York Times* (Aug. 11): A12.

McPhail, Clark. 1991. *The Myth of the Maddening Crowd.* New York: Aldine de Gruyter.

McPhail, Clark, and Ronald T. Wohlstein. 1983. "Individual and Collective Behavior Within Gatherings, Demonstrations, and Riots." In Ralph H. Turner and James F.

Short, Jr. (Eds.), *Annual Review of Sociology,* vol. 9. Palo Alto, CA: Annual Reviews, pp. 579–600.

McPherson, Barry D., James E. Curtis, and John W. Loy. 1989. *The Social Significance of Sport: An Introduction to the Sociology of Sport.* Champaign, IL: Human Kinetics.

Mead, George Herbert. 1934. *Mind, Self, and Society.* Chicago: University of Chicago Press.

Medved, Michael. 1992. *Hollywood vs. America: Popular Culture and the War on Traditional Values.* New York: HarperPerennial.

Mental Medicine. 1994. "Wealth, Health, and Status." *Mental Medicine Update,* 3 (2): 7.

Merchant, Carolyn. 1983. *The Death of Nature: Women, Ecology, and the Scientific Revolution.* San Francisco: Harper & Row.

———. 1992. *Radical Ecology: The Search for a Livable World.* New York: Routledge.

Meredith, Robyn. 1998. "Jobs Out of Reach for Detroiters Without Wheels." *New York Times* (May 26): A12.

Merton, Robert King. 1938. "Social Structure and Anomie." *American Sociological Review,* 3 (6): 672–682.

———. 1949. "Discrimination and the American Creed." In Robert M. MacIver (Ed.), *Discrimination and National Welfare.* New York: Harper & Row, pp. 99–126.

———. 1968. *Social Theory and Social Structure* (enlarged ed.). New York: Free Press.

Messerschmidt, James. 1986. *Capitalism, Patriarchy and Crime.* Totowa, NJ: Rowman and Littlefield.

Messner, Michael A. 2002. *Taking the Field: Women, Men and Sports.* Minneapolis: University of Minnesota Press.

Messner, Michael A., Margaret Carlisle Duncan, and Kerry Jensen. 1993. "Separating the Men from the Girls: The Gendered Language of Televised Sports." *Gender & Society,* 7 (1): 121–137.

Meyer, David S., and Suzanne Staggenborg. 1996. "Movements, Countermovements, and the Structure of Political Opportunity." *American Journal of Sociology,* 101: 1628–1660.

Miall, Charlene. 1986. "The Stigma of Involuntary Childlessness." *Social Problems,* 33 (4): 268–282.

Michael, Robert T., John H. Gagnon, Edward O. Laumann, and Gina Kolata. 1994. *Sex in America.* Boston: Little, Brown.

Michels, Robert. 1949. *Political Parties.* Glencoe, IL: Free Press (orig. pub. 1911).

Mickelson, Roslyn Arlin, and Stephen Samuel Smith. 1995. "Education and the Struggle Against Race, Class, and Gender Inequality." In Margaret L. Andersen and Patricia Hill Collins (Eds.), *Race, Class, and Gender* (2nd ed.). Belmont, CA: Wadsworth, pp. 289–304.

Middleton, Nick. 1999. *The Global Casino: An Introduction to Environmental Issues* (2nd ed.). London: Arnold.

Mies, Maria, and Vandana Shiva. 1993. *Ecofeminism.* Highlands, NJ: Zed.

Miethe, Terance, and Charles Moore. 1987. "Racial Differences in Criminal Processing: The Consequences of Model Selection on Conclusions About Differential Treatment." *Sociological Quarterly,* 27: 217–237.

Milgram, Stanley. 1963. "Behavioral Study of Obedience." *Journal of Abnormal and Social Psychology,* 67: 371–378.

———. 1967. "The Small World Problem." *Psychology Today,* 1: 61–67.

———. 1974. *Obedience to Authority.* New York: Harper & Row.

Miliband, Ralph. 1969. *The State in Capitalist Society.* New York: Basic.

Milkman, Ruth. 1997. *Farewell to the Factory: Auto Workers in the Late Twentieth Century.* Berkeley: University of California Press.

Miller, Casey, and Kate Swift. 1991. *Words and Women: New Language in New Times* (updated ed.). New York: HarperCollins.

Miller, Dan E. 1986. "Milgram Redux: Obedience and Disobedience in Authority Rela-

tions." In Norman K. Denzin (Ed.), *Studies in Symbolic Interaction.* Greenwich, CT: JAI, pp. 77–106.

Miller, L. Scott. 1995. *An American Imperative: Accelerating Minority Educational Advancement.* New Haven, CT: Yale University Press.

Miller, Michele. 2003. "Homeschooling: Tuned to the Individual." *St. Petersburg Times Online.* Retrieved Aug. 7, 2003. Online: http://www.sptimes.com/2003/08/03/news_pf/Pasco/Homeschooling_Tuned_.shtml

Mills, C. Wright. 1956. *White Collar.* New York: Oxford University Press.

———. 1959a. *The Power Elite.* Fair Lawn, NJ: Oxford University Press.

———. 1959b. *The Sociological Imagination.* London: Oxford University Press.

———. 1976. *The Causes of World War Three.* Westport, CT: Greenwood.

Mills, Robert J. 2002. "Health Insurance Coverage: 2001." U.S. Census Bureau, Current Population Reports, P60–220. Washington, DC: U.S. Government Printing Office.

Min, Pyong Gap. 1988. "The Korean American Family." In Charles H. Mindel, Robert W. Habenstein, and Roosevelt Wright, Jr. (Eds.), *Ethnic Families in America: Patterns and Variations* (3rd ed.). New York: Elsevier, pp. 199–229.

Mindel, Charles H., Robert W. Habenstein, and Roosevelt Wright, Jr. (Eds.). 1988. *Ethnic Families in America: Patterns and Variations* (3rd ed.). New York: Elsevier.

Mintz, Beth, and Michael Schwartz. 1985. *The Power Structure of American Business.* Chicago: University of Chicago Press.

Mirowsky, John. 1996. "Age and the Gender Gap in Depression." *Journal of Health and Social Behavior,* 37 (December): 362–380.

Mirowsky, John, and Catherine E. Ross. 1980. "Minority Status, Ethnic Culture, and Distress: A Comparison of Blacks, Whites, Mexicans, and Mexican Americans." *American Journal of Sociology,* 86, 479–495.

Misztal, Barbara A. 1993. "Understanding Political Change in Eastern Europe: A Sociological Perspective." *Sociology,* 27 (3): 451–471.

Monaghan, Peter. 1993. "Facing Jail, a Sociologist Raises Questions About a Scholar's Right to Protect Sources." *Chronicle of Higher Education* (Apr. 7): A10.

Monahan, John. 1992. "Mental Disorder and Violent Behavior: Perceptions and Evidence." *American Psychologist,* 47: 511–521.

Montaner, Carlos Alberto. 1992. "Talk English—You Are in the United States." In James Crawford (Ed.), *Language Loyalties: A Source Book on the Official English Controversy.* Chicago: University of Chicago Press, pp. 163–165.

Moody, Harry R. 2002. *Aging: Concepts and Controversy* (4th ed.). Thousand Oaks, CA: Pine Forge.

Moore, K. A., A. K. Driscoll, and L. D. Lindberg. 1998. *A Statistical Portrait of Adolescent Sex, Contraception, and Childbearing.* Washington, DC: National Campaign to Prevent Teen Pregnancy.

Moore, Patricia, with C. P. Conn. 1985. *Disguised.* Waco, TX: Word.

Moore, R. Laurence. 1995. *Selling God: American Religion in the Marketplace of Culture.* New York: Oxford University Press.

Moore, Robert B. 1992. "Racist Stereotyping in the English Language." In Margaret L. Anderson and Patricia Hill Collins (Eds.), *Race, Class, and Gender.* Belmont, CA: Wadsworth, pp. 317–329.

Moore, Wilbert E. 1968. "Occupational Socialization." In David A. Goslin (Ed.), *Handbook on Socialization Theory and Research.* Chicago: Rand McNally, pp. 861–883.

Moraga, Cherríe. 1994. "From a Long Line of Vendidas: Chicanas and Feminism." In Anne C. Hermann and Abigail J. Stewart (Eds.), *Theorizing Feminism: Parallel Trends in*

Humanities and Social Sciences.* Boulder, CO: Westview, pp. 34–48.

Morgan, Leslie, and Suzanne Kunkel. 1998. *Aging: The Social Context.* Thousand Oaks, CA: Pine Forge.

Morrill, C., and Gary A. Fine. 1997. "Ethnographic Contributions to Organizational Sociology." *Sociological Methods and Research,* 25: 424–451.

Morris, Aldon. 1981. "Black Southern Student Sit-In Movement: An Analysis of Internal Organization." *American Sociological Review,* 46: 744–767.

Morse, Jodie. 2001. "Letting God Back In." *Time* (Oct. 22): 71.

Morselli, Henry. 1975. *Suicide: An Essay on Comparative Moral Statistics.* New York: Arno (orig. pub. 1881).

"Mouse Makes the Man." 1995. *New York Times* (Nov. 19): E2.

Mowbray, Carol T., Sandra E. Herman, and Kelly L. Hazel. 1992. "Gender and Serious Mental Illness." *Psychology of Women Quarterly,* 16 (March): 107–127.

Mucciolo, Louis. 1992. *Eightysomething: Interviews with Octogenarians Who Stay Involved.* New York: Birch Lane.

Mujica, Mauro E. 1993. "Why a Hispanic Heads an Organization Called U.S. English." *The New Yorker* (Dec. 27): 101.

———. 2002. "Why an Immigrant Runs an Organization Called U.S.ENGLISH." Retrieved June 21, 2003. Online: http://www.us-english.org/inc/about/chairman.asp

Munch, Allison, J. Miller McPherson, and Lynn Smith-Lovin. 1997. "Gender, Children, and Social Contact: The Effects of Childrearing for Men and Women." *American Sociological Review,* 62 (August): 509–520.

Murdock, George P. 1945. "The Common Denominator of Cultures." In Ralph Linton (Ed.), *The Science of Man in the World Crisis.* New York: Columbia University Press, pp. 123–142.

Murphy, Robert E., Jessica Scheer, Yolanda Murphy, and Richard Mack. 1988. "Physical Disability and Social Liminality: A Study in the Rituals of Adversity." *Social Science and Medicine,* 26: 235–242.

Mydans, Seth. 1995. "Part-Time College Teaching Rises, as Do Worries." *New York Times* (Jan. 4): B6.

———. 1997a. "Brutal End for an Architect of Cambodian Brutality." *New York Times* (June 14): 5.

———. 1997b. "Its Mood Dark as the Haze, Southeast Asia Aches." *New York Times* (Oct. 26): 3.

Myerhoff, Barbara. 1994. *Number Our Days: Culture and Community Among Elderly Jews in an American Ghetto.* New York: Meridian.

Myerson, Allen R. 1994. "Jeans Makers Flourish on Border: The Two Faces of an Industry Success Story." *New York Times* (Sept. 29): C1, C13.

Myrdal, Gunnar. 1970. *The Challenge of World Poverty: A World Anti-Poverty Program in Outline.* New York: Pantheon/Random House.

Nacos, Brigitte. 2002. "Terrorism, the Mass Media, and the Events of 9–11." *Phi Kappa Phi Forum* (Spring): 13–19.

Naffine, Ngaire. 1987. *Female Crime: The Construction of Women in Criminology.* Boston: Allen & Unwin.

Nagourney, Adam. 2002. "The Battleground Shifts." *New York Times* (June 28): A1, A17.

Nasar, Sylvia. 1999. "Jobless Rate in August Again Dipped to a 29-Year Low: Labor Market Was Tight But Wages Barely Rose." *New York Times* (Sept. 4): B1, B3.

National Campaign to Prevent Teen Pregnancy. 1997. *Whatever Happened to Childhood? The Problem of Teen Pregnancy in the United States.* Washington, DC: National Campaign to Prevent Teen Pregnancy.

National Center for Education Statistics. 2003. "Trends in International Mathematics and

Science Study." Retrieved Aug. 7, 2003. Online: http://nces.ed.gov/timss/results.asp

National Center on Elder Abuse. 2003. "Types of Elder Abuse in Domestic Settings." Retrieved July 26, 2003. Online: http://www.elderabusecenter.org/pdf/basics/fact1.pdf

National Centers for Disease Control. 2001. "43 Percent of First Marriages Break Up Within 15 Years." Retrieved July 14, 2002. Online: http://www.cdc.gov/nchs/releases/01news/firstmarr.htm

National Council on Crime and Delinquency. 1969. *The Infiltration into Legitimate Business by Organized Crime.* Washington, DC: National Council on Crime and Delinquency.

National Home Education Research Institute. 2002. "Facts on Home Schooling." Retrieved Aug. 7, 2003. Online: http://www.nheri.org/content.php?menu=1001&page_id=23

National Opinion Research Center. 1996. *General Social Surveys, 1972–1996: Cumulative Codebook.* Chicago: National Opinion Research Center.

Navarrette, Ruben, Jr. 1997. "A Darker Shade of Crimson." In Diana Kendall (Ed.), *Race, Class, and Gender in a Diverse Society.* Boston: Allyn & Bacon, 1997: 274–279. Reprinted from Ruben Navarrette, Jr., *A Darker Shade of Crimson.* New York: Bantam, 1993.

Neergaard, Lauran. 2003. "Elderly Drivers a Concern." *ContraCostaTimes.com.* Retrieved July 20, 2003. Online: http://www.bayarea.com/mld/cctimes/news/6106194.htm

Neimark, Jill. 1997. "On the Front Lines of Alternative Medicine." *Psychology Today* (January/February): 51–68.

Nelson, Margaret K., and Joan Smith. 1999. *Working Hard and Making Do: Surviving in Small Town America.* Berkeley: University of California Press.

Nelson, Mariah Burton. 1994. *The Stronger Women Get, the More Men Love Football: Sexism and the American Culture of Sports.* New York: Harcourt.

Neuborne, Ellen. 1995. "Imagine My Surprise." In Barbara Findlen (Ed.), *Listen Up: Voices from the Next Feminist Generation.* Seattle: Seal, pp. 29–35.

Neuschler, Edward. 1987. *Medicaid Eligibility for the Elder in Need of Long Term Care.* Congressional Research Service Contract No. 86–26 (September). Washington, DC: National Governors Association.

New York Times. 1996. "The Megacity Summit." (Apr. 8): A14.

___. 1997. "Shift in Schools' Spending." (Dec. 12): A15.

___. 2002a. "The Landscape on Vouchers." (June 28): A17.

___. 2002b. "Text: Senate Judiciary Committee Hearing, June 6, 2002." Retrieved June 9, 2002. Online: http://www.nytimes.com/2002/06/06.../06TEXT-INQ2.html

New York Times Magazine. 1998. "Child-Care Caste System." (Apr. 5): 41.

Newburger, Eric C. 2001. "Home Computers and Internet Use in the United States, August 2000." *Current Population Reports,* P23–207. U.S. Census Bureau. Retrieved Sept. 8, 2001. Online: http://www.census.gov/population/www/socdemo/computer.html

Newitz, Annalee. 1993. "I Was a Credit Card Virgin." *Bad Subjects: Political Education for Everyday Life* (e-magazine). Retrieved May 30, 1999. Online: http://english-www.hss.cmu.edu/bs/08/Newitz-Rubio-Caffrey.html

Newman, Katherine S. 1988. *Falling from Grace: The Experience of Downward Mobility in the American Middle Class.* New York: Free Press.

___. 1993. *Declining Fortunes: The Withering of the American Dream.* New York: Basic.

___. 1999. *No Shame in My Game: The Working Poor in the Inner City.* New York: Knopf and the Russell Sage Foundation.

Newsweek. 1997. "Cult: Now on the Next Level." (Dec. 29/Jan. 5): 17.

___. 1999. "Perils and Promise: Teens by the Numbers." (May 10): 38–39.

Niebuhr, H. Richard. 1929. *The Social Sources of Denominationalism.* New York: Meridian.

Nielsen, Joyce McCarl. 1990. *Sex and Gender in Society: Perspectives on Stratification* (2nd ed.). Prospects Heights, IL: Waveland.

Nisbet, Robert. 1979. "Conservativism." In Tom Bottomore and Robert Nisbet (Eds.), *A History of Sociological Analysis.* London: Heinemann, pp. 81–117.

Noble, Barbara Presley. 1995. "A Level Playing Field, for Just $121." *New York Times* (Mar. 5): F21.

Noel, Donald L. 1972. *The Origins of American Slavery and Racism.* Columbus, OH: Merrill.

Nolan, Patrick, and Gerhard E. Lenski. 1999. *Human Societies: An Introduction to Macrosociology* (8th ed.). New York: McGraw-Hill.

Nolte, Carl. 2003. "Media Join Troops Preparing for War: 600 Journalists 'Embedded' with Military in Kuwait." *SFGATE.com* (Mar. 11). Retrieved July 25, 2003. Online: http://sfgate.com/cgi-bin/article.cgi?file=/chronicle/archive/2003/03/11/MN158132.DTL

NOW (National Organization for Women). 2002. "Stop the Abuse of Women and Girls in Afghanistan!" Retrieved July 14, 2002. Online: http://www.nowfoundation.org/global/taliban.html

Nuland, Sherwin B. 1997. "Heroes of Medicine." *Time* (Fall special edition): 6–10.

Oakes, Jeannie. 1985. *Keeping Track: How High Schools Structure Inequality.* New Haven, CT: Yale University Press.

Oakes, Jeannie, and Martin Lipton. 2003. *Teaching to Change the World* (2nd ed.) New York: McGraw-Hill.

Oberschall, Anthony. 1973. *Social Conflict and Social Movements.* Englewood Cliffs, NJ: Prentice Hall.

Oboler, Suzanne. 1995. *Ethnic Labels, Latino Lives: Identity and the Politics of (Re)presentation in the United States.* Minneapolis: University of Minnesota Press.

O'Connell, Helen. 1994. *Women and the Family. Prepared for the UN-NGO Group on Women and Development.* Atlantic Highlands, NJ: Zed.

O'Connor, James. 1973. *The Fiscal Crisis of the State.* New York: St. Martin's.

Odendahl, Teresa. 1990. *Charity Begins at Home: Generosity and Self-Interest Among the Philanthropic Elite.* New York: Basic.

Office of Juvenile Justice and Delinquency Prevention. 2002. *2000 National Youth Gang Survey.* Retrieved July 12, 2003. Online: http://www.ncjrs.org/pdffiles1/ojjdp/fs200204.pdf

Ogburn, William F. 1966. *Social Change with Respect to Culture and Original Nature.* New York: Dell (orig. pub. 1922).

O'Hearn, Claudine Chiawei (Ed.). 1998. *Half and Half: Writers on Growing Up Biracial and Bicultural.* New York: Pantheon.

Olmos, David R. 1997. "Dr. Robot in the OR." *Austin American-Statesman* (July 21): D1, D8.

Omi, Michael, and Howard Winant. 1994. *Racial Formation in the United States: From the 1960s to the 1990s.* New York: Routledge.

Orbach, Susie. 1978. *Fat Is a Feminist Issue.* New York: Paddington.

Orenstein, Peggy, in association with the American Association of University Women. 1995. *SchoolGirls: Young Women, Self-Esteem, and the Confidence Gap.* New York: Anchor/Doubleday.

Ortner, Sherry B. 1974. "Is Female to Male as Nature Is to Culture?" In Michelle Rosaldo and Louise Lamphere (Eds.), *Women, Culture, and Society.* Stanford, CA: Stanford University Press.

Orum, Anthony M. 1974. "On Participation in Political Protest Movements." *Journal of Applied Behavioral Science,* 10: 181–207.

___. 2001. *Introduction to Political Sociology* (4th ed.). Upper Saddle River, NJ: Prentice Hall.

Orum, Anthony M., and Amy W. Orum. 1968. "The Class and Status Bases of Negro Student Protest." *Social Science Quarterly,* 49 (December): 521–533.

Orzechowski, Shawna, and Peter Sepielli. 2001. "Net Worth and Asset Ownership of Households: 1998 and 2000." U.S. Census Bureau, Current Population Reports, P70–88. Washington, DC: U.S. Government Printing Office.

Ostrower, Francie. 1997. *Why the Wealthy Give: The Culture of Elite Philanthropy.* Princeton, NJ: Princeton University Press.

Ouchi, William. 1981. *Theory Z: How American Business Can Meet the Japanese Challenge.* Reading, MA: Addison-Wesley.

Outhwaite, William, and Tom Bottomore (Eds.). 1994. *The Blackwell Dictionary of Twentieth-Century Social Thought.* Malden, MA: Blackwell.

Oxendine, Joseph B. 2003. *American Indian Sports Heritage* (rev. ed.). Lincoln: University of Nebraska Press.

Oxfam. 2002. "Poverty in the Midst of Wealth: The Democratic Republic of Congo." Retrieved July 12, 2003. Online: http://www.oxfam.org.uk/policy/papers/drc/povertywealth.htm

Padilla, Felix M. 1993. *The Gang as an American Enterprise.* New Brunswick, NJ: Rutgers University Press.

___. 1997. *The Struggle of Latino/Latina University Students.* New York: Routledge.

Page, Charles H. 1946. "Bureaucracy's Other Face." *Social Forces,* 25 (October): 89–94.

Palen, J. John. 1995. *The Suburbs.* New York: McGraw-Hill.

Palmore, Erdman. 1981. *Social Patterns in Normal Aging: Findings from the Duke Longitudinal Study.* Durham, NC: Duke University Press.

Panzarino, Connie. 1994. *The Me in the Mirror.* Seattle: Seal.

The Paper Store Enterprises. 2003. "Need Term Paper Help . . . *FAST?!?*" Retrieved June 14, 2003. Online: http://www.Fastpapers.com

Parenti, Michael. 1994. *Land of Idols: Political Mythology in America.* New York: St. Martin's.

___. 1996. *Democracy for the Few* (5th ed.). New York: St. Martin's.

___. 1998. *America Besieged.* San Francisco: City Lights.

Park, Andrew. 1999. "Health Sites Are Hits—But Are They Reliable?" *Austin American-Statesman* (Sept. 15): A1, A14.

Park, Robert E. 1915. "The City: Suggestions for the Investigation of Human Behavior in the City." *American Journal of Sociology,* 20: 577–612.

___. 1928. "Human Migration and the Marginal Man." *American Journal of Sociology,* 33.

___. 1936. "Human Ecology." *American Journal of Sociology,* 42: 1–15.

Park, Robert E., and Ernest W. Burgess. 1921. *Human Ecology.* Chicago: University of Chicago Press.

Parker, Robert Nash, and Doreen Anderson-Facile. 2000. "Violent Crime Trends." In Joseph F. Sheley (Ed.), *Criminology: A Contemporary Handbook* (3rd ed.). Belmont, CA: Wadsworth, pp. 191–214.

Parkinson, C. Northcote. 1957. *Parkinson's Law and Other Studies in Administration.* New York: Ballantine.

Parrish, Dee Anna. 1990. *Abused: A Guide to Recovery for Adult Survivors of Emotional/Physical Child Abuse.* Barrytown, NY: Station Hill.

Parry, A. 1976. *Terrorism: From Robespierre to Arafat.* New York: Vanguard.

Parsons, Talcott. 1951. *The Social System.* Glencoe, IL: Free Press.

___. 1955. "The American Family: Its Relations to Personality and to the Social Structure." In Talcott Parsons and Robert F. Bales (Eds.), *Family, Socialization and Interaction Process.* Glencoe, IL: Free Press, pp. 3–33.

___. 1960. "Toward a Healthy Maturity." *Journal of Health and Social Behavior,* 1: 163–173.

Patros, Philip G., and Tonia K. Shamoo. 1989. *Depression and Suicide in Children and Adolescents: Prevention, Intervention, and Postvention.* Boston: Allyn & Bacon.

PBS. 1992a. "The Glory and the Power, Part I."

___. 1992b. "Sex, Power, and the Workplace."

Pear, Robert. 1994. "Health Advisers See Peril in Plan to Cut Medicare." *New York Times* (Aug. 31): A1, A10.

___. 1995. "No Cash for Unwed Mothers, G.O.P. Affirms." *New York Times* (Jan. 21): 9.

Pearce, Diana. 1978. "The Feminization of Poverty: Women, Work, and Welfare." *Urban and Social Change Review,* 11 (1/2): 28–36.

Pearson, Judy C. 1985. *Gender and Communication.* Dubuque, IA: Brown.

Peck, Dennis L., and Kenneth Warner. 1995. "Accident or Suicide? Single-Vehicle Car Accidents and the Intent Hypothesis." *Adolescence,* 30: 463–473.

Penhaligon, Greg. 2003. "Post-Boom Job Guide." Retrieved July 23, 2003. Online: http://hotwired.lycos.com/webmonkey/03/09/index2a.html?tw=jobs

People. 1999. "Ready, Set, Go Buy Something." (June 21): 106–114.

Pérez, Ramón ("Tianguis"). 1991. *Diary of an Undocumented Immigrant.* Houston: Arte Publico.

Perrow, Charles. 1986. *Complex Organizations: A Critical Essay* (3rd ed.). New York: Random House.

Perrucci, Robert, and Earl Wysong. 1999. *The New Class Society.* Lanham, MD: Rowman & Littlefield.

Perry, David C., and Alfred J. Watkins (Eds.). 1977. *The Rise of the Sunbelt Cities.* Beverly Hills, CA: Sage.

Peter, Laurence J., and Raymond Hull. 1969. *The Peter Principle: Why Things Always Go Wrong.* New York: Morrow.

Peters, John F. 1985. "Adolescents as Socialization Agents to Parents." *Adolescence,* 20 (Winter): 921–933.

Peters, Linda, and Patricia Fallon. 1994. "The Journey of Recovery: Dimensions of Change." In Patricia Fallon, Melanie A. Katzman, and Susan C. Wooley (Eds.), *Feminist Perspectives on Eating Disorders.* New York: Guilford, pp. 339–354.

Petersen, John L. 1994. *The Road to 2015: Profiles of the Future.* Corte Madera, CA: Waite Group.

Peterson, Robert. 1992. *Only the Ball Was White: A History of Legendary Black Players and All-Black Professional Teams.* New York: Oxford University Press (orig. pub. 1970).

Pettigrew, Thomas. 1981. "The Mental Health Impact." In Benjamin Bowser and Raymond G. Hunt (Eds.), *Impacts of Racism on White Americans.* Beverly Hills, CA: Sage, p. 117 (cited in Feagin and Vera, 1995).

Peyser, Marc. 2002. "The Insiders." *Newsweek* (July 1): 38–43.

Phillips, John C. 1993. *Sociology of Sport.* Boston: Allyn & Bacon.

Phillips, Kevin. 2003. *Wealth and Democracy: A Political History of the American Rich.* New York: Broadway.

Phillips, Peter, and Project Censored. 2002. *Censored 2003: The Top 25 Censored Stories.* New York: Seven Stories.

Piaget, Jean. 1932. *The Moral Judgment of the Child.* London: Routledge and Kegan Paul.

___. 1954. *The Construction of Reality in the Child.* Trans. Margaret Cook. New York: Basic.

Pierre-Pierre, Garry. 1997. "Traditional Church's New Life." *New York Times* (Nov. 15): A11.

Pietilä, Hilkka, and Jeanne Vickers. 1994. *Making Women Matter: The Role of the United Nations.* Atlantic Highlands, NJ: Zed.

Pillemer, Karl A. 1985. "The Dangers of Dependency: New Findings on Domestic Violence Against the Elderly." *Social Problems,* 33 (December): 146–158.

Pillemer, Karl A., and David Finkelhor. 1988. "The Prevalence of Elder Abuse: A Random Sample Survey." *Gerontologist,* 28 (1): 51–57.

Pinderhughes, Dianne M. 1986. "Political Choices: A Realignment in Partisanship

Among Black Voters?" In James D. Williams (Ed.), *The State of Black America 1986.* New York: National Urban League, pp. 85–113.
Pinderhughes, Howard. 1997. *Race in the Hood: Conflict and Violence Among Urban Youth.* Minneapolis: University of Minnesota Press.
Pines, Maya. 1981. "The Civilizing of Genie." *Psychology Today,* 15 (September): 28–29, 31–32, 34.
Pitzer, Ronald. 2003. "Rural Children Under Stress." University of Minnesota Extension Service. Retrieved Aug. 9, 2003. Online: http://www.extension.umn.edu/distribution/familydevelopment/components/7269cm.html
Polakow, Valerie. 1993. *Lives on the Edge: Single Mothers and Their Children in the Other America.* Chicago: University of Chicago Press.
Polanyi, Karl. 1944. *The Great Transformation: The Political and Economic Origins of Our Time.* New York: Beacon.
Pomice, Eva. 1990. "Madison Avenue's Blind Spot." In Karin Swisher (Ed.), *The Elderly: Opposing Viewpoints.* San Diego: Greenhaven, pp. 42–45.
Popenoe, David. 1993. "American Family Decline, 1960–1990: A Review and Appraisal." *Journal of Marriage and the Family,* 55 (3): 527–543.
Popenoe, David, and Barbara Dafoe Whitehead. 1999. "The State of Our Unions: The Social Health of Marriage in America" (June). Retrieved Sept. 21, 1999. Online: http://marriage.rutgers.edu/State.html
Population Reference Bureau. 2001. "Human Population: Fundamentals of Growth Patterns of World Urbanization." Retrieved Nov. 22, 2001. Online: http://www.prb.org/Content/NavigationMenu/PRB/E.../Patterns_of_World_Urbanization.html
Porter, Judith D. R. 1971. *Black Child, White Child.* Cambridge, MA: Harvard University Press.
Portes, Alejandro, and Rubén G. Rumbaut. 1996. *Immigrant America: A Portrait* (2nd ed.). Berkeley: University of California Press.
Postman, Neil, and Steve Powers. 1992. *How to Watch TV News.* New York: Penguin.
Postone, Moishe. 1997. "Rethinking Marx (in a Post-Marxist World)." In Charles Camic (Ed.), *Reclaiming the Sociological Classics: The State of the Scholarship.* Malden, MA: Blackwell, pp. 45–80.
Powell, Brian, and Douglas B. Downey. 1997. "Living in Single-Parent Households: An Investigation of the Same-Sex Hypothesis." *American Sociological Review,* 62 (August): 521–539.
Presthus, Robert. 1978. *The Organizational Society.* New York: St. Martin's.
Proctor, Bernadette D., and Joseph Dalaker. 2002. "Poverty in the United States: 2001." U.S. Census Bureau, Current Population Reports, P60–219. Washington, DC: U.S. Government Printing Office.
Project Censored. 2003. "Censored 2003: The Top 25 Censored Stories of 2001–2003." Retrieved July 29, 2003. Online: http://www.projectcensored.org
Prus, Robert. 1996. *Symbolic Interaction and Ethnographic Research: Intersubjectivity and the Study of Human Lived Experience.* Albany: State University of New York Press.
Pryor, John, and Kathleen McKinney (Eds.). 1991. "Sexual Harassment." *Basic and Applied Social Psychology,* 17 (4). Marketed as a book; Hillsdale, NJ: Erlbaum.
Puffer, J. Adams. 1912. *The Boy and His Gang.* Boston: Houghton Mifflin.
Quadagno, Jill S. 1984. "Welfare Capitalism and the Social Security Act of 1935." *American Sociological Review,* 49: 632–647.
Quarantelli, E. L., and James R. Hundley, Jr. 1993. "A Test of Some Propositions About Crowd Formation and Behavior." In Russell L. Curtis, Jr., and Benigno E. Aguirre (Eds.), *Collective Behavior and Social Movements.* Boston: Allyn & Bacon, pp. 183–193.
Queen, Stuart A., and David B. Carpenter. 1953. *The American City.* New York: McGraw-Hill.
Quinney, Richard. 1979. *Class, State, and Crime.* New York: McKay.
___. 2001. *Critique of the Legal Order.* Piscataway, NJ: Transaction (orig. pub. 1974).
Qvortrup, Jens. 1990. *Childhood as a Social Phenomenon.* Vienna: European Centre for Social Welfare Policy and Research.
Rabinowitz, Fredric E., and Sam V. Cochran. 1994. *Man Alive: A Primer of Men's Issues.* Pacific Grove, CA: Brooks/Cole.
Radcliffe-Brown, A. R. 1952. *Structure and Function in Primitive Society.* New York: Free Press.
Raffalli, Mary. 1994. "Why So Few Women Physicists?" *New York Times Supplement* (January): Sect. 4A, 26–28.
Ramirez, Lisa. 1999. "'Senior Boom' Expected for 100-Year-Olds." *Austin American-Statesman* (Aug. 18): A4.
Ramirez, Marc. 1999. "A Portrait of a Local Muslim Family." *Seattle Times* (Jan. 24). Retrieved Aug. 16, 1999. Online: http://archives.seattletimes.com/cgi-bin/texis.mummy/web/vortex/display?storyID=36d4d218
Ravitch, Diana, and Joseph P. Viteritti. 1997. *New Schools for a New Century: The Redesign of Urban Education.* New Haven, CT: Yale University Press.
Ray, Bipasha. 2003. "Flash Mobs Stage Wacky Pranks." Retrieved Aug. 10, 2003. Online: http://www.siliconvalley.com/mld/siliconvalley/news
Reaves, Brian A. 2001. *Felony Defendants in Large Urban Counties, 1998: State Court Processing Statistics.* Washington, DC: U.S. Government Printing Office.
Reckless, Walter C. 1967. *The Crime Problem.* New York: Meredith.
Reich, Robert. 1993. "Why the Rich Are Getting Richer and the Poor Poorer." In Paul J. Baker, Louis E. Anderson, and Dean S. Dorn (Eds.), *Social Problems: A Critical Thinking Approach* (2nd ed.). Belmont, CA: Wadsworth, pp. 145–149. Adapted from *The New Republic,* May 1, 1989.
Reiman, Jeffrey. 1998. *The Rich Get Richer and the Poor Get Prison: Ideology, Class, and Criminal Justice* (5th ed.). Boston: Allyn & Bacon.
Reinharz, Shulamit. 1992. *Feminist Methods in Social Research.* New York: Oxford University Press.
Reinisch, June. 1990. *The Kinsey Institute New Report on Sex: What You Must Know to Be Sexually Literate.* New York: St. Martin's.
Reissman, Catherine. 1991. *Divorce Talk: Women and Men Make Sense of Personal Relationships.* New Brunswick, NJ: Rutgers University Press.
Relman, Arnold S. 1992. "Self-Referral—What's at Stake?" *New England Journal of Medicine,* 327 (Nov. 19): 1522–1524.
Reskin, Barbara F., and Irene Padavic. 2002. *Women and Men at Work* (2nd ed.). Thousand Oaks, CA: Pine Forge.
Rheingold, Howard. 2003. *Smart Mobs: The Next Social Revolution.* New York: Perseus.
Richardson, Laurel. 1993. "Inequalities of Power, Property, and Prestige." In Virginia Cyrus (Ed.), *Experiencing Race, Class, and Gender in the United States.* Mountain View, CA: Mayfield, pp. 229–236.
Richardson, Lynda. 1994. "Minority Students Languish in Special Education System." *New York Times* (Apr. 6): A1, B8.
Ricketts, Thomas C. III. 1999. "Preface," in Thomas C. Ricketts III (ed.), *Rural Health in the United States.* New York: Oxford University Press, pp. vii–viii.
Riemer, Jeffrey W. 1998. "Durkheim's 'Heroic Suicide' in Military Combat." *Armed Forces & Society: An Interdisciplinary Journal,* 25: 103–118.
Riggs, Robert O., Patricia H. Murrell, and JoAnne C. Cutting. 1993. *Sexual Harassment in Higher Education: From Conflict to Community.* ASHE-ERIC Higher Education Reports, 93–2. Washington, DC: George Washington University.
Rigler, David. 1993. "Letters: A Psychologist Portrayed in a Book About an Abused Child Speaks Out for the First Time in 22 Years." *New York Times Book Review* (June 13): 35.
Riley, Matilda White, and John W. Riley, Jr. 1994. "Age Integration and the Lives of Older People." *Gerontologist,* 34 (1): 110–115.
Risman, Barbara J. 1987. "Intimate Relationships from a Microstructural Perspective: Men Who Mother." *Gender & Society,* 1: 6–32.
Ritzer, George. 1995. *Expressing America: A Critique of the Global Credit Card Society.* Thousand Oaks, CA: Pine Forge.
___. 1996. *Sociological Theory* (4th ed.). New York: McGraw-Hill.
___. 1997. *Postmodern Society Theory.* New York: McGraw-Hill.
___. 1998. *The McDonaldization Thesis.* London: Sage.
___. 1999. *Enchanting a Disenchanted World: Revolutionizing the Means of Consumption.* Thousand Oaks, CA: Pine Forge.
___. 2000a. *The McDonaldization of Society.* Thousand Oaks, CA: Pine Forge.
___. 2000b. *Modern Sociological Theory* (5th ed.). New York: McGraw-Hill.
Riverkeeper.org. 2003. "Our History: The Riverkeeper Story." Retrieved Aug. 11, 2003. Online: ttp://riverkeeper.org/ourstory_history.php
Rizzo, Thomas A., and William A. Corsaro. 1995. "Social Support Processes in Early Childhood Friendships: A Comparative Study of Ecological Congruences in Enacted Support." *American Journal of Community Psychology,* 23: 389–418.
Roberts, Keith A. 2004. *Religion in Sociological Perspective* (4th ed.). Belmont, CA: Wadsworth.
Roberts, Sam. 1994. "Black Women Graduates Outpace Male Counterparts." *New York Times* (Oct. 31): A8.
Robinson, Brian E. 1988. *Teenage Fathers.* Lexington, MA: Lexington.
Robinson, Simon. 2000. "Poverty and War." *Time* (Feb. 21). Retrieved July 12, 2003. Online: http://www.time.com/time/europe/magazine/2000/221/congo.html
Robson, Ruthann. 1992. *Lesbian (Out)law: Survival Under the Rule of Law.* New York: Firebrand.
Rockstargames.com. 2003. "Grand Theft Auto: Vice City." Retrieved July 6, 2003. Online: http://www.rockstargames.com/vicecity
Rockwell, John. 1994. "The New Colossus: American Culture as Power Export." *New York Times* (Jan. 30): Sect. 2, pp. 1, 30.
Rodriguez, Clara E. 1989. *Puerto Ricans: Born in the U.S.A.* New York: Unwin Hyman.
Roethlisberger, Fritz J., and William J. Dickson. 1939. *Management and the Worker.* Cambridge, MA: Harvard University Press.
Rogers, Deborah D. 1995. "Daze of Our Lives: The Soap Opera as Feminine Text." In Gail Dines and Jean M. Humez (Eds.), *Gender, Race and Class in Media: A Text-Reader.* Thousand Oaks, CA: Sage, pp. 325–331.
Rogers, Harrell R. 1986. *Poor Women, Poor Families: The Economic Plight of America's Female-Headed Households.* Armonk, NY: Sharpe.
Rollins, Judith. 1985. *Between Women: Domestics and Their Employers.* Philadelphia: Temple University Press.
Romero, Mary. 1992. *Maid in the U.S.A.* New York: Routledge.
___. 1997. "Introduction." In Mary Romero, Pierrette Hondagneu-Sotelo, and Vilma Ortiz (Eds.), *Challenging Fronteras: Structuring Latina and Latino Lives in the U.S.* New York: Routledge, pp. 3–5.
Romo, Harriett D., and Toni Falbo. 1996. *Latino High School Graduation.* Austin: University of Texas Press.
Roob, Nancy, and Ruth McCambridge. 1992. "Private-Sector Funders: Their Role in Homelessness Projects." In Padraig O'Malley (Ed.), *Homelessness: New England and Beyond: New England Journal of Public Policy* (May special issue): 623–646.
Roof, Wade Clark. 1993. *A Generation of Seekers: The Spiritual Journeys of the Baby Boom Generation.* San Francisco: HarperSanFrancisco.
Root, Maria P. P. 1990. "Disordered Eating in Women of Color." *Sex Roles,* 22 (7/8): 525–536.
Ropers, Richard H. 1991. *Persistent Poverty: The American Dream Turned Nightmare.* New York: Plenum.
Rose, Jerry D. 1982. *Outbreaks.* New York: Free Press.
Rosenblatt, Robert A. 1994. "Entitlement Seen Taking Up Nearly All Taxes by 2012." *Los Angeles Times* (Aug. 9): 1.
Rosenblatt, Roger. 1999. "Let Rivers Run Deep." *Time* (Aug. 2): 74–77.
Rosengarten, Ellen M. 1995. Communication to author.
Rosenthal, Naomi, Meryl Fingrutd, Michele Ethier, Roberta Karant, and David McDonald. 1985. "Social Movements and Network Analysis: A Case Study of Nineteenth-Century Women's Reform in New York State." *American Journal of Sociology,* 90: 1022–1054.
Rosenthal, Robert, and Lenore Jacobson. 1968. *Pygmalion in the Classroom: Teacher Expectation and Student's Intellectual Development.* New York: Holt, Rinehart, and Winston.
Rosnow, Ralph L., and Gary Alan Fine. 1976. *Rumor and Gossip: The Social Psychology of Hearsay.* New York: Elsevier.
Rospenda, Kathleen M., Judith A. Richman, and Stephanie J. Nawyn. 1998. "Doing Power: The Confluence of Gender, Race, and Class in Contrapower Sexual Harassment." *Gender & Society,* 12: 40–61.
Ross, Dorothy. 1991. *The Origins of American Social Science.* Cambridge, England: Cambridge University Press.
Rossi, Alice S. 1980. "Life-Span Theories and Women's Lives." *Signs,* 6 (1): 4–32.
___. 1992. "Transition to Parenthood." In Arlene Skolnick and Jerome Skolnick (Eds.), *Family in Transition.* New York: HarperCollins, pp. 453–463.
Rossi, Peter H. 1989. *Down and Out in America: The Origins of Homelessness.* Chicago: University of Chicago Press.
Rossides, Daniel W. 1986. *The American Class System: An Introduction to Social Stratification.* Boston: Houghton Mifflin.
Rostow, Walt W. 1971. *The Stages of Economic Growth: A Non-Communist Manifesto* (2nd ed.). Cambridge: Cambridge University Press (orig. pub. 1960).
___. 1978. *The World Economy: History and Prospect.* Austin: University of Texas Press.
Roth, Guenther. 1988. "Marianne Weber and Her Circle." In Marianne Weber, *Max Weber.* New Brunswick, NJ: Transaction, p. xv.
Rothchild, John. 1995. "Wealth: Static Wages, Except for the Rich." *Time* (Jan. 30): 60–61.
Rotheram, Mary Jane, and Jean S. Phinney. 1987. "Introduction: Definitions and Perspectives in the Study of Children's Ethnic Socialization." In Jean S. Phinney and Mary Jane Rotheram (Eds.), *Children's Ethnic Socialization.* Newbury Park, CA: Sage, pp. 10–28.
Rothman, Robert A. 2001. *Inequality and Stratification: Class, Color, and Gender* (4th ed.). Upper Saddle River, NJ: Prentice Hall.
Rousseau, Ann Marie. 1981. *Shopping Bag Ladies: Homeless Women Speak About Their Lives.* New York: Pilgrim.
Rubin, Lillian B. 1986. "A Feminist Response to Lasch." *Tikkun,* 1 (2): 89–91.
___. 1994. *Families on the Fault Line.* New York: HarperCollins.
Ruffin, Roy J., and Paul R. Gregory. 2000. *Principles of Economics* (7th ed.). Upper Saddle River, NJ: Pearson Addison-Wesley.
Runnymede Trust, 1997. *Islamophobia: A Challenge for Us All.* London: Commission on British Muslims and Islamophobia.
Rural School and Community Trust. 2003. "Rural Teachers Earn Significantly Less Than

Urban, Suburban Counterparts." Washington, DC: Rural School and Community Trust. Retrieved Aug. 9, 2003. Online: http://www.ruraledu.org/newsroom/teachpay.htm

Russo, Nancy Felipe, and Mary A. Jansen. 1988. "Women, Work, and Disability: Opportunities and Challenges." In Michelle Fine and Adrienne Asch (Eds.), *Women with Disabilities: Essays in Psychology, Culture, and Politics.* Philadelphia: Temple University Press.

Rutstein, Nathan. 1993. *Healing in America.* Springfield, MA: Whitcomb.

Rymer, Russ. 1993. *Genie: An Abused Child's Flight from Silence.* New York: HarperCollins.

Sachs, Jeffrey D. 2003. "A Rich Nation, A Poor Continent." *New York Times* (July 9): A23.

Sadker, David, and Myra Sadker. 1985. "Is the OK Classroom OK?" *Phi Delta Kappan,* 55: 358–367.

___. 1986. "Sexism in the Classroom: From Grade School to Graduate School." *Phi Delta Kappan,* 68: 512–515.

Sadker, Myra, and David Sadker. 1984. *Year 3: Final Report, Promoting Effectiveness in Classroom Instruction.* Washington, DC: National Institute of Education.

___. 1994. *Failing at Fairness: How America's Schools Cheat Girls.* New York: Scribner.

Safilios-Rothschild, Constantina. 1969. "Family Sociology or Wives' Family Sociology? A Cross-Cultural Examination of Decision-Making." *Journal of Marriage and the Family,* 31 (2): 290–301.

Samovar, Larry A., and Richard E. Porter. 1991a. *Communication Between Cultures.* Belmont, CA: Wadsworth.

___. 1991b. *Intercultural Communication: A Reader* (6th ed.). Belmont, CA: Wadsworth.

Sampson, Robert J. 1986. "Effects of Socioeconomic Context on Official Reaction to Juvenile Delinquency." *American Sociological Review,* 51 (December): 876–885.

___. 1997. "Neighborhoods and Violent Crime: A Multilevel Study of Collective Efficacy." *Science,* 277: 18–25.

Sampson, Robert J., and John Laub. 1993. *Crime in the Making: Pathways and Turning Points Through Life.* Cambridge, MA: Harvard University Press.

Samuelson, Paul A., and William D. Nordhaus. 1989. *Economics* (13th ed.). New York: McGraw-Hill.

Sanchez-Ayendez, Melba. 1995. "Puerto Rican Elderly Women: Shared Meanings and Informal Supportive Networks." In Margaret L. Andersen and Patricia Hill Collins (Eds.), *Race, Class, and Gender: An Anthology.* Belmont, CA: Wadsworth.

Sandefur, Gary D., and Arthur Sakamoto. 1988. "American Indian Household Structure and Income." *Demography,* 25 (1): 71–80.

Sanger, David E. 1994. "Cutting Itself Down to Size: Japan's Inferiority Complex." *New York Times* (Feb. 6): E5.

Sapir, Edward. 1961. *Culture, Language and Personality.* Berkeley: University of California Press.

Sargent, Margaret. 1987. *Sociology for Australians* (2nd ed.). Melbourne, Australia: Longman Cheshire.

Sassen, Saskia. 2001. *The Global City: New York, London, Tokyo* (2nd ed.). Princeton, NJ: Princeton University Press.

Savin-Williams, Ritch C. 2004. "Memories of Same-Sex Attractions." In Michael S. Kimmel and Michael A. Messner (Eds.), *Men's Lives* (6th ed.). Boston: Allyn & Bacon, pp. 116–132.

Scarce, Rik. 1990. *Eco-Warriors: Understanding the Radical Environmental Movement.* Chicago: Noble.

Schachter, Jason. 2001. "Geographical Mobility: March 1999 to March 2000." U.S. Census Bureau. Current Population Reports P20–538. Washington, DC: U.S. Government Printing Office.

Schama, Simon. 1989. *Citizens: A Chronicle of the French Revolution.* New York: Knopf.

Schattschneider, Elmer Eric. 1969. *Two Hundred Americans in Search of a Government.* New York: Holt, Rinehart & Winston.

Schemo, Diana Jean. 1994. "Suburban Taxes Are Higher for Blacks, Analysis Shows." *New York Times* (Aug. 17): A1, A16.

Schmidley, Dianne. 2003. "The Foreign-Born Population in the United States: March 2002." Current Population Reports, P20–539, U.S. Census Bureau. Washington, DC: U.S. Government Printing Office.

Schneider, Alison. 1997. "Proportion of Minority Professors Inches Up to About 10%." *Chronicle of Higher Education* (June 20): A12, A13 (a report on the Higher Education Research Institute, University of California, study titled "Race and Ethnicity in the American Professoriate, 1995–96").

Schneider, Donna. 1995. *American Childhood: Risks and Realities.* New Brunswick, NJ: Rutgers University Press.

Schneider, Keith. 1993. "The Regulatory Thickets of Environmental Racism." *New York Times* (Dec. 19): E5.

Schor, Juliet B. 1999. *The Overspent American: Upscaling, Downshifting, and the New Consumer.* New York: HarperPerennial.

Schubert, Hans-Joachim (Ed.). 1998. "Introduction." In *On Self and Social Organization—Charles Horton Cooley.* Chicago: University of Chicago Press, pp. 1–31.

Schur, Edwin M. 1983. *Labeling Women Deviant: Gender, Stigma, and Social Control.* Philadelphia: Temple University Press.

Schutske, John. 2002. "Keeping Farm Children Safe." University of Minnesota Extension Service. Retrieved Aug. 9, 2003. Online: http://www.extension.umn.edu/distribution/youth development/DA6188.html

Schutz, Alfred. 1967. *The Phenomenology of the Social World.* Evanston, IL: Northwestern University Press (orig. pub. 1932).

Schwartz, John. 2002. "Too Much Information, Not Enough Knowledge." *New York Times* (June 9): WK5.

Schwartz, John, and John M. Broder. 2003. "Engineer Warned of Consequences of Liftoff Damage." *New York Times* (Feb. 13): A1–A29.

Schwartz, John (with Matthew L. Wald). 2003. "Costs and Risk Clouding Plans to Fix Shuttles." *New York Times* (June 8): A1–A20.

Schwartz, John, and Matthew L. Wald. 2003. "'Groupthink' Is 30 Years Old, and Still Going Strong." *New York Times* (Mar. 9): WK3.

Schwarz, John E., and Thomas J. Volgy. 1992. *The Forgotten Americans.* New York: Norton.

Scott, Alan. 1990. *Ideology and the New Social Movements.* Boston: Unwin & Hyman.

Scott, Joan W. 1986. "Gender: A Useful Category of Historical Analysis." *American Historical Review,* 91 (December): 1053–1075.

Seccombe, Karen. 1991. "Assessing the Costs and Benefits of Children: Gender Comparisons Among Childfree Husbands and Wives." *Journal of Marriage and the Family,* 53 (1): 191–202.

Seegmiller, B. R., B. Suter, and N. Duviant. 1980. *Personal, Socioeconomic, and Sibling Influences on Sex-Role Differentiation.* Urbana: ERIC Clearinghouse of Elementary and Early Childhood Education, ED 176 895, College of Education, University of Illinois.

Seid, Roberta P. 1994. "Too 'Close to the Bone': The Historical Context for Women's Obsession with Slenderness." In Patricia Fallon, Melanie A. Katzman, and Susan C. Wooley (Eds.), *Feminist Perspectives on Eating Disorders.* New York: Guilford, pp. 3–16.

Segura, Denise A. 1994. "Inside the Work Worlds of Chicana and Mexican Immigrant Workers." In Maxine Baca Zinn and Bonnie Thornton Dill (Eds.), *Women of Color in U.S. Society.* Philadelphia: Temple University Press, pp. 95–111.

Seligman, Martin E. P. 1975. *Helplessness: On Depression, Development and Death.* San Francisco: Freeman.

Sengoku, Tamotsu. 1985. *Willing Workers: The Work Ethic in Japan, England, and the United States.* Westport, CT: Quorum.

Sengupta, Somini. 1997. "At Holidays, Test of Patience of Muslims." *New York Times* (Dec. 25): A12.

Senna, Joseph J., and Larry J. Siegel. 2002. *Introduction to Criminal Justice* (9th ed.). Belmont, CA: Wadsworth.

Seper, Chris. 2003. "Blogging About Your Job? Your Boss Could Take Offense." *Austin American-Statesman* (July 27): J1, J4.

Serbin, Lisa A., Phyllis Zelkowitz, Anna-Beth Doyle, Dolores Gold, and Bill Wheaton. 1990. "The Socialization of Sex-Differentiated Skills and Academic Performance: A Mediational Model." *Sex Roles,* 23: 613–628.

Serrill, Michael S. 1997. "Socialism Dies Again." *Time* (Sept. 22): 44.

Shapiro, Joseph P. 1993. *No Pity: People with Disabilities Forging a New Civil Rights Movement.* New York: Times/Random House.

Sharma, Arvind. 1995. "Hinduism." In Arvind Sharma (Ed.), *Our Religions.* San Francisco: HarperCollins, pp. 3–67.

Shaw, Randy. 1999. *Reclaiming America: Nike, Clean Air, and the New National Activism.* Berkeley: University of California Press.

Shawver, Lois. 1998. "Notes on Reading Foucault's *The Birth of the Clinic.*" Retrieved Oct. 2, 1999. Online: http://www.california.com/~rathbone/foucbc.htm

Sheen, Fulton J. 1995. *From the Angel's Blackboard: The Best of Fulton J. Sheen.* Ligouri, MO: Triumph.

Sheff, David. 1995. "If It's Tuesday, It Must Be Dad's House." *New York Times Magazine* (Mar. 26): 64–65.

Sheff, Nick. 1999. "My Long-Distance Life." *Newsweek* (Feb. 15): 16.

Shenon, Philip. 1994. "China's Mania for Baby Boys Creates Surplus of Bachelors." *New York Times* (Aug. 16): A1, A4.

___. 1997. "Army Puts Blame on Officers." *Austin American-Statesman* (Sept. 12): A1, A6.

Sherman, Suzanne (Ed.). 1992. "Frances Fuchs and Gayle Remick." In *Lesbian and Gay Marriage: Private Commitments, Public Ceremonies.* Philadelphia: Temple University Press, pp. 189–201.

Shevky, Eshref, and Wendell Bell. 1966. *Social Area Analysis: Theory, Illustrative Application and Computational Procedures.* Westport, CT: Greenwood.

Shils, Edward A. 1965. "Charisma, Order, and Status." *American Sociological Review,* 30: 199–213.

Shisslak, Catherine M., and Marjorie Crago. 1992. "Eating Disorders Among Athletes." In Raymond Lemberg (Ed.), *Controlling Eating Disorders with Facts, Advice, and Resources.* Phoenix: Oryx, pp. 29–36.

Shmueli, Sandra. 2003. "'Flash Mob' Craze Spreads." *CNN.com.* Retrieved Aug. 11, 2003. Online: http://www.cnn.com/2003/TECH/internet/08/04/flash.mob

Shorto, Russell. 1997. "Belief by the Numbers." *New York Times Magazine* (Dec. 7): 60.

Shum, Tedd. 1997. "Olympic Gymnast Chow Makes Impact on All Americans." Retrieved Aug. 15, 1999. Online: http://www.dailybruin.ucla.edu/DB/issues/97/05.30/view.shum.html

Shupe, Anson, and Jeffrey K. Hadden. 1989. "Is There Such a Thing as Global Fundamentalism?" In Jeffrey K. Hadden and Anson Shupe (Eds.), *Secularization and Fundamentalism Reconsidered.* New York: Paragon.

Sidel, Ruth. 1986. *Women and Children Last: The Plight of Poor Women in Affluent America.* New York: Viking.

Siegel, Larry J. 1998. *Criminology: Theories, Patterns, and Typologies* (6th ed.). Belmont, CA: West/Wadsworth.

___. 2003. *Criminology* (8th ed.). Belmont, CA: Wadsworth.

Silicon Valley/San Jose Business Journal. 2003. "Silicon Valley Jobless Rate Inches Up as State's Remains Stable." Retrieved July 22, 2003. Online: http://www.bizjournals.com/sanjose/stories/2003/07/07/daily57.html

Silversten, Scott. 1998. "Texas A&M LB Nguyen Overcomes Adversity in Several Areas." SLAM! Sports: College Football Notes. Retrieved Aug. 15, 1999. Online: http://www.canoe.ca/statsFBC/BC-FBC-LGNS-CNNSIALBTS-R.html

Simmel, Georg. 1950. *The Sociology of Georg Simmel.* Trans. Kurt Wolff. Glencoe, IL: Free Press (orig. written in 1902–1917).

___. 1957. "Fashion." *American Journal of Sociology,* 62 (May 1957): 541–558. Orig. pub. 1904.

___. 1990. *The Philosophy of Money.* Ed. David Frisby. New York: Routledge (orig. pub. 1907).

Simon, David R. 1996. *Elite Deviance* (5th ed.). Boston: Allyn & Bacon.

Simons, Marlise. 1993. "Prosecutor Fighting Girl-Mutilation." *New York Times* (Nov. 23): A4.

Simpson, George Eaton, and Milton Yinger. 1972. *Racial and Cultural Minorities: An Analysis of Prejudice and Discrimination* (4th ed.). New York: Harper & Row.

Simpson, Sally S. 1989. "Feminist Theory, Crime, and Justice." *Criminology,* 27: 605–632.

Singer, Margaret Thaler, with Janja Lalich. 1995. *Cults in Our Midst.* San Francisco: Jossey-Bass.

Sivard, Ruth L. 1991. *World Military and Social Expenditures—1991.* Washington, DC: World Priorities.

___. 1993. *World Military and Social Expenditures—1993.* Washington, DC: World Priorities.

Sjoberg, Gideon. 1965. *The Preindustrial City: Past and Present.* New York: Free Press.

Skocpol, Theda, and Edwin Amenta. 1986. "States and Social Policies." In Ralph H. Turner and James F. Short, Jr. (Eds.), *Annual Review of Sociology,* 12: 131–157.

Slugoski, B. F., and G. B. Ginsburg. 1989. "Ego Identity and Explanatory Speech." In John Shotter and Kenneth J. Gergen (Eds.), *Texts of Identity.* London: Sage, pp. 36–55.

Smart Growth America. 2001. "Americans Want Growth and Green; Demand Solutions To Traffic, Haphazard Development." Retrieved Nov. 30, 2001. Online: http://www.smartgrowthamerica.com/release.htm

Smart Growth Network. 2001. "About Smart Growth." Retrieved Nov. 30, 2001. Online: http://www.smartgrowth.org/about/default.asp

Smelser, Neil J. 1963. *Theory of Collective Behavior.* New York: Free Press.

___. 1988. "Social Structure." In Neil J. Smelser (Ed.), *Handbook of Sociology.* Newbury Park, CA: Sage, pp. 103–129.

Smith, Adam. 1976. *An Inquiry into the Nature and Causes of the Wealth of Nations.* Ed. Roy H. Campbell and Andrew S. Skinner. Oxford, England: Clarendon (orig. pub. 1776).

Smith, Allen C., III, and Sheryl Kleinman. 1989. "Managing Emotions in Medical School: Students' Contacts with the Living and the Dead." *Social Science Quarterly,* 52 (1): 56–69.

Smith, Denise. 2003. "The Older Population in the United States: March 2002." U.S. Census Bureau, Current Population Reports, P20–546. Washington, DC: U.S. Government Printing Office.

Smith, Denise I., and Renee T. Spraggins. 2001. "Gender: 2000." U.S. Census Bureau. Retrieved Nov. 28, 2001. Online: http://www.census.gov/prod/2001pubs/c2kbr01-9.pdf

Smith, Dorothy E. 1999. *Writing the Social: Critique, Theory, and Investigations.* Toronto: University of Toronto Press.

Smith, Douglas, Christy Visher, and Laura Davidson. 1984. "Equity and Discretionary Justice: The Influence of Race on Police Arrest Decisions." *Journal of Criminal Law and Criminology,* 75: 234–249.

Smith, Huston. 1991. *The World's Religions.* San Francisco: HarperSanFrancisco.

Smith, Wallace Charles. 1985. *The Church in the Life of the Black Family.* Valley Forge, PA: Judson.

Smith, Wes. 2001. *Hope Meadows: Real-Life Stories of Healing and Caring from an Inspiring Community.* New York: Berkley.

Snow, David A., and Leon Anderson. 1991. "Researching the Homeless: The Characteristic Features and Virtues of the Case Study." In Joe R. Feagin, Anthony M. Orum, and Gideon Sjoberg (Eds.), *A Case for the Case Study.* Chapel Hill: University of North Carolina Press, pp. 148–173.

___. 1993. *Down on Their Luck: A Case Study of Homeless Street People.* Berkeley: University of California Press.

Snow, David A., and Robert Benford. 1988. "Ideology, Frame Resonance, and Participant Mobilization." In Bert Klandermans, Hanspeter Kriesi, and Sidney Tarrow (Eds.), *International Social Movement Research,* Vol. 1, *From Structure to Action.* Greenwich, CT: JAI, pp. 133–155.

Snow, David A., E. Burke Rochford, Jr., Steven K. Worden, and Robert D. Benford. 1986. "Frame Alignment Processes, Micromobilization, and Movement Participation." *American Sociological Review,* 51: 464–481.

Snow, David A., Louis A. Zurcher, and Robert Peters. 1981. "Victory Celebrations as Theater: A Dramaturgical Approach to Crowd Behavior." *Symbolic Interaction,* 4 (1): 21–41.

Snyder, Benson R. 1971. *The Hidden Curriculum.* New York: Knopf.

Solomon, Jay. 2001. "How Mr. Bambang Markets Big Macs in Muslim Indonesia." *Wall Street Journal* (Oct. 26): A1–A7.

Sommers, Ira, and Deborah R. Baskin. 1993. "The Situational Context of Violent Female Offending." *Journal of Research in Crime and Delinquency,* 30 (2): 136–162.

South, Scott J., Charles M. Bonjean, Judy Corder, and William T. Markham. 1982. "Sex and Power in the Federal Bureaucracy." *Work and Occupations,* 9 (2): 233–254.

Specter, Michael. 1994. "Crisis of Bread and Land Afflicts Russian Farming." *New York Times* (Sept. 19): A1, A16.

Spence, Jan. 1997. "Homeless in Russia: A Visit with Valery Sokolov." Share International. Retrieved Sept. 15, 2001. Online: http://www.shareintl.org/archives/homelessness/hl-jsRussia.htm

Spindel, Carol. 2000. *Dancing at Halftime: Sports and the Controversy Over American Indian Mascots.* New York: New York University Press.

Spitzer, Steve. 1975. "Toward a Marxian Theory of Deviance." *Social Problems,* 22: 638–651.

SportsLine.com. 2002. "Study Says Black Head Coaches on Decline in Last 6 Years." Retrieved July 26, 2003. Online: http://register.sportsline.com/u/ce/multi/0,1329,5635433_56,00.html

sportsnetwork.com. 2003. "International Soccer: UEFA Seeks to Eradicate Racism." Retrieved July 17, 2003. Online: http://www.sportnetwork.com/?c=sportsnetwork&page=soc-cup/news/CAN2477113.htm

Sreenivasan, Sreenath. 1996. "Blind Users Add Access on the Web." *New York Times* (Dec. 2): C7.

Stack, Steven. 1998. "Gender, Marriage, and Suicide Acceptability: A Comparative Analysis." *Sex Roles,* 38: 501–521.

Stack, Steven, and I. Wasserman. 1995. "The Effect of Marriage, Family, and Religious Ties on African American Suicide Ideology." *Journal of Marriage and the Family,* 57: 215–222.

Stake, Robert E. 1995. *The Art of Case Study Research.* Thousand Oaks, CA: Sage.

Stannard, David E. 1992. *American Holocaust: Columbus and the Conquest of the New World.* New York: Oxford University Press.

Staples, Robert. 1994. "The Illusion of Racial Equality: The Black American Dilemma." In Gerald Early (Ed.), *Lure and Loathing: Essays on Race, Identity, and the Ambivalence of Assimilation.* New York: Penguin.

Stark, Rodney, and William Sims Bainbridge. 1981. "American-Born Sects: Initial Findings." *Journal for the Scientific Study of Religion,* 20: 130–149.

Starr, Paul. 1982. *The Social Transformation of Medicine: The Rise of a Sovereign Profession and the Making of a Vast Industry.* New York: Basic.

Statham, Anne, Laurel Richardson, and Judith A. Cook. 1991. *Gender and University Teaching: A Negotiated Difference.* Albany: SUNY Press.

Steffensmeier, Darrell, and Emilie Allan. 2000. "Looking for Patterns: Gender, Age, and Crime." In Joseph F. Sheley (Ed.), *Criminology: A Contemporary Handbook* (3rd ed.). Belmont, CA: Wadsworth, pp. 85–128.

Stein, Joel. 2003. "The Singin', Dancin' American Idyll." *Time.com.* Retrieved June 21, 2003. Online: http://www.time.com/time/magazine/printout/0,8816,458781,00.html

Stein, Peter J. 1976. *Single.* Englewood Cliffs, NJ: Prentice Hall.

Stein, Peter J. (Ed.). 1981. *Single Life: Unmarried Adults in Social Context.* New York: St. Martin's.

Steinmetz, Suzanne K. 1987. "Elderly Victims of Domestic Violence." In Carl D. Chambers, John H. Lindquist, O. Z. White, and Michael T. Harter (Eds.), *The Elderly: Victims and Deviants.* Athens: Ohio University Press, pp. 126–141.

Stern, Sharon. 2001. "Americans with Disabilities: 1997." U.S. Census Bureau. Retrieved Nov. 23, 2001. Online: http://www.census.gov/hhes/www.disability.html

Stevenson, Mary Huff. 1988. "Some Economic Approaches to the Persistence of Wage Differences Between Men and Women." In Ann H. Stromberg and Shirley Harkess (Eds.), *Women Working: Theories and Facts in Perspective* (2nd ed.). Mountain View, CA: Mayfield, pp. 87–100.

Stevenson, Richard W. 1997. "World Bank Report Sees Era of Emerging Economies." *New York Times* (Sept. 10): C7.

Stewart, Abigail J. 1994. "Toward a Feminist Strategy for Studying Women's Lives." In Carol E. Franz and Abigail J. Stewart (Eds.), *Women Creating Lives: Identities, Resilience, and Resistance.* Boulder, CO: Westview, pp. 11–35.

Stiehm, Judith Hicks. 1989. *Arms and the Enlisted Woman.* Philadelphia: Temple University Press.

Stoller, Eleanor Palo, and Rose Campbell Gibson. 1997. *Worlds of Difference: Inequalities in the Aging Experience* (2nd ed.). Thousand Oaks, CA: Sage.

Stucki, B. 1992. "The Long Voyage Home: Return Migration Among Aging Cocoa Farmers in Ghana." *Journal of Cross-Cultural Gerontology,* 7: 363–378.

Substance Abuse and Mental Health Services Administration. 2000. *National Household Survey on Drug Abuse, 2000.* Retrieved Nov. 20, 2000. Online: http://www.samhsa.gov/oas/NHSDA/2kNHSDA/chapter2.htm

Sullivan, Teresa A., Elizabeth Warren, and Jay Lawrence Westbrook. 2000. *The Fragile Middle Class: Americans in Debt.* New Haven, CT: Yale University Press.

Sullivan, Thomas J. 2000. *Introduction to Social Problems* (5th ed.). Boston: Allyn & Bacon.

Sumner, William G. 1959. *Folkways.* New York: Dover (orig. pub. 1906).

Sutherland, Edwin H. 1939. *Principles of Criminology.* Philadelphia: Lippincott.

___. 1949. *White Collar Crime.* New York: Dryden.

Swarns, Rachel L. 1997. "Welfare Mothers Prep for Jobs, and Wait." *New York Times* (Aug. 31): 1.

Swerdlow, Marian. 1989. "Men's Accommodations to Women Entering a Nontraditional Occupation: A Case of Rapid Transit Operatives." *Gender & Society,* 3: 373–387.

Swidler, Ann. 1986. "Culture in Action: Symbols and Strategies." *American Sociological Review,* 51 (April): 273–286.

Symonds, William C. 1990. "Is Sex Discrimination Still Par for the Course?" *Business Week* (Dec. 24): 56.

Szasz, Thomas S. 1984. *The Myth of Mental Illness: Foundations of a Theory of Personal Conduct.* New York: HarperCollins.

Tabb, William K., and Larry Sawers. 1984. *Marxism and the Metropolis: New Perspectives in Urban Political Economy* (2nd ed.). New York: Oxford University Press.

Takaki, Ronald. 1989. *Strangers from a Different Shore: A History of Asian Americans.* New York: Penguin.

___. 1993. *A Different Mirror: A History of Multicultural America.* Boston: Little, Brown.

Tannen, Deborah. 1993. "Commencement Address, State University of New York at Binghamton." Reprinted in *Chronicle of Higher Education* (June 9): B5.

Tarbell, Ida M. 1925. *The History of Standard Oil Company.* New York: Macmillan (orig. pub. 1904).

Tavris, Carol. 1993. *The Mismeasure of Woman.* New York: Touchstone.

Tax Foundation. 2003. "Summary of Federal Income Tax Data, 2000." Retrieved July 19, 2003. Online: http://www.taxfoundation.org/prtopincometable.html

Taylor, Carl S. 1993. *Girls, Gangs, Women, and Drugs.* East Lansing: Michigan State University Press.

Taylor, Chris. 1999. "We're Goths and Not Monsters." *Time* (May 3): 45.

Taylor, Howard F. 1995. "Symposium: The Bell Curve." *Contemporary Sociology: A Journal of Reviews,* 24 (2): 153–157.

Taylor, Robert Joseph, Linda M. Chatters, and V. Mays. 1988. "Parents, Children, Siblings, In-Laws, and Non-Kin Sources of Emergency Assistance to Black Americans." *Family Relations,* 37: 298–304.

Taylor, Steve. 1982. *Durkheim and the Study of Suicide.* New York: St. Martin's.

Teir, Robert. 1994. "Homeless Rights v. Public Space." *Austin American-Statesman* (Jan. 22): A23.

Terkel, Studs. 1990. *Working: People Talk About What They Do All Day and How They Feel About What They Do.* New York: Ballantine (orig. pub. 1972).

___. 1996. *Coming of Age: The Story of Our Century by Those Who've Lived It.* New York: St. Martin's Griffin.

thewritemarket.com 2003. "Building an Internet Community: What Is an 'Internet Community'?" Retrieved July 5, 2003. Online: http://www.thewritemarket.com/promotion/community.htm

Thompson, Becky W. 1994. *A Hunger So Wide and So Deep: American Women Speak Out on Eating Problems.* Minneapolis: University of Minnesota Press.

Thompson, Mark. 1997. "Fatal Neglect." *Time* (Oct. 27): 34–38.

Thornberry, Terence P., Marvin D. Krohn, Alan J. Lizotte, and Deborah Chard-Wierschem. 1993. "The Role of Juvenile Gangs in Facilitating Delinquent Behavior." *Journal of Research in Crime and Delinquency,* 30 (1): 55–87.

Thorne, Barrie. 1993. *Gender Play: Girls and Boys in School.* New Brunswick, NJ: Rutgers University Press.

___. 1995. "Girls and Boys Together . . . But Mostly Apart: Gender Arrangements in Elementary Schools." In Michael S. Kimmel and Michael A. Messner (Eds.), *Men's Lives* (3rd ed.). Boston: Allyn & Bacon, pp. 61–73.

Thorne, Barrie, Cheris Kramarae, and Nancy Henley. 1983. *Language, Gender, and Society.* Rowley, MA: Newbury.

Thornton, Arland, and Deborah Freedman. 1983. "The Changing American Family." *Population Bulletin,* 38 (October).

Thornton, Michael C., Linda M. Chatters, Robert Joseph Taylor, and Walter R. Allen. 1990. "Sociodemographic and Environmental Correlates of Racial Socialization by Black Parents." *Child Development,* 61: 401–409.

Thornton, Russell. 1984. "Cherokee Population Losses During the Trail of Tears: A New Perspective and a New Estimate." *Ethnohistory,* 31: 289–300.

Tierney, John. 1993. "Fernando, 16, Finds a Sanctuary in Crime." *New York Times* (Apr. 13): A1, A11.

___. 2003. "Postwar GI Deaths Now Exceed Toll from War." Retrieved Sept. 1, 2003. Online: http://www.nytimes.com/2003/08/27/international/worldspecial/27IRAQ.html

Tilly, Charles. 1973. "Collective Action and Conflict in Large-Scale Social Change: Research Plans, 1974–78." Center for Research on Social Organization. Ann Arbor: University of Michigan, October.

___. 1975. *The Formation of National States in Western Europe.* Princeton, NJ: Princeton University Press.

___. 1978. *From Mobilization to Revolution.* Reading, MA: Addison-Wesley.

Tilly, Chris, and Charles Tilly. 1998. *Work Under Capitalism.* Boulder, CO: Westview.

Tiryakian, Edward A. 1978. "Emile Durkheim." In Tom Bottomore and Robert Nisbet (Eds.), *A History of Sociological Analysis.* New York: Basic, pp. 187–236.

Tittle, Charles, and Robert Meier. 1990. "Specifying the SES/Delinquency Relationship." *Criminology,* 28: 271–299.

Toner, Robin. 1999. "A Majority Over 45 Say Sex Lives Are Just Fine." *New York Times* (Aug. 4): A10.

Tong, Rosemarie. 1989. *Feminist Thought: A Comprehensive Introduction.* Boulder, CO: Westview.

Tönnies, Ferdinand. 1940. *Fundamental Concepts of Sociology* (Gemeinschaft *und* Gesellschaft). Trans. Charles P. Loomis. New York: American Book Company (orig. pub. 1887).

___. 1963. *Community and Society* (Gemeinschaft *and* Gesellschaft). New York: Harper & Row (orig. pub. 1887).

TORAW. 2003. "TORAW: The Organization for the Rights of American Workers—Glossary of Terms." Retrieved July 23, 2003. Online: http://www.toraw.org/GLOSSARY.htm

Tower, Cynthia Crosson. 1996. *Child Abuse and Neglect* (3rd ed.). Boston: Allyn & Bacon.

Tracy, C. 1980. "Race, Crime and Social Policy: The Chinese in Oregon, 1871–1885." *Crime and Social Justice,* 14: 11–25.

Tran, Tranh Van. 1988. "The Vietnamese American Family." In Charles H. Mindel, Robert W. Habenstein, and Roosevelt Wright, Jr. (Eds.), *Ethnic Families in America: Patterns and Variations* (3rd ed.). New York: Elsevier, pp. 276–302.

Trevithick, Alan. 1997. "On a Panhuman Preference for Monandry: Is Polyandry an Exception?" *Journal of Comparative Family Studies* (September): 154–184.

Troeltsch, Ernst. 1960. *The Social Teachings of the Christian Churches,* vols. 1 and 2. Trans. O. Wyon. New York: Harper & Row (orig. pub. 1931).

Tuan, Mia. 1998. *Forever Foreigners or Honorary Whites? The Asian Ethnic Experience Today.* New Brunswick, NJ: Rutgers University Press.

Tumin, Melvin. 1953. "Some Principles of Stratification: A Critical Analysis." *American Sociological Review,* 18 (August): 387–393.

Turk, Austin. 1969. *Criminality and Legal Order.* Chicago: Rand McNally.

___. 1977. "Class, Conflict and Criminology." *Sociological Focus,* 10: 209–220.

Turner, Jonathan, Leonard Beeghley, and Charles H. Powers. 2002. *The Emergence of Sociological Theory* (5th ed.). Belmont, CA: Wadsworth.

Turner, Jonathan H., Royce Singleton, Jr., and David Musick. 1984. *Oppression: A Socio-History of Black–White Relations in America.* Chicago: Nelson-Hall (reprinted 1987).

Turner, Ralph H., and Lewis M. Killian. 1993. "The Field of Collective Behavior." In Russell L. Curtis, Jr., and Benigno E. Aguirre

(Eds.), *Collective Behavior and Social Movements.* Boston: Allyn & Bacon, pp. 5–20.
Twenhofel, Karen. 1993. "Do You Diet?" In Leslea Newman (Ed.), *Eating Our Hearts Out: Personal Accounts of Women's Relationship to Food.* Freedom, CA: Crossing.
Twitchell, James B. 1996. *ADCULTusa: The Triumph of Advertising in American Culture.* New York: Columbia University Press.
___. 1999. *Lead Us into Temptation: The Triumph of American Materialism.* New York: Columbia University Press.
Tyre, Peg, and Daniel McGinn. 2003. "She Works, He Doesn't." *Newsweek* (May 12): 45–52.
uefa.com. 2002. "UEFA Backs Anti-Racism Plan." *uefa.com.* Retrieved July 17, 2003. Online: http://www.uefa.com
UNAIDS/WHO. 2000. "Report on the Global HIV/AIDS Epidemic—June 2000." Retrieved Aug. 5, 2000. Online: http://www.unaids.org/epidemic_update/report/glo_estim.pdf
UNICEF (United Nations Children's Fund). 1991. *Zur Situation der Kinder in der Welt.* Cologne. Quoted in Ingomar Hauchler and Paul M. Kennedy (Eds.). 1994. *Global Trends: The World Almanac of Development and Peace.* New York: Continuum.
United College Marketing Services. 1997. "Credit Strategy Seminars—How a Loan Officer Looks at You." Retrieved May 30, 1999. Online: http://www.college-visa.com
United Nations. 1997. "Global Change and Sustainable Development: Critical Trends." United Nations Department for Policy Coordination and Sustainable Development. Posted online (Jan. 20).
___. 2000. *Long-Range World Population Projections.* New York: United Nations Population Division.
United Nations Conference on Trade and Development. 2002. "The World's 100 Largest Non-Financial TNCs, Ranked by Foreign Assets, 2000." Retrieved Aug. 30, 2003. Online: http://www.unctad.org/sections/dite_dir/docs//Top100WIR2002.pdf
United Nations Development Programme. 1997. *Human Development Report: 1996.* New York: Oxford University Press.
___. 1999. *Human Development Report: 1999.* New York: Oxford University Press.
___. 2002. "The Millennium Development Goals and Human Development." Retrieved July 15, 2003. Online: http://hdr.undp.org/docs/mdg/The%20Millennium%20Development%20Goals%20and%20Human%20Development.pdf
___. 2003. *Human Development Report: 2003.* New York: Oxford University Press.
United Nations DPCSD. 1997. "Report of Commission on Sustainable Development, April 1997." New York: United Nations Department for Policy Coordination and Sustainable Development. Posted online.
United Nations Population Division. 1999. Retrieved Oct. 17, 1999. Online: gopher:/gopher.undp.org/00/ungophers/popin/wdtrends
U.S. Bureau of Justice Statistics. 2003a. "Capital Punishment Statistics." Retrieved July 12, 2003. Online: http://www.ojp.usdoj.gov/bjs/cp.htm
___. 2003b. "Criminal Victimization in the United States, 2001: Statistical Tables." Retrieved July 12, 2003. Online: http://www.ojp.usdoj.gov/bjs/pub/pdf/cvus0101.pdf
U.S. Bureau of Labor Statistics. 2000. "Labor Force Statistics from the Current Population Survey." Retrieved: Sept. 29, 2001. Online: http://www.bls.gov/news.release/work.t04.htm
___. 2002. "Highlights of Women's Earnings in 2001." Retrieved Oct. 12, 2002. Online: http://www.bls.gov/cps/cpswom2001.pdf
___. 2003a. "Consumer Expenditures in 2001." Retrieved June 15, 2003. Online: http://www.bls.gov/cex/csxann01.pdf
___. 2003b. "Unemployed Persons by Marital Status, Race, Age, and Sex." Retrieved July 19, 2003. Online: http://www.bls.gov/cps/cpsaat24.pdf
___. 2003c. "Work Stoppage Data." Retrieved Aug. 30, 2003. Online: http://data.bls.gov/cgi-bin/surveymost
U.S. Census Bureau. 2000. "Employment, Earnings, and Disability." Retrieved Aug. 6, 2000. Online: http://www.census.gov/hhes/www/disable/emperndis.pdf
U.S. Census Bureau. 2001a. "American Housing Survey for the United States: 2001." Retrieved Sept. 6, 2003. Online: http://www.census.gov/hhes/www/housing/ahs/ahs01/tab24.html
___. 2001b. "Overview of Race and Hispanic Origin: Census 2000 Brief." Washington, DC: U.S. Department of Commerce.
___. 2002. *Statistical Abstract of the United States 2002.* Washington, DC: U.S. Government Printing Office.
___. 2003a. "Languages Spoken at Home for the Population 5 Years and Over by State: 2000." Retrieved June 21, 2003. Online: http://www.census.gov/population/cen2000/phc-t20/tab05.pdf
___. 2003b. "Women Edge Men in High School Diplomas, Breaking 13-Year Deadlock." Retrieved July 26, 2003. Online: http://www.census.gov/Press-Release/www/2003/cb03-51.html
U.S. Conference of Mayors. 2002. *A Status Report on Hunger and Homelessness in America's Cities 2002.* Retrieved July 3, 2003. Online: http://usmayors.org/uscm/hungersurvey/2002/onlinereport/HungerAndHomelessReport2002.pdf
U.S. Congress, Office of Technology Assessment. 1979. "The Effects of Nuclear War." Report quoted in Michael E. Howard. 1990. "On Fighting a Nuclear War." In Francesca M. Cancian and James William Gibson (Eds.), *Making War, Making Peace: The Social Foundations of Violent Conflict.* Belmont, CA: Wadsworth, pp. 314–322.
U.S. Department of Education. 1993. *Adult Literacy in America: A First Look at the Results of the National Literacy Survey.* Washington, DC: U.S. Government Printing Office.
___. 2001a. "Homeschoolers Estimated at 850,000." Retrieved Aug. 6, 2003. Online: http://www.ed.gov/PressRelease/08-2001/08012001.html
___. 2001b. "Violent Deaths in or Near Schools Are Rare." Retrieved July 5, 2003. Online: http://www.ed.gov/PressReleases/12-2001/12042001.html
U.S. Department of Health and Human Services. 2000. "Temporary Assistance for Needy Families (TANF) Program: Third Annual Report to Congress." Retrieved Sept. 29, 2001. Online: http://www.acf.dhhs.gov/programs/opre/annual3.doc
___. 2002a. "Facts & Figures: HIV/AIDS Statistics." Retrieved Aug. 7, 2003. Online: http://www.niaid.nih.gov/factsheets/aidsstat.htm
___. 2002b. "A Profile of Older Americans: 2002." Retrieved July 26, 2003. Online: http://www.aoa.dhhs.gov/aoa/stats/profile/profile.pdf
U.S. Department of Justice. 2000a. "Capital Punishment Statistics." Retrieved Oct. 31, 2001. Online: http://www.ojp.usdoj.gov/bjs/cp.htm
___. 2000b. *Intimate Partner Violence.* Retrieved Nov. 3, 2001. Online: http://www.ojp.usdoj.gov/bjs/pub/press/ipv.htm
U.S. Department of Labor. 2002. "Nontraditional Occupations for Women in 2001." Retrieved Oct. 12, 2002. Online: http://www.dol.gov/wb/wb_pubs/nontrad2001.htm
___. 2003. "Business Ownership—Cornerstone of the American Dream." Retrieved July 19, 2003. Online: http://www.dol.gov/odep/pubs/business/business.htm
U.S. Department of Labor, Employment and Training Administration. 1993. *Dictionary of Occupational Titles.* Washington, DC: U.S. Government Printing Office.
U.S. Health Care Financing Administration. 2000. "National Health Expenditure Amounts, and Average Annual Percent Change, by Type of Expenditure: Selected Calendar Years 1970–2008." Retrieved Sept. 3, 2000. Online: http://www.hcfa.gov/stats/NHE-Proj/proj1998/tables/table2.htm
U.S. Immigration and Naturalization Service. 2000. *1998 Statistical Yearbook of the Immigration and Naturalization Service.* Washington, DC: U.S. Department of Justice, Immigration and Naturalization Service.
U.S. Office of Personnel Management. 2001. "Demographic Profile of the Federal Workforce as of September 30, 2000." Retrieved Aug. 9, 2003. Online: http://www.opm.gov/feddata/demograp/00demogr.pdf
___. 2003. "2003 Salary Tables and Related Information." Retrieved Aug. 9, 2003. Online: http://www.opm.gov/oca/PAYRATES
University of Utah. 2003. "U of U Trains Bosnians to Teach Students with Disabilities in Inclusive Settings." Retrieved Aug. 6, 2003. Online: http://www.utah.edu/unews/releases/03/jul/bosnian.html
Van Biema, David. 1993. "But Will It End the Abortion Debate?" *Time* (June 14): 52–54.
___. 1997. "Buddhism in America." *Time* (Oct. 13): 72–81.
Vaughan, Diane. 1985. "Uncoupling: The Social Construction of Divorce." In James M. Henslin (Ed.), *Marriage and Family in a Changing Society* (2nd ed.). New York: Free Press, pp. 429–439.
Vaughan, Ted R., Gideon Sjoberg, and Larry T. Reynolds (Eds.). 1993. *A Critique of Contemporary American Sociology.* Dix Hills, NY: General Hall.
Veblen, Thorstein. 1967. *The Theory of the Leisure Class.* New York: Viking (orig. pub. 1899).
Vetter, Harold J., and Gary R. Perlstein. 1991. *Perspectives on Terrorism.* Pacific Grove, CA: Brooks/Cole.
VirtualSources.com. 2003. "Information and Resources from Countries Around the World." Retrieved Aug. 30, 2003. Online: http://www.virtualsources.com/index.htm
Vissing, Yvonne. 1996. *Out of Sight, Out of Mind: Homeless Children and Families in Small Town America.* Lexington: University Press of Kentucky.
Vito, Gennaro F., and Ronald M. Holmes. 1994. *Criminology: Theory, Research and Policy.* Belmont, CA: Wadsworth.
Volti, Rudi. 1995. *Society and Technological Change* (3rd ed.). New York: St. Martin's.
Voynick, Steve. 1999. "Living with Ozone." *The World & I* (July): 192–199.
Wagner, Elvin, and Allen E. Stearn. 1945. *The Effects of Smallpox on the Destiny of the American Indian.* Boston: Bruce Humphries.
Waldman, Amy. 2001. "Behind the Burka: Women Subtly Fought Taliban." *New York Times* (Nov. 19): A1, B4.
Waldron, Ingrid. 1994. "What Do We Know About Causes of Sex Differences in Mortality? A Review of the Literature." In Peter Conrad and Rachelle Kern (Eds.), *The Sociology of Health and Illness: Critical Perspectives* (4th ed.). New York: St. Martin's.
Walker, Lenore. 1979. *The Battered Woman.* New York: Harper & Row.
Wallace, Harvey. 2002. *Family Violence: Legal, Medical, and Social Perspectives* (3rd ed.). Boston: Allyn & Bacon.
Wallace, Walter L. 1971. *The Logic of Science in Sociology.* New York: Aldine de Gruyter.
Wallerstein, Immanuel. 1979. *The Capitalist World-Economy.* Cambridge, England: Cambridge University Press.
___. 1984. *The Politics of the World Economy.* Cambridge, England: Cambridge University Press.
___. 1991. *Unthinking Social Science: The Limits of Nineteenth-Century Paradigms.* Cambridge, England: Polity.
Warner, W. Lloyd, and Paul S. Lunt. 1941. *The Social Life of a Modern Community.* New Haven, CT: Yale University Press.
Warr, Mark. 1993. "Age, Peers, and Delinquency." *Criminology,* 31 (1): 17–40.
___. 2000. "Public Perceptions and Reactions to Crime." In Joseph F. Sheley (Ed.), *Criminology: A Contemporary Handbook* (3rd ed.). Belmont, CA: Wadsworth, pp. 13–55.
Waters, Malcolm. 1995. *Globalization.* London and New York: Routledge.
Watkins, T. H. 1993. *The Great Depression: America in the 1930s.* Boston: Little, Brown.
Watson, Paul. 1997. "Richer, Poorer." *Toronto Star* (Aug. 10): F1, F5.
Watson, Tracey. 1987. "Women Athletes and Athletic Women: The Dilemmas and Contradictions of Managing Incongruent Identities." *Sociological Inquiry,* 57 (Fall): 431–446.
Waxman, Laura, and Sharon Hinderliter. 1996. *A Status Report on Hunger and Homelessness in America's Cities: 1996.* Washington, DC: U.S. Conference of Mayors.
Webb, Eugene, and others. 1966. *Unobtrusive Measures: Nonreactive Research in the Social Sciences.* Chicago: Rand McNally.
Weber, Max. 1963. *The Sociology of Religion.* Trans. E. Fischoff. Boston: Beacon (orig. pub. 1922).
___. 1968. *Economy and Society: An Outline of Interpretive Sociology.* Trans. G. Roth and G. Wittich. New York: Bedminster (orig. pub. 1922).
___. 1976. *The Protestant Ethic and the Spirit of Capitalism.* Trans. Talcott Parsons. Introduction by Anthony Giddens. New York: Scribner (orig. pub. 1904–1905).
Weeks, John R. 2002. *Population: An Introduction to Concepts and Issues* (8th ed.). Belmont, CA: Wadsworth.
Weeks, Linton. 2003. "Worst Foot Forward: A Guide to Foreign Insults." *Washington Post* (Apr. 11): C1. Retrieved June 16, 2003. Online: http://www.washingtonpost.com/ac2/wp-dyn?pagename=article&node=&contentId=A5449-2003Apr10&notFound=true
Weigel, Russell H., and P. W. Howes. 1985. "Conceptions of Racial Prejudice: Symbolic Racism Revisited." *Journal of Social Issues,* 41: 124–132.
Wei-ming, Tu. 1995. "Confucianism." In Arvind Sharma (Ed.), *Our Religions.* San Francisco: HarperCollins, pp. 141–227.
Weinhouse, Don, and Marilyn Weinhouse. 1994. *Little Children, Big Needs: Parents Discuss Raising Children with Exceptional Needs.* Niwot: University Press of Colorado.
Weinstein, Michael M. 1997. "'The Bell Curve,' Revisited by Scholars." *New York Times* (Oct. 11): A20.
Weisner, Thomas S., Helen Garnier, and James Loucky. 1994. "Domestic Tasks, Gender Egalitarian Values and Children's Gender Typing in Conventional and Nonconventional Families." *Sex Roles* (January): 23–55.
Weiss, Gregory L., and Lynne E. Lonnquist. 2003. *The Sociology of Health, Healing, and Illness* (4th ed.). Upper Saddle River, NJ: Prentice Hall.
Weiss, Meira. 1994. *Conditional Love: Attitudes Toward Handicapped Children.* Westport, CT: Bergin & Garvey.
Weistart, John. 1993. "The 90's University: Reading, Writing, and Shoe Contracts." *New York Times* (Nov. 28): 23.
Weitz, Rose. 1993. "Living with the Stigma of AIDS." In Delos H. Kelly (Ed.), *Deviant Behavior: A Text-Reader in the Sociology of Deviance* (4th ed.). New York: St. Martin's, pp. 222–236.
___. 2004. *The Sociology of Health, Illness, and Health Care* (3rd ed.). Belmont, CA: Wadsworth.
Wellhousen, Karyn, and Zenong Yin. 1997. "Peter Pan Isn't a Girls' Part: An Investigation of Gender Bias in a Kindergarten Classroom." *Women and Language,* 20: 35–40.
Wellman, Barry. 2001. "Physical Place and Cyberplace: The Rise of Personalized Networking." *International Journal of Urban and Regional Research* 22 (2): 227–252.
Welner, Kevin Grant, and Jeannie Oakes. 2000. *Navigating the Politics of Detracking.* Arlington Heights, IL: Skylight.

Westrum, Ron. 1991. *Technologies and Society: The Shaping of People and Things.* Belmont, CA: Wadsworth.

White, Jack E. 1997. "I'm Just Who I Am." *Time* (May 5): 32–36.

White, Merry. 1994. *The Material Child: Coming of Age in Japan and America.* Berkeley: University of California Press.

___. 2002. *Perfectly Japanese: Making Families in an Era of Upheaval.* Berkeley: University of California Press.

Whitney, Craig R. 1997. "Jeanne Calment, World's Elder, Dies at 122." *New York Times* (Aug. 5): B8.

Whorf, Benjamin Lee. 1956. *Language, Thought and Reality.* Ed. John B. Carroll. Cambridge, MA: MIT Press.

Whyte, William Foote. 1988. *Street Corner Society: Social Structure of an Italian Slum.* Chicago: University of Chicago Press (orig. pub. 1943).

___. 1989. "Advancing Scientific Knowledge Through Participatory Action Research." *Sociological Forum,* 4: 367–386.

Whyte, William H., Jr. 1957. *The Organization Man.* Garden City, NY: Anchor.

Wickett, Ann. 1989. *Double Exit: When Aging Couples Commit Suicide Together.* Eugene, OR: Hemlock Society.

Wilkerson, Jamie. 1996. "Thoughts of Internet Friendships." Retrieved July 5, 2003. Online: http://web.cetlink.net/~parrothd/Jamie/Poetry1.htm

Wilkie, Jane Riblett. 1993. "Changes in U.S. Men's Attitudes Toward the Family Provider Role, 1972–1989." *Gender & Society,* 7 (2): 261–279.

Williams, David R., David T. Takeuchi, and Russell K. Adair. 1992. "Socioeconomic Status and Psychiatric Disorders Among Blacks and Whites." *Social Forces,* 71: 179–195.

Williams, Lena. 1995. "A Silk Blouse on the Assembly Lines? (Yes, the Boss's)." *New York Times* (Feb. 5): F7.

Williams, Norma. 1990. *The Mexican American Family: Tradition and Change.* Dix Hills, NY: General Hall.

Williams, Robin M., Jr. 1970. *American Society: A Sociological Interpretation* (3rd ed.). New York: Knopf.

Williamson, Robert C., Alice Duffy Rinehart, and Thomas O. Blank. 1992. *Early Retirement: Promises and Pitfalls.* New York: Plenum.

Willie, Charles V. 1991. *A New Look at Black Families* (4th ed.). Dix Hills, NY: General Hall.

Wilson, David (Ed.). 1997. "Globalization and the Changing U.S. City." *Annals of the American Academy of Political and Social Sciences,* 551 (May special issue).

Wilson, Edward O. 1975. *Sociobiology: A New Synthesis.* Cambridge, MA: Harvard University Press.

Wilson, Elizabeth. 1991. *The Sphinx in the City: Urban Life, the Control of Disorder, and Women.* Berkeley: University of California Press.

Wilson, James Q. 1996. "Foreword." In George L. Kelling and Catherine M. Coles (Eds.), *Fixing Broken Windows: Restoring Order and Reducing Crime in Our Communities.* New York: Touchstone, pp. xiii–xvi.

Wilson, William Julius. 1978. *The Declining Significance of Race: Blacks and Changing American Institutions.* Chicago: University of Chicago Press.

___. 1996. *When Work Disappears: The World of the New Urban Poor.* New York: Knopf.

Winik, Lyric Wallwork. 1997. "Oh Nurse, More Beluga Please." *Forbes FYI: The Good Life* (Winter): 157–166.

Winkleby, Marilyn A., and Catherine Cubbin. 2003. "Influence of Individual and Neighbourhood Socioeconomic Status on Mortality Among Black, Mexican-American, and White Women and Men in the United States." *Journal of Epidemiology and Community Health,* 57: 444–452.

Winn, Maria. 1985. *The Plug-in Drug: Television, Children, and the Family.* New York: Viking.

Wirth, Louis. 1938. "Urbanism as a Way of Life." *American Journal of Sociology,* 40: 1–24.

Wiseman, Jacqueline. 1970. *Stations of the Lost: The Treatment of Skid Row Alcoholics.* Chicago: University of Chicago Press.

Witt, Susan D. 1997. "Parental Influence on Children's Socialization to Gender Roles." *Adolescence,* 32: 253–260.

Wollstonecraft, Mary. 1974. *A Vindication of the Rights of Woman.* New York: Garland (orig. pub. 1797).

Women's Research & Education Institute. 2002. "Women in the Military." Retrieved Aug. 9, 2003. Online: http://www.wrei.org/projects/wiu/index.htm

Wonders, Nancy. 1996. "Determinate Sentencing: A Feminist and Postmodern Story." *Justice Quarterly,* 13: 610–648.

Wood, Daniel B. 2002. "As Homelessness Grows, Even Havens Toughen Up." *Christian Science Monitor* (Nov. 21). Retrieved July 1, 2003. Online: http://www.csmonitor.com/2002/1121/p01s04-ussc.htm

Wood, Julia T. 1994. *Gendered Lives: Communication, Gender, and Culture.* Belmont, CA: Wadsworth.

___. 1999. *Gendered Lives: Communication, Gender, and Culture* (3rd ed.). Belmont, CA: Wadsworth.

Wooley, Susan C. 1994. "Sexual Abuse and Eating Disorders: The Concealed Debate." In Patricia Fallon, Melanie A. Katzman, and Susan C. Wooley (Eds.), *Feminist Perspectives on Eating Disorders.* New York: Guilford, pp. 171–211.

WordSpy.com. "Flash Mob." Retrieved Aug. 11, 2003. Online: http://www.wordspy.com/words/flashmob.asp

World Bank. 2001. *World Development Report 2002: Building Institutions for Markets.* Retrieved Sept. 17, 2001. Online: http: econ.worldbank.org/wdr/subpage.php?pr=2391

___. 2003a. "East Asia Navigates Short-Term Shocks for a Stronger Future." Retrieved July 15, 2003. Online: http://web.worldbank.org/WBSITE/EXTERNAL/NEWS/0,,contentMDK:20106953-menuPK:34463-pagePK:34370-piPK:34424-theSitePK:4607,00.html#

___. 2003b. "PovertyNet: Listen to the Voices." Retrieved July 10, 2003. Online: http://www.worldbank.org/poverty/voices/listen-findings.htm

___. 2003c. "World Development Indicators 2003." Retrieved July 12, 2003. Online: http://www.worldbank.org/data/wdi2003/worldview.pdf

World Health Organization. 1999. *The World Health Report 1999.* Retrieved Oct. 2, 1999. Online: http://www.who.int/whr/1999/en/pdf/burden.pdf

___. 2003. *The World Health Report 2003.* Retrieved Jan. 8, 2004. Online: http://www.who.int/whr/2003/en/overview_en.pdf

Worster, Donald. 1985. *Natures Economy: A History of Ecological Ideas.* New York: Cambridge University Press.

Wouters, Cas. 1989. "The Sociology of Emotions and Flight Attendants: Hochschild's Managed Heart." *Theory, Culture & Society,* 6: 95–123.

Wright, Erik Olin. 1978. "Race, Class, and Income Inequality." *American Journal of Sociology,* 83 (6): 1397.

___. 1979. *Class Structure and Income Determination.* New York: Academic Press.

___. 1985. *Class.* London: Verso.

___. 1997. *Class Counts: Comparative Studies in Class Analysis.* Cambridge, England: Cambridge University Press.

Wright, Erik Olin, Karen Shire, Shu-Ling Hwang, Maureen Dolan, and Janeen Baxter. 1992. "The Non-Effects of Class on the Gender Division of Labor in the Home: A Comparative Study of Sweden and the U.S." *Gender & Society,* 6 (2): 252–282.

Wright, John W. (Ed.). 1997. *The New York Times 1998 Almanac.* New York: Penguin Reference.

WTVM. 2003. "Soldiers Talk About Embedded Media." WTVM, Columbus, GA. Retrieved July 25, 2003. Online: http://www.wtvm.com/Global/story.asp?s=%20%201365415

Yablonsky, Lewis. 1997. *Gangsters: Fifty Years of Madness, Drugs, and Death on the Streets of America.* New York: New York University Press.

Yelin, Edward H. 1992. *Disability and the Displaced Worker.* New Brunswick, NJ: Rutgers University Press.

Yinger, J. Milton. 1960. "Contraculture and Subculture." *American Sociological Review,* 25 (October): 625–635.

___. 1982. *Countercultures: The Promise and Peril of a World Turned Upside Down.* New York: Free Press.

Young, John. 1990. *Sustaining the Earth: The Story of the Environmental Movement—Its Past Efforts and Future Challenges.* Cambridge, MA: Harvard University Press.

Young, Michael Dunlap. 1994. *The Rise of the Meritocracy.* New Brunswick, NJ: Transaction (orig. pub. 1958).

Younge, Gary. 2002. "Everything You Know Is Untrue." *The Guardian* (Aug. 22). Retrieved Aug. 8, 2003. Online: http://www.obv.org.uk/reports/2002/078_2002_0822.html

Zack, Naomi. 1998. *Thinking About Race.* Belmont, CA: Wadsworth.

Zald, Mayer N., and John D. McCarthy (Eds.). 1987. *Social Movements in an Organizational Society.* New Brunswick, NJ: Transaction.

Zavella, Patricia. 1987. *Women's Work and Chicano Families: Cannery Workers of the Santa Clara Valley.* Ithaca, NY: Cornell University Press.

Zeitlin, Irving M. 1997. *Ideology and Development of Sociological Theory* (6th ed.). Upper Saddle River, NJ: Prentice Hall.

Zelizer, Viviana. 1985. *Pricing the Priceless Child: The Changing Social Value of Children.* New Haven, CT: Yale University Press.

Zellner, William M. 1978. Vehicular Suicide: In Search of Incidence. Unpublished M.A. thesis, Western Illinois University, Macomb. Quoted in Richard T. Schaefer and Robert P. Lamm. 1992. *Sociology* (4th ed.). New York: McGraw-Hill, pp. 54–55.

Zill, N., and C. W. Nord. 1994. *Running in Place.* Washington, DC: Child Trends.

Zipp, John F. 1985. "Perceived Representativeness and Voting: An Assessment of the Impact of 'Choices' vs. 'Echoes.'" *American Political Science Review,* 60 (3): 738–759.

Zuboff, Shoshana. 1988. *In the Age of the Smart Machine: The Future of Work and Power.* New York: Basic.

Zweigenhaft, Richard L., and G. William Domhoff. 1998. *Diversity in the Power Elite: Have Women and Minorities Reached the Top?* New Haven, CT: Yale University Press.

PHOTO CREDITS

This page constitutes an extension of the copyright page. We have made every effort to trace the ownership of all copyrighted material and to secure permission from copyright holders. In the event of any question arising as to the use of any material, we will be pleased to make the necessary corrections in future printings. Thanks are due to the following authors, publishers, and agents for permission to use the material indicated.

Chapter 1. 3: © Nancy Richmond/The Image Works **8:** © Lyntha Scott Eiler/Stock Boston **10:** © AP/Wide World Photos **14:** Hulton/Archive/Getty Images **15:** The Granger Collection, New York **16:** Corbis-Bettmann **17:** Corbis-Bettmann **18:** The Granger Collection, New York **19:** © The New Yorker Collection 2001 Edward Koren from cartoonbank.com. All rights reserved. **20:** © Michael Newman/PhotoEdit **22:** The Granger Collection, New York **24:** © Richard Pasley/Stock Boston **25:** © UPI/Corbis **26:** © Bob Daemmrich/The Image Works **28:** © Robert Mora/Getty Images **30:** © JimMcHugh/Corbis/Outline **32:** © Bill Aron/PhotoEdit **Chapter 2. 39:** © Paul Morse/Corbis **40:** © David Keeler/Getty Images **48:** © 2003 AP/Wide World Photos **50:** © Gary Conner/PhotoEdit **52:** © 2003 AP/Wide World Photos **53:** © Spencer Grant/Stock Boston **56:** © HIRB/IndexStock Imagery **58:** © Spencer Grant/PhotoEdit **60:** © Bob Daemmrich/The Image Works **61:** © Flip Chalfant (The Image Bank)/Getty Images **63:** © Addison Geary **65:** © 2003 AP/Wide World Photos **67:** © Rob Crandall/The Image Works **71:** © Sybil Shackman **Chapter 3. 75:** © 2003 AP/Wide World Photos **80:** top right, © James Nelson (Stone)/Getty Images **80:** bottom, © Jean-Marc Truchet (Stone)/Getty Images **80:** top left, © Phil Banka(Stone)/Getty Images **81:** top right, © Mark Richards/PhotoEdit **81:** bottom, © Michael Greenlar/The Image Works **81:** top left, © Spencer Grant/PhotoEdit **82:** © 2002 AP/Wide World Photos **84:** © Joe Polollio Photography **86:** © Michael Newman/PhotoEdit **88:** © Lou Dematteis/ Reuters NewMedia, Inc./Corbis **91:** © Nancy Siesel/The New York Times **93:** © 2003 AP/Wide World Photos **95:** © Gerhard Humer/Contrast/Gamma Press **97:** © Eric Fowke/PhotoEdit **98:** © Scott Barbour/Getty Images **99:** © Vince Bucci/Getty Images **100:** © AP/Wide World Photos **102:** right, © J. Aronovsky/Zuma **102:** left, © Reuters/Corbis **104:** © Bill Bachmann/The Image Works **Chapter 4. 109:** © AP/Wide World Photos **113:** left, © Bill Aron/PhotoEdit **113:** right, © Bob Daemmrich/The Image Works **116:** © Stuart Cohen/Index Stock Imagery **117:** © Tony Freeman/PhotoEdit **118:** © Flash!Light/Stock Boston **119:** © Roberto Soncin Gerometta **122:** top left, © Elizabeth Crews/The Image Works **122:** bottom, © Rachel Epstein/The Image Works **122:** top right, © Richard Hutchings/ Photo Researchers **124:** © David Young-Wolff/PhotoEdit **126:** © Michael Newman/PhotoEdit **127:** © Joseph Schuyler/ Stock Boston **128:** © Michael Newman/PhotoEdit **130:** © Bill Aron/PhotoEdit **131:** © AP/Wide World Photos **133:** © Richard Lord/The Image Works **134:** © Larry Kolvoord/The Image Works **Chapter 5. 139:** © Bob Collins/The Image Works **144:** right, © Yellow Dog Productions(The Image Bank)/Getty Images **144:** left, © Mary Kate Denny/PhotoEdit **146:** © Lisette Le Bon/SuperStock **148:** right, © Jon Feingersh/Corbis **148:** left, © Mug Shots/Corbis **149:** © 2002 AP/Wide World Photos **152:** © Jason Laure/The Image Works **154:** © Michelle Garrett/Corbis **156:** © Alyssa Banta/Getty Images **158:** right, © Alison Wright/The Image Works **158:** left, © Holton Collection/SuperStock **159:** left, © Andy Sacks(Stone)/Getty Images **159:** right, © Lonnie Duka(Stone)/Getty Images **162:** © Tony Freeman/PhotoEdit **163:** left, © 2003 AP/Wide World Photos **163:** right, © Mike Theiler/Getty Images **165:** © 2003 AP/Wide World Photos **168:** left, © Mark Antman/The Image Works **168:** right, © Michael Newman/PhotoEdit **Chapter 6. 175:** © 2001 AP/Wide World Photos **179:** left, © Tony Roberts/Corbis **179:** right, © Bob Daemmrich/The Image Works **180:** © Michael Newman/PhotoEdit **182:** left, © Michael Newman/PhotoEdit **182:** right, © Michael Newman/PhotoEdit **184:** right, © Cindy Charles/PhotoEdit **184:** left, © David R. Frazier (Stone)/Getty Images **189:** top, © 2003 AP/Wide World Photos **189:** bottom, © 2003 AP/Wide World Photos, Tyler Morning Telegraph, Dr. Scott Lieberman **191:** top right, © A. Ramey/Stock Boston **191:** bottom, © Tim Barnwell/Stock Boston **191:** top left, © Tony Savino/The Image Works **194:** © Frank Pedrick/The Image Works **195:** © SuperStock **197:** © Thomas K. Wanstall/The Image Works **199:** right, © Mark Richards/PhotoEdit **199:** left, © Tim Brown(Stone)/Getty Images **203:** © Martin Rogers (Stone)/Getty Images **Chapter 7. 209:** © Mark Richards/PhotoEdit **213:** © Spencer Grant/PhotoEdit **214:** © 2003 AP/Wide World Photos **217:** © AP/Wide World Photos **220:** left, © Bill Bachmann/ PhotoEdit **220:** right, © Michael Newman/PhotoEdit **222:** © 2003 AP/Wide World Photos **225:** © 2003 AP/Wide World Photos **228:** © Dwayne Newton/PhotoEdit **231:** © 2003 AP/Wide World Photos **233:** © 2003 AP/Wide World Photos **240:** © Porter Gifford/Getty Images **Chapter 8. 249:** © Michael Dwyer/Stock Boston **252:** © Ed Kashi/Aurora Photos **255:** bottom left, © Alan Sussman/The Image Works **255:** top, © Stock Montage, Inc. **255:** bottom right, © SuperStock **258:** © Erik Eckholm/The New York Times Pictures **260:** © A. Perigot-Icone/The Image Works **262:** © John Maier, Jr./The Image Works **269:** © Lauren Goodsmith/ The Image Works **270:** © Brenda Prince/Format **271:** © Chris Hondros/Getty Images **272:** © Louise Gubb/The Image Works **275:** © James Marshall/The Image Works **277:** © Todd Bigelow/Aurora Photos **279:** © AP/Wide World Photos **Chapter 9. 285:** © Anthony Barboza **289:** bottom, © Tim Carlson/Stock Boston **289:** top, © William Strode/ Woodfin Camp & Associates **291:** © 2002 AP/Wide World Photos **297:** © Tom Rosenthal/ SuperStock **299:** © Ethan Miller/Reuters **302:** top left, © AP/Wide World Photos **302:** top right, © Bob Daemmrich/ Stock Boston **302:** bottom right, © Gregg Mancuso/Stock Boston **302:** bottom left, © Robin Nelson/Black Star (Stockphoto.com) **310:** © Sonda Dawes/The Image Works **312:** © AP/Wide World Photos **Chapter 10. 319:** © Reuters NewMedia Inc./Corbis **323:** © Jeff Greenberg/PhotoEdit **325:** © Gary Hershorn/ Reuters **327:** © Michael Greenlar/The Image Works **330:** © Bettmann/Corbis **333:** © Shelly Katz/Getty Images **335:** right, © AP/Wide World Photos **335:** left, © Esbin-Anderson/The Image Works **335:** center right, © Richard Heinzen/SuperStock **335:** center left, © Roger Allyn Lee/SuperStock **337:** © Ruth Freeman/The New York Times **338:** © Frank Siteman/Stock Boston **340:** © 2003 AP/Wide World Photos **342:** © Lawrence Migdale/ Stock Boston **344:** © AP/Wide World Photos **346:** © AP/Wide World Photos **348:** © 2003 AP/Wide World Photos **Chapter 11. 355:** © Duncan Smith/Photodisc Green-Getty Images **357:** © Marjorie Farrell/The Image Works **361:** © Bob Dammerich/The Image Works **362:** © Tony Duffy/Getty Images **364:** © Historical Picture Archive/Corbis **368:** left, © Bob Thomas(Stone)/Getty Images **368:** right, © Geri Engberg/The Image Works **371:** © Barbara Campbell **372:** © Mary Kate Denny/PhotoEdit **373:** © 2003 AP/Wide World Photos **379:** right, © Spencer Grant/ PhotoEdit **379:** left, © Tony Freeman/ PhotoEdit **381:** right, © Myrleen Ferguson Cate/PhotoEdit **384:** © Jose Carillo/PhotoEdit **Chapter 12. 389:** © DiMaggio/Kalish/Corbis **390:** © Bob Daemmrich/ Stock Boston **394:** © Jacob Halaska/Index Stock Imagery **397:** © Branson Reynolds/ Index Stock Imagery **399:** © Camille Tokerud(Stone)/ Getty Images **404:** right, © David Young-Wolff/PhotoEdit **404:** left, © Tony Freeman/ PhotoEdit **407:** © AP/Wide World Photos **409:** © Felicia Martinez/PhotoEdit **413:** © Susan Van Etten/PhotoEdit **414:** © Sondra Dawes/The Image Works **Chapter 13. 419:** © 2003 AP/Wide World Photos **423:** left, © Mark Richards/PhotoEdit **423:** right, © Mark Richards/PhotoEdit **425:** © 2003 AP/Wide World Photos **427:** © AFP/Corbis **430:** © 2003 AP/Wide World Photos **436:** © Paul Simcock(Image Bank)/Getty Images **438:** © 2003 AP/Wide World Photos **441:** © Spencer Grant/PhotoEdit **442:** © Charles Gupton(Stone)/Getty Images **443:** © Bill Aron/PhotoEdit **444:** © Rob Crandall/The Image Works **445:** © PhotoEdit **448:** © Mark Newman/PhotoEdit **450:** © Paul A. Souders/Corbis **Chapter 14. 455:** © Richard Ellis/Getty Images **461:** top, © Betty Press/Woodfin Camp & Associates **461:** bottom, © Cary Wolinsky/Stock Boston **461:** center, © SuperStock **462:** bottom, © PressNet/Topham/The Image Works **462:** right, © plainpicture/Warneck/Alamy Images **464:** © AP/Wide World Photos **466:** © Paul Hosefros/The New York Times **467:** © 2003 AP/Wide World Photos **468:** © 2003 AP/Wide World Photos **471:** © Manny Ceneta/ AFP/Corbis **473:** © Jim West Photography **476:** © James Pickerell/The Image Works **480:** © Reuters NewMedia Inc./Corbis **481:** © Mario Villafuerte/Getty Images **Chapter 15. 489:** © Myrleen Cate/IndexStock Imagery **493:** left, © Bob Daemmrich/The Image Works **493:** right, © Steven Rubin/The Image Works **495:** © Tony Howarth/ Woodfin Camp & Associates **498:** © Alan Carey/The Image Works **500:** © Janine Weidel Photolibrary/Alamy Images **503:** © Mary Kate Denny/PhotoEdit **505:** © Jonathan Nourok(The Image Bank)/Getty Images **508:** © Dan Habib/Concord Monitor photo/Corbis-SABA **509:** © Bob Daemmrich/ Stock Boston **513:** © Tony Freeman/ PhotoEdit **514:** © Michelle D. Bridwell/PhotoEdit **517:** top left, © Tony Freeman/ PhotoEdit **517:** top right, © Lawrence Migdale/Stock Boston **517:** bottom left, © Lori Adamski Peek(Stone)/Getty Images **517:** bottom right, © Myrleen Ferguson Cate/PhotoEdit **Chapter 16. 523:** © Larry Martin/Little Big Horn College **526:** © Gabe Palmer/Corbis **528:** © Alan Oddie/ PhotoEdit **529:** © Durand Patrick/ Corbis Sygma **530:** © Donna Day(Stone)/Getty Images **535:** © Spencer Grant/ PhotoEdit **538:** © Courtesy of Baylor University **540:** left, © Mark Newman/PhotoEdit **540:** right, © Dan Habib/Concord Monitor **545:** © Gary Caskey/Reuters/Landov **546:** © Brian Snyder/ Reuters/Landov **549:** © AP/Wide World Photos **550:** © Don Smetzer/PhotoEdit **552:** © Syracuse Newspapers/ Dick Blume/The Image Works **556:** © Bridget Besaw Gorman/Aurora Photos **Chapter 17. 561:** UPI-Corbis **564:** © Tom Carter/PhotoEdit **568:** © Bob Daemmrich/The Image Works **570:** © 2003 AP/Wide World Photos **572:** AP/Wide World Photos **574:** © Mark O'Neill/Canada Wide **575:** © Gary Conner/PhotoEdit **576:** © PhotoEdit **579:** © AP/Wide World Photos **581:** © Tony Freeman/ PhotoEdit **583:** © AP/Wide World Photos **586:** © Michael Newman/PhotoEdit **Chapter 18. 591:** © Geri Engberg/The Image Works **594:** left, © Andrew Holbrooke/The Image Works **594:** right, © Lew Lause/SuperStock **596:** © AP/Wide World Photos **598:** © Mary Kate Denny/PhotoEdit **599:** © Michael Newman/ PhotoEdit **601:** © Bettmann-Corbis **602:** © Flash! Light/ Stock Boston **608:** © Howard Sochurek/Woodfin Camp & Associates **609:** © Michael Newman/PhotoEdit **610:** © George Disario/Corbis **613:** © Jodi Buren/Woodfin Camp & Associates **617:** © Michael Newman/PhotoEdit **Chapter 19. 627:** © A. Ramey/PhotoEdit **629:** © Betty Press/Woodfin Camp & Associates **632:** left, © 2003 AP/Wide World Photos **632:** right, © Paul Conklin/PhotoEdit **638:** © Paul Conklin/PhotoEdit **639:** © Tony Freeman/PhotoEdit **640:** © Bill Varie/Corbis **644:** © Andrew Holbrooke/The Image Works **646:** © Robert Brenner/PhotoEdit **651:** left, © AP/Wide World Photos **651:** right, © Monika Graff/The Image Works **653:** © Spencer Grant/PhotoEdit **Chapter 20. 659:** © Photodisc Blue/Getty Images **664:** © AP/Wide World Photos **665:** © AP/Wide World Photos **666:** © David Young-Wolff/ PhotoEdit **668:** © Jonathan Fickies/Getty Images **671:** © Flip Schulke/Black Star (Stockphoto.com) **674:** left, © A. Ramey/Stock Boston **674:** right, © A. Ramey/Stock Boston **675:** © 2003 AP/Wide World Photos **677:** © Andrew Lichtenstein/The Image Works **682:** ©Jeff Greenberg/ PhotoEdit

NAME INDEX

SUBJECT INDEX